ANNUAL REVIEW
OF PHYSIOLOGY

ANNUAL REVIEW
OF PHYSIOLOGY

VOLUME 64, 2002

JOSEPH F. HOFFMAN, *Editor*
Yale University School of Medicine

PAUL De WEER, *Associate Editor*
University of Pennsylvania School of Medicine

www.annualreviews.org   science@annualreviews.org   650-493-4400

ANNUAL REVIEWS
4139 El Camino Way • P.O. BOX 10139 • Palo Alto, California 94303-0139

ANNUAL REVIEWS
Palo Alto, California, USA

International Standard Serial Number: 0066-4278
International Standard Book Number: 0-8243-0364-4
Library of Congress Catalog Card Number: 39-15404

TYPESET BY TECHBOOKS, FAIRFAX, VA
PRINTED AND BOUND IN THE UNITED STATES OF AMERICA

PREFACE

A primary aim of the *Annual Review of Physiology* series is to track recent advances that have occurred in our field. Our task is daunting, not only because the scope is broad and constantly in flux, but also because we need to limit our overall coverage to provide useful reviews. As our readers are aware, we have approached this task by sectionalizing our subject areas, utilizing an adopted "theme" within each section, with these themes being cycled through different topics in subsequent years, depending on the changes that have taken place in each field. This format, together with the inclusion of special topics, has held us in good stead over the years as judged by the favorable reaction of many readers and by more objective measures such as the Impact Factor. But having said this, we are also on the alert and concerned with how we could improve our publication not just by altering the format but in choosing our various subjects for review. I believe the Editorial Committee would welcome suggestions that would enable us to better serve our constituencies. One reason for raising this question is the almost constant preoccupation of how our fellow physiologists view the current and changing faces of our field. We have seen academic departments of physiology (particularly in the United States) modify their names to include, in various forms, cellular and molecular, biophysics, or cell biology and, more recently, integrative. It is interesting that comparative physiology has long bridged all of these areas. Now come new concerns or at least new catch phrases such as functional genomics, translational physiology, and ecogenomics (see References 1–5). Although the intent and experimental enticements underlying this new lexicon have been in the physiologists arsenal for many years, they do emphasize the need for a broadening of focus, using the tool boxes from the reductionist approach, to study whole animal physiology as well as pathophysiology. It is clear that new horizons are appearing that beckon the discovery of new physiological treasures by drinking from the genomic spring. This subject has been commented on in this space in previous volumes (see *Annual Review of Physiology*, Volumes 54, 55, 58, and 63).

This year's volume contains, in addition to the traditional sections, our prefatory chapter written by Ian Glynn, in which he reflects on his work with the sodium pump, and two special topics. The special topics are aimed at informing our readers of important developments in related fields that lie outside our normal coverage. One special topic, edited by James B. Hurley, is on G protein effector mechanisms. The other special topic, edited by Gerhard Giebisch, concerns transportopathies. As said before, we are open to suggestions and encourage comments about our publication. We can be reached at www.annualreviews.org. It should also be noted that this year's articles, as well as those from previous years, are available on our web site.

Joseph F. Hoffman
Editor

1. Chapman RW. 2001. EcoGenomics—a consilience for comparative immunology? *Dev. Comp. Immunol.* 25:549–51
2. Hochachka PW, Mommsen TP, Walsh PJ, eds. 2001. Knowledge explosion in comparative physiology and biochemistry: its causes and its future. *Comp. Biochem. Physiol. B* 130: 125–26
3. Chapman RW, Almeida J. 2001. Response to the editors Peter W. Hochachka, Thomas P. Mommsen and Patrick J. Walsh. *Comp. Biochem. Physiol.* 130:133–34
4. Hall JE. 2001. The promise of translational physiology. *Am. J. Physiol. Cell Physiol.* 281:C1411–C14
5. Murer H. 2001. From the Editor-in-Chief. *Pflügers Arch.* 443:2

Annual Review of Physiology,
Volume 64, 2002

CONTENTS

ERRATA
An online log of corrections to *Annual Review of Physiology* chapter may be found at http://physiol.annualreviews.org/errata.shtml

OTHER REVIEWS OF INTEREST TO PHYSIOLOGISTS

From the *Annual Review of Neuroscience*, Volume 25 (2002):

The Role of Notch in Promoting Glial and Neural Stem Cell Fates, Nicholas Gaiano and Gord Fishell

AMPA Receptor Trafficking and Synaptic Plasticity, Roberto Malinow and Robert C. Malenka

From the *Annual Review of Pharmacology and Toxicology*, Volume 42 (2002):

Signal Transduction by Cell Adhesion Receptors and the Cytoskeleton: Functions of Integrins, Cadherins, R. L. Juliano

AKAP-Mediated Signal Transduction, Jennifer J. Carlisle Michel and John D. Scott

The Changing Face of the Na^+/H^+ Exchanger, NHE1: Structure, Regulation, and Cellular Actions, L. K. Putney, S. P. Denker, and D. L. Barber

Drug Efficay at G Protein–Coupled Receptors, Terry Kenakin

Dimerization: An Emerging Concept for G Protein–Coupled Receptor Ontogeny and Function, Stephane Angers, Ali Salahpour, and Michel Bouvier

Protein Tyrosine Phosphatases: Structure and Function, Substrate Specificity, and Inhibitor Development, Zhong-Yin Zhang

ANNUAL REVIEWS is a nonprofit scientific publisher established to promote the advancement of the sciences. Beginning in 1932 with the *Annual Review of Biochemistry*, the Company has pursued as its principal function the publication of high-quality, reasonably priced *Annual Review* volumes. The volumes are organized by Editors and Editorial Committees who invite qualified authors to contribute critical articles reviewing significant developments within each major discipline. The Editor-in-Chief invites those interested in serving as future Editorial Committee members to communicate directly with him. Annual Reviews is administered by a Board of Directors, whose members serve without compensation.

Ian Glynn

Annu. Rev. Physiol. 2002. 64:1–18

A HUNDRED YEARS OF SODIUM PUMPING

Ian M. Glynn

Trinity College, Cambridge CB2 1TQ, England; e-mail: img10@cam.ac.uk

Key Words   sodium, potassium, pump, ATPase, Na,K-ATPase

■ **Abstract**   This article gives a history of the evidence (*a*) that animal cell membranes contain pumps that expel sodium ions in exchange for potassium ions; (*b*) that the pump derives energy from the hydrolysis of ATP; (*c*) that it is thermodynamically reversible—artificially steep transmembrane ion gradients make it run backward synthesizing ATP from ADP and orthophosphate; (*d*) that its mechanism is a ping-pong one, in which phosphorylation of the pump by ATP is associated with an efflux of three sodium ions, and hydrolysis of the phosphoenzyme is associated with an influx of two potassium ions; (*e*) that each half of the working cycle involves both the transfer of a phosphate group and a conformational change—the phosphate transfer being associated with the occlusion of ions bound at one surface and the conformational change releasing the occluded ions at the opposite surface.

INTRODUCTION

It has been the convention in recent years for the author of the first chapter in each *Annual Review of Physiology* to take a more historical and more personal view of the topic being discussed than would be appropriate in the rest of the volume. Since it is now seven years since I forsook the sodium pump for the less tractable problems of the origin and machinery of the mind, I am glad to be able to make a virtue of necessity and follow that convention.

THE FIRST FIFTY YEARS

In 1902, Ernest Overton (1) reported a discovery that, he tells us, left him so totally staggered (*vollständig verblüfft*) that he took several hours in the open air to think about it. Overton was an Englishman, and grandson of the man who as a boy had introduced Charles Darwin to beetles. Working in Würzburg, he had been studying the osmotic behavior of frog muscles, and what left him so *verblüfft* was the observation that muscles suspended for some time in an isotonic solution of sucrose were inexcitable. Adding sodium chloride to the solution restored excitability, and although the nature of the anion was not critical, only lithium could substitute for sodium. He concluded that sodium ions must have a specific function, and he suggested that, during a very brief period following stimulation, the muscle membrane becomes permeable to both sodium and potassium ions, leading to an exchange of intracellular potassium for extracellular sodium. There was, he pointed out, a corollary to this hypothesis:

> Consider that, in the course of 70 years, heart muscle cells contract about 24×10^8 times and respiratory muscles about 6×10^8 times. If some sodium ions enter and some potassium ions leave during each contraction, then the differences between internal and external cation concentrations would gradually be levelled out unless there is some mechanism at work which opposes this equilibration. In actual fact, our muscles contain, so far as I am aware, just as much potassium and as little sodium in old age as they do in early youth. (Translation by Bernard Katz.)

What makes this argument for the existence of a sodium pump so remarkable is not the argument itself, which is straightforward, but the premises on which it is based. For Overton was writing before the publication of the famous paper by Bernstein (2), which suggested that the resting potential was roughly a potassium equilibrium potential and that the action potential was the result of a transient loss by the membrane of its selective permeability to potassium ions. How Overton arrived at his hypothesis he doesn't tell us, but I suspect that, like Bernstein's, it owed a great deal to work by Ostwald (3) on so-called precipitation membranes—artificial membranes formed by the precipitation of insoluble material at the interface of two solutions. Ostwald studied the permeability and electrical properties of membranes of copper ferrocyanide, which he formed by putting solutions of copper sulfate and potassium ferrocyanide on opposite sides of thin pieces of parchment. He found that such membranes were permeable to potassium and chloride ions but not to the larger barium or sulfate ions. What is more, when the fluids bathing the membrane contained only a single species of penetrating ion, present in different concentrations on the two sides, he detected transmembrane potentials whose magnitudes fitted the Nernst equation, published the previous year. And Ostwald predicted that analogous behavior in natural membranes would explain not only the electric currents in muscles and nerves but also the mysterious workings of the electric organs of electric fish.

Whether this notion of Overton's debt to Ostwald is right or not, Overton seems to have been equally prescient in his hypothesis and in its corollary. Prescience, though, is recognized only with hindsight, and for Overton's work that hindsight came in the mid-1940s, a decade after his death. Bernstein's paper was published just a month after Overton's and, of course, contains no reference to it; his 1912 book, *Elektrobiologie* (4), has only a brief account of Overton's experiments and no mention of their possible implications. Bayliss's influential *Principles of General Physiology* (5), published in 1915, is similarly inadequate, and Fenn's exhaustive (and exhausting) 1936 review on electrolytes in muscle (6) does not mention the relevant experiments. Andrew Huxley (7) has said that he and Alan Hodgkin were not aware of these experiments when they discovered the overshoot of the action potential in 1939, and he believes that they would have thought of the correct explanation of the overshoot much sooner if they had been.

By 1940, the widely held notion that the high concentration of potassium and the low concentration of sodium within muscle fibers were maintained simply by the impermeability of the membrane to sodium ions was no longer tenable. In experiments on anesthetized rats, Fenn & Cobb (8) had shown that stimulation of the sciatic nerves for 30 min caused the gastrocnemius muscles to lose part of their potassium and take up a roughly equivalent amount of sodium; both changes were largely reversed during a few hours of rest. Using the then newly available radioactive ^{24}Na, Heppel (9) had shown that it took less than an hour for sodium in the muscles of potassium-deprived rats to equilibrate with sodium in the bathing solution. And Steinbach (10) had shown that frog muscles soaked in potassium-free Ringer's solution gradually lost potassium and gained sodium and that the exchange was reversible. Since, in the experiments of both Fenn & Cobb and of Steinbach, the movements of both potassium and sodium during the recovery phase were against the concentration gradients, and these movements were in opposite directions, it seemed to follow either that at least one species of ion can be pumped through the membrane or that much of the potassium inside the fibers is bound to molecules that are unable to penetrate the membrane and prefer potassium to sodium.

Curiously, none of the experimentalists who produced these striking results showed any enthusiasm for the notion of ion pumps, and their papers contain no mention of the relevant work of Overton. Fenn & Cobb suggested that, following excitation, perhaps "only the surface of the fiber breaks down and exchanges its potassium for sodium, the lower layers being still impermeable to sodium"—a suggestion later dismissed by Fenn under the umbrella of "some rather artificial hypotheses." Steinbach has only a brief discussion of the causes of ion movements during the recovery phase, and favors selective binding of potassium to indiffusible organic molecules within the cell. In a slightly later paper (11), he justifies this rejection of the notion of ions being pumped across the membrane, explaining that to postulate the existence of such a pump "removes much of the charm of the old selective permeability idea, since once an auxiliary mechanism must be assumed, the simplicity and clarity of the scheme is destroyed."

Why selective binding is less destructive of the simplicity and clarity of the scheme than ion pumping is not clear, and in the following year, Dean (12), basing his conclusions on much the same experiments, argued strongly for the existence of an outwardly directed sodium pump.

Work on the storage of blood during the Second World War, by Maizels & Patterson in London, and by both Danowski and Harris in the United States, provided independent evidence for the pumping of both sodium and potassium (13–15). During cold-storage, red cells gradually lose potassium and gain sodium, but it was found that these changes could be reversed by replacing the cells in the circulation or by incubating them with glucose at 37°C. Reversal seemed to depend on energy from glycolysis because it was prevented by fluoride or iodoacetate, but not by cyanide or dinitrophenol.

What looked like a serious objection to the idea that the low sodium content of muscle was the result of an outward pumping of sodium was a calculation by Conway (16), in Dublin, suggesting that the energy required would greatly exceed that available. This objection lost its force when Ussing (17), in Copenhagen, suggested that a substantial part of the sodium efflux from muscle might represent a linked one-for-one exchange of intracellular and extracellular sodium ions, which need not require significant energy. Evidence for such an exchange was later found by Keynes & Swan (18) in experiments with frog muscle.

THE LAST FIFTY YEARS

I don't know when I first decided that I wanted to be a doctor. It may have been at the age of eight, when I walked into my grandmother's kitchen and saw one of my aunts sitting at the table dissecting a human brain. (The rules about disposal of body parts were more lax in those days.) She was the second member of the family to "do medicine," her younger brother having led the way, and there was a feeling in the family, as in many Jewish immigrant families, that medicine, both in its worthwhileness and its professional prospects, was the ideal career. We lived in Hackney, and I went to the local elementary school, which had a narrow curriculum but excellent teaching, and where large classes seemed to be compatible with good discipline. From there, through the generosity of the socialist but not too doctrinaire London County Council, I went to the City of London School—one of the few English so-called public (i.e., private) schools that was not a boarding school—where, looking back, I realize that the teaching was always good and that in mathematics, physics, and English it was superb.

When I came up to Cambridge as a medical student in 1946, I was a member of Trinity College. This was not the college my headmaster had suggested, but my mother's greengrocer had a son who was (and is) a distinguished mathematician, and his advice—routed to me through our respective mothers—was "Tell Ian to apply to Trinity because, if he wants to do research later, they have more research Fellowships." This advice was even better than he realized because, as well as research Fellowships, Trinity had Alan Hodgkin as a tutor in physiology. It was only in the previous year that he and Andrew Huxley had returned to Cambridge after their war service and had resumed their work on the nature of conduction in nerves, which they had had to break off at a very exciting stage when Hitler invaded Poland in 1939. By the time I had completed my medical course and returned to Cambridge as a graduate student in 1953, they had essentially solved the nerve problem. The Hodgkin-Huxley theory was as convincing as it was elegant, but it left unanswered an intriguing subsidiary problem. If the energy for each nerve impulse came from the downhill entry of sodium ions and loss of potassium ions, how were the gradients of these ions to be maintained? The question that Overton had raised half a century earlier, almost as an aside in discussing an unsupported and speculative hypothesis, suddenly became pressing. It was the question I decided to work on.

Defining the Problem

Before investigating the mechanism of a pump, it is as well to know just what it pumps and what its immediate source of energy is. Fifty years ago, it was not certain that the same mechanism was responsible for pumping sodium ions outward and potassium ions inward, or that the pumps in nerve and in muscle, which depend on respiration, were the same as the pump in mamalian red cells, which depends on glycolysis.

A possible link between the active movements of sodium and potassium had been suggested by the observation—in human red cells, frog muscle, and invertebrate nerve axons—that sodium efflux was reduced when the extracellular potassium concentration was lowered (19–21). Proof that the linkage not only existed but was tight came from experiments in which both glucose-dependent sodium efflux and glucose-dependent potassium influx were measured in batches of red cells incubated in media with potassium concentrations ranging from 0 to 5 mM (22). When there was no potassium outside the cells, there was no glucose-dependent sodium efflux; as the potassium concentration was increased, there were increases in the glucose-dependent fluxes of both sodium and potassium, and these increases seemed to fit Michaelis curves with similar K_m values. That the linked fluxes were not, as had been thought, $1Na^+:1K^+$ but $3Na^+:2K^+$ was shown by Post & Jolly (23), who monitored the net movements of sodium and potassium when cold-stored human red cells were incubated in media containing concentrations of sodium and potassium similar to those in the cells, so that net passive movements could be assumed to be small.

The identity, or near identity, of the pumps in different tissues became clearer following the discovery by Schatzmann (24) that the cardiac glycoside strophanthin inhibited the reuptake of potassium and expulsion of sodium by cold-stored red cells, without affecting either oxygen consumption or lactic acid production. This strongly suggested that the action was on the pumping mechanism rather than the energy supply; so when others found that active transport of sodium and potassium in other tissues was also sensitive to cardiac glycosides, it seemed likely that a similar mechanism was involved. The potency and specificity of the cardiac glycosides also made it possible to use them to estimate the number of sodium pumps on a cell membrane (25), and they have, of course, been a major tool for investigating the pump mechanism.

The work of Kalckar, Lipmann, and others in the 1930s and 1940s had made it clear that energy from both glycolysis and respiration was made available as the energy-rich phosphate bonds of ATP. The most straightforward interpretation of the dependence of sodium pumping in some tissues on glycolysis and in other tissues on respiration was, therefore, that pumps in both kinds of tissue used ATP as a fuel. Support for this view came from experiments showing that dinitrophenol, which uncouples respiration from ATP synthesis, inhibited pumping in tissues that rely on respiration, whereas arsenate, which uncouples glycolysis from ATP synthesis, inhibited pumping in tissues that rely on glycolysis (21, 26, 27). Direct evidence that ATP fueled the pump came first from experiments by Gárdos (28), working in Budapest, who showed that resealed red cell ghosts containing trapped ATP could accumulate potassium actively in the absence of any other substrate. Although Gárdos's resealed ghosts were heavily contaminated with intact cells, these were unlikely to have been responsible for most of the accumulation since it continued even in the presence of arsenate. Later, Dunham (29) showed that removal of ATP from iodoacetate-poisoned red cells by the addition of glucose prevented cation transport, and Caldwell & Keynes (30) showed that injecting ATP or arginine phosphate into cyanide-poisoned squid axons restored cation transport.

The Identity of the Sodium Pump and the (Na⁺+K⁺)-Activated ATPase—"All Roads Lead to Rome"

In 1956, like others working on the sodium pump, I was thinking mainly in terms of circulating, lipid-soluble, cation-selective carriers. But because the sodium pump seemed so much better at discriminating between sodium and potassium ions than any known cation-binding agent, I wondered whether this discrimination might depend not (or not only) on differences of affinity, but on differences of the reactivity of an enzyme depending on whether sodium or potassium ions were bound to it. (A somewhat analogous hypothesis would be that most of us are dark-haired because, as Anita Loos tells us, "Gentlemen Prefer Blondes," "But-Gentlemen Marry Brunettes.") In my first full paper (22), I therefore listed eight enzymes that showed striking discrimination between sodium and potassium ions, and I noted that several of them catalyzed reactions that involved the transfer of large amounts of energy. The following year, in a long review on red cells written while I was doing my national service as an under-employed doctor in the Royal Air Force, I made the same point, and finished the article by suggesting that investigating the role of the ATPase in the red cell membrane might be fruitful (31).

What I did not know, when I wrote that, was that Jens Christian Skou, in Aarhus, had already done, and was about to publish, experiments showing that fragments of crab nerve membrane possessed Mg-ATPase activity that was stimulated greatly by the joint presence of sodium and potassium ions (32). Skou pointed out in the last paragraph of his paper, "the crab-nerve ATPase ... seems to fulfil a number of conditions that must be imposed on an enzyme which is thought to be involved in the active extrusion of sodium ions from the nerve fibre." The case for a connection between this ATPase and the sodium pump was strengthened when, at the suggestion of Post, Skou tested the effect of ouabain on the ATPase (33, 34). Subsequently, work by Post and his colleagues at Vanderbilt, and by Ned Dunham and me in Cambridge, showed that similar Na,K-ATPase activity was present in the human red cell membrane and that the properties of the ATPase closely paralleled the properties of the pump (35, 36). In particular, experiments on resealed red cell ghosts showed that it was intracellular sodium ions and extracellular potassium ions that activated the ATPase (37). Skou himself found that Na,K-ATPases similar to that in crab nerve could be found in the mammalian brain and kidney (38), and Bonting & Caravaggio, at Nijmegen, found a striking correlation between the fluxes of sodium and potassium and the activity of the membrane Na,K-ATPase in six different tissues of the cat (39). In the late 1960s and early 1970s, even more striking evidence for the identity came from experiments showing that the pump could be driven backward to synthesize ATP (40, 41); that an antibody to partially purified Na,K-ATPase inhibited the pump (42); and that, when ATP was added to a suspension of artificial lipid vesicles with partially purified Na,K-ATPase incorporated into their membranes, movements of sodium and potassium ions (with appropriate stoichiometry) could be detected (43, 44). By then, though, the identity of the pump and the ATPase was hardly in question.

An intriguing feature of Skou's important discovery is that, as he himself makes clear in a commentary written 32 years later (33), he had not been primarily concerned with the sodium pump. He had been investigating the mechanism of local anesthetics and had wanted a lipoprotein enzyme that would sit in a lipid monolayer and whose activity he could follow when the surface pressure in the monolayer was increased by the insertion of anesthetic molecules. Having discovered the properties of the enzyme, he realized that what he was looking at was probably the sodium pump, but he points out that the statement in the introduction to the 1957 paper, "A further study on the ATPase in nerves and its possible role in the active outward transport of sodium ions seems warranted," was "a subsequent rationalization." And in deciding the title of that paper, he rejected the phrase "sodium pump" as "too provocative."

A Peculiar Difficulty

Why was Skou's 1957 paper so important? After all, a devil's advocate could argue that there was already both indirect and direct evidence that the linked Na^+/K^+ pump was fueled by ATP. If the energy in the terminal phosphoryl bond of ATP is to be made available for osmotic work, the end products are likely to be ADP and orthophosphate; so it should not have been surprising that the pumping machinery acted as an ATPase.

The answer is that Skou's work opened up a bundle of new approaches to elucidating the pump mechanism. Transport enzymes in general, and the sodium pump in particular, present a peculiar difficulty to the investigator. Most enzymes convert substrates into products that are chemically different from the substrates, and therefore easily distinguishable. In contrast, the sodium and potassium ions that are pumped across the membrane remain sodium and potassium ions, and the conversion of an intracellular ion into an extracellular ion, or vice versa, can be detected only by working with intact cells. As long as the only way of recognizing the presence of the pump was by observing the movements of the ions across the cell membrane, investigators were restricted to working with intact cells, which is, of course, why resealed red cell ghosts and squid giant axons, whose contents could be regulated to a considerable extent by the experimenter, were so important. But as soon as the pump was identified as an $(Na^+ + K^+)$-activated ATPase, with recognizably different substrates and products (and, as it later turned out, with recognizably different intermediate forms), it could be studied in membrane fragments rich in pumps and, ultimately (45), in more-or-less purified and even solubilized pump preparations.

Working Out the Working Cycle

The reaction catalyzed by an enzyme that requires four substrates (Na_i^+, K_o^+, ATP and H_2O) and produces four products (Na_o^+, K_i^+, ADP and orthophosphate) is unlikely to occur in a single step. Understanding the mechanism of the pump, therefore, involves two stages: discovering the sequence of reaction steps that make up the working cycle and showing how the atomic structure of the pump makes these reactions possible. There is at present a great deal of sophisticated work going on concerned with the second stage, and accounts of this, together with accounts of recent work on the electrical properties of the pump, of the nature and functions of different isoforms of the Na,K-ATPase, and of the possible role of endogenous inhibitors of the pump, can be found in Taniguchi & Kaya's edited volume on Na/K-ATPase and related ATPases (46, see also 47; for recent work on the related calcium-ATPase, see 48). In the very limited space available here, only the history of the first stage is discussed. Even with this restriction, the relevant literature is extensive, and much important work will, inevitably, be omitted. For fuller reviews see 49–51.

By what sequence of steps is the hydrolysis of ATP coupled to the active efflux of three sodium ions and the active influx of two potassium ions? The answer to this question came largely from complementary studies of phosphate-group transfer in membrane fragments and of ion fluxes in intact cells.

The first suggestion that the enzyme becomes phosphorylated during the working cycle came from Skou, who found that membrane fragments from crab nerves catalyzed an ATP-ADP exchange (34). That exchange, however, appeared not to need sodium ions or to be inhibited by ouabain; its relation to the sodium pump was, therefore, uncertain. In the early 1960s, Wayne Albers and his colleagues at the National Institutes of Health, and Post and his colleagues at Vanderbilt, exposed membrane fragments from electric-eel electric organ or from guinea pig kidney to $[\gamma^{32}P]ATP$ and followed the phosphorylation and dephosphorylation of the enzyme under different conditions (52–54). They found that sodium ions were necessary for (or greatly accelerated) phosphorylation, and potassium ions greatly accelerated the hydrolysis of the phosphoenzyme yielding orthophosphate. Later the phosphorylated group was shown to be a β-aspartyl carboxyl (55). Both groups pointed out that the dependence of phosphorylation on sodium ions and of dephosphorylation on potassium ions suggested that the outward movement of sodium involved a phosphorylation step and the inward movement of potassium involved hydrolysis of the phosphoenzyme. There were, though, interesting differences in their interpretations.

Albers and his colleagues thought that the enzyme functioned in turn as a Mg^{2+}-activated kinase, a Na^+-activated transferase, and a K^+-activated phosphatase.

Following phosphorylation of the enzyme, the negatively charged phosphate group was, they supposed, transferred to another site (or more probably a succession of sites) in a channel whose environment of fixed charges was such that only Na^+ ions could act as counter ions. At the outer end of the channel, the K^+-dependent phosphatase action of the enzyme released the phosphate group as orthophosphate, allowing the Na^+ ions to escape. They also suggested, more tentatively, that negatively charged sialic acids at the cell surface might prevent the orthophosphate from leaving and that its diffusion back to the cell interior along a K^+-selective channel could account for the coupled inward transport of potassium.

Post and his colleagues combined the notion of a sodium-phosphoenzyme complex with Trevor Shaw's cyclical carrier model (see 31) to produce a scheme in which both the phosphorylated and unphosphorylated forms of the enzyme could exist in alternative configurations, one with the cation-binding sites facing inward, the other with these sites facing outward. Sodium ions at the inner face of the membrane were supposed to catalyze the formation of the phosphenzyme and be carried outward bound to it. Potassium ions at the outer face of the membrane were supposed to catalyze the hydrolysis of the phosphoenzyme and be carrried inward bound to the dephosphoenzyme.

Despite the differences between the two interpretations, the basic notion that the working cycle of the pump consisted of a phosphorylation step associated with sodium efflux and a hydrolysis step associated with potassium influx became known as the Albers-Post scheme and provided a fertile base for further elucidation of the working cycle.

The reversibility of the pump, discovered in 1966 (40, 41), implied that no individual step in the cycle could be too far from equilibrium to be reversed by appropriate changes in ligand concentrations. It followed that, if the Albers-Post scheme were correct, it should be possible to find conditions in which an inward movement of sodium ions was associated with the transfer of a phosphate group from the phosphoenzyme to ADP and other conditions in which an outward movement of potassium ions was associated with phosphorylation of the enzyme by orthophosphate.

It was already known, from the work of Albers and his colleagues at NIH (52), that when preparations of Na,K-ATPase from the *Electrophorus* electric organ are incubated in high-sodium, potassium-free media containing both ATP and ^{14}C-labeled ADP, they catalyze an ATP-ADP exchange, suggesting the shuttling of a phosphate group between nucleotide and enzyme. Within a few years, experiments on red cells, frog muscle, and squid axons showed that when these cells were incubated in high-sodium potassium-free media there was an ouabain-sensitive exchange of internal and external sodium ions (56–59). In red cells, the exchange was close to one-for-one and showed a marked asymmetry, with a high affinity for sodium at the inner face of the membrane and a low affinity at the outer face. Significantly, it did not occur unless the cells contained both ATP and ADP; and although the ATP was not consumed, replacement of ATP by its non-phosphorylating β,γ-imido analog was ineffective (60, 61). The implication was that the sodium-sodium exchange was accompanied by an ATP-ADP exchange, and this was confirmed later (62, 63). The obvious interpretation was that the outward movement of sodium involved the transfer of a phosphate group from ATP to the enzyme, and the inward movement involved its transfer back to ADP to form ATP.

There was, though, an interesting and important twist to the story. In their 1966 experiments on ATP-ADP exchange by electric organ Na,K-ATPase, Fahn and his colleagues (64) noticed that the exchange was unaffected or slightly faster if the enzyme was pretreated with oligomycin, despite the fact that oligomycin inhibits electric organ Na,K-ATPase. They explained this by supposing (a) that the phosphoenzyme existed in two interconvertible forms (in a later nomenclature, E_1P and E_2P); (b) that only the first of these to be formed (E_1P) could react with ADP, and only the second (E_2P) was hydrolyzed in the presence of potassium; and (c) that oligomycin acted by preventing the conversion of E_1P to E_2P. So when Patricio Garrahan and I (65) found that oligomycin inhibited the ouabain-sensitive sodium:sodium exchange that occurs when red cells are incubated in high-sodium, potassium-free media, the implication was that the outward movement of sodium ions required not just phosphorylation but the conversion of E_1P to E_2P. That made sense if the conversion involved a conformational change in the phosphoenzyme, because such a change seemed to be just what was needed to release the sodium ions to the exterior.

What about the predicted association between an outward movement of potassium ions through the pump and phosphorylation of the pump by orthophosphate? Early experiments on the effects of cardiac glycosides on red cells showed that, under fairly physiological conditions, more than a fifth of the potassium efflux was inhibited by digoxin (25); but at that time there was no way of knowing whether this part of the potassium efflux was the result of a partial failure of discrimination between sodium and potassium by the forward-running pump or of occasional reversal of the part of the cycle concerned with potassium entry. Later experiments, both on intact red cells and on resealed red cell ghosts, showed that the glycoside-sensitive potassium efflux was part of an exchange of internal and external potassium ions (66–69). This exchange had interesting features: It was roughly one-for-one; it had strikingly asymmetric affinities for potassium (high outside, low inside); and it occurred only if the cells contained magnesium ions, orthophosphate, and a high concentration of ATP (or ADP, deoxy-ATP, CTP, or a non-phosphorylating analog of ATP). The need for orthophosphate strongly suggested that the outward movement of potassium involved a reversal of the dephosphorylation step, and studies on Na,K-ATPase preparations from pig kidney and from *Electrophorus* electric organ demonstrated the exchange of ^{18}O between orthophosphate and water that would be expected to result from the alternate formation of phosphoenzyme (from orthophosphate) and its hydrolysis (70). The ability of non-phosphorylating analogs of ATP to support potassium-potassium exchange pointed to a role for ATP in the second half of the pump cycle different from its phosphorylating role in the first half.

Two Conformations of the Dephosphoenzyme

Work in the early 1970s (71, 72) had shown that the affinity of Na,K-ATPase preparations for ATP was high in potassium-free media and dropped dramatically when potassium ions were added. It seemed likely that this drop in affinity reflected a change in conformation, and proof came in 1975, when Peter Jørgensen, at Aarhus, showed that tryptic digestion of kidney Na,K-ATPase yielded different products depending on whether the enzyme was in a sodium medium (E_1 form) or a potassium medium (E_2 form) (73). Later, Jørgensen & Petersen (74) showed that trypsin attacks phosphoenzyme in the E_2P form in the same way as it attacks the E_2 form, implying that, so far as the accessibility of peptide bonds to trypsin is concerned, the conformations must be similar. Comparisons of intrinsic tryptophan fluorescence supported this view and also suggested a similarity between the conformations of E_1 and E_1P (74, 75).

Occlusion of Ions During Their Transport Across the Membrane

It has always seemed possible that, at some stage during their passage across the membrane, transported ions are occluded within the pump molecule, unable to escape to either surface without either a conformational or a chemical change in that molecule. (For a general review of occlusion in enzyme pumps, see 76). The first direct, or almost direct, evidence for occlusion in the sodium pump came from ingenious experiments by Post and his colleagues, published in 1972 (77). Although the sodium pump is rather specific for sodium, it is much less specific for potassium, so various congeners of potassium, including rubidium and lithium, can substitute for potassium. Post and his colleagues found that the rate at which it was possible to rephosphorylate kidney Na,K-ATPase that had just been dephosphorylated differed depending on whether lithium or rubidium had been used to catalyze the hydrolysis. This was true even if the experiment was done in such a way that the conditions during rephosphorylation were identical. In other words, the enzyme appeared to remember which ion had catalyzed the hydrolysis. To explain this memory, they suggested that the catalyzing ions became occluded within the enzyme at the moment of hydrolysis and were released only later after a slow conformational change. Because they found that the enzyme was available for rephosphorylation sooner if higher concentrations of ATP were used, they suggested that the binding of ATP at a low-affinity site accelerated the conformational change that released the occluded ions. Here, then, was the likely role of ATP (or its non-phosphorylating analogs) in supporting potassium-potassium exchange.

At that time there was no reason to suppose that the form of the dephospho-enzyme containing occluded potassium ions could exist more than transiently. In the late 1970s, Steve Karlish, David Yates, and I were measuring the rates of interconversion between the E_1 and E_2 forms of unphosphorylated pig kidney Na,K-ATPase when the sodium or potassium concentrations in the medium were rapidly changed. Because the two forms have such different affinities for ATP, we could monitor the changes in conformation using stopped-flow fluorimetry and formycin triphosphate (FTP), a fluorescent analog of ATP that the enzyme treats much like ATP but which has the convenient property of fluorescing more strongly when it is bound. What we found was that the direction and magnitude of the changes in fluorescence that followed changes in the composition of the suspending medium were much as we expected, but the rates were not (78). In particular, when excess sodium was added to enzyme suspended in a low-potassium, sodium-free, magnesium-free medium containing a very low concentration of FTP, the rise in fluorescence was astonishingly slow, with a rate-constant, at room temperature, of about 0.25 s^{-1}. The rate increased, in a roughly linear fashion, as the concentration of FTP was increased up to 24 μM, a concentration beyond which measurements of binding became inaccurate because the fraction bound was so small.

The combination of a slow conformational change and its acceleration by a nucleotide acting with a low affinity and presumably without phosphorylating (since the medium lacked magnesium ions), was so reminiscent of Post's hypothetical occluded-potassium form that we wondered whether that was what we were looking at. In other words, did dephosphoenzyme in a low-potassium, sodium-free medium contain occluded potassium ions? To answer that question Luis Beaugé, and I suspended pig kidney Na,K-ATPase in a sodium-free solution containing radioactive rubidium and forced it through a small column of cation exchange resin at a rate that was slow enough for the resin to remove nearly all of the free rubidium ions yet fast enough for the enzyme to emerge in less than 2 s, i.e., within a period much smaller than the time constant for the conformational change in the fluorescence experiments (79). The enzyme did indeed carry rubidium ions through the column, about two rubidium ions per phosphorylation site, and a high concentration of ATP or of sodium ions prevented this effect. Later experiments (80), varying the rates of flow through the column and comparing the rates of release of different potassium congeners with the rates of conformational change determined by stopped-flow fluorimetry using several different fluorescent probes, supported the hypothesis that the conformational change releases the occluded ions.

In all these experiments, the occluded form was made directly from the unphosphorylated enzyme, but it was also possible to demonstrate its formation by the normal physiological route (81). Enzyme suspended in a Tris-medium containing magnesium, sodium, and a low concentration of radioactive rubidium was passed first through a thin layer of Sephadex containing 40 μM ATP (or ADP in the controls) and then through the cation exchange resin. With ATP, but not with ADP, roughly two rubidium ions per phosphorylation site were carried through the column. Clearly, there were two routes to the occluded-potassium form: the direct route by reversal of the final step of the normal cycle, picking up potassium ions on low-affinity sites at the inner surface of the membrane, and the physiological route, using the forward-running cycle and picking up potassium ions on high-affinity sites at the outer surface of the membrane (see 82).

The success of the rapid cation-exchange method for detecting occlusion of potassium congeners prompted us to use the same method to test the hypothesis that the E_1P form of the phosphoenzyme contains occluded sodium ions, a hypothesis that Joseph Hoffman and I had proposed 13 years earlier to explain the properties of the sodium-sodium exchange in red cells (60). The idea was to generate E_1P by phosphorylation of the enzyme in the presence of ^{22}Na-labeled sodium ions, to force the phosphorylated enzyme rapidly down a column of cation exchange resin, and to measure the radioactivity of the effluent. Because E_1P changes conformation to E_2P spontaneously, it was necessary to block that change, and this was done by pretreating the enzyme with either N-ethyl maleimide or α-chymotrypsin (83). Dephosphorylation of the E_1P by ADP was prevented by working close to $0°C$ and exposing the enzyme only very briefly to ATP at a low concentration in a thin layer of Sephadex above the resin. Evidence for the occlusion of sodium ions was found with both pretreatments, and the results with the α-chymotrypsin-treated enzyme were accurate enough to show that close to three sodium ions were occluded per phosphorylation site (84).

Later work has shown that the outward release of potassium ions from the occluded-potassium form of the enzyme, when orthophosphate is added, is an ordered release (85–87) and that the outward release of sodium ions from the occluded-sodium form of the enzyme occurs in two stages (88–90). Despite these complications, the working cycle remains an elaboration of the old Albers-Post ping-pong scheme. Another possible complication, the hypothesis that the pump is an $(\alpha\beta)_2$ diprotomer (or even a tetraprotomer) showing half-of-the-sites reactivity, has a long and interesting history, and there is evidence in its favor (for references see 91, 92 and papers in 46). But although it is clear that oligomer formation can occur, obligatory half-of-the-sites reactivity is difficult to reconcile with experiments showing (a) that the pump in the red cell membrane is a monomeric $\alpha\beta$ protomer (93); (b) that monomeric $\alpha\beta$ protomers and $(\alpha\beta)_2$ diprotomers, prepared from the same solubilized dog-kidney Na,K-ATPase, had the same specific activity (94); and (c) that, when a purified preparation of duck salt gland Na,K-ATPase with very high specific activity was exposed to $^{32}P_i$ in the presence of ouabain, close to one phosphate was incorporated per $\alpha\beta$ unit (95).

"Remembrance of Things Past"

Aaron Klug tells us that Rosalind Franklin once said to him, "What is the point of doing all this work if you don't get some fun out of it?" So, was the work I was engaged in fun? The answer is that it sometimes was, and the degree of enjoyment was not simply related to the significance of the results. The early experiments showing a tight link between active sodium efflux and active potassium influx, and the experiment to count the number of pumps in the red cell membrane, were particularly enjoyable because I was a research student and the whole business of discovery was new to me. On the other hand, proving that the pump could be driven backward, synthesizing its own fuel, was less exciting than it ought to have been. This was partly because, initially, the effect was a small one; after a string of successful trials and no failures, Patricio Garrahan and I were convinced, but by then the novelty had worn off. A second reason was that Trevor Shaw had earlier attempted to prove the same point more elegantly by injecting an extract of firefly tails into a squid axon, arranging suitable gradients of sodium and potassium across the axonal membrane, and measuring the light emitted. Although our experiments were successful and his were not, we had the vague feeling that, where he had failed to produce a rabbit from a hat, we had succeeded only in producing a rabbit from a hutch. For me, the most enjoyable experiments were those using stopped-flow fluorimetry to follow the change in enzyme conformation that releases occluded potassium ions. For years I had worked surrounded by electrophysiologists whose exciting results flashed onto their oscilloscope screens during the course of their experiments. My exciting results came, if at all, during sessions with a slide-rule or calculator. So the pleasure of watching the spot on an oscilloscope screen languidly trace a perfect, if tremulous, exponential curve was a pleasure all the greater for being of a kind so long denied. And finally, of course, the occlusion experiments were a pleasure, both because such simple experiments provided such clear-cut results and because in examining the behavior of the occluded-ion forms we felt we were getting down to the nitty gritty of the pump. That is, though, probably the wrong metaphor to use in connection with a pump that is efficient enough to be thermodynamically reversible and that has now kept physiologists happily occupied for a century.

Visit the Annual Reviews home page at www.AnnualReviews.org

LITERATURE CITED

1. Overton E. 1902. Beiträge zur allgemeinen Muskel- und Nerven-Physiologie. II. Über die Unentbehrlichkeit von Natrium- (oder Lithium-) Ionen für den Contractionsact des Muskels. *Pflügers Arch.* 92:346–86

2. Bernstein J. 1902. Untersuchengen zur Thermodynamik der bioelektrische Ströme. *Pflügers Arch.* 92:521–62

3. Ostwald W. 1890. Elektrische Eigenschaften halbdurchlässiger Scheidenwände. *Z. Phys. Chem.* 6:71–96

4. Bernstein J. 1912. *Elektrobiologie.* p. 101. Braunschweig: Vieweg

5. Bayliss WM. 1915. *Principles of General Physiology.* London: Longmans Green

6. Fenn WO. 1936. Electrolytes in muscle. *Physiol. Rev.* 16:450–87

7. Huxley AF. 1999. Overton on the indis-

pensability of sodium ions. *Brain Res. Bull.* 50:307–8

8. Fenn WO, Cobb DM. 1936. Electrolyte changes in muscle during activity. *Am. J. Physiol.* 115:345–56

9. Heppel LA. 1940. The diffusion of radioactive sodium into the muscles of potassium-deprived rats. *Am. J. Physiol.* 128:449–54

10. Steinbach HB. 1940. Sodium and potassium in frog muscle. *J. Biol. Chem.* 133:695–701

11. Steinbach HB. 1940. Electrolyte balance of animal cells. *Cold Spring Harbor Symp. Quant. Biol.* 8:242–54

12. Dean RB. 1941. Theories of electrolyte equilibrium in muscle. *Biol. Symp.* 3:331–48

13. Maizels M, Patterson JH. 1940. Survival

of stored blood after transfusion. *Lancet* 2:417–20

14. Danowski TS. 1941. The transfer of potassium across the human blood cell membrane. *J. Biol. Chem.* 139:693–705

15. Harris JE. 1941. The influence of the metabolism of human erythrocytes on their potassium content. *J. Biol. Chem.* 141:579–95

16. Conway EJ. 1946. Ionic permeability of skeletal muscle fibres. *Nature* 157:715–17

17. Ussing HH. 1947 Interpretation of the exchange of radio-sodium in isolated muscle. *Nature* 160:262–63

18. Keynes RD, Swan RC. 1959. The effect of external sodium concentration on the sodium fluxes in frog skeletal muscle. *J. Physiol.* 147:591–625

19. Harris EJ, Maizels M. 1951. The permeability of human red cells to sodium. *J. Physiol.* 113:506–24

20. Keynes RD. 1954. The ionic fluxes in frog muscle. *Proc. R. Soc. London Ser. B* 142:359–82

21. Hodgkin AL, Keynes RD. 1955. Active transport of cations in giant axons from *Sepia* and *Loligo*. *J. Physiol.* 128:28–60

22. Glynn IM. 1956. Sodium and potassium movements in human red cells. *J. Physiol.* 134:278–310

23. Post RL, Jolly PC. 1957. The linkage of sodium, potassium and ammonium active transport across the human erythrocyte membrane. *Biochim. Biophys. Acta* 25:118–28

24. Schatzmann HJ. 1953. Herzglycoside als Hemmstoffe für den aktiven Kalium und Natrium Transport durch die Erythrocytenmembran. *Helv. Physiol. Acta* 11:346–54

25. Glynn IM. 1957. The action of cardiac glycosides on sodium and potassium movements in human red cells. *J. Physiol.* 136:148–73

26. Maizels M. 1954. Active cation transport in erythrocytes. *Symp. Soc. Exp. Biol.* 8:202–27

27. Straub FB. 1953. Über die Akkumulation der Kaliumionen durch menschliche Blutkörperchen. *Acta Physiol. Acad. Sci. Hung.* 4:235–40

28. Gárdos G. 1954. Akkumulation der Kaliumionen durch menschliche Blutkörperchen. *Acta Physiol. Acad. Sci. Hung.* 6:191–99

29. Dunham ET. 1957. Linkage of active cation transport to ATP utilization. *Physiologist* 1:23

30. Caldwell PC, Keynes RD. 1957. The utilization of phosphate bond energy for sodium extrusion. *J. Physiol.* 137:12–13P

31. Glynn IM. 1957. The ionic permeability of the red cell membrane. *Prog. Biophys.* 8:241–307

32. Skou JC. 1957. The influence of some cations on an adenosinetriphosphatase from peripheral nerves. *Biochim. Biophys. Acta* 23:394–401

33. Skou JC. 1989. The identification of the sodium-pump as the membrane-bound Na$^+$/K$^+$-ATPase: a commentary by J.C. Skou. *Biochim. Biophys. Acta* 1000:435–38

34. Skou JC. 1960. Further investigations on a Mg^{++} + Na$^+$-activated adenosinetriphosphatase, possibly related to the active, linked transport of Na$^+$ and K$^+$ across the nerve membrane. *Biochim. Biophys. Acta* 42:6–23

35. Post RL, Merritt CR, Kinsolving CR, Albright CD. 1960. Membrane adenosine triphosphatase as a participant in the active transport of sodium and potassium in the human erythrocyte. *J. Biol. Chem.* 235:1796–802

36. Dunham ET, Glynn IM. 1961. Adenosine-triphosphatase activity and the active movements of alkali metal ions. *J. Physiol.* 156:274–93

37. Glynn IM. 1961. Activation of adenosine-triphosphatase activity in a cell membrane by external potassium and internal sodium. *J. Physiol.* 160:18–19P

38. Skou JC. 1962. Preparation from mammalian brain and kidney of the enzyme system involved in active transport of Na$^+$ and K$^+$. *Biochim. Biophys. Acta* 58:314–25

39. Bonting SL, Caravaggio LL. 1963. Studies on Na:K activated ATPase. V. Correlation of enzyme activity with cation flux in six tissues. *Arch. Biochem. Biophys.* 101:37–46

40. Garrahan PJ, Glynn IM. 1967. The incorporation of inorganic phosphate into adenosine triphosphate by reversal of the sodium pump. *J. Physiol.* 192:237–56

41. Lew VL, Glynn IM, Ellory JC. 1970. Net synthesis of ATP by reversal of the sodium pump. *Nature* 225:865–66

42. Jorgensen PL, Hansen O, Glynn IM, Cavieres JD. 1973. Antibodies to pig kidney (Na$^+$ + K$^+$)-ATPase inhibit the Na$^+$ pump in human red cells provided they have access to the inner surface of the cell membrane. *Biochim. Biophys. Acta* 291:795–800

43. Sweadner KJ, Goldin SM. 1975. Reconstitution of active ion transport by the sodium and potassium ion-stimulated adenosine triphosphatase from canine brain. *J. Biol. Chem.* 250:4022–24

44. Hilden S, Hokin LE. 1975. Active potassium transport coupled to active sodium transport in vesicles reconstituted from purified sodium and potassium ion-activated adenosine triphosphatase from the rectal gland of *Squalus acanthias*. *J. Biol. Chem.* 250:6296–303

45. Jørgensen PL. 1982. Mechanism of the Na$^+$,K$^+$ pump. Protein structure and conformations of the pure (Na$^+$ + K$^+$)-ATPase. *Biochim. Biophys. Acta* 694:27–68

46. Taniguchi K, Kaya S. 2000. Na/K-ATPase and related ATPases. *Proc. 9th Int. Conf. Na/K-ATPase and Related ATPases. Sapporo, Japan 1999.* Amsterdam: Elsevier. 771 pp.

47. Lingrel JB, Kuntzweiler T. 1994. Na$^+$,K$^+$-ATPase. *J. Biol. Chem.* 269:19659–62

48. Toyoshima C, Nakasako M, Nomura H, Ogawa H. 2000. Crystal structure of the calcium pump of sarcoplasmic reticulum at 2.6 Å resolution. *Nature* 405:647–55

49. Glynn IM. 1985. The Na$^+$,K$^+$-transporting adenosine triphosphatase. In *The Enzymes of Biological Membranes*, ed. AN Martonosi, 3:35–114. New York/London: Plenum. 676 pp. 2nd. ed.

50. Skou JC, Esmann M. 1992. The Na,K-ATPase. *J. Bioenerg. Biomembr.* 24:249–61

51. Robinson JD. 1997. *Moving Questions: A History of Membrane Transport and Bioenergetics.* New York: Oxford. 373 pp.

52. Albers RW, Fahn S, Koval GJ. 1963. The role of sodium ions in the activation of *Electrophorus* electric organ adenosine triphosphatase. *Proc. Natl. Acad. Sci. USA* 50:474–81

53. Charnock JS, Post RL. 1963. Evidence of the mechanism of ouabain inhibition of cation activated adenosine triphosphatase. *Nature* 199:910–11

54. Post RL, Sen AK, Rosenthal AS. 1965. A phosphorylated intermediate in adenosine triphosphate-dependent sodium and potassium transport across kidney membranes. *J. Biol. Chem.* 240:1437–45

55. Post RL, Kume S. 1973. Evidence for an aspartyl phosphate residue at the active site of sodium and potassium ion transport adenosine triphosphatase. *J. Biol. Chem.* 248:6993–7000

56. Garrahan PJ, Glynn IM. 1967. The behaviour of the sodium pump in red cells in the absence of external potassium. *J. Physiol.* 192:159–74

57. Garrahan PJ, Glynn IM. 1967. Factors affecting the relative magnitudes of the sodium:potassium and sodium:sodium exchanges catalysed by the sodium pump. *J. Physiol.* 192:189–216

58. Keynes RD, Steinhardt RA. 1968. The components of the sodium efflux in frog muscle. *J. Physiol.* 198:581–99

59. Baker PF, Blaustein MP, Keynes RD, Manil J, Shaw TI, Steinhardt RA. 1969. The ouabain-sensitive fluxes of sodium and potassium in squid giant axons. *J. Physiol.* 200:459–96

60. Glynn IM, Hoffman JF. 1971. Nucleotide requirements for sodium-sodium exchange

catalysed by the sodium pump in human red cells. *J. Physiol.* 218:239–56

61. Cavieres JD, Glynn IM. 1979. Sodium-sodium exchange through the sodium pump: the roles of ADP and ATP. *J. Physiol.* 297:637–45

62. Cavieres JD. 1980. Extracellular sodium stimulates ATP-ADP exchange by the sodium pump. *J. Physiol.* 308:57P

63. Kaplan JH, Hollis RJ. 1980. External Na dependence of ouabain-sensitive ATP-ADP exchange initiated by photolysis of intracellular caged-ATP in human red cell ghosts. *Nature* 288:587–89

64. Fahn S, Koval GJ, Albers RW. 1966. Sodium-potassium-activated adenosine triphosphatase of *Electrophorus* electric organ. I. An associated sodium-activated transphosphorylation. *J. Biol. Chem.* 241: 1882–89

65. Garrahan PJ, Glynn IM. 1967. The stoichiometry of the sodium pump. *J. Physiol.* 192:217–35

66. Glynn IM, Lew VL, Lüthi U. 1970. Reversal of the potassium entry mechanism in red cells, with and without reversal of the entire pump cycle. *J. Physiol.* 207:371–91

67. Simons TJB. 1974. Potassium-potassium exchange catalysed by the sodium pump in human red cells. *J. Physiol.* 237:123–55

68. Simons TJB. 1975. The interaction of ATP analogs possessing a blocked γ-phosphate group with the sodium pump in human red cells. *J. Physiol.* 244:731–39

69. Sachs JR. 1981. Mechanistic implications of the potassium-potassium exchange carried out by the sodium-potassium pump. *J. Physiol.* 316:263–77

70. Dahms AS, Boyer PD. 1973. Occurrence and characteristics of ^{18}O exchange reactions catalyzed by sodium- and potassium-dependent adenosine triphosphatases. *J. Biol. Chem.* 248:3155–62

71. Hegyvary C, Post RL. 1971. Binding of adenosine triphosphate to sodium and potassium ion-stimulated adenosine triphosphatase. *J. Biol. Chem.* 246:5234–40

72. Nørby JG, Jensen J. 1971. Binding of ATP to brain microsomal ATPase. Determination of the ATP-binding capacity and the dissociation constant of the enzyme-ATP complex as a function of K^+ concentration. *Biochim. Biophys. Acta* 233:104–16

73. Jørgensen PL 1975. Purification and characterization of $(Na^+ + K^+)$-ATPase. V. Conformational changes in the enzyme. Transitions between the Na-form and the K-form studied with tryptic digestion as a tool. *Biochim. Biophys. Acta* 401:399–415

74. Jørgensen PL, Petersen J. 1979. Protein conformations of the phosphorylated intermediates of purified Na,K-ATPase studied with tryptic digestion and intrinsic fluorescence as tools. In *Na,K-ATPase: Structure and Kinetics*, ed. JC Skou, JG Nørby, pp. 143–55. London: Academic

75. Karlish SJD, Yates DW. 1978. Tryptophan fluorescence of $(Na^+ + K^+)$-ATPase as a tool for study of the enzyme mechanism. *Biochim. Biophys. Acta* 527:115–30

76. Glynn IM, Karlish SJD. 1990. Occluded cations in active transport. *Annu. Rev. Biochem.* 59:171–205

77. Post RL, Hegyvary C, Kume S. 1972. Activation by adenosine triphosphate in the phosphorylation kinetics of sodium and potassium ion transport adenosine triphosphatase. *J. Biol. Chem.* 247:6530–40

78. Karlish SJD, Yates DW, Glynn IM. 1978. Conformational transitions between Na^+-bound and K^+-bound forms of $(Na^+ + K^+)$-ATPase, studied with formycin nucleotides. *Biochim. Biophys. Acta* 525:252–64

79. Beaugé LA, Glynn IM. 1979. Occlusion of K ions in the unphosphorylated sodium pump. *Nature* 280:510–12

80. Glynn IM, Hara Y, Richards DE, Steinberg M. 1987. Comparison of rates of cation release and of conformational change in dog kidney Na,K-ATPase. *J. Physiol.* 383:477–85

81. Glynn IM, Richards DE. 1982. Occlusion of rubidium ions by the sodium-potassium pump: its implications for the mechanism

of potassium transport. *J. Physiol.* 330:17–43

82. Blostein R, Chu L. 1977. Sidedness of (sodium, potassium)-adenosine triphosphatase of inside-out red cell membrane vesicles. Interactions with potassium. *J. Biol. Chem.* 252:3035–43

83. Jørgensen PL, Skriver E, Hebert H, Maunsbach AB. 1982. Structure of the Na,K pump: crystallisation of pure membrane-bound Na,K-ATPase and identification of functional domains of the α-subunit. *Ann. NY Acad. Sci.* 402:207–24

84. Glynn IM, Hara Y, Richards DE. 1984. The occlusion of sodium ions within the mammalian sodium-potassium pump: its role in sodium transport. *J. Physiol.* 351:531–47

85. Glynn IM, Howland JL, Richards DE. 1985. Evidence for the ordered release of rubidium ions occluded within the Na,K-ATPase of mammalian kidney. *J. Physiol.* 368:453–69

86. Forbush B. 1987. Rapid release of ^{42}K or ^{86}Rb from two distinct transport sites on the Na,K-pump in the presence of P_i or vanadate. *J. Biol. Chem.* 262:11116–27

87. Glynn IM, Richards DE. 1989. Evidence for the ordered release of rubidium ions occluded within individual protomers of dog kidney Na,K-ATPase. *J. Physiol.* 408:57–66

88. Lee JA, Fortes PAG. 1985. Anthroylouabain binding to different phosphoenzyme forms of Na,K-ATPase. In *The Sodium Pump*, ed. I Glynn, C Ellory, pp. 277–82. Cambridge, UK: Company of Biologists

89. Yoda A, Yoda S. 1987. Two different

phosphorylation-dephosphorylation cycles of Na,K-ATPase proteoliposomes accompanying Na$^+$ transport in the absence of K$^+$. *J. Biol. Chem.* 262:110–15

90. Jørgensen PL. 1991. Conformational transitions in the α-subunit and ion occlusion. In *The Sodium Pump: Structure, Mechanism and Regulation*, Society of General Physiologists Series, ed. JH Kaplan, P De Weer, 46(1):189–200. New York: Rockefeller Univ. Press

91. Taniguchi K, Kaya S, Abe K, Mårdh S. 2001. The oligomeric nature of Na/K-transport ATPase. *J. Biochem.* 129:335–42

92. Donnett C, Arystarkhova E, Sweadner K. 2001. Thermal denaturation of the Na,K-ATPase provides evidence for α-α oligomeric interaction and γ subunit association with the C-terminal domain. *J. Biol Chem.* 276:7357–65

93. Martin DW, Sachs JR. 1992. Cross-linking of the erythrocyte (Na$^+$,K$^+$)-ATPase. *J. Biol. Chem.* 267:23922–29

94. Hayashi Y, Mimura K, Matsui H, Takagi T. 1989. Minimum enzyme unit for Na$^+$/K$^+$-ATPase is the $\alpha\beta$ protomer: determination by low-angle laser light scattering photometry coupled with high-performance gel chromatography for substantially simultaneous measurement of ATPase activity and molecular weight. *Biochim. Biophys. Acta* 983:217–29

95. Martin DW, Sachs JR. 1999. Preparation of Na$^+$,K$^+$-ATPase with near maximal specific activity and phosphorylation capacity: evidence that the reaction mechanism involves all of the sites. *Biochemistry* 38:7485–97

Annu. Rev. Physiol. 2002. 64:19–46

POTASSIUM CHANNEL ONTOGENY

Carol Deutsch

Department of Physiology, University of Pennsylvania, Philadelphia, Pennsylvania 19104-6085; e-mail: cjd@mail.med.upenn.edu

Key Words channel biogenesis, K^+ channel oligomerization, assembly and trafficking, endoplasmic reticulum, membrane integration and topology

■ **Abstract**   Potassium channels are multi-subunit complexes, often composed of several polytopic membrane proteins and cytosolic proteins. The formation of these oligomeric structures, including both biogenesis and trafficking, is the subject of this review. The emphasis is on events in the endoplasmic reticulum (ER), particularly on how, where, and when K^+ channel polypeptides translocate and integrate into the bilayer, oligomerize and fold to form pore-forming units, and associate with auxiliary subunits to create the mature channel complex. Questions are raised with respect to the sequence of these events, when biogenic decisions are made, models for integration of K^+ channel transmembrane segments, crosstalk between the cell surface and ER, and recognition of compatible partner subunits. Also considered are determinants of subunit composition and stoichiometry, their consequence for trafficking, mechanisms for ER retention and export, and sequence motifs that direct channels to the cell surface. It is these mechanistic issues that govern the differential distributions of K^+ conductances at the cell surface, and hence the electrical activity of cells and tissues underlying both the physiology and pathophysiology of an organism.

INTRODUCTION

For the past 50 years, the emphasis of ion channel studies has been the structure/function relationship of mature ion channels. Here I would like to consider what happens between protein synthesis (translation) and these final structures in the plasma membrane. The answers lie in tracking the various stages of ion channel assembly from channel infancy and adolescence to maturity. This ontogeny has two parts: biogenesis and trafficking. We might loosely refer to events in the endoplasmic reticulum (ER) as biogenesis and movement to and from the ER and other compartments as trafficking. The goal of this article is to raise provocative mechanistic questions and, in this context, review what is known regarding K^+ channel ontogeny. I have chosen to focus on biogenesis in the eukaryotic ER, and only highlight some aspects of trafficking and surface expression of channels. Our discussion will be restricted to K^+ channels, with an emphasis on voltage-gated (Kv) channels.

BIOGENESIS

mRNA for the channel protein is made and exported from the nucleus to the cytosol, where it associates in a complex with ribosomes and tRNA, and the nascent peptide is born (Figure 1, see color insert). When a specific series of amino acids, a signal sequence, emerges from the ribosome, this entire complex is targeted to the ER membrane where synthesis continues. Topologically, the inside of the ER is equivalent to the extracellular space and contains a unique group of proteins and factors that modify the nascent channel peptides, as well as regulate the intralumenal milieu. Upon arrival at the membrane, the nascent peptide engages the translocation machinery and is translocated across the membrane through an aqueous, proteinaceous channel. Eventually, the protein integrates laterally into the bilayer, oligomerizes, folds, and associates with auxiliary proteins. Expression of ion channels at the cell surface is governed by all of these steps, as well as those involved in trafficking. Interruption or miscarriage of any of these maturation events will prevent channel expression at the surface.

Although Figure 1 depicts a cartoon representation of sequential biogenic events in K^+ channel assembly, this need not be the case. More likely, these events are overlapping and coordinated, or even cooperative, as depicted in the timeline and discussed herein.

Targeting

Typically, targeting of membrane proteins involves a signal sequence in the N terminus that binds to a cytosolic signal recognition particle (SRP), which subsequently binds to a receptor on the ER membrane. We know something about this targeting step for Kv1.3, a *Shaker*-family member. Several individual transmembrane segments in Kv1.3 are able to target to the ER and translocate without being cleaved. Others also stop translocation across the membrane (1). However, in the full-length native Kv1.3, the second transmembrane segment, S2, functions as the signal sequence (1). This is surprising because more than 200 amino acids are synthesized before the polypeptide is targeted to the ER. However, precedent exists for this scenario (2), which has some bearing on cooperative early events in the ER (see below.). To date, the topogenic determinants in other K^+ channels have not been elucidated.

Translocation

The site to which the targeted ribosome binds is the translocon (Figure 1, illustrated as black ovals), a multi-protein complex that forms an aqueous pore through which the nascent K^+ channel peptide enters the ER lumen (3). This pore only opens to the lumen when a translocating peptide reaches a certain length (4). Two orthogonal events are governed by this complex: translocation across the bilayer and integration into the bilayer. Translocation defines the sidedness (transmembrane topology) of the channel protein and, thus, the location of folding events for cytoplasmic domains and for extracellular/lumenal domains. It is in the translocon that topological fate is determined, but perhaps not irrevocably.

Decades of structure/function studies have supported a six-transmembrane topology for Kv channels, and similar studies, though fewer in number, indicate a two-transmembrane topology for inward rectifiers, six for HERG and KCNQ families, and seven for the large conductance calcium-activated K^+ channels. These conventional models, originally derived from hydropathy analysis, have been verified directly in some cases (e.g., 1, 5–8). How does the channel arrive at these final topologies, and how is this information encoded?

It is encoded by the amino acid sequences in the transmembrane and flanking regions of channel proteins. This encoding been described for only one K^+ channel, Kv1.3 (1). Transmembrane segments S1, S2, S3, S5, and S6 can individually initiate translocation, and only S1 and S2 function as signal anchors and efficiently integrate into the membrane, whereas S4 and S5 show weaker integration into the membrane. S3 and S6 alone lack the ability to integrate into the membrane (1). Moreover, contiguous multi-transmembrane segments alter integration and translocation efficiencies of individual transmembrane segments, indicating that multiple topogenic determinants cooperate during Kv1.3 topogenesis to form the mature subunit in the membrane (1). There is virtually no information deciphering the details of this code, neither the specific amino acids, nor the code-readers/executors.

When are these topogenic decisions made? One possibility is that a transmembrane segment establishes topology before translation of the next C-terminal segment. Alternatively, as is the case for Kv1.3 [and cystic fibrosis transmembrane conductance regulator (CFTR)], the first transmembrane segment achieves its topology only after the second transmembrane segment is synthesized (1, 2). Interestingly, these topogenic decisions do not have to be irrevocable. Although not explored for K^+ channels, this issue has been explored for the aquaporin-1 protein: a four-transmembrane topology initially exhibited in the ER converts to a six-transmembrane topology (9–11).

Modification

While translation and translocation of the nascent peptide across the ER membrane continue, modification of the nascent peptide, including cleavage of signal sequences, glycosylation, and oxidation of the peptide, occurs. Specific enzymes in the ER compartment carry out these functions.

GLYCOSYLATION   N-linked glycosylation in the ER (core glycosylation), entails enzymatic transfer of a block of high mannose chains from a lipid donor to a consensus asparagine. Although Kv channels are core-glycosylated and then undergo higher-order glycosyation in the Golgi (12–15), initial studies suggested that added carbohydrates do not affect cell surface expression, synthesis, turnover, or biophysical function (12, 16–18; however, see 19). Nonetheless, sorting in the Golgi and/or delivery to the surface plasma membrane may be influenced by glycosylation as squid Kv1A channels appear on the surface membrane with a shorter time delay and higher initial rate than do their non-glycosylated counterparts (18). What is the role of glycosylation in K^+ channel ontogeny? One possibility is that carbohydrate-tagging directs specific regions of the K^+ channel polypeptide to the surface of the protein, thus rendering hydrophobic intermediates more soluble (20). The absence of such tagging can target K^+ channel proteins for degradation (see below).

OXIDATION K^+ channel inactivation can be regulated by intracellular reduction or oxidation of cysteine residues (21). The oxidation state of methionine also modulates K^+ channel properties and is regulated by cellular enzymes and molecules (22). Can similar redox reactions in the cytosol modulate biogenic steps in K^+ channel assembly? Although this issue has yet to be explored for the cytosol, enzymatically mediated disulfide bond formation (via protein disulfide isomerase, PDI) in the lumen is well known. It is essential for correct folding of some proteins, including ligand-gated channels (23–27), and limits the number of folded states a protein can visit. The issue is less clear for K^+ channel assembly. In *Shaker*, subunits are not linked by disulfide bonds, nor are conserved cysteines involved in disulfide bonds, nor does the protein contain disulfide bonds essential for protein folding or assembly of functional channels (28). However, when cysteine-free mutant Kv channels are expressed, the levels are low compared with cysteine-containing analogues (29, 30). This may be the result of elimination of a quality control mechanism that occurs in the ER, namely, cysteine residues of unassembled proteins form disulfide-bonded complexes with ER-resident oxidoreductases (e.g., PDI, Erp72), thereby retaining the protein for correct folding (31). Similarly, cysteine oxidation of biogenic intermediates may be important for K^+ channel assembly (32). Alternatively, cysteines may be important in another, as yet undefined, scenario.

OTHER PUTATIVE MODIFIERS   Many other processes may play a role in K^+ channel biogenesis, but they have not been explored. For instance, phosphorylation commonly occurs in the cytosol. Why not during ER assembly of K^+ channels (see below)? Peptidyl-prolyl *cis-trans* isomerization increases the efficiency of protein disulfide isomerase as a protein folding catalyst (33). Is there a role for this enzyme in K^+ channel assembly?

Integration

Eventually, the transmembrane segment of a nascent membrane protein exits the aqueous pore, sequentially folding in association with each protein of the translocon (e.g., Sec61, TRAM), and finally integrates into the hydrophobic interior of the bilayer. Two models may be considered for K^+ channel integration (34). The "spooling" model proposes sequential integration of each transmembrane segment. An alternative model involves an expandable translocon that houses multiple transmembrane segments as a biogenic unit, which is then released into the lipid bilayer while subsequent transmembrane segments continue to be made, oriented, and associated within the translocon. In both models, the lumenal gate and ribosome-membrane junction open and close, modulated by sequences in the nascent peptide, while maintaining the permeability barrier of the membrane.

For some proteins a spooling mechanism makes sense, but for others, it may not. In Kv channels, transmembrane segments S2, S3, and S4 contain negatively and positively charged residues, yet are integrated into the hydrophobic bilayer (1). How is this possible? Interactions between transmembrane regions may actually be required for them to be stable in the membrane bilayer (e.g., neutralizing charges, satisfying H-bonds). In these cases, it makes sense that multi-spanning transmembrane regions should be formed, and even reoriented, within the translocon before integration into the bilayer. The expandable (20–60 Å) translocon can accommodate multiple boarders (35–37). Regardless of charge, we can ask whether transmembrane segments integrate one at a time, independently, or in some coordinated fashion. For some non-channel proteins, transmembrane segments integrate into the bilayer soon after entering the translocon (38), and hydrophobic signal-anchor transmembrane sequences can contact phospholipid while the rest of the polypeptide is still being translated (39). In Kv1.3, translocation and integration efficiencies of S2, S3, S4, and S6 are markedly interdependent (1), suggesting they participate in biogenic units that form, fold, and orient within the translocon. What is it about the membrane-spanning sequences of Kv channels and the translocon that decides whether transmembrane regions traverse the membrane as they are synthesized or following assembly in the translocon? These issues have not been addressed, no less studied, in other K+ channels.

CHAPERONES   Upon exiting the translocon, many polytopic membrane proteins bind to specific chaperone proteins (40, 41), which temporally prevent misfolding of newly formed proteins and help retain unassembled or incorrectly folded proteins in the ER. Aside from calnexin, an ER resident membrane chaperone, little is known about channel chaperone proteins (for AChR channels, see 42, 43). Calnexin transiently associates with glycosylated *Shaker* in the ER but is not required for proper folding and assembly of *Shaker* Kv channels (44).

Although a chaperone-like role has been ascribed to Kvβ subunits, which enhance expression of some Kv channels (45, see below), they are not chaperones in the conventional sense of the term. Classical chaperones only transiently associate, whereas β-subunits remain associated in the mature channel complex. However, a recently cloned and characterized cytoplasmic protein, the so-called K+ channel associated protein, KChAP, which binds to Kv1 and Kv2 family members, may qualify. It belongs to a family of transcription factor binding proteins and binds specifically and transiently to the N terminus, does not modify the biophysical properties of the Kv channels, and increases their surface expression (46, 47). KChAP also binds to the C terminus of Kvβ subunits. Perhaps KChAP modulates Kvα-Kvβ association to increase the stability of the complex. Moreover, it is not known when KChAP exerts its chaperone activity, i.e., during translocation, folding, or oligomerization. Where does it associate, in the ER or later compartments? By what mechanisms does it promote enhanced expression? We do not yet know the answers.

Another even more speculative newcomer on the channel chaperone scene is G protein $\beta\gamma$ ($G\beta\gamma$). Some evidence suggests that $G\beta\gamma$ binds to Kv1.1 and to $Kv\beta1.1$ and stabilizes the $Kv\alpha$-$Kv\beta$ complex, possibly early in biosynthesis in the ER (48). This raises the intriguing possibility of coupling G protein signal transduction initiated by extracellular signals at the cell surface to control of channel assembly in biosynthetic intracellular organelles.

Oligomerization

Several prerequisites must be met in order for subunits to oligomerize. The associating subunits must coexist spatially and temporally, recognize each other, and provide sufficient stabilization energy to form a stable structure.

SITE OF OLIGOMERIZATION   The first evidence that tetrameric K^+ channels form in the ER membrane came in 1992 from the work of Rosenberg & East (49), who showed that *Shaker* translated in vitro in microsomal membranes and reconstituted into bilayers produced functional channels. There is also direct biochemical evidence that Kv subunits associate in the ER (13, 16, 17, 50, 51).

RECOGNITION DOMAINS   Only members of the same Kv subfamilies co-assemble to form channels (52–54). What is responsible for this segregation? The answer is a highly conserved sequence in the cytosolic N terminus of Kv channels, christened the "T1 domain" (first tetramerization domain) (16, 50, 54–59). Although mutations in T1 can create nonfunctional, unassembled protein (18, 60, 61), presumably due to defects early in ER biogenesis (18), T1 can be removed without preventing the formation of functional channels (15, 16, 56–58, 62–64 and references therein). T1-deleted Kv subunits can eventually associate (promiscuously) via transmembrane domains (15) to form stable, functional channels, but both the rates and efficiency of tetramer formation are significantly lower (15, 57).

T1 domains prevent scrambling of Kv channels to homogeneity, thus safeguarding distinct Kv channel repertoires. The likely mechanism is that tetramerization of the cytoplasmic T1 domain promotes transmembrane channel assembly by increasing the effective local subunit concentration for T1 compatible subunits, which kinetically expedites subunit association (1, 65–67). Direct T1-T1 interactions in the full-length channel in the ER have only recently been demonstrated and strongly support this hypothesis (67, see below). A provocative implication of T1's importance comes from squid Kv channels. The mRNA for squid Kv1A (SqKv1A) is extensively edited (68), and eight of these edited sites are in the T1 domain, one of which dramatically decreases expression of the protein and, consequently, the functional channel (18). Why edit T1? No significant biophysical consequences were observed; however, one may speculate that T1 editing regulates some aspects of assembly (J. Rosenthal, personal communication). In 1998, the crystal structure of the soluble T1 domain was determined at 1.55 Å resolution (69) and then corroborated for several other Kv T1 proteins (61, 70, 71). In each case, a tetrameric structure surrounds a narrow pore. At each subunit interface, side-chains of 15 residues are involved in polar intersubunit interactions and are highly conserved in a subfamily-specific manner. The tetrameric structure of T1 is also found in the mature, full-length *Shaker* channel in the plasma membrane (72, 73).

OTHER RECOGNITION DOMAINS   Regions other than the N terminus are involved in K+ channel assembly. These include the transmembrane core regions in Kv channels (15, 16, 62, 63, 74, 75), the proximal C terminus and the second transmembrane segment, M2, in KIR channels (76, 77), and the C-terminal region in both G protein-coupled KIR subunits (78) and the *ether-à-go-go* (*eag*) family of channels (79). In the last case, this does not seem to be true for the human *eag* channel, HERG (80). The N termini of HERG (residues 1–136) alone form tetramers and cause dominant-negative suppression, a criteria for an intersubunit association domain.

A recognition/association role for the cytoplasmic C terminus exists not only in the animal kingdom, but also in the plant world for the inward rectifying AKT1, KAT1, KST1, and SKT1 channels (81, 82). Similar to Kv N-terminal T1 domains, C-terminal recognition domains in plant K+ channels are not required for channel function but may be important for efficient assembly.

In these cases, how does the relatively late translation of recognition domains govern assembly? The mechanism is likely to be distinctly different from that involving N-terminal recognition domains, which are translated and functionally available prior to synthesis of the rest of the nascent subunit. No experiments have addressed this intriguing conundrum.

And finally, multiple intersubunit association sites may contribute to channel formation, some serving as recognition motifs, others as stabilization motifs (15, 77). The detailed assignments and the relative contributions of these motifs have yet to be determined.

In principle, any peptide domain imbued with the necessary properties (e.g., high affinity, specificity) could serve as a recognition and/or stabilization domain. Proof of principle was elegantly demonstrated by Zerangue et al. (66). They replaced the T1 domain of *Shaker* B with a coiled-coil sequence (GCN4-L1) that forms parallel tetramers. This artificial tetramerization domain restored efficient channel formation to a T1-deleted *Shaker* protein.

WHEN DURING BIOGENESIS DOES TETRAMER FORMATION OCCUR?   The first implication came from the observation that Kv1.1 and Kv1.4, when co-translated in isolated microsomal membranes, form co-immunoprecipitated heteromultimers at the earliest time that protein was detected (15 min) (17). However, the antibodies used in this study were directed at the C terminus of the Kv proteins so that only fully synthesized protein was detected, thereby precluding resolution of the protein domains involved and of the timing of the oligomerization steps. These results do not distinguish a series of rapid, separate steps between synthesis and association from a co-translational event.

Papazian and co-workers place *Shaker* tetramerization subsequent to protein synthesis, glycosylation, and insertion into the membrane but prior to folding of the voltage sensor, pore formation, and association with auxiliary subunits (51). This temporal assignment was not based on simultaneous measurement of the time course of these events but rather on logical inference from biochemical assays of mutant *Shaker*. Direct evidence for early tetramerization has only recently been obtained (67). T1 tetramers form between neighboring subunits while the nascent channel peptides are still attached to ribosomes and have translocated across the ER membrane. This association is specific for residues in the T1 domain, perhaps occurring soon after targeting of the nascent chains to the ER membrane and before the monomer is completely synthesized. Thus T1 folding and tetramer formation may be the first assembly event in channel formation.

TETRAMER FORMATION   How do four Kv subunits come together? Tetramers could form via concerted or stepwise mechanisms. Two stepwise possibilities were explored by Tu & Deutsch (65) for Kv1.3. One is by sequential addition of monomers to form dimers, then trimers, and, finally, tetramers. Alternatively, monomers may associate to form dimers, which then dimerize to form tetramers. The dominant pathway in tetramer formation is dimerization of dimers. Moreover, conformational changes accompany dimer formation, thereby creating new interaction sites that are used in the next stage to form tetramers. What specific interactions underlie each stage? It is possible that T1-T1 interactions prevail at the monomer-monomer association stage and that other specific interactions involving the transmembrane core regions govern subsequent association steps to form final tetramers. For Kv1.3, T1 folding and tetramerization are induced by the ER membrane (67). This could be due to surface catalysis by the lipid bilayer or to specific association of the nascent Kv1.3 with translocon machinery. Such mechanisms could protect against premature folding and oligomerization of membrane proteins prior to targeting.

Specific oligomerization mechanisms likely depend on the specific channel type, proceeding by a succession of stepwise folding and oligomerization. There is little information regarding how other K^+ channels tetramerize or what interaction surfaces are created and required for intersubunit interaction. In part, this is due to the lack of good experimental strategies and techniques with which to probe these issues.

When two different monomers containing compatible recognition domains are expressed within the same cell, the resultant compositions and stoichiometries of tetramers can vary extensively. At one extreme, preferential association yields only homotetramers; at the other, random association yields a binomially weighted mixture of homo- and heterotetramers (83, 84). Functional (84, 85) and biochemical strategies have been used (e.g., toxin binding and antibodies) to analyze tetramer stoichiometries both in vivo and in vitro. Although it appears that the degree of temporal overlap and kinetics of subunit expression govern the proportions of homo- and heterotetramers (84), virtually nothing is known about the mechanism for this regulated distribution in vivo.

Folding

Presumably, nascent K^+ channel peptides fold during translocation and integration into the bilayer, most probably with the help of chaperone proteins. Likewise, the oligomeric complex continues to fold into the final functional channel. During folding, the functional units of the channel (e.g., gates, sensors for gating, pores) are created. In some cases, we do not even know the components of these functional units, either in the mature channel or in the immature intermediates.

For Kv channels there is some indication that intra-subunit electrostatic interactions between transmembrane segments mediate folding of Kv monomers. Neutralization of positive charges in S4, part of the Kv voltage sensor, prevents maturation of the protein, suggesting that these charge neutralizations prevent correct folding of the protein (86). These S4 mutants can be rescued by second-site neutralizations of negative charges in S2 and S3. Related experiments producing charge reversals (87) further support the conclusion that electrostatic intra-subunit interactions contribute to proper folding of the protein.

Schulteis et al. (51) used a variety of *Shaker* Kv mutants to infer that folding and assembly steps alternate during channel biogenesis. Folding events to form the pore and sensors occur late in the sequence, after synthesis, insertion, tetramerization, and association with auxiliary subunits, but prior to formation of N- and C-terminal proximity. However, the spatial and temporal details of these folding events are not known. Thus far, the most precisely located folding event is the folding and tetramerization of T1 domains before subunits exit from the translocon (67). Here is a clear call to the bench to elucidate the stepwise mechanism of subunit folding and channel formation. Specifically, when does the voltage-sensor form? When does the permeation pore form? When do the gates and the linkers coupling the gates and sensors form? Finally, when do the mature protein-protein, protein-lipid, and protein-aqueous interfaces indicated by crystallographic- and EPR-derived structures or scanning mutagenesis studies, form?

A channel could conduct K^+ ions once its functional units are formed, but is this when the channel actually first transports K^+ ions? Unfortunately, we remain unenlightened whether K^+ channels function with precise gating behavior, selectivity, and high throughput in the ER.

Auxiliary Subunits

A critical element of channel ontogeny is the association of the pore-forming part of a K^+ channel with auxiliary proteins to ultimately form the mature channel complex. The pore-forming tetramers are catalysts of ion transport, and the auxiliary proteins, either constitutively or intermittently associated with the catalytic unit, serve as regulators of this function. In addition, auxiliary subunits can modulate biogenic and trafficking events, yet except for a few cases (see below), the mechanisms are not known, nor where, when, and how auxiliary subunits associate with their pore-forming counterparts. For instance, all compartments along the path from ER to plasma membrane may host associations between auxiliary subunits and the pore-forming channel polypeptide. Association could be co-translational or post-translational. Auxiliary membrane proteins could associate with pore-forming subunits at their mutual peripheries, or one could intercalate into the other. A single stoichiometry or multiple stoichiometries may predominate. These fundamental assembly issues are both intriguing and, in most cases, unanswered.

CA-ACTIVATED K$^+$ CHANNELS   The large conductance Ca-activated K$^+$ channel (BK for big conductance) associates in a 1:1 complex (88, 89) with an auxiliary membrane protein, which increases the Ca^{2+} and voltage sensitivity of BK and mediates drug sensitivity (90–92). In addition, two cytosolic proteins, slob (Slowpoke-binding protein) and slip1 (Slo-interacting protein 1), regulate BK channels in part by redistributing these channels within cells (93, 94). Moreover, a third protein (14-3-3) interacts with slob and signaling proteins (e.g., kinases) to modulate channel function (95). Thus the pore-forming BK channel associates in a large complex of auxiliary proteins, each with designated functional and signaling tasks, as well as biogenic and trafficking tasks. The interaction sites need to be identified, along with the mechanisms by which these auxiliary proteins direct trafficking and localization of BK channels.

Another type of Ca-activated K$^+$ channel is SK, a small conductance channel, with six transmembrane segments similar to Kv channels, but not activated by voltage. SK channels are also heteromeric complexes. Calmodulin, which constitutively associates with SK subunits in a calcium-independent manner (96), binds calcium and activates the channel. Ca-independent binding of calmodulin to SK requires only the C-terminal domain of calmodulin and is mediated by two noncontiguous subregions (97). Calmodulin also regulates, in a calcium-dependent manner, multimerization of SK subunits, thereby providing a potential link between Ca^{2+} signaling and channel expression (98). It is likely that these Ca-sensing auxiliary proteins associate with the Ca-activated K$^+$ channels during biogenesis.

INWARD RECTIFYING CHANNELS (KIR)   Several KIRs form functional channels as heteromultimers, either with pore-forming subunits or with auxiliary subunits. For example, the K$_{ATP}$ channel is assembled as an octamer with a 1:1 stoichiometry of KIR and an auxiliary subunit, SUR (sulfonyl urea receptor), which modulates function and drug sensitivity of channel activity (99–102). The first transmembrane segment (M1) and the cytosolic N terminus of KIR 6.2 are important for specifying assembly with SUR (103), and subunit association occurs in the ER and early Golgi (104; see below). SUR1 may also recognize a cytoplasmic surface of KIR 6.2 composed of both the N and C termini, and direct interaction between N and C termini of KIR 6.2 has been demonstrated (105).

KV CHANNELS   Co-assembly of KvLQT1 (KCNQ1) and minK (KCNE1) proteins produces the cardiac I_{ks} potassium channel; neither one alone reconstitutes the in vivo current (106, 107). KCNE1 also associates with the pore-forming HERG channel, and a family of minK-related proteins (MiRPS, part of the KCNE family) associates with, and modulates the function of Kv, KCNQ, and HERG channels (108, 109). Association likely occurs co-translationally in the ER (110). However, both the architecture and the stoichiometry of the KCNE:KCNQ complex are highly controversial. KCNE1 could reside at the KCNQ1 channel periphery (111, 112), in the S4 gating pore (113), or in intimate association with the permeation pore (114–116). Moreover, the number of KCNE subunits per channel could be two or more, and variable (112, 117).

The best-known auxiliary subunits are the cytosolic, so-called β-subunits that associate with Kv channels (referred to here as α-subunits). The β-subunits are structurally similar to aldo-keto reductases (118–120), are expressed in both excitable and non-excitable cells, and have diverse modulatory effects (45, 119, 121, 122; see below). Although different β-subunits exist for the different Kv subfamilies, only those pertaining to the Kv1 subfamily are discussed here.

The two major isoforms are Kvβ1 and Kvβ2, each associating through conserved C-terminal regions with the T1 domains of *Shaker* and *Shal*, but not *Shaw* or *Shab* Kv channels (59, 123–126; however, see 122). We do not know what accounts for the specificity.

The two isoforms have different functional effects. Only Kvβ1 speeds inactivation gating, mediated by its N-terminal domain (121, 126). The other prominent effect is increased stability and cell surface expression of Kv channels, mediated by the C-terminal domain of the β-subunit (45, 119, 124, 127). Compared with Kvβ1, Kvβ2 has a higher affinity for both α-subunits and itself, perhaps accounting for the predominance of α-β2 complexes (126).

At least three roles have been suggested for β-subunits: a chaperone-like role early in assembly in the ER (45, 127), a modulator role in gating in the plasma membrane (121), and a transducer role in redox-linked enzymatic control of electrical activity in the plasma membrane (120). There is evidence for the first two cases, but as yet no functional evidence of enzymatic activity, only the structural and genetic similarity to oxidoreductases.

The most detailed work in support of some chaperone-like function was provided by Trimmer and co-workers (45), who showed that α-β interaction in the ER is rapid and permanent. In the case of Kv1.2, surface expression requires β-subunit co-expression and increases in a dose-dependent manner with β DNA. Moreover, the β-subunit decreases Kv1.2 degradation and also promotes ER glycosylation of Kv1.2, thereby suggesting that the β-subunit interacts with the N terminus of Kv1.2 during translocation of Kv1.2 across the ER translocon. The in vivo role for Kvβ-subunits was demonstrated for hyperkinetic (Hk) mutants in

Drosophila, which have decreased neural K^+ channel expression. Co-expression of Hk polypeptide, which is homologous with $Kv\beta$ subunits, with *Shaker* Kv protein in *Xenopus* oocytes increased current amplitudes and modulated the biophysical current properties (119).

The nature of the effect of β-subunits on Kv expression is isoform specific and depends on the composition and stoichiometry of the channel α-subunits. Co-expression of $Kv\beta2$ with Kv1.1, 1.3, 1.4, 1.6, and 1.2 increases surface expression in each case, whereas co-expression with *Shaker* and Kv1.5 does not (13, 127). In fact, Kv1.5 expression is decreased by β-subunit co-expression (127). The rate and extent of *Shaker* maturation (i.e., higher complex glycosylation in the Golgi) is not affected by β-subunit expression (13) and thus differs from Kv1.2, which is predominantly trapped in the ER and is poorly glycosylated in the absence of β-subunits (45). These results suggest that those isoforms existing in the absence of β-subunits predominantly as immature proteins in the ER exhibit the most striking increase in surface expression when β-subunits are co-expressed.

Both $Kv\beta1$ and $Kv\beta2$ subunits also self-associate to form homo- and heterotetramers (120, 126, 128), and the α-β complexes, specifically the T1-$Kv\beta2$ complex, consist of the T1-tetramer (see above) docked onto the β-tetramer to form an octameric structure with a stoichiometry of $T1_4\beta_4$ (71). However, we know nothing about the sequence of events that gives rise to this structure. Although the α:β stoichiometry first reported was 1:1 (129, 130), it now appears that $Kv\beta1$ and $Kv\beta2$ associate with α-subunits in two different functional stoichiometries: $\alpha_4\beta1_n$ (where $n = 0$–4) and $\alpha_4\beta2_4$, which may underlie their different roles in channel function and expression, respectively (131). How do these different stoichiometries arise? Do β-subunits first self-associate to form various oligomeric β_n stoichiometries and then associate with an α-tetramer; do β-subunits individually, and sequentially, associate with α-tetramers; or do α- and β-subunits associate to form a heterodimer, which subsequently associates with other $\alpha\beta$-dimers to form the octameric 4:4 complex? The first scenario is likely to prevail in the case of $\alpha_4\beta2_4$, consistent with the positive cooperativity of $\beta2$ for itself, and would account for the apparent higher avidity of $\beta2$ for Kv channels because a tetrameric $\beta2$ would provide multivalent interactions (131). Why do $\beta1$ and $\beta2$ have different self-association characteristics? Could it be due to differences in post-translational modification, e.g., phosphorylation and/or interaction with other proteins, as suggested by Li and co-workers (131)? There are many post-translational modification sites in β-subunits (e.g., phosphorylation sites), which raises the possibility of modification during assembly.

Among the candidate proteins for such modifications are cytosolic ZIP1 and ZIP2 (PKC-ζ interacting protein), which bind to $Kv\beta2$ subunits and PKC-ζ and serve as physical linkers (132). ZIP1 and ZIP2 differentially stimulate phosphorylation of $Kv\beta2$ by PKC-ζ and may be responsible for specific regulation of PKC-ζ phosphorylation. When does ZIP stimulate PKC-ζ phosphorylation of $\beta2$? Could it be before α and $\beta2$ associate or only after the octameric complex is assembled? Thus phosphorylation may play a biogenic role as well as a functional role.

OTHER AUXILIARY SUBUNITS   Several other proteins may function as auxiliary subunits, including enzymes, cytoskeletal proteins, and scaffolding proteins (see below). In addition, lipids can also serve this purpose. In each case, a precedent exists for a role in channel function in the plasma membrane, but do these auxiliary agents also play a role in assembly? This has yet to be explored.

TRAFFICKING/SORTING

Biogenesis alone is not sufficient to effect physiological channel function. Rather, the appropriate targeting, localization, abundance, and turnover are required. Failure to achieve these prerequisites often leads to pathophysiology. The following discussion highlights a few mechanisms involved in trafficking of K^+ channels.

Retention and Retrieval in the ER

In the octameric K_{ATP} channels, both the KIR6.2 and SUR subunits have cytoplasmic RKR sequences that must be masked before the channel can be transported to the cell surface in a variety of eukaryotic cells (104)[1]. This constitutes the rate-limiting mechanism in K_{ATP} assembly. Similar motifs exist in a cytosolic domain of CFTR (RXT) (133) and in the C terminus of a $GABA_B$ receptor subunit (RSRR) (134). In the latter case, heterodimeric coiled-coil interactions shield the motif, allowing the complex to be expressed at the cell surface. Are there analogous motifs in other K^+ channels? This could be a general mechanism for quality control in K^+ channel assembly. Specific sequence motifs may not be the only way to tag a channel for retention in the ER. Association of wild-type subunits with non-expressing mutant subunits can serve the same purpose (135, 136). The consequence of both mechanisms, i.e., ER trapping, modulates levels of surface expression and distribution of K conductances.

Degradation

Unassembled or misfolded subunits are eventually degraded by proteases. Retrotranslocation from ER to cytosol removes these proteins for degradation by proteasomes in the cytosol. This pathway typically requires modification of the protein (ubiquitination) and chaperone association (e.g., calnexin) and is used by some non-K^+ channels (137–139). A similar pathway is likely involved in the degradation of misfolded mutant Kv channels or unglycosylated Kv channels, which may fold less efficiently (140, 141). Few K^+ channels have been studied, so questions remain as to how K^+ channel subunits avoid destruction or succumb to it.

---

[1]Strategic use of such sequences is a highly innovative approach that has been developed to probe protein-protein interactions between subunits (103).

Cytosolic Export/Targeting Motifs

In addition to motifs that direct retrograde movement, there are motifs that direct export or target channels to the plasma membrane and even to specific domains in the plasma membrane. In KIR 1.1 and 2.1 channels, diacidic motifs (e.g., ENE, EXD) function as direct ER export signals (142). These signals do not act by promoting correct folding or assembly, nor are they context dependent. When inserted into Kv1.2, a poor surface expressor (45), FCYENE promoted Kv1.2 expression, similar to the effects produced by Kvβ2. This raises the intriguing possibility that Kvβ2 may facilitate surface expression because it contains export signals and forms a tight complex with α-subunits. This would suggest a speedy escort function rather than a classical chaperone-folding function for Kvβ2. The flanking residues of diacidic motifs are also critical for ER export and vary among K$^+$ channels, thus providing a mechanism for differential distributions of K$^+$ channels on the cell surface.

A conserved motif in the variable C-terminal region of Kv1 channels, VXXSL, modulates surface expression, the requirement for auxiliary β-subunits and subunit glycosylation, but does not alter β-subunit association (143). However, this motif is not responsible for biogenesis and cell surface expression of the *Shaker* protein (144). C-terminal export signals also exist in the auxiliary subunit, SUR, which regulate K$_{ATP}$ surface expression (145). Together with the retrograde RKR motif (above), these anterograde signals ensure that correctly assembled octameric K$_{ATP}$ arrives at the plasma membrane.

Subtype-specific sorting of Kv channels to different plasma membrane domains is critical for physiological function in polarized cells, e.g., epithelia and neurons. In general, the molecular determinants have not been identified; yet at least one clear example is Kv2.1. In this case, a C-terminal domain is responsible for localization of the channel at the basolateral membrane of mammalian cells (146). There are many examples of Kv localization (e.g., 147–151), but the determinants have not been identified in all cases.

Subunit Composition

Another trafficking determinant is subunit composition. Heteromultimers of different composition/stoichiometry, but overlapping subunits, can be targeted to different subcellular compartments. We do not know what governs this, nor do we know whether such diversity is physiologically important. One intriguing role for composition-dependent or stoichiometry-dependent expression is as a regulator of channel distributions and plasticity. The following Kv and KIR examples illustrate the case.

Co-expression of members of the KIR 3.0 family with KIR 4.1 suppresses KIR 4.1 currents. Shortly after formation, the heteromeric complex is degraded rather than transported to the plasma membrane (152), perhaps serving as a physiological mechanism to eliminate inactive channels or ensure a given type and distribution of channels at the cell surface. KIR5.1 does not produce functional channels in

Xenopus oocytes (153); however, co-expression of KIR 5.1 and KIR 4.1 produces potentiated currents (cf. KIR4.1 alone) with novel electrophysiological properties (154). Moreover, tandem constructs with identical stoichiometries but different relative positions of the subunits give distinctly different properties, perhaps suggesting that KIR subunits do not assemble randomly.

The G protein–coupled KIR channels in brain and heart are heterotetramers (155, 156) of KIR3.1 and KIR3.4; the stoichiometry is most likely 2:2 (157–159). KIR3.4 homotetramers and KIR3.1/KIR3.4 heterotetramers, but not KIR3.1 homotetramers, localize to the plasma membrane (160). It appears that the C-terminal domains are responsible for directing surface versus intracellular localization of KIR channels (78, 160, 161).

Only certain specific Kv1 heterotetramers exist and predominate in mammalian brain, with the notable absence of others (149, 162–165). Both in heterologous transfected mammalian cells and in hippocampal neurons, heteromeric assembly of certain Kv1 members increases cell surface expression of specific Kv subunits, whereas other subunit compositions have a dominant-negative effect (166).

Differential Targeting to Lipid Rafts

In addition to protein-protein interactions, protein-lipid interactions may provide a mechanism for subtype-specific localization of K$^+$ channels. Lipid rafts, which are dynamic microdomains in the plasma membrane, contain tightly packed sphingolipids and cholesterol (167). Kv2.1 channels target to a non-caveolar lipid raft in mammalian cells, whereas Kv4.2 channels do not (168); yet within the same membrane, Kv1.5 channels target to caveolae (169). Such compartmentalization may enable isoform-specific modulation of Kv function, especially as important signaling molecules, including Src-family tyrosine kinases, are concentrated in these caveolae (167).

Clustering

Targeting channels to the cell surface does not itself ensure physiological efficacy. In some cases, it is the density of channel subtypes that is critical for function (e.g., at synapses), and thus mechanisms for clustering and unclustering channels are important. Cell surface clustering of some K$^+$ channels is mediated by the PSD-95 family of membrane-associated guanylate kinases (MAGUK). Specific regions of these proteins (PDZ domains containing GLGF repeats) bind to the last four C-terminal amino acids (X S/T X V) of Kv channel subunits (170). Different Kv1 channels display extremely different clustering efficiencies when co-expressed with PSD-95 (171). Not only do channel isoforms influence the abundance and distribution of ion channels at the membrane surface, but so do the MAGUK isoforms (171). PSD-95 clusters Kv at the cell surface, perhaps the site of initial Kv/PSD-95 interaction. In contrast, SAP97, another MAGUK isoform, leads to accumulation of Kv clusters in ER-derived vesicles, thereby preventing trafficking of Kv channels to the surface (172), consistent with an initial Kv/SAP97 interaction in the ER.

Clustering and unclustering provides a mechanism for plasticity and remodeling of channel distributions in any membrane, but particularly at the synapse. One way to effect unclustering is by promoting signal-linked dissociation of channels from the scaffolding protein. For example, KIR 2.3 co-localizes with PSD-95 in vivo (hippocampus, cerebral cortex) and forms tight complexes. Activation of PKA in intact cells dissociates this complex from PSD-95 owing to phosphorylation of the serine in the C-terminal binding motif (173). KIR2 family members also associate with SAP97 in vivo, and phosphorylation of KIR 2.2 inhibits its association with SAP97 (174). Such mechanisms are likely to be more pervasive than the current literature reveals.

Plasma Membrane Insertion/Retrieval

Channel abundance and distribution at the cell surface are, in part, determined by insertion into, and retrieval from, the plasma membrane. The molecular bases for these processes have yet to be elucidated. However, many membrane proteins are internalized from the plasma membrane via clathrin-coated pits and adaptor proteins that recognize cytosolic signal motifs (e.g., LL and YXXϕ, where X can be any hydrophobic amino acid and ϕ is a bulky hydrophobic amino acid) (175). A clear example of endocytosis of K^+ channels is that of KIR 2.1. Endocytotic retrieval of this channel occurs via interactions between clathrin adaptor proteins and a tyrosine-based sorting motif (YXPL) on the cytosolic C terminus of KIR 2.1 (176). Few examples have been studied, yet retrieval and degradation of channels from the cell surface is critical to regulation of the abundance and distribution of channel repertoires.

An N-terminal domain (amino acids 3–39) in a rabbit Kv1.3 protein appears to modulate channel expression at the cell surface (177). Mutations and/or deletion of this domain increase current amplitudes more than 10-fold, primarily by increasing the amount of channel protein at the cell surface, which has been ascribed to an increased rate of channel synthesis and insertion into the plasma membrane. A more detailed mechanism is not known.

FUTURE DIRECTIONS AND IMPLICATIONS

When we ask what regulates K^+ channel assembly, we now appreciate the complexity of this question. Regulation can occur at the level of the mRNA/polysome complex, the protein, the oligomeric structure, and/or the membrane-protein complex, and in several cellular compartments, including the ER, the Golgi, organelles on the exo- and endocytotic pathways, and/or the plasma membrane.

What Regulates Subunit Composition, Stoichiometry, and Cellular Distribution?

For example, why is the KIR 3.1/3.4 stoichiometry 2:2 and are these heteromers and the KIR 3.4 homotetramers differentially localized in the plasma membrane? KIR

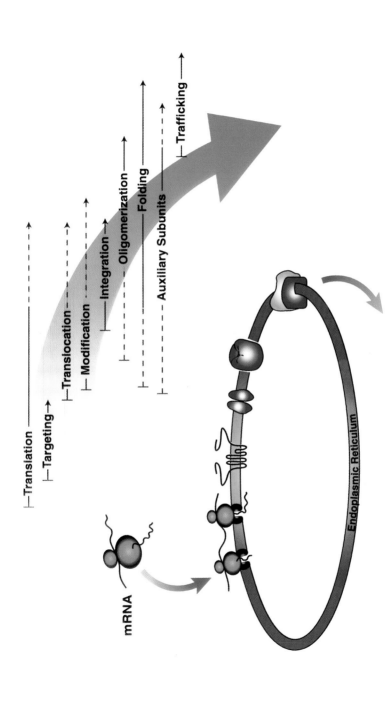

Figure 1 Channel biogenesis in the endoplasmic reticulum. The cartooned events (*from left to right*) illustrate ribosomal translation of a nascent chain, binding of the ribosome/mRNA/protein complex to a translocon, subsequent elongation of the nascent chain, integration of the channel protein into the ER bilayer, oligomerization of subunits, folding to form a channel, and association of the channel with an auxiliary protein. The timeline (*above right*) depicts the cartooned assembly as overlapping events.

1.1 and KIR 4.1 form heterotetramers in vivo in individual auditory hair cells, but the distribution of channel stoichiometries is not binomial. Specifically, the hetero- and homotetramers form with comparable probabilities to yield distinct channel populations with different electrophysiological properties (178). Consequently, a very specific profile of K^+ conductance is established.

What sensing and sorting mechanisms prevail to regulate these distributions? Are there physiological mechanisms at the level of ER biogenesis for regulating K^+ conductance at the plasma membrane, including retrograde and anterograde subunit compositions and stoichiometries (i.e., dominant-negative suppression tagging for degradation versus tagging for plasma membrane targeting)?

Another fundamental issue is whether homotetrameric channels maintain four-fold symmetry throughout assembly. One implication of conformational changes in tetramer formation that accompany the dimerization of Kv monomers (65), is that fourfold symmetry may only be achieved in the final product, but not in the oligomeric intermediates. The identity of the interfaces of oligomeric intermediates during biogenesis remains to be elucidated.

Topogenic and Functional Duality

One of the broadest speculations we may make is that association sites between transmembrane segments are both topogenic and functional determinants, perhaps via the same intramembrane protein-protein interaction. Thus two roles, one functional and one architectural, would be served by the same amino acid sequences. Precedents exist in other non-channel multi-subunit protein complexes. For example, a conserved 20–amino acid sequence in hydrophobic membrane proteins of periplasmic permeases plays a critical role in inter-subunit association interactions and also participates in formation of a functional site for transport activity (179). Another example of the same transmembrane sequence bearing information for assembly and for ER retention/degradation is the T cell receptor α-subunit/CD3-δ (180). Similar dualities may serve to make functional K^+ channels. Candidate cooperative biogenic units, which are also functional units, have already been identified in Kv1.3 (1). It is possible that modular units are first manifest as interacting transmembrane domains that help direct topogenesis, thereby ensuring the requisite structure for their subsequent role as functional units in the channel, e.g., gating, permeation, and voltage-sensing.

Issues of Integration

Do transmembrane segments of K^+ channels integrate into the bilayer shortly after entering the translocon, possibly as biogenic units, or do they remain in the translocon until translation of the entire protein is finished and it detaches from the ribosome? Early T1 tetramerization may, for example, promote integration of some transmembrane segments, possibly as biogenic units, into the bilayer. Virtually nothing is known regarding what governs the accumulation of transmembrane K^+ channel segments as biogenic units in the translocon and whether the resident transmembrane segments interact with each other. One possible mechanism involves intramolecular interactions, such as charge pair interactions (181). Could this also obtain for K^+ channels? Such intramolecular protein-protein interactions have been implicated in the folding, mature structure, and function of the voltage sensor of Kv channels (86, 87, 182–184).

Yet another intriguing implication derives from these considerations. If a channel has large cytosolic loops and integrates as biogenic units, then what governs the final assembly in the membrane? This issue is relatively unexplored in K^+ channels, yet precedents for post-integrational assembly exist, as independently translated or purified fragments can recombine to produce functional, polytopic membrane proteins (185–188).

The Enigma of Late Recognition

A biogenic puzzle is raised by those K^+ channels lacking recognition domains in the N terminus (e.g., inward rectifiers). If C-terminal recognition domains are responsible for tetramer formation and are also the last to be synthesized, then what is the mechanism for oligomerization? Is there an analogous process to that for Kv N-terminal T1 self-association in which early global restriction of subunits shortly after synthesis promotes a higher local concentration and kinetically facilitates tetramerization (66, 67)? Possibly the channel is held in the translocon and released into the bilayer in a coupled, synchronous event with C-terminal tetramerization. Nothing is known about the temporal or spatial sequence of events of this tetramerization with respect to co-translation and/or integration.

Clues from the Temporal Sequence

The sequence of biogenic events may shed light on the role of T1 and β-subunits in assembly (66, 67). Since T1 self-association occurs first, T1 could serve as a scaffold for folding the rest of the channel structure. Likewise, Kvβ tetramer, if it associates early in assembly, could scaffold the tetramerization of T1. Even the reverse may be considered: The early tetramerization of the folded T1 domain may facilitate Kvβ tetramerization. Are these folding events coordinated and/or cooperative events? Detailed time courses of T1-Kvβ interactions have not been determined. The temporal sequence and stoichiometry of association will likely implicate Kvβ's possible role as chaperone, scaffold, modulator, and/or transducer.

Crosstalk with the Plasma Membrane

Finally, a large complex comprised of pore-forming units and regulatory auxiliary units provides a panoply of interactions for cellular crosstalk. For example, ZIP proteins, which themselves can be regulated by growth factors, link PKCζ with Kvβ-subunits (132). These, in turn, modulate Kv function. This raises the intriguing possibility that signaling pathways might also regulate biogenic events in the ER through interaction with Kvβ subunits, especially if Kvβ-subunits have chaperone, scaffolding, or export roles in Kv assembly. A similar role has recently been suggested for $G\beta\gamma$ proteins (48) and for Ca^{2+}calmodulin (98). By extension, we might imagine that the combinations of other signaling molecules, auxiliary subunits, and K^+ channels, govern K^+ channel biogenesis.

Visit the Annual Reviews home page at www.AnnualReviews.org

LITERATURE CITED

1. Tu L, Wang J, Helm A, Skach WR, Deutsch C. 2000. Transmembrane biogenesis of Kv1.3. *Biochemistry* 39:824–36

2. Lu Y, Xiong X, Helm A, Kimani K, Bragin A, Skach WR. 1998. Co- and post-translational translocation mechanisms direct cystic fibrosis transmembrane conductance regulator N terminus transmembrane assembly. *J. Biol. Chem.* 273:568–76

3. Johnson AE, van Waes MA. 1999. The translocon: a dynamic gateway at the ER membrane. *Annu. Rev. Cell Dev. Biol.* 15:799–842

4. Crowley KS, Liao S, Worrell VE, Reinhart GD, Johnson AE. 1994. Secretory proteins move through the endoplasmic reticulum membrane via an aqueous, gated pore. *Cell* 78:461–71

5. Shih TM, Goldin AL. 1998. Topology of the Shaker potassium channel probed with hydrophilic epitope insertions. *J. Cell Biol.* 136:1037–45

6. Ho K, Nichols CG, Lederer WJ, Lytton J, Vassilev PM, et al. 1993. Cloning and expression of an inwardly rectifying ATP-regulated potassium channel. *Nature* 362:31–38

7. Kubo Y, Reuveny E, Slesinger PA, Jan YN, Jan LY. 1993. Primary structure and functional expression of a rat G-protein-coupled muscarinic potassium channel. *Nature* 364:802–6

8. Wallner M, Meera P, Toro L. 1996. Determinant for beta-subunit regulation in high-conductance voltage-activated and $Ca^{(2+)}$-sensitive K^+ channels: an additional transmembrane region at the N ter-

minus. *Proc. Natl. Acad. Sci. USA* 93:14922–27

9. Lu Y, Turnbull IR, Bragin A, Carveth K, Verkman AS, Skach WR. 2000. Reorientation of aquaporin-1 topology during maturation in the endoplasmic reticulum. *Mol. Biol. Cell* 11:2973–85

10. Skach WR, Shi LB, Calayag MC, Frigeri A, Lingappa VR, Verkman AS. 1994. Biogenesis and transmembrane topology of the CHIP28 water channel at the endoplasmic reticulum. *J. Cell Biol.* 125:803–15

11. Jung JS, Preston GM, Smith BL, Guggino WB, Agre P. 1994. Molecular structure of the water channel through aquaporin CHIP. The hourglass model. *J. Biol. Chem.* 269:14648–54

12. Santacruz-Toloza L, Huang Y, John SA, Papazian DM. 1994. Glycosylation of Shaker potassium channel protein in insect cell culture and in *Xenopus* oocytes. *Biochemistry* 33:5607–13

13. Nagaya N, Papazian DM. 1997. Potassium channel alpha and beta subunits assemble in the endoplasmic reticulum. *J. Biol. Chem.* 272:3022–27

14. Schulteis C, Nagaya N, Papazian D. 1996. Intersubunit interaction between amino- and carboxy-terminal cysteine residues in tetrameric Shaker K^+ channels. *Biochemistry* 35:12133–40

15. Tu L, Santarelli V, Sheng Z-F, Skach W, Pain D, Deutsch C. 1996. Voltage-gated K^+ channels contain multiple intersubunit association sites. *J. Biol. Chem.* 271:18904–11

16. Babila T, Moscucci A, Wang H, Weaver FE, Koren G. 1994. Assembly of

mammalian voltage-gated potassium channels: evidence for an important role of the first transmembrane segment. *Neuron* 12:615–26

17. Deal KK, Lovinger DM, Tamkun MM. 1994. The brain Kv1.1 potassium channel: in vitro and in vivo studies on subunit assembly and posttranslational processing. *J. Neurosci.* 14:1666–76

18. Liu TI, Lebaric ZN, Rosenthal JJC, Gilly WF. 2001. Natural substitutions at highly conserved T1-domain residues perturb processing and functional expression of squid Kv1 channels. *J. Neurophysiol.* 85:61–71

19. Thornhill WB, Wu MB, Jiang X, Wu X, Morgan PT, Margiotta JF. 1996. Expression of Kv1.1 delayed rectifier potassium channels in Lec mutant Chinese hamster ovary cell lines reveals a role for sialidation in channel function. *J. Biol. Chem.* 271:19093–98

20. Doms RW, Lamb RA, Rose JK, Helenius A. 1993. Folding and assembly of viral membrane proteins. *Virology* 193:545–62

21. Ruppersberg JP, Stocker M, Pongs O, Heinemann SH, Frank R, Koenen M. 1991. Regulation of fast inactivation of cloned mammalian IK(A) channels by cysteine oxidation. *Nature* 352:711–14

22. Hoshi T, Heinemann S. 2001. Regulation of cell function by methionine oxidation and reduction. *J. Physiol.* 531:1–11

23. Marquardt T, Hebert DN, Helenius A. 1993. Post-translational folding of influenza hemagglutinin in isolated endoplasmic reticulum-derived microsomes. *J. Biol. Chem.* 268:19618–25

24. Braakman I, Helenius J, Helenius A. 1992. Manipulating disulfide bond formation and protein folding in the endoplasmic reticulum. *EMBO J.* 11:1717–22

25. Gelman MS, Prives JM. 1996. Arrest of subunit folding and assembly of nicotinic acetylcholine receptors in cultured muscle cells by dithiothreitol. *J. Biol. Chem.* 271:10709–14

26. Fu DX, Sine SM. 1996. Asymmetric con-

tribution of the conserved disulfide loop to subunit oligomerization and assembly of the nicotinic acetylcholine receptor. *J. Biol. Chem.* 271:31479–84

27. Green WN, Wanamaker CP. 1997. The role of the cysteine loop in acetylcholine receptor assembly. *J. Biol. Chem.* 272:20945–53

28. Schulteis CT, John SA, Huang Y, Tang C-Y, Papazian DM. 1995. Conserved cysteine residues in the *Shaker* K⁺ channel are not linked by a disulfide bond. *Biochemistry* 34:1725–33

29. Boland LM, Jurman ME, Yellen G. 1994. Cysteines in the *Shaker* K⁺ channel are not essential for channel activity or zinc modulation. *Biophys. J.* 66:694–99

30. Lu J, Deutsch C. 2001. Pegylation: a method for assessing topological accessibilities in Kv1.3. *Biochemistry.* In press

31. Reddy PS, Corley RB. 1998. Assembly, sorting, and exit of oligomeric proteins from the endoplasmic reticulum. *BioEssays* 20:546–54

32. Leyland ML, Dart C, Spencer PJ, Sutcliffe MJ, Stanfield PR. 1999. The possible role of a disulphide bond in forming functional Kir2.1 potassium channels. *Pflügers Arch.* 438:778–81

33. Schonbrunner ER, Schmid FX. 1992. Peptidyl-prolyl *cis-trans* isomerase improves the efficiency of protein disulfide isomerase as a catalyst of protein folding. *Proc. Natl. Acad. Sci. USA* 89:4510–13

34. Hegde RS, Lingappa VR. 1997. Membrane protein biogenesis: regulated complexity at the endoplasmic reticulum. *Cell* 91:575–82

35. Thrift RN, Andrews DW, Walter P, Johnson AE. 1991. A nascent membrane protein is located adjacent to ER membrane proteins throughout its integration and translation. *J. Cell Biol.* 112:809–21

36. Do H, Falcone D, Lin J, Andrews DW, Johnson AE. 1996. The cotranslational integration of membrane proteins into the phospholipid bilayer is a multistep process. *Cell* 85:369–78

37. Borel AC, Simon SM. 1996. Biogenesis of polytopic membrane proteins: membrane segments assemble within translocation channels prior to membrane integration. *Cell* 35:379–89

38. Mothes W, Heinrich SU, Graf R, Nilsson I, von Heijne G, et al. 1997. Molecular mechanism of membrane protein integration into the endoplasmic reticulum. *Cell* 89:523–33

39. Martoglio B, Hofmann MW, Brunner J, Dobberstein B. 1995. The protein-conducting channel in the membrane of the endoplasmic reticulum is open laterally toward the lipid bilayer. *Cell* 81:207–14

40. Popot JL, de Vitry C. 1990. On the microassembly of integral membrane proteins. *Annu. Rev. Biophys. Biophys. Chem.* 19:369–403

41. High S. 1995. Protein translocation at the membrane of the endoplasmic reticulum. *Prog. Biophys. Mol. Biol.* 63:233–50

42. Keller SH, Lindstrom J, Taylor P. 1996. Involvement of the chaperone protein calnexin and the acetylcholine receptor beta-subunit in the assembly and cell surface expression of the receptor. *J. Biol. Chem.* 271:22871–77

43. Keller SH, Lindstrom J, Taylor P. 1998. Inhibition of glucose trimming with castanospermine reduces calnexin association and promotes proteasome degradation of the alpha-subunit of the nicotinic acetylcholine receptor. *J. Biol. Chem.* 273:17064–72

44. Nagaya N, Schulteis CT, Papazian DM. 1999. Calnexin associates with Shaker K⁺ channel protein but is not involved in quality control of subunit folding or assembly. *Receptors Channels* 6:229–39

45. Shi G, Nakahira K, Hammond S, Rhodes KJ, Schechter LE, Trimmer JS. 1996. Beta subunits promote K⁺ channel surface expression through effects early in biosynthesis. *Neuron* 16:843–52

46. Kuryshev YA, Gudz TI, Brown AM, Wible BA. 2000. KChAP as a chaperone for specific K⁺ channels. *Am. J. Physiol. Cell Physiol.* 278:C931–C41

47. Wible BA, Yang Q, Kuryshev YA, Accili EA, Brown AM. 1998. Cloning and expression of a novel K⁺ channel regulatory protein, KChAP. *J. Biol. Chem.* 273:11745–51

48. Jing J, Chikvashvili D, Singer-Lahat D, Thornhill WB, Reuveny E, Lotan I. 1999. Fast inactivation of a brain K⁺ channel composed of Kv1.1 and Kvbeta1.1 subunits modulated by G protein beta gamma subunits. *EMBO J.* 18:1245–56

49. Rosenberg RL, East JE. 1992. Cell-free expression of functional Shaker potassium channels. *Nature* 360:166–69

50. Shen NV, Chen X, Boyer MM, Pfaffinger P. 1993. Deletion analysis of K⁺ channel assembly. *Neuron* 11:67–76

51. Schulteis CT, Nagaya N, Papazian DM. 1998. Subunit folding and assembly steps are interspersed during Shaker potassium channel biogenesis. *J. Biol. Chem.* 273:26210–17

52. Covarrubias M, Wei AA, Salkoff L. 1991. Shaker, Shal, Shab, and Shaw express independent K⁺ current systems. *Neuron* 7:763–73

53. Salkoff L, Baker K, Butler A, Covarrubias M, Pak MD, Wei A. 1992. An essential 'set' of K⁺ channels conserved in flies, mice and humans. *Trends Neurosci.* 15:161–66

54. Xu J, Yu W, Jan YN, Jan LY, Li M. 1995. Assembly of voltage-gated potassium channels. Conserved hydrophilic motifs determine subfamily-specific interactions between the alpha-subunits. *J. Biol. Chem.* 270:24761–68

55. Li M, Jan YN, Jan LY. 1992. Specification of subunit assembly by the hydrophilic amino-terminal domain of the Shaker potassium channels. *Science* 257:1225–30

56. Lee TE, Phillipson LH, Kuznetsov A, Nelson DJ. 1994. Structural determinant for assembly of mammalian K⁺ channels. *Biophys. J.* 66:667–73

57. Tu L, Santarelli V, Deutsch C. 1995. Truncated K$^+$ channel DNA sequences specifically suppress lymphocyte K$^+$ channel gene expression. *Biophys. J.* 68: 147–56

58. Hopkins WF, Demas V, Tempel BL. 1994. Both N- and C-terminal regions contribute to the assembly and functional expression of homo- and heteromultimeric voltage-gated K$^+$ channels. *J. Neurosci.* 14:1385–93

59. Yu W, Xu J, Li M. 1996. NAB domain is essential for the subunit assembly of both alpha-alpha and alpha-beta complexes of Shaker-like potassium channels. *Neuron* 16:441–53

60. Minor DL, Lin YF, Mobley BC, Avelar A, Jan YN, et al. 2000. The polar T1 interface is linked to conformational changes that open the voltage-gated potassium channel. *Cell* 102:657–70

61. Cushman SJ, Nanao MH, Jahng AW, DeRubeis D, Choe S, Pfaffinger PJ. 2000. Voltage dependent activation of potassium channels is coupled to T1 domain structure. *Nat. Struct. Biol.* 7:403–7

62. VanDongen AM, Frech GC, Drewe JA, Joho RH, Brown AM. 1990. Alteration and restoration of K$^+$ channel function by deletions at the N- and C-termini. *Neuron* 5:433–43

63. Attali B, Lesage F, Ziliani P, Guillemare E, Honore E, et al. 1993. Multiple mRNA isoforms encoding the mouse cardiac Kv1–5 delayed rectifier K$^+$ channel. *J. Biol. Chem.* 268:24283–89

64. Kobertz WR, Miller C. 1999. K$^+$ channels lacking the 'tetramerization' domain: implications for pore structure. *Nat. Struct. Biol.* 6:1122–25

65. Tu L, Deutsch C. 1999. Evidence for dimerization of dimers in K$^+$ channel assembly. *Biophys. J.* 76:2004–17

66. Zerangue N, Jan YN, Jan LY. 2000. An artificial tetramerization domain restores efficient assembly of functional Shaker channels lacking T1. *Proc. Natl. Acad. Sci. USA* 97:3591–95

67. Lu J, Robinson MR, Edwards D, Deutsch C. 2001. T1-T1 interactions occur in ER membranes while nascent Kv peptides are still attached to ribosomes. *Biochemistry.* 40:10934–46

68. Rosenthal JJC, Bezanilla F. 2000. Editing of delayed rectifier K$^+$ channel mRNA in squid giant axon. *Biophys. J.* 78:214 (Abstr.)

69. Kreusch A, Pfaffinger PJ, Stevens CF, Choe S. 1998. Crystal structure of the tetramerization domain of the Shaker potassium channel. *Nature* 392:945–48

70. Bixby KA, Nanao MH, Shen NV, Kreusch A, Bellamy H, et al. 1999. Zn^{2+}-binding and molecular determinants of tetramerization in voltage-gated K$^+$ channels. *Nat. Struct. Biol.* 6:38–43

71. Gulbis JM, Zhou M, Mann S, MacKinnon R. 2000. Structure of the cytoplasmic beta subunit-T1 assembly of voltage-dependent K$^+$ channels. *Science* 289: 123–27

72. Kobertz WR, Williams C, Miller C. 2000. Hanging gondola structure of the T1 domain in a voltage-gated K$^+$ channel. *Biochemistry* 39:10347–52

73. Sokolova O, Kolmakova-Partensky L, Grigorieff N. 2001. Three-dimensional structure of a voltage-gated potassium channel at 2.5 nm resolution. *Structure.* In press

74. Aiyar J, Grissmer S, Chandy KG. 1993. Full-length and truncated Kv1.3 K$^+$ channels are modulated by 5-HTic receptor activation and independently by PKC. *Am. J. Physiol. Cell Physiol.* 265:C1571–C78

75. Sheng Z, Skach W, Santarelli V, Deutsch C. 1997. Evidence for interaction between transmembrane segments in assembly of Kv1.3. *Biochemistry* 36:15501–13

76. Tinker A, Jan YN, Jan LY. 1996. Regions responsible for the assembly of inwardly rectifying potassium channels. *Cell* 87:857–68

77. Koster JC, Bentle KA, Nichols CG, Ho K. 1998. Assembly of ROMK1 (Kir 1.1a) inward rectifier K$^+$ channel subunits

involves multiple interaction sites. *Biophys. J.* 74:1821–29

78. Woodward R, Stevens EB, Murrell-Lagnado RD. 1997. Molecular determinants for assembly of G-protein-activated inwardly rectifying K$^+$ channels. *J. Biol. Chem.* 272:10823–30

79. Ludwig J, Owen D, Pongs O. 1997. Carboxy-terminal domain mediates assembly of the voltage-gated rat ether-à-go-go potassium channel. *EMBO J.* 16:6337–45

80. Li X, Xu J, Li M. 1997. The human delta1261 mutation of the HERG potassium channel results in a truncated protein that contains a subunit interaction domain and decreases the channel expression. *J. Biol. Chem.* 272:705–8

81. Daram P, Urbach S, Gaymard F, Sentenac H, Cherel I. 1997. Tetramerization of the AKT1 plant potassium channel involves its C-terminal cytoplasmic domain. *EMBO J.* 16:3455–63

82. Ehrhardt T, Zimmermann S, Muller-Rober B. 1997. Association of plant K$^+$(in) channels is mediated by conserved C-termini and does not affect subunit assembly. *FEBS Lett.* 409:166–70

83. MacKinnon R. 1991. Determination of the subunit stoichiometry of a voltage-activated potassium channel. *Nature* 350:232–35

84. Panyi G, Deutsch C. 1996. Assembly and suppression of endogenous Kv1.3 channels in human T cells. *J. Gen. Physiol.* 107:409–20

85. Deutsch C. 1998. The courtship and marriage of K$^+$ channel subunits. *Biol. Skr. Dan. Vid. Selsk.* 49:107–13

86. Papazian DM, Shao XM, Seoh SA, Mock AF, Huang Y, Wainstock DH. 1995. Electrostatic interactions of S4 voltage sensor in Shaker K$^+$ channel. *Neuron* 14:1293–301

87. Tiwari-Woodruff SK, Schulteis CT, Mock AF, Papazian DM. 1997. Electrostatic interactions between transmembrane segments mediate folding of Sha-

ker K$^+$ channel subunits. *Biophys. J.* 72:1489–500

88. Knaus HG, Garcia-Calvo M, Kaczorowski GJ, Garcia ML. 1994. Subunit composition of the high conductance calcium-activated potassium channel from smooth muscle, a representative of the mSlo and slowpoke family of potassium channels. *J. Biol. Chem.* 269:3921–24

89. Knaus HG, Folander K, Garcia-Calvo M, Garcia ML, Kaczorowski GJ, et al. 1994. Primary sequence and immunological characterization of beta-subunit of high conductance Ca^{2+}-activated K$^+$ channel from smooth muscle. *J. Biol. Chem.* 269:17274–78

90. Dworetzky SI, Boissard CG, Lum-Ragan JT, McKay MC, Post-Munson DJ, et al. 1996. Phenotypic alteration of a human BK (hSlo) channel by hSlo beta subunit coexpression: changes in blocker sensitivity, activation/relaxation and inactivation kinetics, and protein kinase A modulation. *J. Neurosci.* 16:4543–50

91. McManus OB, Helms LM, Pallanck L, Ganetzky B, Swanson R, Leonard RJ. 1995. Functional role of the beta subunit of high conductance calcium-activated potassium channels. *Neuron* 14:645–50

92. Wallner M, Meera P, Ottolia M, Kaczorowski GJ, Latorre R, et al. 1995. Characterization of and modulation by a beta-subunit of a human maxi KCa channel cloned from myometrium. *Receptors Channels* 3:185–99

93. Schopperle WM, Holmqvist MH, Zhou Y, Wang J, Wang Z, et al. 1998. Slob, a novel protein that interacts with the Slowpoke calcium-dependent potassium channel. *Neuron* 20:565–73

94. Xia X, Hirschberg B, Smolik S, Forte M, Adelman JP. 1998. dSLo interacting protein 1, a novel protein that interacts with large-conductance calcium-activated potassium channels. *J. Neurosci.* 18:2360–69

95. Zhou Y, Schopperle WM, Murrey H, Jaramillo A, Dagan D, et al. 1999. A

dynamically regulated 14-3-3, Slob, and Slowpoke potassium channel complex in Drosophila presynaptic nerve terminals. *Neuron* 22:809–18

96. Xia XM, Fakler B, Rivard A, Wayman G, Johnson-Pais T, et al. 1998. Mechanism of calcium gating in small-conductance calcium-activated potassium channels. *Nature* 395:503–7

97. Keen JE, Khawaled R, Farrens DL, Neelands T, Rivard A, et al. 1999. Domains responsible for constitutive and Ca^{2+}-dependent interactions between calmodulin and small conductance Ca($^{2+}$)-activated potassium channels. *J. Neurosci.* 19: 8830–38

98. Joiner WJ, Khanna R, Schlichter LC, Kaczmarek LK. 2001. Ca^{2+}/calmodulin regulates assembly of SK channels. *Soc. Neurosci. Abstr.* 27: In press

99. Clement JP, Kunjilwar K, Gonzalez G, Schwanstecher M, Panten U, et al. 1997. Association and stoichiometry of K$_{(ATP)}$ channel subunits. *Neuron* 18:827–38

100. Inagaki N, Gonoi T, Seino S. 1997. Subunit stoichiometry of the pancreatic beta-cell ATP-sensitive K$^+$ channel. *FEBS Lett.* 409:232–36

101. Shyng S, Nichols CG. 1997. Octameric stoichiometry of the K$_{ATP}$ channel complex. *J. Gen. Physiol.* 110:655–64

102. Babenko AP, Aguilar-Bryan L, Bryan J. 1998. A view of sur/KIR6.X, K$_{ATP}$ channels. *Annu. Rev. Physiol.* 60:667–87

103. Schwappach B, Zerangue N, Jan YN, Jan LY. 2000. Molecular basis for K$_{(ATP)}$ assembly: transmembrane interactions mediate association of a K$^+$ channel with an ABC transporter. *Neuron* 26:155–67

104. Zerangue N, Schwappach B, Jan YN, Jan LY. 1999. A new ER trafficking signal regulates the subunit stoichiometry of plasma membrane K$_{(ATP)}$ channels. *Neuron* 22:537–48

105. Tucker SJ, Ashcroft FM. 1999. Mapping of the physical interaction between the intracellular domains of an inwardly rectifying potassium channel, Kir6.2. *J. Biol. Chem.* 274:33393–97

106. Sanguinetti MC, Curran ME, Zou A, Shen J, Spector PS, et al. 1996. Coassembly of K(V)LQT1 and minK (IsK) proteins to form cardiac I(Ks) potassium channel. *Nature* 384:80–83

107. Barhanin J, Lesage F, Guillemare E, Fink M, Lazdunski M, Romey G. 1996. K(V)LQT1 and lsK (minK) proteins associate to form the I(Ks) cardiac potassium current. *Nature* 384:78–80

108. Abbott GW, Butler MH, Bendahhou S, Ptacek LJ, Goldstein SAN. 2001. MiRP2 is associated with periodic paralysis and forms skeletal muscle potassium channels with Kv3.4. *Biophys. J.* 193a (Abstr.)

109. Zhang M, Jiang M, Tseng G. 2001. hMiRP1 coassembles with rKv4.2 and modulates its gating function. *Biophys. J.* 212a (Abstr.)

110. McDonald TV, Yu Z, Ming Z, Palma E, Meyers MB, et al. 1997. A minK-HERG complex regulates the cardiac potassium current I(Kr). *Nature* 388:289–92

111. Romey G, Attali B, Chouabe C, Abitbol I, Guillemare E, et al. 1997. Molecular mechanism and functional significance of the MinK control of the KvLQT1 channel activity. *J. Biol. Chem.* 272:16713–16

112. Wang W Xia J, Kass RS. 1998. MinK-KvLQT1 fusion proteins, evidence for multiple stoichiometries of the assembled IsK channel. *J. Biol. Chem.* 273:34069–74

113. Kurokawa J, Motoike HK, Kass RS. 2001. TEA$^+$-sensitive KCNQ1 constructs reveal pore-independent access to KCNE1 in assembled I$_{Ks}$ channels. *J. Gen. Physiol.* 117:43–52

114. Wang KW, Tai KK, Goldstein SA. 1996. MinK residues line a potassium channel pore. *Neuron* 16:571–77

115. Tai KK, Goldstein SA. 1998. The conduction pore of a cardiac potassium channel. *Nature* 391:605–8

116. Sesti F, Tai KK, Goldstein SA. 2000.

MinK endows the I(Ks) potassium channel pore with sensitivity to internal tetraethylammonium. *Biophys. J.* 79:1369–78

117. Wang KW, Goldstein SA. 1995. Subunit composition of minK potassium channels. *Neuron* 14:1303–9

118. McCormack T, McCormack K. 1994. Shaker K$^+$ channel beta subunits belong to an NAD(P)H-dependent oxidoreductase superfamily. *Cell* 79:1133–35

119. Chouinard SW, Wilson GF, Schlimgen AK, Ganetzky B. 1995. A potassium channel beta subunit related to the aldoketo reductase superfamily is encoded by the *Drosophila* hyperkinetic locus. *Proc. Natl. Acad. Sci. USA* 92:6763–67

120. Gulbis JM, Mann S, MacKinnon R. 1999. Structure of a voltage-dependent K$^+$ channel beta subunit. *Cell* 97:943–52

121. Rettig J, Heinemann SH, Wunder F, Lorra C, Parcej DN, et al. 1994. Inactivation properties of voltage-gated K$^+$ channels altered by presence of beta-subunit. *Nature* 369:289–94

122. Fink M, Duprat F, Lesage F, Heurteaux C, Romey G, et al. 1996. A new K$^+$ channel beta subunit to specifically enhance Kv2.2 (CDRK) expression. *J. Biol. Chem.* 271:26341–48

123. Sewing S, Roeper J, Pongs O. 1996. Kv beta 1 subunit binding specific for Shaker-related potassium channel alpha subunits. *Neuron* 16:455–63

124. Nakahira K, Shi G, Rhodes KJ, Trimmer JS. 1996. Selective interaction of voltage-gated K$^+$ channel beta-subunits with alpha-subunits. *J. Biol. Chem.* 271:7084–89

125. Wang Z, Kiehn J, Yang Q, Brown AM, Wible BA. 1996. Comparison of binding and block produced by alternatively spliced Kvbeta1 subunits. *J. Biol. Chem.* 271:28311–17

126. Xu J, Li M. 1997. Kvbeta2 inhibits the Kvbeta1-mediated inactivation of K$^+$ channels in transfected mammalian cells. *J. Biol. Chem.* 272:11728–35

127. Accili EA, Kiehn J, Yang Q, Wang Z,

Brown AM, Wible BA. 1997. Separable Kvbeta subunit domains alter expression and gating of potassium channels. *J. Biol. Chem.* 272:25824–31

128. van Huizen R, Czajkowsky DM, Shi D, Shao Z, Li M. 1999. Images of oligomeric Kv beta 2, a modulatory subunit of potassium channels. *FEBS Lett.* 457:107–11

129. Parcej DN, Dolly JO. 1989. Dendrotoxin acceptor from bovine synaptic plasma membranes. Binding properties, purification and subunit composition of a putative constituent of certain voltage-activated K$^+$ channels. *Biochem. J.* 257:899–903

130. Parcej DN, Scott VE, Dolly JO. 1992. Oligomeric properties of alpha-dendrotoxin-sensitive potassium ion channels purified from bovine brain. *Biochemistry* 31:11084–88

131. Xu J, Yu W, Wright JM, Raab RW, Li M. 1998. Distinct functional stoichiometry of potassium channel beta subunits. *Proc. Natl. Acad. Sci. USA* 95:1846–51

132. Gong J, Xu J, Bezanilla M, van Huizen R, Derin R, Li M. 1999. Differential stimulation of PKC phosphorylation of potassium channels by ZIP1 and ZIP2. *Science* 285:1565–69

133. Gilbert A, Jadot M, Leontieva E, Wattiaux-De Coninck S, Wattiaux R. 1998. Delta F508 CFTR localizes in the endoplasmic reticulum-Golgi intermediate compartment in cystic fibrosis cells. *Exp. Cell Res.* 242:144–52

134. Margeta-Mitrovic M, Jan YN, Jan LY. 2000. A trafficking checkpoint controls GABA(B) receptor heterodimerization. *Neuron* 27:97–106

135. Ficker EK, Dennis AT, Obejero-Paz CA, Castaldo P, Taglialatela M, Brown AM. 2001. Retention in the endoplasmic reticulum as a mechanism of dominant-negative current suppression in human long QT syndrome. *Biophys. J.* 80:215a (Abstr.)

136. Zarei MM, Zhu N, Alioua A, Eghbali M, Stefani E, Stefani L. 2001. A novel

Maxi-K splice variant exhibits dominant-negative properties. *Biophys. J.* 80:222a (Abstr.)

137. Ward CL, Omura S, Kopito RR. 1995. Degradation of CFTR by the ubiquitin-proteasome pathway. *Cell* 83:121–27

138. Jensen TJ, Loo MA, Pind S, Williams DB, Goldberg AL, Riordan JR. 1995. Multiple proteolytic systems, including the proteasome, contribute to CFTR processing. *Cell* 83:129–35

139. Keller SH, Taylor P. 1999. Determinants responsible for assembly of the nicotinic acetylcholine receptor. *J. Gen. Physiol.* 113:171–76

140. Myers MP, Khanna R, Papazian DM. 2001. Differential targeting of Shaker mutant proteins to proteasomes. *Biophys. J.* 80:219a (Abstr.)

141. Khanna R, Myers MP, Papazian DM. 2001. Unglycosylated Shaker K^+ channel protein is unstable and degraded by cytoplasmic proteasomes. *Biophys. J.* 80:437a (Abstr.)

142. Ma D, Zerangue N, Lin YF, Collins A, Yu M, et al. 2001. Role of ER export signals in controlling surface potassium channel numbers. *Science* 291:316–19

143. Li D, Takimoto K, Levitan ES. 2000. Surface expression of Kv1 channels is governed by a C-terminal motif. *J. Biol. Chem.* 275:11597–602

144. Khanna R, Myers MP, Laine M, Mock AF, Sandoval B, Papazian DM. 2001. Putative cell surface targeting motif unnecessary for expression of functional Shaker channels. *Biophys. J.* 80:218a (Abstr.)

145. Sharma N, Crane A, Clement JP, Gonzalez G, Babenko AP, et al. 1999. The C terminus of SUR1 is required for trafficking of K_{ATP} channels. *J. Biol. Chem.* 274:20628–32

146. Scannevin RH, Murakoshi H, Rhodes KJ, Trimmer JS. 1996. Identification of a cytoplasmic domain important in the polarized expression and clustering of the Kv2.1 K^+ channel. *J. Cell Biol.* 135:1619–32

147. Trimmer JS, Rhodes KJ. 2001. Hetero-multimer formation in native K^+ channels. In *Potassium Channels in Cardiovascular Biology*, ed. S Archer, N Rusch, pp. 163–75. New York: Kluwer Acad./Plenum

148. Wang H, Kunkel DD, Martin TM, Schwartzkroin PA, Tempel BL. 1993. Heteromultimeric K^+ channels in terminal and juxtaparanodal regions of neurons. *Nature* 365:75–79

149. Rhodes KJ, Strassle BW, Monaghan MM, Bekele-Arcuri Z, Matos MF, Trimmer JS. 1997. Association and colocalization of the Kvbeta1 and Kvbeta2 beta-subunits with Kv1 alpha-subunits in mammalian brain K^+ channel complexes. *J. Neurosci.* 17:8246–58

150. Rasband MN, Trimmer JS, Schwarz TL, Levinson SR, Ellisman MH, et al. 1998. Potassium channel distribution, clustering, and function in remyelinating rat axons. *J. Neurosci.* 18:36–47

151. Cooper EC, Milroy A, Jan YN, Jan LY, Lowenstein DH. 1998. Presynaptic localization of Kv1.4-containing A-type potassium channels near excitatory synapses in the hippocampus. *J. Neurosci.* 18:965–74

152. Tucker SJ, Bond CT, Herson P, Pessia M, Adelman JP. 1996. Inhibitory interactions between two inward rectifier K^+ channel subunits mediated by the transmembrane domains. *J. Biol. Chem.* 271:5866–70

153. Bond CT, Pessia M, Xia XM, Lagrutta A, Kavanaugh MP, Adelman JP. 1994. Cloning and expression of a family of inward rectifier potassium channels. *Receptors Channels* 2:183–91

154. Pessia M, Tucker SJ, Lee K, Bond CT, Adelman JP. 1996. Subunit positional effects revealed by novel heteromeric inwardly rectifying K^+ channels. *EMBO J.* 15:2980–87

155. Kofuji P, Davidson N, Lester HA. 1995. Evidence that neuronal G-protein-gated inwardly rectifying K^+ channels are activated by G beta gamma subunits and function as heteromultimers. *Proc. Natl. Acad. Sci. USA* 92:6542–46

156. Krapivinsky G, Gordon EA, Wickman K, Velimirovic B, Krapivinsky L, Clapham DE. 1995. The G-protein-gated atrial K$^+$ channel I(KACh) is a heteromultimer of two inwardly rectifying K$^+$-channel proteins. *Nature* 374:135–41

157. Corey S, Krapivinsky G, Krapivinsky L, Clapham DE. 1998. Number and stoichiometry of subunits in the native atrial G-protein-gated K$^+$ channel, IKACh. *J. Biol. Chem.* 273:5271–78

158. Silverman SK, Lester HA, Dougherty DA. 1996. Subunit stoichiometry of a heteromultimeric G protein-coupled inward-rectifier K$^+$ channel. *J. Biol. Chem.* 271:30524–28

159. Tucker SJ, Pessia M, Adelman JP. 1996. Muscarine-gated K$^+$ channel: subunit stoichiometry and structural domains essential for G protein stimulation. *Am. J. Physiol. Heart Circ. Physiol.* 271:H379–H85

160. Kennedy ME, Nemec J, Corey S, Wickman K, Clapham DE. 1999. GIRK4 confers appropriate processing and cell surface localization to G-protein-gated potassium channels. *J. Biol. Chem.* 274:2571–82

161. Stevens EB, Woodward R, Ho IH, Murrell-Lagnado R. 1997. Identification of regions that regulate the expression and activity of G protein-gated inward rectifier K$^+$ channels in Xenopus oocytes. *J. Physiol.* 503:547–62

162. Koch RO, Wanner SG, Koschak A, Hanner M, Schwarzer C, et al. 1997. Complex subunit assembly of neuronal voltage-gated K$^+$ channels. Basis for high-affinity toxin interactions and pharmacology. *J. Biol. Chem.* 272:27577–81

163. Shamotienko OG, Parcej DN, Dolly JO. 1997. Subunit combinations defined for K$^+$ channel Kv1 subtypes in synaptic membranes from bovine brain. *Biochemistry* 36:8195–201

164. Coleman SK, Newcombe J, Pryke J, Dolly JO. 1999. Subunit composition of Kv1 channels in human CNS. *J. Neurochem.* 73:849–58

165. Felix JP, Bugianesi RM, Schmalhofer WA, Borris R, Goetz MA, et al. 1999. Identification and biochemical characterization of a novel nortriterpene inhibitor of the human lymphocyte voltage-gated potassium channel, Kv1.3. *Biochemistry* 38:4922–30

166. Manganas LN, Trimmer JS. 2000. Subunit composition determines Kv1 potassium channel surface expression. *J. Biol. Chem.* 275:29685–93

167. Simons K, Ikonen E. 1997. Functional rafts in cell membranes. *Nature* 387:569–72

168. Martens JR, Navarro-Polanco R, Coppock EA, Nishiyama A, Parshley L, et al. 2000. Differential targeting of Shaker-like potassium channels to lipid rafts. *J. Biol. Chem.* 275:7443–46

169. Martens JR, Sakamoto N, Sullivan SA, Grobaski TD, Tamkun MM. 2001. Isoform-specific localization of voltage-gated K$^+$ channels to distinct lipid raft populations. *J. Biol. Chem.* 276:8409–14

170. Kim E, Niethammer M, Rothschild A, Jan YN, Sheng M. 1995. Clustering of Shaker-type K$^+$ channels by interaction with a family of membrane-associated guanylate kinases. *Nature* 378:85–88

171. Kim E, Sheng M. 1996. Differential K$^+$ channel clustering activity of PSD-95 and SAP97, two related membrane-associated putative guanylate kinases. *Neuropharmacology* 35:993–1000

172. Tiffany AM, Manganas LN, Kim E, Hsueh YP, Sheng M, Trimmer JS. 2000. PSD-95 and SAP97 exhibit distinct mechanisms for regulating K$^+$ channel surface expression and clustering. *J. Cell Biol.* 148:147–58

173. Cohen NA, Brenman JE, Snyder SH, Bredt DS. 1996. Binding of the inward rectifier K$^+$ channel Kir 2.3 to PSD-95 is regulated by protein kinase A phosphorylation. *Neuron* 17:759–67

174. Leonoudakis D, Mailliard WS, Wingerd

KL, Clegg DO, Vandenberg CA. 2001. Inward rectifier potassium channel KIR2.2 is associated with SAP97. *J. Cell Sci.* 114:987–98

175. Ohno H, Aguilar RC, Yeh D, Taura D, Saito T, Bonifacino JS. 1998. The medium subunits of adaptor complexes recognize distinct but overlapping sets of tyrosine-based sorting signals. *J. Biol. Chem.* 273:25915–21

176. Tong Y, Brandt GS, Li M, Shapovalov G, Slimko E, et al. 2001. Tyrosine decaging leads to substantial membrane trafficking during modulation of an inward rectifier potassium channel. *J. Gen. Physiol.* 117:103–18

177. Yao X, Huang Y, Kwan HY, Chan P, Segal AS, Desir G. 1998. Characterization of a regulatory region in the N-terminus of rabbit kv1.3. *Biochem. Biophys. Res. Commun.* 249:492–98

178. Glowatzki E, Fakler G, Brandle U, Rexhausen U, Zenner HP, et al. 1995. Subunit-dependent assembly of inward-rectifier K+ channels. *Proc. R. Soc. London Ser. B* 261:251–61

179. Mourez M, Hofnung M, Dassa E. 1997. Subunit interactions in ABC transporters: a conserved sequence in hydrophobic membrane proteins of periplasmic permeases defines an important site of interaction with the ATPase subunits. *EMBO J.* 16:3066–77

180. Bonifacino JS, Cosson P, Klausner RD. 1990. Colocalized transmembrane determinants for ER degradation and subunit assembly explain the intracellular fate of TCR chains. *Cell* 63:503–13

181. Dunten RL, Sahin-Toth M, Kaback HR. 1993. Role of the charge pair aspartic acid-237–lysine-358 in the lactose permease of *Escherichia coli. Biochemistry* 32:3139–45

182. Planells-Cases R, Ferrer-Montiel AV, Patten CD, Montal M. 1995. Mutation of conserved negatively charged residues in the S2 and S3 transmembrane segments of a mammalian K+ channel selectively modulates channel gating. *Proc. Natl. Acad. Sci. USA* 92:9422–26

183. Seoh SA, Sigg D, Papazian DM, Bezanilla F. 1996. Voltage-sensing residues in the S2 and S4 segments of the Shaker K+ channel. *Neuron* 16:1159–67

184. Monks SA, Needleman DJ, Miller C. 1999. Helical structure and packing orientation of the S2 segment in the Shaker K+ channel. *J. Gen. Physiol.* 113:415–23

185. Bibi E, Kaback HR. 1990. In vivo expression of the lacY gene in two segments leads to functional lac permease. *Proc. Natl. Acad. Sci. USA* 87:4325–29

186. Yu H, Kono M, McKee TD, Oprian DD. 1995. A general method for mapping tertiary contacts between amino acid residues in membrane-embedded proteins. *Biochemistry* 34:14963–69

187. Naranjo D, Kolmakova-Partensky L, Miller C. 1997. Expression of split Shaker K+ channels in *Xenopus* oocytes. *Biophys. J.* 72 A11 (Abstr.)

188. Popot JL, Trewhella J, Engelman DM. 1986. Reformation of crystalline purple membrane from purified bacteriorhodopsin fragments. *EMBO J.* 5:3039–44

Annu. Rev. Physiol. 2002. 64:47–67

PROLACTIN: The New Biology of an Old Hormone

Vincent Goffin, Nadine Binart, Philippe Touraine, and Paul A. Kelly

*INSERM Unit 344, Faculty of Medicine Necker, Paris Cedex 15, 75730, France;
e-mail: goffin@necker.fr, binart@necker.fr, touraine@necker.fr, kelly@necker.fr*

Key Words   prolactin receptor, animal models, extrapituitary, mammary gland, cancer

■ **Abstract**   Prolactin (PRL) is a paradoxical hormone. Historically known as the pituitary hormone of lactation, it has had attributed to it more than 300 separate actions, which can be correlated to the quasi-ubiquitous distribution of its receptor. Meanwhile, PRL-related knockout models have mainly highlighted its irreplaceable role in functions of lactation and reproduction, which suggests that most of its other reported target tissues are presumably modulated by, rather than strictly dependent on, PRL. The multiplicity of PRL actions in animals is in direct opposition to the paucity of arguments that suggest its involvement in human pathophysiology other than effects on reproduction. Although many experimental data argue for a role of PRL in the progression of some tumors, such as breast and prostate cancers, drugs lowering circulating PRL levels are ineffective. This observation opens new avenues for research into the understanding of whether local production of PRL is involved in tumor growth and, if so, how extrapituitary PRL synthesis is regulated. Finally, the physiological relevance of PRL variants, such as the antiangiogenic 16K-like PRL fragments, needs to be elucidated. This review is aimed at critically discussing how these recent findings have renewed the manner in which PRL should be considered as a multifunctional hormone.

PROLACTIN

Structure and Regulation

Prolactin (PRL) is a polypeptide hormone discovered more than 70 years ago (1) to be a pituitary factor that stimulates mammary gland development and lactation in rabbits, a function from which its name "pro-lactin" originated (2). The gene encoding PRL is unique and is found in all vertebrates. In humans, it is located on chromosome 6 (3). It was initially described as containing five exons and four introns, for an overall length of 10 kb (4); since then, an additional exon 1a has been described. After removal of the signal peptide (28 residues), the mature form of the protein contains 199 residues (23 kDa). Several variants of PRL resulting from posttranslational modifications have been identified

47

(5), some of which considerably enlarge the field of action of the hormone (see below).

PRL is mainly secreted by lactotrophic cells of the anterior pituitary. It is widely accepted that pituitary PRL secretion is positively and negatively regulated, but it is mainly controlled by inhibitory factors originating from the hypothalamus, the most important of which is dopamine, acting through the D2 subclass of dopamine receptors present in lactotrophs. It is interesting to note that mice in which the PRL receptor (PRLR) gene has been invalidated (6) are hyperprolactinemic, reflecting feedback of PRL on its own secretion (7). This negative regulation may be direct on lactotrophs or indirect via an action on neuroendocrine dopaminergic neurons that have been shown to express PRL receptors. The current view of the regulation of PRL synthesis integrates an extremely wide spectrum of molecules (hormones, neurotransmitters, neuropeptides, etc.), the description of which has been recently reviewed (8).

The PRL gene is regulated at the transcriptional level by two distinct promoters. The proximal promoter, also referred to as the pituitary promoter, covers ~5 kb upstream of the transcription site, in which the 250 bp just before the Cap (capping of polymerase on RNA) site (in exon 1b) are necessary and sufficient for transcription (9). The second promoter, referred to as the extrapituitary promoter, includes ~3 kb upstream of exon 1a (itself located ~5.8 kb upstream from the initiation site) and was initially described as directing PRL expression in lymphoid and decidual cells (10, 11). Depending on promoter usage, PRL mRNAs differ in length by 134 bp, but they encode identical mature protein. The dichotomy of promoter usage (pituitary versus extrapituitary) needs to be revisited in view of recent data, which may considerably enlarge our understanding of the mechanism of action of PRL, especially when acting locally (autocrine/paracrine effect).

PRL Is a Cytokine

The amino acid sequence of PRL is similar to that of two other polypeptide hormones, growth hormone (GH) and placental lactogen (PL). Because these homologous proteins share genomic, structural, immunological, and biological features, they have been grouped within a protein family called the PRL/GH/PL family (12, 13). More recently, PRL/PL/GH hormones have been linked to a still more extended family of proteins, referred to as hematopoietic cytokines (14). Considering PRL as a cytokine is based on both molecular and functional evidence. The first, although indirect, argument is that its receptor is a member of the cytokine receptor superfamily. Second, PRL is predicted to adopt the up-up-down-down four α-helix bundle fold (15) characteristic of hematopoietic cytokines (14). Third, similarly to well-recognized cytokines, PRL was shown to act on cells of the immune system, although considerable controversy exists concerning the true immunomodulatory role of PRL (see below).

PROLACTIN RECEPTOR

Structure and Distribution

Nearly three decades ago, PRLR was identified as a specific, high-affinity, saturable membrane-bound protein (16). In addition to the short PRLR isoform whose cDNA was originally cloned (17), other isoforms referred to as long or intermediate, based on their overall length, have been identified. Within a given species, the extracellular (ligand-binding) domain is identical, with only the cytoplasmic tail differing. Extensive descriptions of these various PRLR isoforms have been published (18). The gene encoding the human PRLR is unique and is located on chromosome 5; it contains at least 10 exons, for an overall length >100 kb (19). It is interesting that the genomic organization (coding sequences of exons) closely parallels the functional/folding domains of the mature proteins (20). The 5' UTR of the PRLR gene contains several promoters whose tissue-specific usage in various species has been proposed (20–22). PRLR is virtually expressed in all organs and/or tissues, and the level can be very low in certain cells, such as those of the immune system; depending on cell type, levels vary from $\sim$200 to $\sim$30,000 receptors per cell. The expression of the various isoforms has been shown to vary as a function of the stage of the estrous cycle, pregnancy, and lactation. However, because of the extremely broad distribution of PRLR, it is currently difficult to propose a general overview of its regulation of expression (18).

The receptors for GH (23), PRL (17), and a few other cytokines were the pioneering members of the superfamily of cytokine receptors (24), which currently includes more than 30 members (25). These receptors share typical features, such as two disulfide bonds and a duplicated Trp-Ser sequence named the WS motif within the receptor extracellular domain (18).

Activation of PRLR

The PRLR binds to at least three types of ligands: PRL, PL, and primate GHs (13). This complicates our understanding of the biological effects induced by PRL in vivo because ligands other than PRL activate the prolactin receptor, especially in humans, where the contribution of GH to PRLR-mediated effects occurs. This multiplicity of PRLR ligands may be one of the molecular reasons why the phenotypes observed in PRL knockout (KO) mice are less severe than in PRLR KO mice, because PLs (synthesized during gestation) can still exert their effects via PRLR in PRL-deficient animals (see below). In this context, it is also noteworthy that the receptor for PLs was recently proposed to be a PRLR-GHR heterodimer, although the biological relevance of this observation remains to be demonstrated (26).

Binding of these ligands to PRLR is the first step of receptor activation. Thus far, and in contrast to most cytokine receptors (27), no accessory membrane protein has been shown to be required for effective PRLR signaling. Several studies argue that PRLR is activated by dimerization (13), which is mediated by a single molecule of ligand (28). This involves two regions (so-called binding sites 1 and 2), each interacting with one molecule of PRLR (13). Based on this mechanism of activation, PRLR antagonists have been designed by introducing sterically hindering residues within the second binding site of its various ligands, which consequently maintain the ability to bind to the receptor through their site 1 but are no longer able to induce its dimerization (29, 30).

Signal Transduction

Once bound to one of its ligands, PRLR triggers intracellular signaling cascades; there is currently no evidence that the type of ligand (PRL, GH, or PL) affects the nature of the signal transmitted into the cell. Like all cytokine receptors, PRLR lacks intrinsic enzymatic activity and transduces its message inside the cell via a wide number of associated kinases, which in turn activate downstream effectors. The main and best-known cascades involve the Jak/Stat pathway, the Ras-Raf-MAPK pathway, and the Src tyrosine kinases, but other transducing proteins are also involved that have been extensively detailed elsewhere (18, 31). Recently, we generated a list of the genes that are activated by PRL in rat Nb2 lymphoma cells, which shed new light on the genomic targets of this hormone (32).

Site-directed mutational studies have identified within the PRLR cytoplasmic domain some features specifically linked to certain transducing properties, such as specific tyrosine residues that can be phosphorylated and participate in recruiting Stats, insulin receptor substrates (IRS), and adaptor proteins to the receptor complex (18). Depending on the presence or absence of these features, the various PRLR isoforms are thus expected to exhibit different signaling properties. For example, the short PRLR is not tyrosine-phosphorylated, which prevents this isoform from interacting directly with SH2-containing proteins, such as Stat factors. However, such interactions may occur via indirect mechanisms mediated by adaptor proteins (33), or by other proteins included in the receptor complex, such as Jak2, whose phosphotyrosines can recruit Stat5 (34). These alternative mechanisms of protein recruitment by PRLR probably require classical structure-function interactions to be revisited in part. It is interesting to note that heterodimerization of different PRLR isoforms produces inactive complexes (35, 36), which might also be of importance in the physiological context because PRL target cells usually express more than a single PRLR isoform.

Protein tyrosine phosphatases are believed to be part of the signaling downregulation network, although their mechanism of action is still poorly understood (33, 37). Recently, the SOCS (suppressor of cytokine signaling) gene family was identified as targets of the Jak/Stat pathway and was shown to encode proteins down-regulating this pathway at the level of activation. Although the involvement of individual SOCS proteins in PRLR signaling has been studied mainly using cell transfection approaches (38), the mechanisms by which endogenous SOCS regulate PRLR signaling in more physiological contexts are still to be elucidated

(39). In contrast to SOCS-2 KO mice whose gigantism presumably identifies this SOCS as a down-regulator of GH function (40), linking any phenotype observed in mice lacking a particular SOCS gene to PRL function has proven to be difficult. Obviously, regulation of SOCS expression is a means by which cross talk between cytokine receptors (including PRLR) could occur, which complicates the interpretation of data provided by KO models.

Finally, another emerging field in PRLR signaling is the occurrence of cross talk with members of other receptor families, such as tyrosine kinases (41, 42) or nuclear receptors (43). Interactions of activated Stats with the latter obviously represents a possible molecular mechanism underlying the integrated regulation of multiple hormone-dependent functions known to involve PRL and, for example, sex steroids.

A NEW LOOK AT THE MULTIPLE FUNCTIONS AND MECHANISMS OF ACTIONS OF PRL

PRL was originally isolated based on its ability to stimulate mammary development and lactation in rabbits, and soon thereafter to stimulate the production of crop milk in pigeons. PRL was also shown to promote the formation and action of the corpus luteum (44). As emphasized by the phenotypes of PRL-related KO models, milk production and reproductive properties are functions that cannot be taken over by other hormones or cytokines. However, the biological role of PRL can no longer be restricted to these actions. We recently listed up to 300 separate functions or molecules activated by PRLR, which we organized into categories related to water and electrolyte balance, growth and development, endocrinology and metabolism, brain and behavior, reproduction, and immunoregulation and protection (18). This extremely broad spectrum of activities should probably be regarded as a panel of functions modulated by, rather than unique to, PRL (or its receptor).

In the next part of this chapter, our goal is not to again make an exhaustive list of these biological functions but rather to suggest why PRL should be considered differently. For this purpose, we summarize some recent experimental findings that may open new avenues of PRL research.

Does PRL Act in the Nucleus?

Although internalization of receptor-bound PRL has been clearly demonstrated, whether hormone-receptor complexes are translocated to the nucleus after internalization has long been debated. Obviously, one of the potential areas of concern is the choice of antibodies used for immunocytochemical studies, which may cross-react with other cellular/nuclear proteins and, hence, lead to misinterpretation of experimental observations. Our group failed to detect PRL or its receptor within the nucleus (45). Others reported that significant nuclear translocation of PRL required the presence of costimulatory factors, such as epidermal growth factor (EGF) or interleukins (46), although the molecular mechanisms underlying this observation are unknown. To delineate how nuclear retrotransport of PRL could occur (PRL lacks any consensus nuclear localization sequence), a yeast two-hybrid screen was performed that identified cyclophilin B (which possesses a nuclear localization sequence) as a candidate for interaction with PRL (46). This interaction is thought to occur within the extracellular space, without affecting PRL affinity for its receptor or activation of signaling pathways (46). Exogenous cyclophilin B was shown to enhance not only anti-PRL immunoreactivity in nuclei of target cells, but also cell proliferation, establishing this protein as a chaperone-facilitating nuclear translocation and mitogenic activity of PRL. However, how these molecules/pathways interact to synergistically increase cell responses remains open to investigation, although interactions between PRL–cyclophilin B complex and transcription factors (e.g., Stats) have been cited (46, 47).

Relevance of Extrapituitary PRL

Although the large majority of circulating PRL is of pituitary origin, in the past few years, interest has been raised in locally produced, extrapituitary PRL (48). Recent reports have identified human umbilical vein endothelial cells (49), prostate (50, 51), and myeloid leukemic cells (52) as new PRL sources.

The control of extrapituitary PRL secretion is still poorly understood, both at the trancriptional and at the stimulatory levels. Conventional wisdom has linked the proximal PRL promoter to pituitary gene expression (involving Pit-1 as major activating transcription factor and dopamine as major negative regulator) and the decidual/lymphocyte promoter to extrapituitary gene expression (independent of Pit-1 and dopamine). It is interesting to note that PRL cDNA obtained from various human breast cancer cell lines or biopsies showed that both types of PRL mRNA (differing in their 5′ UTR) are present, reflecting a duality of promoter usage (53). Still more surprising, a recent study has shown that in the SK–BR-3 human mammary tumor cell line, the pituitary and not the decidual/lymphocyte promoter is active, despite the absence of Pit-1 in mammary cells, and a potential stimulatory role has also been proposed for EGF (I. Manfroid, J.A. Martial & M. Muller, personal communication). This pioneering study should shed new light on possible mechanisms of extrapituitary PRL regulation.

Demonstrating the occurrence of an autocrine-paracrine mechanism in PRL target cells is a tricky issue that requires using neutralizing antibodies, hormone antagonists, specific inhibitors, or PRL antisense oligonucleotides to highlight the effects resulting from inhibition of the endogenous hormone. It has been shown that endogenously produced hormones are active at much lower concentrations than those administered exogenously (54, 55). Therefore, although the contribution of local PRL production to circulating PRL levels is presumably low, it may be sufficient to exert significant activity on its local environment. In human breast cancer cells, PRL synthesis has been detected using various approaches (56, 57), and autocrine secretion has been shown to constitutively activate cell proliferation via activation of Jak2, which in turn activates both ErbB-2 (one of the multiple EGF receptors) and downstream pathways (58). Similarly, endometrial PRL has been proposed to be an active player in the morphological and functional changes undergone by the decidua from implantation to delivery (59, 60). More than showing a local PRL secretion, these studies demonstrate the functionality of the autocrine/paracrine loop.

Specific Functions of PRL Isoforms

Posttranslational modifications are not required for the hormone to be fully active (61). In fact, posttranslational modifications are more often detrimental than beneficial to PRL bioactivity (5). For example, glycosylation lowers biological activity, phosphorylation generates PRL antagonists in some species, and proteolytic cleavage of PRL into 16K PRL abolishes PRLR binding. In this section, we discuss recent findings related to two of these variants, with the aim of integrating their functional specificity within the spectrum of PRL functions.

(PSEUDO)PHOSPHORYLATED PRL   To mimic natural PRL phosphorylation (covalent linkage of a phosphate group to serine and/or threonine residues), a recombinant pseudophosphorylated (PP)-PRL has been engineered by substituting an aspartate for serine 179 in the human sequence (62). As anticipated from former reports (63), PP–human (h)PRL was first reported to act as a PRLR antagonist in vitro (62). Accordingly, this analog was shown to reduce tumor incidence of prostate cancer cells injected into Nude mice (64), and to alter maternal behavior in nulliparous female (65) and development of pup tissues (66) in rats. In contrast, PP-hPRL was also reported to promote lobuloalveolar differentiation and casein expression during rat pregnancy (67) and to be even more potent than wild-type hPRL on bone tissue (68), which indicates that this analog also displays agonistic properties in some circumstances. In our hands, using various in vitro bioassays, this PP analog has been shown to be an agonist with no antagonistic properties (69). Hence, using PP-PRL raises more questions than it solves about the true physiological role of phosphorylated PRL, one of which is whether this analog is a true molecular mimic of phosphorylated PRL in vivo. Obviously, further studies using PP-hPRL will need careful interpretation.

16K PRL   16K PRL was discovered more than 20 years ago as the N-terminal 16-kDa fragment resulting from the proteolysis of rat PRL by acidified mammary extracts (70). Since then, this PRL fragment has received considerable interest from the scientific community. The protease responsible for the cleavage of rat PRL into 16K PRL was identified as cathepsin D, whose implication in tumor progression is relevant (71). 16K PRL was shown to have lost PRLR binding ability but otherwise to have acquired the ability to specifically bind another membrane receptor (72) through which it exerts antiangiogenic activity (73). Although this receptor is still not identified, some of its downstream signaling targets have been elucidated (74–77).

However, many questions related to the biology of 16K PRL remain unanswered. First, although the majority of investigations have used rat 16K PRL, results are much less clear for other species, especially humans, in which PRL was recently reported to be resistant to cathepsin D (78). This contrasts with our findings indicating that hPRL yields partial, but reproducible, proteolysis leading to N-terminal 16K-like PRL fragments when incubated with this protease. Second, because it may be generated both centrally (79) and at the periphery, such as in pulmonary fibroblasts (77) and endothelial cells (49), the site(s) of 16K PRL generation remain(s) to be clearly identified. Hence, whether all sites of extrapituitary PRL synthesis can generate 16K PRL from endogenous 23K PRL or, alternatively, whether circulating PRL is internalized before the proteolyzed form is exported (or both) remains open to investigation. Also, the subcellular compartment(s) in which appropriate proteolysis conditions are found remain(s) to be identified, although one can not discard the possibility that the cleavage takes place in the extracellular milieu. These few examples underscore the necessity of posing the appropriate questions to demonstrate the physiological relevance of PRL fragments in vivo. In humans, although various recombinant forms of 16K hPRL were shown to be antiangiogenic (76), they do not provide any insight into the biological relevance of 16K hPRL in vivo. Also, the question is raised whether the effects of PRL on tumors in vivo should be viewed from a new angle, considering a balance between the mitogenic and angiogenic (pro-tumor) activities of full-length PRL versus the antiangiogenic (anti-tumor) activity of 16K-like PRL (76, 80).

Lessons from Animal Models

One of the most important recent advances in the study of mammalian genes has been the development of techniques to obtain defined mutations in mice. Often the deletion of a gene that has accepted functions from biochemical and cell biological experiments results in unexpected phenotypes. Frequently this takes the form of no or mild phenotypes and evokes the possible existence of redundantly functioning genes. Efforts to define the functions of PRL and its receptor from the phenotypes of null mutants illustrate these complexities. In this section, we describe the phenotypes resulting from null mutation of PRL or PRLR genes, which are presented as expected, unexpected, and controversial phenotypes.

EXPECTED PHENOTYPES

Reproductive phenotypes   A large body of literature attests that lactogenic hormones play a role in reproductive function. Accordingly, PRL$^{-/-}$ female mice are completely infertile. After several matings with males of established fertility, no litters were produced. Each female mated repeatedly at irregular intervals, without entering a state of pseudopregnancy. Estrous cycles were irregular, and individual females failed to establish any consistent pattern of cycling. All these observations led to the conclusion that PRL is essential to female reproduction (81).

PRLR$^{-/-}$ females also showed an absence of pseudopregnancy and an arrest of egg development immediately after fertilization, with only a few reaching the stage of blastocysts. The outcome is a complete sterility. Divergent effects of PRL on the rate of implantation and development of mouse embryos have been reported (18). Uterine preparation for embryo implantation is dependent on continued estrogen and progesterone secretion by the corpus luteum, which in rodents is supported by a functional pituitary during the first half of pregnancy (82). PRL has been shown to stimulate progesterone synthesis by dispersed ovarian cells from mice in midpregnancy (83), demonstrating that lactogenic hormones can directly stimulate ovarian progesterone secretion. Thus, whereas PRLR$^{-/-}$ females cannot implant blastocysts, the defect of the preimplantation egg development can be completely rescued by exogenous progesterone. However, although implantation occurs, full-term pregnancy is not achieved (7).

Both PRL and PRLR genes are expressed in the uterus (84), which suggests that a paracrine/autocrine effect might be involved. Our observations indicate that preventing PRL action by disruption of the PRLR gene alters the maternal decidual transformation in response to the implanting blastocyst, demonstrating an essential role of PRL in reproduction. PRLR expression has also been reported in human endometrial tissue. PRL is known to be expressed in the decidualized human endometrium and secreted into amniotic fluid. By using in situ hybridization histochemistry techniques, PRL specific hybridization signals were distributed over the decidual cells in early and term pregnancy. Knowing that PRLR is expressed in the uterus (85), it would be interesting to determine whether production of PRL or the expression of PRLR is altered in pathological conditions associated with female sterility.

Furthermore, the expression pattern of such progesterone-dependent genes as amphiregulin, COX-1, and Hoxa-10 was similar in wild-type and steroid-supplemented PRLR$^{-/-}$ mice (85). These results suggest that the correction of reproductive deficits by progesterone in PRLR$^{-/-}$ mice is accomplished by proper expression of progesterone-dependent genes that are essential in early pregnancy. Thus, the rescue of pregnancy failure by progesterone and the cause of pregnancy loss at a later stage in PRLR$^{-/-}$ mice cannot be ascribed to an aberrant spatial expression of genes that normally contribute to the establishment of pregnancy. At early stages of pregnancy, uterine PRLR expression is restricted to a subpopulation of undecidualized cells adjacent to the uterine crypt and in the antimesometrial stroma. Although the function of PRLR in these cells is unknown, we cannot exclude their contribution to normal decidual function.

Mammary gland development   PRLR$^{+/-}$ mice showed impaired mammary development and alveolar differentiation during pregnancy, which corresponded to reduced phosphorylation levels of Stat5; they also showed impaired expression of milk protein genes. Development of the glands in these mice was arrested at midpregnancy. Although PRL activated Stat5 only in the epithelium, GH and EGF activated Stat5 preferentially in the stroma. Epithelial PRLRs are required for mammary development and milk protein gene expression during pregnancy. Although GH is not required for alveolar development, we were able to demonstrate its lactogenic function in cultured PRLR-null mammary epithelium. However, ductal development in GHR-null mice was impaired, supporting the notion that GH signals through the stromal compartment. Our findings demonstrate that GH, PRL, and EGF activate Stat5 in separate compartments, which in turn reflects their specific role in ductal and alveolar development and differentiation (86). These results demonstrate that two functional alleles of PRLR are required for efficient lactation and that this phenotype in heterozygotes is primarily due to a deficit in the degree of mammary gland development.

The mammary gland undergoes development in utero, at puberty, and during pregnancy. The essential hormonal factors regulating the later two phases in mice have been established as estrogen, glucocorticoids, and growth hormone during puberty, and estrogen, progesterone, and placental lactogen and/or prolactin during pregnancy (87, 88). These hormones produce some development with each estrous cycle and massive development at pregnancy, which following estrus or weaning never fully regresses, resulting in ever-increasing alveolar and ductal development with each episode (89). Our observations suggest that the epithelial cell proliferation during pregnancy and the postpartum period depends on a threshold of PRLR expression that is not achieved with just one functional allele, given that the level of PRLR is closely controlled in mammary gland (90). Heterozygous mice on C57BL/6 pure background never lactate, even after multiple pregnancies. In contrast, on 129Sv background, mammary gland proliferation is insufficient to insure lactation at the first pregnancy, but further estrous cycles or a single pregnancy lead to the development of a mammary gland capable of producing milk. This demonstrates either that continuous hormonal stimulus can overcome the block, or that compensatory mechanisms are established.

Initial histological investigation of virgin glands of mature wild-type, PRL$^{-/-}$, or PRLR$^{-/-}$ animals indicated no dramatic differences with ductal tissue present, confirming that PRL stimulus is not essential for this stage of development (6, 81). Because PRLR$^{-/-}$ females are sterile, the effect of the receptor mutation on mammary development during pregnancy has been analyzed by transplantating PRLR$^{-/-}$ mammary epithelium into PRLR$^{+/+}$ mammary fat pads cleared of endogenous epithelial cells before puberty (91). These results demonstrate that epithelial PRLR is required not for alveolar bud formation but for lobuloalveolar development.

Behavior   A number of experimental behavioral studies have clearly established PRLR as a regulator of maternal behavior. A deficiency in pup-induced maternal behavior was observed in PRLR$^{-/-}$ and PRLR$^{+/-}$ nulliparous females. Moreover, primiparous PRLR$^{+/-}$ females exhibited a profound deficit in maternal care when challenged with foster pups (92). By contrast, the PRL gene mutation did not prevent female mice from manifesting spontaneous maternal behaviors (81). Otherwise, eating, locomotor activity, sexual behavior, configural learning, and olfactory function exploration were normal in PRLR mutant mice (92).

UNEXPECTED PHENOTYPES

Bone   Although no particular growth phenotype was observed in PRL$^{-/-}$ animals, examination of the calvariae of PRLR$^{-/-}$ embryos indicates lower ossification compared with controls. In PRLR$^{-/-}$ adults, histomorphometric analysis showed decreased bone formation rate and reduced bone mineral density. We identified PRLR mRNA encoding the long form in osteoblasts, but not in osteoclast-like cells, which suggests that a direct effect of PRL on osteoblasts could be required for normal bone formation and maintenance of bone mass (93).

Male fertility   PRL was reported to regulate testosterone production by Leydig cells via modulation of the effects of luteinizing hormone and of the level of its receptor. PRL has been also proposed to be involved in sperm capacitation and to enhance in vitro fertilization rates (94, 95), although others failed to confirm these findings (96). PRL can also influence the function of the accessory reproductive glands (97, 98). We thus expected some abnormalities in PRL-related KO models. In contrast, PRLR$^{-/-}$ (N. Binart, C. Pineau, H. Kercret, A.M. Touzalin, P. Imbert-Bolloré, et al., manuscript in preparation) and PRL$^{-/-}$ (99) males are fully fertile, indicating that PRL is not a key player in the control of male fertility in mice. Histological analysis of testes and accessory glands PRLR$^{-/-}$ did not show any abnormality. Moreover, no alterations were detected in plasma testosterone, and the response to exogenous administration of gonadotropins was also completely normal. Although it was previously shown that PRL can repair the reproductive defect in male pituitary dwarf mice, our data are totally consistent with observation, indicating that PRL deficiency alone is not sufficient to cause male infertility and alterations in basal plasma testosterone concentrations (99).

Development of lacrimal and Harderian glands   Analysis of the PRLR KO model suggests that PRL plays a weak role in establishing the sexual dimorphism of male lacrimal glands. In females, hyperprolactinemia causes a hyperfemale morphology, which suggests a role for PRL in dry-eyes syndrome. PRL is required for porphyrin secretion by the Harderian gland but plays no essential role in the secretory immune function of the lacrimal gland (100).

Metabolic status   PRL and PLs are known to play a role in carbohydrate metabolism through effects on pancreatic insulin production and peripheral insulin sensitivity; however, the roles of lactogens in lipid metabolism are poorly understood (101). Progressive reduction in body weight associated with a reduction in total abdominal fat mass and in leptin concentrations was observed in PRLR$^{-/-}$ females but not in males (102). This reduction in abdominal fat reflects in part the absence of lactogen action in the adipocyte because the presence of PRLR mRNA was demonstrated in white adipose tissue. No apparent decrease of weight has been described in PRL$^{-/-}$ mice, which suggests a role for lactogens in adipose tissue growth and metabolism in pregnancy when PLs are present at a very high level.

CONTROVERSIAL PHENOTYPE: IMMUNITY   Among the very broad spectrum of actions attributed to PRL (18), some (103) remain conflictual; one example is its actual role in the immune system. Indeed, an amazing body of literature has described PRL as involved in proliferation, differentiation, and apoptosis in various immune cells, which argues for the immunomodulatory role of PRL (104). However, no clear immune phenotype could be detected in PRL-related KO models. In PRLR$^{-/-}$, thymic or splenic cellularity, composition of the lymphocyte subset present in primary or secondary lymphoid organs, and immune system development and function appeared unaltered (105). In PRL$^{-/-}$, myelopoiesis and primary lymphopoiesis were not affected (81). These data argue that PRL does not play any critical role in primary lymphocyte development and homeostasis, although an immunomodulatory role is not ruled out.

HUMAN PATHOPHYSIOLOGY

To date, and contrary to all other pituitary hormones, no disease has been linked to any genetic abnormality of PRL or its receptor. This would suggest either that mutations have no detectable effect in vivo, thereby preventing their phenotypic detection, or that such mutations might be lethal and thus never detected. PRL/PRLR KO mice are viable, so it is clear that PRL stimulus is not essential for survival. However, the major reproductive defects in females could explain the lack of genetic transmission.

In the current state of the art, the only characterized pathology related to PRL is hyperprolactinemia, which is frequent and is responsible for almost 25% of menstrual cycle disorders in women. It can result from pituitary adenoma or particular physiological circumstances (pregnancy, stress), or it can be secondary to drug intake. Lactotroph adenomas, which arise from monoclonal expansion of a single cell that has presumably undergone somatic mutation (106), can be treated efficiently by surgery or by dopaminergic agonists. At the other extremity, hypoprolactinemia is not a well-described syndrome, except in the few cases resulting from genetic defects of Pit-1 (which alters production of several pituitary hormones). In view of hypoprolactinemia leading to absence of milk secretion in postpartum PRL resistance in related family members, a genetic hypothesis has recently been suggested (107).

Besides these pathologies linked to abnormal levels of circulating PRL, it can be argued that PRL should be included in the list of factors favoring the proliferation of certain tumors. In vitro studies clearly show that PRL stimulates proliferation of many breast cancer cell lines, irrespective of their estrogen/progesterone receptor status (56, 108). Moreover, in contrast to steroid receptors, PRLR expression has been observed in (almost) all human breast tumors analyzed (109–111). In some instances, a higher level of expression could be detected in tumor versus normal

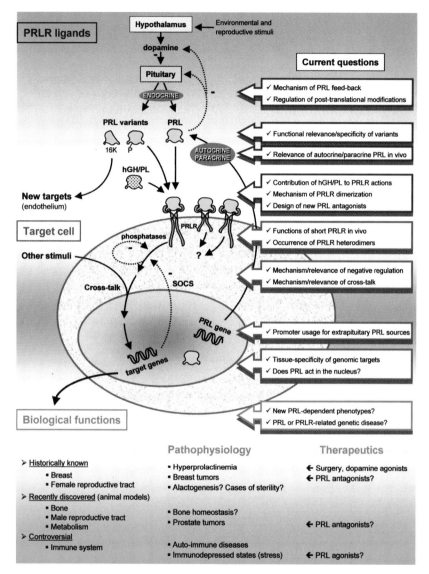

Figure 1 Schematic representation of the prolactin (PRL)/PRL receptor (PRLR) system, which can be divided into three main levels (color coded). The first (*blue*) involves all features concerning PRLR ligands (origin, regulation, nature, posttranslational modification, etc.). The second (*green*) involves the events occurring once PRLR ligands meet their target cells (receptor activation, intracellular signaling, cross talk, gene activation, etc.). The third (*pink*) involves the phenotypic consequences of these molecular/cellular events, i.e., the biological actions of PRLR ligands (including pathophysiological aspects). Some of the major current questions regarding these three levels are indicated on the *right*. Although complex, this figure is not meant to be exhaustive.

breast tissue (111), which could indicate a greater sensitivity of mammary tumors to PRL. In vivo studies using genetically modified mice have shown that elevated PRL levels (PRL transgenic mice) accelerate the rate of spontaneous mammary tumor appearance (112), and that absence of PRL (PRL KO mice) retards the appearance of genetically induced mammary tumors (113). Finally, the recent analysis of the American nurse cohort has clearly shown that breast cancer risk in postmenopausal woman is correlated to high-normal serum PRL levels (114). Despite these convincing reports, the involvement of PRL in the progression of breast cancer has been debated for decades and is still partly disregarded. One of the reasons is that no real benefit was reported when patients suffering from breast cancer were treated with dopamine agonists, although disease stabilization was observed in some cases (115, 116). The failure of dopamine agonist may be linked to the fact that it is local, rather than systemic, PRL that is involved in breast cancer cell proliferation via an autocrine-paracrine mechanism (56, 57, 110). Similar arguments also begin to accumulate for the involvement of PRL in prostate tumor growth (50, 51, 117), the strongest of which resides in the recent observation that transgenic mice overexpressing PRL in the prostate develop dramatic hyperplasia of this organ (118).

As discussed above, there is currently no known inhibitor of local synthesis of PRL. Our group thus turned to the strategy of developing PRL antagonists (29), which may be an alternative to the use of dopamine for the inhibition of PRLR-mediated effects (108). Obviously, in vivo studies will be necessary to confirm whether antagonists complement currently available anti-PRL molecules in clinical use.

Finally, a role of PRL in the pathogenesis of autoimmune diseases has also been suggested (systemic lupus or rheumatoid arthritis). However, experimental more than clinical data suggest such an interaction because most of the therapeutic trials using dopamine agonists in lupic patients have been disappointing (119). Again, because various cells of the immune system express PRL (120), the eventual role of local PRL needs to be addressed. Although the absence of immune phenotype in PRL-related KO models has led to questions about the immunomodulatory role of this hormone, it has recently been suggested that the role of PRL (and GH) in the immune system may be limited to immune responses associated with stress (121). Accordingly, clinical use of PRL in the treatment of immunosuppressed patients suffering from such diseases as AIDS and cancer has been proposed (122).

CONCLUSIONS

Although the main biological functions of PRL are related to its historically described actions on breast and reproductive tract, it can no longer be considered as acting exclusively on these targets. The broad distribution of PRLR and the increasing list of tissues identified as PRL sources are probably correlated to the unusually large number of functions reported for this hormone, some, but obviously not all of which were confirmed by the phenotypes observed in KO models. Until a few years ago, there was a strong discrepancy between the biological versatility of PRL and the paucity of clinical arguments that suggested a role for PRL in human diseases. While awaiting the eventual identification of pathologies resulting from genetic defects of PRL or PRLR, one immediate goal in this field will be to understand how the amazing number of puzzling reports describing targets, mechanisms of actions, or functions of PRL can be linked and integrated into an overall physiological relevance of this old hormone (Figure 1, see color insert). In particular, elucidating the mechanisms of extrapituitary PRL regulation and its real in vivo contribution (especially in tumors), the functional specificity of the numerous PRL variants and PRLR isoforms, and the in vivo impact of signaling cross talks constitute major challenges for the future.

Visit the Annual Reviews home page at www.AnnualReviews.org

LITERATURE CITED

1. Stricker P, Grueter R. 1928. Action du lobe antérieur de l'hypophyse sur la montée laiteuse. *C. R. Soc. Biol.* 99:1978–80

2. Riddle O, Bates RW, Dykshorn SW. 1933. The preparation, identification and assay of prolactin—a hormone of the anterior pituitary. *Am. J. Physiol.* 105:191–216

3. Owerbach D, Rutter WJ, Cooke NE, Martial JA, Shows TB. 1981. The prolactin gene is located on chromosome 6 in humans. *Science* 212:815–16

4. Truong AT, Duez C, Belayew A, Renard A, Pictet R, et al. 1984. Isolation and characterization of the human prolactin gene. *EMBO J.* 3:429–37

5. Sinha YN. 1995. Structural variants of prolactin: occurence and physiological significance. *Endocr. Rev.* 16:354–69

6. Ormandy CJ, Camus A, Barra J, Damotte D, Lucas BK, et al. 1997. Null mutation of the prolactin receptor gene produces multiple reproductive defects in the mouse. *Genes Dev.* 11:167–78

7. Binart N, Helloco C, Ormandy CJ, Barra J, Clement-Lacroix P, et al. 2000. Rescue of preimplantatory egg development and embryo implantation in prolactin receptor-deficient mice after progesterone administration. *Endocrinology* 141:2691–97

8. Freeman ME, Kanyicska B, Lerant A, Nagy G. 2000. Prolactin: structure, function, and regulation of secretion. *Physiol. Rev.* 80:1523–631

9. Peers B, Voz ML, Monget P, Mathy-Hartert M, Berwaer M, et al. 1990. Regulatory elements controlling pituitary-specific expression of the human prolactin gene. *Mol. Cell Biol.* 10:4690–700

10. Gellersen B, Kempf R, Telgmann R, DiMattia GE. 1994. Nonpituitary human prolactin gene transcription is independent of Pit-1 and differentially controlled in lymphocytes and in endometrial stroma. *Mol. Endocrinol.* 8:356–73

11. Berwaer M, Martial JA, Davis JR. 1994. Characterization of an up-stream promoter directing extrapituitary expression of the human prolactin gene. *Mol. Endocrinol.* 8:635–42

12. Miller WL, Eberhardt NL. 1983. Structure and evolution of the growth hormone gene family. *Endocr. Rev.* 4:97–130

13. Goffin V, Shiverick KT, Kelly PA, Martial JA. 1996. Sequence-function relationships within the expanding family of prolactin, growth hormone, placental lactogen and related proteins in mammals. *Endocr. Rev.* 17:385–410

14. Horseman ND, Yu-Lee LY. 1994. Transcriptional regulation by the helix bundle peptide hormones: growth hormone, prolactin, and hematopoietic cytokines. *Endocr. Rev.* 15:627–49

15. Goffin V, Martial JA, Summers NL. 1995. Use of a model to understand prolactin and growth hormone specificities. *Protein Eng.* 8:1215–31

16. Posner BI, Kelly PA, Shiu RP, Friesen HG. 1974. Studies of insulin, growth hormone and prolactin binding: tissue distribution, species variation and characterization. *Endocrinology* 95:521–31

17. Boutin JM, Jolicoeur C, Okamura H, Gagnon J, Edery M, et al. 1988. Cloning and expression of the rat prolactin receptor, a member of the growth hormone/prolactin receptor gene family. *Cell* 53:69–77

18. Bole-Feysot C, Goffin V, Edery M, Binart N, Kelly PA. 1998. Prolactin and its receptor: actions, signal transduction pathways and phenotypes observed in prolactin receptor knockout mice. *Endocr. Rev.* 19:225–68

19. Arden KC, Boutin JM, Djiane J, Kelly PA, Cavenee WK. 1990. The receptors for prolactin and growth hormone are localized in the same region of human chromosome 5. *Cytogenet. Cell Genet.* 53:161–65

20. Ormandy CJ, Binart N, Helloco C, Kelly PA. 1998. Mouse prolactin receptor gene: Genomic organization reveals alternative promoter usage and generation of isoforms via alternative 3′-exon splicing. *DNA Cell Biol.* 17:761–70

21. Moldrup A, Ormandy C, Nagano M, Murthy K, Banville D, et al. 1996. Differential promoter usage in prolactin receptor gene expression: hepatocyte nuclear factor 4 binds to and activates the promoter

preferentially active in the liver. *Mol. Endocrinol.* 10:661–71

22. Hu ZZ, Zhuang L, Meng J, Leondires M, Dufau ML. 1999. The human prolactin receptor gene structure and alternative promoter utilization: the generic promoter hPIII and a novel human promoter hP(N). *J. Clin. Endocrinol. Metab.* 84:1153–56

23. Leung DW, Spencer SA, Cachianes G, Hammonds RG, Collins C, et al. 1987. Growth hormone receptor and serum binding protein: purification, cloning, and expression. *Nature* 330:537–43

24. Bazan F. 1989. A novel family of growth factor receptors: a common binding domain in the growth hormone, prolactin, erythropoietin and IL-6 receptors, and p75 IL-2 receptor β-chain. *Biochem. Biophys. Res. Commun.* 164:788–95

25. Kelly PA, Djiane J, Postel-Vinay MC, Edery M. 1991. The prolactin/growth hormone receptor family. *Endocr. Rev.* 12:235–51

26. Herman A, Bignon C, Daniel N, Grosclaude J, Gertler A, et al. 2000. Functional heterodimerization of prolactin and growth hormone receptors by ovine placental lactogen. *J. Biol. Chem.* 275:6295–301

27. Wells JA, De Vos AM. 1996. Hematopoietic receptor complexes. *Annu. Rev. Biochem.* 65:609–34

28. Elkins PA, Christinger HW, Sandowski Y, Sakal E, Gertler A, et al. 2000. Ternary complex between placental lactogen and the extracellular domain of the prolactin receptor. *Nat. Struct. Biol.* 7:808–15

29. Goffin V, Kinet S, Ferrag F, Binart N, Martial JA, et al. 1996. Antagonistic properties of human prolactin analogs that show paradoxical agonistic activity in the Nb2 bioassay. *J. Biol. Chem.* 271:16573–79

30. Fuh G, Colosi P, Wood WI, Wells JA. 1993. Mechanism-based design of prolactin receptor antagonists. *J. Biol. Chem.* 268:5376–81

31. Yu-Lee LY, Luo G, Book ML, Morris SM. 1998. Lactogenic hormone signal transduction. *Biol. Reprod.* 58:295–301

32. Bole-Feysot C, Perret E, Roustan P, Bouchard B, Kelly PA. 2000. Analysis of prolactin-modulated gene expression profiles during the Nb2 cell cycle using differential screening techniques. *Genome Biol.* 1:0008.1–8.15

33. Ali S, Ali S. 2000. Recruitment of the protein-tyrosine phosphatase SHP-2 to the C-terminal tyrosine of the prolactin receptor and to the adaptor protein Gab2. *J. Biol. Chem.* 275:39073–80

34. Goupille O, Daniel N, Bignon C, Jolivet G, Djiane J. 1997. Prolactin signal transduction to milk protein genes: carboxyterminal part of the prolactin receptor and its tyrosine phosphorylation are not obligatory for Jak2 and Stat5 activation. *Mol. Cell. Endocrinol.* 127:155–69

35. Perrot-Applanat M, Gualillo O, Pezet A, Vincent V, Edery M, et al. 1997. Dominant negative and cooperative effects of mutant forms of prolactin receptor. *Mol. Endocrinol.* 11:1020–32

36. Chang WP, Ye Y, Clevenger CV. 1998. Stoichiometric structure-function analysis of the prolactin receptor signaling domain by receptor chimeras. *Mol. Cell. Biol.* 18:896–905

37. Ali S, Chen Z, Lebrun JJ, Vogel W, Kharitonenkov A, et al. 1996. PTP1D is a positive regulator of the prolactin signal leading to β-casein promoter activation. *EMBO J.* 15:135–42

38. Tomic S, Chughtai N, Ali S. 1999. SOCS-1, -2, -3: selective targets and functions downstream of the prolactin receptor. *Mol. Cell Endocrinol.* 158:45–54

39. Barkai U, Prigent-Tessier A, Tessier C, Gibori GB, Gibori G. 2000. Involvement of SOCS-1, the suppressor of cytokine signaling, in the prevention of prolactin-responsive gene expression in decidual cells. *Mol. Endocrinol.* 14:554–63

40. Metcalf D, Greenhalgh CJ, Viney E, Willson TA, Starr R, et al. 2000. Gigantism in mice lacking suppressor of cytokine signalling-2. *Nature* 405:1069–73

41. Yamauchi T, Ueki K, Tobe K, Tamemoto H, Sekine N, et al. 1997. Tyrosine phosphorylation of the EGF receptor by the kinase Jak2 is induced by growth hormone. *Nature* 390:91–96

42. Fenton SE, Sheffield LG. 1993. Prolactin inhibits epidermal growth factor (EGF)-stimulated signaling events in mouse mammary epithelial cells by altering EGF receptor function. *Mol. Biol. Cell* 4:773–80

43. Stoecklin E, Wissler M, Schaetzle D, Pfitzner E, Groner B. 1999. Interactions in the transcriptional regulation exerted by Stat5 and by members of the steroid hormone receptor family. *J. Steroid Biochem. Mol. Biol* 69:195–204

44. Astwood EB. 1941. The regulation of corpus luteum function by hypophysial luteotrophin. *Endocrinology* 29:309–19

45. Perrot-Applanat M, Gualillo O, Buteau H, Edery M, Kelly PA. 1997. Internalization of prolactin receptor and prolactin in transfected cells does not involve nuclear translocation. *J. Cell. Sci.* 110:1123–32

46. Rycyzyn MA, Reilly SC, O'Malley K, Clevenger CV. 2000. Role of cyclophilin B in prolactin signal transduction and nuclear retrotranslocation. *Mol. Endocrinol.* 14:1175–86

47. Clevenger CV, Rycyzyn MA. 2000. Translocation and action of polypeptide hormones within the nucleus. Relevance to lactogenic transduction. *Adv. Exp. Med. Biol.* 480:77–84

48. Ben-Jonathan N, Mershon JL, Allen DL, Steinmetz RW. 1996. Extrapituitary prolactin: distribution, regulation, functions, and clinical aspects. *Endocr. Rev.* 17:639–69

49. Corbacho AM, Macotela Y, Nava G, Torner L, Duenas Z, et al. 2000. Human umbilical vein endothelial cells express multiple prolactin isoforms. *J. Endocrinol.* 166:53–62

50. Lissoni P, Mandala M, Rovelli F, Casu M, Rocco F, et al. 2000. Paradoxical stimulation of prolactin secretion by L-dopa in metastatic prostate cancer and its possible role in prostate-cancer-related hyperprolactinemia. *Eur. Urol.* 37:569–72

51. Nevalainen MT, Valve EM, Ingleton PM, Nurmi M, Martikainen PM, et al. 1997. Prolactin and prolactin receptors are expressed and functioning in human prostate. *J. Clin. Invest.* 99:618–27

52. Kooijman R, Gerlo S, Coppens A, Hooghe-Peters EL. 2000. Myeloid leukemic cells express and secrete bioactive pituitary-sized 23 kDa prolactin. *J. Neuroimmunol.* 110:252–58

53. Shaw-Bruha CM, Pirrucello SJ, Shull JD. 1997. Expression of the prolactin gene in normal and neoplastic human breast tissues and human mammary cell lines: promoter usage and alternative mRNA splicing. *Breast Cancer Res. Treat.* 44:243–53

54. Kaulsay KK, Zhu T, Bennett W, Lee K, Lobie PE. 2001. The effects of autocrine human growth hormone (hGH) on human mammary carcinoma cell behavior are mediated via the hGH receptor. *Endocrinology* 142:767–77

55. Mertani HC, Zhu T, Goh EL, Lee KO, Morel G, et al. 2001. Autocrine human growth hormone (hGH) regulation of human mammary carcinoma cell gene expression. Identification of CHOP as a mediator of hGH stimulated human mammary carcinoma cell survival. *J. Biol. Chem.* 276:21464–75

56. Ginsburg E, Vonderhaar BK. 1995. Prolactin synthesis and secretion by human breast cancer cells. *Cancer Res.* 55:2591–95

57. Clevenger CV, Plank TL. 1997. Prolactin as an autocrine/paracrine factor in breast cancer. *J. Mammary Gland. Biol. Neopl.* 2:59–68

58. Yamauchi T, Yamauchi N, Ueki K, Sugiyama T, Waki H, et al. 2000. Constitutive tyrosine phosphorylation of ErbB-2 via Jak2 by autocrine secretion of prolactin in human breast cancer. *J. Biol. Chem.* 275:33937–44

59. Jabbour HN, Critchley HO. 2001. Potential roles of decidual prolactin in early pregnancy. *Reproduction* 121:197–205

60. Prigent-Tessier A, Barkai U, Tessier C, Cohen H, Gibori G. 2001. Characterization of a rat uterine cell line, U(III) cells: prolactin (PRL) expression and endogenous regulation of PRL-dependent genes; estrogen receptor beta, alpha(2)-macroglobulin, and decidual PRL involving the Jak2 and Stat5 pathway. *Endocrinology* 142:1242–50

61. Paris N, Rentier-Delrue F, Defontaine A, Goffin V, Lebrun JJ, et al. 1990. Bacterial production and purification of recombinant human prolactin. *Biotechnol. Appl. Biochem.* 12:436–49

62. Chen TJ, Kuo CB, Tsai KF, Liu JW, Chen DY, et al. 1998. Development of recombinant human prolactin receptor antagonists by molecular mimicry of the phosphorylated hormone. *Endocrinology* 139:609–16

63. Wang YF, Walker AM. 1993. Dephosphorylation of standard prolactin produces a more biologically active molecule: evidence for antagonism between nonphosphorylated and phosphorylated prolactin in the stimulation of Nb2 cell proliferation. *Endocrinology* 133:2156–60

64. Xu XL, Kreye E, Kuo CB, Walker AM. 2001. A molecular mimic of phosphorylated prolactin markedly reduced tumor incidence and size when DU145 human prostate cancer cells were grown in nude mice. *Cancer Res.* 61:6098–104

65. Bridges RS, Rigero BA, Byrnes EM, Yang L, Walker AM. 2001. Central infusions of the recombinant human prolactin receptor antagonist, S179D-PRL, delay the onset of maternal behavior in steroid-primed, nulliparous female rats. *Endocrinology* 142:730–39

66. Yang L, Kuo CB, Liu Y, Coss D, Xu X, et al. 2001. Administration of unmodified

prolactin (U-PRL) and a molecular mimic of phosphorylated prolactin (PP-PRL) during rat pregnancy provides evidence that the U-PRL:PP-PRL ratio is crucial to the normal development of pup tissues. *J. Endocrinol.* 168:227–38

67. Chen C, Xu XL, Yang L, Coss D, Walker AM. 2000. Unmodified prolactin promotes ductal and lobulo-alveolar growth, while a molecular mimic of phosphorylated prolactin promotes lobuloalveolus formation and casein gene expression in the pregnant rat mammary gland. *82nd Annu. Meet. Endocr. Soc.*, Toronoto Can. pp. 194–95

68. Coss D, Yang L, Kuo CB, Xu X, Luben RA, et al. 2000. Effects of prolactin on osteoblast alkaline phosphatase and bone formation in the developing rat. *Am. J. Physiol.* 279:1216–25

69. Bernichtein S, Kinet S, Jeay S, Madern M, Martial JA, et al. 2001. S179D-hPRL, a pseudo-phosphorylated human prolactin analog, is an agonist and not an antagonist. *Endocrinology* 142:3950–63

70. Mittra I. 1980. A novel "cleaved prolactin" in the rat pituitary. Part 1. Biosynthesis, characterization and regulatory control. *Biochem. Biophys. Res. Commun.* 95:1750–59

71. Rochefort H, Liaudet-Coopman E. 1999. Cathepsin D in cancer metastasis: a protease and a ligand. *Acta Pathol. Microbiol. Immunol. Scan.* 107:86–95

72. Clapp C, Weiner RI. 1992. A specific, high affinity, saturable binding site for the 16-kilodalton fragment of prolactin on capillary endothelial cells. *Endocrinology* 130:1380–86

73. Clapp C, Martial JA, Guzman RC, Rentier-Delrue F, Weiner RI. 1993. The 16-kilodalton N-terminal fragment of human prolactin is a potent inhibitor of angiogenesis. *Endocrinology* 133:1292–99

74. Martini JF, Piot C, Humeau LM, Struman I, Martial JA, et al. 2000. The antiangiogenic factor 16K PRL induces programmed cell death in endothelial cells

by caspase activation. *Mol. Endocrinol.* 14:1536–49

75. D'Angelo G, Martini JF, Iiri T, Fantl WJ, Martial J, et al. 1999. 16K human prolactin inhibits vascular endothelial growth factor-induced activation of Ras in capillary endothelial cells. *Mol. Endocrinol.* 13:692–704

76. Struman I, Bentzien F, Lee H, Mainfroid V, D'Angelo G, et al. 1999. Opposing actions of intact and N-terminal fragments of the human prolactin/growth hormone family members on angiogenesis: novel mechanism for the regulation of angiogenesis. *Proc. Natl. Acad. Sci. USA* 96:1246–51

77. Corbacho AM, Nava G, Eiserich JP, Noris G, Macotela Y, et al. 2000. Proteolytic cleavage confers nitric oxide synthase inducing activity upon prolactin. *J. Biol. Chem.* 275:13183–86

78. Khurana S, Liby K, Buckley AR, Ben-Jonathan N. 1999. Proteolysis of human prolactin: resistance to cathepsin D and formation of a nonangiostatic, C-terminal 16 K fragment by thrombin. *Endocrinology* 140:4127–32

79. Clapp C, Torner L, Gutiérrez-Ospina G, Alcantara E, Lopez FJ, et al. 1994. Prolactin gene is expressed in the hypothalamo-neurohypophyseal system and the protein is processed into a 14 kDa fragment with 16K prolactin-like activity. *Proc. Natl. Acad. Sci. USA* 91:10384–88

80. Goffin V, Touraine P, Pichard C, Bernichtein S, Kelly PA. 1999. Should prolactin be reconsidered as a therapeutic target in human breast cancer? *Mol. Cell. Endocrinol.* 151:79–87

81. Horseman ND, Zhao W, Montecino-Rodriguez E, Tanaka M, Nakashima K, et al. 1997. Defective mammopoiesis, but normal hematopoiesis, in mice with a targeted disruption of the prolactin gene. *EMBO J.* 16:6926–35

82. Astwood E, Greep R. 1938. A corpus luteum-stimulating substance in the

rat placenta. *Proc. Soc. Exp. Biol. Med.* 38:713–16

83. Galosy S, Talamantes F. 1995. Luteotropic actions of placental lactogens at midpregnancy in the mouse. *Endocrinology* 136:3993–4003

84. Tanaka S, Koibuchi N, Ohtake H, Ohkawa H, Kawatsu T, et al. 1996. Regional comparison of prolactin gene expression in the human decidualized endometrium in early and term pregnancy. *Eur. J. Endocrinol.* 135:177–83

85. Reese J, Binart N, Brown N, Ma WG, Paria BC, et al. 2000. Implantation and decidualization defects in prolactin receptor (PRLR)-deficient mice are mediated by ovarian but not uterine PRLR. *Endocrinology* 141:1872–81

86. Gallego MI, Binart N, Robinson GW, Okagaki R, Coschigano K, et al. 2001. Prolactin, growth hormone and epidermal growth factor activate Stat5 in different compartments of mammary tissue and exert different and overlapping developmental effects. *Dev. Biol.* 229:163–75

87. Neville MC, Daniel CW. 1987. *The Mammary Gland.* New York: Plenum

88. Nandi S. 1958. Endocrine control of mammary gland development and function in the C3H/He Crgl mouse. *J. Natl. Cancer Inst.* 21:1039–63

89. Vonderhaar B. 1988. Regulation of development of the normal mammary gland by hormones and growth factors. In *Breast Cancer: Cellular and Molecular Biology,* ed. M Lippman, R Dickson, pp. 252–66. Norwell, MA: Kluwer Acad.

90. Ormandy CJ, Sutherland RL. 1993. Mechanisms of prolactin receptor regulation in mammary gland. *Mol. Cell. Endocrinol.* 91:C1–6

91. Brisken C, Kaur S, Chavarria TE, Binart N, Sutherland RL, et al. 1999. Prolactin controls mammary gland development via direct and indirect mechanisms. *Dev. Biol.* 210:96–106

92. Lucas BK, Ormandy C, Binart N, Bridges RS, Kelly PA. 1998. Null mutation of prolactin receptor gene produces a defect in maternal behavior. *Endocrinology* 139:4102–7

93. Clement-Lacroix P, Ormandy C, Lepescheux L, Ammann P, Damotte D, et al. 1999. Osteoblasts are a new target for prolactin: analysis of bone formation in prolactin receptor knockout mice. *Endocrinology* 140:96–105

94. Shah GV, Sheth AR. 1979. Is prolactin involved in sperm capacitation? *Med. Hypoth.* 5:909–14

95. Fukuda A, Mori C, Hashimoto H, Noda Y, Mori T, et al. 1989. Effects of prolactin during preincubation of mouse spermatozoa on fertilizing capacity in vitro. *J. In Vitro Fert. Embryo. Transf.* 6:92–97

96. Dodds WG, Fowler J, Peykoff A, Miller KF, Friedman CI, et al. 1990. The effect of prolactin on murine in vitro fertilization and embryo development. *Am. J. Obstet. Gyncecol.* 162:1553–59

97. Costello LC, Franklin RB. 1994. Effect of prolactin on the prostate. *Prostate* 24:162–66

98. Bartke A. 1980. Role of prolactin in reproduction in male mammals. *Fed. Proc.* 39:2577–81

99. Steger RW, Chandrashekar V, Zhao W, Bartke A, Horseman ND. 1998. Neuroendocrine and reproductive functions in male mice with targeted disruption of the prolactin gene. *Endocrinology* 139:3691–95

100. McClellan KA, Robertson FG, Kindblom J, Wennbo H, Tornell J, et al. 2001. Investigation of the role of prolactin in the development and function of the lacrimal and harderian glands using genetically modified mice. *Invest. Ophthalmol. Vis. Sci.* 42:23–30

101. Ling C, Hellgren G, Gebre-Medhin M, Dillner K, Wennbo H, et al. 2000. Prolactin (PRL) receptor gene expression in mouse adipose tissue: increases during lactation and in PRL-transgenic mice. *Endocrinology* 141:3564–72

102. Freemark M, Fleenor D, Driscoll P,

Binart N, Kelly PA. 2001. Body weight and fat deposition in prolactin receptor-deficient mice. *Endocrinology* 142:532–37

103. Nicoll CS. 1980. Ontogeny and evolution of prolactin's functions. *Fed. Proc.* 39:2563–66

104. Matera L. 1996. Endocrine, paracrine and autocrine actions of prolactin on immune cells. *Life Sci.* 59:599–614

105. Bouchard B, Ormandy CJ, Di Santo JP, Kelly PA. 1999. Immune system development and function in prolactin receptor-deficient mice. *J. Immunol.* 163:576–82

106. Herman V, Fagin J, Gonsky R, Kovacs K, Melmed S. 1990. Clonal origin of pituitary adenomas. *J. Clin. Endocrinol. Metab.* 71:1427–33

107. Zargar AH, Masoodi SR, Laway BA, Shah NA, Salahudin M. 1997. Familial puerperal alactogenesis: possibility of a genetically transmitted isolated prolactin deficiency. *Br. J. Obstet. Gynaecol.* 104:629–31

108. Llovera M, Pichard C, Bernichtein S, Jeay S, Touraine P, et al. 2000. Human prolactin (hPRL) antagonists inhibit hPRL-activated signaling pathways involved in breast cancer cell proliferation. *Oncogene* 19:4695–705

109. Mertani HC, Garcia-Caballero T, Lambert A, Gerard F, Palayer C, et al. 1998. Cellular expression of growth hormone and prolactin receptors in human breast disorders. *Int. J. Cancer* 79:202–11

110. Reynolds C, Montone KT, Powell CM, Tomaszewski JE, Clevenger CV. 1997. Expression of prolactin and its receptor in human breast carcinoma. *Endocrinology* 138:5555–60

111. Touraine P, Martini JF, Zafrani B, Durand JC, Labaille F, et al. 1998. Increased expression of prolactin receptor gene assessed by quantitative polymerase chain reaction in human breast tumors versus normal breast tissues. *J. Clin. Endocrinol. Metab.* 83:667–74

112. Wennbo H, Gebre-Medhin M, Gritli-Linde A, Ohlsson C, Isaksson OG, et al. 1997. Activation of the prolactin receptor but not the growth hormone receptor is important for induction of mammary tumors in transgenic mice. *J. Clin. Invest.* 100:2744–51

113. Vomachka AJ, Pratt SL, Lockefeer JA, Horseman ND. 2000. Prolactin gene-disruption arrests mammary gland development and retards T-antigen-induced tumor growth. *Oncogene* 19:1077–84

114. Hankinson SE, Willett WC, Michaud DS, Manson JE, Colditz GA, et al. 1999. Plasma prolactin levels and subsequent risk of breast cancer in postmenopausal women. *J. Natl. Cancer Inst.* 91:629–34

115. Manni A, Boucher AE, Demers LM, Harvey HA, Lipton A, et al. 1989. Endocrine effects of combined somatostatin analog and bromocriptine therapy in women with advanced breast cancer. *Breast Cancer Res. Treat.* 14:289–98

116. Anderson E, Ferguson JE, Morten H, Shalet SM, Robinson EL, et al. 1993. Serum immunoreactive and bioactive lactogenic hormones in advanced breast cancer patients treated with bromocriptine and octreotide. *Eur. J. Cancer* 29A:209–17

117. Leav I, Merk FB, Lee KF, Loda M, Mandoki M, et al. 1999. Prolactin receptor expression in the developing human prostate and in hyperplastic, dysplastic, and neoplastic lesions. *Am. J. Pathol.* 154:863–70

118. Wennbo H, Kindblom J, Isaksson OG, Tornell J. 1997. Transgenic mice overexpressing the prolactin gene develop dramatic enlargement of the prostate gland. *Endocrinology* 138:4410–15

119. Alvarez-Nemegyei J, Cobarrubias-Cobos A, Escalante-Triay F, Sosa-Munoz J, Miranda JM, et al. 1998. Bromocriptine in systemic lupus erythematosus: a double-blind, randomized, placebo-controlled study. *Lupus* 7:414–19

120. Kooijman R, Gerlo S, Coppens A, Hooghe-Peters EL. 2000. Growth hormone and

prolactin expression in the immune system. *Ann. NY Acad. Sci.* 917:534–40

121. Dorshkind K, Horseman ND. 2000. The roles of prolactin, growth hormone, insulin-like growth factor-I, and thyroid hormones in lymphocyte development and function: insights from genetic models of hormone and hormone receptor deficiency. *Endocr. Rev.* 21:292–312

122. Richards SM, Murphy WJ. 2000. Use of human prolactin as a therapeutic protein to potentiate immunohematopoietic function. *J. Neuroimmunol.* 109:56–62

Annu. Rev. Physiol. 2002. 64:69–92

OVULATION: New Dimensions and New Regulators of the Inflammatory-Like Response

JoAnne S. Richards,[1] Darryl L. Russell,[1] Scott Ochsner,[1] and Lawrence L. Espey[2]

[1]Department of Molecular and Cellular Biology, Baylor College of Medicine, Houston, Texas 77030, and [2]Department of Biology, Trinity University, San Antonio Texas 78212; e-mail: joanner@bcm.tmc.edu; rrylr@bmc.tmc.edu; sochsner@bmc.tmc.edu; lespey@trinity.edu

Key Words PR, ADAMTS, TSG-6, cumulus, ovary, follicle

■ **Abstract** Ovulation is a complex process that is initiated by the lutenizing hormone surge and is controlled by the temporal and spatial expression of specific genes. This review focuses on recent endocrine, biochemical, and genetic information that has been derived largely from the identification of new genes that are expressed in the ovary, and from knowledge gained by the targeted deletion of genes that appear to impact the ovulation process. Two main areas are described in most detail. First, because mutant mouse models indicate that appropriate formation of the cumulus matrix is essential for successful ovulation, genes expressed in the cumulus cells and those that control cumulus expansion are discussed. Second, because mice null for the progesterone receptor fail to ovulate and are ideal models for dissecting the critical events downstream of progesterone receptor, genes expressed in mural granulosa cells that regulate the expression of novel proteases are described.

INTRODUCTION

The release of the female germ cell, the ovum, from the ovary is a key event in mammalian reproduction. Termed ovulation, first described elsewhere (1–3), this event has captured the attention of investigators, who have approached the phenomenon in a number of different ways. In simple terms, ovulation is the rupture of follicle at the surface of the ovary, releasing the oocyte. In recent times, ovulation has been likened to an inflammatory response, a description that remains highly germane (4–7). Ovulation is the summation of processes ongoing sequentially as well as simultaneously within several ovarian microenvironments.

To prepare for ovulation, the ovary must undergo a series of closely regulated events. Small follicles must mature to the preovulatory stage, during which the oocyte, granulosa cells, and theca cells acquire specific functional characteristics. The oocyte becomes competent to undergo meiosis, granulosa cells acquire the

0066-4278/02/0315-0069$14.00

ability to produce estrogens and respond to luteinizing hormone (LH) via the LH receptor, and theca cells begin to synthesize increasing amounts of androgens that serve as substrates for the aromatase enzyme in the granulosa cells (Figure 1, see color insert) (for reviews, see 8, 9). The sequence of temporal events that occur during ovulation is initiated in a responsive preovulatory follicle by a surge of LH, which impacts both theca cells and granulosa cells to stimulate cAMP and activate selective protein kinase signaling cascades (8, 10–12). These signaling pathways rapidly induce transcription of specific genes, which are expressed transiently prior to follicle rupture. The induced products initiate or alter additional cell signaling cascades, such as protease-driven cascades, which cause follicular rupture and promote follicular remodeling to form a corpus luteum. Remarkably, many events are spatially restricted to specific microenvironments within the follicle or surrounding interstitial compartments to allow successful expulsion of the cumulus-oocyte complex from the ruptured follicle (13–16).

This review focuses on recent endocrine, biochemical, and genetic information that has been derived largely from the identification of new genes that are expressed in the ovary, and from knowledge gained by the targeted deletion of genes that appear to impact the ovulation process. Although many obligatory events precede cumulus expansion and follicle rupture, less emphasis is placed on new genes expressed prior to the LH surge, or on mutant mice in which ovulation failure is the result of inadequate follicular development that prevents a normal response to LH. Thus, rather than duplicating information reported in a number of earlier review articles (4, 5, 10, 17, 18), this review focuses on knowledge obtained mainly within the past 10 years.

THE PREOVULATORY FOLLICLE: THE GATEWAY TO SUCCESSFUL OVULATION

Maturation of the preovulatory follicle by the combined actions of follicle-stimulating hormone (FSH), estradiol, and various growth factors, such as insulin-like growth factor-1 (IGF-1), sets the stage for subsequent events that are essential for ovulation and luteinization (Figure 1). The preovulatory follicle expresses steroidogenic enzymes necessary for the synthesis of estradiol, which in turn triggers the LH surge. LH then acts on mature follicles to terminate the program of gene expression associated with folliculogenesis. The transcription of genes that control granulosa cell proliferation—IGF-1 (19), FSH receptor (20), estrogen receptor β (ERβ) (21), cyclin D2 (22, 23), and others (12)—is rapidly turned off as a consequence of LH-mediated increases in intracellular cAMP. Expression of genes encoding the steroidogenic enzymes are also rapidly terminated (8). Not surprisingly, the targeted disruption of genes obligatory for follicle maturation precludes ovulation or luteinization. Specifically, in mice null for FSH receptor, FSHβ, LH receptor, LH, IGF-1, IGF-1 receptor, leptin, c-fos, cyclin D2, ERβ, ERα, aromatase, or the corepressor RIP 140, either follicular growth is arrested at a developmentally immature stage or further growth results in the formation of cystic follicles (12, 24–29). An equally severe response (but more remote in mechanistic terms) occurs in mice null for such genes as Wingless/Wnt4 (30) and growth differentiation factor-9 (GDF-9) (31), which are involved in the early organization of the ovary.

In conjunction with the termination of specific gene expression in mature follicles, LH induces genes involved in ovulation (Figures 1 and 2, see color insert). These include the genes for progesterone receptor, (PR) (32, 33), cyclooxygenase-2 (COX-2) (34), CAAT enhancer binding protein beta (C/EBPβ) (35), early growth regulatory factor (Egr-1) (36), and pituitary adenylyl cyclase activating peptide (PACAP) (37, 38). Genes involved in luteinization are also induced rapidly by the LH surge. Some of these include the cell cycle inhibitors, p21CIP and p27KIP, steroidogenic enzymes StAR and P450scc, specific transcription factors Fra2/JunD, protein kinases, and other factors (11, 12, 39).

Although the LH surge simultaneously initiates the processes of ovulation and luteinization, these events are functionally dissociated. In fact, it is critical that the events associated with and controlling ovulation precede those that dictate and finalize the genetic program for luteinization. If the events of luteinization occur too rapidly, as in the PDE4D null mice (40), or if the events associated with ovulation are impaired or delayed, as in the PR null mice (41), oocytes can be trapped within a functional corpus luteum. To this end it has been noted that in species with differing ovulatory time spans following the LH surge, expression of COX-2 occurs at a similar time prior to follicle rupture (42, 43).

The fundamental question is the identity of the principal mediators of the ovulatory process. To clarify this issue, one must consider three functional domains within the follicle: the cumulus cells, the granulosa cells, and the thecal cells.

CUMULUS EXPANSION: THE MICROENVIRONMENT OF AN OOCYTE DURING OVULATION

The pioneering studies of Eppig (9, 44), Salustri et al. (45), and Hess et al. (16) have shown that the matrix on which the cumulus cells move has at least three major components (Figure 3, see color insert). These include hyaluronic acid (HA) (16, 45) and at least two HA binding proteins, tumor necrosis factor (TNF)–stimulated gene (TSG)-6 (46, 47) and the serum-derived inter-α-inhibitor (IαI), also known as inter-α-trypsin inhibitor (ITI) or as serum-derived HA binding protein (SHAP) (13, 14, 16).

HA is a high-molecular-weight (several million daltons), linear, unbranched glycosaminoglycan that consists of alternating glucuronic acid and N-acetyl glucosamine monomers. HA is unique among the glycosaminoglycans in that it is a nonsulfated polymer, it is synthesized without association with a core protein, and its synthesis occurs at the plasma membrane rather than in the Golgi (48, 49).

HA is typically 100–700 times the size of other glycosaminoglycan chains that are attached to proteoglycans. In the ovary, HA is produced by the cumulus cells and granulosa cells adjacent to the antrum and is stabilized in a matrix by covalent coupling to the heavy chain (HC) of IαI (16, 45). This molecular organization allows the phenomenon known as cumulus expansion. In vitro studies show that although HA is produced by cumulus cells, expansion occurs only when IαI enters the follicle or when serum is added to cumulus-oocyte complexes (COC) in vitro. The cumulus-derived matrix also contains other factors, such as the HA binding protein TSG-6 (47) and the proteoglycans brevican and versican (50).

The temporal pattern of cumulus expansion indicates that the LH surge induces the specific genes that are obligatory for expansion to occur (Figures 2 and 3). These include (*a*) COX-2, the rate-limiting enzyme in the synthesis of prostaglandins, such as prostaglandin E2 (PGE2) (34), (*b*) HA synthase-2 (HAS-2), which catalyzes the production of HA (51), and (*c*) TSG-6, which is an HA binding protein (47, 52). Recent studies have shown that ovulation is impaired in mice null for COX-2 (53, 54), and that COCs within the COX-deficient preovulatory follicles fail to undergo cumulus expansion in response to LH (55). Ovulation and cumulus expansion can be restored by exogenous administration of PGE2 or interleukin (IL)-1β (55), indicating that prostaglandins and other signaling pathways are obligatory for both events. EP2, one of the receptors for PGE2, is expressed in the cumulus cells prior to the LH surge, and mice null for EP2 are also infertile and exhibit impaired cumulus expansion (15, 56, 57). Collectively, these observations provide clear evidence that one critical site for PGE2 action in the ovulating follicle is the COC. LH-induced expression of HAS-2 in cumulus cells is altered less in the ovaries of COX-2 or EP2 null mice (S. Ochsner, D.L. Russell & J.S. Richards, submitted for publication). Thus, HAS-2 is a downstream target of LH but not of prostaglandin production (COX-2) or action (PGE2). In contrast, TSG-6 is a confirmed target of prostaglandin action on cumulus cells (S. Ochsner, D.L. Russell & J.S. Richards, submitted for publication). Expression of TSG-6 is selectively reduced in the cumulus cells (but not granulosa cells) of COX-2 and EP2 null mice (S. Ochsner, D.L. Russell & J.S. Richards, submitted for publication). These results indicate that PGE2 regulates expression of TSG-6 within the cumulus microenvironment, and that TSG-6 may play some critical role in expansion of the matrix. Both TSG-6 and HA are expressed several hours prior to any visible physical expansion of the matrix (i.e., dispersion of the cumulus cells away from the oocyte). This suggests that the presence of these molecules is not sufficient for matrix formation or the movement of cumulus cells away from the oocyte.

As mentioned above, a critical requirement for cumulus matrix formation and/or stabilization is the entrance of the serum-derived protein IαI into the follicle. Although IαI is normally excluded from follicular fluid because of its size and the avascular nature of the granulosa cell layer, it enters during ovulation on dissolution of the basal lamina (16). IαI is a complex molecule produced mainly by the liver, and it is present in high concentrations in circulating serum. Of particular relevance to the hypothesis that ovulation is an inflammatory-like response, IαI is known to localize to sites of inflammation (14). IαI is composed of several subunits. One subunit, called the light chain, is also known as bikunin or urinary trypsin inhibitor. Within the IαI molecule, bikunin is covalently (via an ester bond) associated via a chondroitin-sulfate moeity to two additional polypeptides: the HCs of IαI, of which there are three isoforms—HC1, HC2 and HC3, also known as SHAP. In the presence of HA, IαI undergoes a substitution reaction in which the HC (SHAP) is covalently bound to HA, releasing the bikunin-chondroitin-sulfate (urinary trypsin inhibitor) (13). The high degree of covalent linkage between the HCs and HA in the cumulus-oocyte matrix is unprecedented (59) and suggests that the enzyme activity controlling this process is elevated within the follicle. Indeed, studies by Chen et al. (59) have shown high activity of the HC-HA conversion process in mural granulosa cells. Although the enzymatic activity that catalyzes the covalent linkage to HA is essential, the biochemical identity of this converting enzyme is not known. If this putative enzyme is hormonally regulated in the mural granulosa cells, such a condition would also be an important factor in controlling matrix formation.

Mice null for bikunin (the light chain of IαI) can synthesize HCs, but they fail to form the intact IαI complex in serum (13, 14). These mice are infertile and exhibit impaired ovulation, and their COC fail to undergo expansion. Cumulus expansion and ovulation are restored in the bikunin null mice when either IαI or serum is provided. Bikunin itself does not restore cumulus expansion, and it is not found in the matrix. Thus, bikunin is not only a protease inhibitor, it also plays a key role as a "SHAP (HC)-presenting" chaperone that is essential for substitution with HA to occur (13, 14). TSG-6 binds covalently to IαI by mechanisms that are not entirely known, but the reaction appears to involve glycosaminoglycans (60). TSG-6 also binds HA via an HA binding region highly conserved among hyaluron binding proteoglycans, including CD44, the HA cell surface receptor. In this way, one putative role for TSG-6 may be that it serves to bridge HA-HC molecules in the matrix (52, 60, 60a). TSG-6 also has a CUB domain (consisting of the complement subcomponents Clr/Cls, Uegf, Bmp1) that in other proteins is associated with cell differentiation, an indication that TSG-6 may exert other functions as well (61). One possibility is that TSG-6 might be critical for cumulus cell attachment to the matrix.

Collectively, these observations indicate that HA and IαI, as well as COX-2/PGE2/EP2-induced gene products (TSG-6 and others), are critical for cumulus-oocyte matrix formation or cumulus cell differentiation; lack of any one of these factors precludes expansion. However, much more information is needed to validate the role of each factor in cumulus expansion and other aspects of the ovulation process. The putative converting enzyme that links HA to the HC of IαI needs to be identified. One possibility is that the enzyme is regulated in granulosa cells by LH. In addition, it will be important to determine what factor or factors regulate attachment and/or movement of the cumulus cells within the matrix, possibly along the HA chain. Another unresolved question is the identity of the critical cell adhesion molecules. Although one likely candidate is the well-characterized cell surface HA receptor CD44, mice null for CD44 are fertile (62). Another cell surface hyaluronan receptor (HARE) has recently been described, but the main function of this agent appears to be related to endocytosis and it may not be critical for cell attachment (63). Therefore other granulosa cell hyaluron receptors, such as members of the syndecan or glypican families of cell surface proteoglycans, might control cumulus cell attachment to the TSG-6–HA–HC matrix or have other functions related to cumulus cells (64). It is noteworthy that syndecans 1, 3, and 4 are expressed in the ovary. Whereas syndecan 4 appears to be high in atretic follicles, syndecans 1 and 3 are associated with luteinization (65).

The molecular mechanisms by which LH induces expression of HAS-2, COX-2, and TSG-6 genes may be either direct via LH receptors present on cumulus cells (although this idea is controversial) or indirect via the activation of other signaling events in the follicle (Figures 3 and 4, see color insert). The latter has recently gained credence because the previously unknown soluble factor released by the oocyte that is essential for cumulus expansion (9, 45) has been provisionally identified as GDF-9 (66). Purified GDF-9 can induce the expression of both HA and COX-2 in cultures of rodent granulosa cells (66). However, GDF-9 is expressed in oocytes beginning at the small primary follicle stage and continues in the oocyte after ovulation. If GDF-9 is the stimulatory factor, this raises the dilemma of why the expression of HA and COX-2 is restricted to ovulating follicles, i.e., those stimulated by the LH surge. One possible explanation that would link the obligatory requirement of LH action with that of the soluble ooctye-derived factor, GDF-9, is that GDF-9 may need to be modified before it becomes activated. GDF-9, like other members of the TGF-β family, is synthesized as a propeptide, and therefore it is likely present at the surface of the oocyte-cumulus cell junctions, possibly attached to proteoglycans (67) as a latent factor. Induction by LH of a specific protease (factor X) (Figure 4) may be necessary to activate and/or release GDF-9, thereby allowing it to interact with cellular receptors and induce HAS-2 and COX-2 in cumulus cells.

Support for this theory is provided by evidence that a related family member, TGF-β, can enhance cumulus expansion (45). TGF-β can be activated by the matrix metalloproteinase (MMP) MMP-9 and can enhance Ras-mediated induction of COX-2 (68, 69). Within the follicle, MMP-9 (also known as gelatinase B) is induced transiently by the LH surge (70). However, MMP-9 null mice are fertile and exhibit only mild reproductive (ovarian related) defects (71). This suggests that if a protease is essential for activation of GDF-9, it is likely to be a protease other than MMP-9. Because the LH surge induces several proteases (5, 70, 72, 73), these candidates should also be tested for their ability to activate GDF-9, TGF-β, or related TGF-β family members, such as BMP15 (74). Alternatively, LH may lead to the activation of latent TNFα, which is known to induce both COX-2 and TSG-6 in other cell types (52, 75). The TNF-activating protein (ADAM17) is a member of the ADAMs (a disintegrin and metalloproteinase) family of MMPs, which appear to play key roles in reproductive processes (see below). However, mice null for TNFα and TNF receptor type I are fertile (76, 77). In addition, IL-1β as well as PGE2 reverses the infertile phenotype in the COX-2 null mice (55) and facilitates LH-induced ovulation in other models (78, 79). Taken together, these data indicate that although IL-1β can induce COX-2, it may also bypass COX-2 and directly induce TSG-6 (80). It is important to note that in other cell types, IL-1β induces TSG-6 better than TNFα (80). However, a critical role for IL-1β in ovulation seems unlikely because LH induces IL-1β expression selectively in theca cells (79), not cumulus cells, and mice that are null for IL-1 type I receptors are fertile (81). Full resolution of the pathways by which LH induces COX-2 in granulosa cells versus cumulus cells will necessitate stage-specific knockouts (KOs) of GDF-9. Knockouts of TSG-6 are also needed to convincingly show a role for this protein in cumulus expansion in vivo.

FOLLICLE RUPTURE

The Theca Cell Compartment and LH-Regulated Genes

The specific ovulatory role of the theca cells is not clear. For follicle rupture to occur, there must be regulated digestion of the extracellular matrix (ECM) within the theca layers and tunica albuginea at the ovarian surface (5). Although COX-1, a source of ovarian prostaglandins, is expressed in the theca layer (82), mice null for COX-1 are fertile (54), precluding an obligatory role for prostaglandin production within the thecal cell layer. Likewise, thecal cells express a variety of MMPs, including MMP2 (gelatinase A), MMP9 (gelatinase B), MMP14 (MT1-MMP), and MMP19, along with at least one tissue inhibitor of MMPs (TIMP-1) (72, 72a). Whereas MMP2 is expressed exclusively in the theca cells of preantral, preovulatory, and ovulating follicles (83), the other MMPs exhibit more complex patterns of expression. For example, MMP14 is localized in granulosa cells of small and preovulatory follicles but is expressed in theca cells of ovulating follicles. MMP19 is induced by the LH surge and initially appears in theca cells (at 4 h) and then in granulosa cells (at 12 h). The activity of MMP13 (collagenase) is also increased by LH, but the site of its expression and/or activation is not known (84). TIMP-1 has been reported in one study to have an expression pattern similar to that of MMP19 (72). Another study indicates that TIMP-1, TIMP-2, and TIMP-3 are primarily associated with theca cells of luteinizing follicles on proestrous and corpora lutea (85). Although the ovarian expression of MMP-17 (MT-MMP4) is not known, it may be relevant because it is anchored to the surface of cells by glycosylphosphatidylinositol and can be shed by proteolytic activity of members of the ADAMs family (86).

Despite the apparent critical role of MMPs in many biochemical processes (71, 72a, 87, 88), targeted deletion of MMPs has not indicated that individually they exert global effects on embryo development (71) or, of relevance here, ovulation. Mice null for MMP2, TIMP-1, and TIMP-2 are fertile (71, 88–90). Mice null for MMP9 are only slightly subfertile, with defects in placental vascular formation (71). Mice null for MMP14 (MT1-MMP) exhibit severe abnormalities, resulting in death (due to wasting) of most animals between 50–90 days of age (91). This mortality is somewhat surprising because observations of various cell lines show that MMP14 is essential for activation of MMP2 (92). That the MMP14 KO phenotype does not mimic that of the MMP2 KO mice indicates that MMP14 has additional functions, specifically in the breakdown of collagen and possibly the migration of cells on laminin (91). Of note, MMP14 null mice exhibit "no sexual maturation" (91). However, because the cited reference does not provide more specific details, one cannot decipher whether there is an actual defect in ovarian function. Mice null for MMP19 have not yet been reported. Although the expression pattern for this enzyme suggests that it may have some active role in ovulation, there is not sufficient data to conclude that it is essential in ovulation (72). Caution is necessary in view of similar, premature deductions that the temporal and spatial expression of ovarian tissue plasminogen activator (tPA) and urinary palsminogen activator (uPA) indicated that these proteases were important in ovulation. However, more recent data from mice null for these molecules, as well as their substrate plasminogen, now suggest that this is not the case (93). TIMP-2 and TIMP-3 appear to be expressed mostly in corpora lutea (72, 85). That there is a role for TIMP3 at the time of ovulation is especially worthy of consideration because it is the only TIMP that binds tightly to ECM glycosoaminoglycans and inhibits certain members of the ADAM family of MMPs, such as the TNFα converting enzyme (TNF-activating protein) (94) and ADAMTS-4 (95).

The cathepsin proteases (cathepsins B, H, K, L, and S) also have been implicated as regulators of ovarian cell function (96). Of these enzymes only cathepsin H shows strong specific expression in the thecal cell layer, along with moderate expression in granulosa cells. Cathepsin L is expressed in granulosa cells and also corpora lutea (see below). Cathepsin B is localized almost exclusively in the surface epithelium (96).

Based on these observations, the contribution of the thecal cell layer to significant proteolytic events in the ovary at the time of ovulation remains unclear. If ovulation is indeed an inflammatory-like response, perhaps the theca cell layer has some additional role, such as serving as a protective anti-inflammatory barrier to prevent inappropriate release of the oocyte. The reported expression of certain LH-regulated genes involved in metabolizing potentially toxic molecules provides support for such a protective role. Specifically, LH induces distinct members of the aldo-keto reductase (NAD/P oxidoreductases) superfamily. These are carbonyl reductase (97) and the mouse vas deferens protein, MVDP (98). Aldo-keto reductases are important for the detoxification of steroids, prostaglandins, xenobiotic compounds, and other potent lipid molecules that have regulatory functions. It is interesting to note that LH also induces the theca cells to express 3′-hydroxysteroid dehydrogenase, an enzyme that converts active steroids to presumably inactive metabolites (99). Thus, the thecal cell layer acts in part as a protective shield to ensure that toxic levels of compounds do not reach the granulosa cells or the oocyte at an inopportune time. These changes may also serve to protect the theca cells themselves and the ovary in general from exposure to toxic compounds.

The Granulosa Compartment and LH-Regulated Genes

The LH-induced transcription factors in granulosa cells include early growth regulatory factor-1 (Egr-1) (36), CAAT enhancer binding protein beta (C/EBPβ) (35), and progesterone receptor (PR) (32, 33) (Figures 2 and 5, see color insert). Each of these components of the ovulatory process is induced rapidly but is expressed only transiently, with peak levels of message and protein occurring approximately 4 h after the LH surge. Other transcription factors, such as the activator protein-1 family members (e.g., c-Fos, c-Jun, Fra2, and JunD), are induced rapidly and remain elevated during the postovulatory luteal phase (39). Each of these mediators appears to be involved in the functional activity of granulosa cells of ovulating follicles.

Egr-1 is a zinc finger transcription factor that binds regions rich in guanine and cytosine within promoters of numerous genes. It is important to note that Egr-1 often binds overlapping sequences with Sp1, an important transcription factor for several ovarian-expressed genes, such as Sgk, P450scc, and p21CIP (100–102), as well as for MMP14 (103). Egr-1 can exert positive transcriptional events or may exert negative regulation of Sp1, as occurs within its own promoter (D.L. Russell, I. Gonzalez-Robanya, C. Pipaon & J.S. Richards, manuscript in preparation). Mice null for Egr-1 are infertile and exhibit severely impaired synthesis of the β-subunit of the pituitary gonadotropin, LH (105). The ovarian targets of Egr-1 action are not known, but mice null for Egr-1 fail to ovulate and no corpora lutea form. It has also not been resolved whether this is because of the lack of pituitary LH or whether it is due to additional ovarian defects. Although a relationship between Egr-1 and expression of the cell surface hyaluron binding protein CD44 has been suggested (106), mice null for CD44 are normal (62). As discussed below, Egr-1 has been implicated as a regulator of the expression of two proteases thought to impact ovulation (70). Thus, the role of Egr-1 in ovulation-related ovarian events remains to be provided.

C/EBPβ is a member of a family of basic helix-loop-helix transcription factors. Induction of C/EBPβ by LH was first documented in relation to its potential role in the regulation of COX-2 expression in granulosa cells (35). Mice null for C/EBPβ exhibit impaired ovulation and luteinization (107, 108). Similar to the gonads of COX-2 null mice, the ovaries of mice null for C/EBPβ display abnormal vascular morphology and hemorrhagic follicles, indicating that ovulation-related angiogenesis is imparied or altered. By transfection assays and bandshifts, C/EBPβ has been implicated in the regulation not only of COX-2, but also of TSG-6 (80) and StAR (25, 109). However, mice null for C/EBPβ express COX-2 (107) and TSG-6 (S. Ochsner, D.L. Russell & J.S. Richards, submitted for publication), and therefore, neither of these genes appears to directly mediate the actions of C/EPBβ in these cells. Most recently, it has been shown that several genes involved in cholesterol metabolism exhibit altered expression in the C/EBPβ null mice (E. Sterneeck, personal communication). Although these genes may have a role in the process of luteinization, their impact, if any, on ovulation is unclear. Thus, a specific role for C/EBPβ in the ovulation process remains to be determined.

Cyclin D2 regulates cell cycle kinase cascades that are obligatory for entry of cells into the G1 phase of the cell cycle. In the ovary, cyclin D2 is expressed selectively in proliferating granulosa cells of growing follicles (22). Mice null for cyclin D2 fail to ovulate; however, the granulosa cells within the small follicles can be stimulated to differentiate and to express genes associated with ovulation (PR, COX-2) and luteinization (P450scc) (110). How decreased cell number prevents ovulation is not entirely clear but may also relate to appropriate formation of the COC.

PR AND PR-REGULATED GENES: ADAMTS-1 AND CATHEPSIN L   PR is a member of the nuclear receptor superfamily and regulates the numerous functions in reproductive tissues, including the uterus, mammary gland, and ovary. In the ovary, LH rapidly and selectively induces PR in mural granulosa cells of preovulatory follicles (70) (Figure 5). In cultured cells, the effect of LH on PR transcription can be mimicked by cAMP-inducing agonists, such as FSH and forskolin, as well as by the phorbol ester phorbol myristate acetate, but not directly by estradiol (33). Mice null for PR fail to ovulate even when stimulated by exogenous hormones, and these findings support other studies that implicated progesterone as a key player in the ovulatory process (41, 111, 112). Despite the failure of ovulation to occur in PR KO mice, the expression of COX-2, cumulus expansion, and luteinization proceed normally (70). The oocyte remains trapped within the luteinized tissue (70). Thus, molecular targets of PR appear to be those controlling rupture of the follicle, rather than those of luteinization. Recently, two targets of PR action have been identified. These are the ADAMTS-1 (a disintegrin and metalloproteinase with thrombospondin-like repeats) known as METH-1 (in humans) (113, 114) and cathepsin L (70). Expression of both these genes is markedly reduced and/or altered in granulosa cells of PR null mice (Figure 5). Whether or not PR acts directly or indirectly to control the expression of these two distinct proteases remains to be determined.

ADAMTS-1 is a member of the metzincins superfamily of matrix metalloproteinases (MMPs) (115). ADAM proteases are characterized by their multifunctional structure and zinc binding domain (116). ADAMTS-1 was first cloned by differential display reverse transcriptase–polymerase chain reaction from colon 26 cachexigenic tumor cells (117). More recently, using the same procedure to discover differentially expressed genes, ADAMTS-1 was identified as an LH-induced transcript in rat ovary (114). Since the initial discovery in 1997, at least 11 other members of this family have been identified, all of which appear to be highly homologous. With the exception of ADAMTS-1 and ADAMTS-5 (also known as implantin, for its unique expression in placenta), which are colocalized to the same region of human chromosone, 21q21-q22, other members are associated with distinctly different chromosomal regions (118). For example, ADAMTS-9 is localized to a region in which deletions or rearrangements lead to pathogenesis of a number of cancers (119).

In the ovary, ADAMTS-1 is selectively induced by LH in granulosa cells and cumulus cells, with the peak level of mRNA and protein being produced 8–12 h after exposure of ovaries to an ovulatory dose of hCG (a gonadotropin functionally analagous to LH) (70, 114). The peak in ADAMTS-1 transcription occurs after the peak of PR expression but before ovulation, which is usually observed 14–16 h after exposure to ovulatory hormones in mice and rats. It is significant that there are clear data to show that the induction of ADAMTS-1 is drastically reduced in rats when the preovulatory synthesis of progesterone is inhibited with epostane (114) as well as in mice that are null for PR (70). Thus, the temporal pattern of these events indicates that ADAMTS-1 has a critical downstream role in mediating the PR-regulated ovarian activity that culminates in the rupture of a follicle. Because ADAMTS-1 is a multifunctional protein, it is important in future studies to clarify it specific role(s) in ovulation. Three major sites of action are likely (Figure 6, see color insert). First, ADAMTS-1 is a potent active protease that cleaves, among other substrates, the bait region of α2-macroglobulin (120). As an active secreted protease it is likely to initiate one or more proteolytic cascades. Although speculative, one target may be MMP13 (collagenase), which is present in the ECM of the thecal cell layer. Another substrate may be MMP14. Because mice null for this protease fail to exhibit sexual maturity (91), MMP14 may have a function in the ovary. Clearly more studies are needed to identify the downstream MMPs involved in matrix degradation and follicle rupture. These potential preoteolyic actions of ADAMTS-1 relate most closely to observed phenotype of the mice null for PR.

As a protease, ADAMTS-1, like ADAMMTS-4 (also expressed in the ovary), may also control the amount and the cellular location of various proteoglycans. Brevican and versican are present in follicular fluid, perlican is present in the thecal compartment, and such cell surface proteoglycans as syndecan or glycipan may be on either granulosa cells or theca cells (50, 65). Both ADAMTS-1 and ADAMTS-4 have been shown to degrade aggrecan and brevican (121–123). The action of ADAMTS-1 on proteoglycans present in ovarian follicles is highly likely. If ADAMTS-1 also proteolyzes the cell surface ectodomain of the syndecans, the shedding these extracellular domains could release potent biological peptides into the follicular milieu (67). Syndecan 1 and 3 are expressed in healthy follicles, whereas syndecan 4 is expressed in atretic follicles (65). Proteoglycans appear to play a critical role in the physical composition of follicular fluid and are important for cell migration and for other cell functions (45). Proteoglycans can inhibit gonadotropin binding to receptors on the surface of granulosa cells, and they can prevent hormone-mediated induction of cAMP. Thus, ADAMTS-1 could enhance the response of granulosa cells to hormones by reducing the local concentration proteoglycans or by otherwise altering their modulatory function. However, because ovarian ADAMTS-1 is not elevated until 12 h after hCG admdinistration, it is not likely that it influences the initial actions of FSH or LH during ovulation. However, ADAMTS-1 might be regulating the activity of other glycoproteins, such as pituitary adenylyl cyclase activating peptide (PACAP), which is induced by LH in ovulating follicles (124, 125). Alternatively, by altering the local concentrations of proteoglycans, ADAMTS-1 could also regulate the activity of specific growth factors, such as GDF-9, FGF-2 and FGF-7, EGF, TGF-α, or Wnts, whose activity is known to be blocked by proteoglycans (67). Thus, a lack of ADAMTS-1 might prevent the activation of one or more potent bioactive factors in the follicular fluid by preventing their release from the proteoglycans. In addition, anticoagulant heparan sulfate proteoglycans bind antithrombin III and thereby prevent premature clotting of the follicular fluid prior to ovulation (126). Thus, ADAMTS-1 may regulate (i.e., degrade) proteoglycans, thereby allowing clot formation to occur.

A third putative function for ADAMTS-1 in the follicle may be mediated by its ability to interact with specific cellular signaling molecules through disintegrin or thrombospondin motifs at the carboxy terminus of the protein (120). Like some other ADAM proteins, ADAMTS-1 may be a signaling protein that regulates some aspect of granulosa cell function via interactions with specific cell surface G-protein-coupled receptors, integrins, and tetraspan proteins (127). For example (128), one class of protease-activated G-protein-coupled receptors, known as PARs, are activated by proteolytic cleavage of their extracellular domain (129). These PARs are targets of thrombin and cathepsin G (130). ADAMTS-1, like thrombospondin 1 (TS-1), is also a potent antiangiogenic factor that may interact with integrins (113). Although TS-1 and -2 are expressed in ovarian follicles (131), the specific roles of thrombspondins and ADAMTS-1 have not been clearly delineated. Do thrombospondins and ADAMTS-1 act synergistically or individually as antiangiogenic factors in follicular fluid? Because in other situations ADAMTS-1 is a more potent antiangiogenic factor than TS-1, these molecules may have distinct functions. Mutant mice models provide supporting evidence. It is important to note that mice null for TS-1 and -2 are fertile (132, 133), whereas mice null for ADAMTS-1 are severely subfertile, with alterations in ovarian and urogenital morphology (128). In fact, TS-1 null mice exhibit a phenotype distinct from that of ADAMTS-1 but similar to, although less severe than, the phenotype observed in mice null for TGF-β (132). No ovarian phenotypes have been reported for either the TS-1 or TGF-β1 KO mice. As a potent antiangiogenic factor, ADAMTS-1 may regulate blood vessel formation and/or migration in the ruptured follicle. In the absence of ADAMTS-1, the migration of endothelial cells may be altered. How this would impact ovulation is not entirely clear, and evidence for altered vascular growth and organization is lacking in PR KO mice. Thus, which if any of these putative functions of ADAMTS-1 is necessary for ovulation and may contribute to the anovulatory phenotype of the PR KO mice remains to be determined.

Recent studies have shown that in addition to ADAMTS-1, two other family members, ADAMTS-4 and -9, are also expressed in human (119) and rodent (134) ovary. However, the ovarian cell types expressing these two members of the family are not known. The cellular localization of each enzyme is a necessary step in any effort to determine whether these three ADAMTS proteins may have redundant, overlapping, and/or synergistic functions in the ovary. It will be important to assess whether ADAMTS-4 or ADAMTS-9 impact ovulation and whether their expression is regulated by progesterone. That mice null for ADAMTS-1 have severe developmental defects in the kidney, ovary, and urogenital tract indicates that ADAMTS-1 must have some functions in these tissues that are not redundant with other family members (128). The link among these ovarian ADAMTS enzymes to gonadal function is reminiscent of the role of two proteins, GON-1 and MIG-17, that control gonadal formation in *Caenorhabditis elegans* (135, 136). Accordingly it should be noted that GON-1 and MIG-17 are closely related in structure to ADAMTS-1 and ADAMTS-9 (135). Whereas GON-1 is essential for initiating migration of gonadal cells, MIG-17 appears to determine the direction of migration. Also of note, MIG-17 is produced by muscle cells, but it binds to and alters the function of the gonadal cells, which suggests that it localizes to specific matrix or cell surface proteins. Not only is GON-1 synthesized in the muscle cells and in the specialized leader cells, and based on additional defects in gonadogenesis of GON-1 mutants, it is also possible that GON-1 is expressed in gonadal cells. Thus, cell movement and cell orientation appear to be important features of these molecules in *C. elegans*. The biochemical basis for this aspect of cell migration is not known. However, it likely involves both the protease domains of the molecules as well as the ability of their disintegrin- and/or thrombospondin-like domains to bind cell surface receptors, such as specific members of the integrin family as well as proteoglycans. In this manner, ADAMTS-1, ADAMTS-4, and ADAMTS-9 all might regulate specific intracellular signaling events in the mammalian ovary. In this regard, several ADAM proteins are obligatory for other reproductive functions. For example, successful fertilization requires the attachment of sperm to oocytes. This process requires the presence of fertilinβ (ADAM 2) on the surface of the sperm as well as the tetraspan protein CD9 and the integrin proteins $\alpha 6\beta 1$ on the surface of the oocyte (127, 137–139). These examples illustrate complex interactions that, theoretically, could be mediated by ADAMTS-1 in the mammalian ovary. It is also possible that integrin(s) and/or tetraspan proteins are involved. However, based on our limited understanding of ADAMTS-1 in mammalian cells, it is difficult to predict which of its multiple functions might be critical for impacting the process of ovulation. Mutations of ADAMTS-1, ADAMTS-4, and ADAMTS-9 genes are clearly needed to resolve this important area of ovarian cell function and ovulation.

Cathepsin L is another LH and PR-regulated gene in the ovary that was identified by cDNA array technology (70). Cathepsin L is a member of the papain family of enzymes. It is commonly a lysosomal protease, but it is also secreted from certain endocrine cells, such as Sertoli cells of the testis and placental trophoblasts and from certain tumors (140). In cat uterus, cathepsin L is also regulated by progesterone (141). The function of cathepsin L in the ovary appears to be complex. This enzyme is expressed in granulosa cells of follicles at several different stages of development in response to both FSH and LH. In addition, its expression in ovulatory follicles is impaired in PR null mice (70). A functional link between PR-regulated expression of ADAMTS-1 and cathepsin L is not immediately obvious, but this issue will be clarified as more information is gained about the specific roles of these proteases in the ovulation process. Cathepsin L, like cathepsin G, may activate PARs (130).

PACAP and the type-I PACAP receptor (PAC1) are also LH-inducible genes that have been shown to be responsive to regulation by progesterone and PR antagonists in vivo and in vitro (37, 125, 142, 143). Although changes in the expression of PACAP and PAC1 in mice null for PR have not been determined, these results implicate PR in the regulation of cell signaling events during the process of ovulation. PACAP may provide a means for granulosa cells to maintain responsiveness to cAMP under conditions in which the LH receptors have been desensitized by the ovulatory surge of gonadotropins and transcription of the LH receptor gene has been reduced. PACAP has been shown to stimulate progesterone production as well as meiotic maturation in follicle-enclosed, cumulus-enclosed oocytes (142, 143). Thus, PACAP and PAC1 may impact the induction of ADAMTS-1 and cathepsin L that occur later in the ovulation process, as well as other events within the ovulating follicle. Significantly, a potential consensus PR response element has been reported for PACAP promoter but not for known sequences of the ADAMTS-1 or cathepsin L promoters (37). Conversely, PAC1 is expressed in follicles at many stages of development (38).

Other LH-Induced Genes that May Impact the Ovulatory Process: Wnts/Frizzled

Rupture of the ovarian follicle requires cellular signaling events that are appropriately controlled. One class of signaling molecules that has been shown to regulate cell fate, differentiation, and proliferation are the secreted glycoproteins of the Wnt family and their cognate serpentine (some of which are G-protein coupled) receptors of the Frizzled family (144–146). These molecules regulate the morphogenesis of many tissues, including those of the reproductive tract, uterus (147), mammary gland, (145) and ovary (30). Targeted deletion of Wnt4 in mice impacts normal development of the embryonic ovary (30). Female mice null for Wnt4 exhibit a sex-reversed ovarian phenotype that exhibits many characteristics similar to mice null for both ERα and ERβ (148). In each mutant mouse model, ovaries acquire seminiferous tubular-like structures and express genes that control male gonad development, such as Sox-9 and MIS, in an inappropriate temporal pattern. Conversely, in mice null for *Fgf-9* or in mice in which SOX-9 is inactive, gonads of the XY male mice acquire the ovarian program of development (31, 149). Gonads of the *Fgf-9* null mice fail to express SOX-9 and MIS (31). These observations indicate that in the embryonic gonad, Wnt4 may suppress signaling events (FGF-9, Sry, Sox-9, MIS, and others) that dictate development of the embryonic testis and male-specific functions. Of potential interest, SOX9 is a potent regulator of both collagen and proteoglycan (aggrecan) gene expression in bone (150, 151), which suggests that similar activities might be exerted in gonadal development. Concomitantly, Wnt4 supports the female pattern of gene expression. Thus, Wnt4 appears to play an important role in the destiny of female (XX) gonadal cells.

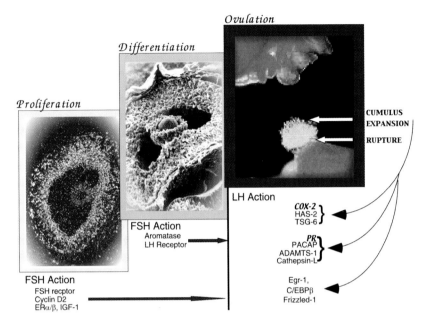

Figure 1 Stage-dependent expression of genes in preovulatory follicles and during ovulation. Granulosa cells of preovulatory follicles respond to FSH, estradiol, and IGF-1; are highly proliferative; and express specific genes that regulate cell cycle progression such as cyclin D2. Granulosa cells also differentiate in response to these hormones and acquire the ability to synthesize estradiol via the aromatase enzyme and to respond to LH via the induction of the LH receptor. The ovulatory surge of LH terminates the follicular program of gene expression (*heavy vertical line*) and induces genes that regulate ovulation. Mice null for the prostaglandin synthesizing enzyme cyclooxygenase-2 (COX-2) and the transcription factor progesterone receptor (PR) have impaired ovulation, indicating that these genes control specific functions during ovulation. The genes regulated by COX-2 and PR are the major topics of this review.

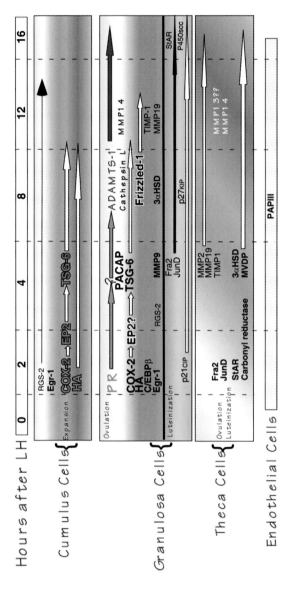

Figure 2 Schematic of LH regulated genes associated with ovulation and luteinization. The LH surge induces specific temporal (*horizontal scale*) and spatial (arranged vertically) patterns of gene expression in various follicular cell types. The cumulus cells and periantral granulosa cells are the major sites of expression of COX-2, EP2, TSG-6, and HA. Along with the serum-derived protein, IαI, HA and TSG-6 assemble to form an extracellular matrix upon which cumulus cells disperse; a process called cumulus expansion. This process is obligatory for successful ovulation. The granulosa cells are the major site of synthesis of potential PR and PR/progesterone-regulated genes such as ADAMTS-1, cathepsin L, and PACAP. Granulosa cells are also the major site of expression of other genes that impact ovulation such as C/EBPβ and Egr-1.

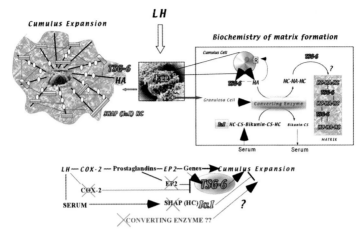

Figure 3 Schematic of cumulus matrix formation: hormonal regulation of the factors and the biochemistry that control this process. Cumulus expansion depends on the synthesis of hyaluronic acid, HA, and the entry of the serum factor, IαI. In the presence of a converting enzyme activity in granulosa cells, the HC of IαI becomes covalently linked to HA, thus forming a complex matrix. Additional components of cumulus cells stabilize this matrix. One such factor is the HA-binding protein TSG-6. Expression of TSG-6 in cumulus cells is dependent on COX-2 and the PGE receptor, EP2. See text for discussion.

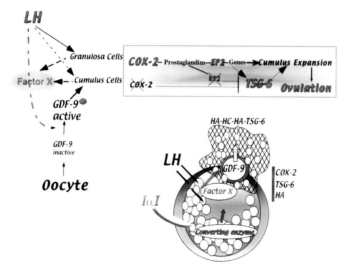

Figure 4 Does induction of COX-2 in cumulus cells by LH require activation of GDF-9 synthesized by the oocytes? GDF-9, a TGF-β family member expressed exclusively in oocytes of growing and ovulating follicles, can induce COX-2 in granulosa cells. However, the LH surge is obligatory for the induction of COX-2 and cumulus expansion. A possible explanation linking the obligatory requirement of LH with that of GDF-9 is that GDF-9 (or another factor) may need to be modified before it becomes activated. GDF-9, like other members of the TGF-β family, is synthesized as a latent, pro-peptide. Therefore, LH may induce and/or activate a protease (Factor X) in mural granulosa cells that can activate and/or release GDF-9 from cumulus cells. Activated GDF-9 could then interact with its cellular receptors and induce HA and COX-2 in cumulus cells. In addition, LH stimulates the dissolution of the basal lamina allowing entry of IαI, which with HA allows the matrix to be formed.

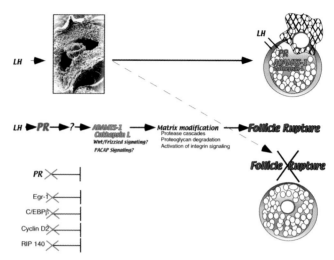

Figure 5 Genes that impact ovulation. Mice null for various transcription factors such as PR, Egr-1, C/EBPβ, and the co-repressor RIP-140 are infertile because the follicles fail to rupture. Mice null for the cell cycle regulator cyclin D2 and for the phosphodiesterase PDE4D also fail to rupture. Of these genes only PR, Egr-1, and C/EBPβ are induced by the LH surge; the others are constitutively present in granulosa cells. Mice null for PR fail to express ADAMTS-1 or cathepsin L in a appropriate manner and also fail to ovulate. Whether PR regulates the expression of ADAMTS-1 and cathepsin L directly or indirectly is not clear and may involve other signaling pathways, as discussed in the text.

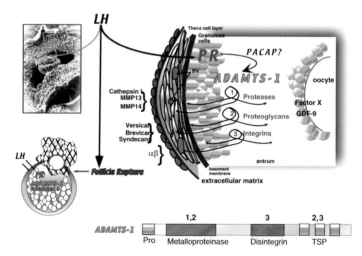

Figure 6 Potential sites of ADAMTS-1 action in the ovulation process. ADAMTS-1 is a multifunctional protein with a metalloproteinase domain, a disintegrin domain, and three thrombospondin(TSP)-like repeats. Which of these functions is most active and most related to the ovulation process is not known. As a protease (1, 2), ADAMTS-1 could alter the structure and function of specific matrix proteins (MMPs) or the proteoglycans. If it binds specific integrins or other cell surface molecules, ADAMTS-1could activate selective cell-signaling pathways. ADAMTS-1 has been shown to be a potent anti-angiogenic factor. Thus one role for this molecule may be to restrict blood vessel formation at a specific time in the ovulation process.

The specific functions of Wnt4 (or other Wnt family members) in the adult ovary are not known. However, recent studies have shown that the adult ovary of rodents expresses several members of the Wnt/Frizzled signaling pathway in a cell- and stage- specific manner (M. Hsieh & J.S. Richards, submitted for publication). For example, Wnt4 is expressed in small, primary follicles in ovaries of immature and adult mice, which suggests that Wnt4 impacts early stages of follicle growth/formation (M. Hsieh & J.S. Richards, submitted for publication) as well as the initial development of the female gonad (30). In addition, Wnt4 is expressed at elevated levels in corpora lutea following the LH surge. At this same time, the Frizzled 4 receptor is also highly expressed, which suggests that it may be a preferred receptor for Wnt4 in this tissue. In the pituitary, Wnt4 appears to control the proliferation of progenitor cells but not the fate of the distinct cell types that eventually develop in the adult pituitary (153). The role of Wnt signaling in the adult pituitary is not known. In mammary epithelial cells, Wnt4 expression colocalizes with that of the progesterone receptor (PR), is regulated in part by progesterone (144), and appears to be involved in ductal branching (i.e., proliferation). Wnt1 and Wnt4 have also been related to oncogenesis in mammary epithelial cells. Likewise, in adipocytes, Wnt10b inhibits differentiation, which suggests that some Wnt signaling is related more to proliferation than to differentiation (154). Based on the temporal expression pattern of Wnt4 in the ovary as well as its normal pattern of expression in ovaries of PR KO mice, this ligand does not appear to be colocalized or a target of PR (or progesterone) in granulosa cells of ovulating follicles (32; M. Hsieh & J.S. Richards, submitted for publication). In contrast, the Frizzled 1 receptor is expressed at high levels in granulosa cells of ovulating follicles between 8–12 h after exposure to LH/hCG (M. Hsieh & J.S. Richards, submitted for publication). This spatial and temporal pattern suggests that Frizzled 1 (and a Wnt ligand yet to be specified) might control unknown events associated with ovulation or effect certain morphogenic changes that occur during the formation of the corpus luteum. The specific functions of Frizzled 1 as well as Wnt4 and Frizzled 4 in the adult ovary need to be determined and will depend on ovarian specific KOs of these genes.

That the Wnt4 null mice share similarities to mice null for ERβ/ERα and inhibin α indicates that these intrafollicular regulators may impact Wnt4 signaling (or vice versa) to control some aspect of follicle formation. Of note, Wnt4 has recently been shown to regulate the expression of DAX-1, a co-repressor of SF-1, thereby impacting gonadal gene expression (154a). The orientation of granulosa cells in a sphere as opposed to a seminiferous tubule is obviously not a random process. What type of molecule could dictate such an orientation? Perhaps it is one or more members of the ADAMTS family that, as mentioned above, are known organizers of gonadal shape and development in *C. elegans* (135). The Wnt/Frizzled pathways and the BMP pathways have been shown to impact and antagonize each other in many aspects of development (155, 156). Thus, Wnts alone or with ADAMTS proteins may modify the spatial actions of various BMP-like molecules (GDF-9, TGF-β, BMP15) or FGF molecules (157) and vice versa in the ovary, as they do in other tissues. These molecules and their interactions will no doubt be topics of investigation for the future.

Pancreatic-associated protein-III (PAPIII) is a member of a group of closely related, lectin-type proteins that are especially abundant in the inflamed pancreas (158, 159). In the ovary, transcripts for PAPIII appear between 4 and 12 h after administration of hCG, with peak levels at 8 h (S. Yoshioka, S. Fujii, J.S. Richards & L.L. Espey, submitted for publication). In situ hybridization shows that PAPIII mRNA is most abundant in endothelial cells that line the inner walls of blood vessels. It is interesting to note that PAPIII is also present in granulosa cells of some small follicles. The presence of PAPIII in the blood vessels, especially those in the hilus of the ovary, provides additional evidence that ovulation is comparable in some ways to an inflammatory reaction. In the ovulatory process, PAPIII may provide a protective role to reduce trauma to the ovarian vasculature during the ovulatory period.

ACKNOWLEDGMENTS

This work was supported, in part, by NIH-HD-16272-16229, SCCPRR-HD07495 (JSR), and National Sciences Foundation Grant 9870793 (LLE).

Visit the Annual Reviews home page at www.AnnualReviews.org

LITERATURE CITED

1. Hartman CG. 1932. Ovulation and the transport and viability of ova and sperm in the female genital tract. In *Sex and Internal Secretions*, ed. E Allen, pp. 647–88. Baltimore, MD: Williams & Wilkins
2. Hisaw FL. 1947. Development of the graafian follicle and ovulation. *Physiol. Rev.* 27:95–119
3. Corner GW. 1963. *The Hormones in Human Reproduction*, pp. 33–75. New York: Athenum
4. Espey LL. 1980. Ovulation as an inflammatory reaction—a hypothesis. *Biol. Reprod.* 22:73–106
5. Espey LL, Lipner H. 1994. Ovulation. In *The Physiology of Reproduction*, ed. E Knobil, JD Neill, pp. 725–80. New York: Raven
6. Parr EL. 1975. Rupture of ovarian follicles at ovulation. *J. Reprod. Fertil.* 22(Suppl.):1–22
7. Parr EL. 1974. Histological examination of the rat ovarian follicle wall prior to ovulation. *Biol. Reprod.* 11:483–503
8. Richards JS. 1994. Hormonal control of gene expression in the ovary. *Endocr. Rev.* 15:725–51
9. Eppig JJ. 1991. Intercommunication between the mammalian oocytes and companion somatic cells. *BioEssays* 13:569–74
10. Richards JS, Robker RL, Russell D, Sharma CS, Espey LE, et al. 2000. Ovulation: a multi-gene, multi-step process. *Steroids* 65:559–70
11. Richards JS. 2001. New signaling pathways for hormones and cyclic adenosine 3′,5′-monophosphate action in endocrine cells. *Mol. Endocrinol.* 15:209–18
12. Richards JS. 2001. Perspective: the ovarian follicle—a perspective in 2001. *Endocrinology* 142:1–10
13. Sato H, Kajikawa S, Kuroda S, Horisawa Y, Nakamura N, et al. 2001.

Impaired fertility in female mice lacking urinary trypsin inhibitor. *Biochem. Biophys. Res. Commun.* 281:1154–60

14. Zhou L, Yoneda M, Zhao M, Yingsung W, Yoshida N, et al. 2001. Defect in SHAP-hyaluronan complex causes severe female infertility: a study by inaction of the bikunin gene in mice. *J. Biol. Chem.* 276:7693–96

15. Hizaki H, Segi E, Sugimoto Y, Hirose M, Saji T, et al. 1999. Abortive expansion of the cumulus and impaired fertility in mice lacking the prostaglandin E receptor subtype EP2. *Proc. Natl. Acad. Sci. USA* 96:10501–6

16. Hess KA, Chen L, Larsen WJ. 1999. Inter-α-inhibitor binding to hyaluronan in the cumulus extracellular matrix is required for optimal ovulation and development of mouse oocytes. *Biol. Reprod.* 61:436–43

17. Espey LL. 1978. See Ref. 5, pp. 503–31

18. Adashi EY, ed. 1998. *Ovulation: Evolving Scientific and Clinical Concepts.* New York: Springer. 335 pp.

19. Zhou J, Kumar TR, Matzuk MM, Bondy C. 1997. Insulin-like growth factor I regulates gonadotropin responsiveness in the murine ovary. *Mol. Endocrinol.* 11:1924–33

20. Richards JS. 1979. Hormonal control of ovarian follicular development: a 1978 perspective. *Recent Prog. Horm. Res.* 35:343–73

21. Sharma SC, Clemens JW, Pisarska MD, Richards JS. 1999. Expression and function of estrogen receptor subtypes in granulosa cells: regulation by estradiol and forskolin. *Endocrinology* 140:4320–34

22. Robker RL, Richards JS. 1998. Hormone-induced proliferation and differentiation of granulosa cells: a coordinated balance of the cell cycle regulators cyclin D2 and p27KIP1. *Mol. Endocrinol.* 12:924–40

23. Robker RL, Richards JS. 1998. Hormonal control of the cell cycle in ovarian cells: proliferation versus differentiation. *Biol. Reprod.* 59:476–82

24. Elvin JA, Matzuk MM. 1998. Mouse models of ovarian failure. *Rev. Reprod.* 3:183–95

25. Orly J. 2000. Molecular events defining follicular developments and steroidogenesis in the ovary. In *Gene Engineering in Endocrinology,* ed. MA Shupkin, pp. 239–75. Totowa, NJ: Humana

26. Couse JF, Korach KS. 1999. Estrogen receptor null mice: What have we learned and where will they lead us? *Endocr. Rev.* 20:358–417

27. Hasegawa T, Zhao L, Caron KM, Majdic G, Suzuki T, et al. 2000. Developmental roles of the steroidogenic acute regulatory protein (StAR) as revealed by StAR knockout mice *Mol. Endocrinol.* 14:1462–71

28. White R, Leonardsson G, Rosewell I, Jacobs MA, Milligan S, Parker M. 2000. The nuclear receptor co-repressor Nrip1 (RIP140) is essential for female fertility. *Nat. Med.* 6:1368–73

29. Johnson RS, Spiegelman BM, Papaioannou V. 1992. Pleiotropic effects of a null mutation in the c-fos proto-oncogene. *Cell* 71:577–86

30. Vainio S, Heikkila M, Kispert A, Chin N, McMahon AP. 1999. Female development in mammals is regulated by Wnt-4 signalling. *Nature* 397:405–9

31. Colvin JS, Green RP, Schmahl J, Capel B, Ornitz DM. 2001. Male-to-female sex reversal in mice lacking fibroblast growth factor 9. *Cell* 104:875–89

32. Park O-K, Mayo K. 1991. Transient expression of progesterone receptor messenger RNA in ovarian granulosa cells after the preovulatory luteinizing hormone surge. *Mol. Endocrinol.* 5:967–78

33. Natraj U, Richards JS. 1993. Hormonal regulation, localization and functional activity of the progesterone receptor in granulosa cells of rat preovulatory follicles. *Endocrinology* 133:761–69

34. Sirois J, Simmons DL, Richards JS. 1992. Hormonal regulation of messenger ribonucleic acid encoding a novel isoform

of prostaglandin endoperoxide H synthase in rat preovulatory follicles. *J. Biol. Chem.* 267:11586–92

35. Sirois J, Richards JS. 1993. Transcriptional regulation of the rat prostaglandin endoperoxide synthase 2 gene in granulosa cells. *J. Biol. Chem.* 268:21931–38

36. Espey LL, Ujoka T, Russell DL, Skelsey M, Vladu B, et al. 2000. Induction of early growth response protein-1 (Egr-1) gene expression in the rat ovary in response to an ovulatory dose of hCG. *Endocrinology* 141:2385–91

37. Park J II, Kim W-J, Wang L, Park H-J, Lee J, et al. 2000. Involvement of progesterone in gonadotropin-induced pituitary adenylate cyclase-activating polypeptide gene expression in preovulatory follicles of rat ovary. *Mol. Hum. Reprod.* 6:238–45

38. Park H-J, Lee J, Wang L, Park J-H, Kwon H-B, et al. 2000. Stage-specific expression of pituitary adenylate cyclase-activating polypeptide type I receptor messenger RNA during ovarian follicle development in the rat. *Endocrinology* 141:702–9

39. Sharma CS, Richards JS. 2000. Regulation of AP1 (Jun/Fos) factor expression and activation in ovarian granulosa cells: relation of JunD and Fra2 to terminal differentiation. *J. Biol. Chem.* 275:33718–28

40. Jin S-LC, Richard FJ, Kuo W-P, D'Ercoel AJ, Conti M. 1999. Imparied growth and fertility of cAMP-specific phosphodiesterase PDE4D-deficient mice. *Proc. Natl. Acad. Sci. USA* 96:11998–2003

41. Lydon JP, DeMayo F, Funk CR, Mani SK, Hughes AR, et al. 1995. Mice lacking progesterone receptor exhibit reproductive abnormalities. *Genes Dev.* 9:2266–78

42. Liu J, Carriere PD, Dore M, Sirois J. 1997. Prostaglandin G/H synthase-2 is expressed in bovine preovulatory follicles after the endogenous LH surge of luteinizing hormone. *Biol. Reprod.* 57:1524–31

43. Boerboom D, Sirois J. 1998. Molecualr characterization of equine prostaglandin

G/H synthase-2 and regulation of its messenger ribonucleic acid in preovulatory follicles. *Endocrinology* 139:1662–70

44. Eppig JJ. 1979. FSH stimulates hyaluronic acid synthesis by oocyte-cumulus cell complexes from mouse preovulatory follicles. *Nature* 281:483–84

45. Salustri A, Camaioni A, Di Giacomo M, Fulop C, Hascall VC. 1999. Hyaluronan and proteoglycans in ovarian follicles. *Hum. Reprod. Update* 5:293–301

46. Fulop C, Kamath RV, Li Y, Otto JM, Salustri A, et al. 1997. Coding sequence, exon-intron structure and chromosomal localization of murine TNF-stimulated gene 6 that is specifically expressed by expanding cumulus cell-oocyte complexes. *Gene* 202:95–102

47. Yoshioka S, Ochsner S, Russell DL, Ujioka T, Fujii S, et al. 2000. Expression of tumor necrosis factor-stimulated gene-6 in the rat ovary in response to an ovulatory dose of gonadotropin. *Endocrinology* 141:4114–19

48. Nandi A, Estess P, Siegleman MH. 2000. Hyaluronan anchoring and regulation on the surface of vascular endothelial cells is mediated through the functionally active form of CD44. *J. Biol. Chem.* 275:14939–48

49. Lesley J, Hascall VC, Tammi M, Hyman R. 2000. Hyaluronan binding by cell surface CD44. *J. Biol. Chem.* 275:26967–75

50. MacArthur ME, Irving-Rodgers HF, Byers S, Rodgers RJ. 2000. Identification and immunolocalization of decorin, versican, perlecan, nidogen, and chondroitin sulfate proteoglycans in bovine small-antral ovarian follicles. *Biol. Reprod.* 63:913–24

51. Weigel PH, Hascall VC, Tammi M. 1997. Hyaluronan synthase. *J. Biol. Chem.* 272:13997–4000

52. Lee TH, Wisiewski H-G, Vilcek J. 1992. A novel secretory tumor necrosis factor-inducible protein (TSG-6) is a member of the family of hyaluronate binding

proteins, closely related to the adhesion receptor CD44. *J. Cell Biol.* 116:545–57

53. Morham SG, Langenback R, Loftin CD, Tiano HF, Vouloumanos N, et al. 1995. Prostaglandin synthase 2 gene disruption causes severe renal pathology in the mouse. *Cell* 83:473–82

54. Dinchuk JE, Car BD, Focht RJ, Johnston JJ, Jaffee BD, et al. 1995. Renal abnormalities and an altered inflammatory response in mice lacking cyclooxygenase II. *Nature* 378:406–9

55. Davis BJ, Lennard DE, Lee CA, Tiano HF, Morham SG, et al. 1999. Anovulation in cyclo-oxygenase-2-deficient mice is restored by prostaglandin E2 and interleukin-1β. *Endocrinology* 140: 2685–96

56. Tilley SL, Audoly LP, Hicks EH, Kim H-S, Flannery PJ, et al. 1999. Reproductive failure and reduced blood pressure in mice lacking the EP2 prostaglandin E2 receptor. *J. Clin. Invest.* 103:1539–45

57. Kennedy CRJ, Zhang Y, Brandon S, Guan Y, Coffee K, et al. 1999. Salt-sensitive hypertension and reduced fertility in mice lacking the prostaglandin EP2 receptor. *Nat. Med.* 5:217–22

58. Deleted in proof

59. Chen L, Zhang H, Powers RW, Russell PT, Larsen WJ. 1996. Covalent linkage between proteins of the inter-α-inhibitor family and hyaluronic acid *cis* mediated by a factor produced by granulosa cells. *J. Biol. Chem.* 271:19409–14

60. Wisniewski H-G, Burgess WH, Oppenheim JD, Vilcek J. 1994. TSG-6, an arthritis-associated hyaluronan binding protein forms a stable complex with the serum protein inter-α-inhibitor. *Biochemistry* 33:7423–29

60a. Carrette O, Nemade RV, Day AJ, Brickner A, Larsen WJ. 2001. TSG-6 is concentrated in the extracellular matrix of mouse cumulus oocyte complexes through hyaluronan and inter-alpha-inhibitor binding. *Biol. Reprod.* 65:301–8

61. Bork P, Beckman G. 1993. The CUB domain: a widespread module in developmentally regulated proteins. *J. Mol. Biol.* 231:539–45

62. Protin U, Schweighoffer T, Jochum W, Hilberg F. 1999. CD44-deficient mice develop normally with changes in subpopulations and recirculation of lymphocyte subsets. *J. Immunol.* 163:4917–23

63. Zhou B, Weigel JA, Fauss L, Weigel PH. 2000. Identification of the hyaluronan receptor for endocytosis (HARE). *J. Biol. Chem.* 275:37733–41

64. Bernfield M, Gotte M, Park PW, Reizes O, Fitzgerald ML, et al. 1999. Functions of cell surface heparan sulfate proteoglycans. *Annu. Rev. Biochem.* 68:729–77

65. Ishiguro K, Kojima T, Taguchi O, Saito H, Muramatsu T, Kadomatsu K. 1999. Syndecan-4 expression is associated with follicular atresia in mouse ovary. *Histochem. Cell Biol.* 112:25–33

66. Elvin JA, Clark AT, Wang P, Wolfman NM, Matzuk MM. 1999. Paracrine actions of growth differentiation factor-9 in the mammalian ovary. *Mol. Endocrinol.* 13:1035–48

67. Park PW, Reizes O, Bernfield M. 2000. Cell surface heparan sulfate proteoglycans: selective regulators of ligand-receptor encounters. *J. Biol. Chem.* 275: 29923–26

68. Yu Q, Stamenkovic I. 2000. Cell surface-localized matrix metalloproteinase-9 proteolytically activates TGF-β and promotes tumor invasion and angiogenesis. *Genes Dev.* 14:163–76

69. Sheng H, Shao J, Dixon DA, Williams CS, Prescott SM, et al. 2000. Transforming growth factor-β1 enhances Ha-ras-induced expression of cyclooygenase-2 in intestinal epithelial cells via stabilitzation of mRNA. *J. Biol. Chem.* 275:6628–35

70. Robker RL, Russell DL, Espey LL, Lydon JP, O'Malley BW, Richards JS. 2000. Progesterone-regulated genes in the ovulation process: ADAMTS-1 and cathepsin

L proteases. *Proc. Natl. Acad. Sci. USA* 97:4689–94

71. Vu TH, Werb Z. 2000. Matrix metalloproteinases: effectors of development and normal physiology. *Genes Dev.* 14:2123–33

72. Hagglund AC, Ny A, Leonardsson G, Ny T. 1999. Regulation and localization of matrix metalloproteinases and tissue inhibitors of metalloproteinases in the mouse ovary during gonadotropin-induced ovulation. *Endocrinology* 140:4351–58

72a. Curry TE Jr, Osteen KG. 2001. Cyclic changes in the matrix metalloproteinase system in the ovary and uterus. *Biol. Reprod.* 64:1285–96

73. Reich R, Tsafriri A, Mechanic GL. 1985. The involvement of collagenase in ovulation in the rat. *Endocrinology* 116:522–27

74. Yan C, Wang P, DeMayo J, DeMayo FJ, Elvin JA, et al. 2001. Synergistic roles of bone morphogenic protein 15 and growth differentiation factor 9 in ovarian function. *Mol. Endocrinol.* 15:854–66

75. Smith WL, DeWitt DL, Garavito RM. 2000. Cyclooxygenases: structural, cellular and molecular biology. *Annu. Rev. Biochem.* 69:145–82

76. Roby KF, Son D-S, Terranova PF. 1999. Alterations of events related to ovarian function in tumor necrosis factor receptor type I knockout mice. *Biol. Reprod.* 61:1616–21

77. Eugster H-P, Muller M, Ryffel B. 1998. *Immunodeficiency of Tumor Necrosis Factor and Lymphotoxin-α Double Deficient Mice.* Totowa, NJ: Humana

78. Ono M, Nakamura Y, Tamura H, Takiguchi S, Sugino N, Kato H. 1997. Role of interleukin-1β in superovulation in rats. *Endocr. J.* 44:797–804

79. Terranova PF, Rice VM. 1997. Review: cytokine involvement in ovarian processes. *Am. J. Reprod. Immunol.* 37:50–63

80. Klampfer L, Lee TH, Hsu W, Vilcek J, Chen-Kiang S. 1994. NF-IL6 and AP1

cooperatively modulate the activation of TSG-6 gene by tumor necrosis factor alpha and interleukin-1. *Mol. Cell. Biol.* 14:6561–69

81. Morrissey PJ, Glaccum M, Maliszewski CR, Peschon J. 1998. *Functional Phenotype of Mice Deficient for the IL-1R Type I Gene.* Totowa, NJ: Humana

82. Wong WYW, Richards JS. 1991. Evidence for two antigenically distinct molecular weight variants of prostaglandin H synthase in the rat ovary. *Mol. Endocrinol.* 5:1269–79

83. Lui K, Wahlberg P, Ny T. 1998. Coordinated and cell-specific regulation of membrane type matrix metalloproteinase 1 (MT1-MMP) and its substrate matrix metalloproteinase 2 (MMP-2) by physiological signals during follicular development and ovulation. *Endocrinology* 139:4735–38

84. Cooke RG, Nothnick WB, Komar C, Burns P, Curry TE Jr. 1999. Collagenase and gelatinase messenger ribonucleic acid expression and activity during follicular development in the rat ovary. *Biol. Reprod.* 61:1309–16

85. Simpson KS, Byers MJ, Curry TE Jr. 2001. Spatiotemporal messenger ribonucleic acid expression of ovarian tissue inhibitors of metalloproteinases throughout the rat estrous cycle. *Endocrinology* 142:2058–69

86. Itoh Y, Kajita M, Kinoh H, Mori H, Okada A, Seiko M. 1999. Membrane type 4 matrix metalloproteinase (MT4-MMP, MMP17) is a glycosylphosphatidylinositol-anchored proteinase. *J. Biol. Chem.* 274:34260–66

87. Nagase H, Woessner JF Jr. 1999. Matrix metalloproteinases. *J. Biol. Chem.* 274:21491–94

88. Shapiro SD. 1998. Matrix metalloproteinase degradation of extracellular matrix: biological consequences. *Curr. Opin. Cell Biol.* 10:602–8

89. Caterina J, Caterina N, Yamada S, Holmback K, Longenecker G, et al. 1998.

Murine TIMP-2 gene-targeted mutation. *Ann. NY Acad. Sci.* 125:528–30

90. Nothnick WB, Soloway P, Curry TE Jr. 1997. Assessment of the role of tissue inhibitor of metalloproteinase-1 (TIMP-1) during the periovulatory period in female mice lacking a functional TIMP-1 gene. *Biol. Reprod.* 56:1181–88

91. Holmbeck K, Bianco P, Caterina J, Yamada S, Kromer M, et al. 1999. MT1-MMP-deficient mice develop dwarfism, osteopenia, arthritis, and connective tissue disease due to inadequate collagen turnover. *Cell* 99:81–92

92. Hernandex-Barrantes S, Toth M, Bernado MM, Yurkova M, Gervasi DC, et al. 2000. Binding of active (57 kDa) membrane type-1-matrix metalloproteinase (MT1-MMP) to tissue inhibitor of metalloproteinase (TIMP)-2 regulates MT1-MMp processing and pro-MMP-2 activation. *J. Biol. Chem.* 275:12080–89

93. Ny A, Leonardsson G, Hagglund A-C, Hagglof P, Ploplis VA, et al. 1999. Ovulation in plasminogen-deficient mice. *Endocrinology* 140:5030–35

94. Yu W-H, Yu S-S, Meng Q, Brew K, Woessner JF Jr. 2000. TIMP-3 binds sulfated glycosaminoglycans of the extracellular matrix. *J. Biol. Chem.* 275:31226–32

95. Hashimoto G, Aoki T, Nakamura H, Tanzawa K, Okada Y. 2001. Inhibition of ADAMTS-4 (aggrecanase) by tissue inhibitors of metalloproteinases (TIMP-1,-2,-3 and 4). *FEBS Lett.* 494:192–95

96. Oksjoki S, Soderstrom M, Vuorio E, Anttila L. 2001. Differential expression patterns of cathepsins B, H, K, L, and S in the mouse ovary. *Mol. Hum. Reprod.* 7:27–34

97. Espey LL, Yoshioka S, Russell DL, Ujioka T, Vladu B, et al. 2000. Characterization of ovarian carbonyl reductase gene expression during ovulation in the gonadotropin-primed immature rat. *Biol. Reprod.* 62:390–97

98. Brockstedt E, Peters-Kottig M, Badock V, Hegele-Hartung C, Lessl M. 2000.

Luteinizing hormone induces mouse vas defers protein expression in the murine ovary. *Endocrinology* 141:2574–81

99. Espey LL, Yoshioka S, Ujioka T, Fujii S, Richards JS. 2001. 3α-Hydroxysteroid dehydrogenase mRNA transcription in the immature rat ovary in response to an ovulatory dose of gonadotropin. *Biol. Reprod.* 65:72–78

100. Pardali K, Kurisaka A, Moren A, ten Dijke P, Kardassis D, Moustakas A. 2000. Role of Smad proteins and transcription factor Sp1 in p21Waf/Cip regulation by transforming growth factor-β. *J. Biol. Chem.* 275:29244–56

101. Alliston TN, Maiyar AC, Buse P, Firestone GL, Richards JS. 1997. Follicle stimulating hormone-regulated expression of serum/glucocorticoid-inducible kinase in rat ovarian granulosa cells: a functional role for the Sp1 family in promoter activity. *Mol. Endocrinol.* 11:1934–49

102. Prowse DM, Bolgan L, Molnar L, Dotto GP. 1997. Involvement of the Sp3 transcription factor in induction of p21CIP1/KIP1 in keratinocyte differentiation. *J. Biol. Chem.* 272:1308–14

103. Haas TL, Stitelman D, Davis SJ, Apte SS, Madri JA. 1999. Egr-1 mediates extracellular matrix-driven transcription of membrane type 1 matrix metalloproteinase in endothelium. *J. Biol. Chem.* 274:22679–85

104. Deleted in proof

105. Lee SL, Sadovsky Y, Swirnoff AH, Polish JA, Goda P, et al. 1996. Luteinizing hormone deficiency and female infertility in mice lacking the transcription factor NGFI-A (Egr-1). *Science* 273:1219–21

106. Fitzgerald KA, O'Neill LA. 1999. Characterization of CD44 induction by IL-1: a critical role for Egr-1. *J. Immunol.* 162:4920–27

107. Sterneck E, Tassarollo L, Johnson PF. 1997. An essential role for C/EBPβ in female reproduction. *Genes Dev.* 11:2153–62

108. Pall M, Hellberg P, Brannstrom M, Mikuni M, Peterson CM, et al. 1997. The transcription factor C/EBPβ and its role in ovarian function: evidence for direct involvement in the ovulatory process. *EMBO J.* 16:2153–62

109. Silverman E, Eimerl S, Orly J. 1999. CCAAT enhancer-binding protein beta and GATA-4 binding regions within the promoter of the steroidogenic acute regulatory protein (StAR) gene are required for transcription in rat granulosa cells. *J. Biol. Chem.* 274:17987–96

110. Sicinski P, Donaher PL, Parker SB, Geng Y, Gardner H, et al. 1996. Cyclin D2 is an FSH-responsive gene involved in gonadal cell proliferation and oncogenesis. *Nature* 384:470–74

111. Rose UM, Hanssen RGJM, Kloosterboer HJ. 1999. Development and characterization of an in vitro ovulation model using mouse ovarian follicles. *Biol. Reprod.* 61:503–11

112. Pall M, Mikuni M, Mitsube K, Brannstrom M. 2000. Time-dependent ovulation inhibition of a selective progesterone-receptor antagonist (Org 31710) and effect on ovulatory mediators in the in vitro perfused rat ovary. *Biol. Reprod.* 63:1642–47

113. Vazquez F, Hastings G, Ortega M-A, Lane TF, Oikemus S, et al. 1999. METH-1, a human ortholog of ADAMTS-1, and METH-2 are members of a new family of proteins with angio-inhibitory activity. *J. Biol. Chem.* 274:23349–57

114. Espey LL, Yoshioka S, Russell DL, Robker RL, Fujii S, Richards JS. 2000. Ovarian expression of a disintegrin metalloproteinase with thrombospondin motifs during ovulation in the gonadotropin-primed immature rat. *Biol. Reprod.* 62: 1090–95

115. Loechel F, Gilip BJ, Engvall E, Albrechtsen R, Wewer UM. 1998. Human ADAM12 (meltrin α) is an active metalloproteinase. *J. Biol. Chem.* 273:16993–97

116. Black RA, White JM. 1998. ADAMs: focus on the protease domain. *Curr. Opin. Cell Biol.* 10:654–59

117. Kuno K, Kanada N, Nakashima E, Fujiki F, Ichimura F, Matsushima K. 1997. Molecular cloning of a gene encoding a new type of metalloproteinase-distintegrin family protein with thromobospondin motifs as an inflammation associated gene. *J. Biol. Chem.* 272:556–62

118. Hurskainen TL, Hirohata S, Seldin MF, Apte SS. 1999. ADAM-TS5, ADAM-TS6, ADAM-TS7, novel members of a new family of zinc metalloproteinases: general features and genomic distribution of the ADAM-TS family. *J. Biol. Chem.* 274:25555–63

119. Clark ME, Kelner GS, Turbeville LA, Boyer A, Arden KC, Maki RA. 2000. ADAMTS9: a novel member of the ADAM-TS/metallospondin gene family. *Genomics* 67:343–50

120. Kuno K, Terashima Y, Matsushima K. 1999. ADAMTS-1 is an active metalloproteinase with the extracellular matrix. *J. Biol. Chem.* 274:18821–26

121. Nakamura H, Fujii Y, Inoka I, Kazuhiko S, Tanzawa K, et al. 2000. Brevican is degraded by matrix metalloproteinases and aggrecanase-1 (ADAMTS-4) at different sites. *J. Biol. Chem.* 275:38885–90

122. Tortorella MD, Pratta M, Liu R-Q, Austin J, Ross OH, et al. 2000. Sites of aggrecan cleavage by recombinant human aggrecanase-1 (ADAMTS-4). *J. Biol. Chem.* 275:18566–73

123. Kuno K, Okada Y, Kawashima H, Nakamura H, Miyasaka M, et al. 2000. ADAMTS-1 cleaves a cartilage proteoglycan, aggrecan. *FEBS Lett.* 478:241–45

124. Lee J, Park H-J, Choi H-S, Kwon H-B, Arimura A, et al. 1999. Gonadotropin stimulation of pituitary adenylate cyclase-activating polypeptide (PACAP) messenger RNA in the rat ovary and the role of PACAP as a follicle survival factor. *Endocrinology* 140:818–26

125. Ko C, In YH, Park-Sarge OK. 1999.

Role of progesterone receptor activation in pituitary adenylate-cyclase activating polypeptide gene expression in rat ovary. *Endocrinology* 140:5185–94

126. Hosseini G, Liu J, de Agostini AI. 1996. Characterization and hormonal modulation of anticoagulant heparan sulfate proteoglycans synthesized by rat ovarian granulosa cells. *J. Biol. Chem.* 271:22090–99

127. Bigler D, Takahashi Y, Chen MS, Almeda EAC, Osbourne L, White JM. 2000. Sequence specific interaction between disintegrin domain of mouse ADAM 2 (fertilin β) and murine eggs. *J. Biol. Chem.* 275:11576–84

128. Shindo T, Kurihara H, Kuno K, Yokoyama H, Wada T, et al. 2000. ADAMTS-1; a metalloproteinase-disintegrin essential for normal growth, fertility and organ morphology and function. *J. Clin. Invest.* 105:1345–52

129. Nakanishi-Matsui M, Zheng Y-W, Sulciner DJ, Weiss EJ, Ludeman MJ, Coughlin SR. 2000. PAR3 is a cofactor for PAR4 activation by thrombin. *Nature* 404:609–13

130. Sambrano GR, Huang W, Faruqi T, Mahrus S, Craik C, Coughlin SR. 2000. Cathepsin G activates protease-activated receptor-4 in human platelets. *J. Biol. Chem.* 275:6819–23

131. Bagavandoss P, Sage EH, Vernon RB. 1998. Secreted protein, acidic and rich in cysteine (SPARC) and thrombospondin in the developing follicle and corpus luteum of the rat. *J. Histochem. Cytochem.* 46:1043–49

132. Crawford SE, Stellmach V, Murphy-Ullrich JE, Ribeiro SMF, Lawler J, et al. 1998. Thrombospondin-1 is a major activator of TGF-β1 in vivo. *Cell* 93:1159–70

133. Kyriakides TR, Leach KJ, Hoffman AS, Ratner BD, Bornstein P. 1999. Mice that lack the angiogenesis inhibitor, thrombospondin 2, mount an altered foreign body reaction characterized by increased

vascularity. *Proc. Natl. Acad. Sci. USA* 96:4449–54

134. Abbaszade I, Liu R-Q, Yang F, Rosenfeld SA, Ross OH, et al. 1999. Cloning and characterization of ADAMTS-11, an aggrecanase form the ADAMTS family. *J. Biol. Chem.* 274:23443–50

135. Blelloch R, Newman C, Kimble J. 1999. Control of cell migration during *Caenorhabditis elegans* development. *Curr. Opin. Cell Biol.* 11:608–13

136. Nishiwaki K, Hisamoto N, Matsumoto K. 2000. A metalloprotease disintegrin that controls cell migration in *Caenorhabditis elegans*. *Science* 288:2205–9

137. Chen MS, Tung KSK, Coonrod SA, Takahashi Y, Bigler D, et al. 1999. Role of the integrin-assoicated protein CD9 in binding between sperm ADAM2 and the egg integrin $\alpha6\beta1$: implications for murine fertilization. *Proc. Natl. Acad. Sci. USA* 96:11830–35

138. Le Naour F, Rubenstein E, Jasmin C, Prenant M, Boucheix C. 2000. Severely reduced female fertility in CD9-deficient mice. *Science* 287:319–24

139. Evans JP, Foster JA, McAvey BA, Gerton GL, Kopf GS, Schultz RM. 2000. Effects of perturbation of cell polarity on molecular markers of sperm-egg binding site on mouse eggs. *Biol. Reprod.* 62:76–84

140. Ishidoh K, Kominami E. 1998. Gene regulation and extracellular functions of procathepsin L. *Biol. Chem.* 379:131–35

141. Jaffe RC, Donnelly KM, Mavrogianis PA, Verhage HG. 1989. Molecular cloning and characterization of a progesterone-dependent cat endometrial secretory protein complementary deoxyribonucleic acid. *Mol. Endocrinol.* 3:1807–14

142. Ko C, Park-Sarge OK. 2000. Progesterone receptor activation mediates LH-induced type-I pituitary adenylate cyclase activating polypeptide receptor (PAC(1)) gene expression in rat granulosa cells. *Biochem. Biophys. Res. Commun.* 277:270–79

143. Gras S, Hannibal J, Fahrenkrug J. 1999. Pituitary adenylate cyclase-activating polypeptide is an auto/paracrine stimulator of acute progesterone accumulation and subsequent luteinization in cultured periovulatory granulosa/lutein cells. *Endocrinology* 140:2199–205

144. Robinson GW, Hennighausen L, Johnson PF. 2000. Side-branching in the mammary gland: the progesterone-Wnt connection. *Genes Dev.* 14:889–94

145. Brisken C, Heineman A, Chavarria T, Elenbaas B, Tan J, et al. 2000. Essential function of Wnt-4 in mammary gland development downstream of progesterone signaling. *Genes Dev.* 14:650–54

146. Parkin NT, Kitajewski J, Varmus HE. 1993. Activity of Wnt-1 as a transmembrane protein. *Genes Dev.* 7:2181–93

147. Parr BA, McMahon AP. 1998. Sexually dimorphic development of the reproductive tract requires Wnt-7a. *Nature* 395:707–10

148. Couse JF, Hewitt SC, Bunch DO, Sar M, Walker VR, et al. 1999. Postnatal sex reversal of the ovaries in mice lacking estrogen receptor α and β. *Science* 286:2328–31

149. Kent J, Wheatley SC, Andrews JE, Sinclair AH, Koopman P. 1996. A male-specific role for SOX9 in vertibrate sex determination. *Development* 122:2813–22

150. Lefebvre V, Huang W, Harley VR, Goodfellow PN, de Crombrugghe B. 1997. SOX9 is a potent activator of the chondrocyte-specific enhancer of the pro alpha(II) collagen gene. *Mol. Cell. Biol.* 17:2336–46

151. Sekiya L, Koopman P, Watanabe H, Ezura Y, Yamada Y, Noda M. 1997. SOX9 enhances aggrecan gene expression via the promoter region containing a single HMG-BOX-sequence in a chondogenic cell line, TC6 J. *Bone Miner. Res.* 12:222–24

152. Deleted in proof

153. Treier M, Gleiberman AS, O'Connell SM, Szeto DP, McMahon JA, et al. 1998. Multiple signaling requirements for pituitary organogenesis in vivo. *Genes Dev.* 12:1691–704

154. Ross SE, Hemati N, Longo KA, Bennett CN, Lucas PC, et al. 2000. Inhibition of adipogenesis by Wnt signaling. *Science* 289:950–53

154a. Jordan BK, Mohammed M, Ching ST, Delot E, Chen X-N, et al. 2001. Up-regulation of WNT-4 signaling and dosage-sensitive sex reversal in humans. *Am. J. Hum. Genet.* 68:1102–9

155. Schneider VA, Mercola M. 2001. Wnt antagonism initiates cardiogenesis in *Xenopus laevis*. *Genes Dev.* 15:304–15

156. Marvin MJ, DiRocco G, Gardiner A, Bush SM, Lassar AB. 2000. Inhibition of Wnt activity induces heart formation from posterior mesoderm. *Genes Dev.* 15:316–27

157. Kawakami Y, Capdevila J, Buscher D, Itoh T, Esteban CR, Belmonte JCI. 2001. WNT signals control FGF-dependent limb initiation and AER induction in the chick embryo. *Cell* 104:891–900

158. Keim V, Rohr G, Stockert HG, Haberich FJ. 1984. An additional secretory protein in the rat pancreas. *Digestion* 29:242–49

159. Keim V, Iovanna JL, Dagorn JC. 1994. The acute phase reaction of the exocrine pancreas. Gene expression and synthesis of pancreatitis-associated proteins. *Digestion* 55:65–72

Annu. Rev. Physiol. 2002. 64:93–127

AROMATASE—A BRIEF OVERVIEW

Evan R. Simpson, Colin Clyne, Gary Rubin,
Wah Chin Boon, Kirsten Robertson, Kara Britt,
Caroline Speed, and Margaret Jones

*Prince Henry's Institute of Medical Research and the Department of Biochemistry
and Molecular Biology, Monash University, Melbourne, Australia;
e-mail: evan.simpson@med.monash.edu.au; colin.clyne@med.monash.edu.au;
gary.rubin@med.monash.edu.au; wah.chin.boon@med.monash.edu.au;
kirsten.robertson@med.monash.edu.au; kara.britt@med.monash.edu.au;
caroline.speed@med.monash.edu.au; margaret.jones@med.monash.edu.au*

Key Words estrogen biosynthesis, gene, tissue-specific expression, extragonadal synthesis, cancer, new therapeutic concepts

■ **Abstract** There is growing awareness that androgens and estrogens have general metabolic roles that are not directly involved in reproductive processes. These include actions on vascular function, lipid and carbohydrate metabolism, as well as bone mineralization and epiphyseal closure in both sexes. In postmenopausal women, as in men, estrogen is no longer solely an endocrine factor but instead is produced in a number of extragonadal sites and acts locally at these sites in a paracrine and intracrine fashion. These sites include breast, bone, vasculature, and brain. Within these sites, aromatase action can generate high levels of estradiol locally without significantly affecting circulating levels. Circulating C_{19} steroid precursors are essential substrates for extragonadal estrogen synthesis. The levels of these androgenic precursors decline markedly with advancing age in women, possible from the mid-to-late reproductive years. This may be a fundamental reason why women are at increased risk for bone mineral loss and fracture, and possibly decline of cognitive function, compared with men. Aromatase expression in these various sites is under the control of tissue-specific promotors regulated by different cohorts of transcription factors. Thus in principle, it should be possible to develop selective aromatase modulators (SAMs) that block aromatase expression, for example, in breast, but allow unimpaired estrogen synthesis in other tissues such as bone.

INTRODUCTION

Models of estrogen insufficiency have revealed new and often unexpected roles for estradiol in both females and males (1, 2). These models include natural mutations in humans of the aromatase gene, of which there are some ten cases known, three of whom are men, as well as one man with a mutation in the estrogen receptor (ER) α. They also include mice with targeted disruptions of ERα and ERβ, the double

0066-4278/02/0315-0093$14.00

93

ERα- and β-knockout mouse (3–6), as well as the aromatase knockout (ArKO) mouse models (7–9). Some of these roles challenge the definitions of the terms estrogen and androgen. For example, the lipid and carbohydrate phenotype of estrogen insufficiency is nonsexually dimorphic and appears to apply equally to males and females (10, 11), as does the bone phenotype of undermineralization and failure of epiphyseal closure. Even more dramatically, the role of estradiol in male germ cell development would indicate that, at least in this local context, estradiol would be more appropriately defined as an androgen (12). The second important point is that, in men and in postmenopausal women when the ovaries cease to produce estrogens, estradiol does not function as a circulating hormone. It is no longer an endocrine factor, instead estradiol is produced in a number of extragonadal sites and acts locally at these sites as a paracrine or even intracrine factor (2, 13). These sites include the mesenchymal cells of adipose tissue, osteoblasts and chondrocytes of bone, numerous sites in the brain, and also the Leydig cells and germ cells of the testes. Thus circulating levels of estrogens in postmenopausal women and in men are not the drivers of estrogen action; they are reactive rather than proactive. This is because circulating estrogens in these individuals originate from the synthesis of estrogen in extragonadal sites where it acts locally, and that which escapes local metabolism then enters the circulation. Thus circulating levels reflect, rather than direct, estrogen action in postmenopausal women and men.

AROMATASE AND ITS GENE

Estrogen biosynthesis is catalyzed by a microsomal member of the cytochrome P450 superfamily, namely aromatase cytochrome P450 (P450arom, the product of the *CYP19* gene). The P450 gene superfamily is very large, containing (as of 1996) over 480 members in 74 families, of which cytochrome P450arom is the sole member of family 19 (14). This heme protein is responsible for binding of the C_{19} androgenic steroid substrate and catalyzing the series of reactions leading to formation of the phenolic A ring characteristic of estrogens. The aromatase reaction employs 3 moles of oxygen and 3 moles of NADPH for every mole of steroid substrate metabolized (15–20). These oxygen molecules are utilized to oxidize the C_{19} angular methyl group to formic acid, which occurs concomitantly with aromatization of the A ring to give the phenolic A ring characteristic of estrogens (21, 22). The reducing equivalents for this reaction are supplied from NADPH via a ubiquitous microsomal flavoprotein, NADPH-cytochrome P450 reductase. In humans, a number of tissues have the capacity to express aromatase and hence synthesize estrogens. These include the ovaries and testes, the placenta and fetal (but not adult) liver, adipose tissue, chondrocytes and osteoblasts of bone, the vasculature smooth muscle, and numerous sites in the brain, including several areas of the hypothalamus, limbic system, and cerebral cortex.

The human *CYP19* gene was cloned some years ago (23–25), when it was shown that the coding region spans 9 exons beginning with exon II. Upstream of exon II are a number of alternative first exon Is that are spliced into the 5′-untranslated region of the transcript in a tissue-specific fashion (Figure 1). Thus placental transcripts contain at their 5′-end a distal exon, I.1, which was originally estimated to be at least 40 kb upstream from the start of translation in exon II. This is because placental expression is driven by a powerful distal promoter upstream of exon I.1 (26, 27). Examination of the Human Genome Project data reveals that exon I.1 is 89 kb upstream of exon II (S. Bulun, personal communication). On the

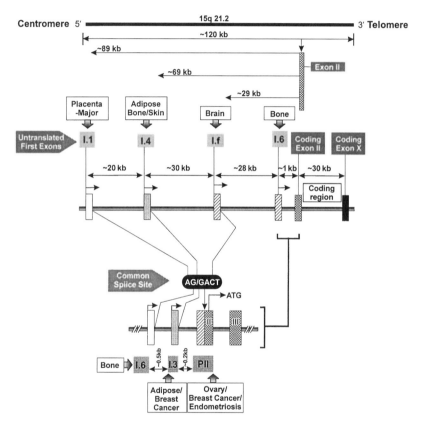

Figure 1   Genomic organization of the human *CYP19* gene. BLAST searches of various promoters and coding region revealed alignment to distinct locations in two overlapping (an end-to-end overlap of 6141 bp) BAC clones of chromosome 15q21.2 region. The distance of each promoter with respect to the first coding exon (exon II) was also determined. The major placental promoter I.1 is the most distally located (approximately 89 kb). Even though each tissue expresses a unique untranslated first exon 5′-UTR, by splicing into a highly promiscuous splice acceptor site (AG/A\GACT) of the exon II, the coding region and the translated protein product are identical in all the tissues. Adapted from unpublished work of S. Sebastian & S. Bulun, with permission.

other hand, transcripts in ovary and testes contain at their 5′-end a sequence that is immediately upstream of the translational start site. This is because expression of the gene in the gonads utilizes a proximal promoter (promoter II) (28). By contrast, transcripts in adipose tissue contain yet another distal exon located 20 kb downstream of exon I.1, namely I.4 (29). Adipose tissue transcripts also contain the promoter II–specific exonic sequence as well as exon I.3, located just upstream of exon II. Exon I.3 actually includes promoter II as an exonic sequence, and exon I.3–containing transcripts originate from promoter I.3. Harada and colleagues use a different nomenclature for the aromatase first exons. Thus exon I.1 is exon Ia, exon I.4 is exon Ib, exon I.3 is Ic, and the gonad-specific exonic sequence is Id. A number of other untranslated exons have been characterized (e.g., I.2, I.5, 2a), including one that is expressed in brain (If) (30) and another that is expressed in bone (I.6) (31). However, apart from these, the significance of the other transcripts is unknown, and generally they are present in tissues at only trace levels. Splicing of these untranslated exons to form the mature transcript occurs at a common 3′-splice junction that is upstream of the translational start site. This means that although transcripts in different tissues have different 5′ termini, the coding region and thus the protein expressed in these various tissue sites is always the same. However, the promoter regions upstream of each of the several untranslated first exons have different cohorts of response elements, and therefore regulation of aromatase expression in each tissue that synthesizes estrogens is different. The gonadal promoter binds the transcription factors CREB and SF1, and thus aromatase expression in gonads is regulated by cAMP and gonadotrophins. On the other hand, the adipose promoter I.4 is regulated by class I cytokines such as IL-6, IL-11, and oncostatin M, as well as by TNFα. Thus the regulation of estrogen biosynthesis in each tissue site of expression is unique (reviewed in 32).

The biosynthesis of estrogens appears to occur throughout the entire vertebrate phylum including mammals, birds, reptiles, amphibians, teleost and elasmobranch fish, and agnatha (hagfish and lampreys) (33–35). It has also been described in the protochordate *Amphioxus*. To our knowledge, estrogen biosynthesis has not been reported in non-chordate animal phyla, although the aromatase P450 gene family appears to be an ancient lineage of P450 gene products, diverging as much as 10^9 years ago (19). In most vertebrate species that have been examined, aromatase expression occurs in the gonads and in the brain. In many species, estrogen biosynthesis in the brain has been implicated in sex-related behavior such as mating responses, and frequently a marked sexually dimorphic difference has been demonstrated. This is true, for example, in avian species in which the song of the male is important in courtship behavior (36). In the case of humans and a number of higher primates, there is a more extensive tissue distribution of estrogen biosynthesis, as mentioned above, because this also occurs in the placenta and liver of the developing fetus, as well as in the adipose tissue of the adult. The ability of the placenta to synthesize estrogens is also the property of a number of ungulate species such as cows, pigs, horses, and sheep.

Analysis of *CYP19* gene structure has been carried out in several fish and mammalian species. Both goldfish and zebrafish contain distinct aromatase isoforms present in brain and ovary that are differentially programmed and regulated during early development (37, 38). These are the products of separate genes that are only about 60% identical, indicating a long evolutionary history as separate genes. On the other hand, each isoform of the zebrafish aromatase is 88% identical to the corresponding A or B isoforms of the goldfish. A phylogenetic tree of the aromatase cDNAs sequenced to date is presented in Figure 2. Although the gene encoding aromatase in humans is a single member of the *CYP19* family, albeit with a number of alternatively spliced first exons, multiple *CYP19* loci and isoforms, each with a different tissue-specific and developmental program, have been identified in other mammalian species. Thus the bovine placenta transcribes and does not translate an aromatase pseudogene consisting of sequences highly homologous to several of the exons of the functional bovine aromatase, but interspersed with a conserved bovine repeat element (39). The functional gene and pseudogene are estimated to lie 24 kb apart. Like the human gene, the bovine gene also expresses a number of

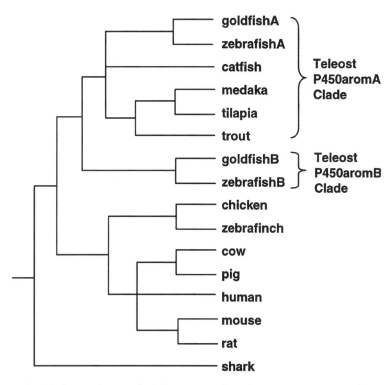

Figure 2   Phylogenetic tree of P450arom proteins. The tree was constructed by maximum parsimony using the BRANCH AND BOUND option of PAUP and represents a consensus of four similar trees. Adapted from (37) with permission.

tissue-specific untranslated first exons. The ovine genome also appears to contain a functional *CYP*19 gene and a pseudogene (40–42). Interestingly, evidence has been presented that in the porcine genome, multiple aromatase transcripts are the products of three distinct genes (paralogs) rather than a consequence of complex splicing mechanisms (43, 44). Thus type I isoform is predominantly expressed in the ovary whereas type II is expressed in the mid- and late-pregnancy endometrium and placenta. Additionally peri-implantation porcine conceptuses synthesize high levels of estrogens, which is a consequence of the high level of expression within the tissue of yet another form of aromatase, the type III isoform. This latter form has been shown to catalyze the formation of 19-nor androgens. The pathway of molecular evolution that has given rise to this complexity of isoforms and alternative splicing is far from clear, but it is evident that different species have chosen different routes in order to achieve tissue-specific regulation of estrogen biosynthesis.

STRUCTURE: FUNCTION RELATIONSHIPS

Aromatase shares a number of structural features common to all cytochrome P450 species. Most notably, toward the carboxy terminus there is the heme-binding region containing a totally conserved cysteine residue that serves as the fifth coordinating ligand of the heme iron. A thiolate ion present at this site, instead of the nitrogenous base most commonly found in other *b*-type cytochromes, is the reason for the unique spectrophotometric as well as catalytic properties of this family of hemeproteins. Upstream of this domain is a region of over 20 amino acids that is highly conserved in all aromatase species, and upstream of this is another region of high conservation, namely the portion of the I helix believed to form the substrate-binding pocket proximal to the heme-prosthetic group (45). The three-dimensional crystal structure of a number of soluble bacterial cytochromes P450 has now been resolved (46–48), and the major structural features believed to be common to all P450s are now apparent.

Application of this methodology to determine the structure of eukaryotic P450s has been hampered because of the difficulty of solubilizing these proteins in the absence of detergent and their intransigence to crystallization. However, recently the crystal structure of a microsomal cytochrome P450 was published (49). Because aromatase is generally present at lower concentrations in the endoplasmic reticulum of cells in which it is expressed than most other P450 species, obtaining a crystal structure of aromatase is unlikely to be achieved in the immediate future. However, the structure of human aromatase has been modeled on the known three-dimensional structures of bacterial P450 species. This model was based on a core structure identified from the structures of the soluble bacterial P450s, rather than by molecular replacement, after which the less-conserved elements and loops were added in a rational fashion. Minimization and dynamic simulations were used to optimize the model, and the reasonableness of the structure was evaluated. On this basis, a membrane-associated hydrophobic region of aliphatic and aromatic residues involved in substrate recognition was postulated, as well as a redox partner–binding region that appears to be unique compared with other P450s, and also residues involved in active site-binding of substrates and inhibitors of aromatase (50). This model has been used to develop a reaction mechanism of aromatase that accounts for the conversion of C_{19} steroids to the corresponding phenolic estrogenic steroids. It also explains how demethylation can occur to form 19-nor steroids without the aromatization step. As indicated above, we are presently unaware of any ongoing efforts to obtain a crystal structure for aromatase, so there may be little further progress in elucidating the relationship of function to structure of this enzyme in the next few years.

TISSUE-SPECIFIC EXPRESSION OF AROMATASE

Expression in Human Ovary

Aromatase P450 is expressed in the pre-ovulatory follicles and corpora lutea of ovulatory women by means of a promoter proximal to the start of translation (PII) (26, 28). Aromatase expression in the granulosa cells of the ovary is primarily under the control of the gonadotropin FSH, whose action is mediated by cAMP. To understand how this transcription is controlled by cAMP, chimeric constructs containing deletion mutations of the proximal promoter 5′-flanking DNA of the human and rat sequences fused to reporter genes were constructed (51, 52). Assay of reporter gene transcription in transfected bovine granulosa and luteal cells and rat granulosa cells revealed that basal- and cAMP-stimulated transcription was lost upon deletion from -278 to -100 bp, indicating the presence of functional response elements in this region. Mutation of a CAAGGTCA motif located at -130 bp in the human sequence revealed that this element is crucial for basal- and cAMP-stimulated reporter gene transcription. When a single copy of this element was placed upstream of a heterologous promoter, it could act as a weak cAMP-response element. An electrophoretic mobility shift assay in the presence of specific antibodies and UV-crosslinking established that Ad4BP/SF-1 binds to this hexameric element. SF-1 (steroidogenic factor-1) is an orphan member of the nuclear receptor gene superfamily that has been shown to be a critical developmental factor for the gonads as well as the adrenals (53, 54). However, deletion mutation analysis revealed the presence of another sequence upstream of the SF-1 site that is also critical for cAMP-responsiveness. This sequence, at $-211/$ -202 bp in the human gene and $-161/-138$ bp in the rat gene, is TGCACGTCA, identical to a canonical CRE except for the extra C. In both human and rat genes, electrophoretic gel mobility shift and antibody super-shift analysis revealed that CREB binds to this site along with several other proteins (52, 55). These results are summarized in Figure 3. Thus aromatase expression in the ovary is transcriptionally regulated by a hexameric sequence binding SF-1 and an imperfect CRE binding CREB and other factors.

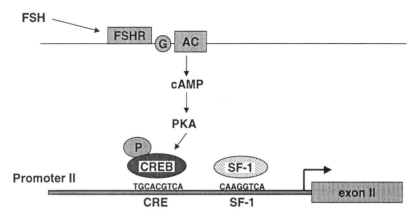

Figure 3   Regulation of aromatase expression in ovary. In ovarian granulosa cells, expression is regulated primarily by promoter II. FSH binds to its receptor and activates adenylylcyclase, which results in formation of cAMP and activation of PKA. PKA is presumed to phosphorylate CREB, which binds to a CREB-like sequence (CLS) on promoter II. Downstream of the CLS is an SF-1 binding site. Binding of both factors to their response elements is presumed to result in recruitment of the cohort of coactivators, which leads to activation of promoter II–specific expression.

The situation with regard to the bovine ovarian aromatase promoter is interesting, because after the ovulatory surge of gonadotropins, aromatase expression is maintained in the human corpus luteum but is lost from that of the bovine. Transfection of the bovine aromatase constructs corresponding to those described above for human and rat revealed no expression of reporter gene activity. Examination of the bovine sequence revealed a 1-bp deletion in the bovine CRE compared with that in human and rat CRE. Mutation of this to the corresponding human sequence led to partial restoration of reporter gene activity (56).

Expression in Adipose Tissue

As mentioned above, aromatase is expressed in adipose tissue in humans. A majority of aromatase transcripts in adipose tissue contain untranslated exon I.4 because their expression is directed by promoter I.4, which, as mentioned previously, is stimulated by class I cytokines and TNFα. In addition, there is an obligatory requirement of glucocorticoids for expression from this promoter (57). Analysis of the sequence upstream of the transcriptional start site of promoter I.4 reveals the presence of the interferon γ activation site (GAS) element followed by a glucocorticoid response element (GRE) that is a SP1-binding site (57). Deletion analysis and mutagenesis experiments have shown that all three of these response elements are required for expression from promoter I.4. Furthermore, the stimulation of expression via class I cytokines is mediated via a JAK1/STAT3 pathway (58) (Figure 4). It should be pointed out that glucocorticoids have also been shown to be required

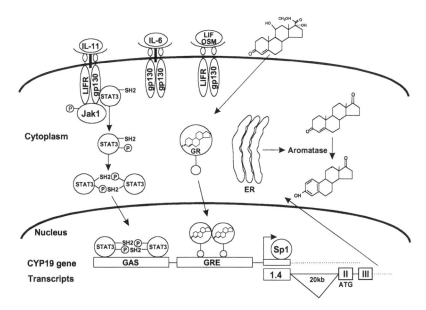

Figure 4 Schematic representation of second messenger signaling pathways whereby class I cytokines stimulate aromatase gene expression in human adipose stromal cells. Jak1 kinase is bound to the common receptor subunit gp130 and activated following ligand binding and receptor dimerization, as a consequence of phosphorylation on tyrosine residues. STAT3 is recruited to binding sites on gp130 and is phosphorylated on tyrosine residues by Jak1. These phosphotyrosine residues are recognized by SH2-homology domains on STAT3, resulting in dimerization followed by translocation to the nucleus and binding to the GAS element of promoter I.4 of the aromatase gene. Following binding of glucocorticoid receptors to the GRE and Sp1 to its site on untranslated exon on I.4, activation of transcription of the aromatase gene from promoter I.4 is initiated. Splicing of the initial transcript results in formation of mature mRNA, which translocates to the ribosomes and is translated to give rise to aromatase protein. GR, glucocorticoid receptor; ER, endoplasmic reticulum.

in other STAT-mediated responses such as the STAT5 mediation of prolactin regulation of casein gene expression (59). The mechanism whereby TNFα regulates promoter I.4 activity appears to involve an imperfect AP1 site upstream of the GAS element, but the details of this signalling pathway have yet to be established. Once again, as is the case with class I cytokines, glucocorticoids are an obligatory requirement for this pathway.

In adipose tissue, aromatase is expressed primarily in the stromal mesenchymal cells or preadipocytes rather than in the lipid-laden adipocytes themselves (60). Aromatase expression also occurs in other cells of mesenchymal origin such as skin fibroblasts and osteoblasts (61–63), so this should more appropriately be termed a mesenchymal promoter. It turns out that aromatase transcripts in adipose tissue contain in addition to exon I.4, exon II (the gonadal exon), as well as exon I.3 (64, 65). Expression from both promoters II and I.3 in adipose stromal cells is stimulated by cAMP, as is the case of ovarian granulosa cells, but in contrast to the latter, in adipose stromal cells, the expression via cAMP is potentiated by phorbol esters. The most potent factor driving aromatase expression via promoter II in these cells appears to be prostaglandin E2 (66), which activates EP1 and EP2 receptors. The latter signals via adenylyl cyclase and increased cAMP, whereas the former signals via IP3, calcium, and PKC. The majority of aromatase transcripts in adipose tissue contain exon I.4 followed by exon I.3 and II. This is true for cells derived from noncancerous breast, abdominal, buttocks, and thigh subcutaneous adipose tissue (67). Because a number of the class 1 cytokines as well as TNFα are produced in adipose tissue, it has been proposed that the regulation is via paracrine and autocrine mechanisms involving locally produced pro-inflammatory cytokines within the adipose tissue. An important feature of aromatase expression in adipose tissue is that it increases with advancing age (68). One possible explanation for this could stem from the observations that levels of pro-inflammatory cytokines such as IL-6 also increase with age, at least in blood (69, 70). The observation that aromatase expression occurs in the pre-adipocyte mesenchymal cells is consistent with the mechanism of stimulation outlined above. TNFα and class 1 cytokines are all known to inhibit adipocyte differentiation in model systems such as 3T3L1 cells (71). Indeed, IL-11 was originally characterized as an adipogenesis inhibitory factor (72), and the previous name for TNFα was cachectin because of its powerful action to induce cachexia.

Consistent with this concept then, it might be anticipated that factors that stimulate adipogenesis and lipid accumulation would be inhibitory of aromatase expression. Such a factor is PPARγ, and ligands for this nuclear receptor include the thiazolidinediones such as troglitazone and rosiglitazone, as well as the so-called endogenous ligand 15-deoxy-Δ^{12-14}-prostaglandin J2 (73). PPARγ acts by binding as a heterodimer with RXR. As anticipated, these PPARγ ligands all inhibit aromatase expression in adipose stromal cells. This has been shown in terms of aromatase activity, transcript expression by semi-quantitative RT-PCR, and expression of luciferase reporter gene constructs driven by promoter I.4 sequence (74). On the other hand, a troglitazone metabolite that is not a ligand for PPARγ did not inhibit aromatase expression. Regional variations in aromatase expression in subcutaneous adipose tissue have been observed, namely that expression in buttocks and thighs is two- to threefold greater than that in subcutaneous abdominal tissue or breast tissue (68). This difference is not reflected in changes in the ratios of promoter-specific transcripts, which are the same in these different region-specific sites. Furthermore, there is no difference in the ratio of lipid-laden adipocytes to adipose stromal cells in these different sites, which indicates that this regional variation in expression does not result from differences in the ratio of these two cell types (75). Furthermore, there appears to be no sexually dimorphic difference in this pattern of aromatase expression, namely it is similar in men and women.

On the other hand, regional differences in aromatase expression within breast adipose tissue have been reported (76). Moreover, this regional variation within the breast does correlate with the ratio of adipocytes to stromal cells, namely that the regions with the highest aromatase expression have the highest proportion of stromal cells relative to lipid-laden adipocytes, consistent with the known cellular pattern of aromatase expression (76). This regional distribution of aromatase expression within the breast also correlates quite well with the known regional distribution of breast cancers. Thus it can be envisioned that in local regions of the breast where the ratio of stromal cells to adipocytes is highest, local aromatase expression and, therefore, local estrogen concentrations are highest, and this would be a favored site for breast tumor growth and development. An important question then arises: Once a breast tumor is established, does it influence estrogen production in the local breast environment in order to optimize its growth and development? It is well established that the presence of a breast tumor enhances the local expression of aromatase three- to fourfold in the surrounding breast adipose tissue to levels equivalent to those within the tumor itself, and that there is a gradient of aromatase expression with the tumor as its focus (68, 77, 78). Breast tumors produce factors that stimulate the proliferation of mesenchymal cells surrounding them—the well-known desmoplastic reaction—but clearly these factors also stimulate aromatase expression within these cells. Three laboratories have shown that this increase in aromatase expression proximal to a breast tumor is accompanied by a switch in promoter utilization such that in tissue proximal to the tumor as well as in the tumor itself, promoters I.3 and II predominate, and transcripts driven from promoter I.4 are now in a minority (64, 65, 78). Thus it would appear that malignant breast epithelial cells secrete factors that induce aromatase expression in adipose fibroblasts and in fibroblasts of the tumor itself via promoter II. One possible factor is prostaglandin E2, which is produced by malignant breast epithelium as well as by macrophages recruited to the tumor site (Figure 5).

Several groups are currently studying the mechanism whereby this promoter switching process takes place. Bulun and colleagues have shown that conditioned medium from T47D breast cancer cell causes a striking induction of C/EBPβ expression without affecting the levels of C/EBPα or δ transcripts (79). This group has postulated that the promoter switch is mediated, at least in part, by the tumor-induced upregulation and enhanced binding of C/EBPβ to a promoter II regulatory element that they identified in this sequence. This group has also shown that TNFα and IL-11 secreted by malignant breast epithelial cells inhibit adipocyte differentiation by selectively downregulating C/EBPα and PPARγ and have proposed that this is a mechanism for the desmoplastic reaction, namely the proliferation of pre-adipocyte fibroblasts (80).

In contrast to ovarian granulosa cells, the expression of SF-1 in adipose stromal cells is extremely low, so an important question arises as to the nature of the factor or factors that bind to the CAAGGTCA regulatory site in adipose stromal cells. Chen and colleagues have identified several potential candidates by means of a yeast one-hybrid screening approach (81). These factors include ERRα1 and

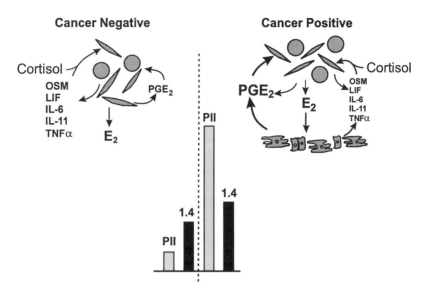

Figure 5 Proposed regulation of aromatase gene expression in breast adipose tissue from cancer-free individuals and from those with breast cancer. In the former case, expression is stimulated primarily by class I cytokines and TNFα produced locally in the presence of systemic glucocorticoids. As a consequence, promoter I.4–specific transcripts of aromatase predominate. In the latter case, PGE_2 produced by the tumorous epithelium, tumor-derived fibroblasts, and/or macrophages recruited to the tumor site appears to be the major factor stimulating aromatase expression, as evidenced by the predominance of promoter II and I.3–specific transcripts of aromatase.

COUP-TF1, which they showed to be expressed in breast tissue and to bind to this site, identified as a silencer element in adipose stromal cells. More recently this group identified a second site upstream of the SF-1 binding site that binds the zinc finger transcription factor Snail (SnaH) (82). SnaH was found to act as a repressor of promoter I.3 activity. RT-PCR analysis from a number of breast cancer cell lines indicates that SnaH is expressed at a higher level in normal breast epithelial cells and stromal fibroblast cell lines than in breast cancer cell lines. Based on this, this group proposed that SnaH acts as a repressor that downregulates the expression of aromatase in normal breast tissue by suppressing the function of promoter I.3. A reduction of the expression of SnaH in breast cancer tissue was proposed to be involved in the increase in expression of aromatase via promoter II and I.3 observed in the presence of breast tumors. Additionally, Goss and colleagues have implicated GATA transcription factors to be involved in promoter II expression in breast cancer cell lines (83). The nature of the factors responsible for the enhanced aromatase expression in breast tissue observed in the presence of a tumor and the mechanism driving the promoter switching are at present under active investigation.

Expression in Human Placenta

Placental expression of aromatase in the human is driven from a powerful distal placental promoter I.1 upstream of untranslated exon I.1, which is located a remarkable 89 kb upstream of the translation start site. Employing various deletion mutations of the upstream flanking region of exon I.1, several groups have examined putative regulatory sequences within this region and the proteins that interact with these sequences to regulate expression of aromatase in choriocarcinoma cells. Toda and colleagues (84) identified a binding site for C/EBP-β that is located between -2141 and -2115 bp relative to the start of transcription in exon I.1. They further identified an element located between -238 and -200 bp that appears to synergize with the C/EBP-β element upstream. Yamada et al. (85) identified two elements within -300 bp upstream of exon I.1 that recognize the same *trans*-acting factor that binds to the trophoblast-specific element previously located in the enhancer region of the human glycoprotein hormone α-subunit gene.

Sun and colleagues (86) identified an imperfect palindromic sequence, 5'-AGGTCATGCCCC-3', located at -183 to -172 bp that is responsible for stimulation of aromatase expression by retinoic acids. This does not function as a binding site for SF-1 because SF-1 is not expressed in placenta; however, it did appear to bind a heterodimer composed of RXRα and VDR. It was reported that levels of RXR and RAR receptor expression increased during the process of cytotrophoblast differentiation into syncytiotrophoblasts in the placenta (87). This is coincident with the increase in aromatase expression. These results suggest that retinoids play an important role in developmental regulation of aromatase gene expression in the placenta. Recently, it was shown that inactivation of the PPARγ gene results in embryonic lethality owing to failure of trophoblast development (88). Because this factor functions as a heterodimer with RXR, it is possible that PPARγ may also be implicated in aromatase expression in placenta. Mendelson's group (89) identified a sequence -42 to -125 bp upstream of the promoter I.1 transcriptional start site. This site was shown to bind the helix-loop-helix factor Mash-2, which is a hypoxia-induced transcription factor shown to inhibit aromatase expression in cultured human syncytiotrophoblastic cells (89). Thus multiple sites on the genomic region upstream of exon I.1 appear to be required for placental regulation of aromatase expression in humans, some acting as stimuli and some as inhibitors.

Although rodents appear not to express aromatase in the placenta, mice that are transgenic for a growth hormone reporter gene driven by the human promoter I.1 sequence express this reporter in the labyrinthine trophoblast, which is likely analogous to the human syncytotrophoblast (90). Thus the murine trophoblast apparently expresses the cohort of transcription factors required to activate promoter I.1. Presumably, therefore, what is lacking in the mouse is a murine equivalent of aromatase promoter I.1. On the other hand, as indicated above, several ungulate species such as cattle, pigs, and sheep do express placental aromatase. The aromatase gene of these species also utilizes a variety of untranslated first exons. However, the predominant exon I species in bovine and sheep placentae are unrelated to that of human and unrelated to each other. Thus whereas there is close homology across species of the ovarian promoter, remarkable differences exist concerning placental-specific transcript expression and presumably promoter usage.

Expression in Bone

Estrogens play an important role in bone physiology of women, especially in maintaining the level of mineralization. Recently, a substantial role for estrogen in bone physiology of men has also been demonstrated with a case of ERα deficiency (91) and two cases of aromatase deficiency (92, 93), showing similar presentations including failure of epiphyseal closure, delayed bone age, osteopenia, and osteoporosis. In men, the main source of estrogen is peripheral conversion, rather than from the testes, similar to the case in postmenopausal women. Eighty-five percent of estradiol in men is estimated to be produced by peripheral conversion of circulating androgens, and a major site of peripheral conversion has been identified in adipose tissue. However, in addition to adipose tissue, bone and cells derived from bone tissue possess aromatase activity (63, 94, 95). Using immunocytochemical and in situ techniques, Sasano and colleagues have located aromatase expression in bone to two major sites, osteoblasts and chondrocytes. No expression in osteoclast cells was detected. Aromatase transcripts have been detected in fresh bone tissue obtained from adults (96) and from fetuses (95). The level of transcripts detected in the bone samples varied greatly among subjects, positively correlating to the degree of osteoporosis (94). It was also reported that the levels of aromatase transcripts in femurs of elderly people were increased at the fracture site in comparison with those distal to the fracture site where aromatase was undetectable (97). These results strongly indicate that expression of aromatase is positively correlated in physiological and pathological conditions with stimulation of osteogenesis.

Nawata et al. (63) reported that 1,25-dihydroxycholecalciferol, IL-1β, Type 1 cytokines such as IL-11, as well as TNFα stimulated aromatase expression in osteoblast-like cells derived from human bone in vitro. Recently, similar studies were reported employing human osteoblast cells derived from fetuses (95). Once again class I cytokines, IL-1, as well as TNFα and TGF-β1 were all shown to be stimulatory of aromatase expression in such cells in the presence of glucocorticoids. On the other hand, dibutyryl cAMP had no effect whatsoever to stimulate expression. RACE analysis revealed that the major promoter employed in the cultured osteoblasts as well as in fetal tibia was promoter I.4, similar to the situation in adipose tissue. Consistent with the failure of dibutyryl cAMP to stimulate expression, promoter II-specific transcripts were present at very low levels.

Aromatase Expression in the Brain

Aromatase is expressed in numerous sites of the brain, and the relative abundance of aromatase expression in these sites is developmentally regulated. A detailed review of aromatase expression in the brain is beyond the scope of this present review; however, aromatase expression based on determination of activity as well as in situ hybridization has been detected in several sites in the hypothalamus including the preoptic nucleus, the sexually dimorphic nucleus, the bed nucleus of the striata terminalis, and the medial amygdala. Low levels have been found in the paraventricular and preoptic nuclei and the ventral hypothalamic nucleus (98, 99). Aromatase is also expressed in the hippocampus and in the subfornicle organ and the pons. Many of these regions are also areas in which one or both of the estrogen receptor isoforms are expressed; for example, ERα tends to be expressed in nuclei of the hypothalamus, whereas ERβ tends to be expressed more commonly in the hippocampus and cortical regions. Various aromatase transcripts have also been reported in the brain including transcripts containing exon If (30). This brain-specific sequence is the major 5′-terminus of transcripts in the rat amygdala and is also present in transcripts in the preoptic area. However, promoter II–specific transcripts have also been detected in amygdala and HPOA regions. I.4–specific transcripts have also been detected in the brain (100). Thus it is likely that different promoters are employed in the various brain loci of expression and that consequently the regulation differs widely in these particular brain sites.

It should also be noted that transcripts derived from promoter If have been detected by RT-PCR in other nonneural cells, namely ovary, placenta, and THP1 cells (101), although these transcripts were present in low abundance. Regulation of aromatase in brain appears to be increased by androgens and either suppressed or not affected by cAMP (102). In cultured cells derived from mouse embryonic hypothalamus, aromatase expression was elevated by α1-adrenergic agonists but not by those selective for α2- or β-adrenergic receptors. Substance P, cholecystokinin, neurotensin, and brain naturetic peptide, as well as phorbol esters and dibutyryl cGMP all increased aromatase expression, suggesting a major role of PKC and PKG pathways in this regulation, which is presumably mediated via the brain-specific promoter (103).

Aromatase Expression in Other Tissues

In humans, aromatase is expressed at high levels in the fetal liver but is undetectable in the adult liver. By use of cultured fetal hepatocytes, the main product of aromatization of androstenedione was found to be estrone sulfate (104). Aromatase activity in human fetal hepatocytes was stimulated by glucocorticoids in the presence of fetal calf serum and was also stimulated by dibutyryl cAMP and cholera toxin. In these respects, the regulation of aromatase activity of human fetal hepatocytes was similar to that of adipose stromal cells, and consistent with this, the primary transcripts found in human fetal liver are those derived from promoter I.4. Promoter I.4–specific transcripts also predominate in skin fibroblasts (61, 105) and have been detected in intestine (106).

Normal endometrium has undetectable levels of aromatase expression; however, in contrast, tissue from endometriotic implants has very high rates of aromatase expression, which is stimulated by PGE_2 via promoter II, similar to the situation in tumorous breast tissue (107). This appears to be so because in eutopic endometrium COUP-TF usually binds to the nuclear receptor half-site on promoter II without competition from SF-1, and SF-1 is not detected in most eutopic endometrial samples. COUP-TF generally functions in an inhibitory fashion. By contrast, SF-1 is present in endometriotic stromal cells and thus competes with COUP-TF. Thus factors that could differentially affect the activity or recruitment of COUP-TF rather than SF-1 in endometriotic tissue would be highly specific inhibitors of aromatase expression in this tissue.

Because of the documented effects of sex steroids on the cardiovascular system, and the reported presence of estrogen receptors in vascular tissues, studies to examine whether aromatase is present in vascular tissue have also been undertaken. Aromatase has been detected by means of in situ hybridization in human vascular smooth muscle cells, but not in endothelial cells of human aorta and pulmonary arteries (108, 109). In human vena cava, aromatase immunoreactivity has been detected both in smooth muscle cells and in endothelial cells. Transcripts derived from promoters II and I.3 appeared to be dominant in tissues expressing high levels of aromatase transcripts, whereas those expressing exon I.4 appear to be present in areas of low expression. These results suggest that the vasculature is yet another site where estrogens have local actions that are paracrine or even intracrine in nature.

Aromatase Expression in Malignant Cells

A number of tumors secrete sufficient quantities of estrogen to cause feminization in male patients. These include a case of hepatocellular carcinoma (110), Sertoli cell tumors obtained from patients with Peutz-Jegher syndrome (111) (a condition characterized by the presence of estrogen-producing bilateral multifocal sex cord tumors), as well as a reported case of adrenocortical tumor (112). A number of endometrial carcinomas also express aromatase (113) as do endometriotic plaques, although healthy endometrium does not express aromatase (114). Interestingly, in all these pathological situations, the major promoter employed to drive aromatase expression is the gonadal-type promoter, promoter II, regardless of the tissue of origin of the tumor. As indicated previously, breast tumors also express predominantly promoter II–specific aromatase transcripts. Likewise, the hepatocellular carcinoma expressed only promoter II–specific aromatase transcripts in spite of the fact that normal adult liver does not express aromatase and fetal liver expresses promoter I.4–specific transcripts. Because promoter II is the proximal promoter of aromatase situated immediately upstream of the start of translation, this may well be the ancestral promoter of aromatase, although why this should make it the candidate of choice for use in malignant cells is not entirely clear.

MODELS OF AROMATASE INSUFFICIENCY

Aromatase Deficiency in Humans

Aromatase deficiency has been reported in some 10 individuals to date, of whom 2 are adult males. With the exception of the first reported case, a Japanese patient (115), all the mutations so far identified are single-base pair changes giving rise to single amino acid substitutions and in one case a premature stop-codon (92, 93, 116–118). The mutation in the Japanese patient is a single base change that destroyed an exon-intron splice junction and gave rise to continued readthrough of an extra 87 bases to a cryptic splice site within the intron, which resulted in an in-frame insertion of 29 amino acids within the coding region. It should be pointed out that the fact that both siblings in the New York family were homozygous for the condition would suggest that the mutation does not cause any diminished likelihood of implantation, nor any serious problem with embryonic or fetal development. Apart from the California patient and the Swiss patient who are compound heterozygotes (93, 117), all the subjects are homozygous for the mutation in question and are the products of consanguineous relationships. In most cases, it was the mother who, during the third trimester, complained of virilization that resulted in facial hair and acne. These symptoms subsided after delivery. In the cases of the female newborns, they had pseudohermaphroditism with clitoromegaly and hypospadias in varying degrees of severity. This virilization of both mother and fetus is a consequence of the inability of DHEA of fetal adrenal origin to be converted to estrogens by the placenta, with its consequent peripheral conversion to androgens. At the time of puberty, these individuals exhibit primary amenorrhea, failure of breast development, hypergonadotrophic-hypogonadism, and cystic ovaries. Subsequent estrogen supplementation leads to regression of these symptoms. Because all the female patients so far studied have been placed on estrogen supplementation, the long-term sequelae of aromatase deficiency into adulthood have not been studied in women.

By contrast, two males have been reported with this condition (92, 118). In each case childhood development was uneventful. However, in their late twenties they were of very tall stature owing to sustained linear growth through puberty as a consequence of failure of epiphyseal fusion. They also had severely delayed bone age, which resulted in osteopenia and undermineralization. Both had a fair degree of android obesity. The New York patient had a lipid and carbohydrate phenotype, i.e., elevated plasma LDL and triglycerides and elevated insulin with normal glucose levels, indicative of insulin resistance. These symptoms subsided upon estradiol administration.

Analysis of the plasma hormone levels of the New York male patient (Table 1) revealed undetectable estrogens and very high circulating androgens (118, 119). Circulating FSH and LH were also elevated, indicative of an important role of estrogens in the negative feedback regulation of gonadotrophins in males as in females. Presumably in the case of males, this estrogen is primarily derived from local aromatization of testosterone within brain sites. The patient was 204-cm tall

TABLE 1   Plasma hormone levels in a male patient with aromatase deficiency

	Result	After estrogen therapy	Control (adult male)	Units
$\Delta_4 A$	335	217	30–263	ng/dl
T	2015	990	200–1200	mg/dl
5α-DHT	125	79	30–85	ng/dl
E_1	<7	49	10–50	pg/ml
E_2	<7	64	10–50	pg/ml
FSH	28.3	12.7	5.0–9.9	mIU/ml
LH	26.1	11.3	2.0–9.9	mIU/ml
Glucose	70	101	70–105	mg/dl
Insulin	52	15	5–25	μU/ml
GH	<0.5		0.5–4.2	ng/ml
IGF-1	203		182–780	ng/ml

Adapted from (119).

at age 24 and had testes that were 35-ml in volume. Unfortunately, no semen sample was available so it was impossible to gauge his sperm count or sperm viability. He was unmarried and had no offspring. However, his hormonal profile would indicate the likelihood that he had functioning testes as far as their capacity to synthesize steroids was concerned.

The second male patient was from a family in southern Italy; he presented at the age of 28 years with tall stature, infertility, and skeletal pain (92). He was found to have open epiphyses and a bone age of 14.8 years. Treatment with testosterone for 8 months failed to result in any improvement of his condition, whereas treatment with transdermal estradiol for 6 months restored his bone density to within the normal range and eliminated his other symptoms. In contrast to the New York patient, this individual had a testicular volume of 8 mls. Testicular biopsy revealed that his seminiferous tubules had little or no sperm present, and gamete development appeared arrested at the spermatocyte level. In contrast to the New York patient again, his circulating testosterone levels were not elevated; however, they were reduced upon administration of estradiol. These results might suggest a relationship between aromatase deficiency and testicular dysfunction. Unfortunately for this concept, this patient has a brother with azoospermia but who was apparently homozygous for the normal active aromatase gene. These observations suggest that the fertility problem in this patient could be independent of his aromatase deficiency. The phenotypes of these men with aromatase deficiency may be compared with that of the one known male with a mutation in the ER α isoform, who also presented with a failure of epiphyseal closure and undermineralized bones, but who has reduced sperm count and sperm motility (91).

The Aromatase Knockout (ArKO)n Mouse

Generation of aromatase knockout (ArKO) mice (7–9) has led to several insights into the multiple roles played by estrogen in the development and maintenance of mammalian physiological systems. By disrupting the *Cyp19* gene, all estrogen production was effectively abolished. Aberrations are seen in both the male and female endocrine profiles: reproductive phenotypes, including reproductive behavior; the skeleton; and in the accumulation of excess adipose tissue. The value of this model is that from conception, the mice never synthesize their own estrogen but are still responsive to the actions of exogenously administered estrogen. These properties differentiate this model from those in which responsiveness to estrogen action is blocked.

The Male ArKO Mouse

The fertility of male ArKO mice is severely compromised by disruption in spermatogenesis and impaired sexual behavior (7, 8, 120–122). Generally, the architecture and germ cell composition of testes from young ArKO mice (12–14 weeks of age) did not differ from those of wild-type (WT) littermates (7, 121, 122). Four out of five ArKO males at 4.5 months of age had normal testis morphology. Testes of a fifth male however, displayed grossly dysmorphic seminiferous tubules and disrupted spermatogenesis. By 1 year of age, all ArKO males examined showed evidence of disrupted spermatogenesis, whereas all wild-type animals had normal testicular morphology. The site of spermatogenic disruption appeared to be early spermiogenesis, with many tubules displaying degenerating round spermatids and multinucleated cells. In these tubules, elongated spermatids were not seen, suggesting that round spermatids did not complete elongation and spermiation. Six of seven 1-year-old ArKO mice examined had tubules displaying spermiogenic arrest, but a few normal tubules were also present. This observation suggests a heterogeneity in the disruption, and such heterogeneity has been reported in other knockout models with testicular phenotypes (123). An interesting feature of the tubules displaying early spermiogenic arrest was that abnormal acrosomes often were observed, in which an uneven spreading over the nuclear membrane was apparent, and more than one acrosomal granule was frequently visible. All 1-year-old ArKO animals showed Leydig cell hyperplasia/hypertrophy. Although the numbers of spermatogonia and spermatocytes were not significantly decreased in 1-year-old ArKO mice, there was a significant decrease in the numbers of both round and elongated spermatids compared with WT. This decrease in round spermatid number appears to result from an increase in apoptotic cell death, as visualized by TUNEL assays (121).

Quantitative histomorphometry revealed that no significant changes in seminiferous tubule lumen volume were seen between WT and ArKO animals. This is in sharp contrast to the report that dilated seminiferous tubule lumens are seen in the ERα KO mouse from an early age (123). Although it cannot be ruled out that ArKO animals have compromised efferent duct function, the current data suggest that the failure of spermatogenesis is not caused by a primary defect in seminiferous tubule fluid resorption. Instead, the phenotype suggests that a lack of aromatase, and thus estrogen, causes a failure of germ cell differentiation, perhaps accompanied by a decrease in seminiferous fluid secretion. Importantly, aromatase expression and immunocytochemical activity has been demonstrated in germ cells from both mice (124) and rats (125). These authors hypothesize that estrogen may be synthesized by germ cells, which act in a paracrine fashion on Sertoli cells causing them to release factors to specifically regulate germ cell development. Alternatively, estrogen could be synthesized by the germ cells and act in a paracrine and an intracrine fashion, providing a local source of estrogen involved in controlling the complex process of spermatogenesis (121). Aromatase activity in the Leydig cells also could contribute to regulation of germ cell development (126–127).

The viability of the spermatozoa located within the epididymides of 15-week-old and 1-year-old ArKO males has been examined using in vitro fertilization (120, 122). At 14–15 weeks of age, ArKO sperm numbers and concentration were comparable to that of WT littermates, although there was a significant decrease in sperm motility (122). Sperm from these young ArKO males was able to fertilize oocytes in vitro (120, 122). Conflicting reports exist regarding sperm viability in older animals. Robertson et al. (122) report significant decreases in both sperm motility and concentration in 1-year-old ArKO males, and the sperm from these individuals was incapable of in vitro fertilization. In contrast, Toda and colleagues (120) report that their 10-month-old ArKO males produced similar numbers of sperm as their WT littermates and were capable of in vitro fertilization.

Appropriate sexual behavior is an integral component of successful matings and hence contributes significantly to fertility. In estrogen-deficient male ArKO mice, sexual behavior has been demonstrated to be severely impaired (8, 120, 122, 128). Adult ArKO males (ranging in age from 10 weeks through to 1 year) were observed to take longer to initiate mounting behavior and had significantly fewer mounts than WT littermates (8, 120), or did not mount at all (8, 122). This is similar to the situation observed with the double ERα- and ERβ-knockout mice but differs from the behavior of the mice with the individual receptor deletions; i.e., the ERα-knockout male mice display mounting behavior but no ejaculation, whereas the ERβ-knockout male mice have normal sexual behavior (129). It is evident that loss of estrogenic stimulation has severely detrimental effects on male sexual behavior. Presumably the estrogen in question is that produced as a consequence of local aromatase action in the brain, utilizing circulating testosterone as substrate.

Prostates of ArKO mice were significantly enlarged compared with those of WT control animals, as measured by wet weight and volume (130). The ventral, anterior, and dorsolateral lobes were up to 60, 40, and 55% larger, respectively, depending upon the age of the animals examined. This significant increase in size is a reflection of hyperplasia of the entire organ, including stroma, epithelia, and luminal compartments, rather than any local dysplastic growth, and is concomitant with upregulation of androgen receptors in epithelial cells. Thus the prostatic phenotype of ArKO mice has some resemblance to that of the ERβKO mouse prostate (131), in which elevated AR levels are reported, as well as multiple hyperplastic foci. Circulating testosterone and 5α-dihydrotestosterone (DHT) levels were elevated in these animals, as expected in an estrogen-deficient animal, but only DHT was elevated in the prostate tissue itself (2.84 $\pm$ 0.39 ng/ml versus 1.48 $\pm$ 0.21 ng/ml, ArKO versus WT, p $<$ 0.02). Serum prolactin levels were elevated approximately threefold above those of WT males. The conclusion is that estrogens exert dual actions in the prostate gland, triggering aberrant growth and/or suppressing androgen-induced hyperplasia. However, it is clear that further work needs to be conducted to establish the precise interactions between estrogens, androgens, and other hormones in the regulation of prostatic growth.

The Female ArKO Mouse

As could be expected, estrogen deficiency resulting from a lack of endogenous aromatase activity had profound effects upon the reproductive system of ArKO females. At about 3 months of age, when mice are young but already mature adults, uteri were severely underdeveloped, weighing less than half the wet weight of uteri of WT littermates (7, 132). Visually ArKO uteri are extremely hypotrophic and are easily discernible from WT uteri upon gross necropsy. Administration of exogenous estrogens in the form of conjugated equine estrogens generated hyperemic ArKO uteri, demonstrating they were still responsive to estrogen (7). The primary reason for ArKO female infertility, however, lies within their ovaries. In a preliminary study, Fisher et al. (7) reported the presence of many large follicles with evidence of antrum formation, but no corpora lutea (CL), indicative of failure to ovulate. Britt and colleagues (132) further investigated this phenotype, performing a detailed examination of the ovary architecture. An age-dependent phenotype was observed in the ArKO ovaries, with a progressive deterioration from the first age group examined (10–12 weeks), until by 1 year of age, the ovaries were dysmorphic and degenerative. At 10–12 weeks of age, ArKO ovaries contained a full complement of follicles, including primordial, primary, secondary, and antral follicles, but no CL. However even at this young age, the follicles appeared unhealthy, and hemorrhagic cysts were present in secondary and antral follicles. The usually uniform layers of granulosa cells were disrupted, and pyknotic nuclei were present, indicative of atretic follicles. By 21–23 weeks of age, the morphology had deteriorated further, and many antral follicles were observed to be cystic and hemorrhagic. All secondary and antral follicles were lost in 1-year-old ovaries, and hemorrhagic cysts made up much of the tissue. Apoptotic granulosa cells were widespread, particularly in larger antral-sized follicles, and extensive deposits of collagen were observed throughout the disorganized interstitial regions. Cells identified as macrophages had infiltrated regions of advanced degeneration.

ArKO ovaries develop but fail to fully mature within an estrogen-free environment. Although it appears, therefore, that estrogen plays an essential role in ovarian growth, other endocrine parameters are also perturbed in this model. Serum estradiol concentrations for all ArKO animals were beneath the level of detection for the assay ($\leq$6–8 pg/ml), but testosterone concentrations were approximately 10 times higher in the ArKO than in the WT females (7). Gonadotropin levels were elevated in the ArKO females; LH and FSH increased by up to 10-fold and 4-fold, respectively (7, 132). Comparison of the ArKO ovarian phenotype with those of the ERα- and ERβ- and double ERα/ERβ knockout mice indicates that each is different and unique. Consequently it is reasonable to interpret the disrupted ovarian morphology of ArKO mice as resulting from a perturbed hormonal environment, in which lack of estrogen per se is only one component. Hormonal manipulation in terms of restoration of estrogen and normalization of gonadotropin levels within this model may provide insights as to the importance each of these endocrine parameters plays in maintaining normal ovarian function.

Adipose Phenotype

ArKO mice accumulate excessive intra-abdominal fat in an age-dependent manner, a reflection of increased adipocyte volume and number in gonadal and infrarenal fat depots (10, 133). Extra total body fat accretion as early as 10 weeks of age was confirmed using magnetic resonance imaging (MRI) (Table 2). Importantly, these parameters were normalized to be comparable with WT animals following the administration of exogenous 17β-estradiol. A concomitant increase in body mass was observed for female mice; however, ArKO males were not heavier than WT siblings until 12 months of age. A parallel increase in circulating lipids was observed in 1-year-old animals where serum cholesterol and HDL were elevated in male and female ArKO mice. Serum triglycerides were also elevated in older male ArKO mice. Extra fat accumulation and elevated serum lipids were not caused by hyperphagia, although this is not surprising because circulating leptin levels were elevated two- to threefold in 4-month-old and 1-year-old male and female ArKO mice compared with those of their WT littermates. Leptin is known to regulate body fat predominantly by decreasing food intake (134). Nor was the obesity generated by reduced resting energy expenditure. Rather, the obese phenotype was associated with decreased lean mass and reduced spontaneous

TABLE 2 Percent adipose tissue

	10 weeks	1 year
Females		
ArKO	17.6 ± 4.4 (5)*	64.3 ± 11.0 (19)*
WT	4.9 ± 1.0 (5)	42.1 ± 6.7 (9)
Males		
ArKO	15.2 ± 2.3 (5)*	40.3 ± 3.8 (13)*
WT	7.3 ± 1.7 (5)	29.5 ± 3.7 (16)

Mean ± S.E.M. (n) * indicates at least p < 0.05 compared to WT.

physical activity. Associated with both increasing age and fat accumulation was a fourfold increase in circulating levels of insulin, although glucose levels remained unchanged. Normoglycemia concomitant with hyperinsulinemia is suggestive of insulin resistance in 1-year-old ArKO mice. A similar phenotype of obesity has been reported in ERα-knockout mice (11), and this phenotype appears to resemble that seen in humans with aromatase deficiency. The accumulation of fat droplets within the livers of ArKO mice has been reported as a corollary to the obese phenotype (10, 135). Nemoto and colleagues (135) reported an impairment to the hepatocellular fatty acid β-oxidation pathway in an aromatase-deficient mouse model they independently generated. These researchers were also able to rescue the fatty liver phenotype by administration of a stringent protocol of 17β-estradiol replacement.

Skeletal Abnormalities

The importance of estrogen in the maintenance of bone mass in postmenopausal women has been well documented. Its role in skeletal metabolism and growth in men has become apparent from analysis of the natural mutations in aromatase and ERα (91, 92, 118). Studies of the ArKO mouse have been able to augment available data on female bone maintenance and also provide data on the role played by estrogen in male bone maintenance. These studies have indicated that estrogen is as important for normal bone metabolism and growth in males as it is in females and that there may exist gender-specific bone responses to estrogen deficiency (136). Oz and collaborators (136) reported that both ArKO males and females exhibited loss of trabecular bone volume and thickness. Although the loss of these parameters in female ArKO mice was consistent with increased bone turnover as seen in postmenopausal women, their loss in males was more closely correlated with decreased osteoblastic activity. Furthermore, femur growth as measured by length was almost 10% less in ArKO males compared with WT littermates (p < 0.001); however, there were no significant differences in femur length between females. Both ArKO males and females demonstrated osteopenia in the lumbar spine. It is apparent from this model that the loss of estrogen and its actions has a significant and detrimental impact upon the maintenance of healthy bones in both males and females but that gender differences in the role of estrogen in bone exist.

CONCLUSIONS—THE CONCEPT OF LOCAL ESTROGEN BIOSYNTHESIS

Our understanding of the role of estrogens in both males and females has expanded greatly in recent years. Considerable emphasis has been focused in this review on the regulation of extragonadal estrogen biosynthesis, in particular that which occurs in adipose tissue, bone, and brain, and its importance in the well-being of the elderly (reviewed in 32).

Whereas the ovaries are the principal source of systemic estrogen in the premenopausal nonpregnant woman, other sites of estrogen biosynthesis become the major sources beyond menopause. These sites include the mesenchymal cells of the adipose tissue and skin, osteoblasts and perhaps chondrocytes in bone, vascular endothelial and aortic smooth muscle cells (109), and a number of sites in the brain including the medial preoptic/anterior hypothalamus, the medial basal hypothalamus, and the amygdala (137). These extragonadal sites of estrogen biosynthesis possess several fundamental features that differ from those of the ovaries. Principally, the estrogen synthesized within these compartments is probably only biologically active at a local tissue level in a paracrine or intracrine fashion. Thus the total amount of estrogen synthesized by these extragonadal sites may be small, but the local tissue concentrations achieved are probably high and exert significant biological influence locally. Thus these sources of estrogen play an important, but hitherto largely unrecognized, physiological and pathophysiological role.

After menopause, the mesenchymal cells of the adipose tissue become the main source of estrogen (32, 138). Therefore in the postreproductive years, the degree of a woman's estrogenization is mainly determined by the extent of her adiposity. This is of clinical importance because corpulent women are relatively protected against osteoporosis (139). Conversely obesity is positively correlated with breast cancer risk (140). Estrogens also play a major role in the regulation of adiposity in both males and females because aromatase- and ERα-deficient humans and mice develop a phenotype of abdominal obesity, elevated blood lipids, insulin resistance, and hepatic steatosis.

In males, it has been estimated that at best the testes can account for 15% of circulating estrogens (141), and local production of estrogens, both intratesticular and extragonadal, is of physiological significance throughout adult life. For example, the Leydig cells (126) and other cells of the testes, including germ cells in various stages of differentiation (124), produce estradiol, which has an important role in spermatogenesis. Estrogen production in bone appears to be as vital for the maintenance of bone mineralization and prevention of osteoporosis in men as it is in women. This is supported by studies of men with either a mutation of the gene encoding the aromatase enzyme (92, 118) or a mutation of the estrogen receptor (91), as well as by mouse models of gene disruption. Similarily, recent evidence indicates that estrogen production in one or more brain sites has a major influence on sexual behavior in males (8, 120, 122).

In this context it is appropriate to reconsider why osteoporosis is more common in women than in men and affects women at a younger age, in terms of fracture incidence. A key factor in the gender difference in the incidence of this and other diseases of estrogen insufficiency appears to be the availability of precursor C_{19} steroids for aromatization to estrogens in extragonadal sites, a concept also advanced by Labrie and colleagues (142). In postmenopausal women, the principal source of C_{19} steroid production is the adrenal cortex, which elaborates androstenedione, dehydroepiandrosterone (DHEA), and DHEA sulfate (DHEAS). However, the secretion of these steroids and their plasma concentrations decrease markedly with advancing age (13). Moreover, DHEA must first be converted to androstenedione prior to aromatization. Another major step is the reduction of the 17-keto group to 17β-hydroxyl, catalyzed by 17β-HSD type I, which is essential for formation of the active estrogen, estradiol. The distribution of this enzyme in the various extragonadal sites of aromatization has not yet been fully established, although it is expressed in tumorous breast epithelium (96) and in bone (94). It should be noted in this context that a recent report indicates that 17β-HSD type III, which converts androstenedione to testosterone, is present in visceral fat (143) together with 17β-HSD type II.

In the male circulation, in contrast, the levels of testosterone are at least an order of magnitude greater than those circulating in the plasma of postmenopausal women. Moreover, it is also two orders of magnitude greater than the mean levels of circulating estradiol in postmenopausal women. Given that much of the circulating estradiol is bound to sex hormone-binding globulin, it is unlikely to have a major impact on transactivation of the estrogen receptor, compared with estrogen produced locally as a consequence of conversion of circulating testosterone. Thus the uninterrupted sufficiency of circulating testosterone in men throughout life supports the local production of estradiol by aromatization of testosterone in estrogen-dependent tissues and thus affords ongoing protection against the so-called estrogen-deficiency diseases. This appears to be important in terms of protecting the bones of men against mineral loss and may contribute to the maintenance of cognitive function and prevention of Alzheimer's disease in men.

As suggested previously (32) (and discussed in this review), the aromatase found in adipose tissue is a marker of the undifferentiated adipose mesenchymal cell phenotype. In support of this, the factors that stimulate expression in adipose tissue are factors that either inhibit or reverse the differentiated phenotype of adipocytes, namely class I cytokines such as IL-6, oncostatin M and IL-11, or TNFα. All these factors act via the mesenchymal promoter I.4 of the aromatase gene and require glucocorticoids as co-stimulators (reviewed in 32). These considerations suggest that factors that stimulate adipocyte differentiation such as ligands of the PPARγ receptor, e.g., troglitazone, rosiglitazone, and 15-deoxy-$\Delta^{12,14}$-PGJ$_2$, would inhibit aromatase expression in adipose tissue, and this has proven to be the case (74). A summary of the interrelationships between adipose metabolism and aromatase expression is presented in Figure 6.

When a breast tumor is present, aromatase activity within the tumor and surrounding adipose tissue is such that intratumoral estradiol levels are at least an order of magnitude greater than those in the circulating plasma of postmenopausal women (144) (possibly one reason why taking HRT carries little increased risk of breast cancer). This is because the tumor produces factors that stimulate aromatase expression locally. The stimulation is associated with switching of the aromatase gene promoter from I.4 to promoter II, the ovarian type promoter, which likely occurs because the tumor-derived factors include PGE$_2$, a powerful stimulator of adenylate cyclase in adipose stromal cells and because promoter II is regulated by cAMP. Promoter II contains a half-site for nuclear receptor binding. In the ovary

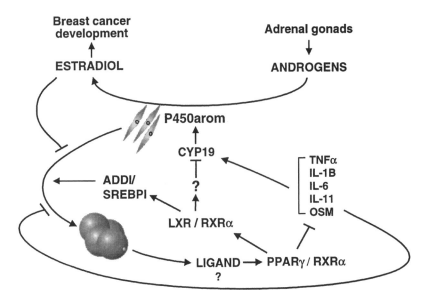

Figure 6   Potential interrelationships between adipose metabolism and aromatase expression in adipose tissue. In this model, the PPARγ/RXR heterodimer activates/induces key factors involved in lipogenesis such as ADD1/SREBP-1c, possibly via induction of LXR. PPARγ/RXR also inhibits aromatase expression and thus estradiol formation. Because estrogen inhibits lipid accumulation (by as yet unknown mechanisms), this double inhibition also results in net lipid accumulation. Class I cytokines and TNFα stimulate aromatase expression. Although not shown in adipose tissue, expression of pro-inflammatory cytokines in macrophages is inhibited by PPARγ. Because these are inhibitory of adipocyte differentiation, at least in the 3T3-L1 model, this double inhibition provides yet another mechanism whereby PPARγ may stimulate lipid accumulation. Thus the pathways of adipocyte differentiation, lipid metabolism, and aromatase expression are linked through PPARγ.

this is occupied by SF-1. In adipose stromal cells, it appears to be occupied by a repressor protein that competes with SF-1 for the site. If this factor proves to be an orphan member of the nuclear receptor family, then agonists of this factor would be good candidates for specific inhibitors of aromatase expression in tumorous breast tissue.

Third generation aromatase inhibitors are finding utility in the treatment of estrogen-dependent diseases such as breast cancer and more recently endometriosis (107). However these have the disadvantage that they inhibit aromatase activity in a global fashion and thus could have a detrimental impact at sites where estrogen is required for normal function, such as the maintenance of bone mineralization and possibly the prevention of hepatic steatosis. The concept of selective aromatase modulators (SAMs) is made possible by three considerations: First, in postmenopausal women and in men, estrogen is not a significant circulating hormone but rather acts at a local level in sites where it is produced in a paracrine or even intracrine fashion. Second, aromatase expresson in these different tissue sites of expression is regulated by the use of tissue-specific promoters. Third, various tissue-specific aromatase promoters employ entirely different signaling pathways and thus different cohorts of transcription factors. It is possible, therefore, to envision tissue-specific inhibition of aromatase expression in a fashion similar to the concept of tissue-specific regulation of estrogen action (the concept of SERMs). Thus SAMs could provide a parallel alternative treatment to SERMs, either as a backup or else as first line modality in their own right.

ACKNOWLEDGMENTS

Work from this laboratory presented in this review was supported, in part, by USPHS Grant R37A908174, by a grant from the Victorian Breast Cancer Research Consortium, and by NH & MRC (Australia) Grant 981126. The authors thank Sue Elger for skilled editorial assistance and Sue Panckridge for expert assistance with figure preparation.

Visit the Annual Reviews home page at www.AnnualReviews.org

LITERATURE CITED

1. Simpson ER, Rubin G, Clyne C, Robertson K, O'Donnell L, et al. 1999. Local estrogen biosynthesis in males and females. *Endocr.-Rel. Cancer* 6:131–37

2. Simpson ER, Rubin G, Clyne C, Robertson K, O'Donnell L, et al. 2000. The role of local estrogen biosynthesis in males and females. *Trends Endocrinol. Metab.* 11:184–88

3. Lubahn DB, Moyer JS, Golding TS, Couse JF, Korach KS, Smithies O. 1993. Alteration of reproductive function but not prenatal sexual development after insertional disruption of the mouse estrogen receptor gene. *Proc. Natl. Acad. Sci. USA* 90:11162–66

4. Krege JH, Hodgin JB, Couse JF, Enmark E, Warner M, et al. 1998. Generation and reproductive phenotypes of mice lacking estrogen receptor-β. *Proc. Natl. Acad. Sci. USA* 95:15677–82

5. Couse JF, Hewitt SC, Bunch DO, Sar M, Walker VR, et al. 1999. Postnatal sex reversal of the ovaries in mice lacking estrogen receptors α and β. *Science* 286:2328–31

6. Dupont S, Krust A, Gansmuller A, Dierich A, Chambon P, Mark M. 2000. Effect of single and compound knockouts of estrogen receptor alpha (ERalpha) and beta (ERbeta) on mouse reproductive phenotypes. *Development* 127:4277–91

7. Fisher CR, Graves KH, Parlow AF, Simpson ER. 1998. Characterization of mice deficient in aromatase (ArKO) because of targeted disruption of the *cyp19* gene. *Proc. Natl. Acad. Sci. USA* 95:6965–70

8. Honda S, Harada N, Takagi Y, Maeda S. 1998. Disruption of sexual behaviour in male aromatase-deficient mice lacking exons 1 and 2 of the *cyp19* gene. *Biochem. Biophys. Res. Commun.* 252:445–49

9. Nemoto Y, Toda K, Ono M, Fujikawa-Adachi K, Saibara T, et al. 2000. Altered expression of fatty acid metabolizing

enzymes in aromatase-deficient mice. *J. Clin. Invest.* 105:1819–25

10. Jones MEE, Thorburn AW, Britt KL, Hewitt KN, Wreford NG, et al. 2000. Aromatase-deficient (ArKO) mice have a phenotype of increased adiposity. *Proc. Natl. Acad. Sci. USA* 97:12735–40

11. Heine PA, Taylor JA, Iwamoto GA, Lubahn DB, Cooke PS. 2000. Increased adipose tissue in male and female estrogen receptor-alpha knockout mice. *Proc. Natl. Acad. Sci. USA* 97:12729–34

12. Simpson ER, Davis SR. 1998. Why do the sequelae of estrogen deficiency affect women more frequently than men? *J. Clin. Endocrinol. Metab.* 83:2214

13. Labrie F, Belanger A, Cusan L, Gomez JL, Candas B. 1997. Marked decline in serum concentrations of adrenal C19 sex steroid precursors and conjugated androgen metabolites during aging. *J. Clin. Endocrinol. Metab.* 82:2396–402

14. Nelson DR, Koymans L, Kamataki T, Stegeman JJ, Feyereisen R, et al. 1996. P450 superfamily: update on new sequences, gene mapping, accession numbers and nomenclature. *Pharmacogenetics* 6:1–42

15. Thompson EA Jr, Siiteri PK. 1974. The involvement of human placental microsomal cytochrome P450 in aromatisation. *J. Biol. Chem.* 249:5373–78

16. Mendelson CR, Wright EE, Porter JC, Evans CT, Simpson ER. 1985. Preparation and characterization of polyclonal and monoclonal antibodies against human aromatase cytochrome P-450 (P-450$_{arom}$), and their use in its purification. *Arch. Biochem. Biophys.* 243:480–91

17. Nakajin S, Shimoda M, Hall PF. 1986. Purification to homogeneity of aromatase from human placenta. *Biochem. Biophys. Res. Commun.* 134:704–10

18. Kellis JT, Vickery LE. 1987. Purification and characterization of human placental aromatase cytochrome P450. *J. Biol. Chem.* 262:4413–20

19. Nelson DR, Kamataki T, Waxman DJ, Guengerich FP, Estabrook RW, et al. 1993. The P450 superfamily: update on new sequences, gene mapping, accession numbers, early trivial names of enzymes, and nomenclature. *DNA Cell Biol.* 12:1–51

20. Osawa Y, Yoshida N, Franckowiak M, Kitawaki J. 1987. Immunoaffinity purification of aromatase cytochrome P450 from human placental microsomes, metabolic switching from aromatization to 1β and 2β-monohydroxylation, and recognition of aromatase isoenzymes. *Steroids* 50:11–28

21. Cole PA, Robinson CH. 1988. A peroxide model reaction for placental aromatase. *J. Am. Chem. Soc.* 110:1284–85

22. Akhtar M, Calder MR, Corina DL, Wright JN. 1982. Mechanistic studies on C19 demethylation in oestrogen biosynthesis. *Biochem. J.* 201:569–80

23. Means GD, Mahendroo M, Corbin CJ, Mathis JM, Powell FE, et al. 1989. Structural analysis of the gene encoding human aromatase cytochrome P-450, the enzyme responsible for estrogen biosynthesis. *J. Biol. Chem.* 264:19385–91

24. Harada N, Yamada K, Saito K, Kibe N, Dohmae S, Takagi Y. 1990. Structural characterization of the human estrogen synthetase (aromatase) gene. *Biochem. Biophys. Res. Commun.* 166:365–72

25. Toda K, Terashima M, Kamamoto T, Sumimoto H, Yamamoto Y, et al. 1990. Structural and functional characterization of human aromatase P450 gene. *Eur. J. Biochem.* 193:559–65

26. Means GD, Kilgore MW, Mahendroo MS, Mendelson CR, Simpson ER. 1991. Tissue-specific promoters regulate aromatase cytochrome P450 gene expression in human ovary and fetal tissues. *Mol. Endocrinol.* 5:2005–13

27. Mahendroo MS, Means GD, Mendelson CR, Simpson ER. 1991. Tissue-specific expression of human P450$_{arom}$: the promoter responsible for expression in

adipose is different from that utilized in placenta. *J. Biol. Chem.* 266:11276–81

28. Jenkins C, Michael D, Mahendroo M, Simpson E. 1993. Exon-specific northern analysis and rapid amplification of cDNA ends (RACE) reveal that the proximal promoter II (PII) is responsible for aromatase cytochrome P450 (CYP19) expression in human ovary. *Mol. Cell. Endocrinol.* 97:R1–R6

29. Mahendroo MS, Mendelson CR, Simpson ER. 1993. Tissue-specific and hormonally controlled alternative promoters regulate aromatase cytochrome P450 gene expression in human adipose tissue. *J. Biol. Chem.* 268:19463–70

30. Honda S, Harada N, Takagi N. 1994. Novel exon I of the aromatase gene specific for aromatase transcripts in human brain. *Biochem Biophys. Res. Commun.* 198:1153–60

31. Shozu M, Zhao Y, Bulun SE, Simpson ER. 1998. Multiple splicing events involved in regulation of human aromatase expression by a novel promoter, I.6. *Endocrinology* 139:1610–17

32. Simpson ER, Zhao Y, Agarwal VR, Michael MD, Bulun SE, et al. 1997. Aromatase expression in health and disease. *Rec. Prog. Horm. Res.* 52:185–214

33. Callard GV, Petro Z, Ryan KJ. 1978. Phylogenetic distribution of aromatase and other androgen-converting enzymes in the central nervous system. *Endocrinology* 103:2283–90

34. Callard GV, Petro Z, Ryan KJ. 1980. Aromatization and 5α-reduction in brain and non-neural tissues of a cyclostome (*Petromyzan marinua*). *Gen. Comp. Endocrinol.* 42:155–59

35. Callard GV. 1981. Aromatization is cyclic AMP-dependent in cultured reptilian brain cells. *Brain Res.* 204:451–54

36. Hutchinson JB. 1991. Hormonal control of behaviour: steroid action in the brain. *Curr. Opin. Neurobiol.* 1:562–70

37. Kishida M, Callard GV. 2001. Distinct cytochrome P450 aromatase isoforms in zebrafish (*D. rerio*) brain and ovary are differentially regulated during early development. *Endocrinology* 142:740–50

38. Chiang EF, Yan YL, Guiguen Y, Postlethwait J, Chung BC. 2001. Two cyp19 (P450 aromatase) genes on duplicated Zebrafish chromosomes are expressed in ovary or brain. *Mol. Biol. Evol.* 18:542–50

39. Brunner RM, Goldammer T, Furbass R, Vanselow J, Schwerin M. 1998. Genomic organization of the bovine aromatase gene and a homologous pseudogene as revealed by DNA fiber FISH. *Cytogenet. Cell Genet.* 82:37–40

40. Hinshelwood MM, Liu Z, Conley AJ, Simpson ER. 1995. Demonstration of tissue-specific promoters in nonprimate species that express aromatase P450 in placentae. *Biol. Reprod.* 53:1151–59

41. Furbass R, Kalbe C, Nanselow J. 1997. Tissue-specific expression of the bovine aromatase-encoding gene uses multiple transcriptional start sites and alternative first exons. *Endocrinology* 138:2813–19

42. Vanselow J, Zsolnai A, Fesus L, Furbass R, Schwerin M. 1999. Placenta-specific transcripts of the aromatase-encoding gene include different untranslated first exons in sheep and cattle. *Eur. J. Biochem.* 265:318–24

43. Graddy LG, Kowalski AA, Simmen FA, Davis SLF, Baumgartner WW, Simmen RCM. 2000. Multiple isoforms of porcine aromatase are encoded by three distinct genes. *J. Steroid Biochem. Mol. Biol.* 73:49–57

44. Corbin CJ, Trant JM, Walters KW, Conley AJ. 1999. Changes in testosterone metabolism associated with the evolution of placental and gonadal isoenzymes of porcine aromatase cytochrome P450. *Endocrinology* 140:5202–10

45. Graham-Lorence S, Khalil MW, Lorence MC, Mendelson CR, Simpson ER. 1991. Structure-function relationships of human aromatase cytochrome P-450 using molecular modeling and site directed mutagenesis. *J. Biol. Chem.* 266:11939–46

46. Poulos TL, Finzel BC, Howard AJ. 1987. High resolution crystal structure of cytochrome P450. *J. Mol. Biol.* 195:687–700

47. Ravichandran KG, Boddupalli SS, Hasemann CA, Peterson JA, Deisenhofer J. 1993. Atomic structure of the hemoprotein domain of cytochrome P450BM3: a prototype for eukaryotic microsomal P450s. *Science* 261:731–36

48. Hasemann CA, Ravichandran KG, Peterson JA, Deisenhofer J. 1994. Crystal structure and refinement of cytochrome P450terp at 2–3 Å resolution. *J. Mol. Biol.* 236:1169–85

49. Williams PA, Cosme J, Sridhar V, Johnson EF, McRee DE. 2000. Mammalian microsomal P450 monooxygenase: structural adaptations for membrane binding and functional diversity. *Mol. Cell* 5:121–31

50. Graham-Lorence S, Amarneh B, White RE, Peterson JA, Simpson ER. 1995. A three-dimensional model of aromatase cytochrome P450. *Protein Sci.* 4:1065–80

51. Michael MD, Kilgore MW, Morohashi K, Simpson ER. 1995. Ad4BP/SF-1 regulates cyclic AMP-induced transcription from the proximal promoter (PII) of the human aromatase P450 (CYP19) gene in the ovary. *J. Biol. Chem.* 270:13561–66

52. Fitzpatrick SL, Richards JS. 1994. Transcriptional regulation of the rat aromatase gene in granulosa cells and R2C cells by SF-1 and CREB. *Proc. Endocr. Soc.* 76:399 (Abstr.)

53. Luo X, Ikeda Y, Parker KL. 1994. A cell-specific nuclear receptor is essential for adrenal and gonadal development. *Cell* 77:481–90

54. Lala DS, Rice DA, Parker KL. 1992. Steroidogenic factor 1, a key regulation of steroidogenic gene expression, is the mouse homolog of fushi tarazu-factor 1. *Mol. Endocrinol.* 6:1249–58

55. Michael MD, Michael LF, Simpson ER. 1997. A CRE-like sequence that binds CREB and contributes to cAMP-dependent regulation of the proximal promoter of the human aromatase P450 (CYP19) gene. *Mol. Cell. Endocrinol.* 134:147–56

56. Hinshelwood MM, Michael MD, Simpson ER. 1997. The 5'-flanking region of the ovarian promoter of the bovine CYP19 gene contains a deletion in a cyclic adenosine 3',5'-monophosphate-like responsive sequence. *Endocrinology* 138:3704–10

57. Zhao Y, Mendelson CR, Simpson ER. 1995. Characterization of the sequences of the human CYP19 (aromatase) gene that mediate regulation by glucocorticoids in adipose stromal cells and fetal hepatocytes. *Mol. Endocrinol.* 9:340–49

58. Zhao Y, Nichols JE, Bulun SE, Mendelson CR, Simpson ER. 1995. Aromatase P450 gene expression in human adipose tissue: role of a Jak/STAT pathway in regulation of the adipose-specific promoter. *J. Biol. Chem.* 270:16449–57

59. Reinhardt HM, Schutz G. 1998. Glucocorticoid signalling—multiple variations of common theme. *Mol. Cell. Endocrinol.* 146:1–6

60. Price T, O'Brien S, Dunaif A, Simpson ER. 1992. Comparison of aromatase cytochrome P450 mRNA levels in adipose tissue from the abdomen and buttock using competitive polymerase chain reaction amplification. *Proc. Soc. Gynecol. Invest.* 179: (Abstr.)

61. Berkovitz GD, Bisat T, Carter KM. 1989. Aromatase activity in microsomal preparations of human genital skin fibroblasts: influence of glucocorticoids. *J. Steroid Biochem.* 33:341–47

62. Bruch HR, Wolf L, Budde R, Romalo G, Schweikert HU. 1992. Androstenedione metabolism in cultured human osteoblast-like cells. *J. Clin. Endocrinol. Metab.* 75:101–5

63. Nawata H, Tanaka S, Tanaka S, Takayanagi R, Sakai Y, et al. 1995. Aromatase

in bone cells: association with osteoporosis in postmenopausal women. *J. Steroid Biochem. Mol. Biol.* 53:165–74

64. Harada N, Utsume T, Takagi Y. 1993. Tissue-specific expression of the human aromatase cytochrome P-450 gene by alternative use of multiple exons 1 and promoters, and switching of tissue-specific exons 1 in carcinogenesis. *Proc. Natl. Acad. Sci. USA* 90:11312–16

65. Agarwal VR, Bulun SE, Leitch M, Rohrich R, Simpson ER. 1996. Use of alternative promoters to express the aromatase cytochrome P450 (CYP19) gene in breast adipose tissues of cancer-free and breast cancer patients. *J. Clin. Endocrinol. Metab.* 81:3843–49

66. Zhao Y, Agarwal VR, Mendelson CR, Simpson ER. 1996. Estrogen biosynthesis proximal to a breast tumor is stimulated by PGE2 via cyclic AMP, leading to activation of promoter II of the CYP19 (aromatase) gene. *Endocrinology* 137:5739–42

67. Agarwal VR, Ashanullah CI, Simpson ER, Bulun SE. 1997. Alternatively spliced transcripts of the aromatase cytochrome P450 (CYP19) gene in adipose tissue of women. *J. Clin. Endocrinol. Metab.* 82:70–74

68. Bulun SE, Simpson ER. 1994. Competitive RT-PCR analysis indicates levels of aromatase cytochrome P450 transcripts in adipose tissue of buttocks, thighs, and abdomen of women increase with advancing age. *J. Clin. Endocrinol. Metab.* 78:428–32

69. Wei J, Xu H, Davies JL, Hemmings JP. 1992. Increase of plasma IL-6 concentration with age in healthy subjects. *Life Sci.* 51:1953–56

70. Daynes RA, Araneo BA, Ershler WB, Maloney C, Li GZ, Ryu SY. 1993. Altered regulation of IL-6 production with normal aging. *J. Immunol.* 150:5219–30

71. Petruschke T, Hauner H. 1993. Tumor necrosis factor-alpha prevents the differentiation of human adipocyte precursor cells and caused delipidation of developed fat cells. *J. Clin. Endocrinol. Metab.* 76:742–47

72. Kawashima I, Ohsumi J, Mita-Honjo K, Shimoda-Takano K, Ishikawa H, et al. 1991. Molecular cloning of cDNA encoding adipogenesis inhibitory factor and identity with interleukin-11. *FEBS Lett.* 283:199–202

73. Kliewer SA, Lenhard JM, Willson TM, Patel I, Morris DC, Lehmann JM. 1995. A prostaglandin J2 metabolite binds peroxisome proliferator-activated receptor gamma and promotes adipocyte differentiation. *Cell* 83:813–19

74. Rubin GL, Zhao Y, Kalus AM, Simpson ER. 2000. Peroxisome proliferator-activated receptor gamma ligands inhibit estrogen biosynthesis in human breast adipose tissue: possible implications for breast cancer therapy. *Cancer Res.* 60:1604–8

75. Rink JD, Simpson ER, Barnard JJ, Bulun SE. 1996. Cellular chacterization of adipose tissue from various body sites of women. *J. Clin. Endocrinol. Metab.* 81:2443–47

76. Bulun SE, Sharda G, Rink J, Sharma S, Simpson ER. 1996. Distribution of aromatase P450 transcripts and adipose fibroblasts in the human breast. *J. Clin. Endocrinol. Metab.* 81:1273–77

77. Harada N. 1997. Aberrant expression of aromatase in breast cancer tissues. *J. Steroid Biochem. Mol. Biol.* 61:175–84

78. Zhou D, Zhou C, Chen S. 1997. Gene regulation studies of aromatase expression in breast cancer and adipose stromal cells. *J. Steroid Biochem. Mol. Biol.* 61:273–80

79. Zhou J, Gurates B, Yang S, Sebastian S, Bulun SE. 2001. Malignant breast epithelial cells stimulate aromatase expression via promoter II in human adipose fibroblasts: an epithelial-stromal interaction in breast tumors mediated by CCAAT/enhancer binding protein beta. *Cancer Res.* 61:2328–34

80. Meng L, Zhou J, Sasano H, Suzuki T,

Zeitoun KM, Bulun SE. 2001. Tumor necrosis factor alpha and interleukin 11 secreted by malignant breast epithelial cells inhibit adipocyte differentiation by selectively down-regulating CCAAT/enhancer binding protein alpha and peroxisome proliferator-activated receptor gamma: mechanism of desmoplastic reaction. *Cancer Res.* 61:2250–55

81. Yang C, Zhou D, Chen S. 1998. Modulation of aromatase expression in the breast tissue by ERR alpha-1 orphan receptor. *Cancer Res.* 58:5695–700

82. Okubo T, Truong TK, Yu B, Itoh T, Zhao J, et al. 2001. Down-regulation of promoter 1.3 activity of the human aromatase gene in breast tissue by zinc-finger protein, snail (SnaH). *Cancer Res.* 61:1338–46

83. Jin T, Zhang X, Li H, Goss PE. 2000. Characterization of a novel silencer element in the human aromatase gene PII promoter. *Breast Cancer Res. Treat.* 62:151–59

84. Toda K, Miyahara K, Kawamoto T, Ikeda H, Sagara Y, Shizuta Y. 1992. Characterization of a *cis*-acting regulatory element involved in human aromatase P450 gene expression. *Eur. J. Biochem.* 205:303–9

85. Yamada K, Harada N, Honda S, Takagi Y. 1995. Regulation of placenta-specific expression of the aromatase cytochrome P450 gene. Involvement of the trophoblast-specific element binding protein. *J. Biol. Chem.* 270:25064–69

86. Sun T, Zhao Y, Fisher CR, Kilgore M, Mendelson CR, Simpson ER. 1996. Characterization of *cis*-acting elements of the human aromatase P450 (CYP19) gene that mediate regulation by retinoids in human choriocarcinoma cells. In *Program & Abstracts of the 10th Int. Congr. Endocrinology.* San Francisco, CA. Abstr. OR38-8

87. Stephanon A, Sarlis NJ, Richards R, Handwerger S. 1994. Expression of retinoic acid receptor subtypes and cellular retinoic acid binding protein II in mRNAs during differentiation of human trophoblast cells. *Biochem. Biophys. Res. Comm.* 202:772–80

88. Barak Y, Nelson MC, Ong ES, Jones YZ, Ruiz-Lozano P, et al. 1999. PPAR gamma is required for placental, cardiac, and adipose tissue development. *Mol. Cell* 4:585–95

89. Jiang B, Kamat A, Mendelson CR. 2000. Hypoxia prevents induction of aromatase expression in human trophoblasts: potential inhibitory role of the hypoxia-inducible transcription factor Mash-2 (mammalian achaete-scute homologous protein-2). *Mol. Endocrinol.* 14:1661–73

90. Kamat A, Graves KH, Smith ME, Richardson JA, Mendelson CR. 1999. A 500-bp region, approximately 40 kb upstream of the human CYP19 (aromatase) gene, mediates placenta-specific expression in transgenic mice. *Proc. Natl. Acad. Sci. USA* 96:4575–80

91. Smith EP, Boyd J, Frank GR, Takahashi H, Cohen RM, et al. 1994. Estrogen resistance caused by a mutation in the estrogen-receptor gene in a man. *N. Engl. J. Med.* 331:1056–61

92. Carani C, Qin K, Simoni M, Faustini-Fustini M, Serpanti S, et al. 1997. Aromatase deficiency in the male: effect of testosterone and estradiol treatment. *N. Engl. J. Med.* 337:91–95

93. Mullis PE, Yoshimura N, Kuhlmann B, Lippuner K, Jaeger P, Harada H. 1997. Aromatase deficiency in a female who is compound heterozygote for two new point mutations in the P450$_{arom}$ gene: impact of estrogens on hypergonadotropic hypogonadism, multicystic ovaries, and bone densitometry in childhood. *J. Clin. Endocrinol. Metab.* 82:1739–45

94. Sasano H, Uzuki M, Sawai T, Nagura H, Matsunaga G, et al. 1997. Aromatase in bone tissue. *J. Bone Min. Res.* 12:1416–23

95. Shozu M, Simpson ER. 1998. Aromatase

expression of human osteoblast-like cells. *Mol. Cell. Endocrinol.* 139:117–29

96. Sasano H, Frost AR, Saitoh R, Harada N, Poutanen M, et al. 1996. Aromatase and 17β-hydroxysteroid dehydrogenase type 1 in human breast carcinoma. *J. Clin. Endocrinol. Metab.* 81:4042–46

97. Lea CK, Ebrahim H, Tennant S, Flanagan AM. 1997. Aromatase cytochrome P450 transcripts are detected in fractured human bone but not in normal skeletal tissue. *Bone* 21:433–40

98. Lauber ME, Lichtensteiger W. 1994. Pre- and postnatal ontogeny of aromatase cytochrome P450 messenger RNA expression in the male rat brain studied by in situ hybridization. *Endocrinology* 135:1661–68

99. Wagner CK, Morrell JI. 1997. Neuroanatomical distribution of aromatase in RNA in the rat brain: indications of regional regulation. *J. Steroid Biochem. Mol. Biol.* 61:307–14

100. Sasano H, Takashashi K, Satoh F, Nagura H, Harada N. 1998. Aromatase in the central nervous system. *Clin. Endocrinol.* 48:325–29

101. Shozu M, Zhao Y, Simpson ER. 1997. Estrogen biosynthesis in THP-1 cells is regulated by promoter switching of the aromatase (CYP19) gene. *Endocrinology* 138:5125–35

102. Abdelgadir SE, Resko JA, Ojeda SR, Lephart ED, McPhaul MJ, Roselli CE. 1994. Androgens regulate aromatase cytochrome P450 messenger ribonucleic acid in the rat brain. *Endocrinology* 135:395–401

103. Abe-Dohmae S, Takagi Y, Harada N. 1997. Autonomous expression of aromatase during development of mouse brain is modulated by neurotransmitters. *J. Steroid Biochem. Mol. Biol.* 61:299–306

104. Lanoux MJ, Cleland WH, Mendelson CR, Carr BR, Simpson ER. 1985. Factors affecting the conversion of androstenedione to estrogens by human fetal hepa-

tocytes in monolayer culture. *Endocrinology* 117:361–67

105. Harada N. 1992. A unique aromatase (P450arom) mRNA formed by alternative use of tissue specific exons I in human skin fibroblasts. *Biochem. Biophys. Res. Commun.* 189:1001–7

106. Toda K, Simpson ER, Mendelson CR, Shizuta Y, Kilgore MW. 1994. Expression of the gene encoding aromatase cytochrome P450 (CYP19) in fetal tissues. *Mol. Endocrinol.* 8:210–17

107. Zeitoun K, Takayama K, Michael MD, Bulun SE. 1999. Stimulation of aromatase promoter (II) activity in endometriosis and its inhibition in endometrium are regulated by competitive binding of steroidogenic factor-1 and chicken ovalbumin upstream promoter transcription factor to the same *cis*-acting element. *Mol. Endocrinol.* 13:239–53

108. Harada N, Sasano H, Murakami H, Ohkuma T, Nagura H, Takagi Y. 1999. Localized expression of aromatase in human vascular tissues. *Circ. Res.* 84: 1285–91

109. Sasano H, Murakami H, Shizawa S, Satomi S, Nagura H, Harada N. 1999. Aromatase and sex steroid receptors in human vena cava. *Endocrine J.* 46:233–42

110. Agarwal VR, Takayama K, van Wyk JJ, Sasano H, Simpson ER, Bulun SE. 1998. Molecular basis of severe gynecomastia associated with aromatase expression in a fibrolamellar hepatocellular carcinoma. *J. Clin. Endocrinol. Metab.* 83:1797–800

111. Bulun SE, Rosenthal IM, Brodie AM, Inkster SE, Zeller WP, et al. 1994. Use of tissue-specific promoters in the regulation of aromatase cytochrome P450 gene expression in human testicular and ovarian sex cord tumors as well as in normal fetal and adult gonads. *J. Clin. Endocrinol. Metab.* 78:1616–21

112. Young J, Bulun SE, Agarwal V, Couzinet B, Mendelson CR. 1996. Aromatase expression in a feminizing adrenocortical

tumor. *J. Clin. Endocrinol. Metab.* 81: 3173–76

113. Bulun SE, Economos K, Miller D, Simpson ER. 1994. CYP19 (aromatase) gene expression in human malignant endometrial tumors. *J. Clin. Endocrinol. Metab.* 79:1831–34

114. Noble LS, Simpson ER, Johns A, Bulun SE. 1996. Aromatase expression in endometriosis. *J. Clin. Endocrinol. Metab.* 81:174–79

115. Shozu M, Akasofu K, Harada T, Kubota Y. 1991. A new cause of female pseudohermaphoditism: placental aromatase deficiency. *J. Clin. Endocrinol. Metab.* 72: 560–66

116. Ito Y, Fisher CR, Conte FA, Grumbach MM, Simpson ER. 1993. Molecular basis of aromatase deficiency in an adult female with sexual infantilism and polycystic ovaries. *Proc. Natl. Acad. Sci. USA* 90:11673–77

117. Conte FA, Grumbach MM, Ito Y, Fisher CR, Simpson ER. 1994. A syndrome of female pseudohermaphrodism, hypergonadotropic hypogonadism, and multicystic ovaries associated with missense mutations in the gene encoding aromatase (P450arom). *J. Clin. Endocrinol. Metab.* 78:1287–92

118. Morishima A, Grumbach MM, Simpson ER, Fisher C, Qin M. 1995. Aromatase deficiency in male and female siblings caused by a novel mutation and the physiological role of oestrogens. *J. Clin. Endocrinol. Metab.* 80:3689–98

119. Bilezikian JP, Morishima A, Bell J, Grumbach MM. 1998. Increased bone mass as a result of estrogen therapy in a man with aromatase deficiency. *N. Engl. J. Med.* 339:599–603

120. Toda K, Okada T, Akira S, Saibara T, Shiraishi M, et al. 2001. Oestrogen at the neonatal stage is critical for reproductive ability of male mice as revealed by supplementation with 17β-oestradiol to aromatase gene (Cyp19) knockout mice. *J. Endocrinol.* 168:455–63

121. Robertson K, O'Donnell L, Jones MEE, Meachem SJ, Boon WC, et al. 2000. Impairment of spermatogenesis in mice lacking a functional aromatase (*cyp 19*) gene. 1999. *Proc. Natl. Acad. Sci. USA* 96:7986–91

122. Robertson KM, Simpson ER, Lacham-Kaplan O, Jones MEE. 2001. Characterization of the fertility of the male aromatase knockout (ArKO) mouse. *J. Androl.* 22:825–30

123. Eddy EM, Washburn TF, Bunch DO, Goulding EH, Gladen BC, et al. 1996. Targeted disruption of the oestrogen receptor gene in male mice causes alteration of spermatogenesis and infertility. *Endocrinology* 137:4796–805

124. Nitta H, Bunick D, Hess RA, Janulis L, Newton SC, et al. 1993. Germ cells of the mouse testis express P450 aromatase. *Endocrinology* 132:1396–401

125. Levallet J, Bilinska B, Mittre H, Genissel C, Fresnel J, Carreau S. 1998. Expression of immunolocalization of functional cytochrome P450 aromatase in mature rat testicular cells. *Biol. Reprod.* 58:919–26

126. Tsai-Morris CH, Aquilana DR, Dufau ML. 1985. Cellular localisation of rat testicular aromatase activity during development. *Endocrinology* 116:38–46

127. Janulis L, Bahr JM, Hess RA, Bunick D. 1996. P450 aromatase messenger ribonucleic acid expression in male rat germ cells: detection by reverse transcription-polymerase chain reaction amplification. *J. Androl.* 17:651–58

128. Simpson ER, Davis SR. 2000. Another role highlighted for estrogens in the male: sexual behaviour. *Proc. Natl. Acad. Sci. USA* 97:14038–40

129. Ogawa S, Chester AE, Curtis Hewitt S, Walker VR, et al. 2000. Abolition of male sexual behaviours in mice lacking estrogen receptors α and β ($\alpha\beta$ERKO). *Proc. Natl. Acad. Sci. USA* 97:14737–41

130. McPherson SJ, Wang H, Jones ME, Pedersen J, Iismaa TP, et al. 2001. Elevated androgens and prolactin in aromatase deficient mice (ArKO) cause enlargement but not malignancy of the prostate gland. *Endocrinology* 142:2458–67

131. Weihua Z, Makela S, Andersson LC, Salmi S, Saji S, et al. 2001. A role for estrogen receptor β in the regulation of growth of the ventral prostate. *Proc. Natl. Acad. Sci. USA* 98:6330–35

132. Britt KL, Drummond AE, Cox VA, Dyson M, Wreford NG, et al. 2000. An age-related ovarian phenotype in mice with targeted disruption of the Cyp 19 (aromatase) gene. *Endocrinology* 141:2614–23

133. Jones MEE, Thorburn AW, Britt KL, Hewitt KN, Misso M, et al. 2001. The obese phenotype of the ArKO mouse. *J. Steroid Biochem. Mol. Biol.* In press

134. Farooqi IS, Jebb SA, Langmack G, Lawrence E, Cheetham CH, et al. 1999. Effects of recombinant leptin therapy in a child with cogenital leptin deficiency. *N. Engl. J. Med.* 341:879–915

135. Nemoto Y, Toda K, Ono M, Fujikawa-Adachi K, Saibara T, et al. 2000. Altered expression of fatty acid-metabolizing enzymes in aromatase-deficient mice. *J. Clin. Invest.* 105:1819–25

136. Oz O, Zerwekh JE, Fisher C, Graves K, Nanu L, et al. 2000. Bone has a sexually dimorphic response to aromatase deficiency. *J. Bone Miner. Res.* 15:507–14

137. Naftolin F, Ryan KJ, Davies IJ, Reddy VV, Flores F, et al. 1975. The formation of estrogens by central neuroendocrine tissues. *Recent Prog. Horm. Res.* 31:295–319

138. Siiteri PK, MacDonald PC. 1973. Role of extraglandular estrogen in human endocrinology. In *Handbook of Physiology*, ed. RO Green, EB Astwood, 2:619–29. Bethesda, MD: Am. Physiol. Soc.

139. Melton LJ. 1997. Epidemiology of spinal osteoporosis. *Spine* 22:2S–11S

140. Huang Z, Hankinson SE, Colditz GA, Stampfer MJ, Hunter DJ, et al. 1997. Dual effects of weight and weight gain on breast cancer risk. *J. Am. Med. Assoc.* 278:1407–11

141. Hemsell DL, Edman CD, Marks JF, Siiteri PK, MacDonald PC. 1974. Plasma precursors of estrogen. II. Correlation of the extent of conversion of plasma androstenedione to estrone with age. *J. Clin. Endocrinol. Metab.* 38:476–79

142. Labrie F, Belanger A, Luu-The V, Labrie C, Simond J, et al. 1998. DHEA and the intracrine formation of androgens and estrogens in peripheral target tissues: its role during aging. *Steroids* 63:322–28

143. Corbould AM, Judd SJ, Rodgers RJ. 1998. Expression of types 1, 2 and 3 β-hydroxysteroid dehydrogenase in subcutaneous abdominal and intra-abdominal adipose tissue of women. *J. Clin. Endocrinol. Metab.* 83:187–94

144. Pasqualini JR, Chetrite G, Blacker C, Feinstein MC, De la Londe L, et al. 1996. Concentrations of estrone, estradiol and estrone sulfate and evaluation of sulfatase and aromatase activities in pre- and postmenopausal breast cancer patients. *J. Clin. Endocrinol. Metab.* 81:1460–64

Annu. Rev. Physiol. 2002. 64:129–52

G PROTEINS AND PHEROMONE SIGNALING

Henrik G. Dohlman

Department of Biochemistry and Biophysics, University of North Carolina School of Medicine, Chapel Hill, North Carolina 27599; e-mail: henrik_dohlman@med.unc.edu

Key Words yeast, desensitization, GPA1, SST2, RGS

■ **Abstract** All cells have the capacity to respond to chemical and sensory stimuli. Central to many such signaling pathways is the heterotrimeric G protein, which transmits a signal from cell surface receptors to intracellular effectors. Recent studies using the yeast *Saccharomyces cerevisiae* have produced important advances in our understanding of G protein activation and inactivation. This review focuses on the mechanisms by which G proteins transmit a signal from peptide pheromone receptors to the mating response in yeast and how mechanisms elucidated in yeast can provide insights to signaling events in more complex organisms.

PERSPECTIVES AND SCOPE

Signal transduction—the process of cell-to-cell communication—is one of the most intensively studied subjects in biology. Among the best-characterized signaling systems are those comprised of a cell surface receptor, a G protein, and an effector enzyme. A recent addition to this list is the regulator of G protein signaling, or RGS protein (Figure 1A). In humans G protein–coupled receptors mediate responses to light, odor, taste, hormones, and neurotransmitters. In the simplest eukaryotes they mediate signals that affect such basic processes as cell division and mating.

In this review I describe factors that regulate G protein activity in the yeast *S. cerevisiae*. A relative newcomer, the yeast model has now emerged as having one of the best-understood signaling systems in any eukaryotic organism. Powerful genetic methods available only in yeast have already told us a great deal about the most fundamental aspects of cell regulation. All of the key signaling proteins, from cell surface to nucleus, have been identified and genetically knocked out. Moreover, principles elucidated in yeast are usually applicable to more complex systems. Among the key proteins discovered in yeast are the first RGS protein (Sst2) and the first kinase-scaffolding protein (Ste5).

I summarize recent advances in the identification and characterization of factors that regulate G protein activity. The focus is on the growing list of proteins that interact with the RGS or G protein and how these interactions affect their assembly, localization, catalytic activity, and stability. Such factors can have both positive or

0066-4278/02/0315-0129$14.00

negative effects on signaling. Negative regulation in the form of desensitization can allow cells to recover from chronic but unproductive stimulation with mating pheromones or other environmental signals. Positive regulation in the form of amplification can transduce a weak external signal to large internal changes in cell physiology. A third type of regulation (not discussed here) involves signal fidelity. Many different pathways use homologous components, which can greatly complicate specificity. Cross talk can be an adaptive process when it allows a single stimulus to trigger multiple responses in a coordinated manner, but it can

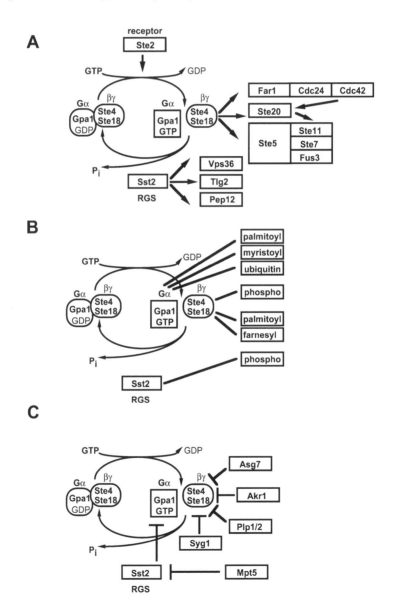

be maladaptive when it leads to the adventitious activation of the wrong target. A recent and comprehensive review of the yeast mating pathway, including a detailed discussion of signal fidelity in yeast, is available elsewhere (1).

G PROTEIN SIGNALING AND REGULATION

The basic mechanism of G protein signaling is now well established (see accompanying articles in this volume). Upon activation, receptors catalyze the exchange of GDP for GTP on the G protein α subunit, which leads to its dissociation from the G protein $\beta\gamma$ subunits. Either Gα, or G$\beta\gamma$, or both are then free to activate downstream effectors. Signaling persists until GTP is hydrolyzed to GDP, and the subunits reassociate, completing the cycle of activation. RGS proteins (and some effectors) act as GTPase accelerating proteins (GAPs), and are thought to contribute to desensitization by shortening the active lifetime of the G protein. Stated differently, RGS proteins act in opposition to the receptor by promoting G protein inactivation. Thus the intensity of the G protein signal depends on (*a*) the rate of nucleotide exchange, (*b*) the rate of GTP hydrolysis, and (*c*) the rate of subunit reassociation. After the G protein, signals can be transmitted through a variety of effector enzymes, but typically converge on a protein kinase cascade and ultimately lead to changes in cellular homeostasis, differentiation, and development. The specificity of the G protein signal depends on its ability to respond to a given receptor and to transmit that signal through the appropriate effector.

G PROTEIN SIGNALING IN YEAST

Signaling in yeast begins with G protein–coupled receptors at the plasma membrane. Two haploid cell types, known as MATα and MAT**a**, each secrete small peptide pheromones, called α-factor and **a**-factor, respectively (2, 3). The α-factor

---

$\leftarrow$

Figure 1   G protein signaling and regulation in yeast. (*A*) The mating factor pheromone binds to a cell surface receptor Ste2 (or Ste3), which promotes GTP binding to the G protein α subunit (Gpa1). GTP triggers dissociation of G$\beta\gamma$ (Ste4, Ste18). Free G$\beta\gamma$ activates multiple effectors including an adaptor protein (Far1) for the Cdc42 exchange factor (Cdc24), a protein kinase (Ste20), and a scaffolding protein (Ste5) for downstream protein kinases (Ste11, Ste7, Fus3). Signaling proceeds until GTP is hydrolyzed and the G protein subunits reassociate. GTP hydrolysis is accelerated by an RGS protein (Sst2). Sst2 appears to activate a separate signaling pathway involving membrane sorting proteins (Vps36, Tlg2, Pep12). (*B*) The RGS and G protein components are post-translationally modified, as indicated. (*C*) The RGS and G protein components are regulated by candidate binding partners, including several novel proteins (Asg7, Akr1, Syg1, Mpt5) and homologs of a known regulatory protein phosducin (Plp1, Plp2).

binds to a specific receptor (Ste2)[1] on MAT**a** cells, whereas the **a**-factor binds to a receptor (Ste3) on MATα cells (4). The pheromone receptors activate a G protein heterotrimer consisting of an α subunit (Gpa1$^{G\alpha}$ or Scg1) and a $\beta\gamma$ dimer (Ste4/Ste18) (5). As described below, the G protein signal is transmitted and amplified via multiple effectors that bind to G$\beta\gamma$. These lead to the sequential activation of the protein kinases Ste20, Ste11, Ste7, and Fus3. Ste20 is a member of the p21-activated protein kinase (PAK) family (6). Fus3 is a member of the mitogen-activated protein kinase (MAPK) family (7). Fus3 phosphorylates and regulates proteins required for pheromone detection (Ste3) (8), G protein inactivation (Sst2) (9), kinase scaffolding and activation (Ste5) (10), component kinases (Ste11, Ste7) (11–15), morphological and cytoskeletal changes (Far1) (16–19), transcriptional activation (Ste12) (16, 20, 21), and transcription inhibition (Dig1, Dig2) (22–24).

Thus signaling in yeast begins with G protein–coupled receptors at the plasma membrane, is followed by activation of a protein kinase cascade, and culminates with the phosphorylation and activation of nuclear proteins that control cell polarity, transcription, and progression through the cell cycle. All of these changes represent a coordinated response to pheromone in preparation for cell fusion. Polarized cell growth is required to establish the site of cell fusion (25). New gene transcription is required to produce (for example) cell adhesion proteins (26–28). Growth arrest is required to synchronize the cell cycles of the two mating partners prior to fusion (29, 30). The following sections will describe each of the known G protein effectors and regulators and their role in coordinating the cell mating process.

G PROTEIN EFFECTORS: CDC24, STE20, STE5

Several proteins have been proposed to act as G$\beta\gamma$ effectors (Figure 1A). One is Cdc24^{GEF}, the guanine nucleotide exchange factor (GEF) for Cdc42 (31). Cdc42 is a member of the Ras superfamily of small GTPases and is specifically involved in actin rearrangements that lead to polarized cell growth and morphogenesis (32, 33). In yeast Cdc42 is required for budding in dividing cells and for projection formation in pheromone-arrested cells (34–37).

In dividing haploid cells each new bud forms next to the previous bud site (33). Upon pheromone stimulation, division stops, but cell growth continues toward the source of pheromone (chemotropism) (38, 39). Thus the bud position is fixed to a predetermined site, whereas projection formation can occur in any direction. Stated differently, an external signal (a pheromone gradient) overrides the internal signal (a previous bud site), leading to polarized cell growth. Only if pheromone concentrations are uniform does the cell revert to using internal cues and form a projection next to the previous bud site (40).

---

[1]Following the standard nomenclature in *S. cerevisiae*, Ste2 is the designation for the protein product of the *STE2* gene. Recessive (usually mutant) alleles are designated *ste2*. Deletion mutations are designated *ste2Δ*. For clarity, protein function is in some cases indicated as a superscript, Ste2^{receptor}.

Regulation of cell polarity may depend in part on the subcellular distribution of $Cdc24^{GEF}$. In naive cells $Cdc24^{GEF}$ forms a complex with Far1, is present in particulate fractions, and localizes to the plasma membrane, as well as to the nucleus (41–43). In pheromone-stimulated cells $Far1-Cdc24^{GEF}$ relocalizes to the tip of the mating projection, through binding of Far1 to $Ste4^{G\beta}$ (41–44). This establishes a landmark for localized activation of Cdc42 and the establishment of cell polarity. Mutations in Cdc42 (45), or mutations in $Cdc24^{GEF}$ that block binding to $Far1-G\beta\gamma$ (44, 46), prevent oriented growth of the projection toward a pheromone source. Remarkably, chemotropic growth does not require components of the MAPK cascade (47). Perhaps $G\beta\gamma$ binding to Far1 and $Cdc24^{GEF}$ is sufficient to modulate cell polarity, whereas additional effectors (Ste5 and $Ste20^{PAK}$) control other aspects of the pheromone response such as cell cycle and transcriptional regulation. These are not separate pathways, however, because $Cdc24^{GEF}$ mutants and Cdc42 temperature-sensitive mutants can also block the growth and transcription response (34, 35, 48). Conversely, cells that overexpress Cdc42, contain a GTPase-deficient form of Cdc42 (34, 35), or lack the Cdc42 GAP Rga1 (49) exhibit an increase in pheromone-dependent gene transcription.

Cdc42 may regulate transcription and cell division through $Ste20^{PAK}$, which has also been implicated in the establishment of cell polarity (50–52). The GTP-bound form of Cdc42 binds $Ste20^{PAK}$ and stimulates its kinase activity (34, 35, 53–56). $Ste20^{PAK}$ in turn initiates the protein kinase cascade comprised of $Ste11^{MAPKKK}$, $Ste7^{MAPKK}$, and $Fus3^{MAPK}$.

Taken together, this evidence suggests that $Cdc24^{GEF}$ and Cdc42 lie between $Ste4^{G\beta}$ and $Ste20^{PAK}$ in the mating response pathway (35). This is not a dedicated pathway, however, because $Ste20^{PAK}$ also functions upstream of $Ste11^{MAPKKK}$ in the invasive growth pathway (through $Ste7^{MAPKK}$ and $Kss1^{MAPK}$) (57, 58) and in the high osmolarity pathway (through $Pbs2^{MAPKK}$ and $Hog1^{MAPK}$) (59, 60). $Ste20^{PAK}$ also regulates targets (largely unidentified) involved in cell adhesion during mating (61), cell elongation (62), bud formation (50), and myosin function (63). Moreover, deletion of *STE20* does not produce complete sterility. This may be due to the action of another PAK family member, Cla4, which is normally required for cell division. An overlap of function is further suggested by the fact that neither *STE20* nor *CLA4* is essential, but a double mutant is inviable (64). To further complicate matters, an activating allele of *STE20* can stimulate the mating response pathway in the absence of exogenous mating pheromone, but this effect cannot be suppressed by deletions of *STE4*, *STE5*, *STE7*, *STE11*, or *STE12* (56). Overexpression of *CDC42*, or an activated allele of *CDC42*, also results in a slight increase in pheromone-induced gene transcription in a variety of sterile mutant strains (65). Thus $Ste20^{PAK}$ can activate multiple pathways, including at least three different MAPKs.

A second putative effector is $Ste20^{PAK}$, which has a high-affinity binding site for $Ste4^{G\beta}$ (66). Upon pheromone stimulation, $Ste20^{PAK}$ becomes concentrated at the tip of the mating projection. The ability of $G\beta\gamma$ to bind to $Ste20^{PAK}$ and $Far1-Cdc24^{GEF}$ may help reinforce signaling, by assembling $Ste11^{MAPKKK}$ with its activator $Ste20^{PAK}$, Ste20 with its activator Cdc42, and Cdc42 with its activator

Cdc24^{GEF} (54, 61, 67, 68). Membrane translocation is not essential for function, however, because mutants of Ste20^{PAK} that cannot bind to Cdc42 fail to concentrate at the tip of the mating projection but still mate and respond normally to pheromone (61, 69). Thus the functional significance of G$\beta\gamma$ binding to Ste20^{PAK} is unclear.

A third putative effector is the scaffolding protein Ste5. This protein binds to Ste4$^{G\beta}$ (70–72), as well as to the three kinases downstream of Ste20: Ste11^{MAPKKK}, Ste7^{MAPKK}, and Fus3^{MAPK} (73–75). One way Ste5 could promote signaling is by assembling all the components of the pathway, thereby enhancing their sequential interaction and efficiency of signaling. In support of this model, point mutations in Ste5 that selectively block binding to G$\beta\gamma$, Ste7^{MAPKK}, or Ste11^{MAPKKK} lead to a dramatic reduction in mating efficiency (72, 73). The binding sites for Ste5 and Ste20^{PAK} on G$\beta\gamma$ partially overlap. It is not known if binding to both effector proteins can occur simultaneously or if they are independent (66, 76).

Another way Ste5 could transmit a signal is through Ste4$^{G\beta}$-mediated translocation to the plasma membrane. This may be important, because many of the components act in different subcellular compartments at different times; indeed, Ste5 has been found at the plasma membrane, cytosol, and nucleus. Pheromone treatment results in the rapid translocation of Ste5 from the nucleus to the tip of the mating projection (77–80). Translocation requires binding to G$\beta\gamma$. If Ste5 lacks the Ste4$^{G\beta}$ binding domain, or if Ste4$^{G\beta}$ is not expressed, Ste5 remains in the nucleus and cannot signal (71, 72, 77–79). Conversely, if Ste5 lacks the nuclear localization signal, it remains in the cytoplasm and still cannot signal. If Ste5 is fused to a membrane-targeting signal, it can signal in the absence of Ste4$^{G\beta}$ (77). Thus plasma membrane localization is sufficient for Ste5 signaling. However, transit through the nucleus may be a prerequisite for Ste5 to reach the plasma membrane. Perhaps Ste5 undergoes some posttranslational modification within the nucleus or associates with a required targeting protein. Alternatively, the nuclear localization signal may also be required for some other function, such as binding and activation by Ste4$^{G\beta}$.

Another mechanism by which Ste5 may transmit the signal is through a G$\beta\gamma$-induced conformational change, possibly leading to enhanced activation by Ste20^{PAK} (79) or direct activation of Ste11^{MAPKKK}. As with other MAPKKK family members, the Ste11^{MAPKKK} N-terminal domain can bind and inhibit its own C-terminal kinase domain (12, 15, 81). Ste5 binds to the N-terminal domain of Ste11^{MAPKKK}, and in this manner could trigger disinhibition of kinase activity and activation of the mating response (74, 81, 82). Regardless of the mechanism, it is clear that the most active pool of cellular Fus3^{MAPK} exists in a complex of 350–500 kDa that contains Ste5, as well as Ste11^{MAPKKK} and Ste7^{MAPKK} (10, 74, 75). These findings suggest that Ste5 is not merely a scaffold but could play an active role in signal transduction and amplification.

Finally, it has been argued that Gpa1$^{G\alpha}$, like many other Gα subunits in other systems, also transmits a signal through some downstream effector, leading to pheromone desensitization (83, 84). The basis for this model is the ability of a dominant mutant, Gpa1^{N388D}, to promote recovery from G1 arrest following prolonged pheromone stimulation. Gpa1^{N388D} is presumed (but not demonstrated) to slow GTPase activity, which if true would lock the protein in a constitutively active state. However, another mutant demonstrated to bind GTP and to lack GTPase activity (Gpa1^{Q327L}) does not promote pheromone adaptation, in the manner of Gpa1^{N388D} (85–87). Thus an alternative possibility is that Gpa1^{N388D} binds more avidly to the receptor or to G$\beta\gamma$. Increased binding to G$\beta\gamma$ has been proposed to explain the ability of two other mutants, Gpa1^{G50V} and Gpa1^{G327S}, to stimulate recovery from G1 arrest in the manner of Gpa1^{N388D} (85, 88). In any case, a Gpa1$^{G\alpha}$ effector, if it exists, has yet to be identified.

G PROTEIN REGULATION: POST-TRANSLATIONAL MODIFICATION

Chronic exposure to pheromone leads to profound changes in cell physiology, in anticipation of mating. Despite these changes, cells that fail to mate eventually become unresponsive to pheromone action, through a process of desensitization. As intermediaries between cell surface receptors and intracellular effectors, G proteins are particularly well positioned to modulate the intensity of the transmembrane signal. G protein activity depends on the rate of GTP binding (accelerated by receptors) and GTP hydrolysis (stimulated by RGS proteins). In addition, there is growing evidence for regulation of G proteins and RGS proteins through posttranslational modifications.

Several types of modifications have been described for the G protein subunits, any one of which may be involved in signal regulation (Figure 1B). All known Gα subunits are myristoylated and/or palmitoylated (89). Myristoylation involves the cotranslational addition of a 14-carbon saturated fatty acid through an amide bond to the N-terminal Gly residue (Gly2) (90, 91). Palmitoylation (also referred to as thioacylation) involves the addition of a 16-carbon saturated fatty acid through a thioester bond to internal Cys residues (usually Cys3) (91). Both modifications have been intensively studied because they are in some cases regulated by extracellular signals. This has been shown for palmitoylation of mammalian G$_s\alpha$ and G$_q\alpha$ (89) and for myristoylation of Gpa1$^{G\alpha}$ (92).

Gpa1$^{G\alpha}$ is normally myristoylated at less than full stoichiometry, and pheromone stimulation appears to improve the efficiency with which newly synthesized Gpa1$^{G\alpha}$ undergoes this modification. Myristoylation appears to be essential for Gpa1$^{G\alpha}$ function, so a pheromone-dependent change in the stoichiometry of myristoylated versus unmyristoylated protein could affect G protein signaling. Mutations in the N-myristoyl-transferase (*NMT1*) gene, or replacement of the myristoylated Gly residue of Gpa1$^{G\alpha}$ (*gpa1^{G2A}*), mimic the *gpa1*Δ null mutant, resulting in sustained release of G$\beta\gamma$ and constitutive signaling (93, 94). Nonmyristoylated Gα (Gpa1^{G2A}) can still form a high affinity complex with G$\beta\gamma$ in vitro but fails to associate with the plasma membrane in vivo. A pool of G$\beta\gamma$ remains at the plasma membrane, however, and this is presumably responsible for the constitutive signaling phenotype of the Gpa1^{G2A} mutant (94). In support of this model, fusion of Gpa1$^{G\alpha}$ to the C terminus of Ste2 confers normal coupling to G$\beta\gamma$, even though the Gpa1$^{G\alpha}$ moiety cannot be myristoylated (95). Gα in yeast is palmitoylated at Cys3, and blocking this modification has consequences similar to (though less severe than) Gpa1^{G2A} with respect to signaling and localization (96, 97). These findings indicate that fatty acylation is primarily needed for proper membrane targeting, rather than for subunit-subunit association.

Madura & Varshavski have demonstrated that Gpa1$^{G\alpha}$ can be covalently modified with ubiquitin (98). Ubiquitin is a highly conserved peptide (73 of 76 residues are identical in human and in yeast) that tags proteins for degradation. Proteolysis is usually carried out by the proteasome; however, some ubiquitinated membrane proteins (including the receptors Ste2 and Ste3) are instead degraded in the vacuole (yeast counterpart to the lysosome) (99, 100). Because overexpression of *GPA1* leads to diminished signaling (101) and deletion of *GPA1* leads to sustained release of G$\beta\gamma$ and constitutive signaling, ubiquitin-dependent degradation of Gpa1$^{G\alpha}$ could alter the pheromone response (102, 103). To test this model, we have purified Gpa1$^{G\alpha}$ and mapped the site of ubiquitination by mass spectrometry (L. Marotti, R. Newitt, Y. Wang, R. Aebersold & H. G. Dohlman, submitted). The ubiquitinated Lys lies within a unique 110–amino acid insert not found in other Gα proteins. Perhaps this insert serves as a proteolytic signal that is regulated by ubiquitin-dependent degradation activity. Ubiquitination has not yet been demonstrated for any other G protein subunit in any organism.

All known G protein γ subunits are isoprenylated. This modification involves a thioether linkage of the 15-carbon farnesyl or 20-carbon geranyl-geranyl isoprenoid to a cysteine residue four amino acids from the carboxyl terminus. Prenylated proteins also undergo endoproteolytic truncation of the last three amino acids and methylation of the new carboxyl-terminal prenylcysteine residue (105). Whereas most Gγ proteins are geranyl-geranylated, Ste18$^{G\gamma}$ is farnesylated (at Cys 107) (106) and palmitoylated (at Cys 106) (97, 107). Substitution of Cys107 results in a loss-of-function (sterile) phenotype, whereas substitution of Cys106 results in a partial loss-of-function phenotype (108, 109). Unmodified Ste18$^{G\gamma}$ is still targeted to the plasma membrane but is readily dissociated following G protein activation. Thus it appears that these modifications are dispensable for G protein activation by the receptor but are required for stable association of G$\beta\gamma$ at the plasma membrane (97, 107).

The Gβ subunit is dynamically phosphorylated at several sites after pheromone stimulation (110). A short deletion mutation in Gβ (*ste4*$^{\Delta310-346}$) prevents pheromone-stimulated phosphorylation of the protein and results in about a sixfold increase in pheromone sensitivity. This phenotype suggests that G protein phosphorylation mediates an adaptive response to pheromone-induced signaling (110). It has also been reported that phosphorylated Ste4$^{G\beta}$ is more likely to be associated with the putative effector Ste5 (72, 73) and more likely to be found in the cytoplasmic fraction (111). However, a more recent study indicates that the adaptive defect exhibited by the $ste4^{\Delta310\text{-}346}$ deletion mutant is due to disruption of the interaction between Ste4$^{G\beta}$ and Gpa1$^{G\alpha}$ (112). Two different Ste4$^{G\beta}$ mutants (Ste4$^{T320A/S335A}$ and Ste4$^{T322A/S335A}$) remain unphosphorylated upon pheromone stimulation, yet have no discernible effect on either signaling or adaptation. These results suggest that pheromone-induced phosphorylation is not an adaptive mechanism but leave open the question of how phosphorylation contributes to Gβ function. Whereas several Gα subtypes are phosphorylated in mammalian cells, phosphorylation has not been demonstrated for Gβ in any other system. Perhaps yeast and mammalian systems have evolved to phosphorylate the subunit that is primarily responsible for effector activation. It is also noteworthy that phosphorylation occurs within a unique insert not found in other Gβ proteins. This follows a pattern in which post-translational modifications occur within insert domains unique to yeast signaling proteins, including the ubiquitination of Gα (described above) and the phosphorylation of the RGS protein (described below).

G PROTEIN REGULATORS: SST2, PLP1/2, MPT5, AKR1, ASG7, SYG1

A number of proteins have been shown to regulate RGS and G protein function (Figure 1C). Sst2^{RGS} plays an especially important role in signal modulation. Sst2^{RGS} was first identified nearly 20 years ago in a screen for mutants that interfere with pheromone adaptation. Deletion of *SST2* allows cells to respond to doses of pheromone at least two orders of magnitude lower than normal and renders the cell completely unable to recover from pheromone-imposed cell cycle arrest (113, 114). The *SST2* gene was cloned in 1987 (115), but the deduced amino acid sequence failed to provide any information about its mechanism of action or its target. Moreover, the existence of G protein heterotrimers in yeast was not yet known, and the relevance of pheromone signaling to hormone signaling in mammalian cells was not appreciated. For these reasons, the discovery of Sst2^{RGS} did not attract much attention at the time.

In the past several years, the function of Sst2^{RGS} has been thoroughly characterized. Initially, a dominant "sterile" allele of *SST2* was used to determine the intracellular target of the Sst2^{RGS} protein. One such dominant gain-of-function mutant (Pro20Leu) could block the response to pheromone but did not prevent downstream activation through overexpression of Ste4$^{G\beta}$, a constitutively active Gβ mutant (Ste4^{Hpl}), or a disruption of the *GPA1* gene (116). These and other genetic arguments implicated Gpa1$^{G\alpha}$ as the direct target of Sst2^{RGS}. It was subsequently shown that Sst2^{RGS} and Gpa1$^{G\alpha}$ colocalize at the plasma membrane and that both proteins copurify as a complex from yeast (87).

A family of proteins homologous to Sst2^{RGS} was since discovered in more complex organisms (117, 118), as reviewed previously (119, 120). Several RGS family members have been purified in recombinant form and shown to be potent GAPs for Gα proteins (121–123). We have shown a similar activity for Sst2^{RGS}

and Gpa1$^{G\alpha}$ in yeast (86). These findings provide a likely mechanism by which Sst2^{RGS} promotes desensitization: Specifically, by stimulating Gpa1$^{G\alpha}$ GTPase activity, Sst2^{RGS} shortens the lifetime of the active GTP-bound species, accelerates reassociation with G$\beta\gamma$, and attenuates the cellular response.

The mechanism of RGS GAP activity was firmly established through an X-ray crystal structure determination of the mammalian G$_{i1}\alpha$-RGS4 complex (124). It is evident from this work that RGS4 does not contribute any "catalytic" residues to the GTP binding pocket of Gα. Rather, RGS proteins act by binding and stabilizing the three "switch" regions (regions that undergo conformational change upon GTP hydrolysis), thereby lowering the energy of activation for the reaction.

In order to demonstrate that Gα GAP activity is responsible for the desensitizing activity of RGS proteins, we screened for mutants of *GPA1* that phenotypically mimic the loss of *SST2*. One such allele was identified (*GPA1^{G302S}*) and shown to be completely resistant to *SST2* action in vivo and completely unresponsive to RGS GAP activity in vitro (125). Analogous mutations in mammalian Gα proteins also confer resistance to RGS action, both in vitro and in transfected cells (125, 126). These mutants are currently being tested in transgenic animals, in lieu of RGS knockouts, to determine which signaling pathways are subject to RGS regulation and how desensitization of G proteins compares with other modes of desensitization (e.g., of receptors) (K. Young, personal communication).

A major question is how Sst2^{RGS} and other regulators of G protein signaling are themselves regulated. One mechanism that may be important is transcriptional induction, as both mRNA (115) and protein (87) levels increase markedly with prolonged receptor stimulation. Sst2^{RGS} has also been shown to undergo posttranslational phosphorylation. One of the sites, Ser-539, is phosphorylated only in response to pheromone stimulation and requires a functional Fus3^{MAPK} or Kss1^{MAPK} (9). Ser-539 lies within a 118–amino acid insert that contains a PEST motif (87), which is often found in proteins that are rapidly degraded (127). Notably, phosphorylation at position 539 appears to slow the rate of Sst2^{RGS} degradation (9). Thus the unique insert in Sst2^{RGS} serves as a proteolytic signal that is regulated by pheromone-dependent MAP kinase activity. Pheromone-dependent phosphorylation and stabilization presumably helps to reinforce pheromone-dependent transcriptional induction and increased expression of Sst2^{RGS}.

A number of proteins appear to negatively signal through binding to Sst2^{RGS}. A two-hybrid screen was used to show that Sst2^{RGS} binds to an RNA-binding protein of the *Drosophila* pumillo repeat family, *Mpt5* (128, 129). Disruption of MPT5 results in pleiotropic effects, including a temperature-sensitive growth defect and a modest increase in pheromone sensitivity. An independent two-hybrid screen using an array of all 6000 yeast genes yielded no positives (130). However, the same array was used to identify 17 proteins that recognize just the N-terminal domain of Sst2 (i.e., lacking the RGS domain). Of these, three (*TLG2*, *PEP12*, and *VPS36*) exhibited a reduction in pheromone-dependent growth arrest and reporter-transcription activity when deleted (131). All three gene products are needed for proper trafficking to the vacuole (yeast counterpart to the lysosome) (see below).

Yeast express two homologues of phosducin, called Plp1 and Plp2 (132). In mammals phosducin binds G protein $\beta\gamma$ subunits in vitro and is postulated to regulate their signaling function in vivo. Plp1 and Plp2 bind to Ste4$^{G\beta}$, and binding is enhanced by pheromone stimulation and by the addition of GTPγS, conditions that favor dissociation of G$\beta\gamma$ from Gα. Cells overexpressing either *PLP1* or *PLP2* exhibit a profound (70–80%) decrease in gene induction, yet they have no effect on pheromone-mediated growth arrest. These data indicate that yeast phosducin can selectively regulate early signaling events following pheromone stimulation. A *plp1*Δ gene disruption mutant is viable and exhibits a very modest increase in pheromone-mediated gene induction. In contrast, the *plp2*Δ mutation is lethal, and cell viability is not restored by a *ste7*Δ^{MAPKK} disruption mutant. These data suggest that phosducin has another essential function in the cell, independent of its regulatory role in cell signaling (132).

Another protein that appears to bind and regulate G$\beta\gamma$ is Syg1 (133), a member of the phosphate permease family of plasma membrane transporters (134, 135). In a screen for high copy suppressors of a *gpa1*$\Delta^{G\alpha}$ mutant, Spain et al. identified a truncated mutant form of *SYG1*, designated *SYG1*-1 (133). This allele is also a potent suppressor of the cell cycle arrest and differentiation phenotypes of cells treated with α-factor or that overexpress wild-type Ste4$^{G\beta}$; *SYG1*-1 does not suppress the lethality of an activated Gβ mutant (Ste4^{Hpl}). This type of allele specificity suggests a direct interaction between Syg1 and Ste4$^{G\beta}$, a supposition supported by interaction in a two-hybrid assay (133). However, the full length Syg1 protein does not appear to regulate $\beta\gamma$ signaling, and no further characterization of Syg1 has been reported.

A third regulator of G$\beta\gamma$ is the ankyrin repeat–containing protein, Akr1. Akr1 binds to G$\beta\gamma$ but not to the G$\alpha\beta\gamma$ heterotrimer (136, 137), leading to inhibition of the mating response. Consistent with this model, *AKR1* overexpression can suppress the cell division arrest resulting from loss of Gpa1$^{G\alpha}$ or overexpression of G$\beta\gamma$. Loss of Akr1 leads to partial activation of downstream kinases Ste20^{PAK}, Ste11^{MAPKKK}, and Ste7^{MAPKK}. Haploid and diploid cells lacking Akr1 grow slowly and develop deformed buds or projections, suggesting that this protein also participates in the control of cell morphology. In contrast to the pheromone-supersensitive phenotype, however, the morphological abnormalities are not rescued by dominant sterile mutations (136, 137). The mechanism of Akr1 action is not known, but one possibility is that it recruits isoprenylated proteins Gγ and Cdc42 to the plasma membrane and in this manner acts to suppress the G$\beta\gamma$ signal (68, 106). This may as well pertain to Ste18$^{G\gamma}$, which is also palmitoylated (97, 107).

More recently, attention has focused on a regulatory process called receptor inhibition and a novel protein called Asg7. The identification of Asg7 stems from the observation that pheromone signaling is blocked when the **a**-factor receptor (Ste3) is expressed inappropriately in **a** cells (138). This inhibition requires Asg7, which apparently acts by redirecting Ste3 and Ste4$^{G\beta}$ from the plasma membrane to an internal compartment where they cannot signal (139, 140). Such action may be appropriate following cell fusion, when Ste3 and Asg7 are at least transiently expressed in the same (diploid) cell, and signaling must stop. Indeed, mating of $asg7\Delta$ mutants produces diploids that continue to form haploid-like mating projections and that are slow to initiate vegetative growth (139, 140).

Of the RGS- and G protein–binding proteins described above, mammalian homologues have only been characterized for Sst2[RGS] and Plp1/2[phosducin]. There has been relatively little biochemical characterization of the novel regulators Mpt5, Akr1, Asg7, or Syg1. Any of these could represent desensitization factors that attenuate the G protein response. Another possibility is that these proteins are not regulators but effectors, which when mutated or overexpressed have the capacity to inhibit $G\beta\gamma$ signaling events. For example, G protein–coupled receptor kinases have the capacity to sequester $G\beta\gamma$ and perturb their function (141). This is not, however, considered to be their main function in the cell. Rather, G protein–coupled receptor kinase binding to $G\beta\gamma$ is needed to facilitate membrane localization as well as phosphorylation and desensitization of receptors (142–144). Similarly, Mpt5, Akr1, Asg7, or Syg1 may be $G\beta\gamma$ effectors that can compete or allosterically modulate binding to known effectors such as Cdc24[GEF], Ste20[PAK], and Ste5. These possibilities remain to be tested directly. No arrestin or G protein–coupled receptor kinase orthologs have been identified in yeast.

A SECOND G PROTEIN IN YEAST

Whereas most attention has focused on signaling through Gpa1[Gα], a second Gα (Gpa2) is present in yeast (145). Gpa2[Gα] is coupled to a putative receptor (Gpr1) (146–149), an RGS protein (Rgs2) (150), and a potential effector phospholipase C (Plc1) (147). All three proteins are required for induction of diploid filamentous growth in response to nitrogen depletion (147). Another candidate effector is Ime2, a protein kinase needed for initiation of meiosis (151a). Gpr1 and Gpa1[Gα] are also required for the activation of adenylyl cyclase upon glucose feeding (146–149). Although the details are unclear, Gpa2[Gα] can activate a kinase cascade that begins with the cAMP-dependent protein kinase. In response to low nitrogen, there is an increase in PKA activity and unipolar budding (148). In response to glucose, there is an increase in cAMP and PKA activity and reduced accumulation of trehalose and glycogen, as well as enhanced heat shock sensitivity. The same pathway also appears to be regulated by the small G proteins Ras1 and Ras2 (58, 146, 151). An intriguing question, as yet unanswered, is how the Gpa1[Gα] and Gpa2[Gα] signaling pathways remain separate, despite their close sequence similarity. Sst2[RGS] does not appear to modulate the glucose/Gpa2[Gα] response, nor does Rgs2 modulate the pheromone/Gpa1[Gα] pathway (150). Another question is, what acts as the $G\beta\gamma$ for Gpa2[Gα]? Despite sequencing of the entire yeast genome, no obvious candidates exist (152).

AN EFFECTOR FUNCTION FOR SST2

All RGS proteins have a common, conserved "RGS core domain" of ~120 amino acids, which is necessary and sufficient for their GTPase accelerating activity. However, many RGS family members also have large N-terminal or C-terminal extensions, suggesting they have other functions as well (119, 120, 153). For instance, the mammalian RGS protein p115RhoGEF has one domain that acts as a GAP for $G_{13}\alpha$ and a second domain that acts as a GDP-GTP exchange factor for RhoA (154, 155). These findings underscore the view that some RGS proteins are not simply GAPs but have separate functions that link them to other signaling pathways.

Among the larger RGS proteins, many have an N-terminal DEP motif (named for Dishevelled, Egl-10, and pleckstrin) (153, 156). Examples include the RGS proteins *FlbA* in *Aspergillus nidulans*; Egl-10 and Eat-16 in *Caenorhabditis elegans*; as well as RGS6, RGS7, RGS9, and RGS11 in mammals. Sst2^{RGS} has two DEP regions, comprised of residues 50–135 and 279–358. On the basis of mutagenesis and NMR structural studies, it was proposed that the DEP domain in Dishevelled is required for membrane targeting (157, 158) and the recognition of upstream molecules (159). Likewise, in the case of Egl-10 and Sst2^{RGS}, the DEP domains appear necessary and sufficient for membrane localization (117, 160).

Recently, we have demonstrated that Sst2^{RGS} undergoes endoproteolytic processing in vivo to yield separate N- and C-terminal fragments (160). Purification and sequencing of the C-Sst2 fragment revealed cleavage sites after Ser-414 and Ser-416, just preceding the region of RGS homology. Although neither fragment alone has RGS activity in vivo, coexpression of N-Sst2 and C-Sst2 does partially restore its ability to regulate the growth arrest response. Whereas the full-length protein is localized to the microsomal and plasma membrane fractions, the N-Sst2 species is predominantly in the microsomal fraction, and C-Sst2 is found in the soluble fraction. Finally, Sst2^{RGS} processing requires expression of other components of the pheromone response pathway, including the receptor and the G protein (160). These results indicate that proteolytic processing of Sst2^{RGS} is regulated, in so far as it requires an intact signaling apparatus.

More recently, we have begun to investigate the physiological function of the N-terminal (non-RGS) domain of Sst2^{RGS}. To this end, we performed genome-wide transcription profiling of cells expressing the N-Sst2 fragment alone and found a number of induced genes, most of which contain a stress-response element in the promoter region.

To identify components of a signaling pathway leading from N-Sst2 to stress-response elements, we performed a genome-wide two-hybrid analysis using N-Sst2 as bait and found 17 interacting proteins (131). Three of these (Vps36, Pep12, Tlg2) are needed for full stress-response signaling. Tlg2 is a syntaxin (t-SNARE) that functions in transport from the endosome to the late Golgi within the endocytic pathway (161, 161a). Vps36 is needed for protein trafficking from the pre-vacuolar compartment to the vacuole (162, 163). Pep12 is a syntaxin that is required for protein sorting between the Golgi and endosome (164, 164a).

The functional interaction between Sst2 and Vps36 has been characterized further (131). The *vps36Δ* mutation diminishes signaling by pheromone, as well as by downstream components, including the transcription factor Ste12. Overexpression of N-Sst2 diminishes signaling in the *vps36Δ* mutant, but not in wild-type cells. Conversely, overexpression of Vps36 enhances the pheromone response in *sst2Δ* cells but not in wild-type. Taken together, these findings indicate that full-length and processed forms of Sst2^{RGS} have different physiological roles in the cell (164). Moreover, the N-terminal domain of Sst2 and Vps36 have opposite and opposing effects on both the pheromone- and stress-response pathways, with Sst2 acting upstream of Vps36, and Vps36 likely acting within the nucleus at the level of transcription.

CONCLUSIONS

Historically, signaling in yeast has been studied largely through in vivo genetic methods, whereas signaling by G proteins has been characterized using mostly biochemical methods. Both approaches are necessary. Genetic studies are needed to establish whether a particular interaction actually affects signaling in vivo. Biochemical studies are needed to determine the molecular basis for any functional changes observed in cells. Moreover, some interactions may not alter protein function but even so might affect protein expression or localization within the cell. I have described some notable examples of how G protein function is regulated in vivo, in a manner that could not easily be discerned biochemically. Prominent examples include (a) Gpa1$^{G\alpha}$ myristoylation, which preserves G$\beta\gamma$ binding but leads to a redistribution of the protein away from the plasma membrane; (b) endoproteolytic processing of Sst2^{RGS}, which appears to uncouple its effector and G protein regulatory functions of the protein; and (c) nucleocytoplasmic shuttling of Ste5 and Far1, which helps to establish polarity of signaling. I have also described some examples of how G protein function could best be studied biochemically. Examples include (a) the GTPase accelerating activity of RGS proteins, which could only have been determined through a detailed analysis of the purified proteins, and (b) the site of Fus3^{MAPK} phosphorylation of Sst2^{RGS}, which could only have been determined with protein purified from the native host cell.

For the future, a number of significant questions remain. One issue is how two G proteins, Gpa1$^{G\alpha}$ and Gpa2$^{G\alpha}$, maintain specific interactions with different receptors, effectors, and RGS proteins. Another challenge will be to determine the biological and biochemical function of accessory proteins that bind to RGS and G proteins. In this regard I believe yeast will continue to be a highly useful and appropriate model for the study of G protein signaling. It remains the only organism in which every gene has been arrayed for the purpose of transcription analysis and nearly every gene has been genetically disrupted. As our attention shifts from genomics to proteomics, much of the pioneering work will continue to be conducted in yeast. For instance, large quantities of protein can be obtained for biochemical or biophysical analysis, through large-scale fermentation and affinity tag purification. More significantly, the functional significance of protein-protein interactions, or protein posttranslational modifications, can easily be tested through gene replacement and gene disruption mutants (165). Undoubtedly, many of these regulatory processes will alter protein expression or localization, even if they leave protein activity unchanged, and can best be studied in vivo. Our ability to integrate such biochemical and genetic information in yeast will remain a unique feature of the system and a continuing source of new insights about signal regulation in eukaryotes.

ACKNOWLEDGMENTS

This work was supported in part by National Institutes of Health grants GM 55316 and GM 59167. The author is an Established Investigator of the American Heart Association.

Visit the Annual Reviews home page at www.AnnualReviews.org

LITERATURE CITED

1. Dohlman HG, Thorner JW. 2001. Regulation of G protein-initiated signal transduction in yeast: paradigms and principles. *Annu. Rev. Biochem.* 70:703–54
2. Fuller RS, Sterne RE, Thorner J. 1988. Enzymes required for yeast prohormone processing. *Annu. Rev. Physiol.* 50:345–62
3. Caldwell GA, Naider F, Becker JM. 1995. Fungal lipopeptide mating pheromones: a model system for the study of protein prenylation. *Microbiol. Rev.* 59:406–22
4. Dohlman HG, Thorner J, Caron MG, Lefkowitz RJ. 1991. Model systems for the study of seven-transmembrane-segment receptors. *Annu. Rev. Biochem.* 60:653–88
5. Blumer KJ, Thorner J. 1991. Receptor-G protein signaling in yeast. *Annu. Rev. Physiol.* 53:37–57
6. Daniels RH, Bokoch GM. 1999. p21-activated protein kinase: a crucial component of morphological signaling? *Trends Biochem. Sci.* 24:350–55
7. Gustin MC, Albertyn J, Alexander M, Davenport K. 1998. MAP kinase pathways in the yeast *Saccharomyces cerevisiae. Microbiol. Mol. Biol. Rev.* 62:1264–300
8. Feng Y, Davis NG. 2000. Feedback phosphorylation of the yeast **a**-factor receptor requires activation of the downstream signaling pathway from G protein through mitogen-activated protein kinase. *Mol. Cell. Biol.* 20:563–74
9. Garrison TR, Zhang Y, Pausch M, Apanovitch D, Aebersold R, et al. 1999. Feedback phosphorylation of an RGS protein by MAP kinase in yeast. *J. Biol. Chem.* 274:36387–91
10. Kranz JE, Satterberg B, Elion EA. 1994. The MAP kinase Fus3 associates with and phosphorylates the upstream signaling component Ste5. *Genes Dev.* 8:313–27
11. Zhou Z, Gartner A, Cade R, Ammerer G, Errede B. 1993. Pheromone-induced signal transduction in *Saccharomyces cerevisiae* requires the sequential function of three protein kinases. *Mol. Cell. Biol.* 13:2069–80
12. Stevenson BJ, Rhodes N, Errede B, Sprague GF Jr. 1992. Constitutive mutants of the protein kinase STE11 activate the yeast pheromone response pathway in the absence of the G protein. *Genes Dev.* 6:1293–304
13. Errede B, Ge QY. 1996. Feedback regulation of MAP kinase signal pathways *Philos. Trans. R. Soc. London Ser. B* 351:143–48; discussion 148–49
14. Bardwell L, Cook JG, Chang EC, Cairns BR, Thorner J. 1996. Signaling in the yeast pheromone response pathway: specific and high-affinity interaction of the mitogen-activated protein (MAP) kinases

Kss1 and Fus3 with the upstream MAP kinase kinase Ste7. *Mol. Cell. Biol.* 16: 3637–50

15. Cairns BR, Ramer SW, Kornberg RD. 1992. Order of action of components in the yeast pheromone response pathway revealed with a dominant allele of the STE11 kinase and the multiple phosphorylation of the STE7 kinase. *Genes Dev.* 6:1305–18

16. Elion EA, Satterberg B, Kranz JE. 1993. FUS3 phosphorylates multiple components of the mating signal transduction cascade: evidence for STE12 and FAR1. *Mol. Biol. Cell* 4:495–510

17. Peter M, Gartner A, Horecka J, Ammerer G, Herskowitz I. 1993. FAR1 links the signal transduction pathway to the cell cycle machinery in yeast. *Cell* 73:747–60

18. Tyers M, Futcher B. 1993. Far1 and Fus3 link the mating pheromone signal transduction pathway to three G1-phase Cdc28 kinase complexes. *Mol. Cell. Biol.* 13:5659–69

19. Gartner A, Jovanovic A, Jeoung DI, Bourlat S, Cross FR, et al. 1998. Pheromone-dependent G1 cell cycle arrest requires Far1 phosphorylation, but may not involve inhibition of Cdc28-Cln2 kinase, in vivo. *Mol. Cell. Biol.* 18:3681–91

20. Song D, Dolan JW, Yuan YL, Fields S. 1991. Pheromone-dependent phosphorylation of the yeast STE12 protein correlates with transcriptional activation. *Genes Dev.* 5:741–50

21. Hung W, Olson KA, Breitkreutz A, Sadowski I. 1997. Characterization of the basal and pheromone-stimulated phosphorylation states of Ste12p. *Eur. J. Biochem.* 245:241–51

22. Cook JG, Bardwell L, Kron SJ, Thorner J. 1996. Two novel targets of the MAP kinase Kss1 are negative regulators of invasive growth in the yeast *Saccharomyces cerevisiae*. *Genes Dev.* 10:2831–48

23. Tedford K, Kim S, Sa D, Stevens K, Tyers M. 1997. Regulation of the mating pheromone and invasive growth responses

in yeast by two MAP kinase substrates. *Curr. Biol.* 7:228–38

24. Olson KA, Nelson C, Tai G, Hung W, Yong C, et al. 2000. Two regulators of Ste12p inhibit pheromone-responsive transcription by separate mechanisms. *Mol. Cell. Biol.* 20:4199–209

25. Dorer R, Pryciak P, Schrick K, Hartwell LH. 1994. The induction of cell polarity by pheromone in *Saccharomyces cerevisiae*. *Harvey Lect.* 90:95–104

26. Elion EA, Trueheart J, Fink GR. 1995. Fus2 localizes near the site of cell fusion and is required for both cell fusion and nuclear alignment during zygote formation. *J. Cell. Biol.* 130:1283–96

27. Guo B, Styles CA, Feng Q, Fink GR. 2000. A *Saccharomyces* gene family involved in invasive growth, cell-cell adhesion, and mating. *Proc. Natl. Acad. Sci. USA* 97:12158–63

28. Heiman MG, Walter P. 2000. Prm1p, a pheromone-regulated multispanning membrane protein, facilitates plasma membrane fusion during yeast mating. *J. Cell. Biol.* 151:719–30

29. Hartwell LH. 1973. Synchronization of haploid yeast cell cycles, a prelude to conjugation. *Exp. Cell Res.* 76:111–17

30. Samokhin GP, Lizlova LV, Bespalova JD, Titov MI, Smirnov VN. 1981. The effect of α-factor on the rate of cell-cycle initiation in *Saccharomyces cerevisiae*: α-factor modulates transition probability in yeast. *Exp. Cell Res.* 131:267–75

31. Zheng Y, Cerione R, Bender A. 1994. Control of the yeast bud-site assembly GTPase Cdc42. Catalysis of guanine nucleotide exchange by Cdc24 and stimulation of GTPase activity by Bem3. *J. Biol. Chem.* 269:2369–72

32. Cabib E, Drgonova J, Drgon T. 1998. Role of small G proteins in yeast cell polarization and wall biosynthesis. *Annu. Rev. Biochem.* 67:307–33

33. Chant J. 1999. Cell polarity in yeast. *Annu. Rev. Cell. Dev. Biol.* 15:365–91

34. Simon MN, De Virgilio C, Souza B,

Pringle JR, Abo A, et al. 1995. Role for the Rho-family GTPase Cdc42 in yeast mating-pheromone signal pathway. *Nature* 376:702–5

35. Zhao ZS, Leung T, Manser E, Lim L. 1995. Pheromone signalling in *Saccharomyces cerevisiae* requires the small GTP-binding protein Cdc42p and its activator *CDC24. Mol. Cell. Biol.* 15:5246–57

36. Oehlen LJ, Cross FR. 1998. Potential regulation of Ste20 function by the Cln1-Cdc28 and Cln2-Cdc28 cyclin-dependent protein kinases. *J. Biol. Chem.* 273:25089–97

37. Chant J, Herskowitz I. 1991. Genetic control of bud site selection in yeast by a set of gene products that constitute a morphogenetic pathway. *Cell* 65:1203–12

38. Valtz N, Peter M, Herskowitz I. 1995. FAR1 is required for oriented polarization of yeast cells in response to mating pheromones. *J. Cell. Biol.* 131:863–73

39. Segall JE. 1993. Polarization of yeast cells in spatial gradients of α mating factor. *Proc. Natl. Acad. Sci. USA* 90:8332–36

40. Madden K, Snyder M. 1992. Specification of sites for polarized growth in *Saccharomyces cerevisiae* and the influence of external factors on site selection. *Mol. Biol. Cell* 3:1025–35

41. Nern A, Arkowitz RA. 1999. A Cdc24p-Far1p-G$\beta\gamma$ protein complex required for yeast orientation during mating. *J. Cell. Biol.* 144:1187–202

42. Shimada Y, Gulli MP, Peter M. 2000. Nuclear sequestration of the exchange factor Cdc24 by Far1 regulates cell polarity during yeast mating. *Nat. Cell Biol.* 2:117–24

43. Toenjes KA, Sawyer MM, Johnson DI. 1999. The guanine-nucleotide-exchange factor Cdc24p is targeted to the nucleus and polarized growth sites. *Curr. Biol.* 9:1183–86

44. Butty AC, Pryciak PM, Huang LS, Herskowitz I, Peter M. 1998. The role of Far1p in linking the heterotrimeric G protein to polarity establishment proteins during yeast mating. *Science* 282:1511–16

45. Kozminski KG, Chen AJ, Rodal AA, Drubin DG. 2000. Functions and functional domains of the GTPase Cdc42p. *Mol. Biol. Cell* 11:339–54

46. Nern A, Arkowitz RA. 1998. A GTP-exchange factor required for cell orientation. *Nature* 391:195–98

47. Schrick K, Garvik B, Hartwell LH. 1997. Mating in *Saccharomyces cerevisiae*: the role of the pheromone signal transduction pathway in the chemotropic response to pheromone. *Genetics* 147:19–32

48. Reid BJ, Hartwell LH. 1977. Regulation of mating in the cell cycle of *Saccharomyces cerevisiae. J. Cell. Biol.* 75:355–65

49. Stevenson BJ, Ferguson B, De Virgilio C, Bi E, Pringle JR, et al. 1995. Mutation of RGA1, which encodes a putative GTPase-activating protein for the polarity-establishment protein Cdc42p, activates the pheromone-response pathway in the yeast *Saccharomyces cerevisiae. Genes Dev.* 9:2949–63

50. Cvrckova F, De Virgilio C, Manser E, Pringle JR, Nasmyth K. 1995. Ste20-like protein kinases are required for normal localization of cell growth and for cytokinesis in budding yeast. *Genes Dev.* 9:1817–30

51. Eby JJ, Holly SP, van Drogen F, Grishin AV, Peter M, et al. 1998. Actin cytoskeleton organization regulated by the PAK family of protein kinases. *Curr. Biol.* 8:967–70

52. Weiss EL, Bishop AC, Shokat KM, Drubin DG. 2000. Chemical genetic analysis of the budding-yeast p21-activated kinase Cla4p. *Nat. Cell Biol.* 2:677–85

53. Leeuw T, Fourest-Lieuvin A, Wu C, Chenevert J, Clark K, et al. 1995. Pheromone response in yeast: association of Bem1p with proteins of the MAP kinase cascade and actin. *Science* 270:1210–13

54. Moskow JJ, Gladfelter AS, Lamson RE, Pryciak PM, Lew DJ. 2000. Role

of Cdc42p in pheromone-stimulated signal transduction in *Saccharomyces cerevisiae. Mol. Cell. Biol.* 20:7559–71

55. Leberer E, Dignard D, Harcus D, Thomas DY, Whiteway M. 1992. The protein kinase homologue Ste20p is required to link the yeast pheromone response G-protein $\beta\gamma$ subunits to downstream signalling components. *EMBO J.* 11:4815–24

56. Ramer SW, Davis RW. 1993. A dominant truncation allele identifies a gene, STE20, that encodes a putative protein kinase necessary for mating in *Saccharomyces cerevisiae. Proc. Natl. Acad. Sci. USA* 90:452–56

57. Liu H, Styles CA, Fink GR. 1993. Elements of the yeast pheromone response pathway required for filamentous growth of diploids. *Science* 262:1741–44

58. Roberts RL, Fink GR. 1994. Elements of a single MAP kinase cascade in *Saccharomyces cerevisiae* mediate two developmental programs in the same cell type: mating and invasive growth. *Genes Dev.* 8:2974–85

59. Raitt DC, Posas F, Saito H. 2000. Yeast Cdc42 GTPase and Ste20 PAK-like kinase regulate Sho1-dependent activation of the Hog1 MAPK pathway. *EMBO J.* 19:4623–31

60. Reiser V, Salah SM, Ammerer G. 2000. Polarized localization of yeast Pbs2 depends on osmostress, the membrane protein Sho1 and Cdc42. *Nat. Cell Biol.* 2:620–27

61. Leberer E, Wu C, Leeuw T, Fourest-Lieuvin A, Segall JE, et al. 1997. Functional characterization of the Cdc42p binding domain of yeast Ste20p protein kinase. *EMBO J.* 16:83–97

62. Roberts RL, Mosch HU, Fink GR. 1997. 14-3-3 proteins are essential for RAS/MAPK cascade signaling during pseudohyphal development in *S. cerevisiae. Cell* 89:1055–65

63. Leberer E, Thomas DY, Whiteway M. 1997. Pheromone signalling and polar-ized morphogenesis in yeast. *Curr. Opin. Genet. Dev.* 7:59–66

64. Cvrckova F, De Virgilio C, Manser E, Pringle JR, Nasmyth K. 1995. Ste20-like protein kinases are required for normal localization of cell growth and for cytokinesis in budding yeast. *Genes Dev.* 9:1817–30

65. Akada R, Kallal L, Johnson DI, Kurjan J. 1996. Genetic relationships between the G protein $\beta\gamma$ complex, Ste5p, Ste20p and Cdc42p: investigation of effector roles in the yeast pheromone response pathway. *Genetics* 143:103–17

66. Leeuw T, Wu C, Schrag JD, Whiteway M, Thomas DY, et al. 1998. Interaction of a G-protein β-subunit with a conserved sequence in Ste20/PAK family protein kinases. *Nature* 391:191–95

67. Ziman M, Preuss D, Mulholland J, O'Brien JM, Botstein D, et al. 1993. Subcellular localization of Cdc42p, a *Saccharomyces cerevisiae* GTP-binding protein involved in the control of cell polarity. *Mol. Biol. Cell* 4:1307–16

68. Ohya Y, Qadota H, Anraku Y, Pringle JR, Botstein D. 1993. Suppression of yeast geranylgeranyl transferase I defect by alternative prenylation of two target GTPases, Rho1p and Cdc42p. *Mol. Biol. Cell* 4:1017–25

69. Peter M, Neiman AM, Park HO, van Lohuizen M, Herskowitz I. 1996. Functional analysis of the interaction between the small GTP binding protein Cdc42 and the Ste20 protein kinase in yeast. *EMBO J.* 15:7046–59

70. Whiteway MS, Wu C, Leeuw T, Clark K, Fourest-Lieuvin A, et al. 1995. Association of the yeast pheromone response G protein $\beta\gamma$ subunits with the MAP kinase scaffold Ste5p. *Science* 269:1572–75

71. Inouye C, Dhillon N, Thorner J. 1997. Ste5 RING-H2 domain: role in Ste4-promoted oligomerization for yeast pheromone signaling. *Science* 278:103–6

72. Feng Y, Song LY, Kincaid E, Mahanty SK, Elion EA. 1998. Functional binding

between Gβ and the LIM domain of Ste5 is required to activate the MEKK Ste11. *Curr. Biol.* 8:267–78

73. Inouye C, Dhillon N, Durfee T, Zambryski PC, Thorner J. 1997. Mutational analysis of STE5 in the yeast *Saccharomyces cerevisiae*: application of a differential interaction trap assay for examining protein-protein interactions. *Genetics* 147:479–92

74. Choi KY, Satterberg B, Lyons DM, Elion EA. 1994. Ste5 tethers multiple protein kinases in the MAP kinase cascade required for mating in *S. cerevisiae. Cell* 78:499–512

75. Choi KY, Kranz JE, Mahanty SK, Park KS, Elion EA. 1999. Characterization of Fus3 localization: active Fus3 localizes in complexes of varying size and specific activity. *Mol. Biol. Cell* 10:1553–68

76. Dowell SJ, Bishop AL, Dyos SL, Brown AJ, Whiteway MS. 1998. Mapping of a yeast G protein $\beta\gamma$ signaling interaction. *Genetics* 150:1407–17

77. Pryciak PM, Huntress FA. 1998. Membrane recruitment of the kinase cascade scaffold protein Ste5 by the G$\beta\gamma$ complex underlies activation of the yeast pheromone response pathway. *Genes Dev.* 12:2684–97

78. Mahanty SK, Wang Y, Farley FW, Elion EA. 1999. Nuclear shuttling of yeast scaffold Ste5 is required for its recruitment to the plasma membrane and activation of the mating MAPK cascade. *Cell* 98:501–12

79. Sette C, Inouye CJ, Stroschein SL, Iaquinta PJ, Thorner J. 2000. Mutational analysis suggests that activation of the yeast pheromone response mitogen-activated protein kinase pathway involves conformational changes in the Ste5 scaffold protein. *Mol. Biol. Cell* 11:4033–49

80. Kunzler M, Trueheart J, Sette C, Hurt E, Thorner J. 2001. Mutations in the YRB1 gene encoding yeast ran-binding-protein-1 that impair nucleocytoplasmic transport

and suppress yeast mating defects. *Genetics* 157:1089–105

81. Marcus S, Polverino A, Barr M, Wigler M. 1994. Complexes between STE5 and components of the pheromone-responsive mitogen-activated protein kinase module. *Proc. Natl. Acad. Sci. USA* 91:7762–66

82. Printen JA, Sprague GF Jr. 1994. Protein-protein interactions in the yeast pheromone response pathway: Ste5p interacts with all members of the MAP kinase cascade. *Genetics* 138:609–19

83. Stratton HF, Zhou J, Reed SI, Stone DE. 1996. The mating-specific Gα protein of *Saccharomyces cerevisiae* downregulates the mating signal by a mechanism that is dependent on pheromone and independent of G$\beta\gamma$ sequestration. *Mol. Cell. Biol.* 16:6325–37

84. Zhou J, Arora M, Stone DE. 1999. The yeast pheromone-responsive Gα protein stimulates recovery from chronic pheromone treatment by two mechanisms that are activated at distinct levels of stimulus. *Cell. Biochem. Biophys.* 30:193–212

85. Apanovitch DM, Iiri T, Karasawa T, Bourne HR, Dohlman HG. 1998. Second site suppressor mutations of a GTPase-deficient G-protein α-subunit. Selective inhibition of G$\beta\gamma$-mediated signaling. *J. Biol. Chem.* 273:28597–602

86. Apanovitch DM, Slep KC, Sigler PB, Dohlman HG. 1998. Sst2 is a GTPase-activating protein for Gpa1: purification and characterization of a cognate RGS-Gα protein pair in yeast. *Biochemistry* 37:4815–22

87. Dohlman HG, Song J, Ma D, Courchesne WE, Thorner J. 1996. Sst2, a negative regulator of pheromone signaling in the yeast *Saccharomyces cerevisiae*: expression, localization, and genetic interaction and physical association with Gpa1 (the G-protein α-subunit). *Mol. Cell. Biol.* 16:5194–209

88. Kallal L, Fishel R. 2000. The GTP hydrolysis defect of the *Saccharomyces*

cerevisiae mutant G-protein Gpa1 (G50V). *Yeast* 16:387–400

89. Wedegaertner PB. 1998. Lipid modifications and membrane targeting of Gα. *Biol. Signals Recept.* 7:125–35

90. Johnson DR, Bhatnagar RS, Knoll LJ, Gordon JI. 1994. Genetic and biochemical studies of protein N-myristoylation. *Annu. Rev. Biochem.* 63:869–914

91. Casey PJ. 1995. Protein lipidation in cell signaling. *Science* 268:221–25

92. Dohlman HG, Goldsmith P, Spiegel AM, Thorner J. 1993. Pheromone action regulates G-protein α-subunit myristoylation in the yeast Saccharomyces cerevisiae. *Proc. Natl. Acad. Sci. USA* 90:9688–92

93. Stone DE, Cole GM, de Barros Lopes M, Goebl M, Reed SI. 1991. N-myristoylation is required for function of the pheromone-responsive G α protein of yeast: conditional activation of the pheromone response by a temperature-sensitive N-myristoyl transferase. *Genes Dev.* 5:1969–81

94. Song J, Hirschman J, Gunn K, Dohlman HG. 1996. Regulation of membrane and subunit interactions by N-myristoylation of a G protein α subunit in yeast. *J. Biol. Chem.* 271:20273–83

95. Medici R, Bianchi E, Di Segni G, Tocchini-Valentini GP. 1997. Efficient signal transduction by a chimeric yeast-mammalian G protein α subunit Gpa1-Gsα covalently fused to the yeast receptor Ste2. *EMBO J.* 16:7241–49

96. Song J, Dohlman HG. 1996. Partial constitutive activation of pheromone responses by a palmitoylation-site mutant of a G protein α subunit in yeast. *Biochemistry* 35:14806–17

97. Manahan CL, Patnana M, Blumer KJ, Linder ME. 2000. Dual lipid modification motifs in Gα and Gγ subunits are required for full activity of the pheromone response pathway in Saccharomyces cerevisiae. *Mol. Biol. Cell* 11:957–68

98. Madura K, Varshavsky A. 1994. Degra-

dation of Gα by the N-end rule pathway. *Science* 265:1454–58

99. Hicke L, Riezman H. 1996. Ubiquitination of a yeast plasma membrane receptor signals its ligand-stimulated endocytosis. *Cell* 84:277–87

100. Roth AF, Davis NG. 1996. Ubiquitination of the yeast **a**-factor receptor. *J. Cell. Biol.* 134:661–74

101. Cole GM, Stone DE, Reed SI. 1990. Stoichiometry of G protein subunits affects the *Saccharomyces cerevisiae* mating pheromone signal transduction pathway. *Mol. Cell. Biol.* 10:510–17

102. Dietzel C, Kurjan J. 1987. The yeast SCG1 gene: a G α-like protein implicated in the **a**- and α-factor response pathway. *Cell* 50:1001–10

103. Miyajima I, Nakafuku M, Nakayama N, Brenner C, Miyajima A, et al. 1987. GPA1, a haploid-specific essential gene, encodes a yeast homolog of mammalian G protein which may be involved in mating factor signal transduction. *Cell* 50:1011–19

104. Deleted in proof

105. Fu HW, Casey PJ. 1999. Enzymology and biology of CaaX protein prenylation. *Recent Progr. Horm. Res.* 54:315–42

106. Finegold AA, Schafer WR, Rine J, Whiteway M, Tamanoi F. 1990. Common modifications of trimeric G proteins and ras protein: involvement of polyisoprenylation. *Science* 249:165–69

107. Hirschman JE, Jenness DD. 1999. Dual lipid modification of the yeast Gγ subunit Ste18p determines membrane localization of Gβγ. *Mol. Cell. Biol.* 19:7705–11

108. Grishin AV, Weiner JL, Blumer KJ. 1994. Biochemical and genetic analysis of dominant-negative mutations affecting a yeast G-protein γ subunit. *Mol. Cell. Biol.* 14:4571–78

109. Whiteway MS, Thomas DY. 1994. Site-directed mutations altering the CAAX box of Ste18, the yeast pheromone-response pathway Gγ subunit. *Genetics* 137:967–76

110. Cole GM, Reed SI. 1991. Pheromone-induced phosphorylation of a G protein β subunit in *S. cerevisiae* is associated with an adaptive response to mating pheromone. *Cell* 64:703–16

111. Hirschman JE, De Zutter GS, Simonds WF, Jenness DD. 1997. The Gβγ complex of the yeast pheromone response pathway. Subcellular fractionation and protein-protein interactions. *J. Biol. Chem.* 272:240–48

112. Li E, Cismowski MJ, Stone DE. 1998. Phosphorylation of the pheromone-responsive Gβ protein of *Saccharomyces cerevisiae* does not affect its mating-specific signaling function. *Mol. Gen. Genet.* 258:608–18

113. Chan RK, Otte CA. 1982. Isolation and genetic analysis of *Saccharomyces cerevisiae* mutants supersensitive to G1 arrest by **a** factor and α factor pheromones. *Mol. Cell Biol.* 2:11–20

114. Chan RK, Otte CA. 1982. Physiological characterization of *Saccharomyces cerevisiae* mutants supersensitive to G1 arrest by **a** factor and α factor pheromones. *Mol. Cell. Biol.* 2:21–29

115. Dietzel C, Kurjan J. 1987. Pheromonal regulation and sequence of the *Saccharomyces cerevisiae* SST2 gene: a model for desensitization to pheromone. *Mol. Cell. Biol.* 7:4169–77

116. Dohlman HG, Apaniesk D, Chen Y, Song J, Nusskern D. 1995. Inhibition of G-protein signaling by dominant gain-of-function mutations in Sst2p, a pheromone desensitization factor in *Saccharomyces cerevisiae. Mol. Cell. Biol.* 15:3635–43

117. Koelle MR, Horvitz HR. 1996. EGL-10 regulates G protein signaling in the *C. elegans* nervous system and shares a conserved domain with many mammalian proteins. *Cell* 84:115–25

118. Siderovski DP, Hessel A, Chung S, Mak TW, Tyers M. 1996. A new family of regulators of G-protein-coupled receptors? *Curr. Biol.* 6:211–12

119. De Vries L, Zheng B, Fischer T, Elenko E, Farquhar MG. 2000. The regulator of G protein signaling family. *Annu. Rev. Pharmacol. Toxicol.* 40:235–71

120. Ross EM, Wilkie TM. 2000. GTPase-activating proteins for heterotrimeric G proteins: regulators of G protein signaling (RGS) and RGS-like proteins. *Annu. Rev. Biochem.* 69:795–827

121. Watson N, Linder ME, Druey KM, Kehrl JH, Blumer KJ. 1996. RGS family members: GTPase-activating proteins for heterotrimeric G-protein α-subunits. *Nature* 383:172–75

122. Hunt TW, Fields TA, Casey PJ, Peralta EG. 1996. RGS10 is a selective activator of Gαi GTPase activity. *Nature* 383:175–77

123. Berman DM, Wilkie TM, Gilman AG. 1996. GAIP and RGS4 are GTPase-activating proteins for the Gi subfamily of G protein α subunits. *Cell* 86:445–52

124. Tesmer JJ, Berman DM, Gilman AG, Sprang SR. 1997. Structure of RGS4 bound to AlF4-activated Giα1: stabilization of the transition state for GTP hydrolysis. *Cell* 89:251–61

125. DiBello PR, Garrison TR, Apanovitch DM, Hoffman G, Shuey DJ, et al. 1998. Selective uncoupling of RGS action by a single point mutation in the G protein α-subunit. *J. Biol. Chem.* 273:5780–84

126. Lan KL, Sarvazyan NA, Taussig R, Mackenzie RG, DiBello PR, et al. 1998. A point mutation in Gαo and Gαi1 blocks interaction with regulator of G protein signaling proteins. *J. Biol. Chem.* 273:12794–97

127. Rechsteiner M, Rogers SW. 1996. PEST sequences and regulation by proteolysis. *Trends Biochem. Sci.* 21:267–71

128. Chen T, Kurjan J. 1997. *Saccharomyces cerevisiae* Mpt5p interacts with Sst2p and plays roles in pheromone sensitivity and recovery from pheromone arrest. *Mol. Cell. Biol.* 17:3429–39

129. Sinclair D, Mills K, Guarente L. 1998.

Aging in *Saccharomyces cerevisiae.* *Annu. Rev. Microbiol.* 52:533–60

130. Uetz P, Giot L, Cagney G, Mansfield TA, Judson RS, et al. 2000. A comprehensive analysis of protein-protein interactions in *Saccharomyces cerevisiae. Nature* 403:623–27

131. Burchett S, Flanary P, Jiang L, Aston C, Young K, et al. 2001. Regulation of stress response signaling by the N-terminal DEP (Dishevelled/EGL-10/Pleckstrin) domain of Sst2, a regulator of G protein signaling in *Saccharomyces cerevisiae. J. Biol. Chem.* Submitted

132. Flanary PL, DiBello PR, Estrada P, Dohlman HG. 2000. Functional analysis of Plp1 and Plp2, two homologues of phosducin in yeast. *J. Biol. Chem.* 275:18462–69

133. Spain BH, Koo D, Ramakrishnan M, Dzudzor B, Colicelli J. 1995. Truncated forms of a novel yeast protein suppress the lethality of a G protein α subunit deficiency by interacting with the β subunit. *J. Biol. Chem.* 270:25435–44

134. Nelissen B, De Wachter R, Goffeau A. 1997. Classification of all putative permeases and other membrane plurispanners of the major facilitator superfamily encoded by the complete genome of *Saccharomyces cerevisiae. FEMS Microbiol. Rev.* 21:113–34

135. Paulsen IT, Sliwinski MK, Nelissen B, Goffeau A, Saier MH Jr. 1998. Unified inventory of established and putative transporters encoded within the complete genome of *Saccharomyces cerevisiae. FEBS Lett.* 430:116–25

136. Pryciak PM, Hartwell LH. 1996. AKR1 encodes a candidate effector of the G $\beta\gamma$ complex in the *Saccharomyces cerevisiae* pheromone response pathway and contributes to control of both cell shape and signal transduction. *Mol. Cell. Biol.* 16:2614–26

137. Kao LR, Peterson J, Ji R, Bender L, Bender A. 1996. Interactions between the ankyrin repeat-containing protein Akr1p

and the pheromone response pathway in *Saccharomyces cerevisiae. Mol. Cell. Biol.* 16:168–78

138. Couve A, Hirsch JP. 1996. Loss of sustained Fus3p kinase activity and the G1 arrest response in cells expressing an inappropriate pheromone receptor. *Mol. Cell. Biol.* 16:4478–85

139. Roth AF, Nelson B, Boone C, Davis NG. 2000. Asg7p-Ste3p inhibition of pheromone signaling: regulation of the zygotic transition to vegetative growth. *Mol. Cell. Biol.* 20:8815–25

140. Kim J, Bortz E, Zhong H, Leeuw T, Leberer E, et al. 2000. Localization and signaling of Gβ subunit Ste4p are controlled by a-factor receptor and the a-specific protein Asg7p. *Mol. Cell. Biol.* 20:8826–35

141. Koch WJ, Hawes BE, Inglese J, Luttrell LM, Lefkowitz RJ. 1994. Cellular expression of the carboxyl terminus of a G protein-coupled receptor kinase attenuates G$\beta\gamma$-mediated signaling. *J. Biol. Chem.* 269:6193–97

142. Krupnick JG, Benovic JL. 1998. The role of receptor kinases and arrestins in G protein-coupled receptor regulation. *Annu. Rev. Pharmacol. Toxicol.* 38:289–319

143. Boekhoff I, Touhara K, Danner S, Inglese J, Lohse MJ, et al. 1997. Phosducin, potential role in modulation of olfactory signaling. *J. Biol. Chem.* 272:4606–12

144. Pitcher JA, Inglese J, Higgins JB, Arriza JL, Casey PJ, et al. 1992. Role of $\beta\gamma$ subunits of G proteins in targeting the β-adrenergic receptor kinase to membrane-bound receptors. *Science* 257:1264–67

145. Nakafuku M, Obara T, Kaibuchi K, Miyajima I, Miyajima A, et al. 1988. Isolation of a second yeast *Saccharomyces cerevisiae* gene (GPA2) coding for guanine nucleotide-binding regulatory protein: studies on its structure and possible functions. *Proc. Natl. Acad. Sci. USA* 85:1374–78

146. Xue Y, Batlle M, Hirsch JP. 1998. GPR1 encodes a putative G protein-coupled receptor that associates with the Gpa2p Gα subunit and functions in a Ras-independent pathway. *EMBO J.* 17:1996–2007

146a. Yun CW, Tamaki H, Nakayama R, Yamamoto K, Kumagai H. 1997. G-protein coupled receptor from yeast *Saccharomyces cerevisiae Biochem. Biophys. Res. Commun.* 240:287–92

147. Ansari K, Martin S, Farkasovsky M, Ehbrecht IM, Kuntzel H. 1999. Phospholipase C binds to the receptor-like GPR1 protein and controls pseudohyphal differentiation in *Saccharomyces cerevisiae. J. Biol. Chem.* 274:30052–58

148. Pan X, Heitman J. 1999. Cyclic AMP-dependent protein kinase regulates pseudohyphal differentiation in *Saccharomyces cerevisiae. Mol. Cell. Biol.* 19:4874–87

149. Pan X, Harashima T, Heitman J. 2000. Signal transduction cascades regulating pseudohyphal differentiation of *Saccharomyces cerevisiae. Curr. Opin. Microbiol.* 3:567–72

150. Versele M, de Winde JH, Thevelein JM. 1999. A novel regulator of G protein signalling in yeast, Rgs2, downregulates glucose-activation of the cAMP pathway through direct inhibition of Gpa2. *EMBO J.* 18:5577–91

151. Kubler E, Mosch HU, Rupp S, Lisanti MP. 1997. Gpa2p, a G-protein α-subunit, regulates growth and pseudohyphal development in *Saccharomyces cerevisiae* via a cAMP-dependent mechanism. *J. Biol. Chem.* 272:20321–23

151a. Donzeau M, Bandlow W. 1999. The yeast trimeric guanine nucleotide-binding protein α subunit, Gpa2p, controls the meiosis-specific kinase Ime2p activity in response to nutrients. *Mol. Cell Biol.* 19:6110–19

152. Lorenz MC, Heitman J. 1997. Yeast pseudohyphal growth is regulated by GPA2, a G protein α homolog. *EMBO J.* 16:7008–18

153. Burchett SA. 2000. Regulators of G protein signaling: a bestiary of modular protein binding domains. *J. Neurochem.* 75:1335–51

154. Kozasa T, Jiang X, Hart MJ, Sternweis PM, Singer WD, et al. 1998. p115 RhoGEF, a GTPase activating protein for Gα12 and Gα13. *Science* 280:2109–11

155. Hart MJ, Jiang X, Kozasa T, Roscoe W, Singer WD, et al. 1998. Direct stimulation of the guanine nucleotide exchange activity of p115 RhoGEF by Gα13. *Science* 280:2112–14

156. Ponting CP, Bork P. 1996. Pleckstrin's repeat performance: a novel domain in G-protein signaling? *Trends Biochem. Sci.* 21:245–46

157. Axelrod JD, Miller JR, Shulman JM, Moon RT, Perrimon N. 1998. Differential recruitment of Disheveled provides signaling specificity in the planar cell polarity and Wingless signaling pathways. *Genes Dev.* 12:2610–22

158. Wong HC, Mao J, Nguyen JT, Srinivas S, Zhang W, et al. 2000. Structural basis of the recognition of the dishevelled DEP domain in the Wnt signaling pathway. *Nat. Struct. Biol.* 7:1178–84

159. Boutros M, Paricio N, Strutt DI, Mlodzik M. 1998. Dishevelled activates JNK and discriminates between JNK pathways in planar polarity and wingless signaling. *Cell* 94:109–18

160. Hoffman GA, Garrison TR, Dohlman HG. 2000. Endoproteolytic processing of Sst2, a multidomain regulator of G protein signaling in yeast. *J. Biol. Chem.* 275:37533–41

161. Holthuis JC, Nichols BJ, Dhruvakumar S, Pelham HR. 1998. Two syntaxin homologues in the TGN/endosomal system of yeast. *EMBO J.* 17:113–26

161a. Abeliovich H, Grote E, Novick P, Ferro-Novick S. 1998. Tlg2p, a yeast syntaxin homolog that resides on the Golgi and endocytic structures. *J. Biol. Chem.* 273:11719–27

162. Luo W, Chang A. 1997. Novel genes involved in endosomal traffic in yeast revealed by suppression of a targeting-defective plasma membrane ATPase mutant. *J. Cell. Biol.* 138:731–46

163. Nothwehr SF, Bryant NJ, Stevens TH. 1996. The newly identified yeast GRD genes are required for retention of late-Golgi membrane proteins. *Mol. Cell. Biol.* 16:2700–71

164. Becherer KA, Rieder SE, Emr SD, Jones EW. 1996. Novel syntaxin homologue, Pep12p, required for the sorting of lumenal hydrolases to the lysosome-like vacuole in yeast. *Mol. Biol. Cell* 7:579–94

164a. Gerrard SR, Levi BP, Stevens TH. 2000. Pep12p is a multifunctional yeast syntaxin that controls entry of biosynthetic, endocytic and retrograde traffic into the prevacuolar compartment. *Traffic* 1:259–69

165. Winzeler EA, Shoemaker DD, Astromoff A, Liang H, Anderson K, et al. 1999. Functional characterization of the *S. cerevisiae* genome by gene deletion and parallel analysis. *Science* 285:901–6

Annu. Rev. Physiol. 2002. 64:153–87

G Proteins and Phototransduction

Vadim Y. Arshavsky[1], Trevor D. Lamb[2], and
Edward N. Pugh, Jr.[3]

[1]*Howe Laboratory of Ophthalmology, Harvard Medical School, Boston,
Massachusetts 02114; e-mail: vadim_arshavsky@meei.harvard.edu*
[2]*Department of Physiology, University of Cambridge, Cambridge CB2 3EG UK;
e-mail: TDL1@cam.ac.uk*
[3]*Department of Ophthalmology, University of Pennsylvania, Philadelphia,
Pennsylvania 19104; e-mail: pugh@mail.med.upenn.edu*

Key Words　transducin, phosphodiesterase, amplification, signal termination

■ **Abstract**　Phototransduction is the process by which a photon of light captured by a molecule of visual pigment generates an electrical response in a photoreceptor cell. Vertebrate rod phototransduction is one of the best-studied G protein signaling pathways. In this pathway the photoreceptor-specific G protein, transducin, mediates between the visual pigment, rhodopsin, and the effector enzyme, cGMP phosphodiesterase. This review focuses on two quantitative features of G protein signaling in phototransduction: signal amplification and response timing. We examine how the interplay between the mechanisms that contribute to amplification and those that govern termination of G protein activity determine the speed and the sensitivity of the cellular response to light.

INTRODUCTION

G Proteins Regulate the Amplification and Timing of Cell Signaling

Heterotrimeric G proteins are the molecular mediators that link G protein–coupled receptors (GPCRs) to their effector proteins — enzymes or ion channels; thus, they form the second stage in a three-step signaling cascade. The first step is the activation of the GPCR by the binding of an agonist molecule. In the second step the activated GPCR activates multiple molecules of G protein by catalyzing the exchange of guanosine triphosphate (GTP) for guanosine diphosphate (GDP) on the G protein α subunit (Gα). The G proteins activated in this way constitute an amplified representation of the activated GPCR. In the third step of the cascade the activated G protein binds to its effector and thereby switches it either on or off in different systems. The timing of the effector's activity is controlled by the G protein, because hydrolysis of the GTP bound to Gα terminates the activity of

0066-4278/02/0315-0153$14.00

both the G protein and the effector. The duration of this signal can be set by the intrinsic GTPase time constant of the G protein or it can be regulated by interaction with a GTPase-activating protein (GAP), either alone or in combination with the G protein effector.

The Vertebrate Rod: A Model System for Investigating G Protein Signaling

Phototransduction is the process by which a photon of light captured by a molecule of visual pigment generates an electrical response in a photoreceptor cell. Visual pigments are members of the superfamily of GPCRs whose natural ligand, the chromophore 11-*cis* retinal, is already covalently attached (1, 2). The 11-*cis* isomer acts as a powerful antagonist, and it is only when a photon of light isomerizes the chromophore to its all-*trans* form, which acts as a powerful agonist, that the visual pigment GPCR becomes active. In all vertebrate and invertebrate photoreceptor cells investigated thus far, phototransduction has been found to be based on a heterotrimeric G protein signaling cascade. The cascade of vertebrate photoreceptors has proven to be an especially productive model system for elucidating general principles of G protein signaling, thanks to several advantages it offers to investigators.

The first advantage arises from the fact that phototransduction takes place in a highly specialized organelle, the outer segment (Figure 1) of the photoreceptor cell. The outer segment contains high concentrations of the proteins involved in the transduction cascade—typically ~500 μM of the G protein, transducin (see Table 3 in Ref. 3)—and, moreover, it lacks most of the proteins involved in other cellular functions. The outer segments of rods, in particular, can easily be detached from the retina and harvested in quantities sufficient for purification of most key protein components, and for conducting biochemical experiments to characterize their properties and interactions.

A second advantage is that the light-evoked electrical responses of vertebrate photoreceptors have been extensively investigated with an array of electrophysiological techniques, providing a rich, quantitative database that has to be explicable in terms of the interactions between the biochemical components of the transduction cascade. Because the electrophysiological recordings are obtained from intact photoreceptors (and even from cells in vivo), it follows that such recordings reflect transduction when the components of the cascade are present in their natural dispositions and concentrations, and are interacting with one another under conditions selected by evolution for their signaling function.

A third advantage of vertebrate photoreceptor preparations, one that applies equally well to in vitro biochemical assays and to electrophysiologically measured responses, is that they allow exquisite precision in quantification of the natural stimulus, measured as the number of photopigment molecules isomerized by light exposure. As a consequence, investigations of rod phototransduction have achieved

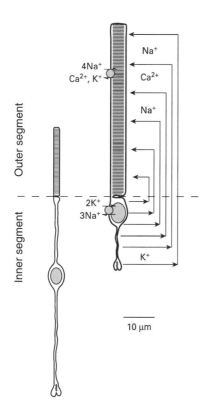

Figure 1 The structure of the mammalian rod (*left*) and a toad rod (*right*) (drawn roughly to scale). In the dark state the cells have a circulating current, characterized by the influx of Na^+ and Ca^{2+} cations into the outer segment through cGMP activated channels, and a balancing outward flux of K^+ through channels in the inner segment. A Na^+/Ca^{2+}-K^+ exchanger in the outer segment membrane, and a conventional Na^+/K^+ exchanger in the inner segment membrane maintain the overall ionic gradients that drive the two limbs of the circulating current.

a level of quantitative analysis that would be difficult to attain with any other second messenger system. Indeed, rod phototransduction provides the unique opportunity to analyze a cellular response evoked by the activation of a single GPCR, i.e., one rhodopsin molecule (4–7).

A number of excellent reviews published over the past decade examine various aspects of vertebrate phototransduction and are recommended to the reader (3, 8–26). This review focuses on recent developments in the understanding of two quantitative features of G protein signaling in phototransduction: signal amplification and response timing. It also focuses entirely on rods, which are the more abundant photoreceptors in the retinas of most vertebrates, with the consequence that the quantitative aspects of G protein signaling have been studied in much greater detail in rods than in cones.

The Major Steps in the Activation Phase of Vertebrate Phototransduction

The major steps of the G protein cascade in vertebrate rods are illustrated schematically in Figure 2 (see color insert).

PHOTOISOMERIZATION   Vision begins when the chromophore, the preattached lig-
and of a single molecule of visual pigment, is isomerized by a captured photon.
Within a millisecond rhodopsin undergoes a series of intramolecular transitions
leading to a conformational state called metarhodopsin II (or R*), which is capable
of activating the photoreceptor-specific G protein, transducin.

TRANSDUCIN ACTIVATION   R* interacts with the GDP-bound form of the trans-
ducin $\alpha\beta\gamma$ trimer; i.e., with $G\alpha_t$-GDP-$G\beta\gamma_t$. The R* activates the transducin by
triggering rapid exchange of bound GDP for GTP on the $G\alpha_t$; this is followed very
rapidly by dissociation of the transducin from the R*, as well as by dissociation of
the active $G\alpha_t$-GTP (or G*) from $G\beta\gamma_t$.

PHOSPHODIESTERASE ACTIVATION   At the next step of the cascade $G\alpha_t$-GTP stimulates the activity of its effector enzyme, the cGMP phosphodiesterase (PDE),
also known as PDE6 (27). The PDE is a heterotetramer consisting of two identical
or nearly identical catalytic subunits ($\alpha\beta$ in rods, $\alpha\alpha$ in cones) and two identical
regulatory γ subunits (PDEγ), which serve as protein inhibitors of PDE activity
and which are responsible for maintaining the activity in the nonactivated state at
its very low basal level. Activation of the PDE results from the binding of $G\alpha_t$
to the γ subunit, thereby removing the inhibitory constraint that the PDEγ had
imposed on the catalytic site of the PDE α or β subunit.

CYCLIC GMP HYDROLYSIS AND cGMP-CHANNEL CLOSURE   Activation of the PDE
causes a reduction in cytoplasmic concentration of cGMP, the second messenger
in phototransduction, and this in turn causes closure of the cation selective cGMP-
gated channels located in the plasma membrane. Closure of these channels reduces
the steady inward current that is normally carried by Na^+ and Ca^{2+} ions in the dark,
resulting in membrane hyperpolarization and decreased release of the synaptic
transmitter glutamate at the photoreceptor terminal.

Timely Termination of Phototransduction

As in all G protein signaling pathways, timely termination of the photoreceptor
signal requires that all the activated intermediates be inactivated rapidly, restoring
the system to its dark, basal state, ready for signaling again. Thus, the three protein
intermediates, R*, $G\alpha_t$-GTP, and activated PDE, must all be inactivated, and the
concentration of cytoplasmic messenger cGMP must be restored to its dark level
by guanylyl cyclase. Because the focus of this review is on transducin, we do
not consider the inactivation of R* or the restoration of cGMP levels by guanylyl
cyclase; these topics are covered in several of the reviews cited above.

In phototransduction the active state of the effector enzyme (PDE) persists until
the GTP bound to its activator ($G\alpha_t$) is hydrolyzed to GDP and P_i, permitting the

dissociation of $G\alpha_t$ from PDE and returning the effector to its inactive state. The two functional features of phototransduction reviewed here, signal amplification and termination of activity, involve distinct molecular and cellular mechanisms, yet are tightly intertwined in determining the response of the photoreceptor. Thus, the activation phase of the signal needs to be very highly amplified to elicit a reliable response to each photon. In addition, the cascade needs to be inactivated rapidly enough to allow the cell to signal reductions in light intensity and repetitive stimulation. A goal of this review is to examine the mechanisms that contribute to signal amplification and those that lead to termination of cascade activity, whose interplay determines the sensitivity of the cell to light.

AMPLIFICATION IN ROD PHOTOTRANSDUCTION IS ACHIEVED IN THREE GAIN STEPS

A Framework for Quantifying Amplification in Phototransduction

The remarkable ability of rods to reliably signal individual photoisomerizations has presented a long-standing challenge to investigators of vertebrate phototransduction (4, 5). The capture of a single photon by a rod results in suppression of 2–5% of the total cGMP-activated current at the peak of the photoresponse (4–7) and a consequent membrane hyperpolarization of about 1 mV. This exquisite signaling capacity is achieved through three steps of amplification, or gain. First, a single R* activates many transducin molecules in the course of its lifetime. Second, each PDE activated by a transducin hydrolyzes many cGMP molecules during its lifetime. Third, the cGMP-gated channels contribute a gain factor corresponding to the cooperativity of their opening (their Hill coefficient).

The contribution of the individual molecular components of the cascade to the overall gain of phototransduction has been quantified in a theoretical framework developed by Lamb & Pugh (28). Details of that LP analysis can be found in (3, 14, 28); here we simply present a summary as a background for key aspects of this review.

The LP analysis predicts, and many studies have confirmed, that the rod's response to a brief flash initially rises as a parabolic function of time (Figure 3):

$$R(t) \approx \frac{1}{2}\Phi A t^2. \qquad\qquad 1.$$

Here $R(t)$ is the normalized response, Φ is the number of photoisomerizations (R*) per rod generated by the flash, and A is a constant characterizing the rate of parabolic rise, and thus the overall signal amplification: The greater the magnitude of A, the more rapidly the response rises per R*. The normalized response $R(t)$ is the fractional suppression of the preexisting circulating current, expressed as the observed response divided by the steady level present before the flash. The initial suppression of current follows a parabolic function of time because the quantity of cGMP hydrolyzed at time t after the stepwise activation of R^* is determined by two cascaded integrating processes: the activation of multiple transducins (and as a consequence, multiple PDEs) by each R^*, followed by the hydrolysis of multiple cGMPs by each activated PDE.

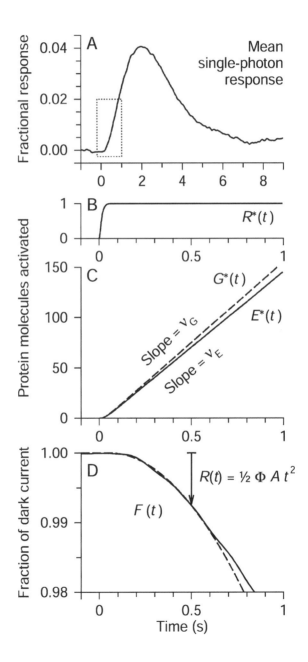

The LP analysis further shows that the parameter A, called the "amplification constant," can be expressed as the product of four gain factors arising at different steps of the cascade:

$$A = \nu_G c_{GE} \beta_{sub} n_{cG}. \qquad 2.$$

Here ν_G is the rate of activation of transducin molecules per R^* and c_{GE} is the coupling efficiency from G^* activation to PDE activation, so that $\nu_G\, c_{GE} = \nu_E$ is the rate of activation of PDE catalytic subunits per R^*; we use the letter E to denote "effector enzyme," which in this cascade is the PDE. β_{sub} is the rate constant of cGMP hydrolysis per activated PDE catalytic subunit, and n_{cG} is the Hill coefficient describing the cooperativity of channel opening by cGMP. The value of n_{cG} is 2–3 (29–31). The parameters ν_G and ν_E, characterizing the rates of activation of transducin and PDE, per R^*, respectively, are examined below.

The parameter β_{sub} captures the gain contributed by the PDE and is defined as the rate of change in cytoplasmic concentration of cGMP elicited by a single activated PDE catalytic subunit. This rate constant can be expressed in terms of the Michaelis parameters of the PDE as

$$\beta_{sub} = \frac{k_{sub}/K_m}{N_{Av} V_{cyto} BP_{cG}}, \qquad 3.$$

where k_{sub} represents the average cGMP turnover rate of an active PDE catalytic subunit (PDE*), given by $\frac{1}{2}k_{cat}$, with k_{cat} denoting the catalytic rate of a fully activated PDE holomer (PDE**), and where K_m is the Michaelis constant of the PDE for cGMP. In addition, N_{Av} is Avogadro's number, V_{cyto} is the cytoplasmic volume of the outer segment, and BP_{cG} is the cell's cytoplasmic buffering power for cGMP.

Figure 3   Kinetics of the rod's response to a single photon of light, as a function of time t after delivery of the photon. (*A*) The mean electrical response of a toad rod to a single photon is plotted on a slow time-base (from 32). (*B, C*) Kinetics of activation of the disc-based proteins, rhodopsin, G protein, and phosphodiesterase (PDE) to their excited forms (R^*, G^*, and E^*, respectively), when inactivation reactions are ignored; note the faster time-base. A single photoisomerization activates a single R^*, which triggers activation of the G protein at a constant rate (of $\nu_G \approx 150\ G^*\ s^{-1}$ at room temperature), and which in turn couples to E^* activation at almost as high a rate, ν_E. (*D*) The steadily increasing quantity of activated PDE, $E^*(t)$ leads to hydrolysis of cGMP at an accelerating rate, which in turn causes the fraction $F(t)$ of cGMP-gated channels remaining open to decline according to a Gaussian function of time. The fractional response, $R(t)$ plotted in *A*, is given by the reduction in $F(t)$ from unity, which can be shown to follow parabolic kinetics (*arrow*) at early times. The noisy trace in *D* is redrawn from the part of panel *A* highlighted by the dotted box on the faster time-base and in terms of fractional circulating current.

The LP analysis provides a framework for quantitative comparison of estimates of the gain factors obtained in biochemical experiments, with the overall signal amplification derived from electrophysiological recordings. In other words, the amplification constant of an intact rod provides a benchmark that must be accounted for by the combined effect of the biochemical parameters, as described by Equations 2 and 3. We examine this issue in detail below, after first discussing an example that illustrates the important role of the volume of the cytoplasm, V_{cyto}, in amplification.

Comparative Analysis of the Amplification of Amphibian and Mammalian Rods

Whereas both amphibian and mammalian rods reliably respond to single photons, those of mammals reach a reliable signal about 10 times faster than those of amphibia, a factor that cannot be explained solely by the temperature dependence of chemical reactions, which would have a Q_{10} no higher than 3. Perhaps even more puzzling, a body of electrophysiological evidence has established that the amplification constant is up to 100 times higher in mammalian rods ($A \approx 5{-}10$ s^{-2}) than in amphibian rods ($A \approx 0.1{-}0.2$ s^{-2}) (reviewed in 3 and 14). In the LP analysis this large difference is accounted for principally by the smaller volume of mammalian rods: Thus, for amphibian rod outer segments $V_{cyto} \approx 1$ pl, whereas for mammalian rods $V_{cyto} \approx 20{-}40$ fl, a factor of 25- to 50-fold smaller. Because V_{cyto} appears in the denominator, Equation 3 predicts that the mammalian rod will have a 25- to 50-fold higher value of β_{sub}, owing to its smaller cytoplasmic volume alone; thus, a given number of activated PDE catalytic subunits can much more rapidly alter the cGMP concentration in the smaller volume. The higher temperature of the mammalian rod is also expected to increase ν_E and k_{sub}, providing, together with the cytoplasmic volume ratio, a full accounting of the 50- to 100-fold higher value of A in two cells that use the same molecular machinery. One can readily see that the higher value of A in the mammalian rod endows it with greater response speed: In Equation 1 an n-fold increase in A causes R to reach a criterion level (e.g., 5%) at a speed $\sqrt{n}$-fold faster, so that a 100-fold higher value of A in the small mammalian rod translates into a 10-fold faster attainment of a criterion response amplitude. Thus, improved temporal resolution is an evolutionary advantage of the small packaging of the mammalian rod photoreceptor.

THE RATES OF ACTIVATION OF TRANSDUCIN AND PHOSPHODIESTERASE BY R*: BIOCHEMICAL MEASUREMENTS

Biochemical assays with radiolabeled nucleotides (usually with the nonhydrolyzable GTP analog, GTPγS) provide a straightforward method of measuring the rate of transducin activation per R* (ν_G). Another, less direct, way to estimate

ν_G is to derive it from the measured rate of PDE activation per $R^*(\nu_E)$ based on the assumption that activation of each catalytic subunit of PDE results from the activation of one $G\alpha_t$. The legitimacy of this assumption is discussed below.

There is great variation among published estimates of ν_G and ν_E derived from biochemical assays, with values ranging from 10 to 200 G^* s^{-1} per R^* (reviewed in 14; their Tables 3, 4). Furthermore, concerns have been expressed regarding the legitimacy of using biochemical assays to determine ν_G and ν_E (reviewed in 14; see 32, 33 for recent updates). Among the concerns are the low time resolution of such assays relative to the speed of normal photoresponses and the relatively dilute suspensions of disrupted photo-receptor membranes that are typically used. Dilution often results in the loss of a substantial fraction of transducin into the aqueous phase, and severe disruption could potentially alter the natural kinetics, for example by altering the lateral diffusion of proteins in or at the membrane surface.

We recently undertook an investigation aimed at addressing a number of issues associated with the estimation of ν_G and ν_E using biochemical assays (32). The experiments first determined, for fresh frog rod outer segment suspensions, the conditions that maximized the activation rates; the factors that were studied included the means of permeabilization of the outer segments, as well as the concentration of membranes and the concentrations of GTP and divalent cations. This investigation found that at 22°C and under optimized conditions ν_E and ν_G were on average 120 $G\alpha_t$ (or subunits of activated PDE) s^{-1} per R^*, with values as high as 150 in individual experiments.

THE RATE OF TRANSDUCIN ACTIVATION BY R^*: MEASUREMENTS WITH INFRARED LIGHT SCATTERING

Mechanisms of Light Scattering by Rod Disc Membranes

Monitoring the light-evoked changes in the scattering of infrared light by suspensions of rod outer segment fragments has played an important role in the investigation of photo-transduction. In this section we review some of the findings obtained by this technique and compare them with related findings obtained using other methods.

To understand the basic idea of light scattering, imagine a red laser beam passing through a dilute suspension of milk in a fish tank: Looking down on the tank, one can see the path of the beam through the tank, because of the light that is scattered by the suspended milk droplets in the direction of the observer's eye. Scattering occurs in all directions (though not equally), and an off-axis detector, such as the eye, will monitor the component of light scattered in that direction; in contrast, a detector on the axis of the laser beam will monitor mainly the light that has not been scattered, and this intensity is usually much higher. Typically one measures the fractional change in intensity, $\Delta I/I$, either off-axis, or axially, or both, using a near-infrared source, and in response to stimulation of the rod particle preparation with a brief flash of visible light that isomerizes a known fraction of the rhodopsin.

The intensity of the scattering in any particular direction depends on a number of factors, including the direction of the detector with respect to the illumination beam, the wavelength of the light, the size distribution and material properties of the suspended particles, and of course, the concentration of particles in the solution. Theoretical analysis shows that, for a dilute suspension of particles whose major dimension is not much greater than the wavelength of the measuring light, the principal factor contributing to a change in the scattering intensity in a particular direction is a change in the mass density of the particles relative to the density of the bulk solution (34, 35). Thus, in a suspension of rod discs, any gain of protein mass onto the membranes from the solution will cause the scattering of light to increase, whereas any loss of mass from the membranes to the solution will cause the scattering to decrease.

Scattering by larger and more organized structures, such as entire rod outer segments, is more complicated to analyze theoretically than that by small fragments, as a number of factors other than the electron density of the particles come into play. In the next section we focus on two signals that are obtained from suspensions of discs or small fragments of outer segment.

The Discovery of the Binding and Dissociation Signals for Near Infrared Light Scattering from Rod Disc Suspensions

Near infrared light scattering by rod outer segments was first reported by Hofmann and colleagues (36) and has subsequently been the subject of numerous investigations by many researchers. Kühn et al. (37) presented the first experiments demonstrating effects of transducin activation on light scattering changes. Using a suspension of outer segment fragments and measuring the intensity of light in the forward direction, they found and analyzed two distinct transducin-dependent infrared scattering signals, which they named the "binding" and "dissociation" signals. The binding signal, which they observed in the absence of GTP, was seen as a negative-going deflection in $\Delta I/I$ in the axial direction. When the membranes were stripped of all proteins but rhodopsin and then reconstituted with G_t, the amplitude of the binding signal was found to saturate when the concentration of R^* was very nearly equal to the total concentration of G_t in the reaction volume. Other investigations had shown that, in the absence of GTP, G_t binds very tightly to R^* (38), and therefore Kühn et al. (37) concluded that the binding signal was likely to represent the formation of an R^*-G_t complex with 1:1 stoichiometry. The second transducin-dependent signal reported by Kühn et al. (37) required the presence of GTP and was seen as an increase in $\Delta I/I$ measured axially; i.e., it corresponded to a reduction in scattering and hence presumably a loss of mass from the membrane. Based on prior work showing that $G\alpha_t$ dissociates from the membrane when rhodopsin is activated (38), it was concluded that this signal most likely reflects the rapid dissociation of $G\alpha_t-GTP$ from the membranes into solution, and so it was called the "dissociation" signal.

The Role of Transducin Redistribution Between Soluble and Membrane-Bound Pools in the Binding and Dissociation Signals

Subsequent research has substantiated the explanations put forward by Kühn et al. (37) for the molecular nature of the two scattering signals they observed but has refined it by focusing attention on the critical role of the soluble and membrane-bound fractions of G_t. Arshavsky and coworkers (39) hypothesized that the binding and dissociation signals both reflected the redistribution of transducin between the soluble and membrane fractions. Because it had been established that G_t dissociates from membranes at low ionic strength (38), they manipulated the ionic strength of the medium to vary the amount of G_t bound to membranes previously stripped of proteins and reconstituted with G_t in the dark. They found that the amplitudes of the binding and dissociation signals varied with the quantities of G_t expected to be in the corresponding fractions: Thus, in the low ionic strength medium (in which most G_t is dissociated) the binding signal was large and the dissociation signal small, whereas the converse was true in normal ionic strength solution. Moreover, the absolute difference between the maximal amplitudes of both signals remained unchanged with the manipulation, consistent with the idea that the two signals reflect the light-dependent redistribution of a fixed pool of transducin.

Comprehensive evidence that both the binding and dissociation signals are generated by a redistribution of the mass of G_t between membrane-bound and soluble phases has recently been presented by Heck & Hofmann (33). Under standard experimental conditions (3 μM rhodopsin–containing membranes reconstituted with 0.5 μM G_t), the saturated amplitudes of the binding and dissociation signals were proportional to the quantities of G_t recovered from the membrane and soluble phases and quantified with SDS-PAGE, at an approximately 50%:50% ratio. In addition, a linear relationship was observed between the maximal amplitude of the dissociation signal and the total added G_t over the concentration range of 0.1–1 μM G_t. These results provided the authors with a basis for using the saturating amplitude of the dissociation signal to estimate the fraction of G_t that was initially membrane bound under the experimental conditions.

Further evidence for the hypothesis that G_t mass redistribution underlies the dissociation signal can be obtained from a calculation of the saturating amplitude of the signal. Thus, the first-order prediction for off-axis scattering by a dilute solution of small rod fragments is that $\Delta I/I = 2\Delta M/M$, where M is the total mass of the fragment; the factor 2 arises because the intensity of light scattered by small particles depends on the square of the number of scattering electrons in the particle (35). Thus, given that (*a*) rhodopsin comprises 31% of the dry mass of rod discs (41); (*b*) that the concentration of rhodopsin was 3 μM, and that of the membrane-bound fraction of transducin ($G_{t,mem}$) was 0.25 μM in the experiments of Heck & Hofmann (33); and that (*c*) the molecular masses of rhodopsin and $G\alpha_t$ are both approximately 38 kDa, the saturated amplitude of the dissociation signal is predicted to be

$$\frac{\Delta I}{I} \approx \frac{2\Delta M}{M} \approx 2\frac{0.25\,\mu\text{M} \times 38\,\text{kDa}}{(3\,\mu\text{M} \times 38\,\text{kDa})/0.31} \approx 0.05, \qquad\qquad 4.$$

whereas the observed magnitude was 0.03. (Possible reasons for the modest discrepancy between prediction and observation are that the rod particles may be too large to undergo coherent electronic oscillation, that the mass of water inside the membrane vesicles is not taken into account in this calculation, and that in addition to $G\alpha_t$ some $G\beta\gamma_t$ may dissociate from the membranes.) Experiments by other investigators (e.g., 37) with similar preparations give saturating amplitudes of comparable magnitude, though the fractions of soluble and membrane-bound G_t were not specified.

The kinetics of the binding signal in dilute suspensions of rod fragments reveal that the rate at which R^* reacts with soluble G_t must be far slower than the rate at which it reacts with membrane-bound G_t. It is therefore important to estimate the fraction of G_t that might be membrane bound in intact rods. The dissociation constant for the binding of holo-G_t to rod outer segment membranes is $K_D \approx 1$–$3\,\mu\text{M}$ (33), in the case of reconstituted membranes, and has been estimated to be an order of magnitude lower for native membranes (42); here K_D is expressed in terms of the bulk concentration of rhodopsin in solution, because rhodopsin is present in a fixed ratio to the membrane lipids (43). Hence, in an intact rod, where the rhodopsin concentration is $R = 6000\,\mu\text{M}$ relative to the cytoplasmic volume and where the concentration ratio G_t to rhodopsin is 1:12 (44, 45), it is expected that the membrane-bound fraction of G_t will be at least $R/(R + K_D) = 6000/(6000 + 1) = 99.98\%$; i.e., essentially unity.

Estimation of ν_G with the Dissociation Signal

The unequivocal assignment of the molecular mechanisms underlying the binding and dissociation signals, together with the finding that the fraction of membrane-bound G_t can be determined from the ratio of the maximal amplitudes of the two signals, provided Heck & Hofmann (33) with a rigorous basis for determining ν_G, the rate of activation of G_t per R^*. Using preparations of bovine rod disc membranes reconstituted with purified transducin, they recorded the kinetics of the dissociation signal over a wide range of GTP and GDP concentrations, at a range of temperatures. They then fitted an analytical model of the interactions and determined the single set of parameters that provided the best global fit to the entire ensemble of results, at each temperature. At 22°C the extrapolated maximal rate of G_t activation was $\nu_{G,\text{max}} \approx 590\ G\alpha_t\ \text{s}^{-1}$ per R^*, and the concentration of membrane-bound G_t required to reach the half-maximal rate (i.e., the K_m for G_t) was 3100–3800 molecules per μm^2 of membrane surface (see Reference 33, tables 1 and 2). At 34°C the corresponding parameters were 1300 s^{-1} and 3000 μm^{-2} (Reference 33, table 2).

From Heck & Hofmann's results and analysis one can derive an estimate of the rate ν_G that would apply in vivo. Given that G_t in intact frog and mammalian rods is present at a molar ratio to rhodopsin of 1:12 (44, 45), and taking the rhodopsin density to be 25,000 molecules μm^{-2} (43), the membrane density of transducin in amphibian rods is calculated to be $G_{t,mem} \approx 2100\ \mu\text{m}^{-2}$. Hence, in the presence of a saturating concentration of GTP and in the absence of GDP, the rate ν_G in rods at 22°C is predicted from the analysis of Heck & Hofmann to be

$$\nu_G = \nu_{G,max} \frac{G_{t,mem}}{G_{t,mem} + K_m} = 220\,G\alpha_t\,\text{s}^{-1}\ \text{per R}^*, \qquad 5.$$

where K_m has been set to the average of the two values measured by Heck & Hofmann, 3500 μm^{-2}. The estimate of ν_G derived with Equation 5 for amphibian rods depends on several assumptions; namely, that frog and bovine membranes at 22°C are functionally equivalent, that the GDP concentration is negligible, and that peripheral proteins (which are not present in the reconstituted system) have negligible effect on ν_G. Despite these uncertainties, the value of 220 $G\alpha_t\,\text{s}^{-1}$ per R* derived from the light-scattering measurements using bovine rod membranes (33) is remarkably close to the value of 120–150 $G\alpha_t\,\text{s}^{-1}$ per R* derived from biochemical assays using frog rod membranes (32).

The Release or Amplified Transient Signal Likely Reflects G_t-Phosphodiesterase Interaction

Another distinct light-scattering signal that has been used to estimate ν_G is the "release" signal, first described by Vuong et al. (46). They used a magnetic field to orient fragments of frog rod outer segments, with normally spaced disc stacks but permeabilized plasma membranes. With the rods oriented at 45° to the incident beam and with two detectors positioned at right angles to the incident beam, they measured infrared scattering simultaneously at the two detectors. Physical analysis showed that changes in scattering recorded by the detector at $-90°$ should reflect mass displacement along the rod's axis, whereas the detector at $+90°$ should reflect mass displacement in a radial direction.

Upon flash illumination of the rod suspension, Vuong et al. (46) measured a small but rapid transient increase in scattering in the "axial" signal, together with a larger and slower decrease in scattering in both directions. They attributed the difference between the two signals to release of $G\alpha_t$ from the membrane into the cytoplasm and termed this the release signal, whereas they attributed the common reduction in scattering to the subsequent leakage of $G\alpha_t$ out of the outer segments, and they termed this the "loss" signal. In analyzing the release signal, they measured a fractional rate of change of scattering of $10^4\ \text{s}^{-1}$ per R*. Then, by assuming that the maximal level of this signal corresponded to the release of the total complement of $G\alpha_t$ from the membrane, they multiplied this value by an assumed ratio of G_t to rhodopsin, 1:10, to obtain a transducin activation rate of $\nu_G = 1000\,G\alpha_t\,\text{s}^{-1}$ per R*. In our view, the assumption by Vuong et al. (46), that the maximal release signal corresponds to release of the entire pool of G_t, is questionable. For the reasons discussed below, we instead take the view that this signal is more likely to arise from the interaction between G_t and PDE.

Subsequent to the investigation by Vuong et al., a number of other studies of light scattering from rod preparations with highly organized structure have been carried out, including further experiments with magnetically oriented rods (47, 48) as well as experiments with isolated retinas (49, 50). In those preparations that most nearly retain their natural structure (despite having been permeabilized), the measured off-axis scattering, $\Delta I/I$, exhibits a simple positive-going form; when ATP is present in addition to GTP, the signal recovers to baseline and can be elicited repeatedly, indicating that the cascade can recover completely (reviewed in 51; see 50). Pepperberg et al. (49) called this simplified, positive-going signal the amplified transient or "AT signal," and it has generally been assumed to have the same molecular origin as the release signal described by Vuong et al. (46). Other investigators using the AT signal have also concluded that $\nu_G \approx 1000$ s^{-1} per R^* (51).

Space limitations do not permit a thorough discussion of the substantial literature on the AT signal. However, a number of observations suggest that the AT signal might originate in an interaction between $G\alpha_t$ and PDE. First, to our knowledge, the AT signal has only been recorded in preparations in which both G_t and PDE were present. Second, using a preparation of magnetically oriented rod outer segments, Kamps et al. (52) tested the hypothesis that the presence of PDE is required for the signal, by preactivating the PDE with protamine, prior to delivery of the light flash. The AT signal was not observed in protamine-treated rods, although a normal dissociation signal was observed. Third, Heck & Hofmann (53), investigating rod outer segment membranes stripped of peripheral proteins and reconstituted with purified G_t and PDE, found (*a*) that an AT-like signal was only seen when native PDE was present in the sample prior to flash activation and (*b*) that preactivation of the PDE with $G\alpha_t$-$GTP\gamma S$ removed this signal stoichiometrically, although a normal dissociation signal was still observed (indicating the competence of the G_t that was not preactivated).

The apparent major conflict between the value of ν_G estimated from the AT signal (~ 1000 s^{-1} $G\alpha_t$ s^{-1} per R^*) and the estimates obtained from the dissociation signal and the $GTP\gamma S$ binding assay (220 and 150 s^{-1} $G\alpha_t$ s^{-1} per R^*, respectively) can be resolved if the AT signal actually represents a binding interaction between G_t and PDE. Because the amount of PDE in rod outer segments is about an order of magnitude smaller than the total amount of G_t (54, 55), the binding interaction of these two proteins will saturate when $\sim 1/10$ of the G_t pool is activated. Thus, if we assume that the AT signal and the axial release signal from magnetically oriented rods are essentially the same, and originate from a $G\alpha_t$-PDE interaction, then the estimate of ν_G derived from these experiments should be scaled by $\sim 1/10$ relative to the estimates derived with the assumption that the maximum signal represents total activation of G_t. This rescaling yields $\nu_G \approx 100$ s^{-1} per R^* for the AT signal and release signals, bringing the estimates from the three types of experiments into reasonable agreement. In summary, we think that the most plausible reconciliation of the estimates of ν_G obtained from the AT/release signal with the estimates from other methods is that the AT/release signal represents the activation of only that fraction of the G_t pool that has interacted with PDE. At a mechanistic level, we think that the rapid (though transient) increase in apparent mass at the membrane might result from the binding of $G\alpha_t$ to the PDE.

THE STOICHIOMETRY BETWEEN ACTIVATED TRANSDUCIN AND PHOSPHODIESTERASE SUBUNITS

Another issue impacting the molecular nature of signal amplification in the phototransduction cascade is the stoichiometry of activation of PDE by $G\alpha_t$ under in vivo conditions. Two factors should be considered in this regard: the fraction of the total PDE activity evoked by the binding of a single $G\alpha_t$-GTP molecule, and a proportionality factor arising from the finite time required for a newly produced $G\alpha_t$-GTP to find, bind to, and activate a PDE.

Given that the PDE consists of two nearly identical functional units, each containing one catalytic α or β subunit and one γ subunit (serving both as the inhibitor of nonactivated PDE and the binding site for transducin), a starting hypothesis would be that each PDE unit is activated independently by a $G\alpha_t$ (cf. 56). However, a variety of other hypotheses have been proposed, ranging from the idea that one $G\alpha_t$ activates both catalytic subunits (57) to the idea that the binding of transducin to the first PDEγ results in only 5% of maximal PDE activity (58).

Determination of the coupling ratio from $G\alpha_t$ to PDE, with assays that measure $G\alpha_t$ production and PDE activity, requires a precise determination of the total PDE amount in the reaction mixture. The most convenient preparation for estimating the PDE content of outer segment membranes is frog rods, because frog PDE has two exchangeable, high affinity noncatalytic bindings sites for cGMP (59, 60). Thus, the total amount of the PDE holoenzyme in a frog rod outer segment preparation can be determined as half the maximal radiolabeled cGMP bound to the membrane fraction. Using this approach to quantify PDE, Leskov et al. (32) found that during the initial phase of activation (up to about one third of the total PDE activity), the ratio of PDE catalytic subunits activated per $G_t\alpha$ was unity. Thus, there is neither gain nor loss of signal amplification in the coupling between $G\alpha_t$ and PDE.

Another finding (61) indicates that in frog rod outer segments $G\alpha_t$ interacts preferentially with the γ subunit associated with one of the PDE catalytic units. Although this would suggest that, under well-stirred conditions, PDE activity would only be driven beyond 50% maximal by a large excess of $G\alpha_t$, it is important to note that the situation in the rod is far from well stirred and that $G\alpha_t$ is produced at high concentration at the location of a single R*. Thus, even though only 100 or so molecules of $G\alpha_t$ may be produced by the time-to-peak of the single-photon response (100–200 ms in a mammalian rod), they will have been at such a locally high concentration that they may well have bound doubly to molecules of PDE in the vicinity of the isomerization.

The delay in interaction between a newly formed $G\alpha_t$ and a PDE can be quantified in terms of the proportion of those $G\alpha_t$s that have been activated and have bound to PDE at any given time relative to the total $G\alpha_t$s that have been activated. This proportion has been expressed as the coupling efficiency c_{GE} in Equation 2, and factors affecting its magnitude have been analyzed by Lamb & Pugh (28) and Lamb (62). Theoretical analysis indicates that this proportionality factor should be quite close to unity, for the measured protein densities and diffusion coefficients, in combination with the measured rate ν_G of transducin activation.

THE KINETIC PARAMETERS OF TRANSDUCIN-ACTIVATED PHOSPHODIESTERASE

Phosphodiesterase is a Nearly Perfect Effector Enzyme

From the measured values of the amplification constant and the rate of transducin activation in amphibian rods, it is possible to use Equations 2 and 3 to obtain a lower limit for the catalytic efficacy of PDE. In Equation 2 we can substitute the values discussed previously, $A \approx 0.1$–0.2 s^{-2} and $\nu_E (= \nu_G c_{GE}) \approx 120$ s^{-1}, together with the accepted cooperativity of channel activation of $n_{cG} \approx 2$–3, to obtain a value of $\beta_{sub} \approx 4 \times 10^{-4}$ s^{-1}. This parameter, β_{sub}, represents the rate constant at which cGMP is hydrolyzed in the intact outer segment by a single activated hydrolytic subunit of PDE and is related to the underlying physical parameters according to Equation 3. Substituting into Equation 3, with a cytoplasmic volume of $V_{cyto} \approx 1$ pl for a frog rod, we obtain a lower bound for the catalytic efficacy of the PDE of $k_{sub}/K_m \geq 2 \times 10^8$ M s^{-1}. For this "$\geq$" relation, the equality applies in the case that there is no buffering of cGMP in the outer segment ($BP_{cG} = 1$), whereas the required value of k_{sub}/K_m must be even higher than this if buffering of cGMP does occur. This value places PDE among the handful of most efficient enzymes known, for which k_{cat}/K_m exceeds 10^8 M^{-1} s^{-1} (63), and qualifies PDE as a nearly perfect effector in fulfilling its function of maximally amplifying the signal during the photoreceptor response to light.

Turnover Rate and Michaelis Constant of Phosphodiesterase

It has been established for many years that the turnover rate of the fully activated PDE** holomer (k_{cat}) exceeds 4000 s^{-1} (3), corresponding to a subunit turnover number (for PDE*) of $k_{sub} \geq 2000$ s^{-1}. On the other hand, the estimated value for the enzyme's Michaelis constant, K_m, has recently been substantially revised. Over the years, the estimates of K_m reported by different investigators for light-activated PDE have varied by almost two orders of magnitude, with many values well into the millimolar range and none below 80 μM (Table V in 14). In 1994 Dumke et al. (55) proposed that this wide variation might result from preparations exhibiting widely different degrees of preservation of the original disc-stacking structure, in conjunction with the phenomenon of cGMP "diffusion with hydrolysis." The idea was that when the stacking of discs resembled that in intact outer segments, and at the same time a great deal of PDE activity was stimulated, then substantial gradients of cGMP concentration would arise along the interdisc diffusional paths. As a result, activated PDE molecules located near the center of the disc stacks would be exposed to much lower concentrations of cGMP than the level set in the bulk solution by the experimenter, and therefore the K_m would be greatly overestimated. Consistent with this analysis, it was found that the K_m measured in suspensions of large rod outer segment fragments was about six times higher than the value observed with severely disrupted membrane fragments. Dumke et al. proposed that the lowest observed value, $K_m \approx 100 \ \mu M$, measured with the most severely disrupted preparations represented the true Michaelis constant of the transducin-activated enzyme.

Subsequently it was discovered that even this value represented a substantial over-estimate (32), because diffusion with hydrolysis was found to occur even in the most severely disrupted preparations of frog rod outer segment membranes. To reduce the effects of this phenomenon, Leskov et al. (32) chose to activate only a small proportion of the PDE, which they accomplished by activating only small amounts of $G\alpha_t$ by using very low concentrations of GTPγS. At very low levels of PDE activation, engaging only 1–2% of the total PDE, the measured K_m stabilized at 10 μM cGMP, which they concluded to be the true K_m of transducin-activated PDE. Importantly, the measurements of the PDE activation rate of $\sim$120–150 s^{-1} obtained in the same study indicate that only 1–2% of the total PDE is activated within an individual interdiscal space during a single-photon response. This means that during the single-photon response the PDE will indeed operate at its true K_m of $\sim$10 μM.

This result reconciled the estimates of the K_m for PDE activated by trypsin proteolysis of the inhibitory γ subunits that is also accompanied by the solubilization of activated PDE from the surface of rod disc membranes to solution (see Table V in 14; see 64 for a more recent update). The previous discrepancy resulted not from any fundamental difference in catalytic properties of PDE when activated by different means, but from what might be described as a geometrical factor in the membrane stack that led to an artificially high value of apparent K_m.

Reconciliation of Biochemical and Electrophysiological Measurements

With the results and insights described above, it is possible to present a unified set of parameters that accounts both for the biochemical and the electrophysiological measurements in the literature. Thus, if we take $k_{sub} = 2200$ s^{-1} and $K_m = 10 \ \mu M$ for PDE, and a cytoplasmic volume of $V_{cyto} = 0.85$ pl (based on frog or toad rod outer segments dimensions of $6 \times 60 \ \mu m$ and the cytoplasm occupying 50% of the envelope volume), and if we assume that $BP_{cG} = 1$ (implying the absence of cGMP buffering), substitution in Equation 3 yields $\beta_{sub} = 4.3 \times 10^{-4}$ s^{-1}. Then, with the mean value of ν_E obtained in biochemical experiments of 120 E* s^{-1} per R* and assuming the most conservative value of the channel Hill coefficient, $n_{cG} = 2$, the overall amplification constant is predicted by Equation 1 to be $A = 0.10$ s^{-2}.

Furthermore, substitution of the best numbers obtained in the biochemical assays (32) or the even higher value of $\nu_G \approx 220 \text{ s}^{-1}$ derived above from analysis of the light-scattering signals (33), together with a channel cooperativity of $n_{cG} = 3$, yields values for A that exceed $A = 0.3 \text{ s}^{-2}$ and therefore allows for the possibility that $BP_{cG} > 1$ (cf. 60).

POSSIBLE CONTRIBUTIONS OF THE G PROTEIN CASCADE TO ADAPTATIONAL CHANGES

The Amplification of Transduction is Unaltered in the Short Term by Exposure to Backgrounds of Moderate Intensity

It has long been known that the sensitivity of the visual system decreases in the presence of steady background illumination, a phenomenon known as light adaptation. In the rod and cone photoreceptors light adaptation is characterized both by desensitization, measured as a reduction in the peak of the incremental response to a dim flash, and by acceleration of the response, manifested by a shortening of the time-to-peak and a more rapid final recovery to the baseline level (e.g., 65).

Although it has been clearly established that photoreceptor light adaptation is mediated to a substantial degree by a light-induced reduction in calcium concentration (66–68), it is important to emphasize that the reduction in sensitivity is not actually caused by the lowered calcium concentration (69, 70). Indeed, the reduced calcium concentration does just the opposite: It rescues the photoreceptor from the massive reduction in sensitivity that would otherwise occur. Such saturation would inevitably accompany the complete closure of channels that would be induced by even quite dim backgrounds, were it not for the occurrence of adaptational changes. The lowered calcium concentration prevents saturation and thereby raises the sensitivity from the very low level that occurs when calcium concentration is "clamped."

As described above, the amplification of transduction can be quantified by measuring the early parabolic rise of the response to a dim flash. Three studies on isolated amphibian rods have reported that during background illumination the amplification constant A is unaltered (70–72), though two other studies have reported considerable reduction (73, 74). The most extensive of these studies (70) found that the reactions that mediate recovery cause deviation of the response from its initial trajectory at much earlier times than previously thought (less than 100 ms for moderate backgrounds), making it essential to restrict analysis to the earliest region of the rising phase. Analysis of the fractional response at these very early times showed the initial rise to be invariant with adaptational state, for backgrounds suppressing up to 75% of the circulating current and applied for periods of several minutes; thus, there was no evidence of any reduction in the amplification constant. Accordingly, the simplest interpretation is that neither the rate of G protein activation, nor any of the other factors that combine to form the amplification constant (Equations 2, 3), are altered during short-term adaptation at moderate intensities.

However, there are no grounds for extending this conclusion beyond the adaptational conditions tested. In particular, it seems possible that exposure to saturating intensities, or to extended durations of illumination (as in the diurnal cycle), might elicit changes in amplification.

Long-Term Changes in Amplification Might be Mediated by Changes in Protein Concentration

Consideration of Equations 2 and 3 indicates that changes in a multitude of different parameters could in principle modulate the gain of phototransduction. One class of possibility that we now consider is whether the rate ν_G of G protein activation might be modulated by alteration of the level of G_t in the disc membranes, either by changes in bulk concentration or by regulation of the competence of the G_t (e.g., by phosducin binding).

An initial question is whether alterations in the concentration of G_t would be expected to alter the rate ν_G of G_t activation or whether transduction operates under conditions in which this rate is saturated. Three lines of evidence have recently emerged in support of the first of these possibilities, that ν_G does depend on the level of G_t.

First, Heck & Hofmann (33) systematically varied the concentration of G_t in their preparation and observed a strong dependence of the rate of activation upon the level of G_t over the physiological range, with a K_m of 3100–3800 molecules of the membrane-bound G_t per μm^2 of membrane. Because the actual level of G_t in amphibian and rodent rods is considerably less than this [$\sim$2000–2500 μm^{-2}, calculated from a G_t:rhodopsin ratio of $1/12$ (44, 45)], the rate ν_G would be expected to depend strongly on the actual amount of G_t present. Second, it has been found that elevation of the level of G_t in suspension of frog rod outer segments above its normal content, elicited by the addition of purified frog transducin, leads to an increase in the activation rate ν_G assayed by GTPγS binding (V. A. Klenchin & M. D. Bownds, unpublished data). Third, it has recently been found in mouse rods that a genetic manipulation that is presumed to increase the frequency of interaction between R* and G_t leads to an increase in amplification constant, A (75). In these rods, which were hemizygous for rhodopsin, the quantity of rhodopsin in the membrane was approximately halved from normal, whereas the level of transducin was unaltered. The observed doubling of A was interpreted to be caused by an increased rate of lateral diffusion of proteins at the disc membrane surface. This result suggests that the rate ν_G in vivo is dependent on the rate at which R* contacts molecules of G_t, and it is expected that this rate will depend on the concentration of G_t in the membrane.

Given that ν_G is affected by the concentration of G_t, two scenarios need to be investigated: whether the effective concentration of G_t is modulated by the binding of other proteins and whether changes in bulk concentration of G_t in the outer segment occur.

PHOSDUCIN   It has been suggested that the effective concentration of G_t might be modulated by interaction with phosducin—a soluble phosphoprotein that complexes tightly with $G_t\beta\gamma$ (76–79). The presumption is that in this bound state G_t would not be competent to be activated by R^*. It has been shown that the unphosphorylated form of phosducin binds to $G_t\beta\gamma$ with much greater affinity than does the phosphorylated form (80–81), and it has therefore been hypothesized that light-induced dephosphorylation of phosducin could mediate a reduction in the gain of transduction, by reducing the concentration of competent G_t. However, a result obtained in three recent studies argues against this possibility. Measurements of the distribution of phosducin throughout the subcellular compartments of rod photoreceptors indicate that the great majority of phosducin is present in the inner segment and that the amount of phosducin in the rod outer segments is less than 10% of the total amount of transducin (82–84). On this basis it would seem that the amount of phosducin in the outer segment would be insufficient to influence the effective level of G_t.

REDISTRIBUTION OF TRANSDUCIN   Another possible mechanism for gain modulation would be a redistribution of transducin between the inner and outer segments. Although such a redistribution of G_t was originally reported in the late 1980s, the method was questioned at the time, but new results confirm the original findings. A redistribution of G_t was reported by four groups (85–87a) when immunohistochemical techniques were applied to rodent retinas prepared under different adaptational conditions, though Roof & Heth (88) argued that these results might have been compromised by an artifact: light-dependent masking of the antibody recognition epitopes. However, a novel technique combining tangential microdissection of the flat-mounted retina with Western blot analysis of protein in the sections has recently confirmed that major movements of transducin do indeed occur in rat rods (89), with a large proportion of G_t translocated into the inner segment during extended light exposure and translocated back to the outer segment during dark adaptation. In those experiments the intensity of light exposure that was used caused closure of all the cGMP-gated channels (i.e., response saturation), so it could not be determined whether the amplification constant had changed during the light exposure. However, experiments in progress (M. Sokolov, A. L. Lyubarsky, K. J. Strissel, A. Savchenko, V. I. Govardovskii, E. N. Pugh, Jr. and V. Y. Arshavsky, submitted) have shown that during dark adaptation there is a window of time during which the rods have recovered sufficiently from saturation for meaningful electrophysiological measurements to be made. The reduction in the rod outer segment levels of G_t observed in those experiments was accompanied by a reduction in the amplification constant measured in the same animals.

GARP   Yet another possible mechanism for regulation of the gain of transduction involves proteins called GARPs (glutamic acid–rich proteins), discovered by Sugimoto et al. (90). At least one of these proteins, GARP2, binds to PDE, thereby preventing its activation, under strongly light-adapted conditions (91). However, electrophysiological measurements to test for alteration of gain have not been conducted yet, so this role of GARP2 is entirely hypothetical.

In closing this section, we draw attention to two points. The first is that the hypothesized mechanisms for the modulation of gain in the cascade may conceivably serve roles not in light adaptation per se, but in protection of the cell's metabolism. In the continued presence of illumination that leads to closure of all the cGMP-activated channels and hence the continued absence of signaling, the rod has little to gain (and perhaps much to lose) by maintaining elevated turnover of substrate in each of the contributing steps. The second point is that the topic of light adaptation is in fact a very complex one, and here we have examined only those phenomena relating to activation of the light response. Most of the mechanisms thought to influence light adaptation under physiological conditions (i.e., when some cGMP channels remain open) instead contribute to the shut-off (or recovery) of the response and are not addressed here. For recent reviews of light adaptation, see (25, 26, 69).

TRANSDUCIN INACTIVATION IN PHOTORECEPTORS IS TIGHTLY CONTROLLED BY SEVERAL REGULATORY PROTEINS

The shape of the rod's response to a single photon embodies an evolutionary compromise between the mutually conflicting factors of speed and sensitivity. Assuming the amplifying mechanisms underlying the activation phase of the response are optimized, the sensitivity, measured by the peak amplitude of the single-photon response, is determined primarily by the lifetimes of the three primary intermediates of the cascade: the lifetime of R^*, the lifetime of $E^*(=G\alpha_t\text{-PDE})$, and the lifetime of cGMP. (The lifetime of cGMP is the reciprocal of the rate constant of steady cGMP hydrolysis.) Because the inactivation of E^* occurs when the terminal phosphate of $G\alpha_t\text{-GTP}$ is hydrolyzed, the GTPase rate of transducin is a determinant of both the time course and the sensitivity of the light response. The lower the GTPase rate, the longer the duration of PDE activity, the larger the response amplitude, and the greater the sensitivity. But such sensitivity comes at the price of a slower response and poorer time resolution.

Rods and cones, the two classes of photoreceptor in the vertebrate retina, represent different compromises between speed and sensitivity: In a given species the rods are more sensitive than the cones, in part because they inactivate more slowly. For example, in an amphibian retina the activation stages of transduction appear to have comparable amplification (as defined by Equations 1–3), and the 10- to 20-fold differences in sensitivity arise in large part from the differences in time-to-peak of the response to a dim flash, which may be around 100 ms for a cone but around 1 s for a rod (at 22°C) (3). Differences in rod and cone response timing reflect differences in evolutionary pressures for speed vs. sensitivity under nighttime and daylight conditions.

Not surprisingly for its critical role in response timing, the rate of transducin GTPase in photoreceptors is tightly regulated by several phototransduction proteins. Working together, a complex of proteins optimize the lifetime of activated transducin to fit the needs of photoreceptors from different species and individual rod and cone photoreceptors of the same retina. In this section we first introduce the proteins known to be involved in regulating the GTPase activity of transducin. We then discuss the roles of these proteins and the contributions of their major structural domains.

Discovery of Transducin GTPase–Regulating Proteins

The early studies of transducin GTPase reaction revealed a paradox: The rate of GTP hydrolysis measured with purified transducin was $\sim$100 times slower than the apparent time constant with which rods recover from flashes (compare, e.g., 92 and 93 with 4 and 5). The resolution of this paradox began when experimental measurements of transducin's GTPase activity were made under the more physiological conditions of either concentrated suspensions of rod outer segments (94, 95) or of rod outer segments with highly preserved disc stack structure (96). These studies found a GTPase rate much higher than had been observed in cell-free systems reconstituted with purified proteins. Subsequent studies revealed that photoreceptors contain a molecular mechanism capable of activating the relatively slow intrinsic GTPase activity of transducin by at least two orders of magnitude. Because many aspects of the regulation of transducin's GTPase activity have been described in detail in two excellent recent reviews (97, 98), we provide only a brief historical outlook and then consider a few of the most recent developments in the field.

The first photoreceptor protein shown to act as a GTPase activating protein (GAP) for transducin was the immediate target of $G\alpha_t$-GTP, the inhibitory γ subunit of PDE (PDEγ). Arshavsky & Bownds (99) found that addition of either purified PDE or recombinant PDEγ to photoreceptor membranes (containing G_t but lacking most other soluble and peripherally bound membrane proteins, including endogenous PDE) caused a several-fold stimulation of transducin's GTPase activity. In the same year Berstein and colleagues (100) showed another G protein effector, the $\beta1$ isoform of phospholipase C (PLC$\beta1$), to be capable of stimulating the GTPase activity of the corresponding G proteins, $G_{q/11}$. Together, these observations raised the possibility that regulation of the lifetime of activated G proteins by their effectors might represent a general principle of operation (101). However, PDEγ is now known not to be the primary GAP for transducin, and the only two conventional effectors known to confer GAP activity on their G proteins are PLC$\beta1$ and type 5 adenylate cyclase (102).

A much more ubiquitous mechanism of regulating the GTPase activity of G proteins became evident with the discovery of a family of GAPs called regulators of G protein signaling, or RGS proteins (see 97, 103–105 for recent reviews of RGS protein classification and properties). All members of the RGS family examined so far have been shown to serve as GAPs for a broad range of G protein α subunits, acting allosterically by stabilizing the transition conformation of the Gα subunit that is most favorable for hydrolysis of GTP.

In the field of phototransduction, evidence that photoreceptors contain a GAP distinct from PDEγ came from experiments in which PDEγ failed to activate transducin's GTPase activity in the absence of photoreceptor membranes (106), experiments in which photoreceptor membranes with PDE removed retained their ability to activate GTPase (107) and experiments showing the GAP effect of PDEγ to be higher with larger concentrations of photoreceptor membranes in the reaction mixture (108–110). Taken together, these results indicated the role of PDEγ in this regard to be one of potentiating the GAP activity of another membrane-associated protein, rather than being the sole GTPase regulator itself.

Wensel and colleagues made a major breakthrough in identifying this membrane-associated GAP, with their discovery that the short splice variant of the ninth member of the RGS family (RGS9) is the photoreceptor-specific GAP responsible for stimulating transducin's GTPase activity (111, 112). Shortly afterwards it was shown that RGS9 exists in photoreceptors as a constitutive complex with the long splice variant type 5 G protein β subunit (Gβ5L) (113). The expression of RGS9 and Gβ5L in photoreceptors occurs under strict reciprocal control. Mice lacking the RGS9 gene do not possess functional Gβ5L in their rods, in spite of the presence of the corresponding mRNA at normal level (114). Likewise, successful expression of RGS9 in both Sf9 cells and in vivo, in the rod and cone photoreceptors of transgenic *Xenopus laevis* tadpoles and in murine cones, requires coexpression of Gβ5 (115, 116).

Recent studies indicate that RGS9-Gβ5L belongs to a subfamily of RGS proteins, including mammalian RGS6, RGS7 and RGS11, whose members share a common structural domain composition, are found predominantly in the nervous system, and appear to exist as complexes with Gβ5 in vivo (117–121). RGS9 from photoreceptors is the only member found to complex with Gβ5L; other members complex with a short splice variant of Gβ5, Gβ5S (117–122). As illustrated in Figure 4, the binding of Gβ5 occurs via a G protein γ subunit–like domain, or GGL, located next to the RGS homology domain in the sequence of all RGS9 subfamily members (123). For further information on the structural and functional aspects of this subfamily we refer the reader to two recent reviews (98, 122).

RGS9 is Necessary for Normal Inactivation of the Phototransduction Cascade in Rods and Cones

Experiments with mice that have the RGS9 gene knocked out have established that this protein is essential for normal inactivation of the cascade in vivo in both rods and cones (114, 116). In single mouse rods the dominant or rate-limiting time constant of recovery of the flash response increases from 0.2 s to about 9 s, a 45-fold increase in the absence of RGS9 (114). Cone-driven responses to strong flashes measured with electroretinographic methods show an ~60-fold increase in their half-time of recovery (116). Combined with histochemical results that show

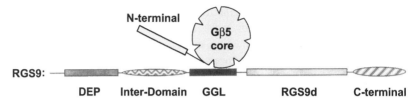

Figure 4   Domain composition of the RGS9-Gβ5L complex. RGS9 contains five distinct structural regions: the ~110–amino acid N-terminal domain called DEP because it is also present in Disheveled, EGL-10 and Pleckstrin (136); the ~80–amino acid sequence (interdomain) with the lowest degree of homology among the members of RGS9 subfamily; the ~80–amino acid G protein γ subunit–like domain (GGL); the RGS homology domain (RGS9d) of ~120 residues; the C-terminal extension of ~55 amino acids, unique among other members of this subfamily. Gβ5L is thought to have a core seven-bladed β-propeller structural domain characteristic of all G protein β subunits and mostly the α helical N-terminal region, including a photoreceptor-specific 42–amino acid extension that is absent in Gβ5S and in other known G protein β subunits (137).

expression in both types of photoreceptors (112), these electrophysiological data unequivocally establish the necessity of RGS9 for normal cascade inactivation.

Cooperation Between RGS9-Gβ5L and PDEγ is Required for Timely Transducin Inactivation In Vivo

The identification of three different proteins (RGS9, Gβ5L, and PDEγ) involved in the regulation of transducin GTPase has raised the issue of whether these proteins perform their GAP function as a coordinated ensemble or simply provide redundancy in an important regulation. Consider first the mechanism of PDEγ action. In principle, there are two possibilities. First, PDEγ could directly contribute to GTP hydrolysis, for example, by further stabilizing the Gα_t transition state beyond the action of the RGS domain. Second, PDEγ could act by increasing the affinity between activated Gα_t and RGS9. Kinetic analysis of the ~15–30-fold PDEγ potentiation of the GAP activity of the native RGS9-Gβ5L complex from bovine rod outer segment membranes has shown that this potentiation consists entirely of an equal magnitude increase in the affinity between RGS9-Gβ5L and the Gα_t-PDEγ complex as compared with activated Gα_t alone (124). The same study revealed another remarkable feature of the native RGS9-Gβ5L complex. At 22°C and saturating Gα_t concentration, RGS9-Gβ5L is able to stimulate transducin GTPase activity to a rate of ~100 turnovers per second, making RGS9-Gβ5L the most efficient GAP known for a heterotrimeric G protein. This rate is sufficiently high to account for photoresponse turnoff in the most rapid photoreceptors, such as human cones, which are capable of resolving the oscillations of light flickering at frequencies of 60 Hz.

The observations that RGS9-Gβ5L plays the central role in stimulating transducin GTPase activity and that PDEγ acts by increasing the affinity between Gα_t and RGS9-Gβ5L raise the hypothesis that PDEγ plays no essential role in regulating the rate of GTP hydrolysis in physiologically intact rods. Thus, in intact rods both Gα_t and RGS9-Gβ5L are present at concentrations higher than in most in vitro experiments, so that their affinity may be sufficient for timely transducin inactivation without any additional impact from PDEγ. A definitive rejection of this hypothesis was provided by the analysis of the responses of the rods of a transgenic mouse whose PDEγ was substituted with a mutant PDEγ containing a single amino acid substitution at position 70 (W70A) (45). The mutation of this residue had been previously shown to reduce the binding affinity between transducin PDEγ (125, 126) and to abolish the ability of PDEγ to activate transducin GTPase in the presence of the RGS9-containing photoreceptor membranes (126). The impaired Gα_t-PDE interactions in the W70A mice resulted in a greatly lowered amplification of the rod photoresponses and slowed kinetics of the response recovery. The lowered amplification was expected from the greatly weakened binding affinity of Gα_t for W70A PDEγ. The slowed recovery is particularly important for this discussion, because it establishes that the binding interaction between RGS9-Gβ5L and wild type PDEγ is essential for a normal rate of inactivation of Gα_t in vivo. The time course of the photoresponse recovery in W70A rods assessed from an exponential fitted to the final response decline ($\tau \approx 1$ s) was about sevenfold slower than that fitting the responses of rods of wild type mice ($\tau \approx 0.15$ s). The factor 7 provides a rough estimate of the degree to which PDEγ activates transducin GTPase in intact rods. Interestingly, rods from knockout mice completely lacking RGS9 recover from responses of similar amplitude only ~2.5-fold slower ($\tau \approx 2.5$ s) than rods from W70A mice (114). Taken together, the data from these two studies argue that PDEγ is essential for the normal regulation of transducin GTPase in vivo and that the relative impact of PDEγ in activating GTP hydrolysis is on the same order of magnitude as the impact of RGS9.

Physiological Utility of the Dual Regulation of Transducin GTPase Activity by an RGS Protein and its Target Enzyme

The physiological utility of a dual regulation of transducin GTPase activity by an RGS protein and its target enzyme can be understood from the following consideration. When a molecule of transducin is activated by an R* it needs to accomplish two competing goals. First, it must transduce the signal from the R* to PDE with high efficiency, i.e., without losing any signal before PDE is activated. Second, it has to inactivate rapidly, within 200 ms in mammalian rods. If Gα_t were to be inactivated by RGS9-Gβ5L before it complexed with PDE, some Gα_t molecules would never activate PDE, and signal amplification would accordingly be smaller. Thus, the dependence of GTPase activation on Gα_t association with PDEγ ensures high efficiency of signal transmission between Gα_t and PDE, because the hydrolysis of GTP does not occur without the binding of Gα_t to PDE. At the same time, the combination of the high affinity of RGS9-Gβ5L for Gα_t-PDEγ together with the high GTPase activity of the complex, leads to rapid termination of the signal. Achieving both efficient transmission and adequate time resolution is a general problem in all signal transduction pathways utilizing G protein α subunits as messenger molecules. In this sense, photoreceptors provide an instructive example of how this problem can be solved at the molecular level by a coordinated action of an RGS protein and a G protein effector.

Targeting Specificity of RGS9-Gβ5L: An Interplay of Contributions from the Catalytic and Noncatalytic Domains of RGS9-Gβ5L

Another way to view the preferential ability of RGS9-Gβ5L to interact with the Gα_t-PDEγ complex as compared with free activated Gα_t is to consider it as an example of specificity in RGS protein action. The problem of RGS specificity has been among the most intensively studied topics in G protein signaling since RGS proteins have been discovered. It is now recognized that the catalytic domains of most RGS proteins accelerate GTPase activity of multiple G protein α subunits and usually show only limited selectivity toward individual Gα subunits (reviewed in 97, 98). Consistent with these observations, the interacting surfaces between RGS and G proteins are among the most conserved regions present in proteins of both families (127, 128).

The promiscuity of the RGS catalytic domains suggests that the large degree of specificity in the interactions between RGS and G proteins is likely to be achieved through a variety of mechanisms including specific expression of individual RGS proteins in appropriate cell types, their precise targeting to specific subcellular compartments, posttranslational modifications, and the action of additional domains and subunits with which many RGS proteins are equipped (reviewed in 97, 98, 104, 105, 122). The ability of RGS9-Gβ5L to distinguish between free activated Gα_t, and Gα_t bound to PDEγ is a particularly interesting example of RGS protein specificity because it cannot be achieved by simple compartmentalization, i.e., localization of a correct RGS–G protein pair in a particular cellular compartment. Free Gα_t and Gα_t-PDEγ must coexist during the light response, so RGS9-Gβ5L needs to be able to discriminate the two species, interacting predominantly with the latter to ensure signal transmission efficiency.

RGS9-Gβ5L is unique among all tested RGS proteins for its ability to cooperate with PDEγ upon stimulating transducin GTPase activity. The GAP activity of other RGS proteins and their recombinant catalytic domains is inhibited by PDEγ (115, 129–132). We now discuss how this property of RGS9-Gβ5L results from contributions of both the catalytic domain of RGS9 (RGS9d) and other structures within the RGS9-Gβ5L complex.

The fact that RGS9d itself is able to positively cooperate with PDEγ upon stimulating the GTPase activity of transducin was first reported by Wensel and colleagues (111) and confirmed later by other investigators (111, 133, 134). More

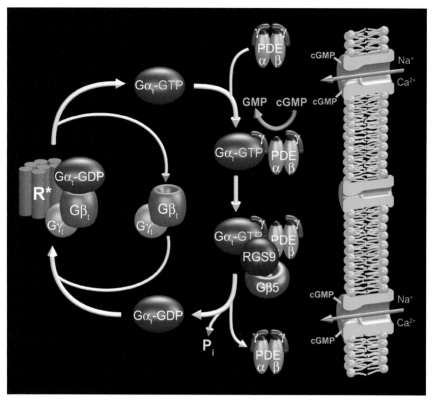

Figure 2 The cycle of G protein activation and inactivation in phototransduction. See text for details.

recently, Sowa et al. (132, 135) provided important structural insights to the effector regulation of RGS9d activity. They conducted site-specific mutagenesis of the RGS homology domain of RGS7, which is the closest relative of RGS9 but whose GAP activity toward transducin is inhibited by PDEγ, and found that as little as three amino acid residues, corresponding to L353, R360, and G367 of RGS9, determine whether PDEγ inhibits or activates the GAP activity of RGS7 (132). They further found that the substitution of the first two residues of RGS7 by the corresponding residues of RGS9 results in a complete reversal of the inhibitory effect of PDEγ into a stimulatory effect.

These results are consistent with an earlier report that the residues required for positive cooperation with PDEγ reside within the $\alpha 3-\alpha 5$ region of RGS9d (133). They are also consistent with information obtained from the crystal structure of RGS9d in complex with G$\alpha_t \cdot$ GDP $\cdot$ AlF$_4^-$ and the PDEγ fragment responsible for regulating transducin GTPase activity (128). In this structure PDEγ and RGS9d are located very close to one another, make extensive interactions with the same region of transducin (switch II), and form one direct contact between V66 of PDEγ and W362 of RGS9. Yet neither one of the three RGS9 residues that determine the direction of the PDEγ effect interacts with PDEγ directly, indicating that the mechanism of their action must be allosteric (132) and remains to be elucidated.

Although RGS9d itself is able to positively cooperate with PDEγ upon stimulating transducin GTPase activity, the degree of this cooperation is small and has not been found to exceed threefold (111, 132–134). This suggests that most of the $\sim$20-fold stimulatory effect of PDEγ on the activity of native RGS9-Gβ5L (124) originates from the action of the structural domains within the RGS9-Gβ5L complex other than RGS9d (Figure 4). Indeed, He et al. (115) and Skiba et al. (139) reported that essentially all structures within RGS9-Gβ5L modulate its catalytic properties and/or enhance its cooperation with PDEγ. The latter study argues that the noncatalytic domains of RGS9-Gβ5L act by modulating its affinities for free Gα_t and the Gα_t-PDEγ complex. The structure, including the seven-bladed β-propeller core of Gβ5 and the GGL domain of RGS9, reduces the affinity between RGS9 and Gα_t, whether Gα_t is present in its free GTP-bound form or is complexed with PDEγ. Several other domains, including DEP and the C-terminus of RGS9 and the N-terminus of Gβ5L, increase the RGS9 affinity for Gα_t-GTP-PDEγ but not for free Gα_t-GTP, thus counteracting the affinity reduction (or inhibition) imposed by GGL-Gβ5. The overall effect of these two affinity shifts of opposite direction is manifested in the physiologically large difference in the RGS9-Gβ5L affinities for free Gα_t-GTP and Gα_t-GTP-PDEγ. What remains to be determined is whether the noncatalytic domains act allosterically by modifying the binding properties of RGS9d or act directly by forming their own contacts with Gα_t and/or PDEγ.

This mechanism by which the noncatalytic domains of RGS9-Gβ5L modulate the affinities of this complex for correct and incorrect targets suggests a general principle by which targeting specificity may be achieved by the members of the RGS9 subfamily. The Gβ5-GGL module may serve as an inhibitor of their GAP activity by reducing the affinity of highly promiscuous RGS homology domains to G protein α subunits (cf. 118). Other domains, for instance DEP present in all RGS9 subfamily members, may restore the affinity specifically toward an appropriate target of each individual RGS protein.

CONCLUDING REMARKS

Phototransduction in vertebrate rods stands as a prototypical example of the operation of a G protein cascade of signal transduction. We understand the molecular steps contributing to activation in great detail and we can account quantitatively for the immense gain of the rising phase of the response to light. In addition, we now appreciate the molecular nature of many of the reactions involved in terminating the light response, though the quantitative details of the interactions are less clear than in the case of activation. Of particular importance to the photoreceptor in achieving both a high gain and a rapid recovery is transducin's ability to resist shut-off until after it has interacted with its effector enzyme, the PDE; central to this ability is the role played by RGS9-Gβ5L. It appears likely that the subtleties of the molecular interactions contributing to the remarkable abilities of G proteins to produce both highly amplified and well-timed signals will be unraveled in the next few years.

ACKNOWLEDGMENTS

We thank Johnathan Hopp for preparing Figure 2. VYA is supported by NIH grants EY10336 and EY12859. ENP is supported by NIH grant EY02660. TDL is supported by Wellcome Trust grant 034792. VYA and ENP are recipients of Jules and Doris Stein Professorships from Research to Prevent Blindness Inc.

Visit the Annual Reviews home page at www.AnnualReviews.org

LITERATURE CITED

1. Koutalos Y, Ebrey TG. 1986. Recent progress in vertebrate photoreception. *Photochem. Photobiol.* 44:809–17
2. Yokoyama S, Yokoyama R. 2000. Comparative molecular biology of visual pigments. See Ref. 138, pp. 257–96
3. Pugh EN Jr, Lamb TD. Phototransduction in vertebrate rods and cones: molecular mechanisms of amplification, recovery and light adaptation. See Ref. 138, pp. 183–255
4. Baylor DA, Lamb TD, Yau K-W. 1979.

Responses of retinal rods to single photons. *J. Physiol.* 288:613–34
5. Baylor DA, Nunn BJ, Schnapf JL. 1984. The photocurrent, noise and spectral sensitivity of rods of the monkey Macaca fascicularis. *J. Physiol.* 357:575–607
6. Rieke F, Baylor DA. 1998. Origin of reproducibility in the responses of retinal rods to single photons. *Biophys. J.* 75:1836–57
7. Whitlock GG, Lamb TD. 1999. Variability in the time course of single photon

responses from toad rods: termination of rhodopsin's activity. *Neuron* 23:337–51

8. Chabre M, Deterre P. 1989. Molecular mechanism of visual transduction. *Eur. J. Biochem.* 179:255–66

9. Stryer L. 1991. Visual excitation and recovery. *J. Biol. Chem.* 266:10711–14

10. Detwiler PB, Gray-Keller MP. 1992. Some unresolved issues in the physiology and biochemistry of phototransduction. *Curr. Opin. Neurobiol.* 2:433–38

11. Hurley JB. 1992. Signal transduction enzymes of vertebrate photoreceptors. *J. Bioenerg. Biomembr.* 24:219–26

12. Lagnado L, Baylor D. 1992. Signal flow in visual transduction. *Neuron* 8:995–1002

13. Pugh EN Jr, Lamb TD. 1990. cGMP and calcium: the internal messengers of excitation and adaptation in vertebrate photoreceptors. *Vis. Res.* 30:1923–48

14. Pugh EN Jr, Lamb TD. 1993. Amplification and kinetics of the activation steps in phototransduction. *Biochim. Biophys. Acta Bio-Energ.* 1141:111–49

15. Yau K-W. 1994. Phototransduction mechanism in retinal rods and cones: the Friedenwald lecture. *Invest. Ophthalmol. Vis. Sci.* 35:9–32

16. Palczewski K. 1994. Is vertebrate phototransduction solved? New insights into the molecular mechanism of phototransduction. *Invest. Ophthalmol. Vis. Sci.* 35:3577–81

17. Hurley JB. 1994. Termination of photoreceptor responses. *Curr. Opin. Neurobiol.* 4:481–87

18. Bownds MD, Arshavsky VY. 1995. What are the mechanisms of photoreceptor adaptation. *Behav. Brain Sci.* 18:415–24

19. Helmreich EJM, Hofmann KP. 1996. Structure and function of proteins in G-protein-coupled signal transfer. *Biochim. Biophys. Acta Rev. Biomembr.* 1286:285–322

20. Koutalos Y, Yau KW. 1996. Regulation of sensitivity in vertebrate rod photoreceptors by calcium. *Trends Neurosci.* 19:73–81

21. Palczewski K, Saari JC. 1997. Activation and inactivation steps in the visual transduction pathway. *Curr. Opin. Neurobiol.* 7:500–4

22. Molday RS. 1998. Photoreceptor membrane proteins, phototransduction, and retinal degenerative diseases—the Friedenwald lecture. *Invest. Ophthalmol. Vis. Sci.* 39:2493–513

23. Pugh EN Jr, Nikonov S, Lamb TD. 1999. Molecular mechanisms of vertebrate photoreceptor light adaptation. *Curr. Opin. Neurobiol.* 9:410–18

24. Dizhoor AM. 2000. Regulation of cGMP synthesis in photoreceptors: role in signal transduction and congenital diseases of the retina. *Cell. Signal.* 12:711–19

25. Fain GL, Matthews HR, Cornwall MC, Koutalos Y. 2001. Adaptation in vertebrate photoreceptors. *Physiol. Rev.* 81:117–51

26. Burns ME, Baylor DA. 2001. Activation, deactivation, and adaptation in vertebrate photoreceptor cells. *Annu. Rev. Neurosci.* 24:779–805

27. Beavo JA. 1995. Cyclic nucleotide phosphodiesterases: functional implications of multiple isoforms. *Physiol. Rev.* 75:725–48

28. Lamb TD, Pugh EN Jr. 1992. A quantitative account of the activation steps involved in phototransduction in amphibian photoreceptors. *J. Physiol.* 449:719–58

29. Fesenko EE, Kolesnikov SS, Lyubarsky AL. 1985. Induction by cyclic GMP of cationic conductance in plasma membrane of retinal rod outer segment. *Nature* 313:310–13

30. Haynes LW, Kay AR, Yau KW. 1986. Single cyclic GMP-activated channel activity in excised patches of rod outer segment membrane. *Nature* 321:66–70

31. Zimmerman AL, Baylor DA. 1986. Cyclic GMP-sensitive conductance of retinal rods consists of aqueous pores. *Nature* 321:70–72

32. Leskov IB, Klenchin VA, Handy JW, Whitlock GG, Govardovskii VI, et al. 2000. The gain of rod phototransduction: reconciliation of biochemical and electrophysiological measurements. *Neuron* 27:525–37

33. Heck M, Hofmann KP. 2001. Maximal rate and nucleotide dependence of rhodopsin-catalyzed transducin activation—initial rate analysis based on a double displacement mechanism. *J. Biol. Chem.* 276:10000–9

34. Hulst HC. 1957. *Light Scattering by Small Particles.* New York: Wiley

35. Harding SE. 1992. Total intensity and quasi-electric light scattering applications in microbiology. In *Laser Light Scattering in Biochemistry*, ed. SE Harding, DB Satelle, VA Bloomfield, pp. 365–86. Cambridge: R. Soc. Chem.

36. Hofmann KP, Uhl R, Hoffmann W, Kreutz W. 1976. Measurements on fast light-induced light-scattering and-absorption changes in outer segments of vertebrate light sensitive rod cells. *Biophys. Struct. Mech.* 2:61–77

37. Kühn H, Bennett N, Michel-Villaz M, Chabre M. 1981. Interactions between photoexcited rhodopsin and GTP-binding protein: kinetic and stoichiometric analyses from light-scattering changes. *Proc. Natl. Acad. Sci. USA* 78:6873–77

38. Kühn H. 1980. Light- and GTP-regulated interaction of GTPase and other proteins with bovine photoreceptor membranes. *Nature* 283:587–89

39. Arshavsky VY, Dizhoor AM, Kaulen AD, Shestakova IK, Philipov PP. 1985. A light-scattering study of the effects of rhodopsin phosphorylation on its interactions with transducin. *Biol. Membr.* 2:5–10 (In Russian)

40. Deleted in proof

41. Daemen FJ. 1973. Vertebrate rod outer segment membranes. *Biochim. Biophys. Acta* 300:255–88

42. Liebman PA, Sitaramayya A. 1984. Role of G-protein-receptor interaction in amplified phosphodiesterase activation of retinal rods. *Adv. Cycl. Nucleotide Protein Phosphorylation Res.* 17:215–25

43. Liebman PA, Parker KR, Dratz EA. 1987. The molecular mechanism of visual excitation and its relation to the structure and composition of the rod outer segment. *Annu. Rev. Physiol.* 49:765–91

44. Gray-Keller MP, Biernbaum MS, Bownds MD. 1990. Transducin activation in electropermeabilized frog rod outer segments is highly amplified, and a portion equivalent to phosphodiesterase remains membrane-bound. *J. Biol. Chem.* 265:15323–32

45. Tsang SH, Burns ME, Calvert PD, Gouras P, Baylor DA, et al. 1998. Role for the target enzyme in deactivation of photoreceptor G protein in vivo. *Science* 282:117–21

46. Vuong RM, Chabre M, Stryer L. 1984. Millisecond activation of transducin in the cyclic nucleotide cascade of vision. *Nature* 311:659–61

47. Bruckert F, Vuong TM, Chabre M. 1988. Light and GTP dependence of transducin solubility in retinal rods. Further analysis by near infra-red light scattering. *Eur. Biophys. J.* 16:207–18

48. Bruckert F, Chabre M, Minh Vuong T. 1992. Kinetic analysis of the activation of transducin by photoexcited rhodopsin. Influence of the lateral diffusion of transducin and competition of guanosine diphosphate and guanosine triphosphate for the nucleotide site. *Biophys. J.* 63:616–29

49. Pepperberg DR, Kahlert M, Krause A, Hofmann KP. 1988. Photic modulation of a highly sensitive, near-infrared light-scattering signal recorded from intact retinal photoreceptors. *Proc. Natl. Acad. Sci. USA* 85:5531–35

50. Kahlert M, Hofmann KP. 1991. Reaction rate and collisional efficiency of the rhodopsin-transducin system in intact retinal rods. *Biophys. J.* 59:375–86

51. Uhl R, Wagner R, Ryba N. 1990. Watching G proteins at work. *Trends Nerosci.* 13:64–70

52. Kamps KM, Reichert J, Hofmann KP. 1985. Light-induced activation of the rod phosphodiesterase leads to a rapid transient increase of near-infrared light scattering. *FEBS Lett.* 188:15–20

53. Heck M, Hofmann KP. 1993. G-protein-effector coupling: a real-time light-scattering assay for transducin-phosphodiesterase interaction. *Biochemistry* 32:8220–27

54. Hamm HE, Bownds MD. 1986. Protein complement of rod outer segments of frog retina. *Biochemistry* 25:4512–23

55. Dumke CL, Arshavsky VY, Calvert PD, Bownds MD, Pugh EN Jr. 1994. Rod outer segment structure influences the apparent kinetic parameters of cyclic GMP phosphodiesterase. *J. Gen. Physiol.* 103:1071–98

56. Wensel TG, Stryer L. 1990. Activation mechanism of retinal rod cyclic GMP phosphodiesterase probed by fluorescein-labeled inhibitory subunit. *Biochemistry* 29:2155–61

57. Melia TJ, Malinski JA, He F, Wensel TG. 2000. Enhancement of phototransduction protein interactions by lipid surfaces. *J. Biol. Chem.* 275:3535–42

58. Whalen MM, Bitensky MW, Takemoto DJ. 1990. The effect of the gamma-subunit of the cyclic GMP phosphodiesterase of bovine and frog (*Rana catesbiana*) retinal rod outer segments on the kinetic parameters of the enzyme. *Biochem. J.* 265:655–58

59. Yamazaki A, Sen I, Bitensky MW, Casnellie JE, Greengard P. 1980. Cyclic GMP specific, high affinity, noncatalytic binding sites on light-activated phosphodiesterase. *J. Biol. Chem.* 255:11619–24

60. Cote RH, Brunnock MA. 1993. Intracellular cGMP concentration in rod photoreceptors is regulated by binding to high and moderate affinity cGMP binding sites. *J. Biol. Chem.* 268:17190–98

61. Norton AW, D'Amours MR, Grazio HJ, Hebert TL, Cote RH. 2000. Mechanism of transducin activation of frog rod photoreceptor phosphodiesterase—allosteric interactions between the inhibitory gamma subunit and the noncatalytic cGMP-binding sites. *J. Biol. Chem.* 275:38611–19

62. Lamb TD. 1994. Stochastic simulation of activation in the G-protein cascade of phototransduction. *Biophys. J.* 67:1439–54

63. Fersht A. 1977. *Enzyme Structure and Mechanism.* Reading/San Francisco: Freeman

64. D'Amours MR, Cote RH. 1999. Regulation of photoreceptor phosphodiesterase catalysis by its non-catalytic cGMP-binding sites. *Biochem. J.* 340:863–69

65. Baylor DA, Hodgkin AL. 1973. Detection and resolution of visual stimuli by turtle photoreceptors. *J. Physiol.* 234:163–98

66. Nakatani K, Yau K-W. 1988. Calcium and light adaptation in retinal rods and cones. *Nature* 334:69–71

67. Matthews HR, Murphy RLW, Fain GL, Lamb TD. 1988. Photoreceptor light adaptation is mediated by cytoplasmic calcium concentration. *Nature* 334:67–69

68. Fain GL, Lamb TD, Matthews HR, Murphy RLW. 1989. Cytoplasmic calcium as the messenger for light adaptation in salamander rods. *J. Physiol.* 416:215–43

69. Pugh EN Jr, Nikonov S, Lamb TD. 1999. Molecular mechanisms of vertebrate photoreceptor light adaptation. *Curr. Opin. Neurobiol.* 9:410–18

70. Nikonov S, Lamb TD, Pugh EN Jr. 2000. The role of steady phosphodiesterase activity in the kinetics and sensitivity of the light-adapted salamander rod photoresponse. *J. Gen. Physiol.* 116:795–824

71. Torre V, Matthews HR, Lamb TD. 1986. Role of calcium in regulating the cyclic GMP cascade of phototransduction in retinal rods. *Proc. Natl. Acad. Sci. USA* 83:7109–13

72. Lamb TD, Whitlock GG. 2001. *Invest. Ophthalmol. Vis. Sci.* 42:S369 (Abstr.)

73. Jones GJ. 1995. Light adaptation and the rising phase of the flash photocurrent of salamander retinal rods. *J. Physiol.* 487: 441–51

74. Gray-Keller MP, Detwiler PB. 1996. Ca^{2+} dependence of dark- and light-adapted flash responses in rod photoreceptors. *Neuron* 17:323–31

75. Calvert PD, Govardovskii VI, Krasnoperova N, Anderson RE, Lem J, Makino CL. 2001. Membrane protein diffusion sets the speed of rod phototransduction. *Nature* 411:90–94

76. Lee RH, Brown BM, Lolley RN. 1984. Light-induced dephosphorylation of a 33K protein in rod outer segments of rat retina. *Biochemistry* 23:1972–77

77. Lee RH, Lieberman BS, Lolley RN. 1987. A novel complex from bovine visual cells of a 33,000-dalton phosphoprotein with beta and gamma transducin: purification and subunit structure. *Biochemistry* 28:3983–90

78. Wilkins JF, Bitensky MW, Willardson BM. 1996. Regulation of the kinetics of phosducin phosphorylation in retinal rods. *J. Biol. Chem.* 271:19232–37

79. Gaudet R, Bohm A, Sigler PB. 1996. Crystal structure at 2.4 Å resolution of the complex of transducin βgamma and its regulator, phosducin. *Cell* 87:577–88

80. Lee RH, Brown BM, Lolley RN. 1990. Protein kinase A phosphorylates retinal phosducin on serine 73 in situ. *J. Biol. Chem.* 265:15860–66

81. Yoshida T, Willardson BM, Wilkins JF, Jensen GJ, Thornton BD, Bitensky MW. 1994. The phosphorylation state of phosducin determines its ability to block transducin subunit interactions and inhibit transducin binding to activated rhodopsin. *J. Biol. Chem.* 269:24050–57

82. Thulin CD, Howes K, Driscoll CD, Savage JR, Rand TA, et al. 1999. The immunolocalization and divergent roles of phosducin and phosducin-like protein in the retina. *Mol. Vis.* 5:40

83. Arshavsky VY, Sokolov M, Govardovskii VI. 2000. Cellular localization of phosducin in the vertebrate retina. *Exp. Eye Res.* 71:S179 (Abstr.)

84. Nakano K, Chen J, Tarr GE, Yoshida T, Flynn JM, Bitensky MW. 2001. Rethinking the role of phosducin: light-regulated binding of phosducin to 14-3-3 in rod inner segments. *Proc. Natl. Acad. Sci. USA* 98:4693–98

85. Philp NJ, Chang W, Long K. 1987. Light-stimulated protein movement in rod photoreceptor cells of the rat retina. *FEBS Lett.* 225:127–32

86. Brann MR, Cohen LV. 1987. Diurnal expression of transducin mRNA and translocation of transducin in rods of rat retina. *Science* 235:585–87

87. Whelan JP, McGinnis JF. 1988. Light-dependent subcellular movement of photoreceptor proteins. *J. Neurosci. Res.* 20:263–70

87a. Organisciak DT, Xie A, Wang HM, Jiang YL, Darrow RM, Donoso LA. 1991. Adaptive changes in visual cell transduction protein levels: effect of light. *Exp. Eye Res.* 53:773–79

88. Roof DJ, Heth CA. 1988. Expression of transducin in retinal rod photoreceptor outer segments. *Science* 241:845–47

89. Sokolov M, Strissel KJ, Govardovskii VI, Arshavsky VY. 2001. *Invest. Ophthalmol. Vis. Sci.* 42:S186 (Abstr.)

90. Sugimoto Y, Yatsunami K, Tsujimoto M, Khorana HG, Ichikawa A. 1991. The amino acid sequence of a glutamic acid-rich protein from bovine retina as deduced from the cDNA sequence. *Proc. Natl. Acad. Sci. USA* 88:3116–19

91. Körschen HG, Beyermann M, Müller F, Heck M, Vantler M, et al. 1999. Interaction of glutamic-acid-rich proteins with the cGMP signalling pathway in rod photoreceptors. *Nature* 400:761–66

92. Fung BBK, Hurley JB, Stryer L. 1981. Flow of information in the light-triggered cyclic nucleotide cascade of

vision. *Proc. Natl. Acad. Sci. USA* 78: 152–56

93. Baehr W, Morita EA, Swanson RJ, Applebury ML. 1982. Characterization of bovine rod outer segment G-protein. *J. Biol. Chem.* 257:6452–60

94. Dratz EA, Lewis JW, Schaechter LE, Parker KR, Kliger DS. 1987. Retinal rod GTPase turnover rate increases with concentration: a key to the control of visual excitation? *Biochem. Biophys. Res. Commun.* 146:379–86

95. Arshavsky VY, Antoch MP, Lukjanov KA, Philippov PP. 1989. Transducin GTPase provides for rapid quenching of the cGMP cascade in rod outer segments. *FEBS Lett.* 250:353–56

96. Wagner R, Ryba N, Uhl R. 1988. Subsecond turnover of transducin GTPase in bovine rod outer segments. *FEBS Lett.* 234:44–48

97. Ross EM, Wilkie TM. 2000. GTPase-activating proteins for heterotrimeric G proteins: regulators of G protein signaling (RGS) and RGS-like proteins. *Annu. Rev. Biochem.* 69:795–827

98. Cowan CW, He W, Wensel TG. 2000. RGS proteins: lessons from the RGS9 subfamily. *Prog. Nucleic Acid Res. Mol. Biol.* 65:341–59

99. Arshavsky VY, Bownds MD. 1992. Regulation of deactivation of photoreceptor G protein by its target enzyme and cGMP. *Nature* 357:416–17

100. Berstein G, Blank JL, Jhon D-Y, Exton JH, Rhee SG, Ross EM. 1992. Phospholipase C-β1 is a GTPase-activating protein for $G_{q/11}$, its physiologic regulator. *Cell* 70:411–18

101. Bourne HR, Stryer L. 1992. G proteins: the target sets the tempo. *Nature* 358:541–43

102. Scholich K, Mullenix JB, Wittpoth C, Poppleton HM, Pierre SC, et al. 1999. Facilitation of signal onset and termination by adenylyl cyclase. *Science* 283:1328–31

103. Siderovski DP, Strockbine B, Behe CI.

1999. Whither goest the RGS proteins? *Crit. Rev. Biochem. Mol. Biol.* 34:215–51

104. Burchett SA. 2000. Regulators of G protein signaling: a bestiary of modular protein binding domains. *J. Neurochem.* 75:1335–51

105. De Vries L, Zheng B, Fischer T, Elenko E, Farquhar MG. 2000. The regulator of G protein signaling family. *Annu. Rev. Pharmacol. Toxicol.* 40:235–71

106. Antonny B, Otto-Bruc A, Chabre M, Minh Vuong T. 1993. GTP hydrolysis by purified α-subunit of transducin and its complex with the cyclic GMP phosphodiesterase inhibitor. *Biochemistry* 32:8646–53

107. Angleson JK, Wensel TG. 1993. A GTPase-accelerating factor for transducin, distinct from its effector cGMP phosphodiesterase, in rod outer segment membranes. *Neuron* 11:939–49

108. Angleson JK, Wensel TG. 1994. Enhancement of rod outer segment GTPase accelerating protein activity by the inhibitory subunit of cGMP phosphodiesterase. *J. Biol. Chem.* 269:16290–96

109. Arshavsky VY, Dumke CL, Zhu Y, Artemyev NO, Skiba NP, et al. 1994. Regulation of transducin GTPase activity in bovine rod outer segments. *J. Biol. Chem.* 269:19882–87

110. Otto-Bruc A, Antonny B, Vuong TM. 1994. Modulation of the GTPase activity of transducin. Kinetic studies of reconstituted systems. *Biochemistry* 33:15215–22

111. He W, Cowan CW, Wensel TG. 1998. RGS9, a GTPase accelerator for phototransduction. *Neuron* 20:95–102

112. Cowan CW, Fariss RN, Sokal I, Palczewski K, Wensel TG. 1998. High expression levels in cones of RGS9, the predominant GTPase accelerating protein of rods. *Proc. Natl. Acad. Sci. USA* 95:5351–56

113. Makino ER, Handy JW, Li TS, Arshavsky VY. 1999. The GTPase activating factor for transducin in rod photoreceptors is the complex between RGS9 and type 5

G protein β subunit. *Proc. Natl. Acad. Sci. USA* 96:1947–52

114. Chen CK, Burns ME, He W, Wensel TG, Baylor DA, Simon MI. 2000. Slowed recovery of rod photoresponse in mice lacking the GTPase accelerating protein RGS9-1. *Nature* 403:557–60

115. He W, Lu LS, Zhang X, El Hodiri HM, Chen CK, et al. 2000. Modules in the photoreceptor RGS9-1 · $G_{\beta 5L}$ GTPase-accelerating protein complex control effector coupling, GTPase acceleration, protein folding, and stability. *J. Biol. Chem.* 275:37093–100

116. Lyubarsky AL, Naarendorp F, Zhang X, Wensel T, Simon MI, Pugh EN Jr. 2001. RGS9-1 is required for normal inactivation of mouse cone phototransduction. *Mol. Vis.* 7:71–78

117. Cabrera JL, De Freitas F, Satpaev DK, Slepak VZ. 1998. Identification of the Gβ5-RGS7 protein complex in the retina. *Biochem. Biophys. Res. Commun.* 249:898–902

118. Levay K, Cabrera JL, Satpaev DK, Slepak VZ. 1999. Gβ5 prevents the RGS7-Gαo interaction through binding to a distinct G gamma-like domain found in RGS7 and other RGS proteins. *Proc. Natl. Acad. Sci. USA* 96:2503–7

119. Zhang JH, Simonds WF. 2000. Co-purification of brain G-protein β5 with RGS6 and RGS7. *J. Neurosci.* 20:RC59–NIL13

120. Liang JJ, Chen HHD, Jones PG, Khawaja XZ. 2000. RGS7 complex formation and co-localization with the Gβ5 subunit in the adult rat brain and influence on Gβ5γ2-mediated PLCγ signaling. *J. Neurosci. Res.* 60:58–64

121. Witherow DS, Wang Q, Levay K, Cabrera JL, Chen J, et al. 2000. Complexes of the G protein subunit Gβ5 with the regulators of G protein signaling RGS7 and RGS9—characterization in native tissues and in transfected cells. *J. Biol. Chem.* 275:24872–80

122. Sondek J, Siderovski DP. 2001. G gamma-like (GGL) domains: new frontiers in G-protein signaling and beta-propeller scaffolding. *Biochem. Pharmacol.* 61:1329–37

123. Snow BE, Krumins AM, Brothers GM, Lee SF, Wall MA, et al. 1998. A G protein gamma subunit-like domain shared between RGS11 and other RGS proteins specifies binding to $G_{\beta 5}$ subunits. *Proc. Natl. Acad. Sci. USA* 95:13307–12

124. Skiba NP, Hopp JA, Arshavsky VY. 2000. The effector enzyme regulates the duration of G protein signaling in vertebrate photoreceptors by increasing the affinity between transducin and RGS protein. *J. Biol. Chem.* 275:32716–20

125. Otto-Bruc A, Antonny B, Minh Vuong T, Chardin P, Chabre M. 1993. Interaction between the retinal cyclic GMP phosphodiesterase inhibitor and transducin. Kinetics and affinity studies. *Biochemistry* 32:8636–45

126. Slepak VZ, Artemyev NO, Zhu Y, Dumke CL, Sabacan L, et al. 1995. An effector site that stimulates G-protein GTPase in photoreceptors. *J. Biol. Chem.* 270:14319–24

127. Tesmer JJG, Berman DM, Gilman AG, Sprang SR. 1997. Structure of RGS4 bound to AlF$_4^-$-activated $G_{i\alpha 1}$: stabilization of the transition state for GTP hydrolysis. *Cell* 89:251–61

128. Slep KC, Kercher MA, He W, Cowan CW, Wensel TG, Sigler PB. 2001. Structural determinants for regulation of phosphodiesterase by a G protein at 2.0 Å. *Nature* 409:1071–77

129. Wieland T, Chen CK, Simon MI. 1997. The retinal specific protein RGS-r competes with the gamma subunit of cGMP phosphodiesterase for the α subunit of transducin and facilitates signal termination. *J. Biol. Chem.* 272:8853–56

130. Nekrasova ER, Berman DM, Rustandi RR, Hamm HE, Gilman AG, Arshavsky VY. 1997. Activation of transducin guanosine triphosphatase by two proteins of the RGS family. *Biochemistry* 36:7638–43

131. Natochin M, Granovsky AE, Artemyev NO. 1997. Regulation of transducin GTPase activity by human retinal RGS. *J. Biol. Chem.* 272:17444–49

132. Sowa ME, He W, Slep KC, Kercher MA, Lichtarge O, Wensel TG. 2001. Prediction and confirmation of a site critical for effector regulation of RGS domain activity. *Nat. Struct. Biol.* 8:234–37

133. McEntaffer RL, Natochin M, Artemyev NO. 1999. Modulation of transducin GTPase activity by chimeric RGS16 and RGS9 regulators of G protein signaling and the effector molecule. *Biochemistry* 38:4931–37

134. Skiba NP, Yang CS, Huang T, Bae H, Hamm HE. 1999. The α-helical domain of Gα_t determines specific interaction with regulator of G protein signaling 9. *J. Biol. Chem.* 274:8770–78

135. Sowa ME, He W, Wensel TG, Lichtarge O. 2000. A regulator of G protein signaling interaction surface linked to effector specificity. *Proc. Natl. Acad. Sci. USA* 97:1483–88

136. Ponting CP, Bork P. 1996. Pleckstrin's repeat performance: a novel domain in G-protein signaling? *Trends Biochem. Sci.* 21:245–46

137. Watson AJ, Aragay AM, Slepak VZ, Simon MI. 1996. A novel form of the G protein β subunit Gβ_5 is specifically expressed in the vertebrate retina. *J. Biol. Chem.* 271:28154–60

138. Stavenga DG, DeGrip WJ, Pugh EN Jr, eds. 2000. *Handbook of Biological Physics*, Vol. 3. *Molecular Mechanisms in Visual Transduction.* Amsterdam: Elsevier

139. Skiba NP, Martemyanov KA, Elfenbein A, Hopp JA, Bohm A, et al. 2001. RGS9-Gβ5 substrate selectivity in photoreceptors: Opposing effects of constituent domains yield high affinity of RGS interaction with G protein-effector complex. *J. Biol. Chem.* 276:37365–72

Annu. Rev. Physiol. 2002. 64:189–222

G Proteins and Olfactory Signal Transduction

Gabriele V. Ronnett[1] and Cheil Moon
Departments of Neuroscience and Neurology[1], The Johns Hopkins University School of Medicine, Baltimore, Maryland 21205; e-mail: gronnett@jhmi.edu; cmoon@jhmi.edu

Key Words olfaction, olfactory receptor neuron, adenylyl cyclase, sensory transduction, second messengers, cross-talk, desensitization

■ **Abstract** The olfactory system sits at the interface of the environment and the nervous system and is responsible for correctly coding sensory information from thousands of odorous stimuli. Many theories existed regarding the signal transduction mechanism that mediates this difficult task. The discovery that odorant transduction utilizes a unique variation (a novel family of G protein–coupled receptors) based upon a very common theme (the G protein–coupled adenylyl cyclase cascade) to accomplish its vital task emphasized the power and versatility of this motif. We now must understand the downstream consequences of this cascade that regulates multiple second messengers and perhaps even gene transcription in response to the initial interaction of ligand with G protein–coupled receptor.

INTRODUCTION

Cell signaling systems have developed to serve the diverse intracellular and intercellular communications needs of complex organisms. Study of these signaling pathways has revealed that cells employ variations of common transduction motifs to generate their responses. This is true even for sensory signaling, in which environmental stimuli must be interpreted as the first step in sensory perception. The correct analysis of sensory input is vital to an organism's survival. The olfactory system has challenged many researchers seeking to understand the molecular aspects of sensory signal transduction and coding mechanisms (1–5).

The olfactory system must discriminate among thousands of odors comprised of chemically divergent structures (odorants). As for other sensory modalities, a combination of molecular, electrophysiological, and cell biological approaches was required to delineate odorant transduction. What has emerged is that odorant transduction combines unique receptive molecules with common G protein–mediated transduction cascades to detect odorants. Many of these features are conserved across phyla, as recently reviewed by Hildebrand & Shepherd (5). Although G protein cascades are involved in the initial events of odor perception,

0066-4278/02/0315-0189$14.00

189

what has emerged is that a variety of signal cascades are activated in addition to a G protein cascade. The roles of these other signals are only beginning to be understood, and many controversial issues remain (6).

CELLULAR COMPOSITION OF THE OLFACTORY EPITHELIUM

The initial events of odor detection occur at the peripheral olfactory system, which is well adapted structurally to perform its function. The primary olfactory sensory receptor neurons are located in the olfactory epithelium, where they are in direct contact with inhaled odorants. There are three principal cell types in the olfactory epithelium: olfactory receptor neurons (ORNs), supporting sustentacular cells, and several types of basal cells (7, 8).

ORNs are bipolar, extending apical dendrites to the surface of the neuroepithelium and sending unmyelinated axons through the basal lamina and cribiform plate (of the ethmoid bone) to terminate in the brain on dendrites of mitral and tufted neurons in the glomeruli of the olfactory bulb. The apical dendrites form dendritic knobs from which arise specialized, nonmotile cilia, where the initial events of olfactory transduction occur (2, 9, 10). Electrophysiological studies indicate that odorant sensitivity and the odorant-induced current are uniformly distributed along the cilia, suggesting that all the components of the immediate responses to odorants are localized to the cilia. Immunoelectron microscopic studies have confirmed the cilial localization of many of these components (11, 12). ORNs comprise 75–80% of the cells in the epithelium (13) and are functionally homogeneous: They all detect odorants. ORNs senesce and die throughout life at a regular rate. They are replenished by the differentiation of globose basal cells (14–16). As they mature, ORNs move apically in the epithelium, permitting determination of neuronal age by position (17), with mature ORNs expressing olfactory marker protein (18, 19). Interestingly, this neurogenesis can be hyperinduced by ablation of the olfactory bulb (termed bulbectomy) (20). Thus, the understanding of the functions of signaling components in signal transduction can be facilitated by studies of the spatial organization and development of ORNs.

Sustentacular cells are in general considered to be supportive cells and share features in common with glia. They stretch from the epithelial surface to the basal lamina, where they maintain foot processes (2, 3). Sustentacular cells electrically isolate ORNs, secrete components into the mucus, and contain detoxifying enzymes (21). The sustentacular cells contain high concentrations of cytochrome P450–like enzymes (22). Regarding odorant transduction, it is thought that these enzymes may modify odorants to make them less membrane permeable or inactivate them. Recent studies indicate that sustentacular cells may produce growth factors important to ORN development (23). Neuropeptide Y (NPY) is an amidated neuropeptide that performs many functions in mammalian physiology (24, 25). NPY mRNA is upregulated following peripheral axotomy and in pheochromocytoma and ganglioneuroblastoma tissue (26). Whereas NPY is expressed in developing ORNs during embryogenesis, it is expressed in sustentacular cells in the adult olfactory epithelium, functioning as a neuroproliferative factor for olfactory neuronal precursors in vivo and in vitro (23). Thus, NPY is the first of possibly many growth factors that maintain ORN homeostasis.

The basal cells underlie the ORNs and serve as precursors for the generation of new ORNs throughout adulthood (7, 8, 16). Basal cells have been divided into two general classes. Horizontal cells are morphologically flat and express cytokeratin (7, 27), and globose basal cells are rounded in shape and express several markers, including GBC-1, GBC-3, and GBC-5 (28, 29). Compared with other neurons, ORNs have a shorter average lifetime, in the range of several months. This may in part be due to the fact that ORNs are exposed to a variety of toxic or infectious agents. Given the turnover of ORNs throughout life, the role of the globose basal cells in providing new ORNs is crucial to the maintenance of the sense of smell.

GENERAL FEATURES OF ODORANT TRANSDUCTION

Odorant signal transduction is initiated when odorants interact with specific receptors on the cilia of ORNs (1, 30–32) (Figure 1). Receptors subsequently couple to a G protein to stimulate adenylyl cyclase (33–35). Electrophysiological and biochemical studies confirm that cAMP is the key messenger in the initial phase of odorant detection (33, 34, 36–41). The concentration of cAMP in the cilia rises, gating open a cyclic nucleotide-gated channel, resulting in an influx of Na^+ and calcium (42, 43). The immediate response is the generation of a graded receptor potential (44, 45).

In addition, several other second messenger cascades that are activated upon odorant detection may regulate secondary events or odorant adaptation. These include pathways activated downstream of the cyclic nucleotide-gated channel and include the consequence of the channels' substantial calcium permeability (46, 47). Odorants also increase phosphoinositide hydrolysis and the production of inositol-1,4,5-trisphosphate (IP_3) (35, 48–50). Cyclic GMP production is also increased with odorant exposure (51, 52). Interestingly, the ORN's response to odorant-induced cGMP production is much slower than the cAMP or IP_3 responses, which normally peak within 500 ms. Thus, the cGMP response does not appear to function in the immediate detection phase of olfaction, such as modulating cyclic nucleotide-gated cation channels or IP_3 receptors, but rather in desensitization or the modulation of the cellular response during longer exposures to odorants (53–56). The relationship of these messengers to the G protein–coupled cascade is discussed in subsequent sections.

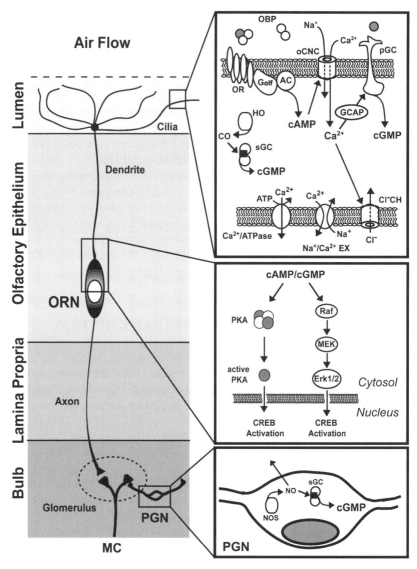

Figure 1 Model of odorant signal transduction. See text for details. There are signaling cascades that mediate the initial phase of odorant detection and that mediate potential long-term responses to odorant detection. Abbreviations: AC, adenylyl cyclase; CO, carbon monoxide; CREB, cAMP-responsive element binding protein; GCAP, guanylyl cyclase activating protein; G_{olf}, olfactory G protein; HO, heme oxygenase; MEK, MAP or ERK kinase; OBP, odorant binding protein; oCNC, olfactory cyclic neucleotide-gated channel; OR, odorant receptor; pGC, particulate guanylyl cyclase; PGN, periglomerular neuron; PKA, cAMP-dependent protein kinase; Raf, MEK kinase; sGC, soluble guanylyl cyclase.

COMPONENTS OF THE ODORANT
TRANSDUCTION CASCADE

Odorant-Binding Proteins

The existence of carrier proteins for odorants resident in the nasal mucus was predicted based upon the fact that hydrophobic odorants must travel through the aqueous mucus barrier toward the cilia of ORNs. In fact, odorant-binding proteins (OBP) were discovered by several laboratories in early attempts to identify odorant receptors using radioactive odorants such as 3-isobutyl-2-methyloxypyrazine (57–59). Native OBP purified from olfactory mucus is a homodimer of two 19-kDa subunits with an affinity for odorants in the micromolar range (60).

The molecular cloning of OBP helped clarifiy its function. OBP is a member of the lipophilic molecule carrier protein family; a well-characterized member of this family is a retinol-binding protein of the retina. This protein conveys retinol from retinal pigment epithelium to rods and cones where it is incorporated into rhodopsin (61). In situ hybridization studies using probes to visualize OBP mRNA revealed its selective concentration in the lateral nasal gland, the largest of 20 discrete nasal glands in mammals (62). OBP thus appears to be secreted from this gland down a long duct to the tip of the nose, where watery secretions are atomized to humidify inspired air. OBP thus localized might trap odorants and carry them with inhalation to ORNs. Alternatively, OBP may function to remove odorants from the sensory epithelium and cilia.

Further studies have revealed that more than one form of OBP is expressed in the nasal epithelium. Rabbitts and colleagues (63) identified a second form of OBP, OBPII. OBPII encodes a secretory protein with significant homology to OBPI, and it is also expressed in the lateral nasal gland, which is the site of OBP expression. Interestingly, the OBPII sequence shows significant homology to the VEG protein, which is thought to be involved in taste transduction (64). Breer and colleagues demonstrated that although OBP proteins appear to share many structural features, recombinant rat OBPI and OBPII each interact with distinct sets of odorants and therefore contain distinct ligand specificities (65). OBPI binds specifically to a pyrazine derivative, 2-isobutyl-3-methoxypyrazine, whereas OBPII binds to the chromophore, 1-anilinonaphthalene 8-sulfonic acid (1,8-ANS), specifically. In other vertebrates multiple forms of OBP have been identified. There are four OBPs in mice (66), three OBPs in rabbit (67), and two OBPs in cow (68, 69). OBP has also been cloned from insects (70, 71).

Odorant Receptors

Mammals perceive a huge variety of environmental odors. The initial step in odor recognition involves the interaction of odorous ligands with specific receptors in the ciliary membrane of ORNs (1, 30–32). Based upon the assumption derived from biochemical evidence that odorant signal transduction involves G proteins and G protein–coupled receptors, a very large gene family of closely related olfactory-specific seven-transmembrane spanning domain receptors was identified by polymerase chain reaction (1, 72, 73). In vertebrates the family of odorant receptors (ORs) encodes as many as 1000 genes, suggesting that some of the steps following odorant recognition occur within the primary sensory neurons themselves. To date, OR genes have been isolated from 12 vertebrate species: rat, mouse, human, catfish, zebrafish, dog, frog, chicken, pig, opossum, mudpuppy, and lamprey (74). In humans estimates for the size of the receptor family range from 500 to 1000 genes. Interestingly, compared with the other species, human OR clones display a high frequency of pseudogenes (74).

The genes encoding ORs may be initially classified as Class I (fish-like) and Class II (tetrapod-specific) ORs. Class I ORs are specific for recognizing water-soluble odorants, whereas Class II ORs bind airborne odorants. All human Class I ORs are localized in a single large cluster, and half of those ORs are apparently functional (75). Expression of Class I ORs has already been reported in rats (76) and in human (77). Class II families are all present in more than one chromosome each, except for a small family of 12 (75). The genes encoding ORs are devoid of introns within their coding regions (72). Mammalian OR genes are typically organized in clusters of 10 or more members and located on many chromosomes. There is a strong correlation between the localization of an OR in a particular chromosomal cluster and its position in a phylogenetic dendrogram derived from comparison of full-length OR protein sequences (78). The repertoire of human OR genes contains a large fraction of pseudogenes, suggesting that olfaction became less important in the course of primate evolution.

Our knowledge of human ORs has been facilitated through the Human Genome Project. The availability of this sequence information indicated that the overall number of human ORs may be in excess of 1000, as previously predicted (79), and that only one third of the human ORs appear to be functional, consistent with previous reports showing a large proportion of pseudogenes (80). One group predicts that 906 human OR genes are present in the human genome, of which approximately 60% appear to be pseudogenes (81), compared with fewer than 5% pseudogenes in rodents or lower primates.

The locations of human ORs have been elucidated in the past several years. Human ORs contain large genomic segments that have been duplicated to many locations in the genome (82), particularly near telomeres (83). The overall localization of human ORs on all chromosomes except 20 and Y is in agreement with previous work based on fluorescence in situ hybridization (80). Human ORs are predominantly localized to the middle of the q arms with several additional genes located near the p telomere. Human chromosome 11 appears to be the origin of the human OR repertoire. This chromosome contains 42% of all human ORs and is the only one containing Class I receptors. Moreover, this chromosome contains the most diverse collection of OR families, 9 out of the 13 Class II families. It also has the two largest clusters in the genome, each with more than 100 ORs. Interestingly, this chromosome shows contiguous conserved synteny in species from humans to the earliest mammals (84).

The expression pattern of ORs in ORNs shows an unusual spatial distribution (85, 86). In situ hybridization studies showed that OR mRNAs are expressed within one of several broad, nonoverlapping zones. Within a zone ORs are expressed in a random manner. Each zone occupies about a quarter of the olfactory epithelium (85) and is represented on the turbinates and on the septum (87). However, the physiological relevance of zonal expression remains unclear.

Although a number of studies have been done on the expression and distribution of ORs at the message level, relatively little is known about the expression of OR proteins. Polyclonal antibodies have been raised against some ORs, permitting visualization of OR proteins. In rats, an OR is expressed as early as E14 in a zonally restricted pattern (88). The expression of ORs appears restricted to the cilia and dendritic knobs of ORNs. The cilia-specific expression of ORs supported a role for ORs in olfactory transduction (12, 89, 90). A concern with studies utilizing antibodies to identify discrete members of the OR family is the specificity of the antibodies, given the large numbers of receptors. Thus, despite the general utility of antisera for immunohistochemical and biochemical studies, the enormous size of the OR repertoire limits the feasibility of proving the specificity of an antibody for a specific receptor.

Significant difficulties with the heterologous expression of ORs have severely limited studies designed to provide functional confirmation of the role of ORs. The most convincing data concerning function have been provided by four approaches. One approach utilized genetic studies in *Caenorhabditis elegans*, which demonstrated that the ODR-10 mutant lacked a seven-transmembrane receptor and was deficient in its ability to detect acetyl (91). Krautwurst et al. (92) achieved similar functional heterologous expression of rodent ORs also using HEK-293 cells. This group generated an expression library of mouse ORs and identified three ORs responding to carvone, (−) citronellal, and limonene using micromolar concentrations of these odorants. Firestein and colleagues also demonstrated functional expression of a cloned OR in rat nasal epithelium by using a recombinant adenovirus containing a putative OR to infect rat nasal epithelium in vivo (93). They demonstrated that this specific OR was overexpressed in the rat olfactory epithelium and, by electro-olfactogram recording, that the expressed OR conferred a response to a small subset of odorants. Malnic et al. (31) performed single-cell polymerase chain reaction on ORNs whose odorant responses had been determined as isolated cells in culture, to demonstrate that a combinatorial code of receptor gene expression exists for odorant perception. These approaches of developing functional expression systems to study ORs may prove to be extremely useful for the screening of ORs on a large scale, as well as understanding the molecular mechanism of odorant recognition.

Besides functioning in odorant transduction at the dendritic end of the bipolar ORN, ORs may be involved in determining or guiding ORN axonal projections to the olfactory bulb and possibly to specific glomeruli (94, 95). In rodents the axons of ORNs that express the same OR message converge to defined glomeruli in the olfactory bulb, suggesting that the rodent olfactory bulb is topographically organized and that an ORN expressing a specific OR projects to and synapses with the representing glomeruli in the olfactory bulb. This type of organizational "wiring" prompts an interesting hypothesis. It suggests that the signals from ORNs expressing a single OR gene (out of the approximately 2000 alleles of the rat OR repertoire) are represented in glomeruli as a topographical map in the olfactory bulb; as a consequence an environmental odor is encoded by activation of a specific set of glomeruli.

Some studies indicate that receptors closely related to OR genes may be expressed in tissues other than the olfactory epithelium. This finding suggests that there may be alternative biological roles for this family of chemosensory receptors. Expression of various ORs was reported in human and murine erythroid cells (77), developing rat heart (96), avian notochord (97), and lingual epithelium (98). The best case for the existence of ORs is the finding that genes related to mammalian ORs are transcribed in testes and expressed on the surface of mature spermatozoa, suggesting a possible role for ORs in sperm chemotaxis (99, 100).

G Proteins

The first evidence for the involvement of G proteins in odorant transduction was obtained through biochemical experiments in which investigators demonstrated that the odorant-induced stimulation of olfactory sensory cilia was dependent upon the presence of GTP (101). Subsequently, a G protein was cloned from an olfactory cDNA library. This clone, termed G_{olf}, was highly and almost exclusively expressed in ORNs (102). G_{olf} was able to stimulate adenylyl cyclase in heterologous systems. Aside from its expression in ORNs of the olfactory epithelium, G_{olf} was expressed in basal ganglia (103). As mentioned, odorants also increased IP_3 production, causing many to postulate that cilia might contain olfactory-specific G_q proteins. To date, an olfactory-specific class of G_q proteins has not been reported.

Mice with targeted disruption of the gene for G_{olf} displayed a striking reduction in the electrophysiological response of ORNs to a wide variety of odors, supporting the hypothesis that G_{olf}, and thus this G protein–mediated cascade, is required for odorant signal transduction (104). Despite this intense attenuation in response to odors, the topographic map of ORN projections to the olfactory bulb was unaltered in G_{olf}-deficient mice. Thus, odorant stimulation may or may not be an essential process in determining the targets of ORN axonal projections to the olfactory bulb. However, for a conclusive answer, these studies may need to be done at higher resolution.

SIGNALING CASCADES AND THEIR COMPONENTS IN ODORANT DETECTION

cAMP

The first direct biochemical studies reported an odorant-induced cAMP response in olfactory sensory cilia isolated from both frog and rat (33, 34). The olfactory sensory cilia were prepared by subcellular fractionation after calcium shock of the olfactory epithelium (101). The odorant-stimulated production of cAMP was tissue specific and occurred only in the presence of GTP, suggesting the involvement of receptors coupled to G proteins.

Electrophysiological studies provided further evidence for the central role of cAMP in odorant detection. Recordings from excised membrane patches of cilia demonstrated a cAMP-gated conductance (42). Investigators proposed that an odorant would increase intracellular cyclic nucleotide concentration to gate a cationic conductance, initiating a depolarizing response. Kinetic studies of odorant-induced currents recorded in the whole-cell configuration (43, 105) suggested that the latency of the odorant response (several hundred milliseconds) indeed supported a role for a second messenger such as cAMP. Further biochemical characterization using isolated rat olfactory sensory cilia showed that cAMP was best produced by fruity, floral, and herbaceous odors (34, 42). Screening many odorants at a single concentration revealed only minimal cAMP production by some, generating the hypothesis that those odorants with small or no cAMP responses employed another cascade, perhaps inositol phosphates (34, 42). These initial measurements were made 15 minutes after the exposure of isolated cilia to odorants.

To demonstrate that the production of cAMP occurs on a relevant time scale, subsecond kinetics of odorant-induced changes were analyzed using a rapid quench-flow device (39, 106). In this device cilia membranes and odorant solutions were reacted together using computer-controlled mixing, with subsequent quenching of samples at intervals from 8–500 ms. cAMP was produced rapidly and transiently in response to odorants, with increases evident as early as 25 ms. Certain odorants, such as fruity odors, were able to stimulate cAMP production at concentrations as low as 10 nM, whereas others, such as putrid odors, had no effect, even at higher concentrations. Those odorants that did not stimulate cAMP production were hypothesized to act through the phosphoinositide cycle. High (millimolar) levels of calcium inhibited the response; intermediate concentration ranges, however, were not tested.

The odorant-induced cAMP response was investigated further using isolated rat olfactory cilia determining the generality of the odorant-induced cAMP response and the calcium dependence of this response (38). Odorants indeed cause rapid and transient elevations of cAMP, as well as the more sustained signal, as seen by Pace et al. (33) and Sklar et al. (34). Different from the observation from Breer's group (39, 106), all odorants stimulated cAMP production. Interestingly, responses were nonlinear. Thus, there was an initial dose-dependent cAMP response with increasing odorant concentration that decreased at higher odorant concentrations, and for some odorants increased at even greater concentrations. Basal and odorant-induced cAMP levels in cilia demonstrated a biphasic calcium dependence, with peak cAMP stimulation in the range of 1–10 μM free calcium. Dose-response curves done at two calcium levels showed that the influence of calcium on odor responses was complex, suggesting the possible involvement of calcium in both signal generation and termination.

To evaluate odorant signal transduction in intact cells, a primary culture system of olfactory epithelium enriched in ORNs was developed (35, 36, 107). Using this primary culture system, cAMP responses to odorant stimulation were monitored in intact ORNs. Odorants were quite potent at producing cAMP, with as little as 0.1 nM isobutylmethoxypyrazine (IBMP) generating a response (35, 36). Responses were multiphasic; cAMP production increased with increasing odorant concentration, decreased at even higher odorant concentrations, and sometimes reappeared at still higher (1–10 μM) concentrations. Signals were calcium dependent, with maximal adenylyl cyclase activity at 10 μM free calcium and inhibition at higher calcium concentrations. Odorant induction of cAMP production was rapid, with peak effects observed at 10–15 sec, but signals continued well above baseline for minutes, confirming results from Sklar et al. (34) and Pace & Lancet (108). The duration of the cAMP response observed in whole cells was significantly longer than that measured using isolated cilia. This may be because of differences between the preparations, or because cAMP functions in more than the initial rapid phase of odor detection. This latter possibility is considered in later sections.

Adenylyl Cyclases

Cyclic AMP is generated by adenylyl cyclases. There are at least nine known isoforms of adenylyl cyclases (109). Bakalyar & Reed (110) cloned a novel adenylyl cyclase, AC3. Northern blot analysis revealed that AC3 mRNA was enriched in the olfactory epithelium and that AC3 message disappeared after bulbectomy. When expressed in HEK293 cells, AC3 had almost no basal activity. In contrast, two other isoforms of adenylyl cyclase, AC1 and AC2, have high basal activities. The low basal activity of AC3 may be relevant to its role in sensory transduction. G_{olf} and AC3 have been ultrastructurally localized to olfactory cilia, indicating that G_{olf} may mediate the activation of AC3 (111).

To evaluate the role of AC3 in the olfactory transduction, the AC3 gene has been disrupted in mice (41). Odorant-induced responses measured by electroolfactogram were completely eliminated in AC3-null mice. Moreover, odor-dependent learning was impaired in these mice. Interestingly, both fruity odors (transduced by cAMP) and putrid odors (transduced by IP$_3$) failed to evoke any response in these animals. This observation was mimicked by a pharmacological study that showed that adenylyl cyclase antagonists reversibly inhibit electroolfactogram responses, even to putrid odors (formerly thought to act through IP$_3$) (112). Taken together, these results confirmed earlier biochemical studies that implicated cAMP as essential for the initial phases of odorant transduction. IP$_3$ was therefore postulated to play more of a modulatory role in the odorant transduction in mammals.

Certain adenylyl cyclases are rather broadly expressed, whereas others are restricted in their distribution (113). Although AC3 is highly enriched in ORNs, especially in cilia, other adenylyl cyclases, such as AC2 or AC4, have also been associated with olfactory neuroepithelium, raising the issue that other adenylyl cyclases may be important in different aspects of olfactory signal transduction.

Adenylyl cyclases are regulated by different mechanisms. Studies by Storm and colleagues (114–116) indicated that the mechanisms of regulation of adenylyl cyclases may not only be dependent upon the specific kind of adenylyl cyclase expressed in a tissue, but by local influences and the expression of regulatory molecules in that specific cell. Thus, whereas ectopically expressed AC3 may be stimulated by calcium, in vivo studies in certain tissues argue for the inhibition of AC3 by calcium. Equally diverse are the effects of protein kinases on adenylyl cyclases. Phorbol esters are used to mimic the effects of protein kinase C (PKC) activation and elicit a stimulatory effect on AC2 but barely stimulate AC1 or AC8. These latter adenylyl cyclases are stimulated up to eightfold by calcium (117). Frings (118) has reported that activation of PKC by phorbol esters increased cAMP in frog olfactory tissue. Calmodulin mediates the stimulation by calcium of AC1, AC3, and AC8 (119); it is unclear how the calcium sensitivity of the calcium inhibition of AC5 and AC6 is achieved. There is also evidence that PKA may affect adenylyl cyclase activity.

Olfactory Phosphodiesterases

The ambient level of cAMP in a cell is dependent upon both the synthesis and degradation of cAMP. Odorants clearly activate adenylyl cyclase, but is there any effect of odorants on phosphodiesterases (PDEs)? There are at least seven different gene families of PDEs whose activities are regulated by calcium, cyclic nucleotides, and phosphorylation (120–123). Thus, odorants could have an indirect effect on the degradation of cAMP, potentially providing a second site of regulation for the odorant-induced cAMP response. Several forms of cAMP-PDE are expressed in rat olfactory cilia (124–125). A novel calcium/calmodulin PDE (CaM-PDE) is selectively found in ORNs, with prominent cilial expression. This novel CaM-PDE has a high affinity (K_m of 1.4 μM) for cAMP and could be activated by odorants in response to intracilial calcium increases. Cloning of the high-affinity PDE revealed it to have a higher affinity for cAMP than any known brain isoform (126). This PDE, designated PDE1C2, is well suited for restoring the submicromolar levels of cAMP after odorant stimulation. In an ectopic expression system, maximum activation by calcium was reached at 10 μM calcium concentration.

A subset of olfactory neurons expresses cGMP-stimulated phosphodiesterase (PDE2) (127). In these specific ORNs, guanylyl cyclase type-D (GC-D) is also expressed, suggesting that it may play an important role in odorant transduction for a specific subset of responses. PDE2 and GC-D are both expressed in olfactory cilia of these neurons; however, only PDE2 is expressed in axons (127). In contrast to most other ORNs, these neurons appear to project to a distinct group of glomeruli in the olfactory bulb similar to the subset that have been termed necklace glomeruli. Furthermore, this subset of neurons are unique in that they do not contain several of the previously identified components of olfactory signal transduction cascades involving cAMP and calcium, including a calcium/calmodulin-dependent PDE (PDE1C2), AC3, and cAMP-specific PDE (PDE4A) (127, 128).

Interestingly, these latter three proteins are expressed in the same neurons; however, their subcellular distributions are distinct. PDE1C2 and AC3 are expressed almost exclusively in the olfactory cilia, whereas PDE4A is present only in the cell bodies and axons. Taken together, these data strongly suggest that selective compartmentalization of different PDEs and cyclases is an important feature for the regulation of signal transduction in ORNs.

A recent study identified some ORNs devoid of G_{olf}, ACIII, PDE1C2, and the cyclic nucleotide-gated channel subunits $\alpha 3$ and $\beta 1b$ that are expressed in prototypical ORNs (127). These particular ORNs express GC-D, PDE2, and cGMP-selective $\alpha 2$ channels (127, 128). GC-D is related to the Ca^{2+}-regulated retinal GC forms GC-E and GC-F, as opposed to the receptor GCs, GC-A, GC-B, and GC-C, which are activated by peptide ligands (129). In particular, GC-D and GC-E/F share characteristic sequence similarity in a regulatory domain that is involved in binding of GCAPs (130). This similarity raises the intriguing possibility that GC-D activity is under the dual control of an unknown extracellular ligand and Ca^{2+} (54). These ORNs project their axons in glomeruli different from ACIII-expressing ORNs and form necklace-shaped synapses in the glomeruli. The necklace glomeruli in the olfactory bulb are spared from the morphological alterations observed in the typical glomeruli of the G_{olf} null-mice, suggesting that a typical glomeruli receives innervation from a subset of receptor neurons that use a pathway independent of cAMP signaling (131).

Channels

In ORNs, ion channels are expressed in ciliary processes of dendritic endings, where they amplify the odor-induced receptor current. Ca^{2+} signals generated by cyclic neucleotide-gated channels (CNCs) are at the heart of sensory transduction in vision and olfaction. The ability of CNCs to conduct Ca^{2+} determines both the rise time and the amplitude of the olfactory receptor current, as well as its termination after the stimulus. Ca^{2+}-gated Cl^- channels are triggered by odor-induced Ca^{2+} influx through CNCs and cause a depolarizing Cl^- efflux that amplifies the receptor current (132). The extrusion of Ca^{2+} ions is mediated by Na^+/Ca^{2+} exchange mechanisms in cilia and probably involves Ca^{2+}-ATPases in knobs, dendrites, and cilia (133–137).

OLFACTORY CYCLIC NUCLEOTIDE-GATED CHANNELS

General

The gating of CNCs accounts for the initial component of the odor-induced electrical response, and this event is crucial to recognizing the environmental signal in the central nervous system. In fact, mice deficient in CNC$\alpha 3$, which is essential for forming functional CNCs, suffer from general anosmia (40, 138). Retinal and olfactory CNCs share a high degree of sequence similarity (over 80% amino acids identity) in the cyclic nucletide binding sites, but they show very different characteristics. First, cyclic nucleotide selectivities of these CNCs are very different (139). The retinal CNCs show much higher apparent affinity for cGMP (140, 141), whereas cAMP and cGMP have very similar effects on the olfactory CNCs (42, 140). cAMP and cGMP are varied only in their purine ring structure. Next, the olfactory CNC (oCNC) has a larger single-channel conductance (55 versus 20 ps), a lower degree of selectivity among monovalent cations, and a larger apparent pore diameter (6.3 versus 5.8 Å, determined from organic cation permeability) than the retinal CNC (142). Notably, in the absence of divalent cations, unit conductance of the oCNC is 25–40 ps (143, 144).

oCNCs have a higher affinity for cyclic nucleotides than visual CNCs and display higher affinity for cGMP than for cAMP (145). However, cGMP may not gate oCNCs in the earlier olfactory signal tansduction because the cGMP response is not large and fast enough to gate oCNC (37, 54).

Composition

The oCNC was identified based on the visual CNC. Initially, the α-subunit was cloned, but the expressed channel showed different characteristics from the wild type. Later the β-subunit was cloned and conferred the affinity for cAMP (146). To date, three subunits that form the oCNCs of ORNs have been identified, and the rod photoreceptor channels have at least two subunits (147–150). In addition, a second type of modulatory subunit is part of the olfactory channels (146, 151, 152). Native oCNCs appear to contain not only α and β subunits but also a splice variant of the β subunit of rod photoreceptor CNC (153). Recent research revealed that native oCNCs comprise a heteromeric channel complex consisting of $\alpha 3$, $\alpha 4$, and $\beta 1b$, which is highly permeable to Ca^{2+} (150, 154).

Exogenous expression of α subunits alone generates functional CNCs in oocytes. In contrast, the β subunit does not yield functional channels when expressed on its own. However, coexpression of the β subunit with the α subunit shifts the $K_{1/2}$ values for cAMP to lower values and yields channels whose properties resemble more closely the native rat olfactory channel (146, 153). However, the $K_{1/2}$ values for cAMP are still threefold higher than those in the native channels (146, 153).

Regulation

The activity of oCNCs is regulated in various ways. The most rapid is the negative feedback inhibition of oCNCs by Ca^{2+}/calmodulin (155, 156). Ca^{2+}/calmodulin causes a decrease in the apparent cyclic nucleotide affinity of both rod (157, 158) and olfactory (155, 159) CNCs. Elevation of Ca^{2+} owing to influx through oCNCs reduces its affinity for cAMP, resulting in a lowering of its open probability (156). This modulation occurs by the direct binding of Ca^{2+}/calmodulin to the amino-terminal region of the CNCs and does not involve the action of a kinase (155). The decrease in apparent cyclic nucleotide affinity arises from a decrease in the stability of the allosteric opening transition, consistent with the effect of mutations in the amino-terminal region (155). Ca^{2+}-mediated desensitization appears to act through an allosteric mechanism with the effect of stabilizing a closed state of the channel.

Divalent cations like Ca^{2+} and Mg^{2+} suppress the conductance of the oCNCs (42). First, Ca^{2+} regulates the activity of oCNCs. In the presence of Ca^{2+}, the open probability of single CNCs was reduced from 0.6 to 0.09 in the presence of 100 μM cAMP, whereas the single channel conductance remained unchanged (160). Kramer & Sieglebaum (161) demonstrated a similar action by intracellular Ca^{2+} in catfish olfactory neurons. The effect of Ca^{2+} is to reduce the open probability by shifting the affinity of the channel for cAMP to higher concentrations; this effect was overcome by application of high concentrations of cAMP (162). Extracellular Ca^{2+} also lowers the affinity of the oCNC for cAMP by a different mechanism (163). Ni^{2+} also produces an inhibition in the oCNCs and may primarily bind to the channel when it is closed (158). This effect is the opposite of the observation in the rCNC. The binding site for the inhibitory effect of Ni^{2+} on the oCNC is localized to a single histidine residue (H396) at a position just three amino acids downstream from the homologous potentiation site in the rod channel.

Phosphorylation is the most common posttranslational regulatory mechanism. The affinity of the oCNCs for cAMP is affected by phosphorylation of oCNC subunits (164). When Ser93 of the oCNC α subunit is phosphorylated, the affinity for cAMP increases. The protein kinases involved in this effect are PKCs (γ, δ, and τ). When the ORNs are activated by phorbol ester, a PKC activator, the ORNs showed larger responses. Interestingly, the oCNC β subunit is a subject of phosphorylation by the protein tyrosine kinase (PTK), whereas the α subunit is not affected by PTK inhibitor (165, 166). Homomeric channel complexes with α subunits are unaffected by genistien, whereas heteromeric channel complex $\alpha + \beta$ channels are inhibited, indicating that the β subunit may interact with the PTK. This is different from the rCNC, in which the α subunit is crucial for allowing interaction with PTK (166). The two subunits that are unable to interact with the PTK (α subunit of oCNC and β subunit of rCNC) share a homologous domain in the NH2 terminus that enables these subunits to interact with Ca^{2+}/calmodulin (155, 159). Expression of oCNC subunits is especially widespread in the nervous system and elsewhere, but channels in different locations may differ in their subunit compositions (167, 168), potentially providing for differential modulation by protein kinases.

Expression of Cyclic Nuceotide-Gated Channels in Other Systems

CNCs have been found elsewhere in the nervous system (153, 169) and have been implicated in processes as diverse as synaptic modulation and axon outgrowth in animals ranging from the nematode to mammals (170, 171). CNCs have also been found in a variety of other cell types including kidney, testis, and heart (47, 172), where they may fulfill various physiological functions.

Transcripts for subunit α of both the rod photoreceptor (rCNCα) and the olfactory receptor cell (oCNCα) subtype of CNC were detected in adult rat hippocampus in most principal neurons, including pyramidal granule cells (169). Two genes are colocalized in individual neurons. Comparison of the patterns of expression of type 1 cGMP–dependent protein kinase and the CNCs suggests that hippocampal neurons can respond to changes in cGMP levels with both rapid changes in CNC activity and slower changes induced by phosphorylation. Recent studies have shown that increased cGMP levels in hippocampal neurons can lead to increased transmitter release and that Ca^{2+} entry into presynaptic terminals plays a critical role. Although some of these effects may be mediated by protein kinases, the CNCs may also contribute to some of the actions of cAMP and cGMP in hippocampus and other regions of the central nervous system. It is possible that they exert their effects at both postsynaptic and presynaptic sites. This will be answered when the subcellular localization of CNCs in hippocampus is determined.

Ca^{2+}-DEPENDENT Cl^- CONDUCTANCE   Many odorants elicit the activation of a Ca^{2+}-permeable nonselective cation conductance (46, 173, 174), followed by a nonlinearly Ca^{2+}-dependent Cl^- conductance (175, 176). Owing to the specific ion concentrations in the mucus, these result in an influx of extracellular Ca^{2+} followed by outflux of intracellular Cl^-, which cooperatively depolarizes the cilia membrane (177). The chloride conductance significantly amplifies the odor-induced depolarization of the cilia, and the nonlinear Ca^{2+} sensitivity of this current is thought to introduce an excitation threshold to improve the signal-to-noise ratio of the transduction process (154, 178). If the stimulus strength is sufficient, the depolarization of the cilia propagates by passive electrotonic spread and finally triggers the generation of action potentials at the initial segment of the axon (179). A Cl^- channel blocker reduces the receptor current by 85% in rat olfactory cells. Cl^- efflux amplifies the cationic current in terrestrial animals. The chloride channel mediating this current remains to be identified.

K^+ CONDUCTANCE   Repolarization of the cells seems to involve Ca^{2+}-dependent and fast inactivation potassium conductances. Repolarization of the action potential is achieved by the activation of various K^+ channels (180–186). A transient, 4-aminopyridine (4-AP)-sensitive K^+ conductance and Ca^{2+}-activated K^+ conductances measured in various species seem to contribute to the repolarization and an increase of the impedance of ORNs (180–189). However, in cultured ORNs of rat, the fast 4-AP-sensitive current and a Ca^{2+}-dependent K^+ current are absent, whereas a large delayed-rectifier K^+ conductance is observed (190).

Na^+ CONDUCTANCE   Excitation of an olfactory neuron generates a receptor potential, and when the membrane potential reaches the firing threshold, Na^+ channels activate and initiate spike generation. The Na^+ currents are increased via cGMP-dependent phosphorylation, whereas the delayed rectifier K^+ currents are not affected by PKG-mediated phosphorylation (191). Cyclic GMP may lower the threshold in olfactory perception by decreasing the current threshold to generate spikes, and also prevent the saturation of odor signals by increasing the maximum spike frequency (191). Variability in this observation could be because these channels are primarily located in the axonal membrane, which is partially lost during preparation (189, 192).

CA^{2+} CONDUCTANCE   High-voltage-activated Ca^{2+} currents in ORNs have been described in various species (180, 181, 185–187, 189, 190). Quantitative ratiometric Ca^{2+} imaging with fluo3 and FuraRed revealed that the high-voltage-activated Ca^{2+} channels in *Xenopus* ORNs are primarily situated on the soma and the proximal dendrite (193).

A low-voltage-activated Ca^{2+} current may also be involved in action potential formation. The low-voltage-activated Ca^{2+} currents play a particular role in relatively large ORNs that have high capacitances and ion membrane time constants (194, 195).

OTHER SIGNALING CASCADES IN OLFACTION

Although targeted deletion studies support the central role of cAMP in odorant detection, many investigators have provided evidence demonstrating that other signaling pathways are activated in response to odor stimulation.

Inositol-1,4,5-trisphosphate (IP_3)

In the brain and peripheral tissues receptor-mediated stimulation of phospholipase C generates IP_3, which releases calcium from endoplasmic reticulum stores by binding to specific IP_3 receptors (196, 197). Plasma membrane IP_3 receptors have been identified in lymphocytes (198) and neurons (199, 200) that permit calcium entry from extracellular sources. Thus, calcium may be made available to affect a variety of targets in response to odor stimulation. There are now five families of IP_3 receptors (201–203).

Studies in several species implicate IP_3 in olfaction. However, electrophysiological experiments have in many cases failed to demonstrate a role for IP_3 in the initial phases of the response to odorants. Huque & Bruch (204) showed phospholipase C activity in isolated catfish olfactory cilia. Restrepo and collaborators (205) showed that amino acids enhanced calcium flux in isolated catfish ORNs. Utilizing the rapid mixing technique, Breer and colleagues (39) demonstrated increases in IP_3 levels in response to some odorants.

Studies in primary cultures of ORNs confirmed that odorants stimulate the production of IP_3. Exposure of cells to low nanomolar concentrations of odorants resulted in IP_3 formation (35, 206). All odorants stimulate cAMP and IP_3 production in primary culture, although with different potencies, suggesting interactions with different receptors. The enhancement by single odors of both cAMP and IP_3 production affords a mechanism for increased specificity of odor detection. However, these studies were only performed at longer (1 s and beyond) times after odor encounter. Ache and coworkers confirmed that odors differentially stimulate dual pathways in isolated lobster antennules (207). Odors elevated cAMP and IP_3 in the outer dendritic membranes of lobster in vitro. IP_3 carried the stimulatory current, while cAMP was inhibitory, providing a mechanism for fine-tuning of the responses. The relevance of IP_3 to mammalian olfaction has been questioned by several groups, whose knock-outs affecting the cAMP signaling cascade resulted in loss of the electro-olfactogram responses, suggesting that cAMP is the sole odorant-generated second messenger (40, 41, 104). These discrepancies may be reconciled if cAMP is indeed the primary second messenger required for the initial events of odor detection and cellular depolarization, whereas IP_3 is involved in other secondary responses, such as adaptation or activity-driven cellular responses, not electro-olfactogram generation.

IP_3 receptors have been localized immunohistochemically to the ciliary surface membrane (208), positioning IP_3 to trigger the influx of extracellular calcium. There is also evidence for plasma membrane IP_3-sensitive channels in lobster ORNs (209, 210). Kalinoski and colleagues have also demonstrated an IP_3-like receptor in isolated catfish cilia, although its micromolar K_d for IP_3 suggests a different type of IP_3 receptor (211). Several phospholipase C isoforms have been demonstrated in olfactory epithelium (209, 212, 213).

Reconciliation of the data thus far obtained for IP_3 will require further work. For some time debate existed as to whether cGMP or calcium was the visual second messenger (214). We now know that, whereas cGMP is central, calcium is the major modulator of cGMP levels (215–218). Additionally, there are striking interspecies differences: Whereas IP_3 is important in amphibian phototransduction, no role has thus far been found in mammals. Olfaction may have similar complexities.

cGMP

Cyclic GMP is well established to be the primary second messenger in visual signal transduction. A number of studies indicate that cGMP may play an important role in olfactory transduction. Odorants increase cGMP levels in olfactory tissues (56) and ORNs (52). When compared with the odorant-induced increase in cAMP and IP_3 levels, the rise in cGMP levels occurred with a slower, sustained time course. This delayed response suggests that cGMP may not be involved in initial signaling events, but rather in long-term cellular events such as desensitization (219) or in the activation of neuronal activity–dependent transcription (55). cGMP levels are regulated by two distinct classes of guanylyl cyclases, soluble guanylyl cyclase and receptor guanylyl cyclase. Soluble guanylyl cyclase is activated by gaseous messengers such as NO or CO, whereas receptor guanylyl cyclase is activated by specific extracellular ligands or calcium. Both guanylyl cyclases are expressed in ORNs, implying a complex regulation of cGMP levels in olfaction (52, 220).

Diffusible gaseous messenger molecules such as NO or CO can stimulate soluble guanylyl cyclase by binding to the heme group in soluble guanylyl cyclases

(221). NO and CO are produced by NO synthase and heme oxygenase (HO), respectively. In ORNs NO synthase is expressed at embryonic stages and is markedly reduced at early postnatal stages, whereas HO is highly expressed after birth (222, 223). These data suggest that NO plays an important role during development, whereas HO functions in mature ORNs. Two forms of HO have been identified: HO-1 and HO-2. HO-1 is a heat shock protein (hsp-32) induced by heme, heavy metals, stress, and hormones (224–229) and is highly expressed in the spleen and liver, where it is responsible for the destruction of heme from red blood cells.

HO-2 is not inducible and is highly expressed in the brain, especially in neurons of the olfactory epithelium and in the neuronal and granule cell layer of the olfactory bulb. In situ hybridization analysis showed that guanylyl cyclase and HO-2 were found in ORNs (52). Incubation of ORNs with the HO inhibitor, zinc protoporphyrin-9 (Zn PP-9), lowered cGMP levels in ORNs (222). In addition, odorants augment cGMP levels in ORNs (52, 222). This odorant-induced cGMP increase could be inhibited by Zn PP-9, but not by an NO synthase inhibitor. Interestingly, the inhibition of HO could not entirely deplete cGMP levels in ORNs, suggesting that particulate guanylyl cyclases may also contribute to cGMP production in ORNs (222). Exposure of isolated cilia derived from OR neurons to various odorants increased cGMP levels (220). Thus, there is a strong suggestion that both soluble and receptor guanylyl cyclases have roles in olfactory signal transduction.

The observation that the inhibition of HO in ORNs could not totally block the cGMP response suggested the involvement of receptor guanylyl cyclases in odorant transduction. The fact that an NO donor and soluble guanylyl cyclase activator, sodium nitroprusside, could not alter the cGMP levels in isolated cilia supported the idea that a receptor guanylyl cyclase might play a role in olfactory cilia. An olfactory-specific receptor guanylyl cyclase, guanylyl cyclase-D (GC-D), has been identified in olfactory epithelium (129). GC-D has been suggested to function as the receptor of sensory neurons to specific odors. The role and regulation of these guanylyl cyclases in the olfactory system are unclear.

Recent studies have identified an odorant-responsive receptor guanylyl cyclase in rat olfactory sensory cilia (220). At least two receptor guanylyl cyclases exist in cilia, a low K_m and a high K_m isoform (220). Odorants were shown to elevate cGMP levels in cultured ORNs (222) and in isolated olfactory cilia (220) in a calcium-dependent manner. A number of experiments suggested that calcium plays a role in odorant transduction and can fluctuate upon odorant exposure (172, 230, 231). Hence, it was hypothesized that an OR guanylyl cyclase could be regulated by a calcium-binding protein, such as guanylyl cyclase–activating protein (GCAP), similar to that found in the visual transduction pathway. Immunohistochemical studies using anti-GCAP1 antibodies revealed that GCAP1 was highly localized to the olfactory cilia (220). Moreover, GCAP1 regulated the odorant-induced cGMP response in isolated rat olfactory cilia in a calcium-dependent manner (220). Thus, ORNs contain multiple cGMP pathways that mediate delayed and sustained cGMP responses to odorants.

Calcium

Calcium regulates diverse cellular functions, and in general these functions are mediated by a variety of calcium-binding proteins (232). Odorant stimulation of ORNs results in a calcium influx, which in turn can modulate a number of transduction pathways. Calmodulin and other calcium-binding proteins may participate in the processing of olfactory information. Therefore, study of the calcium-binding proteins may provide important background information about the complex signal transduction pathway involved in olfaction. Olfactory tissue contains various calcium-binding proteins: calmodulin, calretinin, calbindin-D28k, neurocalcin, and recoverin (233). Another calcium-binding protein, S-100, is restricted to glial cells, primarily around the cribiform plate.

Calmodulin is found in olfactory cilia at a concentration of about 1 mM (234). The odorant-induced intracellular elevation of calcium is thought to promote adaptation because calcium/calmodulin can reduce the affinity of the oCNC for cAMP by 20-fold (155, 235). Extracellular calcium is absolutely required for the decay phase of the odorant-induced whole cell current, which in the absence of extracellular calcium remains at a steady state (236). Calcium/calmodulin can also affect CNC activity (237).

Neurocalcin is a calcium-binding protein with three EF hand motifs and is also expressed in the rat olfactory epithelium (238). Neurocalcin is expressed in ORNs, where it is associated with outer mitochondrial membrane, endoplasmic reticulum, and axon fibers. The intracellular distribution of neurocalcin in ORNs suggested that this protein may participate in cytoskeletal arrangement in ORNs. The expression of neurocalcin in postnatal development was also studied (233, 239). Neurocalcin showed a gradient of expression pattern descending from the central to the lateral areas in the nasal cavity during childhood; this expression pattern became identical to the adult profile after 20 days.

Additional calcium-binding proteins have been localized to the olfactory system. A 26-kDa calcium-binding protein named p26olf was identified in frog olfactory epithelium (240). p26olf contains two S-100-like regions and is localized to the cilia layer of the olfactory epithelium, suggesting that it is a dimeric form of S-100 protein that may be involved in the olfactory transduction or adaptation. Visinin-like protein (VILIP), a member of the neuronal subfamily of EF-hand calcium-sensor proteins, was associated with ORNs of the rat olfactory epithelium (241). VILIP is localized prominently to cilia and dendritic knobs. In vitro recombinant VILIP attenuates odorant-induced cAMP formation in a calcium-dependent manner. The observation that VILIP does not interfere with odorant-induced receptor desensitization and that VILIP inhibits the forskolin-induced cAMP production suggests that it may directly affect adenylyl cyclases and in turn may play a role in adaptation of ORNs.

A GCAP1-like calcium-binding protein is present in rat olfactory cilia (220). GCAP1 was purified and later cloned from bovine retina by Palczewski and colleagues (130). GCAP1 is a 21-kDa cytosolic EF-hand-family protein and has been proposed to function as a photoreceptor-specific calcium-binding protein to activate particulate guanylyl cyclase, thus restoring the cGMP level in light-activated photoreceptor cells. Immunohistochemical studies revealed the presence of GCAP1 in rat olfactory cilia (220). Interestingly, purified GCAP1 potentiated cGMP production at high calcium concentrations in isolated rat olfactory cilia (220). In photoreceptor cells, GCAP1 activates particulate guanylyl cyclase when the intracellular calcium level is low. The size of the olfactory GCAP (19 kDa) was not identical to the retinal GCAP1. Thus, the olfactory GCAP is considered a GCAP1-like protein. The cloning of the olfactory GCAP will answer the precise function and mechanism of the olfactory guanylyl GCAP in olfaction.

Calcium itself mediates Cl^- conductance in ORNs (132, 176, 242). The odor-induced currents show little rectification. It appears that the depolarizing current has two components, an initial inward cationic conductance followed by an inward anionic Cl^- conductance (132, 237, 242). Calcium, which enters the cilia through the cyclic nucleotide-gated channel, triggers a calcium-activated Cl^- channel in olfactory cilia membrane (176). This conductance may serve as a "fail safe" so that cells can depolarize, irrespective of changes in extracellular milieu.

DESENSITIZATION

Desensitization of receptor-mediated responses can occur through a variety of processes, including phosphorylation, internalization, and receptor-effector uncoupling (243–245). The homologous desensitization of G protein–coupled receptors is well established in β2-adrenergic receptor (βAR-2) as a model (246, 247). Phosphorylation of receptors by a specific receptor kinase such as β-adrenergic receptor kinases (β-ARKs) or G protein receptor kinases (GRKs) mediate homologous desensitization. Complete quenching of signal transduction requires the binding of a protein called β-arrestin (βARR) to a phosphorylated receptor (248).

Specific isoforms of GRKs and βARR, βARK-2, and βARR-2 were localized to olfactory neurons, specifically to olfactory cilia and dendritic knobs (249). Other isoforms of βARK or βARR were not present in these cells. Functional studies of βARK-2 and βARR-2 in the olfactory cilia were performed (249, 250). The odorant-induced cAMP production was monitored in the presence or absence of neutralizing antibodies against specific isoforms of βARK and βARR. Preincubation of olfactory cilia with neutralizing antibodies to βARK-2 and βARR-2 increased the absolute levels of odorant-induced cAMP as high as fourfold and completely blocked desensitization. Later mice with targeted disruption of βARK-2 have been available, and cilia preparations derived from the βARK-2-deficient mice showed lack of the agonist-induced desensitization (251). Taken together, the expression of βARK-2 and βARR-2 within the olfactory cilia, the inhibition of desensitization with βARK-2 and βARR-2 neutralizing antibodies, and the lack of the agonist-induced desensitization in the βARK-2 deficient mice suggest that βARK-2 and βARR-2 mediate the odorant-dependent desensitization in olfaction.

In addition to this mechanism for homologous desensitization, it has been suggested that PKA or PKC may play a role in odorant-related heterologous desensitization (252). PKA has been implicated in olfactory desensitization following the increase in cAMP by odorant stimulation, whereas PKC may mediate desensitization following phosphoinositide cycle activation by odorant stimulation. However, these results need to be reexamined, in light of more recent data using knockout animals that indicate that cAMP mediates odorant detection.

Cyclic GMP may also play a role in desensitization. Zufall & Leinders-Zufall (53) showed that cGMP mediated a long-lasting form of odor response adaptation in tiger salamander. The long-lasting adaptation lasted for several minutes and was attributable to cyclic nucleotide-gated channel modulation by cGMP. They showed that this form of long-lasting adaptation was abolished selectively by HO inhibitors (which prevent CO release and cGMP formation), whereas odor excitation was unaffected. The results suggest that endogenous CO/cGMP signals contribute to olfactory desensitization.

LONG-TERM RESPONSES TO ODORANT DETECTION

The theory that extracellular signals, such as hormones, growth factors, and neuronal activity, can modulate transcriptional events to produce long-term changes in cellular activity is well established (253). However, the long-lasting effects of odorant stimulation in ORNs has only recently been studied.

A delayed cAMP response upon odorant stimulation was characterized and was mediated by cGMP via activation of a cGMP-dependent protein kinase (PKG) (220). Based on the kinetics of the delayed cAMP response discussed above, it was suggested that cGMP might mediate a delayed cAMP response to regulate long-term cellular responses to odorant detection, including the regulation of gene expression. Recent studies support this idea. Odorant stimulation could therefore potentially result in transcriptional changes via CREB activation (55). Incubation of ORNs with either 8-Br-cGMP or a soluble guanylyl cyclase activator (sodium nitroprusside) increased CREB activation. Thus, cGMP produced upon odorant stimulation may generate a sustained cAMP signal capable of activating CREB.

Involvement of the Ras-MAPK (mitogen-activated protein kinase) signal transduction pathway in olfaction was recently reported in *C. elegans* (254). The Ras-MAPK pathway plays important roles in cellular proliferation and differentiation in response to extracellular signals. Mutational inactivation and hyperactivation of this pathway impaired efficiency of chemotaxis to a set of odorants. The activation of MAPK upon odorant stimulation was dependent on calcium fluxes via the nucleotide-gated channel and the voltage-activated calcium channel. More recently, Watt & Storm demonstrated that odorants activate MAPK in rodent ORNs (255). The odorant-activation of the MAPK pathway led to the activation of cAMP response element (CRE)-mediated transcription. The odorant stimulation of MAPK activation was ablated by inhibition of CaM-dependent protein kinase II (CaMKII), suggesting that odorant activation of MAPK is mediated through CaMKII. Moreover, discrete populations of ORNs display CRE-mediated gene transcription when stimulated by odorants in mice. These data suggest that ORNs may undergo long-term adaptive changes mediated through CRE-mediated transcription.

CONCLUSIONS

Our understanding of olfactory transduction has advanced rapidly in the past decade with the realization that G proteins and seven-transmembrane spanning domain receptors are involved in odorant detection. It remains to be determined how other transduction cascades interface with these G protein–coupled receptors to either fine-tune the response or mediate other aspects of odor detection. Understanding the olfactory code will allow us to manipulate olfactory perception in both health and disease. Our appreciation of the ability of odor perception to influence long-term neuronal responses, and potentially neuronal survival, may provide clues to understanding this process in other neuronal systems.

Visit the Annual Reviews home page at www.AnnualReviews.org

LITERATURE CITED

1. Buck LB. 1996. Information coding in the vertebrate olfactory system. *Annu. Rev. Neurosci.* 19:517–44
2. Getchell TV. 1986. Functional properties of vertebrate olfactory receptor neurons. *Physiol. Rev.* 66:772–818
3. Getchell TV, Margolis FL, Getchell ML. 1985. Perireceptor and receptor events in vertebrate olfaction. *Prog. Neurobiol.* 23:317–45
4. Firestein S. 1996. Olfaction: scents and sensibility. *Curr. Biol.* 6:666–67
5. Hildebrand JG, Shepherd GM. 1997. Mechanisms of olfactory discrimination: converging evidence for common principles across phyla. *Annu. Rev. Neurosci.* 20:595–631
6. Gold GH. 1999. Controversial issues in vertebrate olfactory transduction. *Annu. Rev. Physiol.* 61:857–71
7. Graziadei PPC, Monti-Graziadei GA. 1979. Neurogenesis and neuron regeneration in the olfactory system of mammals. *J. Neurocytol.* 8:1–18

8. Moulton DG, Beidler LM. 1967. Structure and function in the peripheral olfactory system. *Physiol. Rev.* 47:1–52
9. Labarca P, Bacigalupo J. 1988. Ion channels from chemosensory olfactory neurons. *J. Bioenerg. Biomembr.* 20:551–69
10. Lowe G, Gold GH. 1993. Contribution of the ciliary cyclic nucleotide-gated conductance to olfactory transduction in the salamander. *J. Physiol.* 462:175–96
11. Menco BPM, Brunch RC, Dau B, Danho W. 1992. Ultrastructural localization of olfactory transduction components: the G protein subunit $G_{olf\alpha}$ and type III adenylyl cyclase. *Neuron* 8:441–53
12. Menco BPM. 1997. Ultrastructural aspects of olfactory signaling. *Chem. Senses* 22:295–311
13. Farbman AI. 1992. Development and plasticity. In *Cell Biology of Olfaction*, ed. PW Barlow, D Bray, PB Green, JMW Slack, pp. 167–206. Cambridge, UK: Cambridge Univ. Press

14. Graziadei PPC. 1973. Cell dynamics in the olfactory mucosa. *Tissue Cell* 5:113–31

15. Graziadei PPC, Metcalf JF. 1971. Autoradiographic and ultrastructural observation on the frog's olfactory mucosa. *Zellforsch* 116:305–18

16. Caggiano M, Kauer JS, Hunter DD. 1994. Globose basal cells are neuronal progenitors in the olfactory epithelium: a lineage analysis using a replication-incompetent retrovirus. *Neuron* 13:339–52

17. Roskams AJI, Cai X, Ronnett GV. 1998. Expression of neuron-specific beta-III tubulin during olfactory neurogenesis in the embryonic and adult rat. *Neuroscience* 83:191–200

18. Margolis FL. 1980. A marker protein for the olfactory chemoreceptor neuron. In *Proteins of the Nervous System*, ed. RA Bradshaw, DM Schneider, pp. 59–84. New York: Raven

19. Farbman AI, Margolis FL. 1980. Olfactory marker protein during ontogeny: immunohistochemical localization. *Dev. Biol.* 74:205–15

20. Carr VM, Farbman AI. 1993. The dynamics of cell death in the olfactory epithelium. *Exp. Neurol.* 124:308–14

21. Okano TM. 1974. Secreation and electrogenesis of the supporting cell in the olfactory epithelium. *J. Physiol.* 242:353–70

22. Lazard D, Zupko K, Poria Y, Nef P, Lazarovits J, et al. 1991. Odorant signal termination by olfactory UDP-glucuronosyl transferase. *Nature* 349:790–93

23. Hansel DE, Eipper BA, Ronnett GV. 2001. Neuropeptide Y: functions as a neuroproliferative factor. *Nature* 410:940–44

24. Baraban SC, Hollopeter G, Erickson JC, Schwartzkroin PA, Palmiter RD. 1997. Knock-out mice reveal a critical antiepileptic role for neuropeptide Y. *J. Neurosci.* 17:8927–36

25. Danger JM, Tonon MC, Jenks BG, Saint-Pierre S, Martel JC, et al. 1990. Neuropeptide Y: localization in the central nervous system and neuroendocrine functions. *Fundam. Clin. Pharmacol.* 4:307–40

26. Adrian TE, Allen JM, Bloom SR, Ghatei MA, Rossor MN, et al. 1983. Neuropeptide Y distribution in human brain. *Nature* 306:584–86

27. Calof AL, Chikaraishi DM. 1989. Analysis of neurogenesis in a mammalian neuroepithelium: proliferation and differentiation of an olfactory neuron precursor in vitro. *Neuron* 3:115–27

28. Goldstein BJ, Schwob JE. 1996. Analysis of the globose basal cell compartment in rat olfactory epithelium using GBC-1, a new monoclonal antibody against globose basal cells. *J. Neurosci.* 16:4005–16

29. Huard JM, Youngentob SL, Goldstein BJ, Luskin MB, Schwob JE. 1998. Adult olfactory epithelium contains multipotent progenitors that give rise to neurons and non-neural cells. *J. Comp. Neurol.* 400:469–86

30. Rhein LD, Cagan RH. 1980. Biochemical studies of olfaction: isolation, characterization and odorant binding activity of cilia from rainbow trout olfactory rosettes. *Proc. Natl. Acad. Sci. USA* 77:4412–16

31. Malnic B, Hirono J, Sato T, Buck L. 1999. Combinatorial receptor codes for odors. *Cell* 96:713–23

32. Dwyer ND, Troemel ER, Sengupta P, Bargmann CI. 1998. OR localization to olfactory cilia is mediated by ODR-4, a novel membrane-associated protein. *Cell* 93:455–66

33. Pace U, Hanski E, Salomon Y, Lancet D. 1985. Odorant-sensitive adenylate cyclase may mediate olfactory reception. *Nature* 316:255–58

34. Sklar PB, Anholt RRH, Snyder SH. 1986. The odorant-sensitive adenylate cyclase of olfactory receptor neurons. *J. Biol. Chem.* 261:15538–43

35. Ronnett GV, Cho H, Hester LD, Wood SF, Snyder SH. 1993. Odorants differentially enhance phosphoinositide turnover

and adenylyl cyclase in olfactory receptor neuronal cultures. *J. Neurosci.* 13:1751–58

36. Ronnett GV, Parfitt DJ, Hester LD, Snyder SH. 1991. Odorant-sensitive adenylate cyclase: rapid potent activation and desensitization in primary olfactory neuronal cultures. *Proc. Natl. Acad. Sci. USA* 88:2366–69

37. Ronnett GV, Snyder SH. 1992. Molecular messengers of olfaction. *Trends Neurosci.* 15:508–12

38. Jaworsky DE, Matsuzaki O, Borisy FF, Ronnett GV. 1995. Calcium modulates the rapid kinetics of the odorant-induced cyclic AMP signal in rat olfactory cilia. *J. Neurosci.* 15:310–18

39. Breer H, Boekhoff I, Tareilus E. 1990. Rapid kinetics of second messenger formation in olfactory transduction. *Nature* 345:65–68

40. Brunet LL, Gold GH, Ngai J. 1996. General anosmia caused by a targeted disruption of the mouse olfactory cyclic nucleotide-gated cation channel. *Neuron* 17:681–93

41. Wong ST, Trinh K, Hacker B, Chan GC, Lowe G, et al. 2000. Disruption of the type III adenylyl cyclase gene leads to peripheral and behavioral anosmia in transgenic mice. *Neuron* 27:487–97

42. Nakamura T, Gold GH. 1987. A cyclic nucleotide-gated conductance in olfactory receptor cilia. *Nature* 325:442–44

43. Firestein S, Werblin FS. 1989. Odor-induced membrane currents in vertebrate olfactory receptor neurons. *Science* 244:79–82

44. Getchell TV, Shepherd GM. 1978. Adaptive properties of olfactory receptor analysed with odour pulses of varying durations. *J. Physiol.* 282:541–60

45. Ottoson D. 1956. Analysis of the electrical activity of the olfactory epithelium. *Acta Physiol. Scand.* 122:1–83

46. Frings S, Seifert R, Godde M, Kaupp UB. 1995. Profoundly different calcium permeation and blockage determine the specific function of distinct cyclic nucleotide-gated channels. *Neuron* 15:169–79

47. Kaupp UB. 1991. The cyclic nucleotide-gated channels of vertebrate photoreceptors and olfactory epithelium. *Trends Neurosci.* 14:150–57

48. Breer H, Boekhoff I. 1991. Odorants of the same odor class activate different second messenger pathways. *Chem. Senses* 16:19–29

49. Miyamoto T, Restrepo D, Cragoe EJ, Teeter JH. 1992. IP$_3$ and cAMP-induced responses in isolated olfactory receptor neurons from the channel catfish. *J. Membr. Biol.* 127:173–83

50. Schandar M, Laugwitz KL, Boekhoff I, Kroner C, Gudermann T, et al. 1998. Odorants selectively activate distinct G protein subtypes in olfactory cilia. *J. Biol. Chem.* 273:16669–77

51. Ingi T, Cheng J, Ronnett GV. 1996. Carbon monoxide: an endogenous modulator of the nitric oxide-cyclic GMP signaling system. *Neuron* 16:835–42

52. Verma A, Hirsch DJ, Glatt CE, Ronnett GV, Snyder SH. 1993. Carbon monoxide: a putative neural messenger. *Science* 259:381–84

53. Zufall F, Leinders-Zufall T. 1997. Identification of a long-lasting form of odor adaptation that depends on the carbon monoxide/cGMP second-messenger system. *J. Neurosci.* 17:2703–12

54. Moon C, Jaberi P, Otto-Bruc A, Baehr W, Palczewski K, Ronnett GV. 1998. Calcium-sensitive particulate guanylyl cyclase as a modulator of cAMP in olfactory receptor neurons. *J. Neurosci.* 18:3195–205

55. Moon C, Sung YK, Reddy R, Ronnett GV. 1999. Odorants induce the phosphorylation of the cAMP response element binding protein in olfactory receptor neurons. *Proc. Natl. Acad. Sci. USA* 96:14605–10

56. Breer H, Klemm T, Boekhoff I. 1992. Nitric oxide mediated formation of cyclic

GMP in the olfactory system. *NeuroReport* 3:1030–31

57. Pelosi P, Baldaccini NE, Pisanelli AM. 1982. Identification of a specific olfactory receptor for 2-isobutyl-3-methoxypyrazine. *Biochem. J.* 201:245–48

58. Pevsner J, Sklar PB, Snyder SH. 1986. Odorant-binding protein: localization to nasal gland and secretions. *Proc. Natl. Acad. Sci. USA* 83:4942–46

59. Pevsner J, Trifiletti RR, Strittmatter SS, Snyder SH. 1985. Isolation and characterization of an olfactory receptor protein for odorant pyrazines. *Proc. Natl. Acad. Sci. USA* 82:3050–54

60. Pevsner J, Hou VX, Snyder SH. 1990. Odorant-binding protein: characterization of ligand binding. *J. Biol. Chem.* 265: 6118–25

61. Bok D. 1990. Processing and transport of retinoids by the retinal pigment epithelium. *Eye* 4:326–32

62. Pevsner J, Hwang PM, Sklar PB, Venable JC, Snyder SH. 1988. Odorant-binding protein and its nRNA are localized to lateral nasal gland implying a carrier function. *Proc. Natl. Acad. Sci. USA* 85:2383–87

63. Dear TN, Boehm T, Keverne EB, Rabbitts TH. 1991. Novel genes for potential ligand-binding proteins in subregions of the olfactory mucosa. *EMBO J.* 10:2813–19

64. Burova TV, Rabesona H, Choiset Y, Jankowski CK, Sawyer L, Haertle T. 2000. Why has porcine VEG protein unusually high stability and suppressed binding ability? *Biochim. Biophys. Acta* 1478:267–79

65. Lobel D, Marchese S, Krieger J, Pelosi P, Breer H. 1998. Subtypes of odorant-binding proteins—heterologous expression and ligand binding. *Eur. J. Biochem.* 254: 318–24

66. Pes D, Pelosi P. 1995. Odorant-binding proteins of the mouse. *Comp. Biochem. Physiol. B* 112:471–79

67. Garibotti M, Navarrini A, Pisanelli AM, Pelosi P. 1997. Three odorant-binding proteins from rabbit nasal mucosa. *Chem. Senses* 22:383–90

68. Bianchet MA, Bains G, Pelosi P, Pevsner J, Snyder SH, et al. 1996. The three-dimensional structure of bovine odorant binding protein and its mechanism of odor recognition [see comments]. *Nat. Struct. Biol.* 3:934–39

69. Dal Monte M, Andreini I, Revoltella R, Pelosi P. 1991. Purification and characterization of two odorant-binding proteins from nasal tissue of rabbit and pig. *Comp. Biochem. Physiol. B* 99:445–51

70. Vogt RG, Prestwich GD, Lerner MR. 1990. Odorant-binding-protein subfamilies associate with distinct classes of olfactory receptor neurons in insects. *J. Neurobiol.* 22:74–84

71. Vogt RG, Rybczynski R, Lerner MR. 1991. Molecular cloning and sequencing of general odorant-binding proteins GOBP1 and GOBP2 from the tobacco hawk moth manduca sexta: comparisons with other insect OBPs and their signal peptides. *J. Neurosci.* 11:2972–84

72. Buck L, Axel R. 1991. A novel multigene family may encode ORs: a molecular basis for odor recognition. *Cell* 65:175–87

73. Buck LB. 1992. The olfactory multigene family. *Curr. Biol.* 2:467–73

74. Mombaerts P. 1999. Seven-transmembrane proteins as odorant and chemosensory receptors. *Science* 286:707–11

75. Glusman G, Yanai I, Rubin I, Lancet D. 2001. The complete human olfactory subgenome. *Genome Res.* 11:685–702

76. Raming K, Konzelmann S, Breer H. 1998. Identification of a novel G-protein coupled receptor expressed in distinct brain regions and a defined olfactory zone. *Recept. Channel* 6:141–51

77. Feingold EA, Penny LA, Nienhuis AW, Forget BG. 1999. An olfactory receptor gene is located in the extended human beta-globin gene cluster and is expressed in erythroid cells. *Genomics* 61:15–23

78. Zozulya S, Echeverri F, Nguyen T. 2001.

The human olfactory receptor repertoire. *Genome Biol. Res.* 0018.1–0018.12

79. Lancet D, Ben-Arie N, Cohen S, Gat U, Gross-Isseroff R, et al. 1993. Olfactory receptors: transduction, diversity, human psychophysics and genome analysis. *Ciba Found. Symp.* 179:131–41

80. Rouquier S, Taviaux S, Trask BJ, Brand-Arpon V, van den Engh G, et al. 1998. Distribution of olfactory receptor genes in the human genome. *Nat. Genet.* 18:243–50

81. Venter JC, Adams MD, Myers EW, Li PW, Mural RJ, et al. 2001. The sequence of the human genome. *Science* 291:1304–51

82. Trask BJ, Massa H, Brand-Arpon V, Chan K, Friedman C, et al. 1998. Large multi-chromosomal duplications encompass many members of the olfactory receptor gene family in the human genome. *Hum. Mol. Genet.* 7:2007–20

83. Trask BJ, Friedman C, Martin-Gallardo A, Rowen L, Akinbami C, et al. 1998. Members of the olfactory receptor gene family are contained in large blocks of DNA duplicated polymorphically near the ends of human chromosomes. *Hum. Mol. Genet.* 7:13–26

84. O'Brien SJ, Eisenberg JF, Miyamoto M, Hedges SB, Kumar S, et al. 1999. Genome maps 10. Comparative genomics. Mammalian radiations. Wall chart. *Science* 286:463–78

85. Ressier KJ, Sullivan SL, Buck LB. 1993. A zonal organization of OR gene expression in the olfactory epithelium. *Cell* 73:597–609

86. Vassar R, Ngai J, Axel R. 1993. Spatial segregation of OR expression in the mammalian olfactory epithelium. *Cell* 74:309–18

87. Mombaerts P. 1999. Molecular biology of ORs in vertebrates. *Annu. Rev. Neurosci.* 22:487–509

88. Koshimoto H, Katoh K, Yoshihara Y, Nemoto Y, Mori K. 1994. Immunohistochemical demonstration of embryonic expression of an odor receptor protein and its zonal distribution in the rat olfactory epithelium. *Neurosci. Lett.* 169:73–76

89. Menco BP, Cunningham AM, Qasba P, Levy N, Reed RR. 1997. Putative odour receptors localize in cilia of olfactory receptor cells in rat and mouse: a freeze-substitution ultrastructural study. *J. Neurocytol.* 26:691–706

90. Menco BP, Jackson JE. 1997. A banded topography in the developing rat's olfactory epithelial surface. *J. Comp. Neurol.* 388:293–306

91. Senhupta P, Chou JH, Bargmann CI. 1996. odr-10 encodes a seven transmembrane domain olfactory receptor required for responses to the odorant diacetyl. *Cell* 84:899–909

92. Krautwurst D, Yau KW, Reed RR. 1998. Identification of ligands for olfactory receptors by functional expression of a receptor library. *Cell* 95:917–26

93. Zhao H, Ivic L, Otaki JM, Hashimoto M, Mikoshiba K, Firestein S. 1998. Functional expression of a mammalian OR. *Science* 279:237–41

94. Ressler KJ, Sullivan SL, Buck LB. 1994. Information coding in the olfactory system: evidence for a stereotyped and highly organized epitope map in the olfactory bulb. *Cell* 79:1245–55

95. Mombaerts P, Wang F, Dulac C, Chao SK, Nemes A, et al. 1996. Visualizing an olfactory sensory map. *Cell* 87:675–86

96. Drutel G, Arrang JM, Diaz J, Wisnewsky C, Schwartz K, Schwartz JC. 1995. Cloning of OL1, a putative olfactory receptor and its expression in the developing rat heart. *Recept. Channel* 3:33–40

97. Nef S, Nef P. 1997. Olfaction: transient expression of a putative OR in the avian notochord. *Proc. Natl. Acad. Sci. USA* 94:4766–71

98. Abe K, Kusakabe Y, Tanemura K, Emori Y, Arai S. 1993. Multiple genes for G protein-coupled receptors and their expression in lingual epithelia. *FEBS Lett.* 316:253–56

99. Walensky LD, Ruat M, Bakin RE, Blackshaw S, Ronnett GV, Snyder SH. 1998. Two novel OR families expressed in spermatids undergo 5'-splicing. *J. Biol. Chem.* 273:9378–87

100. Walensky LD, Roskams JA, Lefkowitz RJ, Snyder SH, Ronnett GV. 1995. ORs and desensitization proteins colocalize in mammalian sperm. *Mol. Med.* 1:130–41

101. Rhein LD, Cagan RH. 1983. Biochemical studies of olfaction: binding specificity of odorants to cilia preparation from rainbow trout olfactory rosettes. *J. Neurochem.* 41:569–77

102. Jones DT, Reed RR. 1987. Molecular cloning of five GTP-binding protein cDNA species from rat olfactory neuroepithelium. *J. Biol. Chem.* 262:14241–49

103. Drinnan SL, Hope BT, Snutch TP, Vincent SR. 1991. G_{olf} in the basal ganglia. *Mol. Cell. Neurosci.* 2:66–70

104. Belluscio L, Gold GH, Nemes A, Axel R. 1998. Mice deficient in G(olf) are anosmic. *Neuron* 20:69–81

105. Firestein S, Shepherd GM, Werblin FS. 1990. Time course of the membrane current underlying sensory transduction in salamander olfactory receptor neurones. *J. Physiol.* 430:135–58

106. Boekhoff I, Tareilus E, Strotmann J, Breer H. 1990. Rapid activation of alternative second messenger pathways in olfactory cilia from rats by different odorants. *EMBO J.* 9:2453–58

107. Ronnett GV, Hester LD, Snyder SH. 1991. Primary culture of neonatal rat olfactory neurons. *J. Neurosci.* 11:1243–55

108. Pace U, Lancet D. 1986. Olfactory GTP-binding protein: signal transducing polypeptide of vertebrate chemosensory neurons. *Proc. Natl. Acad. Sci. USA* 83:4947–51

109. Hanoune J, Defer N. 2001. Regulation and role of adenylyl cyclase isoforms. *Annu. Rev. Pharmacol. Toxicol.* 41:145–74

110. Bakalyar HA, Reed RR. 1990. Identification of a specialized adenylyl cyclase that may mediate odorant detection. *Science* 250:1403–6

111. Menco BPM, Bruch RC, Dau B, Danho W. 1992. Ultrastructural localization of olfactory transduction components: the G protein subunit G_{olf} and type III adenylyl cyclase. *Neuron* 8:441–53

112. Chen S, Lane AP, Bock R, Leinders-Zufall T, Zufall F. 2000. Blocking adenylyl cyclase inhibits olfactory generator currents induced by "IP(3)-odors." *J. Neurophysiol.* 84:575–80

113. Mons N, Cooper DMF. 1995. Adenylate cyclases: critical foci in neuronal signaling. *Trends Neurosci.* 18:536–42

114. Choi EJ, Xia Z, Storm DR. 1992. Stimulation of the type III olfactory adenylyl cyclase by calcium and calmodulin. *Biochemistry* 31:6492–98

115. Choi E-J, Wong ST, Dittman AH, Storm DR. 1993. Phorbol ester stimulation of the type I and type III adenylyl cyclases in whole cells. *Biochem. J.* 32:1891–94

116. Wayman GA, Impey S, Storm DR. 1995. Ca^{2+} inhibition of type III adenylyl cyclase *in vivo.* *J. Biol. Chem.* 270:21480–86

117. Cooper DMF, Mons N, Karpen JW. 1995. Adenylyl cyclases and the interaction between calcium and cAMP signaling. *Nature* 374:421–24

118. Frings S. 1993. Protein kinase C sensitizes olfactory adenylate cyclase. *J. Gen. Physiol.* 101:183–205

119. Tang W-J, Gilman AG. 1992. Adenylyl cyclases. *Cell* 70:869–72

120. Beavo JA, Conti M, Heaslip RJ. 1994. Multiple cyclic nucleotide phosphodiesterases. *Mol. Pharmacol.* 46:399–405

121. Beltman J, Sonnenburg WK, Beavo JA. 1993. The role of protein phosphorylation in the regulation of cyclic nucleotide phosphodiesterases. *Mol. Cell Biochem.* 127/128:239–53

122. Beavo JA. 1995. Cyclic nucleotide phosphodiesterases: functional implications of multiple isoforms. *Physiol. Rev.* 75:725–48

123. Burns F, Zhao AZ, Beavo JA. 1996. Cyclic nucleotide phosphodiesterases: gene complexity, regulation by phosphorylation, and physiological implications. *Adv. Pharmacol.* 36:29–48

124. Borisy FF, Hwang PM, Ronnett GV, Snyder SH. 1993. High affinity cyclic AMP phosphodiesterase and adenosine localized in sensory organs. *Brain Res.* 610:199–207

125. Borisy FF, Ronnett GV, Cunningham AM, Juilfs D, Beavo J, Snyder SH. 1991. Calcium/calmodulin activated phosphodiesterase selectively expressed in olfactory receptor neurons. *J. Neurosci.* 12:915–23

126. Yan C, Zhao AZ, Bentley JK, Loughney K, Ferguson K, Beavo JA. 1995. Molecular cloning and characterization of a calmodulin-dependent phosphodiesterase enriched in olfactory sensory neurons. *Proc. Natl. Acad. Sci. USA* 92:9677–81

127. Juilfs DM, Fülle HJ, Zhao AZ, Houslay MD, Garbers DL, Beavo JA. 1997. A subset of olfactory neurons that selectively express cGMP-stimulated phosphodiesterase (PDE2) and guanylyl cyclase-D define a unique olfactory signal transduction pathway. *Proc. Natl. Acad. Sci. USA* 94:3388–95

128. Meyer MR, Angele A, Kremmer E, Kaupp UB, Muller F. 2000. A cGMP-signaling pathway in a subset of olfactory sensory neurons. *Proc. Natl. Acad. Sci. USA* 97:10595–600

129. Fülle H-J, Vassar R, Foster DC, Yang R-B, Axel R, Garbers DL. 1995. A receptor guanylyl cyclase expressed specifically in olfactory sensory neurons. *Proc. Natl. Acad. Sci. USA* 92:3571–75

130. Palczewski K, Subbaraya I, Gorczyca WA, Helekar BS, Ruiz CC, et al. 1994. Molecular cloning and characterization of retinal photoreceptor guanylyl cyclase-activating protein. *Neuron* 13:395–404

131. Baker H, Cummings DM, Munger SD, Margolis JW, Franzen L, et al. 1999. Targeted deletion of a cyclic nucleotide-gated channel subunit (OCNC1): biochemical and morphological consequences in adult mice. *J. Neurosci.* 19:9313–21

132. Lowe G, Gold GH. 1993. Nonlinear amplification by calcium-dependent chloride channels in olfactory receptor cells. *Nature* 366:283–86

133. Dionne VE. 1998. New kid on the block: a role for the Na/Ca exchanger in odor transduction. *J. Gen. Physiol.* 112:527–28

134. Jung A, Lischka FW, Engel J, Schild D. 1994. Sodium/calcium exchanger in olfactory receptor neurones of Xenopus laevis. *NeuroReport* 5:1741–44

135. Menco BP, Birrell GB, Fuller CM, Ezeh PI, Keeton DA, Benos DJ. 1998. Ultrastructural localization of amiloride-sensitive sodium channels and Na^+,K^+-ATPase in the rat's olfactory epithelial surface. *Chem. Senses* 23:137–49

136. Noe J, Tareilus E, Boekhoff I, Breer H. 1997. Sodium/calcium exchanger in rat olfactory neurons. *Neurochem. Int.* 30:523–31

137. Reisert J, Matthews HR. 1998. Na^+-dependent Ca^{2+} extrusion governs response recovery in frog olfactory receptor cells. *J. Gen. Physiol.* 112:529–35

138. Zheng C, Feinstein P, Bozza T, Rodriguez I, Mombaerts P. 2000. Peripheral olfactory projections are differentially affected in mice deficient in a cyclic nucleotide-gated channel subunit. *Neuron* 26:81–91

139. Varnum MD, Black KD, Zagotta WN. 1995. Molecular mechanism for ligand discrimination of cyclic nucleotide-gated channels. *Neuron* 15:619–25

140. Zufall F, Firestein S, Shepherd GM. 1994. Cyclic nucleotide-gated ion channels and sensory transduction in olfactory receptor neurons. *Annu. Rev. Biophys. Biomol. Struct.* 23:577–607

141. Fesenko EE, Kolesnikov SS, Lyubarsky AL. 1985. Induction by cGMP of cationic conductance in plasma membrane of retinal rod outer segment. *Nature* 313:310–13

142. Picco C, Menini A. 1993. The permeability of the cGMP-activated channel to organic cations in retinal rods of the tiger salamander. *J. Physiol.* 460:741–58

143. Firestein S, Zufall F, Shepherd GM. 1991. Single odor-sensitive channels in olfactory receptor neurons are also gated by cyclic nucleotides. *J. Neurosci.* 11:3565–72

144. Kurahashi T, Kaneko A. 1991. High density cAMP-gated channels at the ciliary membrane in the olfactory receptor cell. *NeuroReport* 2:5–8

145. Nakamura T. 2000. Cellular and molecular constituents of olfactory sensation in vertebrates. *Comp. Biochem. Physiol. A* 126:17–32

146. Liman ER, Buck LB. 1994. A second subunit of the olfactory cyclic nucleotide-gated channel confers high sensitivity to cAMP. *Neuron* 13:611–21

147. Chen T-Y, Peng Y-W, Dhallan RS, Ahamed B, Reed RR, Yau K-W. 1993. A new subunit of the cyclic nucleotide-gated cation channel in retinal rods. *Nature* 362:764–67

148. Korschen HG, Illing M, Seifert R, Sesti F, Williams A, et al. 1995. A 240 kDa protein represents the complete beta subunit of the cyclic nucleotide-gated channel from rod photoreceptor. *Neuron* 15:627–36

149. Sautter A, Zong X, Hofmann F, Biel M. 1998. An isoform of the rod photoreceptor cyclic nucleotide-gated channel beta subunit expressed in olfactory neurons. *Proc. Natl. Acad. Sci. USA* 95:4696–701

150. Bonigk W, Bradley J, Muller F, Sesti F, Boekhoff I, et al. 1999. The native rat olfactory cyclic nucleotide-gated channel is composed of three distinct subunits. *J. Neurosci.* 19:5332–47

151. Bradley J, Li J, Davidson N, Lester HS, Zinn K. 1994. Heteromeric olfactory cyclic nucleotide-gated channels: a subunit that confers increased sensitivity to cAMP. *Proc. Natl. Acad. Sci. USA* 91:8890–94

152. Shapiro MS, Zagotta WN. 1998. Stoichiometry and arrangement of heteromeric olfactory cyclic nucleotide-gated ion channels. *Proc. Natl. Acad. Sci. USA* 95:14546–51

153. Bradley J, Zhang Y, Bakin R, Lester HA, Ronnett GV, Zinn K. 1997. Functional expression of the heteromeric "olfactory" cyclic nucleotide-gated channel in the hippocampus: a potential effector of synaptic plasticity in brain neurons. *J. Neurosci.* 17:1993–2005

154. Dzeja C, Hagen V, Kaupp UB, Frings S. 1999. Ca^{2+} permeation in cyclic nucleotide-gated channels. *EMBO J.* 18:131–44

155. Chen T-Y, Yau K-W. 1994. Direct modulation by Ca^{2+}-calmodulin of cyclic nucleotide-activated channel of rat olfactory receptor neurons. *Nature* 368:545–48

156. Kurahashi T, Menini A. 1997. Mechanism of odorant adaptation in the olfactory receptor cell. *Nature* 385:725–29

157. Chen TY, Illing M, Molday LL, Hsu YT, Yau KW, Molday RS. 1994. Subunit 2 (or beta) of retinal rod cGMP-gated cation channel is a component of the 240-kDa channel-associated protein and mediates $Ca(2+)$-calmodulin modulation. *Proc. Natl. Acad. Sci. USA* 91:11757–61

158. Gordon SE, Zagotta WN. 1995. A histidine residue associated with the gate of the cyclic nucleotide-activated channels in rod photoreceptors. *Neuron* 14:177–83

159. Liu M, Chen TY, Ahamed B, Li J, Yau KW. 1994. Calcium-calmodulin modulation of the olfactory cyclic nucleotide-gated cation channel. *Science* 266:1348–54. Erratum. 1994. *Science* 266:1933

160. Zufall F, Shepherd GM, Firestein S. 1991. Inhibition of the olfactory cyclic nucleotide-gated ion channel by intracellular calcium. *Proc. R. Soc. Biol.* 246:225–30

161. Kramer RH, Siegelbaum SA. 1992. Intracellular Ca^{2+} regulates the sensitivity of cyclic nucleotide-gated channels in olfactory receptor neurons. *Neuron* 9:897–906

162. Firestein S, Zufall F. 1994. The cyclic nucleotide gated channel of olfactory receptor neurons. *Semin. Cell Biol.* 5:39–46

163. Kleene SJ. 1999. Both external and internal calcium reduce the sensitivity of the olfactory cyclic-nucleotide-gated channel to cAMP. *J. Neurophysiol.* 81:2675–82

164. Muller F, Bonigk W, Sesti F, Frings S. 1998. Phosphorylation of mammalian olfactory cyclic nucleotide-gated channels increases ligand sensitivity. *J. Neurosci.* 18:164–73

165. Molokanova E, Maddox F, Luetje CW, Kramer RH. 1999. Activity-dependent modulation of rod photoreceptor cyclic nucleotide-gated channels mediated by phosphorylation of a specific tyrosine residue. *J. Neurosci.* 19:4786–95

166. Molokanova E, Savchenko A, Kramer RH. 2000. Interactions of cyclic nucleotide-gated channel subunits and protein tyrosine kinase probed with genistein. *J. Gen. Physiol.* 115:685–96

167. Berghard A, Buck LB. 1996. Sensory transduction in vomeronasal neurons: evidence for G alpha o, G alpha i2, and adenylyl cyclase II as major components of a pheromone signaling cascade. *J. Neurosci.* 16:909–18

168. Wiesner B, Weiner J, Middendorff R, Hagen V, Kaupp UB, Weyand I. 1998. Cyclic nucleotide-gated channels on the flagellum control Ca^{2+} entry into sperm. *J. Cell. Biol.* 142:473–84

169. Kingston PA, Zufall F, Barnstable CJ. 1996. Rat hippocampal neurons express genes for both rod retinal and olfactory cyclic nucleotide-gated channels: novel targets for cAMP/cGMP function. *Proc. Natl. Acad. Sci. USA* 93:10440–45

170. Coburn CM, Bargmann CI. 1996. A putative cyclic nucleotide-gated channel is required for sensory development and function in *C. elegans*. *Neuron* 17:695–706

171. Zufall F, Shepherd GM, Barnstable CJ. 1997. Cyclic nucleotide gated channels as regulators of CNS development and plasticity. *Curr. Opin. Neurobiol.* 7:404–12

172. Yau K-W. 1994. Cyclic nucleotide-gated channels: an expanding new family of ion channels. *Proc. Natl. Acad. Sci. USA* 91:3481–83

173. Frings S. 1999. Tuning Ca^{2+} permeation in cyclic nucleotide-gated channels. *J. Gen. Physiol.* 113:795–98

174. Frings S, Lindemann B. 1991. Current recording from sensory cilia of olfactory receptor cells in situ. *J. Gen. Physiol.* 97:1–15

175. Frings S, Reuter D, Kleene SJ. 2000. Neuronal Ca^{2+}-activated Cl^- channels—homing in on an elusive channel species. *Prog. Neurobiol.* 60:247–89

176. Kleene SJ, Gesteland RC. 1991. Calcium-activated chloride conductance in frog olfactory cilia. *J. Neurosci.* 11:3624–29

177. Reuter D, Zierold K, Schroder WH, Frings S. 1998. A depolarizing chloride current contributes to chemoelectrical transduction in olfactory sensory neurons in situ. *J. Neurosci.* 18:6623–30

178. Kleene SJ. 1997. High-gain, low-noise amplification in olfactory transduction. *Biophys. J.* 73:1110–17

179. Zufall F, Leinders-Zufall T, Greer CA. 2000. Amplification of odor-induced Ca^{2+} transients by store-operated Ca^{2+} release and its role in olfactory signal transduction. *J. Neurophysiol.* 83:501–12

180. Corotto FS, Piper DR, Chen N, Michel WC. 1996. Voltage- and Ca^{2+}-gated currents in zebrafish olfactory receptor neurons. *J. Exp. Biol.* 199:1115–26

181. Firestein S, Werblin FS. 1987. Gated currents in isolated olfactory receptor neurons of the larval tiger salamander. *Proc. Natl. Acad. Sci. USA* 84:6292–96

182. Lynch JW, Barry PH. 1991. Properties of transient K^+ currents and underlying single K^+ channels in rat olfactory receptor neurons. *J. Gen. Physiol.* 97:1043–72

183. Lucero MT, Chen N. 1997. Characterization of voltage- and Ca^{2+}-activated K^+ channels in squid olfactory receptor neurons. *J. Exp. Biol.* 200:1571–86

184. Miyamoto T, Restrepo D, Teeter JH. 1992. Voltage-dependent and odorant-regulated currents in isolated olfactory receptor neurons of the channel catfish. *J. Gen. Physiol.* 99:505–29

185. Nevitt GA, Moody WJ. 1992. An electrophysiological characterization of ciliated olfactory receptor cells of the coho salmon *Oncorhynchus kisutch*. *J. Exp. Biol.* 166:1–17

186. Schild D. 1989. Whole-cell currents in olfactory receptor cells of *Xenopus laevis*. *Exp. Brain Res.* 78:223–32

187. Delgado R, Labarca P. 1993. Properties of whole cell currents in isolated olfactory neurons from the chilean toad *Caudiverbera caudiverbera*. *Am. J. Physiol. Cell Physiol.* 264:C1418–C27

188. Maue RA, Dionne VE. 1987. Patch-clamp studies of isolated mouse olfactory receptor neurons. *J. Gen. Physiol.* 90:95–125

189. Trotier D. 1986. A patch-clamp analysis of membrane currents in salamander olfactory receptor cells. *Pfluegers Arch.* 407:589–95

190. Trombley PQ, Westbrook GL. 1991. Voltage-gated currents in identified rat olfactory receptor neurons. *J. Neurosci.* 11:435–44

191. Kawai F, Miyachi E. 2001. Modulation by cGMP of the voltage-gated currents in newt olfactory receptor cells. *Neurosci. Res.* 39:327–37

192. Restrepo D, Okada Y, Teeter JH, Lowry LD, Cowart B, Brand JG. 1993. Human olfactory neurons respond to odor stimuli with an increase in cytoplasmic Ca^{2+}. *Biophys. J.* 64:1961–66

193. Schild D, Jung A, Schultens HA. 1994. Localization of calcium entry through calcium channels in olfactory receptor neurones using a laser scanning microscope and the calcium indicator dyes Fluo-3 and Fura-Red. *Cell Calcium* 15:341–48

194. Kawai F, Kurahashi T, Kaneko A. 1997. Nonselective suppression of voltage-gated currents by odorants in the newt

olfactory receptor cells. *J. Gen. Physiol.* 109:265–72

195. Trotier D, MacLeod P. 1986. Intracellular recordings from salamander olfactory supporting cells. *Brain Res.* 374:205–11

196. Berridge MJ, Irvine RF. 1984. Inositol trisphosphate, a novel second messenger in cellular signal transduction. *Nature* 312:315–21

197. Berridge MJ, Irvine RF. 1989. Inositol phosphates and cell signalling. *Nature* 341:197–204

198. Kuno M, Gardner P. 1987. Ion channels activated by inositol 1,4,5-trisphosphate in plasma membrane of human T-lymphocytes. *Nature* 326:301–4

199. Fijimoto T, Nakade S, Miyawaki A, Mikoshiba K, Ogawa K. 1992. Localization of inositol 1,4,5-trisphosphate receptor-like protein in plasmalemmal caveolae. *J. Cell Biol.* 119:1507–13

200. Bush KT, Stuart RO, Li SH, Moura LA, Sharp AH, et al. 1994. Epithelial inositol 1,4,5-trisphosphate receptors. Multiplicity of localization, solubility, and isoforms. *J. Biol. Chem.* 269:23694–99

201. Joseph SK. 1996. The inositol triphosphate receptor family. *Cell. Signal.* 8:1–7

202. Taylor CW, Richardson A. 1991. Structure and function of inositol trisphosphate receptors. *Pharmacol. Ther.* 51:97–137

203. Taylor CW, Traynor D. 1995. Calcium and inositol trisphosphate receptor. *J. Membr. Biol.* 145:109–18

204. Huque T, Bruch RC. 1986. Odorant-and guanine nucleotide-stimulated phosphoinositide turnover in olfactory cilia. *Biochem. Biophys. Res. Commun.* 137:36–42

205. Restrepo D, Miyamoto T, Bryant BP. 1990. Odor stimuli trigger influx of Ca^{2+} into olfactory neurons of the channel catfish. *Science* 249:1166–68

206. Ronnett GV, Cho H, Hester LD, Wood SR, Snyder SH. 1993. Odorants differentially enhance phosphoinositide turnover and adenylyl cyclase in olfactory receptor neuronal cultures. *J. Neurosci.* 13:1751–58

207. Boekhoff I, Michel WC, Breer H, Ache BW. 1994. Single odors differentially stimulate dual second messenger pathways in lobster olfactory receptor cells. *J. Neurosci.* 14:3304–9

208. Cunningham AM, Ryugo DK, Sharp AH, Reed RR, Snyder SH, Ronnett GV. 1993. Neuronal inositol 1,4,5-trisphosphate receptor localized to the plasma membrane of olfactory cilia. *Neuroscience* 57:339–52

209. Munger SD, Gleeson RA, Aldrich HC, Rust NC, Ache BW, Greenberg RM. 2000. Characterization of a phosphoinositide-mediated odor transduction pathway reveals plasma membrane localization of an inositol 1,4,5-trisphosphate receptor in lobster olfactory receptor neurons. *J. Biol. Chem.* 275:20450–57

210. Fadool DA, Ache BW. 1992. Plasma membrane inositol 1,4,5-trisphosphate-activated channels mediate signal transduction in lobster olfactory receptor neurons. *Neuron* 9:907–18

211. Kalinoski DL, Aldinger SB, Boyle AG, Huque T, Maracek JF, et al. 1992. Characterization of a novel inositol 1,4,5-trisphosphate receptor in isolated olfactory cilia. *Biochem. J.* 281:449–56

212. Bruch RC, Abogadie FC, Farbman AI. 1995. Identification of three phospholipase C isotypes expressed in rat olfactory epithelium. *NeuroReport* 6:233–37

213. Abogadie FC, Bruch RC, Wurzburger R, Margolis FL, Farbman AI. 1995. Molecular cloning of a phosphoinositide-specific phospholipase C from catfish olfactory rosettes. *Brain Res.* 31:10–16

214. Zuker CS. 1996. The biology of vision of Drosophila. *Proc. Natl. Acad. Sci. USA* 93:571–76

215. Somlyo AV, Walz B. 1995. Ca^{2+} in visual transduction and adaptation in vertebrate and invertebrates. *Cell Calcium* 18:253–55

216. Wang TL, Sterling P, Vardi N. 1999. Localization of type I inositol 1,4,5-trisphosphate receptor in the outer segments of mammalian cones. *J. Neurosci.* 19:4221–28

217. Udovichenko IP, Cunnick J, Gonzalez K, Takemoto DJ. 1994. The visual transduction and the phosphoinositide system: a link. *Cell. Signal.* 6:601–5

218. Coccia VJ, Cote RH. 1994. Regulation of intracellular cyclic GMP concentration by light and calcium in electropermeabilized rod photoreceptors. *J. Gen. Physiol.* 103:67–86

219. Leinders-Zufall T, Shepherd GM, Zufall Z. 1996. Modulation by cyclic GMP of the odour sensitivity of vertebrate olfactory receptor cells. *Proc. R. Soc. Biol.* 263:803–11

220. Moon C, Jaberi P, Otto-Bruc A, Baehr W, Palczewski K, Ronnett G. 1998. Calcium-sensitive particulate guanylyl cyclase as a modulator of cAMP in olfactory neurons. *J. Neurosci.* 18:3195–205

221. Snyder SH. 1994. Nitric oxide and carbon monoxide: unprecedented signalling molecules in the brain. In *Encyclopedia Britannica*, ed. F-Y Sat, D Calhoun, pp. 84–101. Chicago: Britannica

222. Ingi T, Ronnett GV. 1995. Direct demonstration of a physiological role for carbon monoxide in olfactory receptor neurons. *J. Neurosci.* 15:8214–22

223. Roskams JA, Bredt DS, Ronnett GV. 1994. Nitric oxide expression during olfactory neuron development and regeneration. *Am. Chem. Soc.* 16:308

224. Bauer I, Wanner GA, Rensing H, Alte C, Miescher EA, et al. 1998. Expression pattern of heme oxygenase isoenzymes 1 and 2 in normal and stress-exposed rat liver. *Hepatology* 27:829–38

225. Beschorner R, Adjodah D, Schwab JM, Mittelbronn M, Pedal I, et al. 2000. Long-term expression of heme oxygenase-1 (HO-1, HSP-32) following focal cerebral infarctions and traumatic brain injury in humans. *Acta Neuropathol.* 100:377–84

226. Ewing JF, Raju VS, Maines MD. 1994. Induction of heart heme oxygenase-1 (HSP32) by hyperthermia: possible role

in sress-mediated elevation of cyclic 3':5'-guanosine monophosphate. *J. Pharmacol. Exp. Ther.* 271:408–14

227. Hirata K, He JW, Kuraoka A, Omata Y, Hirata M, et al. 2000. Heme oxygenase1 (HSP-32) is induced in myelin-phagocytosing Schwann cells of injured sciatic nerves in the rat. *Eur. J. Neurosci.* 12:4147–52

228. Koistinaho J, Miettinen S, Keinanen R, Vartiainen N, Roivainen R, Laitinen JT. 1996. Long-term induction of haeme oxygenase-1 (HSP-32) in astrocytes and microglia following transient focal brain ischaemia in the rat. *Eur. J. Neurosci.* 8:2265–72

229. Kutty RK, Maines MD. 1989. Selective induction of heme oxygease-1 isozyme in rat testis by human chorionic gonadotropin. *Arch. Biochem. Biophys.* 268:100–7

230. Hatt H, Ache BW. 1994. Cyclic nucletide- and inositol phosphate-gated ion channels in lobster olfactory receptor neurons. *Proc. Natl. Acad. Sci. USA* 91:6264–68

231. Dhallan RS, Yau KW, Schrader KA, Reed RA. 1990. Primary structure and functional expression of a cyclic nucleotide-activated channel from olfactory neurons. *Nature* 347:184–87

232. Baimbridge KG, Celio MR, Rogers JH. 1992. Calcium-binding proteins in the nervous system. *Trends Neurosci.* 15:303–8

233. Bastianelli E, Polans AS, Hidaka H, Pochet R. 1995. Differential distribution of six calcium-binding proteins in the rat olfactory epithelium during postnatal development and adulthood. *J. Comp. Neurol.* 354:395–409

234. Anholt RRH, Rivers AM. 1990. Olfactory transduction: cross-talk between second-messenger systems. *Biochemistry* 29:4049–54

235. Hsu Y-T, Molday RS. 1993. Modulation of the cGMP-gated channel of rod photoreceptor cells by calmodulin. *Nature* 361:76–79

236. Kurahashi T, Shibuya T. 1990. Ca^{2+}-dependent adaptive properties in the solitary olfactory receptor cell of the newt. *Brain Res.* 515:261–68

237. Kurahashi T, Yau K-W. 1993. Co-existence of cationic and chloride components in odorant-induced current of vertebrate olfactory receptor cells. *Nature* 363:71–74

238. Iino S, Kobayashi S, Okazaki K, Hidaka H. 1995. Neurocalcin-immunoreactive receptor cells in the rat olfactory epithelium and vomeronasal organ. *Neurosci. Lett.* 191:91–94

239. Bastianelli E, Pochet R. 1995. Calmodulin, calbindin-D28k, calretinin and neurocalcin in rat olfactory bulb during postnatal development. *Brain Res. Dev. Brain Res.* 87:224–27

240. Miwa N, Kobayashi M, Takamatsu K, Kawamura S. 1998. Purification and molecular cloning of a novel calcium-binding protein, p26olf, in the frog olfactory epithelium. *Biochem. Biophys. Res. Commun.* 251:860–67

241. Boekhoff I, Braunewell KH, Andreini I, Breer H, Gundelfinger E. 1997. The calcium-binding protein VILIP in olfactory neurons: regulation of second messenger signaling. *Eur. J. Cell. Biol.* 72:151–58

242. Kleene SJ. 1993. Origin of the chloride current in olfactory transduction. *Neuron* 11:123–32

243. Sibley DR, Benovic JL, Caron MG, Lefkowitz RJ. 1987. Regulation of transmembrane signaling by receptor phosphorylation. *Cell* 48:913–22

244. Huganir RL, Greengard P. 1990. Regulation of neurotransmitter receptor desensitization by protein phosphorylation. *Neuron* 5:555–67

245. Hausdorff WP, Caron MG, Lefkowitz RJ. 1990. Turning off the signal: desensitization of β-adrenergic receptor function. *FASEB J.* 4:2881–89

246. Benovic JL, Bouvier M, Caron MG, Lefkowitz RJ. 1988. Regulation of adenylyl

cyclase-coupled β-adrenergic receptors. *Annu. Rev. Cell Biol.* 4:405–28

247. Benovic JL, DeBlasi A, Stone WC, Caron MG, Lefkowitz RJ. 1989. β-adrenergic receptor kinase: primary structure delineates a multigene family. *Science* 246:235–40

248. Lohse MJ, Benovic JL, Codina J, Caron MG, Lefkowitz RJ. 1990. β-arrestin: a protein that regulates β-adrenergic receptor function. *Science* 248:1547–50

249. Dawson TM, Arriza JL, Lefkowitz RJ, Jaworsky DE, Ronnett GV. 1993. Beta-adrenergic receptor kinase-2 and beta-arrestin-2: mediators of odorant-induced desensitization. *Science* 259:825–29

250. Schleicher S, Boekoff I, Arriza J, Lefkowitz RJ, Breer H. 1993. A β-adrenergic receptor kinase-like enzyme is involved in olfactory signal termination. *Proc. Natl. Acad. Sci. USA* 90:1420–24

251. Peppel K, Boekhoff I, McDonald P, Breer H, Caron MG, Lefkowitz RJ. 1997. G protein-coupled receptor kinase 3 (GRK3) gene disruption leads to loss of OR desensitization. *J. Biol. Chem.* 272: 25425–28

252. Boekhoff I, Breer H. 1992. Termination of second messenger signaling in olfaction. *Proc. Natl. Acad. Sci. USA* 89:471–74

253. Hill CS, Treisman R. 1995. Transcriptional regulation by extracellular signals: mechanisms and specificity. *Cell* 80:199–211

254. Hirotsu T, Saeki S, Yamamoto M, Iino Y. 2000. The Ras-MAPK pathway is important for olfaction in *Caenorhabditis elegans*. *Nature* 404:289–93

255. Watt WC, Storm DR. 2000. Odorants stimulate the Erk/MAP kinase pathway and activate CRE-mediated transcription in olfactory sensory neurons. *J. Biol. Chem.* 276:2047–52

Annu. Rev. Physiol. 2002. 64:223–62

MYCOSPORINE-LIKE AMINO ACIDS AND RELATED GADUSOLS: Biosynthesis, Accumulation, and UV-Protective Functions in Aquatic Organisms

J. Malcolm Shick

School of Marine Sciences and Department of Biological Sciences, University of Maine, 5751 Murray Hall, Orono, Maine 04469-5751; e-mail: shick@maine.edu

Walter C. Dunlap

Environmental Biochemistry, Australian Institute of Marine Science, PMB No. 3, Townsville MC, Qld. 4810, Australia; e-mail: wdunlap@aims.gov.au

Key Words algae, corals, photooxidative stress, sunscreens, symbiosis

■ **Abstract** Organisms living in clear, shallow water are exposed to the damaging wavelengths of solar ultraviolet radiation (UVR) coincident with the longer wavelengths of photosynthetically available radiation (PAR) also necessary for vision. With the general exception of bacteria, taxonomically diverse marine and freshwater organisms have evolved the capacity to synthesize or accumulate UV-absorbing mycosporine-like amino acids (MAAs), presumably for protection against environmental UVR. This review highlights the evidence for this UV-protective role while also considering other attributed functions, including reproductive and osmotic regulation and vision. Probing the regulation and biosynthesis of MAAs provides insight to the physiological evolution and utility of UV protection and of biochemically associated antioxidant defenses.

INTRODUCTION: BIOCHEMICAL DEFENSES AGAINST UV RADIATION

Solar radiation reaching the Earth consists of infrared (>800 nm), visible or photosynthetically available (PAR, 400–750 nm), ultraviolet-A (UVA, 320–400 nm), and the more energetic ultraviolet-B (UVB, 280–320 nm) wavelengths. Highly energetic ultraviolet-C radiation (UVC, 200–280 nm) does not reach the Earth because it is absorbed by atmospheric ozone and O_2, in the latter case initiating reactions forming the ozone that itself absorbs most solar UVB. Intertidal and epipelagic marine organisms are exposed to the highest levels of ultraviolet radiation (UVR), and even planktonic and benthic organisms may experience harmful levels to depths >20 m (1).

0066-4278/02/0315-0223$14.00

Environmental UVB and the short wavelengths of UVA can be detrimental to marine life (2, 3), and enhancement of UVB (via ozone loss) can disrupt trophic interactions (3, 4). Cellular damage from UV exposure can occur by direct photochemical reaction, e.g., thymine dimerization in DNA (5), or via the photodynamic production of reactive oxygen species (ROS) such as singlet oxygen ($^1O_2$) and superoxide radical ($O_2^{\bullet-}$) (6). Accordingly, shallow-dwelling organisms exposed to high levels of solar UVR have evolved biochemical defenses against such damage, protection that includes the elaboration of natural UV-absorbing sunscreens, the expression and regulation of antioxidant enzymes, the accumulation and cycling of small-molecule antioxidants, and molecular repair. This review focuses on the physiological aspects of UV-absorbing mycosporine-like amino acids (MAAs) in aquatic organisms; antioxidant (2, 7–9) and molecular repair (10, 11) functions are reviewed elsewhere.

Nearly ubiquitous among marine organisms is the ability to synthesize or otherwise acquire MAAs absorbing maximally in the range 310–360 nm (Figure 1). MAAs are transparent to visible light (i.e., they are not pigments) and have high molar absorptivity ($\varepsilon = 28100$–50000 M^{-1} cm^{-1}) for UVA and UVB. MAAs are particularly common in coral-reef algae and animals but also occur in organisms from most shallow-water environments, from tropical coral reefs, to high alpine lakes, to polar seas (8, 12). A protective function for MAAs, inferred from their efficient UV absorption and correlation between their concentrations and ambient levels of UVR such as occur over a bathymetric range, has been verified experimentally. The origin of MAAs as products of photosynthetic organisms necessitates more coverage of algal and cyanobacterial physiology than is usual in this series but is justified because the compounds are important in the photophysiology of invertebrate-microbial symbioses and in metazoans that obtain MAAs from their diet.

HISTORICAL CONTEXT

Discovery of MAAs in Marine Organisms

The abundance of life in marine surface waters is evidence of effective UV protection. Wittenburg (13) was first to report a strong UV-absorbing agent ($\lambda_{max} = 305$ nm) in the gas gland of the epipelagic Portuguese man-of-war, but the isolated substance (λ_{max} corrected to 310 nm) was never fully characterized (14). Soon thereafter, UV-absorbing materials were found to be characteristic of the Rhodophyta (red macroalgae) (15). Similarly, Shibata (16) observed strong UV absorption by the aqueous extracts of several zooxanthellate scleractinian corals and a cyanobacterium from the Great Barrier Reef. This substance (named S-320) in coral reef samples showed broad UV absorption centered at 320 nm, but variations in the exact λ_{max} (315–323 nm) suggested that S-320 is made up of a group of spectrally similar compounds. The first evidence that S-320 is photoprotective was provided by Maragos (17) on finding that S-320 absorbance in colonies of

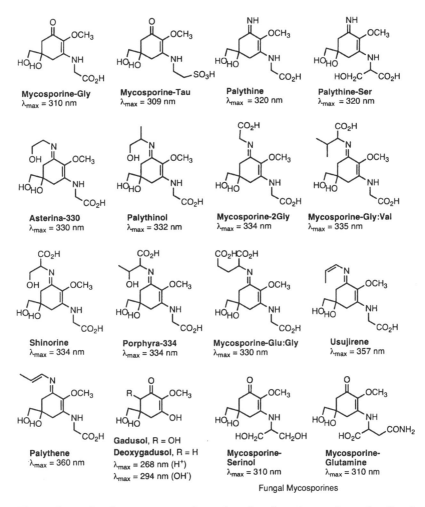

Figure 1   Molecular structures and wavelengths of maximum absorption (λ_{max}) of two fungal mycosporines, and of mycosporine-like amino acids and related gadusols in marine organisms.

Porites lobata varied inversely with depth, presumably compensating for ambient levels of UVR. A decade later, Jokiel & York (18) showed experimentally that S-320 decreased in tissues of *Pocillopora damicornis* on long-term exclusion of UVR, providing the first indication that S-320 was produced in response to UV exposure rather than some other depth-related factor. Contemporaneous reports indicate that UV-absorbing material is almost ubiquitous among marine algae (19, 20).

After the early biological observations of S-320 in coral reef invertebrates, natural products chemists at Nagoya (21), searching for palytoxin, observed a water-soluble metabolite from a tropical zoanthid, *Palythoa tuberculosa*, having a sharp absorption maximum at 310 nm. This metabolite proved to be mycosporine-glycine (Figure 1), having a basic structure identical to a family of metabolites previously described in terrestrial fungi. Following this discovery, Hirata's group isolated a number of imino-mycosporine derivatives (to be grouped later as mycosporine-like amino acids, MAAs), including palythine, palythinol, and palythene, also from *P. tuberculosa* (22); asterina-330 was later characterized from the sea star, *Asterina pectinifera* (23). Rhodophytes yielded several MAAs, including porphyra-334 from *Porphyra tenera* (24), shinorine from *Chondrus yendoi* (25), and usujirene from *Palmaria palmata* (26). There are now 19 known MAAs (see 12 for a complete listing with their taxonomic distribution). Thus S-320 has been identified as a suite of MAAs in corals and other marine organisms (27, 28). Interest in the physiological role of MAAs following the early years of chemical discoveries has intensified, particularly because of environmental concern about global stratospheric ozone depletion (29, 30).

Mycosporines in Terrestrial Fungi

Fungal metabolites strongly absorbing UVB ($\lambda_{max} = 310$ nm), first described by Leach (31) and notionally designated P-310, were present in the mycelia of several genera when sporulation had been induced with near-UVR, but were absent from non-sporulating colonies grown in darkness. The first P-310 substance was isolated from *Stereum hirsutum*, and its structure was elucidated (32) as 2-methoxy-3-bis(hydroxymethyl)methylamino-5-hydroxy-5-hydroxymethyl-2-cyclohexene-1-one (i.e., mycosporine-serinol; Figure 1). An early review (33) indicated that mycosporines are widespread among fungi, with the exception of Agaricales, but their presumed sporogenic activity was questioned on finding mycosporines only in conidiospores within the mycelium during sexual development. These UV-absorbing metabolites were later postulated to provide protection to fungal spores exposed to solar radiation during atmospheric dispersal (34). Although this contention is arguable (35), an evolutionary divide appears between the parent class of mycosporines expressed by fungi exclusively as oxo-carbonyl chromophores (absorbing UVB) and MAAs in aquatic organisms and terrestrial cyanobacteria containing mainly imino-carbonyl constituents (absorbing UVB and principally the short wavelengths of UVA) (Figure 1). Interestingly, only oxo-carbonyl mycosporines occur in the fungal-cyanobacterial symbiosis of terrestrial lichens (36).

OCCURRENCE AND DISTRIBUTION OF MAAs

Their near-ubiquity among marine taxa suggests not only an early origin of MAAs but also a functional importance that has been retained during subsequent evolution. Extensive sampling within some taxa in environments ranging from tropical to polar has discerned relationships between concentrations of MAAs in their tissues and environmental fluences of UVR that the organisms experience on local (bathymetric) and latitudinal scales. The predominantly superficial localization of MAAs in multicellular organisms, and their apparently homogeneous cytoplasmic distribution within single cells, are consistent with optical considerations for a sunscreening role.

MAAs in Marine Micro- and Macroalgae

MAAs occur in all microalgal taxa examined (reviewed in 12), as well as in natural assemblages of phytoplankton (37). When cultured under PAR alone, bloom-forming dinoflagellates tend to have the greatest capacity to accumulate MAAs (38), and when stimulated by changing irradiance, may alter their MAA complements and hence their UV-absorption spectra on the order of hours (39).

Marine, freshwater, and terrestrial cyanobacteria contain multiple MAAs, many being the same as in phylogenetically diverse eukaryotes, but most remaining unidentified (40). As in eukaryotic microalgae, the ability of small (typically $<10\,\mu m$ diameter), seemingly vulnerable, cyanobacteria to inhabit intensely bright environments has prompted study of their UV defenses.

Small cell size itself limits the efficacy of molecular sunscreens because the absorption of UVR is a function both of the concentration of the chromophore and of cell size ($=$ optical path length), with decreases in either reducing the sunscreen factor [S, the fraction of radiation of a given wavelength incident on a cell that will be absorbed by the sunscreen and thus not impinge on other cellular constituents (41)]. In no case examined has S achieved 1.0, i.e., some UVR always reaches other cellular targets and may exert biological effects (42–44), so the term sunscreen rather than sunblock is apt.

Although sometimes listed as sources of MAAs (12, 45), marine bacteria have not been systematically examined for their ability to synthesize these compounds. Only one bacterium, *Micrococcus* sp., is reported to contain an MAA, shinorine (46), although several species in the genera *Pseudoalteromonas* and *Vibrio* can convert the algal MAAs shinorine and porphyra-334 in the medium to mycosporine-glycine or to the related 4-deoxygadusol (8, 47).

The bio-optical model for UV absorption by single cells assumes a homogeneous cytoplasmic distribution of the sunscreen, a condition apparently met by MAAs in most cyanobacteria, where the water-soluble compounds occur free in the cytoplasm and not in cell walls or photosynthetic membranes (48); MAAs covalently linked to oligosaccharides uniquely occur in the extracellular glycan coat of *Nostoc commune* (40, 49), where they yield an S of ~ 0.7. Cytosolic homogeneity has not been checked in eukaryotic microalgae, where localization of MAAs around UV-sensitive organelles might increase their efficacy (50), although the small size of organelles presumably would require both very high local concentrations of MAAs for them to be effective and a more complicated optical model taking into account the cytological density and distribution of the organelles to assess this.

MAAs were early discovered in the Rhodophyta, and it is here among the algae that they achieve their highest concentrations and greatest diversity (12, 51, 52) and where we know most about macroalgal UV photoacclimatization. The extensive surveys by Karsten and colleagues enable the generalizations that Phaeophyta (brown algae such as kelps) either lack MAAs or have only trace amounts of them, presumably relying on other UV absorbers such as phlorotannins (53), and that Chlorophyta (green algae) have a higher proportion of genera containing MAAs, but, again, in minimal concentrations (54). Geographic and bathymetric trends are also evident.

MAAs in Algal-Invertebrate Symbioses

As we have emphasized (2, 8, 9), the UV-intense environment of many coral reefs is rich in MAA-containing symbioses between diverse invertebrates and unicellular phototrophs (including cyanobacteria; Prochlorales, *Prochloron* sp.; and dinoflag-ellates, especially *Symbiodinium* spp., known as zooxanthellae). Such symbioses present a heterogeneous array of MAAs, the concentration of which may vary according to habitat depth, so the symbioses have been studied for environmental determinants of their levels of MAAs.

Reef-building corals (Scleractinia) are particularly well studied and collectively contain at least 13 MAAs, the number in a given zooxanthellate species ranging from 2 (55) to 10 (56). Corals and Indo-Pacific "giant clams" (57) overlap in their MAA complements, seemingly because they form symbioses with the same phylotypes of *Symbiodinium* (58).

MAAs usually are more concentrated in the hosts' tissues than in symbionts freshly isolated from them [(57, 59–61); but for an exception, see (62)], a distribution suggesting that the host's tissues, by virtue of their high concentrations of MAAs and relatively long optical path, afford the first line of defense against UVR. As befit UV sunscreens, MAAs are more concentrated in superficial tissues of tridacnid clams (57) and the ascidian *Lissoclinum patella* (61) than in under-lying layers of the tissues where the algae reside. Likewise, in the sea anemone *Anemonia viridis*, MAAs are more concentrated in the tentacular ectoderm than in the endoderm, where the zooxanthellae are located (D. Allemand & J. M. Shick, unpublished data). The higher concentration of MAAs in the upper surface than in the sides or base of hemispherical colonies of the coral *Montastraea annularis* (63), and in branch tips than in tissues closer to the center of arborescent colonies of *Pocillopora damicornis* (64), probably is a photoadaptive response affording greater protection in the most exposed tissues. Likewise, MAAs are more concen-trated in peripheral, actively growing apical tissues than in older, self-shaded parts of macroalgal thalli (65, 66; K. Wong, N. L. Adams & J. M. Shick, unpublished data).

Unlike the glycosylated MAAs (and scytonemin, a unique cyanobacterial sun-screen) (40, 49) in cyanobacterial sheaths, an extracellular location of MAAs in marine symbioses has been reported only in the mucus of corals (67, 68), wherein they seem to be passively released rather than actively regulated, particularly as the relatively low concentration of MAAs in the mucus (~ 1 μM) would intercept only 7% of the incident solar UVR over the 1 mm thickness of the mucous layer (68).

MAAs in algal-invertebrate symbioses presumably originate in the phototrophic partner via the shikimate pathway. Four MAAs have been found among *Symbiodinium* species in culture: mycosporine-glycine, shinorine, and porphyra-334 (58) and mycosporine-2 glycine (J. M. Shick & C. Ferrier-Pagès, unpublished data), although a given species or phylotype contains only from zero to three of these MAAs. The suites of MAAs in the tridacnid clams and cnidarians from which the algal cultures were derived are similar to those of the algae, but the intact symbioses typically have a greater number of compounds. Differences in their kinetics of biosynthesis suggest that certain algal-type MAAs, such as shinorine and mycosporine-glycine that are synthesized quickly in response to UV exposure or other precursors, may subseqently be converted to different MAAs, perhaps in the host's tissues (56). In extreme cases, the host inhabited by zooxanthellae that produce no MAAs in vitro may have up to seven MAAs (56, 69), so the compounds seemingly must originate other than in the algae.

The finding that zooxanthellae in hospite (within the host) have different patterns of protein expression from the algae in vitro (70) is in keeping with the long-standing notion that the host can alter the biochemistry of its endosymbionts, e.g., there are qualitative differences in photosynthate produced by zooxanthellae in hospite and those in vitro (71). Thus it is possible that similar differences in the MAAs synthesized by cultured algae and by those in hospite explain the foregoing discrepancies. This remains to be tested.

Zooxanthellae generally produce a more restricted suite of MAAs than do free-living species of dinoflagellates in culture, the latter having from four to eight identified MAAs, plus additional, unidentified, presumed MAAs (38, 43, 50, 72–74). Might this reflect phylogenetic differences among dinoflagellates in the complexity of their biosynthetic machinery, and might the animal host harboring zooxanthellae have assumed the function of bioconversion among MAAs (concomitantly broadening the band of UV absorption) during evolution of the symbiosis?

Less is known of other symbioses. For example, a coral-reef sponge, *Dysidea herbacea*, contains four MAAs, one of them the novel mycosporine-glutamate: glycine, as well as the common mycosporine-glycine, and the isomers palythene and usujirene (75), the latter two being relatively rare among marine invertebrates. The unusual MAA complement in this sponge may be related to its symbiosis with *Oscillatoria spongeliae*, a cyanobacterium, a group for which most MAAs remain undescribed (40).

MAAs in Asymbiotic Invertebrates and Vertebrates

MAAs occur in a plethora of asymbiotic animals (see tables in Reference 12), especially in the epidermis (76, 77), as expected for UV sunscreens. An intriguing example is the presence of MAAs in the ocular tissues of fishes. Two of the MAAs found in the eyes of fishes (palythinol and palythene), plus shinorine, also occur in the cornea and lens of the cuttlefish *Sepia officinalis*, a cephalopod mollusc, where the absorption spectra of the ocular MAAs (λ_{max} from 332 to 360 nm) and the single visual pigment ($\lambda_{max} \sim 490$ nm) are distinct; thus the MAAs in *Sepia* probably do not affect its visual sensitivity (78).

There is a clear sexual dichotomy in the occurrence of MAAs: Whereas ovaries and eggs often have the highest concentration of MAAs among the tissues tested, testes and sperm have, at most, trace amounts (55, 76, 79–84). In asymbiotic animals, ovarian MAAs originate in the adult's food (42, 44, 83, 85, 86), and because the sexual difference persists when the dietary availability of MAAs is controlled, the dichotomy probably has an adaptive physiological basis. There are also implications for the specificity of transport mechanisms by which dietary MAAs are sequestered.

As in algal unicells (41), size may determine the occurrence of MAAs in gametes: Relatively large eggs (150 μm in diameter or greater in various species of sea urchins and tunicates, for example) have an optical pathlength sufficiently large so that MAAs in the observed concentrations could have biologically relevant sunscreening effectiveness. As in cyanobacteria, MAAs in eggs of sea urchins appear to occur free in the cytosol and not associated with any specific subcellular fractions (N. L. Adams, A. K. Carroll & J. M. Shick, unpublished data). Conversely, sperm averaging 2–3 μm in diameter would have to accumulate MAAs amounting to $\sim$25% of their dry mass to achieve similar sunscreening factors (42, 69), a physiologically infeasible concentration because of osmotic constraints. The seeming UV-vulnerability of sperm shed to the environment may thus explain why so many marine animals, especially on coral reefs (81), spawn at night. Even the sperm of species spawning in daylight must remain undamaged for only minutes to hours before fertilizing an egg, so their cumulative UV dose is far less than in eggs and in the larvae developing from them (42, 44). The developmental manifestations of UV exposure of embryos having different concentrations of MAAs are considered below.

Bathymetric Distribution of MAAs

The concentration of MAAs in corals is greater in shallow than in deep water (2, 8, 9, 86a). This is apparently an adaptive response to the exponential increase in the fluence of UVR with decreasing depth, and although other depth-dependent variables such as PAR and water movement may contribute (64, 87), the net effect is that corals contain MAAs in positive relation with their need for UV-sunscreen protection. The capacities of some corals to accumulate MAAs are correlated with their depth ranges (88, 89).

Palythine ($\lambda_{max} = 320$ nm) and mycosporine-glycine ($\lambda_{max} = 310$ nm) are the most prevalent MAAs in corals (reviewed in 2, 12), perhaps owing to the transparency of tropical waters to shorter, more damaging wavelengths. Mycosporine-glycine usually shows the greatest increase in shallow water, a correlate that may be related not only to its UVB-absorbing but also to its antioxidant properties (90), for phototrophic corals necessarily experience conditions conducive to oxidative stress (reviewed in 2, 8, 9).

Bathymetric differences in the concentration of MAAs also occur in boreal and polar red macroalgae (51, 66, 86, 91–94). Transplanting deep-growing algae into shallow water often eliminates or reverses such differences, an effect that depends on both PAR and UVR (66, 92, 93), which strengthens the case for bathymetric photoacclimatization of UV defenses. High-intertidal species tend to have higher concentrations of MAAs than do low-shore or subtidal species. Some deep-water species of Rhodophyta seemingly lack the capacity to produce MAAs, whereas some littoral species have constitutively high concentrations (94).

Attempts to correlate the abundance in corals (89) and macroalgae (66) of MAAs having particular absorption maxima with the underwater spectrum or wavelengths stimulating their synthesis have been less convincing. This is partly because of the relatively great breadth of the absorption spectra of some MAAs (44, 88) and because of the apparent multiplicity of interacting signals stimulating the biosynthesis of MAAs, as well as differences in the kinetics of their accumulation, which result in temporal differences in the complement of MAAs and the combined UV-absorption spectra they present.

Data on asymbiotic animals are too few to generalize about bathymetric patterns of their MAA concentrations. Based on very small sample sizes in a broad survey, Shick et al. (76) postulated intergeneric differences among coral reef sea cucumbers (holothuroid echinoderms) associated with their depths of occurrence. Karentz et al. (84) found significantly higher MAA concentrations in the ovaries of the Antarctic sea urchin *Sterechinus neumayeri* collected intertidally and at 8 m than in those from 15 and 24 m depth. Conversely, no depth-related differences in ovarian MAA concentrations occur in the boreal sea urchin *Strongylocentrotus droebachiensis* over a range of 0.5 to 10 m, perhaps because the turbid water at the collection sites itself attenuates UVR (86). Because MAAs in asymbiotic herbivores are derived from their food, including cyanobacterial mats in the case of tropical holothuroids and macroalgae in the case of sea urchins, any depth-related differences in MAAs in the consumers' tissues arise from digesting differentially photoacclimatized phototrophs because UV exposure of the consumers themselves does not affect their accumulation of MAAs (86, 95). The lack of bathymetric trends in the consumers' MAAs may also arise from their ingesting "drift algae" and from their own mobility, which enables them to graze at different depths (76, 86) so that MAA concentrations in consumers' tissues spatially and temporally integrate the UV exposure of their food.

Geographic and Seasonal Occurrence of MAAs

MAAs are ubiquitous, occurring in biomes ranging from polar to tropical (summarized in 12). MAAs occur most frequently and reach their greatest concentrations among tropical species, which may stem from the exposure of tropical organisms to higher levels of UVR owing both to the smaller solar zenith angle and thinner ozone layer there (96). The increase in fluences of UVR with decreasing latitude (97) may help to explain a similar geographic trend in the concentration of MAAs in red algae (51, 52). However, a latitudinal gradient in temperature paralleling that in UVR has not been considered, nor can local variation in temperature be ruled out as a contributor to the variation in MAAs in primary producers, particularly in tropical species under unusually warm conditions associated with the El Niño–Southern Oscillation (98). The primacy of the role of UVR compared with PAR or temperature may be indicated by the higher concentration of MAAs in Antarctic red macrophytes (which rivals those in warm-temperate species) (51, 94), than in antipodal Arctic species, where the UV transparency of seawater is less, owing to higher concentrations of dissolved organic material (99).

Seasonal changes in concentrations of MAAs in a tropical symbiotic sponge were positively related to seawater temperature and PAR (75) and thus presumably to UVR. MAA levels in a scleractinian coral tracked seasonal changes in solar UVB (with a one-week lag) but not temperature, so that UV acclimatization by the symbiotic algae is implicated (100). In comparison, concentrations of MAAs in the soft corals *Lobophytum compactum* and *Sinularia flexibilis* were significantly correlated with annual cycles of both solar irradiation and seawater temperature (101). Conversely, ovarian MAA levels were negatively related to seasonal temperature (and UVR) in a boreal sea urchin (86). In both the tropical sponge and the boreal sea urchin, MAA concentrations are maximal at the time of the annual spawning, which occurs at seasonally high and low fluences of UVR, respectively, so that accumulating UV-protective MAAs in eggs released to the environment is more the determinant than are direct thermal or UV effects on their accumulation.

Despite the bathymetric differences in ovarian concentrations of MAAs in the Antarctic sea urchin *S. neumayeri*, there were no differences attributable to seasonal changes in daylength, or in fluences of PAR or of UVB, even during springtime ozone depletion (84). This is perhaps because MAAs accumulate in ovaries more gradually, in synchrony with gametogenesis and maturation (86) (spanning a wide seasonal range of conditions) and, moreover, depend on diet (42, 44, 83).

In summary, geographic and seasonal trends—probably determined more by differences in solar UVR than by PAR or temperature—in concentrations of MAAs in tissues are more clearly seen in the photosynthetic organisms (notably marine red algae) that produce MAAs than in species that consume them. The lack of clear environmental correlates of MAA accumulation by animals seems related to the tissue-specificity of MAAs sampled and particularly to the tendency for them to be sequestered in ovaries and eggs, which may be released at different seasons by various species in diverse habitats. The strong dietary dependence of MAAs in consumers and the limited knowledge of what they may eat in the field for months before being sampled complicate the matter.

MAAs in Freshwater and Terrestrial Organisms

MAAs are found in cyanobacteria (40) and green microalgae (Chlorophyta) (102) from freshwater, hot spring, and terrestrial habitats. Mycosporine-glycine and several unidentified MAAs absorbing at 309–310 nm occur in terrestrial cyanobacterial lichens, but it is unknown whether the MAAs are produced by the cyanobacterial or fungal partner (36). MAAs occur in phytoplankton in high-alpine lakes and in the copepods that consume them (103, 103a). The MAA concentrations in the copepods are positively correlated with the altitude and the clarity of the water in the lakes, factors that increase the conditions of UV exposure (103a). Benthic cyanobacteria in the same lake have a more diverse suite of MAAs and in higher concentration than do phytoplankton; the unique occurrence of mycosporine-glycine in epilithic cyanobacteria on the lake shore may be related to their greater exposure to UVB (103). Like their marine and brackish-water relatives, freshwater fishes contain MAAs, both in their lenses and skin (104; N. L. Adams & J. M. Shick, unpublished data).

The sampling of freshwater and terrestrial species has not been extensive enough to enable other than the broadest generalizations. MAAs are unreported in higher plants, where the principal protection from UVR is by the multifunctional flavonoids (105, 106). Not only have MAAs not been found in the higher vertebrates (their sunscreening functions assumed by melanins), but unlike diverse invertebrates and fishes, mammals apparently cannot absorb MAAs from their food (77).

PROTECTIVE FUNCTIONS OF MAAs AND RELATED GADUSOLS

The photo-physiochemical properties of a natural "sunscreening" agent are vital to its UV-protective effectiveness. A sunscreen must not only be efficient at absorbing appropriate wavelengths of UVR, but also at dissipating the absorbed energy without transferring it to sensitive biomolecules, or causing the photodynamic production of $^1O_2$ and $O_2^{\bullet-}$ to impose oxidative stress. Sunscreening and antioxidant functions for UV protection are thus closely entwined (107).

Photophysics of UV Suncreening and Related Antioxidant Functions

There are only two reports on the photophysical characteristics of UV dissipation by MAAs, and both examine the excited-state properties and photostability of macroalgal imino-mycosporines in vitro. Shick et al. (69) reported that shinorine (17 μM), despite its high absorptivity for UVA ($\varepsilon = 44,670\,M^{-1}\,cm^{-1}$), showed no detectable fluorescence when excited across its half-maximal absorption waveband (312–348 nm). Furthermore, electron paramagnetic resonance (EPR) spectroscopy revealed that purified shinorine (50 μM) did not produce detectable radicals when irradiated from 305 to 700 nm in the EPR cavity together with the spin-traps 5,5-dimethylpyrroline-*N*-oxide or α-(4-pyridyl-1-oxide)-*N-tert*-butylnitrone. The absence of free radical formation by UV irradiation and a lack of fluorescence are consistent with a high efficiency of thermally dissipating absorbed UV energy.

The photophysical properties of porphyra-334 in aqueous solution were studied in detail (108). Irradiation of porphyra-334 (6 μM) at its absorption maximum produced a weak fluorescence (emission maximum = 395 nm) of extremely low quantum yield ($\Phi_F = 1.6 \times 10^{-3}$). The short lifetime of the excited singlet-state indicated its rapid internal conversion to ground state. Direct excitation of 6 μM porphyra-334 by laser flash photolysis (355 nm) revealed no triplet-state absorption transient in the microsecond range, consistent with a lack of triplet-state reactivities. The excited triplet state, however, could be measured by sensitization with benzophenone, and showed a low quantum efficiency ($\Phi_T < 0.05$) and a strongly exothermic triplet energy level (E_T) ≤ 250 kJ mol^{-1}. Conde et al. (108) conclude, "...the very low quantum yields of fluorescence, intersystem crossing, and photolysis, are in agreement with a photoprotective role of porphyra-334 in living systems. In particular, the very low triplet quantum yield will preclude the action of [this] MAA as a photodynamic agent via singlet oxygen generation." The same holds for the lack of fluorescence and radical production observed for shinorine (69).

The foregoing properties of shinorine and porphyra-334, together with their high degree of photostability in vitro (42, 108) and in vivo (42, 44), are expected attributes of an efficient UV-screening agent elaborated by evolution. Moreover, the MAAs (principally shinorine) in the microalga *Phaeocystis antarctica* do not transfer (directly or by fluorescence) the UV energy that they absorb to chlorophyll *a* and thus do not participate in photosynthesis (109).

As part of their defense against photooxidative stress, marine invertebrates and fishes often contain high levels of gadusols (110, 111), which are related structurally (Figure 1) and biosynthetically to MAAs. The sunscreening and antioxidant roles of these cyclohexenone molecules are not always distinct. Examining the relationship between sunscreening and possible antioxidant functions of MAAs revealed that imino-MAAs are oxidatively robust, whereas the oxo-carbonyl mycosporine-glycine (and mycosporine-taurine, unique to sea anemones of the genus *Anthopleura*) (59) has moderate, concentration-dependent antioxidant activity (90). This may explain why corals growing in shallow water generally have disproportionally greater quantities of mycosporine-glycine than deeper-water conspecifics. Moreover, 4-deoxygadusol (4-DG), presumed to be the immediate precursor of MAAs, has strong antioxidant properties (8), as demonstrated by a comparison of electrochemical properties of water-soluble antioxidants found in marine organisms (Figure 2). 4-DG has been prepared by a bacterial "retrobiosynthetic" pathway (47), yet the biosynthetic relationship between the antioxidant function of 4-DG (and oxo-MAAs) and the sunscreening function of oxo- and imino-MAAs has not been fully explored. No doubt this area of investigation will

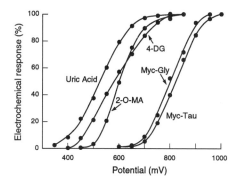

Figure 2   Comparison of voltamperograms for water-soluble antioxidants commonly found in marine organisms: 4-DG, 4-deoxygadusol; 2-O-MA, 2-*O*-methylascorbate, a stable, methylated form of ascorbate (W. C. Dunlap, unpublished data); Myc-Gly, mycosporine-glycine; Myc-Tau, mycosporine-taurine. Antioxidant activities generally increase with decreasing half-wave potential (50% electrochemical response).

flourish as implications of thermal and photooxidative stress in coral bleaching (2, 112–116) gain recognition.

Evolution of Structure and Function of MAAs and Gadusols

The requirement for UV protection in the evolution of phototrophic life on the early Earth has been renewed in scientific discourse (117, 118). The first life developed in the absence of atmospheric O_2, so the early biosphere lacked this defense against the highly energetic wavelengths of solar UVC, and little evidence remains of how early photosynthesizers withstood it. The simpler structure of 4-DG and its intermediate position between MAAs and the shikimate pathway (Figure 3) suggest that gadusols evolved prior to MAAs and may have served originally as a UVB/C screen ($\lambda_{max} = 294$ nm at physiological pH), albeit with lower absorptivity than MAAs (117). The strong antioxidant properties of gadusols concomitantly would have protected early cyanobacteria against oxidative damage at the intracellular sites of oxygenic photosynthesis. More speculatively, gadusols likewise might have detoxified sulfur- and oxygen-centered free radicals in microxic, sulfidic interfacial microhabitats (119). Biochemical evolution involving amine condensation with 4-DG provided MAAs with strong UVB- and UVA-absorbing characteristics but at the expense of moderating or eliminating the antioxidant activity of 4-DG. Oxo-MAAs (absorbing UVB maximally at ~310 nm) might have predominated in cyanobacteria in early evolution, whereas imino-MAAs developed later as rising atmospheric oxygen levels increased the need for protection from UVA and photooxidative stress (117).

The absence of imino-mycosporines from the fungi suggests that these compounds in eukaryotic algae were inherited from the cyanobacterial progenitors

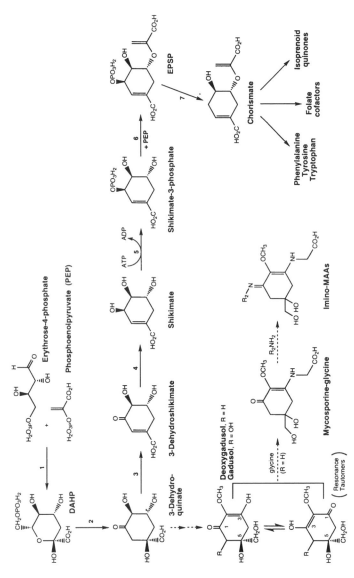

Figure 3   The shikimate pathway, showing intermediates and enzyme-catalyzed steps (*numbered*). DAHP, 3-deoxy-D-arabinoheptulosinate-7-phosphate; EPSP, 5-enolpyruvylshikimate-3-phosphate. Enzymes: 1, DAHP synthase; 2, DHQ synthase; 3, DHQ dehydratase; 4, shikimate dehydrogenase; 5, shikimate kinase; 6, EPSP synthase; 7, chorismate synthase. Broken arrows represent the putative biosynthetic relationship between 3-dehydroquinate (DHQ), gadusols, and MAAs. R_2, amino acids and amino alcohols characterizing individual MAAs. Compiled from various sources.

of their plastids, rather than having a cytosolic, eukaryotic provenance (117). Whether the fungal oxo-mycosporines are redox-active (i.e., have antioxidant activity) apparently has not been examined, but for most this is probable, based on chemical considerations. Their role in sporulation (35) indicates that these mycosporines have assumed additional functions, roles which may be paralleled by mycosporine-glycine in algae and invertebrates and which may involve redox signaling—all intriguing possibilities for future research.

A Critical Look at the Protective Functions of MAAs

Organismal studies have documented MAA concentration-dependent protection of embryonic and larval development (42, 44, 55) and of growth and photosynthesis in free-living algae (43, 50, 92, 93). Also, protection of photosynthesis in symbiotic microalgae in hospite probably is owing to the higher concentration of MAAs (and other UV-absorbing materials) in the hosts' cells (which offer a longer optical path over which UVR is attenuated) because UVR does inhibit photosynthesis in the freshly isolated endosymbionts (60–62, 107, 120). Collectively, these studies also indicate that the protection by MAAs is incomplete, so that MAAs are appropriately seen as part of a suite of defenses against the manifold effects of UVR. Nor is it clear that all of the protection attributed to MAAs indeed derived from them, because MAA levels in some of the test organisms resulted from differential prior exposure to UVR, which might have enhanced other defenses such as antioxidants (69).

Protection against acute effects of UVR is unambiguous in the larvae of a sea urchin (42, 44) and in a free-living dinoflagellate (50), where the concentrations of MAAs were experimentally altered without prior exposure to UVR. In the latter case, protection was evaluated using the biological weighting function (BWF)—a polychromatic action spectrum (121, 122)—showing the wavelength-dependent inhibition of photosynthesis (Figure 4A). In that case, the BWFs for cells grown under high PAR (but no UVR) and rich in MAAs diverged from MAA-depauperate, low-PAR cells in the range of 320 to 360 nm, where cells having higher concentrations of MAAs showed the greatest enhancement of UV absorbance and resistance to acute UV exposure. Moreover, because lowering of the UV effect (biological weight) at 340 nm in MAA-rich cells compared with MAA-depauperate cells exceeded the value of 80% predicted by Garcia-Pichel's (41) optical model for regionally homogeneous intracellular sunscreens, Neale et al. (50) suggested that MAAs may not be uniformly distributed within algal cells; as already noted, there are virtually no data on any such subcellular localization.

Adams & Shick (42) fed omnivorous sea urchins macroalgae having different concentrations of MAAs that were transferred to the ovaries and eggs during gametogenesis and maturation in the absence of UVR. Fertilized eggs acutely exposed to simulated solar UVR showed a delay in cleavage that was inversely and logarithmically related to their concentration of MAAs, a result in keeping with the continuous absorption of UVR along the intracellular optical path, where the highest levels

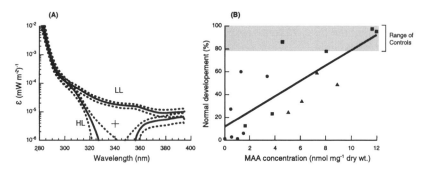

Figure 4   (A) Biological weight (ε, reciprocal of mW m^{-2}) for the inhibition of photosynthesis by acute exposure to UVR in the dinoflagellate *Gymnodinium sanguineum* grown under high PAR (HL) or low PAR (LL) in the absence of UVR. HL cells contained 44.4 and LL cells contained 2.1 nmol total MAAs nmol^{-1} chlorophyll *a*. The broken lines represent the 95% confidence belts for the HL and LL mean curves. The cross indicates the weight predicted by applying a sunscreen factor (*S*) of 0.8 to the biological weight at 340 nm [reproduced from (50) with permission of *J. Phycol.*]. (B) When acutely exposed to PAR + UVR, the percentage of embryos of *Strongylocentrotus droebachiensis* developing normally is correlated with the total MAA-concentration in the eggs ($r^2 = 0.614, P = 0.001$). The data are from separate experiments on embryos from laboratory-maintained adults fed *Laminaria saccharina* (●), which lacks MAAs, or a combination diet of *L. saccharina plus Chondrus crispus* rich in MAAs (■), and field-fresh sea urchins eating a natural diet (▲). The shaded area indicates the range of normal development in control embryos (irradiated with PAR only) from the same batches of eggs at the same time of development (day 4) [reproduced from (44) with permission of *Mar. Biol.*].

of shinorine and porphyra-334 in the eggs gave a calculated sunscreen factor of 0.86. A protective effect extended to later stages when the incidence of normal development during UV irradiation depended on the concentration of MAAs in the eggs and larvae (Figure 4*B*). The nominal UVA- (334 nm) absorbing MAAs protected against UVB- (<320 nm) induced damage because their broad absorption properties have high molar extinction coefficients extending into the UVB.

When endogenous ascorbate (having the lowest electrochemical potential among biological antioxidants and thus easily oxidized to form the ascorbate free radical) was used as a spin-trap for in vivo electron paramagnetic resonance analysis, sea urchin ovaries irradiated with broad-spectrum UVR showed an inverse relationship between their concentration of MAAs and this measure of oxidative stress (69). Thus, although the shinorine and porphyra-334 in the eggs are not themselves antioxidants (90), their sunscreening role includes intercepting and harmlessly dissipating UVR before it reaches more photoreactive biomolecules.

Despite their demonstrated role in protecting developing embryos in laboratory cultures, the ecological importance of UV protection by MAAs in early development (known to be the most sensitive phase of the life cycle) remains largely unexplored. The reported effects of UVR and MAA protection of reproduction in tropical corals are scant and scattered. Early evidence offers that MAA-containing planulae released by shallow-water colonies of *Agaricia agaricites* may be near their limits of tolerance to environmental UVR (55). By extension, this poses the question of UV protection in broadcast spawners: Do deep-water corals provide the same level of UV protection for their buoyant eggs as in those from shallow-water corals acclimatized to UV exposure? The zooxanthellate eggs of *Montipora verrucosa* and the azooxanthellate eggs of *Fungia scutaria* have strikingly different MAA compositions and concentrations, the concentrations being more than eight times greater in the positively buoyant eggs of *M. verrucosa* than in the negatively buoyant eggs of *F. scutaria* (82). A systematic inter- and intraspecific comparison of the UV tolerance and MAA composition of coral eggs and embryos is necessary to establish meaningful correlations.

Although a protective role is clearly demonstrated in dinoflagellates and sea urchin embryos, exactly what targets MAAs protect has not been determined. Such targets are known in principle from studies of UV-induced damage to biological molecules, but demonstration of their protection in vivo is scarce. Damage to DNA, RNA, and proteins is especially well known (5, 123), but in marine organisms, the repair of DNA damage is better documented than its prevention, and there are no published studies showing specifically whether high concentrations of MAAs can reduce UV-related damage to these molecules or impairment of enzyme function. The UV-induced accumulation of MAAs in a dinoflagellate did not prevent a concurrent decrease in its activity of Rubisco (43), the primary CO_2-fixing enzyme.

The pyridine nucleotides NAD(P)H, which absorb maximally at 340 nm, are potential targets of UVR. The metabolic ubiquity of the NAD(P)/NAD(P)H redox couple indicates its early origin (124). The transfer of electrons from NADH to O_2 in the electron transport system is generally familiar to animal physiologists, but NAD(P)/NAD(P)H also maintains redox balance in anaerobic fermentations such as glycolysis and alternate schemes where pyruvate can have fates other than conversion to lactate (125), and in anoxygenic photosynthesis (126). Another crucial role of NAD(P)H is the reduction of coenzyme Q, important in preventing lipid peroxidation in the plasma membrane (127). NADPH is also central to photosynthetic electron flow and important in the oxidative defenses in chloroplasts (128). Thus there might be widespread metabolic consequences of the UV irradiation of NAD(P)H, especially in the presence of O_2, where the potential for the photosensitized production of $O_2^{\bullet-}$ (129) and H_2O_2, and thence $HO^{\bullet}$, would multiply the secondary effects of its irradiation (6).

NAD(P)H has a molar extinction coefficient of $6230\ M^{-1}\ cm^{-1}$ at 340 nm, compared with the values on the order of $40,000\ M^{-1}\ cm^{-1}$ for various UVA-absorbing MAAs (12). Given this within-order-of-magnitude similarity of the molar absorptivities of NAD(P)H and UVA-absorbing MAAs, the latter would have to be present in much greater concentration in order to be protective, and this seems to be the case: MAA concentrations in various invertebrates, including eggs of sea urchins where they demonstrably protect cleavage from UVR, are about 500–2,000 nmol/g of fresh tissue, whereas the total concentration of NAD(P)H is ~50 nmol/ml (1 ml ~1 g) of unfertilized eggs of sea urchins and 135 nmol/ml of fertilized eggs (130, 131). A high molar ratio of MAAs to flavin nucleotides, which are other UVA-targets in cells (6), may also minimize in vivo any O_2-dependent, flavin-mediated photolysis of redox-active oxo-MAAs such as occurs at approximately equimolar concentrations of FAD and fungal mycosporines in vitro (132).

MAAs in Coral Bleaching

The role of MAAs as UV photoprotectants in the synergy of photic and thermal stresses causing coral bleaching has gained prominent attention. Early papers by Lesser et al. (112) and Glynn et al. (133) inferred that MAAs or their biosynthesis is thermally labile, so high temperatures would diminish UV protection, thus providing a molecular link between high UV irradiance and temperature to explain this synergistic stress in coral bleaching. Until recently, there has been no systematic examination of MAA levels in corals during a bleaching event to test this hypothesis.

MAAs in the mucus of the solitary coral *Fungia repanda* (1 m depth), monitored over 18 months, were positively correlated with solar UVR with a lag time of 1 week (100), consistent with the kinetics of UV-stimulated MAA-biosynthesis in corals (56). This significant correlation did not extend to seawater temperature or to the volume of mucus secreted. Although the corals observed during two bleaching events that occurred in this study were pale or partially bleached, the authors did not observe any shift of MAA concentrations or modification of composition in the mucus.

Similarly, over a two-year period, Michalek-Wagner (101) found positive correlations between MAA concentrations in reef-flat colonies of the soft corals *Lobophyton compactum* and *Sinularia flexibilis* and annual cycles in solar radiation and seawater temperature. MAAs in these soft corals during the 1998 mass bleaching event on the Great Barrier Reef clearly were not degraded at bleaching temperatures (134). On the contrary, MAAs were up-regulated under thermal stress, and concentrations were further enhanced during simultaneous exposure to UVR (98). Thermally acclimatized colonies having high MAA levels were not fully protected against solar UV-induced bleaching (loss of zooxanthellae) and, as may be expected, MAAs provided no discernible protection against thermal stress alone.

Another question regards the effects of bleaching and subsequent recovery on UV protection in coral reproduction. Experimental bleaching of the soft coral *Lobophytum compactum* reduced fecundity, fertilization success, and offspring viability in the subsequent breeding season (135). This negative impact on reproduction was associated with lowered levels of protein, lipid, MAA, and carotenoid in bleached adults (136). In contrast, MAAs were not as greatly depleted as were other constituents in the eggs of bleached soft corals (136), but the importance of this conservation of MAA content in UV protection for larval recruitment remains untested. These results provide evidence for Gleason's (86a) hypothesis that the biosynthesis of MAAs is costly and may necessitate trade-offs with other metabolic demands during multiple abiotic stresses.

Other Roles Attributed to MAAs

Because of their UV absorbance, MAAs have been studied primarily in photobiological contexts (sunscreens, vision). Owing to their seasonal changes in concentration that parallel ovarian maturation and their high aqueous solubility and high concentrations in vivo, MAAs have also been considered in reproductive and osmotic contexts.

REPRODUCTIVE REGULATION   Largely by analogy with the case in the fungi, where mycosporines are involved with morphogenesis and sporulation (see review in 35), Bandaranayake proposed that MAAs in marine invertebrates have an undefined but intrinsic involvement with reproduction (75). There is scant empirical evidence to support this hypothesis, which stems mainly from the case of fungal mycosporines and from the observation that the concentrations of individual MAAs, especially in the ovaries, by various marine invertebrates is not necessarily correlated with the seasonal maximum of solar irradiation (which might be expected if the primary role of MAAs is as a UV sunscreen), but in many cases may reach their peak at the time of ovarian reproductive maturity.

Indeed, the seasonal accumulation of MAAs in ovaries of *Strongylocentrotus droebachiensis* occurs in inverse relation to solar irradiation: MAAs increase throughout the autumn and early winter, and peak in late winter just prior to spawning (83). Conversely, MAAs are most concentrated in ovaries and eggs of the crown-of-thorns sea star at its time of spawning in austral summer (35). Likewise, MAAs in the coral reef sponge *Dysidea herbacea* peak in summer, the time of spawning (75). It is a fair generalization that most MAAs reach their maximum concentration in eggs at about the time of spawning, which in the foregoing cases may be near the seasonal minima and maxima, respectively, of solar irradiation.

The synchronous accumulation of MAAs and the development of ovaries does not necessarily mean that MAAs are regulating or controlling such development, and their ovarian sequestration may be among the anabolic processes (e.g., vitellogenesis) that must be coordinated in conjunction with oogenesis and maturation. Thus MAA concentration may be an indicator of ovarian ripeness, not a regulator of it. Sea urchins eating only kelp (lacking MAAs), and in whose ovaries concentrations of previously assimilated MAAs remained constant or declined during gametogenesis, had gonadal indices (percentage of body mass allocated to gonads) as high or higher than those eating MAA-rich diets (83, 86). Importantly, the ovaries in adults on these disparate diets did not differ in their percentage of nutritive or gametic cells in various stages of development despite their very differerent concentrations of MAAs (86).

For MAAs to exert their demonstrated effectiveness as UV sunscreens in spawned eggs in nature, they would ideally reach their maximum intracellular levels at the time of the eggs' release to the environment, regardless of whether UVR is at its seasonal high or low, which indeed is what occurs, but only if adults have eaten diets containing MAAs. That UVR does not enhance the accumulation by adults of dietary MAAs into their ovaries (86) suggests that the preferential concentration of MAAs in the ovaries is under a more general physiological control. It remains to be seen whether the environmental [e.g., photoperiod: (137)] and physiological [e.g., neurochemical and locally secreted chemical messengers (138)] factors that regulate ovarian development also control the sequestration and metabolism of MAAs; however, it does seems clear that MAAs themselves are not controlling oogenesis or maturation in sea urchins. Thus whether MAAs that are so prevalent among marine and freshwater organisms were evolutionarily co-opted to serve as chemical regulators of physiological processes such as reproduction remains speculative, and like so many gaps in our knowledge, testing this experimentally is hindered by the lack of commercial sources of MAAs.

OSMOTIC REGULATION   Perhaps because of their "extremophily," cyanobacteria thriving in harsh environments have been investigated for other functions of MAAs. This includes a postulated role as osmolytes (139), and indeed the $\sim$100 mM intracellular concentration of MAAs in halophilic cyanobacteria approaches that of free amino acids (FAAs) in diverse marine invertebrates, where FAAs account for about 25–75% of the total intracellular osmotic concentration. Such a high concentration of MAAs in cells suggests that they are compatible solutes with respect to their effect on macromolecular function (125), and their zwitterionic or acidic nature is consistent with this hypothesis. Similar concentrations of MAAs ($\sim$60 mM) occur in the ocular lenses of some tropical fishes (W. C. Dunlap & M. Inoue, unpublished data).

As would be expected if they have a role in cellular volume regulation, MAAs are released from cyanobacteria under hypo-osmotic stress (139, 140), although they are not absorbed from a more concentrated medium (139). As is the case for FAAs in many taxa, the steady-state concentration of MAAs is positively related to environmental salinity in *Chlorogloeopsis*, where the osmotically induced biosynthesis of mycosporine-glycine is synergistically enhanced by UVB, whereas shinorine accumulation is more under the control of UVB (140). Nevertheless, in this cyanobacterium, which tolerates salt concentrations only up to 70% of normal seawater, MAAs represent less than 5% of total intracellular osmolytes, so their physiological role here is not in osmotic regulation. It may be fortuitous that osmotic shock induces the biosynthesis of UV-absorbing MAAs where they do not serve as osmolytes; such a dual control by UV and osmotic stress is reminiscent of the activation of the c-Jun amino-terminal protein kinase (JNK) cascade in mammalian cells by these separate stressors (141).

In marine invertebrates, MAAs are far less important than FAAs as organic osmolytes. For example, in sea anemones living in normal seawater of $\sim$1000 milliosmoles, concentrations of FAAs are about 100 mol g^{-1} wet weight of tissue (142); assuming that tissues are 80% water by weight and that 50% of tissue water is intracellular, this gives an intracellular concentration of FAAs of 250 mM. By comparison, concentrations of MAAs in sea anemones (59, 143; J. M. Shick, unpublished data) are about 15 μmol g^{-1} dry weight, or 3 μmol g^{-1} wet weight; this gives an intracellular concentration of MAAs of about 7.5 mM, or only 3% of the FAA concentration in sea anemones and only 7.5% of the MAA concentration in halophilic cyanobacteria.

Intracellular concentrations of MAAs in sea urchin eggs, which are notably stenohaline, are lower, about 2 mM. Concentrations of MAAs in other marine invertebrates are similar to those in sea anemones, whereas FAA concentrations are generally higher, which suggests that MAAs are not major contributors to the intracellular pool of osmolytes in these cases. The extraordinarily high concentrations of MAAs (from 2200 up to 8800 nmol mg^{-1} tissue protein, equivalent to $\sim$220–880 μmol g^{-1} tissue wet weight) reported in some corals (54, 64, 144) would accordingly represent intracellular osmotic concentrations of $\sim$0.55 to 2.2 M, or from about half to double the total osmotic concentration (including organic plus inorganic solutes) of cells in osmotic equilibrium with seawater. Such concentrations are physiologically implausible, unless MAAs in these corals have domains other than free cytosolic dispersal.

MAAs IN MARINE VISION

UVR, even the longer wavelengths of UVA, can damage ocular tissues, and photooxidative damage is the usual consequence of light exposure of the retina (145). Thus the corneal and lenticular MAAs accumulated by many fish may protect their retinas from damage by environmental UVR. Some species having near-UV-sensitive vision, however, have UVA-transparent ocular tissues to allow stimulation of retinal opsin absorbing in the range 360–380 nm. Assuming that UV-sensitive vision imposes a metabolic cost, it follows that vision in the near-UV range is functionally important (146).

Presence of MAAs in Ocular Tissues of Fishes

Kennedy & Milkman (147) and Bon et al. (148) early postulated the existence in the ocular lenses of fishes, amphibians, and cephalopods of UV-absorbing substances having a characteristic absorbance at 320–360 nm. These UV-absorbing pigments (149) were unlike the kynurenine derivatives found in terrestial animals, including humans (150). The biochemical properties reported by Zigman et al. (149) and Zigman (150) suggested they belong to the mycosporine family of metabolites, later confirmed by analysis of ocular tissues from a wide diversity of fishes from the Great Barrier Reef (151). The specific group of MAAs present in the lenses of tropical fishes includes palythine, asterina-330, palythinol, and palythene. Notably absent are the UVB-absorbing mycosporine-glycine (λ_{max} = 310 nm) and the imino-MAAs shinorine and porphyra-334, common to many marine algae and corals. The interspecific comparison of MAAs in fish lenses revealed no clear behavioral or taxonomic trends: e.g., MAA concentrations in the lenses of two diurnal surface-feeders were more than three orders of magnitude lower than in common reef species feeding in deeper water. In the Sciaridae (parrot-fish), concentrations of palythene (λ_{max} = 360 nm) were so great that the lenses were noticeably yellow, the only instance of a pigmented MAA. The presence of MAAs in particular tissues of fishes is attributed to accumulation from the diet and selective sequestration (77); such dietary accumulation of MAAs is treated separately.

Role of MAAs in Vision

Protecting ocular tissues from damaging UVR and preventing the transmission (and focusing) of this energy to the retina would seem adaptive for fishes inhab-iting the shallow photic zone. Accordingly, many species have highly absorbing substances (including MAAs) in their ocular tissues that effectively block trans-mission of wavelengths <400 nm (104). The general presence of UVA-absorbing chromophores in ocular tissues is often attributed to improving visual acuity by reducing chromatic aberration caused by the scattering of short-wavelength radi-ation (152). However, this principle does not appear relevant to all fishes, as many species have tetrachromatic vision (UV opsin λ_{max} = 360–380 nm) with func-tional near-UV perception (reviewed in 153), as do many insects, reptiles, birds, and some mammals (154). Siebeck & Marshall (155) compare light transmittance by the ocular media of 211 species of coral reef fishes, where 50.2% of them strongly absorb light of wavelengths below 400 nm, which is generally consis-tent with known lenticular MAA composition (151), particularly for palythene (λ_{max} = 360 nm). The remaining 49.8% have eyes that transmit wavelengths sufficient to allow UV-sensitive vision. Of the Labridae (wrasses), one of the largest and most diverse families of coral reef fishes, only 5 of 36 species have UV-capable vision (156), a variability again consistent with available data on MAAs (151).

While contemporary research focuses on the ocular transmission of UV wave-lengths to evaluate the constraints of UV vision, visual perception of an object also depends on its reflection or absorption of the relevant wavelengths. Given that MAAs are generally localized in the epidermis of fishes and other marine or-ganisms (8), the occurrence of these UV chromophores in dermal tissues may have relevance in the perception of UV coloration by marine animals having UV-sensitive vision. Notwithstanding evolutionary constraints necessary to achieve UV-sensitive vision, the role of MAAs is unlikely restricted to UV photoprotec-tion, and potential involvement in visual UV perception may implicate MAAs as contributors to the sensory physiology, behavior, and ecology of marine animals.

BIOSYNTHESIS OF MAAs BY ALGAL PRODUCERS

The Shikimate Pathway

Details of the biosynthesis of MAAs in marine algae and phototrophic symbioses remain to be demonstrated, but their origin via the shikimate pathway has been a persistent assumption. Favre-Bonvin et al. (157) showed that the shikimate pathway-intermediate, 3-dehydroquinate (DHQ), is the precursor for the six-membered carbon ring common to fungal mycosporines (Figure 1). Synthesis of fungal mycosporines and of MAAs presumably proceeds from DHQ via gadusols (cyclohexenones) (Figure 3) (see references in 35, 69). Based on this knowledge, mycosporine-glycine and fungal mycosporine-serinol (Figure 1) were prepared starting with natural D-(-)-quinate (158).

The variable kinetics of increase among individual MAAs (when their synthesis is stimulated) in dinoflagellates, where the MAA complement may change on the order of hours (39, 72), and in red macroalgae, where the suite of MAAs varies over several days (51, 92), may also indicate interconversions among MAAs following the initial synthesis of a smaller number of primary compounds, as seems to be the case in the coral *Stylophora pistillata* (56). As in *S. pistillata*, shinorine is among the first MAAs to be synthesized in the red macrophyte *Chondrus crispus* (92), and likewise a compound absorbing at 334 nm (the λ_{max} of shinorine) is the first to increase in the free-living dinoflagellate *Alexandrium excavatum* (72). Reciprocal changes in shinorine and palythine concentrations occur in *C. crispus* as they do in *S. pistillata*, suggesting a precursor-product relationship.

Blockage of the synthesis of MAAs in *S. pistillata* by *N*-phosphonomethylglycine (glyphosate), a specific inhibitor of the shikimate pathway, provides the only direct evidence that MAAs in marine organisms are indeed formed via this route (56). DHQ is formed from 3-deoxy-D-arabinoheptulosinate 7-phosphate (DAHP) at the second step in the shikimate pathway by the enzyme 3-dehydroquinate synthase (DHQ synthase). Both this enzyme and one isozyme of DAHP synthase (which catalyzes the first step in the pathway, the condensation of phosphoenolpyruvate and erythrose 4-phosphate) require Co^{2+}, and chelation of this metal by glyphosate may be the basis for its inhibition (at near-millimolar concentrations) of these enzymes (159, 160). Glyphosate (at 1 μM) is a competitive inhibitor for PEP of 5-enolpyruvylshikimate 3-phosphate (EPSP) synthase, the sixth enzyme in the pathway. Therefore, the use of 1 mM glyphosate on *S. pistillata* would have blocked the first two steps, as well as the penultimate step, in the shikimate pathway. This is important because, if the synthesis of MAAs indeed proceeds via a branchpoint at DHQ, inhibition of EPSP synthase alone would not block accumulation of MAAs (Figure 3).

The biosynthesis of MAAs via the shikimate pathway in a zooxanthellate coral presumably occurs in the algal partner, because animals purportedly lack this pathway, which is thought to be restricted to bacteria, algae, plants, and fungi. However, the oft-repeated dogma that animals lack this pathway apparently stems from the inability of vertebrates (variously given as animals, vertebrates, fish, mammals, and humans in literature accounts) to synthesize essential aromatic amino acids, which they must obtain from their diets (159–162), and is not based on empirical evidence such as failures to detect activities of enzymes of the shikimate pathway or DNA sequences encoding these enzymes.

Nevertheless, which amino acids are essential for most invertebrates is incompletely known, and some these animals apparently can synthesize amino acids that vertebrates cannot (163, 163a). The report that azooxanthellate corals can produce small amounts of tyrosine and phenylalanine (164) seemingly points to the existence of the shikimate pathway and post-chorismate aromatic biosynthesis in these metazoans (although the authors could not rule out the production of the essential amino acids by bacteria associated with the corals). The shikimate pathway does occur exceptionally in a protist, the malaria parasite *Plasmodium falciparum* (165), perhaps in an evolutionarily enigmatic cytosolic form, as in the fungi (166).

Zooxanthellae freshly isolated from their hosts frequently have the same or similar MAA complement as the holobiont (intact symbiosis) or host tissues (60, 62, 143), reinforcing the notion that the algae produce the MAAs. However, in some cases, the freshly isolated zooxanthellae from MAA-containing holobionts lack MAAs (57, 143), which, together with the inability of these and some other zooxanthellae to synthesize MAAs in culture, suggests that the MAAs have another provenance in these cases. If dietary acquisition of MAAs can be excluded, the occurrence in corals and other symbioses of MAAs not synthesized by their endosymbionts demands explanation. Might cnidarians have the enzymes of the early shikimate pathway and the DHQ branchpoint to MAAs, or even the rest of the pre- and post-chorismate pathway, and thus be capable of synthesizing compounds assumed to be essential? We pose this question based on limited but suggestive data and mindful of the different protist ancestry for Cnidaria than for other Metazoa, from which stem the Cnidaria diverged very early (167).

It is noteworthy that cyanobacteria, phototrophic eukaryotes, and the symbioses in which they occur contain imino-MAAs that are lacking from terrestrial fungi, which have only oxo-mycosporines (35). Thus the imino-MAAs probably arose in the plastid line (117), where eukaryotic algae produce a suite of MAAs that both overlaps with and is broader than that in extant cyanobacteria, indicating evolutionary diversification in MAA-biosynthetic pathways in photosynthetic eukaryotes. The occurrence in phototrophic symbioses of MAAs lacking from their endosymbionts in culture and from cyanobacteria may again indicate either bioconversion of primary MAAs or, more speculatively, a de novo synthesis by the hosts themselves.

The metabolic cost of MAAs as a determinant of their concentration in organisms under different photic conditions or in different tissues has been debated (18, 60, 86a, 87, 88, 92, 168, 169). None of the foregoing sources, however, includes a reckoning of the actual costs of synthesizing MAAs, which may or may not affect growth and reproduction, and deposition of protein and lipids in UV-exposed corals and algae (37, 43, 86a, 136, 170–174).

Based on stoichiometry and ATP-coupling coefficients, Haslam (161) quotes a cost of 60 ATP-equivalents for synthesizing one mole of chorismate. Presumably the cost of synthesizing one mole of MAA would be similar, if it proceeds from the DHQ branchpoint prior to investing the final ATP and PEP in synthesizing shikimate-3-phosphate and EPSP in the pathway to chorismate (Figure 3), but with a corresponding ATP-cost of condensing amino acid(s) to the cyclohexenone base structure. This is about double the cost for the more direct synthesis of most non-aromatic amino acids from glycolytic or citrate cycle intermediates. Raven (175) estimates the cost as 300 moles of photons captured in photosynthesis per mole of MAA synthesized, which is about the same as the cost of synthesizing chlorophyll and one-tenth that of producing the light-harvesting complex of chlorophyll, proteins, and accessory pigments (176). Because MAAs are about 5 to 10 times more concentrated than chlorophyll in microalgal (50, 175) and cyanobacterial (177) cells, the cost of MAA synthesis is considerable—perhaps 19% of the total cost of cell production (175). Therefore, although MAA-biosynthesis in phototrophs is coupled to photosynthesis, the frequently observed necessity of a UV stimulus may further check the operation of a costly anabolic process that might compete with other energetic and material demands. In particular, the biosynthesis of MAAs de novo requires nitrogen (scarce in oligotrophic waters) and may divert this limiting nutrient from the competing needs of growth and reproduction (86a).

Stimulation and Regulation of Biosynthesis of MAAs

The concentrations of MAAs in photosynthetic organisms and in symbioses containing them are positively related to the total irradiance they experience in nature, but because solar PAR and UVR co-vary, it is not always clear what wavelengths determine this (reviewed in 2, 8, 12, 69). Experiments using filtered sunlight and sources of artificial light on diverse cyanobacteria, micro- and macroalgae, and corals reveal stimulating effects of UVB, UVA, white light lacking UVR, and blue light, but no effect of red or green light, indicating the presence of specific photoreceptors.

Chemically inhibiting photosynthesis arrests the synthesis of MAAs in cultured dinoflagellates (72), but bright white light stimulates the biosynthesis of MAAs disproportionally more than it enhances photosynthesis (50), so that MAAs do not simply follow photosynthetic carbon, and the upregulation of their biosynthesis in microalgae is part of a suite of responses to high irradiance (50, 109). Non-photic factors such as water flow that enhance photosynthesis also increase the accumulation of MAAs in the coral *Pocillopora damicornis*, but the effect is small compared with that of UVR (64) and is transient in *Porites compressa* (87). The effect of PAR on the levels of MAAs in *Montastraea faveolata* is only 30% of that of UVR when these wavelengths co-vary over a depth range of 3 to 30 m (144).

Studies using controlled spectral irradiance are too few to allow many generalizations about the specific stimuli for biosynthesis of MAAs. UVB but not UVA wavelengths stimulate the synthesis of intracellular and sheath MAAs in diverse cyanobacteria (reviewed in 40, 49, 177a), whereas UVA and blue light are positive effectors in free-living dinoflagellates (72) and in the red macrophyte *Chondrus crispus* (177b). Under the same fluence of PAR, MAAs are synthesized in response to combined UVB + UVA radiation, but not to UVA alone, in the coral *Stylophora pistillata* (56). A differential response occurs for MAA biosynthesis during exposure of Antarctic diatoms to UVB, UVA, and PAR (178), a response also seen in a dinoflagellate (178a). Like cyanobacteria, zooxanthellae isolated from diverse hosts may contain constitutive levels of MAAs in the absence of UVR in culture and may enhance these under UVR (57, 58, 143, 173). Zooxanthellae originally isolated from *S. pistillata* and maintained under the same light source as this species of coral for two weeks scarcely increase their MAAs beyond constitutive levels (J. M. Shick & C. Ferrier-Pagès, unpublished data), although the coral does. The limited complement of MAAs in zooxanthellae may change with the age of the culture (143).

It has not been established whether the different kinetics of increase for various MAAs in *S. pistillata* indicate different UV sensitivities of de novo biosynthesis of these MAAs or whether there is a conversion of some MAAs synthesized early to those accumulating later; moreover, some interconversions may occur in the host's tissues and not in the algae (56, 69). Similar temporal changes in the complement of MAAs occur in red macroalgae under controlled spectral irradiance, where, e.g., only shinorine is synthesized in *C. crispus* under UVR alone but full-spectrum PAR + UVR elicits the additional synthesis of palythine (92, 179), which is also synthesized under blue light alone (177b). The former indicates not only a wavelength-specificity of stimulation, but also that the biosynthesis of shinorine does not directly depend on photosynthesis (at least in experiments of 7 days duration), as no PAR was present.

Little is known of the photoreceptors determining the spectral specificity of MAA biosynthesis. The pronounced effect of blue light and UVA on such biosynthesis in the red-tide dinoflagellate *Alexandrium excavatum* (72) and *C. crispus* (177b) indicates the presence of blue-light/UVA receptors, perhaps the flavoprotein cryptochromes (180, 181), but this has not been ascertained. The biosynthesis of MAAs in response particularly to UVB in cyanobacteria, microalgae, rhodophytes, and zooxanthellate corals suggests the presence of the corresponding receptor, although no receptor specific to UVB has been identified in any organism. Intriguingly, there is a selective induction by UVB of mRNA transcripts for DAHP synthase (the first enzyme in the shikimate pathway) (Figure 3) and for enzymes involved in the biosynthesis of UV-screening flavonoids in higher plants (182), but the signaling mechanism has not been demonstrated.

Cultures of *A. excavatum* kept in logarithmic growth phase increase their MAA content and change its composition within 3 to 6 h of exposure to high irradiance (39). Such rapid UV photoacclimation may be necessary in growing phytoplankton that experience radically different light fields on a daily basis (39). Slower rates of accumulation requiring days occur in batch cultures (presumably in stationary phase) of Antarctic diatoms (45, 178) and natural assemblages of Antarctic phytoplankton (37). Zooxanthellae in stationary-phase culture show smaller changes in MAA content, and increases that do occur require two weeks or more (143, J. M. Shick & C. Ferrier-Pagès, unpublished data). Time-courses of several days to weeks for MAAs to increase are also evident in corals (reviewed in 2, 8), where the steady-state population of zooxanthellae in hospite may have low metabolic rates like those in stationary-phase cultures. Kinetics similar to those in corals are seen in red macrophytes (93).

BIOACCUMULATION OF MAAs BY CONSUMER ORGANISMS

No metazoan has been shown to synthesize MAAs de novo, and the long-standing assumption that consumers acquire them from their diet (39, 76, 183) has been confirmed experimentally (77, 83, 85, 95). Sea cucumbers (Figure 5), sea urchins (83), and fishes (77) remove MAAs from their digesta, whereas hairless mice (which lack MAAs in their tissues) do not absorb or degrade MAAs present in a formulated diet (77).

Uptake of MAAs

The selective uptake of MAAs from food suggests there are specific transporters for them in the gut. Sea urchins accumulate principally shinorine (an acidic MAA) from *Chondrus crispus*, although this red macrophyte contains not only this MAA but also higher concentrations of the uncharged molecules palythine, asterina-330, and usujirene (86). Conversely, the medaka fish *Oryzias latipes* absorbs and accumulates neutral palythine and asterina-330 but not the acidic shinorine from *Mastocarpus stellatus* (77). These results suggest the presence of a transporter for acidic MAAs in sea urchins and one for neutral MAAs in fishes (69). *Sepia officinalis*, a cephalopod mollusc, can assimilate neutral and acidic MAAs, as its eyes contain both classes of MAAs (78), as do eggs and follicle/test cells of the tunicate *Ascidia ceratodes* (184).

The inhibition of the translocation of shinorine across the holothuroid gut by the presence of an equimolar concentration of the structurally similar porphyra-334 (J. M. Shick & W. C. Dunlap, unpublished data) suggests that the passage of MAAs occurs via carrier-mediated mechanisms rather than by paracellular diffusion, as likewise seems the case in the medaka (77). There are also regional differences in the echinoderm gut's capacity to transport MAAs (83; J. M. Shick & W. C. Dunlap, unpublished data). Although the mammalian (murine) gut cannot absorb it, shinorine is taken up in a concentration-dependent fashion by human skin cells in culture (77).

Accumulation of MAAs

The accumulation of MAAs in cultured microalgae, macroalgae, and zooxanthellate corals may occur within hours to days of the appropriate photic stimulation,

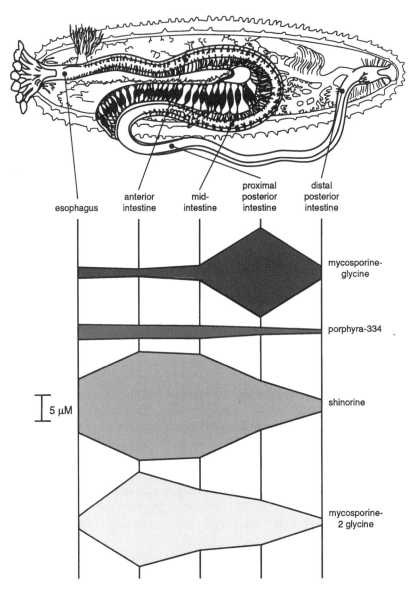

Figure 5   MAAs in the enteric fluid of the holothuroid *Thelenota ananas* feeding on cyanobacterial mats and associated microalgae (W. C. Dunlap & J. M. Shick, unpublished data).

whereas measurable accumulation of dietary MAAs in the consumers' tissues requires weeks to months (77, 83, 86, 95). Although echinoderm larvae can accumulate dissolved FAAs from seawater (163), they do not take up dissolved shinorine (44). The differential occurrence of particular MAAs among organs of metazoan consumers implies a specificity of transport systems, and perhaps tissue-specific interconversions among MAAs, to produce this distribution (8, 76).

Interconversion of Dietary MAAs

The composition of MAAs accumulated by non-symbiotic consumers, notably echinoderms, differs from that in their diets (76, 86). MAAs in sea cucumbers are predominantly localized in their epidermal tissues and gonads, where asterina-330 is usually a major component, yet this MAA is absent from their diet of benthic microalgae and filamentous cyanobacteria, which contain mostly shinorine, mycosporine-2 glycine, and porphyra-334. This conundrum was partially solved by detailed examination of a coral reef sea cucumber, *Thelenota ananas*. The MAA complement in the foregut corresponds with that in dietary microflora, with the lumen of the digestive tract showing decreasing concentrations of algal MAAs and increasing concentrations of mycosporine-glycine (absent from the sediment algae) distally until enteric concentrations of all MAAs decline as they are translocated to body fluids and tissues (Figure 5). Asterina-330 is rarely present in the gut contents but reaches high concentrations in the intestinal tissue and especially in the epidermis. Further examination showed that strains of the ubiquitous marine bacterium *Vibrio harveyi*, isolated from enteric fluids, selectively hydrolyze the hydroxyamino acid substituents of shinorine and porphyra-334, yielding mycosporine-glycine (see 8 for the biochemical scheme), which may explain the complementary changes in these MAAs in the gut fluid (Figure 5). Whereas mycosporine-glycine is the postulated intermediate in the conversion to asterina-330, the pathway to reaminate mycosporine-glycine with endogenous ethanolamine has not been elucidated. Interestingly, this conversion appears not to operate in the medaka fish, which absorbs the small amount of asterina-330 available in a formulated diet but does not absorb or convert the far greater amount of shinorine (77).

CONCLUSION: OUTLOOK AND FUTURE DIRECTIONS

The occurrence and physiological importance of MAAs are becoming standard inclusions in studies of the effects of UVR on aquatic organisms. Most involve environmental correlations between levels of UVR and MAAs, particularly in light of stratospheric ozone depletion, and more intensive and extensive sampling may discern a match between the UV absorption maxima of diverse MAAs and the UV spectral irradiance in the habitats where they occur. Possible localization of MAAs in chloroplasts or other UV-sensitive intracellular sites will be studied.

Cellular targets protected from UVR by MAAs will be elucidated, particularly as measuring damage to DNA becomes more feasible in taxonomically diverse organisms and as techniques of proteomics are applied. Molecular studies of UV signal-transduction eliciting biosynthesis of MAAs can be expected, particularly as researchers on cyanobacteria, algae, and corals extend their collaborations to include workers on higher plants, where UV-signaling is of intense interest. The possibility that MAAs themselves serve as chemical messengers or otherwise regulate physiological function will be examined further. Cooperation between physiologists and photobiochemists has already proved productive, and details of the cellular machinery for synthesizing and interconverting MAAs may yet emerge from interactions between these groups and metabolic biochemists and molecular biologists. Structures, UV-absorbing, and redox properties of unidentified MAAs will be elucidated, particularly in cyanobacteria, perhaps in the context of their protecting biological nitrogen fixation from UVR. The study of symbiotic associations will continue to broaden the physiological horizons of algal, bacterial, fungal, plant, and animal physiologists alike. Comparative genomics might test for the wider occurrence of the shikimate pathway in protists (which are underrepresented in studies of the presence of MAAs) and in Cnidaria, where in some instances the presence of MAAs is otherwise enigmatic. Finally, it may be possible to reintroduce the biosynthetic pathway for MAAs into a higher plant (such as a flavonoid-deficient mutant of *Arabadopsis*) to examine the biomolecular evolution of tolerance to UVR.

Visit the Annual Reviews home page at www.AnnualReviews.org

LITERATURE CITED

1. Booth CR, Morrow JH. 1997. The penetration of UV into natural waters. *Photochem. Photobiol.* 65:254–75
2. Shick JM, Lesser MP, Jokiel PL. 1996. Effects of ultraviolet radiation on corals and other coral reef organisms. *Global Change Biol.* 2:527–45
3. de Mora S, Demers S, Vernet M, eds. 2000. *The Effects of UV Radiation in the Marine Environment.* Cambridge, UK: Cambridge Univ. Press. 324 pp.
4. Worrest RC, Häder D-P. 1997. Overview on the effects of increased solar UV on aquatic microorganisms. *Photochem. Photobiol.* 65:257–59
5. Jagger J. 1985. *Solar UV Actions on Living Cells.* New York: Praeger. 202 pp.
6. Tyrrell RM. 1991. UVA (320–380 nm) as an oxidative stress. In *Oxidative Stress:* *Oxidants and Antioxidants*, ed. H Sies, pp. 57–83. San Diego: Academic
7. DiGiulio RT, Washburn PC, Wennin RJ, Winston GW, Jewell CS. 1989. Biochemical responses in aquatic animals: a review of determinants of oxidative stress. *Environ. Toxicol. Chem.* 8:1103–23
8. Dunlap WC, Shick JM. 1998. Ultraviolet radiation-absorbing mycosporine-like amino acids in coral reef organisms: a biochemical and environmental perspective. *J. Phycol.* 34:418–30
9. Dunlap WC, Shick JM, Yamamoto Y. 2000. UV protection in marine organisms. I. Sunscreens, oxidative stress and antioxidants. In *Free Radicals in Chemistry, Biology and Medicine*, ed. S Yoshikawa, S Toyokuni, Y Yamamoto, Y Naito, pp. 200–14. London: OICA Int.

10. Mitchell DL, Karentz D. 1993. The induction and repair of DNA photodamage in the environment. In *Environmental UV Photobiology*, ed. AR Young, LO Björn, J Moan, W Nultsch, pp. 345–77. New York: Plenum

11. Jeffrey WD, Kase JP, Wilhelm SW. 2000. UV radiation effects on heterotrophic bacterioplankton and viruses in marine ecosystems. See Ref. 3, pp. 206–36

12. Karentz D. 2001. Chemical defenses of marine organisms against solar radiation exposure: UV-absorbing mycosporine-like amino acids and scytonemin. In *Marine Chemical Ecology*, ed. JB McClintock, BJ Baker, pp. 481–520. Boca Raton, FL: CRC Press

13. Wittenburg JB. 1960. The source of carbon monoxide in the float of the Portugese man-of-war *Physalia physalis* L. *J. Exp. Biol.* 37:698–705

14. Price JH, Forrest HS. 1969. 310 μm absorbance in *Physalia physalis*: distribution of the absorbance and isolation of a 310 μm absorbing compound. *Comp. Biochem. Physiol.* 30:879–88

15. Tsujino I. 1961. Studies on the compounds specific for each group of marine algae. II. Extraction and isolation of characteristic ultraviolet absorbing material in Rhodophyta. *Bull. Fac. Fisheries, Hokkaido Univ.* 12:59–65

16. Shibata K. 1969. Pigments and a UV-absorbing substance in corals and a blue-green alga living in the Great Barrier Reef. *Plant Cell Physiol.* 10:325–35

17. Maragos JE. 1972. *A study of the ecology of Hawaiian reef corals*. Ph.D. thesis. Univ. Hawaii, Honolulu. 290 pp.

18. Jokiel PL, York RH Jr. 1982. Solar ultraviolet photobiology of the reef coral *Pocillopora damicornis* and symbiotic zooxanthellae. *Bull. Mar. Sci.* 32:301–15

19. Iwamoto K, Aruga Y. 1973. Distribution of the UV-absorbing substance in algae with reference to the particularity of *Prasiola japonica* Yatabe. *J. Tokyo Univ. Fisheries* 60:43–54

20. Sivalingam PM, Ikawa T, Nisizawa K. 1974. Possible physiological roles of a substance showing characteristic UV-absorbing patterns in some marine algae. *Plant Cell Physiol.* 15:583–86

21. Ito S, Hirata Y. 1977. Isolation and structure of a mycosporine from the zoanthidian *Palythoa tuberculosa*. *Tetrahedron Lett.* 28:2429–30

22. Hirata Y, Uemura D, Ueda K, Takano S. 1979. Several compounds from *Palythoa tuberculosa* (Coelenterata). *Pure Appl. Chem.* 51:1875–83

23. Nakamura H, Kobiashi J, Hirata Y. 1981. Isolation and structure of a 330 nm UV-absorbing substance, asterina–330, from the starfish *Asterina pectinifera*. *Chem. Lett.* 28:1413–14

24. Takano S, Nakanishi A, Uemura D, Hirata Y. 1979. Isolation and structure of a 334 nm UV-absorbing substance Porphyra-334 from the red alga *Porphyra tenera* Kjellman. *Chem. Lett.* 26:419–20

25. Tsujino I, Yabe K, Sekikawa I. 1980. Isolation and structure of a new amino acid, shinorine, from the red alga *Chondrus yendoi* Yamada Mikami. *Bot. Mar.* 23:65–68

26. Sekikawa I, Kubota C, Hiraoki T, Tsujino I. 1980. Isolation and structure of a 357 nm UV-absorbing substance, usujirene, from the red alga *Palmaria palmata* (L) O. Kintzi. *Jpn. J. Phycol.* 34:185–88

27. Nakamura H, Kobiashi J, Hirata Y. 1982. Separation of mycosporine-like amino acids in marine organisms using reverse-phase high-performance liquid chromatography. *J. Chromatogr.* 250:113–18

28. Dunlap WC, Chalker BE. 1986. Identification and quantitation of near-UV absorbing compounds (S-320) in a hermatypic scleractinian. *Coral Reefs* 5:155–59

29. Madronich S, McKenzie RL, Björn LO, Caldwell MM. 1998. Changes in biologically active ultraviolet radiation reaching

the earth's surface. In *Environmental Effects of Ozone Depletion*: 1998 *Assessment*, ed. UNEP, pp. 1–27. Nairobi: United Nations Environ. Progr.

30. Shindell DT, Rind D, Lonergan P. 1998. Increased polar stratospheric ozone losses and delayed recovery owing to increasing green-house gas concentrations. *Nature* 392:589–92

31. Leach CM. 1965. Ultraviolet-absorbing substances associated with light-induced sporulation in fungi. *Can. J. Bot.* 43:185–200

32. Favre-Bonvin J, Arpin N, Brevard C. 1976. Structure de la mycosporine (P310). *Can. J. Chem.* 54:1105–13

33. Arpin N, Curt R, Favre-Bonvin J. 1979. Mycosporines: mise au point et données nouvelles concernant leurs structures, leur distribution, leur localisation et leur biogenèse. *Rev. Mycologie* 43:247–57

34. Young H, Patterson VJ. 1982. A UV protective compound from *Glomerella cingulata*—a mycosporine. *Phytochemistry* 21:1075–77

35. Bandaranayake WM. 1998. Mycosporines: Are they nature's sunscreens? *Natural Prod. Rep.* 15:159–72

36. Büdel B, Karsten U, Garcia-Pichel F. 1997. Ultraviolet-absorbing scytonemin and mycosporine-like amino acid derivatives in exposed, rock-inhabiting cyanobacterial lichens. *Oecologia* 112:165–72

37. Villafañe VE, Helbling EW, Holm-Hansen O, Chalker BE. 1995. Acclimatization of Antarctic natural phytoplankton assemblages when exposed to solar ultraviolet radiation. *J. Plankton Res.* 17:2295–306

38. Jeffrey SW, MacTavish HS, Dunlap WC, Vesk M, Groenewould K. 1999. Occurrence of UVA- and UVB-absorbing compounds in 152 species (206 strains) of marine microalgae. *Mar. Ecol. Prog. Ser.* 189:35–51

39. Carreto JI, Carignan MO, Daleo G, De Marco SG. 1990. Occurrence of mycosporine-like amino acids in the red-tide di-

noflagellate *Alexandrium excavatum*: UV-photoprotective compounds? *J. Plankton Res.* 121:909–21

40. Castenholz RW, Garcia-Pichel F. 2000. Cyanobacterial responses to UV-radiation. In *The Ecology of Cyanobacteria: Their Diversity in Time and Space*, ed. BA Whitton, M Potts, pp. 591–611. Dordrecht: Kluwer

41. Garcia-Pichel F. 1994. A model for internal self-shading in planktonic organisms and its implications for the usefulness of ultraviolet sunscreens. *Limnol. Oceanogr.* 39:1704–17

42. Adams NL, Shick JM. 1996. Mycosporine-like amino acids provide protection against ultraviolet radiation in eggs of the green sea urchin, *Strongylocentrotus droebachiensis*. *Photochem. Photobiol.* 64:149–58

43. Lesser MP. 1996. Acclimation of phytoplankton to UV-B radiation: oxidative stress and photoinhibition of photosynthesis are not prevented by UV-absorbing compounds in the dinoflagellate *Prorocentrum micans*. *Mar. Ecol. Prog. Ser.* 132:287–97

44. Adams NL, Shick JM. 2001. Mycosporine-like amino acids prevent UVB-induced abnormalities during early development of the green sea urchin *Strongylocentrotus droebachiensis*. *Mar. Biol.* 138:267–80

45. Roy S. 2000. Strategies for the minimisation of UV-induced damage. See Ref. 3, pp. 177–205

46. Arai T, Nishijima M, Adachi K, Sano H. 1992. Isolation and structure of a UV absorbing substance from the marine bacterium *Micrococcus* sp. AK-334. pp. 88–94. *MBI Rep. 1992, Marine Biotechnol. Inst. Tokyo*

47. Dunlap WC, Masaki K, Yamamoto Y, Larsen RM, Karube I. 1998. A novel antioxidant derived from seaweed. In *New Developments in Marine Biotechnology*, ed. Y LeGal, H Halvorson, pp. 33–35. New York: Plenum

48. Garcia-Pichel F, Castenholz RW. 1993. Occurrence of UV-absorbing, mycosporine-like compounds among cyanobacterial isolates and an estimate of their screening capacity. *Appl. Environ. Microbiol.* 59:163–69

49. Ehling-Schulz M, Scherer S. 1999. UV protection in cyanobacteria. *Eur. J. Phycol.* 34:329–38

50. Neale PJ, Banaszak AT, Jarriel CR. 1998. Ultraviolet sunscreens in *Gymnodinium sanguineum* (Dinophyceae): mycosporine-like amino acids protect against inhibition of photosynthesis. *J. Phycol.* 34:928–38

51. Karsten U, Sawall T, Hanelt D, Bischof K, Figueroa FL, et al. 1998. An inventory of UV-absorbing mycosporine-like amino acids in macroalgae from polar to warm-temperate regions. *Bot. Mar.* 41:443–53

52. Karsten U, Sawall T, Wiencke C. 1998. A survey of the distribution of UV-absorbing substances in tropical macroalgae. *Phycol. Res.* 46:271–79

53. Pavia H, Cevin G, Lindgren A, Åberg A. 1997. Effects of UV-B radiation and simulated herbivory on phlorotannins in the brown alga *Ascophyllum nodosum*. *Mar. Ecol. Prog. Ser.* 157:139–46

54. Banaszak AT, Lesser MP, Kuffner IB, Ondrusek M. 1998. Relationship between ultraviolet (UV) radiation and mycosporine-like amino acids (MAAs) in marine organisms. *Bull. Mar. Sci.* 63:617–28

55. Gleason DF, Wellington GM. 1995. Variation in UVB sensitivity of planula larvae of the coral *Agaricia agaricites* along a depth gradient. *Mar. Biol.* 123:693–703

56. Shick JM, Romaine-Lioud S, Ferrier-Pagès C, Gattuso J-P. 1999. Ultraviolet-B radiation stimulates shikimate pathway-dependent accumulation of mycosporine-like amino acids in the coral *Stylophora pistillata* despite decreases in its population of symbiotic dinoflagellates. *Limnol. Oceanogr.* 44:1667–82

57. Ishikura M, Kato C, Maruyama T. 1997. UV-absorbing substances in zooxanthellate and azooxanthellate clams. *Mar. Biol.* 128:649–55

58. Banaszak AT, LaJeunesse TC, Trench RK. 2000. The synthesis of mycosporine-like amino acids (MAAs) by cultured, symbiotic dinoflagellates. *J. Exp. Mar. Biol. Ecol.* 249:219–33

59. Stochaj WR, Dunlap WC, Shick JM. 1994. Two new UV-absorbing mycosporine-like amino acids from the sea anemone *Anthopleura elegantissima* and the effects of zooxanthellae and spectral irradiance on chemical composition and content. *Mar. Biol.* 118:149–56

60. Shick JM, Lesser MP, Dunlap WC, Stochaj WR, Chalker BE, Wu Won J. 1995. Depth-dependent responses to solar ultraviolet radiation and oxidative stress in the zooxanthellate coral *Acropora microphthalma*. *Mar. Biol.* 122:41–51

61. Dionisio-Sese ML, Ishikura M, Maruyama T, Miyachi S. 1997. UV-absorbing substances in the tunic of a colonial ascidian protect its symbiont, *Prochloron* sp., from damage by UV-B radiation. *Mar. Biol.* 128:455–61

62. Shick JM, Lesser MP, Stochaj WR. 1991. Ultraviolet radiation and photooxidative stress in zooxanthellate Anthozoa: the sea anemone *Phyllodiscus semoni* and the octocoral *Clavularia* sp. *Symbiosis* 10:145–73

63. Muszynski FZ, Bruckner A, Armstrong RA, Morell JM, Corredor JE. 1998. Within-colony variations of UV absorption in a reef building coral. *Bull. Mar. Sci.* 63:589–94

64. Jokiel PL, Lesser MP, Ondrusek ME. 1997. UV-absorbing compounds in the coral *Pocillopora damicornis*: interactive effects of UV radiation, photosynthetically active radiation, and water flow. *Limnol. Oceanogr.* 42:1468–73

65. Sivalingam PM, Ikawa T, Nisizawa K. 1976. Physiological roles of a substance 334 in algae. *Bot. Mar.* 19:9–21

66. Karsten U, Wiencke C. 1999. Factors

controlling the formation of UV-absorbing mycosporine-like amino acids in the marine red alga *Palmaria palmata* from Spitsbergen (Norway). *J. Plant Physiol.* 155:407–15

67. Drollet JH, Glaziou P, Martin PMV. 1993. A study of mucus from the solitary coral *Fungia fungites* (Scleractinia: Fungiidae) in relation to photobiological UV adaptation. *Mar. Biol.* 115:263–66

68. Teai T, Drollet JH, Bianchini J-P, Cambon A, Martin PMV. 1998. Occurrence of ultraviolet radiation-absorbing mycosporine-like amino acids in coral mucus and whole corals of French Polynesia. *Mar. Freshw. Res.* 49:127–32

69. Shick JM, Dunlap WC, Buettner GR. 2000. Ultraviolet (UV) protection in marine organisms II. Biosynthesis, accumulation, and sunscreening function of mycosporine-like amino acids. In *Free Radicals in Chemistry, Biology and Medicine*, ed. S Yoshikawa, S Toyokuni, Y Yamamoto, Y Naito, pp. 215–28. London: OICA Int.

70. Stochaj WR, Grossman AR. 1997. Differences in the protein profiles of cultured and endosymbiotic *Symbiodinium* sp. (Pyrrophyta) from the anemone *Aiptasia pallida* (Anthozoa). *J. Phycol.* 33:44–53

71. Sutton DC, Hoegh-Guldberg O. 1990. Host-zooxanthella interactions in four temperate marine invertebrate symbioses: assessment of effect of host extracts on symbionts. *Biol. Bull.* 178:175–86

72. Carreto JI, Lutz VA, De Marco SG, Carignan MO. 1990. Fluence and wavelength dependence of mycosporine-like amino acid synthesis in the dinoflagellate *Alexandrium excavatum*. In *Toxic Marine Phytoplankton*, ed. E Graneli, B Sundström, L Edler, DM Anderson, pp. 275–79. Amsterdam: Elsevier

73. Vernet M, Whitehead K. 1996. Release of ultraviolet-absorbing compounds by the red-tide dinoflagellate *Lingulodinium polyedra*. *Mar. Biol.* 127:35–44

74. Sinha RP, Klisch M, Gröninger A, Häder D-P. 1998. Ultraviolet absorbing/screening substances in cyanobacteria, phytoplankton and macroalgae. *J. Photochem. Photobiol. B* 47:83–94

75. Bandaranayake WM, Bourne DJ, Sim RG. 1997. Chemical composition during maturing and spawning of the sponge *Dysidea herbacea* (Porifera:Demospongiae). *Comp. Biochem. Physiol.* 118B: 851–89

76. Shick JM, Dunlap WC, Chalker BE, Banaszak AT, Rosenzweig TK. 1992. Survey of ultraviolet radiation-absorbing mycosporine-like amino acids in organs of coral reef holothuroids. *Mar. Ecol. Prog. Ser.* 90:139–48

77. Mason DS, Schafer F, Shick JM, Dunlap WC. 1998. Ultraviolet radiation-absorbing mycosporine-like amino acids (MAAs) are acquired from their diet by medaka fish (*Oryzias latipes*) but not by SKH−1 hairless mice. *Comp. Biochem. Physiol.* 120A:587–98

78. Shashar N, Hárosi FI, Banaszak AT, Hanlon RT. 1998. UV radiation blocking compounds in the eye of the cuttlefish *Sepia officinalis*. *Biol. Bull.* 195:187–88

79. Karentz D, Mc Euen FS, Land MC, Dunlap WC. 1991. Survey of mycosporine-like amino acid compounds in Antarctic marine organisms: potential protection from ultraviolet exposure. *Mar. Biol.* 129: 157–66

80. Karentz D. 1994. Ultraviolet tolerance mechanisms in Antarctic marine organisms. In *Ultraviolet Radiation and Biological Research in Antarctica*, ed. CS Weiler, PA Penhale, pp. 93–110. Washington, DC: Am. Geophys. Union

81. Gulko D. 1995. Effects of ultraviolet radiation on fertilization and production of planula larvae in the Hawaiian coral *Fungia scutaria*. In *Ultraviolet Radiation and Coral Reefs*, ed. D Gulko, PL Jokiel, pp. 135–47. Honolulu: Univ. Hawaii Inst. Marine Biol.

82. Krupp DA, Blanck J. 1995. Preliminary

report on the occurrence of mycosporine-like amino acids in the eggs of the Hawaiian scleractinian corals *Montipora verrucosa* and *Fungia scutaria*. In *Ultraviolet Radiation and Coral Reefs*, ed. D Gulko, PL Jokiel, pp. 129–34. Honolulu: Univ. Hawaii Inst. Marine Biol.

83. Carroll AK, Shick JM. 1996. Dietary accumulation of mycosporine-like amino acids (MAAs) by the green sea urchin (*Strongylocentrotus droebachiensis*). *Mar. Biol.* 124:561–69

84. Karentz D, Dunlap WC, Bosch I. 1997. Temporal and spatial occurrence of UV-absorbing mycosporine-like amino acids in tissues of the antarctic sea urchin *Sterechinus neumayeri* during springtime ozone-depletion. *Mar. Biol.* 129:343–53

85. Carefoot TH, Harris M, Taylor BE, Donovan D, Karentz D. 1998. Mycosporine-like amino acids: possible UV protection in eggs of the sea hare *Aplysia dactylomela*. *Mar. Biol.* 130:389–96

86. Adams NL, Shick JM, Dunlap WC. 2001. Selective accumulation of mycosporine-like amino acids in ovaries of the green sea urchin *Strongylocentrotus droebachiensis* is not affected by ultraviolet radiation. *Mar. Biol.* 138:281–94

86a. Gleason DF. 2001. Ultraviolet radiation and coral communities. In *Ecosystems, Evolution, and Ultraviolet Radiation*, ed. CS Cockell, AR Blaustein, pp. 118–49. New York: Springer-Verlag

87. Kuffner IB. 2001. Effects of ultraviolet radiation and water motion on the reef coral *Porites compressa* Dana: a flume experiment. *Mar. Biol.* 138:467–76

88. Gleason DF. 1993. Differential effects of ultraviolet radiation on green and brown morphs of the Caribbean coral *Porites astreoides*. *Limnol. Oceanogr.* 38:1452–63

89. Corredor JE, Bruckner AW, Muszynski FZ, Armstrong RA, García R, Morell JM. 2000. UV-absorbing compounds in three species of Caribbean zooxanthellate corals: depth distribution and spectral response. *Bull. Mar. Sci.* 67:821–30

90. Dunlap WC, Yamamoto Y. 1995. Small-molecule antioxidants in marine organisms: antioxidant activity of mycosporine-glycine. *Comp. Biochem. Physiol.* 112B:105–14

91. Maegawa M, Kunieda M, Kida W. 1993. Difference of the amount of UV absorbing substance between shallow- and deep-water red algae. *Jpn. J. Phycol.* 41:351–54

92. Franklin LA, Yakoleva I, Karsten U, Lüning K. 1999. Synthesis of mycosporine-like amino acids in *Chondrus crispus* (Florideophyceae) and the consequences for sensitivity to ultraviolet B radiation. *J. Phycol.* 35:682–93

93. Karsten U, Bischof K, Hanelt D, Tüg H, Wiencke C. 1999. The effect of ultraviolet radiation on photosynthesis and ultraviolet-absorbing substances in the endemic Arctic macroalga *Devaleraea ramentacea* (Rhodophyta). *Physiol. Plant.* 105:58–66

94. Hoyer K, Karsten U, Sawall T, Wiencke C. 2001. Photoprotective substances in Antarctic macroalgae and their variation with respect to depth distribution, different tissues and developmental stages. *Mar. Ecol. Prog. Ser.* 211:117–29

95. Newman SJ, Dunlap WC, Nicol S, Ritz D. 2000. Antarctic krill (*Euphausia superba*) acquire UV-absorbing mycosporine-like amino acids from dietary algae. *J. Exp. Mar. Biol. Ecol.* 255:93–110

96. Cutchis P. 1982. A formula for comparing annual damaging ultraviolet (DUV) radiation doses at tropical and mid-latitude sites. In *The Role of Solar Ultraviolet Radiation in Marine Ecosystems*, ed. J Calkins, pp. 213–28. New York: Plenum

97. Frederick JE, Snell HE, Haywood EK. 1989. Solar ultraviolet radiation at the earth's surface. *Photochem. Photobiol.* 50:443–50

98. Michalek-Wagner K, Dunlap WC. 2001. The effects of elevated irradiance and temperature on the bleaching response and tissue concentrations of UV-absorbing

mycosporine-like amino acids in soft corals. *Coral Reefs*. In review

99. Gibson JAE, Warwick F, Pienitz R. 2000. Control of biological exposure to UV radiation in the Arctic Ocean: comparison of the roles of ozone and riverine dissolved organic matter. *Arctic* 53:372–82

100. Drollet JH, Teai T, Faucon M, Martin PMV. 1997. Field study of compensatory changes in UV-absorbing compounds in the mucus of the solitary coral *Fungia repanda* (Scleractinia:Fungiidae) in relation to solar UV radiation, seawater temperature, and other coincident physico-chemical parameters. *Mar. Freshwater Res.* 48:329–33

101. Michalek-Wagner K. 2001. Seasonal and sex-specific variations in levels of photoprotecting mycosporine-like amino acids (MAAs) in soft corals. *Mar. Biol.* 139:651–60

102. Xiong F, Kopecky J, Nedbal L. 1999. The occurrence of UV-B absorbing mycosporine-like amino acids in freshwater and terrestrial microalgae (Chlorophyta). *Aquat. Bot.* 63:37–49

103. Sommaruga R, Garcia-Pichel F. 1999. UV-absorbing mycosporine-like compounds in planktonic and benthic organisms from a high-mountain lake. *Arch. Hydrobiol.* 144:255–69

103a. Tartarotti B, Laurion I, Sommaruga R. 2001. Large variability in the concentration of mycosporine-like amino acids among zooplankton from lakes located across an altitude gradient. *Limnol. Oceanogr.* 46:1546–52

104. Thorpe A, Douglas RH, Truscott RJW. 1993. Spectral transmission and shortwave absorbing pigments in the fish lens. I. Phylogenetic distribution and identity. *Vision Res.* 33:289–300

105. Caldwell MM, Robberecht R, Flint SD. 1983. Internal filters: prospect for UV-acclimation in higher plants. *Physiol. Plant.* 58:445–50

106. Rozema J, van de Staaij J, Björn LO,

Caldwell M. 1997. UV-B as an environmental factor in plant life: stress and regulation. *Trends Ecol. Evol.* 12:22–28

107. Shick JM. 1993. Solar UV and oxidative stress in algal-animal symbioses. In *Frontiers of Photobiology*, ed. A Shima, M Ichihashi, Y Fujiwara, H Takebe, pp. 561–64. Amsterdam: Excerpta Medica

108. Conde FR, Churio MS, Previtali CM. 2000. The photoprotector mechanism of mycosporine-like amino acids. Excited-state properties and photostability of porphyra-334 in aqueous solution. *J. Photochem. Photobiol. B* 56:139–44

109. Moisan TA, Mitchell BG. 2001. UV absorption by mycosporine-like amino acids in *Phaeocystis antarctica* Karsten induced by photosynthetically available radiation. *Mar. Biol.* 138:217–27

110. Plack PA, Fraser NW, Grant PT, Middleton C, Mitchell AI, Thomson RH. 1981. Gadusol, an enolic derivative of cyclohexane−1,3-dione present in the roes of cod and other marine fish. *Biochem. J.* 199:741–47

111. Grant PT, Middleton C, Plack PA, Thomson RH. 1985. The isolation of four aminocyclohexenimines (mycosporines) and a structurally related derivative of cyclohexane−1:3-dione (gadusol) from the brine shrimp, *Artemia*. *Comp. Biochem. Physiol.* 80B:755–59

112. Lesser MP, Stochaj WR, Tapley DW, Shick JM. 1990. Physiological mechanisms of bleaching in coral reef anthozoans: effects of irradiance, ultraviolet radiation and temperature on the activities of protective enzymes against active oxygen. *Coral Reefs* 8:225–32

113. Gleason DF, Wellington GM. 1993. Ultraviolet radiation and coral bleaching. *Nature* 365:836–37

114. Lesser MP. 1996. Elevated temperature and ultraviolet radiation cause oxidative stress and inhibit photosynthesis in symbiotic dinoflagellates. *Limnol. Oceanogr.* 41:271–83

115. Lesser MP. 1997. Oxidative stress causes coral bleaching during exposure to elevated temperatures. *Coral Reefs* 16:187–92

116. Nii CM, Muscatine L. 1997. Oxidative stress in the symbiotic sea anemone *Aiptasia pulchella* (Carlgren, 1943): contribution of the animal superoxide ion production at elevated temperature. *Biol. Bull.* 192:444–56

117. Garcia-Pichel F. 1998. Solar ultraviolet and evolutionary history of cyanobacteria. *Orig. Life Evol. Biosph.* 28:321–47

118. Cockell CS, Knowland J. 1999. Ultraviolet radiation screening compounds. *Biol. Rev.* 74:311–45

119. Tapley DW, Buettner GR, Shick JM. 1999. Free radicals and chemiluminescence as products of the spontaneous oxidation of sulfide in seawater, and their biological implications. *Biol. Bull.* 196:52–56

120. Masuda K, Goto M, Maruyama T, Miyachi S. 1993. Adaptation of solitary corals and their zooxanthellae to low light and UV radiation. *Mar. Biol.* 117:685–91

121. Coohill TP. 1991. Action spectra again? *Photochem. Photobiol.* 54:859–70

122. Neale PJ. 2000. Spectral weighting functions for quantifying effects of UV radiation in marine ecosystems. See Ref. 3, pp. 72–100

123. Jeffery WR. 1990. A UV-sensitive maternal messenger RNA encoding a cytoskeletal protein may be involved in axis formation in the ascidian embryo. *Dev. Biol.* 141:141–48

124. Broda E. 1978. *The Evolution of the Bioenergetic Processes.* New York: Pergamon. 231 pp.

125. Hochachka PW, Somero GN. 1984. *Biochemical Adaptation.* Princeton, NJ: Princeton Univ. Press. 537 pp.

126. Fenchel T, Finlay BJ. 1995. *Ecology and Evolution in Anoxic Worlds.* Oxford, UK: Oxford Univ. Press. 276 pp.

127. Arroyo A, Kagan VE, Tyurin VA, Burgess JR, de Cabo R, et al. 2000. NADH and NADPH-dependent reduction of coenzyme Q at the plasma membrane. *Antiox. Redox Signal.* 2:251–62

128. Asada K. 1992. Production and scavenging of active oxygen in chloroplasts. In *Molecular Biology of Free Radical Scavenging Systems*, ed. JG Scandalios, pp. 173–92. New York: Cold Spring Harbor Lab. Press

129. Cunningham ML, Johnson JS, Giovanazzi SM, Peak MJ. 1985. Photosensitized production of superoxide anion by monochromatic (290–405 nm) ultraviolet irradiation of NADH and NADPH coenzymes. *Photochem. Photobiol.* 42:125–28

130. Schomer B, Epel D. 1998. Redox changes during fertilization and maturation of marine invertebrate eggs. *Dev. Biol.* 203:1–11

131. Miller BS, Epel D. 1999. The roles of changes in NADPH and pH during fertilization and artificial activation of the sea urchin egg. *Dev. Biol.* 216:394–405

132. Bernillon J, Parussini E, Letoublon R, Favre-Bonvin J, Arpin N. 1990. Flavin-mediated photolysis of mycosporines. *Phytochemistry* 29:81–84

133. Glynn PW, Imai R, Sakai K, Nakano Y, Yamazato K. 1992. Experimental responses of Okinawan (Ryuku Islands, Japan) reef corals to high sea temperature and UV radiation. In *Proc. 7th Int. Coral Reef Symp., Guam*, ed. RH Richmond, pp. 27–37. Mangilao: Univ. Guam Press

134. Michalek-Wagner K, Dunlap WC. 2001. Examination of UV-absorbing mycosporine-like amino acids in soft corals during bleaching and non-bleaching episodes. *Coral Reefs.* In review

135. Michalek-Wagner K, Willis BL. 2001. Impacts of bleaching on the soft coral *Lobophytum compactum.* I. Fecundity, fertilization and offspring viability. *Coral Reefs* 19:231–39

136. Michalek-Wagner K, Willis BL. 2001.

Impacts of bleaching on the soft coral *Lobophytum compactum*. II. Biochemical changes in adults and their eggs. *Coral Reefs* 19:240–46

137. Pearse JS, Pearse VB, Davis KK. 1986. Photoperiodic regulation of gametogenesis and growth in the sea urchin *Strongylocentrotus droebachiensis*. *J. Exp. Zool.* 237:107–18

138. Shirai H, Walker CW. 1988. Chemical control of asexual and sexual reproduction in echinoderms. In *Endocrinology of Selected Invertebrate Types*, ed. H Laufer, GH Downer, pp. 453–76. New York: Liss

139. Oren A. 1997. Mycosporine-like amino acids as osmotic solutes in a community of halophilic cyanobacteria. *Biomicrobiol. J.* 14:231–40

140. Portwich A, Garcia-Pichel F. 1999. Ultraviolet and osmotic stresses induce and regulate the synthesis of mycosporines in the cyanobacterium *Chlorogloeopsis* PCC 6912. *Arch. Microbiol.* 172:187–92

141. Rosette C, Karin M. 1996. Ultraviolet light and osmotic stress: activation of the JNK cascade through multiple growth factor and cytokine receptors. *Science* 274:1194–97

142. Shick JM. 1991. *A Functional Biology of Sea Anemones*. London: Chapman & Hall. 395 pp.

143. Banaszak AT, Trench RK. 1995. Effects of ultraviolet (UV) radiation on marine microalgal-invertebrate symbioses. II. The synthesis of mycosporine-like amino acids in response to exposure to UV in *Anthopleura elegantissima* and *Cassiopeia xamachana*. *J. Exp. Mar. Biol. Ecol.* 94:233–50

144. Lesser MP. 2000. Depth-dependent photoacclimatization to solar ultraviolet radiation in the Caribbean coral *Montastraea faveolata*. *Mar. Ecol. Prog. Ser.* 192:137–51

145. Zigman S. 1993. Ocular damage—yearly review. *Photochem. Photobiol.* 57:1060–68

146. Yokoyama S, Yokoyama R. 1996. Adaptive evolution of photoreceptors and visual pigments in vertebrates. *Annu. Rev. Ecol. Syst.* 27:543–67

147. Kennedy D, Milkman R. 1956. Selective absorption by the lenses of lower vertebrates, and its influences on spectral sensitivity. *Biol. Bull.* 111:375–86

148. Bon WF, Ruttenburg G, Dohrn A, Batnik H. 1968. Comparative physicochemical investigations on the lens proteins of fishes. *Exp. Eye Res.* 7:603–10

149. Zigman S, Paxia T, Waldron W. 1985. Properties and functions of near-UV absorbing pigments in marine animal lenses. *Biol. Bull.* 169:564 (Abstr.)

150. Zigman S. 1987. Biochemical adaptation in vision: lens pigments. *Photochem. Photobiol.* 45s:35S (Abstr.)

151. Dunlap WC, Williams DM, Chalker BE, Banaszak AT. 1989. Biochemical photoadaptations in vision: UV-absorbing pigments in fish eye tissues. *Comp. Biochem. Physiol.* 93B:601–7

152. Muntz WRA. 1973. Yellow filters and the absorption of light by the visual pigments of some Amazonian fishes. *Vision Res.* 13:2235–54

153. Losey GS, Cronin TW, Goldsmith TH, Hyde D, Marshall NJ, McFarland WN. 1999. The visual world of fishes: a review. *J. Fish Biol.* 59:921–43

154. Goldsmith TH. 1994. Ultraviolet receptors and color vision: evolutionary implications and a dissonance of paradigms. *Vision Res.* 34:1479–87

155. Siebeck UE, Marshall NJ. 2000. Ocular media transmission of coral fish—can coral reef fish see ultraviolet light? *Vision Res.* 41:133–49

156. Siebeck UE, Marshall NJ. 2000. Transmission of ocular media in labrid fishes. *Philos. Trans. R. Soc. London Ser. B* 355:1257–61

157. Favre-Bonvin J, Bernillon J, Salin N, Arpin N. 1987. Biosynthesis of mycosporines: mycosporine glutaminol in

Trichothecium roseum. Phytochemistry 29:2509–14

158. White JD, Cammack JH, Sakuma K. 1989. The synthesis and absolute configuration of mycosporins: a novel application of the Staudinger reaction. *J. Am. Chem. Soc.* 111:8970–72

159. Kishore GM, Shah DM. 1988. Amino acid biosynthesis inhibitors as herbicides. *Annu. Rev. Biochem.* 57:627–63

160. Bentley R. 1990. The shikimate pathway—a metabolic tree with many branches. *Crit. Rev. Biochem. Mol. Biol.* 25:307–84

161. Haslam E. 1993. *Shikimic Acid: Metabolism and Metabolites.* New York: Wiley & Sons. 387 pp.

162. Herrmann KM, Weaver LM. 1999. The shikimate pathway. *Annu. Rev. Plant Physiol.* 50:473–503

163. Manahan DT. 1990. Adaptations by invertebrate larvae for nutrient acquisition from seawater. *Am. Zool.* 30:147–60

163a. Wang JT, Douglas AE. 1999. Essential amino acid synthesis and nitrogen recycling in an alga-invertebrate symbiosis. *Mar. Biol.* 135:219–22

164. Fitzgerald LM, Szmant AM. 1997. Biosynthesis of 'essential' amino acids by scleractinian corals. *Biochem. J.* 322:213–21

165. Roberts F, Roberts CW, Johnson JJ, Kyle DE, Krell T, et al. 1998. Evidence for the shikimate pathway in apicomplexan parasites. *Nature* 393:801–5

166. Keeling PJ, Palmer JD, Donald RGK, Roos DS, Waller RF, McFadden GI. 1999. Shikimate pathway in apicomplexan parasites. *Nature* 397:219–20

167. Adouette A, Balavoine G, Lartillot N, de Rosa R. 1999. Animal evolution: the end of the intermediate taxa? *Trends Genet.* 15:104–8

168. Bandaranayake WM, Des Rocher A. 1999. Role of secondary metabolites and pigments in the epidermal tissues, ripe ovaries, viscera, gut contents and diet of the sea cucumber *Holothuria atra. Mar. Biol.* 133:163–69

169. Norris S. 1999. Marine life in the limelight. *BioScience* 49:520–26

170. Wood WF. 1987. Effect of solar ultraviolet radiation on the kelp *Eklonia radiata. Mar. Biol.* 96:143–50

171. Lesser MP, Shick JM. 1989. Effects of irradiance and ultraviolet radiation on photoadaptation in the zooxanthellae of *Aiptasia pallida*: primary production, photoinhibition, and enzymic defenses against oxygen toxicity. *Mar. Biol.* 102:243–55

172. Grottoli-Everett AG. 1995. Bleaching and lipids in the Pacific coral *Montipora verrucosa*. In *Ultraviolet Radiation and Coral Reefs*, ed. D Gulko, PL Jokiel, pp. 107–13. Honolulu: Univ. Hawaii Inst. Marine Biol.

173. Lesser MP. 1996. Elevated temperatures and ultraviolet radiation cause oxidative stress and inhibit photosynthesis in symbiotic dinoflagellates. *Limnol. Oceanogr.* 41:271–83

174. Wängberg S-Å, Persson A, Karlson B. 1997. Effects of UV-B radiation on synthesis of mycosporine-like amino acids and growth in *Heterocapsa triquetra* (Dinophyceae). *J. Photochem. Photobiol. B* 37:141–46

175. Raven JA. 1991. Responses of aquatic photosynthetic organisms to increased solar UVB. *J. Photochem. Photobiol. B* 9:239–44

176. Raven JA. 1984. A cost-benefit analysis of photon absorption by photosynthetic unicells. *New Phytol.* 98:593–625

177. Subramaniam A, Carpenter EJ, Karentz D, Falkowski PG. 1999. Biooptical properties of the marine diazotrophic cyanobacteria *Trichodesmium* spp. I. Absorption and photosynthetic action spectra. *Limnol. Oceanogr.* 44:608–17

177a. Sinha RP, Klisch M, Helbling EW, Häder D-P. 2001. Induction of mycosporine-like amino acids (MAAs) in cyanobacteria by solar ultraviolet-B radiation. *J. Photochem. Photobiol. B* 60:129–35

177b. Franklin LA, Kräbs G, Kuhlenkamp R. 2001. Blue light and UV-A radiation control the synthesis of mycosporine-like amino acids in *Chondrus crispus* (Florideophyceae). *J. Phycol.* 37:257–70

178. Helbling EW, Chalker BE, Dunlap WC, Holm-Hansen O, Villafañe VE. 1996. Photoacclimation of antarctic marine diatoms to solar ultraviolet radiation. *J. Exp. Mar. Biol. Ecol.* 204:85–101

178a. Klisch M, Häder D-P. 2000. Mycosporine-like amino acids in the marine dinoflagellate *Gyrodinium dorsum*: induction by ultraviolet radiation. *J. Photochem. Photobiol. B* 55:178–82

179. Karsten U, Franklin LA, Lüning K, Wiencke C. 1998. Natural ultraviolet radiation and photosynthetically active radiation induce formation of mycosporine-like amino acids in the marine macroalga *Chondrus crispus* (Rhodophyta). *Planta* 205:257–62

180. Rüdiger W, López-Figueroa F. 1992. Photoreceptors in algae. *Photochem. Photobiol.* 55:949–54

181. Cashmore AR, Jarillo JA, Wu Y-J, Liu D. 1999. Cryptochromes: blue light receptors for plants and animals. *Science* 284:760–65

182. Logemann E, Tavernaro A, Schulz W, Somssich IE, Hahlbrock K. 2000. UV light selectively coinduces supply pathways from primary metabolism and flavonoid secondary product formation in parsley. *Proc. Natl. Acad. Sci. USA* 97:1903–7

183. Chalker BE, Dunlap WC, Banaszak AT, Moran PJ. 1988. UV-absorbing pigments in *Acanthaster planci*: photoadaptation during the life history of reef invertebrates. In *Proc. Sixth Int. Coral Reef Symp.*, p. 15

184. Epel D, Hemela K, Shick M, Patton C. 1999. Development in the floating world: defenses of eggs and embryos against damage from UV radiation. *Am. Zool.* 39:271–78

Annu. Rev. Physiol. 2002. 64:263–88

HYPOXIA-INDUCED ANAPYREXIA: Implications and Putative Mediators

Alexandre A. Steiner and Luiz G. S. Branco

Department of Morphology, Estomatology and Physiology, Dental School of Ribeirão Preto and Department of Physiology, Medical School of Ribeirão Preto, University of São Paulo, 14040-904 Ribeirão Preto, SP, Brazil; e-mail: asteiner@rfi.fmrp.usp.br; branco@forp.usp.br

Key Words set point, body temperature, central nervous system, hypothermia, oxygen

■ **Abstract** Hypoxia elicits an array of compensatory responses in animals ranging from protozoa to mammals. Central among these responses is anapyrexia, the regulated decrease of body temperature. The importance of anapyrexia lies in the fact that it reduces oxygen consumption, increases the affinity of hemoglobin for oxygen, and blunts the energetically costly responses to hypoxia. The mechanisms of anapyrexia are of intense interest to physiologists. Several substances, among them lactate, adenosine, opioids, and nitric oxide, have been suggested as putative mediators of anapyrexia, and most appear to act in the central nervous system. Moreover, there is evidence that the drop in body temperature in response to hypoxia, unlike the ventilatory response to hypoxia, does not depend on the activation of peripheral chemoreceptors. The current knowledge of the mechanisms of hypoxia-induced anapyrexia are reviewed.

INTRODUCTION

In 1991, Wood wrote a comprehensive review (1) about the decrease in body temperature (Tb) observed during hypoxia exposure for the *Annual Review of Physiology*. More than 10 years later, the term anapyrexia was introduced, and considerable progress has been made in the understanding of the mechanisms of this thermoregulatory response to hypoxia, which is the focus of this review. A growing number of studies have reported that a lack of oxygen evokes a drop in Tb in a variety of species ranging from protozoa to mammals (1, 2).

Hypoxia elicits a number of compensatory responses, among them an increase in ventilation and a regulated decrease in Tb, i.e., anapyrexia, in order to increase oxygen uptake and decrease oxygen consumption (1–3). It is currently accepted that the hyperventilation induced by hypoxia results from the activation of peripheral chemoreceptors located in the aortic and carotid bodies and the subsequent processing of this information by the central nervous system (CNS), especially by structures in the brainstem (3). However, this active response to hypoxia is oxygen consuming, which sets a limit to its usefulness (1, 2). Therefore, hypoxia-induced anapyrexia may be beneficial owing to a reduction in oxygen consumption, a leftward shift of the oxyhemoglobin dissociation curve with a resulting improvement of oxygen loading in the lungs, and an attenuation in hyperventilation with a resulting blunting in the energetically costly responses to hypoxia (1, 2). In fact, the importance of this anapyrexic response is emphasized by reports that show an increased survival of the tested species if the animals are allowed to reduce their Tb during hypoxia exposure (1, 2, 4–6). A growing body of evidence supports the notion that anapyrexia likely results from a downward resetting of the thermoregulatory set point. Therefore, based on the neuronal theory of the set point for thermoregulation, which accepts that the thermoreguatory set point is given by a balance between warm-sensitive and temperature-insensitive neurons in the preoptic region of the anterior hypothalamus (PO/AH) (7–10), the participation of the CNS in the development of anapyrexia is evident. Indeed, some putative mediators of anapyrexia have been experimentally tested and the site of action identified as the CNS.

HYPOXIA EXPOSURE REDUCES THE THERMOREGULATORY SET POINT

In mammals, thermoregulation is an extremely complex process that uses many thermoreceptors, afferent neural pathways, a number of integratory sites within the CNS, and numerous efferent pathways and effector organs (11). Mammals control Tb by using both autonomic (e.g., thermogenesis, cutaneous vasodilation, panting, piloerection, and sweating) and behavioral (preference temperature selection) mechanisms (11). Therefore, hypoxia could induce a drop in Tb in mammals by interfering with several steps in the thermoregulatory control, including a reduction in the thermoregulatory set point or an impairment in thermoeffector mechanisms, thereby acting on the CNS or directly on the peripheral tissues. In fact, hypoxia reduces the availability of oxygen to thermogenic tissues (1, 2) and has been reported to induce the production of vasoactive factors in superficial vascular beds (12), responses that would cause a drop in Tb. However, evidence indicates that the decrease in Tb evoked by hypoxia is indeed a consequence of a downward resetting of the thermoregulatory set point rather than the result of a direct effect of hypoxia on specific thermoeffector tissues, i.e., the drop in Tb produced by hypoxia is a regulated decrease in Tb, namely anapyrexia (13–17). It has been shown that the ambient temperature threshold of the thermoneutral zone, below which O_2 consumption increases, is lowered in rats exposed to hypoxia (18). Similarly, rats exposed to heat present a lower threshold for panting when exposed to hypoxia (19). These shifts in the thermoneutral zone are in accordance with a shift in the thermoregulatory set point. In agreement with those observations, a recent study (20) has demonstrated that hypoxic hypoxia shifts the thermoneutral zone of the golden mantled squirrel to a lower temperature, confirming that hypoxia reduces the thermoregulatory set point in mammals. Moreover, Gordon & Fogelson (21) demonstrated that rodents (rats, hamsters, and mice) select a lower preferred ambient temperature after they are subjected to hypoxia. Since behavioral thermoregulation has been shown to be directly related to changes in the thermoregulatory set point (22, 23), this evidence reinforces the finding that rodents reduce their thermoregulatory set point in response to hypoxia.

On the basis of this rationale, ectothermic species have emerged as an interesting model to investigate the effect of hypoxia on the thermoregulatory set point because they rely essentially on behavioral mechanisms to regulate Tb. In fact, exposure to hypoxia has been shown to decrease Tb by means of behavior in ectotherms such as lizards (4), alligators (24), and toads (25–27), and also in water breathers such as the crayfish, amphibian larvae, and teleost fish (28–30). Interestingly, even the unicellular organism *Paramecium caudatum* has been shown to choose a lower preferred environmental temperature in a thermal gradient when exposed to hypoxia (5). For amphibians, reptiles, and even mammals, the experimental thermal gradient is usually an elongated chamber with an aluminum floor with one end cooled to approximately 10°C and the other heated to about 40°C, where the animals may choose a preferred temperature. Petri dishes filled with tap water throughout the chamber are essential to provide access to water at all temperatures (4, 5, 21, 24, 27). This is especially important considering that dehydration is known to elicit anapyrexia in ectothermic species (31).

Taking into consideration the definition of anapyrexia, it should be emphasized that anapyrexia is the opposite of fever, which is defined as a regulated increase in Tb due to an increased thermoregulatory set point (32). Functionally, the decreased thermoregulatory set point is expressed as decreases in metabolic heat production (33) and increases in heat loss in mammals (34) or simply as a behavioral response in ectotherms (4, 24, 27). Accordingly, during the onset of anapyrexia, when Tb is above the thermoregulatory set point, animals use the autonomic (when present) and behavioral thermoregulatory mechanisms available to reduce Tb to a level as low as the set point. Once the reduced Tb is reached, the organism is again euthermic, but at a lower set point. It is important to note that anapyrexia is different from hypothermia in the sense that the latter occurs as a deviation of Tb from the respective thermoregulatory set point and as such is accompanied by mechanisms of heat production and heat conservation in order to restore Tb.

ANAPYREXIA IS A PROTECTIVE RESPONSE

As pointed out above, anapyrexia is an important component of the array of physiological responses that occur during exposure to hypoxia. This fact, together with the observation that this response is extremely widespread among taxa, provides evidence that anapyrexia is a protective response because anapyrexia is evolutionarily conserved. Interestingly, anapyrexia has been shown to be elicited not only by hypoxia exposure but also by other oxygen-limiting stimuli such as hemorrage

(35), anemia (36, 37), azide, cyanide (13, 38), and ozone (39). In fact, the protective role of anapyrexia during hypoxia has already been experimentally tested and the importance of anapyrexia is emphasized by reports showing increased survival rates if the tested species are allowed to reduce their Tb (1, 4–6). In this context, rats (1, 6), lizards (4), and even the *Paramecium* (5) presented higher survival rates under hypoxic conditions at lower Tb values, usually close to those chosen by these species when submitted to hypoxia. Moreover, in oxygen-limiting situations such as in mice (40) and rats (41) under hypoxia, rats (42) and pigs (43) with hemorragic shock, anemic rabbits (44), asphyxiated neonates (45), and in clinical studies involving brain and cardiac surgery (2, 46), there is a strong correlation between anapyrexia and survival rate.

Anapyrexia may protect against low oxygen availability, predominantly in oxygen-sensitive tissues such as heart and brain (1, 2), by three main mechanisms. First, it is well documented that a decrease in Tb leads to a reduction in oxygen consumption (1, 47). In mammals, however, this is true for anapyrexia, but not for hypothermia because anapyrexia is not accompanied by a thermogenic and oxygen-consuming response, a fact observed during hypothermia, which reinforces the difference between anapyrexia and hypothermia. Electrophysiological recovery of hippocampal neurons after hypoxia is improved at lower temperatures, a process that appears to be associated with changes in ATP levels (48). Second, a reduction in Tb produces a leftward shift in the oxyhemoglobin dissociation curve with a resulting increase in oxygen loading in the lungs. Third, evidence from ectotherms indicates that both the ventilatory and cardiovascular responses to hypoxia, which are oxygen consuming, are impaired at lower Tb values (49–51). As for mammals, a biphasic ventilatory response to hypoxia has been observed in small animals and neonates, i.e., a rapid increase in pulmonary ventilation followed by a gradual attenuation of this response (52–55). Although the biphasic ventilatory response to hypoxia is accompanied by a decrease in metabolic rate, whether this response is associated with hypoxia-induced anapyrexia remains controversial (52–57).

MECHANISMS OF ANAPYREXIA

Although hypoxia-induced anapyrexia is widespread among taxa, the mechanisms underlying this response remain poorly understood. However, in the past two decades, progress has been made in understanding the mechanisms and mediators of anapyrexia.

Because anapyrexia results from a downward resetting of the thermoregulatory set point, it may require activation of the neural structures controlling the thermoregulatory set point. According to this model, the first step for the development of anapyrexia in response to a lack of oxygen would be activation of oxygen sensors. Potential candidates include the peripheral chemoreceptors, which are located in the carotid bodies and have a role in mediating the ventilatory response to hypoxia (3). However, this idea has not been confirmed experimentally. Numerous studies have reported that conscious peripherally chemodenervated mammals (58–62) and birds (63) consistently reduce their Tb in response to hypoxia, a response which sometimes is even more pronounced than in intact animals. Moreover, Matsuoka et al. (36) have shown that anemic hypoxia, a situation in which there is a decreased arterial oxygen content (C_aO_2) with no change in arterial oxygen partial pressure (P_aO_2) and, consequently, no activation of peripheral chemoreceptors (3), evokes anapyrexia similarly to hypoxic hypoxia, in which a reduction P_aO_2 occurs and peripheral chemoreceptors are activated. Anemic toads have also been shown to reduce their Tb by means of behavior (37). This result also suggests that, unlike the ventilatory response to hypoxia (3), a decrease either in C_aO_2 or in P_aO_2 is capable of inducing anapyrexia.

Another possibility is that limiting oxygen would activate central oxygen sensors, which might themselves be thermoregulatory centers. This suggestion is supported by a study showing that hypoxia affects the thermal sensitivity of pre-optic neurons in vitro (64). This observation may provide the neuronal basis for the reduction in the thermoregulatory set point by hypoxia because, at least in mammals, the thermoregulatory set point is thought to result from a balance between warm-sensitive and temperature-insensitive neurons located in the PO/AH (9). Taken together, these data indicate that the brain is involved in the development of anapyrexia, but the cellular mechanisms responsible for sensing the lowered C_aO_2 or P_aO_2 in the brain to produce anapyrexia remain unexplored.

In this context, the fact that limiting oxygen has direct cellular effects on oxidative metabolism led Malvin et al. (38) to hypothesize that the impairment of oxidative phosphorylation could be a stimulus for anapyrexia. To this end, the effects of sodium azide (NaN_3), which inhibits oxidative phosphorylation by binding to the cytochrome C oxidase complex (13, 38), on the preferred temperature of the protozoan *Paramecium* was tested. It was observed that azide elicits anapyrexia in a dose-dependent manner, a response that increased the survival of the species. However, whether this mechanism was also relevant to vertebrates remained unknown until 1996, when Branco & Malvin (13) reported that subcutaneous administration of NaN_3 and NaCN produces a reduction in preferred Tb of the toad *Bufo marinus*. Moreover, in that study it was shown that intracerebroventricular injection of NaN_3 and NaCN at doses without thermoregulatory effects, when administered systemically, significantly decreases the preferred Tb of toads by about $10°C$, indicating that inhibition of oxidative phosphorylation in the brain produces anapyrexia. In agreement with this finding, it was recently reported that exclusion of glucose from central sites using 2-deoxy-D-glucose, which could impair oxidative phosphorylation by reducing the availability of metabolic substrates, also causes anapyrexia in toads (65). Taken together, these results imply that a reduction in oxidative phosphorylation in the CNS is important for the development of anapyrexia. However, whether reduced oxidative phosphorylation is the main (or sole) mechanism through which decreased availability of oxygen signals the brain to produce anapyrexia remains to be determined.

PUTATIVE MEDIATORS OF ANAPYREXIA

Once the brain perceives the reduced availability of oxygen, some neuronal pathways should be activated and others inactivated in order to produce anapyrexia. Unfortunately, these pathways are almost unknown. The PO/AH is likely to be one region involved because it is probably the thermosensitive and thermointegrative site of the CNS (9, 10). Moreover, other brain regions responsible for the autonomic control of thermoeffector mechanisms, such as the paraventricular nucleus, the nucleus of the tractus solitarius, and the locus coeruleus, are possibly also involved. To our knowledge, the locus coeruleus is the only nucleus whose participation in hypoxia-induced anapyrexia has been experimentally tested. More specifically, electrolytic lesions of the locus coeruleus have been shown to attenuate hypoxia-induced anapyrexia in rats exposed to 7% inspired oxygen (66), indicating that this nucleus plays a major role in the development of anapyrexia.

Because neurons are under the control of numerous neurotransmitters and neuromodulators, it is not surprising that several molecules have been suggested as putative mediators of hypoxia-induced anapyrexia. These putative mediators of hypoxia-induced anapyrexia have been more explored than the neural pathways involved. Most of the data obtained to date result from the study of the effects of pharmacological agents on Tb and anapyrexia. Although many of these mediators had their site of action confirmed as the CNS by the use of intracerebroventricular injections, this technique does not allow the identification of specific sites or nuclei of the CNS where these neurochemical mediators act. The current knowledge about the main putative mediators of anapyrexia proposed to date is reviewed below.

Arginine Vasopressin (AVP)

Aside from the classical effects of AVP on the regulation of water-salt balance and arterial blood pressure (67), this peptide, also has a thermoregulatory role as an endogenous antipyretic molecule. It has been observed that pregnant ewes near term were unable to develop fever (68, 69). Among all hormonal pattern alterations that occur during pregnancy, Naylor et al. (70) found that AVP concentrations were consistently correlated with the suppression of fever. In agreement with this evidence, further studies supported the antipyretic effect of AVP: (*a*) AVP infusion into the ventral septal area (VSA) of the CNS attenuates or prevents febrile increases in Tb in sheeps (71, 72), rabbits (73), and rats (74), whereas in the same experiments AVP infusion into the VSA at the same dose caused no change in the Tb of euthermic animals; (*b*) the nonmammalian AVP analogue, arginine vasotocin (AVT), attenuates behavioral fever in toads (75); (*c*) VSA is a site of AVP-containing nerve terminals and high amounts of AVP receptors in the CNS (76); (*d*) AVP levels increase in plasma and cerebrospinal fluid during fever (77); (*e*) hypertonic saline and hemorrhage, which are potent stimuli for AVP release, cause antipyresis in rats (78); and (*f*) central administration of specific AVP antagonists results in greatly enhanced fevers (79, 80).

In addition to its antipyretic actions, AVP has also been suggested to be a putative mediator of anapyrexia. Rationales for this hypothesis include (*a*) hypoxia increases AVP levels in plasma and cerebrospinal fluid (81, 82); (*b*) AVP injected intracerebroventricularly (83) or systemically (84, 85) reduces the Tb of rats; (*c*) water deprivation, which is known to increase AVP levels in plasma and cerebrospinal fluid in response to increased extracellular tonicity and diminished plasma volume in mammals (67), reduces the preferred Tb of the toad *Bufo paracnemis* (31) and the metabolic rate of camels (86).

The role of AVP in hypoxia-induced anapyrexia was first experimentally tested in 1994 by Clark & Fewell (87). In that study, it was observed that hypoxia-induced anapyrexia is similar in both Brattleboro rats (which lack AVP-containing cells in the CNS) and Long-Evans rats (used as controls), indicating that AVP does not seem to be a mediator of the anapyrexia elicited by hypoxia. However, thermoregulation may be aberrant in Brattleboro rats. It is known that the Brattleboro rat does not consistently develop fever in response to intraperitoneal and intracerebroventricular injections of bacterial pyrogens at doses that produce fever in regular rats (88–90). In order to solve this issue, we have recently tested the participation of AVP in hypoxia-induced anapyrexia in regular rats by using selective AVP V1- and V2-receptor antagonists. Neither intracerebroventricular nor intravenous treatment with the AVP-receptor antagonists altered hypoxia-induced anapyrexia, even though the doses of the AVP antagonists used in that study were shown to be effective in impairing the pressor and antidiuretic effects of AVP, i.e., they were biologically active (16). In summary, it seems that in spite of the evidence in favor of the participation of AVP in hypoxia-induced anapyrexia, this peptide appears not to be involved in the anapyretic response to hypoxia.

Lactate

Lactate has also been suggested as a mediator of hypoxia-induced anapyrexia. The concept that lactate is a classic companion of hypoxic stress in vertebrates is not new. The analysis of the critical P_aO_2 shows that lactate rises in plasma slightly below the critical P_aO_2 (91). Interestingly, Pörtner et al. (92) demonstrated that lactate injected systemically at the dose of 4 mmol/kg elicits behavioral anapyrexia in the toad *Bufo marinus* under normoxic conditions. In addition, we have reported that the site of action of lactate is likely to be the CNS because intracerebroventricular injection of lactate at a 10-fold lower dose (0.4 mmol/kg) than that used systemically in Pörtner's study reduces in a sustained way the preferred Tb of *Bufo paracnemis* toads, whereas no effect on the preferred Tb is observed when the same dose of lactate is injected systemically (14). In agreement with this notion, a recent study has demonstrated that lactate release depends on the uptake of the excitatory neurotransmitter L-glutamate (93), which suggests that lactate might influence neuronal activity by affecting glutaminergic synapses.

Interestingly, these studies used sodium lactate (NaLa) injections instead of acid lactic to investigate the specific function of the lactate anion on Tb because the H^+ has been shown to induce behavioral anapyrexia in toads (94). However, hypoxic animals are actually lactoacidotic (95), a fact suggesting that lactate and H^+ could act in concert in the development of hypoxia-induced anapyrexia.

The data obtained to date indicating lactate as a putative mediator of anapyrexia are based on the facts that lactate may reduce the Tb of vertebrates and that lactate levels are increased under hypoxic conditions. However, to our knowledge, no study has been designed to directly assess if lactate mediates hypoxia-induced anapyrexia. Two approaches should be useful in testing the lactate hypothesis: the use of an inhibitor of acid lactic production such as dichloroacetate (96) and the use of drugs that block the actions of lactate. We are currently performing experiments to determine the effect of dichloroacetate on hypoxia-induced anapyrexia in rats.

Adenosine

Adenosine has been suggested to be an inhibitory neuromodulator and is thought to be a mediator of hypoxia-induced anapyrexia (1, 97). Adenosine is a purine nucleoside synthesized within cells and released into the extracellular fluid during conditions of hypoxic stress (98). Furthermore, changes in the extracellular concentration of adenosine in the brain and carotid bodies appear to be involved in the regulation of pulmonary ventilation (99), a parameter known to be altered by hypoxia (3). In fact, adenosine has been shown to play thermoregulatory actions. It has been demonstrated that systemic injection of adenosine (97), as well as intracerebroventricular injection of an adenosine analogue (2-chloroadenosine), causes a dose-related fall in Tb in rodents (100) that resembles the progressive Tb drop caused by hypoxia exposure. However, available literature shows that administration of adenosine antagonists, such as aminophylline, theophylline or caffeine, to euthermic animals at optimal doses has no effect on the Tb of these mammals (97). Only at higher doses do these antagonists increase heat production and Tb, which are always associated with behavioral excitation (101). Hence, it is tempting to propose that adenosine does not play a tonic role in Tb control but may act as a modulator when its concentration in the CNS is elevated by specific conditions such as hypoxia (99), asphyxia (102), and injection of adenosine analogues (103).

A growing number of studies have demonstrated that systemic administration of adenosine antagonists attenuates hypoxia-induced anapyrexia in a wide variety of animal species, ranging from ectotherms to mammals (26, 57, 97, 98, 104, 105), indicating that adenosine mediates hypoxia-induced anapyrexia. Moreover, recent studies from our group have demonstrated that intracerebroventricular injection of the adenosine antagonist aminophylline, at a dose that does not affect the thermoregulatory response to hypoxia when administered systemically, significantly attenuates hypoxia-induced anapyrexia in toads (26) and rats (57). Taken together, these results indicate that adenosine is a mediator of the anapyrexia evoked by hypoxia and that the CNS is an important site of action. However, more studies are necessary to identify the specific site of the CNS where adenosine acts to evoke anapyrexia as well as the cellular mechanisms responsible for this effect.

Histamine

No consistent evidence exists about the participation of histamine in hypoxia-induced anapyrexia, but evidence that points at histamine as a potential candidate in the mediation of this response deserves comment. First, histamine has been shown to be involved in thermoregulation (106). Accordingly, histamine has been reported to reduce the Tb of rats when infused intracerebroventricularly (107) and to decrease the preferred Tb of fishes (108) and salamanders (109). Moreover, Kandasamy et al. (106), using histamine antagonists, have shown that histamine mediates the drop in Tb caused by radiation in rats. Perhaps histamine could also mediate hypoxia-induced anapyrexia, but, to our knowledge, data regarding whether histamine levels are increased during hypoxia and whether histamine antagonists would affect hypoxia-induced anapyrexia are lacking.

Endogenous Opioids

Endogenous opioids are a class of molecules with numerous effects, many of which are associated with responses to stress (110). Evidence also indicates that endogenous opioids are involved in thermoregulation in *Paramecium* (23) and mammals (111–115). It has been reported that administration of nonselective opioid antagonists to mice impairs hypoxia-induced anapyrexia and also reduces survival (112, 115). Similarly, Malvin (23) recently demonstrated that the nonselective opioid antagonist naloxane completely blocks NaN$_3$-induced anapyrexia in *Paramecium*. Moreover, Spencer et al. (116) have reported that intracerebroventricular injections of the *delta* opioid receptor agonist, cyclic [D-Pen2,D-Pen5]enkephalin (DPDPE), and of the *kappa* opioid receptor agonist, U-50,488H, reduces the preferred Tb of rats in a thermal gradient, suggesting that opioids reduce the thermoregulatory set point by acting on *delta* and *kappa* opioid receptors in the CNS. Conversely, it was also observed in Spencer's study that the selective *mu* opioid receptor agonist, [D-Ala2,MePhe4,Gly5-ol]-enkephalin(DAMGO), administered intracerebroventricularly increased the preferred Tb of rats, indicating that the thermoregulatory actions of opioids depend on the receptor subtypes involved.

Gases as Modulators of Hypoxia-Induced Anapyrexia

In addition to the action of nongaseous mediators of anapyrexia, which are summarized in Table 1, hypoxia-induced anapyrexia has also been shown to be under the control of gaseous molecules.

Recently, a new class of biologically active molecules has been described, the gaseous compounds nitric oxide (NO) and carbon monoxide (CO) (for a review, see 117, 118). These molecules have revolutionized the understanding of physiological systems and have also provided the basis for the development of new drugs. NO and CO have been shown to participate in several physiological and pathophysiological manifestations, including thermoregulation, fever and, more recently, anapyrexia.

TABLE 1   Evidence for characterization of putative nongaseous mediators of hypoxia-induced anapyrexia

Mediator	Tested species	Effect on Tb	Levels during hypoxia	Effect of antagonism on hypoxic anapyrexia	Site of action
AVP (X)	Sheep, rabbit, rat	↓ Tb	Increase	No effect	CNS
Lactate	Toad	↓ Tb	Increase	N/A	CNS
Adenosine	Piglet, cat, rat, toad	↓ Tb	Increase	Impairment	CNS
Histamine	Rat, fish, salamander	↓ Tb	N/A	N/A	CNS
Opioids	Mouse, rat, Paramecium	↓ Tb (*delta* and *kappa* receptors)	Increase	Impairment	CNS

The main characteristics for a substance to be considered a putative mediator of hypoxia-induced anapyrexia: The substance should produce anapyrexia when injected at the supposed site of action; its levels should be increased during hypoxia; and, most importantly, the blockade of the release and/or action of the proposed mediator should impair or block hypoxia-induced anapyrexia. N/A, not assessed; CNS, central nervous system; Tb, body temperature; (X), not a mediator. See text for references.

NITRIC OXIDE (NO)   The description of the endothelium-derived relaxing factor in 1980 by Furchgott & Zawadzki (119) started a revolution in the understanding not only of blood pressure control but also of a number of other physiological systems (117, 118). The endothelium-derived relaxing factor was identified as the labile gas nitric oxide (120, 121) and has been shown to activate the enzyme soluble guanylate cyclase (sGC) and to increase cyclic GMP (cGMP) levels in vascular smooth muscle (122), in the CNS (123), and in several other tissues (117, 118).

The family of NO synthases (NOS), the enzymes that produce NO in vivo, consists of two different classes: the inducible and constitutive forms. At least three isoforms of NOS exist: neuronal NOS (nNOS) and endothelial NOS (eNOS), as the constitutive isoforms, and inducible NOS (iNOS) (17, 118).

Recent evidence indicates that NO has thermoregulatory actions. Accordingly, we and others have provided evidence that NO plays differential thermoregulatory effects by acting at the periphery and in the CNS. This notion is based on the opposite results obtained by injecting pharmacological modifiers of the NO pathway systemically or intracerebroventricularly (for a review, see 124).

Studies on rats in their thermoneutral zone have shown that the systemic non-selective inhibition of NO synthesis, using L-arginine analogues such as L-NAME, elicits a decrease in Tb (56, 85, 125–128), despite the fact that L-NAME should decrease cutaneous heat loss because it causes vasoconstriction of both large and small vessels, including the superficial vascular beds (118). Thus it has been suggested that NO synthesis inhibition likely reduces Tb by causing a failure of thermogenic mechanisms. Actually, evidence has accumulated that NO plays an important role in heat production. It has been demonstrated that the nNOS isoform is mainly found in sympathetic ganglia (129), including those that innervate brown adipose tissue (BAT), where sympathetic stimulation is known to increase non-shivering thermogenesis (130). NO arising from eNOS can also stimulate thermogenesis because NO participates in the noradrenaline-induced increase in blood flow through BAT, an action that is likely to be essential for heat production (131). Additionally, nonselective inhibition of NO synthesis impairs the norepinephrine-induced rise in BAT thermogenesis (131).

In contrast to the data obtained with rats, rabbits treated systemically with L-NAME exhibited a rise instead of a drop in Tb, which was accompanied by decreased respiratory heat dissipation (132). Taken as a whole, these results support the notion that the thermoregulatory effect produced after systemic inhibition of the NO pathway depends on the most prominent thermoregulatory effector mechanism in the tested species. More studies exposing animals to cold and warm environments and, thus, selecting the most manifest effector mechanism, are necessary to firmly establish this hypothesis.

NO acting on the brain also has thermoregulatory effects. Several studies have observed that animals intracerebroventricularly injected with 250 μg of the nonselective NOS inhibitor L-NAME show a slight increase in Tb, indicating that central NO plays a tonic role by reducing Tb (85, 125–127).

Interestingly, hypoxia has been shown to increase NO production in several tissues including blood vessels (133) and brain (134, 135). Moreover, it has been reported in in vitro and in vivo studies that hypoxic stimuli increase both the activity of the constitutive NOS isoforms and the expression of iNOS in several tissues (136–140) including the CNS (136, 138). The regulation of expression of iNOS by hypoxia seems to be dependent on the hypoxia inducible factor-1 (139). It should be remembered that NOS itself can act as an oxygen sensor (141, 142). On the basis of these facts, we then hypothesized that NO plays a role in hypoxia-induced anapyrexia. This hypothesis was tested in rats by using intravenous and intracerebroventricular injections of the nonselective NOS inhibitor L-NAME or its vehicle (saline) under both normoxic and hypoxic conditions. This experimental approach led us to conclude that NO is an important mediator of hypoxia-induced anapyrexia in the CNS because intracerebroventricular administration of L-NAME, at a dose that does not affect hypoxia-induced anapyrexia when given systemically, completely blocks the decrease in Tb evoked by hypoxia (125). This action of NO has recently been extended to other stimuli that reduce Tb, because nonselective inhibition of NO synthesis in the CNS using L-arginine analogues also abolishes the decrease in Tb elicited by different stimuli, including systemic AVP (85), 2-deoxy-D-glucose (143), and insulin (144), pointing at NO in the CNS as a common mediator of decreases in Tb. The role of NO at the central sites is likely to be mediated by activation of sGC and the consequent rise in the intracellular levels of cGMP since intracerebroventricular administration of the sGC inhibitor ODQ elevates Tb similarly to NOS inhibitors (145). This assumption is reinforced by the fact that central inhibition of the heme oxygenase-carbon monoxide (HO-CO) pathway, which may also activate sGC, causes no change in Tb (145, 146).

Efforts have also been made to identify the NOS isoform involved in hypoxia-induced anapyrexia. We recently reported that treatment of rats with the more selective nNOS inhibitor 7-nitroindazole (7-NI), at a dose that does not alter Tb of euthermic animals, significantly attenuates the anapyrexia induced by hypoxia (7% of inspired oxygen), which implies that nNOS is likely to be the NOS isoform involved in the thermoregulatory response to hypoxia (17). On the other hand, there is also evidence against the participation of nNOS in hypoxia-induced anapyrexia. In this context, Gautier & Murariu (147) have demonstrated that 7-NI does not affect hypoxia (11% of inspired oxygen)-induced anapyrexia, even though 7-NI suppresses the hypoxic hypometabolism. These controversies might be related to different experimental protocols, especially with respect to the severity of hypoxia. Another possibility is that 7-NI could also interfere with systems not related to NO, which is the case of 7-NI interaction with brain monoamine oxidase (148). More studies are clearly necessary to identify the NOS isoform involved in hypoxia-induced anapyrexia, a field in which the use of knockout animals for specific NOS isoforms may be of particular importance.

Although NO seems to play a prominent role in the development of hypoxia-induced anapyrexia, to our knowledge no report exists about the mechanisms involved. At least two hypotheses have been proposed about the mechanism by which NO could contribute to anapyrexia. One of them is that NO could be acting on the PO/AH, which is considered to be the thermoregulatory site of the CNS (9), to reduce the thermoregulatory set point. Previous studies using anesthetized animals have provided conflicting results about the thermoregulatory effect of NO in the PO/AH (149, 150). However, because anesthesia itself can disrupt thermo-regulation (22), we recently assessed the thermoregulatory role of NO in the PO/AH in unanesthetized rats (151). Briefly, it was observed that intra-PO/AH microin-jection of the NO donor sodium nitroprusside significantly reduces the Tb of rats, indicating that NO is likely to reduce Tb in the PO/AH (151). Additionally, intra-PO/AH microinjection of the cGMP analogue 8-Br-cGMP has been reported to cause a drop in Tb similarly to sodium nitroprusside (151), suggesting that NO is likely to reduce Tb by acting in the PO/AH through a cGMP-dependent pathway. Because NO seems to reduce Tb by acting in the PO/AH, which is the brain Tc con-troller site (9), it is possible that NO mediates anapyrexia by acting in the PO/AH. Another possibility is that central NO could participate in the genesis of anapyrexia by reducing the sympathetic tonus. In fact, NO has been shown to play a role in reducing sympathetic tonus by acting on several brain regions, including the par-aventricular nucleus, posterior hypothalamus, and nucleus of the tractus solitarius (152, 153). The reduced firing rates of the sympathetic fibers would in turn cause a decrease in thermogenesis (130) and evoke vasodilation of the superficial vascular beds (154), responses that lead to a drop in Tb. Clearly, more studies are necessary to determine the mechanism of action of NO in the genesis of anapyrexia.

CARBON MONOXIDE (CO)   The gaseous compound CO has been shown to play a role as a neurotransmitter and/or neuromodulator, acting similarly to NO (117, 155).

Heme oxygenase (HO), the enzyme responsible for CO synthesis in vivo, catalyses the metabolism of heme to biliverdin, free iron, and CO (156–158). Three distinct HO isoforms have been identified: HO-1, HO-2, and HO-3, among which the isoforms 1 and 2 are the most studied and best known (158). Both isoenzymes, one of them inducible (HO-1) and the other constitutive (HO-2), are expressed in various tissues, including neural tissue (117, 158–160). Studies have suggested that CO arising from heme via metabolism by HO stimulates sGC activity and promotes an increase in cGMP in neural and cardiovascular tissues (157, 158, 161, 162). Subsequent studies have revealed that CO may also exert biological activities via alternative pathways, such as the activation of cyclooxygenase (163).

In spite of the growing evidence showing the participation of CO in several physiological responses, it was only recently that a thermoregulatory effect of CO was assessed. Data (145, 146, 164) indicate that CO is a pyretic molecule in the CNS of rats. In support, intracerebroventricular injection of the nonselective HO inhibitor, zinc deuteroporphyrin 2,4-bis-glycol (ZnDPBG), has been shown to attenuate the rise in Tb evoked by endotoxin (145, 146) but does not affect the Tb of euthermic rats. Moreover, heme overload, which induces the HO pathway (165–170), produces a rapid rise in Tb when applied to the brain, a response that is attenuated by pretreatment with ZnDPBG (145, 146, 164). Corroborating this observation, a recent study (171) reported that intracerebroventricular injection of hemoglobin increases the Tb of rabbits.

In order to identify the HO product with a pyretic action in the CNS, we then tested the thermoregulatory effects of all HO products, i.e., biliverdin, free iron, and CO (145). Intracerebroventricular administration of biliverdin, at the same dose at which heme overload produces a marked increase in Tb, caused no alteration in basal Tb. Moreover, iron overload in the brain with both ferric and ferrous iron did not affect Tb compared with the control group, and intracerebroventricular treatment with deferoxamine, an iron chelator, did not affect basal Tb or the heme-induced rise in Tb. Finally, we observed that intracerebroventricular pretreatment with the sGC inhibitor ODQ completely blocked the elevation in Tb evoked by heme overload, showing that this effect is mediated through a cGMP-dependent pathway. This result confirms that CO is the HO product with thermoregulatory effects because CO activates sGC (157, 158, 161, 162, 172) and because the other HO products, biliverdin and iron, play no thermoregulatory role in the CNS of rats (145). In fact, in animals intracerebroventricularly injected with CO-saturated saline there was an increase in Tb ($\sim$0.5°C), but this effect was slight compared with the group that received a heme overload (1.5°C) (146). This less pronounced effect of CO-saturated saline may have been the result of an insufficient amount of CO delivered to the brain or a rapid diffusion of the gas out of the brain since this gas has a high affinity for hemoglobin.

Interestingly, hypoxia has been shown to induce the enzyme HO-1 in several cell types, including vascular smooth muscle, astrocytes, cardiomyocytes, Chinese hamster ovary cells, and fibroblasts (173, 174). Therefore, we investigated a possible role of centrally acting CO in hypoxia-induced anapyrexia. As previously reported (145, 146), it was observed that intracerebroventricular administration of ZnDPBG caused no change in Tb of euthermic rats, indicating that the central HO-CO pathway plays no tonic role in thermoregulation. However, intracerebroventricular pretreatment with the HO inhibitor ZnDPBG significantly magnified the anapyretic response to hypoxia, whereas its vehicle did not alter the drop in Tb observed during hypoxia exposure compared with the group that received no injection (175). This result implies that, at least in rats, the central HO-CO pathway is an important modulator of hypoxia-induced anapyrexia, which exerts a key function to prevent excessive decreases in Tb.

It is important to point out that CO is the first molecule described to have a pyretic effect during hypoxia-induced anapyrexia. Therefore, we suggest that anapyrexia may result from a balance between the effects of substances that reduce Tb and those that counteract the decrease in Tb and not simply from the action of molecules that reduce Tb. This proposed mechanism is similar to that involved in the febrile response (a regulated rise in Tb), since it is generally accepted that during fever a balance between the production and release of pyrogens and cryogens exists. Endogenous pyrogens are substances capable of increasing the thermoregulatory set point and consequently Tb, whereas endogenous cryogens are substances that counterbalance the effects of pyrogens and reduce the thermoregulatory set point (32). Thus we suggest the introduction of these terms in the context of anapyrexia, where cryogens would be the mediators of anapyrexia and pyrogens would be substances that counterbalance the effect of cryogens on the thermoregulatory set point.

CYCLIC NUCLEOTIDES IN ANAPYREXIA: A WORKING HYPOTHESIS

As reviewed above, several neurochemical mediators and/or modulators have been suggested to be responsible for anapyrexia. However, no attempt has been made to understand the intracellular second messengers involved in this response.

Recently, we have proposed (176) a neural model for the control of the thermoregulatory set point, which involves the action of the cyclic nucleotides cAMP and cGMP in the PO/AH. This model is based on the observations that (a) the microinjection of the cAMP analogue dibutyryl-cAMP into the PO/AH reduces Tb of unanesthetized rats (176) and rabbits (177); (b) cAMP increases thermosensitivity of warm-sensitive preoptic neurons, an effect known to elicit heat loss mechanisms and a decrease in Tb (9); (c) intracerebroventricular administration of prostaglandin E_2, which is known to increase the thermoregulatory set point in the PO/AH (8), reduces cAMP levels in the anteroventral to third ventricular region of rats (176) where the PO/AH is located; and (d) intracerebroventricular treatment with the phosphodiesterase inhibitor aminophylline, which impairs cAMP catabolism, attenuates prostaglandin E_2-induced rise in Tc in rats (176).

Based on these observations, we then proposed that cAMP reduces the thermoregulatory set point in the PO/AH and that a decrease in preoptic cAMP mediates fever.

However, an interesting aspect regarding the thermoregulatory role of cAMP is that treatment with agents that reduce (178) or increase (26, 57) cAMP levels usually does not affect Tb. Therefore, it is likely that a reduction in preoptic cAMP levels alone does not affect Tc, but there should be a synergistic action of another agent to promote changes in Tb. A possible candidate is the NO-cGMP pathway; we recently reported that intra-PO/AH microinjection of sodium nitroprusside, a NO donor (5 μg), or 8-Br-cGMP (40 μg) reduces Tb (151). We then assessed the effect of the co-injection of 8-Br-cGMP at a low dose (10 μg), which does not affect basal Tb, and dibutyryl-cAMP at a dose (40 μg) that produces a drop in Tb of approximately 0.5°C. Interestingly, this treatment produced a rapid drop in Tc similar to that produced by dibutyryl-cAMP alone, but a second drop in Tc (not present in the group that received dibutyryl-cAMP alone) appeared 1 h after injection, indicating that cAMP and cGMP could act synergistically in the PO/AH to reduce Tb (176). However, it should be pointed out that this model was developed in order to explain the cellular mechanisms in the PO/AH responsible for the genesis of fever.

Therefore, because fever, which is the opposite of anapyrexia, seems to result from the simultaneous reduction in the levels of cAMP and cGMP in the PO/AH (176), we propose that anapyrexia could result from a concomitant rise in the levels of cAMP and cGMP in the PO/AH. In support of this notion is the observation that the levels of cAMP (179) and cGMP (134, 135) in the brain are increased during hypoxia. Because the rise in cAMP during hypoxia is under the control of the catecholaminergic system (180, 179), whereas the rise in cGMP is associated with NO (134, 135), it is tempting to speculate that hypoxia-induced anapyrexia results from the simultaneous activation of the catecholamine-cAMP and NO-cGMP systems in the PO/AH. We are currently working to directly test this hypothesis, which is schematically depicted in Figure 1.

FINAL REMARKS

Hypoxia evokes a regulated decrease in Tb, but only recently did the mechanisms responsible for hypoxia-induced anapyrexia begin to be suggested. A route mediating the reduction in Tb may be the impairment of central oxidative phosphorylation because intracerebroventricular injection of inhibitors of oxidative phosphorylation such as azide or cyanide reduces the preferred Tb of toads. Moreover, exclusion of glucose from central sites plays a major role in hypoglycemia-induced anapyrexia. These data, together with the observations that peripherally chemodenervated animals develop anapyrexia similarly to intact animals, imply that the CNS plays a central role in sensing the lack of oxygen and in development of anapyrexia. Several putative mediators of anapyrexia have been proposed. Some

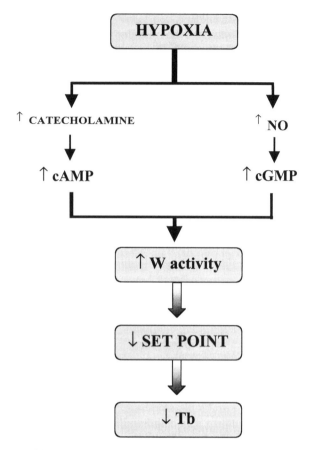

Figure 1   A working hypothesis on the mechanisms involved in the PO/AH to decrease the thermoregulatory set point, i.e., to produce anapyrexia. Hypoxia would increase the activity of the catecholamine-cAMP and nitric oxide (NO)-cGMP systems in the PO/AH, which in turn would increase the activity and thermosensitivity of preoptic warm-sensitive neurons (W). This effect would cause a reduction in the thermoregulatory set point in the PO/AH, leading to an inhibition of thermogenesis and an increase in heat loss, responses that would produce a decrease in body temperature (Tb).

lines of evidence indicate arginine vasopressin (AVP) as one putative mediator. However, studies using Brattleboro rats (which lack AVP-containing neurons in the CNS) and peripherally or centrally injected AVP-receptor antagonists indicate that AVP is not involved in hypoxia-induced anapyrexia. In addition to AVP, many other mediators such as lactate, adenosine, histamine, and opioids have been suggested and experimentally tested, but none of the possible candidates can trigger a full-blown anapyretic response. Recently, the action of NO on the CNS has been shown to play a role in mediating hypoxia-induced anapyrexia, as well as the drop

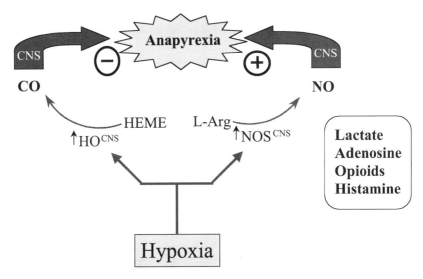

Figure 2   Possible mechanism for hypoxia-induced anapyrexia. Hypoxia activates/ induces both the enzyme nitric oxide synthase (NOS) and heme oxygenase (HO). NOS, in turn, converts the amino acid L-arginine (L-Arg) in the gas nitric oxide (NO), which evokes anapyrexia by acting on the central nervous system (CNS). HO, on the other hand, cleaves the heme group to produce carbon monoxide (CO), which is a pyretic molecule in the CNS and counteracts excessive drops in Tb. Nongaseous mediators such as lactate, adenosine, opioids, and histamine also act on the CNS, together with NO, to evoke anapyrexia.

in Tb evoked by other stimuli, including 2-deoxy-D-glucose, insulin, and systemic AVP. Recent data from our laboratory also indicate that other substances may be formed during hypoxia to counteract the actions of the mediators of anapyrexia and, consequently, to avoid an excessive drop in Tb, similar to what occurs during fever but in an opposite direction. This seems to be the case for endogenously produced CO; inhibition of the enzyme responsible for CO synthesis in the CNS augments hypoxia-induced anapyrexia (Figure 2 illustrates this model). Although progress has been made in understanding the mechanisms of anapyrexia, much remains to be explored.

ACKNOWLEDGMENTS

This work was supported by Fundação de Amparo à Pesquisa do Estado de São Paulo (FAPESP), Conselho Nacional de Desenvolvimento Científico e Tecnológico (CNPq), and PRONEX. A. A. Steiner was the recipient of a FAPESP fellowship. The authors gratefully thank Dr. Clark M. Blatteis for critically reading this manuscript.

Visit the Annual Reviews home page at www.AnnualReviews.org

LITERATURE CITED

1. Wood SC. 1991. Interactions between hypoxia and hypothermia. *Annu. Rev. Physiol.* 53:71–85
2. Wood SC. 1995. Oxygen as a modulator of body temperature. *Braz. J. Med. Biol. Res.* 28:1249–56
3. Taylor EW, Jordan D, Coote JH. 1999. Central control of cardiovascular and respiratory systems and their interactions in vertebrates. *Physiol. Rev.* 79:855–916
4. Hicks JW, Wood SC. 1985. Temperature regulation in lizards: effects of hypoxia. *Am. J. Physiol. Regulatory Integrative Comp. Physiol.* 248:R595–R600
5. Malvin GM, Wood SC. 1992. Behavioral hypothermia and survival of hypoxic protozoans Paramecium caudatum. *Science* 255:1423–25
6. Wood SC, Stabenau EK. 1998. Effect of gender on thermoregulation and survival of hypoxic rats. *Clin. Exp. Pharmacol. Physiol.* 25:155–58
7. Blatteis CM, Sehic E. 1997. Fever: How may circulating pyrogens signal the brain? *News Physiol. Sci.* 12:1–9
8. Blatteis CM, Sehic E. 1997. Prostaglandin E₂: a putative fever mediator. In *Fever: Basic Mechanisms and Management*, ed. PA Mackowiak, pp. 117–45. Philadelphia: Lippincott-Raven
9. Boulant JA. 1998. Hypothalamic neurons: mechanisms of sensitivity to temperature. *Ann. NY Acad. Sci.* 856:108–15
10. Matsuda T, Hori T, Nakashima T. 1992. Thermal and PGE₂ sensitivity of the organum vasculosum laminae terminalis region and preoptic area in rat brain slices. *J. Physiol.* 454:197–212
11. Blatteis CM, ed. 1998. *Physiology and Pathophysiology of Temperature Regulation*. River Edge, NJ: World Sci.
12. Sammut IA, Foresti R, Clark JE, Exon DJ, Vesely MJ, Sarathchandra P, et al. 1998. Carbon monoxide is a major contributor to the regulation of vascular tone in aortas expressing high levels of haeme oxygenase-1. *Br. J. Pharmacol.* 125:1437–44
13. Branco LGS, Malvin GM. 1996. Thermoregulatory effects of cyanide and azide in the toad, *Bufo marinus. Am. J. Physiol. Regulatory Integrative Comp. Physiol.* 270:R169–R73
14. Branco LGS, Steiner AA. 1999. Central thermoregulatory effects of lactate in the toad *Bufo paracnemis. Comp. Biochem. Physiol. A* 122:457–61
15. Gautier H. 1998. Oxygen transport in conscious newborn dogs during hypoxic hypometabolism. *J. Appl. Physiol.* 84:761–62
16. Steiner AA, Carnio EC, Antunes-Rodrigues J, Branco LGS. 1999. Endogenous vasopressin does not mediate hypoxia-induced anapyrexia in rats. *J. Appl. Physiol.* 86:469–73
17. Steiner AA, Carnio EC, Branco LGS. 2000. Role of neuronal nitric oxide synthase in hypoxia-induced anapyrexia in rats. *J. Appl. Physiol.* 89:1131–36
18. Dupré RK, Romero AM, Wood SC. 1988. Thermoregulation and metabolism in hypoxic animals. In *Oxygen Transfer From Atmosphere to Tissues*, ed. NC Gonzalez, MR Fedde, pp. 347–51. New York: Plenum.
19. Bonora M, Gautier H. 1989. Effects of hypoxia on thermal polypnea in intact and carotid body-denervated conscious cats. *J. Appl. Physiol.* 67:578–83
20. Barros RCH, Zimer ME, Branco LGS, Milson WK. 2001. The "hypoxic metabolic response" of the golden mantled ground squirrel. *J. Appl. Physiol.* 91:603–12
21. Gordon CJ, Fogelson L. 1991. Comparative effects of hypoxia on behavioral thermoregulation in rats, hamsters, and mice.

Am. J. Physiol. Regulatory Integrative Comp. Physiol. 260(1):R120–R25

22. Cabanac M. 1998. Thermiatrics and behavior. See Ref. 11, pp. 108–24

23. Malvin GM. 1998. Thermoregulatory changes by hypoxia: lessons from the paramecium. *Clin. Exp. Pharmacol. Physiol.* 25:165–69

24. Branco LGS, Pörtner HO, Wood SC. 1993. Interaction between body temperature and hypoxia in the alligator. *Am. J. Physiol. Regulatory Integrative Comp. Physiol.* 265:R1339–R43

25. Bicego-Nahas KC, Gargaglioni LH, Branco LGS. 2001. Seasonal changes in the preferred body temperature, cardiovascular, and respiratory responses to hypoxia in the toad, *Bufo paracnemis. J. Exp. Zool.* 289:359–65

26. Branco LGS, Steiner AA, Tattersall G, Wood SC. 2000. Role of adenosine in hypoxia-induced hypothermia of toads. *Am. J. Physiol. Regulatory Integrative Comp. Physiol.* 279(1):R196–R201

27. Wood SC, Malvin GM. 1991. Physiological significance of behavioral hypothermia in hypoxic toads (*Bufo marinus*). *J. Exp. Biol.* 159:203–15

28. Bryan JD, Hill LG, Niell WH. 1984. Interdependence of acute temperature preference and respiration in plains minnow. *Trans. Am. Fish Soc.* 113:557–62

29. Dupré RK, Wood SC. 1988. Behavioral temperature regulation by aquatic ectotherms during hypoxia. *Can. J. Zool.* 66: 2649–52

30. Rausch RN, Crawshaw LI. 1990. Effect of hypoxia on behavioral thermoregulation in the goldfish, *Carassius auratus. FASEB J.* 4:A551 (Abstr.)

31. Malvin GM, Wood SC. 1991. Behavioral thermoregulation of the toad, *Bufo marinus*: effects of air humidity. *J. Exp. Zool.* 258(3):322–26

32. Kluger MJ. 1991. Fever: role of pyrogens and cryogens. *Physiol. Rev.* 71(1):93–127

33. Mortola JP, Gautier H. 1995. Interaction between metabolism and ventilation: effects of respiratory gases and temperature. In *Regulation of Breathing*, ed. JÁ Dempsey, pp. 1011–64. New York: Dekker

34. Rohlicek CV, Saik C, Matsuoka T, Mortola JP. 1996. Cardiovascular and respiratory consequences of body warming during hypoxia in conscious newborn cats. *Pediatr. Res.* 40(1):1–5

35. Handerson RA, Whitehurst ME, Morgan KR, Carroll RG. 2000. Reduced oxygen consumption precedes the drop in body core temperature caused by hemorrhage in rats. *Shock* 13(4):320–24

36. Matsuoka T, Saiki C, Mortola JP. 1994. Metabolic and ventilatory responses to anemic hypoxia in conscious rats. *J. Appl. Physiol.* 77(3):1067–72

37. Wood SC. 1990. Effect of hematocrit on behavioral thermoregulation of the toad *Bufo marinus. Am. J. Physiol. Regulatory Integrative Comp. Physiol.* 258(4):R848–R51

38. Malvin GM, Havlen P, Baldwin C. 1994. Interactions between cellular respiration and thermoregulation in the paramecium. *Am. J. Physiol. Regulatory Integrative Comp. Physiol.* 267:R349–R52

39. Slade R, Watkinson WP, Hatch GE. 1997. Mouse strain differences in ozone dosimetry and body temperature changes. *Am. J. Physiol. Lung Cell Mol. Physiol.* 272:L73–L77

40. Artru AA, Michenfelder JD. 1981. Influence of hypothermia or hyperthermia alone or in combination with pentobarbital and phenytoin on survival time in hypoxic mice. *Anesth. Analog.* 60:867–70

41. Sutariya B, Penney D, Barnes J, Helfman C. 1989. Hypothermia protects brain function in acute carbon monoxide poisoning. *Vet. Hum. Toxicol.* 31(5):436–41

42. Takasu A, Stezoski SW, Stezoski J, Safar P, Tisherman AS. 2000. Mild or moderate hypothermia, but not increased oxygen breathing, increases long-term survival after uncontrolled hemorrhagic shock in rats. *Crit. Care Med.* 28(7):2465–74

43. Wladis A, Hjelmqvist H, Brismar B, Kjellstrom BT. 1998. Acute metabolic and endocrine effects of induced hypothermia in hemorrhagic shock: an experimental study in the pig. *J. Trauma* 45(3):527–33

44. Gollan F, Aono M. 1973. The effect of temperature on sanguinous rabbits. *Cryobiology* 10:321–27

45. Dunn JM, Miller JA. 1969. Hypothermia combined with positive pressure ventilation in ressucitation of the asphyxiated neonate. *Am. J. Obstet. Gynecol.* 104:58

46. Beyens T, Biarent D, Bouton JM, Demanet H, Viart P, et al. 1998. Cardiac surgery with extracorporeal circulation in 23 infants weighing 2500 g or less: short and intermediate term outcome. *Eur. J. Cardiothorac. Surg.* 14(2):165–72

47. Krogh A. 1914. The quantitative relation between temperature and standard metabolism in mammals. *Int. Z. Phys. Chem. Biol.* 1:491–508

48. Wang J, Chambers G, Cottrell JE, Kass IS. 2000. Differential fall in ATP accounts for effects of temperature on hypoxic damage in rat hippocampal slices. *J. Neurophysiol.* 83(6):3462–72

49. Branco LGS, Wood SC. 1993. Effect of temperature on central chemical control of ventilation in the alligator *Alligator mississippiensis*. *J. Exp. Biol.* 179:261–72

50. Glass ML, Boutelier RG, Heisler N. 1983. Ventilatory control of arterial PO_2 in the turtle, *Chrysemys picta bellii*: effects of temperature and hypoxia. *J. Comp. Physiol.* 151:145–53

51. Kruhøffer M, Glass ML, Abe AS, Johansen K. 1987. Control of breathing in an amphibian, *Bufo paracnemis*: effects of temperature and hypoxia. *Respir. Physiol.* 69:267–75

52. Burton MD, Kazemi H. 2000. Neurotransmitters in central respiratory control. *Respir. Physiol.* 122:111–21

53. Martin RJ, DiFiore JM, Jana L, Davis RL, Miller MJ, et al. 1998. Persistence of the biphasic ventilatory response to hypoxia in preterm infants. *J. Pediatr.* 132(6):960–64

54. Powell FL, Milsom WK, Mitchell GS. 1998. Time domains of the hypoxic ventilatory response. *Respir. Physiol.* 112:123–34

55. Vizek M, Bonora M. 1998. Diaphragmatic activity during biphasic ventilatory response to hypoxia in rats. *Respir. Physiol.* 111(2):153–62

56. Barros RCH, Branco LGS. 1998. Effect of nitric oxide synthase inhibition on hypercapnia-induced hypothermia and hyperventilation. *J. Appl. Physiol.* 85(3):967–72

57. Barros RCH, Branco LGS. 2000. Role of central adenosine in the respiratory and thermoregulatory responses to hypoxia. *Neuroreport* 11(1):193–97

58. Blatteis CM. 1964. Hypoxia and the metabolic response to cold in newborn rabbits. *J. Physiol.* 172:358–68

59. Gautier H, Bonora M. 1992. Ventilatory and metabolic responses to cold and hypoxia in intact and carotid body-denervated rats. *J. Appl. Physiol.* 73:847–54

60. Gautier H, Bonora M, Schultz SA, Remmers JE. 1987. Hypoxia-induced changes in shivering and body temperature. *J. Appl. Physiol.* 62:2477–84

61. Gautier H, Bonora M, Trinh HC. 1993. Ventilatory and metabolic responses to cold and CO_2 in intact and carotid body-denervated awake rats. *J. Appl. Physiol.* 75:2570–79

62. Matsuoka T, Dotta A, Mortola JP. 1994. Metabolic response to ambient temperature and hypoxia in sinoaortic-denervated rats. *Am. J. Physiol. Regulatory Integrative Comp. Physiol.* 266:R387–R91

63. Gleeson M, Barnas GM, Rautenberg W. 1986. Cardiorespiratory responses to shivering in vagotomized pigeons during normoxia and hypoxia. *Pflügers Arch.* 407:664–69

64. Tamaki Y, Nakayama T. 1987. Effects of air constituents on thermosensitivities of

preoptic neurons: hypoxia versus hypercapnia. *Pfjügers Arch.* 409:1–6

65. Branco LGS. 1997. Effects of 2-deoxy-D-glucose and insulin on plasma glucose levels and behavioral thermoregulation of toads. *Am. J. Physiol. Regulatory Integrative Comp. Physiol.* 272:R1–R5

66. Fabris G, Steiner AA, Anselmo-Franci J, Branco LGS. 2000. Role of nitric oxide in rat locus coeruleus in hypoxia-induced hyperventilation and hypothermia. *NeuroReport* 11(13):2991–95

67. Koeppen BM, Stanton BA, eds. 1997. *Renal Physiology.* St. Louis, MO: Mosby-Year Book

68. Kasting NW, Veale WL, Cooper KE. 1978. Suppression of fever at term of pregnancy. *Nature* 271:245–46

69. Pittman QJ, Cooper KE, Veale WL, Van Petten GR. 1974. Observations on the development of the febrile response to pyrogens in the sheep. *Clin. Sci. Mol. Med.* 46:591–602

70. Naylor AA, Cooper KE, Veale WL. 1987. Vasopressin and fever: evidence supporting the existence of an endogenous antipyretic system in the brain. *Can. J. Physiol. Pharmacol.* 65:1333–38

71. Cooper KE, Kasting NW, Lederis K, Veale WL. 1979. Evidence supporting a role for endogenous vasopressin in natural suppression of fever in the sheep. *J. Physiol.* 295:33–45

72. Kasting NW, Cooper KE, Veale WL. 1979. Antipyresis following perfusion of brain sites with vasopressin. *Experientia* 35:208–9

73. Naylor AM, Ruwe RD, Kohut AF, Veale WL. 1985. Perfusion of vasopressin within the ventral septum of the rabbit suppresses endotoxin fever. *Brain Res. Bull.* 15:209–13

74. Ruwe RD, Naylor AM, Veale WL. 1985. Perfusion of vasopressin within the rat brain suppresses prostaglandin E-hyperthermia. *Brain Res.* 338:219–24

75. Bicego-Nahas KC, Steiner AA, Carnio EC, Antunes-Rodrigues J, Branco LGS.

2000. The antipyretic effect of arginine vasotocin in toads. *Am. J. Physiol. Regulatory Integrative Comp. Physiol.* 278: R1408–R14

76. De Vries GJ, Buijs RM, van Leewen FW, Caffe AR, Swaab DF. 1985. The vasopressinergic innervation of the brain in normal and castrated rats. *J. Comp. Neurol.* 233:236–54

77. Kasting NW, Carr DB, Martin JB, Blume H, Bergland R. 1983. Changes in cerebrospinal fluid and plasma vasopressin in the febrile sheep. *Can. J. Physiol. Pharmacol.* 61:427–31

78. Kasting NW. 1986. Potent stimuli for vasopressin release, hypertonic saline and hemorrhage, cause antipyresis in the rat. *Regul. Pept.* 15(4):293–300

79. Naylor AM, Veale WL, Cooper KE. 1986. Role of endogenous vasopressin in fever suppression in the rat as determined using vasopressin antagonists. Presented at *Program of Thermal Physiology, Satellite Symp. of XXX IUPS.* p. 55. (Abstr.)

80. Wilkinson MF, Kasting NW. 1990. Centrally acting vasopressin contributes to endotoxin tolerance. *Am. J. Physiol. Regulatory Integrative Comp. Physiol.* 258: R443–R49

81. Forsling ML, Rees M. 1975. Effects of hypoxia and hypercapnia on plasma vasopressin concentration. *J. Endocrinol.* 67(2):62P–63P

82. Raff H, Shinsako J, Keil LC, Dallman MF. 1983. Vasopressin, ACTH, and corticosteroids during hypercapnia and graded hypoxia in dogs. *Am. J. Physiol. Endocrinol. Metab.* 244:E453–E58

83. Naylor AM, Ruwe WD, Veale WL. 1986. Thermoregulatory actions of centrally administered vasopressin in the rat. *Neuropharmacology* 25:787–94

84. Shido O, Kifune A, Nagasaka T. 1984. Baroreflexive suppression of heat production and fall in body temperature following peripheral administration of vasopressin in rats. *Jpn. J. Physiol.* 34:397–406

85. Steiner AA, Carnio EC, Antunes-Rodrigues J, Branco LGS. 1998. Role of nitric oxide in systemic vasopressin-induced hypothermia. *Am. J. Physiol. Regulatory Integrative Comp. Physiol.* 275(4):R937–R41

86. Schimidt-Nielsen K, Crawford EC Jr, Newsome AE, Rawson KS, Hammel HT. 1967. Metabolic rate of camels: effect on body temperature and dehydration. *Am. J. Physiol.* 212(2):341–46

87. Clark DJ, Fewell JE. 1994. Body-core temperature decreases during hypoxic hypoxia in Long-Evans and Brattleboro rats. *Can. J. Physiol. Pharmacol.* 72:1528–31

88. Eagan PC, Kasting NW, Veale WL, Cooper KE. 1982. Absence of endotoxin fever but not prostaglandin E2 fever in Brattleboro rat. *Am. J. Physiol. Regulatory Integrative Comp. Physiol.* 242:R116–20

89. Kandasamy SB, Williams BA. 1983. Absence of endotoxin fever but not hyperthermia in Brattleboro rats. *Experientia* 15:1343–44

90. Veale WL, Eagan PC, Cooper KE. 1982. Abnormality of the febrile response of the Brattleboro rat. *Ann. NY Acad. Sci.* 394:776–79

91. Pörtner HO, MacLatchy LM, Toews DP. 1991. Metabolic responses of the toad *Bufo marinus* to environmental hypoxia: an analysis of the critical PO$_2$. *Physiol. Zool.* 64:836–49

92. Pörtner HO, Branco LGS, Malvin GM, Wood SC. 1994. A new role for lactate in the toad, *Bufo marinus. J. Appl. Physiol.* 76:2405–10

93. Demestre M, Boutelle M, Fillenz M. 1997. Stimulated release of lactate in freely moving rats is dependent on the uptake of glutamate. *J. Physiol.* 499(3):825–32

94. Branco LGS, Wood SC. 1994. Role of central chemoreceptors in behavioral thermoregulation of the toad. *Am. J. Physiol. Regulatory Integrative Comp. Physiol.* 266:R1483–R87

95. Talbot CR, Stiffler DF. 1991. Effects of hypoxia on acid-base balance, blood gases, catecholamines, and cutaneous ion exchange in the larval tiger salamander (*Ambystoma tigrinum*). *J. Exp. Zool.* 257:299–305

96. Aaron EA, Foster HV, Lowry TF, Korducki MJ, Ohtake PJ. 1996. Effect of dichloroacetate on Pa$_{CO2}$ responses to hypoxia in awake goats. *J. Appl. Physiol.* 80:176–81

97. Gautier H, Murariu C. 1998. Neuromodulators and hypoxic hypothermia in the rat. *Respir. Physiol.* 112:315–24

98. Murphy DJ, Joran ME, Renninger JE. 1993. Effects of adenosine agonists and antagonists on pulmonary ventilation in conscious rats. *Gen. Pharmacol.* 24:943–54

99. Yan S, Laferrière A, Zhang C, Moss IR. 1995. Microdialyzed adenosine in nucleus tractus solitarii and ventilatory response to hypoxia in piglets. *J. Appl. Physiol.* 79:405–10

100. Yarbrough GG, McGuffin-Clineschimidt JC. 1981. In vivo behavioral assessment of central nervous system purinergic receptors. *Eur. J. Pharmacol.* 76:137–44

101. Lin MT, Chandra A, Liu GG. 1980. The effects of theophylline and caffeine on thermoregulatory functions of rats at different ambient temperatures. *J. Pharm. Pharmacol.* 32:204–8

102. Winn HR, Rubio R, Berne RM. 1981. Brain adenosine concentration during hypoxia in rats. *Am. J. Physiol. Heart Circ. Physiol.* 241:H235–H42

103. Koos BJ, Chau A. 1998. Fetal cardiovascular and breathing responses to an adenosine A2 receptor agonist in sheep. *Am. J. Physiol. Regulatory Integrative Comp. Physiol.* 247:R152–R59

104. Darnall RA, Bruce RD. 1987. Effects of adenosine and xanthine derivatives on breathing control during acute hypoxia in the anesthetized newborn piglet. *Pediatr. Pulmonol.* 3:110–16

105. Long WQ, Anthonisen NR. 1994. Aminophylline partially blocks ventilatory

depression with hypoxia in the awake cat. *Can. J. Physiol. Pharmacol.* 72:673–78

106. Kandasamy SB, Hunt WA, Mickley GA. 1988. Implication of prostaglandin and histamine H1 and H2 receptors in radiation-induced temperature responses of rats. *Radiation Res.* 114:42–53

107. Dey PK, Mukhopadhaya N. 1986. Involvement of histamine receptors in mediation of histamine-induced thermoregulatory response in rat. *Indian J. Physiol. Pharmacol.* 30:300–6

108. Green MD, Lomax P. 1976. Behavioral thermoregulation and neuroamines in fish (*Chromus chromus*). *J. Theor. Biol.* 1:237–40

109. Hutchison VH, Spriesterbach KK. 1986. Histamine and histamine receptors: behavioral thermoregulation in the salamander, *Necturus maculosus*. *Comp. Biochem. Physiol.* 85:199–206

110. Akil H, Watson SJ, Young E, Khachaturian H, Walker JM. 1984. Endogenous opioids: biology and function. *Annu. Rev. Neurosci.* 7:223–55

111. Clark WG. 1979. Influence of opioids on central thermoregulatory mechanisms. *Pharmacol. Biochem. Behav.* 10:609–13

112. Mayfield KP, D'Alecy LG. 1992. Role of endogenous opioids peptides in the acute adaptation to hypoxia. *Brain Res.* 582:226–31

113. Mayfield KP, D'Alecy LG. 1994. Delta-1 opioid receptor dependence of acute hypoxic adaptation. *J. Pharmacol. Exp. Ther.* 268:74–77

114. Mayfield KP, D'Alecy LG. 1994. Delta-1 opioid agonist acutely increases hypoxic tolerance. *J. Pharmacol. Exp. Ther.* 268:683–88

115. Mayfield KP, Hong EJ, Carney KM, D'Alecy LG. 1994. Potential adaptations to acute hypoxia: Hct, stress protein, and set point for temperature regulation. *Am. J. Physiol. Regulatory Integrative Comp. Physiol.* 266:R1615–R22

116. Spencer RL, Hruby VJ, Burks TF. 1990. Alteration of thermoregulatory set point

with opioid agonists. *J. Pharmacol. Exp. Ther.* 252(2):696–705

117. Dawson TM, Snyder SM. 1994. Gases as biological messengers: nitric oxide and carbon monoxide in the brain. *J. Neurosci.* 14(9):5147–59

118. Moncada S, Palmer RMJ, Higgs EA. 1991. Nitric oxide: physiology, pathophysiology and pharmacology. *Pharmacol. Rev.* 43:109–42

119. Furchgott RF, Zawadzki JV. 1980. The obligatory role of endothelial cells in the relaxation of arterial smooth muscle by acetylcholine. *Nature* 288:373–76

120. Ignarro LJ, Buga GM, Wood KS, Byrns RE, Chaudhuri G. 1987. Endothelium-derived relaxing factor produced and released from artery and vein is nitric oxide. *Proc. Natl. Acad. Sci. USA* 84:9265–69

121. Palmer RMJ, Ferrige AG, Moncada S. 1987. Nitric oxide release accounts for the biological activity of endothelium-derived relaxing factor. *Nature* 327:524–26

122. Rapoport RM, Draznin MB, Murad F. 1983. Endothelium-dependent relaxation in rat aorta may be mediated through cyclic GMP-dependent protein phosphorylation. *Nature* 306:174–76

123. Traystman RJ, Moore LE, Helfaer MA, Davis S, Banasiak K, et al. 1995. Nitro-L-arginine analogues: dose- and time-related nitric oxide synthase inhibition in brain. *Stroke* 26:864–69

124. Steiner AA, Branco LGS. 2001. Nitric oxide in the regulation of body temperature and fever. *J. Thermal Biol.* 26:325–30

125. Branco LGS, Carnio EC, Barros RCH. 1997. Role of nitric oxide pathway in hypoxia-induced hypothermia. *Am. J. Physiol. Regulatory Integrative Comp. Physiol.* 273:R967–R71

126. De Luca B, Monda M, Sullo A. 1995. Changes in eating behavior and thermoregulation activity following inhibition of NO formation. *Am. J. Physiol. Regulatory Integrative Comp. Physiol.* 268:R1533–R38

127. De Paula D, Steiner AA, Branco LGS. 2000. The nitric oxide pathway is an important modulator of stress-induced fever in rats. *Physiol. Behav.* 70(5):505–11

128. Scammell TE, Elmquist JK, Saper CB. 1996. Inhibition of nitric oxide synthase produces hypothermia and depresses lipopolysaccharide fever. *Am. J. Physiol. Regulatory Integrative Comp. Physiol.* 271:R333–R38

129. Schmidt HHHW, Gagne GD, Nakane M, Pollock JS, Miller MF, Murad F. 1992. Mapping of nitric oxide synthase in the rat suggests frequent co-localization with NADPH diaphorase but not with soluble guanylyl cyclase, and novel paraneural functions for nitrergic signal transduction. *Histochem. Cytochem.* 40:1439–56

130. Foster DD, Frydman ML. 1979. Tissue distribution of cold-induced thermogenesis in conscious warm or cold acclimated rats reevaluated from changes in tissue blood flow: the dominant role of brown adipose tissue in the replacement of shivering by nonshivering thermogenesis. *Can. J. Physiol. Pharmacol.* 57:257–70

131. Nagashima T, Ohinata H, Kuroshima A. 1994. Involvement of nitric oxide in noradrenaline-induced increase in blood flow through brown adipose tissue. *Life Sci.* 54:17–25

132. Mathai LM, Hjelmqvist H, Keil R, Gerstberger R. 1997. Nitric oxide increases cutaneous and respiratory heat dissipation in conscious rabbits. *Am. J. Physiol. Regulatory Integrative Comp. Physiol.* 272:R1691–R97

133. Renshaw GM, Dyson SE. 1999. Increased nitric oxide synthase in the vasculature of the epaulette shark brain following hypoxia. *NeuroReport* 10(8):1701–12

134. Haxhiu MA, Chang CH, Dreshaj IA, Erokwu B, Prabhakar NR, Cherniack NS. 1995. Nitric oxide and ventilatory response to hypoxia. *Respir. Physiol.* 101(3): 257–66

135. Spanggord H, Sheldon RA, Ferriero DM. 1996. Cysteamine eliminates nitric oxide synthase activity but is not protective to the hypoxic-ischemic neonatal rat brain. *Neurosci. Lett.* 213(1):41–44

136. Cardenas A, Moro MA, Hurtado O, Leza JC, Lorenzo P, et al. 2000. Implications of glutamate in the expression of inducible nitric oxide synthase after oxygen and glucose deprivation in rats forebrain slices. *J. Neurochem.* 74(5):2041–48

137. Hong Y, Suzuki S, Yatoh S, Mizutani M, Nakajima T, et al. 2000. Effect of hypoxia on nitric oxide production and its synthase expression in rat smooth muscle cells. *Biochem. Biophys. Res. Commun.* 268(2):329–32

138. Ikeno S, Nagata N, Yoshida S, Takahashi H, Kigawa J, Terakawa N. 2000. Immature brain injury via peroxynitrite production induced by inducible nitric oxide synthase after hypoxia-ischemia in rats. *J. Obstet. Gynecol. Res.* 26(3):227–34

139. Jung F, Palmer LA, Zhou N, Johns RA. 2000. Hypoxic regulation of inducible nitric oxide synthase via hypoxia inducible factor-1 in cardiac myocytes. *Circ. Res.* 86(3):319–25

140. Vargiu C, Belliardo S, Cravanzola C, Grillo MA, Colombato S. 2000. Oxygen regulation of hepatocyte iNOS gene expression. *J. Hepatol.* 32(4):567–73

141. Abu-Soud HM, Ichimori K, Presta A, Stuehr DJ. 2000. Electron transfer, oxygen binding, and nitric oxide feedback inhibition in endothelial nitric-oxide synthase. *J. Biol. Chem.* 275(23):17349–57

142. Berka V, Tsai AL. 2000. Characterization of interactions among the heme center, tetrahydrobiopterin, and L-arginine binding sites of ferric eNOS using imidazole, cyanide, and nitric oxide as probes. *Biochemistry* 39(31):9373–93

143. Carnio EC, Almeida MC, Fabris G, Branco LGS. 1999. Role of nitric oxide in 2-deoxy-D-glucose-induced hypothermia in rats. *NeuroReport* 10(14):3101–4

144. Almeida MC, Branco LGS. 2001. Role of nitric oxide in insulin-induced hypothermia in rats. *Brain Res. Bull.* 54:49–53

145. Steiner AA, Branco LGS. 2001. Carbon monoxide is the heme oxygenase product with a pyretic action: evidence for a cGMP signaling pathway. *Am. J. Physiol. Regulatory Integrative Comp. Physiol.* 280(2):R448–R57

146. Steiner AA, Colombari E, Branco LGS. 1999. Carbon monoxide as a novel mediator of the febrile response in the central nervous system. *Am. J. Physiol. Regulatory Integrative Comp. Physiol.* 277: R499–R508

147. Gautier H, Murariu C. 1999. Role of nitric oxide in hypoxic hypometabolism in rats. *J. Appl. Physiol.* 87(1):104–10

148. Castagnoli K, Palmer S, Anderson A, Bueters T, Castagnoli NJ. 1997. The neuronal nitric oxide synthase inhibitor 7-nitroindazole also inhibits the monoamine oxidase-B-catalyzed oxidation of 1-methyl-4-phenyl-1,2,3,6-tetrahydropyridine. *Chem. Res. Toxicol.* 10:364–68

149. Amir S, De Blasio E, English AM. 1991. N^G-monomethyl-L-arginine co-injection attenuates the thermogenic and hyperthermic effects of E_2 prostaglandin microinjection into the anterior hypothalamic preoptic area in rats. *Brain Res.* 556(1):157–60

150. Gourine AV, Kulchitshy VA, Gourine VN. 1995. Nitric oxide affects the activity of neurones in the preoptic/anterior hypothalamus of anaesthetized rats: interaction with the effects of centrally administered interleukin-1. *J. Physiol.* 483:72P (Abstr.)

151. Steiner AA, Antunes-Rodrigues J, McCann SM, Branco LGS. 2002. Antipyretic role of the NO-cGMP pathway in the anteroventral preoptic region of the rat brain. *Am. J. Physiol. Regulatory Integrative Comp. Physiol.* In press

152. Goodson AR, Leibold JM, Gutterman DD. 1994. Inhibition of nitric oxide synthesis augments centrally induced sympathetic coronary vasoconstriction in cats. *Am. J. Physiol. Heart Circ. Physiol.* 267: H1272–H78

153. Krukoff TL. 1999. Central nitric oxide in regulation of autonomic functions. *Brain Res. Rev.* 30:52–65

154. Morimoto T. 1998. Heat loss mechanisms. See Ref. 11, pp. 80–90

155. Verma A, Hirsch DJ, Glatt CE, Ronnett GV, Snyder SH. 1993. Carbon monoxide: a putative neural messenger. *Science* 259: 381–84

156. Maines MD. 1988. Heme oxygenase: function, multiplicity, regulatory mechanisms, and clinical applications. *FASEB J.* 2:2557–68

157. Maines MD. 1993. Carbon monoxide: an emerging regulator of cGMP in the brain. *Mol. Cell. Neurosci.* 4:389–97

158. Maines MD. 1997. The heme oxygenase system: a regulator of second messenger gases. *Annu. Rev. Pharmacol. Toxicol.* 37: 517–54

159. Ewing JF, Maines MD. 1992. In situ hybridization immunohistochemical localization of heme oxygenase-2 mRNA and protein in normal rat brain: differential distribution of isoenzyme 1 and 2. *Mol. Cell. Neurosci.* 3:559–70

160. Ewing JF, Haber SN, Maines MD. 1992. Normal and heat-induced patterns of expression of heme oxygenase-1 (HSP32) in rat brain: hyperthermia causes rapid induction of mRNA and protein. *J. Neurochem.* 58:1140–49

161. Johnson RA, Kozma F, Colombari E. 1999. Carbon monoxide: from toxin to endogenous modulator of cardiovascular functions. *Braz. J. Med. Biol. Res.* 32:1–14

162. Morita T, Perrella MA, Lee ME, Kourembanas S. 1995. Smooth muscle cell-derived carbon monoxide is a regulator of cGMP. *Proc. Natl. Acad. Sci. USA* 92: 1475–79

163. Mancuso C, Tringali G, Grossman A, Preziosi P, Navarra P. 1998. The generation of nitric oxide and carbon monoxide produces opposite effects on the release of imminoreactive interleukin-1β from the rat hypothelamus in vitro: evidence for the

involvement of different signaling pathways. *Endocrinol.* 139:1031–37

164. Steiner AA, Branco LGS. 2000. Central CO-heme oxygenase pathway raises body temperature by a prostaglandin-independent way. *J. Appl. Physiol.* 88:1607–13

165. Anning PB, Chen Y, Lamb NJ, Mumby S, Quinlan GJ, et al. 1999. Iron overload upregulates haem oxygenase 1 in the lung more rapidly than in other tissues. *FEBS Lett.* 447:111–14

166. Odaka Y, Takahashi T, Yamasaki A, Suzuki T, Fujiwara T, et al. 2000. Prevention of halothane-induced hepatotoxicity by hemin pretreatment: protective role of heme oxygenase-1 induction. *Biochem. Pharmacol.* 59(7):871–80

167. Ponka P. 1999. Cell biology of heme. *Am. J. Med. Sci.* 318(4):241–56

168. Shibahara S. 1988. Regulation of heme oxygenase gene expression. *Semin. Hematol.* 25:370–76

169. Shibahara S. 1994. Heme oxygenase–regulation of and physiological implication in heme catabolism. In *Regulation of Heme Protein Synthesis*, ed. H Fujita, pp. 103–16. Medina OH: AlphaMed

170. Takahashi K, Hara E, Suzuki H, Sasano H, Shibahara S. 1996. Expression of heme oxygenase isozyme mRNA in the human brain and induction of heme oxygenase-1 by nitric oxide donors. *J. Neurochem.* 67:482–89

171. Frosini M, Sesti C, Valoti M, Palmi M, Fusi F, et al. 1999. Rectal temperature and prostaglandin E_2 increase in the cerebrospinal fluid of conscious rabbits after intracerebroventricular injection of hemoglobin. *Exp. Brain. Res.* 126:252–58

172. Schmidt HHHW. 1992. NO, CO and HO:

endogenous soluble guanylyl cyclase-activating factors. *FEBS Lett.* 307:102–7

173. Murphy BJ, Laderoute KR, Short SM, Sutherland RM. 1991. The identification of heme oxygenase as a major hypoxic stress protein in Chinese hamster ovary cells. *Br. J. Cancer* 64:69–73

174. Panchenko MV, Farber HW, Korn JH. 2000. Induction of heme oxygenase-1 by hypoxia and free radicals in human dermal fibroblasts. *Am. J. Physiol. Cell Physiol.* 278:C92–C101

175. Paro F, Steiner AA, Branco LGS. 2001. Thermoregulatory response to hypoxia after inhibition of the central carbon monoxide-heme oxygenase pathway. *J. Thermal. Biol.* 26:339–43

176. Steiner AA, Antunes-Rodrigues J, Branco LGS. 2002. Role of preoptic second messenger systems (cAMP and cGMP) in thermoregulation and fever. *Am. J. Physiol. Regulatory Integrative Comp. Physiol.* Submitted

177. Dascombe MJ. 1984. Evidence that cyclic nucleotides are not mediators of fever in rabbits. *Br. J. Pharmacol.* 81:583–88

178. Willies GH, Wolf CJ, Rosendorff C. 1976. The effect of an inhibitor of adenylate cyclase on the development of pyrogen, prostaglandin and cyclic AMP fevers in the rabbit. *Pflügers Arch.* 367:177–81

179. Zamboni G, Perez E, Amici R, Parmeggiani PL. 1990. The short-term effects of DL-propanolol on the wake-sleep cycle of the rat are related to selective changes in preoptic cyclic AMP concentration. *Exp. Brain Res.* 81:107–12

180. Gross RA, Ferrendelli JA. 1980. Mechanisms of cyclic AMP regulation in cerebral anoxia and their relationship to glycogenolysis. *J. Neurochem.* 34:1309–18

Annu. Rev. Physiol. 2002. 64:289–311

CALMODULIN AS AN ION CHANNEL SUBUNIT

Yoshiro Saimi[1] and Ching Kung[1,2]

[1]Laboratory of Molecular Biology and [2]Department of Genetics University of Wisconsin, Madison, Wisconsin 53706; e-mail: ysaimi@facstaff.wisc.edu; ckung@facstaff.wisc.edu

Key Words calcium channels, ligand-gated channels, Ca^{2+}-CaM activated and regulated channels, Ca^{2+} sensor, functional bipartition

■ **Abstract** A surprising variety of ion channels found in a wide range of species from *Homo* to *Paramecium* use calmodulin (CaM) as their constitutive or dissociable Ca^{2+}-sensing subunits. The list includes voltage-gated Ca^{2+} channels, various Ca^{2+}- or ligand-gated channels, Trp family channels, and even the Ca^{2+}-induced Ca^{2+} release channels from organelles. Our understanding of CaM chemistry and its relation to enzymes has been instructive in channel research, yet the intense study of CaM regulation of ion channels has also revealed unexpected CaM chemistry. The findings on CaM channel interactions have indicated the existence of secondary interaction sites in addition to the primary CaM-binding peptides and the functional differences between the N- and C-lobes of CaM. The study of CaM in channel biology will figure into our understanding on how this uniform, universal, vital, and ubiquitous Ca^{2+} decoder coordinates the myriad local and global cell physiological transients.

INTRODUCTION

Ca^{2+} is toxic. Prolonged high $[Ca^{2+}]$ presence in the cytoplasm kills cells, but transient rises and falls in $[Ca^{2+}]_{in}$, registered as local or global Ca^{2+} sparks, control numerous physiological events. The signaling pathways in many of these events use calmodulin (CaM) to decode $[Ca^{2+}]_{in}$. These pathways result in protein phosphorylation and dephosphorylation, cyclic-nucleotide formation and breakdown, cytoskeletal rearrangement, and gene transcription, for example (1). CaM also governs two other important parameters: membrane potential and $[Ca^{2+}]_{in}$ itself, by regulating ion pumps and channels. Here, we only review CaM's roles in ion channel regulation. More specifically, we review the direct interaction of Ca^{2+}, CaM, and the pore-forming subunits of ion channels and forgo covering the large literature on regulations through CaM-dependent channel phosphorylation and dephosphorylation. This restriction should focus attention on rapid (millisecond) and local (submicrometer) events near the membrane. We note that disparate studies on individual channels all converge on the notion of CaM being a channel subunit. Much like its original discovery and rediscoveries in enzyme research in the 1970s, CaM has been rediscovered several times in the context of channel research in the 1990s.

CALMODULIN

CaM is a soluble protein, small [148 amino acids (aa), 17 kDa], acidic (pI $\sim$ 4), and highly conserved throughout the eukaryotic evolution. It is abundant, account- ing for $\sim$0.5% of brain proteins and reaching concentrations of 1–10 μM in the cell. As much as half of CaM is associated with membranes, the rest remains soluble in the cytoplasm and inside the nucleus (2, 3). It is known to activate some 40 types of enzymes or channels, and the list will no doubt continue to grow as research continues. CaM functions as a monomer with two pairs of EF hands. In crystal structure, the N-terminal pair (EF hand I and II) forms a lobe as does the C-terminal pair (III and IV), and the two lobes are connected by an eight-turn α helix giving the appearance of a dumbbell (4). CaM in isolation binds Ca^{2+} at the range of physiological [Ca^{2+}]$_{in}$ fluctuations ($K_d = 5 \times 10^{-7}$ to 5×10^{-6} M), but its Ca^{2+} affinity is much higher when in complex with an en- zyme (5). Whereas CaM can bind a maximum of four Ca^{2+} (CaM$_{[Ca:1234]}$), NMR studies have shown that CaM$_{[Ca:34]}$ with Ca^{2+} bound at site III and IV, but not CaM$_{[Ca:12]}$ at I and II, does exist in solution (6). Upon binding Ca^{2+}, CaM$_{[Ca:1234]}$ becomes more extended with each pair of EF hands opening to reveal a hy- drophobic patch (4, 7) that is available to bond with a target peptide(s). Clas- sic co-crystallization and NMR studies show that the two lobes of CaM$_{[Ca:1234]}$ embrace the target peptide of an enzyme by hydrophobic interactions and by salt bridges, with the glutamate residues in the Ca^{2+}-coordinating loop in each of the EF hands (8, 9). Correspondingly, the known primary target peptides of CaM$_{[Ca:1234]}$ are often basic amphiphilic α (Baa) helices with a hydrophobic side and a highly positively charged side. These features do not strictly dic- tate amino acid sequences in the Baa segments, although there are recogniz- able CaM-binding motifs, such as the 1-5-10 and 1-8-14 patterns (5, 10, 11). The 1-5-10 motif found in CaM-dependent kinases I and II and MARCKS proteins denotes a consensus sequence, $X_3 \Phi X_3 \Phi X_4 \Phi$, where X is any amino acid, some of which are positively charged, and Φ is a hydrophobic amino acid. The type A of the 1-8-14 motif in calcineurin and most other CaM-binding proteins comprises $\Phi X_3 \Phi X_2 \Phi X_5 \Phi$, and the type B in fordrin, for example, is $\Phi X_6 \Phi X_5 \Phi$. The long connecting helix in CaM is flexible, allowing the two lobes to position themselves according to the specific sequence of the binding peptide. In several enzymes, the Baa helix is on or near an autoinhibitory domain that obstructs the active site, and upon binding to Ca^{2+}-CaM, the Baa helix moves away to disinhibit the enzyme.

Ca^{2+}-free apocalmodulin (ApoCaM) or partially filled CaM can also bind to certain sites on target proteins. The IQ motif is the best known among the sequences that can bind ApoCaM (10) likely to its C lobe in a semi-open state (12). The motif only loosely defines the amino acid sequence at 5 of 11 possible residues, and different IQ domains bind to CaM at varying Ca^{2+} concentrations or independently of Ca^{2+} (5). The affinity of ApoCaM for Ca^{2+} or its target peptides can also be modulated. In the linked three-party reaction of Ca^{2+}, CaM, and a peptide P, the fact that Ca^{2+}-CaM has higher affinity for P than Ca^{2+}-free ApoCaM conversely implies that P-bound ApoCaM should have higher affinity for Ca^{2+} than unbound ApoCaM. Indeed, binding to an enzyme can increase CaM's affinity for Ca^{2+} a thousandfold (5). Thus we cannot be sure that application of such commonly used Ca^{2+} chelators as EGTA or BAPTA truly leaves CaM free of Ca^{2+} in the channel or enzyme complex.

One needs to consider possible bindings occurring in sites other than the two major hydrophobic patches of CaM and the consensus sequences (the 1-5-10, 1-8-14, IQ motif, etc.) of the target proteins. CaM is known to bind simultaneously to two non-contiguous peptides in the γ subunit of phosphorylase kinase, and the synergy of the two bindings results in an extremely tight complex (13). Interestingly, these two CaM-binding sites are also the kinase's autoinhibitory domains (14). In NO synthase type 2, additional sequences besides the canonical Baa helix are required to bind CaM in a Ca^{2+}-free condition (15). Mutations in the F helices of CaM's EF hand I and III, known as the latch domain, have a disproportionate negative effect of NO synthase activity, indicating that this domain additionally interacts with the enzyme (16). The CaM-binding site of the petunia glutamate decarboxylase contains five negatively charged residues and one CaM seems to bind two such sites, suggesting a role of Ca^{2+}-CaM in dimerizing and therefore activating this decarboxylase (17). Studies of half CaMs (C- or N-lobe alone) have shown that the two lobes are largely independent structures, each capable of wrapping a target peptide in one or the opposite polarity (18 and references therein). Even more radical is the interaction of CaM with basic helix-loop-helix (bHLH) transcription factors to inhibit DNA-binding and transcription activation. Biochemical evidence suggests a model in which two CaMs bind to a bHLH dimer at its pair of DNA-binding sequences by salt bridges or polar interactions (19). Readers are reminded that, in terms of the structures of CaM-target complexes, we have only a few examples at atomic resolution. Recently, Schumacher et al. (20) solved a structure of CaM complexed with the CaM-binding domain of SK2 channels (see below). This astounding crystal structure has revealed several new features of CaM: Ca^{2+} filling of the N but not C lobe, half Ca^{2+}-filled functional states, binding simultaneously to multiple target helices, and tying two target proteins together. At this point, our knowledge of CaM chemistry is extensive but still far from complete (see References 21 and 22 for reviews).

Much less is known about CaM biology. *CAM* genes have not been found in any of some 50 completed bacterial or archael genomes. All examined eukaryotes, on the other hand, have at least one *CAM* gene. However, considering the abundance and wide-ranging involvement of the CaM protein, it seems surprising that there is only one *CAM* gene in the worm *Caenorhabditis elegans* or the fruit fly. Furthermore, the sequence of CaM is extremely conserved: All known vertebrate CaMs are identical, and all metazoan CaMs are essentially the same. CaM of *Paramecium* (a protist) is 94% similar to the mammalian counterpart and that of *Aspergillus* (a fungus) is 92% similar (23). In mammals, each species has three genes that are 80% identical in nucleotide sequence but encode 100% identical

CaM proteins. These genes are flanked differently and presumably under different temporal and tissue-specific transcriptional controls (2).

Given its many roles, it is not surprising that CaM is essential: A *CAM* knockout of the budding yeast *Saccharomyces cerevisiae* cannot grow (24, 25). *CAM*-null *Drosophila* survives, presumably on maternal CaM, but only to the first larval stage, exhibiting an over-excitable terminal phenotype (26). CaMs that are rendered incapacitated in binding Ca^{2+} in vitro (declawed CaM) through multiple substitutions in all four EF hands nonetheless support proliferation of *S. cerevisiae* in the laboratory (25), although not the fission yeast *Schizosaccharomyces pombe* (27). This and other findings reviewed below show that Ca^{2+}-free ApoCaM itself likely has important functions. A hemizygote *Drosophila* with a substitution in the first EF hand of CaM is viable but with structural and functional defects in their synapses (28). Viable behavioral mutants with minor pleiotropic phenotypes have been isolated in *Paramecium* that turned out to bear point substitutions in CaM (29; see below). Budding yeasts with engineered Phe-to-Ala substitutions in CaM are alive at permissive temperatures and can be grouped by their different terminal phenotypes at restrictive temperatures (30). In sum, studies in vivo reveal that CaM plays many roles in life and different parts of the CaM molecule can contribute differently to each role. Studies in silico should complement such in vivo studies in identifying CaM targets (30a).

Ca^{2+} CHEMISTRY OF ION CHANNELS

Although Ca^{2+}-activated K^+ currents had been known earlier (31), it was the work of Brehm & Eckert (32) that first drew attention to the local Ca^{2+} chemistry of ion channels. They found that whereas the voltage-activated Ca^{2+} current of *Paramecium* inactivates, the Ba^{2+} current through the same channel does not. Two-pulse and chelation experiments showed clearly that Ca^{2+} itself is required for inactivation. Ca^{2+}-dependent inactivation was later found in many different Ca^{2+} channels including the cardiac L-type and P/Q-type reviewed below. This early work suggested that Ca^{2+} that had just passed through the channel somehow caused the channel to become inactivated. However, the mechanism of this Ca^{2+}-dependent inactivation was not understood until the advent of patch clamp, gene cloning, and other recombinant-based molecular techniques. As more and more channels were found to be regulated by Ca^{2+}, the search for its mechanisms of action intensified. The ubiquity of CaM and its prominence in enzymology in the 1970s and 1980s made it the likely suspect in any Ca^{2+} action. It was, therefore, common to apply CaM inhibitors such as W7, calmidozolium, or mastoparan to the reaction of interest, and it was presumably the failure to see significant effects on many channel activities that prevented an earlier realization of CaM's role here (33). Nonetheless, by the late 1980s, there was a study of capturing Ca^{2+}-gated K^+ channels from kidney using a CaM-affinity column (34, 35) and of lowering the open probability of a SR Ca^{2+}-release channel upon the addition of CaM (36).

Serendipity played a role in the rediscovery of CaM in channels from a line of hypothesis-free research in vivo. In 1983, Saimi et al. (37) isolated and studied a group of behavioral mutants in *Paramecium* called *pantophobiacs*. They are so named because they overreact to a variety of stimuli including the usual culture media. Voltage-clamp experiments showed that they lacked a Ca^{2+}-dependent K^+ outward current during depolarization. Cytoplasmic transfusion by microinjection from the wild-type to *pantophobiac* was found to restore the current and cure the misbehavior. By fractionating the wild-type cytoplasm, the curing element turned out to be CaM (38), much to the disappointment of one of us (C. Kung) who was hoping for a new find. Even more unexpectedly, genetic analyses showed that *pantophobiac* is allelic to *fast-2*, a group of mutants that have an opposite phenotype and under-react to stimuli owing to a lack of a Ca^{2+}-dependent Na^+ inward current. They too have defects in their CaM, the product of the sole *CAM* gene in *Paramecium* (29). Furthermore, the segregated localization of *pantophobiacs* mutations to the C lobe and the *fast-2* mutations to the N lobe of the CaM molecule was entirely unexpected and informative (see below). Genetics and biochemistry aside, Saimi & Ling in 1990 (39) directly demonstrated the Ca^{2+}-CaM activation of the *Paramecium* Na^+ channel under a patch clamp. CaM and Ca^{2+} added to the inner (bath) side of membrane patches that were excised from *Paramecium* blisters reactivated this Na^+ channel previously inactivated by low $[Ca^{2+}]$. Because the bath solution was free of enzymes or ATP, this result showed that the reactivation of this channel is by its Ca^{2+}-dependent association with Ca^{2+}-CaM and not by a secondary modification such as phosphorylation. Anthropocentric physiologists might question whether this *Paramecium* finding, like the earlier one by Brehm & Eckert, can also be applied to mammals. Indeed, a large variety of unrelated animal ion channels are now found to be directly associated with CaM as reviewed below.

Cyclic Nucleotide-Gated (CNG) Channels

In 1993, Hsu & Molday (40) reported that Ca^{2+}-CaM binding to the CNG channel lowers its affinity for cGMP and reduces the influx of cations into the rod outer segment. CNG channels are oligomers consisting of at least four subunits: $\alpha\alpha\beta\beta$. α and β share 30–50% amino acid identity, but α alone can be expressed heterologously as a functional channel whereas β cannot. Evidence points to binding of Ca^{2+}-CaM to both the N-terminal tail of the β subunit (40–42) as well as the C-terminal tail; the latter does not appear significant in terms of channel regulation (43). Application of Ca^{2+}-CaM to the rod CNG channel in vivo has yielded mixed results. Similarly, although CNG channels from cones also bear a CaM-binding domain, these channels lack crucial molecular determinants in the sequence downstream of the Ca^{2+}-CaM–binding site in order to be regulated by CaM (44), and it is possible that these sites could be used for binding of other similar proteins (45). Thus not all CaM bindings to channel molecules result in modulation of channel activity.

Binding of cyclic nucleotides (cNMP) to the CNG channels of rat olfactory receptors allows Ca^{2+} entry, and in so doing, reduces the channel open probability below its initial peak. This manifests as olfactory adaptation. Chen & Yau (41a) used inside-out patches excised from HEK cells expressing this channel to show that the addition of exogenous CaM reduces the current through the channel to near zero. The Ca^{2+}-CaM–channel association apparently is reversible and can be blocked by the CaM inhibitor mastoparan. Experimentation with chimeras and deletions traced the Ca^{2+}-CaM–binding site to a Baa peptide at the N-terminal tail of the channel, and this peptide was shown to directly bind CaM in gel-shift and fluorescence studies. Further analyses and modeling indicate that this region of the N-tail is required to keep the cNMP-bound channel open constantly and that the Ca^{2+}-CaM binding blocks this N-terminal tail effect. This effect on gating reflects a reduction in the apparent affinity of the channel for cNMP by 20-fold because gating and binding are tightly coupled (46). Varnum & Zagotta (47) showed biochemically that the CaM-binding segment of the N-terminal tail interacts directly with the portion of the C-terminal tail that constitutes the cNMP-binding domain. Apparently, Ca^{2+}-CaM binding to the N-terminal tail then blocks the interaction, removes the autoexcitatory effect of the N-terminal tail, and reduces the open probability. The overall result is an olfactory adaptation: reduction of the channel response to constant stimulus by way of Ca^{2+} that passes through the channel.

NMDA Receptors

NMDA (N-methyl-D-aspartate) receptors are heteropentamers of pore-forming subunits from two families, NR1 and NR2A-D. In a co-immunoprecipitation experiment, Ehlers et al. (48) found CaM to bind NR1 in a Ca^{2+}-dependent manner. They also found that CaM application to excised inside-out patches reduces channel open probability and mean open time. This inactivating effect is reversible and dependent on Ca^{2+}, suggesting that under physiological conditions Ca^{2+}-CaM inactivates NMDA through binding to the NR1 subunit and detaches from it at low $[Ca^{2+}]_{in}$ to relieve inactivation. Two Ca^{2+}-CaM–binding Baa peptides are identified at the C-terminal tail, both having the basic and hydrophobic residues and substantial agreement to the canonical 1-5-10 pattern for Ca^{2+}-CaM binding (10). The downstream site has a higher affinity for Ca^{2+}-CaM than the one upstream. Many splice variants of NR1 are created by including or excluding short exons. A 37-amino acid exon, C1, encompasses the entire downstream Baa and enables generation of NR1 variants with different Ca^{2+}-CaM sensitivities. Therefore, it is surprising that deleting C1 has no significant effect on inactivation, at least not in recombinant NR1/NR2A receptors expressed heterologously (49). All known splice variants contain C0, the upstream 30 amino acid region extending from the end of the last transmembrane helix to C1. Deleting C0 or placing charged substitutions in the upstream Baa abolishes much of the Ca^{2+}-dependent inactivation (49, 50).

The activity of the NMDA receptor can be reduced in a Ca^{2+}- and ATP-dependent manner, but this rundown can be inhibited by phalloidin, an actin depolymerization inhibitor, suggesting that the NMDA channel is attached to the cytoskeleton (51). The interaction of its C-terminal tail with α-actinin-2, an actin-associated protein, was revealed in a yeast two-hybrid complementation experiment (52). C0 is now found to be the site of α-actinin binding as well as the site for other forms of regulation. CaM and α-actinin compete to bind the receptor at C0. Current models, therefore, feature the NMDA channel binding to the actin cytoskeleton through α-actinin-2 at rest. After the opening of NMDA receptors or nearby Ca^{2+} channels, Ca^{2+} arriving at the cytoplasmic side binds CaM, and Ca^{2+}-CaM then binds the Baa in C0, displacing and detaching the actinin-actin fiber while reconfiguring the C-terminal tail to inactivate the channel (49, 50).

Ryanodine Receptors (RyR) and Inositol 1,4,5-Trisphosphate Receptors (IP$_3$R)

Ca^{2+} regulates the triad junction of skeletal muscle in a complex manner. The skeletal-muscle Ca^{2+} channel (the dihydropyridine receptor), like its cardiac counterparts reviewed below, may bind CaM (54). The RyR1 protein of skeletal muscle itself binds Ca^{2+} and functions as a Ca^{2+}-induced Ca^{2+}-release channel, but it contains CaM as its Ca^{2+}-dependent regulator (53). This giant channel is a homotetramer constructed from 565-kDa subunits. Cryomicroscopy and three-dimensional reconstruction reveal a fourfold symmetry with a large cytoplasmic assembly and a smaller transmembrane domain (55). Two other types of proteins are seen to co-assemble with the channel, one of which is CaM. Four CaMs are found to reside strategically in a cleft between adjacent subunits and at the junction between the large and small domains (55).

Experiments on how CaM protects RyR1 from tryptic digestion show that ApoCaM and Ca^{2+}-CaM bind the same or overlapping regions (aa 3630–3637) of RyR1 that included a Baa site of the 1-5-10 motif (56, 57). A similar region has been identified as CaM-binding site through mutational analyses (57a). In a further study, a synthetic peptide of aa 3614–3643 of RyR1 binds both ApoCaM and Ca^{2+}-CaM, and the last 9 amino acids are crucial only for ApoCaM binding, indicating that CaM shifts on the peptide upon binding Ca^{2+} (57b). CaM exhibits opposing effects on RyR1: At nanomolar $[Ca^{2+}]_i$ the bound ApoCaM sensitizes RyR1, but at micromolar $[Ca^{2+}]_i$ the bound Ca^{2+}-CaM inhibits its opening. A mutant CaM that lost its Ca^{2+}-binding sites (CaM) was found to enhance the affinity of RyR1 for Ca^{2+} as well as for single-channel activities at high Ca^{2+}, suggesting that, in the normal situation, Ca^{2+} inhibition of RyR1 is via effects of calcified CaM on RyR1 in vivo (53). The cardiac homolog RyR2, on the other hand, can be inhibited by CaM and has multiple CaM-binding sites (57c), although the ApoCaM sensitization effect is not found for RyR2 (57d).

Oxidation increases RyR1 activity and CaM seems to protect RyR1 from oxidative modification (58). RyR1 is probably under a redox control through S-nitrosylation, and a fraction of its cysteines is apparently naturally nitrosylated. Under a physiologically low pO$_2$, nanomolar NO activates RyR1 in sarcoplasmic reticular vesicles. This effect is reduced when CaM is added, however, and the full effect is restored when competing CaM-binding peptides are also added. In lipid bilayers at a low pO_2, an increase in RyR1's activity by NO is removed by competing CaM-binding peptides. Thus the redox regulation of RyR1 is also dependent on CaM (59).

Inositol 1,4,5-trisphosphate receptors (IP$_3$R) are a second family of Ca^{2+} release channels. RyRs and IP3Rs share structural and functional similarities and have some sequence similarity in the C-terminal domains, although the latter are approximately half the size of the former (60). IP$_3$R-1 can be activated by Ca^{2+} and inhibited by Ca^{2+}-CaM at higher Ca^{2+} concentrations, producing a bell-shaped Ca^{2+} activation curve (62). IP$_3$R-1,2 (Type I and II) have been shown to bind CaM (61) at residues 1564–1585 of IP$_3$R-1 and 1558–1596 of IP$_3$R-2. Some mutations in these residues or reconstitution of IP$_3$R purified away from CaM remove the Ca^{2+}-CaM–dependent inhibition (62). IP$_3$R-1 also has a second CaM-binding site within the first 159 amino acids of the N terminus (63, 64) and CaM binds (EC50 $\approx$ 3 μM) to this site in a Ca^{2+}-independent manner. The exact function of this binding requires further investigation (63–65). IP$_3$R-3, however, does not have a similar putative Ca^{2+}-binding site (61), nor does it exhibit Ca^{2+} inhibition (66).

Ca^{2+}-Activated K^+ Channels of Small or Intermediate Conductance (SK or IK)

Ca^{2+}-activated K^+ channels are usually classified by pharmacological criteria and by the size of their unitary conductances: large (BK > 100 pS), intermediate (IK 10–50 pS), and small (SK < 15 pS). BK apparently uses a "Ca^{2+} bowl" structure in the C-terminal tail of the pore-forming subunit to sense cytoplasmic Ca^{2+}: BK is therefore not reviewed here. Both IK and SK use CaM as their Ca^{2+}-sensing subunit. In 1998, Xia et al. (67) provided the first convincing evidence that CaM is a constitutive element of SK. Beginning with the notion that gating of the SK protein could be controlled by direct Ca^{2+} binding to the pore-forming subunit itself, they replaced all its 21 relevant glutamate or aspartate residues individually but found no marked effects. However, the proximal portion of the C-terminal tail appeared to be crucial for of Ca^{2+} activation because Ca^{2+} dependency was lost upon deletion of this region. They subjected the possible interactions between CaM and this C-terminal region or its sub-regions to four biochemical tests (yeast two-hybrid complementation, GST pulldown, CaM overlay, co-immunoprecipitation) and identified amino acid 390–487 (CaMBD) in the C-terminal tail to be the site of CaM interaction. CaMBD does not include standard ApoCaM or Ca^{2+}-CaM binding motifs (10). In the yeast two-hybrid system, which tests protein-protein interaction in a low $[Ca^{2+}]_{in}$ environment of the yeast nucleoplasm, the interaction between CaM and CaMBD persists even when CaM has been declawed by point substitutions to remove one, two, or three of its EF hands' ability to chelate Ca^{2+}. Clearly, the ability of CaM to bind Ca^{2+} is not required for its association with CaMBD. Whereas co-expressing the clawless CaM$_{1234}$ with SK2 markedly reduced the SK current, co-expression of CaM_{34} (EF hand III and IV no longer bind Ca^{2+}) had little effect, but co-expression of CaM_{12} reduced the current as much as co-expression of CaM_{1234} (68). Co-expressing mutant CaM having the intact C-lobe but only one of the intact N-terminal EF hands (CaM_1 or CaM_2) also resulted in channels whose response to Ca^{2+} was greatly weakened.

The structure of co-crystal CaM with CaMBD of SK2 has recently been solved by Schumacher et al. (20) (Figure 1, see color insert). The is an anti-parallel dimeric structure with a twofold symmetric axis and consists of two CaMBDs and two half Ca^{2+}-filled CaMs. Each CaMBD is a hairpin turn made up of the $\alpha 1$ helix (residues 413–440) and $\alpha 2$ helix (446–489) and a short interconnecting loop. The C-terminal half of $\alpha 2$ (468-488) binds to the primary hydrophobic pocket of the Ca^{2+}-filled N lobe of one CaM in a more-or-less typical manner. The more central part of $\alpha 2$ (aa 458–467), however, interacts with a protrusion of the N lobe of the second CaM. This protrusion, made from the exiting helix of the first EF hand, has not been previously implicated in target binding and constitutes a secondary binding site outside the main hydrophobic pocket. Surprisingly, the C-lobe of CaM appears to be Ca^{2+}-free. Deviations from the basic ApoCaM structure by reorienting the two EF hands and repacking of the hydrophobic core of the C-lobe allow extensive hydrophobic contacts with the $\alpha 1$ helix of CaMBD. Schumacher et al. (20) envision that, at rest, one CaM is consititutively bound to only one CaMBD through the C-lobe-$\alpha 1$ interaction. With a rise in $[Ca^{2+}]_{in}$, Ca^{2+}-filled N-lobes of two neighboring CaMs grab each others $\alpha 2$ helices, thereby generating the force that ultimately opens the channel gate. This work will no doubt stimulate further structural studies as well as site-directed mutagenic dissections.

The intermediate-conductance Ca^{2+}-activated K^+ channel (IK) has no similarity to BK but is 40% identical to SK, and the similarity is striking in the proximal portion of the C-terminal tail, the likely CaM-interaction portion. Fanger et al. (69) investigated human IK1 biochemically and physiologically and arrived at the same conclusion as above, i.e., CaM constitutively binds to this portion of the IK pore-forming subunit in a Ca^{2+}-independent manner, and the binding of Ca^{2+} to this CaM induces the conformation changes toward channel opening.

Trp Family Channels

Trps are now a giant family of Ca^{2+}-permeable cation channels of diverse functions and distributions. The founding member, Trp (*transient receptor potential*), and its homolog, TrpL (*Trp-like*), are the major channels in *Drosophila* phototransduction. Binding between CaM and Trp (1 site) or TrpL (2 sites; CBS-1, CBS-2) has been shown biochemically in vitro (70–72), and a perturbation of the light-sensitive current in a CaM mutant has also been recorded in vivo (73). It is still not clear whether either or both CBS-1 and -2 of TrpL actually bind Ca^{2+} in vivo (74–76). Future work should sort out the relative importance of CBS-1 and CBS-2 in TrpL, as well as the exact mechanism of Ca^{2+}-CaM regulation of these channels in *Drosophila* phototransduction.

More recently, all 7 mammalian Trp channels have been shown to bind CaM at two sites (76a,b). Interestingly, one of these binding sites is close to one of the sites for IP_3R binding that has been shown to gate Trp channels, which may allow store depletion-induced Ca^{2+} entry (76c,d). CaM has also been shown to either inhibit Trp3 (76e) or activate other Trps (76b).

Ca^{2+} Channels

The behavior of Ca^{2+} channels (32) first brought the Ca^{2+} chemistry of ion channels into focus for many of us and has also inspired the current leading-edge research on the complex reactions of Ca^{2+}, CaM, and pore-forming channel proteins. These channels guard the external Ca^{2+} source and determine the amount and timing of local changes in cytoplasmic Ca^{2+}. Given their importance, it is not surprising that Ca^{2+} channels are finely tuned in a complex manner. These channels are clearly not built to only maximize the Ca^{2+} current. The peak current through the cardiac L-type Ca^{2+} channel α_{1C} can be increased 80-fold by simply deleting the N- and C-terminal tails (77). In addition to the tight coupling between the stimulus (depolarization) and the response (conduction), there are several sophisticated feedback modulations of Ca^{2+} channels. At least two rely on Ca^{2+}-CaM: inactivation and facilitation. Inactivation describes the loss of the conductance during stimulation. Facilitation describes the subsequent active enhancement of the conductance after the first stimulus, typically beyond the initial response.

de Leon et al. (78) showed that the 660-aa proximal third of the C-terminal tail of α_{1C} is critical for its Ca^{2+}-dependent inactivation (78, 79). Replacing this Ca^{2+}-dependent inactivation domain (CI) of α_{1C} with the comparable domain of α_{1E}, which shows no inactivation, abolishes the ability of α_{1C} to inactivate. More significantly, grafting the CI of α_{1C} onto the body of α_{1E} confers a Ca^{2+}-dependent inactivation to α_{1E} (78). The CI region has several interesting features including an EF hand and a downstream sequence (80–83), which partially conforms to the consensus IQ sequence (10). This IQ sequence binds CaM, and certain mutations in this sequence block CaM binding and remove inactivation (82, 83, 85).

Peterson et al. (84) found that when co-expressed with wild-type α_{1C} in HEK cells, which presumably express their own wild-type CaM, the completely declawed CaM that no longer binds Ca^{2+} (CaM_{1234}) removes the Ca^{2+}-dependent inactivation. Co-expression of a triply declawed CaM has the same dominant-negative effect on α_{1C} in oocytes (81). Thus the Ca^{2+}-free ApoCaM is apparently tethered to the channel, and calcification of CaM allows it to bind the IQ domain to initiate inactivation.

Surprisingly, Qin et al. showed that a fragment that includes the IQ domain binds CaM even when the key IQEYFRK sequence has been changed into AAAAAAA (83). Pate et al. (2000) identified a 26-amino-acid sequence (called CB), 20 aa upstream of the I of the IQ motif (87), and found that CB binds CaM albeit at a lower affinity than the IQ motif. Zühlke & Reuter (81) previously reported that CB includes a crucial NE dipeptide. They also found that all IQ-like peptides from

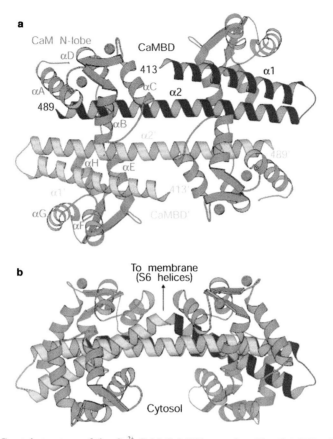

Figure 1 Crystal structure of the Ca^{2+}-CaM-CaMBD complex. The CaMBD of rat SK2 (residues 395–490) was co-crystallized with CaM under a saturating condition for Ca^{2+}. (*a*) Ribbon diagram of the dimeric complex. CaMBD in yellow and blue; CaM in green; Ca^{2+} in red. (*b*) View in (*a*) rotated by 90° showing the orientation of the complex relative to the membrane. The arrow indicates the positions of the first observed residue of each of the CaMBD monomers that are linked to the S6 pore helices. Figure generated with MOLSCRIPT. [Taken from (20), with permission from *Nature*.]

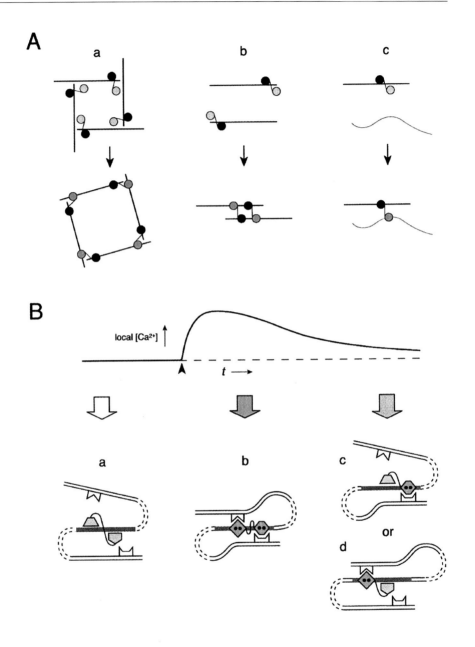

Figure 2 (opposite) Possible CaM-target interactions to be considered in model building including multimerizations (*A*) and functional bipartitions (*B*). Oligomerization of target molecules by Ca^{2+}-CaM binding. (*a*) Sequential and identical bindings of each resident CaM to the target peptide of its neighboring subunit (clockwise here) tie together an oligomer. (*b*) Reciprocal binding leads to a dimer with a twofold symmetry as in SK2 (Figure 1). (*c*) CaM-induced heteromultimers are possible. Different colors represent the N and C lobes or different Ca^{2+}-loaded states. Lobe-specific interactions resulting in CaM's functional bipartition. (*B*) Upper tracing portrays the rise and fall of local $[Ca^{2+}]_i$ upon the opening of a Ca^{2+} source (*arrowhead*). At a location near the source, such as the inner portion of the Ca^{2+} channel itself, Ca^{2+} builds up rapidly and then diffuses or is sequestered away slowly. As a consequence, the channel conformation can be dictated by the fully calcified CaM prior to that dictated by half-filled CaMs. (*B*) The lower portion shows how the two-lobed CaM can become a three-way switch. Depicted is the regulatory domain of a channel or enzyme with CaM (*shaded*), its primary target peptide (*black strip*), and its contiguous or noncontiguous (*dotted lines*) secondary targets from the same or even different proteins. (*a*) At low $[Ca^{2+}]_i$, the ApoCaM (*green*) may be free or already "tethered" near or at the various binding sites. (*b*) At high $[Ca^{2+}]_i$, the fully Ca^{2+}-filled CaM (*red with black circles*) binds both the primary (*black strip*) and all secondary interaction sites (lock-and-key fits). (*c*) At intermediate $[Ca^{2+}]_i$, only the calcified lobe (*red hexagon*) binds one set of secondary sites. (*d*) The opposite lobe can be calcified (*red diamond*) to bind an opposite set of secondary sites in the same regulatory domain at a different conformational state. It also represents a different regulatory domain in the same or different channel or enzyme. Therefore, a-b-c or a-b-d are separate three-way switches. This figure is not meant to explain any set of data of a specific channel or enzyme. We put it forward to stimulate discussions and therefore deliberately incorporated novel features not commonly diagrammed.

different types of Ca^{2+} channels bind Ca^{2+}-CaM (not ApoCaM), but the CB region only of L-type channels, which shows that inactivation and facilitation interact with CaM. CB-containing peptides were found to bind CaM at $[Ca^{2+}]_{in}$ as low as 5×10^{-8} M (87a). The binding between CB and CaM was compared with the binding between IQ and CaM. The results indicate that IQ does not displace CB from CaM when Ca^{2+} is present (87b). In addition to confirming the tethering function of CB, Pitt et al. (87c) identified an upstream region (aa 1565–1578) that conforms to the 1-8-14 of the type A CaM-binding motif (10) and includes an IKTEG sequence known (from previous investigations) to be crucial in inactivation. A peptide (aa 1551–660) encompassing this region can bind to CaM and more importantly to CaM_{1234}. Pitt et al. also discovered apparent Ca^{2+}-dependent conformational changes in this and other peptides independently of CaM and thus accounted for the need of ambient $[Ca^{2+}]_{in}$ (10^{-7} M) for CaM_{1234} binding.

In co-expression experiments, Peterson et al. (84) found that CaM_{12} spares the inactivation of α_{1C} but that CaM_{34} abolishes it, showing a lobe-specific effect. Pate et al. showed that CB of α_{1C} binds CaM_{12} well and CaM_{34} very poorly (87). Using proteolytic hemi-CaMs to interfere with the binding between test peptides and fluorescent CaM, Pitt et al. showed that peptide A (aa 1558–1579) preferentially associates with the N-terminal lobe of CaM, IQ-containing peptide F (aa 1619–638) preferentially associates with the C-terminal lobe, whereas peptide C (aa 1585–1606) shows little preference (87c).

The literature on α_{1C}, taken as a whole, supports the recent model of Pitt et al. (87c) as follows: At rest with an ambient $[Ca^{2+}]_{in}$ of 10^{-7} M, the peptide A region with its 1-8-14 pattern anchors the N-terminal lobe of a constitutive CaM, whereas the CB domain is likely associated with both the N- and the C-lobes of this CaM. Upon channel opening, newly arrived Ca^{2+} binds the C-lobe, and the calcified C-lobe then binds the IQ peptide to effect inactivation. This model of α_{1C} inactivation is therefore similar to that of SK2 activation (Figure 1), both featuring lobe-specific binding of constitutive ApoCaM to nearby but non-contiguous peptides and the binding of a different peptide when one of the lobes is Ca^{2+}-filled to effect activation or inactivation. Interestingly, the anchoring lobe of the constitutive ApoCaM is the C-lobe and the effector lobe that binds Ca^{2+} is the N-lobe in SK2, but the converse is true for α_{1C}. This opposite arrangement recalls the N-lobe–specific effect on the Ca^{2+}-dependent Na^+ channel and the C-lobe–specific effect on the Ca^{2+}-dependent K^+ channel of *Paramecium* (29).

The interaction of IQ and Ca^{2+}-CaM is not sufficient to explain all the CaM effects. R- or N-type Ca^{2+} channels showing little Ca^{2+}-dependent inactivation nonetheless have IQ domains that bind Ca^{2+}-CaM, albeit at reduced affinities. Recall also the complexity of the proximal third of C-terminal tail of α_{1C} that includes an upstream EF hand in addition to the downstream CB and IQ domains. This EF-hand motif may be a necessary part in transducing the CaM binding of IQ into the structural changes toward inactivation of the channel (86). We have certainly not seen the end of α_{1C}'s complexity. Ivanina et al. showed that the N-terminal tail of α_{1C}, like its C-terminal tail, binds CaM in a Ca^{2+}-dependent manner, and N-tail deletion hastens the inactivation of the Ba^{2+} current but retards the inactivation of the Ca^{2+} current (77).

The neuronal P/Q Ca^{2+} channel with its α_{1A} pore-forming subunit also manifests inactivation although the Ca^{2+}-dependent portion of its inactivation is small compared with that of α_{1C} described above. Using the C-terminal tail of α_{1A} as the bait, Lee et al. (88) fished out a *CAM* gene from a rat-brain cDNA library in a yeast two-hybrid screen (88). (Like forward genetics, such a library screen is free of preconceived notions, such as CaM involvement.) Deletion experiments, coupled with His-pulldown and co-immunoprecipitation, delimit the CaM-binding region (CBD) to 32 aa at the C-terminal tail. Interestingly, this sequence does not fall into the known categories of CaM-binding domains (10) but strongly resembles a region in the Ca^{2+}-CaM-stimulated adenylyl cyclase type 8. This region is necessary for both Ca^{2+}-dependent inactivation and facilitation of α_{1A} because its deletion removes both, and co-expression of the peptide with α_{1A} interferes with both. These two robust modulations of the P/Q-type Ca^{2+} channel are dependent on Ca^{2+} and the atypical CaM-binding domain (89). However, when the channel was examined under a Ca^{2+} buffer of intermediate strength, inactivation was lost but facilitation remained. This result again seems to indicate that the peptide-bound CaM has different actions when CaM is empty, partially, or fully loaded with Ca^{2+} (89). The function of CBD in CaM regulation is not yet entirely clear, however (90).

α_{1A} has an IQ-like motif similar to that of α_{1C}. DeMaria et al. (90) found that CaM_{1234} is dominant-negative when co-expressed with α_{1A} as with α_{1C}, leading to loss of inactivation and facilitation. Most interestingly, they found that the expression of the half declawed CaMs (CaM_{34} and CaM_{12}) produced completely selective elimination of facilitation and inactivation, respectively, although the effect of CaM_{12} on α_{1A} seems at odds with that on α_{1C} in regard to inactivation (see above). It, therefore, appears that the two lobes of CaM can somehow split the same local Ca^{2+} signal, "offering custom demodulation of domain Ca^{2+} in the initiation of separate functional sequelae" (91).

In short, we now know that the Ca^{2+}-dependent Ca^{2+} channel inactivation in *Paramecium*, first described by Brehm & Eckert in 1978 (32), is carried out through resident CaM(s). Although the detailed mechanism is not yet entirely clear, it promises to be illuminating in our understanding on how CaM interacts with many of its target proteins in several guises.

Others

A gap junction has two hemichannels (connexons), each with a hexamer of connexin subunits provided by the partners of two connecting cells. Connexin 32, a major connexin in liver, is modeled to traverse the membrane four times, with both its N- and C-terminal tails in the cytoplasm. Two regions are now known to bind to a fluorescent CaM derivative: one within the first 21 amino acids at the N terminus, and the other in a 15-amino acid region in the C-terminal tail immediately after the last transmembrane helix (92). Immunofluorescent data using antibodies against connexin 32 and CaM suggest an interaction between these molecules in HeLa cells (93). Antisense oligonucleotides to CaM messages block CO_2-induced uncoupling of native connexins in oocytes, although a mechanism is yet to be defined (94, 95). These results suggest that CaM regulates gap junction channels, although the exact mechanism for CaM as the Ca^{2+}-sensing gating element of the gap junction channel (93) awaits further study.

Schönherr et al. (96) found that human EAG channels (hEAG1) expressed in oocytes can be inhibited by Ca^{2+} ($IC_{50} = 100$ nM). Upon excision of membrane patches expressing hEAG1, the Ca^{2+} sensitivity is lost but regained upon CaM addition to the bathing solution. One CaM binding per tetrameric channel is apparently sufficient for channel inhibition. CaM overlay assay on C-terminal fragments revealed a CaM-binding domain between amino acid 673 and 770. Mutations created between R711 and R718 result in reduced Ca^{2+} sensitivity of hEAG1. This potential CaM-binding site is similar to those of SK and IK channels and does not conform to the known motifs (10). How widespread Ca^{2+}-CaM is inhibition of other EAG channels remains to be examined, although *Drosophila* EAG channels do not show any Ca^{2+}-CaM inhibition (96).

THE Ca^{2+} SENSOR AND MECHANICAL COUPLING

We have already seen cases where CaMs are detachable (freelance Ca^{2+} reporter) from some channels and are constitutive (resident) in others. The latter are channel subunits in the conventional sense, although here we invoke a looser definition of channel subunits to include ones that are detachable such as the dissociable subunits in G proteins. Unlike many enzymes, channels work on the sub-millisecond scale. Therefore, although we see CaM as freelance reporters to channels in some in vitro cases, resident CaMs might be required for fast responses upon Ca^{2+} signaling in vivo. CaM binding to domains such as IQ domains and CaMBD of SK2 does not require Ca^{2+}, thus allowing CaM to reside on the channel molecules at the resting physiological $[Ca^{2+}]_{in}$ prior to the stimulus. A word of caution: In some of the examples above, not all CaM bindings to channel peptides demonstrated in vitro necessarily correlate with obvious modulation of channel activity in vivo. In those cases, it is not clear whether they are causing changes too subtle to be detected or other CaM-like molecules bind to those sites in vivo. There are other Ca^{2+} sensors, e.g., KChIPs with Ca^{2+}-binding EF hands have been shown to control inactivation of voltage-dependent K channels (Kv4) (96a).

CaM can form complexes with more than one peptide (17, 19, 20). It is now clearly shown that CaM can act as a glue to bring subunits together (Figure 1). For both enzymes and channels, the possibility of CaM-dependent multimerization beyond dimers should also be considered. Figure 2A, *a* (see color insert) illustrates oligomerization of channel peptides through CaM binding upon a surge of Ca^{2+}. Whereas *b* of Figure 2A depicts dimerization, which could be between two identical subunits, as in SK2s CaMBD, *a* schematizes a tetrameric ring formation by CaM "biting one another's tail." Models that maintain the fourfold symmetry of channels such as K^+ channels after Ca^{2+}-CaM binding still remain tenable. Depending on the target architecture, other degrees of oligomerization as well as linking of two different subunits or entirely different molecules (Figure 2A, c) are also possible.

Schumacher et al. (20) postulated that upon binding Ca^{2+} at the N lobe, two CaMs bind two CaMBDs, which would create a rotational force transmitted to the gate area of S6 transmembrane domain to activate the channel. A coupling between Ca^{2+} stimulus and molecular-force generation through CaM-mediated oligomerization of subunits may be widespread among channels and perhaps enzymes. Together with the complex Ca^{2+}-sensing ability of CaM (see below), this may be a crucial feature for gating and modulation of the CaM-regulated channels. Detachable CaM may also provide yet another dimension to channel regulation.

The crystal structure of SK2s CaMBD with CaM (Figure 1) has demonstrated that CaM can indeed bind to peptides without Ca^{2+} binding. Thus SK2 activation requires Ca^{2+} binding only to the N-lobe of CaM. The hydrophobic interaction between the C-lobe and the $\alpha 1$ helix is so strong, that Ca^{2+} may not ever bind to the C-lobe in this case (20). The traditional wisdom that the C-lobe is the first to fill with Ca^{2+} and then the N-lobe upon Ca^{2+} flood does not seem to hold here. Similarly, ApoCaM can be tethered to peptides of L-type Ca^{2+} channels (87a–c) and those of RyR1 (57b). However, it is not clear whether bound CaMs are indeed Ca^{2+} free in vivo and also whether the anchored Ca^{2+}-free lobes can bind Ca^{2+} upon Ca^{2+} flood. Interestingly, although CaM (ApoCaM?) can be bound to P/Q type Ca^{2+} channels in the presence of low Ca^{2+}, CaM does bind Ca^{2+} at both N- and C-lobes, and Ca^{2+} filling is required for channel regulation (90). This illustrates the diversity of CaM channel regulation. We look forward to the solution of crystal structures of CaM bound to other channel peptides and to further methodological development for Ca^{2+}-CaM interactions in the presence of these peptides.

FUNCTIONAL BIPARTITION AND OPPOSING EFFECTS OF CaM

In the 1991 *Paramecium* redux, the most surprising finding is that randomly created mutants of two opposite behaviors both bear mutations in CaM, among other possible proteins (see above). Furthermore, the under-exciters have mutations only in the N-lobe and the over-exciters solely in the C-lobe of CaM. This segregation is absolute and highly statistically significant (29, 97–99). Thus CaM has a functional bipartition, i.e., the integrity of one lobe is important for one set of functions and that of the other lobe for a second set of functions, in particular the opposite functions of the former. The bipartition hypothesis is also supported by yeast research in vivo (100).

Functional bipartition of CaM has also surfaced in other channel research as reviewed above. In the L-type Ca^{2+} channels, CaM_{12}, which is defunct in Ca^{2+}

binding at EF hands 1 and 2, spares the inactivation of α1C but CaM$_{34}$ abolishes it (84). α1C's IQ domain binds both CaM$_{12}$ and CaM$_{34}$, but its CB sequence binds CaM$_{12}$ tightly and CaM$_{34}$ poorly (82). Thus the C-lobe of CaM is the effective switching mechanism for inactivation of the L-type Ca^{2+} channel, whereas the N-lobe may be an anchor at the CB peptide (87c). The P/Q type channel loses its inactivation but not its facilitation under a Ca^{2+}-buffer of intermediate strength (89). Co-expression of CaM$_{12}$ with α1C eliminates inactivation but not facilitation and that of CaM$_{34}$ eliminates facilitation but not inactivation (90). In SK2, both mutational (68) and structural analysis (20) showed incontrovertibly a gating function of the Ca^{2+}-binding N-lobe. The ability of CaM to bind Ca^{2+} into the C-lobe is apparently not required in this case. If we call the Ca^{2+}-free lobe of the resident CaM the anchor lobe and the lobe that binds Ca^{2+} to cause the needed conformational change the effector lobe, it is evident the this functional partition is not fixed with respect to the structural partition of CaM. In the case of α_{1C}, the N-lobe anchors, and the C-lobe is the effector. In SK2, the C-lobe anchors, and the N-lobe is the effector. Thus CaMs mutated in the N-lobe only fail in SK2 activation (68), whereas those mutated in the C-lobe only fail in α_{1C} inactivation (84). This situation is entirely analogous to the serendipitous findings in *Paramecium* where N-lobe CaM mutants fail in their Ca^{2+}-dependent Na^{+} channel, and C-lobe mutants fail in their Ca^{2+}-dependent K^{+} channel (29, 39, 97–99).

Mechanistically, how does CaM manage to actuate two opposing effects, particularly on individual pore-forming channel subunits such as the Ca^{2+} channel and RyR? What happens when the Ca^{2+} concentration in the vicinity of CaM changes in time? In the traditional view, CaM can function in two forms: ApoCaM and the fully calcified CaM$_{[Ca:1234]}$. This two-state model seems insufficient to explain the phenomena described above. Although it is probably unrealistic to have a single general mechanism for all of the CaM actions, we would like to consider two points for a general framework.

First, it seems necessary to consider the role of secondary interaction sites in the target protein. In the structure of CaM$_{[Ca:1234]}$ bound to a Baa helix (8, 9), of the 148 residues of CaM, only approximately 40 residues form bonds with the Baa helix. This leaves open the possibility of bonding elsewhere with the remaining 100 or so residues of CaM, and some apparently do affect enzymatic activities (101, 102). Furthermore, the co-crystal structure (Figure 1) reveals that coordination between SK2's CaMBD and CaM takes place with some residues at the end of the first EF hand of the N-lobe (20), sites for binding previously unknown. Also, it shows that CaM binds three helices of CaMBD from two subunits (20). These findings have broadened our view on how CaM interacts with target peptides beyond the more commonly known binding modes.

Second, it also seems reasonable to invoke functions of half Ca^{2+}-loaded CaMs. Halved CaMs appear to regulate various enzymes (103–106). In the absence of enzymes or channels, stable CaM$_{[Ca:34]}$ does exist (6). It is the form believed to effect α_{1C} inactivation (87c), thus it is natural to assume that it functions in situ. Stable CaM$_{[Ca:12]}$ in a complex with peptides is observed in the crystal structure of

SK2 CaMBD-CaM (Figure 1) and is incorporated into a scheme on the regulation of the P/Q-type Ca^{2+} channel (90).

Figure 2B may be a useful device toward model building. This drawing is meant to provide a general framework to explain the functional bipartition hypothesis and temporal sequences. Here, we envision that the regulatory portion of the channel or enzyme has a primary binding site (*filled bar*) (10, 12). We also assume that secondary bindings (differently notched symbols in lock-and-key fits) can occur on the surface of CaM, as well as at its two major hydrophobic patches exposed upon Ca^{2+} binding and used for primary target binding. Whereas the conventional view is that CaM can be an all-or-none (0 or 4 Ca^{2+}-loaded) two-way switch, we postulate that it is a three-way switch (0, 4, or 2 Ca^{2+}). Upon a local Ca^{2+} flood, the fully loaded $CaM_{[Ca:1234]}$ places this regulatory domain in one configuration (Figure 2B, b). Such a domain can be formed by a single peptide or multiple peptides from the same or different subunits. After local Ca^{2+} is reduced through inactivation or closure of the Ca^{2+} source, CaM in situ becomes half Ca^{2+}-filled and reconfigures the domain (Figure 2B, c). As a result, the domain can exist in three configurations: a, b, or c, and the channel or enzyme in three functional states. Depending on the prior configuration, we postulate that the alternative half Ca^{2+}-filled CaM can dictate another form and a, b, d can form a second three-way switch. The a-b-c and the a-b-d switch in Figure 2B can be in separate proteins of opposite effect, e.g., the depolarizing Ca^{2+}-dependent Na^+ channel and the repolarizing Ca^{2+}-dependent K^+ channel of *Paramecium*. A conglomerate a-b-c-d four-way switch has been proposed to regulate the same domain and thereby afford the P/Q type Ca^{2+} channel four physiological states: resting, open, inactivated, and facilitated (90). The three-way switch is reduced to two-way when one lobe that refuses Ca^{2+} can bind target constitutively in SK2 (modified c-b in Figure 2B) (20).

CONCLUDING REMARKS

How CaM coordinates global as well as local Ca^{2+} actions in the cytoplasm is not understood. Especially challenging is to understand how opposite actions are sequentially timed by Ca^{2+} to produce waves. These actions can take place within a single protein, for instance, of the Ca^{2+} channel or of the ryanodine receptor. Opposite actions can also be carried out by functionally opposing proteins, such as the *Paramecium* Na^+ and K^+ channels. An enigma in calmodulin research has been its activation of opposing enzyme pairs: protein kinase-phosphatase, nucleotide cyclase-phosphodiesterase (21). CaM is not the only Ca^{2+} sensor with EF hands (107, 108). Ultimately, we also need to understand why CaM alone is used in so many important Ca^{2+}-dependent signaling pathways and why CaM uses two (not one or more than two) fairly independent Ca^{2+}-binding lobes. The hallmark of Ca^{2+} actions is their localness and transiency. Modern channel research is uniquely capable of interrogating molecules individually or in a synchronized population and can do so at sub-millisecond time. This research has now been coupled to forward and reverse genetics. With the success of co-crystallization of CaM and its binding peptide of SK2, we look forward to more atomic solutions of how CaM manipulates the cytoplasmic handles of various enzymes and channels and to an understanding of how our two-armed four-handed friend manages to play the concerti of life.

ACKNOWLEDGMENT

We thank Dr. Robin Preston for his critical reading of this review and helpful suggestions. Work in our laboratory has been supported grants from the National Institutes of Health.

Visit the Annual Reviews home page at www.AnnualReviews.org

LITERATURE CITED

1. Cohen P, Klee CB. 1988. *Calmodulin.* Amsterdam: Elsevier. 371 pp.
2. Toutenhoofd SL, Strehler EE. 2000. The calmodulin multigene family as a unique case of genetic redundancy: multiple levels of regulation to provide spatial and temporal control of calmodulin pools? *Cell Calcium* 28:83–96
3. Santella L, Carafoli E. 1997. Calcium signaling in the cell nucleus. *FASEB J.* 11:1091–109. Erratum *FASEB J.* 1997. 11(14)1330
4. Babu YS, Sack JS, Greenhough TJ, Bugg CE, Means AR, Cook WJ. 1985. Three-dimensional structure of calmodulin. *Nature* 315:37–40
5. Jurado LA, Chockalingam PS, Jarrett HW. 1999. Apocalmodulin. *Physiol. Rev.* 79:661–82
6. Klevit RE, Dalgarno DC, Levine BA, Williams RJ. 1984. ^{1}H-NMR studies of calmodulin. The nature of the Ca^{2+}-dependent conformational change. *Eur. J. Biochem.* 139:109–14
7. Wriggers W, Mehler E, Pitici F, Weinstein H, Schulten K. 1998. Structure and dynamics of calmodulin in solution. *Biophys. J.* 74:1622–39
8. Meador WE, Means AR, Quiocho FA. 1992. Target enzyme recognition by

calmodulin: 2.4 Å structure of a calmodulin-peptide complex. *Science* 257:1251–55
9. Ikura M, Clore GM, Gronenborn AM, Zhu G, Klee CB, Bax A. 1992. Solution structure of a calmodulin-target peptide complex by multidimensional NMR. *Science* 256:632–38
10. Rhoads AR, Friedberg F. 1997. Sequence motifs for calmodulin recognition. *FASEB J.* 11:331–40
11. Yap KL, Kim J, Truong K, Sherman M, Yuan T, Ikura M. 2000. Calmodulin target database. *J. Struct. Funct. Genomics* 1:8–14
12. Swindells MB, Ikura M. 1996. Preformation of the semi-open conformation by the apo-calmodulin C-terminal domain and implications for binding IQ-motifs. *Nat. Struct. Biol.* 3:501–4
13. Dasgupta M, Honeycutt T, Blumenthal DK. 1989. The gamma-subunit of skeletal muscle phosphorylase kinase contains two noncontiguous domains that act in concert to bind calmodulin. *J. Biol. Chem.* 264:17156–63
14. Dasgupta M, Blumenthal DK. 1995. Characterization of the regulatory domain of the gamma-subunit of phosphorylase kinase. The two noncontiguous

calmodulin-binding subdomains are also autoinhibitory. *J. Biol. Chem.* 270:22283–89

15. Ruan J, Xie Q, Hutchinson N, Cho H, Wolfe GC, Nathan C. 1996. Inducible nitric oxide synthase requires both the canonical calmodulin-binding domain and additional sequences in order to bind calmodulin and produce nitric oxide in the absence of free Ca^{2+}. *J. Biol. Chem.* 271:22679–86

16. Su Z, Blazing MA, Fan D, George SE. 1995. The calmodulin–nitric oxide synthase interaction. Critical role of the calmodulin latch domain in enzyme activation. *J. Biol. Chem.* 270:29117–22

17. Yuan T, Vogel HJ. 1998. Calcium-calmodulin-induced dimerization of the carboxyl-terminal domain from petunia glutamate decarboxylase. A novel calmodulin-peptide interaction motif. *J. Biol. Chem.* 273:30328–35

18. Barth A, Martin SR, Bayley PM. 1998. Specificity and symmetry in the interaction of calmodulin domains with the skeletal muscle myosin light chain kinase target sequence. *J. Biol. Chem.* 273:2174–83

19. Onions J, Hermann S, Grundstrom T. 2000. A novel type of calmodulin interaction in the inhibition of basic helix-loop-helix transcription factors. *Biochemistry* 39:4366–74

20. Schumacher MA, Rivard AF, Bachinger HP, Adelman JP. 2001. Structure of the gating domain of a Ca^{2+}-activated K$^+$ channel complexed with Ca^{2+}/calmodulin. *Nature* 410:1120–24

21. James P, Vorherr T, Carafoli E. 1995. Calmodulin-binding domains: just two faced or multi-faceted? *Trends Biochem.* 20:38–42

22. Chin D, Means AR. 2000. Calmodulin: a prototypical calcium sensor. *Trends Cell Biol.* 10:322–8

23. Kung C, Preston RR, Maley ME, Ling KY, Kanabrocki JA, et al. 1992. In vivo *Paramecium* mutants show that calmod-

ulin orchestrates membrane responses to stimuli. *Cell Calcium* 13:413–25

24. Davis TN, Thorner J. 1989. Vertebrate and yeast calmodulin, despite significant sequence divergence, are functionally interchangeable. *Proc. Natl. Acad. Sci. USA* 86:7909–13

25. Geiser JR, van Tuinen D, Brockerhoff SE, Neff MM, Davis TN. 1991. Can calmodulin function without binding calcium? *Cell* 65:949–59

26. Heiman RG, Atkinson RC, Andruss BF, Bolduc C, Kovalick GE, Beckingham K. 1996. Spontaneous avoidance behavior in *Drosophila* null for calmodulin expression. *Proc. Natl. Acad. Sci. USA* 93:2420–25

27. Moser MJ, Lee SY, Klevit RE, Davis TN. 1995. Ca^{2+} binding to calmodulin and its role in *Schizosaccharomyces pombe* as revealed by mutagenesis and NMR spectroscopy. *J. Biol. Chem.* 270:20643–52

28. Arredondo L, Nelson HB, Beckingham K, Stern M. 1998. Increased transmitter release and aberrant synapse morphology in a *Drosophila* calmodulin mutant. *Genetics* 150:265–74

29. Kink JA, Maley ME, Preston RR, Ling KY, Wallen-Friedman MA, et al. 1990. Mutations in Paramecium calmodulin indicate functional differences between the C-terminal and N-terminal lobes in vivo. *Cell* 62:165–74

30. Ohya Y, Botstein D. 1994. Structure-based systematic isolation of conditional-lethal mutations in the single yeast calmodulin gene. *Genetics* 138:1041–54

30a. Zhu H, Bilgin M, Banham R, Hall D, Casamayor A, et al. 2001. Global analysis of protein activities using proteome chips. *Science* 293:2101–5

31. Meech RW, Standen NB. 1974. Calcium-mediated potassium activation in *Helix* neurones. *J. Physiol.* 237:43P–44P

32. Brehm P, Eckert R. 1978. Calcium entry leads to inactivation of calcium channel in *Paramecium. Science* 202:1203–6

33. Levitan IB. 1999. It is calmodulin after

all! Mediator of the calcium modulation of multiple ion channels. *Neuron* 22:645–48

34. Klaerke DA, Petersen J, Jorgensen PL. 1987. Purification of Ca^{2+}-activated K^+ channel protein on calmodulin affinity columns after detergent solubilization of luminal membranes from outer renal medulla. *FEBS Lett.* 216:211–16

35. Klaerke DA. 1995. Purification and characterization of epithelial Ca^{2+}-activated K^+ channels. *Kidney Int.* 48:1047–56

36. Smith JS, Rousseau E, Meissner G. 1989. Calmodulin modulation of single sarcoplasmic reticulum Ca^{2+}-release channels from cardiac and skeletal muscle. *Circ. Res.* 64:352–59

37. Saimi Y, Hinrichsen RD, Forte M, Kung C. 1983. Mutant analysis shows that the Ca^{2+}-induced K^+ current shuts off one type of excitation in *Paramecium. Proc. Natl. Acad. Sci. USA* 80:5112–16

38. Hinrichsen RD, Burgess-Cassler A, Soltvedt BC, Hennessey T, Kung C. 1986. Restoration by calmodulin of a Ca^{2+}-dependent K^+ current missing in a mutant of *Paramecium. Science* 232:503–6

39. Saimi Y, Ling KY. 1990. Calmodulin activation of calcium-dependent sodium channels in excised membrane patches of *Paramecium. Science* 249:1441–44

40. Hsu YT, Molday RS. 1993. Modulation of the cGMP-gated channel of rod photoreceptor cells by calmodulin. *Nature* 361:76–79

41. Chen TY, Illing M, Molday LL, Hsu YT, Yau KW, Molday RS. 1994. Subunit 2 (or beta) of retinal rod cGMP-gated cation channel is a component of the 240-kDa channel-associated protein and mediates Ca^{2+}-calmodulin modulation. *Proc. Natl. Acad. Sci. USA* 91:11757–61

42. Grunwald ME, Yu WP, Yu HH, Yau KW. 1998. Identification of a domain on the beta-subunit of the rod cGMP-gated cation channel that mediates inhibition by calcium-calmodulin. *J. Biol. Chem.* 273:9148–57

43. Weitz D, Zoche M, Muller F, Beyermann M, Korschen HG, et al. 1998. Calmodulin controls the rod photoreceptor CNG channel through an unconventional binding site in the N-terminus of the beta-subunit. *EMBO J.* 17:2273–84

44. Grunwald ME, Zhong H, Lai J, Yau KW. 1999. Molecular determinants of the modulation of cyclic nucleotide-activated channels by calmodulin. *Proc. Natl. Acad. Sci. USA* 96:13444–49

45. Hackos DH, Korenbrot JI. 1997. Calcium modulation of ligand affinity in the cyclic GMP-gated ion channels of cone photoreceptors. *J. Gen. Physiol.* 110:515–28

46. Liu M, Chen TY, Ahamed B, Li J, Yau KW. 1994. Calcium-calmodulin modulation of the olfactory cyclic nucleotide-gated cation channel. *Science* 266:1348–54

47. Varnum MD, Zagotta WN. 1997. Interdomain interactions underlying activation of cyclic nucleotide-gated channels. *Science* 278:110–13

48. Ehlers MD, Zhang S, Bernhadt JP, Huganir RL. 1996. Inactivation of NMDA receptors by direct interaction of calmodulin with the NR1 subunit. *Cell* 84:745–55

49. Zhang S, Ehlers MD, Bernhardt JP, Su CT, Huganir RL. 1998. Calmodulin mediates calcium-dependent inactivation of *N*-methyl-D-aspartate receptors. *Neuron* 21:443–53

50. Krupp JJ, Vissel B, Thomas CG, Heinemann SF, Westbrook GL. 1999. Interactions of calmodulin and alpha-actinin with the NR1 subunit modulate Ca^{2+}-dependent inactivation of NMDA receptors. *J. Neurosci.* 19:1165–78

51. Rosenmund C, Westbrook GL. 1993. Calcium-induced actin depolymerization reduces NMDA channel activity. *Neuron* 10:805–14

52. Wyszynski M, Lin J, Rao A, Nigh E, Beggs AH, et al. 1997. Competitive binding of alpha-actinin and calmodulin to

the NMDA receptor. *Nature* 385:439–42

53. Rodney GG, Williams BY, Strasburg GM, Beckingham K, Hamilton SL. 2000. Regulation of RYR1 activity by Ca^{2+} and calmodulin. *Biochemistry* 39:7807–12

54. Meissner G. 1986. Evidence of a role for calmodulin in the regulation of calcium release from skeletal muscle sarcoplasmic reticulum. *Biochemistry* 25:244–51

55. Wagenknecht T, Radermacher M, Grassucci R, Berkowitz J, Xin HB, Fleischer S. 1997. Locations of calmodulin and FK506-binding protein on the three-dimensional architecture of the skeletal muscle ryanodine receptor. *J. Biol. Chem.* 272:32463–71

56. Moore CP, Rodney G, Zhang JZ, Santacruz-Toloza L, Strasburg G, Hamilton SL. 1999. Apocalmodulin and Ca^{2+} calmodulin bind to the same region on the skeletal muscle Ca^{2+} release channel. *Biochemistry* 38:8532–37

57. Takeshima H, Nishimura S, Matsumoto T, Ishida H, Kangawa K, et al. 1989. Primary structure and expression from complementary DNA of skeletal muscle ryanodine receptor. *Nature* 339:439–45

57a. Yamaguchi N, Xin C, Meissner G. 2001. Identification of apocalmodulin and Ca^{2+}-calmodulin regulatory domain in skeletal muscle Ca^{2+} release channel, ryanodine receptor. *J. Biol. Chem.* 276:22579–85

57b. Rodney GG, Moore CP, Williams BY, Zhang JZ, Krol J, et al. 2001. Calcium binding to calmodulin leads to an N-terminal shift in its binding site on the ryanodine receptor. *J. Biol. Chem.* 276:2069–74

57c. Balshaw DM, Xu L, Yamaguchi N, Pasek DA, Meissner G. 2001. Calmodulin binding and inhibition of cardiac muscle calcium release channel (ryanodine receptor). *J. Biol. Chem.* 276:20144–53

57d. Fruen BR, Bardy JM, Byrem TM, Strasburg GM, Louis CF. 2000. Differential Ca^{2+} sensitivity of skeletal and cardiac muscle ryanodine receptors in the pres-

ence of calmodulin. *Am. J. Physiol. Cell Physiol.* 279:C724–C33

58. Zhang JZ, Wu Y, Williams BY, Rodney G, Mandel F, et al. 1999. Oxidation of the skeletal muscle Ca^{2+} release channel alters calmodulin binding. *Am. J. Physiol. Cell Physiol.* 276:C46–C53

59. Eu JP, Sun J, Xu L, Stamler JS, Meissner G. 2000. The skeletal muscle calcium release channel: coupled O_2 sensor and NO signaling functions. *Cell* 102:499–509

60. Berridge MJ. 1993. Inositol trisphosphate and calcium signalling. *Nature* 361:315–25

61. Yamada M, Miyawaki A, Saito K, Nakajima T, Yamamoto-Hino M, et al. 1995. The calmodulin-binding domain in the mouse type 1 inositol 1,4,5–trisphosphate receptor. *Biochem. J.* 308:83–88

62. Michikawa T, Hirota J, Kawano S, Hiraoka M, Yamada M, et al. 1999. Calmodulin mediates calcium-dependent inactivation of the cerebellar type 1 inositol 1,4,5–trisphosphate receptor. *Neuron* 23:799–808

63. Sipma H, De Smet P, Sienaert I, Vanlingen S, Missiaen L, et al. 1999. Modulation of inositol 1,4,5-trisphosphate binding to the recombinant ligand-binding site of the type-1 inositol 1,4, 5-trisphosphate receptor by Ca^{2+} and calmodulin. *J. Biol. Chem.* 274:12157–62

64. Adkins CE, Morris SA, De Smedt H, Sienaert I, Torok K, Taylor CW. 2000. Ca^{2+}-calmodulin inhibits Ca^{2+} release mediated by type-1, -2 and -3 inositol trisphosphate receptors. *Biochem. J.* 345:357–63

65. Patel S, Morris SA, Adkins CE, O'Beirne G, Taylor CW. 1997. Ca^{2+}-independent inhibition of inositol trisphosphate receptors by calmodulin: redistribution of calmodulin as a possible means of regulating Ca^{2+} mobilization. *Proc. Natl. Acad. Sci. USA* 94:11627–32

66. Hagar RE, Burgstahler AD, Nathanson MH, Ehrlich BE. 1998. Type III InsP3 receptor channel stays open in the presence

of increased calcium. *Nature* 396:81–84

67. Xia XM, Fakler B, Rivard A, Wayman G, Johnson-Pais T, et al. 1998. Mechanism of calcium gating in small-conductance calcium-activated potassium channels. *Nature* 395:503–7

68. Keen JE, Khawaled R, Farrens DL, Neelands T, Rivard A, et al. 1999. Domains responsible for constitutive and Ca^{2+}-dependent interactions between calmodulin and small conductance Ca^{2+}-activated potassium channels. *J. Neurosci.* 19:8830–38

69. Fanger CM, Ghanshani S, Logsdon NJ, Rauer H, Kalman K, et al. 1999. Calmodulin mediates calcium-dependent activation of the intermediate conductance KCa channel, IKCa1. *J. Biol. Chem.* 274:5746–54

70. Phillips AM, Bull A, Kelly LE. 1992. Identification of a *Drosophila* gene encoding a calmodulin-binding protein with homology to the trp phototransduction gene. *Neuron* 8:631–42

71. Warr CG, Kelly LE. 1996. Identification and characterization of two distinct calmodulin-binding sites in the Trpl ion-channel protein of *Drosophila melanogaster*. *Biochem. J.* 314:497–503

72. Chevesich J, Kreuz AJ, Montell C. 1997. Requirement for the PDZ domain protein, INAD, for localization of the TRP store-operated channel to a signaling complex. *Neuron* 18:95–105

73. Scott K, Sun Y, Beckingham K, Zuker CS. 1997. Calmodulin regulation of Drosophila light-activated channels and receptor function mediates termination of the light response in vivo. *Cell* 91:375–83

74. Lan L, Bawden MJ, Auld AM, Barritt GJ. 1996. Expression of *Drosophila* trpl cRNA in *Xenopus laevis* oocytes leads to the appearance of a Ca^{2+} channel activated by Ca^{2+} and calmodulin, and by guanosine 5′ [gamma-thio]triphosphate. *Biochem. J.* 316:793–8034

75. Lan L, Brereton H, Barritt GJ. 1998.

The role of calmodulin-binding sites in the regulation of the *Drosophila* TRPL cation channel expressed in *Xenopus laevis* oocytes by Ca^{2+}, inositol 1,4,5-trisphosphate and GTP-binding proteins. *Biochem. J.* 330:1149–58

76. Estacion M, Sinkins WG, Schilling WP. 1999. Stimulation of *Drosophila* TrpL by capacitative Ca^{2+} entry. *Biochem. J.* 341:41–49

76a. Trost C, Bergs C, Himmerkus N, Flockerzi V. 2001. The transient receptor potential, TRP4, cation channel is a novel member of the family of calmodulin binding proteins. *Biochem. J.* 355:663–70

76b. Tang J, Lin Y, Zhang Z, Tikunova S, Birnbaumer L, Zhu MX. 2001. Identification of common binding sites for calmodulin and inositol 1,4,5-trisphosphate receptors on the carboxyl termini of trp channels. *J. Biol. Chem.* 276:21303–10

76c. Kiselyov K, Mignery GA, Zhu MX, Muallem S. 1999. The N-terminal domain of the IP_3 receptor gates store-operated hTrp3 channels. *Mol. Cell* 4:423–29

76d. Boulay G, Brown DM, Qin N, Jiang M, Dietrich A, et al. 1999. Modulation of Ca^{2+} entry by polypeptides of the inositol 1,4, 5-trisphosphate receptor (IP_3R) that bind transient receptor potential (TRP): evidence for roles of TRP and IP_3R in store depletion-activated Ca^{2+} entry. *Proc. Natl. Acad. Sci. USA* 96:14955–60

76e. Zhang Z, Tang J, Tikunova S, Johnson JD, Chen Z, et al. 2001. Activation of Trp3 by inositol 1,4,5-triphosphate receptors through displacement of inhibitory calmodulin from a common binding domain. *Proc. Natl. Acad. Sci. USA* 98:3168–73

77. Ivanina T, Blumenstein Y, Shistik E, Barzilai R, Dascal N. 2000. Modulation of L-type Ca^{2+} channels by beta gamma and calmodulin via interactions with N and C termini of alpha 1C. *J. Biol. Chem.* 275:39846–54

78. de Leon M, Wang Y, Jones L, Perez-Reyes E, Wei X, et al. 1995. Essential

Ca^{2+}-binding motif for Ca^{2+}-sensitive inactivation of L-type Ca^{2+} channels. *Science* 270:1502–6

79. Zhou J, Olcese R, Qin N, Noceti F, Birnbaumer L, Stefani E. 1997. Feedback inhibition of Ca^{2+} channels by Ca^{2+} depends on a short sequence of the C terminus that does not include the Ca^{2+}-binding function of a motif with similarity to Ca^{2+}-binding domains. *Proc. Natl. Acad. Sci. USA* 94:2301–5

80. Soldatov NM, Zühlke RD, Bouron A, Reuter H. 1997. Molecular structures involved in L-type calcium channel inactivation. Role of the carboxyl-terminal region encoded by exons 40–42 in alpha1C subunit in the kinetics and Ca^{2+} dependence of inactivation. *J. Biol. Chem.* 272:3560–66

81. Zühlke RD, Reuter H. 1998. Ca^{2+}-sensitive inactivation of L-type Ca^{2+} channels depends on multiple cytoplasmic amino acid sequences of the alpha1C subunit. *Proc. Natl. Acad. Sci. USA* 95:3287–94

82. Zühlke RD, Pitt GS, Deisseroth K, Tsien RW, Reuter H. 1999. Calmodulin supports both inactivation and facilitation of L-type calcium channels. *Nature* 399:159–62

83. Qin N, Olcese R, Bransby M, Lin T, Birnbaumer L. 1999. Ca^{2+}-induced inhibition of the cardiac Ca^{2+} channel depends on calmodulin. *Proc. Natl. Acad. Sci. USA* 96:2435–38

84. Peterson BZ, DeMaria CD, Adelman JP, Yue DT. 1999. Calmodulin is the Ca^{2+} sensor for Ca^{2+}-dependent inactivation of L-type calcium channels. *Neuron* 22:549–58. Erratum *Neuron* 1999. 22(4):893

85. Zühlke RD, Pitt GS, Tsien RW, Reuter H. 2000. Ca^{2+}-sensitive inactivation and facilitation of L-type Ca^{2+} channels both depend on specific amino acid residues in a consensus calmodulin- binding motif in the(alpha)1C subunit. *J. Biol. Chem.* 275:21121–29

86. Peterson BZ, Lee JS, Mulle JG, Wang Y, de Leon M, Yue DT. 2000. Critical determinants of Ca^{2+}-dependent inactivation within an EF-hand motif of L-type Ca^{2+} channels. *Biophys. J.* 78:1906–20

87. Pate P, Mochca-Morales J, Wu Y, Zhang JZ, Rodney GG, et al. 2000. Determinants for calmodulin binding on voltage-dependent Ca^{2+} channels. *J. Biol. Chem.* 275:39786–92

87a. Romanin C, Gamsjaeger R, Kahr H, Schaufler D, Carlson O, et al. 2000. Ca^{2+} sensors of L-type Ca^{2+} channel. *FEBS Lett.* 487:301–6

87b. Mouton J, Feltz A, Maulet Y. 2001. Interactions of calmodulin with two peptides derived from the C-terminal cytoplasmic domain of the Cav1.2 Ca^{2+} channel provide evidence for a molecular switch involved in Ca^{2+}-induced inactivation. *J. Biol. Chem.* 276:22359–67

87c. Pitt GS, Zühlke RD, Hudmon A, Schulman H, Reuter H, Tsien RW. 2001. Molecular basis of calmodulin tethering and Ca^{2+}-dependent inactivation of L-type Ca^{2+} channels. *J. Biol. Chem.* 276:30794–802

88. Lee A, Wong ST, Gallagher D, Li B, Storm DR, et al. 1999. Ca^{2+}/calmodulin binds to and modulates P/Q-type calcium channels. *Nature* 399:155–59

89. Lee A, Scheuer T, Catterall WA. 2000. Ca^{2+}/calmodulin-dependent facilitation and inactivation of P/Q-type Ca^{2+} channels. *J. Neurosci.* 20:6830–38

90. DeMaria CD, Soong TW, Alseikham BA, Alvania RS, Yue DT. 2001. Calmodulin bifurcates the local Ca^{2+} signal that modulates P/Q-type Ca^{2+} channels. *Nature* 411:484–89

91. DeMaria CD, Soong TW, Alseikham BA, Alvania RS, Yue DT. 2001. Calmodulin bifurcates the local Ca^{2+} signal that modulates P/Q-type Ca channels. *Biophys. J.* 80:197a (Abstr.)

92. Torok K, Stauffer K, Evans WH. 1997. Connexin 32 of gap junctions contains two cytoplasmic calmodulin-binding domains. *Biochem. J.* 326:479–83

93. Peracchia C, Sotkis A, Wang XG, Peracchia LL, Persechini A. 2000. Calmodulin directly gates gap junction channels. *J. Biol. Chem.* 275:26220–24

94. Peracchia C, Wang X, Li L, Peracchia LL. 1996. Inhibition of calmodulin expression prevents low-pH-induced gap junction uncoupling in *Xenopus* oocytes. *Pflügers Arch.* 431:379–87

95. Peracchia C, Wang XG, Peracchia LL. 2000. Slow gating of gap junction channels and calmodulin. *J. Membr. Biol.* 178: 55–70

96. Schönherr R, Lober K, Heinemann SH. 2000. Inhibition of human *ether à go-go* potassium channels by Ca²⁺/calmodulin. *EMBO J.* 19:3263–71

96a. An WF, Bowlby MR, Betty M, Cao J, Ling HP, et al. 2000. Modulation of A-type potassium channels by a family of calcium sensors. *Nature* 403:553–56

97. Ling KY, Preston RR, Burns R, Kink JA, Saimi Y, Kung C. 1992. Primary mutations in calmodulin prevent activation of the Ca²⁺-dependent Na⁺ channel in *Paramecium*. *Proteins* 12:365–71

98. Ling KY, Maley ME, Preston RR, Saimi Y, Kung C. 1994. New non-lethal calmodulin mutations in *Paramecium*. A structural and functional bipartition hypothesis. *Eur. J. Biochem.* 222:433–39

99. Sorenson BR, Eppel J-T, Shea MA. 2001. *Paramecium* calmodulin mutants defective in ion channel regulation associate with melittin in the absence of calcium but require it for tertiary collapse. *Biochemistry* 40:896–903

100. Ohya Y, Botstein D. 1994. Diverse essential functions revealed by complementing yeast calmodulin mutants. *Science* 263:963–66

101. VanBerkum MF, Means AR. 1991. Three amino acid substitutions in domain I of calmodulin prevent the activation of chicken smooth muscle myosin light chain kinase. *J. Biol. Chem.* 266:21488–95

102. Persechini A, Gansz KJ, Paresi RJ. 1996. A role in enzyme activation for the N-terminal leader sequence in calmodulin. *J. Biol. Chem.* 271:19279–82

103. Klee CB. 1988. Interaction of calmodulin with Ca²⁺ and target proteins. In *Calmodulin*, ed. P Cohen, CB Klee, pp. 35–56. Amsterdam: Elsevier

104. Persechini A, McMillan K, Leakey P. 1994. Activation of myosin light chain kinase and nitric oxide synthase activities by calmodulin fragments. *J. Biol. Chem.* 269:16148–54

105. Persechini A, Gansz KJ, Paresi RJ. 1996. Activation of myosin light chain kinase and nitric oxide synthase activities by engineered calmodulins with duplicated or exchanged EF hand pairs. *Biochemistry* 35:224–28

106. Sun H, Squier TC. 2000. Ordered and cooperative binding of opposing globular domains of calmodulin to the plasma membrane Ca-ATPase. *J. Biol. Chem.* 275:1731–38

107. Ikura M. 1996. Calcium binding and conformational response in EF-hand proteins. *Trends Biochem. Sci.* 21:14–17

108. Nakayama S, Kawasaki H, Kretsinger RH. 2000. Evolution of EF-hand proteins. In *Calcium Homeostasis*, ed. E Carafoli, J Krebs, pp. 29–58. Berlin: Springer

Annu. Rev. Physiol. 2002. 64:313–53

STRUCTURE AND FUNCTION OF DENDRITIC SPINES

Esther A. Nimchinsky, Bernardo L. Sabatini, and Karel Svoboda

Howard Hughes Medical Institute, Cold Spring Harbor Laboratory, Cold Spring Harbor, New York 11724; e-mail: nimchins@cshl.org, sabatini@cshl.org, svoboda@cshl.org

Key Words synapse, NMDA receptor, voltage-sensitive calcium channel, fragile X, calcium

■ **Abstract** Spines are neuronal protrusions, each of which receives input typically from one excitatory synapse. They contain neurotransmitter receptors, organelles, and signaling systems essential for synaptic function and plasticity. Numerous brain disorders are associated with abnormal dendritic spines. Spine formation, plasticity, and maintenance depend on synaptic activity and can be modulated by sensory experience. Studies of compartmentalization have shown that spines serve primarily as biochemical, rather than electrical, compartments. In particular, recent work has highlighted that spines are highly specialized compartments for rapid large-amplitude Ca^{2+} signals underlying the induction of synaptic plasticity.

INTRODUCTION

Spines are membranous protrusions from the neuronal surface. They consist of a head (volume $\sim$0.001–1 μm^3) connected to the neuron by a thin (diameter <0.1 μm) spine neck (Figure 1; reviewed in 1). They may arise from the soma, dendrites, or even the axon hillock, and they are found in various neuronal populations in all vertebrates and some invertebrates (2–4). More than 90% of excitatory synapses terminate on spines; the human brain thus contains >10^{13} spines. However, despite their abundance and evident importance, we are in some respects only beginning to appreciate their complexity and understand their function. In this review, we discuss first the structure and structural plasticity of dendritic spines, their changes in certain diseases, and then recent concepts touching on their probable function.

Technical advances have driven our understanding of the morphology and plasticity of dendritic spines. Cajal's early discovery and description (5) was based on material stained with the Golgi method. Most of the subsequent studies of dendritic spines, until rather recently, have been done in tissue stained in this way. Such methods have led to the discovery of abnormalities in dendritic spine morphologies associated with a variety of brain disorders (reviewed in 6). The advent of electron microscopy, and particularly of the ability to reconstruct dendrites and spines from

0066-4278/02/0315-0313$14.00

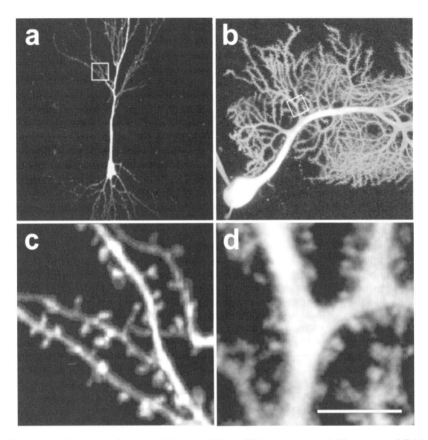

Figure 1   Structure of spines. Spiny dendrites of living neurons. A hippocampal CA1 neuron filled with calcein imaged using 2PLSM is shown at low magnification in (*a*). The region shown at higher magnification in (*b*) is indicated by a box. Note the heterogeneity in spine morphologies apparent at higher magnification. (*c*) A Purkinje neuron loaded with fluorescein dextran in vitro. The electrode is still attached to the soma to the left. Panel (*d*) shows a dendritic segment from this Purkinje cell at high magnification. Note the very-high spine density and the smaller range of spine morphologies. Z-stacks spaced 1 μm apart were collected, and the maximal projection is shown here. Scale bar $= 100 \ \mu$m (*a*), 7 μm (*b*), 40 μm (*c*), 5 μm (*d*).

serial sections, has allowed the analysis of the three-dimensional ultrastructure and actual density of spines on dendrites (1).

Although ultrastructural methods provided the foundation of our knowledge of the spine, they cultivated a rather static view. The development of modern digital microscopy permitted the study of living neurons rendered visible with fluorescent molecules. Most recently, 2-photon laser scanning microscopy (2PLSM) has been especially useful for imaging spines in living slices and in vivo (reviewed in 7). These imaging experiments have also allowed time-lapse studies of spine structural plasticity in vitro and in vivo, which reveals that spine growth and structural plasticity are associated with synaptic plasticity in vitro (reviewed in 8) and experience-dependent plasticity in vivo (9).

Finally, reporters of cellular function, such as Ca^{2+} indicators have permitted a close look at the function of spines in intact living tissues (reviewed in 10). Such studies have directly shown that spines are primarily biochemical compartments and shape a rich repertoire of $[Ca^{2+}]$ signals (reviewed in 11). These signals control the induction of synaptic plasticity.

This review focuses on experiments performed on dendritic spines of Purkinje cells of the cerebellum and pyramidal neurons of the hippocampus and neocortex. This choice is not based only on the fact that these represent the most thoroughly studied spines. Although Purkinje and pyramidal cell spines are similar in some respects in that they are thorn-like protrusions from dendrites that bear excitatory synapses, they differ in size, shape, distribution, development, complement of organelles, the receptors they bear, and the way they handle calcium. It is therefore preferable not to conflate them but to consider the two types of spines separately and to keep in mind that their differences can teach us as much as their similarities.

SPINE STRUCTURE

Structure in the Adult Brain

The spine is a structure specialized for synaptic transmission. Neurotransmitter receptors are largely restricted to the surface of the spine and concentrated close to the presynaptic element (12). This zone is conspicuously indicated by the postsynaptic density (PSD), a membrane-associated disc of electron dense material (13). The PSD consists of the receptors, channels, and signaling systems involved in synaptic transmission and the coupling of synaptic activity to postsynaptic biochemistry. The PSD gives the asymmetric synapse its characteristic ultrastructural appearance and, together with round synaptic vesicles in the presynaptic element, generally indicates an excitatory synapse (Gray's Type 1; 14; see Figure 2a,b). In the absence of a PSD, a synapse has roughly equal dense regions pre- and postsynaptically. This symmetric contact, together with ovoid synaptic vesicles, signifies a nonglutamatergic synapse (Gray's Type 2).

The classical dendritic spine consists of a bulbous head connected to the dendritic shaft by a narrow neck. (Variations on this general scheme exist and are discussed below.) The head contains the PSD and some specialized structures. Other structures appear to be excluded. For example, whereas actin microfilaments are concentrated in spines (15–17), mitochondria and microtubules are excluded (18). Roughly half of spines on hippocampal CA1 cells and virtually all Purkinje cell spines contain some smooth endoplasmic reticulum (SER) (19). Some pyramidal cell spines contain a peculiar structure called the spine apparatus (20; see Figure 2a). This organelle consists of two or more disks of SER separated by an electron-dense material that may consist in part of microtubules (21) or actin filaments (17). The SER is known to play a role in Ca^{2+} handling (22, 23). Small spines are less likely than large spines to contain SER, whereas most of the largest spines contain a spine apparatus (19), which suggests a difference in the way calcium is handled in different-sized spines. Although free ribosomes are rarely

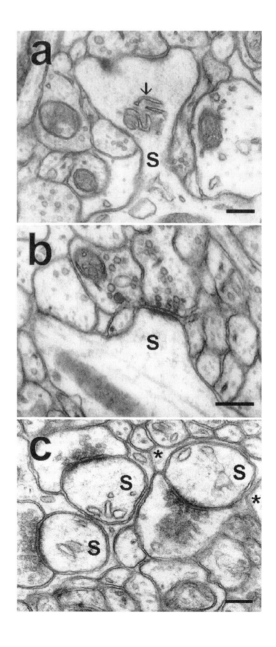

found in spines, polyribosomes are frequently encountered (24), occurring in 82% of spines in the visual cortex, approximately 10% of dentate granule cell spines (25, 26), and in 13% of Purkinje cell spines (27); because of the difficulty detecting polyribosomes even in the electron microscope, these percentages are likely underestimates.

The PSD is the most complex organelle in spines. Recent analysis of the protein composition of PSDs (28, 29) revealed that they contain hundreds of components including receptors, cytoskeletal and adaptor proteins, and associated signaling molecules representing several signaling pathways involved in synaptic plasticity. The PSD can thus be thought of as a collection of signal processing devices associated with the maintenance and plasticity of synaptic function (reviewed in 30). The PSD may assume any of a number of shapes, but it is usually described by ultrastructural criteria as macular, i.e., plaque-like and uninterrupted, or perforated, i.e., appearing annular in three-dimensional reconstructions. Perforated PSDs are invariably associated with puncta adherentia, nonsynaptic cell-to-cell junctions found in all epithelia and abundant in the central nervous system, whereas macular PSDs are not (31). In addition, perforated PSDs tend to occur in the largest spines (32). Recent studies suggest that PSD shape and size are not fixed and may change with alterations in the level of synaptic activity (33–35).

One of the most striking characteristics of dendritic spines, especially in pyramidal neurons, is their morphological diversity. Although the various shapes they assume fall along a continuum from short and squat to long and bulbous, they have been divided into gross morphologic categories. The most commonly used nomenclature, introduced by Peters & Kaiserman-Abramof in 1970, divides spines into three main categories based essentially on the relative sizes of the spine head and neck (24). Mushroom spines have a large head and a narrow neck (Figure 2a); thin spines have a smaller head and a narrow neck; and stubby spines have no obvious constriction between the head and the attachment to the shaft (Figure 2b). Other authors have added another category, the filopodium (Figure 1b; 36), named

---

Figure 2 Ultrastructural appearance of dendritic spines. Characteristic electron microscopic appearance of several types of spines (S). (a) A mushroom spine is shown. Note the constricted spine neck, the prominent irregular PSD, and the well-developed spine apparatus in this spine (arrow). Note also the density of the neuropil in this adult tissue. A stubby spine is illustrated in (b). Purkinje cell spines are very abundant (c) and are often seen in close contact with astrocytic processes (asterisks). The spine necks and their emergence from the parent dendrite are not shown in this photomicrograph. Note the presence of cisternae of SER in virtually all Purkinje cell spines. Micrographs (a) and (b) are from rat hippocampus; (c) is from mouse cerebellum. Scale bars = 300 nm (a, b), 250 nm (c). Electron micrographs from the Atlas of Ultrastructural Neurocytology, J. Spacek (http://www.synapses.bu.edu/atlas/contents.htm), with permission from the author.

for its hairlike morphology, reminiscent of the axonal filopodium on the axonal growth cone, and, like the axonal version, found mostly during development (36). Branched spines (Figure 1*b*) may contain more than one postsynaptic density. In addition, spinelike protrusions may emerge from preexisting spines. These spinules are found adjacent to the synaptic active zone, do not bear a PSD, and invaginate into the presynaptic terminal (37). These categories have been progressively refined, and differences between them have emerged. For instance, mushroom spines are selectively enriched in F-actin (17), are most likely to contain polyribosomes (27) and perforated PSDs (32), and are the almost exclusive bearers of a spine apparatus (19; Figure 2*a*). In contrast, virtually all thin spines have macular PSDs (38). The relative proportions of these classes change with development (see below). Spine morphological criteria are largely qualitative, which makes comparisons between data emerging from different laboratories difficult in some cases.

Cerebellar Purkinje cells bear large numbers of spines, with a mean length of 1.4 μm (39) and a mean volume of approximately 0.12 μm^3 (39, 40), of which over 80% is the spine head (40). The size of the PSD correlates with the spine volume, as well as with the number of presynaptic vesicles. They are filled with SER (Figure 2*c*), which is in continuity with that of the dendritic shaft, so that the Purkinje cell SER constitutes a network linking all spines and dendrites (40, 41). The spines that synapse with climbing fibers, concentrated on the soma and proximal dendrites, are shorter and stubbier than those that contact parallel fibers, which have longer more slender necks (42).

In the CA1 region of the hippocampus, simple and informative relationships have been uncovered between the sizes of dendritic spines and anatomical parameters thought to be related to synaptic strength. Spine volumes are proportional to the areas of PSDs (32). Immunogold labeling studies have shown that the density of AMPA and NMDA receptors is constant within the PSD, and thus the number of receptors per synapse is proportional to PSD area (12, 43, 44) and spine volume. On the presynaptic side, the area of the active zone is proportional to PSD area (and hence spine volume) (45). In addition, the area of the active zone is proportional to the number of docked vesicles (45), which is a good correlate of the quantity of neurotransmitter release per action potential (46). Thus in terms both of transmitter release and postsynaptic sensitivity, large spines are the sites of strong synapses. Consistent with this simple picture are observations that some manipulations that modulate spine size also appear to change measures of synaptic strength, including the amplitude of miniature synaptic currents (47).

The spines on hippocampal pyramidal neurons are somewhat smaller than on Purkinje cells, with a mean volume of 0.062 μm^3, of which, as with Purkinje cells, over 80% is in the spine head (32). However, the coefficient of variation of the measures of spine volume (SD*mean^{-1}) for hippocampal neurons is much greater than that for Purkinje cells (1.29 versus. 0.17, respectively), which reflects great variability in spine shapes in the hippocampus compared with those in the cerebellum. This difference is evident from images of dendrites from both cell types (Figures 1*b,d*).

Relationships with Other Elements

In the cerebellum, Purkinje cell spines are typically contacted by a single presynaptic terminal from a parallel fiber originating from a granule cell. These synapses are glutamatergic and excitatory (48). In one study, approximately one-quarter of axonal varicosities were associated with more than one postsynaptic element, and these were usually spines on the same dendrite (40). In addition, glial Bergmann fibers are in intimate contact with Purkinje cell spines (49; Figure 2c) and may be seen in contact with their tips, even in the absence of a presynaptic element. These processes envelop the synapses in this region and may serve to insulate them from one another, as well as to increase the duration of the glutamate transient in the synaptic cleft by restricting diffusion (50).

In the CA1 field of the hippocampus, spines are also usually contacted by a single presynaptic terminal, although, when more than one spine forms synapses with the same presynaptic bouton, they are usually from different neurons (32). Inhibitory terminals may synapse on pyramidal cell spines in addition to the excitatory terminal (51), an arrangement also seen with dentate granule cells (52, 53). Glial processes are seen in the vicinity of synapses much less regularly than in the cerebellum. Fewer than 60% of excitatory synapses are abutted by a glial process, which, when present, envelops, on average, less than half of the circumference (54).

In neocortex, too, dendritic spines may bear several synaptic arrangements, including a GABAergic (55–57) or dopaminergic (58) symmetric synapse in addition to the asymmetric synapse. Such configurations may permit modulation of excitatory transmission at the level of the spine and could have important consequences for the efficacy of synaptic transmission.

Numbers and Distributions

Accurate estimates of density can only be made using serial section electron microscopy, because it is with this technique that all spines, irrespective of size and orientation, can be identified and counted. With this technique, estimates of dendritic spine density range from two to four spines per micrometer of dendrite in adult hippocampal CA1 pyramidal cells (32, 38, 59) and hippocampal granule cells (60), whereas the density in Purkinje cells is well over ten spines per micrometer (39, 40).

Dendritic spines are distributed throughout the dendritic tree. However, an early observation on the distribution of spines in neocortical pyramidal neurons was their relative absence in the initial portion of the apical dendrite. Valverde demonstrated an exponential increase in apical dendritic spine density with distance from the soma, a peak reached in the middle cortical layers, and a subsequent drop-off in distal dendrites (61). Marin-Padilla later hypothesized that spine distribution could be described by a series of Gaussian curves centered on different cortical layers, reflecting separate afferent systems impinging independently on different parts of the dendritic tree (62). A notable exception to this generalization is the hippocampal CA3 pyramidal neuron. The soma and proximal apical dendrite of this neuron are studded with very large, complex, branched spines called thorny excrescences.

This structure is the postsynaptic target for the mossy fibers of dentate granule cells (63). It is also an exception to other general rules listed above, in that it contains mitochondria, free ribosomes, and microtubules (63).

Spine density varies not only within individual dendritic trees but also across cortical areas. This is most evident in primate neocortex, where many cytoarchitectonically distinct regions exist and can be readily distinguished (64). A recent study by Elston (65) demonstrated that spine density on basal dendrites of layer III neurons in several cortical areas of the frontal pole and orbitofrontal cortex in the macaque monkey (Brodmann areas 10, 11, and 12) was generally threefold greater than in neurons of the primary visual cortex (area 17) and twofold greater than neurons in a parietal visual cortical region (area 7a). Similarly, Jacobs et al. (66), working in the human brain, found that among eight cortical areas sampled, layer III basal dendritic spine density was greatest in areas in the prefrontal and orbitofrontal cortex and up to 40% greater in the frontal polar cortex (area 10) than in the primary somatosensory cortex. Both groups hypothesized a link between spine density and overall number and the level of cortical processing in these regions. For instance, although the frontal polar regions receive highly processed information, they do not receive input from first order sensory areas such as the primary visual and somatosensory cortices. It is believed that these "higher" order areas are involved in a greater degree of convergent processing, which may create a necessity for more synapses and hence more spines. However, not enough is known about these cortical areas to make such an assertion, and data from more cortical areas are needed before these generalizations can be fully justified.

SPINE STRUCTURAL PLASTICITY

Morphological Development

Neonatal mammalian pyramidal cell dendrites are relatively bare (6). Over the first week of life, dendritic protrusions begin to increase in density, and during the second and third weeks their density dramatically increases, as the rate of synaptogenesis reaches a peak (67, 68). There is some evidence that in the juvenile rodent neocortex, a subsequent pruning produces a loss of spines, so that the mature spine density is reached after several months (69). Similar observations have been made in the primate visual cortex (70).

Spine morphology, too, changes with development. Although the most abundant spine type early in development is the stubby spine, early spines are often very long, and filopodia are not infrequently encountered. Over the next few weeks, mean spine length decreases and the incidence of filopodia is dramatically reduced (71). The increase in density and decrease in overall length and the decrease in incidence of dendritic filopodia correspond with the concomitant decrease in spine motility during this time (72; see below). It is worth noting that while patterns may exist for populations of spines at different times during development, spine morphology does not necessarily correlate with spine maturity, and it is probably not accurate to describe a particular spine as immature based on its morphology alone.

Spine development is marked also by ultrastructural changes. Immature spines (under 16 days in the rat) do not contain a well-developed spine apparatus (21, 38). Instead, they may contain a few cisternae of SER that have been termed pre-spine apparatuses (38). In addition, the proportion of dentate granule cell spines containing polyribosomes is greatest when spine formation is at a peak (25), which suggests a role for these structures in the synthesis of proteins needed locally for the formation of new synapses.

A question that is actively debated is how a mature spine is formed. There are at least three general views of this question. In the first model, dendritic filopodia, which are highly motile (73), actively seek out synaptic partners in the developing neuropil. When a partner is found, the filopodium shortens, drawing the axonal element closer to the parent dendrite. Eventually, a fully mature synapse is formed, usually on the spine head, and spine motility decreases as the structure is stabilized. This scheme is based on time-lapse imaging studies in vitro, in which the same protrusions were followed for days, and in a few cases their participation in synaptogenesis was documented (74). This model accounts for both the increased length and motility of immature filopodial spines, but it does not explain the observations by several groups of a much higher density of asymmetric synapses on dendritic shafts than on filopodia early in development (75, 76).

In the second model, dendritic filopodia also seek out synaptic partners, which form synapses on them not necessarily at the tip (76). Here, too, the filopodium retracts, but in this model it retracts completely, which leads to the formation of an asymmetric shaft synapse. Then a spine emerges at that site with a mature synapse at its head (76, 77). This scenario is based on detailed electron microscopic observations involving reconstructions of serially sectioned filopodia and spines, and thus it could take into account the presence of small ultrastructurally identified spines. However, such a static analysis cannot describe either the sequence of events involved in the formation of an individual spine or when the synaptic contact becomes functional. The resolution of this question may require long-term time-lapse functional imaging.

In the third model, the emphasis is on continued spine turnover throughout the lifetime of the animal, a capacity that is greatest during times of enhanced plasticity, such as during critical periods of circuit formation and synaptogenesis (78). In this scheme, spines constantly form by seeking out presynaptic partners and stabilizing into functional spines of any morphology. They can also change or even disappear, depending on the state of the afferent input. The size, not the morphology, is the more relevant characteristic and is related to the strength of the synapse at any point in time, as discussed above. The filopodium is simply a spine in an extreme state of morphologic instability and is not a necessary intermediate for spine formation. The structure and motility of the filopodium are the result of the relative emptiness of the neuropil at early ages and of distances the protrusion must cross to find an axonal element (76).

In any case, it is probably premature to assume that all spines go through the same stages, beginning as filopodia, proceeding to thin or stubby spines, and ending as mushroom spines. Time-lapse imaging in vivo demonstrates that nonmushroom spines may be as stable as mushroom spines and may remain morphologically unchanged for days in vivo, whereas mushroom spines can disappear or change into other morphological types (B. Chen, J. Trachtenberg, and K. Svoboda, unpublished observations).

Deafferentation-Induced Plasticity

Implicit in the 70 years of research from Cajal's description (5) into the 1960s was the idea that once formed, dendritic spines remain in place, forming a synaptic unit with their presynaptic partners for the lifetime of the neuron. As early as 1966, however, there was evidence that spines can be selectively eliminated under certain circumstances without apparent changes in the dendritic tree as a whole. Globus & Scheibel, using the Golgi method, described a decrease in spine density on pyramidal neurons in the visual cortex in response to either contralateral enucleation or ipsilateral lesioning of the lateral geniculate nucleus; both of these interventions deprive the pyramidal cells of a major source of afferent input (79). Nearly concurrently, Valverde (61), also using Golgi-stained material, showed that spine density in portions of layer V pyramidal neurons in the primary visual cortex was decreased in mice raised from birth in total darkness. Several years later, he showed that spine density could recover in some dendrites after only a few days of life in normal lighting (80). Similarly, Parnavelas (81), working in the rat hippocampus, showed that the spine density of dentate granule cells, after an initial decrease due to entorhinal cortex lesions, subsequently returned to baseline levels as a result of reafferentation by sprouting of nearby axons. These findings demonstrated the bidirectional nature of these structural changes and indicated that the initial decrease was not simply a sign of injury to the postsynaptic neuron but was an adaptive response to changes in afferent input.

An interesting exception to the rule that loss of afferents causes loss of postsynaptic spines is the Purkinje cell. In the *weaver*, a spontaneous mouse mutation, granule cells are lost before formation of the parallel fibers. Nontheless, Purkinje cell spines develop normally in the absence of neuronal presynaptic elements (82, 83). Similarly, in cerebellar cultures whose granule cells have been eliminated by poisoning with methylazoxymethanol acetate (84) or with X-irradiation (85), Purkinje cell spines develop normally and even have postsynaptic densities weeks after deafferentation induced by the loss of granule cells. In contrast, pyramidal cell spines are virtually never seen without a presynaptic element (86). It appears that Purkinje cell spines are maintained by entirely different mechanisms from those that maintain spines on hippocampal and neocortical pyramidal neurons.

Spine Motility

The changes described above were all changes in density, i.e., the presence or absence of substantial numbers of spines, taking place over days to months. However, more recent evidence indicates that spines undergo morphologic changes over a large range of spatial and temporal scales and under more physiological conditions.

We now know that dendritic spines are quite motile and exhibit several types of motility under normal circumstances. Enriched with actin (15, 16), they undergo constant small-amplitude changes in shape detectable over the course of seconds (87). This motility is blocked by volatile anesthetics, such as isoflurane and chloroform (88), and by low concentrations of AMPA (α-amino-3-hydroxy-5-methyl-4-isoxazolepropionic acid, a glutamate analog specific for a subtype of ionotropic glutamate receptor) (89). The significance of this motility itself is unclear, but some believe it may be related to changes in glutamate receptor subtypes that occur during certain forms of plasticity (90) (see below).

Spines are also capable of larger movements, especially early during development. The spines of hippocampal pyramidal neurons in slices prepared early in postnatal life and maintained in culture show dramatic structural changes on the order of microns over minutes to hours and even appear and disappear (73). Recently, similar observations were made in neocortical pyramidal cell dendritic spines in the intact brain during periods coincident with accelerated synaptogenesis (78). The extent and prevalence of this motility is developmentally regulated, decreasing dramatically during the first postnatal weeks as synaptic circuits mature (67). Developmental regulation of motility has also been observed in vitro, both in cerebellar Purkinje cells and in hippocampal pyramidal cells (72). In addition, the dendritic filopodia so common early in development are particularly motile structures and capable of changes of several microns in a matter of minutes (78).

A recent study by Lendvai et al. (78) directly probed the relationship between spine motility and sensory experience in the intact cortex. Normally, spine motility decreases with age (72, 73, 78). Pyramidal neurons in the somatosensory cortex were labeled by intracortical injections of a virus encoding enhanced green fluorescent protein (EGFP), and dendritic spines were imaged repeatedly in vivo using a 2PLSM. From these values a measure of spine motility was derived. In animals that had all whiskers on one side of the face trimmed ("deprived" animals), spine motility decreased, but only during the period of peak neocortical synaptogenesis in the barrel cortex. Deprivation before or after this period did not result in a drop in motility. This result demonstrated a tight link between experience-driven synaptogenesis and spine motility.

Synaptic Activity-Dependent Structural Plasticity

The loss of spines upon deafferentation suggests that they are somehow maintained by their afferent input. Because spines bear glutamatergic synapses, one may reason that some aspect of glutamatergic neurotransmission acts as a signal to maintain the spine and that interference with normal synaptic activity may therefore affect spine shape or density.

Numerous studies have been directed at testing this model; varying techniques produced not entirely consistent results. Chronic application of the $GABA_A$ receptor antagonist picrotoxin, which increases neuronal excitability, to dissociated hippocampal cultures resulted in an increase in spine density on second-order dendrites (91), and neocortical organotypic cultures similarly exposed since the day of birth also showed a dose-dependent increase in spine density (92). However, in hippocampal organotypic cultures, chronic exposure to bicuculline (another GABA$_A$ receptor blocker) or picrotoxin caused a dramatic loss of spines (93). This difference may be due to the aberrant recurrent innervation characteristic of these cultures (94), which could give rise to epileptic activity, a situation known to cause spine loss (93, 95). These findings indicate that moderately increased levels of excitatory synaptic activity can induce spine formation, but that excessive and unrestrained activation can cause excitotoxic loss of spines. Consistent with this notion, one study found that application of short pulses of glutamate to dendrites of cultured hippocampal neurons caused spines to elongate, whereas long pulses caused them to shrink (96). Similarly, exposure of cultured hippocampal neurons to a medium that favors NMDA receptor activation induced new spines to form and prompted the pruning of others (97), but acute application of high concentrations of NMDA caused the collapse of spines (98). In addition, release of calcium from internal stores using caffeine, which should generate a moderate (200–400 nM) increase in intracellular [Ca^{2+}] caused a majority of spines to increase in length (99). These findings are consistent with a scheme where moderate increases in intracellular calcium cause spines to form or elongate, but high concentrations have deleterious effects and cause spines to retract. This observation has led some to propose that a major function of dendritic spines is neuroprotective: They spare the dendrite proper the potentially noxious effects of excessive stimulation by spatially confining the calcium rise while still faithfully transmitting synaptic information (100).

As described above, the very-small-amplitude changes in shape that spines constantly undergo can be stopped by application of low concentrations of AMPA (89). In this context it is interesting that a recent study suggests that the spontaneous release of quanta of glutamate, which give rise to miniature excitatory postsynaptic currents (EPSC), is sufficient to maintain dendritic spines (101). Blockade of action potential-induced neurotransmitter release did not affect spine density, but blockade of AMPA receptors or inhibition of vesicular release with botulinum toxin caused a significant decrease in spine density in organotypic cultures (101). Taken together, these studies have led some to speculate that spine maturation and stabilization require the presence of functional AMPA receptor-bearing synapses, in association with functional glutamate-releasing boutons. In this view, immature spines bear immature synapses that contain only NMDA receptors (i.e., silent synapses) (102, 103). These NMDA receptor-only-containing spines are highly motile, reflecting their active role in synapse formation, and are transient unless they meet a presynaptic partner. With synapse formation and maturation, consisting, in part, of AMPA receptor insertion (104) and perhaps recruitment of adhesion molecules to synaptic junctions (105), the synapse is stabilized, and motility decreases (90). This approach also suggests an answer to the question of why deafferentation causes the loss of spines. In this view, it is the loss of the spontaneously released quanta from the presynaptic partner that signals its absence and therefore the superfluity of the postsynaptic spine.

Structural Plasticity Associated with Long-Term Potentiation

It is perhaps not surprising that the remarkable capacity of dendritic spines to change shape so rapidly has enlivened the longstanding debate over the possible structural basis of learning and memory. This subject has recently been exhaustively reviewed (8), and only selected studies are considered below.

Numerous studies have attempted to link morphologic changes with increases in synaptic efficacy such as long-term potentiation (LTP), a well-established synaptic model for memory (106). Van Harreveld & Fifkova (107), only a few years following the first descriptions of LTP (108), described an increase in spine volume in mouse hippocampal dentate granule cells following stimulation in vivo that was capable of inducing LTP. This increase was greatest (nearly 40%) between 10 and 60 min after stimulation and decreased somewhat with time, but was still evident for as long as 23 h after stimulation (109), strengthening the correlation of this morphologic change with the physiologic characteristics of LTP. Also, it is pathway-specific because the change was found only in the outer molecular layer, which receives input from the stimulated perforant path, and not in the inner molecular layer, which does not receive this input. Later, the same group described the shortening of spines and enlargement of spine neck width using the same paradigm (110). Subsequent support for these findings came from a series of ultrastructural studies by Desmond & Levy, in which they documented a more subtle morphologic change consisting of a 48% increase in concave spine heads and a concomitant increase in PSD area (33, 34) and non–PSD-associated spine membrane (111). In a similar vein, Geinisman et al. described a selective increase in the ratio of perforated to nonperforated PSDs when LTP was induced in vivo, but they reported no overall change in the numbers of axospinous synapses (112). Because perforated PSDs tend to occur in mushroom spines (32), this change could indicate an alteration in spine morphology, although this issue was not explicitly addressed. Lee (113) reported a 33% increase in shaft synapses in the hippocampal CA1 field 10 min after induction of LTP. It was not specified whether these synapses were symmetric or asymmetric (and thus presumably inhibitory or excitatory, respectively), but they tended to occur on dendrites that were relatively devoid of spines and therefore very likely belonging to inhibitory interneurons. Chang & Greenough replicated this finding, reporting a 175% increase in shaft synapses but also a near-doubling of synapses on stubby spines (114). They also found no change in overall synapse density (114). Because they were counting synapses, not spines, it is unclear whether there was actually an increase in stubby spine density or in the frequency of synapses on stubby spines. Unfortunately, they also did not distinguish between symmetric and asymmetric synapses. If symmetric synapses were being counted, their results could be consistent with an increase in inhibitory synapses on preexisting stubby spines already bearing excitatory synapses. Because each synapse was not analyzed serially throughout its extent, this possibility cannot be excluded, and recent work showing changes in GABAergic innervation of spines with whisker deprivation makes it very real (115). It should be noted that none of the aforementioned studies described a change in spine density. More recently, Sorra & Harris (59), addressing this question directly, found no changes in synapse or spine density in hippocampal slices 2 h after LTP had been induced.

Many of these studies probably suffered from the fact that the number of synapses stimulated in these paradigms was a very small proportion of the total, and therefore the chance of detecting a subtle change would be small (the familiar needle-in-a-haystack problem). To circumvent this complication, several approaches have been used. One study in CA1 neurons in hippocampal slices used a chemically induced form of LTP, which would be expected to potentiate all synapses in the slice. Dendritic segments were imaged repeatedly over time and monitored for changes in spine density. This approach yielded no overall change in spine density or length, a relatively small increase in small spines, and an increase in their tendency to have changed in their angular position relative to the dendritic shaft (116), which suggests some motility as a result of synaptic plasticity. Two more recent studies utilized 2PLSM to monitor dendritic segments targeted for potentiation-induced changes either (*a*) by their proximity to the focal stimulation electrode (117) or (*b*) by the fact that they were the only locations in the slice able to experience changes, because the rest of the slice was inhibited by the absence of extracellular calcium (118). In both these studies, new spines appeared with greater frequency in the targeted regions than in other more distant or inhibited regions, respectively. In one, filopodia-like protrusions increased in number by 145% (117). In addition, in one study, new spines failed to appear in cases where LTP did not occur (118), which further emphasizes the relationship between LTP and structural change.

Yet another approach has been used by Muller and his colleagues, who have developed a histochemical procedure that marks synapses recently activated by reacting with intracellular calcium and forming an electron-dense reaction product (119). This method permits the analysis of synaptic features of labeled spines, but not the determination of whether entirely new spines have been induced as a result of the stimulation. Nonetheless, these studies demonstrated a transient increase in the proportion of labeled spines with complex perforated PSDs among activated spines relative to the general spine population (35), in agreement with the findings of Geinisman et al. (112). A longer-lasting increase in the incidence of boutons forming synapses with multiple spines has also been reported (120). These boutons are a special case of the multisynaptic bouton, whose postsynaptic elements are not necessarily identified as spines. Notably, the likelihood that the spines synapsing with these multiple-spine boutons were from the same dendrite (66%) was much greater than the proportion found under normal conditions (11%), which suggested that LTP induced the duplication of existing axon-dendrite contacts. It should be noted, however, that the change in the proportion of perforated PSDs on labeled spines is also consistent with the fact (mentioned above) that SER or a spine apparatus, and therefore, a well-developed Ca^{2+}-handling machinery, is found mostly in large mushroom spines, which are also most likely to bear perforated PSDs under normal circumstances. To address this concern, the authors showed that the frequency of perforated PSDs was similar in a set of unstimulated unlabeled synapses and in the labeled synapses that occur occasionally in unstimulated preparations, indicating that they were not preferentially selecting synapses that a priori bear perforated PSDs (35). Nonetheless, the alteration that is the subject of the study is such that it can only be recorded in a subset of dendritic spines. The effects of LTP on other types of spines or on spines in general cannot be described using this technique and neither can the possible effects of LTP on overall spine number or density. The complete set of structural changes associated with LTP is still not well-defined, but it appears likely that the effect is focal and that changes in a tissue will be correlated with the numbers of stimulated axons and dendrites—values that vary widely from study to study. Nonetheless, there is evidence for the remodeling of the PSD and for the formation of novel spines in response to LTP-inducing stimuli, which could form, in part, the basis of long-lasting changes in synaptic strength. It is unlikely, however, that the growth of new spines explains LTP, since it is a relatively slow process, while the onset of LTP is rapid.

Structural Plasticity Associated with Experience and Learning

If the complexity of LTP in vitro makes finding a structural correlate very difficult, the search for structural correlates of learning is even more challenging. Nonetheless, several studies have directly attempted to link changes in dendritic spine morphology with learning. Moser et al. (121), using confocal microscopy, described an increase in spine density on basal dendrites in CA1 pyramidal neurons after rats underwent spatial training. The variability within the control and experimental groups, however, made this result difficult to evaluate, since there was considerable overlap between groups. A smaller increase (15%) in spine density has been observed in piriform cortex following olfactory learning (122). Rusakov et al. (123) performed a study similar to Moser's, using electron microscopy and stereologic methods. Although they described an apparent increase in the clustering of synaptic active zones in CA1, they were unable to replicate Moser et al.'s findings, seeing no changes in synapse density or size. Similarly, Geinisman (124) examined synapses in rabbit hippocampus following trace eyeblink conditioning, a form of hippocampus-dependent associative learning. They described an increase in PSD area reminiscent of the same group's observations of PSD changes after LTP (112). However, they, too, found no change in total synapse number. In the rat dentate gyrus, O'Malley et al. (125) described a transient synapse increase in a passive avoidance training paradigm. Six hours after training there was a twofold increase in the number of axospinous synapses in the middle molecular layer. This change subsided by 72 h. Later, they described a similar transient increase following spatial learning (126), which indicated that the earlier-described change was not restricted to one type of learning. The transience of these changes suggests that they reflected a net rearrangement of synapses, rather than a lasting increase in their number. Thus in most studies the consensus is that even if there are transient increases, there is no lasting increase in synapse number following learning. Synapse and spine morphology may change, but it is clear that further studies will be necessary to show a definitive link between changes in spines and learning.

Hormonal Control of Structural Plasticity

An interesting form of physiologic plasticity of dendritic spines is the variation in spine density in hippocampal CA1 cells over the five-day estrus cycle of the rat (127). During proestrus, when estrogen levels are at their highest, spine density is at its maximum. Between proestrus and estrus, when estrogen levels fall, there is a 30% decrease in spine density; a comparable decrease in synapse density at the electron microscope level is the ultrastructural reflection of the spine density change (128). A similar effect was seen when rats were ovariectomized and had their levels of estrogen and progesterone artificially manipulated (129) and also appeared in neurons in vitro (130), which confirms the link between the sex steroids and spine density. No comparable change was found in CA3 or the dentate gyrus. Estrogen also increased the occurrence of multisynaptic boutons, which are presynaptic elements that make more than a single synapse at the apparent expense of single-synaptic boutons (131). This suggests that during the portion of the estrous cycle when estrogen levels are at their highest, new dendritic spines make synapses with preexisting boutons. More recently, the synapses of a multisynaptic bouton are made with different postsynaptic neurons (132), which indicates not only an increase in spine and synapse number, but also in divergence of inputs. This change was blocked by progesterone (133) and by NMDA receptor antagonists (134), and in vitro the change was mimicked by an inhibitor of GABA synthesis (135), which indicates that estrogen may act indirectly by inhibiting GABAergic neurotransmission, thereby increasing excitatory neurotransmission, particularly through NMDA receptors. This decrease in GABAergic tone, in turn, may be the result of estrogen-induced down-regulation of BDNF (136). It is interesting that secreted BDNF has been found to destabilize dendritic spines and may encourage remodeling (137). Thus this neurotrophin may play an important role in physiologically meaningful long-term changes in spine density.

Thyroid hormone also influences spine density. Exogenous administration of thyroid hormone to neonatal rats caused an increase in spine density in CA1 and CA3 neurons (138) and an acceleration in spine development in visual cortical neurons (139). On the other hand, thyroidectomy performed as early as possible in postnatal life prevented the normal development of visual cortical spines, such that at 30 days, the spine density in thyroidectomized rats was only 60% of that found in control animals (140). Furthermore, the spine density reached a plateau at 30 days, whereas in the control rats, the density continued to increase (140). This abnormality could not be reversed by administration of thyroid hormone unless it was instituted within 2 days of thyroidectomy and maintained (141). Thyroidectomy of juvenile (40-day-old) or adult (120-day-old) mice, however, caused a decrease in spine density that could be reversed by hormone replacement, even if treatment was delayed by 25 to 30 days. Thus thyroid hormone is critical for the initial phase of spinogenesis to occur, but it may serve a somewhat different role in spine maintenance later in life. It is interesting to note that hyperthyroidism in adult animals caused a 26% loss of spines on CA1 (but not CA3) pyramidal neurons (142), in marked contrast to the increase seen in neonates (138), which suggests that thyroid hormone must be kept within a certain range of levels for normal spine maintenance. The mechanism by which thyroid hormone regulates spine density is not known. However, thyroid hormone has such wide-ranging metabolic effects that a direct effect on pyramidal neurons need not be invoked. For instance, hyperthyroid animals open their eyes up to 4 days earlier than control animals (138), a change that could partially explain the accelerated spine development described above.

SPINE STRUCTURE IN BRAIN DISORDERS

In view of the central role occupied by dendritic spines in synaptic transmission, it is not surprising that a number of human disease states are associated with alterations with spine morphology or density.

Schizophrenia

Two recent studies using the Golgi technique have shown decreases in spine density in neocortical pyramidal neurons in patients with schizophrenia. The first study found 59 and 66% decreases in spine density in temporal and frontal cortical regions, respectively (143). The effects of age and postmortem interval were controlled, but the effect of long-term neuroleptic use could not be excluded. The other study (144) showed a 21% decrease in spine density in schizophrenic patients compared with age-matched controls. However, a statistically comparable decrease was also found in age-matched nonschizophrenic psychiatric patients, which casts doubt on the specificity of the decrease to the disease. As in other studies of dendritic spines in human tissue, these results must be evaluated with caution. Given the remarkable plasticity of dendritic spines and their ability to change in density and in shape within minutes, the fact that human material is subject not only to autolysis but also to changes due to the events preceding death makes interpretation of changes in dendritic spine distribution and morphology difficult.

Aging

It is well known that the cognitive functions are impaired during normal aging. Most studies have focused on neuronal loss to account for this change. However, stereologic studies have suggested that the cell loss associated with normal aging may be insufficient to account for the observed cognitive changes (145). The most comprehensive studies of the effect of age have focused on the morphology of neocortical pyramidal neurons (146, 147). The total dendritic length, mean segment length, dendritic segment count, dendritic spine number, and dendritic spine density on basal dendrites of layer III pyramidal cells in prefrontal area 10 and occipital area 18 were examined (147). Tissue was obtained from 26 neurologically intact individuals aged 14 to 106 years. A 9–11% decrease in total dendritic length and about 50% decrease in spine numbers in both areas was reported when comparing the older group (>50 years) with the younger group (≤50 years).

Similar findings have been reported in aged monkeys, which, because they can be perfused, are free of both dying-related and autolytic changes. Cupp & Uemura (148), using the Golgi method, examined the layer III and IV pyramidal cells in the prefrontal cortex of nine rhesus monkeys aged from 7 to 28 years old. They reported significant decreases in dendritic branch order, number of branches, total dendritic length, and spine density for both the apical and basal dendritic trees in monkeys 27–28 years of age (148, 149). In a more recent EM study, Peters et al. (150) noticed a loss of apical dendritic tufts of pyramidal cells in layer I of area 46 of old monkeys (27–32 years of age) compared with young monkeys (6–9 years of age). They estimated that 50% of spines were lost in layer I of area 46 of the old monkeys. A parallel reduction of synaptic density with aging has been observed in quantitative EM studies (149, 150). These age-related changes on dendrites and dendritic spines of neocortical pyramidal cells in aged individuals may lead to a disruption in cortical circuits during normal aging.

Mental Retardation

A link between abnormal dendritic spines and mental retardation was first suggested by Purpura (151). Although isolated reports exist that indicate an abnormality of dendritic spines in numerous mental retardation disorders, including tuberous sclerosis type I (152), fetal alcohol syndrome (153), and nonsyndromic mental retardation (151), the best studied to date have been trisomy 21 (Down syndrome) and the fragile X syndrome.

Patients with Down syndrome have a decreased spine density in neocortex (154–156) and hippocampus (155, 157). This is observed both in young patients (154, 156) and in adults (156, 157). In addition, in a case described by Marin-Padilla, there was an increased incidence of abnormally long and short spines (154). Because many of the critical genes on chromosome 21 implicated in the human disease are found on the mouse chromosome 16, two trisomy 16 mouse models have been developed. Unfortunately, neither the fully trisomic mouse, which has very poor survival, nor the more restricted segmental trisomy 16 mouse has been analyzed for dendritic spine abnormalities. The only suggestive morphologic finding has been a 30% decrease in asymmetric synapse density in the temporal neocortex in the absence of a change in symmetric synapse density (158) and an overall decrease in the number of hippocampal pyramidal cells in the partial trisomics (159). However, since neither the number nor the identity of the genes directly responsible for mental retardation in human trisomy 21 has been identified conclusively, and because there are significant differences in the genes involved in the human and murine trisomies, it is not clear how closely the murine trisomy

a

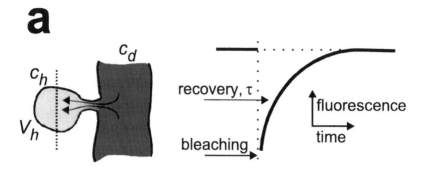

b

c

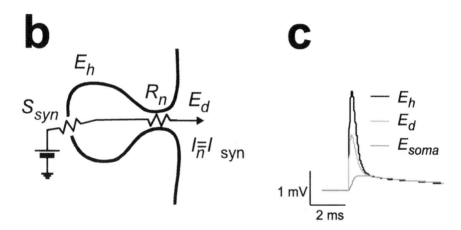

See text page C-2

Figure 3 Electrical and diffusional compartmentalization in spines. The electrical and diffusional properties of spines have been explored with fluorescence recovery after photobleaching (FRAP) measurements. (*a*) On the right is a schematic of a typical experiment. Spine heads, filled with a fluorescent molecule, are bleached by a laser scan (*dotted line*), transiently reducing the dye concentration in the spine head (c_h) but not the dendrite (c_d). Because spine heads are small (< 1 μm), dye concentration within the spine (c_h) will equilibrate rapidly and c_h is therefore uniform. Because the parent dendrites are large compared with spine heads, c_d is hardly perturbed by bleaching in the spine, and c_d is constant. (*a*) On the left is the time course of diffusional equilibration between spine and dendrite, monitored as fluorescence recovery in the spine. The time course of dye concentration in the spine head is then governed by

$$\frac{dc_h}{dt} = \tau^{-1}(c_h - c_d),\qquad\qquad 1.$$

where τ is the measured time constant of FRAP, the time over which the fluorescent probe is compartmentalized in the spine (range: 20–200 ms). τ is inversely proportional to the diffusion coefficient, D. Therefore, a measurement performed with a particular τ having a particular D provides information about diffusional exchange of other molecules with different diffusion coefficients. (*b, c*), Electrical properties of spines. Based on measurements of diffusional transport it is possible to make inferences about the electrical properties of the spine neck (182). The time constant of FRAP can be written as $\tau = V_h\ W_n/D$, where V_h is the spine head volume, and W_n is the diffusional resistance of the spine neck. For a cylindrical spine neck, W_n is given by the spine neck length over the cross-sectional area. A close analogy exists between diffusional currents driven by concentration gradients and electrical currents driven by potential gradients (191), as can be seen by comparing Equation 1 with Ohm's law:

$$C\frac{dE_h}{dt} = I_n = r_i W_n(E_h - E_d),\qquad\qquad 2.$$

where C is the spine head capacitance, I_n is the current through the spine neck, E_h, E_d are the electrical potentials in spine head and dendrite, and r_i is the cytoplasmic resistivity. The spine neck electrical resistance can thus be expressed in terms of measurable quantities as $R_n = r_i\ W_n = r_i\ \tau\ D/V_h$ (182). Taking account of uncertainties in estimating these parameters gives an upper bound of $R_n < 200$ MΩ. The spine neck conductance ($1/R_n > 7000$ pS) is much larger than unitary synaptic conductances ($S_{syn} < 200$ pS in CA1; < 600 pS in Purkinje neurons), and therefore the spine neck geometry is unlikely to restrict current flow. The voltage drop across a spine neck is likely to be small ($E_h - E_d < 4$ mV) for CA1 pyramidal cells (assuming unitary currents $I_{syn} < 20$ pA) (195). Larger voltage drops are possible for Purkinje cells spines ($E_h - E_d < 12$ mV; assuming $I_{syn} < 60$ pA) (194). (*c*) A simulation showing the attenuation of the membrane potential due to cable attenuation. Parameters in Neuron notation (www.neuron.yale.edu). Passive parameters: g_pas, .0001 micro mho; e_pas, -70; cm = 1 microF/cm²; Ri = 200 Ωcm. All compartments are cylindrical. Soma, length, and diameter are 10 μm each. Dendrite: 10 dendritic segments, total length, 200 μm; diameter, 1 μm. Spine attached to the middle of the dendrite, consisting of five compartments in the neck: diameter, 0.1 μm; total length, 1 μm; spine head, 0.5 μm; length, 0.5 μm synapse on spine head; synaptic conductance: alpha function with tau, 0.2 ms; g_{max}, 200 pS.

models the human disease. Resolution of this issue may have to await the complete genetic description of Down syndrome in humans, at which time a precise gene-by-gene trisomic animal model can be constructed, and the relationship, if any, between this form of mental retardation and synaptic and spine morphology will be in reach.

An example of a mental retardation disorder uncomplicated by the involvement of multiple genes is the fragile X syndrome. The fragile X syndrome (Martin-Bell syndrome) is an X-linked mental retardation syndrome that consists of a constellation of signs in addition to the cognitive deficit, including macroorchidism (enlarged testes), certain facial features (such as protruberant ears), and abnormalities in attention and short term memory (160, 161). It is remarkable in being caused by a mutation in a single gene, FMR1. The mutation giving rise to the syndrome, a CGG repeat expansion in the 5′ untranslated region of the FMR1 gene, interferes with transcription, and patients do not have measurable amounts of the gene product, FMRP (162). The function of this protein is unclear. It appears to act as an RNA binding protein (163, 164) and is localized to neurons and to dendrites in particular (163, 165, 166). Interestingly, mRNA for FMRP is found in dendrites, and its expression is increased by activation of metabotropic glutamate receptors (167), which links FMRP to synaptic function. One of the only consistently reported neuropathologic findings in patients with this disorder is an abnormality in dendritic spines, which were described as being unusually long (168–170) and of increased density. These observations have led some to speculate that the absence of FMRP causes a defect in spine maturation and pruning (171). We recently tested this hypothesis directly in a mouse model of the fragile X syndrome, the FMR1 knockout mouse (71). This study consisted of an analysis of the dendritic spines of layer V neocortical pyramidal neurons at 1, 2, and 4 weeks of postnatal life, using 2PLSM imaging of neurons virally transfected with EGFP and fixed in situ. We found that dendritic spines were, indeed, longer in the knockout animals than in the controls and that the abnormality is greatest at the earliest time point studied (25% longer at 1 week). At 2 weeks, the difference was still significant but smaller (10%), and at four weeks, smaller still (3%). Spine density, too, was abnormally high in the knockouts, but only at one week (33% higher). We concluded that the dendritic spine abnormality caused by the absence of the FMR1 gene is most significant very early in postnatal life, coincident with a period of massive synaptogenesis in these animals (39) (68). These results are consistent with recent data in the same mouse model from the group that first published a murine spine abnormality (W. T. Greenough, personal communication) but inconsistent with observations in humans that show persistence of abnormally long spines well into adulthood (170). This discrepancy could be related to species differences or to the considerable difficulties in interpreting spine changes in postmortem human tissue. It could also be due to the fact that the nature of the abnormality in the human brain is still far from clear, since most accounts have been purely qualitative and based on the same one to three cases (168, 169). In addition, the one quantitative spine study (170) is difficult to interpret because some of the "control" tissue may have been abnormal, as evidenced by the fact that nearly all spines were 0.5 μm or less in length. Even in the absence of unambiguous human data, however, a developmentally regulated abnormality in the mouse is an encouraging finding and increases confidence in the animal model.

Other Disorders

Spine abnormalities have also been described in several other neurologic disorders. For instance, decreases in spine density have been reported in neocortical neurons in HIV encephalitis (172), subacute sclerosing panencephalitis (173), and tuberous sclerosis Type I (152); hippocampal pyramidal neurons in murine scrapie (174); Purkinje cells in olivopontocerebellar atrophy (175); and in a mouse model of Menkes kinky hair disease (176).

The list of neurologic disorders associated with changes in dendritic spines is long and growing. The highly plastic nature of dendritic spines, however, and their ability and tendency to change even in response to subtle changes in afferent input, makes these alterations difficult to interpret, because it is very difficult to state with any certainty that a given abnormality is due to a primary defect of the spine rather than a response of the spine to some other abnormality, to changes induced by the process of dying, or to postmortem changes. It is probably only in those abnormalities for which an animal model exists that such a causal link may eventually be made.

COMPARTMENTAL MODELS OF SPINE FUNCTION

Models of spine function have focused on the properties conferred by the spine neck most prominent in mushroom-shaped spines. Because it presents a thin and tortuous path between spine head and parent dendrite, the spine neck could serve as a resistive element and impede synaptic currents (Figure 3, see color insert). Changes in spine neck structure could then regulate the amplitudes of synaptic currents in synaptic plasticity. With the discovery of actin in spines, this model of synaptic plasticity became associated with the term twitching spine (177). Apart from a role in electrical signaling, the spine neck could also serve to restrict diffusional exchange of signaling molecules between spine head and parent dendrite (178–180); this could be important to localize biochemical changes to a particular synapse.

Spines as Diffusional Compartments

A number of groups have investigated diffusional compartmentalization by spine necks. Early studies were mostly computational (178–180) and based on EM reconstructions of spine membrane geometries (1). However, since it is impossible to infer transport properties of the cytoplasmic space from static images, direct measurements of compartmentalization were performed. Transport through the spine neck can be directly probed by measuring diffusion using fluorescence recovery after photobleaching (FRAP) (181) in the spine (Figure 3) (182–184). In these studies a freely diffusible fluorophore was bleached in the spine head and the time for fluorescence recovery by diffusion from the parent dendrite was measured (Figure 3a) (182–184). Since photobleaching always carries the risk of photodamage, we have recently developed a fluctuation analysis as an alternative technique to estimate the time of compartmentalization (B. L. Sabatini, T. Oertner & K. Svoboda, submitted). These studies have produced largely consistent results, which demonstrate that diffusional exchange between spine head and dendrite is in the range 20–200 ms, about a factor of 100 slower than expected for free diffusion over the small distances between spine and dendrite (182–184; B. L. Sabatini, T. Oertner & K. Svoboda, submitted). Other diffusible molecules with comparable diffusion coefficients ($\sim 1 \times 10^{-6}$ cm^2/s), such as some second messengers (186) including free Ca^{2+} and small proteins (187), would be compartmentalized over similar times, while larger signaling and/or more slowly diffusing molecules would be compartmentalized over longer times. These measurements show that spine necks act as diffusion barriers, isolating spine heads from their parent dendrites for durations that are long on time scales of biochemical reactions. Diffusional compartmentalization of second messengers and activated enzymes could underlie synapse specificity in synaptic plasticity (106).

Spines as Electrical Compartments

The geometric constriction of the spine neck could also act as a resistive element limiting synaptic currents; changes in spine neck geometry could then modulate synaptic strength (177, 188, 189) (Figure 3b). However, compartmental models based on electron microscopic reconstructions of spine geometries (1) indicated that spine necks are probably not sufficiently restrictive to pose a significant impediment to synaptic currents (180, 190). Because of the close relationship between diffusion and electrical conduction (191), measurements of diffusional transport through spine necks can be used to compute the electrical resistance of spine necks (182). These measurements have confirmed the suspicion that spine neck resistances are too small to modulate synaptic currents (182, 184). (See the legend of Figure 3 for details.)

But even without influencing the sizes of synaptic currents, spine neck resistances could still produce compartmentalized elevations of membrane potential; synaptic currents passing through the spine neck produce a voltage difference between spine head and dendrite that may selectively activate voltage-sensitive conductances in the spine head (192, 193). These differences are proportional to spine neck resistances and the amplitudes of single-synapse synaptic currents. Unitary synaptic currents generated at parallel fiber (PF) to Purkinje cell synapses (194) are severalfold larger than those generated at Schaffer collateral to CA1 synapses (195), but the spine neck resistances at these synapses are comparable (182, 183). Thus the degree of voltage compartmentalization may differ between cell and synapse types. At CA1 synapses voltage drops across the spine neck are expected to be at most several millivolts. Although these potential differences are large compared with unitary synaptic potentials measured in the soma (Figure 3c), they probably do not elevate spine head potentials sufficiently to selectively activate voltage-gated conductances in the spine head (182). It is therefore unlikely that CA1 spines function as electrical compartments. On the other hand, in Purkinje cell spines the voltage drops across the spine neck could be larger than 10 mV, leaving open the possibility that unitary currents open voltage sensitive currents selectively in spine heads (192, 196).

Ca^{2+} SIGNALING IN SPINES

The entry of Ca^{2+} into neurons is known to activate many cellular pathways that lead to the regulation of synaptic transmission (197). Because the spine neck serves as a barrier to Ca^{2+} exchange between the spine head and the dendrite (see above), spine Ca^{2+} may play an important role in activating synapse-specific regulatory mechanisms. $[Ca^{2+}]$ transients in spines and dendrites following two types of stimuli (back-propagating action potentials and synaptic stimulation) and from three sources (voltage-sensitive Ca^{2+} channels, VSCCs; Ca^{2+}-permeable ligand-gated channels, such as the NMDA-type glutamate receptor; and intracellular Ca^{2+} stores) are considered. A general framework is discussed in this section. Details about the specifics of Ca^{2+} signaling in pyramidal neurons and cerebellar Purkinje neurons are presented in separate sections below. This topic has recently been reviewed with emphasis on technology (10) and Ca^{2+} handling (11, 198; see Figure 4).

In many neuronal types action potentials triggered in the axon hillock can invade the proximal dendrite and open VSCCs (reviewed in 199, 200). This results in a relatively uniform $[Ca^{2+}]$ elevation throughout the proximal portion of the dendritic tree (201). In more distal dendrites, the amplitude of the action potential (AP) decreases with distance from soma, resulting in progressively smaller AP–evoked Ca^{2+} influx (202). Typically action potentials alone do not result in Ca^{2+} influx through ligand-gated channels or in the release of Ca^{2+} from intracellular stores (185, 203, 204).

Synaptic stimulation can lead to Ca^{2+} influx from all three sources. VSCCs are, in certain cell types, opened by the depolarization produced by synaptic stimulation. In some cell types, released neurotransmitter opens Ca^{2+}-permeable receptors, such as NMDA receptors or certain subclasses of AMPA receptors. Lastly, release of Ca^{2+} from intracellular stores can be triggered by metabotropic neurotransmitter receptors or by Ca^{2+}-induced Ca^{2+}-release (CICR). The sources of synaptic Ca^{2+} have been controversial because it is difficult to selectively inhibit each of these three components. For example, blocking neurotransmission not only disrupts the direct Ca^{2+} influx through ligand-gated channels but also prevents the depolarization that leads to VSCCs opening and the initial Ca^{2+} influx that triggers CICR. Similarly bath application of VSCC blockers disrupts not only postsynaptic VSCCs but also prevents presynaptic Ca^{2+} influx thereby reducing or abolishing release of neurotransmitter.

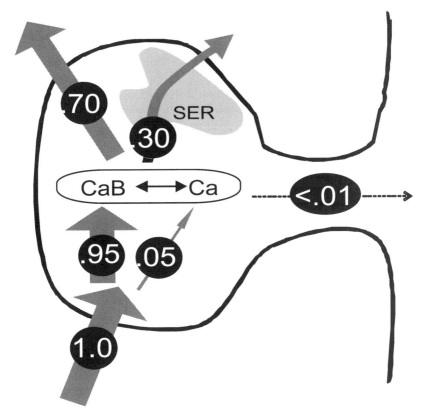

Figure 4   Sinks and sources of Ca^{2+}. Arrows show pathways of Ca^{2+} flux and the associated numbers show the fraction of Ca^{2+} handled by a particular pathway. Of the Ca^{2+} entering through Ca^{2+}-permeable channels, 95% binds to endogenous buffers and 5% stays free. Seventy percent of Ca^{2+} extrusion occurs directly across the plasma membrane, whereas 30% passes through the SER. A negligible fraction diffuses from the spine head to the dendrite.

Ca^{2+} Signaling in Pyramidal Cell Spines

In pyramidal neurons, back-propagating action potentials invade the proximal dendritic arbor (<200 μm) and trigger Ca^{2+} influx through VSCCs located on both the spine head and the dendritic shaft. The presence of VSCCs on spine heads is evident from the rapid rise time (<2 ms) of AP-evoked Ca^{2+} signals in diffusionally isolated spines (203, 205). The number of VSCCs opened per spine by a back-propagating action potential is small (usually <10), which leads to large trial-to-trial variability in the amplitude of AP-evoked [Ca^{2+}] transients, including failures (203). Pharmacological analysis has shown that action potential–evoked Ca^{2+} influx in CA1 spines is mostly through R-type VSCCs (203), whereas N/P/Q-type channels may contribute in neocortical pyramidal spines (206). Differences in channel subtype distribution may permit the differential regulation of Ca^{2+} channels in the spine versus the dendrite or soma, such as occurs in CA1 pyramidal neurons where spine but not dendritic VSCCs are inhibited by the activation of metabotropic GABA receptors (203). However, the functional role of Ca^{2+} influx through VSCCs in regulation of neuronal function is unclear. The inhibition of spine motility with low concentrations of AMPA is due to elevations of intracellular calcium (89). It is therefore possible that calcium entry through VSCCs could contribute to the stabilization of spines. Ca^{2+} released from internal stores does not appear to contribute to AP-evoked Ca^{2+} influx in spines (185, 204, 207).

Synaptic stimulation also elevates spine Ca^{2+}, but the sources of these transients are somewhat more controversial. Stimulation of a small number of axonal fibers allows the measurement of Ca^{2+} accumulations in an isolated spine due to neurotransmitter release at a single synapse (208). With these subthreshold stimuli, $[Ca^{2+}]$ transients are limited to the spine head and are abolished by blocking NMDA receptors (204–207, 209–211). Kovalchuk and colleagues were able to test individually the contributions of VSCCs, CICR, and NMDA receptor to synaptic $[Ca^{2+}]$ transients in acute brain slices and convincingly demonstrated that Ca^{2+} influx through NMDA receptors accounts for most of synaptic spine Ca^{2+} (209). They showed that blockade of CICR or of VSCCs in the postsynaptic cell does not reduce synaptic $[Ca^{2+}]$ transients. Furthermore, synaptic $[Ca^{2+}]$ transients were largely unaffected by blocking AMPA receptors and hence most of the synaptic potential, again ruling out significant contributions from VSCCs. Thus NMDA receptors, which at resting membrane potentials have weak voltage dependence and open relatively readily ($\sim$15% of peak) even in the presence of Mg^{2+} (212), are the main Ca^{2+} source of subthreshold synaptic $[Ca^{2+}]$ transients. This conclusion is in agreement with most studies (204–207, 209–211), although a few point to a contribution of CICR (204) or Ca^{2+} influx through VSCCs (206) downstream of glutamate receptor opening.

A large contribution of CICR was postulated by Emptage et al. who found that, in cultured hippocampal brain slices and using sharp microelectrodes, NMDA receptor-dependent synaptic $[Ca^{2+}]$ transients were nearly completely abolished following disruption of intracellular stores (204). The discrepancies between these studies and those discussed above may be largely due to differences in preparation (cultured versus acute brain slices and intracellular versus whole-cell recording). Indeed, as mentioned above, fewer than half of CA1 spines in intact brain tissue contain SER (19), the organelle responsible for CICR. A number of other studies, without directly investigating spine calcium, have found evidence for postsynaptic Ca^{2+} release from stores in response to synaptic activation when paired with APs (see below), which points to an important function of store-released Ca^{2+} (213). However, the precise conditions producing Ca^{2+} release from stores in spines and its role in synaptic plasticity remain obscure (reviewed in 214).

Lastly we consider Ca^{2+} influx during patterns of synaptic stimulation that lead to AP firing (i.e., suprathreshold) or in which APs have been experimentally paired with synaptic input. Pairing protocols in which a single AP is repeatedly fired within ~10 ms of a synaptic input effectively lead to LTP or long-term depression (LTD) of the synapse depending on whether the synaptic input precedes or lags the AP, respectively. Several studies have shown that pairing of short trains of APs and synaptic stimuli leads to Ca^{2+} accumulations larger than the sum of those generated by each stimulus alone (i.e., supralinear) (206, 210, 215). Koester & Sakmann examined the effects of pairing single APs and EPSCs and found that an AP followed closely by a synaptic input leads to sublinear Ca^{2+} accumulations in spines whereas, when the temporal order is reversed, supralinear Ca^{2+} accumulations are seen (211). It is interesting that trains of suprathreshold synaptic stimulation, which generate large Ca^{2+} accumulations ($>10 \ \mu M$) in both spines and dendrites (216), raise the Na^+ concentration in the spine by tens of millimolar (217). This accumulation is due to Na^+ flux through NMDA receptors whose Mg^{2+} block has been relieved by depolarization and is of sufficient magnitude to alter the reversal potential for current flow through Na^+ channels and glutamate receptors during subsequent stimuli. Although both the Na^+ accumulation and the nonlinear Ca^{2+} signals described above are all associative, their role in the induction of plasticity is as yet unproven.

Ca^{2+} Signaling in Purkinje Cell Spines

The synaptic structure of cerebellar Purkinje cells suggests immediately that the sources of synaptic Ca^{2+} are different from those in pyramidal neurons. Mature Purkinje cell spine synapses lack NMDA receptors, and the dendrites and spines are filled with an intricate ER that is studded with inositol triphosphate (IP3) receptors (218). The contribution of these intracellular Ca^{2+} stores to the calcium signal varies with the stimulation paradigm. Weak PF stimulation produces rapid-onset $[Ca^{2+}]$ transients localized to individual spines (196). Although blocking AMPA receptors uniformly abolishes these signals, the underlying Ca^{2+} source varies from spine to spine; hyperpolarizing the cell eliminates Ca^{2+} accumulations in some spines but accentuates them in others. Thus subthreshold synaptic stimulation results in Ca^{2+} influx through VSCCs in most spines with contributions from Ca^{2+}-permeable glutamate receptors in others (196). Short trains of subthreshold parallel fiber stimuli result in a biphasic $[Ca^{2+}]$ signal that is localized to individual spines (219, 220). The rapid component has the same properties as described above, whereas the late response can be blocked with antagonists of mGluRs as well as by drugs that interfere with Ca^{2+} release from IP3-sensitive stores. Stronger stimuli excite local regenerative spikes in fine dendritic branches that produce large Ca^{2+} accumulations mediated by VSCCs (221).

A further complexity is introduced when parallel fiber synaptic input is paired with climbing fiber input. Wang et al. paired trains of weak PF stimuli and single climbing fiber (CF) stimuli and, using low-affinity Ca^{2+} indicators to accurately measure large Ca^{2+} signals, found that CF input could potentiate PF-induced $[Ca^{2+}]$ transients severalfold (222). This potentiation was dependent on intact intracellular Ca^{2+} stores and, as is true for LTD, is largest when the CF stimulus occurs ~100 ms after the PF stimulus. The causal relationship between IP3-mediated Ca^{2+} release in spines and LTD was nicely demonstrated by Miyata et al. using mutant mice and rats deficient in myosin Va (223). These animals lack ER and associated IP3 receptors in spines but have normal synaptic structure and function including IP3-mediated synaptic Ca^{2+} elevations in dendrites. As expected, synaptic stimulation in these animals produces diminished IP3-mediated $[Ca^{2+}]$ transients in spines, and LTD is absent at their PF synapses. However, LTD can still be elicited by uncaging of Ca^{2+}, which shows convincingly that IP3-mediated Ca^{2+} release from intracellular stores in spines is a key mediator of synaptic plasticity.

Calcium Handling in Spines and Synaptic Plasticity

How can Ca^{2+}, a seemingly ubiquitous second messenger, carry information that will trigger a specific cellular response? For example, what properties of a particular spine Ca^{2+} signal determine whether LTD, LTP, or no change in synaptic efficacy will result? Clearly the details of the Ca^{2+} signal, including its amplitude, kinetics, and location, as well as its source, must be important. Unfortunately the relevant parameters are difficult to uncover because Ca^{2+} indicators, which are necessary to monitor Ca^{2+} levels, severely perturb $[Ca^{2+}]$ transients (see 11 for review). Ca^{2+} indicators are by necessity Ca^{2+} buffers and therefore act to counteract changes in intracellular Ca^{2+}, which makes stimulus-evoked changes in Ca^{2+} entry smaller and more prolonged. In addition, because indicators are small molecules that, unlike Ca^{2+} ions, can diffuse relatively freely within the cytoplasm, the presence of an indicator accelerates the diffusion of Ca^{2+} by a factor of 10–100 (186). Because of the dim signals from dendritic spines, large quantities (>100 μM) of high affinity ($K_d < 1$ μM) fluorescent Ca^{2+} indicators are often used for studying spine Ca^{2+}, thus dominating the endogenous buffering capacity of the spine and significantly perturbing Ca^{2+} signaling.

Fortunately, a framework for taking these effects into account has been developed (224, 225) and has been used to calculate the amplitude and timecourse of AP-evoked $[Ca^{2+}]$ transients and the endogenous Ca^{2+} buffering capacity in the main apical dendrite of unperturbed pyramidal neurons (226, 227). This method relies on using AP-evoked $[Ca^{2+}]$ transients to measure the impulse response of a cellular compartment under conditions of varying indicator concentrations and extrapolating back to the zero added indicator (i.e., unperturbed) case. Recently, this approach has been used to study Ca^{2+} transients and handling in spines of hippocampal pyramidal neurons (B. L. Sabatini, T. Oertner & K. Svoboda, submitted). In the absence of exogenous Ca^{2+} buffers, $[Ca^{2+}]$ transients in spines evoked by a single AP reach >1 μM and decay within <20 ms (Figure 5c). Since the diffusion of Ca^{2+} across the spine neck is slow (>100 ms) (182, 185) in unperturbed neurons, the spine head operates as a completely isolated compartment for the duration of AP-evoked transients. In contrast, in neurons filled with Ca^{2+} indicators the accelerated Ca^{2+} diffusion and prolonged $[Ca^{2+}]$ transients permit Ca^{2+} from the dendrite and spine head to mix, thus allowing the details of spine neck geometry to affect the time course of clearance of spine Ca^{2+} (184, 228).

Synaptically evoked $[Ca^{2+}]$ transients are also large in neurons without added Ca^{2+} buffers; NMDA receptor-mediated calcium transients following release of a single vesicle of glutamate reach ~ 1 μM and ~ 10 μM in spines held at resting and depolarized potentials, respectively (B. L. Sabatini, T. Oertner & K. Svoboda, submitted). The time course of NMDA receptor-mediated $[Ca^{2+}]$ transients are well matched to the convolution of the time course of the NMDA receptor EPSC and the impulse response of the spine measured with APs (Figure 5d). Thus Ca^{2+} handling in the spine is largely linear, i.e., the high levels of Ca^{2+} reached during synaptic stimulation do not saturate Ca^{2+} buffers and pumps and do not trigger CICR. Furthermore, correcting for the accelerated diffusion of Ca^{2+} in the presence of Ca^{2+} indicator, the average spine head is isolated on time scales of >1 s in unperturbed neurons, which allows for compartmentalized Ca^{2+} accumulation during trains of synaptic stimulation. The fast and large spine $[Ca^{2+}]$ transients are a consequence of the low endogenous Ca^{2+} buffer capacity (~ 25) and the high surface-to-volume ratio of the distal dendrite and spine. Thus the spine is a cellular compartment specialized for spatially restricted Ca^{2+} signaling in which $[Ca^{2+}]$ transients closely follow the kinetics of Ca^{2+} sources (B. L. Sabatini, T. Oertner & K. Svoboda, submitted).

What are the implications for synaptic plasticity? First, because spines operate as independent compartments on long time scales, local Ca^{2+} signaling necessary for synapse-specific plasticity is feasible. Secondly, because of the fast clearance of Ca^{2+} from the spine, the kinetics of Ca^{2+} sources are as important as the amplitude of $[Ca^{2+}]$ transients in determining the activation of Ca^{2+}-dependent processes. A useful test case is calmodulin wherein the Ca^{2+} sites are half occupied by ~ 1 μM Ca^{2+}, which binds Ca^{2+} with a rate-limiting time constant of ~ 15 ms in this concentration range [adjusted from (229) for $37°C$]. Because of this slow Ca^{2+}-binding rate, large but brief AP-evoked Ca^{2+} signals are relatively ineffective at activating calmodulin, whereas the smaller but prolonged NMDA receptor-mediated transients generated by subthreshold synaptic stimulation activate calmodulin and trigger LTD (Figure 5e). Larger $[Ca^{2+}]$ transients during suprathreshold synaptic stimulation activate calmodulin more robustly and trigger LTP. Such Ca^{2+} levels cannot be reached during AP trains as the rapid clearance of Ca^{2+} prevents its accumulation even during high frequency during trains (Figure 5f) (B. L. Sabatini, T. Oertner & K. Svoboda, submitted). Thus the kinetics of Ca^{2+} sources may explain the inability of AP-evoked $[Ca^{2+}]$ transients to evoke LTP and LTD, whereas the differences in amplitude of synaptic $[Ca^{2+}]$ transients in depolarized or resting neurons may explain the differential activation of LTP or LTD, respectively.

SOME OPEN QUESTIONS

Although our understanding of the structure and function of dendritic spines has increased dramatically in recent years, many key issues have yet to be addressed. For instance, we are only beginning to appreciate the remarkable complexity of the PSD as one of the richest molecular signal processing machines known. The

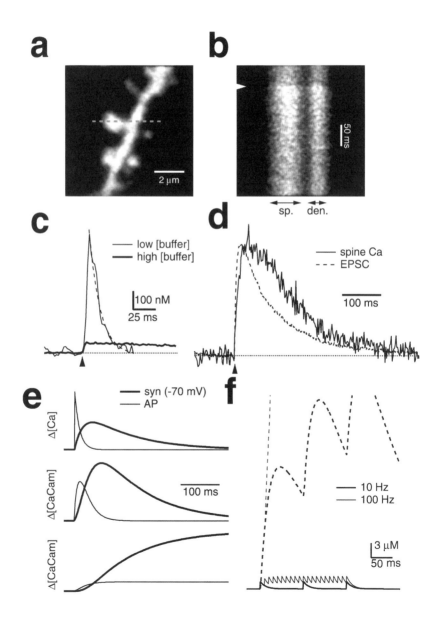

correspondence of one PSD per spine suggests that the spine serves to biochemically isolate PSDs from one another. We need to measure signal transduction in individual PSDs, how it is shaped by synaptic transmission, and how it feeds back to the regulation of synaptic strength. Similarly, the role of polyribosomal complexes in spines remains unexplored. Can translation be turned on in a synapse-specific manner in individual spines? Now that we know that dendritic spines are motile and that they turn over, it should be possible to determine the lifetime of a spine in the intact experimental animal brain. This would be an indication of the lifetime of synapses in the adult human brain, a critical parameter for models of long-term memory storage. A growing number of human disorders have been associated with abnormalities in dendritic spines, but the connection between morphological and clinical phenotype has yet to be established in a single instance. Most disorders are associated with subtle spine malformations, such as shifts in the length

---

Figure 5 Ca^{2+} handling in dendritic spines. (*a*) Image of an apical spine of a CA1 pyramidal neuron filled with Ca^{2+} indicator (100 μM OGB1) and collected with 2PSLM. (*b*) Repeated line scan (500 Hz) over the segment indicated in (*a*). At the time indicated by the arrowhead, current was injected into the soma to elicit a back-propagating action potential. Fluorescence increases quickly and then decays slowly in both the spine and dendrite. The fluorescence from the spatial extent of the spine, as indicated by the arrows, is averaged to produce a fluorescence transient, $F_{spine}(t)$, which is then used to calculate $[Ca^{2+}]_{spine}(t)$. (*c*) AP-evoked $[Ca^{2+}]$ transient in an apical spine of a neuron filled with a small amount (20 μM Fluo-4, *thin trace*) or large amount (100 μM OGB1, *thick trace*) of exogenous Ca^{2+} buffer. The arrowhead shows the time of current injection and AP firing at the soma. The large amount of added buffer reduces the amplitude and prolongs the time course of the $[Ca^{2+}]$ transient. The dashed line shows an exponential fit ($\tau = 12$ ms) to the rapid $[Ca^{2+}]$ transient. (*d*) Time course of NMDA receptor mediated fluorescence increases (*solid trace*) in an apical spine from a neuron containing a small amount of exogenous Ca^{2+} buffer (100 μM MgGreen) and held at 0 mV. Under these conditions, fluorescence is linearly related to intracellular $[Ca^{2+}]$. The arrowhead shows the time of stimulation of the axonal fiber that synapses onto the imaged spine. Owing to the rapid clearance of Ca^{2+} from the spine, the time course of the $[Ca^{2+}]$ transient only slightly lags that of the EPSC (*inverted, dashed line*). (*e*) The kinetics of Ca^{2+} sources can differentially affect the activation of Ca^{2+}-binding proteins including calmodulin. Model showing spine $[Ca^{2+}]$ (*top*), activated calmodulin (*middle*), and total calmodulin activation (*bottom*) following an AP (*thin trace*) or synaptic (*thick trace*) stimulation at resting potentials. Calmodulin bound Ca^{2+} with an equilibration time constant of 15 ms. (*f*) Ca^{2+} accumulation in a dendritic spine during 200 ms of 10 Hz (*thick trace*) and 100 Hz (*thin trace*) trains of synaptic stimuli in a depolarized neuron (*dashed traces*) or APs (*solid traces*). The Ca^{2+} accumulations during the synaptic trains continue off the upper range of the graph. The rapid clearance of Ca^{2+} from the spine prevents its accumulation during AP trains.

distributions of spines or changes in their morphologies. Detailed ultrastructural, physiological, and biochemical studies will likely be necessary to discover the associated functional abnormalities in synaptic connectivity and function.

ACKNOWLEDGMENTS

We thank Drs. Dimitri Chklovskii, Hollis Cline, Patrick R. Hof, Alla Karpova, and Karen Zito for a critical reading of the manuscript and Dr. Josef Spacek for the use of electron micrographs from his online atlas. Work reviewed here was supported by fellowships from the National Institutes of Health (E. A. N.) and Helen Hay Whitney Foundations (B. L. S.); by grants from the Pew, Klingenstein, Whitaker, and Mathers Foundations; and by the National Institutes of Health.

Visit the Annual Reviews home page at www.AnnualReviews.org

LITERATURE CITED

1. Harris KM, Kater SB. 1994. Dendritic spines: cellular specializations imparting both stability and flexibility to synaptic function. *Annu. Rev. Neurosci.* 17:341–71

2. Coss RG, Brandon JG, Globus A. 1980. Changes in morphology of dendritic spines on honeybee calycal interneurons associated with cumulative nursing and foraging experiences. *Brain Res.* 192:49–59

3. Sarnat HB, Netsky MG. 1985. The brain of the planarian as the ancestor of the human brain. *Can. J. Neurol. Sci.* 12:296–302

4. Nassel DR, Strausfeld NJ. 1982. A pair of descending neurons with dendrites in the optic lobes projecting directly to thoracic ganglia of dipterous insects. *Cell Tissue Res.* 226:355–62

5. Ramón y Cajal S. 1891. Sur la structure de l'écorce cérébrale de quelques mamifères. *La Cellule* 7:125–76

6. Purpura DP. 1975. Normal and aberrant neuronal development in the cerebral cortex of human fetus and young infant. *UCLA Forum Med. Sci.* 18:141–69

7. Denk W, Svoboda K. 1997. Photon upmanship: why multiphoton imaging is more than a gimmick. *Neuron* 18:351–57

8. Yuste R, Bonhoeffer T. 2001. Morphological changes in dendritic spines associated with long-term synaptic plasticity. *Annu. Rev. Neurosci.* 24:1071–89

9. Jontes JD, Smith SJ. 2000. Filopodia, spines, and the generation of synaptic diversity. *Neuron* 27:11–14

10. Denk W, Yuste R, Svoboda K, Tank DW. 1996. Imaging calcium dynamics in dendritic spines. *Curr. Opin. Neurobiol.* 6:372–78

11. Sabatini BL, Maravall M, Svoboda K. 2001. Ca^{2+} signaling in spines. *Curr. Opin. Neurobiol.* 11:349–56

12. Nusser Z, Lujan R, Laube G, Roberts JD, Molnar E, Somogyi P. 1998. Cell type and pathway dependence of synaptic AMPA receptor number and variability in the hippocampus. *Neuron* 21:545–59

13. Peters A, Palay SL, Webster HD. 1991. Synapses. In *The Fine Structure of the Nervous System: Neurons and Their Supporting Cells*, pp. 138–211. New York: Oxford Univ. Press

14. Gray EG. 1959. Axo-somatic and axo-dendritic synapses of the cerebral cortex: an electron microscope study. *J. Anat.* 93:420–33

15. Fifkova E, Delay RJ. 1982. Cytoplasmic

actin in neuronal processes as a possible mediator of synaptic plasticity. *J. Cell Biol.* 95:345–50

16. Matus A, Ackermann M, Pehling G, Byers HR, Fujiwara K. 1982. High actin concentrations in brain dendritic spines and postsynaptic densities. *Proc. Natl. Acad. Sci. USA* 79:7590–94

17. Capani F, Martone ME, Deerinck TJ, Ellisman MH. 2001. Selective localization of high concentrations of F-actin in subpopulations of dendritic spines in rat central nervous system: a three-dimensional electron microscopic study. *J. Comp. Neurol.* 435:156–70

18. Peters A, Palay SL, Webster HD. 1991. Dendrites. In *The Fine Structure of the Nervous System: Neurons and Their Supporting Cells*, pp. 70–100. New York: Oxford Univ. Press

19. Spacek J, Harris KM. 1997. Three-dimensional organization of smooth endoplasmic reticulum in hippocampal CA1 dendrites and dendritic spines of the immature and mature rat. *J. Neurosci.* 17:190–203

20. Gray EG, Guillery RW. 1963. A note on the dendritic spine apparatus. *J. Anat. Lond.* 97:389–92

21. Westrum LE, Jones DH, Gray EG, Barron J. 1980. Microtubules, dendritic spines and spine appratuses. *Cell Tissue Res.* 208:171–81

22. Andrews SB, Leapman RD, Landis DM, Reese TS. 1988. Activity-dependent accumulation of calcium in Purkinje cell dendritic spines. *Proc. Natl. Acad. Sci. USA* 85:1682–85

23. Burgoyne RD, Gray EG, Barron J. 1983. Cytochemical localization of calcium in the dendritic spine apparatus of the cerebral cortex and at synaptic sites in the cerebellar cortex. *J. Anat.* 136:634–35

24. Peters A, Kaiserman-Abramof IR. 1970. The small pyramidal neuron of the rat cerebral cortex. The perikaryon, dendrites and spines. *Am. J. Anat.* 127:321–55

25. Steward O, Falk PM. 1985. Polyribosomes under developing spine synapses: growth specializations of dendrites at sites of synaptogenesis. *J. Neurosci. Res.* 13:75–88

26. Steward O, Reeves TM. 1988. Protein-synthetic machinery beneath postsynaptic sites on CNS neurons: association between polyribosomes and other organelles at the synaptic site. *J. Neurosci.* 8:176–84

27. Spacek J. 1985. Three-dimensional analysis of dendritic spines. II. Spine apparatus and other cytoplasmic components. *Anat. Embryol.* 171:235–43

28. Husi H, Ward MA, Choudhary JS, Blackstock WP, Grant SG. 2000. Proteomic analysis of NMDA receptor-adhesion protein signaling complexes. *Nat. Neurosci.* 3:661–69

29. Walikonis RS, Jensen ON, Mann M, Provance DW Jr, Mercer JA, Kennedy MB. 2000. Identification of proteins in the postsynaptic density fraction by mass spectrometry. *J. Neurosci.* 20:4069–80

30. Kennedy MB. 2000. Signal-processing machines at the postsynaptic density. *Science* 290:750–54

31. Spacek J. 1985. Relationships between synaptic junctions, puncta adhaerentia and the spine apparatus at neocortical axospinous synapses. A serial section study. *Anat. Embryol.* 173:129–35

32. Harris KM, Stevens JK. 1989. Dendritic spines of CA1 pyramidal cells in the rat hippocampus: serial electron microscopy with reference to their biophysical characteristics. *J. Neurosci.* 9:2982–97

33. Desmond NL, Levy WB. 1986. Changes in the numerical density of synaptic contacts with long-term potentiation in the hippocampal dentate gyrus. *J. Comp. Neurol.* 253:466–75

34. Desmond NL, Levy WB. 1986. Changes in the postsynaptic density with long-term potentiation in the dentate gyrus. *J. Comp. Neurol.* 253:476–82

35. Buchs PA, Muller D. 1996. Induction of long-term potentiation is associated with major ultrastructural changes of activated

synapses. *Proc. Natl. Acad. Sci. USA* 93: 8040–45

36. Skoff RP, Hamburger V. 1974. Fine structure of dendritic and axonal growth cones in embryonic chick spinal cord. *J. Comp. Neurol.* 153:107–47

37. Tarrant SB, Routtenberg A. 1977. The synaptic spinule in the dendritic spine: electron microscopic study of the hippocampal dentate gyrus. *Tissue Cell* 9: 461–73

38. Harris KM, Jensen FE, Tsao B. 1992. Three-dimensional structure of dendritic spines and synapses in rat hippocampus (CA1) at postnatal day 15 and adult ages: implications for the maturation of synaptic physiology and long-term potentiation. *J. Neurosci.* 12:2685–705

39. Napper RM, Harvey RJ. 1988. Quantitative study of the Purkinje cell dendritic spines in the rat cerebellum. *J. Comp. Neurol.* 274:158–67

40. Harris KM, Stevens JK. 1988. Dendritic spines of rat cerebellar Purkinje cells: serial electron microscopy with reference to their biophysical characteristics. *J. Neurosci.* 8:4455–69

41. Terasaki M, Slater NT, Fein A, Schmidek A, Reese TS. 1994. Continuous network of endoplasmic reticulum in cerebellar Purkinje neurons. *Proc. Natl. Acad. Sci. USA* 91:7510–14

42. Palay SL, Chan-Palay V. 1974. The Purkinje cell. In *Cerebellar Cortex: Cytology and Organization*, pp. 11–62. Berlin: Springer-Verlag

43. Racca C, Stephenson FA, Streit P, Roberts JD, Somogyi P. 2000. NMDA receptor content of synapses in stratum radiatum of the hippocampal CA1 area. *J. Neurosci.* 20:2512–22

44. Takumi Y, Ramirez-Leon V, Laake P, Rinvik E, Ottersen OP. 1999. Different modes of expression of AMPA and NMDA receptors in hippocampal synapses. *Nat. Neurosci.* 2:618–24

45. Schikorski T, Stevens CF. 1997. Quantitative ultrastructural analysis of hippo-

campal excitatory synapses. *J. Neurosci.* 17:5858–67

46. Murthy VN, Sejnowski TJ, Stevens CF. 1997. Heterogeneous release properties of visualized individual hippocampal synapses. *Neuron* 18:599–612

47. El-Husseini AE, Schnell E, Chetkovich DM, Nicoll RA, Bredt DS. 2000. PSD-95 involvement in maturation of excitatory synapses. *Science* 290:1364–68

48. Crepel F, Dhanjal SS, Garthwaite J. 1981. Morphological and electrophysiological characteristics of rat cerebellar slices maintained in vitro. *J. Physiol.* 316:127–38

49. Spacek J. 1985. Three-dimensional analysis of dendritic spines. III. Glial sheath. *Anat. Embryol.* 171:245–52

50. Barbour B, Keller BU, Llano I, Marty A. 1994. Prolonged presence of glutamate during excitatory synaptic transmission to cerebellar Purkinje cells. *Neuron* 12: 1331–43

51. Katona I, Acsady L, Freund TF. 1999. Postsynaptic targets of somatostatin-immunoreactive interneurons in the rat hippocampus. *Neuroscience* 88:37–55

52. Fifkova E, Eason H, Schaner P. 1992. Inhibitory contacts on dendritic spines of the dentate fascia. *Brain Res.* 577:331–36

53. Halasy K, Somogyi P. 1993. Distribution of GABAergic synapses and their targets in the dentate gyrus of rat: a quantitative immunoelectron microscopic analysis. *J. Hirnforsch.* 34:299–308

54. Ventura R, Harris KM. 1999. Three-dimensional relationships between hippocampal synapses and astrocytes. *J. Neurosci.* 19:6897–906

55. Beaulieu C, Kisvarday Z, Somogyi P, Cynader M, Cowey A. 1992. Quantitative distribution of GABA-immunopositive and -immunonegative neurons and synapses in the monkey striate cortex (area 17). *Cereb. Cortex* 2:295–309

56. Jakab RL, Goldman-Rakic P, Leranth C. 1997. Dual role of substance P/GABA

axons in cortical neurotransmission: synaptic triads on pyramidal cell spines and basket-like innervation of layer II–III calbindin interneurons in primate prefrontal cortex. *Cereb. Cortex* 7:359–73

57. Carr DB, Sesack SR. 1998. Callosal terminals in the rat prefrontal cortex: synaptic targets and association with GABA-immunoreactive structures. *Synapse* 29:193–205

58. Goldman-Rakic PS, Leranth C, Williams SM, Mons N, Geffard M. 1989. Dopamine synaptic complex with pyramidal neurons in primate cerebral cortex. *Proc. Natl. Acad. Sci. USA* 86:9015–19

59. Sorra KE, Harris KM. 1998. Stability in synapse number and size at 2 hr after long-term potentiation in hippocampal area CA1. *J. Neurosci.* 18:658–71

60. Trommald M, Hulleberg G. 1997. Dimensions and density of dendritic spines from rat dentate granule cells based on reconstructions from serial electron micrographs. *J. Comp. Neurol.* 377:15–28

61. Valverde F. 1967. Apical dendritic spines of the visual cortex and light deprivation in the mouse. *Exp. Brain Res.* 3:337–52

62. Marin-Padilla M, Stibitz GR. 1968. Distribution of the apical dendritic spines of the layer V pyramidal cells of the hamster neocortex. *Brain Res.* 11:580–92

63. Chicurel ME, Harris KM. 1992. Three-dimensional analysis of the structure and composition of CA3 branched dendritic spines and their synaptic relationships with mossy fiber boutons in the rat hippocampus. *J. Comp. Neurol.* 325:169–82

64. Brodmann K. 1909. *Vergleichende Lokalisationslehre der Grosshirnrinde in ihren Prinzipien dargestellt auf Grund des Zellenbaues.* Leipzig: Barth

65. Elston GN. 2000. Pyramidal cells of the frontal lobe: all the more spinous to think with. *J. Neurosci.* 20:RC95

66. Jacobs B, Schall M, Prather M, Kapler E, Driscoll L, et al. 2001. Regional dendritic and spine variation in human cerebral cortex: a quantitative Golgi study. *Cereb. Cortex* 11:558–71

67. Micheva KD, Beaulieu C. 1996. Quantitative aspects of synaptogenesis in the rat barrel field cortex with special reference to GABA circuitry. *J. Comp. Neurol.* 373:340–54

68. White EL, Weinfeld L, Lev DL. 1997. A survey of morphogenesis during the early postnatal period in PMBSF barrels of mouse SmI cortex with emphasis on barrel D4. *Somatosens. Motil. Res.* 14:34–55

69. Wise SP, Fleshman JW Jr, Jones EG. 1979. Maturation of pyramidal cell form in relation to developing afferent and efferent connections of rat somatic sensory cortex. *Neuroscience* 4:1275–97

70. Boothe RG, Greenough WT, Lund JS, Wrege K. 1979. A quantitative investigation of spine and dendrite development of neurons in visual cortex (area 17) of *Macaca nemestrina* monkeys. *J. Comp. Neurol.* 186:473–89

71. Nimchinsky EA, Oberlander AM, Svoboda K. 2001. Abnormal development of dendritic spines in FMR1 knock-out mice. *J. Neurosci.* 21:5139–46

72. Dunaevsky A, Tashiro A, Majewska A, Mason C, Yuste R. 1999. Developmental regulation of spine motility in the mammalian central nervous system. *Proc. Natl. Acad. Sci. USA* 96:13,438–43

73. Dailey ME, Smith SJ. 1996. The dynamics of dendritic structure in developing hippocampal slices. *J. Neurosci.* 16:2983–94

74. Ziv NE, Smith SJ. 1996. Evidence for a role of dendritic filopodia in synaptogenesis and spine formation. *Neuron* 17:91–102

75. Adams I, Jones DG. 1982. Quantitative ultrastructural changes in rat cortical synapses during early-, mid- and late-adulthood. *Brain Res.* 239:349–63

76. Fiala JC, Feinberg M, Popov V, Harris KM. 1998. Synaptogenesis via dendritic

filopodia in developing hippocampal area CA1. *J. Neurosci.* 18:8900–11

77. Harris KM. 1999. Structure, development, and plasticity of dendritic spines. *Curr. Opin. Neurobiol.* 9:343–48

78. Lendvai B, Stern E, Chen B, Svoboda K. 2000. Experience-dependent plasticity of dendritic spines in the developing rat barrel cortex in vivo. *Nature* 404:876–81

79. Globus A, Scheibel AB. 1966. Loss of dendrite spines as an index of pre-synaptic terminal patterns. *Nature* 212:463–65

80. Valverde F. 1971. Rate and extent of recovery from dark rearing in the visual cortex of the mouse. *Brain Res.* 33:1–11

81. Parnavelas JG, Lynch G, Brecha N, Cotman CW. 1974. Spine loss and regrowth in hippocampus following deafferentiation. *Nature* 248:71–73

82. Hirano A, Dembitzer HM, Yoon CH. 1977. Development of Purkinje cell somatic spines in the weaver mouse. *Acta Neuropathol.* 40:85–90

83. Hirano A. 1983. The normal and aberrant development of synaptic structures between parallel fibers and Purkinje cell dendritic spines. *J. Neural Transm. Suppl.* 18:1–8

84. Takacs J, Gombos G, Gorcs T, Becker T, de Barry J, Hamori J. 1997. Distribution of metabotropic glutamate receptor type 1a in Purkinje cell dendritic spines is independent of the presence of presynaptic parallel fibers. *J. Neurosci. Res.* 50:433–42

85. Baloyannis SJ, Kim SU. 1979. Experimental modification of cerebellar development in tissue culture: X-irradiation induces granular degeneration and unattached Purkinje cell dendritic spines. *Neurosci. Lett.* 12:283–88

86. Mates SL, Lund JS. 1983. Spine formation and maturation of type 1 synapses on spiny stellate neurons in primate visual cortex. *J. Comp. Neurol.* 221:91–97

87. Fischer M, Kaech S, Knutti D, Matus A. 1998. Rapid actin-based plasticity in dendritic spines. *Neuron* 20:847–54

88. Kaech S, Brinkhaus H, Matus A. 1999. Volatile anesthetics block actin-based motility in dendritic spines. *Proc. Natl. Acad. Sci. USA* 96:10,433–37

89. Fischer M, Kaech S, Wagner U, Brinkhaus H, Matus A. 2000. Glutamate receptors regulate actin-based plasticity in dendritic spines. *Nat. Neurosci.* 3:887–94

90. Matus A. 2000. Actin-based plasticity in dendritic spines. *Science* 290:754–58

91. Papa M, Segal M. 1996. Morphological plasticity in dendritic spines of cultured hippocampal neurons. *Neuroscience* 71:1005–11

92. Annis CM, O'Dowd DK, Robertson RT. 1994. Activity-dependent regulation of dendritic spine density on cortical pyramidal neurons in organotypic slice cultures. *J. Neurobiol.* 25:1483–93

93. Muller M, Gähwiler BH, Rietschin L, Thompson SM. 1993. Reversible loss of dendritic spines and altered excitability after chronic epilepsy in hippocampal slice cultures. *Proc. Natl. Acad. Sci. USA* 90:257–61

94. Gutierrez R, Heinemann U. 1999. Synaptic reorganization in explanted cultures of rat hippocampus. *Brain Res.* 815:304–16

95. Drakew A, Muller M, Gähwiler BH, Thompson SM, Frotscher M. 1996. Spine loss in experimental epilepsy: quantitative light and electron microscopic analysis of intracellularly stained CA3 pyramidal cells in hippocampal slice cultures. *Neuroscience* 70:31–45

96. Korkotian E, Segal M. 1999. Bidirectional regulation of dendritic spine dimensions by glutamate receptors. *NeuroReport* 10:2875–77

97. Goldin M, Segal M, Avignone E. 2001. Functional plasticity triggers formation and pruning of dendritic spines in cultured hippocampal networks. *J. Neurosci.* 21:186–93

98. Halpain S, Hipolito A, Saffer L. 1998. Regulation of F-actin stability in dendritic spines by glutamate receptors and calcineurin. *J. Neurosci.* 18:9835–44

99. Korkotian E, Segal M. 1999. Release of calcium from stores alters the morphology of dendritic spines in cultured hippocampal neurons. *Proc. Natl. Acad. Sci. USA* 96:12,068–72

100. Segal M. 1995. Dendritic spines for neuroprotection: a hypothesis. *Trends Neurosci.* 18:468–71

101. McKinney RA, Capogna M, Durr R, Gähwiler BH, Thompson SM. 1999. Miniature synaptic events maintain dendritic spines via AMPA receptor activation. *Nat. Neurosci.* 2:44–49

102. Liao D, Hessler NA, Malinow R. 1995. Activation of postsynaptically silent synapses during pairing-induced LTP in CA1 region of hippocampal slice. *Nature* 375:400–4

103. Isaac JT, Nicoll RA, Malenka RC. 1995. Evidence for silent synapses: implications for the expression of LTP. *Neuron* 15:427–34

104. Shi SH, Hayashi Y, Petralia RS, Zaman SH, Wenthold RJ, et al. 1999. Rapid spine delivery and redistribution of AMPA receptors after synaptic NMDA receptor activation. *Science* 284:1811–16

105. Bozdagi O, Shan W, Tanaka H, Benson DL, Huntley GW. 2000. Increasing numbers of synaptic puncta during late-phase LTP: N-cadherin is synthesized, recruited to synaptic sites, and required for potentiation. *Neuron* 28:245–59

106. Bliss TVP, Collingridge GL. 1993. A synaptic model of memory: long-term potentiation in the hippocampus. *Nature* 361:31–39

107. Van Harreveld A, Fifkova E. 1975. Swelling of dendritic spines in the fascia dentata after stimulation of the perforant fibers as a mechanism of post-tetanic potentiation. *Exp. Neurol.* 49:736–49

108. Bliss TV, Lomo T. 1973. Long-lasting potentiation of synaptic transmission in the dentate area of the anaesthetized rabbit following stimulation of the perforant path. *J. Physiol.* 232:331–56

109. Fifkova E, Van Harreveld A. 1977. Long-lasting morphological changes in dendritic spines of dentate granular cells following stimulation of the entorhinal area. *J. Neurocytol.* 6:211–30

110. Fifkova E, Anderson CL. 1981. Stimulation-induced changes in dimensions of stalks of dendritic spines in the dentate molecular layer. *Exp. Neurol.* 74:621–27

111. Desmond NL, Levy WB. 1988. Synaptic interface surface area increases with long-term potentiation in the hippocampal dentate gyrus. *Brain Res.* 453:308–14

112. Geinisman Y, deToledo-Morrell L, Morrell F. 1991. Induction of long-term potentiation is associated with an increase in the number of axospinous synapses with segmented postsynaptic densities. *Brain Res.* 566:77–88

113. Lee KS, Schottler F, Oliver M, Lynch G. 1980. Brief bursts of high-frequency stimulation produce two types of structural change in rat hippocampus. *J. Neurophysiol.* 44:247–58

114. Chang FL, Greenough WT. 1984. Transient and enduring morphological correlates of synaptic activity and efficacy change in the rat hippocampal slice. *Brain Res.* 309:35–46

115. Vees AM, Micheva KD, Beaulieu C, Descarries L. 1998. Increased number and size of dendritic spines in ipsilateral barrel field cortex following unilateral whisker trimming in postnatal rat. *J. Comp. Neurol.* 400:110–24

116. Hosokawa T, Rusakov DA, Bliss TVP, Fine A. 1995. Repeated confocal imaging of individual dendritic spines in the living hippocampal slice: evidence for changes in length and orientation associated with chemically induced LTP. *J. Neurosci.* 15:5560–73

117. Maletic-Savatic M, Malinow R, Svoboda K. 1999. Rapid dendritic morphogenesis in CA1 hippocampal dendrites induced by synaptic activity. *Science* 283:1923–27

118. Engert F, Bonhoeffer T. 1999. Dendritic

spine changes associated with hippocampal long-term synaptic plasticity. *Nature* 399:66–70

119. Buchs PA, Stoppini L, Parducz A, Siklos L, Muller D. 1994. A new cytochemical method for the ultrastructural localization of calcium in the central nervous system. *J. Neurosci. Methods* 54:83–93

120. Toni N, Buchs PA, Nikonenko I, Bron CR, Muller D. 1999. LTP promotes formation of multiple spine synapses between a single axon terminal and a dendrite. *Nature* 402:421–25

121. Moser MB, Trommald M, Andersen P. 1994. An increase in dendritic spine density on hippocampal CA1 cells following spatial-learning in adult rats suggests the formation of new synapses. *Proc. Natl. Acad. Sci. USA* 91:12,673–75

122. Knafo S, Grossman Y, Barkai E, Benshalom G. 2001. Olfactory learning is associated with increased spine density along apical dendrites of pyramidal neurons in the rat piriform cortex. *Eur. J. Neurosci.* 13:633–38

123. Rusakov DA, Davies HA, Harrison E, Diana G, Richter-Levin G, et al. 1997. Ultrastructural synaptic correlates of spatial learning in rat hippocampus. *Neuroscience* 80:69–77

124. Geinisman Y. 2000. Structural synaptic modifications associated with hippocampal LTP and behavioral learning. *Cereb. Cortex* 10:952–62

125. O'Malley A, O'Connell C, Regan CM. 1998. Ultrastructural analysis reveals avoidance conditioning to induce a transient increase in hippocampal dentate spine density in the 6 hour post-training period of consolidation. *Neuroscience* 87:607–13

126. O'Malley A, O'Connell C, Murphy KJ, Regan CM. 2000. Transient spine density increases in the mid-molecular layer of hippocampal dentate gyrus accompany consolidation of a spatial learning task in the rodent. *Neuroscience* 99:229–32

127. Woolley CS, Gould E, Frankfurt M, McEwen BS. 1990. Naturally occurring fluctuation in dendritic spine density on adult hippocampal pyramidal neurons. *J. Neurosci.* 10:4035–39

128. Woolley CS, McEwen BS. 1992. Estradiol mediates fluctuation in hippocampal synapse density during the estrous cycle in the adult rat. *J. Neurosci.* 12:2549–54

129. Gould E, Woolley CS, Frankfurt M, McEwen BS. 1990. Gonadal steroids regulate dendritic spine density in hippocampal pyramidal cells in adulthood. *J. Neurosci.* 10:1286–91

130. Murphy DD, Segal M. 1996. Regulation of dendritic spine density in cultured rat hippocampal neurons by steroid hormones. *J. Neurosci.* 16:4059–68

131. Woolley CS, Wenzel HJ, Schwartzkroin PA. 1996. Estradiol increases the frequency of multiple synapse boutons in the hippocampal CA1 region of the adult female rat. *J. Comp. Neurol.* 373:108–17

132. Yankova M, Hart SA, Woolley CS. 2001. From the cover: estrogen increases synaptic connectivity between single presynaptic inputs and multiple postsynaptic CA1 pyramidal cells: a serial electron-microscopic study. *Proc. Natl. Acad. Sci. USA* 98:3525–30

133. Murphy DD, Segal M. 2000. Progesterone prevents estradiol-induced dendritic spine formation in cultured hippocampal neurons. *Neuroendocrinology* 72:133–43

134. Wooley CS, McEwen BS. 1994. Estradiol regulates hippocampal dendritic spine density via an N-methyl-D-aspartate receptor-dependent mechanism. *J. Neurosci.* 14:7680–87

135. Murphy DD, Cole NB, Greenberger V, Segal M. 1998. Estradiol increases dendritic spine density by reducing GABA neurotransmission in hippocampal neurons. *J. Neurosci.* 18:2550–59

136. Murphy DD, Cole NB, Segal M. 1998. Brain-derived neurotrophic factor mediates estradiol-induced dendritic spine

formation in hippocampal neurons. *Proc. Natl. Acad. Sci. USA* 95:11,412–17

137. Horch HW, Kruttgen A, Portbury SD, Katz LC. 1999. Destabilization of cortical dendrites and spines by BDNF. *Neuron* 23:353–64

138. Gould E, Westlind-Danielsson A, Frankfurt M, McEwen BS. 1990. Sex differences and thyroid hormone sensitivity of hippocampal pyramidal cells. *J. Neurosci.* 10:996–1003

139. Schapiro S, Vukovich K, Globus A. 1973. Effects of neonatal thyroxine and hydrocortisone administration on the development of dendritic spines in the visual cortex of rats. *Exp. Neurol.* 40:286–96

140. Ruiz-Marcos A, Sanchez-Toscano F, Escobar del Rey F, Morreale de Escobar G. 1979. Severe hypothyroidism and the maturation of the rat cerebral cortex. *Brain Res.* 162:315–29

141. Ruiz-Marcos A, Sanchez-Toscano F, Obregon MJ, Escobar del Rey F, Morreale de Escobar G. 1982. Thyroxine treatment and recovery of hypothyroidism-induced pyramidal cell damage. *Brain Res.* 239:559–74

142. Gould E, Allan MD, McEwen BS. 1990. Dendritic spine density of adult hippocampal pyramidal cells is sensitive to thyroid hormone. *Brain Res.* 525:327–29

143. Garey LJ, Ong WY, Patel TS, Kanani M, Davis A, et al. 1998. Reduced dendritic spine density on cerebral cortical pyramidal neurons in schizophrenia. *J. Neurol. Neurosurg. Psychiatry* 65:446–53

144. Glantz LA, Lewis DA. 2000. Decreased dendritic spine density on prefrontal cortical pyramidal neurons in schizophrenia. *Arch. Gen. Psychiatry* 57:65–73

145. Morrison JH, Hof PR. 1997. Life and death of neurons in the aging brain. *Science* 278:412–19

146. Jacobs B, Batal HA, Lynch B, Ojemann G, Ojemann LM, Scheibel AB. 1993. Quantitative dendritic and spine analyses of speech cortices: a case study. *Brain Lang.* 44:239–53

147. Jacobs B, Driscoll L, Schall M. 1997. Life-span dendritic and spine changes in areas 10 and 18 of human cortex: a quantitative Golgi study. *J. Comp. Neurol.* 386:661–80

148. Cupp CJ, Uemura E. 1980. Age-related changes in prefrontal cortex of *Macaca mulatta*: quantitative analysis of dendritic branching patterns. *Exp. Neurol.* 69:143–63

149. Uemura E. 1980. Age-related changes in prefrontal cortex of *Macaca mulatta*: synaptic density. *Exp. Neurol.* 69:164–72

150. Peters A, Sethares C, Moss MB. 1998. The effects of aging on layer 1 in area 46 of prefrontal cortex in the rhesus monkey. *Cereb. Cortex* 8:671–84

151. Purpura DP. 1974. Dendritic spine "dysgenesis" and mental retardation. *Science* 186:1126–28

152. Machado-Salas JP. 1984. Abnormal dendritic patterns and aberrant spine development in Bourneville's disease—a Golgi survey. *Clin. Neuropathol.* 3:52–58

153. Stoltenburg-Didinger G, Spohr HL. 1983. Fetal alcohol syndrome and mental retardation: spine distribution of pyramidal cells in prenatal alcohol-exposed rat cerebral cortex; a Golgi study. *Brain Res.* 313:119–23

154. Marin-Padilla M. 1976. Pyramidal cell abnormalities in the motor cortex of a child with Down's syndrome. A Golgi study. *J. Comp. Neurol.* 167:63–81

155. Suetsugu M, Mehraein P. 1980. Spine distribution along the apical dendrites of the pyramidal neurons in Down's syndrome. A quantitative Golgi study. *Acta Neuropathol.* 50:207–10

156. Takashima S, Ieshima A, Nakamura H, Becker LE. 1989. Dendrites, dementia and the Down syndrome. *Brain Dev.* 11:131–33

157. Ferrer I, Gullotta F. 1990. Down's syndrome and Alzheimer's disease: dendritic spine counts in the hippocampus. *Acta Neuropathol.* 79:680–85

158. Kurt MA, Davies DC, Kidd M, Dierssen

M, Florez J. 2000. Synaptic deficit in the temporal cortex of partial trisomy 16 (Ts65Dn) mice. *Brain Res.* 858:191–97

159. Insausti AM, Megias M, Crespo D, Cruz-Orive LM, Dierssen M, et al. 1998. Hippocampal volume and neuronal number in Ts65Dn mice: a murine model of Down syndrome. *Neurosci. Lett.* 253:175–78

160. Schapiro MB, Murphy DG, Hagerman RJ, Azari NP, Alexander GE, et al. 1995. Adult fragile X syndrome: neuropsychology, brain anatomy, and metabolism. *Am. J. Med. Genet.* 60:480–93

161. de Vries BB, Halley DJ, Oostra BA, Niermeijer MF. 1998. The fragile X syndrome. *J. Med. Genet.* 35:579–89

162. Pieretti M, Zhang FP, Fu YH, Warren ST, Oostra BA, et al. 1991. Absence of expression of the FMR-1 gene in fragile X syndrome. *Cell* 66:817–22

163. Feng Y, Gutekunst CA, Eberhart DE, Yi H, Warren ST, Hersch SM. 1997. Fragile X mental retardation protein: nucleocytoplasmic shuttling and association with somatodendritic ribosomes. *J. Neurosci.* 17:1539–47

164. Brown V, Small K, Lakkis L, Feng Y, Gunter C, et al. 1998. Purified recombinant Fmrp exhibits selective RNA binding as an intrinsic property of the fragile X mental retardation protein. *J. Biol. Chem.* 273:15,521–27

165. Devys D, Lutz Y, Rouyer N, Bellocq JP, Mandel JL. 1993. The FMR-1 protein is cytoplasmic, most abundant in neurons and appears normal in carriers of a fragile X premutation. *Nat. Genet.* 4:335–40

166. Verheij C, Bakker CE, de Graaff E, Keulemans J, Willemsen R, et al. 1993. Characterization and localization of the FMR-1 gene product associated with fragile X syndrome. *Nature* 363:722–24

167. Weiler IJ, Irwin SA, Klintsova AY, Spencer CM, Brazelton AD, et al. 1997. Fragile X mental retardation protein is translated near synapses in response to neurotransmitter activation. *Proc. Natl. Acad. Sci. USA* 94:5395–400

168. Rudelli RD, Brown WT, Wisniewski K, Jenkins EC, Laure-Kamionowska M, et al. 1985. Adult fragile X syndrome. Clinico-neuropathologic findings. *Acta Neuropathol.* 67:289–95

169. Hinton VJ, Brown WT, Wisniewski K, Rudelli RD. 1991. Analysis of neocortex in three males with fragile X syndrome. *Am. J. Med. Genet.* 41:289–94

170. Irwin SA, Patel B, Idupulapati M, Harris JB, Crisostomo RA, et al. 2001. Abnormal dendritic spine characteristics in the temporal and visual cortices of patients with fragile-X syndrome: a quantitative examination. *Am. J. Med. Genet.* 98:161–67

171. Comery TA, Harris JB, Willems PJ, Oostra BA, Irwin SA, et al. 1997. Abnormal dendritic spines in fragile X knockout mice: maturation and pruning deficits. *Proc. Natl. Acad. Sci. USA* 94:5401–4

172. Masliah E, Ge N, Morey M, DeTeresa R, Terry RD, Wiley CA. 1992. Cortical dendritic pathology in human immunodeficiency virus encephalitis. *Lab. Invest.* 66:285–91

173. Paula-Barbosa MM, Tavares MA, Saraiva AA. 1980. Dendritic abnormalities in patients with subacute sclerosing panencephalitis (SSPE). A Golgi study. *Acta Neuropathol.* 52:77–80

174. Brown D, Belichenko P, Sales J, Jeffrey M, Fraser JR. 2001. Early loss of dendritic spines in murine scrapie revealed by confocal analysis. *NeuroReport* 12:179–83

175. Ferrer I, Genis D, Davalos A, Bernado L, Sant F, Serrano T. 1994. The Purkinje cell in olivopontocerebellar atrophy. A Golgi and immunocytochemical study. *Neuropathol. Appl. Neurobiol.* 20:38–46

176. Iwane S, Kawasaki H, Yamano T, Shimada M. 1989. Golgi study on the homozygote (Ml/Ml) of macular mutant mouse. *Brain Dev.* 11:154–60

177. Crick F. 1982. Do dendritic spines twitch? *Trends Neurosci.* 5:44–46

178. Gamble E, Koch C. 1987. The dynamics of free calcium in dendritic spines in

response to repetitive synaptic input. *Science* 236:1311–15

179. Holmes WR. 1990. Is the function of spines to concentrate calcium? *Brain Res.* 519:338–42

180. Koch C, Zador A. 1993. The function of dendritic spines: devices subserving biochemical rather than electrical compartmentalization. *J. Neurosci.* 13:413–22

181. Axelrod D, Koppel DE, Schlessinger J, Elson E, Webb WW. 1976. Mobility measurement by analysis of fluorescence photobleaching recovery kinetics. *Biophys. J.* 16:1055–69

182. Svoboda K, Tank DW, Denk W. 1996. Direct measurement of coupling between dendritic spines and shafts. *Science* 272: 716–19

183. Hausser M, Paresys G, Denk W. 1997. Coupling between dendritic spines and shafts in cerebellar Purkinje cells. *Soc. Neurosci. Abstr.* 23:781.5

184. Majewska A, Tashiro A, Yuste R. 2000. Regulation of spine calcium dynamics by rapid spine motility. *J. Neurosci.* 20: 8262–68

185. Deleted in proof

186. Allbritton NL, Meyer T, Streyer L. 1992. Range of messenger action of calcium ion and inositol 1,4,5-trisphosphate. *Science* 258:1812–15

187. Swaminathan R, Hoang CP, Verkman AS. 1997. Photobleaching recovery and anisotropy decay of green fluorescent protein GFP-S65T in solution and cells: cytoplasmic viscosity probed by green fluorescent protein translational and rotational diffusion. *Biophys. J.* 72:1900–7

188. Rall W. 1978. Dendritic spines and synaptic potency. In *Studies in Neurophysiology*, ed. R Porter, pp. 203–9. Cambridge, UK: Cambridge Univ. Press

189. Koch C, Poggio T. 1983. A theoretical analysis of electrical properties of spines. *Proc. R. Soc. London Ser. B* 218:455–77

190. Wickens J. 1988. Electrically coupled but chemically isolated synapses: dendritic spines and calcium in a rule for synaptic modification. *Prog. Neurobiol.* 31:507–28

191. Berg HC. 1993. *Random Walks in Biology.* Princeton, NJ: Princeton Univ. Press. 152 pp.

192. Rall W, Segev I. 1987. Functional possibilities for synapses on dendrites and dendritic spines. In *Synaptic Function*, ed. GM Edelman, WE Gall, WM Cowan, pp. 603–36. New York: Wiley

193. Wilson CJ. 1984. Passive cable properties of dendritic spines and spiny neurons. *J. Neurosci.* 4:281–97

194. Barbour B. 1993. Synaptic currents evoked in Purkinje cells by stimulating individual granule cells. *Neuron* 11:759–69

195. Manabe T, Renner P, Nicoll RA. 1992. Postsynaptic contribution to long-term potentiation revealed by the analysis of miniature synaptic currents. *Nature* 355: 50–55

196. Denk W, Sugimori M, Llinas R. 1995. Two types of calcium response limited to single spines in cerebellar Purkinje cells. *Proc. Natl. Acad. Sci. USA* 92:8279–82

197. Zucker RS. 1999. Calcium- and activity-dependent synaptic plasticity. *Curr. Opin. Neurobiol.* 9:305–13

198. Yuste R, Majewska A, Holthoff K. 2000. From form to function: calcium compartmentalization in dendritic spines. *Nat. Neurosci.* 3:653–59

199. Johnston D, Magee JC, Colbert CM, Christie BR. 1996. Active properties of neuronal dendrites. *Annu. Rev. Neurosci.* 19:165–86

200. Johnston D, Hoffman DA, Colbert CM, Magee JC. 1999. Regulation of back-propagating action potentials in hippocampal neurons. *Curr. Opin. Neurobiol.* 9:288–92

201. Callaway JC, Ross WN. 1995. Frequency-dependent propagation of sodium action potentials in dendrites of hippocampal CA1 pyramidal neurons. *J. Neurophysiol.* 74:1395–403

202. Regehr WG, Tank DW. 1992. Calcium concentration dynamics produced by

synaptic activation of CA1 hippocampal pyramidal cells. *J. Neurosci.* 12:4202–23

203. Sabatini BL, Svoboda K. 2000. Analysis of calcium channels in single spines using optical fluctuation analysis. *Nature* 408:589–93

204. Emptage N, Bliss TVP, Fine A. 1999. Single synaptic events evoke NMDA receptor-mediated release of calcium from internal stores in hippocampal dendritic spines. *Neuron* 22:115–24

205. Yuste R, Denk W. 1995. Dendritic spines as basic functional units of neuronal integration. *Nature* 375:682–84

206. Schiller J, Schiller Y, Clapham DE. 1998. Amplification of calcium influx into dendritic spines during associative pre- and postsynaptic activation: the role of direct calcium influx through the NMDA receptor. *Nat. Neurosci.* 1:114–18

207. Mainen ZF, Malinow R, Svoboda K. 1999. Synaptic calcium transients in single spines indicate that NMDA receptors are not saturated. *Nature* 399:151–55

208. Muller W, Connor JA. 1991. Dendritic spines as individual neuronal compartments for synaptic Ca^{2+} responses. *Nature* 354:73–76

209. Kovalchuk Y, Eilers J, Lisman J, Konnerth A. 2000. NMDA receptor-mediated subthreshold $Ca^{(2+)}$ signals in spines of hippocampal neurons. *J. Neurosci.* 20:1791–99

210. Yuste R, Majewska A, Cash SS, Denk W. 1999. Mechanisms of calcium influx into hippocampal spines: heterogeneity among spines, coincidence detection by NMDA receptors, and optical quantal analysis. *J. Neurosci.* 19:1976–87

211. Koester HJ, Sakmann B. 1998. Calcium dynamics in single spines during coincident pre- and postsynaptic activity depend on relative timing of back-propagating action potentials and subthreshold excitatory postsynaptic potentials. *Proc. Natl. Acad. Sci. USA* 95:9596–601

212. Wollmuth LP, Kuner T, Sakmann B. 1998. Adjacent asparagines in the NR2-subunit of the NMDA receptor channel control the voltage-dependent block by extracellular Mg^{2+}. *J. Physiol.* 506:13–32

213. Nakamura T, Barbara JG, Nakamura K, Ross WN. 1999. Synergistic release of Ca^{2+} from IP3-sensitive stores evoked by synaptic activation of mGluRs paired with backpropagating action potentials. *Neuron* 24:727–37

214. Svoboda K, Mainen ZF. 1999. Synaptic $[Ca^{2+}]$: intracellular stores spill their guts. *Neuron* 22:427–30

215. Magee JC, Johnston D. 1997. A synaptically controlled, associative signal for Hebbian synaptic plasticity in hippocampal neurons. *Science* 275:209–13

216. Petrozzino JJ, Miller LDP, Connor JA. 1995. Micromolar Ca^{2+} transients in dendritic spines of hippocampal pyramidal neurons in brain slice. *Neuron* 14:1223–31

217. Rose CR, Konnerth A. 2001. NMDA receptor-mediated Na^+ signals in spines and dendrites. *J. Neurosci.* 21:4207–14

218. Satoh T, Ross CA, Villa A, Supattapone S, Pozzan T, et al. 1990. The inositol 1,4,5,-trisphosphate receptor in cerebellar Purkinje cells: quantitative immunogold labeling reveals concentration in an ER subcompartment. *J. Cell Biol.* 111:615–24

219. Takechi H, Eilers J, Konnerth A. 1998. A new class of synaptic response involving calcium release in dendritic spines. *Nature* 396:757–60

220. Finch EA, Augustine GJ. 1998. Local calcium signalling by inositol-1,4,5-trisphosphate in Purkinje cell dendrites. *Nature* 396:753–56

221. Eilers J, Augustine GJ, Konnerth A. 1995. Subthreshold synaptic Ca^{2+} signalling in fine dendrites and spines of cerebellar Purkinje neurons. *Nature* 373:155–58

222. Wang SS, Denk W, Hausser M. 2000. Coincidence detection in single dendritic

spines mediated by calcium release. *Nat. Neurosci.* 3:1266–73

223. Miyata M, Finch EA, Khiroug L, Hashimoto K, Hayasaka S, et al. 2000. Local calcium release in dendritic spines required for long-term synaptic depression. *Neuron* 28:233–44

224. Neher E, Augustine GJ. 1992. Calcium gradients and buffers in bovine chromaffin cells. *J. Physiol.* 450:273–301

225. Tank DW, Regehr WG, Delaney KR. 1995. A quantitative analysis of presynaptic calcium dynamics that contribute to short term enhancement. *J. Neurosci.* 15:7940–52

226. Maravall M, Mainen ZM, Sabatini B, Svoboda K. 2000. Estimating intracellular calcium concentrations and buffering without wavelength ratioing. *Biophys. J.* 78:2655–67

227. Helmchen F, Imoto K, Sakmann B. 1996. Ca^{2+} buffering and action potential-evoked Ca^{2+} signaling in dendrites of pyramidal neurons. *Biophys. J.* 70:1069–81

228. Majewska A, Brown E, Ross J, Yuste R. 2000. Mechanisms of calcium decay kinetics in hippocampal spines: role of spine calcium pumps and calcium diffusion through the spine neck in biochemical compartmentalization. *J. Neurosci.* 20:1722–34

229. Holmes WR. 2000. Models of calmodulin trapping and CaM kinase II activation in a dendritic spine. *J. Comput. Neurosci.* 8:65–85

Annu. Rev. Physiol. 2002. 64:355–405

SHORT-TERM SYNAPTIC PLASTICITY

Robert S. Zucker

Department of Molecular and Cell Biology, University of California, Berkeley,
California 94720; e-mail: zucker@socrates.berkeley.edu

Wade G. Regehr

Department of Neurobiology, Harvard Medical School, Boston, Massachusetts 02115;
e-mail: wregehr@hms.harvard.edu

Key Words synapse, facilitation, post-tetanic potentiation, depression, augmentation, calcium

■ **Abstract** Synaptic transmission is a dynamic process. Postsynaptic responses wax and wane as presynaptic activity evolves. This prominent characteristic of chemical synaptic transmission is a crucial determinant of the response properties of synapses and, in turn, of the stimulus properties selected by neural networks and of the patterns of activity generated by those networks. This review focuses on synaptic changes that result from prior activity in the synapse under study, and is restricted to short-term effects that last for at most a few minutes. Forms of synaptic enhancement, such as facilitation, augmentation, and post-tetanic potentiation, are usually attributed to effects of a residual elevation in presynaptic $[Ca^{2+}]_i$, acting on one or more molecular targets that appear to be distinct from the secretory trigger responsible for fast exocytosis and phasic release of transmitter to single action potentials. We discuss the evidence for this hypothesis, and the origins of the different kinetic phases of synaptic enhancement, as well as the interpretation of statistical changes in transmitter release and roles played by other factors such as alterations in presynaptic Ca^{2+} influx or postsynaptic levels of $[Ca^{2+}]_i$. Synaptic depression dominates enhancement at many synapses. Depression is usually attributed to depletion of some pool of readily releasable vesicles, and various forms of the depletion model are discussed. Depression can also arise from feedback activation of presynaptic receptors and from postsynaptic processes such as receptor desensitization. In addition, glial-neuronal interactions can contribute to short-term synaptic plasticity. Finally, we summarize the recent literature on putative molecular players in synaptic plasticity and the effects of genetic manipulations and other modulatory influences.

INTRODUCTION

Neurons communicate with each other primarily through fast chemical synapses. At such synapses an action potential generated near the cell body propagates down the axon where it opens voltage-gated Ca^{2+} channels. Ca^{2+} ions entering nerve terminals trigger the rapid release of vesicles containing neurotransmitter, which is ultimately detected by receptors on the postsynaptic cell. A dynamic

0066-4278/02/0315-0355$14.00

enhancement of such synaptic transmission has been recognized for over 60 years (1, 2). Virtually all types of synapses are regulated by a variety of short-lived and long-lasting processes, some of which lead to a decrease in synaptic strength and others that lead to synaptic enhancement. At some synapses with repeated use, synaptic enhancement occurs and facilitatory processes dominate; at others the result is a decrease in synaptic strength and depression prevails. In most cases, it is apparent that multiple processes are present, and the result can be a combination of facilitation and depression in which synaptic strength is highly dependent on the details of the timing of synaptic activation (3–6). Here we are concerned with the properties and mechanisms of use-dependent plasticity on the tens of milliseconds to several minutes time scale.

ENHANCEMENT OF TRANSMISSION

Many chemical synapses show a multi-component increase in synaptic efficacy or a growth in the amplitude of individual postsynaptic potentials (PSPs) or postsynaptic currents (PSCs) on repetitive activation. This enhancement of transmission comes in several flavors, with quite distinct lifetimes (7).

Facilitation

Synaptic enhancement that is prominent on the hundreds of milliseconds time scale is referred to as facilitation. It can be seen with pairs of stimuli, in which the second PSP can be up to five times the size of the first (Figure 1). During brief trains of action potentials (APs), successive PSPs grow within about a second to a size that can easily reach several times—and in some synapses several dozen times—the original PSP. Facilitation often builds and decays with a time course that can be approximated with an exponential of ∼100 ms. At some synapses, facilitation can be further subdivided into a rapid phase lasting tens of milliseconds (F1) and a slower phase lasting hundreds of milliseconds (F2).

Post-Tetanic Potentiation

There are also processes that become increasingly important as the number of stimuli in a train is increased. For these processes, each AP enhances synaptic strength by 1–15%, but because they last for five seconds to several minutes, the integrated effect of a train of hundreds of pulses can lead to a many-fold enhancement (Figure 2). There is considerable variability in synaptic plasticity exhibited by synapses on these time scales. Sometimes processes such as augmentation, which grows and decays with a time constant of ∼5–10 s, can be distinguished from post-tetanic potentiation (PTP), which lasts for 30 s to several minutes. At other synapses, these components are not easily separable, and they are often lumped together and referred to as PTP. During realistic stimulus trains, multiple processes are often present. This is illustrated in Figure 2 where both facilitation and PTP are important during the prolonged train. This is further complicated by the fact that

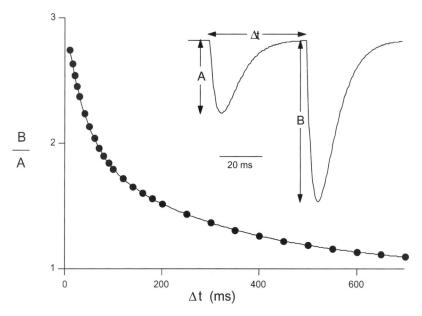

Figure 1 Simulated experiment showing paired-pulse facilitation of the sort that occurs at many synapses. As shown in the inset, activation with pairs of stimuli separated by time Δt evokes synaptic currents with the second response (B) larger than the first (A). As shown in the plot of B/A versus Δt, the magnitude of facilitation decreases as the interpulse interval is increased. In this experiment the amplitude of facilitation can be approximated by a double exponential decay of the form $1 + C_1 \exp(-t/\tau_1) + C_2 \exp(-t/\tau_2)$ (*solid line*). Based on this fit, this synapse would have two components of facilitation: F1 facilitation with $\tau_1 = 40$ ms, and F2 facilitation with $\tau_2 = 300$ ms. At many synapses the distinction between F1 and F2 is not clear, and the duration of facilitation is well approximated by a single exponential fit.

prolonged stimulation is usually accompanied by depression, and potentiation is often only observed after a tetanus, following recovery from depression.

MECHANISMS OF ENHANCEMENT

Statistics of Release Indicate Changes are Presynaptic

In all synapses studied, facilitation, augmentation, and PTP have all been shown by quantal analysis to be presynaptic in origin—to involve specifically an increase in the number of transmitter quanta released by an AP without any change in quantal size or postsynaptic effectiveness (reviewed in 7). Much additional effort has gone into analysis of the changes in statistics of transmitter release using a binomial model of release from a pool of available quanta n, with release probability p. Enhanced release is accompanied by increases in p, n or both (8, 9). The parameter

n corresponds most closely to the number of release sites or active zones that contain clusters of vesicles, some of which appear docked near the presynaptic membrane immediately opposing postsynaptic receptors (10). Interpretation of these results is complicated by the fact that a simple binomial model, assuming uniform p at all release sites, ignores the likely variability in p (9, 11). This can lead to underestimation of changes in p and spurious increases in n.

It seems clear, then, that short-term synaptic enhancement reflects an increase in the probability of release of available quanta, with perhaps also an increase in the

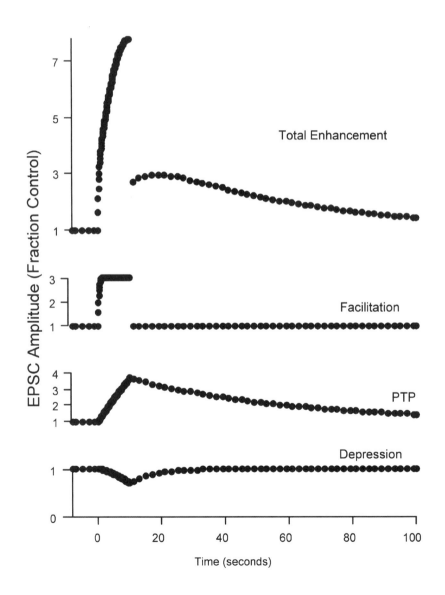

number of release sites capable of releasing a quantum. Either statistical change could be due in turn to an increase in the probability of activating exocytosis of a docked vesicle or an increase in the probability that a release site is occupied by a docked vesicle ready for release (12). The latter could occur if the pool of vesicles available to rapidly occupy release sites is increased. This pool, often called the readily releasable pool, is released within a few seconds by hyperosmotic shock (13, 14), which provides a measure of its size similar to that obtained by measures of synaptic depression (see below). Augmentation in hippocampal synapses is unaccompanied by an increase in the size of this readily releasable pool, suggesting instead an increase in the probability of release from this pool (15).

In correlated ultrastructural and statistical studies (16), control values of n appear to correspond to active zones with multiple dense bodies, whereas the increased n in PTP seems to include the number of active zones with single dense bodies. Thus increase in n may reflect a real recruitment of release from previously dormant or silent active zones with fewer dense bodies.

THE CRUCIAL ROLE OF Ca^{2+} IONS

Early attempts at explaining short-term synaptic enhancement focused on electrical events in presynaptic terminals. Possibilities such as increased invasion of nerve terminals by APs, or broadening APs, or effects of afterpotentials, or increased

---

Figure 2   Simulated experiment showing synaptic plasticity during and following high-frequency stimulation. In this experiment, a synaptic input was stimulated at 0.5 Hz and the amplitude of the PSC remained constant. Tetanic stimulation at 10 Hz for 10 s resulted in an eight-fold synaptic enhancement by the end of the train (*top, closed circles*). Some synaptic enhancement persisted upon returning to 0.5 Hz stimulation. In this example, the total synaptic enhancement (*top, closed circles*) was a result of facilitation, PTP, and depression. Facilitation, which is relatively short-lived with a time constant of 400 ms, built up rapidly during tetanic stimulation, but did not persist after commencing low-frequency stimulation. A slower process resulted in enhancement that increased gradually during the train. After returning to low-frequency stimulation this form of enhancement persists and is known as post-tetanic potentiation (PTP). At many synapses, a form of enhancement (augmentation) intermediate between facilitation and potentiation also exists. This experiment illustrates that at most synapses multiple processes contribute to synaptic enhancement. It also shows that PTP can be studied in isolation from facilitation following a train, but that during the train, both slow and fast forms of enhancement contribute to enhancement. Because longer-lasting forms of enhancement typically result in a small enhancement per pulse, paired-pulse experiments as in Figure 1 are suited to studying facilitation with little contamination from PTP.

Ca^{2+} influx due to facilitation of Ca^{2+} channels, were eliminated in a variety of preparations (8, 17, 18, although see below). Facilitation can even be evoked by constant depolarizing pulses under voltage clamp that activates an invariant Ca^{2+} influx and constant presynaptic $[Ca^{2+}]_i$ change (19). Although, as described below, facilitation of Ca^{2+} influx can contribute to synaptic facilitation, in most cases other mechanisms make more prominent contributions.

The Residual Ca^{2+} Hypothesis

At neuromuscular junctions, an AP can cause facilitation equally well, even if transmitter release does not occur (20, 21); therefore, facilitation seems to arise from some process following AP invasion but preceding secretion. This led naturally to the possibility that facilitation was somehow a consequence of the influx of Ca^{2+} ions during conditioning stimulation. Strong evidence for this idea came from the seminal experiments of Katz & Miledi (22). They used a focal extracellular pipette to provide Ca^{2+} ions to neuromuscular junctions in a Ca^{2+}-free medium and showed that a conditioning impulse not only failed to release transmitter in the absence of external Ca^{2+}, it also failed to facilitate release. Subsequently it was shown that augmentation and potentiation also depend, at least in part, on the presence of external Ca^{2+} during conditioning stimulation (23–25). Such results led to the "residual Ca^{2+} hypothesis": Facilitation is caused by an action of Ca^{2+} remaining in the nerve terminals after the conditioning stimulus. In the past 25 years, substantial evidence has accumulated in support of the residual Ca^{2+} hypothesis: There is a correlation between elevations in $[Ca^{2+}]_i$ and synaptic enhancement; elevating $[Ca^{2+}]_i$ enhances synaptic strength, and preventing increases in $[Ca^{2+}]_i$ eliminates short-term enhancement.

PRESYNAPTIC $[Ca^{2+}]_i$ CORRELATES WITH FACILITATION, AUGMENTATION, AND POTENTIATION   The first attempts to correlate $[Ca^{2+}]_i$ with synaptic plasticity compared post-tetanic Ca^{2+}-activated K^+ current in the cell body of a presynaptic neuron in *Aplysia* to the decay of PTP in a postsynaptic cell (26). It can be difficult to determine the time course of $[Ca^{2+}]_i$ with Ca^{2+}-dependent K^+ channels (27) because of the voltage- and Ca^{2+} dependence of these channels and their degree of co-localization with voltage-gated Ca^{2+} channels (28–30). Subsequently the kinetics of presynaptic post-tetanic $[Ca^{2+}]_i$ changes were measured with the Ca^{2+}-sensitive metallochromic dye arsenazo III and compared with PTP (31). With the advent of fluorescent dyes allowing ratiometric measurement of $[Ca^{2+}]_i$ without knowledge of dye concentration, more accurate estimation of $[Ca^{2+}]_i$ became possible. Numerous studies have demonstrated an apparently linear relationship between magnitude of potentiation, augmentation, or F2 facilitation, and residual $[Ca^{2+}]_i$ concentration in vertebrate and invertebrate neuromuscular junctions and mammalian central and peripheral synapses (18, 32–42). Facilitation triggered by conditioning pulses under voltage clamp also correlates linearly with the measured (43) or inferred (44) magnitude of Ca^{2+} influx during those pulses. That this residual

Ca^{2+} is capable of influencing transmitter release is suggested by the fact that all phases of enhancement of evoked release are accompanied by an increase in the frequency of spontaneously released quanta (miniature PSPs or mPSPs) (42, 45–54).

ELEVATING PRESYNAPTIC $[Ca^{2+}]_i$ ENHANCES AP-EVOKED RELEASE   Various manipulations have been used to mimic the effect of residual Ca^{2+}: fusion of Ca^{2+}-containing liposomes with nerve terminals (55), exposure to Ca^{2+} ionophores (56, 57), release of Ca^{2+} from poisoned mitochondria (58), presynaptic Ca^{2+} injection by iontophoresis (19), or release of Ca^{2+} by photolysis of presynaptic caged Ca^{2+} chelators (59–62). In all cases, AP-induced PSPs were dramatically increased. At calyx of Held synapses, small conditioning Ca^{2+} influx could facilitate release to a later influx (63).

BUFFERING PRESYNAPTIC $[Ca^{2+}]_i$ REDUCES SHORT-TERM ENHANCEMENT   The Ca^{2+} chelators EGTA and BAPTA can be loaded into nerve terminals in the acetoxymethylester form, where they can be de-esterified by endogenous esterases. The de-esterified buffer can accumulate to millimolar levels (as opposed to the micromolar concentration of the AM-ester used to bathe the preparation). In many studies, presynaptic loading of such exogenous Ca^{2+} buffers strongly reduced both components of facilitation (34–36, 40, 41, 45, 52, 64–69) and augmentation (15, 69). Facilitation and augmentation have also been reduced by presynaptic injection of Ca^{2+} buffers into squid and crayfish terminals (66, 70–72), and PTP is reduced in hippocampal synapses expressing excess amounts of the native Ca^{2+}-binding protein calbindin (73), whereas deletion of the gene for the Ca^{2+}-binding protein parvalbumin increases facilitation at cerebellar synapses (74). The finding that the slowly acting buffer EGTA is effective in reducing synaptic enhancement suggests that the target(s) of Ca^{2+} action cannot be very close to Ca^{2+} channels and affected only by the transient local micro-domains of high $[Ca^{2+}]_i$ because these would be affected little by EGTA. In a few instances, authors have reported difficulty in reducing facilitation or augmentation by use of exogenous buffers (65, 75, 76), which may be due to inadequate buffer concentrations or saturation of the buffer by repeated activity.

Another way of rapidly reducing Ca^{2+} concentrations is to use caged Ca^{2+} chelators that increase their affinity for Ca^{2+} upon exposure to ultraviolet light. Zucker and colleagues (61, 77) injected the caged BAPTA diazo-2 or diazo-4 into *Aplysia* central nerve terminals and crayfish peripheral nerve terminals and photolyzed it to increase presynaptic Ca^{2+} buffering after conditioning stimuli and post-tetanic waiting periods designed to select only facilitation, augmentation, or potentiation. All three forms of enhancement were reduced on diazo photolysis and reduction of residual $[Ca^{2+}]_i$.

REDUCING Ca^{2+} INFLUX REDUCES SHORT-TERM ENHANCEMENT   Another way to probe the role of $[Ca^{2+}]_i$ in synaptic enhancement is to assess the effects of altering Ca^{2+} influx. This is done by changing extracellular Ca^{2+}, blocking Ca^{2+} channels with toxins or divalent ions, or by altering the presynaptic waveform, which changes Ca^{2+} entry. Experiments of this sort are, however, difficult to interpret. At most synapses, reducing Ca^{2+} entry not only decreases $[Ca^{2+}]_i$ but also reduces depression by decreasing the initial release of neurotransmitter (see below), which can resemble an increase in facilitation. Facilitation can also be genuinely increased by a desaturation of the release process. In studies of synapses under conditions of little depression and far from saturation of release, it was shown that reducing Ca^{2+} influx decreased facilitation and augmentation phases of enhanced release (78–81). Elevating external $[Ca^{2+}]$ was often, but not always, able to reverse the effects of Ca^{2+} channel blockers.

Facilitation is sometimes stronger following a successful conditioning stimulus than a failure (82, 83). This may reflect the fact that the number of Ca^{2+} channels opening and the amount of Ca^{2+} entry in an active zone are stochastic processes, and both the probability of secretion and the magnitude of residual Ca^{2+} and, consequently, of facilitation depend on this random variable.

PSEUDOFACILITATION   In some counterintuitive experiments, it was found that presynaptic perfusion of the Ca^{2+} buffer BAPTA not only reduced transmission, but concurrently increased facilitation, at some cortical synapses (64). The additional facilitation was shown not to reflect a reduction in depression, but rather an artificial saturation of BAPTA by the first AP, leaving less buffer to capture Ca^{2+} in a conditioned response. Unlike genuine facilitation, this pseudofacilitation could not be blocked by EGTA perfusion because slow-binding EGTA cannot steal entering Ca^{2+} from fast-binding BAPTA. Moreover, pseudofacilitation increased with elevation of external $[Ca^{2+}]$, unlike real facilitation. Although the results suggest that natural facilitation does not work in the same way as pseudofacilitation, they emphasize the possibility of modulating facilitation by altering the Ca^{2+} buffering properties of cytoplasm.

The Single-Site Hypothesis

In the original formulation of the residual Ca^{2+} hypothesis (22), it was proposed that the peak incremental $[Ca^{2+}]_i$ elevation following an AP acts at some presynaptic site to trigger phasic release, while residual Ca^{2+} from that event summates with the incremental Ca^{2+} rise in a subsequent test AP to produce short-term synaptic enhancement. It is now recognized that neurotransmitter release is triggered by an increase in Ca^{2+} levels near open voltage-gated Ca^{2+} channels (Ca_{local}). Within tens of milliseconds, Ca^{2+} then diffuses and equilibrates throughout the presynaptic bouton giving rise to a residual Ca^{2+} signal (Ca_{res}). Thus for a bouton with a resting Ca^{2+} level of Ca_{rest}, the Ca^{2+} level available to trigger release in response to the first stimulus is ($Ca_{rest} + Ca_{local}$) and for a second closely spaced stimulus is ($Ca_{rest} + Ca_{local} + Ca_{res}$).

Independent estimates of the relative magnitudes of $[Ca^{2+}]_i$ triggering phasic secretion and synaptic enhancement suggest that a single type of Ca^{2+}-binding site cannot account for both phasic transmitter release and short-term plasticity. Ca_{rest} is $\sim$100 nM, Ca_{local} is tens of micromolar or higher (84–89), whereas Ca_{res} at times of substantial facilitation, augmentation, or potentiation reaches only $\sim$1 μM (32–34, 41). For PSC $\propto Ca^4$ (90), $Ca_{rest} = 100$ nM, and $Ca_{local} = 20$ μM, a Ca_{res} of 1 μM would produce a synaptic enhancement of only $\sim$20% if there were a single type of Ca^{2+} binding site, far smaller than the observed enhancement, which can be 10-fold.

Another test of this model is whether short-term synaptic enhancement accumulates as predicted if each AP in a train adds a constant increment to residual Ca^{2+}, which decays according to kinetics necessary to account for the phases of enhancement following a single AP. These sorts of calculations have almost always failed to account for the accumulation of facilitation, augmentation, and potentiation of evoked and spontaneous release (48, 91–97; but see 40). Instead, augmentation and potentiation appear to multiply the effects of facilitation (48, 98), and the effect of Ca^{2+} on facilitation appears to multiply its effect on secretion (52). These quantitative studies thus suggest that facilitation is a separate process from both phasic secretion and the slower processes of augmentation and potentiation.

A comparison of the enhancement of evoked synaptic responses and spontaneous neurotransmitter release provides another test of the single-site hypothesis. A conditioning stimulus also increases mPSP frequency, a phenomenon often called delayed release and also thought to reflect increases in $[Ca^{2+}]_i$ (49). According to the single-site model, if $Ca_{res}(t)$, expressed as a fraction of Ca_{local}, is represented as $\varepsilon(t)$, where t is time from the end of the conditioning stimulus and Ca_{rest} is ignored, then the fourth power dependence of transmitter release on $[Ca^{2+}]_i$ (90) predicts that enhanced mPSP frequency decays as $\varepsilon^4(t)$, while facilitation, $f(t)$ (the fractional increase in enhanced evoked release compared with un-enhanced release), should decay as $f(t) = (1 + \varepsilon(t))^4 - 1 \approx 4\varepsilon(t)$, for small $\varepsilon(t)$. Comparison of post-tetanic decays in mPSP frequency and evoked PSP seemed roughly consistent with this prediction, at least for facilitation, in some preparations (46, 53, 99). Elevation of presynaptic Ca^{2+} at crayfish neuromuscular junctions by photolysis of caged Ca^{2+} also increased evoked release and mPSP frequency in ways reasonably consistent with predictions (60). However, other studies have found serious discrepancies between observations and quantitative predictions of the single-site hypothesis. The enhancement of evoked release was much smaller then predicted by the single-site hypothesis from the rate of spontaneous release for facilitation at the mouse neuromuscular junctions (52); for facilitation, augmentation, and potentiation at frog neuromuscular junctions (47); and for facilitation at parallel fiber synapses onto stellate cells in the cerebellum (95, 100).

Facilitation seems to reach a peak immediately after each AP at normal temperature (48). However, near 0°C a delay of a few milliseconds appears (101), so that facilitation is maximal long after phasic secretion has terminated. If this apparent delay in facilitation is real, and not because of a superimposed very fast phase of depression (82), it is further evidence that facilitation and phasic secretion result from distinct Ca^{2+} actions.

A different clue that short-term enhancement involves one or more processes distinct from phasic release is its diversity. Different synapses in the same species, and even different terminals from the same presynaptic neuron, can show vastly different magnitudes of facilitation (64, 69, 102–107). Normally, synapses with a high output to low-frequency stimulation show less facilitation than low-output synapses, which could result from a saturation of release (93) from high-output synapses or from a concurrent depression masking facilitation (see below). Either action would be alleviated by reducing external $[Ca^{2+}]$. However, in one study it was found that reducing $[Ca^{2+}]$ did not erase the differences in amount of facilitation expressed by different types of synapses (108). Synaptic enhancement is also separable from baseline transmission in *Aplysia* neurons, where heterosynaptic activity was found to specifically reduce augmentation and PTP (109). A use-dependent reduction of paired-pulse facilitation has also been reported at *Aplysia* synapses (72). Thus facilitation and augmentation/PTP appear to be independently regulated properties of synaptic transmission.

Multiple Site Hypotheses

These considerations force the notion that synaptic enhancement is due to Ca^{2+} acting at a site or sites different from the fast low-affinity site triggering secretion (32–35, 45, 48, 52, 89, 97, 110–112). There are many unresolved questions regarding the properties of the Ca^{2+}-binding sites and the factors governing the time course and magnitude of synaptic enhancement.

WHAT DETERMINES THE TIME COURSE OF ENHANCEMENT? Does enhancement arise from the continuing action of residual Ca^{2+} due to the extended presence of free Ca^{2+} ions acting in equilibrium with the sites causing enhancement? If so, the kinetics (accumulation and decay) of facilitation, augmentation, and potentiation depend on the kinetics of residual Ca^{2+}. Alternatively, enhancement may decay with its own intrinsic kinetics, reflecting slow unbinding of Ca^{2+} from enhancement sites or, alternatively, aftereffects of Ca^{2+} binding owing to subsequent reactions.

Initially, the bound Ca^{2+} idea was favored, especially for facilitation, which seemed to last longer than would be expected from calculations of diffusion of Ca^{2+} away from the region where secretion is triggered near Ca^{2+} channel mouths (89). Also, a lower apparent Ca^{2+} cooperativity in triggering facilitated release (110) was interpreted as meaning that some Ca^{2+} ions were already bound, although the results could be explained by effects of depression and saturation. Persistent binding was also suggested by measures of residual Ca^{2+} based on Ca^{2+}-dependent K^+ current (27), but these results were confounded by the voltage dependence of the K^+ channels, and bound Ca^{2+} models were favored by investigators who could not block enhancement with exogenous Ca^{2+} buffers (see above).

Many studies indicate that the time course of residual Ca^{2+} contributes to the duration of synaptic enhancement. Recent Ca^{2+} diffusion simulations suggest sluggish residual Ca^{2+} kinetics at a diffusional distance of ~100 nm from Ca^{2+} channels where facilitation could be activated (70). Ca^{2+} accumulation at such sites can account roughly, but with some significant imperfections, for accumulation of facilitation in a train. At many synapses there is a close correlation between the time course of residual Ca^{2+} and the durations of facilitation, augmentation, and potentiation (18, 32–42). At crayfish neuromuscular junctions and *Aplysia* central synapses, elimination of all phases of enhancement by flash photolysis of diazo supports a causal relation between residual Ca^{2+} and these forms of plasticity (61, 77). Further support for facilitation arising from residual Ca^{2+} comes from studies of cerebellar parallel fiber synapses onto Purkinje cells (35). Buffering presynaptic residual Ca^{2+} with EGTA reduces both the magnitude and the duration of facilitation from 200 ms to a process decaying with a 40-ms time constant. The remaining facilitation appears to reflect the intrinsic kinetics of facilitation, when chelation of residual Ca^{2+} allows Ca^{2+} ions to bind to the facilitation site only during the AP.

HOW MANY SITES OF Ca^{2+} ACTION?   Although there is strong evidence in support of at least one high-affinity Ca^{2+} binding site responsible for synaptic enhancement, it is not clear how many types of Ca^{2+} binding sites are involved in synaptic enhancement. The existence of four phases of enhancement—F1, F2, augmentation and PTP—suggests that four distinct processes are involved, each, perhaps, with its own unique binding site. But all of these phases could in principle arise from Ca^{2+} acting at one site but decaying (and accumulating) with multiple phases controlled by diffusion, buffering, extrusion, and uptake. Numerous studies have examined the issue of the number of Ca^{2+} binding sites involved in enhancement, but no consensus has been reached.

The effects of rapid reductions in $[Ca^{2+}]_i$ with flash photolysis of diazo suggest that there are two distinct sites involved in synaptic enhancement at the crayfish neuromuscular junction (61). F1 and F2 facilitation were eliminated within 10 ms. Only part of augmentation and potentiation disappeared this rapidly; most decayed with a time constant of $\sim$0.4 s after photolysis. Augmentation and potentiation also show similar dependence on residual $[Ca^{2+}]_i$, increasing transmission about 10-fold per micromolar (33). These results are consistent with three types of Ca^{2+} binding sites: a low-affinity rapid site involved in phasic release, a moderate-affinity site with relatively rapid kinetics involved in the F1 and F2 components of facilitation, and a high-affinity site with slow kinetics that contributes to PTP and augmentation.

A multiplicative relationship between augmentation and potentiation on the one hand, and facilitation on the other (48, 98), suggests that at vertebrate neuromuscular junctions there are at least two distinct processes governing short-term synaptic enhancement. As argued previously, these two sites are separate from the site triggering phasic secretion. However, at lobster neuromuscular junctions, a model in which fast and slow facilitation components and augmentation summate in accordance with Ca^{2+} accumulating at a single site, describes the accumulation of synaptic enhancement better than multiplicative models (40).

Differential effects of Sr^{2+} or Ba^{2+} ions on synaptic enhancement suggest multiple sites of enhancement. Ba^{2+} selectively increases augmentation, whereas Sr^{2+} selectively increases and prolongs F2 facilitation of both evoked and post-tetanic delayed release (47, 113–115). However, measurements of presynaptic Sr^{2+} dynamics show that the effects of Sr^{2+} are due to differences in buffering and removal between presynaptic Sr^{2+} and Ca^{2+} (100, 116), leading specifically to an increase in $[Sr^{2+}]_i$ during the F2 facilitation phase. Thus the Sr^{2+} results do not necessarily imply different sites of action.

Thus further experiments are required to determine how many types of Ca^{2+}-binding sites are involved in synaptic enhancement at different synapses. Based on the evidence above, it seems likely that for some synapses synaptic enhancement is best described by at least a fast site involved in facilitation and a slower site in augmentation and potentiation.

KINETICS OF POTENTIATION AND AUGMENTATION   Augmentation arises from the phase of $[Ca^{2+}]_i$ decay that follows diffusional equilibration in nerve boutons and appears to be regulated by two plasma membrane extrusion pumps—a Ca^{2+}-ATPase and Na^+/Ca^{2+} exchange (117–119). Prolonged stimulation loads nerve terminals with both Na^+ and Ca^{2+}. The reduction in the Na^+ gradient reduces removal of Ca^{2+} by Na^+/Ca^{2+} exchange and can even reverse this process, which results in influx through this system during and after a long tetanus (118, 119). This results in amplification of residual $[Ca^{2+}]_i$ and prolongation of its removal, contributing to PTP, the slowest phase of synaptic enhancement (120–124). Prolonged stimulation also results in Ca^{2+}-loading of presynaptic mitochondria at crayfish and lizard neuromuscular junctions; leakage of this stored Ca^{2+} provides a major source of the long-lasting residual Ca^{2+} underlying PTP (39, 125–127). A similar role is played by endoplasmic reticulum at frog neuromuscular junctions (128), while both mitochondria and endoplasmic reticulum appear to be involved in the regulation of residual Ca^{2+} at peptidergic nerve terminals (129–131). Thus augmentation and potentiation result from the different dynamics of Ca^{2+} removal, dependent on extrusion and uptake processes, that dominate, respectively, after short and long tetani. Residual Ca^{2+} acts on both rapid (<10 to 40 ms) and somewhat slower-acting (~0.4 s) targets to regulate transmission.

KINETICS OF FACILITATION   Facilitation appears to arise from Ca^{2+} acting at a site with intrinsic kinetics of <10 ms at crayfish neuromuscular junctions (61) and ~40 ms at cerebellar synapses (35). The longer duration of normal facilitation apparently reflects the duration of $[Ca^{2+}]_i$ at its site(s) of generation. The two components of facilitation observed at some synapses may reflect two separate sites of action in F1 and F2 facilitation. Alternatively, $[Ca^{2+}]_i$ may decay non-exponentially, for example by diffusion away from active zones or clusters of active zones, with non-exponential kinetics (70, 132, 133). At hippocampal synapses, residual Ca^{2+} underlying facilitation can also be affected by release from intracellular stores such as the endoplasmic reticulum (134), although this contribution of release from internal stores to facilitation is controversial (135). Treatments such as Sr^{2+} may affect the two sites differentially (47, 113), or may have different effects on the processes controlling diffusion, such as buffer mobility or saturation (100, 116). The number of sites involved in facilitation remains unclear.

The quantitative relationship between decay of mPSP frequency and enhanced evoked release is not simple (95). However, it is possible that the two processes may be controlled by Ca^{2+} acting at two sites, the facilitation site and the secretory trigger, in ways that reflect different degrees of saturation and interaction in controlling spontaneous release and evoked release.

SYNAPTIC ENHANCEMENT, SLOW RELEASE, AND MOBILIZATION   When presynaptic $[Ca^{2+}]_i$ is suddenly elevated, either by strong depolarization or photolysis of caged Ca^{2+}, transmitter release often exhibits a biphasic time course: an initial intense rapid phase of secretion, called the secretory burst, followed by a slower phase with a time constant of hundreds of milliseconds to seconds (136–142). This phase is intermediate in duration between F2 facilitation and augmentation and shares with them a Ca^{2+} dependence that is easily blocked by intracellular EGTA. It is usually interpreted as a Ca^{2+}-dependent mobilization of vesicles from a reserve to a readily releasable pool. After exhaustion of the readily releasable pool, recovery also proceeds with time constants of hundreds of milliseconds to seconds, and this process can also be Ca^{2+} dependent (138, 140, 142–145). Finally, recovery of secreted vesicles can occur by slow and fast pathways, and Ca^{2+} can favor a fast pathway occurring within seconds (146, 147, but see 148) or a fraction of a second (137).

It is tempting to suppose that one or both of these processes (mobilization and Ca^{2+}-dependent recovery of released vesicles) are related to enhancement of synaptic transmission on repetitive stimulation. This is not likely, however, especially for facilitation. The reason is that facilitation has intrinsic kinetics of tens of milliseconds or less (35, 61), much faster than mobilization of vesicles into or recovery from depletion of the releasable pool.

Augmentation may resemble a seconds-long component of recovery from depletion of the readily releasable pool (138, 142, 145). However, augmentation's intrinsic rate constant is ~ 0.4 sec (61) and could only be related to the fastest forms of replenishment of the readily releasable pool. But augmentation appears to occur without any increase in the size of that pool (15). PTP is certainly long enough that it could involve slow recovery and mobilization processes. However, its intrinsic time constant is also a fraction of a second (61) and normally PTP is governed by the slow removal of residual Ca^{2+}. Any relationship between synaptic enhancement and vesicle recovery processes is therefore unlikely.

USE-DEPENDENT CHANGES IN Ca^{2+} ENTRY

The picture painted here of residual Ca^{2+}-dependent short-term plasticity describes the situation at the great majority of chemical synapses. However, a few examples have arisen where there is a change during repetitive activation in the Ca^{2+} influx evoked by an AP. In most cases the contributions of such changes in Ca^{2+} entry make a relatively small contribution to the overall plasticity of the synapse.

Ca^{2+} influx during a train can change by virtue of the properties of the Ca^{2+} channels. At the brainstem calyx of Held giant synapse, a form of Ca^{2+} channel facilitation to repeated depolarizations dependent on the build-up of presynaptic Ca^{2+} has been observed (149, 150). For other experimental conditions there can be a reduction in presynaptic Ca^{2+} current (151). It is possible that this behavior contributes to synaptic facilitation or depression at some synapses. In hippocampal neurons, facilitation was attributable partly to a potential-dependent relief of G-protein–mediated Ca^{2+} channel inhibition, resulting in increased influx through P/Q-type Ca^{2+} channels in repeated APs (152).

Another possibility is that during trains of activity Ca^{2+} entry into cells is sufficient to reduce $[Ca^{2+}]$ in extracellular space. At the calyx of Held, depolarization of the postsynaptic cell to 0 mV for 100 ms resulted in a 35% reduction in the PSC (153). This depression recovered within half a second and was accompanied by an inhibition of postsynaptic Ca^{2+} entry. Such a decrease in Ca^{2+} entry is consistent with the small volume of extracellular space and the magnitude of Ca^{2+} influx that can occur during prolonged depolarization (154). For typical levels of activity, the depletion of extracellular Ca^{2+} and its contribution to short-term plasticity are likely to be small.

At a few synapses, presynaptic spike broadening owing to cumulative K^+ channel inactivation contributes to facilitation by increasing Ca^{2+} entry during later APs in a train. Whole-cell recording from pituitary terminals reveals that repeated stimulation produces presynaptic APs that broaden, which in turn produces more Ca^{2+} influx per spike and facilitation of hormone release (155). Similarly, whole-cell voltage clamping of hippocampal mossy fibers revealed that spike broadening occurs during trains of presynaptic activity and that this broadening results in increased Ca^{2+} entry evoked by APs late in a train (156). This broadening arises from rapid inactivation of a K^+ channel involved in AP repolarization.

At some synapses, reductions in spike amplitude or duration, or failures of spike invasion of terminals seem to play a role in synaptic depression, especially to long trains of stimuli (157–160). In one instance, AP failure was attributable to the gradual activation of Ca^{2+}-dependent K^+ current (161).

Consideration of the magnitude of synaptic plasticity present at synapses where changes in Ca^{2+} entry has been observed suggests that often other mechanisms dominate the overall plasticity of the synapse. For example, at the calyx of Held, facilitation of Ca^{2+} entry occurs even while depression dominates the overall behavior of the synapse (150). The inaccessibility of presynaptic terminals has made it difficult to assess the importance of changes in Ca^{2+} entry to plasticity at synapses where the presynaptic terminal cannot be voltage clamped. Sometimes it is possible to use optical techniques to measure presynaptic AP waveforms or Ca^{2+} influx. Such an approach at the parallel fiber synapses in the cerebellum suggests that neither waveform changes nor changes in Ca^{2+} entry make important contributions to plasticity during trains of presynaptic APs (18, 162–164).

DEPRESSION OF TRANSMITTER RELEASE

At many synapses, periods of elevated activity lead to a decrease in synaptic strength. Multiple mechanisms can contribute to such synaptic plasticity. The most widespread mechanism appears to be a presynaptic decrease in the release of neurotransmitter that likely reflects a depletion of a release-ready pool of vesicles. In addition, a decrease in synaptic strength can arise from the release of modulatory substances from the activated presynaptic terminals, postsynaptic cells, or neighboring cells. Finally, postsynaptic properties such as desensitization of ligand-gated receptors can make the target neuron less sensitive to neurotransmitter.

Often a presynaptic mechanism contributes to a decline in PSC amplitude during repeated stimulation and takes seconds to minutes to recover after stimulation (4, 20, 165, 166). Although such synaptic depression was described nearly 60 years ago (1, 2), the mechanisms responsible for it are still poorly understood.

Statistical analysis of changes in quantal parameters during depression almost always reveals a reduction in the average number of quanta released, but binomial models of release may indicate reductions either in p, the probability of release of releasable quanta, or n, the number of releasable quanta, or both (reviewed in 8). Both changes could result from a reduction in AP effectiveness or Ca^{2+} influx, from feedback inhibitory actions of released transmitter on presynaptic autoreceptors, or from a reduction in the occupation of release sites by docked vesicles.

Depletion Models of Depression

A key characteristic of depression at many synapses is use dependence. Higher levels of transmission are associated with larger depression, and reduction of baseline transmission (for example by reducing external $[Ca^{2+}]$), relieves depression. Sometimes there is even a negative correlation between statistical fluctuations in the first of paired responses and the magnitude of the second (166, 167). Most models of short-term synaptic depression are based on the idea that it reflects depletion of a pool of vesicles that are poised and release ready. A number of such depletion models exist that vary in their assumptions and level of complexity.

According to the depletion model in its simplest form (165, 168), a synaptic connection contains a store of S releasable vesicles, and an AP releases a fraction, F, of this store. If each vesicle released produces a synaptic current i, stimulation produces a response equal to FSi. In this model, the store is transiently depleted of FS vesicles, so immediately following a stimulation, only $S-FS$ vesicles are available for release. If the fraction of available vesicles released by an AP remains unchanged, then a second stimulus will produce a response that is proportional to the number of remaining release-competent vesicles. The amplitude of the second response is then $S(1-F)Fi$, and the ratio of the amplitude of two closely spaced responses is $(1-F)$. It is usually assumed that there is a mono-exponential recovery of the release-ready store or pool of vesicles.

This model provides a simple explanation of some basic features of depression that are apparent at many synapses. For example, the larger the initial probability of release, the more pronounced is depression for two closely spaced stimuli. If F is very low initially, there is no depression, if F is 0.5, the second PSC is half as large as the first, and if F is 1, the second stimulus evokes no response at all because no releasable vesicles are available. This dependence of the magnitude of depression on the baseline probability of release has been described at many synapses. The exponential recovery from depression has also been observed at many synapses for a variety of experimental conditions.

It is important to distinguish measures of store size S and fraction of release F, estimated from the magnitude of depression, from measures of the statistical parameters of transmitter release n and p, estimated from a statistical quantal analysis of the variance of transmitter release (8, 10, 169). Typically, F is greater than p and S is less than n (170). This is because the statistical parameter n apparently corresponds to the total number of release sites, each of which may be able to release only one quantum (vesicle) to an AP (see below). The statistical parameter p has two components, one (p_{occ}) corresponding to the probability that a release site is occupied by a docked and releasable vesicle, and one (p_{eff}) the probability that an AP releases docked vesicles. The pool of immediately releasable vesicles (S) then corresponds to $p_{occ}n$, which is less than n, and the fraction of this pool released by an AP (F) corresponds to p_{eff}, which is greater than $p = p_{occ} p_{eff}$.

In spite of its success, this form of the depletion model is inadequate in several ways. Some of the basic assumptions are likely incorrect, such as F being constant. For example, different release sites are likely to release quanta with different initial probabilities, p_{eff} (107): The sites with highest release probability are depleted of more "willing" vesicles first, and the sites containing more "reluctant" docked vesicles after some stimulation have a lower probability of release. This results in a reduction in the average value of F, as well as in S during depression (63, 165, 171, 172). The simple model can also either underpredict or overpredict steady-state depression produced by trains of activity, and the assumption of a mono-exponential recovery has been questioned (4, 173–177).

In trying to improve upon such a model it is instructive to consider the anatomy of synapses. Studies at the electron microscope level reveal that synapses usually contain hundreds of vesicles in the vicinity of release sites, but only a small number of these vesicles are in contact with the membrane. These docked vesicles are thought to be poised and nearly ready to fuse in response to AP invasion of the presynaptic bouton or are at least available for immediate replenishment of release sites that are vacated by exocytosis. These vesicles are referred to as the readily releasable pool, whereas the more distant vesicles that are unable to respond rapidly are referred to as the reserve pool. In some synapses, the immediately releasable vesicles may be restricted to a fraction of the apparently docked and release-ready vesicles, with perhaps only one per active zone (see below). These immediately releasable vesicles are sometimes referred to conceptually as docked and primed for release. Following exocytosis of docked vesicles, recovery appears to occur by multiple processes into both reserve and readily releasable pools (Figure 3) (146, 178–180).

The concept of pools of vesicles in different states has provided a useful framework for thinking about synaptic depression (3, 4, 171, 175–177, 181–185). Repeated stimulation would result in the depletion of the readily releasable pool, which in turn would lead to a decrease in the number of vesicles that could be released by an AP and depression. According to depletion models that consider pools of vesicles, the extent of depression will depend on the number of vesicles in the reserve and release-ready pools and the transition rates between these pools. Other factors such as maximum sizes of the immediately releasable pool (limited by number of active zones), the release-ready pool (limited by the number of docking sites), the reserve pool (limited by the availability of vesicle membrane recoverable by endocytosis from recently secreted vesicles), and use-dependent changes in the movements of vesicles between pools impose additional restrictions not envisioned in the classical depletion models that treated vesicle movements between pools as a simple mass action without any limits on pool size. The mathematical description of such models can be quite complex, and several unanswered questions remain that are important to developing and assessing such models.

A curious aspect of synaptic depression is that it is not accompanied by a reduction in frequency of spontaneous release of quanta (186–188). This is hard to reconcile with depletion models of depression unless spontaneous release comes from a different pool of vesicles than evoked release or is subject to different limitations.

In a few instances, deep depression on extensive stimulation is accompanied by a presynaptic reduction in quantal amplitude (189–191). This may reflect the release of newly recovered and incompletely filled vesicles.

HOW BIG IS THE READILY RELEASABLE POOL AT INDIVIDUAL RELEASE SITES?   A number of approaches have been taken to measure the size of this pool of vesicles, including a large depolarizing pulse in the presynaptic terminal, caged Ca^{2+} in the presynaptic terminal, application of high-osmolarity solution, and integration of fully depressing responses to brief stimulus trains (13, 14, 107, 171). The challenge with all of these approaches is to measure the full size of the release-ready pool without it being replenished from the reserve pool. Failure to release all of the vesicles will lead to an underestimation of the size of the pool, whereas contributions from the reserve pool can lead to an overestimation (192). Serial electron microcopy has also been used to determine the number of morphologically docked vesicles, which may correspond to the readily releasable pool. Estimates of the functional size of the readily releasable pool per active zone are 7–8 at the parallel fiber and climbing fiber synapses, 22 in goldfish bipolar neurons (193), 32 in frog saccular hair cells (194), 130 in cat rod photoreceptors (195), an average of about 10 for the CA1 region of the hippocampus (196, 197), and in layers 1a and 1b of pyriform cortex, the average number of docked vesicles is 16 and 27, respectively (198). There is considerable variability in the number of docked vesicles at individual synapses, even for the same type of synapse.

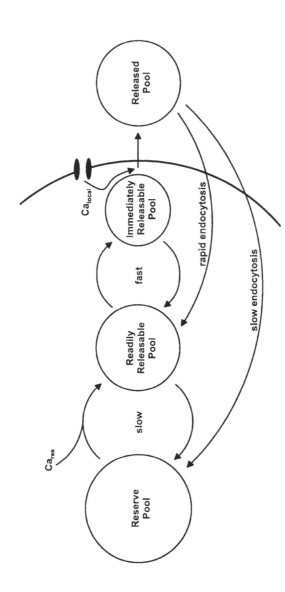

Figure 3   Functional anatomy of an active zone. Each active zone contains one, or at most a very few, vesicles attached to release sites, docked and primed at the plasma membrane, and immediately releasable by the local high $[Ca^{2+}]_i$ near Ca^{2+} channels opened by APs. These vesicles are rapidly replaced by other docked vesicles in a readily releasable pool. Most vesicles are stored in a large reserve cluster behind the plasma membrane. These can be used to refill the readily releasable pool after it is depleted in depression by a Ca^{2+}-dependent mobilization process. Released vesicles are recovered by fast and slow endocytic processes into readily releasable and reserve pools.

AT A SINGLE RELEASE SITE, CAN ONLY A SINGLE VESICLE BE RELEASED IN RESPONSE TO AN AP, OR IS IT POSSIBLE TO HAVE MULTIVESICULAR RELEASE?   The observation that the number of release sites determined by quantal analysis is very similar to the number of active zones or synaptic contacts gives rise to the idea that some mechanism restricts the number of vesicle fusions to a single one per synaptic contact per impulse, so that a single release site is a single active zone (10, 197, 199, 200). More recent experiments support this hypothesis and also propose that following neurotransmitter release at an individual release site the inability to trigger additional release events persists for ~10 ms (13). However, other studies at hippocampal synapses (201) and inhibitory (202) and excitatory (203) cerebellar synapses suggest that individual release sites may release multiple vesicles.

The issue has important implications for the mechanisms that give rise to depression. This is illustrated by considering the cerebellar climbing fiber synapse, which is depressed to 50% of control amplitudes by a single conditioning pulse. If depression reflects a depletion of the readily releasable pool, and there is an average of 7–8 morphologically docked vesicles at each release site (204), how can a single conditioning pulse result in so much depression? This suggests that either each morphological docked vesicle is not release ready, or that multivesicular release must occur (203).

Ca^{2+}-DEPENDENT RECOVERY FROM DEPRESSION   Recent studies have provided new insight into depression and recovery from depression. Whereas recovery from depression usually can be approximated by an exponential with a time constant of several seconds, for some experimental conditions an elevation of presynaptic Ca^{2+} levels accelerates the recovery from depression (143–145). At the climbing fiber synapse, when external $[Ca^{2+}]$ is elevated to 4 mM, significant recovery from depression occurs in less than 100 ms. This finding suggested the hypothesis that high levels of residual Ca^{2+} accelerate recovery from depression. To test the involvement of residual Ca^{2+}, EGTA was introduced into the presynaptic terminal to accelerate the decay of residual Ca^{2+}, which did not affect the initial probability of release but eliminated rapid recovery from depression. Residual Ca^{2+} also accelerated recovery from depression at the calyx of Held.

These studies help to resolve a long-standing deficiency of depletion models of depression in that they failed to predict the magnitude of synaptic responses during high-frequency presynaptic activity. Based on the magnitude and time course of depression evoked by a brief conditioning train, depletion models greatly underestimated synaptic efficacy during prolonged trains of activity. This led to the hypothesis that the rate of recovery from depression is accelerated during trains. This can now be understood in terms of the build-up of presynaptic Ca^{2+} levels accelerating recovery from depression, and allowing the presynaptic terminal to meet the increased demand for neurotransmitter during a train.

One way of accounting for Ca^{2+}-dependent recovery from depression is to update the depletion model such that increases in residual $[Ca^{2+}]_i$ accelerate the mobilization of vesicles from a reserve pool (143–145). Another possibility, put forth by Wu & Borst (171), suggests that "during repetitive firing, accumulation of intracellular calcium may facilitate release of the rapidly replenished but reluctant vesicles, making them available for sustaining synaptic transmission." This study suggests that recovery from depression at the calyx of Held does not result from the Ca^{2+} dependence of recruitment of vesicles from a reserve pool to a ready releasable pool.

Other studies have also questioned a role for residual Ca^{2+} in the acceleration of recovery from depression. Weis et al. (205) manipulated presynaptic levels at the calyx of Held with the rapid Ca^{2+} chelator fura-2 and observed the effects on presynaptic $[Ca^{2+}]_i$ and on synaptic transmission. Based on these experiments and the behavior of several models, they concluded that at the calyx of Held "$[Ca^{2+}]_i$ in the range 50–500 nM does not significantly affect the rate of vesicle filling at this synapse," and that recovery from synaptic depression is "governed by localized, near membrane Ca^{2+} signals not visible to the indicator dye, or else by an altogether different mechanism."

A recent study by Sakaba & Neher of the calyx of Held (206) resolves some of the questions regarding Ca^{2+}-dependent recovery from depression. They manipulated resting presynaptic $[Ca^{2+}]_i$ levels and found that higher initial $[Ca^{2+}]_i$ accelerates recovery from depression. These findings argue against the study by Weis et al. (205) and confirm the importance of residual Ca^{2+} (143–145).

WHY DO SOME SYNAPSES SHOW NO DEPRESSION?   Some synapses, for example at crayfish opener muscles (207) and chick ciliary ganglion (208), show virtually no sign of synaptic depression. Prolonged high-frequency activity eventually depletes the entire or reserve pool of transmitter, but no immediately releasable and rapidly depletable pool is evident. Presumably, this is because at such synapses the readily releasable pool is very rapidly replenished between APs and thus is effectively never depleted.

DEPRESSION INCONSISTENT WITH DEPLETION MODELS   At crayfish fast flexor neuromuscular junctions (209), locust motor neuron synapses (160), and at *Aplysia* synapses between sensory and motor neurons (210–212), depression arising presynaptically occurs independently of changes in the initial level of transmission, and the steady-state degree of depression is nearly independent of stimulation frequency. At the *Aplysia* synapses, buildup of residual $[Ca^{2+}]_i$ and Ca^{2+} channel inactivation also play no roles in depression (212), which is probably caused by switching off of release sites (213). Similarly, for both an excitatory synapse in the goldfish (214) and inhibitory synapses in the rat (164), the extent of depression does not depend on the magnitude of the first release, and at giant cochlear nucleus synapses, a Ca^{2+}-dependent form of depression appears to occur independently of changes in initial release level (215). Strangely, cyclothiazide (normally a blocker of glutamate receptor desensitization) appears to eliminate this depression by some presynaptic action. These studies argue for a mechanism of presynaptic depression that does not reflect depletion of neurotransmitter-containing vesicles. One possible explanation is that there is an activity-dependent gating mechanism limiting vesicle fusion through a refractory process and that this actually prevents vesicle depletion. At the crayfish synapse, depression is largely relieved by inhibition of NO synthase (216), suggesting that at this synapse NO generation is somehow responsible for synaptic depression.

At the squid giant synapse, a step elevation in $[Ca^{2+}]_i$ by caged Ca^{2+} photolysis activated secretion that decayed with a 30-ms time constant (217). Although resembling depletion of a readily releasable store, this explanation appears inconsistent with the property that further step increases in $[Ca^{2+}]_i$ evoked additional bouts of secretion and that the rate of fatigue is independent of $[Ca^{2+}]_i$ step magnitude. The authors characterized this behavior as a form of adaptation in Ca^{2+} sensitivity of release that could contribute to depression, although its kinetics seem too slow to be involved in AP-evoked release. Some of these properties could be explained by the existence of multiple pools of vesicles with different Ca^{2+} sensitivities, as has been shown for secretion of cortical granules by sea urchin eggs (218, 219). Other properties can be explained by invoking a Ca^{2+}-sensitive mobilization of vesicles from reserve to readily releasable pools (R. S. Zucker, unpublished calculations).

SYNAPTIC DEPRESSION VIA ACTIVATION OF METABOTROPIC RECEPTORS

Many presynaptic terminals in the mammalian CNS possess high-affinity metabotropic receptors that can be activated by chemical messengers such as GABA, glutamate or adenosine. Synaptic strength is controlled in part by the occupancy of these receptors, which in turn is set by the extracellular concentrations of their agonists. In some cases, tonic levels are sufficient to partially activate the receptors, but synaptic activity can further increase receptor occupancy by transiently elevating neuromodulator concentration. Following release, transmitter molecules can act either homosynaptically and bind to presynaptic autoreceptors or heterosynaptically by diffusing to nearby terminals. Examples of such signaling are given below and reviews of such modulation (220, 221) should be consulted for a more comprehensive treatment of synaptic modulation by activation of metabotropic receptors.

Homosynaptic Inhibition

The contents of a vesicle can act on the presynaptic terminal from which they were released. This is called homosynaptic modulation, and it is usually inhibitory. In most cases, when neurotransmitter builds up sufficiently to activate presynaptic receptors, the end result is negative feedback and a reduction in the future release of neurotransmitter. For example, vesicles contain ATP at relatively high concentrations, and when a vesicle fuses, ATP is released (222). In extracellular space, ATP is broken down into adenosine, which can activate presynaptic adenosine receptors and lead to homosynaptic inhibition (223). At neuromuscular junctions, this process contributes to depression to prolonged stimulation, but it is not responsible for the major part of synaptic depression (224). At some GABAergic synapses (225, 226), but not at others (227, 228), depression arises partly from a retrograde action of GABA on presynaptic $GABA_B$ receptors, presumably reducing Ca^{2+} influx to subsequent APs. A similar process involving presynaptic metabotropic glutamate receptors appears responsible for only a very tiny proportion of synaptic depression at calyx of Held synapses (229).

Heterosynaptic Inhibition

Activation of synaptic inputs can also affect neighboring synapses. For example, activation of excitatory inputs to a region can activate GABAergic interneurons causing a widespread increase in extracellular GABA levels (230–232). This can activate presynaptic $GABA_B$ receptors and inhibit synaptic strength for seconds following periods of elevated activity. Presynaptic boutons contain many types of receptors that sense a variety of extracellular chemical messengers, all potentially involved in heterosynaptic depression.

Retrograde Control of Neurotransmitter Release

It is also possible for the postsynaptic cell to influence release from the presynaptic terminal. Different types of dendrites can release a variety of messengers that can act through G-protein–coupled receptors located on presynaptic terminals to influence neurotransmitter release. Neuromodulators such as dopamine, dynorphin, glutamate, GABA, and oxytocin are released by fusion of vesicles that are located within the dendrites and cell bodies (233–239). Retrograde messengers are also released by non-vesicular mechanisms. Endogenous cannabinoids such as anandimide and 2-AG are produced by cleavage of phospholipids and are sensed by CB1 receptors on presynaptic terminals. Retrograde signaling by endogenous non-vesicular release of cannabinoids has been shown to suppress inhibitory synapses in the hippocampus and both excitatory and inhibitory synapses in the cerebellum (240–244). Vesicular and non-vesicular release of retrograde messengers are both Ca^{2+} dependent and provide a way for activity of the postsynaptic cell to influence release from its synaptic inputs.

PRESYNAPTIC IONOTROPIC RECEPTORS

Many presynaptic terminals contain ionotropic receptors that can also contribute to short-term synaptic plasticity (245). A variety of ionotropic receptors can be present, although the extent to which these receptors contribute to synaptic plasticity is not known. These receptors include Ca^{2+} permeable receptors such as NMDA receptors and $\alpha7$ nicotinic receptors, as well as Ca^{2+} impermeable receptors such as $GABA_A$ and glycine receptors, which are coupled to Cl^--permeable channels.

Activation of presynaptic ionotropic receptors can either increase or decrease neurotransmitter release by several mechanisms.

At brain stem synapses, glycine opens presynaptic Cl^- channels, which in turn depolarizes the terminal, opens voltage-gated Ca^{2+} channels, elevates presynaptic $[Ca^{2+}]_i$, and facilitates subsequent release of neurotransmitter (246). At the mossy fiber synapse on hippocampal CA3 pyramidal cells, glutamate activation of presynaptic kainate autoreceptors can contribute to synaptic enhancement (247). Blockade of these autoreceptors, as well as their genetic deletion (248), reduces the magnitude of synaptic enhancement during trains. It is hypothesized that this synaptic enhancement is caused by presynaptic depolarization induced by kainate receptor activation by previously released glutamate. High levels of kainate can also contribute to synaptic depression at this synapse (249); thus it appears that a bi-directional short-term regulation of synaptic transmission by autoreceptors is possible.

INVOLVEMENT OF GLIA IN SHORT-TERM PLASTICITY

There is growing realization that glia may be involved in some forms of short-term plasticity (250, 251). With their intimate association with synapses, astrocytes and perisynaptic Schwann cells are well positioned to regulate synapses. They have an established role in clearance of neurotransmitter and may participate in synaptic plasticity by controlling the speed and extent of such clearance (252, 253). This can in turn impact the degree of postsynaptic receptor activation and desensitization.

Another way that glia may be involved in synaptic plasticity is by sensing extracellular messengers and then releasing substances that can affect synaptic efficacy (250, 251). Glia have receptors for many neurotransmitters such as glutamate, GABA, acetylcholine, and ATP. Appropriate signaling molecules elevate glial $[Ca^{2+}]_i$ levels either through Ca^{2+}-permeable channels or by release from internal stores. The resulting increases in $[Ca^{2+}]_i$ can trigger vesicular release of substances from astrocytes, which can then act on presynaptic terminals to regulate neurotransmitter release. This signaling system provides a way for glia to sense release from presynaptic terminals and provide feedback to that terminal.

This interaction between neurons and glia and a potential role for glia in short-term synaptic plasticity have been observed at several synapses. In hippocampal cell culture, stimulation of astrocytes can depress the strength of synaptic contacts between neurons (254). This is the result of Ca^{2+}-evoked glutamate release that inhibits release by activating presynaptic metabotropic glutamate receptors (254–256). At the frog neuromuscular junction, perisynaptic Schwann cells can either potentiate or depress transmission. During prolonged repetitive stimulation of motor neurons, perisynaptic Schwann cells show a rise in $[Ca^{2+}]_i$ levels owing to ATP- and acetylcholine-activated IP_3-dependent release from endoplasmic reticulum (257–259). This increase in $[Ca^{2+}]_i$ can enhance the release of neurotransmitter from the presynaptic terminal (260). In addition, disruption of G protein and NO signaling reduced the extent of depression induced by high frequency stimulation, suggesting that glia can also contribute to depression at this synapse (261, 262). In hippocampal slices, glia are involved in the enhancement of transmission between interneurons and pyramidal cells (263). During repetitive stimulation, inhibitory neurons activate $GABA_B$ receptors on astrocytes, which raises their internal $[Ca^{2+}]_i$ and somehow feeds back to the presynaptic terminal to enhance transmission.

These studies suggest that astrocytes contribute to multiple forms of synaptic plasticity at many synapses in the brain. Clarification of the role of glia in short-term plasticity is in its early stages and promises to remain an exciting area of research in coming years.

POSTSYNAPTIC MECHANISMS OF SYNAPTIC DEPRESSION

Desensitization

Another mechanism that can lead to use-dependent decreases in synaptic strength is through desensitization of postsynaptic receptors (reviewed in 264). Ligand-gated channels undergo a process called desensitization that is analogous to inactivation of voltage-gated channels. Exposure of ligand-gated channels to an agonist can lead to channel opening and can also put some of the channels into a nonresponsive state. It can take tens of milliseconds or even minutes for channels to recover from such a desensitized state.

In *Aplysia*, early work revealed cholinergic excitatory and inhibitory synapses in which desensitization of postsynaptic receptors is the major process responsible for depression of a component of the postsynaptic potential (265, 266). Desensitization of AMPA receptors has been shown to play a role in synaptic transmission at a number of synapses. Desensitization contributes to plasticity at the calyceal synapse between the auditory nerve and the nucleus magnocellularis of the chick (267, 268). At this synapse, inhibitors of desensitization partially relieve synaptic depression to pairs of pulses and trains (269). Blockade of glutamate transporters enhanced depression (270), suggesting that they normally limit depression by reducing the extracellular accumulation of glutamate. Desensitization also occurs at the synapses between cones and bipolar cells in the retina (271), at giant cochlear nucleus synapses (272), and at retinogeniculate synapses, where AMPA receptor desensitization contributes to a reduction in AMPA receptor ePSCs during realistic patterns of activity (C. Chen, D.M. Blitz & W.G. Regehr, in review). Desensitization also contributes to depression of NMDA receptor responses in hippocampal cultures, where the amount of desensitization is regulated by postsynaptic potential (274). Desensitization may also play a role in the apparent reduction in quantal amplitude inferred at hippocampal glutamatergic synapses (275).

The extent of AMPA receptor desensitization to short-term plasticity is likely dictated by the structure of the synapse, the probability of release, and the time course of transmitter clearance. Receptor desensitization does not play a widespread role in short-term plasticity (162, 276, 277). The chick calyceal synapse and the retinogeniculate synapse both have many closely spaced release sites that are not well isolated from one another. This allows glutamate release at one site to diffuse, bind to, and desensitize nearby receptors. The glial sheath that encompasses aggregates of synaptic contacts onto the geniculate neuron may allow glutamate to pool and may contribute to the occurrence of AMPA receptor desensitization. In contrast, even though climbing fibers make hundreds of synaptic contacts with Purkinje cells, because each synapse is ensheathed by glia, the release sites are well isolated from one another, and there is no opportunity for glutamate released from one site to desensitize AMPA receptors at nearby release sites (204, 278). Another consideration is that if vesicle fusion occurs at a given release site and leads to receptor desensitization of the corresponding receptors, presynaptic depression could prevent the release of transmitter at that site and obscure receptor desensitization. Desensitization occurs at other types of receptors where it may contribute to synaptic plasticity (279). Low concentrations of GABA reduce synaptic currents at GABAergic hippocampal synapses, and it has been proposed that desensitization can regulate the availability of GABA receptors (280). NMDA receptors can also be desensitized by a variety of mechanisms (264, 281–283) that may contribute to synaptic plasticity of the NMDA responses. For acetylcholine receptors, although they can also desensitize, at the neuromuscular junction desensitization has little effect on the amplitude of synaptic responses during trains of activity (284).

Relief of Polyamine Block

Another postsynaptic mechanism leading to short-term synaptic plasticity was revealed at synaptic connections between pyramidal neurons and postsynaptic multipolar interneurons in layer 2/3 of rat neocortex (285). Synaptic currents at this synapse are mediated by AMPA receptors, which are blocked by endogenous intracellular polyamines that are present in almost all cells. In excised patches, this polyamine block is relieved by depolarization (286). At the synapses onto layer 2/3 interneurons, this relief of polyamine block enhances the responses of AMPA receptors and helps to offset a synaptic depression produced by another mechanism. Thus a facilitation of AMPA receptors due to potential-dependent relief of polyamine block can lead to a postsynaptic form of facilitation.

MOLECULAR TARGETS IN SYNAPTIC ENHANCEMENT

The search for molecular mediators of short-term synaptic plasticity has been intense. In the case of postsynaptic mechanisms, presynaptic ionotropic receptors, and metabotropic receptors, there has been good success in identification of the molecules involved in synaptic plasticity. It has proven to be more difficult to identify the molecules responsible for facilitation, PTP, and depression, but many possibilities remain to be explored.

Difficulties In Identifying Specific Effects on Short-Term Plasticity

One of the primary difficulties in the identification of molecules directly involved in short-term plasticity is that manipulations affecting the baseline level of transmission indirectly influence the magnitude of short-term plasticity. At most synapses, an increase in the initial probability of transmitter release decreases the magnitude of synaptic enhancement, and, conversely, a decrease in the probability of release results in larger synaptic enhancement or reduced synaptic depression. This is such a widely recognized relationship that it has become a standard means of gauging whether a neuromodulator has a presynaptic or postsynaptic site of action.

Thus the interpretation of effects on short-term plasticity often requires a measure of the effects on the baseline level of transmission. There are several possible ways of making such measurements: (*a*) Ideally, an analysis of quantal content should be performed. Often neurons are connected by a large number of synaptic contacts and determining the quantal content of baseline recordings is not straightforward. In that case, measures of PSP or PSC amplitude should be made, along with tests for postsynaptic effects such as measuring miniature PSP or PSC amplitudes. (*b*) At synapses where NMDA receptors are present, use-dependent block of NMDA receptors by MK801 can be used to detect changes in the probability of release. (*c*) With use of styryl dyes such as FM1-43, vesicle turnover rates to low-frequency stimulation can be assessed.

Often it is experimentally difficult to measure the baseline level of synaptic transmission. For example, transmission may be examined in brain slices where extracellular stimulation is used to evoke responses that are detected with extracellular electrodes. The amplitude of the evoked response does not provide a measure of baseline transmission because it reflects the activation of many presynaptic fibers and depends upon the positioning of the stimulus and recording electrodes, the intensity of stimulation, and the overall health of the slice. Although NMDA receptors may be present, they are difficult to measure with extracellular methods because of their slow time course. Thus it is difficult to quantify the initial magnitude of release with extracellular methods, and many studies do not determine whether changes in plasticity are direct effects on facilitation or depression, or if they are secondary consequences of changes in the initial probability of release.

Another potential complication arises in situations where synapses are manipulated for a prolonged time before the effects on plasticity are assessed. Particularly in the case of genetic mutations, a variety of homeostatic mechanisms allow neuronal systems to respond to the initial effect of a manipulation with secondary changes. This makes it difficult to exclude the possibility that changes in short-term plasticity arise as a secondary consequence through a homeostatic mechanism.

Despite these potential complications, important advances have been made both in the elimination of certain molecular schemes and in the identification of molecules with potential roles in synaptic transmission.

Ca^{2+}-Binding Proteins

An early hypothesis (287) for longer forms of enhancement, particularly PTP, posed that it was caused by Ca^{2+} acting via calmodulin and Ca^{2+}/calmodulin-dependent protein kinase II (CaMKII) to phosphorylate synapsin I, a vesicle-associated protein that binds actin filaments. Phosphorylation was shown to reduce synapsin I cross-linking to actin, and this could liberate vesicles and make them more available for release.

TESTS OF THE SYNAPSIN HYPOTHESIS   Unfortunately, most data from genetic mutants of CaMKII and synapsin I failed to corroborate predictions of this elegant hypothesis. Synapsin I null mutants display normal PTP in hippocampal CA1 neurons, as do CaMKII heterozygotes expressing reduced CaMKII (288–291). PTP was reduced in synapsin II mouse knockouts (292), but this form of the protein is not phosphorylated by CaMKII. CaMKII inhibitors failed to block PTP in hippocampal neurons (293), augmentation at CA3 association/commissural synapses (69), or any form of short-term enhancement at crayfish neuromuscular junctions (61). In contrast, at *Aplysia* synapses PTP was reduced by presynaptic injection of synapsin I/II antibodies and was enhanced by synapsin I injection (294).

Synapsins do appear to play an important role in the maintenance of a reserve pool of vesicles and the mobilization of vesicles from that pool to the readily releasable pool (292, 295–297). Genetic disruption of synapsins leads to a deepening of synaptic depression and loss of undocked vesicles from active zone clusters. The association of vesicle-associated synapsin with actin filaments may also explain why actin depolymerization deepened synaptic depression and slowed its recovery in snake motor neurons (298) and disrupted the reserve pool of vesicles and deepened depression at *Drosophila* motor neurons (299).

CaMKII also seems to influence synaptic plasticity at some synapses. CaMKII inhibition reduced PTP at several fish electroreceptor lateral line synapses, but not others (300), and it reduced augmentation at CA3 mossy fiber synapses (69). Mutations in CaMKII targeted to *Drosophila* presynaptic sensory terminals reduced depression without necessarily affecting baseline transmission (301). Facilitation was increased in the synapsin I null mutants and decreased in CaMKII mutants (290), but unaffected in *Aplysia* synapses (72); these effects occurred in the absence of effects on baseline transmission. It seems likely that synaptic plasticity can be regulated by, if not mediated by, the phosphorylation state of presynaptic proteins.

OTHER SECRETORY PROTEINS   A lot of attention is focused now on a number of Ca^{2+}-binding proteins found in synaptic terminals, especially proteins associated with the secretory apparatus. The discovery that facilitation is accompanied by a reduction in synaptic delay or speeding of release kinetics (302, 303) encourages interest in such proteins. So does the inference from modeling studies that facilitation is caused by Ca^{2+} acting fairly close to Ca^{2+} channel mouths, although not as close as the secretory trigger (70).

Synaptotagmin III and VII are isoforms of synaptotagmin I, the low-affinity Ca^{2+}-binding vesicle protein that is the most popular candidate for the Ca^{2+} trigger of phasic exocytosis (304, 305). These plasma membrane isoforms (306) have higher Ca^{2+}-binding affinity to their C2 domains (307), and interfering with their Ca^{2+} binding blocked exocytosis (308). Because a high-affinity Ca^{2+} receptor involved in generating facilitation could also be required for normal baseline secretion (70), these synaptotagmins are good candidates as Ca^{2+} targets involved in facilitation.

Doc2α (for double C2 domain) is a presynaptic Ca^{2+}-binding protein involved in synaptic transmission (57). Overexpression of Doc2 enhanced growth hormone secretion from PC12 cells, and peptides disrupting its interaction with Munc13 (another presynaptic protein; see below) inhibited transmitter release from sympathetic neurons (309). Doc2 was a good candidate Ca^{2+} target in synaptic enhancement until recently, when it was found that in Doc2α null mutants paired-pulse facilitation and PTP were normal (310). Steady stimulation for 30 s resulted in even more enhancement in Doc2α null mutants, possibly reflecting an enhancement in augmentation. These results suggest that Doc2α may under some circumstances modulate synaptic enhancement, but it is unlikely to mediate its activation by Ca^{2+}.

RIM (for rab-3 interacting molecule) is a third Ca^{2+}-binding protein localized to the plasma membrane (311), which binds to rab-3, a vesicle protein thought to be involved in vesicle trafficking and targeting. Transfection of PC12 cells with RIM enhances secretion, so this protein remains another putative Ca^{2+} target for activating synaptic enhancement.

An additional rab-3 and Ca^{2+}-binding protein called rabphilin is located in synaptic vesicle membranes. Reducing normal rabphilin expression inhibited secretion in PC12 cells (312), while injecting rabphilin into squid giant presynaptic terminals also inhibited secretion (313). Rabphilin knockout mice, on the other hand, showed normal hippocampal baseline synaptic transmission, as well as normal facilitation and augmentation (314). Thus rabphilin is a less-favored candidate as a Ca^{2+} sensor mediating short-term synaptic enhancement.

Munc13-1 is a syntaxin-binding presynaptic protein containing C2 domains, although its Ca^{2+} binding could not be established (315). In mouse knockouts, most hippocampal excitatory synapses were defective, apparently due to interference with a vesicle priming step (316), but in the few synapses where transmission remained, augmentation was enhanced by increase in the store of readily releasable vesicles (A. Sigler, J.-S. Rhee, I. Augustin, N. Brose & C. Rosenmund, in review). This contrasts with normal augmentation where no increase in store size is observed (15). Thus rather than mediating normal synaptic enhancement, this protein may modulate its expression.

CAPS (Ca^{2+}-dependent activator protein for secretion), a protein restricted to dense core vesicles, binds Ca^{2+} with low affinity (270 μM) (318). CAPS antibodies inhibit norepinephrine release from synaptosomes, which suggests a role in triggering phasic secretion. But in melanotrophs, CAPS antibodies specifically block a phase of secretion that is highly sensitive to $[Ca^{2+}]_i$ levels below 10 μM, leaving open the possibility that this site may play a role in synaptic enhancement (319).

Another presynaptic protein with Ca^{2+}-binding C2 domains is called piccolo or aczonin (320, 321). This protein binds Ca^{2+} with low affinity, more consistent with a role in triggering phasic vesicle fusion than in mediating synaptic enhancement (322). At present, physiological evidence for any role of this protein in synaptic transmission is lacking.

Scinderin is a Ca^{2+}-dependent actin-binding protein that appears to regulate exocytosis in chromaffin cells (323). Although no neuronal homologue is known, such a protein could easily mediate Ca^{2+}-dependent changes in the size of the readily releasable store or the probability of release of quanta by APs.

Frequenin is a presynaptic Ca^{2+}-binding protein whose overexpression can greatly enhance synaptic facilitation (324). However, depolarizing pulses without Na^+ influx (in tetrodotoxin) do not exhibit excessive facilitation. Frequenin may therefore affect the accumulation of $[Na^+]_i$, and consequently of $[Ca^{2+}]_i$ under regulation of Na^+/Ca^{2+} exchange and lead to an increase in synaptic enhancement.

There are two other Ca^{2+}-dependent presynaptic protein interactions. Ca^{2+} acting at low micromolar concentrations can trigger binding of the presynaptic Ca^{2+} channels to the plasma membrane SNARE proteins syntaxin and SNAP-25 (325). A potential cation-binding site coordinated with synaptobrevin and syntaxin within the SNARE assembly has also been inferred from structural studies (326). Effects of Ca^{2+} binding to either of these sites on synaptic transmission remain to be studied.

Other Genetic Manipulations

In addition to Ca^{2+}-binding proteins, a number of other presynaptic proteins have been identified that play important roles in synaptic transmission. Mutations or knockouts of genes encoding such proteins have been used to probe their roles in synaptic plasticity, particularly in mice and fruit flies.

MOUSE MUTANTS   Rab-3A is a GTP-binding protein found in synaptic vesicle membranes, believed to function in the targeting of vesicles to active zones. Mouse knockouts of this gene show enhanced facilitation, even in the presence of increased baseline transmission (327); PTP was unaffected (304). Synapses showed an unusual ability to release more than one vesicle per active zone, suggesting that a limit on secretion had been removed, thereby allowing more facilitation to be expressed. Curiously, presynaptic injection of rab-3 into *Aplysia* neurons also increased facilitation (but not PTP), but now baseline transmission was reduced (81); thus in these experiments release could have been desaturated. It appears that Rab-3A can modulate synaptic transmission in a way that indirectly affects facilitation.

Synaptogyrin I and synaptophysin I are synaptic vesicle membrane proteins of unknown function. Double knockouts of both genes showed modest reductions in facilitation and PTP, whereas facilitation was normal in single knockouts of either gene, and PTP was only slightly depressed in synaptogyrin mutants (328). Baseline transmission was unaffected. These proteins can influence, especially in concert, the magnitude of short-term enhancement.

Facilitation was impaired in null mutants of the gene encoding neurotrophin-3, a nerve growth factor expressed mainly in the hippocampal dentate gyrus (329). Baseline transmission was normal. Application of exogenous neutrophilin-3 also reduces facilitation in normal rat hippocampal CA1 cells, but this is probably a consequence of a strengthening of synaptic transmission (330).

Mutations in the gene encoding the brain protein ataxin-1 showed a modest impairment in paired-pulse facilitation, although PTP remained normal (331). However, reduced facilitation could be an indirect consequence of altered baseline transmission, which was not recorded.

Mouse knockouts of the mGluR4 subtype of metabotropic glutamate receptor show reduced facilitation and PTP (332). The results suggest a role for glutamatergic feedback via presynaptic metabotropic receptors in enhancing transmitter release. However, effects on baseline transmission were not assessed, so effects of saturation or countervailing depression cannot be excluded.

DROSOPHILA MUTANTS   A number of genetic manipulations in *Drosophila* have interesting effects on short-term synaptic enhancement. *Rutabaga* and *dunce* are mutants defective in an adenylyl cyclase and phosphodiesterase, respectively. The former has reduced capability for cAMP synthesis; the latter has enhanced levels of cAMP owing to diminished degradative activity. Both mutants displayed remarkably diminished facilitation and PTP under low-$[Ca^{2+}]$ conditions in which basal transmission was also reduced in *rutabaga* but was normal in the *dunce* mutants (333). The latter also showed reduced Ca^{2+} cooperativity in triggering secretion. Surprisingly, some boutons showed enhanced augmentation. Without knowledge of possible changes in presynaptic $[Ca^{2+}]_i$ regulation, the results are difficult to interpret. In a different study, *rutabaga* mutants and inhibitors of adenylate cyclase and protein kinase A showed a diminished mobilization of vesicles from reserve to readily releasable pools (334). Certainly a role or roles, although perhaps indirect, are implicated for cAMP in the regulation of synaptic plasticity.

Shibire is a temperature-sensitive mutant of the gene for dynamin, a protein essential for vesicle endocytosis. At non-permissive temperatures, in addition to the expected gradual decline in transmission owing to the inability to restore vesicles to the reserve pool, a rapid form of depression is seen that is inconsistent with the delays involved in recovering vesicles by endocytosis, which suggests an additional role for dynamin in maintaining an intact readily releasable pool (335).

Leonardo is a *Drosophila* cytosolic protein also located in synaptic vesicles and involved in regulation of a number of second messenger pathways. Null mutants displayed enhanced facilitation but reduced augmentation and PTP (336).

However, baseline transmission was reduced, which could account for the effect on facilitation, whereas excessive depression could have been mistaken for effects on augmentation and PTP.

Latheo is a DNA-regulating protein in *Drosophila* also concentrated presynaptically and able to affect synaptic transmission. Null mutants of the *latheo* gene show seriously impaired facilitation, augmentation, and PTP at neuromuscular junctions (337). Although transmission is strongly increased, the effects on synaptic plasticity persist even when baseline transmission is restored to normal in low-$[Ca^{2+}]$ medium. Transgenic overexpression of latheo rescues synaptic enhancement in the mutants. Not known to be a Ca^{2+}-binding protein, latheo may not be the target of Ca^{2+} action in regulating synaptic transmission, but it certainly seems to play important roles in short-term synaptic enhancement.

Finally, αPS3-integrin is a *Drosophila* cell adhesion protein localized to presynaptic boutons. Its genetic inactivation results in enlarged presynaptic arborizations, enhanced transmission, and reduced facilitation and PTP (338). Integrin inhibitory peptides have similar effects. Compensating the effect on baseline transmission by reducing $[Ca^{2+}]$ leaves PTP impaired, but surprisingly restores augmentation to greater than control levels (facilitation was not tested under these conditions). Deficits in synaptic enhancement are conditionally rescued by heat shock when mutants are transfected with an integrin gene under control of a heat shock promoter. Integrins thus also appear capable of strongly influencing the expression of short-term synaptic enhancement.

SUMMARY OF MOLECULAR MEDIATORS   There are indications that the presynaptic proteins synapsin I and II, synaptogyrin, synaptophysin, munc13-1, latheo, integrin, and native Ca^{2+}-binding proteins all influence the expression of facilitation, augmentation, and/or potentiation. Neutrophilin-3 and metabotropic glutamate receptors can also influence synaptic plasticity. However, present evidence does not point to any of these as the Ca^{2+} target in activating any particular phase of short-term synaptic plasticity.

Other Modulatory Influences

Many factors can influence synaptic plasticity, but the effects may be indirect, merely a consequence of changes in strength of synaptic transmission. For example, numerous aminergic and peptidergic neuromodulators either enhance or reduce facilitation at crustacean neuromuscular junctions, but the reported effects are always inversely related to effects on baseline transmission (339). At crayfish neuromuscular junctions, serotonin enhances synaptic transmission and prolongs facilitation without prolonging the time course of residual Ca^{2+} (340). This curious result is hard to reconcile with the evidence that the duration of facilitation reflects that of residual Ca^{2+}.

The identity of postsynaptic targets can influence the magnitude or dominance of facilitation or depression (341, 342), as can developmental age (343) and the removal of synaptic competition by sensory deprivation (344, 345). These findings support the possibility of independent developmental regulation of short-term synaptic plasticity.

Disruption of the tight association of presynaptic Ca^{2+} channels with the SNARE complex can be achieved by injecting peptides that compete with this binding. This increases synaptic facilitation, but also decreases baseline transmission (346). A low-$[Ca^{2+}]$ medium had similar effects, so the modulation of facilitation is probably a consequence of the reduction in basal transmission.

Perfusion of calyx of Held terminals with inhibitors of G proteins increased synaptic depression and slowed its recovery (347). It is not clear what proteins were affected by this treatment, but because rab3 mutants show increased facilitation (327), they may not be the only targets affected by G protein inhibitors.

Calmodulin inhibition slowed recovery from depression at the calyx of Held and prevented refilling of the readily releasable pool of vesicles (206). It is possible that calmodulin is required for the Ca^{2+}-dependent mobilization of transmitter.

The phosphatase inhibitor okadaic acid depressed facilitation at frog neuromuscular junctions (45). But effects on basal transmission and depression were not reported, so the effect may be only indirect or apparent. Another phosphatase inhibitor with a similar spectrum of action, calyculin A, was without effect.

Overexpression of synaptotagmin I or II and treatment with cytochalasin (an actin depolymerizer) can appear to increase facilitation at neuromuscular junctions, but this may be a consequence of reduced short-term depression (348, 349).

Brain-derived neurotrophic factor reduced facilitation at these junctions as well as at cortical synapses, but this could be a consequence of increased basal transmission (330, 350).

Induction of status epilepticus by kainate reduces facilitation and PTP, probably by increasing the basal probability of transmitter release (351).

Deletion of the gene for the Ca^{2+}-binding protein calbindin-D_{28k} reduced facilitation and PTP (352), a surprising result because overexpression of calbindin has similar effects (73); however, the effects on synaptic plasticity in the deletion mutants might have been caused by effects on basal transmission levels, which were not monitored.

Bacterial lipopolysaccharide endotoxin increased paired-pulse facilitation without affecting baseline transmission (353), but a possible reduction in concurrent depression was not eliminated.

Finally, tetanic stimulation of cortical synapses has been found to reduce post-tetanic paired-pulse facilitation in a way dependent upon a postsynaptic $[Ca^{2+}]_i$ elevation and protein kinase activation (354, 355). The apparent reduction in facilitation was actually caused, however, by a concurrent desensitization in postsynaptic AMPA-type glutamate receptors and so was really a modulation of depression (see below).

A rise in postsynaptic $[Ca^{2+}]_i$ has been reported to be essential for PTP in some *Aplysia* central synapses (356). However, it appears that it is a particularly long-lasting component of PTP that is affected (357); therefore this may more

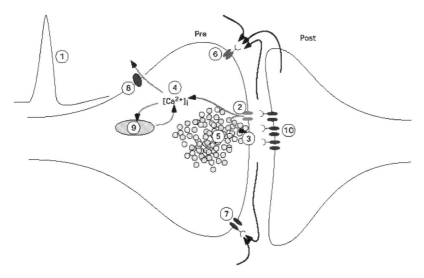

Figure 4 Sites of regulation of short-term synaptic plasticity. (1) AP waveform, (2) Ca^{2+} channel activation, (3) facilitation trigger and the readily releasable pool, (4) residual $[Ca^{2+}]_i$, (5) reserve pool, (6) metabotropic autoreceptors, (7) ionotropic autoreceptors, (8) Ca^{2+}-ATPase, regulating residual $[Ca^{2+}]_i$ in augmentation, (9) mitochondrial regulation of residual $[Ca^{2+}]_i$ in PTP, (10) postsynaptic receptor desensitization.

properly be considered an example of a decremental form of long-term potentiation, sometimes called short-term or slowly-decrementing potentiation, known to depend on postsynaptic Ca^{2+} (358, 359).

At cricket central synapses, a form of short-term synaptic depression appears to be mediated by a rise in postsynaptic $[Ca^{2+}]_i$ (360). The mechanism of this effect remains to be elucidated.

This summary of the search for molecular targets and regulatory influences indicates how important it is to distinguish effects on synaptic plasticity from effects on baseline transmission (Figure 4). If baseline transmission is in fact affected, alterations in external $[Ca^{2+}]_i$ or $[Mg^{2+}]_i$ can be used to restore transmission to comparable levels in experimental and control preparations. Effects on depression can usually be distinguished from short-term enhancement by their kinetic differences and by the relief of depression by reducing external $[Ca^{2+}]$.

Why is reducing external $[Ca^{2+}]_i$ usually so much more effective in relieving depression than in reducing synaptic enhancement? The reason is that baseline transmission normally depends on about the third power of external $[Ca^{2+}]$ near normal $[Ca^{2+}]$ levels. Thus a reduction of $[Ca^{2+}]$ to half will reduce transmission to about one tenth. Because depression by depletion is proportional to transmission strength, depression is also reduced to 10%. However, because synaptic enhancements all seem to be approximately linearly dependent on $[Ca^{2+}]_i$ accumulation, they should be reduced to only half and can now easily be studied in relative isolation from depression.

CONCLUSIONS, PERSPECTIVES

An impressive array of mechanisms have been identified that contribute to short-term synaptic plasticity (Figure 4). In coming years advances will be made on several fronts. First, it is likely that additional mechanisms of short-term synaptic plasticity will be discovered. The past decade saw a new appreciation of retrograde control of synapses, glial contributions to short-term plasticity, the discovery of Ca^{2+}-dependent recovery from depression, and contributions of presynaptic metabotropic ionotropic receptors. It therefore seems likely that new, perhaps unanticipated, forms of short-term plasticity remain to be discovered. A second area of interest is the clarification of the various mechanisms that give rise to different forms of short-term plasticity. Short-term presynaptic depression is one of the least understood forms of plasticity, despite its prevalence. Although more is known about the mechanisms that give rise to facilitation and posttetanic potentiation, a more detailed understanding awaits the identification of the proteins that give rise to these forms of synaptic enhancement. A third challenging area of investigation is the determination of the physiological roles of these various forms of short-term plasticity.

ACKNOWLEDGMENTS

Our research is this area is supported by grants from the National Institutes of Health and the National Science Foundation.

Visit the Annual Reviews home page at www.AnnualReviews.org

LITERATURE CITED

1. Feng TP. 1941. The changes in the endplate potential during and after prolonged stimulation. *Chin. J. Physiol.* 13:79–107
2. Eccles JC, Katz B, Kuffler SW. 1941. Nature of the "endplate potential" in curarized muscle. *J. Neurophysiol.* 4:362–87
3. Tsodyks MV, Markram H. 1997. The neural code between neocortical pyramidal neurons depends on neurotransmitter release probability. *Proc. Natl. Acad. Sci. USA* 94:719–23
4. Varela JA, Sen K, Gibson J, Fost J, Abbott LF, Nelson SB. 1997. A quantitative description of short-term plasticity at excitatory synapses in layer 2/3 of rat primary visual cortex. *J. Neurosci.* 17:7926–40
5. Magleby KL. 1987. Short-term changes in synaptic efficacy. In *Synaptic Function*, ed. GM Edelman, WE Gall, WM Cowan, pp. 21–56. New York: Wiley
6. Dittman JS, Kreitzer AC, Regehr WG. 2000. Interplay between facilitation, depression, and residual calcium at three presynaptic terminals. *J. Neurosci.* 20:1374–85

7. Fisher SA, Fischer TM, Carew TJ. 1997. Multiple overlapping processes underlying short-term synaptic enhancement. *Trends Neuosci.* 20:170–77

8. Zucker RS. 1989. Short-term synaptic plasticity. *Annu. Rev. Neurosci.* 12:13–31

9. McLachlan EM. 1978. The statistics of transmitter release at chemical synapses. *Int. Rev. Physiol.* 17:49–117

10. Zucker RS. 1973. Changes in the statistics of transmitter release during facilitation. *J. Physiol.* 229:787–810

11. Zucker RS. 1977. Synaptic plasticity at crayfish neuromuscular junctions. In *Identified Neurons and Behavior of Arthropods*, ed. G Hoyle, pp. 49–69. New York: Plenum

12. Worden MK, Bykhovskaia M, Hackett JT. 1997. Facilitation at the lobster neuromuscular junction: a stimulus-dependent mobilization model. *J. Neurophysiol.* 78:417–28

13. Stevens CF, Tsujimoto T. 1995. Estimates for the pool size of releasable quanta at a single central synapse and for the time required to refill the pool. *Proc. Natl. Acad. Sci. USA* 92:846–49

14. Rosenmund C, Stevens CF. 1996. Definition of the readily releasable pool of vesicles at hippocampal synapses. *Neuron* 16:1197–207

15. Stevens CF, Wesseling JF. 1999. Augmentation is a potentiation of the exocytotic process. *Neuron* 22:139–46

16. Wojtowicz JM, Marin L, Atwood HL. 1994. Activity-induced changes in synaptic release sites at the crayfish neuromuscular junction. *J. Neurosci.* 14:3688–703

17. Atwood HL. 1976. Organization and synaptic physiology of crustacean neuromuscular systems. *Prog. Neurobiol.* 7:291–391

18. Kreitzer AC, Regehr WG. 2000. Modulation of transmission during trains at a cerebellar synapse. *J. Neurosci.* 20:1348–57

19. Charlton MP, Smith SJ, Zucker RS. 1982. Role of presynaptic calcium ions and channels in synaptic facilitation and depression at the squid giant synapse. *J. Physiol.* 323:173–93

20. Del Castillo J, Katz B. 1954. Statistical factors involved in neuromuscular facilitation and depression. *J. Physiol.* 124:574–85

21. Dudel J, Kuffler SW. 1961. Mechanism of facilitation at the crayfish neuromuscular junction. *J. Physiol.* 155:530–42

22. Katz B, Miledi R. 1968. The role of calcium in neuromuscular facilitation. *J. Physiol.* 195:481–92

23. Rosenthal J. 1969. Post-tetanic potentiation at the neuromuscular junction of the frog. *J. Physiol.* 203:121–33

24. Weinreich D. 1971. Ionic mechanism of post-tetanic potentiation at the neuromuscular junction of the frog. *J. Physiol.* 212:431–46

25. Erulkar SD, Rahamimoff R. 1978. The role of calcium ions in tetanic and post-tetanic increase of miniature end-plate potential frequency. *J. Physiol.* 278:501–11

26. Kretz R, Shapiro E, Kandel ER. 1982. Post-tetanic potentiation at an identified synapse in *Aplysia* is correlated with a Ca^{2+}-activated K^+ current in the presynaptic neuron: evidence for Ca^{2+} accumulation. *Proc. Natl. Acad. Sci. USA* 79:5430–34

27. Blundon JA, Wright SN, Brodwick MS, Bittner GD. 1993. Residual free calcium is not responsible for facilitation of neurotransmitter release. *Proc. Natl. Acad. Sci. USA* 90:9388–92

28. Roberts WM. 1993. Spatial calcium buffering in saccular hair cells. *Nature* 363:74–76

29. Robitaille R, Garcia ML, Kaczorowski GJ, Charlton MP. 1993. Functional colocalization of calcium and calcium-gated potassium channels in control of transmitter release. *Neuron* 11:645–55

30. Yazejian B, Sun XP, Grinnell AD. 2000. Tracking presynaptic Ca^{2+} dynamics during neurotransmitter release with Ca^{2+}-

activated K^+ channels. *Nat. Neurosci.* 3:566–71

31. Connor JA, Kretz R, Shapiro E. 1986. Calcium levels measured in a presynaptic neurone of *Aplysia* under conditions that modulate transmitter release. *J. Physiol.* 375:625–42

32. Delaney KR, Zucker RS, Tank DW. 1989. Calcium in motor nerve terminals associated with posttetanic potentiation. *J. Neurosci.* 9:3558–67

33. Delaney KR, Tank DW. 1994. A quantitative measurement of the dependence of short-term synaptic enhancement on presynaptic residual calcium. *J. Neurosci.* 14:5885–902

34. Regehr WG, Delaney KR, Tank DW. 1994. The role of presynaptic calcium in short-term enhancement at the hippocampal mossy fiber synapse. *J. Neurosci.* 14:523–37

35. Atluri PP, Regehr WG. 1996. Determinants of the time course of facilitation at the granule cell to Purkinje cell synapse. *J. Neurosci.* 16:5661–71

36. Feller MB, Delaney KR, Tank DW. 1996. Presynaptic calcium dynamics at the frog retinotectal synapse. *J. Neurophysiol.* 76:381–400

37. Brain KL, Bennett MR. 1995. Calcium in the nerve terminals of chick ciliary ganglia during facilitation, augmentation and potentiation. *J. Physiol.* 489:637–48

38. Brain KL, Bennett MR. 1997. Calcium in sympathetic varicosities of mouse vas deferens during facilitation, augmentation and autoinhibition. *J. Physiol.* 502:521–36

39. Lin YQ, Brain KL, Bennett MR. 1998. Calcium in sympathetic boutons of rat superior cervical ganglion during facilitation, augmentation and potentiation. *J. Auton. Nerv. Syst.* 73:26–37

40. Ogawa S, Takeuchi T, Ohnuma K, Suzuki N, Miwa A, et al. 2000. Facilitation of neurotransmitter release at the spiny lobster neuromuscular junction. *Neurosci. Res.* 37:33–48

41. Suzuki S, Osanai M, Murase M, Suzuki N, Ito K, et al. 2000. Ca^{2+} dynamics at the frog motor nerve terminal. *Pflügers Arch.* 440:351–65

42. Ravin R, Parnas H, Spira ME, Volfovsky N, Parnas I. 1999. Simultaneous measurement of evoked release and $[Ca^{2+}]_i$ in a crayfish release bouton reveals high affinity of release to Ca^{2+}. *J. Neurophysiol.* 81:634–42

43. Wright SN, Brodwick MS, Bittner GD. 1996. Calcium currents, transmitter release and facilitation of release at voltage-clamped crayfish nerve terminals. *J. Physiol.* 496:363–78

44. Vyshedskiy A, Lin JW. 1997. Activation and detection of facilitation as studied by presynaptic voltage control at the inhibitor of the crayfish opener muscle. *J. Neurophysiol.* 77:2300–15

45. Van der Kloot W, Molgó J. 1993. Facilitation and delayed release at about $0°C$ at the frog neuromuscular junction: effects of calcium chelators, calcium transport inhibitors, and okadaic acid. *J. Neurophysiol.* 69:717–29

46. Zucker RS, Lara-Estrella LO. 1983. Posttetanic decay of evoked and spontaneous transmitter release and a residual-calcium model of synaptic facilitation at crayfish neuromuscular junctions. *J. Gen. Physiol.* 81:355–72

47. Zengel JE, Magleby KL. 1981. Changes in miniature endplate potential frequency during repetitive nerve stimulation in the presence of Ca^{2+}, Ba^{2+}, and Sr^{2+} at the frog neuromuscular junction. *J. Gen. Physiol.* 77:503–29

48. Magleby KL, Zengel JE. 1982. A quantitative description of stimulation-induced changes in transmitter release at the frog neuromuscular junction. *J. Gen. Physiol.* 80:613–38

49. Miledi R, Thies R. 1971. Tetanic and posttetanic rise in frequency of miniature endplate potentials in low-calcium solutions. *J. Physiol.* 212:245–57

50. Liley AW. 1956. The quantal components

of the mammalian end-plate potential. *J. Physiol.* 133:571–87

51. Hubbard JI. 1963. Repetitive stimulation at the mammalian neuromuscular junction, and the mobilization of transmitter. *J. Physiol.* 169:641–62

52. Bain AI, Quastel DM. 1992. Multiplicative and additive Ca^{2+}-dependent components of facilitation at mouse endplates. *J. Physiol.* 455:383–405

53. Barrett EF, Stevens CF. 1972. The kinetics of transmitter release at the frog neuromuscular junction. *J. Physiol.* 227:691–708

54. Ravin R, Spira ME, Parnas H, Parnas I. 1997. Simultaneous measurement of intracellular Ca^{2+} and asynchronous transmitter release from the same crayfish bouton. *J. Physiol.* 501:251–62

55. Rahamimoff R, Meiri H, Erulkar SD, Barenholz Y. 1978. Changes in transmitter release induced by ion-containing liposomes. *Proc. Natl. Acad. Sci. USA* 75:5214–16

56. Kita H, Van der Kloot W. 1974. Calcium ionophore X-537A increases spontaneous and phasic quantal release of acetylcholine at frog neuromuscular junction. *Nature* 250:658–60

57. Duncan RR, Shipston MJ, Chow RH. 2000. Double C2 protein. A review. *Biochimie* 82:421–26

58. Alnaes E, Rahamimoff R. 1975. On the role of mitochondria in transmitter release from motor nerve terminals. *J. Physiol.* 248:285–306

59. Mulkey RM, Zucker RS. 1991. Action potentials must admit calcium to evoke transmitter release. *Nature* 350:153–55

60. Mulkey RM, Zucker RS. 1993. Calcium released by photolysis of DM-nitrophen triggers transmitter release at the crayfish neuromuscular junction. *J. Physiol.* 462:243–60

61. Kamiya H, Zucker RS. 1994. Residual Ca^{2+} and short-term synaptic plasticity. *Nature* 371:603–6

62. Delaney KR, Zucker RS. 1990. Calcium released by photolysis of DM-nitrophen stimulates transmitter release at squid giant synapse. *J. Physiol.* 426:473–98

63. Sakaba T, Neher E. 2001. Quantitative relationship between transmitter release and calcium current at the calyx of held synapse. *J. Neurosci.* 21:462–76

64. Rozov A, Burnashev N, Sakmann B, Neher E. 2001. Transmitter release modulation by intracellular Ca^{2+} buffers in facilitating and depressing nerve terminals of pyramidal cells in layer 2/3 of the rat neocortex indicates a target cell-specific difference in presynaptic calcium dynamics. *J. Physiol.* 531:807–26

65. Tanabe N, Kijima H. 1992. Ca^{2+}-dependent and -independent components of transmitter release at the frog neuromuscular junction. *J. Physiol.* 455:271–89

66. Delaney K, Tank DW, Zucker RS. 1991. Presynaptic calcium and serotonin-mediated enhancement of transmitter release at crayfish neuromuscular junction. *J. Neurosci.* 11:2631–43

67. Hochner B, Parnas H, Parnas I. 1991. Effects of intra-axonal injection of Ca^{2+} buffers on evoked release and on facilitation in the crayfish neuromuscular junction. *Neurosci. Lett.* 125:215–18

68. Robitaille R, Charlton MP. 1991. Frequency facilitation is not caused by residual ionized calcium at the frog neuromuscular junction. *Ann. NY Acad. Sci.* 635:492–94

69. Salin PA, Scanziani M, Malenka RC, Nicoll RA. 1996. Distinct short-term plasticity at two excitatory synapses in the hippocampus. *Proc. Natl. Acad. Sci. USA* 93:13304–9

70. Tang Y, Schlumpberger T, Kim T, Lueker M, Zucker RS. 2000. Effects of mobile buffers on facilitation: experimental and computational studies. *Biophys. J.* 78:2735–51

71. Swandulla D, Hans M, Zipser K, Augustine GJ. 1991. Role of residual calcium in synaptic depression and posttetanic

potentiation: fast and slow calcium signaling in nerve terminals. *Neuron* 7:915–26

72. Jiang XY, Abrams TW. 1998. Use-dependent decline of paired-pulse facilitation at *Aplysia* sensory neuron synapses suggests a distinct vesicle pool or release mechanism. *J. Neurosci.* 18:10310–19

73. Chard PS, Jordan J, Marcuccilli CJ, Miller RJ, Leiden JM, et al. 1995. Regulation of excitatory transmission at hippocampal synapses by calbindin D_{28k}. *Proc. Natl. Acad. Sci. USA* 92:5144–48

74. Caillard O, Moreno H, Schwaller B, Llano I, Celio MR, Marty A. 2000. Role of the calcium-binding protein parvalbumin in short-term synaptic plasticity. *Proc. Natl. Acad. Sci. USA* 97:13372–77

75. Blundon JA, Wright SN, Brodwick MS, Bittner GD. 1995. Presynaptic calcium-activated potassium channels and calcium channels at a crayfish neuromuscular junction. *J. Neurophysiol.* 73:178–89

76. Winslow JL, Duffy SN, Charlton MP. 1994. Homosynaptic facilitation of transmitter release in crayfish is not affected by mobile calcium chelators: implications for the residual ionized calcium hypothesis from electrophysiological and computational analyses. *J. Neurophysiol.* 72:1769–93

77. Fischer TM, Zucker RS, Carew TJ. 1997. Activity-dependent potentiation of synaptic transmission from L30 inhibitory interneurons of *Aplysia* depends on residual presynaptic Ca^{2+} but not on postsynaptic Ca^{2+}. *J. Neurophysiol.* 78:2061–71

78. Wang YX, Quastel DM. 1991. Actions of lead on transmitter release at mouse motor nerve terminals. *Pflügers Arch.* 419:274–80

79. Zengel JE, Sosa MA, Poage RE. 1993. ω-Conotoxin reduces facilitation of transmitter release at the frog neuromuscular junction. *Brain. Res.* 611:25–30

80. Zengel JE, Lee DT, Sosa MA, Mosier DR. 1993. Effects of calcium channel blockers on stimulation-induced changes

in transmitter release at the frog neuromuscular junction. *Synapse* 15:251–62

81. Doussau F, Clabecq A, Henry JP, Darchen F, Poulain B. 1998. Calcium-dependent regulation of rab3 in short-term plasticity. *J. Neurosci.* 18:3147–57

82. Dobrunz LE, Huang EP, Stevens CF. 1997. Very short-term plasticity in hippocampal synapses. *Proc. Natl. Acad. Sci. USA* 94:14843–47

83. Chen Y, Chad JE, Wheal HV. 1996. Synaptic release rather than failure in the conditioning pulse results in paired-pulse facilitation during minimal synaptic stimulation in the rat hippocampal CA1 neurones. *Neurosci. Lett.* 218:204–8

84. Adler EM, Augustine GJ, Duffy SN, Charlton MP. 1991. Alien intracellular calcium chelators attenuate neurotransmitter release at the squid giant synapse. *J. Neurosci.* 11:1496–507

85. Heidelberger R, Heinemann C, Neher E, Matthews G. 1994. Calcium dependence of the rate of exocytosis in a synaptic terminal. *Nature* 371:513–15

86. Bollmann JH, Sakmann B, Borst JG. 2000. Calcium sensitivity of glutamate release in a calyx-type terminal. *Science* 289:953–57

87. Schneggenburger R, Neher E. 2000. Intracellular calcium dependence of transmitter release rates at a fast central synapse. *Nature* 406:889–93

88. Roberts WM. 1994. Localization of calcium signals by a mobile calcium buffer in frog saccular hair cells. *J. Neurosci.* 14:3246–62

89. Yamada WM, Zucker RS. 1992. Time course of transmitter release calculated from simulations of a calcium diffusion model. *Biophys. J.* 61:671–82

90. Dodge FA Jr, Rahamimoff R. 1967. Cooperative action of calcium ions in transmitter release at the neuromuscular junction. *J. Physiol.* 193:419–32

91. Linder TM. 1974. The accumulative properties of facilitation at crayfish neuromuscular synapses. *J. Physiol.* 238:223–34

92. Bittner GD, Schatz RA. 1981. An examination of the residual calcium theory for facilitation of transmitter release. *Brain. Res.* 210:431–36

93. Zucker RS. 1974. Characteristics of crayfish neuromuscular facilitation and their calcium dependence. *J. Physiol.* 241:91–110

94. Bittner GD, Sewell VL. 1976. Facilitation at crayfish neuromuscular junction. *J. Comp. Physiol.* 109:287–308

95. Atluri PP, Regehr WG. 1998. Delayed release of neurotransmitter from cerebellar granule cells. *J. Neurosci.* 18:8214–27

96. Younkin SG. 1974. An analysis of the role of calcium in facilitation at the frog neuromuscular junction. *J. Physiol.* 237:1–14

97. Quastel DMJ. 1984. Calcium cooperativity in coupling neurotransmitter release to nerve terminal depolarization: facilitation. *J. Theoret. Neurobiol.* 3:79–90

98. Landau EM, Smolinsky A, Lass Y. 1973. Post-tetanic potentiation and facilitation do not share a common calcium-dependent mechanism. *Nat. New Biol.* 244:155–57

99. Bain AI, Quastel DM. 1992. Quantal transmitter release mediated by strontium at the mouse motor nerve terminal. *J. Physiol.* 450:63–87

100. Xu-Friedman MA, Regehr WG. 2000. Probing fundamental aspects of synaptic transmission with strontium. *J. Neurosci.* 20:4414–22

101. Van der Kloot W. 1994. Facilitation of transmission at the frog neuromuscular junction at 0°C is not maximal at time zero. *J. Neurosci.* 14:5722–24

102. Katz PS, Kirk MD, Govind CK. 1993. Facilitation and depression at different branches of the same motor axon: evidence for presynaptic differences in release. *J. Neurosci.* 13:3075–89

103. Cooper RL, Marin L, Atwood HL. 1995. Synaptic differentiation of a single motor neuron: conjoint definition of transmitter release, presynaptic calcium signals, and ultrastructure. *J. Neurosci.* 15:4209–22

104. Mallart A, Martin AR. 1968. The relation between quantum content and facilitation at the neuromuscular junction of the frog. *J. Physiol.* 196:593–604

105. Atwood HL, Wojtowicz JM. 1986. Short-term and long-term plasticity and physiological differentiation of crustacean motor synapses. *Int. Rev. Neurobiol.* 28:275–362

106. Dobrunz LE, Stevens CF. 1997. Heterogeneity of release probability, facilitation, and depletion at central synapses. *Neuron* 18:995–1008

107. Murthy VN, Sejnowski TJ, Stevens CF. 1997. Heterogeneous release properties of visualized individual hippocampal synapses. *Neuron* 18:599–612

108. Parnas I, Parnas H, Dudel J. 1982. Neurotransmitter release and its facilitation in crayfish muscle. V. Basis for synapse differentiation of the fast and slow type in one axon. *Pflügers Arch.* 395:261–70

109. Fischer TM, Blazis DE, Priver NA, Carew TJ. 1997. Metaplasticity at identified inhibitory synapses in *Aplysia*. *Nature* 389:860–65

110. Stanley EF. 1986. Decline in calcium cooperativity as the basis of facilitation at the squid giant synapse. *J. Neurosci.* 6:782–89

111. Bertram R, Sherman A, Stanley EF. 1996. Single-domain/bound calcium hypothesis of transmitter release and facilitation. *J. Neurophysiol.* 75:1919–31

112. Chen C, Regehr WG. 1999. Contributions of residual calcium to fast synaptic transmission. *J. Neurosci.* 19:6257–66

113. Zengel JE, Magleby KL, Horn JP, McAfee DA, Yarowsky PJ. 1980. Facilitation, augmentation, and potentiation of synaptic transmission at the superior cervical ganglion of the rabbit. *J. Gen. Physiol.* 76:213–31

114. Zengel JE, Magleby KL. 1980. Differential effects of Ba^{2+}, Sr^{2+}, and Ca^{2+} on stimulation-induced changes in transmitter release at the frog neuromuscular junction. *J. Gen. Physiol.* 76:175–211

115. Goda Y, Stevens CF. 1994. Two components of transmitter release at a central synapse. *Proc. Natl. Acad. Sci. USA* 91:12942–46

116. Xu-Friedman MA, Regehr WG. 1999. Presynaptic strontium dynamics and synaptic transmission. *Biophys. J.* 76:2029–42

117. Zenisek D, Matthews G. 2000. The role of mitochondria in presynaptic calcium handling at a ribbon synapse. *Neuron* 25:229–37

118. Zhong N, Beaumont V, Zucker RS. 2001. Roles for mitochondrial and reverse mode Na^+/Ca^{2+} exchange and the plasmalemma Ca^{2+} ATPase in post-tetanic potentiation at crayfish neuromuscular junctions. *J. Neurosci.* In press

119. Regehr WG. 1997. Interplay between sodium and calcium dynamics in granule cell presynaptic terminals. *Biophys. J.* 73:2476–88

120. Wojtowicz JM, Atwood HL. 1985. Correlation of presynaptic and postsynaptic events during establishment of long-term facilitation at crayfish neuromuscular junction. *J. Neurophysiol.* 54:220–30

121. Birks RI, Cohen MW. 1968. The action of sodium pump inhibitors on neuromuscular transmission. *Proc. R. Soc. London Ser. B* 170:381–99

122. Birks RI, Cohen MW. 1968. The influence of internal sodium on the behaviour of motor nerve endings. *Proc. R. Soc. London Ser. B* 170:401–21

123. Parnas I, Parnas H, Dudel J. 1982. Neurotransmitter release and its facilitation in crayfish. II. Duration of facilitation and removal processes of calcium from the terminal. *Pflügers Arch.* 393:232–36

124. Mulkey RM, Zucker RS. 1992. Posttetanic potentiation at the crayfish neuromuscular junction is dependent on both intracellular calcium and sodium ion accumulation. *J. Neurosci.* 12:4327–36

125. Tang Y, Zucker RS. 1997. Mitochondrial involvement in post-tetanic potentiation of synaptic transmission. *Neuron* 18:483–91

126. David G, Barrett EF. 2000. Stimulation-evoked increases in cytosolic $[Ca^{2+}]$ in mouse motor nerve terminals are limited by mitochondrial uptake and are temperature-dependent. *J. Neurosci.* 20:7290–96

127. David G, Barrett JN, Barrett EF. 1998. Evidence that mitochondria buffer physiological Ca^{2+} loads in lizard motor nerve terminals. *J. Physiol.* 509:59–65

128. Narita K, Akita T, Hachisuka J, Huang S, Ochi K, Kuba K. 2000. Functional coupling of Ca^{2+} channels to ryanodine receptors at presynaptic terminals. Amplification of exocytosis and plasticity. *J. Gen. Physiol.* 115:519–32

129. Peng Y. 1996. Ryanodine-sensitive component of calcium transients evoked by nerve firing at presynaptic nerve terminals. *J. Neurosci.* 16:6703–12

130. Peng YY. 1998. Effects of mitochondrion on calcium transients at intact presynaptic terminals depend on frequency of nerve firing. *J. Neurophysiol.* 80:186–95

131. Muschol M, Salzberg BM. 2000. Dependence of transient and residual calcium dynamics on action-potential patterning during neuropeptide secretion. *J. Neurosci.* 20:6773–80

132. Cooper RL, Winslow JL, Govind CK, Atwood HL. 1996. Synaptic structural complexity as a factor enhancing probability of calcium-mediated transmitter release. *J. Neurophysiol.* 75:2451–66

133. Issa NP, Hudspeth AJ. 1996. The entry and clearance of Ca^{2+} at individual presynaptic active zones of hair cells from the bullfrog's sacculus. *Proc. Natl. Acad. Sci. USA* 93:9527–32

134. Emptage NJ, Reid CA, Fine A. 2001. Calcium stores in hippocampal synaptic boutons mediate short-term plasticity, store-operated Ca^{2+} entry, and spontaneous transmitter release. *Neuron* 29:197–208

135. Carter AG, Vogt KE, Foster KA, Regehr

WG. 2001. Assessing the role of calcium-induced calcium release in short-term presynaptic plasticity at excitatory central synapses. *J. Neurosci.* In press

136. Borges S, Gleason E, Turelli M, Wilson M. 1995. The kinetics of quantal transmitter release from retinal amacrine cells. *Proc. Natl. Acad. Sci. USA* 92:6896–900

137. Beutner D, Voets T, Neher E, Moser T. 2001. Calcium dependence of exocytosis and endocytosis at the cochlear inner hair cell afferent synapse. *Neuron* 29:681–90

138. Gomis A, Burrone J, Lagnado L. 1999. Two actions of calcium regulate the supply of releasable vesicles at the ribbon synapse of retinal bipolar cells. *J. Neurosci.* 19:6309–17

139. Hsu SF, Jackson MB. 1996. Rapid exocytosis and endocytosis in nerve terminals of the rat posterior pituitary. *J. Physiol.* 494:539–53

140. Moser T, Beutner D. 2000. Kinetics of exocytosis and endocytosis at the cochlear inner hair cell afferent synapse of the mouse. *Proc. Natl. Acad. Sci. USA* 97:883–88

141. Mennerick S, Matthews G. 1996. Ultra-fast exocytosis elicited by calcium current in synaptic terminals of retinal bipolar neurons. *Neuron* 17:1241–49

142. Neves G, Lagnado L. 1999. The kinetics of exocytosis and endocytosis in the synaptic terminal of goldfish retinal bipolar cells. *J. Physiol.* 515:181–202

143. Dittman JS, Regehr WG. 1998. Calcium dependence and recovery kinetics of presynaptic depression at the climbing fiber to Purkinje cell synapse. *J. Neurosci.* 18:6147–62

144. Wang LY, Kaczmarek LK. 1998. High-frequency firing helps replenish the readily releasable pool of synaptic vesicles. *Nature* 394:384–88

145. Stevens CF, Wesseling JF. 1998. Activity-dependent modulation of the rate at which synaptic vesicles become available to undergo exocytosis. *Neuron* 21:415–24

146. Pyle JL, Kavalali ET, Piedras-Renteria

ES, Tsien RW. 2000. Rapid reuse of readily releasable pool vesicles at hippocampal synapses. *Neuron* 28:221–31

147. Sankaranarayanan S, Ryan TA. 2001. Calcium accelerates endocytosis of vSNAREs at hippocampal synapses. *Nat. Neurosci.* 4:129–36

148. Stevens CF, Williams JH. 2000 "Kiss and run" exocytosis at hippocampal synapses. *Proc. Natl. Acad. Sci. USA* 97:12828–33

149. Borst JG, Sakmann B. 1998. Facilitation of presynaptic calcium currents in the rat brainstem. *J. Physiol.* 513:149–55

150. Cuttle MF, Tsujimoto T, Forsythe ID, Takahashi T. 1998. Facilitation of the presynaptic calcium current at an auditory synapse in rat brainstem. *J. Physiol.* 512:723–29

151. Forsythe ID, Tsujimoto T, Barnes-Davies M, Cuttle MF, Takahashi T. 1998. Inactivation of presynaptic calcium current contributes to synaptic depression at a fast central synapse. *Neuron* 20:797–807

152. Brody DL, Yue DT. 2000. Relief of G-protein inhibition of calcium channels and short-term synaptic facilitation in cultured hippocampal neurons. *J. Neurosci.* 20:889–98

153. Borst JG, Sakmann B. 1999. Depletion of calcium in the synaptic cleft of a calyx-type synapse in the rat brainstem. *J. Physiol.* 521:123–33

154. King RD, Wiest MC, Montague PR. 2001. Extracellular calcium depletion as a mechanism of short-term synaptic depression. *J. Neurophysiol.* 85:1952–59

155. Jackson MB, Konnerth A, Augustine GJ. 1991. Action potential broadening and frequency-dependent facilitation of calcium signals in pituitary nerve terminals. *Proc. Natl. Acad. Sci. USA* 88:380–84

156. Geiger JR, Jonas P. 2000. Dynamic control of presynaptic Ca^{2+} inflow by fast-inactivating K^+ channels in hippocampal mossy fiber boutons. *Neuron* 28:927–39

157. Brody DL, Yue DT. 2000. Release-independent short-term synaptic depression in

cultured hippocampal neurons. *J. Neurosci.* 20:2480–94

158. Hatt H, Smith DO. 1976. Synaptic depression related to presynaptic axon conduction block. *J. Physiol.* 259:367–93

159. Parnas I. 1972. Differential block at high frequency of branches of a single axon innervating two muscles. *J. Neurophysiol.* 35:903–14

160. Parker D. 1995. Depression of synaptic connections between identified motor neurons in the locust. *J. Neurophysiol.* 74:529–38

161. Bielefeldt K, Jackson MB. 1993. A calcium-activated potassium channel causes frequency-dependent action potential failures in a mammalian nerve terminal. *J. Neurophysiol.* 70:284–98

162. Silver RA, Momiyama A, Cull-Candy SG. 1998. Locus of frequency-dependent depression identified with multiple-probability fluctuation analysis in rat climbing fibre-Purkinje cell synapses. *J. Physiol.* 510:881–902

163. Mennerick S, Zorumski CF. 1995. Paired-pulse modulation of fast excitatory synaptic currents in microcultures of rat hippocampal neurons. *J. Physiol.* 488:85–101

164. Kraushaar U, Jonas P. 2000. Efficacy and stability of quantal GABA release at a hippocampal interneuron-principal neuron synapse. *J. Neurosci.* 20:5594–607

165. Betz WJ. 1970. Depression of transmitter release at the neuromuscular junction of the frog. *J. Physiol.* 206:629–44

166. Thomson AM, Deuchars J, West DC. 1993. Large, deep layer pyramid-pyramid single axon EPSPs in slices of rat motor cortex display paired pulse and frequency-dependent depression, mediated presynaptically and self-facilitation, mediated postsynaptically. *J. Neurophysiol.* 70:2354–69

167. Debanne D, Guerineau NC, Gahwiler BH, Thompson SM. 1996. Paired-pulse facilitation and depression at unitary synapses in rat hippocampus: quantal fluctuation affects subsequent release. *J. Physiol.* 491:163–76

168. Liley AW, North KAK. 1952. An electrical investigation of effects of repetitive stimulation on mammalian neuromuscular junction. *J. Neurophysiol.* 16:509–27

169. Vere-Jones D. 1966. Simple stochastic models for the release of quanta of transmitter from a nerve terminal. *Aust. J. Stat.* 8:53–63

170. Christensen BN, Martin AR. 1970. Estimates of probability of transmitter release at the mammalian neuromuscular junction. *J. Physiol.* 210:933–45

171. Wu LG, Borst JG. 1999. The reduced release probability of releasable vesicles during recovery from short-term synaptic depression. *Neuron* 23:821–32

172. Burrone J, Lagnado L. 2000. Synaptic depression and the kinetics of exocytosis in retinal bipolar cells. *J. Neurosci.* 20:568–78

173. Gingrich KJ, Byrne JH. 1985. Simulation of synaptic depression, posttetanic potentiation, and presynaptic facilitation of synaptic potentials from sensory neurons mediating gill-withdrawal reflex in *Aplysia. J. Neurophysiol.* 53:652–69

174. Kusano K, Landau EM. 1975. Depression and recovery of transmission at the squid giant synapse. *J. Physiol.* 245:13–32

175. Wu LG, Betz WJ. 1998. Kinetics of synaptic depression and vesicle recycling after tetanic stimulation of frog motor nerve terminals. *Biophys. J.* 74:3003–9

176. Stevens CF, Wesseling JF. 1999. Identification of a novel process limiting the rate of synaptic vesicle cycling at hippocampal synapses. *Neuron* 24:1017–28

177. von Gersdorff H, Matthews G. 1997. Depletion and replenishment of vesicle pools at a ribbon-type synaptic terminal. *J. Neurosci.* 17:1919–27

178. Koenig JH, Ikeda K. 1996. Synaptic vesicles have two distinct recycling pathways. *J. Cell Biol.* 135:797–808

179. Klingauf J, Kavalali ET, Tsien RW. 1998. Kinetics and regulation of fast

endocytosis at hippocampal synapses. *Nature* 394:581–85

180. Richards DA, Guatimosim C, Betz WJ. 2000. Two endocytic recycling routes selectively fill two vesicle pools in frog motor nerve terminals. *Neuron* 27:551–59

181. Matveev V, Wang XJ. 2000. Implications of all-or-none synaptic transmission and short-term depression beyond vesicle depletion: a computational study. *J. Neurosci.* 20:1575–88

182. Delgado R, Maureira C, Oliva C, Kidokoro Y, Labarca P. 2000. Size of vesicle pools, rates of mobilization, and recycling at neuromuscular synapses of a *Drosophila* mutant, *shibire*. *Neuron* 28:941–53

183. Glavinovic MI, Narahashi T. 1988. Depression, recovery and facilitation of neuromuscular transmission during prolonged tetanic stimulation. *Neuroscience* 25:271–81

184. Li Y, Burke RE. 2001. Short-term synaptic depression in the neonatal mouse spinal cord: effects of calcium and temperature. *J. Neurophysiol.* 85:2047–62

185. O'Donovan MJ, Rinzel J. 1997. Synaptic depression: a dynamic regulator of synaptic communication with varied functional roles. *Trends Neurosci.* 20:431–33

186. Cummings DD, Wilcox KS, Dichter MA. 1996. Calcium-dependent paired-pulse facilitation of miniature EPSC frequency accompanies depression of EP-SCs at hippocampal synapses in culture. *J. Neurosci.* 16:5312–23

187. Eliot LS, Kandel ER, Hawkins RD. 1994. Modulation of spontaneous transmitter release during depression and post-tetanic potentiation of *Aplysia* sensory-motor neuron synapses isolated in culture. *J. Neurosci.* 14:3280–92

188. Zengel JE, Sosa MA. 1994. Changes in MEPP frequency during depression of evoked release at the frog neuromuscular junction. *J. Physiol.* 477:267–77

189. Highstein SM, Bennett MV. 1975. Fatigue and recovery of transmission at the Mauthner fiber-giant fiber synapse of the hatchetfish. *Brain. Res.* 98:229–42

190. Glavinovic MI. 1995. Decrease of quantal size and quantal content during tetanic stimulation detected by focal recording. *Neuroscience* 69:271–81

191. Naves LA, Van der Kloot W. 2001. Repetitive nerve stimulation decreases the acetylcholine content of quanta at the frog neuromuscular junction. *J. Physiol.* 532:637–47

192. Liu G, Tsien RW. 1995. Properties of synaptic transmission at single hippocampal synaptic boutons. *Nature* 375:404–8

193. von Gersdorff H, Vardi E, Matthews G, Sterling P. 1996. Evidence that vesicles on the synaptic ribbon of retinal bipolar neurons can be rapidly released. *Neuron* 16:1221–27

194. Lenzi D, Runyeon JW, Crum J, Ellisman MH, Roberts WM. 1999. Synaptic vesicle populations in saccular hair cells reconstructed by electron tomography. *J. Neurosci.* 19:119–32

195. Rao-Mirotznik R, Harkins AB, Buchsbaum G, Sterling P. 1995. Mammalian rod terminal: architecture of a binary synapse. *Neuron* 14:561–69

196. Harris KM, Sultan P. 1995. Variation in the number, location and size of synaptic vesicles provides an anatomical basis for the nonuniform probability of release at hippocampal CA1 synapses. *Neuropharmacology* 34:1387–95

197. Schikorski T, Stevens CF. 1997. Quantitative ultrastructural analysis of hippocampal excitatory synapses. *J. Neurosci.* 17:5858–67

198. Schikorski T, Stevens CF. 1999. Quantitative fine-structural analysis of olfactory cortical synapses. *Proc. Natl. Acad. Sci. USA* 96:4107–12

199. Redman S. 1990. Quantal analysis of synaptic potentials in neurons of the central nervous system. *Physiol. Rev.* 70:165–98

200. Korn H, Faber DS. 1991. Quantal analysis and synaptic efficacy in the CNS. *Trends Neurosci.* 14:439–45

201. Tong G, Jahr CE. 1994. Multivesicular release from excitatory synapses of cultured hippocampal neurons. *Neuron* 12:51–59

202. Auger C, Kondo S, Marty A. 1998. Multivesicular release at single functional synaptic sites in cerebellar stellate and basket cells. *J. Neurosci.* 18:4532–47

203. Wadiche J, Jahr C. 2000. Elevated levels of glutamate at cerebellar synapses. *Soc. Neurosci. Abstr.* 26:1385

204. Xu-Friedman MA, Harris KM, Regehr WG. 2001. Three-dimensional comparison of ultrastructural characteristics at depressing and facilitating synapses onto cerebellar Purkinje cells. *J. Neurosci.* 21:6666–72

205. Weis S, Schneggenburger R, Neher E. 1999. Properties of a model of Ca(++)-dependent vesicle pool dynamics and short term synaptic depression. *Biophys. J.* 77:2418–29

206. Sakaba T, Neher E. 2001. Calmodulin mediates rapid recruitment of fast-releasing synaptic vesicles at a calyx-type synapse. *Neuron.* In press

207. Wang C, Zucker RS. 1998. Regulation of synaptic vesicle recycling by calcium and serotonin. *Neuron* 21:155–67

208. Martin AR, Pilar G. 1964. Presynaptic and postsynaptic events during posttetanic potentiation and facilitation in the avian ciliary ganglion. *J. Physiol.* 175:17–30

209. Zucker RS, Bruner J. 1977. Long-lasting depression and the depletion hypothesis at crayfish neuromuscular junctions. *J. Comp. Physiol.* 121:223–40

210. Castellucci VF, Kandel ER. 1974. A quantal analysis of the synaptic depression underlying habituation of the gill-withdrawal reflex in *Aplysia. Proc. Natl. Acad. Sci. USA* 71:5004–8

211. Byrne JH. 1982. Analysis of synaptic depression contributing to habituation of gill withdrawal reflex in *Aplysia californica. J. Neurophysiol.* 48:431–38

212. Armitage BA, Siegelbaum SA. 1998. Presynaptic induction and expression of homosynaptic depression at *Aplysia* sensorimotor neuron synapses. *J. Neurosci.* 18:8770–79

213. Royer S, Coulson RL, Klein M. 2000. Switching off and on of synaptic sites at *Aplysia* sensorimotor synapses. *J. Neurosci.* 20:626–38

214. Waldeck RF, Pereda A, Faber DS. 2000. Properties and plasticity of paired-pulse depression at a central synapse. *J. Neurosci.* 20:5312–20

215. Bellingham MC, Walmsley B. 1999. A novel presynaptic inhibitory mechanism underlies paired pulse depression at a fast central synapse. *Neuron* 23:159–70

216. Aonuma H, Nagayama T, Takahata M. 2000. Modulatory effects of nitric oxide on synaptic depression in the crayfish neuromuscular system. *J. Exp. Biol.* 203:3595–602

217. Hsu SF, Augustine GJ, Jackson MB. 1996. Adaptation of Ca^{2+}-triggered exocytosis in presynaptic terminals. *Neuron* 17:501–12

218. Blank PS, Vogel SS, Cho MS, Kaplan D, Bhuva D, et al. 1998. The calcium sensitivity of individual secretory vesicles is invariant with the rate of calcium delivery. *J. Gen. Physiol.* 112:569–76

219. Blank PS, Cho MS, Vogel SS, Kaplan D, Kang A, et al. 1998. Submaximal responses in calcium-triggered exocytosis are explained by differences in the calcium sensitivity of individual secretory vesicles. *J. Gen. Physiol.* 112:559–67

220. Miller RJ. 1998. Presynaptic receptors. *Annu. Rev. Pharmacol. Toxicol.* 38:201–27

221. Wu LG, Saggau P. 1997. Presynaptic inhibition of elicited neurotransmitter release. *Trends Neurosci.* 20:204–12

222. Greene RW, Haas HL. 1991. The electrophysiology of adenosine in the mammalian central nervous system. *Prog. Neurobiol.* 36:329–41

223. Redman RS, Silinsky EM. 1994. ATP released together with acetylcholine as the mediator of neuromuscular depression

at frog motor nerve endings. *J. Physiol.* 477:117–27

224. Malinowski MN, Cannady SB, Schmit KV, Barr PM, Schrock JW, Wilson DF. 1997. Adenosine depresses transmitter release but is not the basis for 'tetanic fade' at the neuromuscular junction of the rat. *Neurosci. Lett.* 230:81–84

225. Davies CH, Davies SN, Collingridge GL. 1990. Paired-pulse depression of monosynaptic GABA-mediated inhibitory postsynaptic responses in rat hippocampus. *J. Physiol.* 424:513–31

226. Lambert NA, Wilson WA. 1994. Temporally distinct mechanisms of use-dependent depression at inhibitory synapses in the rat hippocampus in vitro. *J. Neurophysiol.* 72:121–30

227. Jiang L, Sun S, Nedergaard M, Kang J. 2000. Paired-pulse modulation at individual GABAergic synapses in rat hippocampus. *J. Physiol.* 523:425–39

228. Wilcox KS, Dichter MA. 1994. Paired pulse depression in cultured hippocampal neurons is due to a presynaptic mechanism independent of $GABA_B$ autoreceptor activation. *J. Neurosci.* 14:1775–88

229. von Gersdorff H, Schneggenburger R, Weis S, Neher E. 1997. Presynaptic depression at a calyx synapse: the small contribution of metabotropic glutamate receptors. *J. Neurosci.* 17:8137–46

230. Isaacson JS, Solis JM, Nicoll RA. 1993. Local and diffuse synaptic actions of GABA in the hippocampus. *Neuron* 10:165–75

231. Dittman JS, Regehr WG. 1997. Mechanism and kinetics of heterosynaptic depression at a cerebellar synapse. *J. Neurosci.* 17:9048–59

232. Mitchell SJ, Silver RA. 2000. GABA spillover from single inhibitory axons suppresses low-frequency excitatory transmission at the cerebellar glomerulus. *J. Neurosci.* 20:8651–58

233. Drake CT, Terman GW, Simmons ML, Milner TA, Kunkel DD, et al. 1994. Dynorphin opioids present in dentate granule cells may function as retrograde inhibitory neurotransmitters. *J. Neurosci.* 14:3736–50

234. Zilberter Y, Kaiser KM, Sakmann B. 1999. Dendritic GABA release depresses excitatory transmission between layer 2/3 pyramidal and bitufted neurons in rat neocortex. *Neuron* 24:979–88

235. Zilberter Y. 2000. Dendritic release of glutamate suppresses synaptic inhibition of pyramidal neurons in rat neocortex. *J. Physiol.* 528.3:489–96

236. Kombian SB, Mouginot D, Pittman QJ. 1997. Dendritically released peptides act as retrograde modulators of afferent excitation in the supraoptic nucleus in vitro. *Neuron* 19:903–12

237. Llano I, Leresche N, Marty A. 1991. Calcium entry increases the sensitivity of cerebellar Purkinje cells to applied GABA and decreases inhibitory synaptic currents. *Neuron* 6:565–74

238. Pitler TA, Alger BE. 1992. Postsynaptic spike firing reduces synaptic $GABA_A$ responses in hippocampal pyramidal cells. *J. Neurosci.* 12:4122–32

239. Feigenspan A, Gustincich S, Bean BP, Raviola E. 1998. Spontaneous activity of solitary dopaminergic cells of the retina. *J. Neurosci.* 18:6776–89

240. Wilson RI, Nicoll RA. 2001. Endogenous cannabinoids mediate retrograde signalling at hippocampal synapses. *Nature* 410:588–92

241. Kreitzer AC, Regehr WG. 2001. Retrograde inhibition of presynaptic calcium influx by endogenous cannabinoids at excitatory synapses onto Purkinje cells. *Neuron* 29:717–27

242. Ohno-Shosaku T, Maejima T, Kano M. 2001. Endogenous cannabinoids mediate retrograde signals from depolarized postsynaptic neurons to presynaptic terminals. *Neuron* 29:729–38

243. Kreitzer AC, Regehr WG. 2001. Retrograde inhibition of inhibitory synapses onto Purkinje cells by endogenous cannabinoids. *J. Neurosci.* In press

244. Wang J, Zucker RS. 2001. Photolysis-induced suppression of inhibition in rat hippocampal CA1 pyramidal neurons. *J. Physiol.* 533:757–63

245. MacDermott AB, Role LW, Siegelbaum SA. 1999. Presynaptic ionotropic receptors and the control of transmitter release. *Annu. Rev. Neurosci.* 22:443–85

246. Turecek R, Trussell LO. 2001. Presynaptic glycine receptors enhance transmitter release at a mammalian central synapse. *Nature* 411:587–90

247. Schmitz D, Mellor J, Nicoll RA. 2001. Presynaptic kainate receptor mediation of frequency facilitation at hippocampal mossy fiber synapses. *Science* 291:1972–76

248. Contractor A, Swanson G, Heinemann SF. 2001. Kainate receptors are involved in short- and long-term plasticity at mossy fiber synapses in the hippocampus. *Neuron* 29:209–16

249. Schmitz D, Frerking M, Nicoll RA. 2000. Synaptic activation of presynaptic kainate receptors on hippocampal mossy fiber synapses. *Neuron* 27:327–38

250. Araque A, Carmignoto G, Haydon PG. 2001. Dynamic signaling between astrocytes and neurons. *Annu. Rev. Physiol.* 63:795–813

251. Haydon PG. 2001. GLIA: listening and talking to the synapse. *Nat. Rev. Neurosci.* 2:185–93

252. Bergles DE, Diamond JS, Jahr CE. 1999. Clearance of glutamate inside the synapse and beyond. *Curr. Opin. Neurobiol.* 9:293–98

253. Danbolt NC. 2001. Glutamate uptake. *Prog. Neurobiol.* 65:1–105

254. Araque A, Parpura V, Sanzgiri RP, Haydon PG. 1998. Glutamate-dependent astrocyte modulation of synaptic transmission between cultured hippocampal neurons. *Eur. J. Neurosci.* 10:2129–42

255. Araque A, Li N, Doyle RT, Haydon PG. 2000. SNARE protein-dependent glutamate release from astrocytes. *J. Neurosci.* 20:666–73

256. Araque A, Sanzgiri RP, Parpura V, Haydon PG. 1998. Calcium elevation in astrocytes causes an NMDA receptor-dependent increase in the frequency of miniature synaptic currents in cultured hippocampal neurons. *J. Neurosci.* 18:6822–29

257. Robitaille R. 1995. Purinergic receptors and their activation by endogenous purines at perisynaptic glial cells of the frog neuromuscular junction. *J. Neurosci.* 15:7121–31

258. Jahromi BS, Robitaille R, Charlton MP. 1992. Transmitter release increases intracellular calcium in perisynaptic Schwann cells in situ. *Neuron* 8:1069–77

259. Robitaille R, Jahromi BS, Charlton MP. 1997. Muscarinic Ca^{2+} responses resistant to muscarinic antagonists at perisynaptic Schwann cells of the frog neuromuscular junction. *J. Physiol.* 504:337–47

260. Castonguay A, Robitaille R. 2001 Differential regulation of transmitter release by presynaptic and glial Ca^{2+} internal stores at the neuromuscular synapse. *J. Neurosci.* 21:1911–22

261. Robitaille R. 1998. Modulation of synaptic efficacy and synaptic depression by glial cells at the frog neuromuscular junction. *Neuron* 21:847–55

262. Thomas S, Robitaille R. 2001. Differential frequency-dependent regulation of transmitter release by endogenous nitric oxide at the amphibian neuromuscular synapse. *J. Neurosci.* 21:1087–95

263. Kang J, Jiang L, Goldman SA, Nedergaard M. 1998. Astrocyte-mediated potentiation of inhibitory synaptic transmission. *Nat. Neurosci.* 1:683–92

264. Jones MV, Westbrook GL. 1996. The impact of receptor desensitization on fast synaptic transmission. *Trends Neurosci.* 19:96–101

265. Wachtel H, Kandel ER. 1971. Conversion of synaptic excitation to inhibition at a dual chemical synapse. *J. Neurophysiol.* 34:56–68

266. Gardner D, Kandel ER. 1977. Physiological and kinetic properties of cholinergic receptors activated by multiaction interneurons in buccal ganglia of *Aplysia. J. Neurophysiol.* 40:333–48

267. Otis T, Zhang S, Trussell LO. 1996. Direct measurement of AMPA receptor desensitization induced by glutamatergic synaptic transmission. *J. Neurosci.* 16:7496–504

268. Trussell LO, Zhang S, Raman IM. 1993. Desensitization of AMPA receptors upon multiquantal neurotransmitter release. *Neuron* 10:1185–96

269. Brenowitz S, Trussell LO. 2001. Minimizing synaptic depression by control of release probability. *J. Neurosci.* 21:1857–67

270. Turecek R, Trussell LO. 2000. Control of synaptic depression by glutamate transporters. *J. Neurosci.* 20:2054–63

271. DeVries SH. 2000. Bipolar cells use kainate and AMPA receptors to filter visual information into separate channels. *Neuron* 28:847–56

272. Oleskevich S, Clements J, Walmsley B. 2000. Release probability modulates short-term plasticity at a rat giant terminal. *J. Physiol.* 524:513–23

273. Deleted in proof

274. Mennerick S, Zorumski CF. 1996. Postsynaptic modulation of NMDA synaptic currents in rat hippocampal microcultures by paired-pulse stimulation. *J. Physiol.* 490:405–7

275. Larkman AU, Jack JJ, Stratford KJ. 1997. Quantal analysis of excitatory synapses in rat hippocampal CA1 in vitro during low-frequency depression. *J. Physiol.* 505:457–71

276. Hashimoto K, Kano M. 1998. Presynaptic origin of paired-pulse depression at climbing fibre-Purkinje cell synapses in the rat cerebellum. *J. Physiol.* 506:391–405

277. Hjelmstad GO, Nicoll RA, Malenka RC. 1997. Synaptic refractory period provides a measure of probability of release in the hippocampus. *Neuron* 19:1309–18

278. Palay SL, Chan-Palay V. 1974. *Cerebellar Cortex.* New York: Springer. 348 pp.

279. Jones MV, Westbrook GL. 1995. Desensitized states prolong GABA$_A$ channel responses to brief agonist pulses. *Neuron* 15:181–91

280. Overstreet LS, Jones MV, Westbrook GL. 2000. Slow desensitization regulates the availability of synaptic GABA$_A$ receptors. *J. Neurosci.* 20:7914–21

281. Tong G, Shepherd D, Jahr CE. 1995. Synaptic desensitization of NMDA receptors by calcineurin. *Science* 267:1510–12

282. Sather W, Dieudonne S, MacDonald JF, Ascher P. 1992. Activation and desensitization of *N*-methyl-D-aspartate receptors in nucleated outside-out patches from mouse neurones. *J. Physiol.* 450:643–72

283. Mayer ML, Vyklicky L, Clements J. 1989. Regulation of NMDA receptor desensitization in mouse hippocampal neurons by glycine. *Nature* 338:425–27

284. Magleby KL, Pallotta BS. 1981. A study of desensitization of acetylcholine receptors using nerve-released transmitter in the frog. *J. Physiol.* 316:225–50

285. Rozov A, Burnashev N. 1999. Polyamine-dependent facilitation of postsynaptic AMPA receptors counteracts paired-pulse depression. *Nature* 401:594–98

286. Rozov A, Zilberter Y, Wollmuth LP, Burnashev N. 1998. Facilitation of currents through rat Ca^{2+}-permeable AMPA receptor channels by activity-dependent relief from polyamine block. *J. Physiol.* 511:361–77

287. Greengard P, Valtorta F, Czernik AJ, Benfenati F. 1993. Synaptic vesicle phosphoproteins and regulation of synaptic function. *Science* 259:780–85

288. Stevens CF, Tonegawa S, Wang Y. 1994. The role of calcium-calmodulin kinase II in three forms of synaptic plasticity. *Curr. Biol.* 4:687–93

289. Chapman PF, Frenguelli BG, Smith A, Chen CM, Silva AJ. 1995. The α-Ca^{2+}/calmodulin kinase II: a bidirectional

modulator of presynaptic plasticity. *Neuron* 14:591–97

290. Silva AJ, Rosahl TW, Chapman PF, Marowitz Z, Friedman E, et al. 1996. Impaired learning in mice with abnormal short-lived plasticity. *Curr. Biol.* 6:1509–18

291. Rosahl TW, Geppert M, Spillane D, Herz J, Hammer RE, et al. 1993. Short-term synaptic plasticity is altered in mice lacking synapsin I. *Cell* 75:661–70

292. Rosahl TW, Spillane D, Missler M, Herz J, Selig DK, et al. 1995. Essential functions of synapsins I and II in synaptic vesicle regulation. *Nature* 375:488–93

293. Malinow R, Madison DV, Tsien RW. 1988. Persistent protein kinase activity underlying long-term potentiation. *Nature* 335:820–24

294. Humeau Y, Doussau F, Vitiello F, Greengard P, Benfenati F, Poulain B. 2001. Synapsin controls both reserve and releasable synaptic vesicle pools during neuronal activity and short-term plasticity in *Aplysia. J. Neurosci.* 21:4195–206

295. Pieribone VA, Shupliakov O, Brodin L, Hilfiker-Rothenfluh S, Czernik AJ, Greengard P. 1995. Distinct pools of synaptic vesicles in neurotransmitter release. *Nature* 375:493–97

296. Hilfiker S, Schweizer FE, Kao HT, Czernik AJ, Greengard P, Augustine GJ. 1998. Two sites of action for synapsin domain E in regulating neurotransmitter release. *Nat. Neurosci.* 1:29–35

297. Ryan TA, Li L, Chin LS, Greengard P, Smith SJ. 1996. Synaptic vesicle recycling in synapsin I knock-out mice. *J. Cell Biol.* 134:1219–27

298. Cole JC, Villa BR, Wilkinson RS. 2000. Disruption of actin impedes transmitter release in snake motor terminals. *J. Physiol.* 525:579–86

299. Kuromi H, Kidokoro Y. 1998. Two distinct pools of synaptic vesicles in single presynaptic boutons in a temperature-sensitive *Drosophila* mutant, shibire. *Neuron* 20:917–25

300. Wang D, Maler L. 1998. Differential roles of Ca^{2+}/calmodulin-dependent kinases in posttetanic potentiation at input selective glutamatergic pathways. *Proc. Natl. Acad. Sci. USA* 95:7133–38

301. Jin P, Griffith LC, Murphey RK. 1998. Presynaptic calcium/calmodulin-dependent protein kinase II regulates habituation of a simple reflex in adult *Drosophila. J. Neurosci.* 18:8955–64

302. Vyshedskiy A, Lin JW. 1997. Change of transmitter release kinetics during facilitation revealed by prolonged test pulses at the inhibitor of the crayfish opener muscle. *J. Neurophysiol.* 78:1791–99

303. Vyshedskiy A, Allana T, Lin JW. 2000. Analysis of presynaptic Ca^{2+} influx and transmitter release kinetics during facilitation at the inhibitor of the crayfish neuromuscular junction. *J. Neurosci.* 20:6326–32

304. Geppert M, Goda Y, Hammer RE, Li C, Rosahl TW, et al. 1994. Synaptotagmin I: a major Ca^{2+} sensor for transmitter release at a central synapse. *Cell* 79:717–27

305. Fernandez-Chacon R, Königstorfer A, Gerber SH, Garcia J, Matos MF, et al. 2001. Synaptotagmin I functions as a calcium regulator of release probability. *Nature* 410:41–49

306. Butz S, Fernandez-Chacon R, Schmitz F, Jahn R, Südhof TC. 1999. The subcellular localizations of atypical synaptotagmins III and VI. Synaptotagmin III is enriched in synapses and synaptic plasma membranes but not in synaptic vesicles. *J. Biol. Chem.* 274:18290–96

307. Li C, Ullrich B, Zhang JZ, Anderson RG, Brose N, Südhof TC. 1995. Ca^{2+}-dependent and -independent activities of neural and non-neural synaptotagmins. *Nature* 375:594–99

308. Sugita S, Han W, Butz S, Liu X, Fernandez-Chacon R, et al. 2001. Synaptotagmin VII as a plasma membrane Ca^{2+}-sensor in exocytosis. *Neuron* 30:459–73

309. Mochida S, Orita S, Sakaguchi G, Sasaki

T, Takai Y. 1998. Role of the Doc2a-Munc13-1 interaction in the neurotransmitter release process. *Proc. Natl. Acad. Sci. USA* 95:11418–22

310. Sakaguchi G, Manabe T, Kobayashi K, Orita S, Sasaki T, et al. 1999. Doc2α is an activity-dependent modulator of excitatory synaptic transmission. *Eur. J. Neurosci.* 11:4262–68

311. Wang Y, Okamoto M, Schmitz F, Hofmann K, Südhof TC. 1997. Rim is a putative Rab3 effector in regulating synaptic-vesicle fusion. *Nature* 388:593–98

312. Komuro R, Sasaki T, Orita S, Maeda M, Takai Y. 1996. Involvement of rabphilin-3A in Ca^{2+}-dependent exocytosis from PC12 cells. *Biochem. Biophys. Res. Commun.* 219:435–40

313. Burns ME, Sasaki T, Takai Y, Augustine GJ. 1998. Rabphilin-3A: a multifunctional regulator of synaptic vesicle traffic. *J. Gen. Physiol.* 111:243–55

314. Schlüter OM, Schnell E, Verhage M, Tzonopoulos T, Nicoll RA, et al. 1999. Rabphilin knock-out mice reveal that rabphilin is not required for rab3 function in regulating neurotransmitter release. *J. Neurosci.* 19:5834–46

315. Brose N, Hofmann K, Hata Y, Südhof TC. 1995. Mammalian homologues of *Caenorhabditis elegans unc-13* gene define novel family of C$_2$-domain proteins. *J. Biol. Chem.* 270:25273–80

316. Augustin I, Rosenmund C, Südhof TC, Brose N. 1999. Munc13-1 is essential for fusion competence of glutamatergic synaptic vesicles. *Nature* 400:457–61

317. Deleted in proof

318. Berwin B, Floor E, Martin TF. 1998. CAPS (mammalian UNC-31) protein localizes to membranes involved in dense-core vesicle exocytosis. *Neuron* 21:137–45

319. Rupnik M, Kreft M, Sikdar SK, Grilc S, Romih R, et al. 2000. Rapid regulated dense-core vesicle exocytosis requires the CAPS protein. *Proc. Natl. Acad. Sci. USA* 97:5627–32

320. Fenster SD, Chung WJ, Zhai R, Cases-Langhoff C, Voss B, et al. 2000. Piccolo, a presynaptic zinc finger protein structurally related to bassoon. *Neuron* 25:203–14

321. Wang X, Kibschull M, Laue MM, Lichte B, Petrasch-Parwez E, Kilimann MW. 1999. Aczonin, a 550-kD putative scaffolding protein of presynaptic active zones, shares homology regions with Rim and Bassoon and binds profilin. *J. Cell Biol.* 147:151–62

322. Gerber SH, Garcia J, Rizo J, Südhof TC. 2001. An unusual C$_2$-domain in the active-zone protein piccolo: implications for Ca^{2+} regulation of neurotransmitter release. *EMBO J.* 20:1605–19

323. Trifaro J, Rose SD, Lejen T, Elzagallaai A. 2000. Two pathways control chromaffin cell cortical F-actin dynamics during exocytosis. *Biochimie* 82:339–52

324. Rivosecchi R, Pongs O, Theil T, Mallart A. 1994. Implication of frequenin in the facilitation of transmitter release in *Drosophila. J. Physiol.* 474:223–32

325. Sheng ZH, Rettig J, Cook T, Catterall WA. 1996. Calcium-dependent interaction of N-type calcium channels with the synaptic core complex. *Nature* 379:451–54

326. Fasshauer D, Sutton RB, Brunger AT, Jahn R. 1998. Conserved structural features of the synaptic fusion complex: SNARE proteins reclassified as Q- and R-SNAREs. *Proc. Natl. Acad. Sci. USA* 95:15781–86

327. Geppert M, Goda Y, Stevens CF, Südhof TC. 1997. The small GTP-binding protein Rab3A regulates a late step in synaptic vesicle fusion. *Nature* 387:810–14

328. Janz R, Südhof TC, Hammer RE, Unni V, Siegelbaum SA, Bolshakov VY. 1999. Essential roles in synaptic plasticity for synaptogyrin I and synaptophysin I. *Neuron* 24:687–700

329. Kokaia M, Asztely F, Olofsdotter K, Sindreu CB, Kullmann DM, Lindvall O. 1998. Endogenous neurotrophin-3

regulates short-term plasticity at lateral perforant path-granule cell synapses. *J. Neurosci.* 18:8730–39

330. Kang H, Schuman EM. 1995. Long-lasting neurotrophin-induced enhancement of synaptic transmission in the adult hippocampus. *Science* 267:1658–62

331. Matilla A, Roberson ED, Banfi S, Morales J, Armstrong DL, et al. 1998. Mice lacking ataxin-1 display learning deficits and decreased hippocampal paired-pulse facilitation. *J. Neurosci.* 18:5508–16

332. Pekhletski R, Gerlai R, Overstreet LS, Huang XP, Agopyan N, et al. 1996. Impaired cerebellar synaptic plasticity and motor performance in mice lacking the mGluR4 subtype of metabotropic glutamate receptor. *J. Neurosci.* 16:6364–73

333. Zhong Y, Wu CF. 1991. Altered synaptic plasticity in *Drosophila* memory mutants with a defective cyclic AMP cascade. *Science* 251:198–201

334. Kuromi H, Kidokoro Y. 2000. Tetanic stimulation recruits vesicles from reserve pool via a cAMP-mediated process in *Drosophila* synapses. *Neuron* 27:133–43

335. Kawasaki F, Hazen M, Ordway RW. 2000. Fast synaptic fatigue in *shibire* mutants reveals a rapid requirement for dynamin in synaptic vesicle membrane trafficking. *Nat. Neurosci.* 3:859–60

336. Broadie K, Rushton E, Skoulakis EM, Davis RL. 1997. Leonardo, a *Drosophila* 14-3-3 protein involved in learning, regulates presynaptic function. *Neuron* 19:391–402

337. Rohrbough J, Pinto S, Mihalek RM, Tully T, Broadie K. 1999. *latheo*, a *Drosophila* gene involved in learning, regulates functional synaptic plasticity. *Neuron* 23:55–70

338. Rohrbough J, Grotewiel MS, Davis RL, Broadie K. 2000. Integrin-mediated regulation of synaptic morphology, transmission, and plasticity. *J. Neurosci.* 20:6868–78

339. Jorge-Rivera JC, Sen K, Birmingham JT, Abbott LF, Marder E. 1998. Temporal dynamics of convergent modulation at a crustacean neuromuscular junction. *J. Neurophysiol.* 80:2559–70

340. Qian SM, Delaney KR. 1997. Neuromodulation of activity-dependent synaptic enhancement at crayfish neuromuscular junction. *Brain. Res.* 771:259–70

341. Bittner GD. 1968. Differentiation of nerve terminals in the crayfish opener muscle and its functional significance. *J. Gen. Physiol.* 51:731–58

342. Markram H, Wang Y, Tsodyks M. 1998. Differential signaling via the same axon of neocortical pyramidal neurons. *Proc. Natl. Acad. Sci. USA* 95:5323–28

343. Reyes A, Sakmann B. 1999. Developmental switch in the short-term modification of unitary EPSPs evoked in layer 2/3 and layer 5 pyramidal neurons of rat neocortex. *J. Neurosci.* 19:3827–35

344. Finnerty GT, Roberts LS, Connors BW. 1999. Sensory experience modifies the short-term dynamics of neocortical synapses. *Nature* 400:367–71

345. Finnerty GT, Connors BW. 2000. Sensory deprivation without competition yields modest alterations of short-term synaptic dynamics. *Proc. Natl. Acad. Sci. USA* 97:12864–68

346. Mochida S, Sheng ZH, Baker C, Kobayashi H, Catterall WA. 1996. Inhibition of neurotransmission by peptides containing the synaptic protein interaction site of N-type Ca^{2+} channels. *Neuron* 17:781–88

347. Takahashi T, Hori T, Kajikawa Y, Tsujimoto T. 2000. The role of GTP-binding protein activity in fast central synaptic transmission. *Science* 289:460–63

348. Morimoto T, Wang XH, Poo MM. 1998. Overexpression of synaptotagmin modulates short-term synaptic plasticity at developing neuromuscular junctions. *Neuroscience* 82:969–78

349. Wang XH, Zheng JQ, Poo MM. 1996. Effects of cytochalasin treatment on short-term synaptic plasticity at developing

neuromuscular junctions in frogs. *J. Physiol.* 491:187–95

350. Stoop R, Poo MM. 1996. Synaptic modulation by neurotrophic factors: differential and synergistic effects of brain-derived neurotrophic factor and ciliary neurotrophic factor. *J. Neurosci.* 16:3256–64

351. Goussakov IV, Fink K, Elger CE, Beck H. 2000. Metaplasticity of mossy fiber synaptic transmission involves altered release probability. *J. Neurosci.* 20:3434–41

352. Klapstein GJ, Vietla S, Lieberman DN, Gray PA, Airaksinen MS, et al. 1998. Calbindin-D28k fails to protect hippocampal neurons against ischemia in spite of its cytoplasmic calcium buffering properties: evidence from calbindin-D28k knockout mice. *Neuroscience* 85:361–73

353. Commins S, O'Neill LA, O'Mara SM. 2001. The effects of the bacterial endotoxin lipopolysaccharide on synaptic transmission and plasticity in the CA1-subiculum pathway in vivo. *Neuroscience* 102:273–80

354. Wang JH, Kelly PT. 1997. Attenuation of paired-pulse facilitation associated with synaptic potentiation mediated by postsynaptic mechanisms. *J. Neurophysiol.* 78:2707–16

355. Wang JH, Kelly PT. 1996. Regulation of synaptic facilitation by postsynaptic Ca^{2+}/CaM pathways in hippocampal CA1 neurons. *J. Neurophysiol.* 76:276–86

356. Schaffhausen JH, Fischer TM, Carew TJ. 2001. Contribution of postsynaptic Ca^{2+} to the induction of post-tetanic potentiation in the neural circuit for siphon withdrawal in *Aplysia*. *J. Neurosci.* 21:1739–49

357. Bao JX, Kandel ER, Hawkins RD. 1997. Involvement of pre- and postsynaptic mechanisms in posttetanic potentiation at *Aplysia* synapses. *Science* 275:969–73

358. Malenka RC. 1991. Postsynaptic factors control the duration of synaptic enhancement in area CA1 of the hippocampus. *Neuron* 6:53–60

359. Malenka RC, Lancaster B, Zucker RS. 1992. Temporal limits on the rise in postsynaptic calcium required for the induction of long-term potentiation. *Neuron* 9:121–28

360. Ogawa H, Baba Y, Oka K. 2001. Dendritic calcium accumulation regulates wind sensitivity via short-term depression at cercal sensory-to-giant interneuron synapses in the cricket. *J. Neurobiol.* 46:301–13

Annu. Rev. Physiol. 2002. 64:407–29

Intracellular Transport Mechanisms of Signal Transducers

Gerald W. Dorn, II
*Department of Medicine, University of Cincinnati, Cincinnati, Ohio 45267-0542;
e-mail: dorngw@ucmail.uc.edu*

Daria Mochly-Rosen
*Department of Molecular Pharmacology, Stanford University School of Medicine,
Stanford, California 94305; e-mail: mochly@stanford.edu*

Key Words   protein kinase C, protein kinase A, translocation, anchoring proteins, Rab GTPase

■ **Abstract**   Recent discoveries have revolutionized our conceptions of enzyme-substrate specificity in signal transduction pathways. Protein kinases A and C are localized to discreet subcellular regions, and this localization changes in an isozyme-specific manner upon activation, a process referred to as translocation. The mechanisms for translocation involve interactions of soluble kinases with membrane-bound anchor proteins that recognize individual kinase isoenzymes and their state of activation. Recently, modulation of kinase-anchor protein interactions has been used to specifically regulate, positively or negatively, the activity of C kinase isozymes. Also described in this review is a role for the Rab family of small G proteins in regulating subcellular protein trafficking. The pathophysiological significance of disrupted subcellular protein transport in cell signaling and the potential therapeutic utility of targeted regulation of these events are in the process of being characterized.

LOCATION, LOCATION . . . TRANSLOCATION

Function is inextricably linked with location. This was aptly demonstrated by the cardiac investigator who was to present at a conference in Portland, Maine, but who flew instead to Portland, Oregon. There were two consequences resulting from this miscommunication. First, the conference attendees failed to benefit from a lecture targeted to their interests; i.e., a proper function was not performed. Second, assuming that the wayward scientist presented instead to the unsuspecting Oregon natives, it would have been regarded as eccentric or worse; i.e., an improper function was performed. The analogous situation in cellular systems is for an effector protein (receptor, signal transducer, or enzyme) to act where it should not. Take, for example, a hypothetical signaling enzyme X. Like all enzymes, X

0066-4278/02/0315-0407$14.00   **407**

has a unique function in each different type of cell in which it is expressed. This function is determined by three characteristics: enzymatic activity (an intrinsic feature of the enzyme), the particular conditions of activation, and the identity of substrates. Normally, X is activated by the appropriate stimulus, finds its proper substrate, and acts. If, however, inactive X is not in its proper location, it may not be activated appropriately. On the other hand, if activated X is not in its proper location, it cannot act on its normal substrates and may therefore act promiscuously on whatever substrate is available. Furthermore, activation of X may require that it be located in one cellular locale, while its proper substrates are elsewhere. Under these conditions, X must move, or translocate, within the cell upon activation. Specific targeting of enzymes to compartmentalized substrates is therefore not only necessary for proper enzyme function but also suggests a mechanism for substrate specificity of closely related enzymes with similar catalytic activity. Thus, if enzyme X translocates to the nucleus upon activation, whereas related enzyme Y translocates to the plasma membrane, they will act upon different substrate proteins and have vastly different cellular effects.

This paper explores the mechanisms by which subcellular location and trafficking of signaling molecules is organized and regulated. Because kinase anchoring and translocation were initially described for protein kinase C (PKC), the first and most detailed section reviews the mechanisms of translocation for this ubiquitous group of at least 11 related calcium and/or phospholipid activated kinases (1–3). Protein kinase Cs are the terminal effector molecules of seven transmembrane-spanning receptors coupled to the Gq class of heterotrimeric G proteins (4). In the heart, various PKC members or isozymes have been implicated in development of cardiac hypertrophy (5, 6), heart failure (7), and in the transient protection afforded the myocardium by brief periods of ischemia (ischemic preconditioning) (8, 9). Initial studies of these PKC functions described associations between various conditions and an increase in membrane-associated PKC, measured as PKC isozyme immunoreactivity in subcellular particulates (10). This type of assay, which compares the relative amounts of soluble ($100,000 \times g$ supernatant) and particulate (Triton X100 soluble $100,000 \times g$ sediments) PKC, is the biochemical analogue of PKC translocation from cytosolic to membranous cellular structures (11). In this context, PKC isozyme translocation is generally considered to be synonymous with activation (12). Whereas standard in vitro transfection and in vivo transgenic overexpression approaches have been used to characterize PKC isozyme-specific effects (see below), the possibility for interactions of highly expressed wild-type or mutationally activated PKCs (abnormally located in unusual subcellular locales) with atypical substrates has complicated such studies because a consequence of such opportunistic interactions would be atypical effects. Improved understanding of the mechanisms for differential subcellular localization and activation/translocation of PKC isozymes and identification of PKC tethering molecules has prompted a new approach of specifically modulating PKC isozyme translocation.

The second section reviews the current knowledge of protein kinase A (PKA) activation and translocation. Protein kinase A, the cAMP-dependent kinase, is the terminal effector molecule of membrane receptors coupled to the Gs or Gi heterotrimeric G proteins (4). Receptor-mediated activation of Gs, such as by β-adrenergic receptors, activates PKA by stimulating adenylyl cyclase and consequently increasing cAMP formation. Conversely, receptor-mediated activation of Gi, as by cholinergic muscarinic receptors, inhibits PKA by diminishing adenylyl cyclase activity. These are ubiquitous signal transduction systems that, in the heart, have critical functions of modulating minute-by-minute heart rate, vascular tone, and cardiac output (the fight or flight response). Although PKA translocation has not traditionally been equated with activation, recent identification and delineation of possible pathological roles for PKA-specific membrane anchor proteins suggest that maintenance of proper subcellular localization is important for this kinase (13).

The third section of this review briefly examines the Rab GTPases. This largest family of ras-related G proteins are not signaling molecules per se, but instead direct subcellular trafficking of select proteins. The Rab GDP/GTP exchange cycle regulates Rab translocation between cytosol and intracellular organelles. This dynamic association of Rab GTPases with different subcellular compartments, in turn, regulates protein secretion, endocytosis, and recycling by controlling vesicle docking and fusion at all steps, from endoplasmic reticulum, through the Golgi stack, to the plasma membrane (14). These functions are obviously required for cellular growth as well as maintenance of normal cellular homeostasis, and disordered subcellular trafficking caused by mutations in Rab effectors has recently been identified as the cause of a variety of genetic diseases (15).

PKC TRANSLOCATION AND ANCHORING

Subcellular Transport and Substrate Specificity

The first evidence for distinct localization of individual PKC isozymes was obtained over 10 years ago (16). Using isozyme-selective antibodies to localize individual PKC isozymes in non-stimulated neonatal cardiac myocytes, it was evident that each isozyme has a unique subcellular localization. Moreover, within seconds of stimulation with norepinephrine (which stimulates PKC via the α1-adrenergic receptor) or with phorbol myristate acetate (PMA), a diacylglycerol (DAG) analogue that directly activates PKC, each isozyme relocated to a new subcellular site. Localization was not restricted to the plasma membrane; rather, activated PKC isozymes were also found on cytoskeletal elements and at perinuclear structures. Subcellular localization to structures other than plasma membrane were also observed in other cell types (17–19). Subsequent studies using commercially available isozyme-specific anti-PKC antibodies (20) confirmed these earlier findings by examining the localization of five different DAG-sensitive PKC isozymes present in cardiac myocytes: α, βI, βII, δ, and εPKC (20, 21). Activated βIPKC translocated from a punctuated structure in the cell body into the nucleus, whereas βIIPKC translocated from fibrillar structures in the cytosol to the perinucleus and plasma membrane. Cell stimulation also caused translocation of δPKC from the nucleus to fibrillar structures in the cell body, and translocation of εPKC from the nucleus to cross-striated structures and cell-cell contact areas (intercalated disks). Finally, αPKC translocated from diffuse cytosolic locations to the perinucleus following activation with hormone or PMA. These studies demonstrated that each PKC isozyme is localized to distinct subcellular sites both before and after activation.

At the time that these observations were made, activated PKC isozymes were all assumed to be associated with plasma membranes. The rationale for that notion was that DAG, which is necessary for activation of most PKCs, is derived from membrane phospholipids (22). Activation of PKC in cells is accompanied by translocation of the enzyme from the cell soluble fraction to the cell particulate fraction, which is enriched with plasma membranes (10). It was, therefore, natural to assume that activation of PKCs induces translocation of cytosolic soluble enzymes to the plasma membrane. It also followed that localization of activated PKC isozymes would be determined solely by lipid-protein interactions. However, because of the observed distinct subcellular location of PKC isozymes, it was suggested that localization of each activated PKC isozyme must also require protein-protein interactions between each activated isozyme and a specific, isozyme-selective anchoring protein, which we collectively termed receptors for activated C-kinase (RACKs) (23). The existence of a different RACK for each PKC isozyme was hypothesized. The anchored RACK binds its specific activated PKC isozyme in a locale containing a defined subset of protein substrates and away from others, which is the determining factor for substrate specificity of that isozyme (see Figure 1 for a schematic presentation; see color insert). In addition, based on immunofluorescence studies, it was suggested that inactive PKCs must also be anchored via protein-protein interactions. The anchoring proteins for the inactive PKC isozymes were collectively termed RICKs for receptors for inactive C-kinases (24). However, to date, the hypothetical RICKs have not been identified.

Proof of the RACK hypothesis required identification of isozyme-specific RACKs and demonstration of the importance of the interaction between activated PKC isozymes and their corresponding RACKs. Four simple predictions were made:

1. Introduction of an unanchored RACK into cells should inhibit PKC function.

2. Introduction of peptides that contain the interaction sites on a RACK or its corresponding PKC isozyme should inhibit translocation and function of that isozyme without affecting the translocation and function of any other isozyme.

3. Introduction of peptides that induce an interaction between a particular PKC isozyme and its RACK should selectively activate that isozyme.

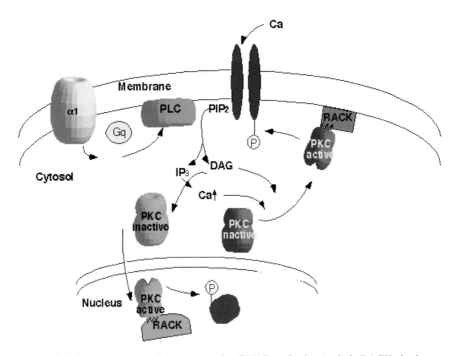

Figure 1 Schematic presentation on the role of PKC anchoring to their RACKs in determining substrate specificity. Shown are two PKC isozymes (*red and light blue*) and their corresponding RACKs. On activation of α1-adrenergic receptor that couples to phospholipase C (PLC) via the Gq protein, both diacylglycerol (DAG) and inositol triphosphate (IP$_3$) are generated. IP$_3$ increases the intracellular calcium ions in the cytosol and, together with DAG, causes translocation of the PKC isozymes to distinct subcellular sites. One PKC isozyme (*red*) translocates from the cell body to the plasma membrane, whereas the other (*light blue*) translocates into the nucleus. The distinct localization of each activated isozyme to different subcellular sites brings them close to different set of substrates; whereas the red isozyme is located nearby the L-type calcium channel (*green*) and thus regulates its function, the activated blue isozyme is localized near nuclear proteins and thus may regulate gene expression by phosphorylation of, for example, a transcription factor (*purple*). (See text for details and references.)

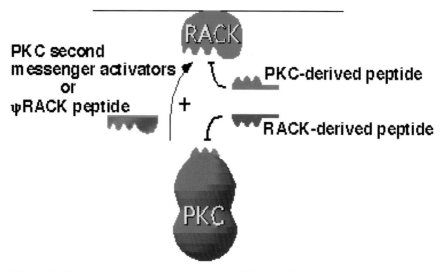

Figure 2 Schematic presentation of peptide inhibitors and activators of PKC transloca-tion. Shown is a PKC isozyme (*red*) and its RACK (*green*). A peptide derived from the RACK (*green*) that contains the PKC-binding site can bind to the activated PKC and thus compete with its binding to its RACK. Peptide derived from PKC that contains the RACK-binding site (*red*) can also act as a competitive inhibitor of PKC translocation to its RACK. Finally, a peptide derived from PKC that is homologous to the RACK (*red and green*), col-lectively termed pseudoRACK peptide (ψRACK), can mimic second messenger-induced translocation and activation of PKC. (See text for details and references.)

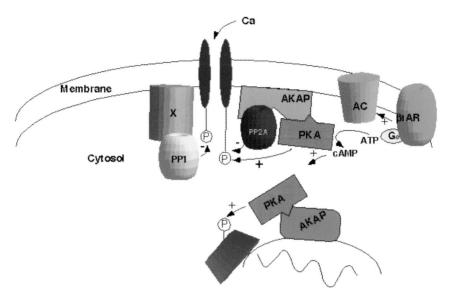

Figure 3 Schematic presentation on the role of PKA anchoring to their AKAPs in determining substrate specificity. Shown PKA (*red*) and two AKAPS (*light blue and green*).
Upon activation of the β1-adrenergic receptor, which couples to adenylyl cyclase via the
Gs protein, cAMP is generated. cAMP causes the departure of the catalytic subunit of PKA
from the AKAP at two distinct subcellular sites. The distinct localization of each activated
PKA by anchoring two AKAPS at different subcellular sites renders them proximal to different substrates; the first AKAP is located near the L-type calcium channel (*blue*) and thus
regulates its function, whereas the other is localized on the mitochondrial membrane adjacent to a different substrate (*purple*). In addition, AKAP can anchor a selective phosphatase, PP2A (*dark purple*) to the same site, leading to synchronized phosphorylation and
dephosphorylation of the channel. (See text for details and references.)

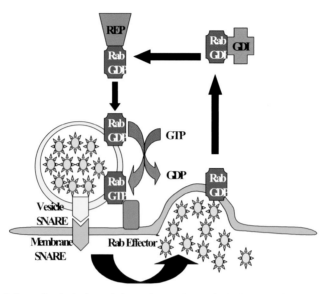

Figure 4 Schematic depiction of the Rab GTP/GDP hydrolysis cycle. Free GDP-Rab, bound to one or more Rab-effector proteins (REP, *upper left*), exhanges GDP for GTP, thus becoming active. GTP-Rab binds to vesicle membranes, thus targeting the vesicle to its specific acceptor organelle via the interactions of docking proteins or SNAREs. Hydrolysis of GTP to GDP is accompanied by dissociation of Rab from membrane, and its free cytosolic inactive form is stablized through maintenance of GDP binding by one or more GDP dissociation inhibitors (GDI).

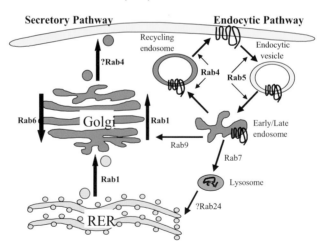

Figure 5 Rab-family member effects on different stages of vesicular transport. Rab1 mediates antegrade transport and Rab6 retrograde transport through the Golgi. Rab4 targets vesicles to plasma membranes, whereas Rab5 mediates internalization of membrane proteins via clathrin-coated endocytic vesicles; these Rabs constitute a recycling pathway. Rabs 7 and 9 play roles in protein degradation.

4. Together, the isozyme-selective inhibitors and activators should identify the function of each isozyme in normal and diseased states.

Work in the past 12 years confirmed these predictions. The first prediction was confirmed in non-cardiac cells; purified RACKs (25) or a peptide derived from them (23, 26) inhibited βPKC function in *Xenopus* oocytes. Subsequent work that confirmed the other three predictions was carried out in cardiac myocytes and focused mainly on elucidating the role of PKC isozymes in normal cardiac function and in cardiac disease. Using peptides that interfered with or promoted anchoring of individual PKC isozymes to their site of action, the role of each isozyme was elucidated. This approach not only validated the hypothesis that functional specificity of individual isozymes is determined by subcellular location, but also demonstrated a need for isozyme-selective reagents to adequately assess PKC functions because different PKC isozymes may exert opposing effects, no effect, or the same effect on different organ functions (see below).

The use of PKC inhibitors and activators in combination, i.e., antithetic loss and gain of function, helped to unambiguously confirm specific roles for individual PKC isozymes. Function induced by a PKC isozyme translocation activator peptide was reversed only by the peptide translocation inhibitor of that isozyme, and not by inhibitors of other isozymes (e.g., see 27). Furthermore, the effect of the peptide translocation activator could be abolished by an inhibitor of its catalytic activity (e.g., see 27). These results demonstrated that the function of each PKC isozyme requires both localization to the correct subcellular site (where its RACK is present) and ability of that isozyme to phosphorylate protein substrates at the site of anchoring.

The isozyme-selective inhibitors used in these studies were mainly derived from the RACK-binding site on individual PKCs (Figure 2, see color insert) (11). The isozyme-selective activator was derived from the short sequence of homology between PKC and its RACK (28). These regulators of PKC binding to RACKs are short peptides identified by a rational search (recently reviewed in 28) (see Figure 2 for scheme). As predicted, following their introduction into cardiac myocytes, each peptide selectively inhibited or induced the translocation of only its corresponding isozyme (e.g., see 27). The biological activities of the peptides are obtained at an intracellular concentration of 5–50 nM, which makes them potent regulators of the function of individual PKC isozymes. Tables 1 and 2 provide a summary of the PKC inhibitor and activator peptides that were identified, their isozyme specificity, the rationale that led to their identification, and the PKC function that was elucidated by their use in cardiac cells.

Before discussing the specific biological activities of these peptides in cardiac myocytes and the PKC functions they helped to define, the means by which these peptides were introduced into cells should be described. Peptides cannot cross biological membranes. Therefore, for initial studies the peptides were microinjected into cells (25, 26). In subsequent studies, cells were transiently permeabilized using saponin in the presence of ATP (29). Cell viability was not affected by this transient

TABLE 1 Peptide translocation inhibitors of specific PKC isozymes

Peptide or fragment	βC2-1, -2, -4	βIV5-3, βIIV5-3	εV1-2 (or εV1)	δV1-1
Specificity	cPKC, possibly βPKC	βI and βIIKC, respectively	εPKC	δPKC
Identified by	Conserved sequences in homologous domains of otherwise unrelated proteins	The least conserved sequences between βI and βIIPKCs	Conserved sequences in εPKC in evolutionarily remote species (*Aplysia* and rat)	The least conserved sequences between δ and θPKCs
Biological activity of the peptide in heart cells	Inhibits PMA-regulation of L-type Ca channel in cardiac myocytes Inhibits hormone or PMA-induced hypertrophy of neonatal cardiac myocytes	Inhibits hormone or PMA-induced hypertrophy of neonatal cardiac myocytes	Inhibits hormone or PMA-induced reduction in contraction rate of neonatal cardiac myocytes Inhibits protection of cardiac myocytes from ischemia-induced cell death in isolated myocytes and in vivo Increases hormone- or PMA-induced hypertrophy of neonatal cardiac myocytes. Causes dilated cardiomyopathy in vivo	Enhances hormone or PMA-induced reduction in contraction rate of neonatal cardiac myocytes Increases protection of cardiac myocytes from ischemia-induced cell death in isolated myocytes
Reference	(105)	(105)	(27, 44)	(27, 44, 32)

TABLE 2   Peptide translocation activators of specific PKC isozymes

Peptide	$\psi\beta$**RACK**	$\psi\varepsilon$**RACK**	$\psi\delta$**RACK**
Specificity	cPKC, possibly βPKC	εPKC	δPKC
Identified by	Homology between βPKC and RACK1, the βPKC-selective RACK	Homology between εPKC and RACK2, the εPKC-selective RACK	The least conserved sequences between δ and θPKCs and in the homologous position of $\psi\varepsilon$RACK in εPKC
Biological activity of the peptide in heart cells	None studied in heart	Enhances contraction rate of neonatal cardiac myocytes	
		Inhibits cardiac myocyte death by ischemia in isolated myocytes, in intact heart, and in vivo	Increases cardiac myocyte death by ischemia in isolated myocytes, in intact heart, and in vivo
		Causes concentric hypertrophy without changes in cardiac function	Causes concentric hypertrophy without changes in cardiac function
		Inhibits hormone- or PMA-induced hypertrophy of neonatal cardiac myocytes	
Reference		(27, 64)	(32)

permeabilization protocol and approximately 5% of the applied peptide entered 95% of the cells (29). However, both microinjection and transient permeabilization are technically challenging and have experimental drawbacks. Therefore, peptide transporters were developed that could cross biological membranes and carry peptide regulators of PKC translocation into cells. Antenna pedia-derived peptide (27, 30), the TAT-derived peptide, or poly-arginine (31) were used as transporters to deliver translocation modifier peptides as cargo into cardiac myocytes. Peptides were conjugated to these transporters via a Cys-Cys bond that is reduced in the cell, enabling the exit of the transporter but trapping the cargo peptide inside the cells. A recent study comparing the available three transporters demonstrated that the poly-arginine transporter was superior to the other transporters by a factor of ~5 (31). Nevertheless, similar results on cardiac myocyte function were obtained with any one of these transporters if sufficient peptide conjugate was used. A fourth approach to deliver PKC-regulating peptides into the heart is by generating transgenic mice, which is described in greater detail below. Finally, and most relevant to the effort to model drugs from the translocation modulating peptides, is the ability to deliver such peptides into intact hearts by coronary perfusion. Recent ex vivo studies using Langendorff rat and mouse hearts (31, 32), and in vitro studies using catheter-based intracoronary delivery in pigs (F. Lee & D. Mochly-Rosen, unpublished data) or intravenous delivery in mice (R. Bolli & D. Mochly-Rosen, unpublished data), have demonstrated that peptides conjugated to the aforementioned transporters are effectively delivered into the heart and thus are potential therapeutic agents.

Effects of PKC Translocation Modifiers on Isolated Cardiac Myocytes

PKC AND CARDIAC CHANNEL ACTIVATION   A variety of signaling events involving G protein–coupled receptors in the heart regulate the L-type calcium channel (33–35). Here, the role of PKC anchoring and function on this channel is described (Figure 1); the role of PKA anchoring is also discussed (see Figure 3).

α1-Adrenergic or PMA-mediated activation of PKC inhibits β-adrenergic stimulation of L-type calcium current in adult rat cardiac myocytes (36). To determine which PKC isozyme(s) mediate this effect, Boutjdir and collaborators introduced inhibitors of conventional PKC (cPKC; i.e., α, β, and γ isozymes) binding to their RACKs into adult cardiac myocytes. These peptide inhibitors (βC2-1, 2, and 4; Table 1) are derived from the C2 domain of βPKC, which is common to all the cPKCs, and therefore should affect all the C2-containing isozymes in heart; i.e., α, βI, and βIIPKC (37 and Table 1). The βC2 peptides, which inhibit cPKC binding to their RACKs (37), blocked ~75% of isoproterenol-induced inhibition of L-type calcium channel activity (36), whereas scrambled peptides with the same amino acid composition had no effect. Of the cPKC isozymes in heart, only activated βIIPKC and its RACK, RACK1, are found at diffuse sites on the plasma membrane (20, 37) where L-type calcium channels may be present; activated βIPKC is inside the nucleus and activated αPKC is at the perinuclear membrane (20). Therefore, it is likely that βIIPKC is the cPKC isozyme that regulates this channel (see Figure 1).

More recently, the εPKC-specific translocation agonist $\psi\varepsilon$RACK (Table 2) also suppressed L-type calcium channel activity in adult rat cardiac myocytes, which was reversed by the εPKC translocation antagonist, εV1-2 (38). Importantly, activated εPKC and its RACK, β'COP, are localized at cardiomyocyte cross-striated structures (20, 39), placing it near T tubules where L-type calcium channels are found. Thus it is possible that, depending on the location of the channel, different PKC isozymes can regulate calcium currents; L-type calcium channels at plasma membranes may be regulated by βIIPKC, whereas channels restricted to T tubules may be regulated by εPKC. It will be interesting to determine if selective disruption of L-type calcium channel localization could help address these possibilities.

Finally, recent work by McHugh and collaborators identified two sites of PKC phosphorylation at the N-terminal domain of the channel (40). However, it is not known which PKC isozyme phosphorylates these sites. Moreover, conflicting data on the role of PKC activation in regulating this channel were reported; both decreases and increases followed by decrease of calcium currents were found (41). Those conflicting observations are attributed to different experimental conditions (40), presumably resulting in activation of different PKC isozymes. For example, application of PKC translocation inhibitor peptides showed that hypoxia-induced inhibition of the L-type calcium current sensitivity to β-adrenergic receptor stimulation in adult guinea pig is mediated through cPKC, not through εPKC (42). However, the effects of these peptides on the regulation of the channel under normoxic conditions have not been examined. It is anticipated that future studies targeting PKC translocation in an isozyme-selective manner under different experimental conditions will determine the role of PKC isozymes in the complex regulation of this key channel.

εPKC AND MYOCYTE CONTRACTION RATE   Isolated neonatal cardiac myocytes (unlike isolated adult cardiac myocytes) contract spontaneously at a rate of 300 beats per min, and the rate of spontaneous contraction is a variable that is subject to regulation by hormonal influences. Yuan and collaborators first showed that activation of PKC inhibited contraction rate (43). A specific role for εPKC as a regulator of contraction rate was subsequently suggested by studies examining the time course and relative sensitivity of individual PKC isozymes to PMA exposure (21). In these studies, maximal εPKC translocation correlated with maximal inhibition of contraction rate (21). Isozyme-selective inhibitors and activators of εPKC were later identified and used to conclusively demonstrate that εPKC mediates PMA-and hormone-induced inhibition of contraction rate. Inhibition of εPKC blocked PMA- or α1-adrenergic-induced depression of contraction rate (44), whereas activation of εPKC enhanced the diminished rate of contraction stimulated by sub-optimal amounts of PMA (T. Liron & D. Mochly-Rosen, unpublished data). Therefore, εPKC activation was proven to be both necessary and sufficient to depress spontaneous contraction rate in isolated cardiac myocytes.

As a specificity control in the aforementioned studies, δPKC-selective translocation inhibitors were also applied to cardiac myocytes, demonstrating that δPKC did not inhibit PMA- or α1-adrenergic-induced depression of contraction rates (44). Subsequent analysis of these data suggested, in fact, that δPKC and εPKC may have opposing regulating effects on the rate of myocyte contraction (45). Future studies will examine these potentially antithetic effects of δPKC and εPKC and determine the significance of these findings on regulated function in the adult heart.

εPKC AND ISCHEMIC INJURY   Short periods of ischemia just prior to prolonged ischemia reduce cardiac damage by prolonged ischemia in vivo and in vitro [see recent review (9)]. The role of PKC in this protection (also termed preconditioning) has been implicated by many studies (8, 46, 47) using both inhibitors and activators of PKC. However, whereas some studies suggest that PKC activation protects the heart, others have suggested that PKC may contribute to the damage induced by the ischemic insult (48). To examine this question, a model of simulated ischemia was developed using neonatal rat cardiac myocytes to determine the role of PKC in cardioprotection (49). Subjecting isolated cardiac myocytes to short bouts of hypoxia prior to prolonged hypoxia, or activating PKC using low concentrations of PMA, reduced hypoxic damage by ~60%. This protection was lost, however, if the isozyme-selective inhibitor of εPKC, εV1-2, was included during the preconditioning treatments of isolated neonatal rat cardiac myocytes (49). Similar results were obtained using isolated adult rabbit (50) or rat cardiac myocytes (27). Using a conscious instrumented rabbit model, Bolli et al. demonstrated that brief ischemia causes a selective translocation of η and εPKC in vitro (51, 52), implicating one or both isozymes in cardioprotection from ischemia. Subsequent studies using the εPKC activator peptide, $\psi\varepsilon$RACK, supported the findings on the role of εPKC in this process; activation of εPKC during ischemia protected both isolated neonatal rat cardiac myocytes and adult rat cardiac myocytes from simulated ischemia-induced damage (27). Fewer than half of the cells were damaged by ischemia if they were pretreated with ~10 nM of $\psi\varepsilon$RACK, compared with untreated cells or cells treated with transporter peptide only or transporter carrying scrambled peptides into the cardiac myocytes (27). The protective effect of εPKC activation observed using isolated myocytes in a model of cardiac ischemia was also confirmed in vivo by transgenic expression of $\psi\varepsilon$RACK in mouse cardiac myocytes (27). $\psi\varepsilon$RACK transgenic mouse hearts were resistant to prolonged no-flow ischemia, with accelerated recovery of contractile function and a cytoprotective effect, indicated by decreased release of creatine phosphokinase. Together, these studies suggest that potential medical benefit could accrue from cardiac administration of an εPKC-selective agonist.

δPKC AND ISCHEMIC INJURY εPKC and δPKC are the only PKC isozymes that are activated during ischemia (49). Indeed, overexpression of constitutively active δPKC, and not constitutively αPKC or wild-type δPKC, in cultured neonatal cardiac myocytes increased resistance to simulated ischemic insult (53). However, the restriction of the protective effect to the constitutively active δPKC suggests a non-isozyme selective effect. Indeed, we have found that δV1-1, a δPKC-selective inhibitor (Table 1), did not affect preconditioning-induced protection in simulated ischemia using isolated rabbit cardiac myocytes (L.E. Chen & D. Mochly-Rosen, unpublished data). Yet, inhibition of δPKC by δV1-1 during the ischemic period reduced the number of cardiomyocytes that died because of the ischemic insult (32). Moreover, δPKC activation with $\psi\varepsilon$RACK, a δPKC-selective translocation activator, increased damage from the simulated ischemia twofold (32). Therefore, although both δPKC and εPKC are activated by simulated ischemia in neonatal and adult cardiac myocytes, εPKC activation has a cardioprotective action, whereas δPKC activation mediates myocyte death induced by ischemia. These opposing roles of individual PKC isozymes illustrate the importance of using isozyme-selective tools to study the role of PKC. Moreover, cardioprotection by the δPKC-selective inhibitor peptide or εPKC-selective translocation activator was obtained after perfusion of these peptides through the coronary arteries (32). Therefore, these studies raise the intriguing possibility that either the inhibitory peptide of δPKC translocation and/or the activating peptide of εPKC translocation (or compounds that mimic their biological effects) could be useful therapeutic agents for ischemic heart disease.

In Vivo Studies of PKC Translocation Modifiers

TRANSGENIC STUDIES OF CARDIAC PKC   Whereas causal relationships between PKC isozyme activation and specific cardiomyocyte responses are beginning to emerge from in vitro studies, most in vivo observations of PKC isozyme effects have tended to be correlative. For example, εPKC has been variously described as selectively translocated to particulate ventricular fractions during acute or chronic pressure overload (5, 6), after angiotensin II stimulation (6), and after chronic ethanol consumption in guinea pigs, which is therefore postulated to be the mechanism for ethanol-induced cardioprotection from ischemic reperfusion injury (30).

In addition to observational and correlative experiments, gain and loss of activity studies of several PKC isozymes have been achieved through genetic manipulation of the mouse. Loss of PKC isozyme activity has been accomplished by individually ablating the genes for βPKC, εPKC, and γPKC (54–56). Surprisingly, none of the PKC gene knockout mice created thus far has manifested a cardiac phenotype, despite abundant evidence that PKC β and ε isozymes can have critical functions in this organ. Instead, a variety of central nervous system aberrations have been observed for these three PKC isozyme knockout mice, including memory or learning deficits and deficient pain perception (55–59). The βPKC-deficient mouse also has an immunodeficiency syndrome (54). Lack of cardiac phenotypes in the existing PKC-deficient mouse models suggests that endogenous cardiac PKC isozymes can opportunistically compensate for absent enzymes. Thus, experimentally, in vitro targeting of PKC isozyme activity requires a more precise approach than simple loss-of-function gene ablation.

The opposite approach, that of overexpressing PKC isozymes in the mouse heart, has been more successful in creating mouse models with cardiac phenotypes, but this has also generated controversial and apparently conflicting results. Overexpression of wild-type (i.e., not mutationally activated) βIIPKC in mouse hearts was initially reported to cause a dilated, fibrotic cardiomyopathic phenotype, with pathologic up-regulation of fetal cardiac genes and progression to cardiac failure (60). Another group simultaneously described differing effects of expressing a truncated βIIPKC, with constitutive activity in adult compared with neonatal hearts (61). Induced expression of activated βIIPKC in adult mouse hearts caused a mild hypertrophic phenotype without molecular or functional markers of heart failure. In contrast, expression of the same gene in neonatal mouse hearts caused premature lethality, apparently from arrhythmic disturbances. Thus βIIPKC expressed at different levels, with different intrinsic activities, and at different times during the development of the mouse results in three different syndromes. Although these data have all been valuable in helping to delineate the possible consequences of increased βIIPKC activity, the question, "What is the function of physiological activation of βIIPKC in the in vitro heart?" remains unanswered.

Transgenic overexpression of mutationally activated εPKC in the mouse heart has generated similarly confusing results but also suggests a mechanism for the plethora of phenotypes seen with standard overexpression strategies. The initial report of εPKC overexpression, in which the cardiomyocytes contained levels that were six times greater than those in nontransgenic cardiomyocytes, described concentric hypertrophy with normal cardiac function but, surprisingly, with abnormal cardiomyocyte function and calcium cycling (62). A follow-up report of another line of mice with higher expression levels described a cardiomyopathic syndrome closer to that originally described for wild-type βIIPKC overexpression (62, see above). The mechanism for the pathological cardiac phenotype at very high levels of εPKC expression was determined to be promiscuous interactions of the activated transgenic εPKC with RACK1, the membrane anchor protein for βIIPKC. Although εPKC is highly specific for its own RACK, RACK2 or β'-COP, this specificity is achieved because of an affinity 10 times higher than for RACK1. However, at the levels of activated εPKC achieved in the cardiomyopathic mouse line (30 times the normal endogenous level) (63), nonspecific interaction with the βIIPKC RACK occurred, and the consequence was mimicry of the βIIPKC overexpression phenotype. Thus both gene ablation and overexpression strategies have inherent problems when applied to the study of PKC isozyme function in the in vitro heart. The ideal approach would selectively activate or inhibit PKC isozyme function in the target organ, e.g., in the heart. This has been achieved by applying the translocation modification paradigm to in vitro mouse models.

Transgenic techniques were employed to express εPKC translocation inhibitory and facilitory peptides in mouse hearts. In this application, transgenesis is not synonymous with overexpression because no active protein or enzyme is being expressed. Rather, transgenesis was employed as a method of cardiac-specific gene delivery for drugs that selectively target εPKC translocation and, hence, activation. To date, transgenic models have been described that express εPKC activating or inhibiting peptides (27, 64), and δPKC activating peptides (32). In all cases, endogenous PKC isozyme expression was unchanged from control. Hence, proper stoichiometry was maintained between the target and non-targeted PKC isozymes, their RACKs, and presumably their respective substrates. Peptide translocation activators for εPKC and δPKC increased their basal partitioning to cardiac membranes by a modest 15 to 20%, indicating that the increase in PKC isozyme activity achieved was similar to that seen with physiological stimuli and substantially less than that which results from treatment with phorbol ester or overexpression (62, 63). Likewise, the εPKC translocation inhibitor peptide decreased εPKC membrane partitioning by approximately 20%, indicating a similarly physiological decline in basal εPKC activation (64, 65).

Based on prior in vivo and in vitro associations of PKC activation with cardiac hypertrophy/failure and ischemic preconditioning (see above), the translocation modifier mice were examined for altered growth, function, and response to ischemic insults. εPKC and δPKC activation were found to have essentially identical effects on myocardial hypertrophy; i.e., they stimulated a mild hypertrophy with preserved ventricular and cardiomyocyte function (32, 64). Ventricular mass increased by 15%, and this was a cumulative effect over at least 12 weeks. However, two of the hallmark molecular markers of pathological cardiac hypertrophy, increased atrial natriuretic factor and α-skeletal actin gene expression, were absent in these two hypertrophy models, whereas β-myosin heavy chain expression, an additional molecular marker of hypertrophy, was increased in the expected fashion. Thus at the morphological, functional, cellular, and molecular levels, εPKC and δPKC activation results in identical physiological hypertrophy responses.

What then was the effect of εPKC and δPKC translocation inhibition? Interestingly, at higher levels of inhibitor peptide expression, the result in both cases was early lethality. This was especially surprising for the εPKC inhibitor mouse because ablation of the εPKC gene reportedly had no detectable effect on cardiac function, and the εPKC knockout mice had normal longevity (55). Cardiac-specific inhibition of εPKC translocation resulted in development of an aggressive and early dilated cardiomyopathy, apparently from insufficient growth of cardiac myocytes in the early postnatal period, i.e., cardiac hypotrophy (64). This appeared to be the opposite of εPKC activation, which caused a benign myocardial hypertrophy owing to increased numbers of cardiomyocytes in the early postnatal period (64). Taken together, these findings suggested that εPKC activity was, at least in the context of myocardial growth and ventricular function, a generally favorable adaptation, which was contrary to previous conclusions and accepted dogma (5–7). Therefore, to determine whether εPKC activity might have different effects in the context of a pathological hypertrophy associated with εPKC translocation, the εPKC activator and (lower expressor) inhibitor mice were crossed with transgenic mice expressing the α subunit of the Gαq heterotrimeric G protein. Transgenic Gαq mice develop eccentric cardiac hypertrophy with decreased ventricular and cardiomyocyte contractile function and an aggressive recapitulation of the embryonic cardiac gene expression program (66–68), and εPKC translocation is increased in Gαq transgenic hearts (66). If εPKC activation by Gαq contributes to the cardiac pathology, then crossing with the εPKC inhibitor mice should salvage the phenotype, and crossing with the εPKC activator mice should worsen the phenotype. Strikingly, just the opposite was seen, with εPKC activation rescuing Gαq hypertrophy and contractile depression, but εPKC inhibition causing lethal decompensation (65). Thus available evidence from in vitro εPKC-specific translocation modulation (64, 65) and from a lower-expressing line of εPKC overexpressing mice (62) indicate that εPKC activation is necessary and sufficient for normal and physiologically compensatory myocardial growth.

PKA AND A KINASE ANCHORING PROTEINS (AKAPS)

Elevation of cAMP and activation of the cAMP-dependent protein kinase, PKA, results in variety of cellular responses. A family of anchoring proteins, termed AKAPs for a kinase anchoring proteins, bind inactive PKA via its regulatory subunit and localize it to distinct subcellular sites (13) (see scheme in Figure 3; see color insert). There are multiple AKAPs in each cell type, each localized to different subcellular sites (13). Upon elevation of cAMP, the catalytic subunit is released and phosphorylates nearby substrates. Work by Scott and collaborators demonstrated that disruption of PKA anchoring to AKAP via a peptide derived from one of the AKAPs inhibits the anchoring of PKA and its function in the cell (69). This peptide, Ht31, is not selective for individual AKAPs. However, it was effective in demonstrating the role of anchoring of PKA to AKAP in PKA-mediating signaling. Most of the work on the role of PKA anchoring and cardiac function has focused on the regulation of L-type calcium channels, and therefore our discussion focuses only on this function.

PKA and Regulation of L-type Ca^{2+} Channels

L-type calcium channels are multi-subunit proteins containing an $\alpha 1$ subunit, which is the ion channel, and two modulatory subunits, β and $\alpha 2\delta$ (35). The major form of the channel in heart is called class C L-type calcium channel, and recent work by Hosey and collaborators demonstrated that the $\beta 2$ subunit is phosphorylated in cardiac myocytes at three loose-consensus phosphorylation sites for PKA (70). Moreover, these sites do not overlap with PKC sites on this subunit (70), indicating that PKA and PKC have distinct roles in regulating this channel. Direct phosphorylation of the channel by PKA, and possibly phosphorylation of other proteins associated with it, stimulates calcium flux through this channel (reviewed in 35). Importantly, association of PKA with AKAP was shown to be required for β-adrenergic receptor-mediated regulation of the channel (71). Introduction of 100 μM Ht31 peptide through a recording pipette inhibited isoproterenol-induced increases in calcium current through this channel. In contrast, a control peptide in which the α-helix structure of Ht31 was disrupted by introduction of a proline residue did not affect L-type calcium channel activity (71). The effect of the peptide was relatively slow to commence (5–10 min), supporting the interpretation that Ht31 peptide competes with AKAP for binding of PKA, gradually replacing PKA from its anchoring site. Together with previous work of Catterall and collaborators (72), these results indicate that anchoring of PKA on AKAP is essential in the cAMP-mediated regulation of this cardiac calcium channel. However, the AKAP required for this association in heart is not known. In cardiac tissue and other tissues—MAP2B, AKAP79, AKAP100, and/or AKAP—15 have been implicated in this function (35). Nevertheless, these observations strongly support the notion that anchored PKA in close proximity to this channel plays an important role in providing specificity in signaling (Figure 3).

Because L-type calcium channels are dynamically controlled, regulation by phosphatases anchored in close proximity to the channels has long been suggested. Indeed, Coghlan and collaborators demonstrated that at least some AKAPs can bind concomitantly both PKA and the Ser/Thr phosphatase 2B (73). Moreover, when using membrane patches of rabbit cardiac myocytes, more than one phosphatase found in close proximity to this channel was shown to regulate channel activity. Although studies examining the consequence on channel activity of removal of phosphates from anchoring sites have not been reported, these observations strongly suggest that anchored phosphatase next to the channels also provides a means for select temporal and spatial regulation of their signaling events (Figure 3).

Rab GTPases—TRAFFIC COPS FOR CELLULAR PROTEINS

As discussed above for signaling molecules, virtually all cellular functions are dependent upon proper compartmentalization of proteins. This requirement mandates that protein transport between specialized intracellular organelles be rigidly controlled in order to prevent potentially harmful mixing or improper targeting of organelle proteins. Transcompartmental movement of cellular proteins involves a cascade of vesicular budding, docking, and fusion events between donor and acceptor organelle membranes (74). Thus proper localization of molecules within cells relies on formation of membranous carrier structures (vesicles) that bud off of donor compartments, move along filamentous cytoskeletal tracks, and then fuse with recipient target membrane (tethering and docking) in order to deliver membrane-associated or soluble luminal contents to the target organelle. Coordination of vesicle docking and fusion is achieved in large measure by a subfamily of Ras-like small G proteins, the 40 member large Rab GTPases (14). In general, the initial interaction between a vesicle and its target membrane occurs through assembly of a protein complex linking the two membranes together. Mutual recognition of a transport vesicle and its target organelle, which involves interactions of membrane anchor proteins termed SNAREs (14, 74, 75), is the critical process controlled by Rab GTP/GDP exchange (Figure 4, see color insert).

As initially described for the prototypical Rab protein, the *Saccharomyces cerevisiae* Ypt1 GTPase discovered in 1983 (1, 76, 77), Rab GDP/GTP nucleotide exchange modulates a dynamic equilibrium between GDP-bound cytosolic (inactive) and GTP-bound organelle-associated (active) forms. The intrinsic GTPase activity of Rab proteins is very low but is subject to acceleration by Rab GTPase-activating proteins (GAPs), several of which have been identified using cross-linking or co-precipitation studies and yeast two-hybrid screening (78–81). The hallmark characteristic of Rab proteins is that a cycle of GDP/GTP nucleotide exchange is superimposed on a cycle of association with and dissociation from subcellular membranes (14, 74). In the cytosol, inactive Rabs are maintained in the GDP-bound state by GDP dissociation inhibitor proteins (GDI), which probably also act as Rab chaperones during translocation between membrane and cytosol. As with other small G proteins, guanine nucleotide exchange factors (GEF) have also been identified as Rab family members. Thus Rab GTPase activity and, hence, Rab function in intracellular protein transport are highly regulated processes that serve to fine tune protein trafficking. Importantly, specific mutations have been identified in the Rabs, analogous to those in Ras proteins, which either abolish or greatly enhance intrinsic GTPase activity and therefore result in constitutively active (GTP-Rab) or dominant-negative (GDP-Rab) forms, respectively (82–87). These Rab mutants have been valuable experimental tools for dissecting in vivo and in vitro Rab functions.

Rab structure and function are remarkably conserved across different tissues and species. The closest mammalian relative to yeast Ypt1 is Rab1, which is over 70% identical at the amino acid level and can fully replace Ypt1 function in the yeast protein secretory pathway (88, 89). Thus the general mechanism of Rab GTPases regulating vesicular transport appears to be an essential biological function among eukaryotic cells. However, the specific function for most of the approximately 40 mammalian Rab family members is not known. In yeast, which has a total of 11 Ypt GTPases, 3 appear to be sufficient to regulate protein transport from the endoplasmic reticulum (ER) to plasma membrane, and another 3 regulate endocytosis. This leaves 5 Ypts for which no crucial function has been identified and which may even be redundant (75). Mammalian Rab proteins, by analogy to the more thoroughly studied yeast system, presumably serve similar functions, and the multiplicity of Rab family members is explained by the observation that different Rab proteins localize to distinct intracellular organelles involved in protein secretion, recycling, and endocytosis (14, 74, 75). For example, Rabs1 and 6 are localized to the *cis*-Golgi network (90–93), Rab4 is found in early endosomes (94–96), and Rab5 is found at the plasma membrane and clathrin-coated vesicles (85, 97). These distinct patterns of localization are consistent with observed diversity in function (Figure 5, see color insert). Rabs1 and 6 both play pivotal roles in the secretory pathway but regulate antegrade and retrograde *trans*-Golgi transport, respectively. Rabs4 and 5 also play opposing roles for protein transport and recycling of membrane proteins: Relevant to the cellular signaling cascades, Rab4 regulates β2-adrenergic receptor insertion into plasma membranes, whereas Rab5 controls β2-adrenergic receptor internalization via the endocytic pathway (98).

Recently, the possibility that disordered Rab-mediated protein transport could contribute to cardiac disease was demonstrated through detection of increased Rab1, 4, and 6 expression in a murine model of dilated cardiomyopathy (G. Wu & G. W. Dorn, unpublished results). Indeed, transgenic overexpression of Rab1 proved sufficient to cause development of a progressive cardiomyopathy phenotype associated with the ultrastructural abnormalities of Golgi vesicles that are characteristic of disturbed Rab1 function (G. Wu & G. W. Dorn, unpublished results). Whether simple increases in expresssion of other Rab family members can similarily contribute to aspects of cardiac disease in which regulated expression is observed remains to be tested.

The molecular mechanisms governing membrane transport and involvement of Rab GTPases have only begun to be elucidated within the last decade. An important consequence of the increased understanding of vesicular transport processes has been identification of a growing list of human diseases caused by defective intracellular membrane transport proteins, including mucolipidosis II (99), Chediak-Higashi syndrome (100), combined deficiency of coagulation factors V and VIII (101), and oculocerebrorenal syndrome (102). Two human diseases are now specifically known to be the consequence of abnormal Rab function caused by genetic defects in Rab escort protein 1 (X-linked choroideremia) (103) or Rab GDP-dissociation inhibitor (X-linked nonspecific mental retardation) (104, 105). Identification of these vesicular "traffic cops" and the obscure membrane anchor proteins provides specific targets for translocation and transport modifiers that may have therapeutic value.

PERSPECTIVES–OF PROTEIN ANCHORS AND SAFE HARBORS

Why do signaling proteins need to be anchored? For PKC, translocation and binding of activated enzymes to RACKs is the mechanism for determining substrate specificity of different isozymes, by enforcing proximity to a particular set of substrates (Figure 1). Furthermore, because PKC substrates co-localized to specific subcellular compartments (such as plasma membrane, nuclear membrane, cytoskeletal filaments, or myofilaments) are likely to have similar or related functions (such as outside-in signaling, nuclear signaling, cellular structure, and contraction, respectively), binding to RACKs determines the distinct functional effects of PKC isozymes with essentially identical catalytic activity. For PKA, anchoring to AKAPs similarly localizes the (inactive) enzyme in proximity to a distinct subset of substrates, which will tend to be the first phosphorylated upon PKA activation and release (Figure 3). However, non-anchored PKA has the potential, through simple diffusion, to phosphorylate more distant substrates, for example in the nucleus, if cAMP levels are sufficiently high. In other words, the strength of the cAMP stimulus could determine which substrates are phosphorylated: weak stimulus, first tier substrates near the AKAP only; strong stimulus, first tier and second tier substrates farther away from the AKAP. Binding to anchoring proteins would not only regulate access to substrates under different conditions of activation, but would also segregate the activities of different kinases. Finally, subcellular compartmentalization has the potential to determine the access of phosphorylated proteins to cellular phosphatases in distinct signaling complexes. These issues will benefit from additional study using modifiers of kinase binding to anchor proteins.

Visit the Annual Reviews home page at www.AnnualReviews.org

LITERATURE CITED

1. Nishizuka Y. 1986. Studies and perspectives of protein kinase C. *Science* 233:305–12
2. Nishizuka Y. 1995. Protein kinase C and lipid signaling for sustained cellular responses. *FASEB J.* 9:484–96
3. Newton AC. 1995. Protein kinase C: structure, function, and regulation. *J. Biol. Chem.* 270:28495–98
4. Simon MI, Strathmann MP, Gautam N. 1991. Diversity of G proteins in signal transduction. *Science* 252:802–8
5. Gu X, Bishop SP. 1994. Increased protein kinase C and isozyme redistribution in pressure-overload cardiac hypertrophy in the rat. *Circ. Res.* 75:926–31
6. Schunkert H, Sadoshima J, Cornelius T, Kagaya Y, Weinberg EO, et al. 1995. Angiotensin II-induced growth responses in isolated adult rat hearts. Evidence for load-independent induction of cardiac protein synthesis by angiotensin II. *Circ. Res.* 76:489–97
7. Bowling N, Walsh RA, Song G, Estridge T, Sandusky GE, et al. 1999. Increased protein kinase C activity and expression of Ca^{2+}-sensitive isoforms in the failing human heart. *Circulation* 99:384–91
8. Speechly-Dick ME, Mocanu MM, Yellon DM. 1994. Protein kinase C. Its role in ischemic preconditioning in the rat. *Circ. Res.* 75:586–90
9. Cohen MV, Baines CP, Downey JM. 2000. Ischemic preconditioning: from adenosine receptor of K_{ATP} channel. *Annu. Rev. Physiol.* 62:79–109
10. Kraft AS, Anderson WB, Cooper HL, Sando JJ. 1982. Decrease in cytosolic calcium/phospholipid-dependent protein kinase activity following phorbol ester treatment of EL4 thymoma cells. *J. Biol. Chem.* 257:13193–96
11. Mochly-Rosen D. 1995. Localization of

protein kinases by anchoring proteins: a theme in signal transduction. *Science* 268:247–51
12. Blackshear PJ, Nairn AC, Kuo JF. 1988. Protein kinases 1988: a current perspective. *FASEB J.* 2:2957–69
13. Dell'Acqua ML, Scott JD. 1997. Protein kinase A anchoring. *J. Biol. Chem.* 272:12881–84
14. Martinez O, Goud B. 1998. Rab proteins. *Biochim. Biophys. Acta* 1404:101–12
15. Olkkonen VM, Ikonen E. 2000. Genetic defects of intracellular-membrane transport. *N. Engl. J. Med.* 343:1095–104
16. Mochly-Rosen D, Henrich CJ, Cheever L, Khaner H, Simpson PC. 1990. A protein kinase C isozyme is translocated to cytoskeletal elements on activation. *Cell Regul.* 1:693–706
17. Gopalakrishna R, Barsky SH, Thomas TP, Anderson WB. 1986. Factors influencing chelator-stable, detergent-extractable, phorbol diester-induced membrane association of protein kinase C. Differences between Ca^{2+}-induced and phorbol ester-stabilized membrane bindings of protein kinase C. *J. Biol. Chem.* 261:16438–45
18. Papadopoulos V, Hall PF. 1989. Isolation and characterization of protein kinase C from Y-1 adrenal cell cytoskeleton. *J. Cell Biol.* 108:553–67
19. Kiley SC, Jaken S. 1990. Activation of alpha-protein kinase C leads to association with detergent-insoluble components of GH4C1 cells. *Mol. Endocrinol.* 4:59–68
20. Disatnik MH, Jones SN, Mochly-Rosen D. 1995. Stimulus-dependent subcellular localization of activated protein kinase C; a study with acidic fibroblast growth factor and transforming growth factor-beta 1 in cardiac myocytes. *J. Mol. Cell Cardiol.* 27:2473–81

21. Johnson JA, Mochly-Rosen D. 1995. Inhibition of the spontaneous rate of contraction of neonatal cardiac myocytes by protein kinase C isozymes. A putative role for the epsilon isozyme. *Circ. Res.* 76:654–63

22. Takai Y, Kishimoto A, Kikkawa U, Mori T, Nishizuka Y. 1979. Unsaturated diacylglycerol as a possible messenger for the activation of calcium-activated, phospholipid-dependent protein kinase system. *Biochem. Biophys. Res. Commun.* 91:1218–24

23. Mochly-Rosen D, Khaner H, Lopez J. 1991. Identification of intracellular receptor proteins for activated protein kinase C. *Proc. Natl. Acad. Sci. USA* 88:3997–4000

24. Mochly-Rosen D, Gordon AS. 1998. Anchoring proteins for protein kinase C: a means for isozyme selectivity. *FASEB J.* 12:35–42

25. Smith BL, Mochly-Rosen D. 1992. Inhibition of protein kinase C function by injection of intracellular receptors for the enzyme. *Biochem. Biophys. Res. Commun.* 188:1235–40

26. Ron D, Mochly-Rosen D. 1994. Agonists and antagonists of protein kinase C function, derived from its binding proteins. *J. Biol. Chem.* 269:21395–98

27. Dorn GW II, Souroujon MC, Liron T, Chen CH, Gray MO, et al. 1999. Sustained in vivo cardiac protection by a rationally designed peptide that causes epsilon protein kinase C translocation. *Proc. Natl. Acad. Sci. USA* 96:12798–803

28. Souroujon MC, Mochly-Rosen D. 1998. Peptide modulators of protein-protein interactions in intracellular signaling. *Nat. Biotechnol.* 16:919–24

29. Johnson JA, Gray MO, Karliner JS, Chen CH, Mochly-Rosen D. 1996. An improved permeabilization protocol for the introduction of peptides into cardiac myocytes. Application to protein kinase C research. *Circ. Res.* 79:1086–99

30. Chen CH, Gray MO, Mochly-Rosen D. 1999. Cardioprotection from ischemia by a brief exposure to physiological levels of ethanol: role of epsilon protein kinase C. *Proc. Natl. Acad. Sci. USA* 96:12784–89

31. Chen LE, Wright LR, Chen C-H, Oliver SF, Wender PA, Mochly-Rosen D. 2001. Molecular transporter for peptides: delivery of a cardioprotective εPKC agonist peptide into cells and intact ischemic heart using a transporter system, R7. *Chem. Biol.* In press

32. Chen L, Hahn H, Wu G, Chen C-H, Liron T, et al. 2001. Opposing cardioprotective actions and parallel hypertrophc effects of δ and εPKC. *Proc. Natl. Acad. Sci. USA.* In press

33. Reuter H. 1983. Calcium channel modulation by neurotransmitters, enzymes and drugs. *Nature* 301:569–74

34. Yatani A, Codina J, Imoto Y, Reeves JP, Birnbaumer L, et al. 1987. A G protein directly regulates mammalian cardiac calcium channels. *Science* 238:1288–92

35. Kamp TJ, Hell JW. 2000. Regulation of cardiac L-type calcium channels by protein kinase A and protein kinase C. *Circ. Res.* 87:1095–102

36. Zhang ZH, Johnson JA, Chen L, El Sherif N, Mochly-Rosen D, et al. 1997. C2 region-derived peptides of beta-protein kinase C regulate cardiac Ca^{2+} channels. *Circ. Res.* 80:720–29

37. Ron D, Luo J, Mochly-Rosen D. 1995. C2 region-derived peptides inhibit translocation and function of beta protein kinase C in vivo. *J. Biol. Chem.* 270:24180–87

38. Hu K, Mochly-Rosen D, Boutjdir M. 2000. Evidence for functional role of epsilonPKC isozyme in the regulation of cardiac Ca(2+) channels. *Am. J. Physiol. Heart Circ. Physiol.* 279:H2658–H64

39. Csukai M, Chen CH, De Matteis MA, Mochly-Rosen D. 1997. The coatomer protein beta'-COP, a selective binding protein (RACK) for protein kinase C epsilon. *J. Biol. Chem.* 272:29200–6

40. McHugh D, Sharp EM, Scheuer T, Catterall WA. 2000. Inhibition of cardiac L-type calcium channels by protein kinase

C phosphorylation of two sites in the N-terminal domain. *Proc. Natl. Acad. Sci. USA* 97:12334–38

41. Lacerda AE, Rampe D, Brown AM. 1988. Effects of protein kinase C activators on cardiac Ca^{2+} channels. *Nature* 335:249–51

42. Hool LC. 2000. Hypoxia increases the sensitivity of the L-type Ca(2+) current to beta-adrenergic receptor stimulation via a C2 region-containing protein kinase C isoform. *Circ. Res.* 87:1164–71

43. Yuan SH, Sunahara FA, Sen AK. 1987. Tumor-promoting phorbol esters inhibit cardiac functions and induce redistribution of protein kinase C in perfused beating rat heart. *Circ. Res.* 61:372–78

44. Johnson JA, Gray MO, Chen CH, Mochly-Rosen D. 1996. A protein kinase C translocation inhibitor as an isozyme-selective antagonist of cardiac function. *J. Biol. Chem.* 271:24962–66

45. Csukai M, Mochly-Rosen D. 1999. Pharmacologic modulation of protein kinase C isozymes: the role of RACKs and subcellular localisation. *Pharmacol. Res.* 39:253–59

46. Ytrehus K, Liu Y, Downey JM. 1994. Preconditioning protects ischemic rabbit heart by protein kinase C activation. *Am. J. Physiol. Heart Circ. Physiol.* 266:H1145–H52

47. Mitchell MB, Meng X, Ao L, Brown JM, Harken AH, et al. 1995. Preconditioning of isolated rat heart is mediated by protein kinase C. *Circ. Res.* 76:73–81

48. Brooks G, Hearse DJ. 1996. Role of protein kinase C in ischemic preconditioning: player or spectator? *Circ. Res.* 79:627–30

49. Gray MO, Karliner JS, Mochly-Rosen D. 1997. A selective epsilon-protein kinase C antagonist inhibits protection of cardiac myocytes from hypoxia-induced cell death. *J. Biol. Chem.* 272:30945–51

50. Liu GS, Cohen MV, Mochly-Rosen D, Downey JM. 1999. Protein kinase C-epsilon is responsible for the protection of preconditioning in rabbit cardiomyocytes. *J. Mol. Cell Cardiol.* 31:1937–48

51. Ping P, Zhang J, Qiu Y, Tang XL, Manchikalapudi S, et al. 1997. Ischemic preconditioning induces selective translocation of protein kinase C isoforms epsilon and eta in the heart of conscious rabbits without subcellular redistribution of total protein kinase C activity. *Circ. Res.* 81:404–14

52. Qiu Y, Ping P, Tang XL, Manchikalapudi S, Rizvi A, et al. 1998. Direct evidence that protein kinase C plays an essential role in the development of late preconditioning against myocardial stunning in conscious rabbits and that epsilon is the isoform involved. *J. Clin. Invest.* 101:2182–98

53. Zhao J, Renner O, Wightman L, Sugden PH, Stewart L, et al. 1998. The expression of constitutively active isotypes of protein kinase C to investigate preconditioning. *J. Biol. Chem.* 273:23072–79

54. Leitges M, Schmedt C, Guinamard R, Davoust J, Schaal S, et al. 1996. Immunodeficiency in protein kinase c beta-deficient mice. *Science* 273:788–91

55. Khasar SG, Lin YH, Martin A, Dadgar J, McMahon T, et al. 1999. A novel nociceptor signaling pathway revealed in protein kinase C epsilon mutant mice. *Neuron* 24:253–60

56. Malmberg AB, Chen C, Tonegawa S, Basbaum AI. 1997. Preserved acute pain and reduced neuropathic pain in mice lacking PKCgamma. *Science* 278:279–83

57. Weeber EJ, Atkins CM, Selcher JC, Varga AW, Mirnikjoo B, et al. 2000. A role for the beta isoform of protein kinase C in fear conditioning. *J. Neurosci.* 20:5906–14

58. Abeliovich A, Paylor R, Chen C, Kim JJ, Wehner JM, et al. 1993. PKC gamma mutant mice exhibit mild deficits in spatial and contextual learning. *Cell* 75:1263–71

59. Abeliovich A, Chen C, Goda Y, Silva AJ, Stevens CF, et al. 1993. Modified hippocampal long-term potentiation in PKC gamma-mutant mice. *Cell* 75:1253–62

60. Wakasaki H, Koya D, Schoen FJ, Jirousek MR, Ways DK, et al. 1997. Targeted overexpression of protein kinase C beta2

isoform in myocardium causes cardiomyopathy. *Proc. Natl. Acad. Sci. USA* 94:
9320–25

61. Bowman JC, Steinberg SF, Jiang T, Gee-
nen DL, Fishman GI, et al. 1997. Expression of protein kinase C beta in the
heart causes hypertrophy in adult mice and
sudden death in neonates. *J. Clin. Invest*
100:2189–95

62. Takeishi Y, Ping P, Bolli R, Kirkpatrick
DL, Hoit BD, et al. 2000. Transgenic overexpression of constitutively active protein
kinase C epsilon causes concentric cardiac
hypertrophy. *Circ. Res.* 86:1218–23

63. Pass JM, Zheng Y, Wead WB, Zhang J,
Li RC, et al. 2001. PKC epsilon activation
induces dichotomous cardiac phenotypes
and modulates PKC epsilon-RACK interactions and RACK expression. *Am. J. Physiol. Heart Circ. Physiol.* 280:H946–H55

64. Mochly-Rosen D, Wu G, Hahn H, Osin-
ska H, Liron T, et al. 2000. Cardiotrophic
effects of protein kinase C epsilon: analysis by in vivo modulation of PKC epsilon
translocation. *Circ. Res.* 86:1173–79

65. Wu G, Toyokawa T, Hahn H, Dorn GW.
2000. Epsilon protein kinase C in pathological myocardial hypertrophy. Analysis by
combined transgenic expression of translocation modifiers and G alpha q. *J. Biol.
Chem.* 275:29927–30

66. D'Angelo DD, Sakata Y, Lorenz JN,
Boivin GP, Walsh RA, et al. 1997. Transgenic G alpha q overexpression induces
cardiac contractile failure in mice. *Proc.
Natl. Acad. Sci. USA* 94:8121–26

67. Adams JW, Sakata Y, Davis MG, Sah
VP, Wang Y, et al. 1998. Enhanced G alpha q signaling: a common pathway mediates cardiac hypertrophy and apoptotic
heart failure. *Proc. Natl. Acad. Sci. USA*
95:10140–45

68. Sakata Y, Hoit BD, Liggett SB, Walsh
RA, Dorn GW. 1998. Decompensation of
pressure-overload hypertrophy in G alpha
q-overexpressing mice. *Circulation* 97:
1488–95

69. Rosenmund C, Carr DW, Bergeson SE,

Nilaver G, Scott JD, et al. 1994. Anchoring
of protein kinase A is required for modulation of AMPA/kainate receptors on hippocampal neurons. *Nature* 368:853–56

70. Gerhardstein BL, Puri TS, Chien AJ,
Hosey MM. 1999. Identification of the sites
phosphorylated by cyclic AMP-dependent
protein kinase on the beta 2 subunit of L-
type voltage-dependent calcium channels.
Biochemistry 38:10361–70

71. Gao T, Yatani A, Dell'Acqua ML, Sako
H, Green SA, et al. 1997. cAMP-dependent
regulation of cardiac L-type Ca^{2+} channels
requires membrane targeting of PKA and
phosphorylation of channel subunits. *Neuron* 19:185–96

72. Johnson BD, Scheuer T, Catterall WA.
1994. Voltage-dependent potentiation of L-
type Ca^{2+} channels in skeletal muscle cells
requires anchored cAMP-dependent protein kinase. *Proc. Natl. Acad. Sci. USA*
91:11492–96

73. Coghlan VM, Perrino BA, Howard M,
Langeberg LK, Hicks JB, et al. 1995. Association of protein kinase A and protein
phosphatase 2B with a common anchoring
protein. *Science* 267:108–11

74. Rodman J, Wandinger-Ness A. 2000. Rab
GTPases coordinate endocytosis. *J. Cell
Sci.* 113:183–92

75. Lazar T, Gotte M, Gallwitz D. 1997.
Vesicular transport: how many Ypt/Rab-
GTPases make a eukaryotic cell? *Trends
Biochem. Sci.* 22:468–72

76. Schmitt HD, Wagner P, Pfaff E, Gallwitz
D. 1986. The ras-related YPT1 gene product in yeast: a GTP-binding protein that
might be involved in microtubule organization. *Cell* 47:401–12

77. Gallwitz D, Donath C, Sander C. 1983. A
yeast gene encoding a protein homologous
to the human c-has/bas proto-oncogene
product. *Nature* 306:704–7

78. Stenmark H, Vitale G, Ullrich O, Zerial
M. 1995. Rabaptin-5 is a direct effector of
the small GTPase Rab5 in endocytic membrane fusion. *Cell* 83:423–32

79. Shirataki H, Kaibuchi K, Yamaguchi T,

Wada K, Horiuchi H, et al. 1992. A possible target protein for smg-25A/rab3A small GTP-binding protein. *J. Biol. Chem.* 267:10946–49

80. Allan BB, Moyer BD, Balch WE. 2000. Rab1 recruitment of p115 into a *cis*-SNARE complex: programming budding COPII vesicles for fusion. *Science* 289:444–48

81. Janoueix-Lerosey I, Goud B. 2000. Use of the two-hybrid system to identify Rab-interacting proteins. *Methods* 20:399–402

82. Wagner P, Molenaar CM, Rauh AJ, Brokel R, Schmitt HD, et al. 1987. Biochemical properties of the ras-related YPT protein in yeast: a mutational analysis. *EMBO J.* 6:2373–79

83. Walworth NC, Goud B, Kabcenell AK, Novick PJ. 1989. Mutational analysis of SEC4 suggests a cyclical mechanism for the regulation of vesicular traffic. *EMBO J.* 8:1685–93

84. Gorvel JP, Chavrier P, Zerial M, Gruenberg J. 1991. rab5 controls early endosome fusion in vitro. *Cell* 64:915–25

85. Bucci C, Parton RG, Mather IH, Stunnenberg H, Simons K, et al. 1992. The small GTPase rab5 functions as a regulatory factor in the early endocytic pathway. *Cell* 70:715–28

86. Tisdale EJ, Bourne JR, Khosravi-Far R, Der CJ, Balch WE. 1992. GTP-binding mutants of rab1 and rab2 are potent inhibitors of vesicular transport from the endoplasmic reticulum to the Golgi complex. *J. Cell Biol.* 119:749–61

87. Brondyk WH, McKiernan CJ, Burstein ES, Macara IG. 1993. Mutants of Rab3A analogous to oncogenic Ras mutants. Sensitivity to Rab3A-GTPase activating protein and Rab3A-guanine nucleotide releasing factor. *J. Biol. Chem.* 268:9410–15

88. Haubruck H, Disela C, Wagner P, Gallwitz D. 1987. The ras-related ypt protein is an ubiquitous eukaryotic protein: isolation and sequence analysis of mouse cDNA clones highly homologous to the yeast YPT1 gene. *EMBO J.* 6:4049–53

89. Haubruck H, Prange R, Vorgias C, Gallwitz D. 1989. The ras-related mouse ypt1 protein can functionally replace the YPT1 gene product in yeast. *EMBO J.* 8:1427–32

90. Plutner H, Cox AD, Pind S, Khosravi-Far R, Bourne JR, et al. 1991. Rab1b regulates vesicular transport between the endoplasmic reticulum and successive Golgi compartments. *J. Cell Biol.* 115:31–43

91. Nuoffer C, Davidson HW, Matteson J, Meinkoth J, Balch WE. 1994. A GDP-bound of rab1 inhibits protein export from the endoplasmic reticulum and transport between Golgi compartments. *J. Cell Biol.* 125:225–37

92. Martinez O, Schmidt A, Salamero J, Hoflack B, Roa M, et al. 1994. The small GTP-binding protein rab6 functions in intra-Golgi transport. *J. Cell Biol.* 127:1575–88

93. Martinez O, Antony C, Pehau-Arnaudet G, Berger EG, Salamero J, et al. 1997. GTP-bound forms of rab6 induce the redistribution of Golgi proteins into the endoplasmic reticulum. *Proc. Natl. Acad. Sci. USA* 94:1828–33

94. vander Slvijs P, Hull M, Webster P, Male P, Goud B, et al. 1992. The small GTP-binding protein rab4 controls an early sorting event on the endocytic pathway. *Cell* 70:729–40

95. Daro E, vander Slvijs P, Galli T, Mellman I. 1996. Rab4 and cellubrevin define different early endosome populations on the pathway of transferrin receptor recycling. *Proc. Natl. Acad. Sci. USA* 93:9559–64

96. Mohrmann K, van der SP. 1999. Regulation of membrane transport through the endocytic pathway by rabGTPases. *Mol. Membr. Biol.* 16:81–87

97. Stenmark H, Parton RG, Steele-Mortimer O, Lutcke A, Gruenberg J, et al. 1994. Inhibition of rab5 GTPase activity stimulates membrane fusion in endocytosis. *EMBO J.* 13:1287–96

98. Seachrist JL, Anborgh PH, Ferguson SS. 2000. beta 2-adrenergic receptor internalization, endosomal sorting, and plasma

membrane recycling are regulated by rab GTPases. *J. Biol. Chem.* 275:27221–28

99. Kornfeld S. 1986. Trafficking of lysosomal enzymes in normal and disease states. *J. Clin. Invest* 77:1–6

100. Faigle W, Raposo G, Tenza D, Pinet V, Vogt AB, et al. 1998. Deficient peptide loading and MHC class II endosomal sorting in a human genetic immunodeficiency disease: the Chediak-Higashi syndrome. *J. Cell Biol.* 141:1121–34

101. Nichols WC, Seligsohn U, Zivelin A, Terry VH, Hertel CE, et al. 1998. Mutations in the ER-Golgi intermediate compartment protein ERGIC-53 cause combined deficiency of coagulation factors V and VIII. *Cell* 93:61–70

102. Zhang X, Jefferson AB, Auethavekiat V, Majerus PW. 1995. The protein deficient in Lowe syndrome is a phosphatidylinositol-4,5-bisphosphate 5-phosphatase. *Proc. Natl. Acad. Sci. USA* 92: 4853–56

103. Seabra MC, Brown MS, Goldstein JL. 1993. Retinal degeneration in choroideremia: deficiency of rab geranylgeranyl transferase. *Science* 259:377–81

104. D'Adamo P, Menegon A, Lo NC, Grasso M, Gulisano M, et al. 1998. Mutations in GDI1 are responsible for X-linked nonspecific mental retardation. *Nat. Genet.* 19:134–39

105. Stebbins L, Mochly-Rosen D. 2001. Binding specificity for RACK1 resides in the V5 region of βII protein kinsase C. *J. Biol. Chem.* In press

Annu. Rev. Physiol. 2002. 64:431–75

CARDIAC ION CHANNELS

Dan M. Roden,[1,2] Jeffrey R. Balser,[2,3] Alfred L. George Jr.,[1,2] and Mark E. Anderson[1,2]

Departments of [1]Medicine, [2]Pharmacology, and [3]Anesthesiology, Vanderbilt University School of Medicine, Nashville, Tennessee 37232; e-mail: dan.roden@mcmail.vanderbilt.edu; jeff.balser@mcmail.vanderbilt.edu; al.george@mcmail.vanderbilt.edu; mark.anderson@mcmail.vanderbilt.edu

Key Words   heart, arrhythmias, ion channel structure, ion channel function

■ **Abstract**   The normal electrophysiologic behavior of the heart is determined by ordered propagation of excitatory stimuli that result in rapid depolarization and slow repolarization, thereby generating action potentials in individual myocytes. Abnormalities of impulse generation, propagation, or the duration and configuration of individual cardiac action potentials form the basis of disorders of cardiac rhythm, a continuing major public health problem for which available drugs are incompletetly effective and often dangerous. The integrated activity of specific ionic currents generates action potentials, and the genes whose expression results in the molecular components underlying individual ion currents in heart have been cloned. This review discusses these new tools and how their application to the problem of arrhythmias is generating new mechanistic insights to identify patients at risk for this condition and developing improved antiarrhythmic therapies.

INTRODUCTION

Abnormalities of cardiac rhythm are a major public health burden: 300,000 Americans suffer sudden death due to arrhythmias each year (1), and many more require therapy for other symptomatic arrhythmias. The normal electrophysiologic behavior of the heart is determined by ordered propagation of excitatory stimuli resulting in rapid depolarization and slow repolarization, generating action potentials in individual myocytes. At the most generic level, abnormalities of impulse generation, propagation, or the duration and configuration of individual cardiac action potentials form the basis of disordered cardiac rhythm. These concepts evolved during the twentieth century from clinical descriptions of arrhythmias, to descriptions of action potentials in specific regions of cardiac tissue, and then to identification of specific whole-cell and single-channel ionic currents whose integrated activity generates action potentials. In the past decade, cloning efforts have defined genes whose expression generates specific molecular components, including pore-forming ion channel proteins, underlying individual ion currents in

0066-4278/02/0315-0431$14.00

cardiac myocytes. Thus, new tools are becoming available that will help explain the molecular basis of arrhythmias and, hence, develop improved therapies.

One common theme is that although expression of a single gene, encoding a pore-forming α subunit, is often sufficient to generate an ion current, recapitulation of all the physiologic features of a current in myocytes frequently requires function-modifying accessory (often termed β) subunits (Table 1). Extending this story further, it is now apparent that the generation of ion currents in cardiac myocytes requires coordinated function of not only of α and β subunits, but also multiple other gene products that determine such functions as trafficking, phosphorylation and dephosphorylation, posttranslational modifications, assembly, and targeting and anchoring to specific subcellular domains. The mechanisms underlying these very important aspects of channel physiology are only now being delineated; moving information about individual gene products to a molecular view of cardiac physiology that incorporates multiple cell types, the extracellular milieu, and cell-cell communication continues to be a major experimental challenge.

TABLE 1   Genes encoding cardiac ion channel α and β subunits[a]

| Current | α Subunits | | β Subunits | |
	Gene	Human chromosomal location	Gene	Human chromosomal location
Inward currents				
I_{Na}	SCN5A	3p21	β_1 (*SCN1B*)	19q13.1-q13.2
			β_2 (*SCN2B*)	11q23
I_{Ca-L}	α_1C (*CACNL1A1*)	12pter-p13.2	β_1 (*CACNB1*)	17q21-q22
			β_2 (*CACNB2*)	10p12
			$\alpha_2\delta$ (*CACNA2D1*)	7q21-22
I_{Ca-T}	α_1H (*CACNA1H*)	16p13.3		
Outward currents				
I_{Ks}	*KvLQT1* (*KCNQ1*)	11p15.5	*minK/IsK* (*KCNE1*)	21q22.12
I_{Kr}	*HERG* (*KCNH2*)	7q36-q36	*minK/IsK* (*KCNE1*)	21q22.12
			MiRP1 (*KCNE2*)	21q22.12
I_{Kur}	*Kv1.5* (*KCNA5*)		*Kvβ1* (*KCNAB1*)	3q26.1
			Kvβ2 (*KCNAB2 ??*)	1p36.3
I_{K1}	*Kir2.1* (*KCNJ2*)	17q		
	Kir2.2 (KCNJ12)	17p11.1		
I_{K-Ach}	*GIRK1, Kir3.1* (*KCNJ3*)	2q24.1		
	$^+$ *GIRK4, Kir3.4* (*KCNJ5*)	11q24		
I_{K-ATP}	*Kir6.2, BIR* (*KCNJ11*)	11p15.1	*SUR2* (*ABCC9*)	12p12.1
I_{TO}	*Kv4.3* (*KCND3*)	1p13.2		
	Kv1.4 (*KCNA4*)	11p14		
I_f, I_h	*BCNG2, HCN2*	19p13.3		
(pacemaker current)	HCN4	15q24-q25		
I_{Kp}	*TWIK1*(*KCNK1*)	1q42-q43		
	CFTR (*ABCC7*)	7q31.2		
	KvLQT1 (*KCNQ1*)	11p15.5	*MiRP1* (*KCNE2*)	21q22.12

[a]Common names (usually first assigned) are given first, with the currently designated gene name in parentheses. Information obtained from NCBI website (http://www2.ncbi.nlm.nih.gov/), including LocusLink and OMIM, as well the Human Genome Nomenclature database (http://www.gene.ucl.ac.uk/nomenclature/).

ION CURRENTS AND THE CARDIAC ACTION POTENTIAL

In a prototypical fast response cell (i.e., from atrium, ventricle, or the Purkinje system), the membrane is highly permeable to K^+, as demonstrated by the fact that the reversal potential for K^+ is very close to the resting membrane potential, i.e., there is no substantial electrochemical gradient for K^+ to enter or exit cells. This permeability reflects the fact that inward rectifier K^+ channels in the membrane are open at rest. By contrast, the resting membrane is Na^+ impermeant despite the large electrochemical gradient favoring Na^+ entry; this reflects the fact that in resting cells, cardiac Na^+ channels, which provide the route for Na^+ to enter cells, are closed. A change in the potential across the cell (due to a propagating impulse or an experimentalist's stimulus) is sensed by the Na^+ channel protein, which alters its conformation to open, allowing a large, rapid Na^+ flux, producing the typical rapid phase 0 depolarization (Figure 1). In some cells, a rapid phase 1 repolarization then ensues, because of outward movement of K^+ via transient outward channels. During phase 0 and phase 1, Ca^{2+} channels open. Phase 2, the characteristically long (hundreds of milliseconds) plateau phase of the action potential, reflects a balance between inward current, largely through L-type Ca^{2+} channels, and outward current, largely through delayed rectifier K^+ channels. The net outward current during phase 3 repolarization is provided by delayed rectifier K^+ channels, along with inactivation of Ca^{2+} channels. Final repolarization is accomplished by outward movement of K^+ through inward rectifier channels. Slow response cells, those in the sinus node and in the atrioventricular node, demonstrate slow depolarization during phase 4, a manifestation of pacemaker channel activity. Furthermore, a rapid phase 1 upstroke is absent, and initial depolarization is accomplished by opening of L-type (and perhaps T-type) Ca^{2+} channels. Other electrogenic behaviors (due to exchangers and pumps) are readily demonstrated in cardiac tissue and are crucial in maintaining intracellular ionic homeostasis in the face of large ion fluxes accompanying each action potential.

Action potential configuration and durations vary in specific regions (e.g., atrium versus ventricle) as well as in specific areas within those regions. Epicardial cells in the ventricle demonstrate a prominent phase 1 notch, which is much less prominent in the endocardium (2). Purkinje and midmyocardial cells display a phase 1 notch and action potentials that are much longer than those in epicardium. Such physiologic heterogeneities likely reflect variations in expression or function of the repertoire of ion channels and other proteins that constitute cardiac ion currents. Exaggeration of these heterogeneities, by changes in rate, ion channel mutations, or drug exposures, promote reentrant excitation, a common mechanism for many cardiac arrhythmias. The acute electrophysiologic response of a myocyte to exogenous stressors such as drugs, myocardial ischemia, or autonomic activation likely reflects changes in function of individual ion channels, including channels activated by specific stimuli such as ATP depletion, muscarinic stimulation, or stretch (3). More chronic responses to such exogenous stressors may also include changes in gene expression. Collectively, such changes in

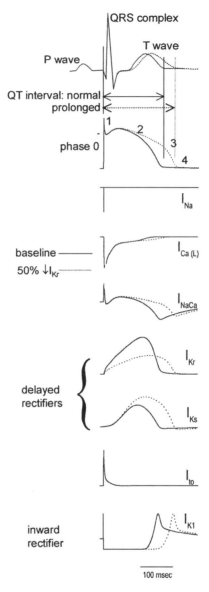

Figure 1   Relationship between the electrocardiogram (*top*), the action potential in the ventricle (*second panel*), and individual ion currents computed using the Luo-Rudy formalism (171). The amplitudes of the currents are not on the same scales. The solid lines represent the baseline; the dotted lines the computation when I_{Kr} is reduced by 50%. Note that this change not only prolongs the action potential duration (APD) (as expected), it also generates changes in the time course of I_{Ca-L}, I_{Ks}, and the sodium-calcium exchange current, each one of which thus also modulates the effect of reduced I_{Kr} on the APD (figure and computation courtesy of P. Viswanathan, Vanderbilt University).

response to pathophysiologic stimuli have been termed electrophysiologic remodeling (4, 5).

CLONING CARDIAC ION CHANNEL GENES

The first Na^+ channel was cloned from electric eel (6), and the predicted structure of the encoded protein includes four roughly homologous domains (I–IV), each containing six membrane-spanning segments, thought to be organized as α helices (Figure 2). This architecture is preserved across other voltage-gated Na^+ and Ca^{2+} channels. The first K^+ channel was cloned from the *Drosophila* mutant

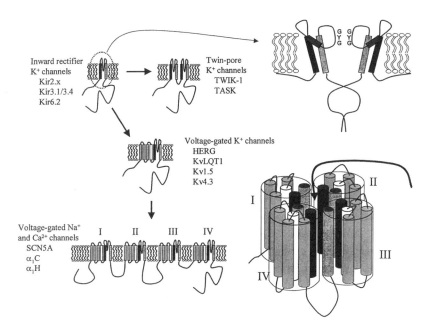

Figure 2   An evolutionary view of ion channel phylogeny. The most primitive ion channel structure is represented by the inward rectifiers, which include a pore region between the two membrane-spanning segments. Although the structure is drawn as a series of membrane-spanning segments perpendicular to the membrane, structural studies indicate considerable detail (suggested at *top right*) that is important for function; this is discussed further by Bezanilla & Perozo (278). Gene duplication of the inward rectifier structure likely led to the twin-pore channels, whereas addition of the S1–S4 segments likely generated voltage-gated K^+ channels, from which gene reduplication produced Ca^{2+} and then Na^+ channels. Voltage-gated Na^+, Ca^{2+}, and K^+ channels form structures with fourfold symmetry to generate ion-permeant pores (*heavy arrow*), as indicated at the lower right (the pore loops are omitted for clarity). Members of each family expressed in heart are listed.

Shaker (7). The molecular architecture of *Shaker* K^+ channels is homologous to a single domain of Na^+ channels, and so it was immediately hypothesized [and subsequently confirmed (8)] that *Shaker* and other voltage-gated K^+ channels assemble as tetramers to generate the pore-forming structure. Papazian (8a) discusses specific domains and mechanisms underlying K^+ channel tetramerization. The first cardiac channels were identified by screening cardiac cDNA with Na^+ (9), Ca^{2+} (10), and *Shaker*-derived probes (11). Cloning from noncardiac cDNA libraries identified the β subunit *minK* (12), members of the inward rectifier family (13), and pacemaker channels (14). The increasing availability of large databases of genomic and expressed sequence tags (EST) sequences has allowed identification of probes with which to screen cDNA libraries for channels, or indeed *in silico* cloning of entire α or β subunits. This approach as been successful in identification of genes encoding pacemaker channels (15, 16), T-type Ca^{2+} channels (17, 18), minK-related proteins (19), and twin-pore K^+ channels (20, 21).

Human molecular genetics has been crucial to identifying genes encoding key cardiac ion channels. Candidate gene (22–24), positional candidate gene (25, 26), and positional cloning (27) approaches have been used to identify disease genes in the long QT (LQTS) and other congenital arrhythmia syndromes (Table 2). Investigation of mechanisms in these diseases has provided important data on the relationship between individual ion channel proteins and integrated normal and abnormal cardiac electrophysiology. This information not only is useful in relatively uncommon individuals with the congenital syndromes, increasingly it can be viewed as a new starting point for developing therapies for common acquired arrhythmias. Recent comprehensive reviews of this area (28, 29) have emphasized the following:

1. Mutations in multiple genes can cause a range of congenital arrhythmia syndromes (30). To date, coding-region missense, deletions, and insertions have been described; these result in alterations of one or more amino acids or in prematurely truncated proteins.

2. These mutations produce diverse functional defects in vitro.

3. Multiple mechanisms (e.g., defective gating or trafficking) underlying these functional defects have been identified (31, 32).

4. Clinical features [specific electrocardiogram (ECG) characteristics, response to drug therapies] vary by genotype and perhaps by specific mutation (33–35).

5. Although defects in the function of a single ion channel protein can be sufficient to explain such specific ECG changes as QT interval prolongation, further mechanisms must generally be invoked to explain why these changes produce arrhythmias. The action potential represents the net effect of multiple time- and voltage-dependent events; thus, one common arrhythmogenic mechanism is that lesions altering one component of such a complex system result in altered behavior of other genetically normal components (Figure 1),

TABLE 2   Monogenic cardiac arrhythmia syndromes[a]

Syndrome	Disease gene	Inheritance	Functional consequences of mutations	Key clinical features
Long QT[b,c]				
LQT1	KvLQT1 (KCNQ1)	Autosomal dominant	$\downarrow I_{Ks}$	TdP with exertion or emotional stress; often β-blocker responsive
LQT2	HERG (KCNH2)	Autosomal dominant	$\downarrow I_{Kr}$	TdP with stress, at rest, or with abrupt auditory stimulus
LQT3	SCN5A	Autosomal dominant	↑ plateau I_{Na}	Syncope/death at rest/sleep
LQT4	Unknown	Autosomal dominant	—	Bradycardia, atrial fibrillation
LQT5	MinK/IsK (KCNE1)	Autosomal dominant	$\downarrow I_{Ks}$	
LQT6	MiRP1 (KCNE2)	Autosomal dominant	$\downarrow I_{Kr}$	
JLN1	KvLQT1 (KCNQ1)	Autosomal recessive	$\downarrow\downarrow I_{Ks}$	Congenital deafness, severe arrhythmia symptoms in childhood
JLN2	MinK/IsK (KCNE1)	Autosomal recessive	$\downarrow\downarrow I_{Ks}$	Congenital deafness, severe arrhythmia symptoms in childhood
Idiopathic ventricular fibrillation				
Brugada[b]	SCN5A	Autosomal dominant	$\downarrow I_{Na}$	Normal QT; right precordial ECG abnormalities; sudden death due to VF
Catecholamine-triggered VT with short QT	Cardiac ryanodine release channel (RYR2)	Autosomal dominant	Unknown; altered (↑) $[Ca^{2+}]_i$ likely	Normal/short QT; bidirectional VT, polymorphic VT VF with stress
Polymorphic VT with short QT (275)	Unknown	Autosomal recessive		
Familial conduction system disease	SCN5A	Autosomal dominant	Multiple Na^+ channel gating defects (see text)	Atrioventricular block, often at a young age
Andersen's	Kir2.1 (KCNJ2)			Bidirectional VT

[a]ECG, electrocardiogram; TdP, Torsades de Pointes; VF, ventricular fibrillation; VT, ventricular tachycardia.

[b]Linkage has been reported to further loci, at which disease genes have not yet been identified (276, 277).

[c]The Romano-Ward syndrome describes the autosomal dominant forms of LQTS (LQT1-6) and Jervell-Lange-Neilsen (JNL) syndrome, the autosomal recessive form with congenital deafness.

which then generate arrhythmias. Examples include action potential prolongation owing to decreased I_K (36) or altered I_{Na} (37), which leads to increased I_{Ca-L}, and then to arrhythmias. $[Ca^{2+}]_i$ often plays a key role in mediating these secondary effects because it is increased by action potential prolongation (38), and elevated $[Ca^{2+}]_i$ in turn directly and indirectly modulates activity of multiple ion currents, as described below.

6. Cases of incomplete penetrance (i.e., mutation carriers lacking an ECG phenotype) have now been recognized (39) and raise the possibility that such

individuals may have enhanced susceptibility to arrhythmias under drug or other stress.

MOLECULAR STRUCTURE-FUNCTION RELATIONSHIPS

Voltage-gated ion channels, the focus of this review, share a number of common structural and functional features that are discussed here, prior to consideration of the key ionic currents in heart. The pioneering experiments of Hodgkin & Huxley identified the concept that membrane currents represent the integrated activity of unitary pore-forming structures, now recognized as ion channels (40, 41). Macroscopic current recordings under voltage clamp have provided the starting point for extraordinarily fruitful hypothesis generation with respect to the underlying function of individual channels. With cloning and mutagenesis of individual ion channel genes, it has been possible to further test these hypotheses. Most recently, the reported structures of the bacterial inward rectifier K^+ channel KcsA (42) and of the eel Na^+ channel (43) provide powerful starting points for inferring molecular structure of cardiac channels.

Permeation

The segment between the fifth and sixth membrane-spanning α helices (S5 and S6) reenters the membrane to form the P-loop, which generates the outer pore. The pore region of virtually all K^+-selective channels includes a signature GXG [GYG, GFG, or GLG (21)] motif that the KcsA structure identifies as a critical element of the selectivity filter, i.e., the region of the pore that allows the channel to discriminate between cations. The KcsA structure also indicates that the four S6 segments form an inverted teepee-like structure that makes up the inner lining of the pore. In Na^+ and Ca^{2+} channels, the selectivity filter is thought to reside in a ring of four amino acids, each contributed by one of the P-loops. In Ca^{2+} channels, this ring is made up of four glutamates (44) whereas the putative selectivity filter in Na^+ channels is generated by four different amino acids (45). Exchange of a glutamate for a lysine in the P-loop of domain III renders Na^+ channels permeant to Ca^{2+} (46). The KcsA structure supports the idea of multiple cations occupying the permeation pathway of an open channel, with the result that repulsive forces would accelerate ion throughput (47), although the extent to which this structure can be applied to other channels (that lack the GYG and other distinctive motifs) has not been established (48).

Voltage Gating

Na^+ and Ca^{2+} channels, transient outward and delayed rectifier K^+ channels, and pacemaker channels are voltage gated, i.e., they display changes in physiologic behavior in response to changes in applied voltage. Such changes (e.g., channel opening) are readily recognized at the macroscopic level and indicate that the channel complex must include region(s) that sense the voltage, i.e., that move in response to a change in applied voltage, thereby changing the conformation, and function, of the ion channel complex. When the eel Na^+ channel was cloned, the fourth membrane-spanning segment (S4) of each six-membrane–spanning segment domain was noted to include a positive charge, arginine or lysine, at every third residue, an ideal arrangement for a voltage sensor. This regularly spaced arrangement of positive charges in S4 is repeated in all other voltage-gated cardiac channels. Cysteine-scanning mutagenesis has indicated that accessibility of the charges in S4 is altered on channel opening or closing (49, 50), reinforcing the role of S4 as the voltage sensor and indicating that it likely moves within a water-filled crevice to accomplish its voltage-sensing function (Figure 3). Whether these channels include circumscribed activation gates (e.g., S6 or the S4–S5 linker) or whether movement of S4 results in a much more complex rearrangement of the entire channel protein complex is not settled.

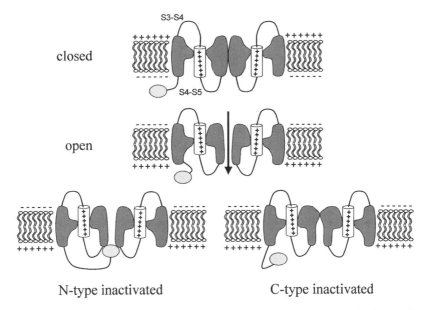

Figure 3 Voltage-gated ion channel state transitions reflect changes in the conformation of the channel protein complex. Channels in the closed state are impermeant. Movement of the voltage sensor (S4) within a water-accessible crevice in the protein complex initiates a protein rearrangement (perhaps involving the S4–S5 linker) to open the channel, i.e., allow it to pass ions along their electrochemical gradients. Two major types of inactivation are shown: N-type (ball and chain), and C-type, involving rearrangements of the pore. Inactivation can occur from closed or open states and can be rapid or slow compared with opening.

Inactivation

Many voltage-gated ion channels display the characteristic of inactivation, entry into a nonconducting state during depolarization. Although the hallmark of inactivation is a slow decline of activated macroscopic current during a square wave depolarization (i.e., inactivation of open channels, as seen with I_{Na}, I_{Ca}, I_{TO}), it is now clear that multiple types of inactivation, with varying time courses, play a role in gating of cardiac ion channels, and that a single channel can exhibit entry into multiple inactivated states. Two major mechanisms of inactivation have been proposed. Elegant experiments using *Shaker* K^+ channels (51, 52) have validated Armstrong's original "ball and chain" model for one type of inactivation, termed N-type inactivation (53). The other type of inactivation, termed C-type (54), appears to involve a rearrangement of residues in or near the pore, analogous to closing the shutter on a camera (Figure 3).

Drug and Toxin Block

Specific blockers are available for most classes of ion channels expressed in heart. Mutagenesis and sidedness experiments have identified two general mechanisms of block. In the first, typified by many toxins, the blocking agent interacts with site(s) on the extracellular face of the channel, thereby disrupting permeation either by physically blocking the pore or by altering protein conformation to prevent channel opening (an allosteric effect). On the other hand, block by available pharmaceutical agents is generally time and voltage dependent, leading to the frequent finding of use-dependent (or frequency-dependent) channel block. One conventional interpretation of use-dependence findings is that drugs bind to a specific receptor site on the channel and that drug accessibility to that site is determined by channel state, i.e., channel protein conformation (55, 56). The S6 segment lining the inner vestibule of the pore is a drug receptor site for many channels, including the Na^+ channel (57), the Ca^{2+} channel (58), and a number of K^+ channels (59, 60). A difficulty with clearly implicating a region such as S6 as the drug receptor is that this and adjacent regions may also play a role in gating of the channel (61). Thus, an alternative mechanism for drug block is that drug binding to the channel protein induces an allosteric change in gating (62).

THE CARDIAC Na^+ CURRENT

The Celera human genome reports 11 separate genes encoding distinct, but highly homologous, Na^+ channel α subunits (63). *SCN5A* (Table 1) encodes the cardiac isoform, a 2016 amino acid polypeptide whose expression in *Xenopus* oocytes or mammalian cells recapitulates cardiac I_{Na} (9). Two ancillary subunits, $\beta1$ and $\beta2$, can be coimmunoprecipitated with brain Na^+ channel α subunits and do appear to enhance surface expression and accelerate activation and inactivation (64). Although the subunits are also expressed in heart, their role in modifying cardiac α subunit function is less clear (65, 66). Most experimentalists coexpress $\beta1$ with wild-type or mutant *SCN5A* in heterologous expression systems because this substantially increases current amplitude.

Myocytes are long and narrow cells, and the cardiac Na^+ channel protein is localized to the intercalated disk at the end of the cell (67). The mechanisms underlying this localization, and its functional significance, have not been elucidated. Sodium channel protein has been detected by in situ hybridization in the limbic region of rat and human brain (68). The role that *SCN5A* might play in an extracardiac locus is unknown. Since the most frequent presentation of patients with mutations in *SCN5A* is abrupt episodes of loss of consciousness, at least some of these patients may have seizure activity rather than (or in addition to) arrhythmias.

Na^+ Channel Structure/Function

Movement of the four S4 regions opens the channel and also initiates the process of fast inactivation; it is now thought that activation and fast inactivation are coupled, rather than sequential (69). Under such a scheme, a channel may actually enter an inactivated state without transitioning through the open state, a process termed closed-state inactivation (70). Mutation of adjacent isoleucine-phenylalanine-methionine residues in the III–IV linker of the rat neuronal channel to glutamines (QQQ) nearly eliminates fast inactivation, i.e., the channel opens and then remains opened during a prolonged depolarization (71). These and other data provide compelling evidence for the role of the III–IV linker in fast inactivation. One view is that the III–IV linker acts as a lid to occlude the permeation pore (an N-type–like mechanism); in this scheme, the site to which the lid docks has not been definitively identified, but it may involve S4–S5 linkers.

Channels that open and then fast inactivate can be reopened following even brief hyperpolarizations that allow the channel to reassume the closed (activate-able) conformation. However, after prolonged depolarizations, recovery does not occur with brief hyperpolarizations, indicating the presence of other (slow) inactivated state(s). Several lines of evidence suggest that this may be analogous to C-type inactivation, involving the pore. In particular, channels engineered to include dual cysteines in the outer pore can be shown to generate disulfide bonds with prolonged, but not brief, depolarizations, and the accessibility of a cysteine placed deep in the pore is reduced during long depolarizations. These results support the idea that the structural rearrangement required to enter intermediate-slow inactivated states involves the outer pore (72). Furthermore, outer pore cysteine accessibility is further reduced by pulse-dependent lidocaine action, implicating C-type inactivation in drug-induced use-dependent block.

The prototypical Na^+ channel toxin tetrodotoxin (TTX) blocks neuronal Na^+ channels in the nanomolar range, but a much higher concentration, 1–10 μM, is required to block the cardiac isoform. Much, but not all, of this difference can be attributed to a single residue, cysteine 373 in the cardiac isoform (and phenylalanine or tyrosine in TTX-sensitive isoforms) in the P-loop/S6 region of domain I (73, 74).

The μ-conotoxins have been used to establish that the arrangement of domains I–IV is clockwise, when viewed from the extracellular face (75). Sea anemone toxins (anthopleurins and ATXII) and scorpion toxins all inhibit Na^+ channel activation (76) and have been used in experimental models to mimic the *SCN5A*-linked form of LQTS (77). One site at which these toxins bind is the S3–S4 extracellular loop of domain IV (78), possibly modifying S4 function.

Regulation of I_{Na}

A silencing element, termed REST, is widely expressed in extraneuronal tissues and is responsible for suppression of neuronal Na^+ channel expression in extraneuronal tissues (79). The molecular basis for cardiac Na^+ channel expression is unexplored. In vivo experiments suggest that chronic therapy with Na^+ channel blockers increases Na^+ channel synthesis (80). Follow-up in vitro studies indicate that elevated $[Ca^{2+}]_i$ may be a common pathway whereby this apparent regulation of Na^+ channel expression occurs (81). On the other hand, in experimental atrial fibrillation (where rapid rates may be accompanied by increased $[Ca^{2+}]_i$), Na^+ currents are generally depressed (82).

The cardiac channel can undergo posttranslational modification, including glycosylation (although not to the same extent as brain or muscle channels) and phosphorylation (83). Both protein kinase (PK)C-mediated and PKA-mediated phosphorylation can alter function (84). PKC effects appear to be mediated by phosphorylation of a serine in the III–IV linker, which is highly conserved among Na^+ channel isoforms (85). Multiple PKA consensus sites are located in the I–II linker, and a role for PKA-mediated phosphorylation of the channel in mediating its trafficking to and from the cell surface has been proposed (86, 87). A sodium channel–specific A kinase anchoring protein (AKAP) has been described in brain but not in heart (88).

Cardiac Na^+ Channel Dysfunction in Disease

CONGENITAL DISEASE   Mutations in *SCN5A* cause the LQT3 variant of the LQTS (26). A common mechanism is failure of fast inactivation, resulting in a population of channels entering a gating mode with recurrent openings throughout the plateau (89, 90). The small net inward pedestal current through these channels is sufficient to upset the balance between inward and outward currents in the plateau and, hence, to prolong action potential duration (APD). The first such defect to be studied was a nine-nucleotide deletion, resulting in deletion of three amino acids (ΔKPQ), in the III–IV linker of the channel. Preliminary reports indicate that knock-in mice engineered to express ΔKPQ do exhibit marked QT prolongation and susceptibility to arrhythmias (91). These findings are consistent with a prominent role of this region in normal fast inactivation. The next two mutations to be described, R1644H and N1325S, are both located in the S4–S5 linkers, further supporting a role for these regions as docking sites for the inactivation lid (92). However, further studies in additional kindreds showed that mutations in other regions of the channel protein also result in defective inactivation (30). This, in turn, reinforces the notion that primary sequence changes in this large channel protein can exert prominent allosteric effects on distant regions of the channel protein. One group has reported that the D1790G mutation produces action potential prolongation not by disrupting fast inactivation but by altering Na^+ channel gating (in the presence of the β_1 subunit) (93). Others, however, dispute this finding and report a persistent plateau Na^+ current as in other mutants (94).

Mutations in *SCN5A* cause two other congenital arrhythmia syndromes (Table 2), idiopathic ventricular fibrillation and conduction block, which unlike LQT3 appear to arise from reduced I_{Na}. Some patients with idiopathic ventricular fibrillation, i.e., sudden death in the absence of any detectable heart disease (and a normal QT interval), do display an unusual electrocardiographic feature at rest, J-point elevation in the right precordial leads. This distinctive electrocardiogram was first popularized by Brugada & Brugada (96) and is termed the Brugada syndrome (95, 96). Some Brugada syndrome mutations result in a highly truncated, and therefore nonfunctional, protein (23). Others augment components of slow inactivation (97, 98); in this situation, Na^+ current is reduced because interdiastolic intervals are insufficiently long to allow complete recovery from the enhanced slow inactivated state. The ECG manifestations of the Brugada syndrome likely reflect marked shortening of epicardial but not endocardial action potentials by loss of Na^+ channel function (99); the rate dependence of recovery from slow inactivation explains why these ECG findings are exaggerated at fast rates. At least one mutation, insertion of a glutamate at position 1795 (1795insD) produces defects in both fast and slow inactivation and results in both the LQT3 and Brugada syndrome phenotypes (98, 100). Similarly, closely adjacent loci in the III–IV linker, and in the domain IV S3–S4 linker, produce LQT3 or the Brugada syndrome (101). Thus, mutations have provided important insights into the function of this large protein and point to the need for further understanding of the structural determinants of gating.

Conduction system disease is the third phenotype associated with Na^+ channel mutations (102). One such mutation, G514C in the I–II linker, has been studied in detail (103). The voltage dependence of activation was shifted positive, and inactivation rate was enhanced; both of these effects would reduce I_{Na} (and therefore might be associated with the Brugada phenotype). However, G514C also shifted steady-state channel availability, consistent with reduced closed-state inactivation. Computational modeling of these complex effects indicated conduction slowing, but without development of action potential heterogeneities associated with LQT3 or the Brugada syndrome.

ACQUIRED DISEASE   Up to the end of the 1980s, Na^+ channel blocking drugs were widely used for the treatment, and indeed were the prophylaxis against, serious arrhythmias. However, in 1989, the Cardiac Arrhythmia Suppression Trial (CAST) investigators reported that treatment with the potent Na^+ channel blocking drugs flecainide and encainide enhanced mortality (likely due to arrhythmias) in patients convalescing from acute myocardial infarction (104). Analysis of the CAST database indicated that the group at highest risk for an increase in drug-associated mortality were those patients at greatest risk for recurrent myocardial ischemia (105), thus implicating an arrhythmogenic interaction between sodium channel block and myocardial ischemia. One possibility is that by slowing conduction, Na^+ channel blocking drugs enhance the likelihood of reentry through a scarred myocardium (106). An alternative explanation comes from experiments in isolated canine ventricular slabs. In this preparation, epicardial cells display a distinctive "spike and dome" action potential configuration due to a prominent I_{TO}, absent in endocardial cells. A number of experimental interventions, including simulated ischemia and Na^+ channel block with flecainide, disrupt the balance between I_{TO} and I_{Na} at the end of phase 1 of the action potential, producing marked abbreviation of epicardial but not endocardial action potentials (107, 108). Thus, these interventions produce marked heterogeneity of APDs, which can result in reentrant excitation. Such epicardial-endocardial gradients may also underlie the distinctive ECG changes present in the Brugada syndrome. Indeed, the observation that flecainide produces similar findings in vitro was one major clue to identification of *SCN5A* as a candidate gene for the Brugada syndrome (23).

One prominent characteristic of ischemically damaged cardiac myocytes is a shift of steady-state Na^+ channel availability to negative potentials (109). This, in turn, reduces Na^+ channel availability at any point during the action potential. Thus, studies of rare congenital syndromes and more common acquired disease converge on loss of Na^+ channel availability, through a variety of mechanisms, as a final common pathway leading to sudden death. Determining the extent to which DNA variants in *SCN5A* or other ion channels may produce little or no baseline clinical phenotype but nevertheless increase an individual patient's risk for sudden death on exposure to a range of stressors, including myocardial ischemia or drugs, is an interesting challenge to contemporary medical genetics.

CARDIAC Ca^{2+} CURRENT

The major Ca^{2+} currents in heart are the L-type (for long lasting) and T-type (for high threshold or tiny). The L-type Ca^{2+} current determines two important and interrelated features of cardiac muscle. First, it contributes inward current to sustain the characteristically long action potentials in heart. Second, the Ca^{2+} it introduces into the cell acts as a trigger to release myofilament-activating Ca^{2+} from sarcoplasmic reticulum (SR) stores, by promoting opening of the SR Ca^{2+} release channel, the ryanodine receptor (RYR). T-type channels are found in the sinus node, atrioventrical (AV) node, and atrium, and in a specialized conducting system, but they are thought to be absent from normal ventricle. T-type current may play a role in determining automaticity.

Multiple calcium channel α subunits have been cloned: $\alpha_1 C$ encodes the L-type α-subunit (10), and $\alpha_1 H$ (or perhaps $\alpha_1 G$) the T-type (18). Two major L-type Ca^{2+} channel ancillary subunits have been identified. β subunits, which may represent multiple gene products (110), are intracellular: α-β interactions increase Ca^{2+} current, accelerate activation and inactivation, and shift activation to more hyperpolarized potentials (110, 111). Some β subunits enhance prepulse facilitation (112), and β subunits may be involved in trafficking α channels. The other major Ca^{2+} channel subunit is a large polypeptide generated by expression of a single gene, with two protein products, α_2 and δ subunits linked by a disulfide bridge (113). The α_2 protein is large and located extracellularly, whereas the δ subunit is small and includes a single membrane-spanning segment and an extracellular domain that links to the α_2 subunit. The $\alpha_2\delta$ subunit may increase cell surface α subunit membrane expression. Function-modifying γ subunits have been described and linked to hereditary epilepsy (114) but have not been identified in heart.

Ca^{2+} Channel Structure/Function

A highly conserved glutamate in each of the four P-loops generates a ring in the pore that confers Ca^{2+} selectivity (44). Multiple Ca^{2+} binding sites may be present within the permeation pathway, with repulsion between adjacent Ca^{2+} ions facilitating high ionic throughput (47). Other cations permeate or block calcium current: L-type Ca^{2+} channels are blocked by cadmium and zinc, and T-type current is more sensitive than L-type to block by Ni^+. L-type channels are more permeable to barium than to Ca^{2+}, and barium currents through L-type Ca^{2+} channels consistently inactivate much more slowly than do Ca^{2+} currents. This finding, as well as experiments with flash release of caged Ca^{2+} (115), implicate Ca^{2+} in mediating L-type Ca^{2+} current inactivation. Initial studies implicated an EF hand-containing domain in the intracellular C terminus of the channel as the Ca^{2+} sensor (116). More recent evidence suggests that this EF hand domain may aid in transmission of signals from two calmodulin-binding domains located nearby on the C terminus (117, 118) and that inactivating Ca^{2+} signals are mediated through calmodulin binding rather than by a direct interaction of Ca^{2+} with the channel protein. Calmodulin may also act as the Ca^{2+} sensor for facilitation of I_{Ca-L}, perhaps by directly binding to the Ca^{2+} channel α subunit C terminus or through activation of the Ca^{2+}/calmodulin–dependent protein kinase CaM kinase II.

Ca^{2+} channels are blocked by three major classes of drugs: dihydropyridines (nifedipine and others), phenylalkylamines (verapamil and others), and benzothiazepines (diltiazem). Multiple binding sites for dihydropyridines have been described, including domains III and IV S6 (119) as well as the domain IV P-loop (120). There are conflicting data with regard to whether the binding site is accessible from the extracellular face (121, 122). Similarly, although the evidence is not conclusive, S6 in domain IV appears important for binding of phenylalkylamines and benzothiazepines to the channel (120, 123). Ca^{2+} channel blockers shift the steady-state inactivation toward more negative voltages.

Regulation of I_{Ca}

Multiple gating modes of the Ca^{2+} channel have been described in single-channel experiments. Sweeps displaying mode 0 gating have very rare brief openings, whereas those showing mode 1 gating show frequent brief openings. Mode 2 is characterized by very long openings and is promoted by dihydropyridine agonists (such as BayK8644) and by PKA- or CaM kinase-mediated phosphorylation (124, 125). Beta-adrenergic stimulation markedly increases Ca^{2+} current through the traditional Gs-related second messenger pathway (126). A direct membrane-delimited stimulatory action of G proteins has also been proposed, but it is controversial (126, 127). Of the multiple consensus PKA sites in the L-type Ca^{2+} channel, S1928 in the C terminus appears to be functionally important (128). Cardiac channel–specific AKAPs appear important for mediating these PKA effects (129, 130). Ca^{2+} channels can be also PKC phosphorylated, but the physiologic significance of this event is not clear.

Thus, increased $[Ca^{2+}]_i$, and in particular the high free Ca^{2+} concentrations achieved in the immediate vicinity of an open Ca^{2+} channel, are likely to have multiple effects on Ca^{2+} channel function, decreasing the current by promoting Ca^{2+}-mediated inactivation and increasing the current through Ca^{2+} stimulation of CaM kinase II (and possibly PKA). In contrast to L-type Ca^{2+} channels, T-type Ca^{2+} channels are not regulated by β-adrenergic signaling. However, their amplitude is increased by α-agonists, extracellular ATP, and endothelin (131, 132).

Chronic stimulation of cAMP-dependent signaling in myocytes increases L-type Ca^{2+} current, and increased transcription of α_1C, β, and $\alpha_2\delta$ subunits has been reported (133, 134). Alternatively spliced forms of both the L- and the T-type Ca^{2+} channel have been described in cardiac preparations, and differences in drug sensitivity and in isoform expression in heart failure have been described (135–139).

Cardiac Ca^{2+} Channel Dysfunction in Disease

I_{Ca-L} and I_{Na} are reduced in atrial myocytes from animals with atrial fibrillation, and it is possible that these changes help perpetuate the arrhythmia (82). A consistent finding in hypertrophy is action potential prolongation (140), and reduction in Ca^{2+} current has been implicated in some studies. On the other hand, L-type Ca^{2+} channels have been implicated as a carrier of arrhythmogenic inward current when action potentials are prolonged. Indeed, Ca^{2+} channel blockers inhibit arrhythmogenic early afterdepolarizations (EADs) under these conditions (141). Prolongation of the action potential also increases the amplitude of the $[Ca^{2+}]_i$ transient, activating CaM kinase (and thus I_{Ca-L} and other arrhythmogenic inward currents), thereby promoting arrhythmogenic EADs and delayed afterdepolarizations (142, 143). CaM kinase inhibition suppresses experimental long QT-related EADs, which suggests a role for this intervention in arrhythmia suppression (144). Although it is tempting, on the basis of these data, to speculate that mutations in L-type Ca^{2+} channel genes will result in congenital arrhythmia syndromes, such cases have not been reported to date. However, mutations have recently been recognized in the cardiac ryanodine release channel RYR2 in patients with idiopathic familial polymorphic ventricular tachycardia, unassociated with QT prolongation (24, 145). The way in which these mutations alter RYR2 function to result in arrhythmias has not yet been established, but presumably elevated $[Ca^{2+}]_i$ may well form part of such a pathway. Whether elevated $[Ca^{2+}]_i$ engages L-type Ca^{2+} current or other arrhythmogenic currents (such as Na^+-Ca^{2+} exchange or the non-selective cationic arrhythmogenic transient inward current I_{TI}) is not established.

The amplitude of T-type currents is increased in animal models of hypertrophy (146). This increase is reminiscent of the electrophysiologic phenotype observed in fetal cells (which exhibit large T-type currents) and may therefore represent reversion to the fetal phenotype, a common finding in hypertrophy.

THE TRANSIENT OUTWARD CURRENT, I_{TO}

I_{TO} is readily identified in most, but not all, human ventricular epicardial and atrial cells and is the major repolarizing current in adult mouse and rat atrium and ventricle, accounting for the extraordinarily short action potentials in these species. One of the first members of the *Shaker* superfamily cloned from human heart, Kv1.4, displays rapid inactivation, superficially typical of I_{TO}, in heterologous expression systems. However, human I_{TO} recovers from inactivation much more rapidly than does Kv1.4-mediated current (147), and Kv1.4 mRNA and protein have been difficult to detect in human and rat heart. By contrast, heterologous expression of Kv4.3 results in a transient outward current with very rapid recovery from inactivation, i.e., a current that would continue to contribute a rapid repolarization "notch" in phase 1, even at rapid rates. Kv4.3 also displays drug sensitivity similar to human I_{TO}, reinforcing the idea that it is a reasonable candidate for human I_{TO}. In rat, I_{TO} likely represents expression of the closely related isoform, Kv4.2, or heteromeric Kv4.2/4.3 assembly. As well, there is a gradient of Kv4.2 expression across the ventricular wall in rat, with greatest transcript abundance in epicardium, corresponding to the larger I_{TO} in epicardium (148); it is not yet known if a similar Kv4.3 gradient is present in humans. Although human endocardium does not generally display a prominent phase 1 notch during stimulation of physiologic rates, I_{TO} with very slow kinetics of recovery from inactivation can be recorded (149); this current would be expected to remain largely inactivated at usual rates. These kinetics, and detection of Kv1.4 in some cardiac preparations (including animals in which Kv4.2 has been suppressed by transgenic expression of a dominant negative subunit), suggests that Kv1.4 may mediate I_{TO} in some regions (e.g., endocardium) of human heart. Kv4-specific function-modifying subunits, termed Kv channel-interacting proteins (KchIPs), have been described and modulate Kv4-mediated current in brain. The extent to which they modulate cardiac I_{TO} is uncertain. KchIPs appear to interact with Kv4 N termini and may act as Ca^{2+} sensors (150). In some species, a calcium-activated I_{TO} (termed I_{TO2}) can be recorded; this may be carried by K^+ or by Cl^- (151, 152).

Regulation of I_{TO}

I_{TO} is blocked by many antiarrhythmic agents, (quinidine, flecainide, propafenone), most of which are viewed as having other primary cardiac ion channel targets

(153, 154). Thus, the extent to which I_{TO}/Kv4 block contributes to the actions of these drugs is uncertain. Kv4.3 undergoes alternative splicing in rats, and evidence has been presented that an alternatively spliced variant is actually "the" cardiac isoform (155). A common finding in human heart failure is action potential prolongation, an effect that has been attributed in part to reduction of I_{TO} (156). A remarkably good correlation has been found between the extent of reduction of I_{TO} and reduction in the abundance of mRNA transcripts encoding Kv4.3 in human heart failure (157). The mechanisms underlying this transcriptional regulation have not been elucidated. In mice expressing a dominant negative Kv4.2 construct in heart, action potentials are markedly prolonged, reflecting the primacy of I_{TO} in rodent heart (158). Electrophysiologic studies in wild-type mice identify up to four components of I_{TO} that vary by region; these patterns are altered in the transgenic animals, reflecting both suppression of I_{TO} and increased expression of Kv1.4 (158–160).

DELAYED RECTIFIER K$^+$ CURRENTS

With a square wave depolarization typically used in a voltage clamp experiment, these currents activate with some delay following initial depolarization. Thus, they generally do not play a major role in determining the balance of inward and outward current during phases 0 and 1, but rather they contribute outward current to later phases of the action potential. Three major delayed rectifier currents (I_{Kr}, I_{Ks}, and I_{Kur}), representing distinct gene products, are present in human heart, distinguishable by their kinetics of activation and deactivation and by their highly variable sensitivity to blocking drugs.

THE RAPID COMPONENT OF DELAYED RECTIFIER CURRENT, I_{Kr}

HERG encodes the α subunit underlying I_{Kr}. *HERG* is expressed most abundantly in heart and to only a minor extent in hippocampus (161), where its function remains obscure. There is some evidence that HERG is associated with subunits that modify its function. Antisense suppression of *minK* expression in mouse atrial tumor cells (AT-1 cells) reduced I_{Kr} amplitude without altering its gating, which suggests a minK-HERG interaction (162). Similarly, coexpression of *minK* with *HERG* in COS cells increased I_{Kr} amplitude compared with *HERG* alone (163). The *minK*-related gene *KCNE2* (also termed *MiRP1*) has been implicated as a HERG interactor. When *MiRP1* and *HERG* are coexpressed, macroscopic HERG current is reduced, reflecting reduction of single-channel amplitude. In addition, both I_{Kr} deactivation and the rate of onset of drug block are accelerated (19). The extent of *KCNE1* expression in human heart and its localization are sources of controversy. Hence, although these interactions, and in particular the effect of *MiRP1* mutations on I_{Kr} pharmacology described further below, are interesting, the extent to which they occur in the human heart remains somewhat uncertain.

HERG Channel Structure Function

HERG sequence is very highly conserved across mammals. In some species, such as guinea pigs, humans, and dogs, I_K is composed of both I_{Ks} and I_{Kr}. In others (cats, rabbits, fetal mice), the current is almost exclusively I_{Kr}. Indeed, it was in rabbit myocytes that the first extensive physiologic characterization of the current was performed, even antedating recognition of separate I_{Ks} and I_{Kr} components in guinea pigs (164). I_{Kr} is the major repolarizing current in fetal mouse and rat heart, but it is rapidly supplanted postnatally by a very large I_{TO}, resulting in the characteristically short action potential in small rodents (165, 166).

Although a cursory inspection of typical HERG currents suggests a gating model with transitions between closed and open states only, I_{Kr} displays a number of unusual gating features that are best explained by prominent and very rapid inactivation during depolarization (Figure 4). The amplitude of activating I_{Kr} decreases markedly as depolarization potential is more positive, i.e., the current displays apparent striking inward rectification. Increasing extracellular K^+ increases the amplitude of outward current (contrary to what is expected by simple Nernstian considerations). Finally, close inspection of deactivating tail currents reveals a rapid hook of outward current preceding relatively slow deactivation (Figure 4).

These observations are all consistent with a model in which transitions between closed and open states are slow and those between open and inactivated states are much more rapid (164, 167). Thus, with maintained depolarization, the channel distributes into open and inactivated states, and at more positive potentials, distribution into the inactivated state is favored, thereby generating the appearance of inward rectification. This is most readily appreciated (Figure 4) if, during maintained depolarization, a cell is briefly hyperpolarized. This allows inactivated channels to move to the open state (a rapid process), but if the hyperpolarization is sufficiently brief, open channels do not move to the closed state. Therefore, on return to depolarized potentials, channels are distributed primarily into open states, and rapid inactivation can be readily appreciated. When this experiment is conducted over a range of potentials, I_{Kr} current-voltage relations are linear, indicating that inward rectification is, indeed, attributable to voltage-dependent fast inactivation (168). Similarly, the hook preceding slow tail-current decay represents rapid recovery of inactivated channels to the open state prior to slower deactivation (transitions from open to closed). The unusual $[K^+]_o$ dependence of activating I_{Kr} is also corrected when inactivation is transiently removed by a brief hyperpolarization, indicating that transitions to the inactivated state are K^+ dependent (168, 169). A second mechanism for the unusual $[K^+]_o$ dependence of I_{Kr} involves $[Na^+]_o$-$[K^+]_o$ interactions in the pore of the channel. When $[K^+]_o$ is removed, $[Na^+]_o$ can be shown to be a potent blocker of outward I_{Kr}, with a 50% inhibitory concentration of 3.1 mM, and $[K^+]_o$ competes with $[Na^+]_o$ to relieve this block (170). Inactivation is removed by mutations within the P-loops, supporting a pore-mediated (non–N-type) inactivation mechanism.

This unusual balance between slow opening and closing and fast inactivation of HERG current position I_{Kr} is an especially important contributor to repolarization

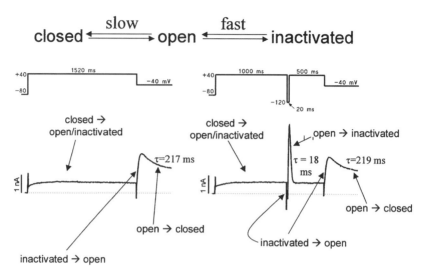

Figure 4 Demonstration of I_{Kr} fast inactivation. A 1520-ms depolarizing voltage clamp step to +40 mV generates a small outward current followed on repolarization to −40 mV by a larger tail current that then decays slowly (*left*). Interruption of the depolarizing pulse by a very brief (20 ms) hyperpolarizing pulse allows inactivated channels to move to the open state but is not long enough for open channels to close. Hence, at the end of the hyperpolarizing pulse, return to +40 mV generates a very large outward current (because so many channels are open), which decays very rapidly (τ 18 ms) as open channels inactivate (tracings courtesy of T. Yang, Vanderbilt University).

in normal heart. This is well demonstrated by experiments in which the time course of HERG current during an action potential is assessed, either in computer models (171) or with action potential clamp of cells expressing *HERG* (172). Despite the name (rapidly activating), the current does not rapidly activate during a depolarizing pulse; indeed, I_{Kr} activation during an action potential can appear slower than that of I_{Ks}, the slow component of I_K (Figure 1). Rather, during the plateau phase of the action potential, little current flows through HERG channels, reflecting extensive inactivation. As the cell begins to repolarize, channels recover from inactivation, producing a large outward current, which further promotes rapid repolarization. Loss of this rapid repolarizing function of I_{Kr}, by mutations or by drug block, therefore, prolongs APD in the plateau range (Figure 1).

HERG current is blocked by many antiarrhythmics and noncardiovascular drugs that share a potential to produce marked QT prolongation and a distinctive ventricular tachycardia, Torsades de Pointes (Figure 5), also characteristic of the congenital LQTS (173). Regulatory agencies have recognized that it may be possible to predict that a given drug will cause Torsades de Pointes on exposure to large numbers of patients, if it is a HERG blocker. However, the extent to which

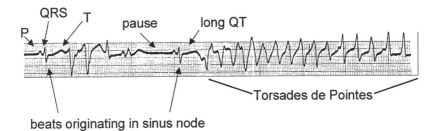

Figure 5 Torsades de Pointes. This is the arrhythmia that occurs in LQTS as well as on challenge with drugs that prolong the QT interval. The changes in cardiac cycle lengths prior to the initiation of the arrhythmia are typical and are consistent with suggested roles for early afterdepolarizations in initiation and heterogeneity of repolarization times in maintenance of the arrhythmia.

such in vitro testing can quantify risk in large numbers of patients exposed, and the way in which such risks should be balanced against any potential benefit of a given therapeutic agent, is very uncertain. Nevertheless, the recognition that HERG is promiscuous in its ability to be blocked by drugs, and that this may have important implications for drug development, has led naturally to the question of structural determinants of HERG block. Drug block of HERG current washes out very slowly, a finding that has been attributed to block at the inner vestibule, followed by channel closing with the drug trapped at that site (174). Block is also enhanced by inactivation through as-yet incompletely understood mechanisms. Modeling based on the KcsA structure reveals two unusual features of the HERG inner vestibule (the site of drug block) that are not predicted to be shared by any other K^+ channel (60). The first of these features is the absence of two highly conserved prolines in S6. The presence of these prolines is predicted to kink the S6 segment, and therefore limit the size of the inner vestibule, in other K^+ channels. In contrast, their absence in HERG enlarges the inner vestibule, allowing relatively bulky drugs to enter the inner vestibule. Second, HERG channels have two aromatic residues predicted to face the inner pore whereas other K^+ channels have none or one (KvLQT1). Alanine mutagenesis of these residues markedly decreases drug block; hence, it is inferred that any one or more of the eight aromatic residues lining the tetrameric pore structure provides the drug-blocking site, likely by bonding to charged residues or aromatics (by π bonding) on the blocking drugs. Whether these evolving structural insights will allow *in silico* screening of drugs to eliminate the possibility of incorporating HERG block is an open question.

Regulation of HERG Channels

The structure of the HERG N terminus has been solved and resembles a PAS domain, a motif that is frequently involved in signal transduction (175). The role of this region, however, in HERG function remains poorly understood. Two

N-terminal splice variants have been described, one of which is not abundantly expressed and the other one of which appears to be cardiac specific (176, 177). Heterologous expression of the latter isoform results in a channel that displays deactivation kinetics that are faster than those of canonical HERG and are thought to resemble those of human I_{Kr}, implicating this isoform, or a heterotetramer that includes the canonical isoform, in the human current. A C-terminal splice variant is expressed more abundantly in human heart than is the canonical variant but does not, itself, generate current (178). The C-terminal variant does, however, confer a dominant negative effect on canonical HERG when they are coexpressed.

Following translation, HERG undergoes extensive glycosylation (in the S1–S2 linker), and antibodies directed against the HERG C terminus have been used to discriminate between core-glycosylated and fully glycosylated forms of the channel (31). This tool was used to identify defective trafficking to the cell surface of some LQTS-associated *HERG* variants, a mechanism that appears to be relatively common in LQT2. Defective trafficking of HERG (and other molecules such as the cystic fibrosis transport regulator) has been corrected in vitro by culturing transfected cells at low temperatures or in the presence of glycerol. In addition, exposure of transfected cells to the I_{Kr} blocker E4031 also corrects the defect, through unknown mechanisms (179).

HERG is a member of a K^+-channel family that includes a cyclic nucleotide-binding domain in its C terminus. Isoproterenol is reported to exert no effect on I_{Kr} in guinea pig cells (180), although a recent report suggests cAMP does modulate the function of heterologously expressed HERG (181).

HERG Channel Function in Disease

Drugs that block HERG current tend to do so with "reverse use-dependent" action potential prolongation (i.e., greatest action potential prolongation at slow rates), and this is one important contributor to Torsades de Pointes occurring with I_{Kr} block or *HERG* mutations (182). Two mechanisms have been proposed. The first arises from the observation that I_{Ks} deactivation is incomplete at rapid rates, and so current may accumulate at fast rates. Under this scenario, I_{Kr} is thus a relatively more important component of repolarizing current at slow rates, and hence I_{Kr} block produces marked prolongation of action potentials, and the QT interval, at slow rates or after a long diastolic interval (183). A second postulated mechanism is that block itself is reverse frequency dependent, an effect that may be promoted by *MiRP1* coexpression (19).

The development of Torsades de Pointes during exposure to an I_{Kr}-blocking drug is infrequent and, although certain clinical risk factors can be identified, remains incompletely predictable. Similarly, patients with LQTS are life-long mutation carriers yet have relatively rare arrhythmia events. Multiple risk factors have been identified, including hypokalemia, slow heart rates, presence of blocking drug, and other demographic features (e.g., female gender, recent conversion from atrial fibrillation, left ventricular hypertrophy). It is postulated that such risk factors come together in an individual patient to culminate in Torsades de Pointes. Thus, patients with LQTS may be viewed as having a defect in repolarization that may be subclinical and reduces what we have termed repolarization reserve (184). Only when repolarization reserve is sufficiently impaired does Torsades de Pointes occur. Reports of patients with LQTS mutations who only present after drug challenge support this concept. These reports implicate two distinct mechanisms: (*a*) increased sensitivity of the channel to drug block (an effect reported with MiRP1 variants) (185); and (*b*) reduction of I_{Ks} or other repolarizing current that is well tolerated until the superposition of I_{Kr} block (186–189).

THE SLOW DELAYED RECTIFIER, I_{Ks}

In guinea pig myocytes, long depolarizing pulses give rise to a slowly activating outward current. Exposure to specific I_{Kr} blockers, such as La^{3+}, E4031, MK499, dofetilide, *d*-sotalol, or almokalant, leaves a large, slowly activating, outwardly rectifying, methanesulfonanilide-resistant component, I_{Ks} (190–192).

Expression of *minK* in *Xenopus* oocytes (but not mammalian cell models) gives rise to a slowly activating, I_{Ks}-like current (12). The mechanism whereby this small protein, predicted to have a single membrane-spanning segment, could give rise to a channel remained uncertain until *KvLQT1*, the disease gene in LQT1, was identified by positional cloning (27). Expression of *KvLQT1* alone gives rise to rapidly activating, slowly deactivating current, which does not have a readily identifiable correlate in cardiac myocytes. Coexpression of *minK* and *KvLQT1*, on the other hand, recapitulates I_{Ks}; the observation of I_{Ks}-like currents in *Xenopus* oocytes reflects low-level constitutive expression of a *Xenopus* homolog of *KvLQT1* (193, 194).

Structure-Function Relationships in the I_{Ks} Channel Complex

When *minK*-mediated I_{Ks} was studied using cysteine-scanning mutagenesis in *Xenopus* oocytes, two adjacent residues in the putative membrane-spanning region, G56 and F57, were shown to be accessible from the outside and inside, respectively (195). These data have been interpreted as showing that minK forms part of the ion-conducting pore in the I_{Ks} channel complex. However, KvLQT1 retains the molecular architecture of a typical voltage-gated K^+ channel; thus, it is difficult to reconcile the idea that minK could be part of the pore itself while maintaining the architecture of the selectivity filter. The accessibility data are also consistent with the idea that minK resides in a crevice that is not part of the pore but nevertheless determines cysteine accessibility, and movement of the minK-KvLQT1 channel complex, with gating (196). In addition to expression in heart, *KvLQT1* and *minK* are expressed in extracardiac tissues, notably inner ear (197, 198).

Regulation of I_{Ks}

The *KvLQT1* gene is very large (>700 kb), and regulatory elements have not been identified (27). An alternatively spliced variant of the channel (isoform 2), with deletion of nearly the whole N terminus, is expressed in heart (199). When isoform 2 is expressed alone, no current is generated, and coexpression of the canonical form and isoform 2 results in dominant negative suppression of I_{KvLQT1}. Some evidence has been presented that the abundance of isoform 2 is greatest in the midmyocardium (200); this is consistent with other studies indicating that I_{Ks} is reduced in this region, a finding that may underlie longer action potentials in this region than in endocardium and epicardium (201).

APD in the ventricle shortens with adrenergic stimulation, even controlling for heart rate. Because Ca^{2+} current increases with adrenergic stimulation (an effect that would prolong APD), a counterbalancing increase in outward current must occur, and two sources of outward current have been suggested: I_{Ks}, and a background cAMP-activated chloride current that can be readily demonstrated in guinea pig myocytes (202) but whose presence in human heart has not been clarified. In nonhuman mammalian studies, I_{Ks} amplitude is increased by adrenergic stimulation as well as by increases in $[Ca^{2+}]_i$ or by PKC stimulation (203). An exception is PKC stimulation of guinea pig I_{Ks}, which reduces the current (204). A consensus PKC phosphorylation site is present in the C terminus of minK in humans (and most other mammalian species) but is absent in guinea pigs. Heterologous expression studies suggest that this sequence variation in *minK* accounts for the difference in sensitivity to PKC stimulation of I_{Ks} among species. One recent report suggested that target residues for the PKA effect reside in the C terminus of KvLQT1, possibly reflecting a physical association between KvLQT1 and elements of the signaling pathway (205). Furthermore, coexpression of an AKAP with *KvLQT1* + *minK* is necessary for the I_{Ks} response to PKA stimulation (206).

I_{Ks}-specific blockers have been developed but not commercialized because of a perceived risk of Torsades de Pointes (207). Chromanol-293B has been reported to be an I_{Ks}-specific blocker, targeting the KvLQT1 protein (208, 209). It has been used in multicellular in vitro experiments to elucidate the whole heart effect of I_{Ks} block (210). In such experiments, I_{Ks} block tends to produce homogeneous action potential prolongation in ventricular tissue; however, with addition of a β-adrenergic agonist such as isoproterenol, marked heterogeneities of APD and arrhythmias emerge. This experimental finding is consistent with the sensitivity of I_{Ks} to adrenergic stimulation and with the finding that arrhythmias in patients with *KvLQT1* or *minK* mutations almost always arise during periods of adrenergic stress (33).

KvLQT1-minK Channel Dysfunction in Disease

Examination of function of LQTS-associated *KvLQT1* mutations has yielded interesting new data on channel function. A 31–amino acid domain in the C terminus of KvLQT1 has been implicated in subunit assembly (211). Examination of currents generated by heterologous expression of KvLQT1 channels with mutations in widely divergent areas of the protein shows currents decreased in amplitude or with altered gating (32). At least one mutation has been identified that results in apparent failure of the mutant channel to traffic to the cell surface (212). When *minK* mutations causing LQTS have been studied in heterologous expression systems, a range of defects has been observed. One mutant (L51H, in the putative membrane-spanning region) did not result in functional channels at the cell surface, a finding that suggests a defect in normal trafficking of channel protein from the endoplasmic reticulum to the cell surface (213). Other mutations did result in I_{Ks}, but with a reduced amplitude (and altered gating) compared with wild type. It is interesting to note that *minK* LQT5 mutations also alter amplitude and/or gating of *HERG*-mediated currents, providing further evidence that minK interacts with other K^+ channel α subunits.

The rare Jervell-Lange-Neilson (JLN) variant of the LQTS (Table 2) associated with congenital deafness arises in children who inherit abnormal *KvLQT1* or *minK* alleles from both parents (214, 215). This situation arises through consanguinity (i.e., the child inherits alleles each encoding the same abnormal protein) or occasionally by chance (in which case the child inherits alleles that encode different abnormal proteins). Thus, both parents are obligate gene carriers for LQT1 or LQT5, except in the unusual instance of a de novo mutation. Although QT intervals are generally normal in these parents, it is now recognized that occasional cases of sudden unexpected death, presumably due to long QT-related arrhythmias, can occur in the parents, an observation that has obvious implications for family screening of JLN patients (215).

Mice with *minK* deletions display a characteristic tail-chasing movement disorder (shaker/waltzer) and are deaf (197). The KvLQT1-minK complex is responsible for endolymph secretion in the inner ear, and its absence in JLN patients, and in *minK* $(-/-)$ mice, results in collapse of the endolymphatic space in the inner ear. Some investigators report QT prolongation at slow rates in *minK* $(-/-)$ mice (216), whereas others find no such effect (217). *MinK* $(-/-)$ mice also exhibit paroxysmal atrial fibrillation (218). This may relate to regional expression of *minK*, determined in *minK* $(-/-)$ mice engineered to incorporate the lacZ reporter in the *minK* locus; in these animals, lacZ expression is confined to the sinus and AV nodes, proximal conducting system, and heterogeneously within the atria (217). Mice with deletion of the *KvLQT1* gene also display shaker/waltzer behavior and are deaf. One group of investigators reports no effect on repolarization in these mice, consistent with the concept that I_{Ks} is not an important modulator of electrophysiology in adult mice (198). Others do report morphologic ECG changes in telemetered unrestrained animals but no changes in the isolated heart (219). Transgenic animals with overexpression of the *KvLQT1* dominant negative isoform 2 display abnormalities of sinus and AV nodal function, implicating *KvLQT1* in normal automaticity (220).

THE ULTRARAPID DELAYED RECTIFIER, I_{Kur}

This current activates even more rapidly than I_{Kr} (i.e., ultrarapidly and hence the name of the current, I_{Kur}) and is present in human and dog atrium and throughout rat and mouse heart. I_{Kur} displays outward rectification and very slow inactivation during strong depolarizations. Heterologous expression of Kv1.5 results in a delayed rectifier-type current with gating kinetics essentially identical to those of I_{Kur} (221, 222). Inactivation is temperature dependent and accelerates at experiments done closer to body temperature. Furthermore, native I_{Kur} and heterologously expressed Kv1.5 are both sensitive to quinidine and to 4-aminopyridine and resistant to dendrotoxin, tetraethylammonium (TEA), and flecainide (154, 223). Antisense inhibition of Kv1.5 in cultured cardiomyocytes reduces I_{Kur} (224). Kv1.5 is not expressed in dog atrium, but another member of the *Shaker* superfamily (Kv3.1) is present and probably underlies I_{Kur} in that species (225). This finding complicates the use of dog models in evaluating I_{Kur}/Kv1.5 blockers, as they are developed for the treatment of atrial fibrillation.

Kv1.5 transcripts are also detected in extracardiac tissues, including pituitary, brain, and pancreas (226–228). Multiple members of a $K_v\beta_1$ subunit family for Kv1.5 have been cloned and multiple N-terminal splice isoforms detected (229, 230). Coexpression of β subunits with Kv1.5 accelerates inactivation (making the channel appear more like I_{TO}) and may alter sensitivity to phosphorylation by PKC or PKA (231, 232).

Structure Function

Drug binding (quinidine has been best studied) likely involves residues in the (intracellular portion) of S6 (59). β subunits appear to be almost exclusively cytosolic and may mediate inactivation through a ball and chain mechanism whereby the β subunit acts as "the ball" (233). The crystal structure of $K_v\beta$ has been solved (in the absence of an α subunit) and displays fourfold symmetry. The structure is very similar to oxidoreductase enzymes; thus, it has been speculated the $K_v\beta$ subunit may act to couple the redox state to channel function (42, 234). The C termini of a range of $K_v\beta$ subunits are highly homologous, and it is in this region that the subunit interacts with the α subunit (235).

I_{Kur} Regulation

In rats, Kv1.5 displays striking regulation by endogenous or exogenous glucocorticoids, increasing with dexamethasone administration by more than 20-fold in ventricle but not in the atrium (236). Thyroid hormone also upregulates Kv1.5 in rats (237). In adult myocytes, Kv1.5 protein (like the sodium channel) is regionally restricted to intercalated disks, a pattern that is absent in neonatal cells and in cells that have undergone ischemic damage (238). The mechanism of this targeting is uncertain; Kv1.5 does include a C-terminal PDZ-binding motif (ETDL). As well, the channel contains SH3 domains that interact with tyrosine kinase, and Kv1.5 is tyrosine phosphorylated in human heart (239). Kv1.5 has been reported to target to lipid rafts, specific membrane regions enriched in certain subtypes of lipids, such as cholesterol and lipids. Kv1.5 appears to target a specific subpopulation of lipid rafts, those in caveola (240, 241). Localization to lipid rafts is thought to be a mechanism for concentrating signaling molecules and substrates within the cell. Another mechanism whereby association of such molecules may be accomplished is via adaptor proteins (ZIP1 and ZIP2) that appear to generate a physical link between the $K_v\alpha/\beta$ complex and protein kinase C (242). The extent to which these interesting biochemical associations are operative in the heart, versus other tissues in which the Kv1.5 complex is expressed, is unknown.

I_{Kur} Dysfunction in Disease

Because Kv1.5 is abundant in human atrium but is absent from normal human ventricle, a drug targeting Kv1.5 would be expected to prolong atrial APD and yet not run the risk of inducing Torsades de Pointes. Although such a chamber-specific antiarrhythmic drug development strategy is appealing, side effects due to Kv1.5 expression in extracardiac tissues could still be a problem. Furthermore, in chronic atrial fibrillation, Kv1.5 mRNA and protein (like other ion currents, including I_{Ca-L} and I_{Na}) are downregulated (243), which might render a drug less effective. Transgenic animals with ventricular overexpression of a Kv1.1 construct that exerts dominant negative effects on Kv1.5 current in vitro do exhibit action potential prolongation in ventricle and ventricular tachyarrhythmias likely related to heterogeneous APDs (244, 245).

CARDIAC INWARD RECTIFIERS

Cardiac inward rectifier K^+ channels, encoded by the Kir superfamily, subserve diverse physiologic functions, including setting the resting potential, hyperpolar-izing depolarized membranes, and contributing to repolarization. The channels do not display intrinsic voltage sensitivity, presumably because of the absence of S4 (Figure 2).

Structure Function

The inward rectifier current I_{K1} likely represents expression of members of the Kir2.x family (246). Further evidence supporting a role for Kir2.1/I_{K1} in termi-nal repolarization (e.g., Figure 1) has come from the linkage of mutations in Kir2.1 with Andersen's syndrome, a neuromuscular disease whose manifestations include bidirectional and other unusual forms of ventricular tachycardia (247). The acetylcholine-gated channel requires coexpression of Kir3.1 and Kir3.4 (also termed GIRK1 and GIRK4) in heart (248). The ATP-inhibited channel results from coexpression of Kir6.x with an ancillary protein, the sulfonylurea receptor (SUR) (249). SUR is a member of the ATP-binding-cassette superfamily, which includes the cystic fibrosis transport regulator (CFTR) and the MDR1 gene, encoding the drug efflux pump P-glycoprotein. Members of this family share a common structural motif, consisting of a putative 12-membrane–spanning segment and two intracellular ATP-binding cassettes (Walker motifs).

Inward rectification of all these channels reflects block of outward current by intracellular constituents. For I_{K1}, Mg^{2+} alone appears sufficient, because when I_{K1} is studied in inside-out patches in the absence of intracellular magnesium, the current-voltage relationship is linear (250, 251). With addition of magnesium to the intracellular face, outward current through the channel becomes progressively occluded with no outward current at physiologic magnesium concentrations. Inward rectification of other members of this group is only partially relieved by removing intracellular magnesium, and polyamines (such as spermidine) have been implicated as further mediators of rectification (252–254).

Acetylcholine activates I_{K-Ach} via a $G\beta\gamma$-delineated pathway (255), and direct binding of $G\beta\gamma$ subunits to both C and N termini of Kir3.1 and Kir3.4 has been demonstrated (256). Kir3.4 knockout results in decreased Kir3.1 protein and the absence of I_{K-ACh} in the heart (257). These mice have normal resting heart rates but do not develop bradycardia with vagal stimulation. Moreover, heart-rate variability (which in the normal human heart is thought to be determined by vagal tone) is markedly reduced in Kir3.4(−/−) mice. The RGS proteins (regulators of G-protein signaling) increase GTPase activity and appear to be involved in $G\alpha$ subunit specification of the duration of G-protein action on I_{K-ACh} (258).

Neither Kir6.x nor SUR alone generates ion currents or proteins at the cell surface when expressed in heterologous systems. Both proteins have endoplasmic reticulum retention motifs that are thought to prevent surface expression, and coexpression of the two likely shields these motifs, allowing the channel complex to traffic appropriately to the cell surface (259). Tissue diversity in the I_{K-ATP} family is created by expression of different members of the Kir and SUR families; the cardiac channel is recapitulated by coexpression of Kir6.2 with SUR2A. Drugs that activate this channel complex (K^+-channel openers such as nicorandil or pinacidil) likely reduce affinity of the channel complex for ATP (260). ATP-sensitive K^+ channels provide a link between the metabolic state of the cell and electrophysiologic activity; given the recent recognition of their unique subunit composition, cardiac-specific agents to protect against the deleterious effects of ischemia (including arrhythmias) are now a consideration. ATP-sensitive channels are expressed not only on the cell surface but also in mitochondria, where they may be particularly attractive targets for cardioprotection (261).

PACEMAKER CURRENT

Rhythmic firing is the sine qua non characteristic of the heart. The ionic basis for the slow depolarization underlying pacemaker activity, particularly evident in sinus node cells, was uncertain until the 1980s, when an increasing body of knowledge identified a hyperpolarization-activated cation current (262). Because the current was (unlike most other ion currents in heart) activated by hyperpolarization, its behavior was termed funny (I_f) or hyperpolarization activated (I_h). In heart, and in other tissues, pacemaker activity is regulated by cyclic nucleotide-related signaling pathways, and the pacemaker channel superfamily does indeed include cyclic nucleotide-binding domains in its C terminus (making this family related to *eag* and *HERG*). Four members of this superfamily have been identified in mice, and HCN1, HCN2, and HCN4 have been found in the heart (15, 16, 263, 264). Heterologous expression of these channels does result in hyperpolarization-activated current that, like I_h, displays only very weak selectivity for K^+ over Na^+ (and with a reversal potential of -35 mV). I_h activation kinetics are generally not well fit with a single exponential term. The zebrafish mutant *Smo* displays marked heart-rate slowing and loss of the fast component, which suggests that multiple components may represent different gene products (265). I_h can be elicited by hyperpolarization to very negative potentials even in nonpacemaker cells, such as those in the ventricle (266). Whether such pacemaker activity might underlie arrhythmias (e.g., if the voltage dependence of activation were shifted positive) is an intriguing possibility. The identification of the molecular basis for pacemaker activity in heart opens the way to the development of new therapeutic agents for disorders related to excess pacemaker activity.

There are a number of differences between the gating and the voltage dependence of I_h and of HCN channels in heterologous expression systems (267). The molecular basis for these differences (such as coexpression of other proteins) has not been elucidated. Similarly, although HCN channels display the typical six-membrane-segment-spanning architecture of voltage-gated K^+ channels, the mechanisms for their unusual hyperpolarization-activated gating and lack of selectivity has not been fully determined. Comparisons of the HCN sequence with that of other K^+ channels reveals retention of the S4 voltage sensor and a GYG selectivity filter. Changes in residues adjacent to the GYG may underlie altered selectivity of the channel.

One possible mechanism for hyperpolarization-induced activation has been provided by studies in which individual residues within S4 have been mutated in *Shaker*, resulting in striking shifts to hyperpolarized potentials of voltage dependence of activation (268). If it is assumed that I_h channels, like I_{Kr}, undergo fast inactivation from an open state, then the unusual voltage dependence of I_h gating could arise from channels that occupy the inactivated state at depolarized potentials and, with hyperpolarization, move slowly to the open state, i.e., maintain the gating scheme depicted in Figure 4, but move the voltage dependence such that it is the inactivated state that is occupied at rest. The molecular basis of regulation of I_h by intracellular signaling systems has not been elucidated, nor have the mechanisms underlying its subunit assembly, stoichiometry, or trafficking been defined. Pacemaker channels are further discussed by Siegelbaum.

BACKGROUND CURRENT

Voltage clamp of heart cells after elimination of time-dependent current continues to elicit a time-independent, voltage-dependent current, usually termed background current. The background current can confound experimental results (particularly if it displays nonlinear current-voltage relationships). Whether any background currents are appropriate targets for drug therapy is not yet known.

At least three types of channels have been postulated as underlying background current in heart. (*a*) I_{Kp} (for plateau) is a time-independent current recorded in guinea pig myocytes. The current is barium sensitive, is not modulated by changes intracellular in extracellular chloride, is insensitive to chloride channel blockers, and persists in the absence of extracellular K^+; these characteristics separate it from chloride current, I_K components, and I_{K1} (269, 270). It has been suggested that expression of the twin-pore K^+ channel TWIK1 may underlie I_{Kp} (21). It is not known whether I_{Kp} is detected in other species. (*b*) In guinea pig myocytes, application of isoproterenol or other maneuvers to active cAMP-dependent signaling increases an outwardly rectifying chloride current (202). Some data suggest that this may represent CFTR activity because CFTR transcripts can be identified in heart (271). More generally, chloride channels activated by changes in cell volume and by changes in $[Ca^{2+}]_i$ have been identified (272). Isoform-specific antagonists have not been available, so the role of individual chloride channels in cardiac physiology and pathophysiology remains undetermined. (*c*) Coexpression of *KvLQT1* with *KCNE2* (*MiRP1*) in heterologous expression systems results in a time-independent (i.e., background) current (273). The general question of the extent of *MiRP1* expression in heart, its localization, and the specific subunits whose function it may modify (HERG, KvLQT1, and Kv4 have all been suggested) remains undetermined.

FUTURE CONSIDERATIONS

Over the past decade, the genes encoding α subunits for the major ionic currents in heart have been cloned. At the same time, however, a number of lines of evidence continue to point to substantial cell-cell and region-region heterogeneity in the electrophysiologic properties of the heart. One precedent comes from studies with *Caenorhabditis elegans*, an organism that demonstrates an astounding number of ion channels, many of which are expressed only in small regions (as small as a single cell), and at specific times during development (274). This precedent suggests that a similar pattern may emerge in the human heart. Second, studies of specific regions of mammalian heart continue to identify variability in electrophysiologic properties, and such variability is increasingly appreciated as an important contributor to arrhythmias. Third, although identification of α subunits subserving ion currents has been a crucial and important step in advancing our understanding of molecular electrophysiology, it is becoming increasingly clear that normal function of a multitude of other gene products is also required to maintain a normal cardiac rhythm: connexins, pumps, exchangers, elements of the extracellular matrix, β subunits, elements of the cytoskeleton, trafficking machinery, and targeting molecules, to name but a few.

Cardiac arrhythmias arise when function or expression of one or more of these elements is perturbed. This can occur in monogenic syndromes, such as the congenital LQTS or Brugada syndrome, or in the course of adaptive or maladaptive changes to other cardiac disease, such as acute and chronic ischemia, hypertrophy, and heart failure. Currently available drugs to treat arrhythmias have very incomplete efficacy and substantial risks of toxicity. The challenge, and opportunity, of an emerging picture of perturbed myocyte biology in arrhythmogenesis is to identify new targets, which will provide safe and effective drug therapies. Furthermore, it seems possible that interindividual variability in cardiac electrophysiology, and in particular in its response to stressors such as acute myocardial ischemia or drug challenge, may have a genetic basis. Defining these genetic factors may ultimately allow mechanism-specific and patient-specific prescription of effective and well tolerated new drug therapies.

Visit the Annual Reviews home page at www.AnnualReviews.org

LITERATURE CITED

1. Zipes DP, Wellens HJJ. 1998. Sudden cardiac death. *Circulation* 98:2334–51
2. Antzelevitch C. 2000. Electrical heterogeneity, cardiac arrhythmias, and the sodium channel. *Circ. Res.* 87:964–65
3. Zeng T, Bett GC, Sachs F. 2000. Stretch-activated whole cell currents in adult rat cardiac myocytes. *Am. J. Physiol. Heart Circ. Physiol.* 278:H548–H57
4. Nattel S, Li D. 2000. Ionic remodeling in the heart: pathophysiological significance and new therapeutic opportunities for atrial fibrillation. *Circ. Res.* 87:440–47
5. Pinto JM, Boyden PA. 1999. Electrical remodeling in ischemia and infarction. *Cardiovasc. Res.* 42:284–97
6. Noda M, Shimizu S, Tanabe T, Takai T, Kayano T, et al. 1984. Primary structure of *Electrophorus electricus* sodium channel deduced from cDNA sequence. *Nature* 312:121–27
7. Papazian DM, Schwarz TL, Tempel BL, Jan YN, Jan LY. 1987. Cloning of genomic and complementary DNA from *Shaker*, a putative potassium channel gene from Drosophila. *Science* 237:749–53
8. MacKinnon R. 1991. Determination of the subunit stoichiometry of a voltage-activated potassium channel. *Nature* 350:232–35
9. Gellens ME, George AL Jr, Chen LQ, Chahine M, Horn R, et al. 1992. Primary structure and functional expression of the human cardiac tetrodotoxin-insensitive voltage-dependent sodium channel. *Proc. Natl. Acad. Sci. USA* 89:554–58
10. Schultz D, Mikala G, Yatani A, Engle DB, Iles DE, et al. 1993. Cloning, chromosomal localization, and functional expression of the alpha 1 subunit of the L-type voltage-dependent calcium channel from normal human heart. *Proc. Natl. Acad. Sci. USA* 90:6228–32
11. Tamkun MM, Knoth KM, Walbridge JA, Kroemer H, Roden DM, Glover DM. 1991. Molecular cloning and characterization of two voltage-gated K$^+$ channel

cDNAs from human ventricle. *FASEB J.* 5:331–37

12. Takumi T, Ohkubo H, Nakanishi S. 1988. Cloning of a membrane protein that induces a slow voltage-gated potassium current. *Science* 242:1042–45

13. Kubo Y, Baldwin TJ, Jan YN, Jan LY. 1993. Primary structure and functional expression of a mouse inward rectifier potassium channel. *Nature* 362:127–33

14. Santoro B, Grant SG, Bartsch D, Kandel ER. 1997. Interactive cloning with the SH3 domain of N-src identifies a new brain specific ion channel protein, with homology to eag and cyclic nucleotide-gated channels. *Proc. Natl. Acad. Sci. USA* 94:14815–20

15. Ludwig A, Zong X, Jeglitsch M, Hofmann F, Biel M. 1998. A family of hyperpolarization-activated mammalian cation channels. *Nature* 393:587–91

16. Santoro B, Liu DT, Yao H, Bartsch D, Kandel ER, et al. 1998. Identification of a gene encoding a hyperpolarization-activated pacemaker channel of brain. *Cell* 93:717–29

17. Cribbs LL, Lee J, Yang J, Satin J, Zhang Y, et al. 1998. Cloning and characterization of α1H from human heart, a member of the T-type Ca^{2+} channel gene family. *Circ. Res.* 83:103–9

18. Perez-Reyes E, Cribbs LL, Daud A, Lacerda AE, Barclay J, et al. 1998. Molecular characterization of a neuronal low-voltage-activated T-type calcium channel. *Nature* 391:896–900

19. Abbott GW, Sesti F, Splawski I, Buck ME, Lehmann MH, et al. 1999. MiRP1 forms I_{Kr} potassium channels with HERG and is associated with cardiac arrhythmia. *Cell* 97:175–87

20. Ketchum KA, Joiner WJ, Sellers AJ, Kaczmarek LK, Goldstein SA. 1995. A new family of outwardly rectifying potassium channel proteins with two pore domains in tandem. *Nature* 376:690–95

21. Lesage F, Guillemare E, Fink M, Duprat F, Lazdunski M, et al. 1996. TWIK-1, a ubiquitous human weakly inward rectifying K^+ channel with a novel structure. *EMBO J.* 15:1004–11

22. Splawski I, Tristani-Firouzi M, Lehmann MH, Sanguinetti MC, Keating MT. 1997. Mutations in the *hminK* gene cause long QT syndrome and suppress I_{Ks} function. *Nat. Genet.* 17:338–40

23. Chen QY, Kirsch GE, Zhang DM, Brugada R, Brugada J, et al. 1998. Genetic basis and molecular mechanism for idiopathic-ventricular fibrillation. *Nature* 392:293–96

24. Priori SG, Napolitano C, Tiso N, Memmi M, Vignati G, et al. 2001. Mutations in the cardiac ryanodine receptor gene (hRyR2) underlie catecholaminergic polymorphic ventricular tachycardia. *Circulation* 103:196–200

25. Curran ME, Splawski I, Timothy KW, Vincent GM, Green ED, Keating MT. 1995. A molecular basis for cardiac arrhythmia: *HERG* mutations cause long QT syndrome. *Cell* 80:795–803

26. Wang Q, Shen J, Splawski I, Atkinson D, Li Z, et al. 1995. *SCN5A* mutations associated with an inherited cardiac arrhythmia, long QT syndrome. *Cell* 80:805–11

27. Wang Q, Curran ME, Splawski I, Burn TC, Millholland JM, et al. 1996. Positional cloning of a novel potassium channel gene: *KVLQT1* mutations cause cardiac arrhythmias. *Nat. Genet.* 12:17–23

28. Keating MT, Sanguinetti MC. 2001. Molecular and cellular mechanisms of cardiac arrhythmias. *Cell* 104:569–80

29. Roden DM, Spooner PM. 1999. Inherited Long QT syndromes: a paradigm for understanding arrhythmogenesis. *J. Cardiovasc. Electrophysiol.* 10:1664–83

30. Splawski I, Shen J, Timothy KW, Lehmann MH, Priori S, et al. 2000. Spectrum of mutations in long-QT syndrome genes. KVLQT1, HERG, SCN5A, KCNE1, and KCNE2. *Circulation* 102:1178–85

31. Zhou ZF, Gong QM, Epstein ML,

January CT. 1998. HERG channel dysfunction in human long QT syndrome—intracellular transport and functional defects. *J. Biol. Chem.* 273:21061–66

32. Bianchi L, Priori SG, Napolitano C, Surewicz KA, Dennis AT, et al. 2000. Mechanisms of I(Ks) suppression in LQT1 mutants. *Am. J. Physiol. Heart Circ. Physiol.* 279:H3003–H11

33. Schwartz PJ, Priori SG, Spazzolini C, Moss AJ, Vincent GM, et al. 2001. Genotype-phenotype correlation in the long-QT syndrome: gene-specific triggers for life-threatening arrhythmias. *Circulation* 103:89–95

34. Zhang L, Timothy KW, Vincent GM, Lehmann MH, Fox J, et al. 2000. Spectrum of ST-T-wave patterns and repolarization parameters in congenital long-QT syndrome: ECG findings identify genotypes. *Circulation* 102:2849–55

35. Zareba W, Moss AJ, Schwartz PJ, Vincent GM, Robinson JL, et al. 1998. Influence of the genotype on the clinical course of the long-QT syndrome. *N. Engl. J. Med.* 339:960–65

36. Wu Y, Macmillan LB, McNeill RB, Colbran RJ, Anderson ME. 1999. CaM kinase augments cardiac L-type Ca^{2+} current: a cellular mechanism for long Q-T arrhythmias. *Am. J. Physiol. Heart Circ. Physiol.* 276:H2168–H78

37. Abriel H, Cabo C, Wehrens XH, Rivolta I, Motoike HK, et al. 2001. Novel arrhythmogenic mechanism revealed by a long-qt syndrome mutation in the cardiac Na(+) channel. *Circ. Res.* 88:740–45

38. Wu Y, Roden DM, Anderson ME. 1999. CaM kinase inhibition prevents development of the arrhythmogenic transient inward current. *Circ. Res.* 84:906–12

39. Priori SG, Napolitano C, Schwartz PJ. 1999. Low penetrance in the long-QT syndrome: clinical impact. *Circulation* 99:529–33

40. Hodgkin AL, Huxley AF. 1952. A quantitative description of membrane current and its application to conduction and excitation in nerve. *J. Physiol.* 117:500–44

41. Hodgkin AL, Huxley AF. 1952. Currents carried by sodium and potassium ions through the membrane of the giant axon of *Loligo. J. Physiol.* 116:449–72

42. Doyle DA, Cabral JM, Pfuetzner RA, Kuo AL, Gulbis JM, et al. 1998. The structure of the potassium channel—molecular basis of K^+ conduction and selectivity. *Science* 280:69–77

43. Sato C, Ueno Y, Asai K, Takahashi K, Sato M, et al. 2001. The voltage-sensitive sodium channel is a bell-shaped molecule with several cavities. *Nature* 409:1047–51

44. Yang J, Ellinor PT, Sather WA, Zhang JF, Tsien RW. 1993. Molecular determinants of Ca^{2+} selectivity and ion permeation in L-type Ca^{2+} channels. *Nature* 366:158–61

45. Chiamvimonvat N, Perez-Garcia MT, Ranjan R, Marban E, Tomaselli GF. 1996. Depth asymmetries of the pore-lining segments of the Na^+ channel revealed by cysteine mutagenesis *Neuron* 16:1037–47

46. Heinemann SH, Terlau H, Stuhmer W, Imoto K, Numa S. 1992. Calcium channel characteristics conferred on the sodium channel by single mutations. *Nature* 356:441–43

47. Hess P, Tsien RW. 1984. Mechanism of ion permeation through calcium channels. *Nature* 309:453–56

48. Lipkind GM, Fozzard HA. 2000. KcsA crystal structure as framework for a molecular model of the Na(+) channel pore. *Biochemistry* 39:8161–70

49. Goldstein SAN. 1996. A structural vignette common to voltage sensors and conduction pores—canaliculi. *Neuron* 16:717–22

50. Yang N, George AL Jr, Horn R. 1996. Molecular basis of charge movement in voltage-gated sodium channels. *Neuron* 16:113–22

51. Zagotta WN, Hoshi T, Aldrich RW. 1990. Restoration of inactivation in mutants of Shaker potassium channels by a peptide derived from ShB. *Science* 250:568–71

52. Hoshi T, Zagotta WN, Aldrich RW. 1990. Biophysical and molecular mechanisms of Shaker potassium channel inactivation. *Science* 250:533–38

53. Armstrong CM, Bezanilla F. 1977. Inactivation of the sodium channel. II. Gating current experiments. *J. Gen. Physiol.* 70:567–90

54. Hoshi T, Zagotta WN, Aldrich RW. 1991. Two types of inactivation in Shaker K^+ channels: effects of alterations in the carboxy-terminal region. *Neuron* 7:547–56

55. Hille B. 1977. Local anesthetics: hydrophilic and hydrophobic pathways for the drug-receptor reaction. *J. Gen. Physiol.* 69:497–515

56. Hondeghem LM, Katzung BG. 1977. Time- and voltage-dependent interactions of antiarrhythmic drugs with cardiac sodium channels. *Biochim. Biophys. Acta* 472:373–98

57. Ragsdale DS, McPhee JC, Scheuer T, Catterall WA. 1994. Molecular determinants of state-dependent block of Na^+ channels by local anesthetics. *Science* 265:1724–28

58. Kraus R, Reichl B, Kimball SD, Grabner M, Murphy BJ, et al. 1996. Identification of benz(othi)azepine-binding regions within L-type calcium channel alpha-1 subunits. *J. Biol. Chem.* 271:20113–18

59. Yeola SW, Rich TC, Uebele VN, Tamkun MM, Snyders DJ. 1996. Molecular analysis of a binding site for quinidine in a human cardiac delayed rectifier K^+ channel—role of S6 in antiarrhythmic drug binding. *Circ. Res.* 78:1105–14

60. Mitcheson JS, Chen J, Lin M, Culberson C, Sanguinetti MC. 2000. A structural basis for drug-induced long QT syndrome. *Proc. Natl. Acad. Sci. USA* 97:12329–33

61. McPhee JC, Ragsdale DS, Scheuer T, Catterall WA. 1995. A critical role for transmembrane segment IVS6 of the sodium channel alpha subunit in fast inactivation. *J. Biol. Chem.* 270:12025–34

62. Balser JR, Nuss HB, Orias DW, Johns DC, Marban E, et al. 1996. Local anesthetics as effectors of allosteric gating. Lidocaine effects on inactivation-deficient rat skeletal muscle Na channels. *J. Clin. Invest.* 98:2874–86

63. Venter JC, Adams MD, Myers EW, Li PW, Mural RJ, et al. 2001. The sequence of the human genome. *Science* 291:1304–51

64. Isom LL, De Jongh KS, Catterall WA. 1994. Auxiliary subunits of voltage-gated ion channels. *Neuron* 12:1183–94

65. Nuss HB, Chiamvimonvat N, Pérez-García MT, Tomaselli GF, Marban E. 1995. Functional association of the $\beta 1$ subunit with human cardiac (hH1) and rat skeletal muscle ($\mu 1$) sodium channel α subunits expressed in *Xenopus* oocytes. *J. Gen. Physiol.* 106:1171–91

66. Makita N, Bennett PB, George AL Jr. 1996. Molecular determinants of $\beta 1$ subunit-induced gating modulation in voltage-dependent Na^+ channels. *J. Neurosci.* 16:7117–27

67. Cohen SA. 1996. Immunocytochemical localization of rH1 sodium channel in adult rat heart atria and ventricle: presence in terminal intercalated disks. *Circulation* 94:3083–86

68. Donahue LM, Coates PW, Lee VH, Ippensen DC, Arze SE, Poduslo SE. 2000. The cardiac sodium channel mRNA is expressed in the developing and adult rat and human brain. *Brain Res.* 887:335–43

69. Aldrich RW, Corey DP, Stevens CF. 1983. A reinterpretation of mammalian sodium channel gating based on single channel recording. *Nature* 306:436–41

70. Horn R, Patlak J, Stevens CF. 1981. Sodium channels need not open before they inactivate. *Nature* 291:426–27

71. Patton DE, West JW, Catterall WA, Goldin AL. 1992. Amino acid residues required for fast Na^+-channel inactivation: charge neutralizations and deletions in the III-IV linker. *Proc. Natl. Acad. Sci. USA* 89:10905–9

72. Benitah JP, Tomaselli GF, Marban E.

1996. Adjacent pore-lining residues within sodium channels identified by paired cysteine mutagenesis. *Proc. Natl. Acad. Sci. USA* 93:7392–96

73. Satin J, Kyle JW, Chen M, Bell P, Cribbs LL, et al. 1992. A mutant of TTX-resistant cardiac sodium channels with TTX-sensitive properties. *Science* 256:1202–5

74. Backx PH, Yue DT, Lawrence JH, Marban E, Tomaselli GF. 1992. Molecular localization of an ion-binding site within the pore of mammalian sodium channels. *Science* 257:248–51

75. Li RA, Ennis IL, French RJ, Dudley SC, Tomaselli GF, Marban E. 2001. Clockwise domain arrangement of the sodium channel revealed by μ-conotoxin (GI-IIA) docking orientation. *J. Biol. Chem.* 276:11072–77

76. El-Sherif N, Fozzard HA, Hanck DA. 1992. Dose-dependent modulation of the cardiac sodium channel by sea anemone toxin ATXII. *Circ. Res.* 70:285–301

77. Shimizu W, Antzelevitch C. 1997. Sodium channel block with mexiletine is effective in reducing dispersion of repolarization and preventing torsade des pointes in LQT2 and LQT3 models of the long-QT syndrome. *Circulation* 96:2038–47

78. Rogers JC, Qu Y, Tanada TN, Scheuer T, Catterall WA. 1996. Molecular determinants of high affinity binding of alpha-scorpion toxin and sea anemone toxin in the S3-S4 extracellular loop in domain IV of the Na$^+$ channel alpha subunit. *J. Biol. Chem.* 271:15950–62

79. Chong JA, Tapia-Ramirez J, Kim S, Toledo-Aral JJ, Zheng Y, et al. 1995. REST: a mammalian silencer protein that restricts sodium channel gene expression to neurons. *Cell* 80:949–57

80. Taouis M, Sheldon RS, Duff HJ. 1991. Upregulation of the rat cardiac sodium channel by in vivo treatment with a class I antiarrhythmic drug. *J. Clin. Invest.* 88:375–78

81. Duff HJ, Offord J, West J, Catterall WA. 1992. Class I and IV antiarrhythmic drugs and cytosolic calcium regulate mRNA encoding the sodium channel alpha subunit in rat cardiac muscle. *Mol. Pharmacol.* 42:570–74

82. Yue L, Melnyk P, Gaspo R, Wang Z, Nattel S. 1999. Molecular mechanisms underlying ionic remodeling in a dog model of atrial fibrillation. *Circ. Res.* 84:776–84

83. Murphy BJ, Rogers J, Perdichizzi AP, Colvin AA, Catterall WA. 1996. cAMP-dependent phosphorylation of two sites in the alpha subunit of the cardiac sodium channel. *J. Biol. Chem.* 271:28837–43

84. Qu YS, Rogers JC, Tanada TN, Catterall WA, Scheuer T. 1996. Phosphorylation of S1505 in the cardiac Na$^+$ channel inactivation gate is required for modulation by protein kinase C. *J. Gen. Physiol.* 108:375–79

85. Li M, West JW, Numann R, Murphy BJ, Scheuer T, Catterall WA. 1993. Convergent regulation of sodium channels by protein kinase C and cAMP-dependent protein kinase. *Science* 261:1439–42

86. Smith RD, Goldin AL. 2000. Potentiation of rat brain sodium channel currents by PKA in Xenopus oocytes involves the I-II linker. *Am. J. Physiol. Cell Physiol.* 278:C638–C45

87. Zhou J, Yi J, Hu N, George AL, Murray KT. 2000. Activation of protein kinase A modulates trafficking of the human cardiac sodium channel in Xenopus oocytes. *Circ. Res.* 87:33–38

88. Tibbs VC, Gray PC, Catterall WA, Murphy BJ. 1998. AKAP15 anchors cAMP-dependent protein kinase to brain sodium channels. *J. Biol. Chem.* 273:25783–88

89. Bennett PB, Yazawa K, Makita N, George AL Jr. 1995. Molecular mechanism for an inherited cardiac arrhythmia. *Nature* 376:683–85

90. Dumaine R, Wang Q, Keating MT, Hartmann HA, Schwartz PJ, et al. 1996. Multiple mechanisms of sodium channel-linked long QT syndrome. *Circ. Res.* 78:916–24

91. Nuyens D. 2001. Electrophysiological mechanisms of the long QT syndrome in mice with targeted deletion of ^{1505}KPQ^{1507} in the cardiac sodium channel. *PACE* 24:598 (Abstr.)

92. Smith MR, Goldin AL. 1997. Interaction between the sodium channel inactivation linker and domain III S4-S5. *Biophys. J.* 73:1885–95

93. An RH, Wang XL, Kerem B, Benhorin J, Medina A, et al. 1998. Novel LQT-3 mutation affects Na$^+$ channel activity through interactions between alpha- and beta1-subunits. *Circ. Res.* 83:141–46

94. Baroudi G, Chahine M. 2000. Biophysical phenotypes of SCN5A mutations causing long QT and Brugada syndromes. *FEBS Lett.* 487:224–28

95. Martini B, Nava A, Thiene G, Buja GF, Canciani B, et al. 1989. Ventricular fibrillation without apparent heart disease: description of six cases. *Am. Heart J.* 118:1203–9

96. Brugada P, Brugada J. 1992. Right bundle branch block, persistent ST segment elevation and sudden cardiac death: a distinct clinical and electrocardiographic syndrome. A multicenter report. *J. Am. Coll. Cardiol.* 20:1391–96

97. Wang DW, Makita N, Kitabatake A, Balser JR, George AL Jr. 2000. Enhanced Na(+) channel intermediate inactivation in Brugada syndrome. *Circ. Res.* 87:E37–43

98. Veldkamp MW, Viswanathan PC, Bezzina C, Baartscheer A, Wilde AA, Balser JR. 2000. Two distinct congenital arrhythmias evoked by a multidysfunctional Na(+) channel. *Circ. Res.* 86:E91–97

99. Gussak I, Antzelevitch C, Bjerregaard P, Towbin JA, Chaitman BR. 1999. The Brugada syndrome: clinical, electrophysiologic and genetic aspects. *J. Am. Coll. Cardiol.* 33:5–15

100. Bezzina C, Veldkamp MW, van Den Berg MP, Postma AV, Rook MB, et al. 1999. A single Na(+) channel mutation causing both long-QT and Brugada syndromes. *Circ. Res.* 85:1206–13

101. Balser JR. 1999. Structure and function of the cardiac sodium channels. *Cardiovasc. Res.* 42:327–38

102. Schott JJ, Alshinawi C, Kyndt F, Probst V, Hoorntje TM, et al. 1999. Cardiac conduction defects associate with mutations in SCN5A. *Nat. Genet.* 23:20–21

103. Tan HL, Bink-Boelkens MT, Bezzina CR, Viswanathan PC, Beaufort-Krol GC, et al. 2001. A sodium-channel mutation causes isolated cardiac conduction disease. *Nature* 409:1043–47

104. Cardiac Arrhythmia Suppression Trial (CAST) Invest. 1989. Increased mortality due to encainide or flecainide in a randomized trial of arrhythmia suppression after myocardial infarction. *N. Engl. J. Med.* 321:406–12

105. Akiyama T, Pawitan Y, Greenberg H, Kuo CS, Reynolds-Haertle RA, CAST Invest. 1991. Increased risk of death and cardiac arrest from encainide and flecainide in patients after nonQ-wave acute myocardial infarction in the Cardiac Arrhythmia Suppression Trial. *Am. J. Cardiol.* 68:1551–55

106. Coromilas J, Saltman AE, Waldecker B, Dillon SM, Wit AL. 1995. Electrophysiological effects of flecainide on anisotropic conduction and reentry in infarcted canine hearts. *Circulation* 91:2245–63

107. Lukas A, Antzelevitch C. 1993. Differences in the electrophysiological response of canine ventricular epicardium and endocardium to ischemia. Role of the transient outward current. *Circulation* 88:2903–15

108. Krishnan SC, Antzelevitch C. 1993. Flecainide-induced arrhythmia in canine ventricular epicardium. Phase 2 reentry? *Circulation* 87:562–72

109. Lue WM, Boyden PA. 1992. Abnormal electrical properties of myocytes from chronically infarcted canine heart. Alterations in V_{max} and the transient outward current. *Circulation* 85:1175–88

110. Chien AJ, Zhao X, Shirokov RE, Puri TS, Chang CF, et al. 1995. Roles of a membrane-localized beta subunit in the formation and targeting of functional L-type Ca^{2+} channels. *J. Biol. Chem.* 270:30036–44

111. Perez-Reyes E, Castellano A, Kim HS, Bertrand P, Baggstrom E, et al. 1992. Cloning and expression of a cardiac/brain beta subunit of the L-type calcium channel. *J. Biol. Chem.* 267:1792–97

112. Cens T, Mangoni ME, Richard S, Nargeot J, Charnet P. 1996. Coexpression of the beta2 subunit does not induce voltage-dependent facilitation of the class C L-type Ca channel. *Pflügers Arch.* 431:771–74

113. De Jongh KS, Warner C, Catterall WA. 1990. Subunits of purified calcium channels. Alpha 2 and delta are encoded by the same gene. *J. Biol. Chem.* 265:14738–41

114. Letts VA, Felix R, Biddlecome GH, Arikkath J, Mahaffey CL, et al. 1998. The mouse stargazer gene encodes a neuronal Ca^{2+}-channel gamma subunit. *Nat. Genet.* 19:340–47

115. Hadley RW, Lederer WJ. 1991. Ca^{2+} and voltage inactivate Ca^{2+} channels in guinea-pig ventricular myocytes through independent mechanisms. *J. Physiol.* 444:257–68

116. de Leon M, Wang Y, Jones L, Perez-Reyes E, Wei X, et al. 1995. Essential Ca(2+)-binding motif for Ca(2+)-sensitive inactivation of L-type Ca^{2+} channels. *Science* 270:1502–6

117. Zuhlke RD, Pitt GS, Deisseroth K, Tsien RW, Reuter H. 1999. Calmodulin supports both inactivation and facilitation of L-type calcium channels. *Nature* 399:159–62

118. Peterson BZ, Lee JS, Mulle JG, Wang Y, de Leon M, Yue DT. 2000. Critical determinants of Ca(2+)-dependent inactivation within an EF-hand motif of L-type Ca(2+) channels. *Biophys. J.* 78:1906–20

119. Striessnig J, Murphy BJ, Catterall WA. 1991. Dihydropyridine receptor of L-type Ca^{2+} channels: identification of binding domains for [^{3}H](+)-PN200–110 and [^{3}H]azidopine within the alpha 1 subunit. *Proc. Natl. Acad. Sci. USA* 88:10769–73

120. Schuster A, Lacinova L, Klugbauer N, Ito H, Birnbaumer L, Hofmann F. 1996. The IVS6 segment of the L-type calcium channel is critical for the action of dihydropyridines and phenylalkylamines. *EMBO J.* 15:2365–70

121. Uehara A, Hume JR. 1985. Interactions of organic calcium channel antagonists with calcium channels in single frog atrial cells. *J. Gen. Physiol.* 85:621–47

122. Kass RS, Arena JP, Chin S. 1991. Block of L-type calcium channels by charged dihydropyridines. Sensitivity to side of application and calcium. *J. Gen. Physiol.* 98:63–75

123. Hering S, Aczel S, Grabner M, Doring F, Berjukow S, et al. 1996. Transfer of high sensitivity for benzothiazepines from L-type to class A (BI) calcium channels. *J. Biol. Chem.* 271:24471–75

124. Yue DT, Herzig S, Marban E. 1990. Beta-adrenergic stimulation of calcium channels occurs by potentiation of high-activity gating modes. *Proc. Natl. Acad. Sci. USA* 87:753–57

125. Dzhura I, Wu Y, Colbran RJ, Balser JR, Anderson ME. 2000. Calmodulin kinase determines calcium-dependent facilitation of L-type calcium channels. *Nat. Cell Biol.* 2:173–77

126. Hartzell HC, Mery PF, Fischmeister R, Szabo G. 1991. Sympathetic regulation of cardiac calcium current is due exclusively to cAMP-dependent phosphorylation. *Nature* 351:573–76

127. Cavalie A, Allen TJ, Trautwein W. 1991. Role of the GTP-binding protein Gs in the beta-adrenergic modulation of cardiac Ca channels. *Pflügers Arch.* 419:433–43

128. Perets T, Blumenstein Y, Shistik E, Lotan I, Dascal N. 1996. A potential site of functional modulation by protein kinase A in the cardiac Ca^{2+} channel alpha 1C subunit. *FEBS Lett.* 384:189–92

129. Gao T, Yatani A, Dell'Acqua ML, Sako H, Green SA, et al. 1997. cAMP-dependent regulation of cardiac L-type Ca^{2+} channels requires membrane targeting of PKA and phosphorylation of channel subunits. *Neuron* 19:185–96

130. Gray PC, Scott JD, Catterall WA. 1998. Regulation of ion channels by cAMP-dependent protein kinase and A-kinase anchoring proteins. *Curr. Opin. Neurobiol.* 8:330–34

131. Tseng GN, Boyden PA. 1989. Multiple types of Ca^{2+} currents in single canine Purkinje cells. *Circ. Res.* 65:1735–50

132. Alvarez JL, Vassort G. 1992. Properties of the low threshold Ca current in single frog atrial cardiomyocytes. A comparison with the high threshold Ca current. *J. Gen. Physiol.* 100:519–45

133. Fan IQ, Chen B, Marsh JD. 2000. Transcriptional regulation of L-type calcium channel expression in cardiac myocytes. *J. Mol. Cell. Cardiol.* 32:1841–49

134. Maki T, Gruver EJ, Davidoff AJ, Izzo N, Toupin D, et al. 1996. Regulation of calcium channel expression in neonatal myocytes by catecholamines. *J. Clin. Invest.* 97:656–63

135. Welling A, Ludwig A, Zimmer S, Klugbauer N, Flockerzi V, Hofmann F. 1997. Alternatively spliced IS6 segments of the alpha 1C gene determine the tissue-specific dihydropyridine sensitivity of cardiac and vascular smooth muscle L-type Ca^{2+} channels. *Circ. Res.* 81:526–32

136. Klockner U, Mikala G, Eisfeld J, Iles DE, Strobeck M, et al. 1997. Properties of three COOH-terminal splice variants of a human cardiac L-type Ca^{2+}-channel alpha(1)-subunit. *Am. J. Physiol. Heart Circ. Physiol.* 41:H1372–H81

137. Soldatov NM, Bouron A, Reuter H. 1995. Different voltage-dependent inhibition by dihydropyridines of human Ca^{2+} channel splice variants. *J. Biol. Chem.* 270:10540–43

138. Satin J, Cribbs LL. 2000. Identification of a T-type Ca(2+) channel isoform in murine atrial myocytes (AT-1 cells) *Circ. Res.* 86:636–42

139. Yang Y, Chen X, Margulies K, Jeevanandam V, Pollack P, et al. 2000. L-type Ca^{2+} channel alpha 1c subunit isoform switching in failing human ventricular myocardium. *J. Mol. Cell. Cardiol.* 32:973–84

140. Hart G. 1994. Cellular electrophysiology in cardiac hypertrophy and failure. *Cardiovasc. Res.* 28:933–46

141. Marban E, Robinson SW, Weir WG. 1986. Mechanisms of arrhythmogenic delayed and early after depolarizations and triggered activity in vivo. *J. Clin. Invest.* 78:1185–92

142. Anderson ME Braun AP, Schulman H, Premack BA. 1994. Multifunctional Ca^{2+}/calmodulin-dependent protein kinase mediates $Ca^{(2+)}$-induced enhancement of the L-type Ca^{2+} current in rabbit ventricular myocytes. *Circ. Res.* 75:854–61

143. Wu Y, Roden DM, Anderson ME. 1999. Calmodulin kinase inhibition prevents development of the arrhythmogenic transient inward current. *Circ. Res.* 84:906–12

144. Mazur A, Roden DM, Anderson ME. 1999. Systemic administration of calmodulin antagonist W-7 or protein kinase A inhibitor H-8 prevents Torsade de Pointes in rabbits. *Circulation* 100:2437–42

145. Laitinen PJ, Brown KM, Piippo K, Swan H, Devaney JM, et al. 2001. Mutations of the cardiac ryanodine receptor (RyR2) gene in familial polymorphic ventricular tachycardia. *Circulation* 103:485–90

146. Nuss HB, Houser SR. 1993. T-type Ca^{2+} current is expressed in hypertrophied adult feline left ventricular myocytes. *Circ. Res.* 73:777–82

147. Po SS, Snyders DJ, Baker R, Tamkun MM, Bennett PB. 1992. Functional expression of an inactivating potassium channel cloned from human heart. *Circ. Res.* 71:732–36

148. Dixon JE, McKinnon D. 1994. Quantitative analysis of potassium channel mRNA expression in atrial and ventricular muscle of rats. *Circ. Res.* 75:252–60

149. Näbauer M, Beuckelmann DJ, Überfuhr P, Steinbeck G. 1996. Regional differences in current density and rate-dependent properties of the transient outward current in subepicardial and subendocardial myocytes of human left ventricle. *Circulation* 93:168–77

150. An WF, Bowlby MR, Betty M, Cao J, Ling HP, et al. 2000. Modulation of A-type potassium channels by a family of calcium sensors. *Nature* 403:553–56

151. Tseng GN, Hoffman BF. 1989. Two components of transient outward current in canine ventricular myocytes. *Circ. Res.* 64:633–47

152. Zygmunt AC, Gibbons WR. 1991. Calcium-activated chloride current in rabbit ventricular mycoytes. *Circ. Res.* 68:424–37

153. Slawsky MT, Castle NA. 1994. K^+ channel blocking actions of flecainide compared with those of propafenone and quinidine in adult rat ventricular myocytes. *J. Pharmacol. Exp. Ther.* 269:66–74

154. Wang Z, Fermini B, Nattel S. 1995. Effects of flecainide, quinidine, and 4-aminopyridine on transient outward and ultrarapid delayed rectifier currents in human atrial myocytes. *J. Pharmacol. Exp. Ther.* 272:184–96

155. Takimoto K, Li DQ, Hershman KM, Li P, Jackson EK, Levitan ES. 1997. Decreased expression of Kv4.2 and novel Kv4.3 K+ channel subunit mRNAs in ventricles of renovascular hypertensive rats. *Circ. Res.* 81:533–39

156. Beuckelmann DJ, Näbauer M, Erdmann E. 1993. Alterations of K^+ currents in isolated human ventricular myocytes from patients with terminal heart failure. *Circ. Res.* 73:379–85

157. Kaab S, Dixon J, Duc J, Ashen D, Nabauer M, et al. 1998. Molecular basis of transient outward potassium current downregulation in human heart failure: a decrease in Kv4.3 mRNA correlates with a reduction in current density. *Circulation* 98:1383–93

158. Barry DM, Xu H, Schuessler RB, Nerbonne JM. 1998. Functional knockout of the transient outward current, long-QT syndrome, and cardiac remodeling in mice expressing a dominant-negative Kv4 alpha subunit. *Circ. Res.* 83:560–67

159. Xu H, Li H, Nerbonne JM. 1999. Elimination of the transient outward current and action potential prolongation in mouse atrial myocytes expressing a dominant negative Kv4 alpha subunit. *J. Physiol.* 519:11–21

160. Xu H, Guo W, Nerbonne JM. 1999. Four kinetically distinct depolarization-activated K^+ currents in adult mouse ventricular myocytes. *J. Gen. Physiol.* 113:661–78

161. Warmke JW, Ganetzky B. 1994. A family of potassium channel genes related to *eag* in *Drosophila* in mammals. *Proc. Natl. Acad. Sci. USA* 91:3438–42

162. Yang T, Kupershmidt S, Roden DM. 1995. Anti-minK antisense decreases the amplitude of the rapidly-activating cardiac delayed rectifier K^+ current. *Circ. Res.* 77:1246–53

163. McDonald TV, Yu Z, Ming Z, Palma E, Meyers MB, et al. 1997. A minK-HERG complex regulates the cardiac potassium current I(Kr). *Nature* 388:289–92

164. Shibasaki T. 1987. Conductance and kinetics of delayed rectifier potassium channels in nodal cells of the rabbit heart. *J. Physiol.* 387:227–50

165. Wang L, Duff HJ. 1997. Developmental changes in transient outward current in mouse ventricle. *Circ. Res.* 81:120–27

166. Wang L, Feng ZP, Kondo CS, Sheldon RS, Duff HJ. 1996. Developmental changes in the delayed rectifier K^+ channels in mouse heart. *Circ. Res.* 79:79–85

167. Smith PL, Baukrowitz T, Yellen G. 1996. The inward rectification mechanism of the

HERG cardiac potassium channel. *Nature* 379:833–36

168. Yang T, Snyders DJ, Roden DM. 1997. Rapid inactivation determines the rectification and $[K^+]_o$ dependence of the rapid component of the delayed rectifier K^+ current in cardiac cells. *Circ. Res.* 80:782–89

169. Wang S, Morales MJ, Liu S, Strauss HC, Rasmusson RL. 1997. Modulation of HERG affinity for E-4031 by [K+]o and C-type inactivation. *FEBS Lett.* 417:43–47

170. Numaguchi H, Johnson JP Jr, Petersen CI, Balser JR. 2000. A sensitive mechanism for cation modulation of potassium current. *Nat. Neurosci.* 3:429–30

171. Viswanathan PC, Shaw RM, Rudy Y. 1999. Effects of IKr and IKs heterogeneity on action potential duration and its rate dependence: a simulation study. *Circulation* 99:2466–74

172. Zhou Z, Gong Q, Ye B, Fan Z, Makielski JC, et al. 1998. Properties of HERG channels stably expressed in HEK 293 cells studied at physiological temperature. *Biophys. J.* 74:230–41

173. Roden DM. 2000. Point of view: acquired long QT syndromes and the risk of proarrhythmia. *J. Cardiovasc. Electrophysiol.* 11:938–40

174. Mitcheson JS, Chen J, Sanguinetti MC. 2000. Trapping of a methanesulfonanilide by closure of the HERG potassium channel activation gate. *J. Gen. Physiol.* 115:229–40

175. Morais JH, Lee A, Cohen SL, Chait BT, Li M, MacKinnon R. 1998. Crystal structure and functional analysis of the HERG potassium channel N terminus: a eukaryotic PAS domain. *Cell* 95:649–55

176. London B, Trudeau MC, Newton KP, Beyer AK, Copeland NG, et al. 1997. Two isoforms of the mouse ether-a-go-go-related gene coassemble to form channels with properties similar to the rapidly activating component of the cardiac delayed rectifier K^+ current. *Circ. Res.* 81:870–78

177. Lees-Miller JP, Kondo C, Wang L, Duff HJ. 1997. Electrophysiological characterization of an alternatively processed ERG K^+ channel in mouse and human hearts. *Circ. Res.* 81:719–28

178. Kupershmidt S, Snyders DJ, Raes A, Roden DM. 1998. A K^+ channel splice variant common in human heart lacks a C-terminal domain required for expression of rapidly-activating delayed rectifier current. *J. Biol. Chem.* 273:27231–35

179. Zhou Z, Gong Q, January CT. 1999. Correction of defective protein trafficking of a mutant HERG potassium channel in human long QT syndrome. Pharmacological and temperature effects. *J. Biol. Chem.* 274:31123–26

180. Sanguinetti MC, Jurkiewicz NK, Scott A, Siegl PKS. 1991. Isoproterenol antagonizes prolongation of refractory period by the class III antiarrhythmic agent E-4031 in guinea pig myocytes: mechanism of action. *Circ. Res.* 68:77–84

181. Cui J, Melman Y, Palma E, Fishman GI, McDonald TV. 2000. Cyclic AMP regulates the HERG K(+) channel by dual pathways. *Curr. Biol.* 10:671–74

182. Hondeghem LM, Snyders DJ. 1990. Class III antiarrhythmic agents have a lot of potential, but a long way to go: reduced effectiveness and dangers of reverse use-dependence. *Circulation* 81:686–90

183. Jurkiewicz NK, Sanguinetti MC. 1993. Rate-dependent prolongation of cardiac action potentials by a methanesulfonanilide class III antiarrhythmic agent: specific block of rapidly activating delayed rectifier K^+ current by dofetilide. *Circ. Res.* 72:75–83

184. Roden DM. 1998. Taking the idio out of idiosyncratic—predicting Torsades de Pointes. *Pace Pac. Clin. Electrophysiol.* 21:1029–34

185. Sesti F, Abbott GW, Wei J, Murray KT, Saksena S, et al. 2000. A common polymorphism associated with antibiotic-induced cardiac arrhythmia. *Proc. Natl. Acad. Sci. USA* 97:10613–18

186. Donger C, Denjoy I, Berthet M, Neyroud N, Cruaud C, et al. 1997. KVLQT1 C-terminal missense mutation causes a forme fruste long-QT syndrome. *Circulation* 96:2778–81

187. Yang P, Wei J, Murray KT, Hohnloser S, Shimizu W, et al. 2001. Frequency of ion channel mutations and polymorphisms in a large population of patients with drug-associated long QT Syndrome. *PACE* 4(Pt. 2):539–786

188. Napolitano C, Schwartz PJ, Brown AM, Ronchetti E, Bianchi L, et al. 2000. Evidence for a cardiac ion channel mutation underlying drug-induced QT prolongation and life-threatening arrhythmias. *J. Cardiovasc. Electrophysiol.* 11:691–96

189. Piippo K, Holmstrom S, Swan H, Viitasalo M, Raatikka M, et al. 2001. Effect of the antimalarial drug halofantrine in the long QT syndrome due to a mutation of the cardiac sodium channel gene SCN5A. *Am. J. Cardiol.* 87:909–11

190. Sanguinetti MC, Jurkiewicz NK. 1990. Two components of cardiac delayed rectifier K^+ current: differential sensitivity to block by class III antiarrhythmic agents. *J. Gen. Physiol.* 96:195–215

191. Sanguinetti MC, Jurkiewicz NK. 1990. Lanthanum blocks a specific component of I_K and screens membrane surface change in cardiac cells. *Am. J. Physiol. Heart Circ. Physiol.* 259:H1881–H89

192. Balser JR, Bennett PB, Roden DM. 1990. Time-dependent outward current in guinea pig ventricular myocytes. Gating kinetics of the delayed rectifier. *J. Gen. Physiol.* 96:835–63

193. Sanguinetti MC, Curran ME, Zou A, Shen J, Spector PS, et al. 1996. Coassembly of KvLQT1 and *minK* (IsK) proteins to form cardiac I_{Ks} potassium channel. *Nature* 384:80–83

194. Barhanin J, Lesage F, Guillemare E, Fink M, Lazdunski M, Romey G. 1996. KvLQT1 and IsK (*minK*) proteins associate to form the I_{Ks} cardiac potassium current. *Nature* 384:78–80

195. Tai KK, Goldstein SN. 1998. The conduction pore of a cardiac potassium channel. *Nature* 391:605–8

196. Tapper AR, George AL Jr. 2001. The KVLQT1 S6 transmembrane seqment is a structural requirment for minK-mediated gating modulation. *Biophys. J.* 80:192a (Abstr.)

197. Vetter DE, Mann JR, Wangemann P, Liu J, McLaughlin KJ, et al. 1996. Inner ear defects induced by null mutation of the *isk* gene. *Neuron* 17:1251–64

198. Lee MP, Ravenel JD, Hu RJ, Lustig LR, Tomaselli G, et al. 2000. Targeted disruption of the Kvlqt1 gene causes deafness and gastric hyperplasia in mice. *J. Clin. Invest.* 106:1447–55

199. Demolombe S, Baro I, Pereon Y, Bliek J, Mohammadpanah R, et al. 1998. A dominant negative isoform of the long QT syndrome 1 gene product. *J. Biol. Chem.* 273:6837–43

200. Pereon Y, Demolombe S, Baro I, Drouin E, Charpentier F, Escande D. 2000. Differential expression of KvLQT1 isoforms across the human ventricular wall. *Am. J. Physiol. Heart Circ. Physiol.* 278:H1908–H15

201. Liu DW, Antzelevitch C. 1995. Characteristics of the delayed rectifier current (I_{Kr} and I_{Ks}) in canine ventricular epicardial, midmyocardial, and endocardial myocytes: a weaker I_{Ks} contributes to the longer action potential of the M cell. *Circ. Res.* 76:351–65

202. Harvey RD, Hume JR. 1989. Isoproterenol activates a chloride current, not the transient outward current, in rabbit ventricular myocytes. *Am. J. Physiol. Heart Circ. Physiol.* 257:C1177–H81

203. Tohse N. 1990. Calcium-sensitive delayed rectifier potassium current in guinea pig ventricular cells. *Am. J. Physiol. Heart Circ. Physiol.* 258:H1200–H7

204. Varnum MD, Busch AE, Bond CT, Maylie J, Adelman JP. 1993. The minK channel underlies the cardiac potassium current I_{Ks} and mediates species-specific

responses to protein kinase C. *Proc. Natl. Acad. Sci. USA* 90:11528–32

205. Motoike HK, Marx SO, Reiken S, Kurokawa J, Kass RS, Marks AR. 2001. Modulation of the cardiac delayed rectifier K+ channel (I_{Ks}) by macromolecular signaling complex. *Biophys. J.* 80:507a (Abstr.)

206. Potet F, Scott JD, Mohammad-Panah R, Escande D, Baro I. 2001. AKAP proteins anchor cAMP-dependent protein kinase to KvLQT1/IsK channel complex. *Am. J. Physiol. Heart Circ. Physiol.* 280:H2038–H45

207. Salata JJ, Jurkiewicz NK, Sanguinetti MC, Siegl PKS, Claremon DA, et al. 1996. The novel class III antiarrhythmic agent, L-735,821 is a potent and selective blocker of IKs in guinea pig ventricular myocytes. *Circulation* 94(8):I529 (Abstr.)

208. Busch AE, Suessbrich H, Waldegger S, Sailer E, Greger R, et al. 1996. Inhibition of IKs in guinea pig cardiac myocytes and guinea pig IsK channels by the chromanol 293B. *Pflügers Arch.* 432:1094–96

209. Loussouarn G, Charpentier F, Mohammad-Panah R, Kunzelmann K, Baro I, Escande D. 1997. KvLQT1 potassium channel but not IsK is the molecular target for *trans*-6-cyano-4-(*N*-ethylsulfonyl-*N*-methylamino)-3-hydroxy-2,2-dimethyl-chromane. *Mol. Pharmacol.* 52:1131–36

210. Shimizu W, Antzelevitch C. 1998. Cellular basis for the ECG features of the LQTt1 form of the long-QT syndrome—effects of beta-adrenergic agonists and antagonists and sodium channel blockers on transmural dispersion of repolarization and torsade de pointes. *Circulation* 98:2314–22

211. Schmitt N, Schwarz M, Peretz A, Abitbol I, Attali B, Pongs O. 2000. A recessive C-terminal Jervell and Lange-Nielsen mutation of the KCNQ1 channel impairs subunit assembly. *EMBO J.* 19:332–40

212. Yamashita F, Horie M, Kubota T, Yoshida H, Yumoto Y, et al. 2000. Characterization and subcellular localization of KCNQ1 with a heterozygous mutation in the C terminus. *J. Mol. Cell. Cardiol.* 33:197–207

213. Bianchi L, Shen Z, Dennis AT, Priori SG, Napolitano C, et al. 1999. Cellular dysfunction of LQT5-minK mutants: abnormalities of IKs, IKr and trafficking in long QT syndrome. *Hum. Mol. Genet.* 8:1499–507

214. Schulze-Bahr E, Wang Q, Wedekind H, Haverkamp W, Chen Q, Sun Y. 1997. KCNE1 mutations cause Jervell and Lange-Nielsen syndrome. *Nat. Genet.* 17:267–68

215. Splawski I, Timothy KW, Vincent GM, Atkinson DL, Keating MT. 1997. Molecular basis of the long QT syndrome associated with deafness. *N. Engl. J. Med.* 336:1562–67

216. Drici MD, Arrighi I, Chouabe C, Mann JR, Lazdunski M, et al. 1998. Involvement of IsK-associated K^+ channel in heart rate control of repolarization in a murine engineered model of Jervell and Lange-Nielsen syndrome. *Circ. Res.* 83:95–102

217. Kupershmidt S, Yang T, Anderson ME, Wessels A, Niswender KD, et al. 1999. Replacement by homologous recombination of the *minK* gene with *lacZ* reveals restriction of *minK* expression to the mouse cardiac conduction system. *Circ. Res.* 84:146–52

218. Temple J, Frias PA, Kupershmidt S, Gbadebo D, Zhang W, et al. 2000. Spontaneous and induced atrial fibrillation in minK knockout mice. *Circulation* 102:153 (Abstr.)

219. Casimiro MC, Knollmann BC, Ebert SN, Vary JC Jr, Greene AE, et al. 2001. Targeted disruption of the Kcnq1 gene produces a mouse model of Jervell and Lange-Nielsen syndrome. *Proc. Natl. Acad. Sci. USA* 98:2526–31

220. Demolombe S, Lande G, Charpentier F, van Roon MA, van den Hoff MJ, et al. 2001. Transgenic mice overexpressing human KvLQT1 dominant-negative

isoform. Part I: Phenotypic characterisation. *Cardiovasc. Res.* 50:314–27

221. Snyders DJ, Tamkun MM, Bennett PB. 1993. A rapidly-activating and slowly-inactivating potassium channel cloned from human heart. *J. Gen. Physiol.* 101:513–43

222. Fedida D, Wible B, Wang Z, Fermini B, Faust F, et al. 1993. Identity of a novel delayed rectifier current from human heart with a cloned K^+ channel current. *Circ. Res.* 73:210–16

223. Wang Z, Fermini B, Nattel S. 1993. Sustained depolarization-induced outward current in human atrial myocytes. Evidence for a novel delayed rectifier K^+ current similar to Kv1.5 cloned channel currents. *Circ. Res.* 73:1061–76

224. Feng JL, Wible B, Li GR, Wang ZG, Nattel S. 1997. Antisense oligodeoxynucleotides directed against Kv1.5 mRNA specifically inhibit ultrarapid delayed rectifier K^+ current in cultured adult human atrial myocytes. *Circ. Res.* 80:572–79

225. Yue L, Wang Z, Rindt H, Nattel S. 2000. Molecular evidence for a role of Shaw (Kv3) potassium channel subunits in potassium currents of dog atrium. *J. Physiol.* 527(3):467–78

226. Takimoto K, Fomina AF, Gealy R, Trimmer JS, Levitan ES. 1993. Dexamethasone rapidly induces Kv1.5 K^+ channel gene transcription and expression in clonal pituitary cells. *Neuron* 11:359–69

227. Roy ML, Saal D, Perney T, Sontheimer H, Waxman SG, Kaczmarek LK. 1996. Manipulation of the delayed rectifier Kv1.5 potassium channel in glial cells by antisense oligodeoxynucleotides. *GLIA* 18:177–84

228. Philipson LH, Hice RE, Schaefer K, LaMendola J, Bell GI, et al. 1991. Sequence and functional expression in Xenopus oocytes of a human insulinoma and islet potassium channel. *Proc. Natl. Acad. Sci. USA* 88:53–57

229. England SK, Uebele VN, Kodali J, Bennett PB, Tamkun MM. 1995. A novel K^+ channel beta-subunit (hKvβ1.3) is produced via alternative mRNA splicings. *J. Biol. Chem.* 270:28531–34

230. England SK, Uebele VN, Shear H, Kodali J, Bennett PB, Tamkun MM. 1995. Characterization of a voltage-gated K^+ channel beta subunit expressed in human heart. *Proc. Natl. Acad. Sci. USA* 6309–13

231. Kwak YG, Hu N, Wei J, George ALJ, Grobaski TD, et al. 1999. Protein kinase A phosphorylation alters Kvβ1.3 subunit-mediated inactivation of the Kv1.5 potassium channel. *J. Biol. Chem.* 274:13928–32

232. Murray KT, Hu N, England SK, Mashburn A, Watson M, Tamkun MM. 1996. Coexpression of a beta subunit enhances the effect of a phorbol ester on the Kv1.5 channel. *Circulation* 94:I473 (Abstr.)

233. Wissmann R, Baukrowitz T, Kalbacher H, Kalbitzer HR, Ruppersberg JP, et al. 1999. NMR structure and functional characteristics of the hydrophilic N terminus of the potassium channel beta-subunit Kvβ1.1. *J. Biol. Chem.* 274:35521–25

234. Bahring R, Milligan CJ, Vardanyan V, Engeland B, Young BA, et al. 2001. Coupling of voltage-dependent potassium channel inactivation and oxidoreductase active site of Kvβ subunits. *J. Biol. Chem.*

235. Wang Z, Kiehn J, Yang Q, Brown AM, Wible BA. 1996. Comparison of binding and block produced by alternatively spliced Kvβ1 subunits. *J. Biol. Chem.* 271:28311–17

236. Levitan ES, Hershman KM, Sherman TG, Takimoto K. 1996. Dexamethasone and stress upregulate Kv1.5 K^+ channel gene expression in rat ventricular myocytes. *Neuropharmacology* 35:1001–6

237. Nishiyama A, Kambe F, Kamiya K, Seo H, Toyama J. 1998. Effects of thyroid status on expression of voltage-gated potassium channels in rat left ventricle. *Cardiovasc. Res.* 40:343–51

238. Mays DJ, Boyden PA, Tamkun MM. 1997. Redistribution of the Kv1.5 potassium channel protein on the surface of

myocytes from the epicardial border zone of the infarcted canine heart. *Cardiovasc. Pathobiol.* 2:79–87

239. Sobko A, Peretz A, Attali B. 1998. Constitutive activation of delayed-rectifier potassium channels by a Src family tyrosine kinase in Schwann cells. *EMBO J.* 17:4723–34

240. Martens JR, Navarro-Polanco R, Coppock EA, Nishiyama A, Parshley L, et al. 2000. Differential targeting of Shaker-like potassium channels to lipid rafts. *J. Biol. Chem.* 275:7443–46

241. Martens JR, Sakamoto N, Sullivan SA, Grobaski TD, Tamkun MM. 2000. Isoform-specific localization of voltage-gated K^+ channels to distinct lipid raft populations: targeting of Kv1.5 to caveolae. *J. Biol. Chem.* 276:8409–14

242. Gong J, Xu J, Bezanilla M, van Huizen R, Derin R, Li M. 1999. Differential stimulation of PKC phosphorylation of potassium channels by ZIP1 and ZIP2. *Science* 285:1565–69

243. Van Wagoner DR, Pond AL, Mccarthy PM, Trimmer JS, Nerbonne JM. 1997. Outward K^+ current densities and Kv1.5 expression are reduced in chronic human atrial fibrillation. *Circ. Res.* 80:772–81

244. London B, Jeron A, Zhou J, Buckett P, Han X, et al. 1998. Long QT and ventricular arrhythmias in transgenic mice expressing the N terminus and first transmembrane segment of a voltage-gated potassium channel. *Proc. Natl. Acad. Sci. USA* 95:2926–31

245. Folco E, Mathur R, Mori Y, Buckett P, Koren G. 1997. A cellular model for long QT syndrome—trapping of heteromultimeric complexes consisting of truncated Kv1.1 potassium channel polypeptides and native kv1.4 and kv1.5 channels in the endoplasmic reticulum. *J. Biol. Chem.* 272:26505–10

246. Wible BA, De Biasi M, Majumder K, Taglialatela M, Brown AM. 1995. Cloning and functional expression of an inwardly rectifying K^+ channel from human atrium. *Circ. Res.* 76:343–50

247. Plaster NM, Tawil R, Tristani-Firouzi M, Canun S, Bendahhou S, et al. 2001. Mutations in Kir2.1 cause the development and episodic electrical phenotypes of Andersen/s syndrome. *Cell* 105:511–19

248. Krapivinsky G, Gordon EA, Wickman K, Velimirovic B, Krapivinsky L, Clapham DE. 1995. The G-protein-gated atrial K^+ channel I_{KACh} is a heteromultimer of two inwardly rectifying K^+-channel proteins. *Nature* 374:135–41

249. Inagaki N, Gonoi T, Clement JP, Namba N, Inazawa J, et al. 1995. Reconstitution of I_{KATP}: an inward rectifier subunit plus the sulfonylurea receptor. *Science* 270:1166–70

250. Matsuda H, Saigusa A, Irisawa H. 1987. Ohmic conductance through the inwardly rectifying K channel and blocking by internal Mg^{2+}. *Nature* 325:156–58

251. Vandenberg CA. 1987. Inward rectification of a potassium channel in cardiac ventricular cells depends on internal magnesium ions. *Proc. Natl. Acad. Sci. USA* 84:2560–64

252. Lopatin AN, Makhina EN, Nichols CG. 1994. Potassium channel block by cytoplasmic polyamines as the mechanism of intrinsic rectification. *Nature* 372:366–69

253. Fakler B, Braendle U, Glowatzki E, Weidemann S, Zenner HP, Ruppersberg JP. 1995. Strong voltage-dependent inward rectification of inward rectifier K^+ channels is caused by intracellular spermine. *Cell* 80:149–54

254. Ficker E, Taglialatela M, Wible BA, Henley CM, Brown AM. 1994. Spermine and spermidine as gating molecules for inward rectifier K^+ channels. *Science* 266:1068–72

255. Kurachi Y, Nakajima T, Sugimoto T. 1986. Acetylcholine activation of K^+ channels in cell-free membrane of atrial cells. *Am. J. Physiol. Heart Circ. Physiol.* 251:H681–H84

256. Huang CL, Jan YN, Jan LY. 1997.

Binding of the G protein betagamma subunit to multiple regions of G protein-gated inward-rectifying K$^+$ channels. *FEBS Lett.* 405:291–98

257. Wickman K, Nemec J, Gendler SJ, Clapham DE. 1998. Abnormal heart rate regulation in GIRK4 knockout mice. *Neuron* 20:103–14

258. Mark MD, Herlitze S. 2000. G-protein mediated gating of inward-rectifier K+ channels. *Eur. J. Biochem.* 267:5830–36

259. Zerangue N, Schwappach B, Jan YN, Jan LY. 1999. A new ER trafficking signal regulates the subunit stoichiometry of plasma membrane K(ATP) channels. *Neuron* 22:537–48

260. Grover GJ, Garlid KD. 2000. ATP-sensitive potassium channels: a review of their cardioprotective pharmacology. *J. Mol. Cell. Cardiol.* 32:677–95

261. Sato T, Sasaki N, Seharaseyon J, O'Rourke B, Marban E. 2000. Selective pharmacological agents implicate mitochondrial but not sarcolemmal K(ATP) channels in ischemic cardioprotection. *Circulation* 101:2418–23

262. DiFrancesco D. 1993. Pacemaker mechanisms in cardiac tissue. *Annu. Rev. Physiol.* 55:455–72

263. Moroni A, Gorza L, Beltrame M, Gravante B, Vaccari T, et al. 2001. HCN1 is a molecular determinant of the cardiac pacemaker current If. *J. Biol. Chem.*

264. Shi W, Wymore R, Yu H, Wu J, Wymore RT, et al. 1999. Distribution and prevalence of hyperpolarization-activated cation channel (HCN) mRNA expression in cardiac tissues. *Circ. Res.* 85:E1–6

265. Baker K, Warren KS, Yellen G, Fishman MC. 1997. Defective "pacemaker" current (I_h) in a zebrafish mutant with a slow heart rate. *Proc. Natl. Acad. Sci. USA* 94:4554–59

266. Yu H, Chang F, Cohen IS. 1993. Pacemaker current exists in ventricular myocytes. *Circ. Res.* 72:232–36

267. Santoro B, Tibbs GR. 1999. The HCN gene family: molecular basis

of the hyperpolarization-activated pacemaker channels. *Ann. NY Acad. Sci.* 868:741–64

268. Miller AG, Aldrich RW. 1996. Conversion of a delayed rectifier K$^+$ channel to a voltage-gated inward rectifier K$^+$ channel by three amino acid substitutions. *Neuron* 16:853–58

269. Backx PH, Marban E. 1993. Background potassium current active during the plateau of the action potential in guinea pig ventricular myocytes. *Circ. Res.* 72:890–900

270. Yue DT, Marban E. 1988. A novel cardiac potassium channel that is active and conductive at depolarized potentials. *Pflügers Arch.* 413:127–33

271. Levesque PC, Hart PJ, Hume JR, Kenyon JL, Horowitz B. 1992. Expression of cystic fibrosis transmembrane regulator Cl$^-$ channels in heart. *Circ. Res.* 71:1002–7

272. Hume JR, Duan D, Collier ML, Yamazaki J, Horowitz B. 2000. Anion transport in heart. *Physiol. Rev.* 80:31–81

273. Tinel N, Diochot S, Borsotto M, Lazdunski M, Barhanin J. 2000. KCNE2 confers background current characteristics to the cardiac KCNQ1 potassium channel. *EMBO J.* 19:6326–30

274. Salkoff L, Butler A, Fawcett G, Kunkel M, McArdle C, et al. 2001. Evolution tunes the excitability of individual neurons. *Neuroscience* 103:853–59

275. Lahat H, Eldar M, Levy-Nissenbaum E, Bahan T, Friedman E, et al. 2001. Autosomal recessive catecholamine- or exercise-induced polymorphic ventricular tachycardia: clinical features and assigment of the disease gene to chromosome 1p13-21. *Circulation.* 103:2822–27

276. McNamara DM, Weiss R, Seibel J, Barmada M, Wang J, et al. 1998. Molecular heterogeneity in the Brugada syndrome. *Circulation* 98:I456 (Abstr.)

277. Marks ML, Trippel DL, Keating MT. 1995. Long QT syndrome associated with syndactyly identified in females. *Am. J. Cardiol.* 76:744–45

Annu. Rev. Physiol. 2002. 64:477–502

FATTY ACID OXIDATION DISORDERS

Piero Rinaldo and Dietrich Matern

*Biochemical Genetics Laboratory, Department of Laboratory Medicine and
Pathology, Mayo Clinic and Foundation, Rochester, Minnesota 55905;
e-mail: rinaldo@mayo.edu; matern@mayo.edu*

Michael J. Bennett

*Departments of Pathology and Pediatrics, University of Texas Southwestern Medical
Center, Dallas, Texas 75390–9073; e-mail: michaelj.bennett@email.swmed.edu*

Key Words   carnitine, inborn errors, ketogenesis, membrane transport,
tissue specificity

■ **Abstract**   Genetic disorders of mitochondrial fatty acid β-oxidation have been
recognized within the last 20 years as important causes of morbidity and mortality,
highlighting the physiological significance of fatty acids as an energy source. Although the mammalian mitochondrial fatty acid-oxidizing system was recognized at
the beginning of the last century, our understanding of its exact nature remains incomplete, and new components are being identified frequently. Originally described
as a four-step enzymatic process located exclusively in the mitochondrial matrix, we
now recognize that long-chain-specific enzymes are bound to the inner mitochondrial
membrane, and some enzymes are expressed in a tissue-specific manner.
   Much of our new knowledge of fatty acid metabolism has come from the study of
patients who were diagnosed with single-gene autosomal recessive defects, a situation
that seems to be further evolving with the emergence of phenotypes determined by
combinations of multiple genetic and environmental factors. This review addresses
the normal process of mitochondrial fatty acid β-oxidation and discusses the clinical,
metabolic, and molecular aspects of more than 20 known inherited diseases of this
pathway that have been described to date.

INTRODUCTION

Mitochondrial fatty acid β-oxidation (FAO) represents a physiological response
to tissue energy depletion when fasting, during febrile illness, and increased mus-
cular activity. It provides as much as 80% of energy for heart and liver functions at
all times (1). In the liver, the oxidation of fatty acids fuels the synthesis of ketone
bodies, 3-hydroxy butyrate and acetoacetate, which are utilized as an alternative
energy source by extrahepatic organs, particularly the brain (2). The oxidation of
long-chain fatty acids also provides the energy required for nonshivering thermogenesis by brown adipose tissue (3).

0066-4278/02/0315-0477$14.00

The first genetic defect of FAO in humans was recognized in 1973 as a disorder of skeletal muscle presenting with exercise-induced rhabdomyolysis and myoglobinuria (4). In the early 1980s, hepatic presentations of FAO defects were reported in individuals who were eventually diagnosed with medium-chain acyl-CoA dehydrogenase (MCAD) deficiency (5). In the ensuing years, new disorders have been discovered at a rapid pace. The clinical phenotypes have expanded considerably to include association with a growing number of clinical entities such as Reye syndrome (6), sudden infant death syndrome (7, 8), cyclic vomiting syndrome (9), fulminant liver disease (10), and maternal complications of pregnancy (11). Advances in the field of FAO disorders have been so rapid that recent reviews of the topic (12, 13) already require updating.

In this review we describe the components of normal FAO, the known genetic defects of the pathway, and future directions for this expanding field including the rationale for neonatal and postmortem screening programs.

COMPONENTS OF FATTY ACID METABOLISM

Lipid Mobilization and Plasma Membrane Fatty Acid Transport and Binding

Long-chain fatty acids (LCFA) are critically important in cellular homeostasis as they are involved in a wide variety of processes including phospholipid synthesis, protein post-translational modifications, cell signaling, membrane permeability, and transcription control (14). LCFA are also key substrates in the response to increased energy demands. The physiological response to fasting is linked to the endocrine-mediated utilization of stored hepatic and muscle glycogen to maintain normoglycemia. When glycogen reserves are depleted, triglycerides are mobilized from lipid stores, and fatty acids are released at the endothelial walls of capillaries and at the hepatocyte surface by the action of lipoprotein lipase and hepatic lipase, respectively. The physiologically available fatty acids are mostly C_{16} and C_{18} species and include saturated and both mono- and di-unsaturated species. The free fatty acids (FFA) released by lipolysis are transported across the plasma membrane of liver and muscle cells by tissue-specific fatty acid transporters (15). At least five different long-chain fatty acid transporters have been characterized in different species, and a murine gene family has been recently described, providing evidence that long-chain fatty acid transporters are a large evolutionarily conserved family of proteins with considerable tissue specificity (16).

The multiple roles of LCFA suggest that cells tightly regulate all aspects of LCFA utilization beginning from cellular uptake and retention. Although some investigators have found evidence in favor of passive diffusion, others believe in an active involvement of plasma membrane proteins in LCFA uptake in mammalian cells. Evidence includes saturability of LCFA uptake by mammalian cells with respect to the concentration of nonesterified LCFA (14, 15), competition of LCFA uptake by structurally related LCFA analogues, inhibition of LCFA uptake by nonpermeant reagents, and increased uptake activity after overexpression of genes coding for LCFA transporters.

Plasma Membrane Uptake of Carnitine

Carnitine (β-hydroxy γ-trimethylaminobutyric acid) is a critical factor for LCFA intramitochondrial transport. Tissue-specific plasma membrane carnitine transporters actively transport carnitine from the blood up a tissue gradient that concentrates intracellular carnitine levels to 50 times the plasma level (17). The best-characterized plasma membrane carnitine transporter, which is encoded by the *OCTN2* gene, a member of the organic cation transporter family, is the one present in muscle, heart, and renal tubule cells (18). A distinct carnitine transporter is expressed in the liver. The carnitine transporters differ considerably with regard to their physiological properties. The product of the *OCTN2* gene is a low K_m (2–6 μmol) high-affinity, sodium-dependent active transporter, whereas the hepatic transporter is a high K_m (500 μmol), low-affinity transporter.

Mitochondrial Transport of Fatty Acids

At the outer mitochondrial membrane, nonesterified fatty acids are activated to acyl-CoA esters by ATP-dependent acyl-CoA synthetases, which are probably involved also in the transport of long-chain acyl-CoA species into the transmembrane space (19). When intracellular levels of malonyl-CoA decline in response to fasting, the activity of carnitine palmitoyl transferase I (CPT I) is upregulated. This drives the CPT I reaction toward the formation of long-chain acyl-carnitine from acyl-CoA and free carnitine (Figure 1). The resulting long-chain acyl-carnitine is transported across the inner mitochondrial membrane by carnitine:acylcarnitine translocase (CACT). Once inside the mitochondrial matrix, the acyl-CoA is reformed by the action of the inner mitochondrial membrane-bound CPT II using free CoA. The carnitine released by this reaction is recycled for subsequent reutilization via CACT.

The step catalyzed by CPT I is rate limiting for entry of fatty acids into the mitochondria. During the postprandial carbohydrate load, intracellular levels of malonyl-CoA increase and inhibit the activity of CPT I, driving metabolism in the direction of fatty acid and triacylglycerol synthesis. Fasting ends the synthesis of malonyl-CoA and redirects acetyl-CoA toward ketogenesis. Medium- and short-chain fatty acids (chain length 12 and lower) enter the mitochondria independently of the carnitine shuttle and are presumably activated within the mitochondrial matrix by different acyl-CoA synthetases.

CPT I was the first enzyme of FAO shown to exist in distinct tissue-specific isoforms (20). Liver CPT I (L-CPT I) is encoded on chromosome 11q13; the locus of muscle CPT I (M-CPT I) is on chromosome 22q13.3 (21). Each of the CPT I genes encodes an 88-kDa polypeptide with 63% identity. However, they have different kinetic properties: L-CPT I has a K_m to carnitine of approximately 30 μmol, the level that is present as free carnitine in serum, whereas M-CPT I has

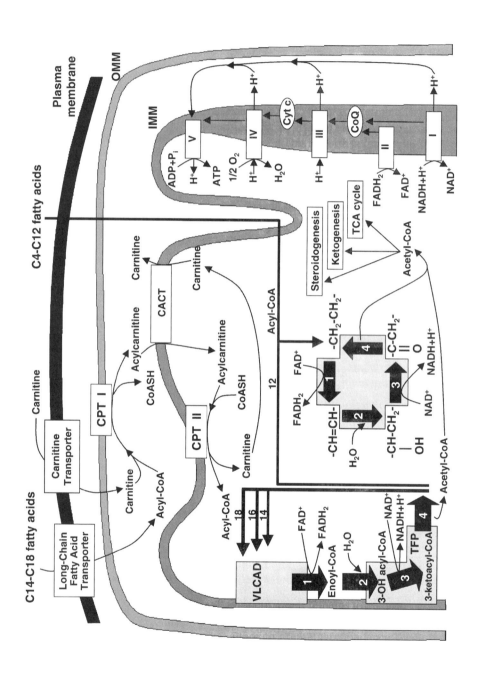

a K_m of 500 μmol, which approximates to intracellular carnitine concentrations. The CACT gene has recently been cloned (22); it encodes a 33-kDa membrane-associated polypeptide related to the mitochondrial transporter family. The gene encoding CPT II is found on chromosome 1p32 (23). CACT and CPT II are not known to have differential tissue expression.

Fatty Acid β-Oxidation and Hepatic Ketogenesis

The β-oxidation process involves the concerted action of a series of chain-length-specific enzymes, which sequentially remove a molecule of acetyl-CoA per cycle from the initial fatty acid substrate (Figure 1). The first reaction involves a member of the acyl-CoA dehydrogenase family of FAD-requiring oxidoreductases. Very-long-chain acyl-CoA dehydrogenase (VLCAD) is a membrane-associated enzyme responsible for reduction across the 2,3 position of the acyl-CoA moiety to produce a 2,3-enoyl-CoA (24). The electrons are transferred via electron transfer flavoprotein (ETF) to ETF:coenzyme Q oxidoreductase (ETF-QO) and ultimately to the oxidative phosphorylation pathway from which energy of oxidation is derived. ETF and ETF-QO are mitochondrial enzymes encoded by nuclear genes (25). ETF, a heterodimer consisting of an α- and a β-subunit, is located within the mitochondrial matrix; ETF-QO is a mitochondrial membrane protein. Both enzymes are involved in FAO only indirectly by way of accepting electrons from FAD-dependent acyl-CoA dehydrogenases.

VLCAD is responsible for reducing acyl-CoAs of chain lengths $C_{12}-C_{18}$. Two additional members of the acyl-CoA dehydrogenase family, medium-chain and short-chain acyl-CoA dehydrogenases, respectively (MCAD, SCAD), are found in the mitochondrial matrix. MCAD is responsible for the metabolism of acyl-CoAs

---

Figure 1 Pathway of fatty acid metabolism from the transport of long-chain fatty acids and carnitine across the plasma membrane to the production of acetyl-CoA. Medium- and short-chain fatty acids do not require an active transport mechanism to reach the mitochondrial matrix. Enzymes of the carnitine cycle (CPT I, CACT, and CPT II) shuttle long-chain fatty acids across the mitochondrial membranes. The fatty acid β-oxidation spiral includes an FAD-dependent acyl-CoA dehydrogenase step (1) followed by a 2,3-enoyl-CoA hydratase reaction (2), the NAD-dependent 3-hydroxyacyl-CoA dehydrogenase step (3), and the thiolase cleavage reaction (4). Oxidation of long-chain fatty acids (membrane-bound enzymes) and medium-/short-chain fatty acids (matrix) are shown separately. Reducing equivalents (FADH$_2$ and NADH+H$^+$) are directed to the oxidative phosphorylation pathway (*right*). Abbreviations: CACT, carnitine:acylcarnitine translocase; CPT, carnitine palmitoyltransferase; IMM, inner mithocondrial membrane; OMM, outer mitochondrial membrane; TCA, tricarboxylic acid; TFP, trifunctional protein; VLCAD, very-long chain acyl-CoA dehydrogenase. Figure designed by Joshua J. Jacobson (Mayo High School Gifted and Talented Education Program, Rochester, MN).

of chain lengths C_6–C_{10}, and SCAD is specific for C_4 and C_6 substrates in the two final rounds of the β-oxidation cycle in the mitochondrial matrix (Figure 1). MCAD and SCAD are catalytically active as homotetramers of 44-kDa polypeptides (26).

Long-chain acyl-CoA dehydrogenase (LCAD) is also a member of this family of enzymes and has substrate chain length specificity between VLCAD and MCAD. Although no human case of LCAD deficiency has been described, a mouse model of this disorder results in gestational loss, lipidosis, hypoglycemia, and myocardial degeneration, a biochemical phenotype similar to human VLCAD deficiency (27). The role of LCAD in maintaining fasting energy homeostasis may be of greater importance in mice than in humans. However, its function and potential clinical significance require elucidation, particularly in human embryonic development (28).

The second reaction involves hydration of the double bond in the 2,3 position to produce a stereo-specific L-3-hydroxyacyl-CoA species. Two enzymes able to carry out this reaction have been described to date: Long-chain enoyl-CoA hydratase (LHYD) is responsible for hydrating long-chain species and is part of a membrane-associated trifunctional enzyme complex called the mitochondrial trifunctional protein (TFP). Active TFP is an heterooctamer of $\alpha_4\beta_4$ subunits (29, 30). The α-subunit encodes LHYD activity and the subsequent enzyme in the sequence, long-chain L-3-hydroxyacyl-CoA dehydrogenase (LCHAD). The β-subunit has long-chain 3-ketoacyl-CoA thiolase (LKAT) activity, the fourth enzyme in the cycle. Association of the complex is necessary for membrane translocation and for catalytic stability of the individual enzymes (31, 32). The genes for both subunits are contiguous on chromosome 2p23.3–p24.1 and are coordinately regulated (33, 34). The second hydratase is crotonase (short-chain enoyl-CoA hydratase), which is responsible for the hydration of medium- and short-chain enoyl-CoA species (35). Although a medium-chain hydratase has also been reported (36), it was later reclassified by Hashimoto and his group as a peroxisomal protein (37), a fitting example of the degree of uncertainty still to be found in our understanding of the catalytic and regulatory components of this pathway.

The third step of FAO involves the reduction at the L-3-hydroxy position to yield a 3-ketoacyl-CoA species. In this reaction, NAD^+ is reduced to NADH. LCHAD is an integral part of the TFP and is responsible for the reduction of long-chain substrates, while a medium- short-chain L-3-hydroxyacyl-CoA dehydrogenase (M/SCHAD) with broad-chain-length specificity is responsible for medium- and short-chain species (38). M/SCHAD is a soluble homodimer located in the mitochondrial matrix, which was originally designated as short-chain 3-hydroxyacyl-CoA dehydrogenase based on its increased ratio of C_4: C_{16} activity in comparison with the homologous enzyme, LCHAD. A third enzyme with 3-hydroxyacyl-CoA dehydrogenase activity, classified as HAD-II, has been purified from bovine liver based on its in vitro activity, and the cDNA was cloned (39). HAD-II does not share homology with M/SCHAD and LCHAD, and its role in mitochondrial β-oxidation is unclear. The gene for SCHAD (M/SCHAD) is located on chromosome 4q22–q26 and encodes a 27-kDa polypeptide (40).

The final step of β-oxidation involves thiolytic cleavage of the 3-ketoacyl-CoA to yield acetyl-CoA and a two-carbon chain-shortened acyl-CoA. The enzyme responsible for long-chain species is LKAT, which is also part of the TFP. A purified protein with medium-chain 3-ketoacyl-CoA thiolase (MKAT) activity has been described (41), and a patient with a defect at this level has been diagnosed (42), although a MKAT gene has not yet been isolated. Short-chain 3-ketoacyl-CoA thiolase activity (SKAT) represents the final step of β-oxidation. SKAT is also known as β-ketothiolase and represents a common step in the metabolism of fatty acids and isoleucine.

In the liver, acetyl-CoA generated by this process under fasting conditions is targeted toward ketogenesis. 3-hydroxy 3-methylglutaryl-CoA (HMG-CoA) synthetase takes one molecule of acetyl-CoA and one of acetoacetyl-CoA to form HMG-CoA, which is then cleaved by HMG-CoA lyase to yield a molecule of acetoacetate, which is in redox equilibrium with D-3-hydroxy butyrate, the quantitatively more abundant ketone body species. Ketone bodies enter the bloodstream and are taken up by tissues with a limited capacity to carry out β-oxidation, in particular the brain. Acetoacetate is converted back to its active form, acetoacetyl-CoA, by acetoacetate:succinyl-CoA transferase. The acetoacetyl-CoA is then thiolytically cleaved into two molecules of acetyl-CoA and fed into the Krebs cycle (2).

FATTY ACID OXIDATION DISORDERS

Disorders of Plasma Membrane Functions

LONG-CHAIN FATTY ACID TRANSPORT/BINDING DEFECT   Two patients with in vitro evidence of a defect of long-chain fatty acid transport at the plasma membrane level have been reported (10). At five years of age, one patient underwent liver transplantation following his seventh life-threatening episode of acute liver failure, which evolved into chronic hepatic insufficiency. The second patient was in apparently good health before presenting with a single episode of fulminant liver failure at four years of age. He also underwent successful liver transplantation.

In cultured skin fibroblasts, localization of the defect at the plasma membrane level was inferred by observing reduced intracellular concentration of C_{14}–C_{18} fatty acids and reduced cellular oxidation of palmitate and oleate. The latter impairment was normalized by permeabilization of the plasma membrane with digitonin. Uptake of 2-deoxy-D-glucose, carnitine, and palmitoyl-L-carnitine were normal, supportive of the specificity of the defect to fatty acids. The exact mechanism (transport, intracellular-binding, channeling) remains to be elucidated. Similar cases have been reported, but quite possibly more than one genetic defect is involved, a hypothesis supported by the outcome of in vitro complementation studies (43; A. Al Odaib & P. Rinaldo, unpublished results). At least one patient has been found to have a complex genotype with a defect of fatty acid transport associated with heterozygosity for two other disorders of fatty acid metabolism (44).

Sequencing and immunoblot experiments targeting several candidate genes have been uninformative to date (A. Al Odaib & P. Rinaldo, unpublished results). In these patients, an underlying defect of the myocardial fatty acid translocase (FAT/CD36) is unlikely considering the clinical presentation with predominant liver disease without cardiac involvement (45). Limited utilization of fatty acids in normal skin fibroblasts makes the study of premitochondrial steps of this pathway particularly difficult and probably not a good representation of the situation in much more metabolically active cells such as hepatocytes, myocytes, and adipocytes.

CARNITINE UPTAKE DEFECT   A genetic defect of the plasma membrane carnitine transporter represents the only true entity of primary carnitine deficiency (46). Carnitine losses owing to defective renal reabsorption are substantial and deplete tissue levels of carnitine to the point of compromising the mitochondrial transport of long-chain fatty acids. Typical manifestations in the first year of life are fasting intolerance, hypoglycemia, and sudden death. In older patients, progressive cardiomyopathy and muscle weakness are observed (47). Characteristically low plasma carnitine concentrations (typically less than 10% of normal) are promptly corrected by oral carnitine and are sufficient to restore intracellular levels to allow for normal function. Treatment is for life because discontinuation is rapidly followed by clinical relapse. Diagnosis is based either on uptake studies in skin fibroblasts or by direct molecular analysis of the *OCTN2* gene (18).

Disorders of Mitochondrial Fatty Acid Transport

CARNITINE PALMITOYLTRANSFERASE I DEFICIENCY   Liver CPT I deficiency was first described in 1981 in a patient with fasting-induced nonketotic hypoglycemia and absence of characteristic fatty acid metabolites in blood and urine (48). However, high plasma concentrations of total and free carnitine frequently offer an important clue in the differential diagnosis of this disorder (49). Approximately 30 cases have been diagnosed worldwide, with a presentation that is primarily hepatic (50) and evolving, in some cases, to a potentially fatal hepatic encephalopathy mimicking Reye syndrome. Recently, a myopathic form of the disorder has been described in some patients, including an infant with concurrent hepatic and skeletal abnormalities and an adult with a progressive myopathy without liver disease (51, 52). In the latter case, residual CPT-I activity in fibroblasts was 15% of controls, a level that was apparently sufficient to maintain function in the liver but, for reasons not yet elucidated, not in muscle.

To date, gene mutations have been reported in at least eight patients (53). The majority of the mutations appear to be private, that is, a particular mutation restricted to a single family, and only offspring of consanguineous marriages were homozygous for a given mutation. The mutation described in a Hutterite infant may have implications for genetic analysis of that population (54).

CARNITINE: ACYLCARNITINE TRANSLOCASE DEFICIENCY   First described in 1992, many patients affected with this disorder died with a chronic progressive liver failure and persistent nonreversible hyperammonemia or because of hypertrophic cardiomyopathy and septal heart defects (55). Skeletal myopathy is also a common feature; hypoglycemia has rarely been described, which makes this a rather difficult disorder to diagnose. Recently, a few patients have been described with a milder clinical course, anecdotally responsive to treatment (56). Multiple private mutations (often splicing site mutations) in the CACT gene have been identified since the initial report (57), and no genotype/phenotype correlations have been recognized. With more CACT-deficient patients identified by expanded newborn screening using tandem mass spectrometry (58), a better understanding of the natural history of this disorder can be expected in the future.

CARNITINE PALMITOYLTRANSFERASE II DEFICIENCY   CPT II deficiency was the first defect of mitochondrial FAO to be described in a patient with exercise-induced myopathy and myoglobinuria (4). A common point mutation (439C → T, S113L) has been found among cases who present with the myopathic phenotype (59). The S113L mutant peptide is translocated normally into the mitochondrion and is processed appropriately. However, homozygosity for this common mutation results in a mature enzyme with 15–25% of residual activity. This appears to be sufficient to spare the liver and heart from disease involvement and affects only skeletal muscle during prolonged exercise, especially when combined with exposure and fasting.

Neonatal presentation of CPT II deficiency, resulting from mutations that either compromise translocation and processing or yield a mature enzyme with no measurable residual activity, has also been described. The phenotype for this severe disorder involves multiple organs and congenital anomalies and often results in neonatal death (60). An intermediate form of CPT II deficiency has also been described with a somewhat milder course owing to compound heterozygosity for a severe and a mild mutation.

Disorders of Long-Chain Fatty Acid β-Oxidation

VERY-LONG-CHAIN ACYL-CoA DEHYDROGENASE DEFICIENCY   There has been some confusion in the literature with regard to defects of palmitoyl-CoA dehydrogenation, initially thought to be a deficiency of the mitochondrial matrix enzyme LCAD. All cases diagnosed with LCAD deficiency before 1992 were subsequently shown to have a defect of the membrane-bound enzyme VLCAD (61). The early reports of (V)LCAD deficiency were of a severe, often lethal presentation in infancy, with manifestations ranging from dilated cardiomyopathy to fasting hypoglycemia, Reye-like disease, or sudden unexpected death (62).

Molecular genetic studies of the severe cases revealed mutations causing total loss of enzymatic activity (63). There are, however, other mutations that conserve some residual activity causing a milder, later onset disease similar to the adult form of CPT II deficiency (64). For both forms, it is clear that early recognition and appropriate clinical management have resulted in reduced morbidity and mortality, including patients with the severe phenotype for whom long-term survival has been documented.

TRIFUNCTIONAL PROTEIN DEFICIENCY AND ISOLATED LONG-CHAIN L-3–HYDROXY-ACYL-CoA DEHYDROGENASE DEFICIENCY   Although LCHAD is one of the three enzymatic components of the TFP, there are many clinical and molecular genetic distinctions between isolated LCHAD deficiency and combined defects of all three catalytic activities (TFP deficiency). Isolated LCHAD deficiency was first described in 1989 and is recognized as one of the most severe FAO disorders. The first cases presented with profound liver disease, varying from acute hepatic failure in the newborn period to a more insidious chronic progressive liver disease in later infancy, evolving into fulminant liver failure with cirrhosis (65). As more cases were recognized, the phenotypic spectra expanded to include cardiomyopathy, skeletal myopathy, retinal pigmentary changes, peripheral neuropathy, and sudden death. Isolated LCHAD deficiency appears to be a much more common entity than TFP deficiency. This can probably be attributed to the existence of a common mutation in the LCHAD coding region of the α-subunit (1528G → C). This mutation results in an amino acid change from glutamine to glutamate in position 474 (E474Q) of the mature polypeptide and accounts for approximately 60% of alleles in LCHAD deficiency (66). Mutations elsewhere in both the α- and β-subunit genes invariably comprise all three catalytic activities and result in TFP deficiency, presenting primarily with heart disease and/or skeletal myopathy (67). Only a few patients with TFP deficiency had a predominant liver disease. Although no patients with isolated deficiency of LHYD or LKAT have been found to date, we reason that point mutations at the catalytic sites for LHYD or LKAT activities, which do not comprise the assembly of the $\alpha_4\beta_4$ complex, are likely to occur.

Since 1990, more than 60 patients with LCHAD deficiency have been reported who were born following pregnancies complicated with severe maternal liver disease, in particular acute fatty liver of pregnancy (AFLP) and HELLP syndrome (hemolysis, elevated liver function tests, low platelets). In 21 pregnancies complicated by third-trimester AFLP, the common LCHAD gene mutation was present on at least one allele of the TFP α-subunit gene in all 12 of the affected infants for whom molecular genetic studies were performed (11).

During pregnancy, increased activity of hormone-sensitive lipase in combination with gestational insulin resistance cause an increase in maternal plasma FFA levels (68). The maternal liver responds to this increase by synthesizing triglycerides, which are secreted as VLDL and LDL (69). In the last trimester, however, the greater energy demands of fetal growth shifts maternal metabolism toward ketogenesis (70), and the fetus uses maternal ketone bodies for lipogenesis as well as for fuel (71).

FFAs serve as a significant energy source in the placenta. This energy is generated via FAO, implying that disruption of the FAO pathway is likely to have detrimental effects on placental function (72). It can be speculated that ketogenesis is impaired under conditions of greater and sustained demand, for instance in a pregnant woman who is heterozygous for a FAO disorder. If the fetus is affected with LCHAD deficiency, the metabolic defect disrupts placental FFA metabolism, leading to progressive accumulation of fatty acid metabolites. Although the fetus may initially use these metabolites to sustain lipogenesis, they eventually re-enter the maternal circulation and ultimately overwhelm an already challenged metabolic balance, resulting in steatosis, liver failure, and potentially life-threatening clinical symptoms (73). At the same time, the placenta will suffer from energy depletion, which may lead to intrauterine growth retardation of the fetus and premature delivery. Both complications are relatively common in newborns with LCHAD deficiency.

Vulnerability to maternal complications of pregnancy is not limited to LCHAD deficiency (74), suggesting that a defect at different levels of the FAO pathway may cause AFLP, the HELLP syndrome, or additional phenotypes where the most prominent prenatal manifestation is intrauterine growth retardation (75). Despite the possibly coincidental nature of some anecdotal observations, a detailed inquiry of the past obstetric history should be a required component of the clinical evaluation of new patients with any FAO disorder. Infants born following such complicated pregnancies should undergo further evaluation by plasma acylcarnitine analysis and molecular genetic analysis at least for the common LCHAD gene mutation. Furthermore, routine monitoring of subsequent pregnancies in known families should include measures to achieve early detection and management of maternal complications.

Disorders of Medium-Chain Fatty Acid β-Oxidation

MEDIUM-CHAIN ACYL-CoA DEHYDROGENASE DEFICIENCY   Since the initial clinical and biochemical reports almost 20 years ago (5), MCAD deficiency has emerged as the most frequently encountered disorder of the FAO pathway and overall as one of the most recognizable inborn errors of metabolism (76). MCAD deficiency is a disease that is prevalent in Caucasians, especially of Northern European descent, with a carrier frequency for the common K304E mutation of the *ACADM* gene of approximately 1:40. The overall frequency of the disease has been estimated to range between 1:6500 to 1:17,000.

Patients with MCAD deficiency are regarded as normal at birth even when transient episodes of "benign" hypoglycemia are observed. Later, usually between 3 and 24 months of age, acute decompensation occurs in response to either prolonged fasting (e.g., weaning the infant from nighttime feedings) or intercurrent and common infections (e.g., viral gastrointestinal or upper respiratory tract infections) associated with reduced food intake and increased energy requirements. Unexpected death during the first metabolic decompensation is common and may occur as late as in adulthood (77). Although metabolic stress may quickly progress to a life-threatening situation, the prognosis is excellent once the diagnosis is established, especially if detected by newborn screening before the onset of symptoms (78).

The frequent observation of intrafamilial differences of the phenotypic expression of MCAD deficiency is inconsistent with a possible genotype-phenotype correlation. Some authors have nevertheless suggested that a subset of patients still recognizable by a "mild" biochemical phenotype could have a benign form of the disease (79–81). This claim implies the existence of a genotype/phenotype correlation and, more importantly, raises a concern that patients with no risk of metabolic decompensation could be detected by newborn screening. These reports, however, were focused on genotype analysis, relied on short-term follow-up and failed to provide convincing biochemical evidence that such a milder phenotype actually exists. In one report (78), for example, an apparent lack of phenylpropionylglycine excretion after an oral phenylpropionic acid load was hardly consistent with a deficient MCAD activity in vitro ascertained using phenylpropionyl-CoA as substrate!

To prove that a benign form of MCAD deficiency truly exists, a normal clinical and laboratory response to prolonged fasting (up to 24 h) should be documented (82). Until then, MCAD deficiency must be regarded as a disorder in which environmental factors (fasting, body temperature, response to illness) play a paramount role in determining the outcome of all patients and are far more important than genotype characterization (76).

MEDIUM- AND SHORT-CHAIN L-3–HYDROXYACYL-CoA DEHYDROGENASE DEFICIENCY
In earlier reviews of FAO disorders, the clinical entity now known as M/SCHAD deficiency was described as a short-chain defect (13). Our present understanding of the protein encoded by the SCHAD gene suggests a much broader chain length specificity (38). At the clinical and biochemical levels, a deficiency of this enzyme has been described with three distinct phenotypes. The first patient presented with cardiomyopathy and recurrent rhabdomyolysis, which culminated in death as a teenager (83). Enzyme activity was reduced in skeletal muscle but not in skin fibroblasts, other tissues were not studied enzymatically. A second phenotype was described with a presentation similar to that seen in ketotic hypoglycemia, and enzyme deficiency was described in isolated mitochondria from cultured skin fibroblasts (84). Other tissues were not analyzed in these patients. A third group of patients had hepatic involvement and steatosis, the catalytic activity with the C_4 substrate was deficient in liver but normal in muscle and fibroblasts (85). Mutation analysis of the SCHAD gene in these patients has been inconclusive, suggesting that other proteins in addition to M/SCHAD may be necessary for enzymatic activity and that there may be a tissue-specific distribution of some of these proteins. A mouse knockout model of M/SCHAD deficiency expresses systemic disease with activities reduced in all tissues examined (86). Apparent disease-causing mutations have been recently described in a patient with hypoketotic hypoglycemia (87) and in another patient with fulminant liver failure that required transplantation (88). Further molecular information and

detailed studies of M/SCHAD protein interactions should improve our understanding of the function, substrate specificity, and tissue expression of this peculiar enzyme.

MEDIUM-CHAIN 3-KETOACYL-CoA THIOLASE DEFICIENCY   This disorder was first described in a Japanese male neonate who died shortly after presentation at two days of age with vomiting, dehydration, metabolic acidosis, liver dysfunction, and terminal rhabdomyolysis with myoglobinuria (42). Urine organic acid analysis revealed ketotic lactic aciduria and significant C_6-C_{12} dicarboxylic aciduria, with strikingly elevated C_{10} and C_{12} species. In skin fibroblasts, palmitate oxidation was normal, octanoate oxidation was reduced to 31% of controls, and there was an isolated deficiency of MKAT activity, a result supported by the finding of a reduced protein signal by immunoprecipitation. A better understanding of this disorder will require additional patients to be identified and the gene to be isolated.

Disorders of Short-Chain Fatty Acid β-Oxidation

SHORT-CHAIN ACYL-CoA DEHYDROGENASE DEFICIENCY   SCAD deficiency was first described in 1987 (89). The number of unequivocally confirmed cases is still small, and the natural history remains poorly defined. The spectrum of manifestations in patients with SCAD deficiency ranges from those who are lifetime asymptomatic to others who had fatal outcomes (90). Muscle hypotonia and developmental delay are relatively common symptoms. The identification of the SCAD gene and subsequent genotyping of patients thought to have SCAD deficiency led to the recognition of two common SCAD gene variants (625G $\rightarrow$ A and 511C $\rightarrow$ T). These alleles are not regarded as true disease-causing mutations nor are they polymorphisms, but rather mutations that confer disease susceptibility (91). The current approach to confirming a diagnosis of SCAD deficiency is based on a systematic use of biochemical studies on body fluids, enzymatic studies in skin fibroblasts and muscle, and complete sequencing of the SCAD gene. Once this information is gathered in a sufficient number of patients, along with clinical parameters, it will be possible to establish consensus criteria and determine the influence of additional genetic, cellular, and/or environmental factors on the phenotypic expression of SCAD deficiency.

Other FAO Disorders

A single patient with biochemical and enzymatic evidence of 2,4-dienoyl-CoA reductase deficiency was described more than 10 years ago (92). The diagnosis originated from the identification of 2-*trans*,4-*cis*-decadienoylcarnitine in plasma and urine. It is surprising that no additional cases have been found after the initial report despite the routine application of acylcarnitine analysis in the laboratory evaluation of metabolic patients.

Disorders of Ketogenesis

3-HYDROXY 3-METHYLGLUTARYL-CoA SYNTHASE DEFICIENCY   Mitochondrial HMG-CoA synthase deficiency has been described in only two patients, both of whom had recurrent fasting-induced hypoketotic-hypoglycemia (93, 94). A recent report has questioned the reported absence of a biochemical phenotype, suggesting that a combination of C_6-C_{12} hypoketotic dicarboxylic aciduria and a normal plasma acylcarnitine profile under acute conditions should be regarded as a specific indicator of this disorder (95). Confirmation of the enzymatic defect requires a liver biopsy, but the gene has been isolated and disease-causing mutations have been identified (96), making the molecular approach preferable. Clearly, it is too early to provide a definitive description of this rare disorder, but it seems to be a mild disorder that is relatively easy to treat by prevention of fasting.

3-HYDROXY 3-METHYLGLUTARYL-CoA LYASE DEFICIENCY   HMG-CoA lyase deficiency was first described in a patient with nonketotic hypoglycemia in 1971 and more than 60 patients have subsequently been described in the literature (2). The biochemical presentation differs from that of HMG-CoA synthase deficiency in that there is a characteristic and large urinary excretion of 3-hydroxy-3-methylglutaric acid and other related metabolites, a pathognomonic pattern for the disease. The enzyme is expressed in cultured skin fibroblasts, a variety of private mutations have been described in the HMG-CoA lyase gene (97).

Primary Disorders of Respiration Effecting FAO

Glutaric acidemia type 2 (GA2) has been linked to mutations in three genes, two coding for ETF (α-ETF, β-ETF) and one for ETF-QO (25). The loci are on chromosomes 15q23, 19q13, and 4q32, respectively.

Since the first description of GA2 in 1976, three different clinical phenotypes have been described, but a correlation to the genotype is not apparent (98). The most severely affected patients present in the neonatal period with congenital anomalies, dysmorphic features reminiscent of peroxisomal disorders, and acute metabolic decompensation. The second group of patients also presents in the newborn period, but the absence of severe congenital anomalies allows for survival beyond infancy. Then, the patients succumb either during an acute metabolic decompensation and/or the onset of progressive cardiomyopathy. The third group has a milder phenotype, usually after the neonatal period, with recurrent episodes of vomiting, hypoglycemia, hepatomegaly, and myopathy. These episodes are potentially life-threatening and sudden unexpected death has been reported (7). Some patients with muscle weakness were not diagnosed until early adulthood. A few patients are clinically responsive to pharmacological doses of riboflavin (99).

Plasma acylcarnitines and urine organic acid and acylglycine profiles provide strong biochemical evidence for a diagnosis of GA2 but do not distinguish between the three specific defects. As for most other FAO disorders, the diagnosis of GA2 is best confirmed by specific enzyme assays in fibroblast cultures (25). Molecular genetic analyses are not as practical because of the involvement of three different genes and the absence of common mutations.

METHODS FOR THE INVESTIGATION
OF FAO DISORDERS

Prenatal Diagnosis

To date, all known FAO disorders are inherited as autosomal recessive traits, with a recurrence risk of 25%. Prenatal diagnosis by biochemical, enzymatic, and molecular methods following chorionic villi sampling and/or amniocentesis is feasible for most of the FAO disorders when there has been appropriate characterization of an index case (74). Even when conventional prerequisites to prenatal testing have been met, seeking a prenatal diagnosis of treatable FAO disorders may still raise ethical issues. On the other hand, performing a prenatal diagnosis could also be a medical necessity to protect the mother's health when previous pregnancies were complicated by maternal liver disease (11).

Newborn Screening

Since the early 1990s, tandem mass spectrometry (MS/MS) has made screening possible for most FAO disorders based on the profiling of acylcarnitines in blood spots (58). The inclusion of FAO disorders in screening programs is highly desirable to achieve significant prevention of morbidity and mortality, implement simple and inexpensive treatment strategies, and contain the cost of care of affected patients (100). However, the specificity and sensitivity of screening by MS/MS has not yet been proven conclusively based on the prospective detection of a significant number of affected cases.

Biochemical Evaluation of Symptomatic Patients

Quantitative profiling of carnitine, acylcarnitines, and fatty acids in plasma, organic acids, and acylglycines in urine are the methods of choice to pursue a biochemical diagnosis for most of the FAO disorders (12, 13, 49, 58, 101). Recognition of an abnormal pattern of metabolite accumulation, ideally within a few hours following onset of symptoms, will allow the implementation of appropriate therapeutic measures pending confirmation of the preliminary diagnosis by a specialized laboratory.

Tables 1 and 2 summarize the characteristic biochemical findings in plasma and urine of patients with individual disorders of membrane-bound and mitochondrial matrix enzymes, respectively. A conclusive diagnosis can rarely be established by a single test and often requires the application of multiple analyses and their integrated interpretation. Uninformative metabolite profiles are frequently observed in patients with FAO defects when clinically and nutritionally stable, and therefore

TABLE 1   Laboratory evaluation of FAO disorders: defects of membrane-bound enzymes and transporters

	Reported prenatal diagnosis	Diagnosed by newborn screening	Laboratory investigations					Diagnosed by postmortem screening
			Plasma			Urine		
			Total, free carnitine	Acyl-carnitines	Fatty acids	Organic acids	Acyl-glycines	
Plasma membrane								
Carnitine uptake defect	+	−	+	−	−	−	−	+
LCFA transport/binding defect	−	−	−	−	−	−	−	+
FAP defect	−	−	−	−	−	−	−	−
Mitochondrial membranes								
CPT I deficiency (liver)	+	(+)	+	±	−	−	−	+
CACT deficiency	+	+	+	+	+	−	−	+
CPT II deficiency (neonatal onset)	+	+	±	+	−	−	−	+
CPT II deficiency (late onset)	+	+	±	+	−	−	−	−
VLCAD deficiency	+	+	±	+	+	+	−	+
ETF-QO deficiency (GA2)	+	+	±	+	+	+	+	+
LCHAD deficiency	+	(+)	±	±	+	±	−	+
TFP deficiency	+	(+)	±	±	+	±	−	+

+, reported; (+), not yet reported but expected to be possible; ±, uninformative findings possible according to the clinical status of the patient; −, not reported (negative). Abbreviations are as follows: CACT, carnitine: acylcarnitine translocase; CPT, carnitine palmitoyltransferase; FAP, fatty acid translocase; ETF-QO, electron transfer flavoprotein-ubiquinone oxidoreductase; LCFA, long-chain fatty acids; LCHAD, long-chain L-3-hydroxyacyl-CoA dehydrogenase; TFP, trifunctional protein; VLCAD, very-long-chain acyl-CoA dehydrogenase. Prenatal diagnosis and postmortem screening are performed by a combination of metabolite analysis, enzyme assays, and mutation analysis (74, 58). Newborn screening is performed by tandem mass spectrometry analysis of blood spot acylcarnitines (78).

TABLE 2 Laboratory evaluation of FAO disorders: defects of mitochondrial matrix enzymes

| | Reported prenatal diagnosis | Diagnosed by newborn screening | Laboratory investigations | | | | | Diagnosed by postmortem screening |
| | | | Plasma | | | Urine | | |
			Total, free carnitine	Acyl-carnitines	Fatty acids	Organic acids	Acyl-glycines	
MCAD deficiency	+	+	±	+	+	+	+	+
SCAD deficiency	+	+	−	+	−	+	+	−
α-ETF deficiency (GA2)	+	+	±	+	+	+	+	+
β-ETF deficiency (GA2)	+	+	±	+	+	+	+	+
Riboflavin responsive form(s) (GA2)	−	−	±	+	+	+	+	−
SCHAD deficiency (muscle)	−	−	±	±	−	−	−	−
SCHAD deficiency (fibroblasts)	+	−	±	±	+	+	−	−
SCHAD deficiency (liver)	−	−	−	−	+	−	−	+
M/SCHAD deficiency	−	−	±	±	+	±	−	−
MKAT deficiency	−	−	±	±	−	+	−	−
2,4-Dienoyl-CoA reductase deficiency	−	−	±	±	−	−	−	−
HMG-CoA synthase deficiency	−	−	−	±	−	±	−	−
HMG-CoA lyase deficiency	+	+	−	±	−	+	−	−

+, reported; (+), not reported but expected to be possible; ±, uninformative findings possible according to the clinical status of the patient; −, not reported (negative). Abbreviations are as follows: ETF, electron transfer flavoprotein; GA2, glutaric acidemia type II; MCAD, medium-chain acyl-CoA dehydrogenase; MKAT, medium chain ketoacyl-CoA dehydrogenase; HMG, 3-hydroxy 3-methylglutaryl-CoA dehydrogenase; M/SCHAD, medium- and short-chain L-3-hydroxyacyl-CoA dehydrogenase; SCAD, short-chain acyl-CoA dehydrogenase; SCHAD, short-chain L-3-hydroxyacyl-CoA dehydrogenase. Prenatal diagnosis and postmortem screening are performed by a combination of metabolite analysis, enzyme assays, and mutation analysis (74,58). Newborn screening is performed by tandem mass spectrometry analysis of blood spot acylcarnitines (78).

a noninformative profile should not be taken as evidence sufficient to rule out an underlying FAO disorder in a patient with fasting intolerance or other suggestive clinical evidence.

Postmortem Screening

FAO disorders frequently manifest with sudden and unexpected death (7, 8). We strongly advocate a complete postmortem investigation and genetic counseling of parents who lose a child suddenly and/or unexpectedly. Figure 2 shows a flow chart for the postmortem evaluation of sudden death cases that focuses on the analysis of acylcarnitines in blood and bile spots (8). Blood and bile could be conveniently collected on the same filter paper card, one identical to those used for newborn screening, which can be shipped at room temperature. Both specimens should be collected in order to provide a better chance to detect and independently confirm the largest possible number of disorders. In cases with a higher level of suspicion, an effort should be made to collect a frozen specimen of liver and a skin biopsy (102). Although fatty infiltration of the liver and/or other organs (heart, kidneys) is a common finding in FAO disorders, caution should be exercised not to use steatosis as the sole criterion to indicate a possible underlying FAO disorder during the postmortem evaluation of a case of sudden death (7). Additional risk

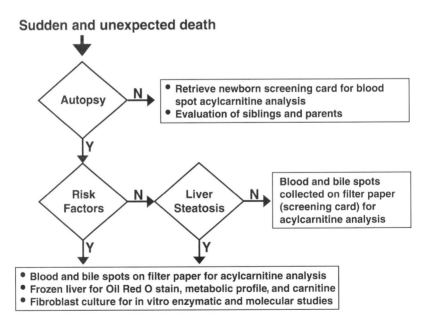

Figure 2 Protocol for the postmortem screening of FAO disorders. Risk factors include evidence of lethargy, vomiting, and/or fasting in the 48 h prior to death; family history of sudden death, Reye syndrome, or maternal pregnancy complication; and alleged child abuse.

factors include a family history of sudden death, Reye syndrome, or maternal pregnancy complication, and evidence of lethargy, vomiting, and/or fasting in the 48 h prior to death. To ensure all possible avenues are covered, cases of alleged child abuse should also be fully investigated. The frozen liver and skin biopsy could be discarded at a later time without further testing when a credible cause of death has been established but could otherwise be crucial to reach a proper diagnosis and conclusive confirmation in vitro.

ANIMAL MODELS OF FAO DISORDERS

In recent years, a key contribution to our understanding of the metabolic pathogenesis of FAO disorders has come from the study of either spontaneous or gene knockout models. Mouse models in which specific FAO genes have been disrupted by homologous recombination can now be used to study the clinical, biochemical, and histologic responses to dietary and environmental stress. The increasing availability of mice knockout models paves the way to a better understanding of disease mechanism, anticipating the long-term natural history of human disorders, and predicting the response to therapies in vivo and the feasibility of gene therapy (103).

Currently, the known spontaneous models of this group of diseases are the juvenile visceral steatosis mouse, a model of the carnitine uptake defect (104), and the BALB/cByJ mouse model of SCAD deficiency (105). Knockout mice models include VLCAD deficiency (29), LCAD deficiency (28), TFP deficiency (106), and M/SCHAD deficiency (88). The authors are aware of work in progress to generate new gene knockout mutants for MCAD, CPT-I, and CPT-II deficiencies (P. A. Wood, personal communication).

In the near future, it will also be possible to investigate mouse models with multiple defects to test the hypothesis that combinations of complete or even partial (double heterozygote) FAO defects could induce recognizable clinical and biochemical phenotypes. In particular, it will be exciting to see whether double-heterozygous enzyme-deficient states could trigger a phenotype either spontaneously or in response to environmental stressors. This evidence would substantiate the hypothesis that mutations in one of the alleles in two or more functionally related genes of the FAO pathway could cause a phenotype that resembles the homozygous deficiency of a single gene (44). Based on the frequencies of known disorders of energy metabolism, this could represent a relatively common mechanism of disease of potentially great clinical relevance, possibly playing a causative role in combination with susceptibility variants and environmental factors (107).

UNSPECIFIED FAO DISORDERS AND FUTURE TRENDS

Our understanding of FAO disorders is still incomplete. Many patients are identified with a history of fasting intolerance and/or abnormal but nonspecific laboratory findings, including inconclusive in vitro testing. These patients often carry a tentative diagnosis of an unspecified FAO disorder and are treated empirically with fasting avoidance, dietary changes, and aggressive preventive measures in response to common illnesses.

Clearly, there are a number of issues that remain unsolved concerning the organization and tissue distribution of the components of this pathway (108). With regard to the organization within the mitochondria, it seems logical that the various enzymes should be physically related to form functional units wherein metabolites pass rapidly from the active site of one enzyme to the next, rather than diffusing randomly to the next enzyme. To date, however, no such structural organization has been conclusively described for FAO enzymes.

ACKNOWLEDGMENTS

We warmly express our gratitude to our mentors and many collaborators for their efforts to elucidate what has become a pathway far more complex than ever imagined. We are particularly indebted to Y. T. Chen, Stephen I. Goodman, Daniel E. Hale, Willi Lehnert, Helmut Niederhoff, Rodney J. Pollitt, Charles A. Stanley, Arnold W. Strauss, and Kay Tanaka.

Joshua J. Jacobson of the Mayo High School Gifted and Talented Education Program designed Figure 1 with exemplary effort and dedication.

Visit the Annual Reviews home page at www.AnnualReviews.org

LITERATURE CITED

1. Eaton S, Bartlett K, Pourfarzam M. 1996. Mammalian mitochondrial β-oxidation. *Biochem. J.* 320:345–57

1a. Scriver CR, Beaudet AL, Sly WS, Valle D, Childs B, et al., eds. 2001. *The Metabolic and Molecular Bases of Inherited Diseases.* New York: McGraw-Hill. 8th ed. 6338 pp.

2. Mitchell GA, Fukao T. 2001. Inborn errors of ketone body metabolism. See Ref. 1a, pp. 2327–56

3. Nicholls DG, Locke RM. 1984. Thermogenic mechanisms in brown fat. *Physiol. Rev.* 64:1–64

4. DiMauro S, DiMauro PMM. 1973. Muscle carnitine palmitoyltransferase deficiency and myoglobinuria. *Science* 182:929–31

5. Stanley CA, Hale DE, Coates PM, Hall CL, Corkey BE, et al. 1983. Medium-chain acyl-CoA dehydrogenase deficiency in children with non-ketotic hypoglycemia

and low carnitine levels. *Pediatr. Res.* 17: 877–84

6. Belay ED, Bresee JS, Holman RC, Khan AS, Shahriari A, Schonberger LB. 1999. Reye's syndrome in the United States from 1981 through 1997. *N. Engl. J. Med.* 340:1377–82

7. Boles RG, Buck EA, Blitzer MG, Platt MS, Martin SK, et al. 1998. Retrospective biochemical screening of fatty acid oxidation disorders in postmortem liver of 418 cases of sudden unexpected death in the first year of life. *J. Pediatr.* 132:924–33

8. Chace DH, DiPerna JC, Mitchell BL, Sgroi B, Hofman LF, Naylor EW. 2001. Electrospray tandem mass spectrometry for analysis of acylcarnitines in dried postmortem blood specimens collected at autopsy from infants with unexplained cause of death. *Clin. Chem.* 47:1166–82

9. Rinaldo P. 1999. Mitochondrial fatty acid

oxidation disorders and cyclic vomiting syndrome. *Dig. Dis. Sci.* 44:97S–102S

10. Al Odaib A, Shneider BL, Bennett MJ, Pober BR, Reyes-Mugica M, et al. 1998. A defect in the transport of long-chain fatty acids associated with acute liver failure. *N. Engl. J. Med.* 339:1752–57

11. Ibdah JA, Bennett MJ, Rinaldo P, Zhao Y, Gibson B, et al. 1999. A fetal fatty acid oxidation disorder causes maternal liver disease of pregnancy. *N. Engl. J. Med.* 340:1723–31

12. Wanders RJ, Vreken P, den Boer ME, Wijburg FA, van Gennip AH, IJlst L. 1999. Disorders of mitochondrial fatty acyl-CoA β-oxidation. *J. Inher. Metab. Dis.* 22:442–87

13. Bennett MJ, Rinaldo P, Strauss AW. 2000. Inborn errors of mitochondrial fatty acid oxidation. *Crit. Rev. Clin. Lab. Sci.* 37:1–44

14. Dutta-Roy AK. 2000. Cellular uptake of long-chain fatty acids: role of membrane-associated fatty-acid-binding/transport proteins. *Cell Mol. Life Sci.* 57:1360–72

15. Berk PD, Stump DD. 1999. Mechanisms of cellular uptake of long chain free fatty acids. *Mol. Cell. Biochem.* 192:17–31

16. Hirsch D, Stahl A, Lodish HF. 1998. A family of fatty acid transporters conserved from mycobacterium to man. *Proc. Natl. Acad. Sci. USA* 95:8625–29

17. Bremer J. 1983. Carnitine: metabolism and functions. *Physiol. Rev.* 63:1420–80

18. Nezu J, Tamai I, Oku A, Ohashi R, Yabuuchi H, et al. 1999. Primary systemic carnitine deficiency is caused by mutations in a gene encoding sodium ion-dependent carnitine transporter. *Nat. Genet.* 21:91–94

19. McGarry JD, Brown NF. 1997. The mitochondrial carnitine palmitoyltransferase system. From concept to molecular analysis. *Eur. J. Biochem.* 244:1–14

20. Britton CH, Schultz RA, Zhang B, Esser V, Foster DW, McGarry JD. 1995. Human liver mitochondrial carnitine palmi-

toyltransferase I: characterization of its cDNA and chromosomal localization and partial analysis of the gene. *Proc. Natl. Acad. Sci. USA* 92:1984–88

21. Britton CH, Mackay DW, Esser V, Foster DW, Burns DK, et al. 1997. Fine chromosome mapping of the genes for human liver and muscle carnitine palmitoyltransferase I (CPT IA and CPT IB). *Genomics* 40:209–11

22. Huizing M, Iacobazzi V, IJlst L, Savelkoul P, Ruitenbeek W, et al. 1997. Cloning of the human carnitine-acylcarnitine carrier cDNA and identification of the molecular defect in a patient. *Am. J. Hum. Genet.* 61:1239–45

23. Finocchiaro G, Taroni F, Rocchi M, Liras Martin A, Colombo I, et al. 1991. cDNA cloning, sequence analysis, and chromosomal localization of the gene for human carnitine palmitoyltransferase. *Proc. Natl. Acad. Sci. USA* 88:661–65

24. Izai K, Uchida Y, Orii T, Yamamoto S, Hashimoto T. 1992. Novel fatty acid β-oxidation enzymes in rat liver mitochondria. I: purification and properties of very-long-chain acyl coenzyme A dehydrogenase. *J. Biol. Chem.* 267:1027–33

25. Frerman FE, Goodman SI. 2001. Defects of electron transfer flavoprotein and electron transfer flavoprotein-ubiquinone oxidoreductase: glutaric acidemia type II. See Ref. 1a, pp. 2357–65

26. Ikeda Y, Okamura-Ikeda K, Tanaka K. 1985. Purification and characterization of short-chain, medium-chain, and long-chain acyl-CoA dehydrogenases from rat liver mitochondria. Isolation of the holo- and apoenzymes and conversion of the apoenzyme to the holoenzyme. *J. Biol. Chem.* 260:1311–25

27. Kurtz D, Rinaldo P, Rhead WJ, Tian L, Millington DS, et al. 1998. Targeted disruption of mouse long-chain acyl-CoA dehydrogenase reveals crucial role in fatty acid oxidation. *Proc. Natl. Acad. Sci. USA* 95:15592–97

28. Cox KB, Hamm DA, Millington DS,

Matern D, Vockley J, et al. 2001. Very long-chain acyl-CoA dehydrogenase deficiency is distinct from long-chain acyl-CoA dehydrogenase deficiency in the mouse. *Hum. Mol. Genet.* In press

29. Uchida Y, Izai K, Orii T, Hashimoto T. 1992. Novel fatty acid β-oxidation enzymes in rat liver mitochondria. II: purification and properties of enoyl-CoA hydratase/3-hydroxyacyl-CoA dehydrogenase/3-ketoacyl-CoA thiolase trifunctional protein. *J. Biol. Chem.* 267:1034–41

30. Carpenter K, Pollitt RJ, Middleton B. 1992. Human long-chain 3-hydroxyacyl-CoA dehydrogenase is a multifunctional membrane-bound β-oxidation enzyme of mitochondria. *Biochem. Biophys. Res. Commun.* 183:443–48

31. Ushikubo S, Aoyama T, Kamijo T, Wanders RJA, Rinaldo P, et al. 1996. Molecular characterization of mitochondrial trifunctional protein deficiency: formation of the enzyme complex is important for stabilization of both alpha and beta subunits. *Am. J. Hum. Genet.* 58:979–88

32. Weinberger MJ, Rinaldo P, Strauss AW, Bennett MJ. 1995. Intact α-subunit is required for membrane-binding of human mitochondrial trifunctional β-oxidation protein, but is not necessary for conferring 3-ketoacyl-CoA thiolase activity to the β-subunit. *Biochem. Biophys. Res. Commun.* 209:47–52

33. Ijlst L, Ruiter JPN, Hoovers JMN, Jakobs ME, Wanders RJA. 1996. Common missense mutation G1528C in long-chain 3-hydroxyacyl-CoA dehydrogenase deficiency-characterization and expression of the mutant protein, mutation analysis on genomic DNA and chromosomal localization of the mitochondrial trifunctional protein alpha subunit gene. *J. Clin. Invest.* 98:1028–33

34. Ushikubo S, Aoyama T, Kamijo T, Wanders RJA, Rinaldo P, et al. 1996. Molecular characterization of mitochondrial trifunctional protein deficiency: formation of the enzyme complex is important for stabiliza-

tion of both α- and β-subunits. *Am. J. Hum. Genet.* 58:979–88

35. Kanazawa M, Ohtake A, Abe H, Yamamoto S, Satoh Y, et al. 1993. Molecular cloning and sequence analysis of the cDNA for human mitochondrial short-chain enoyl-CoA hydratase. *Enzyme Protein* 47:9–13

36. Jackson S, Schaefer J, Middleton B, Turnbull DM. 1995. Characterisation of a novel enzyme of human fatty acid β-oxidation: a matrix-associated, mitochondrial 2-enoyl-CoA hydratase. *Biochem. Biophys. Res. Comm.* 214:247–53

37. Jiang LL, Kobayashi A, Matsuura H, Fukushima H, Hashimoto T. 1996. Purification and properties of human D-3-hydroxyacyl-CoA dehydratase: medium-chain enoyl-CoA hydratase is D-3-hydroxyacyl-CoA dehydratase. *J. Biochem.* 120:624–32

38. Kobayashi A, Jiang LL, Hashimoto T. 1996. Two mitochondrial 3-hydroxyacyl-CoA dehydrogenases in bovine liver. *J. Biochem.* 119:775–82

39. He XY, Schulz H, Yang SY. 1998. A human brain L-3-hydroxyacyl-coenzyme A dehydrogenase is identical to an amyloid beta-peptide-binding protein involved in Alzheimer's disease. *J. Biol. Chem.* 273:10741–46

40. Vredendaal PJCM, van den Berg IET, Malingre HEM, Stroobants AK, Olde Weghuis DE, Berger R. 1996. Human short-chain L-3-hydroxyacyl-CoA dehydrogenase: cloning and characterization of the coding sequence. *Biochem. Biophys. Res. Commun.* 223:718–23

41. Miyazawa S, Osumi T, Hashimoto T. 1980. The presence of a new 3-oxoacyl-CoA thiolase in rat liver peroxisomes. *Eur. J. Biochem.* 103:589–96

42. Kamijo T, Indo I, Souri M, Aoyama T, Hara T, et al. 1997. Medium-chain 3-ketoacyl-CoA thiolase deficiency: a new disorder of mitochondrial fatty acid β-oxidation. *Pediatr. Res.* 42:569–76

43. Rinaldo P, Al Odaib A, Bennett MJ.

1999. Complementation analysis of six patients with a defect of long-chain fatty acid transport in fibroblasts. *J. Inher. Metab. Dis.* 22(Suppl.):15 (Abstr.)

44. Vockley J, Rinaldo P, Bennett MJ, Matern D, Vladutiu GD. 2000. Synergistic heterozygosity: disease resulting from multiple partial defects in one or more metabolic pathways. *Mol. Genet. Metab.* 71:10–18

45. Tanaka T, Nakata T, Oka T, Ogawa T, Okamoto F, et al. 2001. Defect in human myocardial long-chain fatty acid uptake is caused by FAT/CD36 mutations. *J. Lipid Res.* 42:751–59

46. Treem WR, Stanley CA, Finegold DN, Hale DE, Coates PM. 1988. Primary carnitine deficiency due to a failure of carnitine transport in kidney, muscle, and fibroblasts. *N. Engl. J. Med.* 319:1331–36

47. Stanley CA, DeLeeuw S, Coates PM, Vianey-Liaud C, Divry P, et al. 1991. Chronic cardiomyopathy and weakness or acute coma in children with a defect in carnitine uptake. *Ann. Neurol.* 30:709–16

48. Bougneres P-F, Saudubray J-M, Marsac C, Bernard O, Odievre M, Girard J. 1981. Fasting hypoglycemia resulting from hepatic carnitine palmitoyltransferase deficiency. *J. Pediatr.* 98:742–46

49. Rinaldo P. 2001. Laboratory diagnosis of inborn errors of metabolism. In *Liver Disease in Children*, ed. FJ Suchy, RJ Sokol, WF Balistreri, pp. 171–84. Philadelphia: Lippincott, Williams & Wilkins. 2nd ed.

50. Bonnefont J-P, Demaugre F, Prip-Buus C, Saudubray J-M, Brivet M, et al. 1999. Carnitine palmitoyltransferase deficiencies. *Mol. Genet. Metab.* 68:424–40

51. Olpin SE, Allen J, Bonham JR, Clark S, Clayton PT, et al. 2001. Features of carnitine palmitoyltransferase type I deficiency. *J. Inher. Metab. Dis.* 24:35–42

52. Brown NF, Mullur RS, Subramanian I, Esser V, Bennett MJ, et al. 2001. Molecular characterization of liver-type carnitine palmitoyltransferase 1 (L-CPT-I) deficiency in six patients: insights into function of the native enzyme. *J. Lipid Res.*, 42:1134–42

53. Ijlst L, Mandel HW, Oostheim W, Ruiter PN, Gutman A, Wanders RJA. 1998. Molecular basis of hepatic carnitine palmitoyltransferase I deficiency. *J. Clin. Invest.* 102:527–31

54. Abadi N, Thuillier L, Prasad C, Dilling L, Demaugre F, et al. 1999. Molecular resolution of carnitine palmitoyltransferase I deficiency in a Hutterite family. *Am. J. Hum. Genet.* 65:A230 (Abstr.)

55. Stanley CA, Hale DE, Berry GT, DeLeeuw S, Boxer J, Bonnefont J-P. 1992. A deficiency of carnitine-acylcarnitine translocase in the inner mitochondrial membrane. *N. Engl. J. Med.* 327:19–23

56. Morris AAM, Olpin SE, Brivet M, Turnbull DM, Jones RAK. Leonard JV. 1998. A patient with carnitine-acylcarnitine translocase deficiency with a mild phenotype. *J. Pediatr.* 132:514–16

57. Huizing M, Iacobazzi V, Ijlst L, Savelkoul P, Ruitenbeek W, et al. 1997. Cloning of the human carnitine-acylcarnitine carrier cDNA and identification of the molecular defect in a patient. *Am. J. Hum. Genet.* 61:1239–45

58. Rinaldo P, Matern D. 2000. Disorders of fatty acid transport and mitochondrial oxidation: challenges and dilemmas of metabolic evaluation. *Genet. Med.* 2:338–44

59. Taroni F, Verderio E, Dworzak F, Willems PJ, Cavadini P, DiDonato S. 1993. Identification of a common mutation in the carnitine palmitoyltransferase II gene in familial recurrent myoglobinuria patients. *Nat. Genet.* 4:314–20

60. Demaugre F, Bonnefont J-P, Colonna M, Cepenec C, Leroux J-P, Saudubray J-M. 1991. Infantile form of carnitine palmitoyltransferase II deficiency with hepatomuscular symptoms and sudden death. Physiopathological approach to carnitine palmitoyltransferase II deficiencies. *J. Clin. Invest.* 87:859–64

61. Yamaguchi S, Indo Y, Coates PM, Hashimoto T, Tanaka K. 1993. Identification of very-long-chain acyl-CoA dehydrogenase deficiency in three patients previously diagnosed with long-chain acyl-CoA dehydrogenase deficiency. *Pediatr. Res.* 34:111–13

62. Hale DE, Stanley CA, Coates PM. 1990. The long chain acyl-CoA dehydrogenase deficiency. *Progr. Clin. Biol. Res.* 321:411–18

63. Mathur A, Sims HF, Gopalakrishnan D, Gibson B, Rinaldo P, et al. 1999. Molecular heterogeneity in very long-chain acyl-CoA dehydrogenase deficiency causing pediatric cardiomyopathy and sudden death. *Circulation* 99:1337–43

64. Andresen BS, Olpin S, Poorthuis BJ, Scholte HR, Vianey-Saban C, et al. 1999. Clear correlation of genotype with disease phenotype in very-long-chain acyl-CoA dehydrogenase deficiency. *Am. J. Hum. Genet.* 64:479–94

65. Wanders RJA, Ijlst L, van Gennip AH, Jakobs C, de Jager JP, et al. 1990. Long-chain 3-hydroxyacyl-CoA dehydrogenase deficiency: identification of a new inborn error of mitochondrial fatty acid β-oxidation. *J. Inher. Metab. Dis.* 13:311–14

66. Ijlst L, Ruiter JPN, Hoovers JMN, Jakobs ME, Wanders RJA. 1996. Common missense mutation G1528C in long-chain 3-hydroxyacyl-CoA dehydrogenase deficiency: characterization and expression of the mutant protein, mutation analysis on genomic DNA and chromosomal localization of the mitochondrial trifunctional protein β-subunit gene. *J. Clin. Invest.* 98:1028–33

67. Jackson S, Singh Kler R, Bartlett K, Briggs H, Bindoff LA, et al. 1992. Combined enzyme defect of mitochondrial fatty acid oxidation. *J. Clin. Invest.* 90:1219–25

68. Sattar N, Gaw A, Packard CJ, Greer IA. 1996. Potential pathogenic roles of aberrant lipoprotein and fatty acid metabolism in pre-eclampsia. *Br. J. Obstet. Gynaecol.* 103:614–20

69. Sattar N, Bendomir A, Berry C, Shepherd J, Greer IA, Packard CJ. 1997. Lipoprotein subfraction concentrations in preeclampsia: pathogenic parallels to atherosclerosis. *Obstet. Gynecol.* 89:403–8

70. Shambaugh GE 3rd. 1985. Ketone body metabolism in the mother and fetus. *Fed. Proc.* 44:2347–51

71. Herrera E, Amusquivar E. 2000. Lipid metabolism in the fetus and the newborn. *Diabetes Metab. Res. Rev.* 16:202–10

72. Shekawat PS, Bennett MJ, Rakheja D, Strauss AW. 2001. Fatty acid oxidation (FAO) in normal human placenta: developmental expression and activity of enzymes of mitochondrial β-oxidation. *Pediatr. Res.* 49:55A (Abstr.)

73. Tein I. 2000. Metabolic disease in the fetus predispose to maternal complications of pregnancy. *Pediatr. Res.* 47:6–8

74. Rinaldo P, Studinski A, Matern D. 2001. Prenatal diagnosis of disorders of fatty acid transport and mitochondrial oxidation. *Prenat. Diagn.* 21:52–54

75. Matern D, Shehata BM, Shekhawat P, Strauss AW, Bennett MJ, Rinaldo P. 2001. Placental floor infarction complicating the pregnancy of a fetus with long-chain L-3-hydroxy acyl-CoA dehydrogenase deficiency. *Mol. Genet. Metab.* 72:265–68

76. Matern D, Rinaldo P. 2000. Medium chain acyl-coenzyme A (MCAD) deficiency. In *GeneClinics: Medical Genetics Knowledge Base* [database online]. © Univ. Washington, Seattle. http://www.geneclinics.org/profiles/mcad.

77. Raymond K, Bale AE, Barnes CA, Rinaldo P. 1999. Sudden adult death and medium-chain acyl-CoA dehydrogenase deficiency. *Genet. Med.* 1:293–94

78. Chace DH, Hillman SL, Van Hove JL, Naylor EW. 1997. Rapid diagnosis of MCAD deficiency: quantitatively analysis of octanoylcarnitine and other acylcarnitines in newborn blood spots by tandem mass spectrometry. *Clin. Chem.* 43:2106–13

79. Andresen BS, Dobrowolski SF, O'Reilly L, Muenzer J, McCandless SE, et al. 2001.

Medium-chain acyl-CoA dehydrogenase (MCAD) mutations identified by MS/MS-based prospective screening of newborns differ from those observed in patients with clinical symptoms: identification and characterization of a new, prevalent mutation that results in mild MCAD deficiency. *Am. J. Hum. Genet.* 68:1408–18

80. Zschocke J, Schulze A, Lindner M, Fiesel S, Olgemoller K, et al. 2001. Molecular and functional characterisation of mild MCAD deficiency. *Hum. Genet.* 108:404–8

81. Albers S, Levy HL, Irons M, Strauss AW, Marsden D. 2001. Compound heterozygosity in four asymptomatic siblings with medium-chain acyl-CoA dehydrogenase deficiency. *J. Inher. Metab. Dis.* 24:417–18

82. Treem WR, Stanley CA, Goodman SI. 1989. Medium-chain acyl-CoA dehydrogenase deficiency: metabolic effects and therapeutic efficacy of long-term L-carnitine supplementation. *J. Inher. Metab. Dis.* 12:112–19

83. Tein I, De Vivo DC, Hale DE, Clarke JTR, Zinman H, et al. 1991. Short-chain L-3-hydroxyacyl-CoA dehydrogenase deficiency in muscle: a new cause of recurrent myoglobinuria and encephalopathy. *Ann. Neurol.* 30:415–19

84. Bennett MJ, Weinberger MJ, Kobori JA, Rinaldo P, Burlina AB. 1996. Mitochondrial short-chain L-3-hydroxybutyryl-CoA dehydrogenase deficiency: a new defect of fatty acid oxidation. *Pediatr. Res.* 39:185–88

85. Bennett MJ, Spotswood SD, Ross KF, Comfort S, Koonce R, et al. 1999. Fatal hepatic short-chain L-3-hydroxyacyl-coenzyme A dehydrogenase deficiency: clinical, biochemical, and pathological studies on three subjects with this recently identified disorder of mitochondrial beta-oxidation. *Pediatr. Dev. Pathol.* 2:337–45

86. O'Brien LK, Bennett MJ, Rinaldo P, Sims H, Strauss AW. 2001. A mouse model for medium and short chain L-3-hydroxyacyl-CoA dehydrogenase deficiency. *Pediatr. Res.* 49:181A (Abstr.)

87. Clayton PT. 2001. Applications of mass spectrometry in the study of inborn errors of metabolism. *J. Inher. Metab. Dis.* 24:139–50

88. O'Brien LK, Rinaldo P, Sims HF, Alonso EM, Charrow J, et al. 2000. Fulminant hepatic failure associated with mutations in the medium and short chain L-3-hydroxyacyl-CoA dehydrogenase gene. *J. Inher. Metab. Dis.* 23(Suppl.):127 (Abstr.)

89. Coates PM, Hale DE, Finocchiaro G, Tanaka K, Winter SC. 1988. Genetic deficiency of short-chain acyl-coenzyme A dehydrogenase in cultured fibroblasts from a patient with muscle carnitine deficiency and severe skeletal muscle weakness. *J. Clin. Invest.* 81:171–75

90. Bhala A, Willi SM, Rinaldo P, Bennett MJ, Schmidt-Sommerfeld E, Hale DE. 1995. The emerging clinical and biochemical picture of short-chain acyl-CoA dehydrogenase deficiency. *J. Pediatr.* 126:910–15

91. Corydon MJ, Vockley G, Rinaldo P, Rhead WJ, Kjeldsen M, et al. 2001. Role of common variant alleles in the molecular basis of short-chain acyl-CoA dehydrogenase deficiency. *Pediatr. Res.* 49:18–23

92. Roe CR, Millington DS, Norwood DL, Kodo N, Sprecher H, et al. 1990. 2,4-Dienoyl-coenzyme A reductase deficiency: a possible new disorder of fatty acid oxidation. *J. Clin. Invest.* 85:1703–7

93. Thompson GN, Hsu BYL, Pitt JJ, Treacy E, Stanley CA. 1997. Fasting hypoketotic coma in a child with deficiency of mitochondrial 3-hydroxy-3-methylglutaryl-CoA synthase. *N. Engl. J. Med.* 337:1203–7

94. Morris AA, Lascelles CV, Olpin SE, Lake BD, Leonard JV, Quant PA. 1998. Hepatic mitochondrial 3-hydroxy-3-methylglutaryl-coenzyme A synthase deficiency. *Pediatr. Res.* 44:392–96

95. Zschocke J, Hegardt FG, Casals N, Penzien JM, Aledo R, et al. 2000. Clinical, biochemical and molecular characterization of 3-hydroxy 3-methylglutaryl-CoA synthase deficiency. *J. Inher. Metab. Dis.* 23(Suppl.):107 (Abstr.)

96. Bouchard L, Robert MF, Vinarov D, Stanley CA, Thompson GN, et al. 2001. Mitochondrial 3-hydroxy-3-methylgluta-ryl-CoA synthase deficiency: clinical course and description of causal mutations in two patients. *Pediatr. Res.* 49:326–31

97. Mitchell GA, Robert M-F, Hruz PW, Wang S, Fontaine G, et al. 1993. 3-Hydroxy-3-methylglutaryl coenzyme A lyase (HL): cloning of human and chicken liver HL cDNAs and characterization of a mutation causing human HL deficiency. *J. Biol. Chem.* 268:4376–81

98. Loehr JP, Goodman SI, Frerman FE. 1990. Glutaric acidemia type II: heterogeneity of clinical and biochemical phenotypes. *Pediatr. Res.* 27:311–15

99. Gregersen N, Rhead W, Christensen E. 1900. Riboflavin responsive glutaric aciduria type II. *Progr. Clin. Biol. Res.* 321:477–94

100. Wood J, Seashore MR, Magera MJ, Rinaldo P, Strauss AW, Friedman A. 2001. Retrospective diagnosis of very long chain acyl-CoA dehydrogenase deficiency from an infant's newborn screening card. *Pediatrics* 108:e19

101. Rinaldo P, Raymond K, Al Odaib A, Bennett MJ. 1998. Fatty acid oxidation disorders: clinical and biochemical features. *Curr. Opin. Pediatr.* 10:615–21

102. Bennett MJ, Rinaldo P. 2001. The metabolic autopsy comes of age. *Clin. Chem.* 47:1145–46

103. Kelly CL, Rhead WJ, Kutschke WK, Brix AE, Hamm DA, et al. 1997. Functional correction of short-chain acyl-CoA dehydrogenase deficiency in transgenic mice: implications for gene therapy of human mitochondrial enzyme deficiencies. *Hum. Mol. Genet.* 6:1451–55

104. Koizumi T, Nikaido H, Hayakawa J, Nonomura A, Yoneda T. 1988. Infantile disease with microvesicular fatty infiltration of viscera spontaneously occurring in the C3H-H-20 strain of mouse with similarities to Reye's syndrome. *Lab. Animals* 22:83–87

105. Wood PA, Amendt BA, Rhead WJ, Millington DS, Inoue F, Armstrong D. 1989. Short-chain acyl-coenzyme A dehydrogenase deficiency in mice. *Pediatr. Res.* 25:38–43

106. Ibdah JA, Paul H, Zhao Y, Binford S, Cline M, et al. 2001. Lack of mitochondrial trifunctional protein in mice causes neonatal hypoglycemia and sudden death. *J. Clin. Invest.* 107:1403–9

107. Dipple KM, McCabe ERB. 2000. Phenotypes of patients with "simple" Mendelian disorders are complex traits: thresholds, modifiers, and systems dynamics. *Am. J. Hum. Genet.* 66:1729–35

108. Parker A, Engel PC. 2000. Preliminary evidence for the existence of specific functional assemblies between enzymes of the beta-oxidation pathway and the respiratory chain. *Biochem. J.* 345:429–35

Annu. Rev. Physiol. 2002. 64:503–27

MOLECULAR CHAPERONES IN THE KIDNEY

Steven C. Borkan

Evans Biomedical Research Center, Boston Medical Center, Renal Section,
650 Albany Street, Boston, Massachusetts 02118-2518; e-mail: sborkan@bu.edu

Steven R. Gullans

Harvard Institutes of Medicine, Brigham and Women's Hospital, Department
of Medicine, 77 Avenue Louis Pasteur, Boston, Massachusetts 02115;
e-mail: sgullans@rics.bwh.harvard.edu

Key Words stress proteins, heat shock, ischemia, apoptosis, nephrotoxins

■ **Abstract** The normal milieu of the kidney includes hypoxia, large osmotic fluxes, and an enormous amount of fluid/solute reabsorption. Renal adaptation to these conditions requires a host of molecular chaperones that stabilize protein conformation, target nascent proteins to their final intracellular destination, and prevent protein aggregation. Under physiologic or pharmacologic stress, inducible molecular chaperones provide additional mechanisms for repairing or degrading non-native proteins and for inhibiting stress-induced apoptosis. In contrast to intracellular chaperones, chaperones present on the cell surface regulate the immune system and have cytokine-like effects. A diverse range of chaperones and chaperone functions provide the renal cell with an armamentarium of responses to improve the chances of survival.

INTRODUCTION

The normal kidney offers a challenging milieu. Despite a filtration volume of 160–180 liters each day, only 1–2 liters of urine are excreted. Over 98% of the glomerular filtrate must be reabsorbed. The work of solute and fluid reabsorption requires substantial aerobic energy production and extracts a large amount of oxygen from the blood stream. Due in part to the high oxygen extraction ratio and the countercurrent mechanism required for urinary concentration, the medulla operates on the brink of hypoxia (1). In humans the renal inner medullary cells are chronically exposed to a relatively hyperosmotic interstitium, and during changes in hydration state, these cells can experience wide swings in the osmolality ranging from 50 to 1200 mOsm. Thus even under normal circumstances, the kidney experiences a substantial level of stress. The kidney is also exposed to a variety of disease states and adverse challenges such as hypoxia, energy deprivation, and toxins that can compromise cell survival. To adapt to stressful conditions, kidney cells, like cells of all living organisms, utilize

0066-4278/02/0315-0503$14.00

503

inducible cytoprotective mechanisms involving molecular chaperones or stress proteins.

The stress response has been extensively reviewed elsewhere (2–6). Using information obtained from recent investigations, the present review focuses on the role of molecular chaperones in mediating common physiologic and chemical insults that alter renal epithelial cell function. These insults include ischemia, nephrotoxin exposure, glomerulonephritis, kidney transplantation, and osmotic stress. Individual molecular chaperones that modulate cell injury during stress and/or recovery are discussed with an emphasis on the their mechanism(s) of action.

STRESS PROTEINS: A HISTORICAL PERSPECTIVE

In 1962, Ferruccio Ritossoa observed a characteristic puffing of the specific chromosomes after subjecting isolated *Drosophila* salivary glands to temperature shock (7). Twelve years later, Tissieres and colleagues noted that the de novo synthesis of only six prominent proteins (i.e., heat shock proteins) accounted for 30% of the total new protein synthesis in the salivary glands of heated larvae (8). Subsequent studies characterized many heat shock proteins that are divided into families based primarily on their molecular mass. These include HSP 20–30, HSP 50–60, HSP 70 (including HSP 68–78 kDa), HSP 90, and HSP 100–110 kDa. Stress proteins are arbitrarily designated as constitutively expressed (cognate stress proteins) or stress-inducible, although constitutively expressed HSPs can be modestly up-regulated and inducible HSPs may be constitutively expressed (9). Constitutively expressed stress proteins participate in normal cell maintenance. In contrast, accumulation of inducible stress proteins requires an acute stress.

ROLE OF MOLECULAR CHAPERONES IN NORMAL RENAL FUNCTION

Stress proteins serve diverse functions (Table 1). In the kidney, their best-characterized role is that as a molecular chaperone (Figure 1). As the term implies, a molecular chaperone binds various protein substrates, particularly nascent polypeptides and proteins, in a non-native (denatured) conformation. It is now recognized that the primary amino acid sequence is not sufficient to determine tertiary protein structure. A host of molecular chaperones, present in multiple cellular compartments, work in a coordinated and substrate-specific fashion to produce a protein in its final conformation (10). Chaperones present in the endoplasmic reticulum stabilize and fold peptide intermediates (11). Cytosolic chaperones exert quality control in assuring that nascent proteins achieve their mature conformation. Molecular chaperones also collaborate in the maturation of signal transducing proteins including the steroid receptor (12) and protein kinases (13). To deliver newly

TABLE 1   Stress proteins and their diverse functions

Housekeeping functions
    Fold proteins into native conformation
    Assemble multimeric complexes
    Deliver proteins across organelle membranes ("unfoldase")
    Facilitate degradation of malformed proteins
    Uncoat clathrin-coated vesicles (after endocytosis)

Stress functions
    Repair denatured proteins
    Prevent protein aggregate formation
    Facilitate degradation of severely damaged proteins
    Deliver replacement proteins to target organelles
    Inhibit cell death (apoptosis)
    Stabilize the cytoskeleton
    Repair nuclear DNA damage

Immune modulating functions
    Stimulate immune system (cytokine-like function)
    Act as target for generating auto-antibodies
    Induce apoptosis

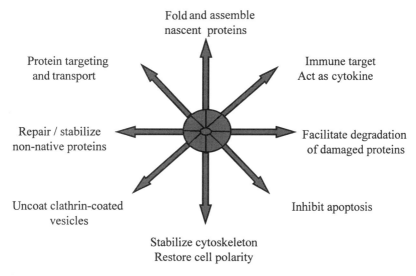

Figure 1   Functional roles of renal molecular chaperones. The roles of stress proteins in the kidney include both chaperone and non-chaperone functions. Constitutively expressed stress proteins fulfill most of these functions under normal circumstances. To enhance resistance to injury and improve cell survival during physiologic or pharmacologic stressors, inducible stress proteins accumulate in the kidney.

synthesized proteins destined for organelles, chaperones maintain peptides in an unfolded state, traffic the peptides across biologic membranes, and then orchestrate re-folding within the lumen (14). Protein complexes made up of multimers are assembled in the cytosol with the assistance of chaperones (10). Most organelles possess unique sets of molecular chaperones. Many of these chaperones coordinate the final re-folding of peptides that cross their membranes (15). This assembly line process assures final delivery of the mature protein to its destination. Importantly, these chaperones also prevent newly synthesized proteins from forming large and potentially toxic aggregates (14, 16–18). The burden of unfolded or denatured proteins appears to be an important sensor for regulating the synthesis of molecular chaperones. In the kidney, chaperones mediate effective binding of mineralocorticoid to its receptor. HSP 90 has been directly implicated in this process (19). Members of the HSP 70 family assist in re-cycling of clathrin-coated vesicles involved in the endocytosis and transport of hormone receptors, glucose, and other molecules (20).

When proteins exist in a non-native state due to an error in synthesis or an adverse stress, molecular chaperones attempt either to refold and repair the nonnative protein or to facilitate its degradation (10, 21, 22). The function of stress proteins extends beyond those of a protein chaperone. Intracellular accumulation of stress proteins inhibits apoptosis (23–28). In contrast, expression of the same stress proteins on the cell surface can precipitate cell death (29, 30), activate the immune system (3, 31), or serve as a target for the generation of autoantibodies in systemic autoimmune disease (32–34). These observations demonstrate that the effects of stress proteins are highly variable and may be stress and cell-type specific.

Induction of the Cellular Stress Response

Hundreds of diverse stimuli have been shown to stimulate the induction of molecular chaperones (35). Changes in intracellular pH, temperature, reduced ATP content, hypoxia, infection, cancer, exposure to radiation or toxins, or even non-toxic drugs have all been shown to illicit the stress response. Surprisingly, cold-shock also induces molecular chaperones (4). Although each stimulus could operate through a unique biochemical pathway, several investigators conceptualize protein denaturation as a unifying model for the cellular stress response (16, 36). In this model, misfolded proteins tend to self-aggregate into larger, nonfunctional protein complex proteins or aggresomes (17). This model system appears to be highly consistent with the mechanism for inducing HSP 70 (72 kDa), the best characterized of the inducible stress proteins. Denaturation of intracellular proteins, especially those proteins that tend to form large complexes, is a potent stimulus for up-regulating HSP 72 (37). Constitutively expressed molecular chaperones (e.g., HSP 73 kDa or HSC 70) appear to act as a quality control mechanism for recognizing nascent proteins with normal conformation (10, 14, 38). Several chaperones cooperate with HSP 70 members to stabilize intrinsically unstable folding intermediates. These co-chaperones include HSP 40, HSP 60, and HSP 90 (10, 39). The presence of both constitutive and inducible chaperones permits a flexible response that exhibits substrate, cell, and stress specificity (40).

When confronted with an increased burden of non-native proteins, the cellular stress response is activated (41). Once induced, molecular chaperones have the onerous task of distinguishing irreparably damaged proteins from those that can be refolded (21). Substantial evidence suggests that selected enzymes can be successfully refolded by molecular chaperones after partial denaturation (21, 42). Proteins that cannot be refolded are destined to be degraded either by lysosomes (38) or by a ubiquitin-mediated proteolysis pathway (22, 43). HSP 70 members have been shown to facilitate both processes of protein degradation (43). Although the consequences of accumulating denatured intracellular proteins is poorly characterized, denatured proteins have the potential to cause injury, referred to as proteotoxicity (16).

The cytoprotection afforded by molecular chaperones represents a potentially elegant and complex system. Renal cells, like the cells of other organs, possess an array of compartment-specific chaperones capable of responding to alterations in protein conformation (44). Both the inducible and constitutively expressed chaperones have numerous isoforms, demonstrate distinct peptide-binding preferences (45), and exhibit stimulus specificity (46). Furthermore, different regions of a single chaperone (e.g., HSP 72) may be critical for mediating cytoprotection against a specific insult (25, 47). In some circumstances, the ATPase region of the chaperone is required to permit high-affinity substrate binding and to mediate release of the repaired peptide, a process referred to as the chaperone function (25). In other cases, non-chaperone domains are equally effective cytoprotectants (25).

In sum, the essential purpose of constitutively expressed molecular chaperones is to assure that nascent polypeptides achieve their native conformation, to target delivery of newly synthesized proteins across biologic membranes and to facilitate the timely degradation of misfolded proteins. In contrast, the inducible molecular chaperones improve the likelihood of cell survival following a noxious insult. A host of diverse renal diseases have been associated with the induction of molecular chaperones (Table 2). The protective effects of molecular chaperones in renal disease are mediated by one or more potential mechanisms (Table 3).

Renal Ischemia/ATP Depletion

Ischemia, often resulting from a transient decrease in blood flow, is a common cause of acquired renal failure. Ischemic renal injury frequently occurs in the presence of hypovolemia, hypotension, sepsis, cardiopulmonary bypass, and during renal transplantation. Both ischemia of the intact kidney and ATP depletion in vitro (a surrogate model for ischemia) result in characteristic changes in renal epithelial cell morphology (48–50). In vivo, the most dramatic changes are observed in epithelial cells within the S_3 segment of the proximal tubule (51). Alone or in combination, collapse of the actin cytoskeleton, loss of cell-cell contact,

TABLE 2   Role of molecular chaperones in renal disease

Disease state	Reference
Glomerulonephritis	(3, 122, 123, 125)
Obstruction	(113, 116, 117)
Interstitial nephritis	(123)
Ischemia	
Nephrotoxins	(111)
Heavy metals	
Cadmium	(73, 153, 154)
Mercury	(106, 109)
Cisplatin	
Iron	(112)
Antibiotics	
Gentamicin	(104, 105, 155)
Hypoxia	(156)
Renal Transplantation	
Ischemia	(87)
Rejection	(136, 137)

and disruption of cell adhesion to the substratum cause epithelial cell dysfunction (52, 53). Cell polarity is disrupted and as a result, Na^+,K^+-ATPase, normally restricted to the basolateral membrane, is re-distributed (54, 55), thereby compromising vectoral solute transport (54). Loss of cell-cell contact permits paracellular backleak, one cause of decreased renal function (53). Intact and/or necrotic cells form casts that obstruct tubular flow and exacerbate renal dysfunction (48, 50). In addition to cytoskeletal and cell contact sites, mitochondria are a primary target for ischemic or ATP depletion-mediated injury (49, 56). Both mitochondrial injury (57) and loss of cell contact sites (58) precipitate apoptosis, a primary form of cell death after transient renal ischemia (28, 59).

TABLE 3   Potential cytoprotective pathways mediated by chaperones

Site of action	Reference
Endoplasmic Reticulum	
Glucose-regulated proteins (GRP78 or BiP)	
Bind misfolded ER proteins	(11, 24)
Promote degradation of misfolded proteins	(38, 69)
Prevent calcium redistribution	(24)
Inhibit oxidant stress	(24)

TABLE 3   (*Continued*)

Site of action	Reference
Calnexin/calreticulin	
Protein folding/assembly	(10)
Maturation of glycoproteins	(10)
Cytosol	
HSP 90	
Translocate cell surface receptors	(10)
Regulate apoptosis	(157)
Assemble enzyme complexes	(10)
HSP 70 (72 and 73 kDa)	
Re-fold damaged proteins	(10)
Prevent protein aggregation	(21)
Facilitate protein degradation	(22)
Replace damaged organelle proteins	(15)
Stabilize actin cytoskeleton	(77, 78)
Stabilize centrosome, microtubules	(158)
Re-establish cell polarity	(55, 159)
Regulate protein kinase(s) (e.g., c-Src)	(234)
Act as a cytokine-like molecule	(31)
Regulate the cell cycle	(160, 161)
Mediate inflammatory response	(111)
Suppress apoptosis	
Inhibit caspase activation	(162)
Prevent apoptosome assembly	(163)
Inhibit JNK activation	(23)
Augment Bcl-2 effect	(28)
Inhibit TNF action	(164)
Inhibit FAS-Fas-ligand effect	(164)
Augment apoptosis in damaged cells	(137)
HSP 47	
Regulate collagen processing	(104)
and deposition	(125)
HSP 27	
Stabilize cytoskeleton (microfilaments,	
intermediate filaments)	(158)
Prevent nuclear protein aggregation	(165)
Inhibit apoptosis	(166)
Osmotic stress protein 94	
Molecular protein chaperone?	(149)
Mitochondria	
Mitochondrial HSP 70 (mt70)	(44)
Protein translocation into mitochondria	(167)
Chaperonin 60 (cpn60)	
Protein translocation into mitochondria	(15, 158)
Fold proteins into native conformation	(15)
Re-fold damaged proteins	(42)

Substantial evidence suggests that molecular chaperones participate in ischemic injury and repair. Within hours after even brief renal ischemia, enhanced expression of several molecular chaperones, including the cytoprotectant protein HSP 72, is observed (60). HSP 72 is the major inducible molecular chaperone found in virtually all mammalian cells. HSP 72 localizes to the cytosol in quiescent cells but rapidly accumulates in nucleoli after stress (61), a pattern observed in most eukaryotic cells (62). During recovery from ATP depletion, HSP 72 translocates to the cytosol where its content exceeds the pre-stress level (63). Expression of HSP 72 is most robust in the inner medulla (60), the region of the kidney least susceptible to ischemic injury (51). Increased expression of HSP 72 is likely to be precipitated by stress-induced aggregation of cytoskeletal proteins including actin (36). ATP depletion (perhaps by perturbing protein conformation) has been shown to stimulate HSF-1, a major transcriptional factor that regulates HSP 72 expression (64). Akcetin and colleagues recently demonstrated that two HSP-70 genes coding for HSP 72 respond to renal ischemia. HSP 70-1 increased after brief renal ischemia, whereas HSP 70-2 increased after more prolonged ischemic stress (65), suggesting that the kidney has a sophisticated stress response system. In order to respond to cell stress, it is likely that HSP 72 requires co-chaperones and regulatory factors including HSP 40 (66), BAG-1 (67), Hip-Hop (12), and others. Although less robust than the HSP 72 response, renal ischemia alters the expression and/or intracellular distribution of other chaperones such as HSP 25, HSP 73, HSP 90, and GRP 78 (BiP).

A host of resident molecular chaperones in the endoplasmic reticulum (ER) assist with routine protein folding (68). After stress, some of these chaperones are up-regulated. GRP 78, an ER member of the HSP 70 family, is induced by stressors that precipitate protein denaturation (69). Interestingly, selective induction of ER chaperones using tunicamycin or A23187 protects against subsequent cell membrane injury caused by ATP depletion (70). These important observations demonstrate that perturbations in protein conformation accompany ischemic stress and that molecular chaperones in the ER are key mediators of protein repair.

HSP 25, an actin-stabilizing protein, normally localizes to the brush border of cortical renal tubules. After transient ischemia, HSP 25 accumulates in a cell fraction of kidney homogenates that also contains cytoskeletal proteins (71). This finding was substantiated by the observation that HSP 25 is constitutively expressed in the kidney in a detergent-soluble protein fraction but reversibly accumulates in the detergent-insoluble fraction during 45 min of ischemia (72). Under basal conditions, HSP 25 co-localized with microfilamentous actin in the brush border but exhibited a diffuse cytosolic pattern similar to the distribution of actin aggregates during ischemia and recovery, supporting the hypothesis that HSP 25 participates in stabilizing or repairing the actin cytoskeleton after an ischemic insult (71, 72). After ischemia in vivo, HSP 25 and α-B crystallin, another small molecular weight HSP, accumulate in the cytosol of the proximal tubule (73).

In contrast to the rapid and dramatic changes in HSP 25 and HSP 72 that accompany renal ischemia, HSP 73 and HSP 90 do not exhibit marked alterations in distribution or expression. No acute changes in HSP 73 or 90 were observed in ATP-depleted renal epithelial cells (63). In the intact kidney, HSP 73 modestly increased 3–6 days after ischemia, whereas HSP 90 accumulation peaked during days 5–7, suggesting that these HSPs may participate in the renal repair process (74). HSP 90 operates much like HSP 72 with regard to its dependence upon ATP hydrolysis to function as a chaperone (75). HSP 90 is often found in complexes that contain HSP 70, suggesting that the two chaperones cooperate in repairing non-native proteins (39, 40). HSP 90 also exerts an important role in the maturation process of protein tyrosine kinases including c-Src (13). c-Src regulates the assembly of key structural proteins that maintain cell contact sites (13). The role of HSP 90 in modulating protein kinase activity after ischemic stress has not been reported.

Elegant studies showed that HSP 70 members have three distinct functional domains: a C-terminal peptide recognition and binding site, a short linking sequence, and an N-terminal ATPase domain (76). Tsang generated a novel hypothesis that the ATPase domain of HSP 70 binds to and stabilizes actin (77). Proximity between HSP 70 and actin in oxidant-stressed cells has been shown using electron microscopy (78). In renal and non-renal cells, a variety of stressors precipitate the formation of large cytosolic aggregates that contain actin, structural proteins, and HSP 72 (36, 79). In renal epithelial cells, prior heat stress, sufficient to induce HSP 72, ameliorates collapse of the actin cytoskeleton during ATP depletion (53). Either as a result of improved actin stability or via an independent mechanism, accumulation of HSP 72 is also associated with improved integrity of the tight junction (53). A recent observation suggests that HSP 72 may interact with c-Src in ATP-depleted renal cells (80). By mediating the activity or distribution of protein kinases such as c-Src, HSP 72 could alter protein tyrosine phosphorylation of key regulatory proteins that mediate cell contact sites (81). Changes in cell-cell and cell-substrate adhesion have important implications in the pathogenesis of acute renal failure. HSP 72 may assist in the restoration of epithelial cell polarity after an ischemic insult. Na^+,K^+-ATPase, a membrane protein responsible for vectoral solute transport, is lost from the basolateral cell surface when integrity of the junctional complex is compromised (55). HSP 72 co-localizes with the Na^+,K^+-ATPase and appears to facilitate its re-insertion into the proper target membrane in an ATP-dependent manner (79). Together, these studies suggest that HSP 72 is important for stabilizing and repairing cell structures that are critical for maintaining epithelial cell function. A recent study has also implicated HSP 25 in preserving microfilamentous actin in the post-ischemic rat renal cortex (72).

Renal cells subjected to ischemia or ATP depletion undergo apoptosis (57–59). Maneuvers that increase HSP 72 content inhibit apoptotic cell death after diverse stresses (23, 25, 28, 41, 82, 83). HSP 72 inhibits Jun N-terminal kinase (JNK), a stress-activated kinase, from initiating the apoptotic pathway in non-renal cells subjected to heat stress (23). A similar effect of HSP 72 was shown after proteasome inhibition, another cause of apoptosis (26). The ATPase domain appears to be unnecessary for prevention of JNK-mediated apoptosis after UV irradiation or exposure to interleukin-1, suggesting that cytoprotection is independent of protein re-folding (47). Recent work by Meriin et al. showed that stress-induced activation of JNK is caused by the inhibition of the phosphatases that dephosphorylate and de-activate JNK (41). HSP 72 antagonizes the effect of stress on phosphatase activity, thereby promoting the deactivation of JNK (41). The anti-apoptotic effects of HSP 72 are not limited to JNK inhibition; however, the chaperone function of HSP 72 is required for inhibiting the activation of pro-apoptotic proteases (procaspases 9 and 3) after lethal heat stress (25). In vitro, HSP 70 also interferes with the assembly of the apoptosome, a complex comprised of cytochrome c, apoptosis-activating factor (APAF-1), and procaspase 9 (83). In ATP-depleted renal epithelial cells, HSP 72 binds Bcl-2, an anti-apoptotic protein and increases the Bcl-2:BAX ratio (28), a determinant of the apoptotic setpoint or rheostat in stressed cells. Importantly, the cellular distribution of HSPs (such as HSP 70) could determine a cell's fate. Up-regulation of HSP 70 in the cytosol or nucleus appears to be cytoprotective, whereas expression on the cell's surface may initiate cell destruction by the immune system (3, 29). In addition to HSP 70, the potential contribution of the small HSPs (e.g., HSP 25–27) as regulators of the apoptotic pathway has recently been recognized (84).

To date, attempts to enhance resistance to ischemic injury in the intact kidney by up-regulating HSP 72 in situ have been inconsistent. Joannidis and colleagues failed to show protection from tubular necrosis in the isolated perfused kidney after subjecting the intact rat to whole-body hyperthermia (85). Although renal ischemia in the intact rat protected against subsequent hypoxic damage in a suspension of proximal tubule, the resistance to injury did not correlate with HSP 72 content (86). In contrast, others showed that subjecting the isolated kidney to hyperthermia prior to transplantation increases HSP 72 and decreases ischemic injury (87). In myocardial cells, HSP 72 accumulation affords consistent cytoprotection against ATP depletion and ischemia. In addition, selective overexpression of HSP 72 protected against cardiac ischemia in a transgenic mouse (88). Pharmacologic approaches have been attempted to induce stress proteins and improve cytoresistance to ischemic injury. Bimoclomol, an hydroxylamine derivative, increases expression of HSPs without significant toxicity and protects mice against ischemic tissue injury (89). Nontoxic methods for inducing molecular chaperones in the kidney hold promise for improving renal epithelial resistance to anticipated episodes of ischemia.

Nephrotoxin Exposure

Acute renal failure is a well-recognized complication of exposure to intravenous contrast agents, nephrotoxic antibiotics, heavy metals, or a variety of chemotherapeutic agents (e.g., cisplatin). The mechanisms of cytotoxicity appear to differ for these insults. Intravenous contrast agents may be directly toxic to renal epithelial cells (90) and can precipitate vasoconstriction, an important cause of cellular ischemia (91). Heavy metals target the mitochondria (92) and cause protein denaturation (93). Cisplatin induces both renal epithelial cell apoptosis and necrosis in a dose-dependent manner (94, 95).

Although cisplatin-induced apoptosis has been assumed to be the direct result of DNA damage (96, 97), other pathways may contribute to cell death. Activation of the Fas/Fas-L system by cisplatin can promote apoptotic renal cell death (98). Cisplatin activates interleukin-1 β-converting enzyme (ICE) proteases in the apoptotic cascade (99). Recently, cisplatin-mediated phosphorylation of α-adducin, an actin-capping protein, was shown to precede caspase activation (100). Once activated, however, caspase 3 led to irreversible cleavage of α-adducin. This observation suggests that the actin stress fibers and focal adhesions might be additional target sites for cisplatin-mediated apoptosis in renal cells (100). Although HSP 72 content did not increase after exposure to cisplatin, the distribution to the nucleus was dramatically altered in HeLa cells (97). Several studies demonstrated that molecular chaperones increase cytoresistance to nephrotoxic injury. In noncancerous renal cells, resistance to cisplatin-induced cell death was observed in transfected epithelial LLC-PK1 cells that overexpressed HSP 72 (103). The ability of HSP 72 to regulate cisplatin-induced apoptosis is not surprising given that HSP 72 inhibits multiple steps in the apoptotic pathway (27, 28, 41). An increase in HSP 70 content may alter the prognosis for renal cell carcinomas (101), perhaps by altering the sensitivity of cancer cells to chemotherapeutic drugs (102).

Exposure to gentamicin, a nephrotoxic antibiotic known to cause acute tubular necrosis, induces the expression of renal molecular chaperones. Increased expression of HSP 47 and HSP 73 was observed in rat kidneys after subcutaneous injection of gentamicin (104). Accumulation of HSP 47 was maximal at day 3, several days after the appearance of ATN. In this same study, HSP 47 immunostaining was most abundant in the tubular epithelial cells and interstitial cells in the regions of collagen III deposition, suggesting that HSP 47 may participate in interstitial repair or fibrosis (104). In contrast, HSP 73 rapidly accumulated within lysosomes of damaged proximal tubular epithelial cells after gentamicin exposure (105), suggesting that this chaperone may facilitate lysosomal protein degradation.

Heavy metals such as mercuric ($HgCl_2$) or cadmium chloride ($CdCl_2$) cause marked toxicity to renal cells. Heavy metals also induce the synthesis of molecular chaperones. Although heavy metals cause protein denaturation (93), a potent stimulus for chaperone induction (37), other events may contribute. Exposure of LLC-PK1 cells to $HgCl_2$ generates a substantial oxidant stress from endogenous hydrogen peroxide that originates, at least in part, from damaged mitochondria (106). In addition to mitochondrial injury, the lysosomal proton gradient required for normal protein degradation is disrupted (106). The protective effect afforded by overexpressing anti-apoptotic genes of the Bcl-2 family against $CdCl_2$ injury suggests a central role of mitochondrial injury caused by $CdCl_2$ (106, 107). As a response to either mitochondrial injury or the protein-damaging effects of hydrogen peroxide, molecular chaperones are induced. An increase in the de novo synthesis of both HSP 72 and HSP 90 was observed in slices of rat kidney after heavy metal exposure (108). Even a single dose of $HgCl_2$, sufficient to induce only single epithelial cell necrosis, induced HSP 72 in the rat renal cortex (109).

However, accumulation of HSP 72 was most abundant in undamaged distal convoluted cells, making its role in proximal tubular cell injury less clear in this study. Other investigators detected HSP 65 and HSP 72 in cortical tubules with the most overt histologic injury (110). In the later study, immunoelectron microscopy of severely damaged cells showed abundant HSP 65 in mitochondria and nucleoli, whereas HSP 72 was overexpressed in the cytoplasm, mitochondria, lysosomes, the cytoskeleton, and in the nucleus (110). Activation of T lymphocytes may also contribute to the inflammatory process that causes chronic interstitial nephritis. After prolonged exposure to cadmium, kidney-derived T cells were capable of inducing interstitial nephritis after passive transfer to cadmium-exposed mice before the onset of overt nephritis (111). These investigators suggested that HSP 70 is an important target for T cell–mediated inflammation and chronic renal injury. Increases in HSP 90 within the renal proximal renal tubule have also been reported after exposure to toxic doses of iron (112).

Obstructive Nephropathy

Urinary tract obstruction is a common cause of both acute and chronic renal dysfunction. Obstruction produces interstitial renal injury with inflammation and promotes chronic apoptosis with progressive renal failure and interstitial fibrosis (113–115). Unilateral obstruction in the rat is associated with an increase in HSP 72 only in the obstructed kidney (113, 116, 117), possibly resulting from localized renal oxidant stress kidney (116). In addition, HSP 47, a collagen-binding stress protein, accumulates in the mouse kidney after unilateral obstruction (118). HSP 47 mRNA expression is increased within 12 h of acute obstruction (118). In this study, administration of either an AII receptor antagonist or an ACE inhibitor decreased HSP 47 and type I collagen mRNA levels by ∼60% and prevented interstitial fibrosis. These observations support the hypothesis that interstitial fibrosis after urinary tract obstruction involves molecular chaperones and may be amenable to therapeutic treatment.

Glomerulonephritis (GN) and Immune-Mediated Injury

Glomerular injury represents a balance between insults delivered by infiltrating leukocytes and platelets that elaborate damaging cytokines, eicosanoids, complement, and oxygen radical species and the ability of the glomerulus to resist injury (119). Noxious stimuli result in vascular injury, alterations in basement membrane composition, damage the glomerular epithelial cell and podocytes, stimulate cell proliferation and, ultimately, precipitate glomerular fibrosis.

Stress proteins may be an important component of resistance to glomerular injury. The presence of constitutive and inducible stress proteins has been demonstrated in animal kidneys with experimental GN (120–122). An increase in stress proteins was detected in human kidneys with various forms of acute GN (120, 123–125). Proteinuria, a potential cause of renal tubular injury that often accompanies GN, increases the expression of HSP 72 (123). The presence of protein in the tubular lumen was increased with pro-apoptotic stimuli, including tumor necrosis factor (TNF), a potent initiator of apoptosis (126). In non-renal cells, the selective overexpression of HSP 72 blocks TNF-induced cell death (127).

Stress proteins have also been implicated in the pathogenesis of autoimmune diseases including systemic lupus erythematosis. In the kidney, molecular chaperones may serve as antigenic targets for immune-mediated injury (111, 128, 129). An allele of HSP70 has been linked to the human leukocyte antigen (HLA) haplotypes that are associated with increased susceptibility to SLE in a Spanish population (34). The pathogenic significance of this finding awaits further clarification.

Increasing evidence implicates HSP 70, HSP 90, and HSP 110 members as potential antigen targets for cell-mediated inflammation (3, 111, 129). Recent investigations also support a possible role for HSP 47, 65, 70, and 90 in the pathogenesis of some forms of glomerulopathy (122, 123, 125). Work by Warr and colleagues suggests that cross reactivity between microbial and human HSPs may stimulate a T cell subset that has been implicated in the pathogenesis of progressive IgA nephritis in humans (124). In this study, a peptide derived from mycobacterial HSP 65 stimulated T cell proliferation. In experimental nephritis caused by injection of Thy1.1 antigen (a form of mesangioproliferative GN), HSP 90 appears to regulate the response to pro-mitogenic signals required for mesangial cells to enter G1 or to progress through the S-phase (120). Mesangial cell proliferation is an important precursor of glomerular dysfunction.

In sum, molecular chaperones may mediate glomerular injury via immune and non-immune mediated mechanisms. Some renal chaperones appear likely to alter the acute infiltration of pro-inflammatory cells, perhaps by inadvertently mimicking an invading micro-organism (124). Other chaperones modulate cell proliferation, determine the susceptibility to pro-apoptotic stimuli, or the propensity to undergo irreversible fibrosis.

Chronic, Progressive Renal Failure

Although the mechanism is debated, many forms of renal injury can lead to progressive renal failure even in the absence of the initial insult (130). Many factors have been proposed to cause progressive renal injury including hemodynamic stress, phosphate deposition, hyperlipidemia, reactive oxygen species, immune-mediated injury resulting from increased ammoniagenesis, and dysregulation of apoptosis leading to unrelenting cellular dropout (49). One or more of these insults could be responsible for the progressive interstitial and glomerular fibrosis that are important hallmarks of chronic, progressive renal failure. In aging rats, increased accumulation of both HSP 47 (131) and HSP 72 (132) have been observed, suggesting that progressive loss of renal function is associated with an increased burden of non-native proteins.

The expression of HSP 47, a collagen-binding protein, in glomeruli with segmental sclerosis paralleled the expression of type I, III, and IV collagen in rats subjected to subtotal nephrectomy (131). These rats showed glomerulosclerosis with marked tubulointerstitial damage, as well as interstitial fibrosis with increased collagen and HSP 47 deposition in glomeruli, tubular epithelial cells, and interstitial cells of remnant rat kidneys (131). Administration of HSP 47 antisense oligodeoxynucleotides suppressed the collagen deposition and attenuated the histologic manifestations of the glomerular fibrosis (131). HSP 47 is unlikely to act alone in processing collagen. GRP 78 and GRP 94 cooperate with the HSP 47-procollagen complex in metabolically stressed cells, preventing the proper folding and release of procollagen (133).

Renal Transplantation

Removal, storage, and re-implantation of a kidney are accompanied by a variable degree of ischemic damage (87, 134). Ischemic renal injury is an important cause of early graft failure (135) and may increase the likelihood of subsequent episodes of acute or chronic rejection (134). Although therapy for acute allograft rejection with active inflammation is often effective, no treatment presently exists for chronic progressive rejection associated with progressive interstitial fibrosis. Accumulation of HSP 47 in the interstitium positively correlates with interstitial fibrosis in allografts with chronic progressive dysfunction (136). A novel HSP (45 kDa or HJD-2) was recently identified in human kidney biopsies that exhibited either acute or chronic rejection. In contrast, normal, pretransplant kidneys and transplanted kidneys without histologic evidence of rejection (acute or chronic) showed no HJD-2 (137). These investigators suggest that this novel HSP might be an antigen against which cytotoxic T cells that mediate acute rejection are directed. However, HSP 60 was also increased in allografts with rejection (137), making it difficult to ascribe causality to a specific HSP. Although expression of molecular chaperones after a nonlethal ischemic insult is assumed to confer cytoprotection, these same chaperones expressed on the cell surface may mark cells for apoptosis (29) or precipitate cell infiltration and inflammation (3, 129).

Osmotic Stress

Over a decade ago, Cohen and colleagues observed a brisk induction of HSP 72 in renal MDCK cells subjected to physiologic, hyperosmolar stress (138). The transcriptional response to hyperosmotic stress has been recently reviewed (139, 140). Renal medullary stress is a consequence of physiological changes in extracellular osmolality associated with normal fluctuations in water, urea, and ion excretion. Cells acutely exposed to hypo- or hyperosmolar stress exhibit changes in cytoskeletal organization (141), membrane transporter activities (142), and stimulation of the cell death pathway (139, 143). To compensate for this adverse physiologic situation, kidney medullary cells constitutively overexpress molecular chaperones, with an increasing gradient of molecular chaperones evident from cortex to medulla (139, 144).

Intracellular cytoprotective osmolytes (i.e., sorbitol, betaine, inositol, taurine, and glycerophosphorylcholine) are accumulated during hypertonic stress (142), although the rate of their accumulation is relatively slow (145). Similarly, cells exposed to a hypo-osmolar environment respond by dumping organic osmolytes, inhibiting intracellular osmolyte production, and increasing their degradation (141). These adaptive mechanisms are likely inadequate to protect renal cells from rapid changes in extracellular osmolality. Thus constitutive expression of inducible HSPs or their rapid induction with severe osmotic stress could fill an acute cytoprotective gap (146, 147).

Several osmotic-sensitive molecular chaperones have been implicated as potential cytoprotectants, including HSP 25, 60, 72, 78, 110, 200, and the osmotic stress protein OSP 94. Expression of HSP 25 and HSP 72 is enhanced in vivo in rat inner medulla in response to dehydration-mediated hyperosmolar stress (139, 143, 148). In addition, molecular chaperones of 46, 60, 78, and 200 kDa have been identified in renal epithelial cells subjected to hypertonic NaCl (147). More recently, OSP 94, a member of the HSP 110 family, was found to be highly expressed in murine inner medulla in vivo (149). In the dehydrated mouse, increased medullary expression of OSP 94 and HSP 110 were observed, emphasizing the importance of these proteins in the adaptive response to hyperosmotic stress (144).

Studies of renal cells in culture yielded comparable results. After exposure to hypertonic NaCl, murine inner medullary collecting duct (IMCD3) cells increased HSP 72 mRNA levels prior to increases in either OSP 94 or HSP 110 (144). The relatively rapid induction of HSP 70 in response to increased osmolality suggests that this chaperone is a critical, but not sufficient, adaptive response to perturbations in protein conformation. An assay for ATP-dependent binding to an unfolded protein identified additional hyperosmotic stress-inducible proteins including mitochondrial HSP 70, as well as 60 and 200 kDa proteins (147). Furthermore, preconditioning with heat stress, a well-established cause of protein denaturation and inducer of heat shock proteins, protects IMCD3 cells against subsequent osmotic stress (144). The existence of this cross tolerance suggests that hyperosmolality and thermal stress share many key features such as protein denaturation and refolding.

Recent work has sought to identify specific molecular chaperones responsible for cytoprotection during hyperosmotic stress. In MDCK cells, a priming osmotic stress correlated with protection against a more severe osmotic shock. MDCK cells exposed to 600 mOsm NaCl were more likely to survive a subsequent urea stress (143). Improved cell survival correlated with the accumulation of HSP 72 (as well as increased levels of betaine and glycerophosphorylcholine) but not HSP 25, suggesting that HSP 72 may mediate cytoprotection (143). Other investigators detected constitutive expression of HSP 72 in vivo, especially in the regions of the medulla exposed to the highest osmolality. However, in vivo water restriction failed to elicit an induction of HSP 72 above constitutively high levels in mice or rats in some studies (144, 145). In fact, water-restricted mice showed increased expression of medullary OSP 94 and HSP 110, suggesting that HSP 72 may not be the only cytoprotectant during osmolar stress (144, 145).

What is the primary stimulus for molecular chaperone induction in medullary cells exposed to increased osmolality? In vivo, urea may be a primary stimulus for inducing molecular chaperones (139, 148). This is an attractive hypothesis because high concentrations of urea cause protein denaturation in vitro (42). Stress kinases appear to signal the osmotic stress response, and induction of HSP 72 appears dependent upon p38 kinase-mediated regulation of Jun N-terminal kinase (JNK) (46). In contrast, thermal induction of HSP 72 is independent of p38 kinase (46). Another in vitro study suggests that vasopressin activates heat shock transcription factor (HSF), a precursor for the induction of HSP 70 (150). In this study, blockade of the V_2 receptor prevented HSP 70 induction, suggesting that the response to osmotic stress is renal specific and may be mediated by through the cyclic-AMP pathway. Ultimately, osmotic stress has the potential to cause cell death. Given that acute hyperosmolality activates stress kinases, it is not surprising that exposure to hypertonic NaCl or urea precipitates apoptosis in inner medullary collecting duct cells (151, 152), a process that could be modulated by the presence of HSPs.

FUTURE DIRECTIONS

In kidney, accumulation of molecular chaperones is a critical step in inducing cytoprotection under adverse circumstances. Induction of chaperones can be achieved experimentally using drugs, heat exposure, or molecular strategies that stimulate overexpression of heat shock transcription factor (HSF)-specific stress proteins. Although this preemptive approach might appear limited in most clinical settings, many episodes of acute renal injury can be anticipated. Renal injury caused by exposure to intravenous contrast agents or nephrotoxins (e.g., aminoglycosides, cisplatin), surgical procedures involving cardiopulmonary bypass, and cold storage of kidneys prior to organ transplantation could potentially be prevented by pre-induction of molecular chaperones. Ultimately, induction of molecular chaperones or the administration of cytoprotective domains of specific chaperones could prevent or ameliorate acute renal failure in high-risk situations. In the meantime, appreciation of the impact of molecular chaperones on cell function has provided a better understanding of the pathways that mediate cellular injury and survival.

ACKNOWLEDGMENTS

This review was supported by the National Institutes of Health grants DK-53387 to S.C.B. and DK-36031 and DK-51606 to S.R.G.

Visit the Annual Reviews home page at www.AnnualReviews.org

LITERATURE CITED

1. Brezis M, Rosen S. 1995. Hypoxia of the renal medulla—its implications for disease. *N. Engl. J. Med.* 332:647–55
2. Feder ME, Hofmann GE. 1999. Heat-shock proteins, molecular chaperones, and the stress response: evolutionary and ecological physiology. *Annu. Rev. Physiol.* 61:243–82

3. Moseley P. 2000. Stress proteins and the immune response. *Immunopharmacology* 48:299–302

4. Cullen KE, Sarge KD. 1997. Characterization of hypothermia-induced cellular stress response in mouse tissues. *J. Biol. Chem.* 272:1742–46

5. Van Why SK, Siegel NJ. 1998. Heat shock proteins in renal injury and recovery. *Curr. Opin. Nephrol. Hypertens.* 7:407–12

6. Beck FX, Neuhofer W, Muller E. 2000. Molecular chaperones in the kidney: distribution, putative roles, and regulation. *Am. J. Physiol. Renal Physiol.* 279:F203–F15

7. Ritossa F. 1962. A new puffing pattern induced by temperature shock and DNP in Drosophila. *Experientia* 18:571–73

8. Tissieres A, Mitchell HK, Tracy UM. 1974. Protein synthesis in salivary glands of *Drosophila melanogaster*: relation to chromosome puffs. *J. Mol. Biol.* 84:389–98

9. Knowlton AA. 1995. The role of heat shock proteins in the heart. *J. Mol. Cell. Cardiol.* 27:121–31

10. Fink AL. 1999. Chaperone-mediated protein folding. *Physiol. Rev.* 79:425–49

11. Nigam SK, Goldberg AL, Ho S, Rohde MF, Bush KT, Sherman M. 1994. A set of endoplasmic reticulum proteins possessing properties of molecular chaperones includes Ca^{2+}-binding proteins and members of the thioredoxin superfamily. *J. Biol. Chem.* 269:1744–49

12. Frydman J, Hohfeld J. 1997. Chaperones get in touch: the hip-hop connection. *Trends Biochem.* 22:87–92

13. Xu Y, Singer M, Lindquist S. 1999. Maturation of the tyrosine kinase c-Src as a kinase and as a substrate depends on the molecular chaperone HSP 90. *Proc. Natl. Acad. Sci. USA* 96:109–14

14. Becker J, Craig EA. 1994. Heat-shock proteins as molecular chaperones. *Eur. J. Biochem.* 219:11–23

15. Stuart RA, Cyr DM, Craig EA, Neupert W. 1994. Mitochondrial molecular chaperones: their role in protein translocation. *Trends Biochem. Sci.* 19:87–92

16. Hightower LE. 1991. Heat shock, stress proteins, chaperones, and proteotoxicity. *Cell* 66:191–97

17. Garcia-Mata R, Bebok Z, Sorscher EJ, Sztul ES. 1999. Characterization and dynamics of aggresome formation by a cytosolic GFP-chimera. *J. Cell Biol.* 146:1239–54

18. Leppa S, Sistonen L. 1997. Heat shock response—pathophysiological implications. *Ann. Med.* 29:73–78

19. Couette B, Jalaguier S, Hellal-Levy C, Lupo B, Fagart J, et al. 1998. Folding requirements of the ligand-binding domain of the human mineralocorticoid receptor. *Mol. Endocrinol.* 12:855–63

20. Wakeham DE, Ybe JA, Brodsky FM, Hwang PK. 2000. Molecular structures of proteins involved in vesicle coat formation. *Traffic* 1:393–98

21. McClellan AJ, Frydman J. 2001. Molecular chaperones and the art of recognizing a lost cause. *Nat. Cell Biol.* 3:E51–53

22. Hayes SA, Dice JF. 1996. Roles of molecular chaperones in protein degradation. *J. Cell Biol.* 132:255–58

23. Gabai VL, Meriin AB, Mosser DD, Caron AW, Rits S, et al. 1997. Hsp70 prevents activation of stress kinases. A novel pathway of cellular thermotolerance. *J. Biol. Chem.* 272:18033–37

24. Liu H, Bowes RC 3rd, van de Water B, Sillence C, Nagelkerke JF, Stevens JL. 1997. Endoplasmic reticulum chaperones GRP78 and calreticulin prevent oxidative stress, Ca^{2+} disturbances, and cell death in renal epithelial cells. *J. Biol. Chem.* 272:21751–59

25. Mosser DD, Caron AW, Bourget L, Meriin AB, Sherman MY, et al. 2000. The chaperone function of hsp70 is required for protection against stress-induced apoptosis. *Mol. Cell. Biol.* 20:7146–59

26. Meriin AB, Gabai VL, Yaglom J, Shifrin VI, Sherman MY. 1998. Proteasome inhibitors activate stress kinases and induce

Hsp72. Diverse effects on apoptosis. *J. Biol. Chem.* 273:6373–79

27. Volloch V, Gabai VL, Rits S, Force T, Sherman MY. 2000. HSP72 can protect cells from heat-induced apoptosis by accelerating the inactivation of stress kinase JNK. *Cell Stress Chaperones* 5:139–47

28. Wang Y, Knowlton AA, Christensen TG, Shih T, Borkan SC. 1999. Prior heat stress inhibits apoptosis in adenosine triphosphate-depleted renal tubular cells. *Kidney Int.* 55:2224–35

29. Poccia F, Piselli P, Vendetti S, Bach S, Amendola A, et al. 1996. Heat-shock protein expression on the membrane of T cells undergoing apoptosis. *Immunology* 88:6–12

30. Sapozhnikov AM, Ponomarev ED, Tarasenko TN, Telford WG. 1999. Spontaneous apoptosis and expression of cell surface heat-shock proteins in cultured EL-4 lymphoma cells. *Cell Prolif.* 32:363–78

31. Asea A, Kraeft SK, Kurt-Jones EA, Stevenson MA, Chen LB, et al. 2000. HSP70 stimulates cytokine production through a CD14-dependent pathway, demonstrating its dual role as a chaperone and cytokine. *Nat. Med.* 6:435–42

32. Minota S, Cameron B, Welch WJ, Winfield JB. 1988. Autoantibodies to the constitutive 73-kD member of the hsp70 family of heat shock proteins in systemic lupus erythematosus. *J. Exp. Med.* 168:1475–80

33. Minota S, Koyasu S, Yahara I, Winfield J. 1988. Autoantibodies to the heat-shock protein hsp90 in systemic lupus erythematosus. *J. Clin. Invest.* 81:106–9

34. Pablos JL, Carreira PE, Martin-Villa JM, Montalvo G, Arnaiz-Villena A, Gomez-Reino JJ. 1995. Polymorphism of the heat-shock protein gene HSP70-2 in systemic lupus erythematosus. *Br. J. Rheumatol.* 34:721–23

35. Nover L. 1991. Inducers of HSP synthesis: heat shock and chemical inducers. In *Heat Shock Response*, ed. L Nover, pp. 5–40. Boca Raton, FL: CRC Press

36. Kabakov AE, Gabai VL. 1993. Protein aggregation as primary and characteristic cell reaction to various stresses. *Experientia* 49:706–13

37. Mifflin L, Cohen R. 1994. Characterization of denatured protein inducers of the heat shock (stress) response in *Xenopus laevis oocytes*. *J. Biol. Chem.* 269:15710–17

38. Bush KT, Goldberg AL, Nigam SK. 1997. Proteasome inhibition leads to a heat-shock response, induction of endoplasmic reticulum chaperones, and thermotolerance. *J. Biol. Chem.* 272:9086–92

39. Buchner J. 1999. HSP 90 & Co-a holding for folding. *Trends Biochem. Sci.* 24:136–41

40. James P, Pfund C, Craig E. 1997. Functional specificity among HSP 70 molecular chaperones. *Science* 275:387–89

41. Meriin AB, Yaglom JA, Gabai VL, Zon L, Ganiatsas S, et al. 1999. Protein-damaging stresses activate c-Jun N-terminal kinase via inhibition of its dephosphorylation: a novel pathway controlled by HSP72. *Mol. Cell. Biol.* 19:2547–55

42. Mendoza JA, Lorimer GH, Horowitz PM. 1992. Chaperonin cpn60 from *Escherichia coli* protects the mitochondrial enzyme rhodanese against heat inactivation and supports folding at elevated temperatures. *J. Biol. Chem.* 267:17631–34

43. Sherman M, Goldberg A. 1996. Involvement of molecular chaperones in intracellular protein breakdown. *Exp. Suppl.* 77:57–78

44. Schmitt M, Neupert W, Langer T. 1996. The molecular chaperone Hsp78 confers compartment-specific thermotolerance to mitochondria. *J. Cell Biol.* 134:1375–86

45. Gierasch LM. 1994. Molecular chaperones. Panning for chaperone-binding peptides. *Curr. Biol.* 4:173–74

46. Sheikh-Hamad D, Di Mari J, Suki WN, Safirstein R, Watts BA 3rd, Rouse D. 1998. p38 kinase activity is essential for osmotic induction of mRNAs for HSP70 and transporter for organic solute betaine

in Madin-Darby canine kidney cells. *J. Biol. Chem.* 273:1832–37

47. Volloch V, Gabai VL, Rits S, Sherman MY. 1999. ATPase activity of the heat shock protein hsp72 is dispensable for its effects on dephosphorylation of stress kinase JNK and on heat-induced apoptosis. *FEBS Lett.* 461:73–76

48. Racusen LC. 1998. Epithelial cell shedding in acute renal injury. *Clin. Exp. Pharmacol. Physiol.* 25:273–75

49. Lieberthal W, Levine JS. 1996. Mechanisms of apoptosis and its potential role in renal tubular epithelial cell injury. *Am. J. Physiol. Renal Physiol.* 271:F477–F88

50. Edelstein CL, Ling H, Schrier RW. 1997. The nature of renal cell injury. *Kidney Int.* 51:1341–51

51. Venkatachalam MA, Bernard DB, Donohoe JF, Levinsky NG. 1978. Ischemic damage and repair in the rat proximal tubule: differences among the S1, S2, and S3 segments. *Kidney Int.* 14:31–49

52. Fish EM, Molitoris BA. 1994. Alterations in epithelial polarity and the pathogenesis of disease states. *N. Engl. J. Med.* 330:1580–88

53. Borkan SC, Wang YH, Lieberthal W, Burke PR, Schwartz JH. 1997. Heat stress ameliorates ATP depletion-induced sublethal injury in mouse proximal tubule cells. *Am. J. Physiol. Renal Physiol.* 272:F347–F55

54. Molitoris BA, Geerdes A, McIntosh JR. 1991. Dissociation and redistribution of $Na^+,K(+)$-ATPase from its surface membrane actin cytoskeletal complex during cellular ATP depletion. *J. Clin. Invest.* 88:462–69

55. Van Why SK, Mann AS, Ardito T, Siegel NJ, Kashgarian M. 1994. Expression and molecular regulation of $Na(+)$-$K(+)$-ATPase after renal ischemia. *Am. J. Physiol. Renal Physiol.* 267:F75–F85

56. Wang YH, Borkan SC. 1996. Prior heat stress enhances survival of renal epithelial cells after ATP depletion. *Am. J. Physiol. Renal Physiol.* 270:F1057–F65

57. Green D, Reed J. 1998. Mitochondria and apoptosis. *Science* 81:1309–12

58. Bergin E, Levine JS, Koh JS, Lieberthal W. 2000. Mouse proximal tubular cell-cell adhesion inhibits apoptosis by a cadherin-dependent mechanism. *Am. J. Physiol. Renal Physiol* 278:F758–F68

59. Lieberthal W, Menza SA, Levine JS. 1998. Graded ATP depletion can cause necrosis or apoptosis of cultured mouse proximal tubular cells. *Am. J. Physiol. Renal Physiol.* 274:F315–F27

60. Emami A, Schwartz JH, Borkan SC. 1991. Transient ischemia or heat stress induces a cytoprotectant protein in rat kidney. *Am. J. Physiol. Renal Physiol.* 260:F479–F85

61. Borkan SC, Emami A, Schwartz JH. 1993. Heat stress protein-associated cytoprotection of inner medullary collecting duct cells from rat kidney. *Am. J. Physiol. Renal Physiol.* 265:F333–F41

62. Lewis MJ, Pelham HR. 1985. Involvement of ATP in the nuclear and nucleolar functions of the 70 kd heat shock protein. *EMBO J.* 4:3137–43

63. Kumar Y, Tatu U. 2000. Induced hsp70 is in small, cytoplasmic complexes in a cell culture model of renal ischemia: a comparative study with heat shock. *Cell Stress Chaperones* 5:314–27

64. Van Why SK, Mann AS, Thulin G, Zhu XH, Kashgarian M, Siegel NJ. 1994. Activation of heat-shock transcription factor by graded reductions in renal ATP, in vivo, in the rat. *J. Clin. Invest.* 94:1518–23

65. Akcetin Z, Pregla R, Darmer D, Heynemann H, Haerting J, et al. 1999. Differential expression of heat shock proteins 70-1 and 70-2 mRNA after ischemia-reperfusion injury of rat kidney. *Urol. Res.* 27:306–11

66. Muller E, Neuhofer W, Burger-Kentischer A, Ohno A, Thurau K, Beck F. 1998. Effects of long-term changes in medullary osmolality on heat shock proteins HSp25, HSP60, HSP72 and HSP73

in the rat kidney. *Pflügers Arch.* 435:705–12

67. Stuart JK, Myszka DG, Joss L, Mitchell RS, McDonald SM, et al. 1998. Characterization of interactions between the anti-apoptotic protein BAG-1 and Hsc70 molecular chaperones. *J. Biol. Chem.* 273:22506–14

68. Kuznetsov G, Nigam SK. 1998. Folding of secretory and membrane proteins. *N. Engl. J. Med.* 339:1688–95

69. Haas I. 1994. Bip (GRP78), an essential hsp70 resident protein in the endoplasmic reticulum. *Experientia* 50:1012–20

70. Bush KT, George SK, Zhang PL, Nigam SK. 1999. Pretreatment with inducers of ER molecular chaperones protects epithelial cells subjected to ATP depletion. *Am. J. Physiol. Renal Physiol.* 277:F211–F18

71. Schober A, Burger-Kentischer A, Muller E, Beck FX. 1998. Effect of ischemia on localization of heat shock protein 25 in kidney. *Kidney Int. Suppl.* 67:S174–76

72. Aufricht C, Ardito T, Thulin G, Kashgarian M, Siegel NJ, Van Why SK. 1998. Heat-shock protein 25 induction and redistribution during actin reorganization after renal ischemia. *Am. J. Physiol. Renal Physiol.* 274:F215–F22

73. Somji S, Sens DA, Garrett SH, Sens MA, Todd JH. 1999. Heat shock protein 27 expression in human proximal tubule cells exposed to lethal and sublethal concentrations of CdCl2. *Environ. Health Perspect.* 107:545–52

74. Morita K, Wakui H, Komatsuda A, Ohtani H, Miura AB, et al. 1995. Induction of heat-shock proteins HSP73 and HSP90 in rat kidneys after ischemia. *Renal Fail.* 17:405–19

75. Obermann WM, Sondermann H, Russo AA, Pavletich NP, Hartl FU. 1998. In vivo function of Hsp90 is dependent on ATP binding and ATP hydrolysis. *J. Cell Biol.* 143:901–10

76. Freeman BC, Myers MP, Schumacher R, Morimoto RI. 1995. Identification of a regulatory motif in Hsp70 that affects AT-Pase activity, substrate binding and interaction with HDJ-1. *EMBO J.* 14:2281–92

77. Tsang T. 1993. New model for 70 kDa heat-shock proteins' potential mechanism of function. *FEBS Lett.* 323:1–3

78. Hinshaw D, Armstrong BC, Burger J, Beals T, Hyslop P. 1988. ATP and microfilaments in cellular oxidant injury. *Am. J. Pathol.* 132:479–88

79. Aufricht C, Lu E, Thulin G, Kashgarian M, Siegel NJ, Van Why SK. 1998. ATP releases HSP-72 from protein aggregates after renal ischemia. *Am. J. Physiol. Renal Physiol.* 274:F268–F74

80. Wang Y, Li F, Schwartz J, Flint P, Borkan S. 2001. c-Src and HSP 72 interact in ATP depleted renal epithelial cells. *Am. J. Physiol. Cell Physiol.* 281:In press

81. Tsukamoto T, Nigam SK. 1999. Role of tyrosine phosphorylation in the reassembly of occludin and other tight junction proteins. *Am. J. Physiol. Renal Physiol.* 276:F737–F50

82. Gabai VL, Yaglom JA, Volloch V, Meriin AB, Force T, et al. 2000. Hsp72-mediated suppression of c-Jun N-terminal kinase is implicated in development of tolerance to caspase-independent cell death. *Mol. Cell. Biol.* 20:6826–36

83. Beere HM, Wolf BB, Cain K, Mosser DD, Mahboubi A, et al. 2000. Heat-shock protein 70 inhibits apoptosis by preventing recruitment of procaspase-9 to the Apaf-1 apoptosome. *Nat. Cell Biol.* 2:469–75

84. Mehlen P, Schulze-Osthoff K, Arrigo A. 1996. Small stress proteins as novel regulators of apoptosis. *J. Biol. Chem.* 271:16510–14

85. Joannidis M, Cantley LG, Spokes K, Medina R, Pullman J, et al. 1995. Induction of heat-shock proteins does not prevent renal tubular injury following ischemia. *Kidney Int.* 47:1752–59

86. Zager RA, Iwata M, Burkhart KM, Schimpf BA. 1994. Post-ischemic acute renal failure protects proximal tubules

from O_2 deprivation injury, possibly by inducing uremia. *Kidney Int.* 45:1760–68

87. Perdrizet GA, Kaneko H, Buckley TM, Fishman MS, Pleau M, et al. 1993. Heat shock and recovery protects renal allografts from warm ischemic injury and enhances HSP72 production. *Transplant Proc.* 25:1670–73

88. Marber MS, Mestril R, Chi SH, Sayen MR, Yellon DM, Dillmann WH. 1995. Overexpression of the rat inducible 70-kD heat stress protein in a transgenic mouse increases the resistance of the heart to ischemic injury. *J. Clin. Invest.* 95:1446–56

89. Vigh L, Literati PN, Horvath I, Torok Z, Balogh G, et al. 1997. Bimoclomol: a nontoxic, hydroxylamine derivative with stress protein-inducing activity and cytoprotective effects. *Nat. Med.* 3:1150–54

90. Messana JM, Cieslinski DA, Humes HD. 1990. Comparison of toxicity of radiocontrast agents to renal tubule cells in vitro. *Renal Fail.* 12:75–82

91. Bakris GL, Lass NA, Glock D. 1999. Renal hemodynamics in radiocontrast medium-induced renal dysfunction: a role for dopamine-1 receptors. *Kidney Int.* 56:206–10

92. Zazueta C, Sanchez C, Garcia N, Correa F. 2000. Possible involvement of the adenine nucleotide translocase in the activation of the permeability transition pore induced by cadmium. *Int. J. Biochem. Cell Biol.* 32:1093–101

93. DalleDonne I, Milzani A, Colombo R. 1997. Actin assembly by cadmium ions. *Biochim. Biophys. Acta* 1357:5–17

94. Lieberthal W, Triaca V, Levine J. 1996. Mechanisms of death induced by cisplatin in proximal tubular epithelial cells: apoptosis vs. necrosis. *Am. J. Physiol. Renal Physiol.* 270:F700–F8

95. Lau AH. 1999. Apoptosis induced by cisplatin nephrotoxic injury. *Kidney Int.* 56:1295–98

96. Aubrecht J, Narla RK, Ghosh P, Stanek J, Uckun FM. 1999. Molecular genotoxicity profiles of apoptosis-inducing vanadocene complexes. *Toxicol. Appl. Pharmacol.* 154:228–35

97. Melendez-Zajgla J, Garcia C, Maldonado V. 1996. Subcellular redistribution of HSP72 protein during cisplatin-induced apoptosis in HeLa cells. *Biochem. Mol. Biol. Int.* 40:253–61

98. Razzaque MS, Koji T, Kumatori A, Taguchi T. 1999. Cisplatin-induced apoptosis in human proximal tubular epithelial cells is associated with the activation of the Fas/Fas ligand system. *Histochem. Cell Biol.* 111:359–65

99. Okuda M, Masaki K, Fukatsu S, Hashimoto Y, Inui K. 2000. Role of apoptosis in cisplatin-induced toxicity in the renal epithelial cell line LLC-PK1. Implication of the functions of apical membranes. *Biochem. Pharmacol.* 59:195–201

100. van de Water B, Tijdens IB, Verbrugge A, Huigsloot M, Dihal AA, et al. 2000. Cleavage of the actin-capping protein alpha-adducin at Asp-Asp-Ser-Asp633-Ala by caspase-3 is preceded by its phosphorylation on serine 726 in cisplatin-induced apoptosis of renal epithelial cells. *J. Biol. Chem.* 275:25805–13

101. Santarosa M, Favaro D, Quaia M, Galligioni E. 1997. Expression of heat shock protein 72 in renal cell carcinoma: possible role and prognostic implications in cancer patients. *Eur. J. Cancer* 33:873–77

102. Abe T, Gotoh S, Higashi K. 1999. Higher induction of heat shock protein 72 by heat stress in cisplatin-resistant than in cisplatin-sensitive cancer cells. *Biochim. Biophys. Acta* 1445:123–33

103. Komatsuda A, Wakui H, Oyama Y, Imai H, Miura AB, et al. 1999. Overexpression of the human 72 kDa heat shock protein in renal tubular cells confers resistance against oxidative injury and cisplatin toxicity. *Nephrol. Dialysis Transplant.* 14:1385–90

104. Cheng M, Razzaque MS, Nazneen A, Taguchi T. 1998. Expression of the heat

shock protein 47 in gentamicin-treated rat kidneys. *Int. J. Exp. Pathol.* 79:125–32

105. Komatsuda A, Wakui H, Satoh K, Yasuda T, Imai H, et al. 1993. Altered localization of 73-kilodalton heat-shock protein in rat kidneys with gentamicin-induced acute tubular injury. *Lab. Invest.* 68:687–95

106. Nath KA, Croatt AJ, Likely S, Behrens TW, Warden D. 1996. Renal oxidant injury and oxidant response induced by mercury. *Kidney Int.* 50:1032–43

107. Kim MS, Kim BJ, Woo HN, Kim KW, Kim KB, et al. 2000. Cadmium induces caspase-mediated cell death: suppression by Bcl-2. *Toxicology* 145:27–37

108. Goering PL, Fisher BR, Chaudhary PP, Dick CA. 1992. Relationship between stress protein induction in rat kidney by mercuric chloride and nephrotoxicity. *Toxicol. Appl. Pharmacol.* 113:184–91

109. Goering PL, Fisher BR, Noren BT, Papaconstantinou A, Rojko JL, Marler RJ. 2000. Mercury induces regional and cell-specific stress protein expression in rat kidney. *Toxicol. Sci.* 53:447–57

110. Hernandez-Pando R, Pedraza-Chaverri J, Orozco-Estevez H, Silva-Serna P, Moreno I, et al. 1995. Histological and subcellular distribution of 65 and 70 kD heat shock proteins in experimental nephrotoxic injury. *Exp. Toxicol. Pathol.* 47:501–8

111. Weiss RA, Madaio MP, Tomaszewski JE, Kelly CJ. 1994. T cells reactive to an inducible heat shock protein induce disease in toxin-induced interstitial nephritis. *J. Exp. Med.* 180:2239–50

112. Fukuda A, Osawa T, Oda H, Tanaka T, Toyokuni S, Uchida K. 1996. Oxidative stress response in iron-induced acute nephrotoxicity: enhanced expression of heat shock protein 90. *Biochem. Biophys. Res. Commun.* 219:76–81

113. Chan W, Krieg RJ Jr, Ward K, Santos F Jr, Lin KC, Chan JC. 2001. Progression after release of obstructive nephropathy. *Pediatr. Nephrol.* 16:238–44

114. Klahr S. 2001. Urinary tract obstruction. *Semin. Nephrol.* 21:133–45

115. Choi YJ, Baranowska-Daca E, Nguyen V, Koji T, Ballantyne CM, et al. 2000. Mechanism of chronic obstructive uropathy: increased expression of apoptosis-promoting molecules. *Kidney Int.* 58:1481–91

116. Lin KC, Krieg RJ Jr, Saborio P, Chan JC. 1998. Increased heat shock protein-70 in unilateral ureteral obstruction in rats. *Mol. Genet. Metab.* 65:303–10

117. Sawczuk IS, Hoke G, Olsson CA, Connor J, Buttyan R. 1989. Gene expression in response to acute unilateral ureteral obstruction. *Kidney Int.* 35:1315–19

118. Moriyama T, Kawada N, Ando A, Yamauchi A, Horio M, et al. 1998. Upregulation of HSP47 in the mouse kidneys with unilateral ureteral obstruction. *Kidney Int.* 54:110–19

119. Kitamura M, Fine LG. 1999. The concept of glomerular self-defense. *Kidney Int.* 55:1639–71

120. Pieper M, Rupprecht HD, Bruch KM, De Heer E, Schocklmann HO. 2000. Requirement of heat shock protein 90 in mesangial cell mitogenesis. *Kidney Int.* 58:2377–89

121. Smoyer WE, Gupta A, Mundel P, Ballew JD, Welsh MJ. 1996. Altered expression of glomerular heat shock protein 27 in experimental nephrotic syndrome. *J. Clin. Invest.* 97:2697–704

122. Sunamoto M, Kuze K, Tsuji H, Ohishi N, Yagi K, et al. 1998. Antisense oligonucleotides against collagen-binding stress protein HSP47 suppress collagen accumulation in experimental glomerulonephritis. *Lab. Invest.* 78:967–72

123. Venkataseshan VS, Marquet E. 1996. Heat shock protein 72/73 in normal and diseased kidneys. *Nephron* 73:442–49

124. Warr K, Fortune F, Namie S, Wilson A, Shinnick T, et al. 1997. T-cell epitopes recognized within the 65,000 MW hsp in patients with IgA nephropathy. *Immunology* 91:399–405

125. Razzaque MS, Kumatori A, Harada T, Taguchi T. 1998. Coexpression of collagens and collagen-binding heat shock protein 47 in human diabetic nephropathy and IgA nephropathy. *Nephron* 80:434–43

126. Kumar A, Jasmin A, Eby MT, Chaudhary PM. 2001. Cytotoxicity of tumor necrosis factor related apoptosis-inducing ligand towards Ewing's sarcoma cell lines. *Oncogene* 20:1010–14

127. Jaattela M, Wissing D, Bauer P, Li G. 1992. Major heat shock protein HSP 70 protects tumor cells from tumor necrosis factor cytotoxicity. *EMBO J.* 11:3507–12

128. Cardenas ME, Zhu D, Heitman J. 1995. Molecular mechanisms of immunosuppression by cyclosporine, FK506, and rapamycin. *Curr. Opin. Nephrol. Hypertens.* 4:472–77

129. Trieb K, Grubeck-Loebenstein B, Eberl T, Margreiter R. 1996. T cells from rejected human kidney allografts respond to heat shock protein 72. *Transplant. Immunol.* 4:43–45

130. Mimran A, Ribstein J. 1999. Angiotensin receptor blockers: pharmacology and clinical significance. *J. Am. Soc. Nephrol.* 10:S273–77

131. Sunamoto M, Kuze K, Iehara N, Takeoka H, Nagata K, et al. 1998. Expression of heat shock protein 47 is increased in remnant kidney and correlates with disease progression. *Int. J. Exp. Pathol.* 79:133–40

132. Maiello M, Boeri D, Sampietro L, Pronzato MA, Odetti P, Marinari UM. 1998. Basal synthesis of heat shock protein 70 increases with age in rat kidneys. *Gerontology* 44:15–20

133. Ferreira LR, Norris K, Smith T, Hebert C, Sauk JJ. 1994. Association of Hsp47, Grp78, and Grp94 with procollagen supports the successive or coupled action of molecular chaperones. *J. Cell Biochem.* 56:518–26

134. Womer KL, Vella JP, Sayegh MH. 2000. Chronic allograft dysfunction: mecha-nisms and new approaches to therapy. *Semin. Nephrol.* 20:126–47

135. Dragun D, Hoff U, Park JK, Qun Y, Schneider W, et al. 2000. Ischemia-reperfusion injury in renal transplantation is independent of the immunologic background. *Kidney Int.* 58:2166–77

136. Abe K, Ozono Y, Miyazaki M, Koji T, Shioshita K, et al. 2000. Interstitial expression of heat shock protein 47 and alpha-smooth muscle actin in renal allograft failure. *Nephrol. Dialysis Transplant.* 15:529–35

137. Alevy YG, Brennan D, Durriya S, Howard T, Mohanakumar T. 1996. Increased expression of the HDJ-2 heat shock protein in biopsies of human rejected kidney. *Transplantation* 61:963–67

138. Cohen DM, Wasserman JC, Gullans SR. 1991. Immediate early gene and HSP70 expression in hyperosmotic stress in MDCK cells. *Am. J. Physiol. Cell Physiol.* 261:C594–C601

139. Beck FX, Grunbein R, Lugmayr K, Neuhofer W. 2000. Heat shock proteins and the cellular response to osmotic stress. *Cell Physiol. Biochem.* 10:303–6

140. Stears RL, Gullans SR. 2000. Transcriptional response to hyperosmotic stress. In *Environmental Stressors and Gene Responses*, ed. K Storey, J Storey, pp. 129–39. London: Elsevier

141. Beck FX, Burger-Kentischer A, Muller E. 1998. Cellular response to osmotic stress in the renal medulla. *Pflügers Arch.* 436:814–27

142. Burg MB. 1995. Molecular basis of osmotic regulation. *Am. J. Physiol. Renal Physiol.* 268:F983–F96

143. Neuhofer W, Muller E, Burger-Kentischer A, Fraek ML, Thurau K, Beck F. 1998. Pretreatment with hypertonic NaCl protects MDCK cells against high urea concentrations. *Pflügers Arch.* 435:407–14

144. Santos BC, Chevaile A, Kojima R, Gullans SR. 1998. Characterization of the Hsp110/SSE gene family response to

hyperosmolality and other stresses. *Am. J. Physiol. Renal Physiol.* 274:F1054–F61

145. Martial S, Price SR, Sands JM. 1995. Regulation of aldose reductase, sorbitol dehydrogenase, and taurine cotransporter mRNA in rat medulla. *J. Am. Soc. Nephrol.* 5:1971–78

146. Sheikh-Hamad D, Garcia-Perez A, Ferraris JD, Peters EM, Burg MB. 1994. Induction of gene expression by heat shock versus osmotic stress. *Am. J. Physiol. Renal Physiol.* 267:F28–F34

147. Rauchman MI, Pullman J, Gullans SR. 1997. Induction of molecular chaperones by hyperosmotic stress in mouse inner medullary collecting duct cells. *Am. J. Physiol. Renal Physiol.* 273:F9–F17

148. Ohno A, Muller E, Fraek ML, Thurau K, Beck F. 1997. Solute composition and heat shock proteins in rat renal medulla. *Pflügers Arch.* 434:117–22

149. Kojima R, Randall J, Brenner BM, Gullans SR. 1996. Osmotic stress protein 94 (Osp94). A new member of the Hsp110/SSE gene subfamily. *J. Biol. Chem.* 271:12327–32

150. Xu Q, Ganju L, Fawcett T, Holbrook N. 1996. Vasopressin-induced heat shock protein expression in renal tubular cells. *Lab. Invest.* 74:178–87

151. Zhang Z, Tian W, Cohen DM. 2000. Urea protects from the proapoptotic effect of NaCl in renal medullary cells. *Am. J. Physiol. Renal Physiol* 279:F345–F52

152. Michea L, Ferguson DR, Peters EM, Andrews PM, Kirby MR, Burg MB. 2000. Cell cycle delay and apoptosis are induced by high salt and urea in renal medullary cells. *Am. J. Physiol. Renal Physiol.* 278:F209–18

153. Somji S, Todd JH, Sens MA, Garrett SH, Sens DA. 1999. Expression of the constitutive and inducible forms of heat shock protein 70 in human proximal tubule cells exposed to heat, sodium arsenite, and CdCl(2). *Environ. Health Perspect.* 107:887–93

154. Liu J, Squibb KS, Akkerman M, Nord-

berg GF, Lipsky M, Fowler BA. 1996. Cytotoxicity, zinc protection, and stress protein induction in rat proximal tubule cells exposed to cadmium chloride in primary cell culture. *Renal Fail.* 18:867–82

155. Ohtani H, Wakui H, Komatsuda A, Satoh K, Miura AB, et al. 1995. Induction and intracellular localization of 90-kilodalton heat-shock protein in rat kidneys with acute gentamicin nephropathy. *Lab. Invest.* 72:161–65

156. Turman MA, Kahn DA, Rosenfeld SL, Apple CA, Bates CM. 1997. Characterization of human proximal tubular cells after hypoxic preconditioning: constitutive and hypoxia-induced expression of heat shock proteins HSP70 (A, B, and C), HSC70, and HSP90. *Biochem. Mol. Med.* 60:49–58

157. Galea-Lauri J, Richardson A, Latchman D, Katz D. 1996. Increased heat shock protein 90 (hsp90) expression leads to increased apoptosis in the monoblastoid cell line U937 following induction of TNF-alpha and cycloheximide: a possible role in immunopathology. *J. Immunol.* 157:4109–18

158. Liang P, MacRae TH. 1997. Molecular chaperones and the cytoskeleton. *J. Cell Sci.* 110:1431–40

159. Bidmon B, Endemann M, Muller T, Arbeiter K, Herkner K, Aufricht C. 2000. Heat shock protein-70 repairs proximal tubule structure after renal ischemia. *Kidney Int.* 58:2400–7

160. Suzuki K, Watanabe M. 1994. Modulation of cell growth and mutation induction by introduction of the expression vector of human hsp70 gene. *Exp. Cell. Res.* 215:75–81

161. Kuhl N, Kunz J, Rensing L. 2000. Heat shock-induced arrests in different cell cycle phases of rat C6-glioma cells are attenuated in heat shock-primed thermotolerant cells. *Cell Prolif.* 33:147–66

162. Li C, Lee J, Ko Y, Kim J, Seo J. 2000. Heat shock protein 70 inhibits apoptosis downstream of cytochrome c release and

upstream of caspase 3 activation. *J. Biol. Chem.* 275:25665–71

163. Saleh A, Srinivasula S, Balkir L, Robbins P, Alnemri E. 2000. Negative regulation of the Apaf-1 aspoptosome by HSP 70. *Nat. Cell Biol.* 2:476–83

164. Klosterhalfen B, Tons C, Hauptmann S, Tietze L, Offner FA, et al. 1996. Influence of heat shock protein 70 and metallothionein induction by zinc-bis-(DL-hydrogenaspartate) on the release of inflammatory mediators in a porcine model of recurrent endotoxemia. *Biochem. Pharmacol.* 52:1201–10

165. Kampinga HH, Brunsting JF, Stege GJ, Konings AW, Landry J. 1994. Cells over-expressing Hsp27 show accelerated recovery from heat-induced nuclear protein aggregation. *Biochem. Biophys. Res. Commun.* 204:1170–77

166. Mehlen P, Schulze-Osthoff K, Arrigo A. 1996. Small stress proteins as novel regulators pf apoptosis. Heat shock protein 27 blocks FAS/APO-1 and staurosporine-induced cell death. *J. Biol. Chem.* 271:16510–14

167. Lim JH, Martin F, Guiard B, Pfanner N, Voos W. 2001. The mitochondrial Hsp70-dependent import system actively unfolds preproteins and shortens the lag phase of translocation. *EMBO J.* 20:941–50

Annu. Rev. Physiol. 2002. 64:529–49

MOLECULAR MECHANISM OF ACTIVE Ca^{2+} REABSORPTION IN THE DISTAL NEPHRON

Joost G. J. Hoenderop,[1] Bernd Nilius,[2] and René J. M. Bindels[1]

[1]Department of Cell Physiology, Institute of Cellular Signalling, University Medical Centre Nijmegen, The Netherlands; e-mail: reneb@sci.kun.nl
[2]Department of Physiology, Campus Gasthuisberg, KU Leuven, Belgium

Key Words ECaC, CaT1, vitamin D, thiazide diuretics, calbindin

■ **Abstract** The identification of the epithelial Ca^{2+} channel (ECaC) complements the group of Ca^{2+} transport proteins including calbindin-D$_{28K}$, Na$^+$/Ca^{2+} exchanger and plasma membrane Ca^{2+}-ATPase, which are co-expressed in 1,25(OH)$_2$D$_3$-responsive nephron segments. ECaC constitutes the rate-limiting apical entry step in the process of active transcellular Ca^{2+} transport and belongs to a superfamily of Ca^{2+} channels that includes the vanilloid receptor and transient receptor potential channels. This new Ca^{2+} channel consists of six transmembrane-spanning domains, including a pore-forming hydrophobic stretch between domain 5 and 6. The C- and N-terminal tails contain several conserved regulatory sites, implying that the channel function is modulated by regulatory proteins. The distinctive functional properties of ECaC include a constitutively activated Ca^{2+} permeability, a high selectivity for Ca^{2+}, hyperpolarization-stimulated and Ca^{2+}-dependent feedback regulation of channel activity, and 1,25(OH)$_2$D$_3$-induced gene activation. This review covers the distinctive properties of this new highly Ca^{2+}-selective channel and highlights the implications for active transcellular Ca^{2+} reabsorption in health and disease.

INTRODUCTION

The kidney plays an eminent role in the maintenance of the Ca^{2+} balance by regulating the Ca^{2+} excretion of the body. On a daily basis, $\sim$8 g Ca^{2+} is filtered at the glomerulus of which less than 2% is excreted into the urine. The majority of Ca^{2+} reabsorption along the nephron occurs via passive paracellular Ca^{2+} reabsorption in the proximal tubules and the thick ascending limb of Henle, whereas approximately 20% is reabsorbed via an active transcellular Ca^{2+} transport route in the distal part of the nephron (1–4). Based on functional and anatomical features, this distal part can be divided in two segments of the distal convoluted tubule (DCT1 and DCT2) and the connecting tubule (CNT) (5). Paracellular Ca^{2+} transport is driven by the transepithelial electrochemical gradient that is generated by

0066-4278/02/0315-0529$14.00

sodium and water reabsorption and is, therefore, only indirectly regulated by hormones such as angiotensin II and dopamine and by α-adrenergic stimulation (1). In contrast, only the transcellular pathway allows the body to regulate Ca^{2+} reabsorption independent of the Na^+ balance. This route is controlled specifically by the calciotropic hormones including PTH, calcitonin, and 1,25-dihydroxyvitamin D_3 $(1, 25(OH)_2D_3)$ (1–4). Due to this active process, the organism can respond to dietary fluctuations of nutritional Ca^{2+} and adapt to the body's demand during processes such as growth, pregnancy, lactation, and aging (1). Disturbances in active Ca^{2+} reabsorption are most likely accompanied by significant alterations in the overall Ca^{2+} homeostasis.

At the cellular level, transcellular Ca^{2+} reabsorption proceeds through a well-controlled sequence of events consisting of luminal Ca^{2+} entry via the epithelial Ca^{2+} channel (ECaC), cytosolic diffusion of Ca^{2+} bound to calbindin-D_{28K}, and basolateral extrusion of Ca^{2+} through Na^+/Ca^{2+} exchanger (NCX) and plasma membrane Ca^{2+}-ATPase (PMCA) (6). The Ca^{2+} entry pathway has remained elusive for a long time, but the recent molecular identification of the responsible protein now makes it possible to study the characteristics of this rate-limiting step in active Ca^{2+} reabsorption in health and disease. This review covers the individual molecular and functional features of the four aforementioned Ca^{2+} transport proteins, presents a comprehensive mechanism of transcellular Ca^{2+} transport, and subsequently outlines the clinical implications of these recent advances.

MOLECULAR FEATURES OF Ca^{2+} TRANSPORT PROTEINS

Epithelial Ca^{2+} Channel (ECaC)

The identification of ECaC provides new opportunities to study the process of Ca^{2+} reabsorption in the kidney. It is the postulated prime target for hormonal control of active Ca^{2+} flux from the urine space to the blood compartment. This section provides a detailed picture of the distinctive molecular and functional features of this Ca^{2+}-selective channel to further substantiate its role under (patho)physiological conditions.

IDENTIFICATION OF ECaC   Despite numerous attempts to identify the molecular nature of the apical Ca^{2+} entry step using multiple approaches, including protein fractionation based on binding to intracellular Ca^{2+}-binding proteins (7), homology cloning based on previously identified voltage-operated Ca^{2+} channels (8–10), and pharmacological characterization of Ca^{2+} influx in cultured renal cells (11), this protein (ECaC) was not conclusively identified. Our group applied, therefore, a different approach in which a cDNA library was generated from a Ca^{2+}-transporting primary culture of rabbit connecting tubules. Subsequently, a functional expression cloning strategy was used to screen cRNA pools for radioactive Ca^{2+} uptake. A single cDNA was isolated that encoded a unique protein named epithelial Ca^{2+}

channel or ECaC (12). The rabbit ECaC cDNA contains an open reading frame of 2190 nucleotides encoding a protein of 730 amino acids with a corresponding molecular mass of 83 kDa. A Kyte-Doolittle hydrophobicity plot predicts a topology of six transmembrane segments (S1–S6), including a putative loop pore structure between S5 and S6 that is similar to the core structure of the pore-forming subunits of transient receptor potential (TRP) channels, vanilloid receptors (VR), polycystins, and voltage-gated Ca^{2+} channels (3, 13–15). By analogy with voltage-gated Ca^{2+} channels, which have four linked domains of six transmembrane segments, it is likely that ECaC channels form tetramers of four single subunits. Although the predicted core structure is conserved within these ion channels, ECaC is a distant relative of the above mentioned ion channels (6). A phylogenetic tree calculated for ECaC-homologous proteins shows that these channels can be clustered in three groups: TRP channels, the group of vanilloid receptors, and ECaC. Detailed sequence comparison reveals a significant amino acid similarity among these three groups that is restricted to the pore loop and the adjacent sixth transmembrane segment (16). However, the overall homology of ECaC to these channels is as low as 30%, suggesting that ECaC is the first member of a new class of Ca^{2+} channels.

REGULATORY SITES OF ECaC   To date, ECaC has been identified from several species including rabbit, rat, mouse, and human (6). Sequence alignment of these different species allows the identification of conserved regulatory sites. Because the amino- and carboxy-terminal tails of ECaC have putative motifs for protein kinase C (PKC) phosphorylation, binding to PDZ domain-containing proteins, and binding to proteins that interact with ankyrin repeats (Figure 1, see color insert), it is anticipated that multiple intracellular partner proteins are involved in the regulation of ECaC activity. Currently, cytosolic proteins that interact with ECaC have not been identified. Previous studies indicated that PKC isoforms play an important regulatory role in the hormonal stimulation of transcellular Ca^{2+} reabsorption (3, 4). Thus it is possible that PKC directly phosphorylates the channel to regulate its activity. In general, PDZ motifs are part of a molecular scaffold that contains multiprotein signaling complexes. PDZ-mediated interactions between proteins can facilitate cell biological processes including linkage of ion channels to the cytoskeleton, phosphorylation of ion channels, and targeting of ion channels in correct spatial arangement in relation to each other and to specialized regions of the cell (17). Interestingly, several PDZ domain-containing proteins have recently been identified in the kidney that could interact with apical transporters such as the Na^{+}-H^{+} exchanger (NHE), renal outer medullary potassium (ROMK) channel, cystic fibrosis transmembrane regulator (CFTR), the Na^{+}-phosphate (NaPi) transporter, and ECaC (18–21). The ankyrin repeat is a common protein sequence motif present in a large family of membrane-associated proteins that bind via their membrane-binding domains to diverse proteins including proteins involved in Ca^{2+} homeostasis such as inositol triphosphate (IP$_3$) and ryanodine receptors (22, 23). Little is known about how this latter group of proteins is targeted to specialized sites within the cell; presumably, accessory proteins play an important role in this sorting process. Detailed molecular studies are now feasible and necessary to delineate the function of these putative regulatory sites in ECaC.

LOCALIZATION OF ECaC IN THE KIDNEY   In rabbit kidney, mRNA analysis on microdissected tubular segments and detailed immunohistochemical analysis demonstrated that ECaC is predominantly present in the connecting tubule (CNT), whereas in rat and mouse, ECaC is also present in the second part of the distal convoluted tubule (DCT2) (24–26). Both segments have frequently been implicated in active Ca^{2+} reabsorption (1, 3, 5). In line with its postulated function as a Ca^{2+} influx protein, ECaC is definitely present along the apical membrane within these distal tubular segments (Figure 2, see color insert). In addition, a diffuse subapical immunoreactivity was also observed, suggesting the presence of ECaC channels in vesicles that could be involved in the shuttling of ECaC to and from the plasma membrane to regulate the Ca^{2+} reabsorptive capacity of the cell. ECaC is predominantly expressed in principle cells of the distal part of the nephron, where it co-localizes with the associated Ca^{2+} transport proteins, i.e., calbindin-D_{28K}, NCX1, and PMCA (24). These findings underline the putative function of ECaC as the gatekeeper of active Ca^{2+} reabsorption in the distal part of the nephron.

FUNCTIONAL HALLMARKS OF ECaC   The electrophysiological properties of ECaC have been extensively documented in several cell types heterologously expressing ECaC (16, 27–33) and are summarized in Table 1. This new calcium channel exhibits distinctive features consistent with its putative role as apical Ca^{2+} influx channel that controls transcellular Ca^{2+} transport, including a constitutively

TABLE 1   Electrophysiological hallmarks of ECaC

Parameter	ECaC	Ref.
Current-voltage relationship	Inward rectification	(28)
Inactivation time constant (-140 mV)	481 ± 45 ms	(28)
Block of monovalent currents by extracellular Ca^{2+} affinity	161 ± 30 nM	(32)
Inhibition by intracellular Ca^{2+}	~ 120 nM	(32)
Divalent permeation	$Ca^{2+} > Ba^{2+} > Sr^{2+} > Mn^{2+}$	(28)
Monovalent permeation	$Na^+ > Li^+ > K^+ > Cs^+ \gg NMDG^+$	(31)
$P_{Ca}^{2+} : P_{Na^+}$	$100 : 1$	(28)
Mg^{2+} block monovalent current	$IC_{50} = 63 \pm 9$ μM	(29)
Mg^{2+} block 100 μM Ca^{2+} current	$IC_{50} = 329 \pm 50 \mu$M	(29)
Single channel conductance in the absence of divalents	78 ± 5 pS	(31)

activated Ca^{2+} permeability at physiological membrane potentials, a high Ca^{2+} selectivity, voltage- and Ca^{2+}-dependent block by extracellular Mg^{2+}, and Ca^{2+}-dependent feedback regulation of channel activity.

Transient expression of ECaC in *Xenopus laevis* oocytes and human embryonic kidney (HEK) cells considerably increased the plasma membrane permeability for Ca^{2+}, which is reflected by a large Ca^{2+}-induced Cl$^-$ current and an elevated cytosolic Ca^{2+} concentration directly proportional to the electrochemical driving force for Ca^{2+} entry, respectively (27, 28). In ECaC-expressing HEK cells, elevation of extracellular Ca^{2+} induced a transient inward current that dramatically elevates [Ca^{2+}]$_i$ (Figure 3*A*, *B*). In the presence of BAPTA in the patch pipette solution, large inward currents were recorded. The amplitude of these latter currents showed a strong Ca^{2+} dependence and a positive reversal potential that shifted by +21 mV per 10 mM change in extracellular Ca^{2+} concentrations (Figure 3*C*). This is close to the predicted Nernstian shift of +29 mV for an ideal Ca^{2+} permeable channel. Using different divalent cations as charge carriers, a conductance sequence of Ca^{2+} > Ba^{2+} > Sr^{2+} > Mn^{2+} was determined (28). These findings are in line with the operation of a constitutively activated Ca^{2+}-selective channel.

Previously, it was demonstrated that the molecular determinants of the Ca^{2+} selectivity and permeation of ECaC reside at a single aspartate residue (D542) present in the putative pore-forming region (16). The pore region of ECaC contains two additional negatively charged amino acids that have only minor effects on the Ca^{2+} permeation properties. It is tempting to speculate that the selectivity filter for Ca^{2+} in ECaC consists of a ring of four negatively charged aspartate residues in a tetrameric pore molecule, as has been suggested for the glutamate residues present in the pore of voltage-operated Ca^{2+} channels (34).

ECaC is approximately 100 times more permeable for Ca^{2+} than for Na$^+$ (28). This permeation ratio is, however, smaller than that of voltage-dependent Ca^{2+} channels but higher than that of the homologues TRP and vanilloid channels. It implies that at physiological Na$^+$ concentrations in the distal tubule, Ca^{2+} influx dominates over Na$^+$ influx.

It has been conclusively established that the activity of ECaC is subject to autoregulatory feedback control mechanisms as shown in Figures 3 and 4 (27, 28, 32). First, ECaC-mediated currents rapidly inactivate during hyperpolarizing voltage steps. This inactivation is eliminated when Ba^{2+} or Sr^{2+} are used as charge carriers, suggesting that ECaC activity is controlled by Ca^{2+}-dependent feedback. Fast-binding and high-affinity Ca^{2+} buffers such as BAPTA attenuate this feedback action of permeating Ca^{2+} ions but are unable to completely abolish it. Second, the current response slowly vanishes during repetitive activation (Figure 4*D*). This current decay is significantly decreased when Ca^{2+} is replaced by Ba^{2+} as charge carrier and abolished when extracellular Ca^{2+} is lowered to 1 nM. This demonstrates that the Ca^{2+}-dependent process inhibits ECaC activity. Third, these regulatory processes are strongly influenced by the surrounding Ca^{2+} concentrations. Elevation of the extracellular Ca^{2+} concentration significantly increases the rate of current decay, whereas Ca^{2+} currents through ECaC are significantly inhibited by elevated intracellular Ca^{2+} concentrations (Figure 3D) (31). Fourth, Ca^{2+} influx is a prerequisite for this phenomenon because the Ca^{2+} impermeable D542A mutant lacks a monovalent current decay in response to repetitive stimulation (32). Together, these data suggest that the ECaC activity is downregulated by Ca^{2+} influx through

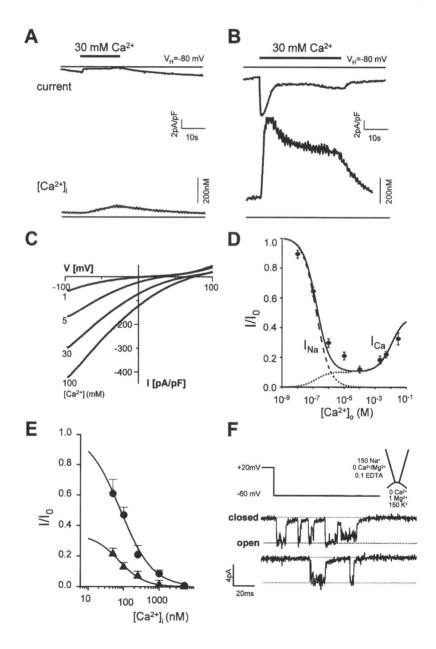

the channel and thus increases the Ca^{2+} concentration in a microdomain near the pore region. This is a prime example of feedback inhibition of ECaC. The operation of this high-affinity mechanism of Ca^{2+}-dependent ECaC inhibition suggests that the presence of intracellular Ca^{2+} buffers such as calbindin-D$_{28K}$ is important in regulating Ca^{2+} influx via ECaC.

ECaC becomes permeable to monovalent cations by lowering extracellular [Ca^{2+}]. Monovalent currents through ECaC are much larger, activate slower than Ca^{2+} currents, and do not inactivate (Figure 4E, F). The permeation sequence for monovalent cations matches the Eisenman X pattern indicating a high field strength–binding site. Elevation of extracellular Ca^{2+} first inhibits the monovalent current through ECaC, but the current is increased at higher Ca^{2+} concentrations (Figure 3E) (28, 32). This feature indicates multiple binding sites for Ca^{2+} within the pore and also reflects anomalous fraction behavior that is well described for high Ca^{2+}-selective cation channels. As far as single channel properties are concerned, only in divalent-free solution could unitary currents be measured (Figure 3F). Single channel conductance under these conditions ranged between 55 and 100 pS (31). In the presence of Ca^{2+}, the single channel conductance decreased dramatically and could not be measured. Likely, open probability is increased at more negative potentials that reflects the inward rectification properties of ECaC.

Several divalent cations readily permeate ECaC, but many others, including Mg^{2+}, Cd^{2+}, La^{3+}, and Gd^{3+}, block the channel at micromolar concentrations

---

←

Figure 3   (A) Currents and Ca^{2+} transients recorded at -80 mV in non-transfected HEK cells during application of 30 mM Ca^{2+} in the presence of NMDG$^+$. The registrations were obtained in the absence of BAPTA in the patch pipette. (B) Recordings in ECaC-expressing HEK cells [see (A)]. (C) Currents recorded in different Ca^{2+} concentrations in the presence of NMDG$^+$, in response to 50 ms voltage ramps from -100 to $+100$ mV, with a holding potential of $+20$ mV. ECaC-expressing HEK cells were loaded with 10 mM BAPTA via the pipette (29). (D) Density of ECaC currents at different intracellular Ca^{2+} concentrations at -80 mV. ECaC-expressing HEK cells were loaded with 10 mM BAPTA via the pipette. Currents were measured in nominal Ca^{2+}-free extracelluar solutions (*circles*) and 1 mM Ca^{2+} (*triangles*) and were normalized to the Na$^+$ currents obtained in 0 mM [Ca^{2+}]$_e$ (29). (E) Current values measured at -80 mV during linear voltage ramps in various extracellular Ca^{2+} concentrations. Currents were normalized to the current values for the same cell in buffered divalent-free solutions. The continuous line represents the current densities as predicted by a model with one high-affinity binding site flanked by a low-affinity binding at each side. The dashed and dotted lines represent the fractions of the currents carried by Ca^{2+} and Na$^+$, respectively (32). (F) Current traces from an inside-out patch in response to a hyperpolarizing step from $+20$ to -60 mV. Pipette and intracellular solutions are depicted.

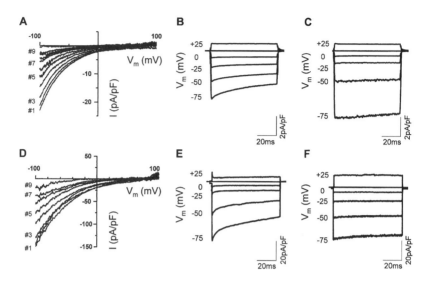

Figure 4   (A, D) Current decay through ECaC in the presence of 30 mM Ca²⁺. Voltage ramps were applied every 5 s (ramps of 400 ms from −100 mV, holding potential +20 mV). (B, E) Current responses to 400 ms voltage steps applied from a holding potential of +20 mV. Steps ranged from −75 to +25 mV, voltage increment is +25 mV, extracellular Ca²⁺ concentration is 1 mM, extracellular Na⁺ is 150 mM. (C, F) Current responses to 400 ms voltage steps applied from a holding potential of +20 mV. Steps ranged from −75 to +25 mV, voltage increment is +25 mV, extracellular Ca²⁺ concentration is buffered to 1 nM, extracellular Na⁺ is 150 mM. Recordings were obtained in primary cultures of rabbit connecting tubules (A–C) and ECaC-expressing HEK cells (D–F) with 10 mM BAPTA in the patch pipette.

(12). We have investigated in detail the concentration dependence of the block by extracellular Mg²⁺ of currents carried by Ca²⁺ or by monovalent ions in the absence of Ca²⁺ (29). Block of ECaC by extracellular Mg²⁺ depends on both membrane voltage and extracellular Ca²⁺ concentration. Block of monovalent currents by Mg²⁺ was one order of magnitude more sensitive than that of Ca²⁺ currents. In addition, block of ECaC by Mg²⁺ is increased fivefold by hyperpolarization of 40 mV. The data strongly support the finding that Mg²⁺ blocks the pore of ECaC at a site approximately 30% inside the transmembrane electrical field. This is further supported by a complete loss of Mg²⁺ block by neutralizing the crucial charge in the pore, D542.

Furthermore, the outlined functional properties of ECaC, including a strong inward rectifying current in response to a voltage ramp, current decay during repetitive stimulation, and a fast inactivating current in response to a hyperpolarizing voltage step, are identical to those determined in primary cultures of rabbit connecting tubules that endogenously express ECaC (Figure 4). Taken together, the

TABLE 2 Family of epithelial Ca^{2+} channels

Gene name[a]	Name	Source	Accession number	Ref.
ECAC1	ECaC	Rabbit kidney	AJ133128[b]	(12)
	ECaC1	Human kidney	AJ271207	(37)
	CaT2	Rat kidney	AF209196	(38)
	ECaC	Rat kidney	AB032019	(6)
ECAC2	CaT1	Rat duodenum	AF160798[b]	(35)
	ECaC2	Human duodenum	AJ277909	(40)
	CaT-like	Human placenta	AJ243500	(39)

[a]HGMW-approved human gene symbol.
[b]Original publication.

experimental data provide evidence that ECaC exhibits the defining characteristics for being the apical Ca^{2+} channel in transepithelial Ca^{2+} reabsorption.

THE FAMILY OF ECaC CHANNELS   After the original cloning of ECaC, several groups identified homologue channels as summarized in Table 2. Hediger and co-workers cloned Ca^{2+} transporter 1 (CaT1) from rat intestine, which shares an 80% amino acid identity with ECaC (35). Patch-clamp studies demonstrate that the characteristics of CaT1 are comparable to those measured for ECaC, but its expression pattern is different (33, 36, 37). Initial Northern blot analysis demonstrated that CaT1 is predominantly expressed in small intestine but not in kidney (35). Additionally, robust expression was detected in human placenta and pancreas in a subsequent study (36). Furthermore, the same investigators re-cloned ECaC, confusingly called CaT2, from rat kidney 38). Recently, CaT-like cloned from human placenta is almost identical to the human orthologue of CaT1 (39). Detailed sequence comparison demonstrates that CaT1 and CaT-like share more than 99% identity, differing by only five amino acids. At present it is unclear whether this can be explained by single nucleotide polymorphisms in the CaT1 gene or by the existence of two separate genes. DNA analysis demonstrated that the conserved pore sequence of the ECaC family is restricted to the previously identified genes encoding ECaC and CaT1, making this latter option less likely (33). ECaC and CaT1 originate from two distinct genes juxtaposed with a distance of approximately 20 kb on chromosome 7q35, suggesting an evolutionary gene duplication. The HGMW approved nomenclature of the corresponding human genes ECAC1 and ECAC2 for the previously reported transcripts ECaC and CaT1, respectively (37, 40).

Calbindin

A well-documented physiological function of calbindins is their eminent Ca^{2+}-binding capacity, the result of high cytosol concentrations in the millimolar range (41, 42). Perhaps the best-defined molecular expression of $1,25(OH)_2D_3$ action is the induction of calbindins. Calbindin is virtually absent in vitamin D–deficient animals and appears within 2–3 h after injection of $1,25(OH)_2D_3$ (41). In addition to its vitamin D dependency, the concentration of this protein varies directly with the efficiency of Ca^{2+} (re)absorption under a wide variety of physiological and nutritional circumstances. A hypothesis proposed by Kretsinger et al. (43) and supported by Feher et al. (44–46) and Bronner et al. (47) is that calbindins facilitate the diffusion of Ca^{2+} through the cytosol and simultaneously serve as an intracellular Ca^{2+} buffer to keep the free cytoplasmic $[Ca^{2+}]$ below toxic levels during periods of stimulated transcellular Ca^{2+} transport. Importantly, because of the relatively slow Ca^{2+} binding kinetics of calbindin-D_{28K}, hormone-induced Ca^{2+} signaling can occur independently of the rate of transcellular Ca^{2+} transport (48).

The calbindin-D_{28K} family contains calbindin-D_{28K}, which is mainly present in kidney, and calbindin-D_{9K}, which is expressed primarily in the gut (41, 42, 49, 50). In fact, nature has produced a wide variety of EF-hand calcium-binding proteins (calbindin-D_{9K}, calbindin-D_{28K}, calmodulin, parvalbumin, S100, troponin C) that have little overall sequence homologies. An important functional feature of calmodulin and troponin is their ability to interact with and regulate the function of voltage-gated Ca^{2+} channels in a calcium-dependent fashion (51, 52). Calmodulin plays a critical role in the Ca^{2+}-dependent (in)activation process of voltage-operated Ca^{2+} channels. This ubiquitously expressed cytosolic Ca^{2+}-binding protein directly interacts with an IQ motif present in the C-tail of these channels where it functions as a Ca^{2+} sensor. This IQ motif is, however, not present in ECaC. Interestingly, Niemeyer et al. identified a new calmodulin-binding site in the C terminus of the ECaC homologue CaT-like (53). These investigators showed that Ca^{2+}-dependent calmodulin binding to CaT-like facilitates channel inactivation, which was counteracted by PKC-mediated phosphorylation of this putative calmodulin-binding site. Although ECaC homologues display an identity of approximately 75% at the amino acid level, the proposed novel calmodulin-binding and PKC phosphorylation sites are not conserved among these channels, suggesting that this regulatory mechanism is not operating in ECaC.

It is not known whether calbindin-D_{28K} fulfills a similar Ca^{2+} sensor function for which a specific interaction with ECaC would be required. Christakos and co-workers reported an interaction of calbindin-D_{28K} with microsomal membranes in kidney (54), whereas, Shimura & Wasserman showed that calbindin-D_{28K} is associated with purified chicken intestinal brush border membranes (55). Together with the striking co-localization of calbindin in all ECaC-expressing tissues, these findings suggest a functional interaction between these two proteins. Future experiments are warranted to delineate the function of calbindin: either being restricted to a buffering function thus maintaining low Ca^{2+} concentrations close to the channel mouth in order to maximally activate the channel or as a molecular interaction between calbindin and ECaC that exerts a direct regulatory effect.

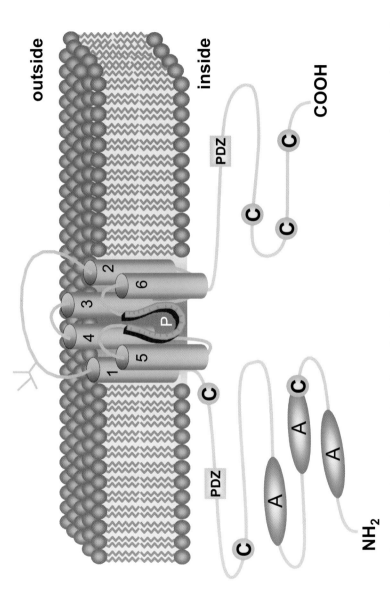

Figure 1 Schematic representation of proposed topology of ECaC protein consisting of six transmembrane-spanning domains with a short hydrophobic stretch between transmembrane segment 5 and 6. Cytosolic C- and N-terminal tails containing several potential regulatory domains, including ankyrin repeats, PDZ motifs, and sites for PKC phosphorylation.

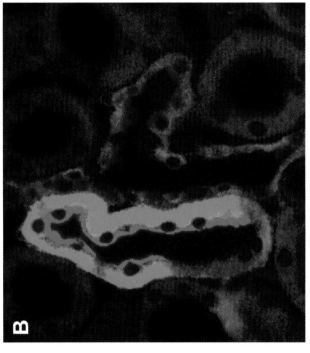

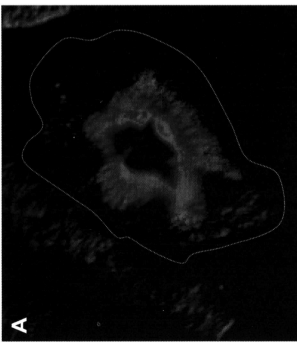

Figure 2 (*A*) Superimposed picture of a Nomarski light microscopic image of a distal tubule and immuno-positive fluorescence staining for ECaC, which was predominantly found along the apical membrane. (*B*) Immuno-positive co-localization of ECaC (*red*) and Na$^+$/Ca^{2+} exchanger (*green*) in a distal tubule (26).

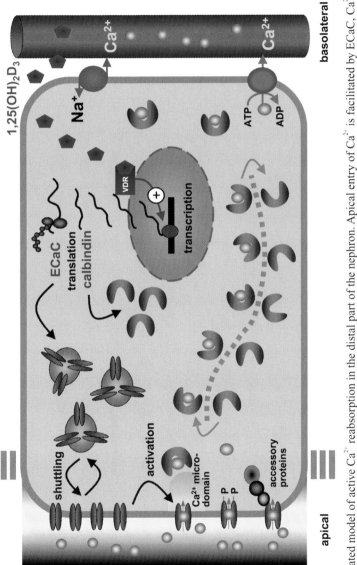

Figure 5 Integrated model of active Ca^{2+} reabsorption in the distal part of the nephron. Apical entry of Ca^{2+} is facilitated by ECaC, Ca^{2+} then binds to calbindin-D_{28K}, and this complex diffuses through the cytosol to the basolateral membrane, where Ca^{2+} is extruded by a Na^+/Ca^{2+} exchanger and a plasma membrane Ca^{2+}-ATPase. The individually controlled steps in the activation process of the rate-limiting Ca^{2+} entry channel include $1,25(OH)_2D_3$-mediated transcriptional and translational activation, shuttling to the apical membrane, and subsequent activation of apically located channels by ambient Ca^{2+} concentration, direct phosphorylation and/or accessory proteins.

Ca^{2+} Extrusion Mechanisms

The extrusion of Ca^{2+} from the distal tubule cell occurs against a steep electrochemical gradient, and two active Ca^{2+} transporters are present in the basolateral membrane: NCX and PMCA (3, 56). All segments of the kidney contain a PMCA pump with the highest expression in cells lining the distal part of the nephron (56, 57). PCR analysis demonstrated the expression of all four presently known isoforms in the kidney. On the basis of the fact that PMCA1 and PMCA4 are ubiquitously expressed, whereas PMCA2 and PMCA3 are more tissue specific, it has been suggested that PMCA1 and PMCA4 are housekeeping isoforms involved in the maintenance of cellular Ca^{2+} homeostasis (58). At variance with this conclusion is the observation that PMCA1b transcripts were detected in rabbit CNT and CCD (56). Furthermore, PMCA1b is the predominant isoform expressed in rabbit small intestine, a site also involved in transcellular Ca^{2+} transport (3). These data suggest that PMCA1 isoforms take part in the process of transcellular Ca^{2+} transport. However, the capacity of this PMCA pump is apparently insufficient to keep pace with the absorptive flux of Ca^{2+} in the distal nephron (56). Thus there is a clear need for another Ca^{2+} transport system. The Na^{+}/Ca^{2+} exchanger has been shown to be a prerequisite for transcellular Ca^{2+} transport, which suggests that Ca^{2+} extrusion across the basolateral membrane is mediated primarily by NCX (56, 59). The Na^{+}/Ca^{2+} exchangers contain a family of three genes: NCX1, NCX2, and NCX3 (60). These transporters have similar hydropathy patterns, suggesting a comparable overall structure, and moderate sequence identity. However, only NCX1 is expressed in kidney. Functional and immunohistochemical evidence indicate that this Na^{+}/Ca^{2+} exchanger is mainly expressed in the connecting tubule and extends to the flanking segments (24, 61, 62). At present, there is no evidence suggesting that the expression level of these Ca^{2+} efflux transporters is regulated by hormones (56).

PHYSIOLOGICAL REGULATION OF ECaC ACTIVITY

Regulation by 1,25(OH)$_2$D$_3$

Vitamin D is one of the most important regulators of the body Ca^{2+} homeostasis and is required for proper development and maintenance of bone mass. The distal part of the nephron determines the final excretion of Ca^{2+} into the urine and is a target for the regulation of the extracellular Ca^{2+} concentration. Several studies reported a stimulatory effect of vitamin D, via its active metabolite 1,25(OH)$_2$D$_3$, on Ca^{2+} reabsorption in the distal nephron (1, 3, 56, 63). This stimulatory effect is accompanied by a well-documented increased expression of calbindin-D$_{28K}$ (41, 42, 56). Recently, the effect of this secosteroid hormone on the expression of ECaC in the kidney was investigated for the first time. In this study, vitamin D–depleted hypocalcemic rats were used and the renal expression of ECaC was determined at the mRNA and protein level. Injections of 1,25(OH)$_2$D$_3$ in the depleted rats normalized the plasma Ca^{2+} concentration within 48 h and significantly increased the renal ECaC mRNA abundance, but the stimulatory effect on protein levels was more pronounced. ECaC protein was barely detectable in the depleted animals but was completely restored after the injection of $1,25(OH)_2D_3$. This observation suggests that in addition to an increase in mRNA, post-transcriptional regulation of ECaC may also take place, which ultimately should result in increased channel activity at the apical membrane (25). Irrespective of this major post-transcriptional effect, this study demonstrates that the ECaC gene is under the control of $1,25(OH)_2D_3$. This is further strengthened by the recent elucidation of the human ECaC promoter that contains four putative vitamin D–responsive elements (25, 64). Promoter reporter analysis will be necessary to confirm the functional role of these elements in the transcriptional regulation of ECaC.

Is ECaC Involved in the Hypocalciuric Effect of Thiazides?

Thiazide diuretics act in the distal convoluted tubule to inhibit the apical Na^+-Cl^- co-transporter (NCC). The resulting natriuretic effect is accompanied by a decreased Ca^{2+} excretion in the urine (65). In a pioneering study by Costanzo et al., an inverse relationship between the rates of Na^+ and Ca^{2+} reabsorption was demonstrated in microperfused rat distal tubules (66). This provided a rationale for the clinical effects of thiazides. Although many studies describe this hypocalciuric effect, there is no definitive clue about the molecular mechanism involved in the differential regulation of Na^+ and Ca^{2+} transport. Two hypotheses have been proposed to clarify this phenomenon. One explanation is that inhibition of NCC results in a reduction of the intracellular Cl^- activity leading to a hyperpolarization of the plasma membrane thus favoring apical Ca^{2+} entry (67). This hypothesis is attractive because the inward Ca^{2+} current through ECaC is maximal at hyperpolarizing membrane potentials (27, 28), but it is not convincingly supported by experimental evidence. The thiazide-induced hyperpolarization has been demonstrated only in the basolateral membrane, and it is difficult to reconcile how this basolateral effect can stimulate apical Ca^{2+} entry (68). In addition, the stimulated Ca^{2+} entry has only been measured as a dihydropyridine-sensitive Ca^{2+} uptake in a suspension of cells with DCT characteristics, which disqualifies ECaC as mediator of this latter effect (67). Alternatively, the basolateral Na^+/Ca^{2+} exchanger is implicated as primarily responsible for the dissociation of Na^+ and Ca^{2+} reabsorption (59, 67). The thiazide-induced inhibition of apical Na^+ influx results in a decrease of $[Na^+]_i$ and thereby enhances the Ca^{2+} efflux through the Na^+/Ca^{2+} exchanger. A prerequisite to support both hypotheses is a substantial co-localization between NCC and the Ca^{2+} transport proteins including ECaC in the distal tubule. Extensive immunohistochemical studies in rats, rabbits, and mice (24–26) demonstrate that the Ca^{2+} transporters completely co-localize but show surprisingly little overlap with NCC, which is restricted to the late part of the distal convoluted tubule (DCT2). The majority of NCC, however, is observed in DCT1, whereas the expression of the Ca^{2+} transport proteins starts at the DCT2 and extends to the connecting tubule. Thus the Ca^{2+}-sparing and natriuretic action of thiazides do not necessarily reside in a single cell type but could be located in subsequent nephron segments and operative via an unknown mechanism. Alternatively, the NCC-containing tubular cells could exhibit an unidentified Ca^{2+} transport mechanism distinct from presently known Ca^{2+} transport proteins. This latter hypothesis seems highly unlikely. Microperfusion studies of distal tubular segments in which the contribution of the individual segments is detailed are extremely difficult to perform if not at all feasible, but they are mandatory to elucidate the molecular mechanism of the hypocalciuric effect of thiazides.

pH Sensitivity

There is ample experimental evidence that the activity of ECaC is affected by the ambient proton concentration. The first piece of evidence was provided by ^{45}Ca^{2+} uptake measurements in ECaC-expressing *Xenopus* oocytes (12). Extracellular acidification of the medium to pH 5.9 diminished the Ca^{2+} uptake by these oocytes. Subsequently, detailed whole-cell patch-clamp measurements on HEK cells heterogenously expressing ECaC demonstrated that extracellular acidification indeed reduced the ECaC current amplitude in the presence of mono- or divalent ions as charge carriers. Finally, acidification of the luminal compartment lessens transcellular Ca^{2+} absorption across primary cultures of rabbit kidney connecting tubules (11). This inhibitory effect was attributed to a pH-dependent reduction in apical Ca^{2+} entry. An important question is whether the effect on the ECaC channel activity is from changes in intra- or extracellular pH. The observation that lowering the pH resulted immediately in reduced channel activity, together with the previous report that a selective reduction of the intracellular pH from 7.4 to 6.9 only moderately reduced Ca^{2+} reabsorption in primary cultures of connecting tubules (11), supports the conclusion that the inhibition of ECaC activity is mainly due to an extracellular pH effect (11, 30). For cyclic nucleotide-gated channels, it was suggested that protonation of a single glutamate residue in the pore region is responsible for the dramatic changes in divalent cation affinity for the channel with respect to variations in the extracellular pH (69). In this respect, it would be interesting to explore the role of the existing negatively charged residues in the ECaC pore of which a single aspartate residue determines the Ca^{2+} permeation and Mg^{2+} block of the channel. Extrapolating this pH influence to the in vivo situation, this effect could at least in part provide the molecular basis of acidosis-induced calciuresis. It is well known that chronic metabolic acidosis induces an increase in urinary calcium excretion (1). The reason for the hypercalciuric effect of acidosis has not been conclusively elucidated, but a stimulated release of Ca^{2+} from bone and a direct inhibition of distal tubule Ca^{2+} reabsorption may account for the observed hypercalciuria (70). During metabolic acidosis, the kidney has a increased capacity to excrete acid. The accumulation of acid in the tubular system will have a direct effect on ECaC channel activity resulting in a reduction of transepithelial Ca^{2+} transport in the distal part of the tubule.

Potential Regulators

In addition to vitamin D, several other hormones, including PTH, vasopressin, prostaglandines, nucleotides, and atrial natriuretic peptide, have been implicated as players in renal Ca^{2+} homeostasis (1, 63, 71–76). A prominent pathway involves the stimulation of active Ca^{2+} reabsorption in DCT and CNT by PTH (3, 4, 63). It has been shown that the PTH receptor couples to both adenylyl cyclase and phospholipase C and that the stimulatory effects of PTH are mediated by their respective cAMP and PKC signaling cascades (4, 75). In addition, evidence for the involvement of a cGMP-dependent kinase (cGK) pathway activated by atrial natriuretic peptide emerged from studies performed with primary cultures of rabbit connecting tubules (73). The means by which these pathways act together to produce an adequate cellular response to these hormones is incompletely understood, and the downstream effectors are largely unknown. ECaC, however, contains several conserved sites for PKC phosphorylation, suggesting that the entry channel could be a direct effector of hormones utilizing PKC to exert a regulatory effect on Ca^{2+} reabsorption. On the other hand, ECaC lacks conserved PKA/cGK sites, indicating that the hormones utilizing these pathways affect intermediate processes, for example shuttling of ECaC to the apical membrane or stimulating ECaC gene expression or apical membrane potential. Further studies should delineate the molecular mechanisms of the activation of ECaC by stimulatory and inhibitory calciotropic hormones.

INTEGRATED MODEL OF ACTIVE Ca^{2+} REABSORPTION

The active transport of Ca^{2+} from the tubular lumen to the bloodstream is a multistep process involving movement of Ca^{2+} into, across, and out of the cell. The Ca^{2+} influx from the apical compartment is likely the rate-limiting step in this process. This conclusion is based on previous observations that the Ca^{2+} influx rate at the apical membrane of renal distal cells is tightly coupled to transepithelial Ca^{2+} flux over a wide range of transport rates (77). Furthermore, the functional and pharmacological properties of ECaC are identical to those of transepithelial Ca^{2+} transport (6, 12) (Figure 4), providing evidence that the protein is the anticipated gatekeeper in active Ca^{2+} reabsorption. Conceivably, this implicates that hormonal regulation of a single influx pathway, i.e., ECaC, controls the rate of transcellular Ca^{2+} transport. In this respect it is important to consider multiple potential mechanisms of ECaC activation, including synthesis of ECaC, activation of existing ECaC channels by regulatory factors, and shuttling of ECaC between intracellular vesicles and the apical membrane. At present evidence is accumulating that regulation takes place at each of these individual levels as shown in Figure 5 (see color insert). Vitamin D stimulates the synthesis of ECaC by transcriptional and possibly post-transcriptional action (25). These newly synthesized channels are then targeted to the apical membrane via still unspecified actions. The subsequent activation of these apically routed channels is controlled by various means that possibly involve the calbindin-D$_{28K}$-buffered Ca^{2+} concentration in a micro do-main near the inner mouth of the channel, direct phosporylation of the channel, the membrane potential, and interacting accessory proteins. Finally, the activity of ECaC should be tightly coordinated with that of the other Ca^{2+} transporters to maintain low cytosolic Ca^{2+} concentrations during the stimulation of the process of transepithelial Ca^{2+} transport.

Clinical and Future Implications

ECaC represents the molecular basis of the apical Ca^{2+} entry step in transcel-lular Ca^{2+} reabsorption. Because this pathway allows the body to actively reg-ulate the net amount of Ca^{2+} leaving the body, the delineation of the molecular mechanisms involved in its regulation could provide new insights in disturbances in Ca^{2+} homeostasis. Secondary hyperparathyroidism (SHPT) is frequently ob-served in patients with end-stage renal disease. Although not entirely understood, one of the main contributing factors is the decreased activity of the renal enzyme, 25-hydroxyvitamin D$_3$-1α-hydroxylase, converting 25-hydroxyvitamin D$_3$ to its active metabolite 1,25(OH)$_2$D$_3$ (78). In order to overcome this situation, the ad-ministration of 1,25(OH)$_2$D$_3$ has gained a place in the therapeutic regimen of these patients. 1,25(OH)$_2$D$_3$ efficiently suppresses the transcriptional rate of PTH messenger in the parathyroid glands but at the same time increases intestinal ab-sorption of Ca^{2+} leading to hypercalcemia with all its adverse effects. Recently, new analogues of 1,25(OH)$_2$D$_3$ were introduced that have been shown to suppress PTH secretion and exhibit significantly less calcemic activity when compared with 1,25(OH)$_2$D$_3$ (79–81). The identification of ECaC provides a new target in study-ing the effects of 1,25(OH)$_2$D$_3$ and its analogues on renal Ca^{2+} handling in more detail than was previously possible. The four putative vitamin D response elements in the promoter region of ECaC, together with the observed stimulatory effect of 1,25(OH)$_2$D$_3$ on the renal expression of ECaC, strongly suggest that calcitriol plays a major regulatory role (25, 64). Furthermore, disturbed Ca^{2+} homeostasis in SHPT and in many other diseases is flanked by different symptoms frequently encountered in end-stage renal disease, for example, metabolic acidosis, which further maintains the vicious circle of Ca^{2+} wasting and increasing PTH levels and increases the risk for severe complications during the course of the disease (82, 83). Finally, elevated excretion of Ca^{2+} in the urine is correlated with an increased risk for the development of kidney stones and nephrocalcinosis (84). Although it ap-pears that hypercalciuria is in many situations a secondary event, idiopathic forms are also known. In both forms, reduction of urinary Ca^{2+} excretion, for exam-ple, by using the Ca^{2+} retaining effect of thiazides, has been shown to reduce the risk of developing kidney stones. The side effects, however, limit their ther-apeutic use (85). A drug specifically designed to stimulate ECaC function might be able to increase Ca^{2+} reabsorption. Based on the data obtained so far, it is of interest to determine whether defects in ECaC can be the primary cause of inher-ited forms of increased renal Ca^{2+} loss. Alternatively, the latter forms could also be caused by increased intestinal activity of ECaC, leading via augmented absorption of Ca^{2+}, to surpassing the tubular threshold for Ca^{2+} reabsorption and thus to hypercalciuria.

Future investigations should address these important clinical issues by considering ECaC as candidate gene in Ca^{2+} homeostasis-related disorders, by unraveling in detail the molecular mechanisms of hormone-regulated Ca^{2+} reabsorption, by characterization of conditional ECaC knockout mice, and by the development of pharmaceutics to influence the activity of ECaC.

ACKNOWLEDGMENTS

The authors acknowledge Dr. C. van Os for critical reading of this review and stimulating discussions. The work in the Nijmegen laboratory was supported in part by grants from EMBO (ALTL 160-2000), the Dutch Organization of Scientific Research (NWO-ALW 805.09.042 and 810.38.004), Deutsche Forschungsgemeinschaft (MU 1497 2-1), and the Dutch Kidney Foundation (C.00.1881). The work in the Leuven laboratory was supported by the Belgian Federal Government, the Flemish Government, and the Onderzoeksraad KU Leuven (GOA 99/07; F.W.O. G.0237.95; F.W.O. G.0214.99; F.W.O. G. 0136.00), and a grant from the Alphonse and Jean Forton-Koning Boudewijn Stichting R7115 B0 (B.N.).

Visit the Annual Reviews home page at www.AnnualReviews.org

LITERATURE CITED

1. Suki WN, Rose D. 1996. Renal transport of calcium, magnesium and phosphate. In *The Kidney*, ed. BM Brenner, I:472–15. Philadelphia: Saunders. 2702 pp.

2. Bindels RJM. 1993. Calcium handling by the mammalian kidney. *J. Exp. Biol.* 184:89–104

3. Hoenderop JGJ, Willems PHGM, Bindels RJM. 2000. Toward a comprehensive molecular model of active calcium reabsorption. *Am. J. Physiol. Renal Physiol.* 278:F352–F60

4. Friedman PA, Gesek FA. 1995. Cellular calcium transport in renal epithelia: measurement, mechanisms, and regulation. *Physiol. Rev.* 75:429–71

5. Reilly RF, Ellison DH. 2000. Mammalian distal tubule: physiology, pathophysiology, and molecular anatomy. *Physiol. Rev.* 80:277–13

6. Hoenderop JGJ, Müller D, Suzuki M, van Os CH, Bindels RJM. 2000. Epithelial calcium channel: gate-keeper of active calcium reabsorption. *Curr. Opin. Nephrol. Hypertens.* 9:335–40

7. Huo T, Lytton J. 1995. Potential calbindin-D_{28K} associated proteins in rat kidney cortex. *J. Am. Soc. Nephrol.* 6:964 (Abstr.)

8. Yu AS, Boim M, Hebert SC, Castellano A, Perez-Reyes E, et al. 1995. Molecular characterization of renal calcium channel beta-subunit transcripts. *Am. J. Physiol. Renal Physiol.* 268:F525–F31

9. Yu AS, Hebert SC, Brenner BM, Lytton J. 1992. Molecular characterization and nephron distribution of a family of transcripts encoding the pore-forming subunit of Ca^{2+} channels in the kidney. *Proc. Natl. Acad. Sci. USA* 89:10494–98

10. Van Kuijck MA, van Aubel RA, Busch AE, Lang F, Russel FG, et al. 1996. Molecular

cloning and expression of a cyclic AMP-activated chloride conductance regulator: a novel ATP-binding cassette transporter. *Proc. Natl. Acad. Sci. USA* 93:5401–6

11. Bindels RJM, Hartog A, Abrahamse SL, van Os CH. 1994. Effects of pH on apical calcium entry and active calcium transport in rabbit cortical collecting system. *Am. J. Physiol. Renal Physiol.* 266:F620–F27

12. Hoenderop JGJ, van der Kemp AW, Hartog A, van de Graaf SFJ, van Os CH, et al. 1999. Molecular identification of the apical Ca²⁺ channel in 1,25-dihydroxyvitamin D₃-responsive epithelia. *J. Biol. Chem.* 274:8375–78

13. Nilius B, Vennekens R, Prenen J, Hoenderop JGJ, Droogmans G, et al. 2001. The single pore residue Asp542 determines Ca²⁺ permeation and Mg²⁺ block of the epithelial Ca²⁺ channel. *J. Biol. Chem.* 276:1020–25

14. Caterina MJ, Rosen TA, Tominaga M, Brake AJ, Julius D. 1999. A capsaicin-receptor homologue with a high threshold for noxious heat. *Nature* 398:436–41

15. Birnbaumer L, Zhu X, Jiang M, Boulay G, Peyton M, et al. 1996. On the molecular basis and regulation of cellular capacitative calcium entry: roles for Trp proteins. *Proc. Natl. Acad. Sci. USA* 93:15195–202

16. Chen XZ, Vassilev PM, Basora N, Peng JB, Nomura H, et al. 1999. Polycystin-L is a calcium-regulated cation channel permeable to calcium ions. *Nature* 401:383–86

17. Shenolikar S, Weinman EJ. 2001. NHERF: targeting and trafficking membrane proteins. *Am. J. Physiol. Renal Physiol.* 280: F389–F95

18. Weinman EJ, Steplock D, Wang Y, Shenolikar S. 1995. Characterization of a protein cofactor that mediates protein kinase A regulation of the renal brush border membrane Na⁺-H⁺ exchanger. *J. Clin. Invest.* 95:2143–49

19. Wade JB, Welling PA, Donowitz M, Shenolikar S, Weinman EJ. 2001. Differential renal distribution of NHERF isoforms and their colocalization with NHE3, ezrin, and ROMK. *Am. J. Physiol. Cell Physiol.* 280:C192–C98

20. Gisler SM, Stagljar I, Traebert M, Bacic D, Biber J, et al. 2001. Interaction of the type IIa Na/Pᵢ cotransporter with PDZ proteins. *J. Biol. Chem.* 276:9206–13

21. Moyer BD, Denton J, Karlson KH, Reynolds D, Wang S, et al. 1999. A PDZ-interacting domain in CFTR is an apical membrane polarization signal. *J. Clin. Invest.* 104:1353–61

22. Bourguignon LY, Chu A, Jin H, Brandt NR. 1995. Ryanodine receptor-ankyrin interaction regulates internal Ca²⁺ release in mouse T-lymphoma cells. *J. Biol. Chem.* 270:17917–22

23. Bourguignon LY, Jin H, Iida N, Brandt NR, et al. 1993. The involvement of ankyrin in the regulation of inositol 1,4,5 trisphosphate receptor-mediated internal Ca²⁺ release from Ca²⁺ storage vesicles in mouse T-lymphoma cells. *J. Biol. Chem.* 268:7290–97

24. Hoenderop JGJ, Hartog A, Stuiver M, Doucet A, Willems PHGM, et al. 2000. Localization of the epithelial Ca²⁺ channel in rabbit kidney and intestine. *J. Am. Soc. Nephrol.* 11:1171–78

25. Hoenderop JGJ, Müller D, van der Kemp AWCM, Hartog A, Suzuki M, et al. 2001. Calcitriol controls the epithelial calcium channel in kidney. *J. Am. Soc. Nephrol.* 12:1342–49

26. Loffing J, Loffing-Cueni D, Valderrabano V, Kläusli L, Hebert SC, et al. 2001. The mouse distal nephron: distributions of sodium-, calcium- and water-transport pathways. *Am. J. Physiol. Renal Physiol.* In press

27. Hoenderop JGJ, van der Kemp AWCM, Hartog A, van Os CH, Willems PHGM, et al. 1999. The epithelial calcium channel, ECaC, is activated by hyperpolarization and regulated by cytosolic calcium. *Biochem. Biophys. Res. Commun.* 261: 488–92

28. Vennekens R, Hoenderop JGJ, Prenen

J, Stuiver M, Willems PHGM, et al. 2000. Permeation and gating properties of the novel epithelial Ca^{2+} channel. *J. Biol. Chem.* 275:3963–69

29. Vennekens R, Prenen J, Hoenderop JGJ, Bindels RJM, Droogmans G, et al. 2001. Pore properties and ionic block of the rabbit epithelial calcium channel expressed in HEK 293 cells. *J. Physiol.* 530:183–91

30. Vennekens R, Prenen J, Hoenderop JGJ, Bindels RJM, Droogmans G, et al. 2001. Modulation of the epithelial calcium channel ECaC by intracellular Ca^{2+}. *Pflügers Arch.* 442:237–42

31. Nilius B, Vennekens R, Prenen J, Hoenderop JGJ, Bindels RJM, et al. 2000. Whole-cell and single channel monovalent cation currents through the novel rabbit epithelial Ca^{2+} channel ECaC. *J. Physiol.* 527:239–48

32. Nilius B, Prenen J, Vennekens R, Hoenderop JGJ, Bindels RJM, et al. 2001. Modulation of the epithelial calcium channel, ECaC, by intracellular Ca^{2+}. *Cell Calcium* 29:417–28

33. Hoenderop JGJ, Vennekens R, Müller D, Prenen J. 2001. Function and expression of the epithelial Ca^{2+} channel family: comparison of the epithelial Ca^{2+} channel 1 and 2. *J. Physiol.* 537:747–61

34. Ugate G, Pérez F, Latorre R. 1998. How do calcium channels transport calcium ions? *Biol. Res.* 31:17–32

35. Peng JB, Chen XZ, Berger UV, Vassilev PM, Tsukaguchi H, et al. 1999. Molecular cloning and characterization of a channel-like transporter mediating intestinal calcium absorption. *J. Biol. Chem.* 274:22739–46

36. Peng JB, Chen XZ, Berger UV, Weremowicz S, Morton CC, et al. 2000. Human calcium transport protein CaT1. *Biochem. Biophys. Res. Commun.* 278:326–32

37. Müller D, Hoenderop JGJ, Meij IC, van den Heuvel LP, Knoers NV, et al. 2000. The human epithelial calcium channel (ECAC1): cloning, tissue distribution and chromosomal mapping. *Genomics* 67:48–53

38. Peng JB, Chen XZ, Berger UV, Vassilev PM, Brown EM, et al. 2000. A rat kidney-specific calcium transporter in the distal nephron. *J. Biol. Chem.* 275:28186–94

39. Wissenbach U, Niemeyer BA, Fixemer T, Schneidewind A, Trost C, et al. 2001. Expression of Cat-like, a novel calcium-selective channel, correlates with the malignancy of prostate cancer. *J. Biol. Chem.* 276:19461–68

40. Barley NF, Howard A, O'Callaghan D, Legon S, Walters JR. 2001. Epithelial calcium transporter expression in human duodenum. *Am. J. Physiol. Gastrointest. Liver Physiol.* 280:G285–G90

41. Christakos S, Gabrielides C, Rhoton WB. 1989. Vitamin D-dependent calcium binding proteins: chemistry, distribution, functional considerations, and molecular biology. *Endocr. Rev.* 10:3–26

42. Hunziker W, Schrickel S. 1988. Rat brain calbindin-D$_{28}$: six domain structure and extensive amino acid homology with chicken calbindin-D$_{28}$. *Mol. Endocrinol.* 2:465–73

43. Kretsinger RH, Mann JE, Simmonds JG. 1982. Model of facilitated diffusion of calcium by the intestinal calcium binding protein. In *Vitamin D, Chemical, Biochemical and Clinical Endocrinology of Calcium Metabolism.* ed. AW Norman, K Schaefer, HG Grigoleit, DV Herrath, pp. 234–48. Berlin/New York: de Gruyter

44. Feher JJ. 1983. Facilitated calcium diffusion by intestinal calcium-binding protein. *Am. J. Physiol. Cell Physiol.* 244:C303–C7

45. Feher JJ, Fullmer CS, Fritzsch GK. 1989. Comparison of the enhanced steady-state diffusion of calcium by calbindin-D9K and calmodulin: possible importance in intestinal calcium absorption. *Cell Calcium* 10:189–203

46. Feher JJ, Fullmer CS, Wasserman RH. 1992. Role of facilitated diffusion of calcium by calbindin in intestinal calcium absorption. *Am. J. Physiol. Cell Physiol.* 262:C517–C26

47. Bronner F, Pansu D, Stein WD. 1986. An analysis of intestinal calcium-transport across the rat intestine. *Am. J. Physiol. Gastrointest. Liver Physiol.* 250:G561–G69

48. Koster HPG, Hartog A, van Os CH, Bindels RJM. 1995. Calbindin-D$_{28K}$ facilitates cytosolic calcium transport without interfering with calcium signaling. *Cell Calcium* 18:187–96

49. Bindels RJM, Timmermans JAH, Hartog A, Coers W, van Os CH. 1991. Calbindin-D$_{9K}$ and parvalbumin are exclusively located along basolateral membranes in rat distal nephron. *J. Am. Soc. Nephrol.* 2:1122–29

50. Bindels RJM, Hartog A, Timmermans JAH, van Os CH. 1991. Immunocytochemical localization of calbindin-D$_{28K}$, calbindin-D$_{9k}$ and parvalbumin in rat kidney. *Contrib. Nephrol.* 91:7–13

51. Lee A, Wong ST, Gallagher D, Li B, Storm DR, et al. 1999. Ca^{2+}/calmodulin binds to and modulates P/Q-type calcium channels. *Nature* 399:155–59

52. Zuhlke RD, Pitt GS, Deisseroth K, Tsien RW, Reuter H. 1999. Calmodulin supports both inactivation and facilitation of L-type calcium channels. *Nature* 399:159–62

53. Niemeyer BA, Bergs C, Wissenbach U, Flockerzi V, Trost C. 2001. Competitive regulation of CaT-like-mediated Ca^{2+} entry by protein kinase C and calmodulin. *Proc. Natl. Acad. Sci. USA* 98:3600–5

54. Freud TS, Christakos S. 1985. Enzyme modification by renal calcium-binding proteins. In *Vitamin D, Chemical, Biochemical and Clinical Endocrinology of Calcium Metabolism.* ed. AW Norman, K Schaefer, HG Grigoleit, DV Herrath, pp. 369–70. Berlin/New York: de Gruyter

55. Shimura F, Wasserman RH. 1984. Membrane-associated vitamin D-induced calcium-binding protein (CaBP): quantification by a radioimmunoassay and evidence for a specific CaBP in purified intestinal brush borders. *Endocrinology* 115:1964–72

56. Van Baal J, Yu A, Hartog A, Fransen JAM, Willems PHGM, et al. 1996. Localization and regulation by vitamin D of calcium transport proteins in rabbit cortical collecting system. *Am. J. Physiol. Renal Physiol.* 271:F985–F93

57. Magocsi M, Yamaki M, Penniston JT, Dousa TP. 1992. Localization of messenger RNAs coding for isozymes of plasma membrane Ca^{2+}-ATPase pump in rat kidney. *Am. J. Physiol. Renal Physiol.* 263:F7–F14

58. Stauffer TP, Guerini D, Carafoli E. 1995. Tissue distribution of the four gene products of the plasma membrane Ca^{2+} pump—a study using specific antibodies. *J. Biol. Chem.* 270:12184–90

59. Bindels RJM, Ramakers PLM, Dempster JA, Hartog A, van Os CH. 1992. Role of Na^{+}-Ca^{2+} exchange in transcellular Ca^{2+} transport across primary cultures of rabbit kidney collecting system. *Pflügers Arch.* 420:566–72

60. Philipson KD, Nicoll DA, Matsuoka S, Hryshko LV, Levitsky DO, et al. 1996. Molecular regulation of the Na^{+}-Ca^{2+} exchanger. *Ann. NY Acad. Sci.* 779:20–28

61. Lytton J, Lee SL, Lee WS, van Baal J, Bindels RJM, et al. 1996. The kidney sodium-calcium exchanger. *Ann. NY Acad. Sci.* 779:58–72

62. White KE, Gesek FA, Reilly RF, Friedman PA. 1998. NCX1 Na/Ca exchanger inhibition by antisense oligonucleotide in mouse distal convoluted tubule cells. *Kidney Int.* 54:896–906

63. Bindels RJM, Hartog A, Timmermans JAH, van Os CH. 1991. Active Ca^{2+} transport in primary cultures of rabbit kidney CCD: stimulation by 1,25-dihydroxyvitamin D$_3$ and PTH. *Am. J. Physiol. Renal Physiol.* 261:F799–F807

64. Müller D, Hoenderop JGJ, Merkx GF, van Os CH, Bindels RJM. 2000. Gene structure and chromosomal mapping of human epithelial calcium channel. *Biochem. Biophys. Res. Commun.* 275:47–52

65. Sutton RAL, Wong NLM, Dirks JH. 1979. Effects of metabolic acidosis and alkalosis on sodium and calcium transport in the dog kidney. *Kidney Int.* 15:520–33

66. Costanzo LS. 1984. Comparison of calcium and sodium transport in early and late rat distal tubules: effect of amiloride. *Am. J. Physiol. Renal Physiol.* 246:F937–F45

67. Friedman PA. 1998. Codependence of renal calcium and sodium transport. *Annu. Rev. Physiol.* 60:179–97

68. Stanton BA. 1990. Cellular actions of thiazide diuretics in the distal tubule. *J. Am. Soc. Nephrol.* 1:832–36

69. Rho SH, Park CS. 1998. Extracellular proton alters the divalent cation binding affinity in a cyclic nucleotide-gated channel pore. *FEBS Lett.* 440:199–202

70. Bushinsky DA, Chabala JM, Gavrilov KL, Levi-Setti R. 1999. Effects of in vivo metabolic acidosis on midcortical bone ion composition. *Am. J. Physiol. Renal Physiol.* 277:F813–F19

71. Van Baal J, de Jong MD, Zijlstra FJ, Willems PHGM, Bindels RJM. 1996. Endogenously produced prostanoids stimulate calcium reabsorption in the rabbit cortical collecting system. *J. Physiol.* 497:229–39

72. Van Baal J, Raber G, de Slegte J, Pieters R, Bindels RJM, et al. 1996. Vasopressin-stimulated Ca^{2+} reabsorption in rabbit cortical collecting system: effects on cAMP and cytosolic Ca^{2+}. *Pflügers Arch.* 433:109–15

73. Hoenderop JGJ, Vaandrager AB, Dijkink L, Smolenski A, Gambaryan S, et al. 1999. ANP-stimulated Ca^{2+} reabsorption in rabbit kidney requires membrane-targeted cGMP-dependent protein kinase type. *Proc. Natl. Acad. Sci. USA* 96:6084–89

74. Hoenderop JGJ, Hartog A, Willems PHGM, Bindels RJM. 1998. Adenosine-stimulated Ca^{2+} reabsorption is mediated by apical A1 receptors in rabbit cortical collecting system. *Am. J. Physiol. Renal Physiol.* 274:F736–F43

75. Hoenderop JGJ, De Pont JJHHM, Bindels RJM, Willems PHGM. 1999. Hormone-stimulated Ca^{2+} reabsorption in rabbit kidney cortical collecting system is cAMP-independent and involves a phorbol ester-insensitive PKC isotype. *Kidney Int.* 55:225–33

76. Koster HPG, Hartog A, van Os CH, Bindels RJM. 1996. Inhibition of Na^+ and Ca^{2+} reabsorption by P_{2u}-purinoceptors requires protein kinase C but not Ca^{2+} signaling. *Am. J. Physiol. Renal Physiol.* 270:F53–F60

77. Raber G, Willems PHGM, Lang F, Nitschke R, van Os CH, et al. 1997. Coordinated control of apical calcium influx and basolateral calcium efflux in rabbit cortical collecting system. *Cell Calcium* 22:157–66

78. Slatopolsky E, Brown A, Dusso A. 1999. Pathogenesis of secondary hyperparathyroidism. *Kidney Int.* 73:14–19

79. Martin KJ, Gonzales EA, Gellens M, Hamm LL, Abboud H, et al. 1998. 19-Nor-1α,25–dihydroxyvitamin D_2 (Paricalcitol) safely and effectively reduces the levels of intact parathyroid hormone in patients on hemodialysis. *J. Am. Soc. Nephrol.* 9:1427–32

80. Kurokawa K, Akizawa T, Suzuki M, Akiba T, Ogata E, et al. 1996. Effect of 22-oxacalcitriol on hyperparathyroidism of dialysis patients: results of a preliminary study. *Nephrol. Dial. Transplant.* 11:121–24

81. Tan AU Jr, Levine BS, Mazess RB, Kyllo DM, Bishop CW, et al. 1997. Effective suppression of parathyroid hormone by 1 alpha-hydroxy-vitamin D_2 in haemodialysis patients with moderate to severe secondary hyperparathyroidism. *Kidney Int.* 51:317–23

82. Kraut JA. 1995. The role of metabolic acidosis in the pathogenesis of renal osteodystrophy. *Adv. Ren. Replace Ther.* 2:40–51

83. Lu KC, Lin SH, Yu FC, Chyr SH, Shieh SD. 1995. Influence of metabolic

acidosis on serum 1,25(OH)$_2$D$_3$ levels in chronic renal failure. *Miner. Electrolyte Metab.* 21:398–02

84. Lloyd SE, Pearce SH, Fisher SE, Steinmeyer K, Schwappach B, et al. 1996. A common molecular basis for three inherited kidney stone diseases. *Nature* 379:445–49

85. Loffing J, Loffing-Cueni D, Hegyi I, Kaplan MR, Hebert SC, et al. 1996. Thiazide treatment of rats provokes apoptosis in distal tubular cells. *Kidney Int.* 50:1180–90

Annu. Rev. Physiol. 2002. 64:551–61

THE RENIN ANGIOTENSIN SYSTEM AND KIDNEY DEVELOPMENT

Taiji Matsusaka[1,2,3], Yoichi Miyazaki[1], and Iekuni Ichikawa[1,2,4]

Department of [1]Pediatrics and [2]Medicine, Vanderbilt University School of Medicine, Nashville, Tennessee 37232; [3]Institute of Medical Science, Molecular and Cellular Nephrology and [4]Department of Pediatrics, Tokai University, Isehara, Kanagawa 259-1193, Japan; e-mail: iekuni.i@is.icc.u-tokai.ac.jp

Key Words AT1 receptor, ureter, pelvis, vascular hypertrophy, knockout mice

■ **Abstract** When angiotensin II or AT1 receptor is experimentally inhibited during the perinatal period, either by pharmacological intervention or genetic manipulation, the kidney develops with profound structural abnormalities. Most prominent are hypertrophy of arterial vasculatures and atrophy of the papilla. Although the mechanism by which the vascular hypertrophy occurs remains unknown, study of the atrophic papilla gives us a new clue for understanding the physiological role of angiotensin. Mutant mice completely devoid of AT1 receptor fail to develop the renal pelvis and the ureteral peristaltic movement. Normally, angiotensin and AT1 receptor are transiently up-regulated around the renal outlet at birth. Thus angiotensin II induces the peristaltic machinery during the perinatal period in a timely fashion to accommodate the dramatic increase in urine production that occurs during the transition from intra- to extra-uterine life. Further studies revealed that in adult animals angiotensin augments the peristaltic movement when the urinary tract is partially obstructed, thereby protecting the kidney from hydronephrosis. This newly discovered function of angiotensin to protect kidney architecture at the time of urine outflow obstruction is reminiscent of its similar kidney structure-protecting function that is active during arterial blood flow obstruction.

INTRODUCTION

Angiotensin-converting enzyme (ACE) inhibitor became available in the late 1970s and was immediately impressive with its potent antihypertensive efficacy in a wide variety of hypertension, including essential hypertension. ACE inhibitors were shown to protect several vital organs, including the heart (1, 2), kidney (3), and retina (4), from tissue injury in diseases. Furthermore, a potential for even broader benefit has been suggested in animal studies in which daily administration of ACE inhibitors led to an increase in life span (5, 6). Although ACE inhibitors are known to be relatively safe compared with other antihypertensive drugs, they are

0066-4278/02/0315-0551$14.00

contraindicated for use in pregnant women. There are a few reports of fetopathy or kidney anomalies in human babies whose mothers took ACE inhibitor during their late pregnancy (7–9). These babies showed immature kidneys and renal failure, suggesting that angiotensin has a role in the kidney development. Genetically manipulated mice that lack angiotensin type 1 (AT1) receptor or other components of the renin angiotensin system revealed that angiotensin has a critically important role in the development of normal kidney architecture. This review focuses on the physiological effect of angiotensin (Ang) II on the renal architecture that is mediated through the AT1 receptor based on the findings from gene targeting studies and related experiments.

Maturation of the Kidney and Urinary Tract During the Perinatal Period

Birth of placental mammals involves an establishment of two new channels for eliminating respiratory and metabolic wastes. Thus as the placental circulation gradually closes, the lung and the kidney become fully matured by birth. In humans, new nephron formation is completed between the 28th and 36th gestational weeks, and at birth, all glomeruli are matured and have capillaries (10). Not only the nephron within the kidney but also the urinary tract play an important role in this process. The ureteral smooth muscle, through its peristaltic movement and the resultant pulsatile pressure generated, expels urine and protects the renal parenchyma from physical stress. When this complex muscle movement is absent, the urine is not efficiently transferred downward, and the pressure thus generated is directly transmitted to the renal parenchyma owing to the absence of the partition that isolates the parenchyma from the downstream pressure. In humans, smooth muscle, as shown by histological staining, first appears at the 12th gestational week and gradually develops in the ureter and pelvis. Full maturation of calcial musculatures occurs between 27 weeks and term (11). In contrast to humans, who are born with fully developed kidneys and urinary tracts, mice and rats are born with immature kidneys and urinary tracts. In rats, nephrogenesis continues until the 18th postnatal day (12), and the formation of pelvic smooth muscle cells starts after birth and finishes at the 10th postnatal day (13). Mice have a similar course of the development of the kidney and urinary tract (14, 15).

Interestingly, corresponding to the period of the maturation of the kidney and urinary tract, i.e., late gestational period in humans and early postnatal days in rodents, a marked increase in urine output occurs (16, 17). Therefore, the above maturation of the renal pelvis appears timely so that urine is removed from the kidney.

Perinatal Ang II/AT1 Expression

The renin-angiotensin system is transiently up-regulated during the perinatal period (18, 15), the significance of which has been long debated because pharmacological blockade has little effect as far as acute hemodynamic changes are concerned (19).

In various mammalian embryonic kidneys, AT1 protein and mRNA are intensely expressed in immature and mature glomeruli, but not in S-shaped or comma bodies, in temporal correlation with the appearance of mesangial cells (20–24). Moderate expression of AT1 protein and mRNA is observed in maturing renal tubules. In most species other than the mouse, AT1 protein and mRNA are intensely expressed in the inner stripe of the outer medulla. Of note, AT1 is also expressed in the urinary tract. In fetal rats at the 17th gestational day, AT1A and AT1B mRNAs are expressed in the renal pelvis and the superior part of the ureter (21). In mice, AT1 receptor is detectable in the renal hilum at birth by binding autoradiography and further increased in the smooth muscle cell layer of the renal pelvis (Figure 1, see color insert) (15).

Abnormal Kidney Phenotypes of Ang II- and AT1-Deficient Mutants

Null-mutant mice for the angiotensinogen gene ($Agt-/-$) showed severe hypotension and renin overexpression (25, 26). In addition, $Agt-/-$ mice show a temporal delay in the development of renal glomerulus (14). Thus $Agt-/-$ kidneys contain more immature glomeruli at 1 week than wild-type kidneys. The glomerular maturity in $Agt-/-$ mice soon catches up with that of wild-type mice, and at 3 weeks, all glomeruli in both strains are matured. The observed glomerular immaturity in $Agt-/-$ mice is consistent with the earlier observation that Ang II exerts potent mitogenic actions in mesangial cells isolated from the fetus (27). However, the lack of major defect in glomerular morphology in $Agt-/-$ mice indicates that Ang II has little effect at the onset of nephrogenesis, a notion consistent with the fact that AT1 receptors appear only after the commitment of glomerulogenesis.

In addition to these somewhat anticipated phenotypes, $Agt-/-$ mice have two impressive abnormal phenotypes in the kidney. One is medial hypertrophy of renal small arteries (Figure 2*b*, see color insert). The lesion develops in renal afferent arterioles and interlobular arteries but not in vessels outside the kidney. Continuous to the hypertrophic afferent arterioles, mesangial matrix is often expanded. The other abnormal phenotype is atrophy of the papilla and the inner medulla with dilated calyx (Figure 3*b*, see color insert) (14, 26, 28).

Similar abnormal phenotypes are commonly observed in null-mutant mice for the *Ace* gene ($Ace-/-$) (29, 30). Esther et al. disrupted the C-terminal half of the *Ace* gene including the portion encoding the transmembrane domain (31). Mice homozygous for this mutation (ACE.2 strain) have plasma ACE activity 34% of that in wild-type mice but completely lack tissue ACE. Interestingly, these animals developed renal vascular hypertrophic lesions similar to those of $Ace-/-$ or $Agt-/-$ mice, in contrast to the normal renal morphology of $Ace+/-$ mice with reduced plasma ACE activity. These findings indicate that the majority of Ang II necessary for maintaining the arterial morphology is generated by tissue-bound ACE activity. On the other hand, papillary hypoplasia was observed only in some ACE.2 strains, suggesting that Ang II generated in the circulation serves to minimize this abnormal phenotype.

Mutant mice completely devoid of AT1 receptor (designated as *Agtr1−/−*), namely mice dual null-mutant for AT1 subtype A and B genes (*Atgr1a, Atgr1b*), develop abnormal renal phenotypes identical to those observed in *Agt−/−* mice (Figure 3*c*) (32, 33). In contrast, mice null-mutant for the AT2 receptor gene do not show these phenotypes (34, 35), which indicates that the lack of AT1, but not AT2, function is responsible for the development of the abnormal phenotypes seen in *Agt−/−* and *Ace−/−* mice. Although 99% of AT1 mRNA is generated from *Agtr1a* in the mouse kidney (36, 37), the phenotypes of *Agtr1a−/−* mice are remarkably mild compared with those of *Agt−/−* or *Agtr1−/−* mice (Figure 2*f*) (38). Mutation of *Agtr1a* causes hypotension and stimulation of renin, leading to a profound increase in Ang II. This up-regulation of the ligand may account for the surprisingly great compensatory function of AT1B prevailing in *Agtr1a−/−* mice, which can be demonstrated when they are given an ACE inhibitor (39).

Of note, the above-described abnormal phenotypes in *Agt−/−*, *Agtr1−/−*, or *Ace−/−* mice remain largely unrecognized at birth, but become evident 1 to 3 weeks after birth. This suggests that Ang II is most important for the maturation of the kidney during the perinatal or postnatal period. This notion is supported by findings in studies performed on rats immediately after birth. Newborn rats treated with an ACE inhibitor or AT1 antagonist developed renal vascular hypertrophy and papillary atrophy (40, 41). In another study, rats born from mothers treated with losartan, an AT1 antagonist, starting at gestation day 15 (full term = 20–21 days), developed vascular hypertrophy and papillary atrophy by the 21st postnatal day. At 90th day, the vascular change was resolved, but the atrophic papilla was still present (42). These findings collectively indicate that the lack of Ang II in the perinatal period causes irreversible changes in the kidney structure. As discussed above, species differences between rodents and humans exist in terms of the rate of maturation of the kidney and urinary tract. Despite these differences, the abnormal phenotypes can be produced by Ang II at a final phase of the development of the kidney and urinary tract, namely at 1 to 3 weeks in rodents and at birth in humans.

In view of the well-known phenomenon that activation of Ang II action induces vascular hypertrophy, the hypertrophic vascular lesion seen in *Agt−/−*, *Agtr1−/−*, and *Ace−/−* mutants appears odd. The pathogenesis of this phenotype remains unknown. When the function of Ang II or AT1 is blocked genetically or pharmacologically, the hypertrophy of the renal artery is uniformly accompanied by renin accumulation. Of note, within the renal vasculature, hypertrophy is distinguished from mere renin accumulation. The hypertrophy is most prominent in the interlobular artery, whereas the renin accumulation occurs mainly in the afferent arterioles. Severe arterial hypertrophy is irreversible, whereas renin accumulation is reversible (43). In addition, not all hypertrophic arteries in the mutants contain renin (14).

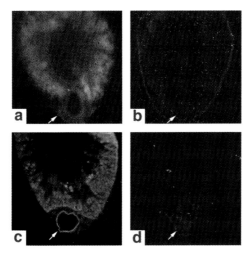

Figure 1 AT1 receptor expression in the kidney and pelvis. AT1 receptor is depicted by binding autoradiography using ^{125}I-labeled [Sar[1], Ile[8]] Ang with excess AT2 antagonist. (*a*) Newborn wild-type mice, (*b*) newborn *Agtr1-/-* mice, (*c*) wild-type mice at 4 weeks postnatal, (*d*) *Agtr1-/-* mice at 4 weeks postnatal (reproduced with a modification with permission from Reference 15).

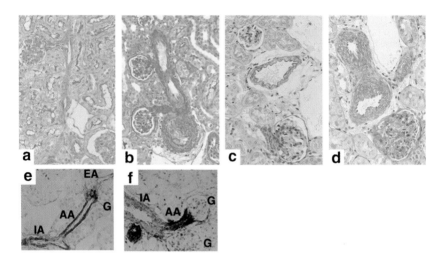

Figure 2 Renal vascular hypertrophy in Ang II-deficient mice. Arterial walls are markedly hypertrophic in *Agt-/-* mice at 4 weeks (*b*) and in renin-deficient mice at 7 weeks (*d*), but not in wild-type litter mates at 4 weeks (*a*) or at 7 weeks (*c*). Arteries are mildly hypertrophic in *Agtr1a-/-* mice at 6 weeks (*f*) but not in *Agtr1a+/-* mice (*e*). (*a–d*) PAS staining; (*e–f*) staining for α-SMA (*brown*) and lacZ (*green*). The latter represents the *Agtr1a* promoter activity. IA, interlobular artery; AA, afferent arteriole; EA, efferent arteriole; G, glomerulus.

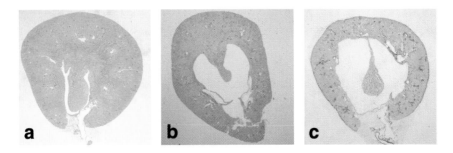

Figure 3 Atrophy of the papilla in Ang II-deficient mice. (*a*) Wild-type mice, (*b*) *Agt-/-* mice, (*c*) *Agtr1-/-* mice, all at 5 weeks. PAS staining (reproduced with a modification with permission from Reference 15).

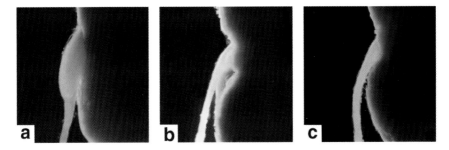

Figure 4 Defective pelvis formation in Ang II-deficient mice. Triangular pelvis is formed at 4 weeks in wild-type mice (*a*) but it is absent in *Agt-/-* mice (*b*) and *Agtr1a-/-* mice (*c*) at the same age (reproduced with a modification with permission from Reference 15).

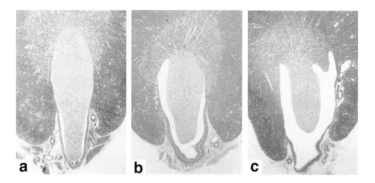

Figure 6 An AT1 antagonist augments hydronephrosis in rats with partial urinary tract obstruction. Coronal sections of kidneys 7 days after sham operation (*a*), partial urinary obstruction with (*c*) or without (*b*) an AT1 antagonist. PAS staining (reproduced with a modification with permission from Reference 54).

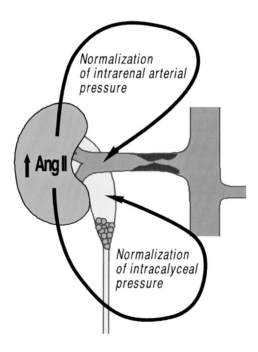

Figure 7 Adaptive roles of Ang II in both vascular and urinary tract obstruction. Renovascular stenosis up-regulates Ang II, which normalizes intrarenal arterial pressure by increasing systemic blood pressure. Urinary tract occlusion also up-regulates Ang II. Ang II normalizes intracalyceal pressure by augmenting ureteral peristalsis during the perinatal period and later in life.

One possible mechanism for the vascular hypertrophy is that renin per se may play an intermediary role in inducing the hypertrophy independently of angiotensin production. However, this possibility has been ruled out by the observations in renin-deficient mutant mice that developed phenotypes similar to those of *Atg—/—* mice, which include renal vascular hypertrophy (Figure 3, see color insert) (44, 45).

In the adventitia of hypertrophic lesions, mononuclear cell infiltration is often observed, suggesting that factor(s) with both hypertrophic effects on vascular smooth muscle cells and chemotactic effects on mononuclear cells are expressed in the lesions. In situ hybridization showed that TGF-β is up-regulated in the lesions of *Atg—/—* mice (14). However, it remains unclear whether TGF-β, indeed, has an intermediary role for the development of the vascular lesion and what mechanism is involved in the TGF-β up-regulation. In chimeric mice made with *Agtr1a—/—* and *Agtr1a+/+* cells, hypertrophic lesion is absent from the renal artery, suggesting that systemic factor(s) or locally produced diffusible factor(s) activated by the blockade of Ang II are involved in the development of renal vascular hypertrophy in the mutants (38).

The atrophic papilla observed in *Agtr1—/—* mice are morphologically, histologically, and histochemically indistinguishable from hydronephrosis induced by surgical ligation of the ureter (46). Thus in both, the calyx is enlarged whereas the papilla is atrophic; tubulointerstitial cells are not only abnormally proliferative but are also apoptotic. Both mice models are also characterized by interstitial macrophage infiltration and fibrosis, and, within local lesions, TGF-β1, PDGF-A, and IGF-1 are up-regulated, whereas EGF is down-regulated. Overall, a qualitatively remarkable similarity exists between *Agtr1—/—* kidneys and wild-type kidneys with ureteral obstruction.

Angiotensin Induces Urinary Peristaltic Machinery at Birth

The similarity between *Agtr1—/—* kidneys and hydronephrosis, together with the intense expression of AT1 receptor along the urinary tract near the end of gestation, collectively suggested that *Agtr1—/—* mice may have a defective development of the urinary tract as a primary phenotype, which secondarily leads to hydronephrotic injury in the renal parenchyma. Indeed, anatomical and histological studies revealed that the ampulla-shape structure of the renal pelvis that develops shortly after birth in wild-type mice is never acquired by *Agtr1—/—* mice (Figure 4, see color insert) (15). The renal pelvis, located at the most cranial part of the ureter, consists of a smooth muscle layer extending from the renal parenchyma to the downstream ureter.

The proliferative activity of ureteral smooth muscle cells was examined immediately after birth with bromodeoxyuridine (BrdU) incorporation. In wild-type mice, BrdU-positive nuclei were observed in great abundance within the rapidly growing smooth muscle layer, whereas, in similarly young *Agtr1—/—* pups, there were only few positive nuclei in a poorly developing smooth muscle layer along the pelvis. The same phenomenon prevailed in mutant mice lacking angiotensinogen and in wild-type mice subjected to pharmacological ACE inhibition. Clearly, therefore, when Ang II action is inhibited, ureteral smooth muscle cells do not proliferate normally.

The capacity of Ang II to directly induce ureteral smooth muscle cells was demonstrated by ureteric organ culture (15). In embryonic ureter specimens taken from the wild-type mice, Ang II caused a remarkable increase in the number of α-smooth muscle actin (SMA) -positive cells, whereas in *Agtr1*−/− tissues, Ang II treatment had no effect on the expression of α-SMA. Moreover, Ang II content in the kidney is transiently and markedly up-regulated during the perinatal period in wild-type kidneys. Collectively, therefore, Ang II directly induces differentiation and/or proliferation of ureteral smooth muscle cells.

The function of the renal pelvis is to eliminate urine from the calyx by generating pressure, directed only forward by a constant unidirectional peristaltic movement initiated by the pacemaker cells within the pelvis (47). This movement allows removal of urine from the calyx without generating backward pressure, which is injurious to the renal parenchyma. The function of the renal pelvis was assessed by monitoring the pelvic pressure using a survo-nulling micropipette method (15). In wild-type mice, the renal pelvic pressure exhibited rhythmic pulsatile changes with constant frequency (Figure 5). In mutant mice, however, the regular contraction of the renal pelvis was absent, and the baseline of intrapelvic pressure was higher than that in the wild- type animals (Figure 5), although the urine flow rate was similar. The renal parenchyma of mutant mice, therefore, is directly and constantly exposed to abnormally high pressure in the absence of the pelvis, the normal

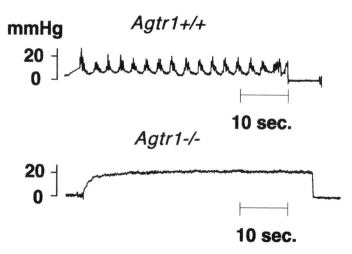

Figure 5   Ureteric peristalsis in Ang II-deficient mice. In wild-type (*Agtr1*+/+) mice, the intrapelvic pressure exhibits rhythmic and constant pulsatile contraction. In *Agtr1*−/− mice, the normal pulsatile pattern is absent (reproduced with a modification with permission from Reference 15).

function of which is to isolate the renal parenchyma from the high downstream pressure. It appears, therefore, that in *Agtr1*−/− mice, an exposure of the renal parenchyma to the high pelvic pressure causes atrophy of the papilla and enlarged calyx.

Overall, these studies revealed that Ang II induces the ureteral peristaltic machinery in the renal pelvis, which is critical for protecting the architecture of renal parenchyma at the time when urine output increases markedly as the mammals lose placental circulation.

Role of AT1 in the Adult Urinary Tract

In adulthood, the kidney may again be exposed to the risk of developing hydronephrosis as a result of a mechanical hindrance produced by stone formation, prostate hypertrophy, neurogenic bladder, etc. Of note, mechanical hindrance of urine flow causes a marked up-regulation of the renin-angiotensin system in the kidney (48–51). Because the AT1 receptor is present on the smooth muscle cells of the ureter (52) and the renal pelvis in adults (53), it also appears plausible that in adult animals Ang II facilitates peristalsis during ureteral obstruction and by so doing protects the kidney from hydronephrosis.

Therefore, we examined whether Ang II modifies the function and/or structure of the urinary tract in animals with partial ureteral obstruction, i.e., a setting where, unlike in complete obstruction, enhancement in the upper-stream ureteral peristalsis can dampen the intrapelvic pressure (54). In rats, after 2 weeks of partial ureteral obstruction, hydronephrosis developed as the calyx expanded and the papilla became atrophic, and hydronephrosis advanced as the obstruction continued. Notably, when these animals were given an Ang II type 1 receptor antagonist, the severity of hydronephrosis, quantified by the calyx-to-papilla area ratio, was augmented (Figure 6, see color insert), indicating that Ang II plays an adaptive role in protecting the kidney from hydronephrosis. Detailed morphometrical analyses revealed no appreciable effects of Ang II on the structure of the urinary tract. However, the antagonist decreased both peak pressure and frequency of ureteral peristalsis in vivo. In addition, Ang II was found to induce contractile responses in isolated ureteral strips via AT1 receptor. Thus Ang II strengthens the function of the peristaltic machinery of the urinary tract without affecting its structure. The frequency of ureteral peristalsis is primarily controlled by the pacemaker system in the proximal portion of the renal pelvis, which induces constant regular pelvic contraction through spontaneous action potential (47, 55). It is likely, therefore, that Ang II stimulates the activity of the pacemaker system during partial ureteral obstruction, in addition to the direct effect on the contractility of ureteral smooth muscles.

In the renal parenchyma, the magnitude of interstitial injury, quantified by collagen fractional volume and α-SMA score, was not affected by the antagonist. However, the detrimental effect of losartan in partial ureteral obstruction via suppressing ureteral peristalsis can be appreciated when one considers that Ang II inhibition has a direct protective effect on the renal parenchyma, which was evident in complete obstruction (56, 57, 51, 58).

SUMMARY

In summary, in both the perinatal period and adulthood, Ang II has distinctive effects on the renal pelvis and the ureter to protect the renal parenchyma from obstructive injury by promoting peristalsis of the urinary tract. There is a remarkable parallelism between this effect and the established role of Ang II in renovascular hypertension (Figure 7, see color insert). Thus in the former, hindrance of the urinary outflow from the kidney leads to renin release, which is geared toward the normalization of the pressure imposed on the kidney, and in the latter, hindrance of arterial blood inflow to the kidney, likewise, stimulates renal renin release, which is likewise geared toward the normalization of blood pressure imposed on the kidney. Both pressure-normalizing effects serve to prevent the kidney from otherwise incurring renal parenchymal damage, i.e., hydronephrosis in the former and atrophy in the latter. In this context, the renin-angiotensin system is viewed as a highly effective tool for the kidney to guard its own structural integrity.

ACKNOWLEDGMENTS

This work was supported by National Institutes of Health grants DK-44757 and DK-37868, Research for the Future Program of Japan Society for the Promotion of Science, Japan Heart Foundation Research Grant, and the Mochida Memorial Foundation for Medical and Pharmaceutical Research.

Visit the Annual Reviews home page at www.AnnualReviews.org

LITERATURE CITED

1. Pfeffer MA, Lamas GA, Vaughan DE, Parisi AF, Braunwald E. 1988. Effect of captopril on progressive ventricular dilatation after anterior myocardial infarction. *N. Engl. J. Med.* 319:80–86

2. Pfeffer MA, Braunwald E, Moye LA, Basta L, Brown EJ Jr, et al. 1992. Effect of captopril on mortality and morbidity in patients with left ventricular dysfunction after myocardial infarction. Results of the survival and ventricular enlargement trial. The SAVE Investigators. *N. Engl. J. Med.* 327:669–77

3. Lewis EJ, Hunsicker LG, Bain RP, Rohde RD. 1993. The effect of angiotensin-converting-enzyme inhibition on diabetic nephropathy. The Collaborative Study Group. *N. Engl. J. Med.* 329:1456–62

4. Chaturvedi N, Sjolie AK, Stephenson JM, Abrahamian H, Keipes M, et al. 1998. Effect of lisinopril on progression of retinopathy in normotensive people with type 1 diabetes. The EUCLID Study Group. EURODIAB Controlled Trial of Lisinopril in Insulin-Dependent Diabetes Mellitus. *Lancet* 351:28–31

5. Heudes D, Michel O, Chevalier J, Scalbert E, Ezan E, et al. 1994. Effect of

chronic ANG I-converting enzyme inhibition on aging processes. I. Kidney structure and function. *Am. J. Physiol. Regulatory Integrative Comp. Physiol.* 266:R1038–R51

6. Inserra F, Romano L, Ercole L, de Cavanagh EM, Ferder L. 1995. Cardiovascular changes by long-term inhibition of the renin-angiotensin system in aging. *Hypertension* 25:437–42

7. Nightingale SL. 1992. From the Food and Drug Administration. *J. Am. Med. Assoc.* 267:2445

8. Pryde PG, Sedman AB, Nugent CE, Barr M Jr. 1993. Angiotensin-converting enzyme inhibitor fetopathy. *J. Am. Soc. Nephrol.* 3:1575–82

9. Sedman AB, Kershaw DB, Bunchman TE. 1995. Recognition and management of angiotensin converting enzyme inhibitor fetopathy. *Pediatr. Nephrol.* 9:382–85

10. Ewan AP, Larsson L. 1992. Morphologic development of the nephron. In *Pediatric Kidney Disease*, ed. CM Edelmann Jr, pp. 19–48. Boston/Toronto/London: Little, Brown. 2nd ed.

11. Itatani H, Koide T, Okuyama A, Mizutani S, Sonoda T. 1979. Development of the calyceal system in the human fetus. *Invest. Urol.* 16:388–94

12. Nigam SK, Aperia AC, Brenner BM. 1996. Development and maturation of the kidney. In *The Kidney*, ed. BM Brenner, pp. 72–98. Philadelphia/London/Toronto/Montreal/Sydney/Tokyo: Saunders. 5th ed.

13. Baker LA, Gomez RA. 1998. Embryonic development of the ureter and bladder: acquisition of smooth muscle. *J. Urol.* 160:545–50

14. Niimura F, Labosky PA, Kakuchi J, Okubo S, Yoshida H, et al. 1995. Gene targeting in mice reveals a requirement for angiotensin in the development and maintenance of kidney morphology and growth factor regulation. *J. Clin. Invest.* 96:2947–54

15. Miyazaki Y, Tsuchida S, Nishimura H, Pope JCT, Harris RC, et al. 1998. Angiotensin induces the urinary peristaltic machinery during the perinatal period. *J. Clin. Invest.* 102:1489–97

16. Boylan BJW, Colbourn EP, McCance RA. 1958. Renal function in the foetal and newborn guinea-pig. *J. Physiol.* 141:323–31

17. Rabinowitz R, Peters MT, Vyas S, Campbell S, Nicolaides KH. 1989. Measurement of fetal urine production in normal pregnancy by real-time ultrasonography. *Am. J. Obstet. Gynecol.* 161:1264–66

18. Taylor GM, Peart WS, Porter KA, Zondek LH, Zondek T. 1986. Concentration and molecular forms of active and inactive renin in human fetal kidney, amniotic fluid and adrenal gland: evidence for renin-angiotensin system hyperactivity in 2nd trimester of pregnancy. *J. Hypertens.* 4:121–29

19. Jose PA, Slotkoff LM, Montgomery S, Calcagno PL, Eisner G. 1975. Autoregulation of renal blood flow in the puppy. *Am. J. Physiol.* 229:983–88

20. Ciuffo GM, Viswanathan M, Seltzer AM, Tsutsumi K, Saavedra JM. 1993. Glomerular angiotensin II receptor subtypes during development of rat kidney. *Am. J. Physiol. Renal Physiol.* 265:F264–F71

21. Shanmugam S, Corvol P, Gasc JM. 1994. Ontogeny of the two angiotensin II type 1 receptor subtypes in rats. *Am. J. Physiol. Endocrinol. Metab.* 267:E828–E36

22. Kakuchi J, Ichiki T, Kiyama S, Hogan BL, Fogo A, et al. 1995. Developmental expression of renal angiotensin II receptor genes in the mouse. *Kidney Int.* 47:140–47

23. Schutz S, Le Moullec JM, Corvol P, Gasc JM. 1996. Early expression of all the components of the renin-angiotensin system in human development. *Am. J. Pathol.* 149:2067–79

24. Butkus A, Albiston A, Alcorn D, Giles M, McCausland J, et al. 1997. Ontogeny of angiotensin II receptors, types 1 and 2, in ovine mesonephros and metanephros. *Kidney Int.* 52:628–36

25. Tanimoto K, Sugiyama F, Goto Y, Ishida J,

Takimoto E, et al. 1994. Angiotensinogen-deficient mice with hypotension. *J. Biol. Chem.* 269:31334–37

26. Kim HS, Krege JH, Kluckman KD, Hagaman JR, Hodgin JB, et al. 1995. Genetic control of blood pressure and the angiotensinogen locus. *Proc. Natl. Acad. Sci. USA* 92:2735–39

27. Ray PE, Aguilera G, Kopp JB, Horikoshi S, Klotman PE. 1991. Angiotensin II receptor-mediated proliferation of cultured human fetal mesangial cells. *Kidney Int.* 40:764–71

28. Nagata M, Tanimoto K, Fukamizu A, Kon Y, Sugiyama F, et al. 1996. Nephrogenesis and renovascular development in angiotensinogen-deficient mice. *Lab. Invest.* 75:745–53

29. Krege JH, John SW, Langenbach LL, Hodgin JB, Hagaman JR, et al. 1995. Male-female differences in fertility and blood pressure in ACE-deficient mice. *Nature* 375:146–48

30. Esther CR Jr, Howard TE, Marino EM, Goddard JM, Capecchi MR, Bernstein KE. 1996. Mice lacking angiotensin-converting enzyme have low blood pressure, renal pathology, and reduced male fertility. *Lab. Invest.* 74:953–65

31. Esther CR, Marino EM, Howard TE, Machaud A, Corvol P, et al. 1997. The critical role of tissue angiotensin-converting enzyme as revealed by gene targeting in mice. *J. Clin. Invest.* 99:2375–85

32. Oliverio MI, Kim HS, Ito M, Le T, Audoly L, et al. 1998. Reduced growth, abnormal kidney structure, and type 2 (AT2) angiotensin receptor-mediated blood pressure regulation in mice lacking both AT1A and AT1B receptors for angiotensin II. *Proc. Natl. Acad. Sci. USA* 95:15496–501

33. Tsuchida S, Matsusaka T, Chen X, Okubo S, Niimura F, et al. 1998. Murine double null zygotes of the angiotensin type 1A and 1B receptor genes duplicate severe abnormal phenotypes of angiotensinogen null zygotes. *J. Clin. Invest.* 101:755–60

34. Hein L, Barsh GS, Pratt RE, Dzau VJ, Kobilka BK. 1995. Behavioural and cardiovascular effects of disrupting the angiotensin II type-2 receptor in mice. *Nature* 377:744–47

35. Ichiki T, Labosky PA Shiota C, Okuyama S, Imagawa Y, et al. 1995. Effects on blood pressure and exploratory behaviour of mice lacking angiotensin II type-2 receptor. *Nature* 377:748–50

36. Burson JM, Aguilera G, Gross KW, Sigmund CD. 1994. Differential expression of angiotensin receptor 1A and 1B in mouse. *Am. J. Physiol. Endocrinol. Metab.* 267:E260–E67

37. Nishimura H, Matsusaka T, Fogo A, Kon V, Ichikawa I. 1997. A novel in vivo mechanism for angiotensin type 1 receptor regulation. *Kidney Int.* 52:345–55

38. Matsusaka T, Nishimura H, Utsunomiya H, Kakuchi J, Niimura F, et al. 1996. Chimeric mice carrying 'regional' targeted deletion of the angiotensin type 1A receptor gene. Evidence against the role for local angiotensin in the in vivo feedback regulation of renin synthesis in juxtaglomerular cells. *J. Clin. Invest.* 98:1867–77

39. Oliverio MI, Best CF, Kim HS, Arendshorst WJ, Smithies O, Coffman TM. 1997. Angiotensin II responses in AT1A receptor-deficient mice: a role for AT1B receptors in blood pressure regulation. *Am. J. Physiol. Renal Physiol.* 272:F515–F20

40. Friberg P, Sundelin B, Bohman SO, Bobik A, Nilsson H, et al. 1994. Renin-angiotensin system in neonatal rats: induction of a renal abnormality in response to ACE inhibition or angiotensin II antagonism. *Kidney Int.* 45:485–92

41. Tufro-McReddie A, Romano LM, Harris JM, Ferder L, Gomez RA. 1995. Angiotensin II regulates nephrogenesis and renal vascular development. *Am. J. Physiol. Renal Physiol.* 269:F110–F15

42. Spence SG, Allen HL, Cukierski MA, Manson JM, Robertson RT, Eydelloth RS. 1995. Defining the susceptible period of

developmental toxicity for the AT1-selective angiotensin II receptor antagonist losartan in rats. *Teratology* 51:367–82

43. Hashimoto K, Imai K, Yoshimura S, Ohtaki T. 1981. Twelve month studies on the chronic toxicity of captopril in rats. *J. Toxicol. Sci.* 6(Suppl.)2:215–46

44. Matsusaka T, Kon V, Takaya J, Katori H, Chen X, et al. 2000. Dual renin gene targeting by Cre-mediated interchromosomal recombination. *Genomics* 64:127–31

45. Yanai K, Saito T, Kakinuma Y, Kon Y, Hirota K, et al. 2000. Renin-dependent cardiovascular functions and renin-independent blood-brain barrier functions revealed by renin-deficient mice. *J. Biol. Chem.* 275:5–8

46. Miyazaki Y, Tsuchida S, Fogo A, Ichikawa I. 1999. The renal lesions that develop in neonatal mice during angiotensin inhibition mimic obstructive nephropathy. *Kidney Int.* 55:1683–95

47. Constantinou CE. 1974. Renal pelvic pacemaker control of ureteral peristaltic rate. *Am. J. Physiol.* 226:1413–19

48. el-Dahr SS, Gee J, Dipp S, Hanss BG, Vari RC, Chao J. 1993. Upregulation of renin-angiotensin system and downregulation of kallikrein in obstructive nephropathy. *Am. J. Physiol. Renal Physiol.* 264:F874–F81

49. Pimentel JL Jr, Martinez-Maldonado M, Wilcox JN, Wang S, Luo C. 1993. Regulation of renin-angiotensin system in unilateral ureteral obstruction. *Kidney Int.* 44:390–400

50. Pimentel JL Jr, Montero A, Wang S, Yosipiv I, el-Dahr S, Martinez-Maldonado M. 1995. Sequential changes in renal expression of renin-angiotensin system genes in acute unilateral ureteral obstruction. *Kidney Int.* 48:1247–53

51. Pimentel JL Jr, Sundell CL, Wang S, Kopp JB, Montero A, Martinez-Maldonado M. 1995. Role of angiotensin II in the expression and regulation of transforming growth factor-beta in obstructive nephropathy. *Kidney Int.* 48:1233–46

52. Paxton WG, Runge M, Horaist C, Cohen C, Alexander RW, Bernstein KE. 1993. Immunohistochemical localization of rat angiotensin II AT1 receptor. *Am. J. Physiol. Renal Physiol.* 264:F989–F95

53. Gasc JM, Shanmugam S, Sibony M, Corvol P. 1994. Tissue-specific expression of type 1 angiotensin II receptor subtypes. An in situ hybridization study. *Hypertension* 24:531–37

54. Fujinaka H, Miyazaki Y, Matsusaka T, Yoshida H, Fogo AB, et al. 2000. Salutary role for angiotensin in partial urinary tract obstruction. *Kidney Int.* 58:2018–27

55. Constantinou CE, Hrynczuk JR. 1976. Urodynamics of the upper urinary tract. *Invest. Urol.* 14:233–40

56. Kaneto H, Morrissey J, McCracken R, Reyes A, Klahr S. 1994. Enalapril reduces collagen type IV synthesis and expansion of the interstitium in the obstructed rat kidney. *Kidney Int.* 45:1637–47

57. Ishidoya S, Morrissey J, McCracken R, Reyes A, Klahr S. 1995. Angiotensin II receptor antagonist ameliorates renal tubulointerstitial fibrosis caused by unilateral ureteral obstruction. *Kidney Int.* 47:1285–94

58. Morrissey JJ, Klahr S. 1997. Rapid communication. Enalapril decreases nuclear factor kappa B activation in the kidney with ureteral obstruction. *Kidney Int.* 52:926–33

Annu. Rev. Physiol. 2002. 64:563–94

MOLECULAR ASPECTS OF RENAL ANIONIC DRUG TRANSPORT

Frans G. M. Russel, Rosalinde Masereeuw, and Rémon A. M. H. van Aubel

Department of Pharmacology and Toxicology, Nijmegen Center for Molecular Life Sciences, University Medical Center Nijmegen, The Netherlands; e-mail: F.Russel@ncmls.kun.nl

Key Words   drug excretion, organic anion transporter, multidrug resistance protein, organic anion transporting polypeptide

■ **Abstract**   Multiple organic anion transporters in the proximal tubule of the kidney are involved in the secretion of drugs, toxic compounds, and their metabolites. Many of these compounds are potentially hazardous on accumulation, and it is therefore not surprising that the proximal tubule is also an important target for toxicity. In the past few years, considerable progress has been made in the cloning of these transporters and their functional characterization following heterologous expression. Members of the organic anion transporter (OAT), organic anion transporting polypeptide (OATP), multidrug resistance protein (MRP), sodium-phosphate transporter (NPT), and peptide transporter (PEPT) families have been identified in the kidney. In this review, we summarize our current knowledge on their localization, molecular and functional characteristics, and substrate and inhibitor specificity. A major challenge for the future will be to understand how these transporters work in concert to accomplish the renal secretion of specific anionic substrates.

INTRODUCTION

The study of the mechanisms by which the kidney excretes foreign anions has traditionally belonged to the field of renal physiology. It was only many years after Marshall and coworkers in 1923 obtained the first conclusive proof of active secretion with phenol red (1) that pharmacologists and toxicologists recognized the clinical importance of renal organic anion transport. The practical significance of this secretory system became clear during the Second World War, when penicillin was first introduced. The antibiotic was in short supply, and its rapid renal excretion was thus an important problem. This urged Beyer and associates to search for an organic anion that would compete with the tubular secretion of penicillin, thereby preventing its rapid loss from the body and prolonging its antibiotic activity. The discovery in 1951 of the benzoic acid derivative, probenecid, was the successful result of this quest (2). As the years went by, probenecid became obsolete

0066-4278/02/0315-0563$14.00

as a penicillin-sparing drug because large-scale production procedures and new semisynthetic congeners became available. In 1988, probenecid reappeared on the scene, but now in a bad light, in sports. During the Tour de France of that year the Spanish cyclist, Pedro Delgado, was caught having probenecid in his urine. The intention seemed obvious, just as probenecid inhibits the renal excretion of penicillin, it also reduces the transport of metabolites of the forbidden anabolic steroids into urine. Thanks to this cunning maneuver, Delgado managed to escape from a positive doping test, and he was allowed to wear the yellow championship jersey up to the podium in Paris, as the International Cycling Union had failed to put probenecid as a masking agent on the doping list. Recently, a new application for probenecid emerged as a nephroprotectant in the therapy with the antiviral drug cidofovir. Patients treated with cidofovir may develop nephrotoxicity, probably as a result of extensive accumulation in tubular cells associated with the process of active secretion. Probenecid is coadministered with cidofovir to inhibit tubular organic anion transport, thus reducing its potential for nephrotoxicity (3).

These few examples illustrate the importance of the renal organic anion transport system in pharmacology and toxicology. Its primary function appears to be the active elimination of endogenous and exogenous substances and their metabolites that are not easily degraded by the body and that might be toxic if allowed to accumulate. In accordance with this vast task, a complex array of membrane transport proteins exists that accepts a wide variety of organic anions, having only in common a negative charge and a hydrophobic backbone. Most of these transporters are confined to the proximal tubule, together with the transport proteins belonging to the organic cation secretory system. Although these cation transporters are distinct proteins with separate specificities, at the molecular level they show many structural similarities with organic anion transporters, and they even share a number of substrates (4, 5). The effectivity and toxicity of many clinically important drugs and harmful environmental compounds in the body is determined by their interaction with the renal organic anion system. Examples include antibiotics, chemotherapeutics, diuretics, nonsteroidal anti-inflammatory drugs (NSAIDs), angiotensin-converting enzyme inhibitors, radiocontrast agents, cytostatics, drug metabolites (especially glutathione, glucuronide, glycine, sulfate, and acetate conjugates), and toxicants and their metabolites, such as mycotoxins, herbicides, plasticizers, glutathione S-conjugates of polyhaloalkanes, polyhaloalkenes, hydroquinones, aminophenols. As the proximal tubule is the primary site of excretion, this nephron segment is often an important target for toxicity because of its rich transport function and concentrative capacity.

This review aims at updating the molecular aspects of organic anion transport in the kidney, particularly from a pharmacological and toxicological point of view. Occasionally, some information is derived from other tissues, but emphasis is placed on the molecular and functional characteristics of cloned renal organic anion transporters that contribute to the excretion of xenobiotics and their metabolites. Several reviews covering part of the present theme have been published in the last few years (5–10). During this period, remarkable progress has been made in the molecular identification of the transport proteins involved in tubular organic anion secretion. Clearly, the list is not yet complete, and there is still much to be learned about the functional characteristics of the various transporters, but our understanding of their role in tubular drug secretion is rapidly increasing. Our current knowledge on renal organic anion transporters is summarized in Figure 1 (see color insert) and Tables 1 and 2 and is discussed below with a special emphasis on nephron and membrane localization, regulation, transport properties, and substrate and inhibitor specificity.

FUNCTIONAL MODEL OF RENAL ORGANIC ANION TRANSPORT

The systems involved in organic anion secretion can be functionally subdivided in the well-characterized, sodium-dependent p-aminohippurate (PAH) system (OAT1) and a recently discovered sodium-independent system (11–13). Both systems mediate two membrane translocation steps arranged in series: uptake from blood across the basolateral membrane of renal epithelial cells followed by efflux into urine across the apical membrane. Although transported through the cytoplasm, substrates for both systems are subject to sequestration within intracellular compartments (Figure 1).

Basolateral Transport

Because cells maintain an inside negative potential, uptake across the basolateral membrane takes place against an electrochemical gradient and requires energy input. Uptake of PAH appears to be a tertiary active process coupled indirectly to the Na^+ gradient (13). This gradient, which is maintained by the Na^+/K^+ ATPase, drives Na^+/dicarboxylate cotransport into the cell and enables uptake of PAH in exchange for a dicarboxylate ion. Based on similarities with transport characteristics found in membrane vesicles, SDCT2/NaDC3 (Slc13A3) has been proposed as the basolateral Na^+/dicarboxylate cotransporter (14). α-ketoglutarate is the most abundant potential dicarboxylate counterion within the proximal tubular cell; its outwardly directed gradient is sustained by the activity of the Na^+/dicarboxylate exchanger and intracellular metabolic generation. Apart from the classical PAH transporter, an additional uptake system has been characterized by using the bulky organic anion fluorescein-methotrexate as a substrate (12). Uptake of fluorescein-methotrexate is independent of Na^+ and is not inhibited by PAH or the dicarboxylate ion glutarate. The molecular identity of this transporter is still elusive, but a possible candidate may be OAT3.

Apical Transport

Efflux across the apical membrane of organic anions is thought to occur by low-affinity anion exchange and/or facilitated diffusion, and a Na^+-independent

TABLE 1 Molecular characteristics of renal organic anion transporters[a,b]

Gene symbol	Gene product		Chromosome localization	mRNA (kb)	M_r (kDa)	Nephron distribution	Membrane localization	Carboxy-terminus	Accession numbers
SLC22A6	OAT1-1	human	11q13.1-13.2	2.4	62	PT	BLM	NGL	NM_004790
	OAT1-2				60	PT	BLM		AB009698
	OAT1-3			56	?	?	?		AJ251529
	OAT1-4				57	?	?		AJ271205
Slc22a6	Oat1	mouse/rat	19		PT (S2)	BLM	?		U52842/ AB004559
SLC22A7	OAT2	human	6p21.2-21.1	?	?	?	?	VQN	NM_006672
Slc22a7	Oat2	rat		2.1	60			EDV	L27651
SLC22A8	OAT3	human	11q11.7	2.2	62	PT	BLM	GSS	NM_004254
Slc22a8	Oat3	mouse/rat	19					GGP	AF078869/ AB017446
	OAT4	human	11cen-q12.3	2.7	60	?	?	TSL[c]	NM_018484
SLC17A1	NPT1	human	6p23-p21.3		51	PT	BBM	TRL[c]	NM_005074
Slc17a1	Npt1	mouse							NM_009198
ABCC1	MRP1	human	16p13.1	6.5	171	CCD	BLM	GLV	L05628
Abcc1	Mrp1	mouse							NM_008576
ABCC2	MRP2	human	10q24	6.5	175	PT	BBM	TKF[c]	XM_000392
Abcc2	Mrp2	mouse/rat	19						NM_013806/ L49379
ABCC3	MRP3	human	17q21.3	6.5	170	CCD	BLM	GLA	XM_003786
Abcc3	Mrp3	rat							AF072816
ABCC4	MRP4	human	13q31-32	6.0	150	PT	BBM	TAL[c]	AF071202
Abcc4	Mrp4	rat							AF376781

Gene	Protein	Species	Chromosome	kb	kDa	Localization	Membrane	PDZ motif	Accession
ABCC5	MRP5	human	3q27	6.0	160	?	BLM	VKG	U83661
Abcc5	Mrp5	mouse/rat							AB019003/ AB020209
ABCC6	MRP6	human	16p13.1	6.0	165	PT	BLM	GLV	NM_001171
Abcc6	Mrp6	mouse/rat						GLA	NM_018795/ U73038
Slc21a1	Oatp1	rat	?	2.7	74	PT (S3)	BBM	TKL[c]	L19031
		mouse							AB031813
Slc21a4	Oat-k1	rat	?	3.0	74	PT (S3)	BBM	TKL[c]	D79981
	Oat-k2	rat	?	2.5	55	PT+CCD	BBM	TKL[c]	AB012662
Slc21a7	Oatp3	rat	?	2.8	74	?	?	TRL[c]	AF041105
		mouse	6						AF240694
SLC21A11	OATP-D	human	15q26.1	?	76	?	?	SVL[c]	NM_013272
	Pgt2	rat							AF239219
SLC21A12	OATP-E	human	20	?	77	?	?	SSV[c]	NM_016354
	Oatp-E	rat							AF239262
Slc21a13	Oatp5	rat	?	?	74	?	?	TKL[c]	AF053317
		mouse							AF213260
SLC15A1	PEPT1	human	13q33-34	3.3	78	PT (S1)	BBM	KQM	NM_005073
Slc15a1	Pept1	mouse/rat						TNM	AF205540/ D50664
SLC15A2	PEPT2	human	3q13.3-q21	4.2	78	PT (S3)	BBM	TKL[c]	NM_021082
Slc15a2	Pept2	mouse/rat						TRL[c]	AF257711/ NM_031672

[a]Abbreviations: BLM: basolateral membrane; BBM: brush-border membrane; PT: proximal tubule; CCD: cortical collecting duct.
[b]Only differences for rat/mouse orthologs compared with human are indicated.
[c]Matches the PDZ-motif consensus sequence T/S-X-V/I/L.

TABLE 2   Substrate specificity of renal organic anion transporters[a]

Name		Transport mechanism	Substrates (K_m)[b]	Inhibitors (K_i)[b]	References
OAT1	human	OA/dicarboxylate antiport	PAH (5 μM); PMEA (30 μM); cidofovir (46 μM); PMEG; PMEDAP	Probenecid (12 μM); furosemide; indomethacin; urate; α-KG; glutarate; betamipron (24 μM); cilastatin (1.5 mM)	(25, 27, 28, 155)
Oat1	rat		PAH (14/70 μM); salicylate; MTX; cAMP; acetylsalicylate; indomethacin; folate; cGMP; prostaglandin E_2; urate; α-KG; ochratoxin A (2 μM); cephaloridine; benzylpenicillin; AZT (68 μM); acyclovir (242 μM); cidofovir (238 μM); PMEA (270 μM); zalcitabine; lamivudine; stavudine; trifluridine; PMEG; PMEDAP	Probenecid; naproxen (2 μM); ibuprofen (3.5 μM); salicylurate (11 μM); piroxicam (52 μM); salicylate (341 μM); acetylsalicylate (428 μM); phenacetin (488 μM); paracetamol (2 mM); benzylpenicillin (1.7 mM); carbenicillin (500 μM); cephaloridine (2.3 mM); cephalothin (290 μM); cefazolin (450 μM); cephalexin (2.3 mM); furosemide; indomethacin (10 μM); urate; α-KG; glutarate; MTX; prostaglandin E_2; cAMP; cGMP	(18, 19, 21–25, 156)
OAT2	human	?	PAH; MTX; cAMP; α-KG		(30)
Oat2	rat	?	PAH; MTX; prostaglandin E_2; α-KG (18); salicylate (90 μM); acetylsalicylate	BSP; ketoprofen; rifampicin; bumetanide; enalapril; cefoperazone; cholate	(29)
OAT3	human	?	PAH (87 μM); MTX (11 μM); cimetidine (57 μM); estrone-sulfate (3 μM); DHEA-sulfate; $E_2$17βG; glutarate; prostaglandin E_2; ochratoxin A; cAMP; salicylate; urate	Probenecid (9 μM); cholate; BSP; betamipron (48 μM); cilastatin (231 μM); diclofenac; ibuprofen; indomethacin; bumetanide; furosemide; benzylpenicillin; corticosterone; quinidine; tetraethylammonium	(31, 155)
Oat3	rat	?	PAH (65 μM); ochratoxin A (0.74 μM); estrone-sulfate (2.3 μM); cimetidine	Probenecid; BSP; indocyanine green; bumetanide; piroxicam; furosemide; AZT; DIDS; melatonin	(32)
OAT4	human	?	PAH; ochratoxin A; DHEA-sulfate (0.6 μM); estrone-sulfate (1 μM)	Probenecid; BSP; indomethacin; ibuprofen; diclofenac; furosemide; bumetanide; corticosterone	(33)

			Substrates	Inhibitors	
NPT1	human	?	PAH (3 mM); faropenem (1 mM); urate; benzylpenicillin (0.5 mM); indomethacin; E217βG	Probenecid; salicylate; DIDS; indomethacin; ampicillin; furosemide	(36)
Npt1	mouse		Benzylpenicillin; faropenem; foscarnet; mevalonate	Probenecid; ampicillin; cephalexin; cephradine; cephaloridine; cyclacillin; apalcillin; cloxacillin; dicloxacillin; nafcillin; cefixime; cefoperazone; cefpiramide; ceftizoxime; cephalothin	(38)
MRP1	human	primary active	LTC4 (0.1 μM); E217βG (1.5 μM); DNP-SG (3.6 μM); bilirubin-glucuronide; GSSG (93 μM); AFB1-SG (0.2 μM); PGA1-SG; PGA2-SG (1 μM); etoposide-glucuronide; S-(ethacrynic acid)-glutathione; GSH; PAH (1 mM); MTX (1 mM)	Probenecid; MK571 (0.6 μM); cyclosporin A (5 μM); PSC833 (27 μM); S-(decyl)-glutathione (0.7 μM); sulfinpyrazone; V-104; indomethacin	(44, 61, 157)
MRP2	human	primary active	LTC4 (1 μM); E217βG (7 μM); DNP-SG; PGA1-SG; bilirubin-glucuronide; S-(ethacrynic acid)-glutathione; GSH; MTX (1 mM); PAH (1 mM); ochratoxin A	MK571; indomethacin; furosemide; benzbromarone	(44, 61)
Mrp2	rat		LTC4 (1 μM); E217βG (7 μM); LTD4(1.5 μM); NAc-LTE4 (5.2 μM); DNP-SG (0.18 μM); bilirubin-glucuronide; GSSG (111 μM); p-nitrophenyl-glucuronide (20 μM); α-naphthyl-β-D-glucuronide; E3040-glucuronide; GSH; MTX (100 μM); folate; pravastatin (170 μM); BSP (31 μM); Fluo-3 (4 μM); BQ123 (100 μM); temocaprilat (100 μM); cefpiramide; ceftriaxone; endothelin-I	Probenecid; MK571; cyclosporin A; GSH (20 mM)	(44)
MRP3	human	primary active	LTC4 (5 μM); DNP-SG (6 μM); E217βG (26 μM); MTX (1 mM)		(70, 72)

(*Continued*)

TABLE 2 (*Continued*)

Name		Transport mechanism	Substrates (K_m)[b]	Inhibitors (K_i)[b]	References
Mrp3	rat		MTX; $E_217\beta G$ (70 μM); E3040-glucuronide	E3040-glucuronide; α-naphthyl-β-D-glucuronide; MTX	(158)
MRP4	human	primary active	PMEA; PMEG; AZTMP; MTX (1 mM); $E_217\beta G$; cAMP; cGMP	Probenecid; α-naphthyl-β-D-glucuronide; p-nitrophenyl-glucuronide; DNP-SG; NAc-DNP-Cys; dipyridamole	(73a, 74, 76a)
MRP5	human	primary active	DNP-SG; PMEA; 6-MP; cAMP (380 μM); cGMP (2 μM); GSH; CMFDA; BCECF; FDA	Probenecid; sulfinpyrazone; FDA; 8-bromo-cGMP; zaprinast; trequinsin; sildenafil	(81–83)
Mrp6	rat	primary active	BQ123		(87)
Oatp1	rat	OA/GSH antiport	LTC_4 (0.27 μM); DNP-SG (408 μM); $E_217\beta G$ (3 μM); estrone-sulfate (4.5 μM); $E_217\beta G$ (27 μM); taurocholate (32 μM); BSP (1.5 μM); aldosterone (15 nM); cortisol (13 nM); ouabain (17 mM); ochratoxin A (17 μM); thyroxine; triiodo-L-thyronine; enalapril (214 μM); temocrapilat (47 μM); DHEA-sulfate; prostaglandin E_2	Probenecid; furosemide; corticosterone (5 μM); corticosterone 21-sulfate (2 μM); aldosterone; hydrocortisone; androsterone (11 μM); androsterone-3-acetate (76 μM); androsterone-3-sulfate (2.4 μM); testosterone-17β-D-glucuronide (4 μM)	(96, 159–164)
Oatp3	rat	?	Thyroxine (5 μM); triiodo-L-thyroxine (7 μM); bile acids		(103, 104)
Oat-k1	rat	?	MTX (1 μM); folate	Probenecid; MTX (2 μM); BSP; PAH; furosemide; valproate; DIDS; taurocholate; ibuprofen; flufenamate; phenylbutzone; indomethacin (1 mM); ketoprofen (2 mM); aminopterin (0.5 μM); 5-methyl-tetrahydrofolic acid (1 μM); leucovorin (8 μM); folate (14 μM)	(99–101, 165)

			Substrates	Inhibitors	Ref.
Oat-k2	rat	?	MTX; prostaglandin E2; folate	Probenecid; MTX; BSP; PAH; furosemide; valproate; DIDS; taurocholate; levofloxacin; indomethacin; testosteron; dexamethasone; 17β-estradiol; ouabain	(102)
OATP-D	human	?	Estrone-sulfate; prostaglandin E2; benzylpenicillin		(95)
OATP-E	human	?	Estrone-sulfate; prostaglandin E2; benzylpenicillin; triiodo-L-thyroxine (1 μM); thyronine		(95, 166)
Oatp-E	rat		Triiodo-L-thyroxine		(166)
PEPT1 PepT1 PepT1	human rat rabbit	peptide/H$^+$ symport	Glycylsarcosine (1 mM); ALA (280 μM); ceftibuten (300 μM); cyclacillin (500 μM); cephalexin; cefixime; cefadroxil; valacyclovir (5 mM); Val-AZT; L-dopa-L-Phe; *formyl*-Met-Leu-Phe	Ampicillin (50 mM); amoxicillin (13 mM); cyclacillin (170 μM); cephalexin (5 mM); cefadroxil (2 mM); cephradine (9 mM); cefdinir (12 mM); ceftibuten (0.6 mM); cefixime (7 mM); bestatin (500 μM); enalapril (4 mM); captopril (9 mM)	(107–115, 167)
PEPT2 PepT2 PepT2	human rat rabbit	peptide/H$^+$ symport	Glycylsarcosine (110 μM); ALA (230 μM); cephalexin; bestatin; valacyclovir	Ampicillin (670 μM); amoxicillin (180 μM); cyclacillin (27 μM); cephalexin (50 μM); cefadroxil (3 μM); cephradine (47 μM); cefdinir (20 mM); ceftibuten (1 mM); cefixime (12 mM); bestatin (20 μM)	(107, 110, 113, 114, 167)

[a] Abbreviations: PAH: *p*-aminohippurate; OA: organic anion; α-KG: α-ketoglutarate; MTX: methotrexate; BSP: bromosulfophthalein; DIDS: 4,4′-diisothiocyanostilbene-2,2′-disulfonic acid; LTC$_4$/LTD$_4$: leukotriene C$_4$/D$_4$; E$_2$17βG: estradiol-17β-D-glucuronide; DNP-SG: S-(dinitrophenyl)-glutathione; GSSG: oxidized glutathione; GSH: reduced glutathione; AFB$_1$-SG, S-(aflatoxin B$_1$)-glutathione; E3040-glucuronide: 6-hydroxy-5,7-dimethyl-2-methylamino-4-(3-pyridylmethylbenzothiazole glucuronide; PMEA/PMEG/PMEDAP: 9-(2-phosphonylmethoxyethyl)adenine/-guanine/-diaminopurine; AZTMP: azidothymidine monophosphate; CMFDA, 5-chloromethylfluorescein; FDA: fluorescein-diacetate; BCECF: 2′,7′-bis-(2-carboxyethyl)-5 (and-6)-carboxyfluorescein acetoxymethyl ester; AZT: azidothymidine; ALA: delta-aminolevulinic acid; PGA$_2$-SG: S-(prostaglandin A$_2$)-glutathione; BQ123: (cyclo [Trp-Asp-Pro-Val-Leu]).

[b] Substrates and inhibitors mentioned for some transporters are indicative but not complete.

ATP-driven system (7). In isolated brush-border membrane vesicles, small organic anions, such as PAH and fluorescein, are transported via a potential difference-driven facilitated transporter. In addition, an anion exchange mechanism has been identified in some species, such as dog, rat, and human, but not in rabbit. It is not clear yet whether one or more of the recently cloned apical organic anion transporters represent these efflux systems at the molecular level. Possible candidates are OATP1, OAT-K1/K2, OAT4, and NPT1. Studies with intact killifish proximal tubules have indicated that next to Na^+-dependent efflux of small organic anions, bulky organic anions, such as fluorescein-methotrexate and lucifer yellow, are excreted via an energy-dependent efflux mechanism (11, 12, 15). This system is insensitive to depletion of Na^+ but inhibited by leukotriene C_4 and S-(dinitrophenyl)-glutathione. The specificity and localization of this carrier closely resembles the apical ATP-dependent anionic conjugate transporter, multidrug resistance protein 2 (MRP2). Although MRP2 primarily transports large organic anions, PAH appears to be a low-affinity substrate in membrane vesicles from cells overexpressing rabbit or human MRP2 (16, 17). MRP2 may be a third low-affinity pathway mediating urinary PAH excretion. Apart from efflux mechanisms, the brush-border membrane also contains transport systems for reabsorption of compounds from primary urine. Small anionic peptides and peptide-like drugs are actively taken up via peptide cotransporters (PEPT1/2) coupled to the H^+-gradient.

MOLECULAR AND FUNCTIONAL CHARACTERISTICS OF CLONED ORGANIC ANION TRANSPORTERS

Organic Anion Transporters OAT1–OAT4, Solute Carrier Family 22A

OAT1 (*SLC22A6*) represents the classical PAH/dicarboxylate exchanger located at the proximal tubule basolateral membrane (18–21). Well before its molecular identification, the transport properties of OAT1 were predicted using basolateral membrane vesicles and intact proximal tubules (7). OAT1 mediates high-affinity uptake of PAH by a sodium-dependent tertiary active process. Functional expression of a rat *Oat1* cDNA in *Xenopus* oocytes has identified a wide variety of endogenous organic anions, such as prostaglandins, cyclic nucleotides, folates (19), and xenobiotics that include β-lactam antibiotics (22), NSAIDs (23), and antiviral drugs (24, 25). The substrate specificity of human OAT1 is far more complex to interpret, as four splice variants of the human OAT1 gene (*OAT1-1*, *-2*, *-3* and *-4*) have been identified (26). OAT1-1 (1689 bp open reading frame; 563 amino acids) and *OAT1-2* (deletion of amino acids 523–535) both mediate PAH/dicarboxylate exchange and exhibit a comparable affinity for PAH as rat Oat1 (27, 28). In contrast to rat Oat1, human OAT1-2 does not transport methotrexate, prostaglandin E_2, or urate (28). On the other hand, human OAT1-2 exhibits a five- and ninefold higher affinity for the antiviral drugs cidofovir and adefovir (or PMEA), respectively (25).

OAT1-3 and OAT1-4 completely lack putative transmembrane-spanning domains 11 and part of 12 and do not seem to transport PAH (26).

Several homologs of OAT1 have been identified and named: OAT2 (*SLC22A7*), OAT3 (*SLC22A8*), and OAT4, which have amino acid identity to OAT1 of 40, 51, and 44%, respectively. Human *OAT2* and rat *Oat2* mRNA is expressed in kidney, but its expression level in liver is higher (29, 30). Human OAT2 and rat Oat2 exhibit a substrate specificity comparable to rat Oat1 (29, 30). Expression of human *OAT3* mRNA is exclusively found in kidney (30, 31), whereas in rats, *Oat3* transcript is most abundantly expressed in liver and to a lesser extent in kidney and brain (32). Similar to OAT1, human OAT3 is expressed at the proximal tubule basolateral membrane (31). Distribution of OAT1 along the tubule seems limited to the S2 segment (20, 27), whereas OAT3 shows immunoreactivity in all segments (S1 > S2 = S3) (31). OAT3 exhibits a 20-fold lower affinity for PAH compared with OAT1 but exhibits a high affinity for anionic conjugates such as estrone sulfate, dihydroepiandrosterone (DHEA) sulfate, and estradiol-17β-D-glucuronide ($E_2$17βG) (31, 32). OAT3 also mediates uptake of cimetidine, and both tetraethylammonium and quinidine were able to inhibit OAT3-mediated estrone sulfate uptake. This raises the interesting possibility that OAT3 may also recognize cations as substrates. Human *OAT4* mRNA is highly expressed in kidney and to a lesser extent in placenta (30, 33). *Xenopus* oocytes expressing OAT4 exhibited only a marginally increased PAH uptake above background (33). However, OAT4 transports estrone sulfate and DHEA sulfate with high affinity similar to OAT3 (33). Although the localization of OAT4 is unknown, it is speculated that its expression is confined to the apical membrane (9). The major difference between OAT1, on the one hand, and OAT2, OAT3, and OAT4, on the other hand, is that the latter three mediate uptake of organic anions independent of the Na^+-gradient. This suggests that an endogenous counter ion different from dicarboxylates drives uphill transport into the proximal tubule. Possible candidates may include reduced glutathione (GSH) or sulfate.

Sodium/Phosphate Cotransporter Type I (NPT1), Solute Carrier Family 17A

NPT1 (*SLC17A1*), or NaPi-I cotransporter, belongs to a family of several sodium-phosphate transporters (NPTs) expressed at the renal proximal brush-border membrane (34). Whereas the NaPi-IIa transport characteristics correspond well to that observed in the brush-border membrane, the contribution of NPT1 to phosphate reabsorption appears to be minimal (34). Expression of rabbit Npt1 in *Xenopus* oocytes generated a chloride conductance sensitive to chloride channel blockers, such as DIDS, and organic anions, such as benzylpenicillin and probenecid (35). Furthermore, mouse Npt1 and human NPT1 were shown to mediate transport of various organic anions in a chloride-dependent fashion (36–38). Human NPT1 exhibits an affinity for PAH, which corresponds to previous reports using brush-border membrane vesicles and suggests that NPT1 may represent either the classical voltage-dependent PAH transporter or the anion exchanger (36). Rabbit Npt1 does not seem to transport PAH, suggesting that species differences may be involved.

Multidrug Resistance Proteins MRP1–MRP6, ATP-Binding Cassette Subfamily ABCC

MRP1 (*ABCC1*) is the first member identified of a group of multidrug resistance proteins now comprising a total of nine members. Besides MRP1–MRP9, the ABCC subfamily within the ATP-binding cassette (ABC) transporter superfamily also includes the cystic fibrosis transmembrane conductance regulator (CFTR/*ABCC7*) and the sulphonyl urea receptors SUR1 (*ABCC8*) and SUR2 (*ABCC9*) (39). MRP1 was initially identified as a novel drug transporter mediating resistance to anti-cancer agents in drug-selected cancer cell lines, which did not overexpress the drug transporter P-glycoprotein (*ABCB1*) (40). Expression of *MRP1* mRNA is found in numerous tissues, including kidney (41). Immunolocalization in mouse kidney has indicated expression of MRP1 at the basolateral membrane of cells of Henle's loop and in the cortical collecting duct (42, 43). MRP1 mediates ATP-dependent transport of a variety of conjugates to glucuronate, sulfate, and GSH and transports GSH itself (44–46). ATP-dependent transport of certain unconjugated drugs by MRP1 requires GSH as a cofactor (47–49), and MRP1-mediated drug resistance is reversed by an inhibitor of GSH synthesis (50). Mice homozygous for a disrupted *Mrp1* gene are hypersensitive to drugs and display an increased GSH tissue level (42, 51, 52). The localization in the collecting duct suggests that MRP1 may play a role in protecting this nephron part against exposure to drugs and toxic compounds, which are locally concentrated to high levels as a result of hormone-regulated water reabsorption. Furthermore, MRP1 might be involved in controlling the redox state of collecting duct cells.

MRP2 (*ABCC2*, 48% amino acid identity to MRP1) is the ATP-dependent anionic conjugate transporter of the liver canalicular (apical) membrane, previously described as cMOAT (53), and is the major biliary excretion route of a broad variety of organic anions (44). MRP2 is also expressed at the brush-border membrane of renal proximal tubules and small intestinal villi (54–56). The Wistar TR⁻ and Sprague Dawley EHBR strain do not express Mrp2 as a result of a mutation introducing a premature stop codon (57, 58). These rats are hyperbilirubinemic as a result of their inability to excrete anionic conjugates, such as bilirubin glucuronides (via Mrp2), into bile (53). The substrate specificity of rat Mrp2 has been characterized in comparative studies using wild-type and mutant rats (44). Substrates have also been identified by functional expression of human, rat, and rabbit *Mrp2* cDNAs (59–63). Similar to MRP1, human MRP2 is able to confer resistance to some anti-cancer drugs (59, 60, 64), requires GSH to transport these compounds (45, 59, 65), and transports GSH itself (45, 46). In contrast to MRP1, MRP2 confers resistance to cisplatin, which presumably is able to form a conjugate with GSH (66). Kidney perfusion studies and transport studies with isolated proximal tubules have indicated that the absence of Mrp2 from TR-/EHBR kidneys does not affect renal excretion of certain organic anions (7). This suggests the involvement of additional transporters compensating for the loss of Mrp2. However, under certain pathological conditions or during continuous drug exposure, MRP2 may serve an important protective role in renal proximal tubules. For instance, the expression level of Mrp2 in the kidney, but not the liver, is elevated up to eight days after administration of cisplatin (67). Furthermore, subtotal nephrectomy increases *Mrp2* mRNA and Mrp2 protein in remnant kidney up to 200% (68).

Human MRP3 (*ABCC3*) is most closely related to MRP1 (58% amino acid identity) and is expressed in the liver, kidney, small intestine, and colon (69). In human liver, MRP3 is localized to the basolateral membrane of cholangiocytes and hepatocytes surrounding the portal tracts (70, 71). In kidney, MRP3 expression is found at the basolateral membrane of distal cells (G. Scheffer, Free Univ. Amsterdam, personal communication). The substrate specificity of human MRP3 resembles that of MRP1 and MRP2, but its affinity for the anionic conjugates tested so far seems lower, and MRP3 does not transport GSH (70, 72). Furthermore, MRP3 exhibits a narrow drug resistance profile as it confers resistance to etoposide, teniposide, and methotrexate but not to many other anti-cancer drugs and heavy metals (70). The expression level of MRP3/Mrp3, which is low in normal human and rat liver, is increased in EHBR liver and bile duct-ligated (cholestatic) human/rat liver (44, 70, 73). Whether the expression level of MRP3 in the kidney is subjected to regulation under pathophysiological conditions remains to be investigated.

Expression of *MRP4* (*ABCC4*) mRNA has been found in a few tissues including kidney (69). We have recently found that MRP4 is localized to the brush-border membrane of renal proximal tubules. Similar to MRP1–MRP3, MRP4 mediates probenecid-sensitive ATP-dependent transport of methotrexate and $E_217\beta G$ but in contrast does not seem to transport leukotriene C_4 (73a). Based on inhibition experiments, GS-DNP, NAc-DNP-Cys, and the glucuronate conjugates of α-naphthol and *p*-nitrophenol may also be MRP4 substrates (73a). MRP4 mediates resistance to the antiviral drugs PMEA (or adefovir), PMEG, and AZT (74, 75). Pharmacokinetic studies in humans have indicated that the kidney is the main excretory organ for antiviral drugs such as PMEA (76). Furthermore, these studies have shown that the excretory pathway rather than the uptake pathway (presumably mediated by OAT1; 25) is rate limiting, thus explaining the observed nephrotoxic side-effects of antiviral drug treatment. MRP4 may play a crucial role in the renal elimination of antiviral drugs, and it will be of great interest to see how *mrp4*(−/−) knockout mice respond to the administration of such drugs. MRP4 also mediates ATP-dependent transport of cAMP and cGMP (73a, 76a) and might represent the previously identified probenecid-sensitive efflux pump for cyclic nucleotides in renal epithelial cells (77–79). Urinary cAMP and cGMP are regulators of phosphate and sodium transport, respectively (78, 80). Whether MRP4 contributes to renal fluid homeostasis remains to be investigated.

Expression of MRP5 (*ABCC5*) has been detected in numerous tissues, including kidney, but immunohistochemistry has yet to be performed (69, 81). Similar to human MRP1, MRP5 is targeted to the basolateral membrane upon expression in MDCKII cells and transports DNP-SG and GSH (82). MRP5 does not confer resistance to many of the anti-cancer agents previously identified for MRP1 (81). In contrast, resistance in two independent MRP5-transfected HEK293 clones was found for PMEA and the anti-cancer drugs 6-mercaptopurine and thioguanine (82). One clone also displayed resistance to etoposide and tenoposide (82). Recently, MRP5 was shown to mediate ATP-dependent cAMP and cGMP transport (83). The transporter might represent the basolateral cyclic nucleotide export pump previously identified in airway and urinary bladder epithelial cells (84–86).

MRP6 (*ABCC6*) is an unusual member of the ABCC subfamily. Rat Mrp6 does not transport any of the MRP-prototypic substrates, such as DNP-SG, LTC_4, or $E_217\beta G$ (87). The only substrate identified so far is the endothelin antagonist BQ123 (87). RNase protection assay has shown that MRP6 is exclusively expressed in kidney and liver (88). Immunolocalization has detected MRP6 at the basolateral membrane of renal proximal tubules and hepatocytes (87; G. Scheffer, Free Univ. Amsterdam, personal communication). Recently, mutations have been identified in the *MRP6* gene of patients with the connective tissue disorder Pseudoxanthoma elasticum (PXE) (89, 90). The PXE phenotype is characterized by alteration (calcification) of elastic structures in skin, eye, and the cardiovascular system, but expression of MRP6 in the PXE-affected tissues has not been found (90). It has therefore been suggested that defective function of MRP6 in kidney and liver is the source for the observed phenotype in the PXE tissues. Whether MRP6 functions as an ATP-dependent drug transporter or perhaps a channel or channel regulator, such as CFTR and SUR1/SUR2, remains to be investigated. MRP7 (*ABCC10*), MRP8 (*ABCC11*), and MRP9 (*ABCC12*) have recently been identified but appear not to be expressed in the kidney (39).

Organic Anion Transporting Polypeptides, Solute Carrier Family SLC22A

Rat Oatp1 (*Slc21a1*) is the first member characterized of the fast-growing SLC22A subfamily, which currently consists of 14 organic anion polypeptides (OATP) in human and rat/mouse (91, 92). Recently, analysis of a full-length mouse cDNA collection resulted in the discovery of at least three unknown Oatps, which may belong to a new branch of the subfamily (93). Expression of *Oatp1* mRNA is found in various tissues, including kidney and liver (91, 94). A mouse ortholog, with an amino acid identity of 80% to rat Oatp1, has also been cloned. Based on differences in their tissue distribution and low amino acid homology, none of the cloned human OATPs appears to represent orthologs of rat/mouse Oatp1 (92, 95). Expression of Oatp1 in rat kidney is found at the brush-border membrane of proximal tubules, whereas in liver, expression is found at the hepatocyte basolateral membrane (94). The range of Oatp1 substrates is extremely broad and contains negatively, positively, and uncharged endogenous compounds and xenobiotics (Table 2). Its transport mechanism is unclear; however, substrates can be exchanged against either GSH or HCO_3^- (96, 97). This raises the possibility that liver Oatp1 mediates uptake of compounds into the hepatocyte, whereas renal Oatp1 mediates reabsorption into the proximal tubule.

Rat Oat-k1 (*Slc21a4*) shows highest identity to rat and mouse Oatp1 (72%) and is localized to the renal proximal tubule brush-border membrane (98). The Oat-k1 substrate specificity is narrow and includes only methotrexate and folate (99). Surprisingly, a large number of compounds act as inhibitors but are not transported by Oat-k1 (100). Methotrexate efflux from MDCK cells expressing Oat-k1 was shown to be enhanced by an inwardly directed gradient of folic acid derivatives but not by taurocholate, uric acid, or GSH (101). Folic acid derivatives are excreted by the kidney via glomerular filtration and, once present in the pre-urine, enhance urinary excretion of methotrexate, thus contributing to the so-called folinic acid rescue. Recently, a smaller Oat-k1 homolog, Oat-k2, was isolated (102). The *Oat-k2* cDNA is almost identical to the 3′ end of the *Oat-k1* cDNA, and the 5′ UTR of the *Oat-k2* cDNA still matches the Oat-k1 coding sequence up to 180 nucleotides. Northern blot analysis to determine *Oat-k2* tissue distribution has been conducted with a probe identical to the corresponding *Oat-k1* sequence but did not reveal two messengers (102). Furthermore, immunoblot analysis of rat proximal tubule membrane vesicles using an antibody that should detect both Oat-k1 and the smaller 19-kDa Oat-k2 revealed only one gene product (98). Therefore, it seems unlikely that Oat-k2 represents a novel family homolog. Alignment of all rat *Slc21a* family members identified so far has indicated that Oat-k1 lacks a conserved cysteine residue. In this respect, Oat-k2 transports taurocholate and prostaglandin E_2, which are not substrates for Oat-k1 (102). Recently, our laboratory has isolated an *Oat-k1* cDNA (K. Kooiman, R. Masereeuw & F. Russel, unpublished data) that encodes a gene product including the conserved cysteine residue. Additional amino acid differences were also found, and functional analysis should reveal whether Oat-k1 has a broader substrate specificity than documented so far.

Rat Oatp3 (*Slc21a7*) has been cloned independently by two groups. Abe et al. (103) found expression of *Oatp3* exclusively in eye and kidney, whereas Walters et al. (104) detected *Oatp3* in small intestine, brain, and lung of Sprague Dawley rats, but not in eye or kidney. Recently, we were able to isolate an *Oatp3* cDNA from a kidney of the Wistar mutant TR⁻ strain, supporting the possibility that tissue distribution may vary between rat strains. In MDCK cells Oatp3 is targeted to the apical membrane, and in rat intestine the transporter is localized to the brush-border membrane (104). Its substrate specificity suggests that in intestine Oatp3 mediates uptake of bile acids via an exchange mechanism (104). Whether Oatp3 transports other organic anions important to the kidney remains to be elucidated. Similar to rat Oatp1 and Oat-k1, there does not seem to exist a human ortholog of rat/mouse Oatp3 (95). Based on chromosome mapping, OATP-A (*SLC22A3*) is proposed as the human ortholog of rat/mouse Oatp3 (104); however, its identity to rat/mouse Oatp3 (72%) is much lower compared with the identity shared between mouse and rat Oatp3 (90%). Furthermore, OATP-A mRNA is expressed in brain and liver, but not in small intestine or kidney.

Human OATP-D (*SLC22A11*) and OATP-E (*SLC22A12*) are ubiquitously expressed (95). The prostaglandin transporter type 2 (Pgt2) may represent the rat ortholog of human OATP-D (86% amino acid identity to OATP-D). Based on a rather low amino acid identity (69%), Oatp-E may not be the rat ortholog of human OATP-E (166). Estrone-sulfate, which is a substrate for rat Oatp1 and many of the human OATPs (92), is also transported by OATP-D and OATP-E (95). Furthermore, OATP-E transports $E_217\beta G$ presumably with a low affinity (95). The membrane localization of either OATP-D or OATP-E has not yet been resolved.

Mouse Oatp5 is exclusively expressed in kidney (105). A rat ortholog has also been cloned (78% amino acid identity to mouse Oatp5), whereas none of the OATPs exhibits an amino acid identity high enough to be considered as the human ortholog. No data are yet available on the Oatp5 substrate specificity nor its membrane localization.

Peptide Transporters PEPT1–PEPT2, Solute Carrier Family SCL15A

Peptide transporters mediate the uptake of di-and tripeptides into renal and intestinal epithelial cells. Studies with brush-border membrane vesicles from small intestine and renal proximal tubule indicate that uptake of peptides is a H^+-gradient-driven transport mechanism. These studies also showed that the intestinal and renal peptide uptake pathways differ in their kinetics. Two peptide transporters have been cloned from various species. PEPT1 is the small intestine low-affinity peptide transporter and is also expressed at low levels in kidney. PEPT2 is the high-affinity renal peptide transporter and is also expressed in lung and brain but not in small intestine. Immunohistochemistry on rat kidney sections has indicated that PepT2 is expressed at the S3 section of the proximal tubule, whereas PepT1 is localized to parts S1 and S2 (106). Substrates of PepT1 and PepT2, apart from peptides, include anti-cancer drugs (107), pro-drugs (108–111), inhibitors of angiotensin-converting enzyme (112), and β-lactam antibiotics (113–115). Based on their localization and substrate specificity, PEPT1 may contribute to the intestinal absorption of orally administered xenobiotics, whereas in the kidney, glomerular-filtered xenobiotics may be accumulated in the proximal tubule via PEPT2 (or PEPT1). How PEPT1 and PEPT2 contribute to drug accumulation in the presence of renal and intestinal ATP-dependent drug-efflux pumps, such as MRP2, MRP4, and P-glycoprotein, warrants further investigation.

PDZ DOMAIN–CONTAINING PROTEINS IN THE KIDNEY

PDZ Domains

PDZ domains are named after the proteins in which characteristic 70–90 amino acid domains initially have been found: postsynaptic density protein PSD-95, *Drosophila* disk large tumor suppressor Dlg-1, and the tight-junction protein ZO-1

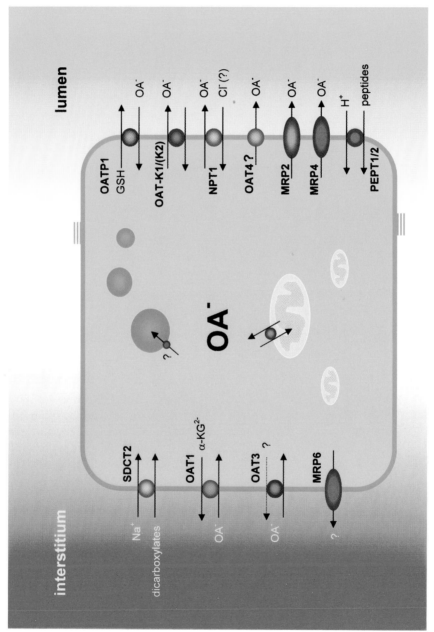

See text page C-2

Figure 1 Schematic model of organic anion transporters in kidney proximal tubule. Uptake of organic anions (OA⁻) across the basolateral membrane is mediated by the classic Na⁺-dependent organic anion transport system, which includes α-ketoglutarate (α-KG2²⁻)/OA⁻ exchange via the organic anion transporter, OAT1 and Na⁺-ketoglutarate transport via the Na⁺/dicarboxylate cotransporter (SDCT2). A second Na⁺-independent uptake system for bulky OA⁻ has been identified, but its molecular identity and driving force are unknown. OAT3 may be a possible candidate. The role of the multidrug resistance transporter, MRP6, as a putative ATP-dependent extrusion mechanism to the interstitium is still unclear. The apical (luminal) membrane contains various transport systems for efflux of OA⁻ into the lumen. The multidrug resistance proteins MRP2 and MRP4 mediate primary active transport. The organic anion-transporting polypeptide, OATP1, the kidney-specific OAT-K1, and the supposed isoforms OAT-K2 and OAT4 may mediate facilitated OA⁻ efflux, but they could also be involved in reabsorption via an exchange mechanism. Finally, PEPT1 and PEPT2 are responsible for the active reabsorption of peptide-like drugs.

(116). PDZ domains have the ability to interact with short peptide sequences at the C terminus of other proteins, such as transmembrane proteins, and PDZ domain–containing proteins are suggested to be involved in stabilization, targeting, and regulation of their binding partner. In the kidney proximal tubule, at least three PDZ domain–containing proteins have been identified (Table 3).

Immunolocalization studies have indicated that Na^+/H^+ exchanger regulatory factor 1 (NHERF-1) and the type IIa Na^+/P_i (NaPi-IIa) cotransporter C-terminal-associated proteins 1 and 2 (NaPi-Cap1 and -Cap2) are localized close to the brush-border membrane of the proximal tubule (117, 118). Expression of a NHERF-1 homolog, called E3KARP (or NHERF-2), has been confined to apical membrane of distal convoluted tubules and the collecting duct (118, 119). Fouassier et al. (119) have also localized E3KARP to the proximal tubule brush-border membrane.

The PDZ domain–containing protein NHERF-1 contains two PDZ domains (PDZ-I and -II) and interacts with the Na^+/H^+ exchanger 3 (NHE3) preferentially through PDZ-II (120). In addition, NHERF-1 forms a complex with ezrin, which binds cAMP-dependent protein kinase II and actin filaments (121). This signaling complex links NHE3 with the cytoskeleton and enables cAMP-dependent phosphorylation of NHE3, which inhibits Na^+/H^+ exchange (121). NHERF-1 also binds to MRP4 (R. van Aubel & F. Russel, unpublished) and CFTR (122, 123). Mouse NaPi-Cap1 and NaPi-Cap2 contain four PDZ domains and have recently been identified for their interaction with the NaPi-IIa cotransporter preferentially through PDZ-III (117). The human ortholog of mouse NaPi-Cap1, called PDZK1, has been shown to bind to MRP2 through PDZ-I (124).

Alignment of NHERF-1 binding partners has suggested a consensus PDZ-motif sequence for the three C-terminal amino acids (S/T-X-L/V/I, where X is any amino acid) in order to bind to NHERF-1 (122). The CFTR mutant *S1455X*, which lacks the 26 C-terminal amino acids, no longer binds to NHERF-1 and is mislocated to the lateral membrane upon expression in kidney (MDCK) and airway (16HBE14o-) epithelial cells (125). Comparable results are found for CFTR-ΔTRL and MRP2-ΔTKL (125, 126). In agreement with the proposed concensus sequence, mutation of either the leucine (position 0) or threonine (position -2) of the CFTR or MRP2 C terminus abrogates binding to NHERF-1 and targeting of CFTR and MRP2 to the apical membrane (126, 127). Recently, with identification of the crystal structure of the first PDZ domain of NHERF-1; further insight has been gained in the importance of a leucine at position 0 of the CFTR C terminus for a NHERF-1/CFTR interaction (128).

Predicting the Targeting of Renal Organic Anion Transporters

The current knowledge of the targeting of transmembrane proteins mediated by PDZ domain–containing proteins allows for speculation that in kidney the presence of a PDZ motif is predictive for an apical localization. Indeed, the organic anion transporters MRP2, MRP4, NPT1, PEPT2, Oatp1, and Oat-k1/k2 are localized to the brush-border membrane, and their C terminus matches the S/T-X-L/V/I

TABLE 3   Characteristics of PDZ domain–containing proteins in the kidney[a]

Gene product	Alternative name	Species	Accession numbers	Mr (kDa)	Tissue distribution	Nephron distribution	Interaction			
							PDZ-I	PDZ-II	PDZ-III	PDZ-IV
NHERF-1	EBP50/ NHERF	human mouse rat rabbit	NM.004252 NM.012030 NM.021594 U19815	39	k, l, si, pl	PT	CFTR, NaPi-IIa	NHE3	np	np
NHERF-2	E3KARP	human mouse rat	NM.04785 NM.023055 AF259898	50		PT(?), DCT, CD	?	NHE3, CFTR	np	np
PDZK1	diphor-1 NaPi-Cap1	human rat mouse	NM.002614 AF013145 AF220100	63	k, l, si	PT	MRP2	?	NaPi-IIa	?
NaPi-Cap2	FLJ22756	mouse human	AF334612 NM.024791		k, si	PT	?	?	NaPi-IIa	?

[a]Abbreviations: np, not present; k, kidney; l, liver; si, small intestine; p, placenta; NHERF: Na$^+$,H$^+$ exchanger regulatory factor; EBP50: ERM-binding phosphoprotein 50; ERM: ezrin-radixin-moesin; E3KARP: Na$^+$,H$^+$ exchanger type 3 kinase A regulatory protein; NaPi-Cap: Na$^+$,P$_i$ type IIa (or NTP2a) cotransporter C-terminal-associated protein; PDZK1: PDZ domain–containing kidney protein.

PDZ-motif consensus sequence (Table 1). On the other hand, MRP1, MRP3, MRP5, MRP6, and OAT1–OAT3, which are localized to the basolateral membrane, lack a potential C terminal PDZ motif (Table 1). Immunolocalization of human OAT4, OATP-D, and OATP-E and rat Oatp3 and Oatp5 has not yet been conducted, but these transporters contain a PDZ motif at the C terminus that favors apical targeting (Table 1). On the other hand, the apical transporters NHE3 and PEPT1 lack the S/T-X-L/V/I PDZ motif at their C terminus; however, PDZ domain–containing proteins may also bind internal regions near the C terminus, as shown for binding of NHERF-1 to NHE3 (129).

REGULATION OF ORGANIC ANION TRANSPORT

Basolateral Transport

Transport of PAH by OAT1 is inhibited by phorbol 12-myristate 13-acetate, and this inhibition is reversed by staurosporine, indicating that organic anion uptake is negatively correlated with protein kinase C (PKC) activity (21, 28). These results are in close agreement with previous reports on PKC-regulated inhibition of organic anion uptake into the opossum kidney (OK) cell line and proximal tubules of rabbit, killifish, and flounder, by using different substrates (130–132). Because Na^+/K^+-ATPase, the putative basolateral Na^+/dicarboxylate exchanger SDCT2/NaDC3, and Oat1 have multiple PKC phosphorylation sites, it remains to be established whether PKC controls basolateral organic anion transport directly or indirectly. Experiments using mouse Oat1 showed no direct phosphorylation of the transporter protein, indicating that an indirect regulation is the likely cause of PKC-induced inhibition of PAH transport (133). In addition, mouse Oat1 is regulated, also by direct phosphorylation of serine residues through protein phosphatase 1 and 2A (PP1/PP2A), which results in a diminished transporter activity (133). Furthermore, human OAT1 was found to be downregulated after tubular injury induced by hydrogen peroxide and gentamicin, which was a result of a reduction in maximum transport velocity (134). Whether the decrease in organic anion transport is a result of activation or inhibition of a signal transduction machinery upon nephrotoxicant exposure needs to be investigated. Numerous neural and humoral agents activate signaling pathways in renal proximal tubules thereby affecting cellular uptake of organic compounds; this has been clearly described by Berkhin & Humphreys (10). Recent data indicate that the uptake of PAH in OK cells is stimulated by epidermal growth factor through phosphorylation of mitogen-activated protein kinases (MAPKs), mitogen-activated/extracellular signal regulated kinase (MEK), and extracellular-regulated kinase isoforms 1 and 2 (ERK1/2) (135), indicating that cell viability and transport activity are highly controlled and correlated. In agreement with OAT1, OAT3 may be regulated by PKC (136), and the transporter is downregulated after exposure of proximal tubules to the nephrotoxic hydrogen peroxide (134).

Apical Transport

Regulation of Mrp2-mediated drug excretion in killifish proximal tubules by the vasoactive hormone, endothelin-1 (ET-1), was recently described (137). ETs are important regulators in kidney functioning and are involved in the control of renal blood flow, glomerular haemodynamics, and sodium and water homeostasis. Furthermore, several disorders associated with dysfunctional endothelium, including acute ischemic renal failure, cyclosporin, cisplatin, and radiocontrast agent toxicity, and vascular rejection of the transplanted kidney, are mediated through ETs (138–140). It was shown that subnanomolar to nanomolar concentrations of ET-1 rapidly reduced the cell-to-tubular lumen transport of fluorescein-methotrexate, whereas the tubular uptake of this compound was unaffected. The effects were prevented by a B-type receptor antagonist, but not by an ET_A-receptor antagonist, and by inhibitors of the second messenger PKC. Furthermore, when proximal tubules were exposed to nephrotoxic drugs, an influx of calcium into the tubule cells was induced, which resulted in a release of ET-1. Subsequently, ET-1 decreased Mrp2 transport via activation of the B-receptor, activation of nitric oxide synthase (NOS), and production of nitric oxide (NO) and PKC, indicating that ET-1 acts via an autocrine/paracrine mechanism in response to injury (141, 141a). In addition to this short-term regulatory effect after nephrotoxicant exposure, an induction in Mrp2 expression and function was found after long-term exposure (68; S. Terlouw, F. Russel, R. Masereeuw & D. Miller, unpublished data).

Intracellular Disposition

In vitro studies in isolated proximal tubule cells revealed that, at low medium concentrations, PAH and fluorescein accumulated within the cells. Intracellular concentrations of three to five times the medium concentration were determined (142). Fluorescein accumulation in proximal tubules during secretory transport has been confirmed in rat kidney in vivo (143). Confocal microscopic images of rat proximal tubule cells showed that fluorescein was compartmentalized within subcellular organelles, predominantly in mitochondria. Uptake of anionic compounds in mitochondria is energetically uphill (inside negative membrane potential of -220 mV) and must therefore be driven by active transport. Interaction studies revealed that metabolite anion transporters appear to mediate the intramitochondrial concentration of fluorescein (142, 144, 145). Miller et al. showed a nocodazole-sensitive vesicular compartmentation in teleost kidneys (146–148). Nocodazole disrupts the Golgi apparatus and, subsequently, microtubules, which are involved in the movement of transporting vesicles through the cells (149). Although the species used in these studies differ, the findings suggest that at least two different compartments are involved in the intracellular accumulation of this anion. The involvement of microtubules in vesicular compartmentations suggests a role in transcellular transport and/or secretion. Another possibility is that endosomal membranes contain carrier proteins for the regulation of transport activity on plasma membranes, which are coincidentally involved in the accumulation of organic anions into these vesicles. However, an argument against this possibility is that immunogold labeling showed no staining for OAT1 in cytoplasmic vesicles (20).

Anion-binding proteins, such as ligandin or glutathione *S*-transferase B, may also determine the intracellular fate of anionic drugs (150–152). PAH, penicillin, probenecid, phlorizin, and diatrizoate were found to interact with ligandin, suggesting that this protein plays a role in controlling free cytoplasmic drug concentrations and in transcellular drug trafficking (151, 153, 154). However, binding to glutathione *S*-transferase does not seem to be a determinant in the rate of luminal secretion (152). Unfortunately, more recent data on the role of binding proteins are not available.

CONCLUSIONS

Renal organic anion transport systems play a critical role in the defense against a wide variety of potentially harmful compounds to which we are continually exposed via food, drugs, occupation, and environment. The proximal tubule is responsible for the carrier-mediated transport from blood to urine of numerous xenobiotics and their metabolites. As a result of its rich transport function and concentrative capacity, this nephron segement is often the victim of toxicity by these compounds.

As the number of cloned transporters grows, it is now appreciated that the organic anion system is a complex organization of multiple plasma membrane proteins, many of which are isoforms that use ATP and transmembrane ion gradients to drive active drug secretion into urine. A striking feature of these transporters is their broad substrate specificity and susceptibility to inhibitors. There is not a single transport protein that recognizes all substrates, and it is difficult, if not impossible, to find substrates that are specific for one transporter. The large number of organic anion transporters with overlapping substrate specificity appears to be the evolutionary answer to the enormous diversity of chemicals and toxins in nature against which the human body has learned to defend itself. The importance of this class of proteins is underlined by their wide distribution in other tissues and organs. An exciting area of future research will be to identify the regulatory sites (e.g., posphorylation sites, PDZ domains, and other protein-protein interaction sites) in the cloned transporters and to study the molecular mechanisms that modulate their function and govern their subcellular distribution. Finally, an important approach to study the in vivo function of transporters and their mutual interaction is to generate knockout mice of the genes encoding these proteins. As there seems to be a large redundancy in the transporter families, conditional or multiple null mutants may be required to overcome the problem of compensatory mechanisms.

Soon the list of transporter genes will be completed, although much work remains to identify and characterize all the isoforms and splice variants at the proteome level. An important next step will be to elucidate the relative participation of the cloned transporters in the membrane steps involved in tubular secretion of specific anionic substrates. However, the most prodigious challenge for the future will be the integration of molecular knowledge on separate genes and proteins into the complex processes that determine the transport function of renal and other mammalian cells. Knowledge on number, activity, specificity, and regulation in health and disease of all different transport proteins in the proximal tubular cell should be brought together in a comprehensive dynamic model of active anionic drug secretion. Such a model should also include the genetic variations in transporter genes throughout the patient population and their significance for drug excretion. For many years this has been a central topic in drug metabolism, but knowledge on the functional implications of single-nucleotide polymorphisms of drug transporters is only in its infancy. Molecular-based knowledge of the renal mechanisms that govern xenobiotic excretion will advance our understanding of renal drug-drug interactions, nephrotoxicity, and interindividual variability in renal clearance. This is essential for rational drug use, development of safer drugs, and a better assessment of the risks associated with the exposure to toxic compounds.

Visit the Annual Reviews home page at www.AnnualReviews.org

LITERATURE CITED

1. Marshall EK, Vickers JL. 1923. The mechanism of the elimination of phenol-sulphonphtalein by the kidney—a proof of secretion by the convoluted tubules. *Bull. Johns Hopkins Hosp.* 34:1–6
2. Beyer KH, Russo HF, Tillson EK, Miller AK, Verwey WF, Gass SR. 1951. 'Benemid' p-(di-n-propylsulfamyl)-benzoic acid: its renal affinity and its elimination. *Am. J. Physiol.* 166:625–40
3. Lacy SA, Hitchcock MJ, Lee WA, Tellier P, Cundy KC. 1998. Effect of oral probenecid coadministration on the chronic toxicity and pharmacokinetics of intravenous cidofovir in cynomolgus monkeys. *Toxicol. Sci.* 44:97–106
4. Koepsell H, Gorboulev V, Arndt P. 1999. Molecular pharmacology of organic cation transporters in kidney. *J. Membr. Biol.* 167:103–17
5. Burckhardt G, Wolff NA. 2000. Structure of renal organic anion and cation transporters. *Am. J. Physiol. Renal Physiol.* 278:F853–F66
6. Sweet DH, Pritchard JB. 1999. The molecular biology of renal organic anion and

organic cation transporters. *Cell Biochem. Biophys.* 31:89–118
7. van Aubel RAMH, Masereeuw R, Russel FGM. 2000. Molecular pharmacology of renal organic anion transporters. *Am J. Physiol. Renal Physiol.* 279:F216–F32
8. Inui KI, Masuda S, Saito H. 2000. Cellular and molecular aspects of drug transport in the kidney. *Kidney Int.* 58:944–58
9. Sekine T, Cha SH, Endou H. 2000. The multispecific organic anion transporter (OAT) family. *Pflügers Arch.* 440:337–50
10. Berkhin EB, Humphreys MH. 2001. Regulation of renal tubular secretion of organic compounds. *Kidney Int.* 59:17–30
11. Masereeuw R, Moons MM, Toomey BH, Russel FGM, Miller DS. 1999. Active Lucifer Yellow secretion in renal proximal tubule: evidence for organic anion transport system crossover. *J. Pharmacol. Exp. Ther.* 289:1104–11
12. Masereeuw R, Russel FGM, Miller DS. 1996. Multiple pathways of organic anion secretion in renal proximal tubule

revealed by confocal microscopy. *Am. J. Physiol. Renal Physiol.* 271:F1173–F82

13. Pritchard JB, Miller DS. 1993. Mechanisms mediating renal secretion of organic anions and cations. *Physiol. Rev.* 73:765–96

14. Chen X, Tsukaguchi H, Chen X-Z, Berger UV, Hediger MA. 1999. Molecular and functional analysis of SDCT2, a novel rat sodium-dependent dicarboxylate transporter. *J. Clin. Invest.* 103:1159–68

15. Miller DS, Letcher S, Barnes DM. 1996. Fluorescence imaging study of organic anion transport from renal proximal tubule cell to lumen. *Am. J. Physiol. Renal Physiol.* 271:F508–F20

16. van Aubel RAMH, Peters JGP, Masereeuw R, van Os CH, Russel FGM. 2000. Multidrug resistance protein mrp2 mediates ATP-dependent transport of classic renal organic anion *p*-aminohippurate. *Am. J. Physiol. Renal Physiol.* 279:F713–F17

17. Leier I, Hummel-Eisenbeiss J, Cui Y, Keppler D. 2000. ATP-dependent para-aminohippurate transport by apical multidrug resistance protein MRP2. *Kidney Int.* 57:1636–42

18. Sweet DH, Wolff NA, Pritchard JB. 1997. Expression cloning and characterization of ROAT1. The basolateral organic anion transporter in rat kidney. *J. Biol. Chem.* 272:30088–95

19. Sekine T, Watanabe N, Hosoyamada M, Kanai Y, Endou H. 1997. Expression cloning and characterization of a novel multispecific organic anion transporter. *J. Biol. Chem.* 272:18526–29

20. Tojo A, Sekine T, Nakajima N, Hosoyamada M, Kanai Y, et al. 1999. Immunohistochemical localization of multispecific renal organic anion transporter 1 in rat kidney. *J. Am. Soc. Nephrol.* 10:464–71

21. Uwai Y, Okuda M, Takami K, Hashimoto Y, Inui K-I. 1998. Functional characterization of the rat multispecific organic anion transporter Oat1 mediating basolateral uptake of anionic drugs in the kidney. *FEBS Lett.* 438:321–24

22. Jariyawat S, Sekine T, Takeda M, Apiwattanakul N, Kanai Y, et al. 1999. The interaction and transport of beta-lactam antibiotics with the cloned rat renal organic anion transporter 1. *J. Pharmacol. Exp. Ther.* 290:672–77

23. Apiwattanakul N, Sekine T, Chairoungdua A, Kanai Y, Nakajima N, et al. 1999. Transport properties of nonsteroidal anti-inflammatory drugs by organic anion transporter 1 expressed in *Xenopus laevis* oocytes. *Mol. Pharmacol.* 55:847–54

24. Wada S, Tsuda M, Sekine T, Cha SH, Kimura M, et al. 2000. Rat multispecific organic anion transporter 1 (rOAT1) transports zidovudine, acyclovir, and other antiviral nucleoside analogs. *J. Pharmacol. Exp. Ther.* 294:844–49

25. Cihlar T, Lin DC, Pritchard JB, Fuller MD, Mendel DB, Sweet DH. 1999. The antiviral nucleotide analogs cidofovir and adefovir are novel substrates for human and rat renal organic anion transporter 1. *Mol. Pharmacol.* 56:570–80

26. Bahn A, Prawitt D, Buttler D, Reid G, Enklaar T, et al. G. 2000. Genomic structure and in vivo expression of the human organic anion transporter 1 (hOAT1) gene. *Biochem. Biophys. Res. Commun.* 275:623–30

27. Hosoyamada M, Sekine T, Kanai Y, Endou H. 1999. Molecular cloning and functional expression of a multispecific organic anion transporter from human kidney. *Am. J. Physiol. Renal Physiol.* 276:F122–F28

28. Lu R, Chan BS, Schuster VL. 1999. Cloning of the human kidney PAH transporter: narrow substrate specificity and regulation by protein kinase C. *Am. J. Physiol. Renal Physiol.* 276:F295–F303

29. Sekine T, Cha SH, Tsuda M, Apiwattanakul N, Nakajima N, et al. 1998. Identification of multispecific organic anion

transporter 2 expressed predominantly in the liver. *FEBS Lett.* 429:179–82

30. Sun W, Wu RR, van Poelje PD, Erion MD. 2001. Isolation of a family of organic anion transporters from human liver and kidney. *Biochem. Biophys. Res. Commun.* 283:417–22

31. Cha SH, Sekine T, Fukushima JI, Kanai Y, Kobayashi Y, et al. 2001. Identification and characterization of human organic anion transporter 3 expressing predominantly in the kidney. *Mol. Pharmacol.* 59:1277–86

32. Kusuhara H, Sekine T, Utsunomiya-Tata N, Tsuda M, Kojima R, et al. 1999. Molecular cloning and characterization of a new multispecific organic anion transporter from rat brain. *J. Biol. Chem.* 274:13675–80

33. Cha SH, Sekine T, Kusuhara H, Yu E, Kim JY, et al. 2000. Molecular cloning and characterization of multispecific organic anion transporter 4 expressed in the placenta. *J. Biol. Chem.* 275:4507–12

34. Murer H, Hernando N, Forster I, Biber J. 2000. Proximal tubular phosphate reabsorption: molecular mechanisms. *Physiol. Rev.* 80:1373–409

35. Busch AE, Schuster A, Waldegger S, Wagner CA, Zempel G, et al. 1996. Expression of a renal type I sodium/phosphate transporter (NaPi-1) induces a conductance in *Xenopus* oocytes permeable for organic and inorganic anions. *Proc. Natl. Acad. Sci. USA* 93:5347–51

36. Uchino H, Tamai I, Yamashita K, Minemoto Y, Sai Y, et al. 2000. *p*-aminohippuric acid transport at renal apical membrane mediated by human inorganic phosphate transporter NPT1. *Biochem. Biophys. Res. Commun.* 270:254–59

37. Uchino H, Tamai I, Yabuuchi H, China K, Miyamoto K, et al. A. 2000. Faropenem transport across the renal epithelial luminal membrane via inorganic phosphate transporter Npt1. *Antimicrob. Agents Chemother.* 44:574–77

38. Yabuuchi H, Tamai I, Morita K, Kouda T, Miyamoto K, et al. 1998. Hepatic sinusoidal membrane transport of anionic drugs mediated by anion transporter Npt1. *J. Pharmacol. Exp. Ther.* 286:1391–96

39. Dean M. Rzhetsky A, Allikmets R. 2001. The human ATP-binding cassette (ABC) transporter superfamily. *Genome Res.* 11:1156–66

40. Cole SPC, Bhardwaj G, Gerlach JH, Mackie JE, Grant CE, et al. 1992. Overexpression of a transporter gene in a multidrug-resistant human lung cancer cell line. *Science* 258:1650–5

41. Zaman GJR, Versantvoort CH, Smit JJ, Eijdems EW, de Haas M, et al. 1993. Analysis of the expression of MRP, the gene for a new putative transmembrane drug transporter, in human multidrug resistant lung cancer cell lines. *Cancer Res.* 53:1747–50

42. Wijnholds J, Scheffer GL, vander Valk M, vander Valk P, Beijnen JH, et al. 1998. Multidrug resistance protein 1 protects the oropharyngeal mucosal layer and the testicular tubules against drug-induced damage. *J. Exp. Med.* 188:797–808

43. Peng K-C, Cluzeaud F, Bens M, Duong van Huyen J-P, Wioland MA, et al. 1999. Tissue and cell distribution of the multidrug resistance-associated protein (MRP) on mouse intestine and kidney. *J. Histochem. Cytochem.* 47:757–67

44. König J, Nies AT, Cui Y, Leier I, Keppler D. 1999. Conjugate export pumps of the multidrug resistance protein (MRP) family: localization, substrate specificity, and MRP2-mediated drug resistance. *Biochim. Biophys. Acta* 1461:377–94

45. Evers R, de Haas M, Sparidans R, Beijnen J, Wielinga PR, et al. 2000. Vinblastine and sulfinpyrazone export by the multidrug resistance protein MRP2 is associated with glutathione export. *Br. J. Cancer* 83:375–83

46. Paulusma CC, van Geer M, Evers R, Heijn M, Ottenhoff R, et al. 1999. Canalicular multispecific organic anion transporter/multidrug resistance protein 2

mediates low-affinity transport of reduced glutathione. *Biochem. J.* 338:393–401

47. Rappa G, Lorico A, Flavell RA, Sartorelli AC. 1997. Evidence that the multidrug resistance protein (MRP) functions as a co-transporter of glutathione and natural product toxins. *Cancer Res.* 57:5232–37

48. Loe DW, Stewart RK, Massey TE, Deeley RG, Cole SPC. 1997. ATP-dependent transport of aflatoxin B_1 and its glutathione conjugates by the product of the multidrug resistance protein (MRP) gene. *Mol. Pharmacol.* 51:1034–41

49. Loe DW, Deeley RG, Cole SPC. 1998. Characterization of vincristine transport by the M(r) 190,000 multidrug resistance protein (MRP): evidence for cotransport with reduced glutathione. *Cancer Res.* 58:5130–36

50. Zaman GJR, Lankelma J, Beijnen J, van Tellingen O, Dekker H, et al. 1995. Role of glutathione in the export of compounds from cells by the multidrug resistance-associated protein. *Proc. Natl. Acad. Sci. USA* 92:7690–94

51. Lorico A, Rappa G, Finch RA, Yang D, Flavell RA, Sartorelli AC. 1997. Disruption of the murine MRP (multidrug resistance protein) gene leads to increased sensitivity to etoposide (VP-16) and increased levels of glutathione. *Cancer Res.* 57:5238–42

52. Wijnholds J, Evers R, Leusden MR, Mol CA, Zaman GJR, et al. 1997. Increased sensitivity to anticancer drugs and decreased inflammatory responce in mice lacking the multidrug resistance-associated protein. *Nat. Med.* 3:1275–79

53. Oude Elferink RPJ, Jansen PLM. 1994. The role of the canalicular multispecific organic anion transporter in the disposal of endo- and xenobiotics. *Pharmacol. Ther.* 64:77–97

54. Keppler D, König J. 1997. Expression and localisation of the conjugate export pump encoded by the MRP2 (cMRP/cMOAT) gene in liver. *FASEB J.* 11:509–16

55. Mottino AD, Hoffman T, Jennes L, Vore M. 2000. Expression and localization of multidrug resistant protein mrp2 in rat small intestine. *J. Pharmacol. Exp. Ther.* 293:717–23

56. van Aubel RAMH, Hartog A, Bindels RJM, van Os CH, Russel FGM. 2000. Expression and immunolocalization of multidrug resistance protein 2 in rabbit small intestine. *Eur. J. Pharmacol.* 400:195–98

57. Ito K, Suzuki H, Hirohashi T, Kume K, Shimizu T, Sugiyama Y. 1997. Molecular cloning of canalicular multispecific organic anion transporter defective in EHBR. *Am. J. Physiol. Gastrointest. Liver Physiol.* 272:G16–G22

58. Paulusma CC, Bosma PJ, Zaman GJR, Bakker CTM, Otter M, et al. 1996. Congenital jaundice in rats with a mutation in a multidrug resistance-associated protein gene. *Science* 271:1126–28

59. Cui Y, König J, Buchholz U, Spring H, Leier I, Keppler D. 1999. Drug resistance and ATP-dependent conjugate transport mediated by the apical multidrug resistance protein, MRP2, permanently expressed in human and canine cells. *Mol. Pharmacol.* 55:929–37

60. Evers R, Kool M, van Deemter L, Janssen H, Calafat J, et al. 1998. Drug export activity of the human canalicular multispecific organic anion transporter in polarized kidney MDCK cells expressing cMOAT (MRP2) cDNA. *J. Clin. Invest.* 101:1310–19

61. Bakos E, Evers R, Sinko E, Varadi A, Borst P, Sarkadi B. 2000. Interactions of the human multidrug resistance proteins MRP1 and MRP2 with organic anions. *Mol. Pharmacol.* 57:760–68

62. Ito K, Suzuki H, Hirohashi T, Kume K, Shimizu T, Sugiyama Y. 1998. Functional analysis of a canalicular multispecific organic anion transporter cloned from rat liver. *J. Biol. Chem.* 273:1684–88

63. van Aubel RAMH, van Kuijck MA, Koenderink JB, Deen PMT, van Os CH,

Russel FGM. 1998. Adenosine triphosphate-dependent transport of anionic conjugates by the rabbit multidrug resistance-associated protein Mrp2 expressed in insect cells. *Mol. Pharmacol.* 53:1062–67

64. Hooijberg JH, Broxterman HJ, Kool M, Assaraf YG, Peters GJ, et al. 1999. Antifolate resistance mediated by the multidrug resistance proteins MRP1 and MRP2. *Cancer Res.* 59:2532–35

65. van Aubel RAMH, Koenderink JB, Peters JGP, van Os CH, Russel FGM. 1999. Mechanisms and interaction of vinblastine and reduced glutathione transport in membrane vesicles by the rabbit multidrug resistance protein Mrp2 expressed in insect cells. *Mol. Pharmacol.* 56:714–19

66. Kawabe T, Chen ZS, Wada M, Uchiumi T, Ono M, et al. 1999. Enhanced transport of anticancer agents and leukotriene C4 by the human canalicular multispecific organic anion transporter (cMOAT/MRP2). *FEBS Lett.* 456:327–31

67. Demeule M, Brossard M, Beliveau R. 1999. Cisplatin induces renal expression of P-glycoprotein and canalicular multispecific organic anion transporter. *Am. J. Physiol. Renal Physiol.* 277:F832–F40

68. Laouari D, Yang R, Veau C, Blanke I, Friedlander G. 2001. Two apical multidrug transporters, P-gp and MRP2, are differently altered in chronic renal failure. *Am. J. Physiol. Renal Physiol.* 280:F636–F45

69. Kool M, de Haas M, Scheffer GL, Scheper RJ, van Eijk MJT, et al. 1997. Analysis of expression of cMOAT (MRP2), MRP3, MRP4, and MRP5, homologous of the multidrug resistance-associated protein gene (MRP1) in human cancer cell lines. *Cancer Res.* 57:3537–47

70. Kool M, van der Linden M, de Haas M, Scheffer GL, de Vree JML, et al. 1999. MRP3, an organic anion transporter able to transport anti-cancer drugs. *Proc. Natl. Acad. Sci. USA* 96:6914–19

71. König J, Rost D, Cui Y, Keppler D. 1999. Characterization of the human multidrug resistance protein isoform MRP3 localized to the basolateral hepatocyte membrane. *Hepatology* 29:1156–63

72. Zeng H, Liu G, Rea PA, Kruh GD. 2000. Transport of amphipathic anions by human multidrug resistance protein 3. *Cancer Res.* 60:4779–84

73. Hirohashi T, Suzuki H, Ito K, Ogawa K, Kume K, et al. 1998. Hepatic expression of multidrug resistance-associated protein-like proteins maintained in Eisai hyperbilirubinemic rats. *Mol. Pharmacol.* 53:1068–75

73a. van Aubel RAMH, Smeets PHE, Peters JGP, Bindels RJM, Russel FGM. 2002. The *MRP4/ABCC4* gene encodes a novel apical organic anion transporter in human kidney proximal tubule: the putative efflux pump for urinary cAMP and cGMP. *J. Am. Soc. Nephrol.* In press

74. Schuetz JD, Connelly MC, Sun D, Paibir SG, Flynn PM, et al. 1999. MRP4: a previously unidentified factor in resistance to nucleoside-based antiviral drugs. *Nat. Med.* 5:1048–51

75. Lee K, Klein-Szanto AJ, Kruh GD. 2000. Analysis of the MRP4 drug resistance profile in transfected NIH3T3 cells. *J. Natl. Cancer Inst.* 92:1934–40

76. Cundy KC. 1999. Clinical pharmacokinetics of the antiviral nucleotide analogues cidofovir and adefovir. *Clin. Pharmacokinet.* 36:127–43

76a. Chen Z-S, Lee K, Kruh GD. 2001. Transport of cyclic nucleotides and estradiol-17β-D-glucuronide by multidrug resistance protein 4. *J. Biol. Chem.* 276:33747–54

77. Strewler GJ. 1984. Release of cAMP from a renal epithelial cell line. *Am. J. Physiol. Cell Physiol.* 246:C224–C30

78. Chevalier RL, Fang GD, Garmey M. 1996. Extracellular cGMP inhibits transepithelial sodium transport by LLC-PK1 renal tubular cells. *Am. J. Physiol. Renal Physiol.* 270:F283–F88

79. Broadus AE, Kaminsky NI, Hardman JG, Sutherland EW, Liddle GW. 1970. Kinetic parameters and renal clearances of plasma adenosine 3′,5′-monophosphate and guanosine 3′,5′-monophosphate in man. *J. Clin. Invest.* 49:2222–36

80. Butlen D, Jard S. 1972. Renal handling of 3′-5′-cyclic AMP in the rat. The possible role of luminal 3′-5′-cyclic AMP in the tubular reabsorption of phosphate. *Pflügers Arch.* 331:172–90

81. McAleer MA, Breen MA, White NL, Matthews N. 1999. pABC11 (also known as MOAT-C and MRP5), a member of the ABC family of proteins, has anion transporter activity but does not confer multidrug resistance when overexpressed in human embryonic kidney 293 cells. *J. Biol. Chem.* 274:23541–48

82. Wijnholds J, Mol CA, van Deemter L, de Haas M, Scheffer GL, et al. 2000. Multidrug-resistance protein 5 is a multispecific organic anion transporter able to transport nucleotide analogs. *Proc. Natl. Acad. Sci. USA* 97:7476–81

83. Jedlitschky G, Burchell B, Keppler D. 2000. The multidrug resistance protein 5 functions as an ATP-dependent export pump for cyclic nucleotides. *J. Biol. Chem.* 275:30069–74

84. Urakabe S, Handler JS, Orloff J. 1975. Release of cyclic AMP by toad urinary bladder. *Am. J. Physiol.* 228:954–58

85. Geary CA, Goy MF, Boucher RC. 1993. Synthesis and vectorial export of cGMP in airway epithelium: expression of soluble and CNP-specific guanylate cyclases. *Am. J. Physiol. Lung Cell Mol. Physiol.* 265:L598–L605

86. Boom A, Golstein PE, Frerotte M, Sande JV, Beauwens R. 2000. Inhibition of basolateral cAMP permeability in the toad urinary bladder. *J. Physiol* 528(Pt.1):189–98

87. Madon J, Hagenbuch B, Landmann L, Meier PJ, Stieger B. 2000. Transport function and hepatocellular localization of mrp6 in rat liver. *Mol. Pharmacol.* 57:634–41

88. Kool M, van der Linden M, de Haas M, Baas F, Borst P. 1999. Expression of human MRP6, a homologue of the multidrug resistance protein gene MRP1, in tissues and cancer cells. *Cancer Res.* 59:175–82

89. Bergen AA, Plomp AS, Schuurman EJ, Terry S, Breuning M, et al. 2000. Mutations in ABCC6 cause pseudoxanthoma elasticum. *Nat. Genet.* 25:228–31

90. Ringpfeil F, Lebwohl MG, Christiano AM, Uitto J. 2000. Pseudoxanthoma elasticum: mutations in the MRP6 gene encoding a transmembrane ATP-binding cassette (ABC) transporter. *Proc. Natl. Acad. Sci. USA* 97:6001–6

91. Jacquemin E, Hagenbuch B, Stieger B, Wolkoff AW, Meier PJ. 1994. Expression cloning of a rat liver Na(+)-independent organic anion transporter. *Proc. Natl. Acad. Sci. USA* 91:133–37

92. Kullak-Ublick GA, Ismair MG, Stieger B, Landmann L, Huber R, et al. 2001. Organic anion-transporting polypeptide B (OATP-B) and its functional comparison with three other OATPs of human liver. *Gastroenterology* 120:525–33

93. Kawai J, Shinagawa A, Shibata K, Yoshino M, Itoh M, et al. 2001. Functional annotation of a full-length mouse cDNA collection. *Nature* 409:685–90

94. Bergwerk AJ, Shi X, Ford AC, Kanai N, Jacquemin E, et al. 1996. Immunologic distribution of an organic anion transport protein in rat liver and kidney. *Am. J. Physiol. Gastrointest. Liver Physiol.* 271:G231–G38

95. Tamai I, Nezu J, Uchino H, Sai Y, Oku A, et al. 2000. Molecular identification and characterization of novel members of the human organic anion transporter (OATP) family. *Biochem. Biophys. Res. Commun.* 273:251–60

96. Li LQ, Lee TK, Meier PJ, Ballatori N. 1998. Identification of glutathione as a driving force and leukotriene C_4 as a

substrate for oatp1, the hepatic sinusoidal organic anion solute transporter. *J. Biol. Chem.* 273:16184–91

97. Satlin LM, Amin V, Wolkoff AW. 1997. Organic anion transporting polypeptide mediates organic anion/ HCO$_3$-exchange. *J. Biol. Chem.* 272:26340–45

98. Masuda S, Saito H, Nonoguchi H, Tomita K, Inui K-I. 1997. mRNA distribution and membrane localization of the OAT-K1 organic anion transporter in rat renal tubules. *FEBS Lett.* 407:127–31

99. Saito H, Masuda S, Inui K-I. 1996. Cloning and functional characterization of a novel rat organic anion transporter mediating basolateral uptake of methotrexate in the kidney. *J. Biol. Chem.* 271:20719–25

100. Masuda S, Saito H, Inui KI. 1997. Interactions of nonsteroidal anti-inflammatory drugs with rat renal organic anion transporter, OAT-K1. *J. Pharmacol. Exp. Ther.* 283:1039–42

101. Takeuchi A, Masuda S, Saito H, Hashimoto Y, Inui K. 2000. Trans-stimulation effects of folic acid derivatives on methotrexate transport by rat renal organic anion transporter, OAT-K1. *J. Pharmacol. Exp. Ther.* 293:1034–39

102. Masuda M, Ibaramoto K, Takeuchi A, Saito H, Hashimoto Y, Inui KI. 1999. Cloning and functional characterization of a new multispecific organic anion transporter, OAT-K2, in rat kidney. *Mol. Pharmacol.* 55:743–53

103. Abe T, Kakyo M, Sakagami H, Tokui T, Nishio T, et al. 1998. Molecular characterization and tissue distribution of a new organic anion transporter subtype (oatp3) that transports thyroid hormones and taurocholate and comparison with oatp2. *J. Biol. Chem.* 273:22395–401

104. Walters HC, Craddock AL, Fusegawa H, Willingham MC, Dawson PA. 2000. Expression, transport properties, and chromosomal location of organic anion transporter subtype 3. *Am. J. Physiol. Gastrointest. Liver Physiol.* 279:G1188–G200

105. Choudhuri S, Ogura K, Klaassen CD. 2001. Cloning, expression, and ontogeny of mouse organic anion-transporting polypeptide-5, a kidney-specific organic anion transporter. *Biochem. Biophys. Res. Commun.* 280:92–98

106. Shen H, Smith DE, Yang T, Huang YG, Schnermann JB, Brosius FC III. 1999. Localization of PEPT1 and PEPT2 proton-coupled oligopeptide transporter mRNA and protein in rat kidney. *Am. J. Physiol. Renal Physiol.* 276:F658–F65

107. Doring F, Walter J, Will J, Focking M, Boll M, et al. 1998. Delta-aminolevulinic acid transport by intestinal and renal peptide transporters and its physiological and clinical implications. *J. Clin. Invest.* 101:2761–67

108. Han H, de Vrueh RL, Rhie JK, Covitz KM, Smith PL, et al. 1998. 5′-amino acid esters of antiviral nucleosides, acyclovir, and AZT are absorbed by the intestinal PEPT1 peptide transporter. *Pharm. Res.* 15:1154–59

109. Balimane PV, Tamai I, Guo A, Nakanishi T, Kitada H, et al. 1998. Direct evidence for peptide transporter (PepT1)-mediated uptake of a nonpeptide prodrug, valacyclovir. *Biochem. Biophys. Res. Commun.* 250:246–51

110. Ganapathy ME, Huang W, Wang H, Ganapathy V, Leibach FH. 1998. Valacyclovir: a substrate for the intestinal and renal peptide transporters PEPT1 and PEPT2. *Biochem. Biophys. Res. Commun.* 246:470–75

111. Tamai I, Nakanishi T, Nakahara H, Sai Y, Ganapathy V, et al. 1998. Improvement of L-dopa absorption by dipeptidyl derivation, utilizing peptide transporter PepT1. *J. Pharm. Sci.* 87:1542–46

112. Temple CS, Boyd CA. 1998. Proton-coupled oligopeptide transport by rat renal cortical brush border membrane vesicles: a functional analysis using ACE inhibitors to determine the isoform of the transporter. *Biochim. Biophys. Acta* 1373:277–81

113. Ganapathy ME, Brandsch M, Prasad PD, Ganapathy V, Leibach FH. 1995. Differential recognition of beta-lactam antibiotics by intestinal and renal peptide transporters, PEPT 1 and PEPT 2. *J. Biol. Chem.* 20:25672–77

114. Ganapathy ME, Prasad PD, Mackenzie B, Ganapathy V, Leibach FH. 1997. Interaction of anionic cephalosporins with the intestinal and renal peptide transporters PEPT 1 and PEPT 2. *Biochim. Biophys. Acta* 1324:296–308

115. Wenzel U, Gebert I, Weintraut H, Weber WM, Clauss W, Daniel H. 1996. Transport characteristics of differently charged cephalosporin antibiotics in oocytes expressing the cloned intestinal peptide transporter PepT1 and in human intestinal Caco-2 cells. *J. Pharmacol. Exp. Ther.* 277:831–39

116. Fanning AS, Anderson JM. 1999. PDZ domains: fundamental building blocks in the organization of protein complexes at the plasma membrane. *J. Clin. Invest* 103:767–72

117. Gisler SM, Stagljar I, Traebert M, Bacic D, Biber J, Murer H. 2001. Interaction of the type IIa Na/Pi cotransporter with PDZ proteins. *J. Biol. Chem.* 276:9206–13

118. Wade JB, Welling PA, Donowitz M, Shenolikar S, Weinman EJ. 2001. Differential renal distribution of NHERF isoforms and their colocalization with NHE3, ezrin, and ROMK. *Am. J. Physiol. Cell Physiol.* 280:C192–C98

119. Fouassier L, Duan CY, Feranchak AP, Yun CH, Sutherland E, et al. 2001. Ezrin-radixin-moesin-binding phosphoprotein 50 is expressed at the apical membrane of rat liver epithelia. *Hepatology* 33:166–76

120. Shenolikar S, Weinman EJ. 2001. NHERF: targeting and trafficking membrane proteins *Am. J. Physiol. Renal Physiol.* 280:F389–F95

121. Weinman EJ, Minkoff C, Shenolikar S. 2000. Signal complex regulation of renal transport proteins: NHERF and regulation of NHE3 by PKA. *Am. J. Physiol. Renal Physiol.* 279:F393–F99

122. Wang S, Raab RW, Schatz PJ, Guggino WB, Li M. 1998. Peptide binding consensus of the NHE-RF-PDZ1 domain matches the C- terminal sequence of cystic fibrosis transmembrane conductance regulator (CFTR). *FEBS Lett.* 427:103–8

123. Short DB, Trotter KW, Reczek D, Kreda SM, Bretscher A, et al. 1998. An apical PDZ protein anchors the cystic fibrosis transmembrane conductance regulator to the cytoskeleton. *J. Biol. Chem.* 273:19797–801

124. Kocher O, Comella N, Gilchrist A, Pal R, Tognazzi K, et al. 1999. PDZK1, a novel PDZ domain-containing protein upregulated in carcinomas and mapped to chromosome 1q21, interacts with cMOAT (MRP2), the multidrug resistance-associated protein. *Lab. Invest.* 79:1161–70

125. Moyer BD, Denton J, Karlson KH, Reynolds D, Wang S, et al. 1999. A PDZ-interacting domain in CFTR is an apical membrane polarization signal. *J. Clin. Invest.* 104:1353–61

126. Harris MJ, Kuwano M, Webb M, Board PG. 2001. Identification of the apical membrane-targeting signal of the multidrug resistance-associated protein 2 (MRP2/MOAT). *J. Biol. Chem.* 276:20876–81

127. Moyer BD, Duhaime M, Shaw C, Denton J, Reynolds D, et al. 2000. The PDZ-interacting domain of cystic fibrosis transmembrane conductance regulator is required for functional expression in the apical plasma membrane. *J. Biol. Chem.* 275:27069–74

128. Karthikeyan S, Leung T, Birrane G, Webster G, Ladias JA. 2001. Crystal structure of the PDZ1 domain of human Na(+)/H(+) exchanger regulatory factor provides insights into the mechanism of carboxyl-terminal leucine recognition by class I PDZ domains. *J. Mol. Biol.* 308:963–73

129. Weinman EJ, Steplock D, Tate K, Hall RA, Spurney RF, Shenolikar S. 1998. Structure-function of recombinant Na/H exchanger regulatory factor (NHE- RF). *J. Clin. Invest.* 101:2199–206

130. Gekle M, Mildenberger S, Sauvant C, Bednarczyk D, Wright SH, Dantzler WH. 1999. Inhibition of initial transport rate of basolateral organic anion carrier in renal PT by BK and phenylephrine. *Am. J. Physiol. Renal Physiol.* 277:F251–F56

131. Halpin PA, Renfro JL. 1996. Renal organic anion secretion: evidence for dopaminergic and adrenergic regulation. *Am. J. Physiol. Regulatory Integrative Comp. Physiol.* 271:R1372–R79

132. Takano M, Nagai J, Yasuhara M, Inui K. 1996. Regulation of *p*-aminohippurate transport by protein kinase C in OK kidney epithelial cells. *Am. J. Physiol. Renal Physiol.* 271:F469–F75

133. You G, Kuze K, Kohanski RA, Amsler K, Henderson S. 2000. Regulation of mOAT-mediated organic anion transport by okadaic acid and protein kinase C in LLC-PK(1) cells. *J. Biol. Chem.* 275: 10278–84

134. Takeda M, Hosoyamada M, Cha SH, Sekine T, Endou H. 2000. Hydrogen peroxide downregulates human organic anion transporters in the basolateral membrane of the proximal tubule. *Life Sci.* 68:679–87

135. Sauvant C, Holzinger H, Gekle M. 2001. Modulation of the basolateral and apical step of transepithelial organic anion secretion in proximal tubular opossum kidney cells. Acute effects of epidermal growth factor and mitogen-activated protein kinase. *J. Biol. Chem.* 276:14695–703

136. Takeda M, Sekine T, Endou H. 2000. Regulation by protein kinase C of organic anion transport driven by rat organic anion transporter 3 (rOAT3). *Life Sci.* 67:1087–93

137. Masereeuw R, Terlouw SA, van Aubel RAMH, Russel FGM, Miller DS. 2000. Endothelin B receptor-mediated regulation of ATP-driven drug secretion in renal proximal tubule. *Mol. Pharmacol.* 57:59–67

138. Bruzzi I, Remuzzi G, Benigni A. 1997. Endothelin: a mediator of renal disease progression. *J. Nephrol.* 10:179–83

139. Hocher B, Thone Reineke C, Bauer C, Raschack M, Neumayer HH. 1997. The paracrine endothelin system: pathophysiology and implications in clinical medicine. *Eur. J. Clin. Chem. Clin. Biochem.* 35:175–89

140. Clavell AL, Burnett JC Jr. 1994. Physiologic and pathophysiologic roles of endothelin in the kidney. *Curr. Opin. Nephrol. Hypertens.* 3:66–72

141. Terlouw SA, Masereeuw R, Russel FGM, Miller DS. 2001. Nephrotoxicants induce endothelin release and signaling in renal proximal tubules: effect on drug efflux. *Mol. Pharmacol.* 59:1433–40

141a. Notenboom S, Miller DS, Smits P, Russel FGM, Masereeuw R. 2001. Role of NO in endothelin regulated drug transport in the renal proximal tubule. *Am. J. Physiol. Renal Physiol.* In press

142. Masereeuw R, van den Bergh EJ, Bindels RJ, Russel FG. 1994. Characterization of fluorescein transport in isolated proximal tubular cells of the rat: evidence for mitochondrial accumulation. *J. Pharmacol. Exp. Ther.* 269:1261–67

143. Boyde A, Capasso G, Unwin RJ. 1998. Conventional and confocal epi-reflection and fluorescence microscopy of the rat kidney in vivo. *Exp. Nephrol.* 6:398–408

144. Masereeuw R, Saleming WC, Miller DS, Russel FGM. 1996. Interaction of fluorescein with the dicarboxylate carrier in rat kidney cortex mitochondria. *J. Pharmacol. Exp. Ther.* 279:1559–65

145. Terlouw SA, Russel FGM, Masereeuw R. 2000. Metabolite anion carriers mediate the uptake of the anionic drug fluorescein in renal cortical mitochondria.

J. Pharmacol. Exp. Ther. 292:968–73

146. Miller DS, Stewart DE, Pritchard JB. 1993. Intracellular compartmentation of organic anions within renal cells. *Am. J. Physiol. Regulatory Integrative Comp. Physiol.* 264:R882–R90

147. Miller DS, Barnes DM, Pritchard JB. 1994. Confocal microscopic analysis of fluorescein compartmentation within crab urinary bladder cells. *Am. J. Physiol. Regulatory Integrative Comp. Physiol.* 267:R16–R25

148. Miller DS, Pritchard JB. 1994. Nocodazole inhibition of organic anion secretion in teleost renal proximal tubules. *Am. J. Physiol. Regulatory Integrative Comp. Physiol.* 267:R695–R704

149. Brown D. 1995. Epithelial cell polarity: molecular mechanisms. In *Molecular Nephrology, Kidney Function in Health and Disease*, ed. D Sclöndorff, JV Bonventre, pp. 21–32. New York: Dekker

150. Campbell JA, Bass NM, Kirsch RE. 1980. Immunohistological localization of ligandin in human tissues. *Cancer* 45:503–10

151. Litwack G, Ketterer B, Arias IM. 1971. Ligandin: a hepatic protein which binds steroids, bilirubin, carcinogens and a number of exogenous organic anions. *Nature* 234:466–67

152. Sheehan D, Mantle TJ. 1984. Evidence for two forms of ligandin (YaYa dimers of glutathione *S*-transferase) in rat liver and kidney. *Biochem. J.* 218:893–97

153. Goldstein EJ, Arias IM. 1976. Interaction of ligandin with radiographic contrast media. *Invest. Radiol.* 11:594–97

154. Kirsch R, Fleischner G, Kamisaka K, Arias IM. 1975. Structural and functional studies of ligandin, a major renal organic anion-binding protein. *J. Clin. Invest.* 55:1009–19

155. Takeda M, Narikawa S, Hosoyamada M, Cha SH, Sekine T, Endou H. 2001.

156. Tsuda M, Sekine T, Takeda M, Cha SH, Kanai Y, et al. 1999. Transport of ochratoxin A by renal multispecific organic anion transporter 1. *J. Pharmacol. Exp. Ther.* 289:1301–5

157. Evers R, Kool M, Smith AJ, van Deemter L, de Haas M, Borst P. 2000. Inhibitory effect of the reversal agents V-104, GF120918 and Pluronic L61 on MDR1 Pgp- MRP1- and MRP2- mediated transport. *Br. J. Cancer* 83:366–74

158. Hirohashi T, Suzuki H, Sugiyama Y. 1999. Characterization of the transport properties of cloned rat multidrug resistance-associated protein 3 (MRP3). *J. Biol. Chem.* 274:15181–85

159. Kanai N, Lu R, Bao Y, Wolkoff AW, Vore M, Schuster VL. 1996. Estradiol 17 beta-D-glucuronide is a high-affinity substrate for oatp organic anion transporter. *Am. J. Physiol. Renal Physiol.* 270:F326–F31

160. Eckhardt U, Schroeder A, Stieger B, Hochli M, Landmann L, et al. 1999. Polyspecific substrate uptake by the hepatic organic anion transporter Oatp1 in stably transfected CHO cells. *Am. J. Physiol. Gastrointest. Liver Physiol.* 276:G1037–G42

161. Kanai N, Lu R, Bao Y, Wolkoff AW, Schuster VL. 1996. Transient expression of oatp organic anion transporter in mammalian cells: identification of candidate substrates. *Am. J. Physiol. Renal Physiol.* 270:F319–F25

162. Ishizuka H, Konno K, Naganuma H, Nishimura K, Kouzuki H, et al. 1998. Transport of temocaprilat into rat hepatocytes: role of organic anion transporting polypeptide. *J. Pharmacol. Exp. Ther.* 287: 37–42

163. Pang KS, Wang PJ, Chung AYK, Wolkoff AW. 1998. The modified dipeptide,

enalapril, an angiotensin-converting enzyme inhibitor, is transported by the rat liver organic anion transport protein. *Hepatology* 28:1341–46

164. Friesema ECH, Docter R, Moerings EPCM, Stieger B, Hagenbuch B, et al. 1999. Identification of thyroid hormone transporters. *Biochem. Biophys. Res. Commun.* 254:497–50

165. Masuda M, Takeuchi A, Saito H, Hashimoto Y, Inui K-I. 1999. Functional analysis of rat renal organic anion transporter OAT-K1: bidirectional methotrex-ate transport in apical membrane. *FEBS Lett.* 459:128–32

166. Fujiwara K, Adachi H, Nishio T, Unno M, Tokui T, et al. 2001. Identification of thyroid hormone transporters in humans: different molecules are involved in a tissue-specific manner. *Endocrinology* 142:2005–12

167. Terada T, Saito H, Mukai M, Inui K. 1997. Recognition of beta-lactam antibiotics by rat peptide transporters, PEPT1 and PEPT2, in LLC-PK1 cells. *Am. J. Physiol. Renal Physiol.* 273:F706–F11

Annu. Rev. Physiol. 2002. 64:595–608

Trafficking of Canalicular ABC Transporters in Hepatocytes

Helmut Kipp
Max-Planck-Institut für Molekulare Physiologie, 44227 Dortmund, Germany

Irwin M. Arias
Department of Physiology, Tufts University School of Medicine, Boston, Massachusetts 02111; e-mail: Irwin.Arias@tufts.edu

Key Words bile secretion, cholestasis, PI3-kinase, cAMP

■ **Abstract** ATP-binding cassette (ABC) transporters located in the hepatocyte canalicular membrane of mammalian liver are critical players in bile formation and detoxification. Although ABC transporters have been well characterized functionally, only recently have several canalicular ABC transporters been cloned and their molecular nature revealed. Subsequently, development of specific antibodies has permitted a detailed investigation of ABC transporter intrahepatic distribution under varying physiological conditions. It is now apparent that there is a complex array of ABC transporters in hepatocytes. ABC transporter molecules reside in intrahepatic compartments and are delivered to the canalicular domain following increased physiological demand to secrete bile. Insufficient amounts of ABC transporters in the bile canalicular membrane result in cholestasis (i.e., bile secretory failure). Therefore, elucidation of the intrahepatic pathways and regulation of ABC transporters may help to understand the cause of cholestasis at a molecular level and provide clues for novel therapies.

INTRODUCTION

The bile canalicular membrane of the mammalian hepatocyte contains several primary active transporters that couple ATP hydrolysis to the transport of specific substrates into the bile canaliculus (1–4). These transporters are members of the superfamily of ATP-binding cassette (ABC) membrane transport proteins (5) and currently include p-glycoprotein (MDR1) for organic cations (6); MDR2 for phosphatidylcholine translocation (7, 8); sister of p-glycoprotein (SPGP), the bile salt export pump (9); and MRP2 (or cMOAT) for non-bile acid organic anions (10). Thus canalicular ABC transporters play a key role in bile formation in mammalian liver. Insufficient amounts of ABC transporters or their malfunction in the canalicular membrane most likely impair bile formation, thereby resulting in cholestasis.

0066-4278/02/0315-0595$14.00

Recent studies indicate that the amount of each ABC transporter in the canalicular membrane is regulated by the physiological demand to secrete bile acids. Intravenous administration to rats of taurocholate or dibutyryl-cAMP rapidly and selectively increased the functional activity and amount of each ABC transporter in the canalicular membrane (11). This effect was inhibited by prior administration of colchicine, which disrupts microtubules (11), and Wortmannin, which inhibits PI3-kinase (12). These observations indicate that an intracellular microtubule-dependent transport mechanism, which is sensitive to active PI3-kinase, is required for the traffic of ABC transporters to the canalicular membrane. Therefore, intracellular traffic and regulation of canalicular ABC transporters are critical for bile formation under physiological conditions (13). This review summarizes recent knowledge regarding intrahepatic distribution, regulation, and traffic of canalicular ABC transporters.

Traffic of Newly Synthesized ABC Transporters in Hepatocytes

Membrane targeting of the newly synthesized canalicular ectoenzymes dipeptidylpeptidase IV, aminopeptidase N and 5′-nucleotidase, and the canalicular cell adhesion molecule cCAM105 (also known as HA4) has been studied in rat liver by in vivo metabolic pulse chase labeling. After biosynthesis, these canalicular proteins are transferred from Golgi to the basolateral membrane and subsequently reach the bile canaliculus only by transcytosis (14, 15). Based on these results, it was proposed that all newly synthesized canalicular proteins, including canalicular ABC transporters, are targeted via this indirect route (16, 17).

An important observation from previous studies was that the canalicular cell adhesion molecule cCAM105, but not canalicular ABC transporters, was readily detected in highly purified sinosoidal/basolateral membrane vesicles (SMV) from rat liver using Western blots (12, 18). The presence of cCAM105 in SMV can be explained by the fact that cCAM105 is initially transferred to the basolateral membrane after biosynthesis and subsequently reaches the apical pole by transcytosis. This scenario is in good accordance with detectable steady-state levels of cCAM105 in SMV. These observations suggest that canalicular ABC transporters may not undergo transcytosis after biosynthesis. The hypothesis of direct apical targeting of canalicular ABC transporters in rat hepatocytes was subsequently tested using metabolic pulse chase labeling (18).

Rats were metabolically labeled with ^{35}S-methionine for 15 min, and the content of newly synthesized cCAM105, MDR1, MDR2, and SPGP was determined after 15 min, 30 min, 1 h, 2 h, and 3 h in purified canalicular membrane vesicles (CMV), SMV, and Golgi membranes from rat liver by immunoprecipitation with specific antibodies. These studies (18) confirmed the transcytotic pathway for apical targeting of newly synthesized cCAM105 (HA4), as described above (14). In contrast, at no time between passage through Golgi and arrival at the bile canaliculus were the apical ABC transporters MDR1, MDR2, and SPGP detected in SMVs, indicating a direct Golgi-to-bile canaliculus pathway for their membrane targeting. Also, newly synthesized MDR1, MDR2, and SPGP were not initially transferred to the basolateral membrane, i.e., their post-Golgi trafficking differed. After passage through Golgi, MDR1 and MDR2 were rapidly delivered directly to the bile canaliculus, whereas Golgi-to-bile canaliculus trafficking of SPGP involved additional intermediate steps. At 1 h after metabolic labeling, only the mature form of SPGP was detected in the homogenate, indicating that processing and passage through the Golgi were complete at this point. This is also supported by decreased radioactivity to background levels in SPGP immunoprecipitates from Golgi membranes after 1 h. At this time point, SPGP was not detected in SMVs and CMVs and, therefore, had not reached the cell surface, which occurred 2 h after metabolic labeling. The most likely explanation is that SPGP is sequestered in an intracellular pool prior to delivery to the canalicular membrane. Intrahepatic sequestering of newly synthesized SPGP was demonstrated in a later study, which included a combined endosomal fraction in metabolic labeling experiments (19).

The membrane targeting pathways of newly synthesized canalicular proteins discovered by in vivo labeling studies are depicted in Figure 1. These studies had two implications. They provide direct biochemical evidence for intrahepatic pools of ABC transporters. The characteristics of these intrahepatic ABC transporter pools are described below. Furthermore, these studies promoted investigation of

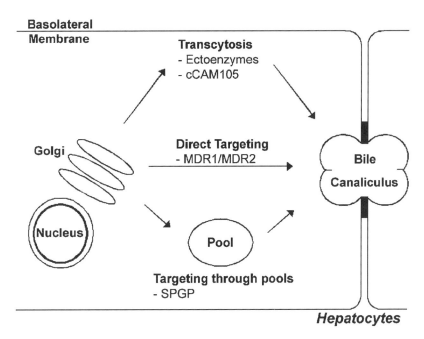

Figure 1   Membrane targeting of newly synthesized canalicular proteins in rat hepatocytes.

direct Golgi-to-bile canaliculus trafficking of MDR1-green fluorescent protein (GFP) in WIF-B cells, a polarized hepatocyte cell culture model.

WIF-B cells are a hybrid of rat hepatoma cells and human fibroblasts, have functional bile canaliculi, and are a useful model for hepatocytes (20, 21). Functional features of hepatocytes are also observed in WIF-B cells: i.e., basolateral to apical membrane transcytosis of canalicular ectoenzymes (22); secretion of fluorescent bile acids and substrates for MDR1 (20, 23) and MRP2 (24); inhibition of the secretion of fluorescent substrates by the phosphoinositide 3-kinase (PI3K) inhibitor, Wortmannin (23); and enhancement of the secretion of fluorescent substrates by taurocholate and PI3K-activating synthetic peptide (23).

Intracellular distribution and trafficking of MDR1-GFP was recently studied in stably transfected WIF-B cells (23). Fluorescence of the MDR1-GFP chimera was exclusively detected in Golgi and bile canalicular membranes; no labeling of basolateral plasma membranes was observed. To visualize movement of MDR1-GFP chimeric protein between Golgi and bile canaliculi, fluorescent images of stably transfected WIF-B cells were serially examined by confocal microscopy. Digital fluorescent images were collected for 20 min and converted into a QuickTime movie (see complete movie at http://www.healthsci.tufts.edu/LABS/IMArias/Sai_F9.htm). Selected sequences from the movie are depicted in Figure 2. The upper six panels of Figure 2A show a time sequence over 2–6 min, in which MDR1-GFP moved rapidly from the Golgi (G) directly along straight or curvilinear paths and merged with the bile canaliculus (BC). In Figure 2A, middle right nine panels, tubulovesicular movement of MDR1-GFP was also observed between the bile canaliculus and pericanalicular region (arrows). Single, long tubules shrank, formed vesicles, and subsequently fused with the canalicular membrane. Other tubular structures extended from the canalicular membrane into the subapical region and retracted to the canalicular membrane (arrowhead in Figure 2A, left middle panel). Figure 2B shows a series of confocal images with tubular structures reaching directly from Golgi to the bile canaliculus (arrows). Movement of MDR1-GFP was not synchronous. Individual tubulovesicular structures frequently changed shape during translocation. Frequently, there was a brief delay following which tubular vesicles fused with the canalicular membrane. This event appeared distinct from movements of tubules in other directions (i.e., those presumably not fusing). Multiple confocal examinations of many cells indicate that the tubule is not moving in and out of focus but fuses with the canalicular membrane.

In addition, incubation of MDR1-GFP stably transfected WIFB9 cells at 15°C for 20 h revealed only colocalization of MDR1-GFP with Golgi markers. Following an increase in incubation temperature to 37°C, MDR1-GFP progressively moved to the canalicular plasma membrane within 30–60 min; this process was accelerated on incubation of cells with taurocholate, and the entire process, including release from Golgi, was prevented by preincubation with Wortmannin. At no time was basolateral membrane localization of MDR1-GFP observed. This model

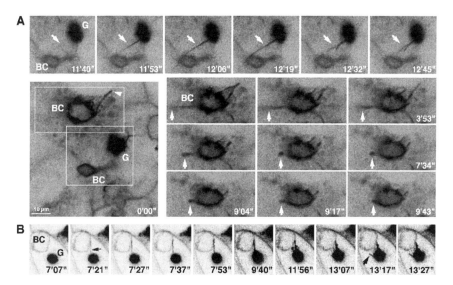

Figure 2 Visualization of intracellular movement of MDR1-GFP protein between Golgi and canalicular membrane in stably transfected WIF-B9 cells. WIF-B9 cells stably transfected with MDR1-GFP were grown on glass cover slips, mounted on the microstage, and maintained at 37°C. *A* and *B* represent independent experiments using independent cell cultures. The cells that express MDR1-GFP both in canaliculi and the perinuclear Golgi region were identified under phase and epifluorescence microscope. Digital images were collected using confocal microscope at 3.24-s intervals for about 20 min. Every four images were averaged to reduce background noise. See QuickTime movie at http://www.healthsci.tufts.edu/LABS/IMArias/Sai_F9.htm. Smaller panels on top and right were time sequences of clipped images (actual time in observation was indicated) from white rectangles in the large middle-left panel. G, Golgi compartment; BC, canalicular membranes. See text for explanation (23).

provides further opportunity to examine the role of specific candidate participants in intracellular trafficking and membrane localization.

The observed direct Golgi-to-bile canalicular trafficking of MDR1-GFP in WIFB 9 cells is consistent with the membrane targeting detected using C219 antibody (MDR1, MDR2) in rat metabolic labeling studies in vivo (18). Furthermore, the movement of MDR1-GFP from Golgi to the canalicular membrane was tubulovesicular in appearance and intermittent (occurring every 5–20 min). A speed of 0.02–0.6 μm/s was slightly lower than values obtained from other studies, which ranged from 0.3 to 1 μm/s (25–27). The process closely resembles the movement of VSV-G protein from ER to Golgi in nonpolarized cells (25, 28) and previously described Golgi-to-plasma-membrane trafficking involving large tubular-vesicular structures (26, 29–31) rather than discrete vesicles, which has been the conventional interpretation (32).

Characteristics of Intrahepatic ABC Transporter Pools

Gatmaitan et al. (11) and others (12) have observed that bile secretion is significantly enhanced in isolated perfused rat liver after treatment of these animals with the second messenger, cAMP, or the bile salt taurocholate. Increased bile secretion resulted from increased amounts of ABC transporters in the bile canalicular membrane after administration of cAMP or taurocholate. Because the increase in canalicular ABC transporter amount was dependent on an intact microtubule system and occurred within minutes of administration, it was proposed that intrahepatic pools provide additional transporters that could be rapidly recruited to the canalicular membrane. Previous morphological studies in rats rendered cholestatic by bile duct ligation (33) or by administration of phalloidin (34) or lipopolysaccharide (35) suggested that MRP2 and SPGP traffic from the bile canaliculus to intracellular sites. In addition, MRP2 (35) and SPGP (9) were observed by immunogold staining and electron microscopy in undefined vesicular structures that were distinct from the bile canalicular membrane.

ABC transporter trafficking from intracellular sites to the hepatocyte apical domain was induced by cAMP and taurocholate and recently used to confirm the existence and define the properties of potential intrahepatic ABC-transporter pools in an in vivo study in rats (19). Administration of cAMP or taurocholate increased amounts of MDR1, MDR2, and SPGP in the bile canalicular membrane by approximately threefold. These effects abated after 6 h, and the bile canalicular content of MDR1, MDR2, and SPGP returned to basal levels. Pretreatment of rats with cycloheximide, an agent that inhibits protein biosynthesis, did not prevent the increase of canalicular ABC transporter amounts upon application of cAMP or taurocholate. These data clearly demonstrate that additional ABC transporters in the bile canalicular membrane do not result from enhanced transcription or translation but indicate recruitment from existing intracellular pools. Using ^{35}S-methionine metabolic labeling, the overall half life of MDR1, MDR2, and SPGP was 5 days in rat liver, suggesting that ABC transporters cycle between intracellular pools and the bile canalicular membrane prior to degradation.

The kinetics of the intrahepatic distribution of SPGP has been investigated in detail (Figure 3) after metabolic labeling of rats with ^{35}S-methionine and immunoprecipitation of SPGP from Golgi membranes, as well as from a combined endosomal fraction (CEF) and CMV. It was observed in a previous study that newly synthesized SPGP was never detected in the sinusoidal/basolateral plasma membrane of the rat hepatocyte (18), indicating non-transcytotic apical targeting of newly synthesized SPGP. Radiolabeled SPGP peaked in Golgi membranes after a chase time of 30 min, and thereafter it was virtually absent from the Golgi, indicating that processing and passage of SPGP through Golgi is complete after 30–60 min (Figure 3). SPGP peaked in CEF at 1 h and, after a chase time of 2 h, first appeared in CMVs. These experiments demonstrated that newly synthesized SPGP is targeted through an endosomal compartment before reaching the bile canalicular membrane. Furthermore, SPGP was not completely transferred

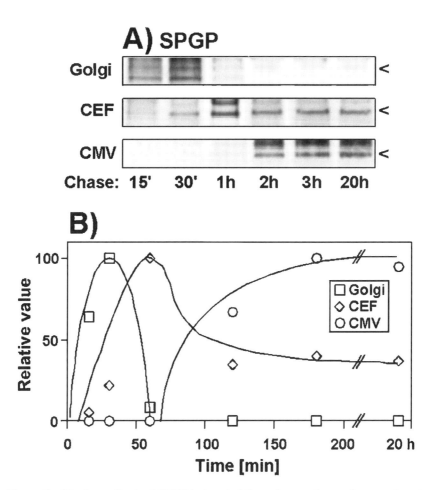

Figure 3   Newly synthesized SPGP is targeted through an endosomal compartment. Rats were pulse-labeled for 15 min with ^{35}S-methionine (5 mCi) and then chased with unlabeled methionine for 15 min, 30 min, 1 h, 2 h, 3 h, and 20 h, respectively. SPGP was then immunoprecipitated from Golgi membranes, CEF and CMVs. Immunoprecipitates were separated by SDS-PAGE and [^{35}S]SPGP was detected in a PhosphorImager. *Panel A* shows representative results observed in three independent sets of rats; arrowheads indicate the position of mature antigens. Panel B: To establish the kinetics of newly synthesized SPGP trafficking through cellular compartments, intensities of [^{35}S]SPGP-bands were quantified with a PhosphorImager. The relative intensity (highest reading in each fraction equals 100) was plotted versus labeling time; mean values $\pm$S.D., $n = 3$ (19).

from CEF, and a significant amount of SPGP remained in the endosomal fraction. These results suggested distribution of SPGP between canalicular membrane and intracellular pools (2 and 3 h chase), which was also observed after a chase time of 20 h and presumably represents steady-state distribution of SPGP between the bile canaliculus and intracellular pools under basal conditions (Figure 3).

Previous studies (11) indicate that the effects of cAMP and taurocholate on bile canalicular ABC transporter amount are additive rather than alternative, which suggests the presence of at least two distinct intrahepatic pools of ABC transporters: one that is mobilized to the canalicular membrane by cAMP (cAMP-pool) and the other by taurocholate (TC-pool). The hypothesis of two distinct intrahepatic pools of ABC transporters was further supported by the observation that targeting of newly synthesized SPGP through intrahepatic sites to the bile canalicular membrane is accelerated by cAMP but not by taurocholate (19). A tentative model for the intrahepatic pathways of ABC transporters is shown in Figure 4. After passage through Golgi, SPGP accumulates in an intrahepatic cAMP-pool and later equlibrates with the TC-pool. Whether equilibration of newly synthesized SPGP with the TC-pool occurs from the bile canalicular membrane or the cAMP-pool remains unclear. Newly synthesized MDR1 and MDR2 bypass the intracellular pools on their journey to the bile canaliculus (18, 23). However, at steady-state

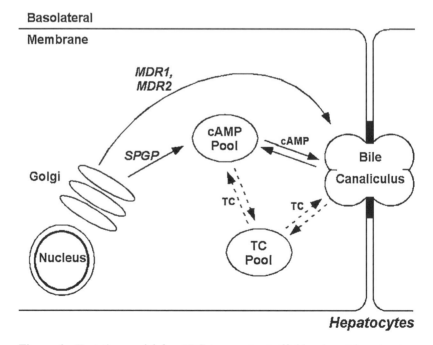

Figure 4 Tentative model for ABC-transporter trafficking in rat hepatocytes. Dashed arrows indicate possible pathways of ABC transporters from the TC-pool. See text for explanation (19).

levels, MDR1 and MDR2 are also mobilized to the bile canalicular membrane by cAMP and taurocholate, suggesting that these ABC transporters also equilibrate with intrahepatic pools after reaching the bile canalicular membrane.

Upon stimulation with cAMP, trafficking of membrane transporters from intracellular sites to the plasma membrane has been described in several systems: (a) cystic fibrosis transmembrane regulator (CFTR) channel into the apical surface of rat duodenal villous epithelia (36); (b) Na^+-taurocholate cotransport protein (ntcp) in the basolateral membrane of rat hepatocytes (37); (c) aquaporin-2 water channel into the apical membrane of LLC-PK1 cells, a polarized renal cell line (38); (d) H^+/K^+-ATPase into the apical membrane of gastric parietal cells (39); and (e) insulin-responsive glucose transporter 4 (GLUT4) into the plasma membrane of rat adipocytes (40). In each of these examples, recruitment of transporters to the plasma membrane from a recycling endosome has been suggested. In particular, trafficking of GLUT4 in rat adipocytes parallels that of ABC transporter trafficking in rat hepatocytes. GLUT4 traffics from distinct intracellular sites to the plasma membrane in response to cAMP and insulin (40). In analogy to these other systems, we propose that cAMP recruits ABC transporters to the bile canalicular membrane from a recycling endosome, whereas the effect of taurocholate appears to be hepatocyte specific and involves a different mechanism yet to be defined.

An interesting question is the distribution ratio of ABC transporters between the bile canalicular membrane and intrahepatic sites. For MRP2, the canalicular- to-intracellular ratio was calculated to be 1:1 by quantitation of immunogold-stained MRP2 in electron microscopy (35) in rat liver under basal conditions. Investigation of ABC transporters by immunoblots at steady-state levels revealed that the amounts of MDR1, MDR2, and SPGP significantly increased in the bile canalicular membrane after stimulation with cAMP or taurocholate, whereas a change in ABC transporter amount in a combined endosomal fraction prepared from the same rat liver remained below detection limit (19). This observation is in accordance with the presence of large intrahepatic pools. Under basal conditions, most MDR1, MDR2, and SPGP appear to reside in intrahepatic pools rather than in the bile canalicular membrane. However, the increase in the amount of ABC transporter in the canalicular membrane after stimulation with cAMP or taurocholate is approximately threefold for each effector. Taking into account that cAMP and taurocholate recruit transporters from different intracellular sources, the intrahepatic pool of ABC transporters is at least six times greater than the amount present in the bile canalicular membrane. Because this calculation presumes that all intracellular ABC transporters are translocated to the bile canalicular membrane upon stimulation, this number represents a lower limit. Thus the intrahepatic/canalicular ratio of MDR1, MDR2, and SPGP probably exceeds 6:1.

Intrahepatic Regulation of ABC Transporters

The mechanism of formation of the tubular vesicles and factors that control their movement to the canalicular membrane are unknown. Based on studies of vesicular transport in other polarized cells, the process probably requires ATP, Ca^{2+}, dynein and other motors, many specific GTPases including dynamin and rab family members, and other factors yet to be identified. A prominent regulator of hepatic ABC transporters has been recently identified to be PI3K (12, 41).

PI3Ks are ubiquitous lipid kinases that function as signal transducers downstream of cell surface receptors and are essential for cell proliferation, adhesion, survival, and cytoskeletal rearrangement (42, 43). Furthermore, PI3K is required for vesicle traffic in plant, yeast, and animal cells (44). The products of PI3K-catalyzed reactions serve as second messengers in many signal transduction pathways. Studies of the function of PI3K were facilitated by Wortmannin, which, when used at the appropriate concentration, is a specific inhibitor for PI3K in isolated systems and cells (45).

Misra and coworkers (12) investigated the combined effects of Wortmannin and taurocholate on bile secretion in isolated perfused rat liver (46) and on the amount of ABC transporters in the bile canalicular membrane. Different schedules of taurocholate and Wortmannin administration in perfused liver and comparison with ABC transporter amount in the bile canalicular membrane revealed two effects of Wortmannin on taurocholate-induced bile secretion. When compared with a control experiment, taurocholate perfusion increased bile secretion, which resulted from an increased amount of canalicular ABC transporters (11). Administration of Wortmannin before taurocholate prevented increase in bile secretion and simultaneously prevented increase in the amount of canalicular ABC transporters. These observations indicate that active PI3K is required for recruitment and vesicular trafficking of additional ABC transporters from intracellular pools to the bile canalicular membrane after stimulation by taurocholate. When Wortmannin was administered following taurocholate administration, the levels of canalicular ABC transporters remained elevated; however, bile secretion decreased by 50% following addition of Wortmannin to the perfusion. In these experiments, the amount of canalicular ABC transporters remained elevated; however, transport activity and bile secretion were impaired, which suggests that active PI3K is not only required for taurocholate-induced vesicular trafficking but may also play a role in regulation of ABC transporter activity in the canalicular membrane. These findings resulted in investigation of the effects of PI3K modulators on ABC transporter function in CMVs (41).

That PI3K lipid products are sufficient to enhance ATP-dependent transport activity of canalicular ABC transporters was proven by adding the $3'$-phosphoinositide lipid products to CMVs, which were prepared from rats that had received various pretreatments. ATP-dependent transport of [^{3}H]-taurocholate (by SPGP) or [^{3}H]-dinitrophenyl glutathione (by MRP2) into CMVs was enhanced by prior treatment of rats with taurocholate and inhibited by addition of Wortmannin. Addition of PI3K lipid product PtdIns$(3, 4)P_2$ to CMVs not only rescued Wortmannin inhibition but restored ATP-dependent transport of substrates above the level induced following taurocholate administration. A similar effect was observed on addition of the PI3K lipid products PtdIns(3, 4, 5)P_3 and PtdIns(3)P, but not for the structurally related PtdIns(4, 5)P_2, which lacks 3-hydroxyl phophorylation. These studies reveal a specific effect of PI3K lipid products on ABC transporter activation. Further evidence for involvement of PI3K lipid products in maximal ABC transporter function was gained by investigation of the effect of a synthetic peptide, which specifically activates PI3K. This rhodamine-linked decapeptide (47) selectively increases PtdIns(3, 4)P_2 and PtdIns(3–5)P_3 (48, 49). Addition of PI3K-activating peptide to CMV doubled ATP-dependent [^{3}H]-taurocholate transport into CMVs in a dose-dependent and saturable manner. Studies of the kinetics at ATP-dependent transport of taurocholate by SPGP and dinitrophenyl glutathione by MRP2 in CMV were previously performed at saturating levels of all substrates, and kinetic constants K_m and V_{max} were determined. Addition of the PI3K-stimulating peptide increased ATP-dependent transport of taurocholate and dinitrophenyl glutathione 1.5- to 3-fold (41). The mechanism responsible for this unexpected result is uncertain; however, we speculate that an active PI3K signal transduction system within the canalicular membrane resulted in lipid kinase products that alter and may physiologically regulate transporter activity and overall bile secretion.

These studies indicate not only that active PI3K is required for intracellular vesicular trafficking of ABC transporters in rat liver but that the lipid products of PI3K are necessary for maximal ATP-dependent transport by canalicular ABC transporters in the canalicular membrane. The mechanism by which PI3K regulates ATP-dependent transporters is not known. Direct interaction with phospholipids has been proposed for MDR1 (50, 51). Recent studies on the regulation of the K_{ATP} channel by PtdIns(4)P and PtdIns(4, 5)P_2 suggest that these negatively charged lipids bind to positive charges at the protein and thereby open the channel (52, 53). However, because the effect of the PI3K lipid products was observed on ATP-dependent transport of [^{3}H]-taurocholate (SPGP), ATP-dependent transport of [^{3}H]-dinitrophenyl glutathione (MRP2), and ATP-dependent NBD-phosphatidylcholine translocation (MDR2), but not on ATP-dependent daunomycin transport (MDR1) (S. Misra & M. Arias, unpublished observation), some, but not all, canalicular ABC transporters are regulated by PI3K lipid products.

CLOSING REMARKS

The demand to secrete bile varies with the nutritional state of the individual, and the hepatocytes cope by shifting ABC transporters between intracellular and plasma-membrane pools. However, physiological bile secretion not only depends on the static presence of ABC transporters in the appropriate location and at the appropriate time, but also involves a complex interaction with regulated traffic and transporter activation. It is a challenge to identify each of the players in this concert and, of course, the conductor.

ACKNOWLEDGMENTS

This work was supported by Deutsche Forschungsgemeinschaft research grant Ki 640 (to H. K.) and by National Institutes of Health grants DK35652 (NIDDK) and 30DK34928 (Digestive Disease Center, NIDDK) (to I. M. A.).

Visit the Annual Reviews home page at www.AnnualReviews.org

LITERATURE CITED

1. Nishida T, Gatmaitan Z, Che M, Arias IM. 1991. Rat liver canalicular membrane vesicles contain an ATP-dependent bile acid transport system. *Proc. Natl. Acad. Sci. USA* 88:6590–94
2. Nishida T, Hardenbrook C, Gatmaitan Z, Arias IM. 1992. ATP-dependent organic anion transport system in normal and TR-rat liver canalicular membranes. *Am. J. Physiol. Gastrointest. Liver Physiol.* 262:G629–G35
3. Stieger B, O'Neill B, Meier PJ. 1992. ATP-dependent bile-salt transport in canalicular rat liver plasma-membrane vesicles. *Biochem. J.* 284:67–74
4. Gatmaitan ZC, Arias IM. 1995. ATP-dependent transport systems in the canalicular membrane of the hepatocyte. *Physiol. Rev.* 75:261–75
5. Higgins CF. 1992. ABC transporters: from microorganisms to man. *Annu. Rev. Cell Biol.* 8:67–113
6. Kamimoto Y, Gatmaitan Z, Hsu J, Arias IM. 1989. The function of Gp170, the multidrug resistance gene product, in rat liver canalicular membrane vesicles. *J. Biol. Chem.* 264:11693–98
7. Ruetz S, Gros P. 1994. Phosphatidylcholine translocase: a physiological role for the mdr2 gene. *Cell* 77:1071–81
8. Nies AT, Gatmaitan Z, Arias IM. 1996. ATP-dependent phosphatidylcholine translocation in rat liver canalicular plasma membrane vesicles. *J. Lipid Res.* 37:1125–36
9. Gerloff T, Stieger B, Hagenbuch B, Madon J, Landmann L, et al. 1998. The sister of P-glycoprotein represents the canalicular bile salt export pump of mammalian liver. *J. Biol. Chem.* 273:10046–50
10. Büchler M, König J, Brom M, Kartenbeck J, Spring H, et al. 1996. cDNA cloning of the hepatocyte canalicular isoform of the multidrug resistance protein, cMrp, reveals a novel conjugate export pump deficient in hyperbilirubinemic mutant rats. *J. Biol. Chem.* 271:15091–98
11. Gatmaitan ZC, Nies AT, Arias IM. 1997. Regulation and translocation of ATP-dependent apical membrane proteins in rat liver. *Am. J. Physiol. Gastrointest. Liver Physiol.* 272:G1041–G49
12. Misra S, Ujhazy P, Gatmaitan Z, Varticovski L, Arias IM. 1998. The role of phosphoinositide 3-kinase in taurocholate-induced trafficking of ATP-dependent canalicular transporters in rat liver. *J. Biol. Chem.* 273:26638–44
13. Kipp H, Arias IM. 2000. Intracellular trafficking and regulation of canalicular ABC-transporters. *Semin. Liver Dis.* 20:339–51
14. Bartles JR, Feracci HM, Stieger B, Hubbard AL. 1987. Biogenesis of the rat hepatocyte plasma membrane in vivo: comparison of the pathways taken by apical and basolateral proteins using subcellular fractionation. *J. Cell Biol.* 105:1241–51
15. Schell MJ, Maurice M, Stieger B, Hubbard AL. 1992. 5'nucleotidase is sorted to the apical domain of hepatocytes via an indirect route. *J. Cell Biol.* 119:1173–82
16. Bartles JR, Hubbard AL. 1988. Plasma membrane protein sorting in epithelial

cells: Do secretory pathways hold the key? *Trends Biochem. Sci.* 13:181–84

17. Roelofsen H, Soroka CJ, Keppler D, Boyer JL. 1998. Cyclic AMP stimulates sorting of the canalicular organic anion transporter (Mrp2/cMoat) to the apical domain in hepatocyte couplets. *J. Cell Sci.* 111:1137–45

18. Kipp H, Arias IM. 2000. Newly synthesized canalicular ABC-transporters are directly targeted from Golgi to the hepatocyte apical domain in rat liver. *J. Biol. Chem.* 275:15917–25

19. Kipp H, Pichetshote N, Arias IM. 2001. Transporters on demand: intrahepatic pools of canalicular ATP binding cassette transporters in rat liver. *J. Biol. Chem.* 276: 7218–24

20. Ihrke G, Neufeld EB, Meads T, Shanks MR, Cassio D, et al. 1993. WIF-B cells: an in vitro model for studies of hepatocyte polarity. *J. Cell Biol.* 123:1761–75

21. Shanks MR, Cassio D, Lecoq O, Hubbard AL. 1994. An improved polarized rat hepatoma hybrid cell line. Generation and comparison with its hepatoma relatives and hepatocytes in vivo. *J. Cell Sci.* 107:813–25

22. Ihrke G, Martin GV, Shanks MR, Schrader M, Schroer TA, et al. 1998. Apical plasma membrane proteins and endolyn-78 travel through a subapical compartment in polarized WIF-B hepatocytes. *J. Cell Biol.* 141:115–33

23. Sai Y, Nies AT, Arias IM. 1999. Bile acid secretion and direct targeting of mdr1–green fluorescent protein from Golgi to the canalicular membrane in polarized WIF-B cells. *J. Cell Sci.* 112:4535–45

24. Nies AT, Cantz T, Brom M, Leier I, Keppler D. 1998. Expression of the apical conjugate export pump, Mrp2, in the polarized hepatoma cell line, WIF-B. *Hepatology* 28:1332–40

25. Presley JF, Cole NB, Schroer TA, Hirschberg K, Zaal KJ, et al. 1997. ER-to-Golgi transport visualized in living cells. *Nature* 389:81–85

26. Toomre D, Keller P, White J, Olivo JC,

Simons K. 1999. Dual-color visualization of *trans*-Golgi network to plasma membrane traffic along microtubules in living cells. *J. Cell Sci.* 112:21–33

27. Nakata T, Terada S, Hirokawa N. 1998. Visualization of the dynamics of synaptic vesicle and plasma membrane proteins in living axons. *J. Cell Biol.* 140:659–74

28. Hirschberg K, Miller CM, Ellenberg J, Presley JF, Siggia ED, et al. 1998. Kinetic analysis of secretory protein traffic and characterization of Golgi to plasma membrane transport intermediates in living cells. *J. Cell Biol.* 143:1485–93

29. McNiven MA. 1998. Dynamin: a molecular motor with pinchase action. *Cell* 94: 151–54

30. Keller P, Simons K. 1997. Post-Golgi biosynthetic trafficking. *J. Cell Sci.* 110: 3001–9

31. Lippincott-Schwartz J. 1998. Cytoskeletal proteins and Golgi dynamics. *Curr. Opin. Cell Biol.* 10:52–59

32. Traub LM, Kornfeld S. 1997. The *trans*-Golgi network: a late secretory sorting station. *Curr. Opin. Cell Biol.* 9:527–33

33. Paulusma CC, Kothe MJ, Bakker CT, Bosma PJ, van Bokhoven I, et al. 2000. Zonal down-regulation and redistribution of the multidrug resistance protein 2 during bile duct ligation in rat liver. *Hepatology* 31:684–93

34. Rost D, Kartenbeck J, Keppler D. 1999. Changes in the localization of the rat canalicular conjugate export pump Mrp2 in phalloidin-induced cholestasis. *Hepatology* 29:814–21

35. Dombrowski F, Kubitz R, Chittattu A, Wettstein M, Saha N, et al. 2000. Electron-microscopic demonstration of multidrug resistance protein 2 (Mrp2) retrieval from the canalicular membrane in response to hyperosmolarity and lipopolysaccharide. *Biochem. J.* 348:183–88

36. Ameen NA, Martensson B, Bourguinon L, Marino C, Isenberg J, et al. 1999. CFTR channel insertion to the apical surface in

rat duodenal villus epithelial cells is upregulated by VIP in vivo. *J. Cell Sci.* 112:887–94

37. Mukhopadhayay S, Ananthanarayanan M, Stieger B, Meier PJ, Suchy FJ, et al. 1997. cAMP increases liver Na^+-taurocholate cotransport by translocating transporter to plasma membranes. *Am. J. Physiol. Gastrointest. Liver Physiol.* 273:G842–G48

38. Fushimi K, Sasaki S, Marumo F. 1997. Phosphorylation of serine 256 is required for cAMP-dependent regulatory exocytosis of the aquaporin-2 water channel. *J. Biol. Chem.* 272:14800–4

39. Yao X, Karam SM, Ramilo M, Rong Q, Thibodeau A, et al. 1996. Stimulation of gastric acid secretion by cAMP in a novel alpha-toxin-permeabilized gland model. *Am. J. Physiol. Cell Physiol.* 271:C61–C73

40. Pessin JE, Thurmond DC, Elmendorf JS, Coker KJ, Okada S. 1999. Molecular basis of insulin-stimulated GLUT4 vesicle trafficking. Location! Location! Location! *J. Biol. Chem.* 274:2593–96

41. Misra S, Ujhazy P, Varticovski L, Arias IM. 1999. Phosphoinositide 3-kinase lipid products regulate ATP-dependent transport by sister of P-glycoprotein and multidrug resistance associated protein 2 in bile canalicular membrane vesicles. *Proc. Natl. Acad. Sci. USA* 96:5814–19

42. Toker A, Cantley LC. 1997. Signalling through the lipid products of phosphoinositide-3-OH kinase. *Nature* 387: 673–76

43. Domin J, Waterfield MD. 1997. Using structure to define the function of phosphoinositide 3-kinase family members. *FEBS Lett.* 410:91–95

44. Fruman DA, Meyers RE, Cantley LC. 1998. Phophoinositide kinases. *Annu. Rev. Biochem.* 67:481–507

45. Arcaro A, Wymann MP. 1993. Wortmannin is a potent phosphatidylinositol 3-kinase inhibitor: the role of phosphatidylinositol 3,4,5–trisphosphate in neutrophil responses. *Biochem. J.* 296:297–301

46. Hems R, Ross BD, Berry MN, Krebs HA. 1966. Glucogenesis in the perfused rat liver. *Biochem. J.* 101:284–92

47. Janmey PA, Cunningham CC, Stossel TP, Vegner R. U.S. Patent No. 5,846,743

48. Hartwig JH, Bokoch GM, Carpenter CL, Janmey PA, Taylor LA, et al. 1995. Thrombin receptor ligation and activated Rac uncap actin filament barbed ends through phosphoinositide synthesis in permeabilized human platelets. *Cell* 82:643–53

49. Lu PJ, Shieh WR, Rhee SG, Yin HL, Chen CS. 1996. Lipid products of phosphoinositide 3-kinase bind human profilin with high affinity. *Biochemistry* 35:14027–34

50. Dogie CA, Yu X, Sharom FJ. 1993. The effect of lipids and detergents on ATPase-active P-glycoprotein. *Biochim. Biophys. Acta* 1146:65–72

51. Sharom FJ. 1997. The P-glycoprotein multidrug transporter: interactions with membrane lipids, and their modulation of activity. *Biochem. Soc. Trans.* 25:1088–96

52. Shyng SL, Nichols CG. 1998. Membrane phospholipid control of nucleotide sensitivity of K_{ATP} channels. *Science* 282:1138–41

53. Baukrowitz T, Schulte U, Oliver D, Herlitze S, Krauter T, et al. 1998. PIP_2 and PIP as determinants for ATP inhibition of K_{ATP} channels. *Science* 282:1141–44

Annu. Rev. Physiol. 2002. 64:609–33

CHLORIDE CHANNELS AND HEPATOCELLULAR FUNCTION: Prospects for Molecular Identification

Xinhua Li[1] and Steven A. Weinman[1,2]

[1]*Department of Physiology and Biophysics and* [2]*Department of Internal Medicine, University of Texas Medical Branch, Galveston, Texas 77555-0641; e-mail: xinli@utmb.edu, sweinman@utmb.edu*

Key Words ClC channels, liver, cell volume regulation, apoptosis

■ **Abstract** Hepatocytes possess chloride channels at the plasma membrane and in multiple intracellular compartments. These channels are required for cell volume regulation and acidification of intracellular organelles. Evidence also supports a role of chloride channels in modulation of apoptosis and cell growth. Swelling- and Ca^{2+}-activated chloride channels have been identified in hepatocyte plasma membranes, and chloride channels have been observed in the membranes of lysosomes, endosomes, Golgi, endoplasmic reticulum, mitochondria, and the nucleus. This review summarizes the functions of these channels and discusses the specific channel molecules they may represent. Chloride channel molecules shown to be expressed in hepatocytes include members of the ClC channel family (ClC-2, ClC-3, ClC-5, and ClC-7), members of the newly identified CLIC family of intracellular chloride channels (CLIC-1 and CLIC-4), the mitochondrial voltage-dependent anion channel, and a newly identified intracellular channel, MCLC (Mid-1 related chloride channel). Current understanding does not include a molecular identification of most of the observed channel functions, but details of the molecular properties of these channel molecules should allow future identification and further understanding of chloride channel function in hepatocytes.

INTRODUCTION

Hepatocytes, the parenchymal cells of the liver, are multifunctional epithelial cells that engage in transcellular solute transport, processing of metabolites, and synthesis and export of numerous important proteins. Each of these processes requires the participation of anion channels. Hepatocytes express an array of anion channels that allow them to carry out these functions. Plasma membrane chloride channels are important in cell volume regulation in hepatocytes (1), as in most other cells (2, 3). Other roles for hepatocyte plasma membrane chloride channels include control of membrane potential, transcellular transport, and mediating the cell shrinkage associated with apoptosis. Chloride channels in the endocytic and secretory compartments play a critical role in acidification and function of these organelles (4), and chloride channels present in mitochondria are responsible for changes in mitochondrial permeability that are essential for apoptosis (5).

In spite of functional identification of some of these channels, in most cases there has been no molecular identification of the channel molecules themselves. This contrasts dramatically with the situation for hepatic organic ion transporters, where there has been molecular identification of proteins responsible for almost all known transport functions (6–8). There are several reasons for this difficulty in identifying chloride channels. The functional properties of the different channels are quite similar, there are no expression systems that lack endogenous chloride channels, and endogenous channels are frequently activated in response to protein overexpression (9, 10).

Nonetheless, the past decade has brought great progress in the molecular identification of chloride channels. There are now six classes of chloride channels that have been identified and members of four of these [the ClC family, the CLCA family, the CLIC family, and the mitochondrial voltage-dependent anion channel (VDAC)] appear to be expressed in hepatocytes. Two other channel types, the cystic fibrosis transmembrane conductance regulator (CFTR) and the neuronal ligand-gated chloride channels, do not appear to be expressed in hepatocytes. In spite of this information, there is no clear correspondence of channel function to channel identity for any of these channels except VDAC (5).

The central challenge in understanding the role of chloride channels in hepatocytes is to reconcile the functional data with the emerging molecular identification of channels. This review describes the functional aspects of hepatocyte chloride channels at the plasma membrane and intracellular sites and discusses the molecular candidates for these channels. Although absolute identifications cannot yet be made, there has been rapid progress in this field, and it is likely that the next several years will see further reconciliation of these two aspects of channel biology.

PLASMA MEMBRANE CHLORIDE CHANNELS

Classical electrophysiological studies of hepatocytes have demonstrated that the hepatocyte plasma membrane is highly permeable to chloride. Under basal physiological conditions, chloride accounts for 25%–80% of membrane conductance, and intracellular chloride is at electrochemical equilibrium (11). The high basal chloride conductance is the primary reason hepatocyte membrane potential is only in the range of -30 to -40 mV (12).

Swelling-Activated Chloride Channels

In response to cell swelling, hepatocytes increase their chloride conductance by as much as 30- to 100-fold (13, 14). Although a similar phenomenon is present in almost all cell types, both the magnitude of this effect and its physiological importance are particularly large in hepatocytes. Unlike most other cells, hepatocyte function requires cycles of swelling and shrinkage. Hepatocytes take up large quantities of amino acids and other nutrients and consequently swell in response to meals (15). Volume regulation is dependent on a rapid increase in chloride conductance (1, 16).

Volume regulatory decrease is chloride dependent and blocked by chloride channel inhibitors or Cl^- substitution in perfused liver (16), isolated hepatocytes (1), and hepatoma cells (17). Studies in both primary hepatocytes (13, 18) and a rat hepatoma cell line (HTC cells) (17) have observed outwardly rectifying chloride conductances that activate with a lag time of minutes, have an $I^- > Cl^-$ selectivity sequence, and are inhibited by several channel blockers including 4,4'-diisothiocyanostilbene-2,2'-disulfonic acid (DIDS) and 5-nitro-2-(3-phenylpropylamino)benzoic acid (NPPB) (see Table 1). The chloride channels are also permeable to nonelectrolytes such as taurine (19). This situation is similar to other models of swelling-activated chloride channels, but in hepatocytes these channels can be activated by alanine uptake as well as osmotic gradients (20).

The mechanism of activation of the swelling-activated chloride channel has been the subject of considerable investigation, but no definitive model is applicable to all situations. Several kinases can be activated by cell swelling, and blocking their activity impairs channel activation. In H4IIE rat hepatoma cells, cell swelling activates the mitogen-activated protein kinases Erk-1 and Erk-2 (21), and this activation is dependent on the action of a tyrosine kinase. In lymphocytes the p56lck tyrosine kinase is activated by cell swelling and is required for chloride channel activation (22), but a direct role of this kinase has not been demonstrated in hepatocytes. Protein kinase C (PKC)-dependent events play a role in channel activation,

TABLE 1 Plasma membrane chloride currents in hepatocytes and hepatoma cells

Cell type	Activated by	Rectification[a]	Inactivation at positive voltage	Ca^{2+}-dependent	Inhibitors[b]	Selectivity	Reference
Rat hepatocytes	Swelling	Outward	Yes	No	NPPB, DIDS	I > Br > Cl	13
Human hepatocytes	Swelling	Outward	Yes		NPPB		34
	ATP	Outward	Yes		NPPB		34
Guinea pig hepatocytes	Ca^{2+}	Outward	No	Yes			52
Skate hepatocytes	Swelling	Outward	No		DIDS		14, 130
HTC cells	Swelling	Outward	Yes		NPPB		17, 49
	ATP	Outward	No	No			23, 30, 32
	Alanine	Outward	No		NPPB	I > Br > Cl	20
AML12 cells	Swelling	Outward	Yes		NPPB, DIDS	I > Br > Cl	79

[a]With symmetric Cl^--containing solutions.

[b]NPPB, 5-nitro-2-(3-phenylpropylamino)benzoic acid; DIDS, 4,4'-diisothiocyanostilbene-2,2'-disulfonic acid.

and PKC inhibitors block swelling-induced chloride currents in HTC cells without inhibiting the currents once they have been activated (23). Phosphatidylinositol 3-kinase is activated by cell swelling and plays a role in a number of vesicular trafficking steps in hepatocytes (24, 25). Blockers of this process inhibit chloride channel activation (26), but the precise relationship of this kinase to the chloride channels is unknown.

Several other second messenger systems have been investigated but do not appear to be involved. Although cell swelling does cause a transient increase in intracellular Ca^{2+}, the swelling-activated channels are not Ca^{2+} dependent. Cyclic AMP dependent processes alter the set point of the channel but do not activate chloride channels when cell volume is reduced (27). Endogenous cysteinyl leukotriene production induced by cell swelling has been suggested to be responsible for chloride channel activation (28), but isolated hepatocytes are able to activate this channel even though they do not express 5-lipoxygenase, the enzyme required for endogenous leukotriene production (29).

An interesting aspect of hepatocyte swelling-activated anion channels is the close relationship with extracellular ATP release. It has been clearly established that hepatocytes, like many other cells, release ATP in response to mechanical stimuli (30, 31). In addition, HTC hepatoma cells respond to micromolar concentrations of extracellular ATP by developing chloride conductances that are indistinguishable from swelling-activated chloride currents (32). In a series of elegant experiments, Wang et al. (30) demonstrated that removal of extracellular ATP with the phosphatase apyrase or with hexokinase/glucose abolished both volume regulation and swelling-induced currents. Furthermore, addition of extracellular ATP under this condition reconstituted both (30). Further studies have shown that inhibition of ATP release by gadolinium blocks volume regulation, but this can be circumvented with extrinsic ATP (33). These observations suggest a model in which the primary event in cell swelling is channel-mediated ATP release, which then binds to a P_2 receptor and directly increases chloride conductance in a Ca^{2+}-independent manner. Both the identity of the P_2 receptor and the mechanisms by which it is coupled to the Cl^- channel are unknown.

Extracellular ATP can clearly trigger Cl^- channel activation in response to cell swelling, but it may not be the sole mechanism of swelling-induced chloride channel activation. In other cell types, osmotic cell swelling also causes ATP release, but this ATP is not responsible for chloride channel activation. In human intestine 407 cells, osmotic cell swelling results in release of micromolar concentrations of ATP at the cell surface. However, an antibody that completely blocked ATP release had no effect on swelling-activated chloride conductance (35). Furthermore, although extracellular ATP potentiated chloride channel activation by hyposmolality, it did not directly activate the channels by itself (35a). These data show that autocrine ATP signaling is not necessary for swelling-induced chloride channel activation in intestinal 407 cells. In hepatocytes and hepatoma cells, extracellular ATP does directly activate chloride channels (30, 34), but other activation pathways may also play a role.

A related issue is the possibility of multiple different chloride channels involved in the swelling response. Unfortunately, there is no molecular identification of the channel itself, and the biophysical signature of this channel (Table 1) is not unique. It is possible there are two different channels, a direct cell swelling-activated channel not dependent on ATP release, as well as an extracellular ATP-activated channel. Depending on exact experimental conditions, one or the other of these may not be activated.

Single-channel studies to define the swelling-activated channel in hepatocytes have not been reported, although a plasma membrane chloride-selective single channel displaying outward rectification has been observed in rat hepatocytes (36) (see Table 2). In other cell types, including lymphocytes (22), rabbit atrial myocytes (37), lung cancer cells (35), and glioma cells (38), the swelling-activated channels have been tentatively identified as intermediate conductance (20–80 pS) channels that display single-channel outward rectification [see Nilius et al. (3) for a detailed review]. There is still no convincing molecular identity of these channels. Potential molecular candidates have been proposed, such as pI_{Cln} (39) and P-glycoprotein (40), but in each case further investigation has shown that the proposed channel was not responsible for the observed currents (41). This issue has recently been discussed (9, 10).

Two candidates have specifically been proposed to represent the swelling-activated channel in hepatocytes, ClC-3 and ClC-2. Duan et al. proposed that ClC-3 is the swelling-activated channel in cardiac myocytes (42). Pursuant to this idea, Shimada et al. (43) demonstrated that ClC-3 was expressed in hepatocytes in both an intracellular and canalicular membrane location. Canalicular localization could contribute to volume regulatory chloride flux into the biliary space. However, further examination demonstrated that ClC-3 does not have the biophysical or pharmacological properties of the swelling-activated chloride channel (44). When expressed in either CHO-K1 cells or the human hepatoma cell line Huh-7, it is primarily intracellular, but a small fraction of expressed ClC-3 does appear on the plasma membrane. In this expression system, ClC-3 is an extremely outward-rectifying channel that is insensitive to DIDS and NPPB and that is not activated by cell swelling. These properties are identical to those of ClC-4 and ClC-5 (45). ClC-3, therefore, is not the swelling-activated channel. This conclusion is definitively supported by the finding that swelling-activated channels in hepatocytes are normal in ClC-3 knockout mice (46).

ClC-2 has also been proposed to be a swelling-activated chloride channel of hepatocytes. It is an inward-rectifying plasma membrane channel that has previously been shown to be activated by cell swelling (47, 48). It is expressed in hepatocytes and localizes in or around the plasma membrane of HTC hepatoma cells (49). Several lines of evidence suggest that it could be responsible for a component of the swelling-activated channels in these cells. Although ClC-2 is an inward-rectifying channel and swelling-activated currents are primarily outward rectifying, there is an inward component to swelling-activated current, which could result from ClC-2. Because ClC-2 currents are DIDS insensitive and DIDS inhibits only the outward

TABLE 2 Single chloride channels in hepatocytes

Cell type	Site	Method	Conductance[a]	Inhibitors[b]	Other	Reference
Guinea pig hepatocytes	Plasma membrane	Patch clamp	7.4 pS	DIDS	Slight OR[c], Ca^{2+}- dependent	52
Rat hepatocytes	Plasma membrane	Patch clamp	30 pS[d]		OR	36
Rat hepatocytes	Canalicular membrane	Planar lipid bilayer	30 pS	No	Voltage-dependent Po	115
Rat hepatocytes	Canalicular membrane	Planar lipid bilayer	90 pS	No		115
Rat liver	Lysosomes	Planar lipid bilayer	120 pS (300 mM KCl)	DIDS	ATP or GTP activated	85
Rat liver	Golgi fraction	Planar lipid bilayer	130 pS		Activated by low pH, six subconductances	50
Rat hepatocytes	Rough endoplasmic reticulum	Planar lipid bilayer	160 pS (450/50 mM KCl)[a]		Double-barreled gating	86
Rat hepatocytes	Rough endoplasmic reticulum	Planar lipid bilayer	164 pS	NPPB	Anion selectivity: Br > Cl > 1	87
Rat liver	Isolated nuclei	Patch clamp	58 pS, 150 pS	ATP		89
Canine liver	Nuclear membrane vesicles	Planar lipid bilayers or proteoliposomes	171 pS (150 mM Cl)			90

[a]Conductances were measured in symmetric (approximately 150 mM) Cl^- solutions except where noted.

[b]DIDS, 4,4'-diisothiocyanostilbene-2,2'-disulfonic acid; NPPB, 5-nitro-2-(3-phenylpropylamino)benzoic acid.

[c]OR, outward rectification.

[d]Single-channel conductance is outwardly rectifying: 30 pS in outward direction; 10 pS in inward direction.

direction of the swelling-activated current, it is possible that ClC-2 accounts for the inward component of the swelling current. Most important, intracellular dialysis with an antibody against ClC-2 inhibits ClC-2 currents in human embryonic kidney (HEK) cells, and it also inhibits swelling-activated chloride currents in HTC cells (49).

Nonetheless, ClC-2 is probably not an important component of the swelling-activated chloride currents in hepatocytes. First, DIDS inhibition of outward but not inward currents is a common observation with many different chloride currents and has been seen even on the single-channel level (50). It may be a consequence of voltage-dependent binding of external DIDS to the channel and thus would not require the postulation of separate DIDS-sensitive and -insensitive channels. Furthermore, single channels associated with cell swelling are generally outwardly rectifying, but these tend to rectify only weakly and thus conduct inward currents as well (Tables 1, 2). Therefore, there is no need to postulate separate inwardly and outwardly conducting channels to account for the whole cell voltage dependence. Finally, although the ClC-2 antibody clearly inhibits the ClC-2 currents in transfected cells, it completely inhibited both inward and outward components of the swelling-activated chloride current in hepatocytes. Because the outward component cannot result from ClC-2, the antibody must be inhibiting other channels as well. Therefore, although ClC-2 is a swelling-activated channel and is present in hepatocytes, it does not account for the observed currents. Antibodies against ClC-2 seem to prevent activation of other channels. This suggests that ClC-2 may play a role in channel activation, as has been suggested for other ClC channels (51).

Ca^{2+}-Activated Chloride Channels

Ca^{2+}-activated chloride channels have been identified in hepatocytes both by whole cell and single-channel patch clamp. Koumi et al. (52) demonstrated a distinct Ca^{2+}-dependent whole cell chloride current in guinea pig hepatocytes, which shows outward rectification and voltage-dependent activation at positive voltages. We have observed a similar current in rat hepatocytes, which can be distinguished from the swelling-activated current by its dependence on Ca^{2+}, its positive voltage-dependent activation, and its sensitivity to low concentrations of niflumic acid (53). These properties are similar to those of Ca^{2+}-activated chloride currents seen in many other cells (54).

Single-channel studies in guinea pig hepatocytes have identified a small, 7-pS channel with Ca^{2+}-dependent open probability and half maximal activation at a free Ca^{2+} concentration of 0. 5 μM (52). These channels are likely to underlie the norepinephrine-induced increase in chloride conductance seen in guinea pig hepatocytes (55).

There has been recent progress in the identification of a family of Ca^{2+}-activated chloride channels (the CLCA family), which are expressed in a broad array of cells and tissues (56, 57). There are currently 12 members of this family identified from bovine, human, mouse, pig, and rat tissues. The prototype, bovine CLCA1, is expressed as a 130-kDa precursor cleaved into a 90-kDa transmembrane component and a 32- to 38-kDa component. One family member, hCLCA-3, is a secreted protein and probably does not function as a chloride channel (58).

When expressed in HEK cells, the CLCA channels have very similar biophysical properties, including Ca^{2+} dependence, outward rectification, and sensitivity to DIDS and niflumic acid. In expression systems they do not show the time- and voltage-dependent activation that characterizes Ca^{2+}-dependent chloride currents in hepatocytes and other cells (59, 60). Single-channel studies have shown a single-channel conductance of approximately 13 pS (61), close to that seen for the Ca^{2+}-dependent chloride in guinea pig hepatocytes (Table 2). Most of the members of the family are expressed in chloride-secreting epithelial cells, such as the trachea, small intestine, colon, kidney, and mammary gland. Immunohistochemistry demonstrates an association with the apical membrane and has led to the suggestion that these channels may serve as an alternate route for Cl^- secretion in CFTR-expressing cells (57).

Of the members so far examined, only mouse mCLCA1 mRNA is present in liver by Northern blotting (60). However, in this study, gallbladder expression was present, and it is possible that the expression in liver could be associated with bile ducts and not hepatocytes. If this is the case, then there is currently no molecular candidate for the hepatocyte Ca^{2+}-activated chloride channel.

cAMP-Activated Chloride Channels

Unlike the situation in chloride-secreting epithelia, intracellular chloride in hepatocytes is at equilibrium, and therefore opening a chloride channel does not result in vectorial chloride secretion. In chloride-secreting epithelia, CFTR functions as a cAMP-dependent chloride channel. Several studies have shown that although CFTR can be readily detected in bile duct epithelial cells, it is not present in normal hepatocytes (62, 63). This contrasts with the situation in some hepatoma cells in which CFTR is present (64).

cAMP-stimulated whole cell chloride currents have been observed in primary rat hepatocytes (13); however, these whole cell currents result from a cAMP-dependent shift in the volume set point of swelling-activated currents and not from a volume-independent cAMP current. There is thus no compelling evidence of a direct cAMP- or PKA-activated chloride channel in hepatocytes.

CHLORIDE CHANNELS AND CELL INJURY

The only definitely established function of hepatocyte plasma membrane chloride channels is the mediation of anion efflux necessary for the regulatory decrease in cell volume. Nonetheless, evidence suggests a role for chloride channels in initiation of apoptosis, cell injury, and cell proliferation as well.

Cell shrinkage is associated with apoptosis in nearly all cell types. Cidlowski and colleagues first suggested that cell shrinkage is an early event in apoptosis and results primarily from the activation of a K^+ efflux pathway (65, 66). In order for K^+ efflux to result in cell shrinkage, there must be a parallel anion efflux pathway. In volume regulatory decrease in hepatocytes, cell shrinkage results from activation of both K^+ and Cl^- channels. This results in a change in both cell volume and cell ionic strength. It is precisely this decrease in ionic strength that is necessary for activity of many of the effector enzymes of apoptosis (67). A dependence of apoptosis on channel-mediated K^+ and Cl^- efflux has been seen in neurons (68), HeLa cells (69), and lymphocytes (70). In Jurkat lymphocytes, induction of apoptosis with an anti-Fas antibody results in the activation of a 40-pS, outwardly rectifying chloride channel (70). The characteristics of this channel are similar to those of the swelling-activated chloride channel, and like that channel, activation was dependent on the tyrosine kinase p56lck (22).

The possibility that chloride channel activation in hepatocytes may precede apoptosis and injury is supported by several observations. Meng et al. (71) observed that hepatocytes isolated from lipopolysaccharide-treated rats have greatly increased chloride currents. The biophysical properties of this current were identical to those of the swelling-activated current but were not explained by changes in cell volume (71). Nietsch et al. (72) similarly observed activation of both K^+ and Cl^- channels in tumor necrosis factor (TNF)-treated HTC hepatoma cells. These authors extended this observation by demonstrating that the chloride currents could be inhibited by NPPB or N-phenylanthranilic acid, and both of these inhibitors delayed cell death induced by TNF/actinomycin D (72). Similar results were obtained by Maeno et al. (69), who observed that either staurosporin- or TNF/cycloheximide-induced apoptosis was preceded by early cell shrinkage, and that both the change in cell volume and eventual apoptosis could be prevented by the chloride channel blockers NPPB or DIDS (69).

The mechanisms by which hepatocellular chloride channels are activated in inflammation have not been defined. Intracellular cysteinyl leukotrienes can activate these channels (71) and leukotrienes accumulate in hepatocytes during inflammation (29). Channel activation by leukotrienes, however, is not a result of ligand binding to leukotriene receptors. Rather it requires leukotriene interaction with the canalicular membrane transport protein MRP2 and can be duplicated by other substrates of MRP2. MRP2 appears to function as a substrate-dependent channel activator, which accelerates the activation of the swelling-activated channel (73).

Several other observations have also associated chloride channels with hepatocyte growth and death. Glycine prevents hepatocyte death from anoxia, ATP depletion, or ischemia/reperfusion (74, 75). Similar observations have been made in renal tubule cells, where it has been proposed that glycine cytoprotection is due to its ability, at high concentrations, to block glycine-gated chloride channels (76). This is supported by the demonstration that both glycine and strychnine, another inhibitor of glycine-gated chloride channels, prevent a rise of hepatocellular chloride content induced by hypoxia and that both are cytoprotective (77). However, although glycine is clearly cytoprotective, the mechanism of this effect may not involve chloride channels. Glycine cytoprotection is unaffected by complete chloride replacement (78), and hepatocytes have never been shown to express the ligand-gated chloride channels expected to be inhibited by excess glycine or strychnine. There are thus no compelling data to link chloride channels to the glycine effect.

Wondergem et al. observed an association of chloride conductance with cell proliferation in a transformed hepatocyte cell line (79). In this study, dividing cells have a high membrane chloride conductance that has similar properties to the swelling-activated conductance. Nondividing cells did not have this conductance, and two chloride channel blockers, NPPB and DIDS, inhibited proliferation. Of course, nonspecific effects of the channel blockers cannot be excluded, but this is an additional piece of circumstantial evidence that suggests chloride channel activation is involved in cell growth and cell death.

INTRACELLULAR CHLORIDE CHANNELS

Intracellular chloride channels are present in membranes of various organelles in hepatocytes, where they facilitate electrogenic cation transport by dissipating membrane potential gradients. This process is important for transport of H^+ (80), Ca^{2+} (81), and Cu^+ (82). Acidification of intracellular organelles, mediated by the vacuolar V-type H^+-ATPase, is particularly dependent on chloride channels. Electrogenic proton transport produces an interior positive electrochemical potential, which opposes further transport (80, 83). Chloride conductance is necessary to allow proton translocation to proceed, and rates of ATP-dependent acidification vary depending on the presence of chloride in endosomes and lysosomes (80, 84). Properties of the channel activities observed in several intracellular preparations from liver are summarized in Table 2.

Functional Identification of Hepatocyte
Intracellular Chloride Channels

LYSOSOMES   Lysosomes maintain a low intraluminal pH by a chloride-dependent process. Tilly et al. (85) studied chloride conductance in purified liver lysosomal membranes. They showed that the vesicles possessed a high anion permeability, as assessed by chloride-dependent quenching of the chloride-sensitive fluorescent dye 6-methoxy-N-(3-sulfopropyl quinolinium) (SPQ) fluorescence. They also observed single chloride channel activity in lipid bilayer fusion experiments. These channels were strongly activated by ATP, GTP, and ATPγS, but not by AMP, ADP, cAMP, CTP, or UTP. They were inhibited by relatively low concentrations of DIDS. Single-channel conductance was 120 pS in symmetrical 300 mM KCl solutions, and the channel rectified slightly. Because of uncertainty about the orientation of the vesicles, it cannot be determined whether this rectification is physiologically relevant.

GOLGI   Nordeen et al. (50) studied an ion channel from rat liver Golgi membrane, which they named GOLAC (Golgi anion channel). To exclude the possibility of contamination by channel molecules transiting through the Golgi, they first cleared the Golgi with cycloheximide treatment. A predominantly single-channel species was observed that had a large open-state conductance of 130 pS and that displayed six partial-conductance states. The channel activity was not changed by nucleotides. The channel is selective for $I^- > Br^- > Cl^-$, has an open probability that is voltage independent, and is inhibited by DIDS. The channel was activated by pH reduction from 7.2 to 5.2 on the side of the bilayer corresponding to the Golgi lumen.

ENDOPLASMIC RETICULUM   Chloride channels in the endoplasmic reticulum (ER) have been proposed to serve the function of charge neutralization during Ca^{2+} efflux (81). Two studies have explicitly examined the nature of rat liver ER channels in bilayer fusion experiments. Morier & Sauvé (86) isolated rat hepatocyte rough ER membrane vesicles and fused them with lipid bilayers. They primarily observed a chloride channel that had two current levels, i.e., transitions of 160 and 320 pS. They interpreted this to represent double-barrel channel behavior. This channel displayed voltage-dependent open probability, which increased at negative potentials. Ionic selectivity and inhibitor sensitivity of this channel were not determined.

Eliassi et al. (87) similarly examined channels resulting from fusion of rat hepatocyte rough endoplasmic reticulum (RER) vesicles with lipid bilayers. They predominantly observed a single chloride channel type with a single-channel conductance of 164 pS, Br > Cl > I selectivity, sensitivity to NPPB, and resistance to 4,4′-dinitrostilbene-2,2′-disulfonic acid. Similar to the findings of Morier & Sauvé, their channel had a voltage-dependent open probability that increased at negative potentials. Although this latter study did not identify a double-barrel behavior of this channel, the other properties are quite similar and suggest that both studies identified the same molecule.

The voltage dependence of this channel is significant. It has a near unity open probability at negative voltages, and this is decreased at least twofold at positive voltages. The orientation of the vesicles in the bilayer experiments suggests that the channels would have the higher open probability when the cytoplasm is negative to the vesicle lumen. This would make the channels most active in the physiological state, i.e., when the lumen is positive. A sudden change to a lumen negative potential, for example by opening of a Ca^{2+} channel, would tend to reduce chloride conductance and could be a factor limiting the quantity of Ca^{2+} released.

NUCLEAR MEMBRANE   The nuclear envelope consists of an outer membrane, which is continuous with the ER, and an inner nuclear membrane. These fuse at points to form the nuclear pore complexes (88). The outer membrane has been shown to have several types of chloride channels. Tabares et al. (89) performed patch clamp on isolated rat liver nuclei and observed two chloride channels: a 150-pS channel with an outwardly rectifying single-channel conductance, and a 58-pS channel with a linear conductance. Both channels were inhibited by DIDS and niflumic acid and were Ca^{2+} independent. A similar channel was identified by Guihard et al. (90), who performed patch clamp studies of giant proteoliposomes reconstituted with nuclear membranes from hepatocytes. They observed a 171-pS chloride channel, which had a voltage-dependent open probability and a very high P_{Cl}/P_K ratio. Inhibitor sensitivity of this channel was not determined.

Molecular Identification of Hepatocyte Intracellular Chloride Channels

There has been great progress in the identification of molecules that may function as intracellular chloride channels. A summary of the channel molecules known to be expressed in hepatocytes or liver is presented in Table 3.

CLC CHLORIDE CHANNEL FAMILY   The ClC channel family represents a group of voltage-dependent chloride channels that are all closely related. The prototype member of this group, ClC-0, was first cloned from the electric organ of *Torpedo marmorata*. The ClC family can be divided based on sequence homology into three major groups (Figure 1). The first group, consisting of ClC-0, ClC-1, ClC-2, ClC-Ka/1, and ClC-Kb/2, are plasma membrane channels. The second group, consisting of ClC-3, ClC-4, and ClC-5, are likely intracellular channels, as are the members of the third group, ClC-6 and ClC-7. The properties of the ClC channel family have been reviewed extensively in the past few years (91–95).

The ClC channels are widely expressed. ClC-2 (49), ClC-3 (43), ClC-5 (96), and ClC-7 (97) have specifically been reported to be present in hepatocytes, and preliminary reports have identified mRNA for ClC-4, as well as for ClC-6 (98, 99). The channel properties of the ClC channels observed by single-channel studies show that they are small channels of single-channel conductance, 5–10 pS, and they are not inhibited by DIDS or NPPB (94). The ClC-3/4/5 branch shows extreme outward rectification (44, 45), and ClC-2 is inward rectifying (47).

Several members of the ClC family have been shown to be intracellular channels and thus are candidates for intracellular channels in hepatocytes. The most extensively documented situation is for ClC-5. Mutations in this channel produce a syndrome of nephrocalcinosis and low-molecular-weight proteinuria known as Dent's disease (100). The proteinuria results from a defect in endocytic uptake of protein from the tubular fluid. ClC-5 is localized to recycling early endosomes of proximal tubule and intercalated cells of the collecting duct (101–103). Mice with a disruption of the gene for ClC-5 have defective endocytosis and acidification of endosomes in proximal tubule. However, endocytosis in hepatocytes is normal, demonstrating that ClC-5 does not perform this function in the liver (96).

Of the ClC chloride channels, the best candidates for an intracellular function in hepatocytes are ClC-3 and ClC-7. Both molecules have been demonstrated in an intracellular localization in hepatocytes (43, 97). Currently, however, there

TABLE 3 Properties of chloride channel molecules expressed in liver[a]

Molecule	Amino acids	Whole cell currents[b]	Single-channel currents	Inhibitors	Anion selectivity	Other	Accession number	Reference
ClC-2	898	Yes	ND		I > Cl	Strong IR	P51788	49
ClC-3	716	Yes	ND	Not blocked by NPPB or DIDS	I > Cl	Strong OR	D17521	44
ClC-4	760	Yes	ND		I > Cl	Strong OR	Q61418	45
ClC-5	746	Yes	ND	Not blocked by DIDS	I > Cl	Strong OR	P51795	45
ClC-7	805	ND	ND				NP001278	131
P64	437	ND	42 pS (140 mM KCl/CsCl)	DNDS, TS-TM calix(4)arene		OR, activated by alkaline phosphatase	L16547, U31302	132
CLIC1	241	ND	67.5 pS (150 mM KCl)	IAA-94	Br ~ Cl > I		U93205	114
CLIC4	253	ND	43 pS (50 mM KCl)	Not blocked by NPPB, DIDS, or IAA-94			AF109196 AF104119	121
VDAC	283	ND	480 pS (100 mM KCl)	Low pH facilitates channel close		Closed at high +/− potential	NM003374	126
MCLC	541	ND	70 pS (100 mM KCl)		Br ~ Cl > F > SO_4	Not blocked by DIDS, ATP, or Gd	AB052922, AB052915	129
MCLCA1	902	Yes	ND	DIDS, niflumic acid, DTT		OR, activated by Ca^{2+}	AF047838	60

[a]ND, not determined; IR, inward rectification; NPPB, 5-nitro-2-(3-phenylpropylamino)benzoic acid; DIDS, 4,4'-diisothiocyanostilbene-2,2'-disulfonic acid; OR, outward rectification; DNDS, 4,4'-dinitrostilbene-2,2'-disulfonic acid; TS-TM, p-tetra-sulfonato-tetra-methoxy-calix(4)arene; IAA, indanyloxyacetic acid; VDAC, voltage-dependent anion channel; DTT, dithiothreitol.
[b]In a heterologous expression system.

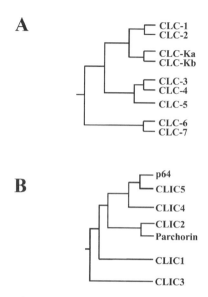

Figure 1   Dendrogram showing evolutionary relationships of the ClC family (*A*) and CLIC family (*B*) of chloride channels. Sequences were analyzed by Clustal W software.

are no functional data to define the role of either of these channels in the liver. Given the very low single-channel conductance of ClC channels, it is unlikely that they represent any of the observed intracellular channels described in the previous section. It is important to note, however, that the very small conductance of ClC channels would have made them quite difficult to observe in the bilayer studies, and the failure to identify them does not preclude an important role of these molecules as hepatocellular intracellular channels.

CLIC CHLORIDE CHANNEL FAMILY   The CLIC family represents a separate class of chloride channels that function in intracellular compartments. The initially identified family member, p64, was isolated and affinity purified taking advantage of its binding to the chloride channel blocker indanyloxyacetic acid 94 (IAA-94) (104, 105). The other six members of the family were identified by sequence homology to p64 (106, 107). Sequence homology of the members of this family is shown in Figure 1.

The p64 channel is a 438–amino acid protein predicted to be 48 kDa. On sodium dodecyl sulfate–polyacrylamide gel electrophoresis, it migrates at 64 kDa. It was purified from bovine kidney and trachea on an IAA-94 affinity column, and its incorporation into lipid bilayers revealed three distinct single-channel conductances, 26, 100, and 400 pS (104, 105). Hydropathy analysis suggests that p64 has two transmembrane segments close to the C terminus and a presumed localization of both N and C termini in the cytoplasm. It has consensus phosphorylation sites for

PKA, PKC, and casein kinase II, and tyrosine kinase–mediated phosphorylation has been shown to enhance channel activity (108).

P64 is enriched in Golgi-like microsomal membranes, and immunocytochemistry has localized it to intracellular organelles and the apical plasma membrane in pancreatic carcinoma cells. Heterologous expression produced a protein that localized to microsomes but not plasma membrane. These observations demonstrate that p64 is an intracellular channel (104, 109). Its expression is abundant in kidney cortex and skeletal muscle, and it is expressed at lower levels in liver.

CLIC-1 was initially designated NCC27 for nuclear chloride channel-27. It is a 241–amino acid protein homologous to the C-terminal half of p64 (110). It is a transmembrane protein and has only one or at most two transmembrane domains. The N terminus is oriented outward and the C terminus inward (111). In human tissues, CLIC-1 is abundant in liver, heart, kidney, and pancreas (107). Although CLIC1 was initially observed to be a nuclear channel (110), it is found in both soluble and membrane-associated fractions of kidney brush borders (112). Expression of CLIC1 correlates with cell cycle stage, and CLIC1 inhibition arrests CHO-K1 cells in the G2/M stage (113)

Electrophysiological studies confirm that CLIC1 functions as a chloride channel, but somewhat disparate results of its channel properties have been obtained. Valenzuela et al. (110) examined CLIC-1 channels in stably transfected CHO-K1 cells. Two distinct single channels were observed, a 22-pS channel in the plasma membrane and a 33-pS channel on the nuclear membrane. The nuclear channel had an ionic selectivity of $F^- > Cl^- > I^-$ (110). Tulk et al. (114) also demonstrated that CLIC-1 is capable of conducting chloride in the absence of other subunits or proteins. Similar to p64, it produces an IAA-94-sensitive Cl efflux and demonstrates single chloride channel activity with a conductance of 67.5 pS in symmetric 150 mM KCl solutions. The single-channel current had a selectivity sequence of $Br^- \sim Cl^- > I^-$, was weakly inward rectifying, showed voltage-dependent inactivation, and was inhibited by IAA-94 (114). This channel rarely closed at voltages between ± 50 mV, but beyond that range, closures occurred more frequently. A similar phenomenon was observed in rat liver canalicular plasma membranes (115). In another study using stably transfected CHO-K1 cells, Tonini et al. (111) found an 8-pS single-channel current. The explanation for the discrepancies may lie in the different experimental conditions, possibly including activation of endogenous channels in some studies.

CLIC-2 (or XAP121) has been identified solely by genomic sequencing and has not been studied at the protein level (116). The gene is on the telomeric region of Xq28 and is encoded in five exons. It predicts a 243–amino acid peptide that is 62% identical to p64 at its C-terminal half. Reverse transcriptase–polymerase chain reaction demonstrated expression in fetal liver and adult skeletal muscle tissue. No information regarding subcellular distribution or function is available.

CLIC-3 was identified in a yeast two-hybrid screening experiment using the mitogen-activated protein kinase ERK7 (extracellular signal-regulated kinase 7) as bait (117). It consists of 207 amino acids and has approximately 46% identity to other members of the CLIC family. The hydrophobic region of CLIC3 represents a very conserved putative transmembrane segment. It has two potential phosphorylation sites for casein kinase II and PKC, respectively, and one N-myristoylation site. CLIC3 mRNA is abundant in placenta, lung, and heart and is present at a lower level in skeletal muscle, kidney, and liver. It was not detected in brain. Although the protein shows a prominent nuclear localization (117), a novel Cl⁻-selective current has been detected in mouse fibroblast LTK cells transfected with CLIC-3. This is presumably due to some of the overexpressed protein appearing on the plasma membrane. The CLIC-3 current showed weak inward rectification and is selective for chloride over gluconate.

CLIC-4 (also known as RS43, p64H1, and mtCLIC) was initially identified as the first p64 homologous protein. Based on its homology with p64, Howell et al. (118) detected, by immunoblotting, a protein they called RS43. CLIC-4 is abundantly expressed in microsomes from rat liver, kidney, heart, and brain (118). Homologs of CLIC-4 in different species are about 95% identical at the amino acid level, indicating the strong conservation of this molecule in evolution. It is a 253–amino acid protein that is structurally similar to p64. Northern blotting demonstrated expression in various human organs, including liver, kidney, pancreas, heart, and skeletal muscle (106, 107, 119).

The subcellular localization of CLIC-4 varies according to the species and cell types examined. In human pancreatic carcinoma cells (Panc-1), it colocalizes with caveolae and *trans*-Golgi network. In the proximal tubule it was found in a subapical location (106). In rat hippocampal neurons, CLIC-4 was distributed in association with large, dense-core vesicles (120). CLIC-4 has been shown to reside in the mitochondrial and cytoplasmic compartments of mouse keratinocytes and is enriched in mitochondrial fractions from rat liver homogenates (119). In transfected cells, rat CLIC-4 localizes to the ER (121). There are limited data available on the channel properties of CLIC-4. In bilayer fusion studies a single channel of 43-pS conductance in 50 mM Cl was observed. This channel lacked voltage dependence or rectification (121).

CLIC-5 was first isolated as a component of cytoskeletal proteins using ezrin as an affinity bait. It has 251 amino acids and is 90% identical to the C terminus of p64. CLIC-5 appears to be expressed primarily in heart, skeletal muscle, and placenta, with little or no expression in liver (107). In placental microvilli, the immunostaining is primarily near the apical plasma membrane. To date there is no functional information on its channel activity.

Another member of the CLIC/p64 family was recently reported and has been named parchorin (122). It is a protein of 637 amino acids, 120 kDa, and is expressed primarily in the cytosol but is recruited to the apical membrane of rabbit gastric glands when secretion is activated (123). It was designated parchorin (for parietal choroid) because of its enrichment in the parietal cells of the gastric glands and cytosolic fractions of choroid plexus (122). Parchorin is 45% homologous to p64 and 64% homologous to CLIC-2. It is the largest member of the family. The highly conserved fragment is the C-terminal region, which is 75% identical to p64. Immunoblotting has shown it to be distributed exclusively in tissues involved in fluid secretion, such as chorioretinal epithelia, lacrimal glands, submandibular glands, airway epithelium, kidney, and gastric mucosa. It does not appear to be expressed in muscle, liver, or lung. Studies in parchorin-transfected pig renal epithelial cells (LLC-PK1) with the chloride indicator dye SPQ show an increased Cl^- flux, which suggests that parchorin functions as a channel.

OTHER INTRACELLULAR CHLORIDE CHANNELS   The voltage-dependent anion channel (VDAC) (or mitochondrial porin) is a family of closely related proteins that form large conductance anion channels in the mitochondrial outer membrane. It serves as a voltage-dependent permeability pathway for NADH and metabolites. VDAC is of special interest because it is a component of the mitochondrial permeability transition pore, and its activity is controled by several members of the Bcl family of proteins. The role of VDAC in apoptosis has been reviewed extensively (124, 125).

The VDACs of mammalian liver form large channels of 200–500 pS conductance, with a voltage dependence that results in channel opening at near zero voltages but with closing at hyperpolarized or depolarized voltages (126, 127). Although it has been suggested that VDAC may be present and functional in intracellular sites other than the mitochondria and also the plasma membrane, these results are likely to reflect nonspecificity of the antibody reagents used because studies with epitope-tagged VDAC derivatives identify them exclusively in association with mitochondria (128).

MCLC (Mid-1 related chloride channel) is a recently discovered intracellular chloride channel that represents a distinct channel molecule unrelated to the channels described above (129). Nagasawa et al. (129) first identified MCLC as an expressed sequence tag with sequence homology to a yeast stretch-activated Ca^{2+} channel, Mid-1. A full-length rat cDNA was subsequently obtained from a rat brain cDNA library. The protein has four putative transmembrane domains and is expressed by Northern and Western blotting in multiple rat tissues, including testis, liver, lung, muscle, and brain. When MCLC was transfected into CHO-K1 cells, it localized in the ER and Golgi. In the testis, it targeted to the nucleus, as well as to cytoplasmic sites. Fusion of microsomal vesicles from MCLC-expressing cells with planar lipid bilayers resulted in a chloride-selective single channel with a linear conductance of 70 pS in symmetric 150 mM KCl solutions. The anion selectivity sequence was $Br^- > Cl^- > F^- > SO_4^-$ (129). The role of this channel in hepatocytes, if any, remains unknown.

CONCLUSIONS

Chloride channels are critical for hepatocyte function and some channel molecules, for example ClC-3, are more than 99% identical between humans and mice. This makes them among the most highly conserved of proteins. At the hepatocyte plasma membrane, chloride channels are absolutely required for volume regulation, and they may play a necessary role in apoptosis. Intracellularly they are necessary for acidification of endosomes, lysomes, ER, and Golgi. Furthermore, a large anion channel is responsible for influx of metabolites into mitochondria and can be a part of the trigger for apoptosis.

Other than VDAC, there is no known correspondence between molecular identity and function for any of the hepatocyte chloride channels. The swelling-activated chloride channel in the plasma membrane has not been identified, and there do not seem to be any promising candidates on the immediate horizon. For the intracellular channels, identification is difficult because the most easily identifiable channel characteristic, single-channel conductance, is highly dependent on ionic composition as well as the presence of other proteins. Nonetheless, extrapolating for expected conductance in 150 mM Cl^- solutions predicts that ER, lysosomes, and Golgi all have a channel of approximately 120 pS. This is much too high to be one of the ClC channels, but it could be CLIC-4 or MCLC. In nuclei, 150-pS channels have also been observed. CLIC-1 is known to be a nuclear membrane protein, but unfortunately the reports of its channel properties are so divergent as to make positive identification impossible.

ClC-3 and ClC-7 are both intracellular channels in hepatocytes and are strong candidates for intracellular channel functions. However, this family produces very small conductance channels, which are not likely be resolvable as single-channel events in bilayer studies. Therefore, although ClC channels may be quite important physiologically, they are unlikely to represent any of the intermediate-to-large conductance channels seen in the bilayer experiments. However, recent advances have identified a number of candidate molecules and further studies will be required to match these molecules with specific functions.

Visit the Annual Reviews home page at www.AnnualReviews.org

LITERATURE CITED

1. Graf J, Haussinger D. 1996. Ion transport in hepatocytes: mechanisms and correlations to cell volume, hormone actions and metabolism. *J. Hepatol.* 24 (Suppl. 1): 53–77

2. Strange K, Emma F, Jackson PS. 1996. Cellular and molecular physiology of volume-sensitive anion channels. *Am. J. Physiol. Cell Physiol.* 270:C711–C30

3. Nilius B, Eggermont J, Voets T, Buyse G, Manolopoulos V, Droogmans G. 1997. Properties of volume-regulated anion channels in mammalian cells. *Prog. Biophys. Mol. Biol.* 68:69–119

4. Al-Awqati Q. 1995. Chloride channels of intracellular organelles. *Curr. Opin. Cell Biol.* 7:504–8

5. Shimizu S, Matsuoka Y, Shinohara Y, Yoneda Y, Tsujimoto Y. 2001. Essential role of voltage-dependent anion channel in various forms of apoptosis in mammalian cells. *J. Cell Biol.* 152:237–50

6. Kullak-Ublick GA, Stieger B, Hagenbuch B, Meier PJ. 2000. Hepatic transport of bile salts. *Semin. Liver Dis.* 20:273–92

7. Suzuki H, Sugiyama Y. 2000. Transport of drugs across the hepatic sinusoidal membrane: sinusoidal drug influx and

efflux in the liver. *Semin. Liver Dis.* 20:251–63

8. Keppler D, Konig J. 2000. Hepatic secretion of conjugated drugs and endogenous substances. *Semin. Liver Dis.* 20:265–72

9. Strange K. 1998. Molecular identity of the outwardly rectifying, swelling-activated anion channel: time to reevaluate pICln. *J. Gen. Physiol.* 111:617–22

10. Clapham DE. 1998. The list of potential volume-sensitive chloride currents continues to swell (and shrink). *J. Gen. Physiol.* 111:623–24

11. Graf J, Henderson RM, Krumpholz B, Boyer JL. 1987. Cell membrane and transepithelial voltages and resistances in isolated rat hepatocyte couplets. *J. Membr. Biol.* 95:241–54

12. Moule SK, McGivan JD. 1990. Regulation of the plasma membrane potential in hepatocytes—mechanism and physiological significance. *Biochim. Biophys. Acta* 1031:383–97

13. Meng X-J, Weinman SA. 1996. Cyclic AMP and swelling activated chloride conductance in rat hepatocytes. *Am. J. Physiol. Cell Physiol.* 271:C112–C20

14. Jackson PS, Churchwell K, Ballatori N, Boyer JL, Strange K. 1996. Swelling-activated anion conductance in skate hepatocytes: regulation by cell Cl⁻ and ATP. *Am. J. Physiol. Cell Physiol.* 270:C57–C66

15. Wondergem R, Wang K. 1993. Redistribution of hepatocyte chloride during L-alanine uptake. *J. Membr. Biol.* 135:237–44

16. Haddad P, Beck JS, Boyer JL, Graf J. 1991. Role of chloride ions in liver cell volume regulation. *Am. J. Physiol. Gastrointest. Liver Physiol.* 261:G340–G48

17. Bodily K, Wang Y, Roman R, Sostman A, Fitz JG. 1997. Characterization of a swelling-activated anion conductance in homozygous typing cell hepatoma cells. *Hepatology* 25:403–10

18. Graf J, Rupnik M, Zupancic G, Zorec R. 1995. Osmotic swelling of hepatocytes increases membrane conductance but not membrane capacitance. *Biochem. J.* 68:1359–63

19. Ballatori N, Simmons TW, Boyer JL. 1994. A volume-activated taurine channel in skate hepatocytes: membrane polarity and role of intracellular ATP. *Am. J. Physiol. Gastrointest. Liver Physiol.* 30:G285–G91

20. Lidofsky SD, Roman RM. 1997. Alanine uptake activates hepatocellular chloride channels. *Am. J. Physiol. Gastrointest. Liver Physiol.* 273:G849–G53

21. Schliess F, Schreiber R, Haussinger D. 1995. Activation of extracellular signal-regulated kinases Erk-1 and Erk-2 by cell swelling in H4IIE hepatoma cells. *Biochem. J.* 309:13–17

22. Lepple-Wienhues A, Szabo I, Laun T, Kaba NK, Gulbins E, Lang F. 1998. The tyrosine kinase p56lck mediates activation of swelling-induced chloride channels in lymphocytes. *J. Cell Biol.* 141:281–86

23. Roman RM, Bodily KO, Wang Y, Raymond JR, Fitz JG. 1998. Activation of protein kinase Cα couples cell volume to membrane Cl⁻ permeability in HTC hepatoma and Mz-ChA-1 cholangiocarcinoma cells. *Hepatology* 28:1073–80

24. Krause U, Rider MH, Hue L. 1996. Protein kinase signaling pathway triggered by cell swelling and involved in the activation of glycogen synthase and acetyl-CoA carboxylase in isolated rat hepatocytes. *J. Biol. Chem.* 271:16668–73

25. Kipp H, Arias IM. 2000. Intracellular trafficking and regulation of canalicular ATP-binding cassette transporters. *Semin. Liver Dis.* 20:339–51

26. Feranchak AP, Roman RM, Schwiebert EM, Fitz JG. 1998. Phosphatidylinositol 3-kinase contributes to cell volume regulation through effects on ATP release. *J. Biol. Chem.* 273:14906–11

27. Meng X-J, Weinman SA. 1995. cAMP and volume activated chloride conductance in rat hepatocytes. *Hepatology* 22:308A (Abstr.)

28. Jorgensen NK, Lambert IH, Hoffmann EK. 1996. Role of LTD_4 in the regulatory volume decrease response in Ehrlich ascites tumor cells. *J. Membr. Biol.* 151:159–73

29. Shimada K, Navarro J, Goeger DE, Mustafa SB, Weigel PH, Weinman SA. 1998. Expression and regulation of leukotriene-synthesis enzymes in rat liver cells. *Hepatology* 28:1275–81

30. Wang Y, Roman R, Lidofsky SD, Fitz JG. 1996. Autocrine signaling through ATP release represents a novel mechanism for cell volume regulation. *Proc. Natl. Acad. Sci. USA* 93:12020–25

31. Braunstein GM, Roman RM, Clancy JP, Kudlow BA, Taylor AL, et al. 2001. CFTR facilitates ATP release by stimulating a separate ATP release channel for autocrine control of cell volume regulation. *J. Biol. Chem.* 276:6621–30

32. Fitz JG, Sostman AH. 1994. Nucleotide receptors activate cation, potassium, and chloride currents in a liver cell line. *Am. J. Physiol. Gastrointest. Liver Physiol.* 29:G544–G53

33. Roman RM, Feranchak AP, Davison AK, Schwiebert EM, Fitz JG. 1999. Evidence for Gd(3+) inhibition of membrane ATP permeability and purinergic signaling. *Am. J. Physiol. Gastrointest. Liver Physiol.* 277:G1222–G30

34. Feranchak AP, Fitz JG, Roman RM. 2000. Volume-sensitive purinergic signaling in human hepatocytes. *J. Hepatol.* 33:174–82

35. Hazama A, Shimizu T, Ando-Akatsuka Y, Hayashi S, Tanaka S, et al. 1999. Swelling-induced, CFTR-independent ATP release from a human epithelial cell line. Lack of correlation with volume-sensitive Cl^- channels. *J. Gen. Physiol.* 114:525–33

35a. van der Wijk T, De Jonge HR, Tilly BC. 1999. Osmotic cell swelling-induced ATP release mediates the activation of extracellular signal-regulated protein kinase (Erk)-1/2 but not the activation of osmosensitive anion channels. *Biochem. J.* 343:579–86

36. Breit S, Kolb H, Apfel H, Haberland C, Schmitt M, et al. 1998. Regulation of ion channels in rat hepatocytes. *Pflügers Arch.* 435:203–10

37. Duan D, Hume J, Nattel S. 1997. Evidence that outwardly rectifying Cl^- channels underlie volume-regulated Cl^- currents in heart. *Clin. Res.* 80:103–13

38. Jackson P, Strange K. 1995. Single-channel properties of a volume-sensitive anion conductance. Current activation occurs by abrupt switching of closed channels to an open state. *J. Gen. Physiol.* 105:643–60

39. Laich A, Gschwentner M, Krick W, Nagl U, Fürst J, et al. 1997. I_{Cln}, a chloride channel cloned from kidney cells, is activated during regulatory volume decrease. *Kidney Int.* 51:477–78

40. Higgins CF. 1995. P-glycoprotein and cell volume-activated chloride channels. *J. Bioenerg. Biomembr.* 27:63–70

41. Vanoye CG, Altenberg GA, Reuss L. 1997. P-glycoprotein is not a swelling-activated Cl^- channel; possible role as a Cl^- channel regulator. *J. Physiol.* 502:249–58

42. Duan D, Winter C, Cowley S, Hume JR, Horowitz B. 1997. Molecular identification of a volume-regulated chloride channel. *Nature* 390:417–21

43. Shimada K, Li X-H, Xu G-Y, Nowak DE, Showalter LA, Weinman SA. 2000. Expression and canalicular localization of two isoforms of the ClC-3 chloride channel from rat hepatocytes. *Am. J. Physiol. Gastrointest. Liver Physiol.* 279:G268–G76

44. Li X, Shimada K, Showalter LA, Weinman SA. 2000. Biophysical properties of ClC-3 differentiate it from swelling-activated chloride channels in

CHO-K1 cells. *J. Biol. Chem.* 275:35994–98

45. Friedrich T, Breiderhoff T, Jentsch TJ. 1999. Mutational analysis demonstrates that ClC-4 and ClC-5 directly mediate plasma membrane currents. *J. Biol. Chem.* 274:896–902

46. Stobrawa SM, Breiderhoff T, Takamori S, Engel D, Schweizer M, et al. 2001. Disruption of ClC-3, a chloride channel expressed on synaptic vesicles, leads to a loss of the hippocampus. *Neuron* 29:185–96

47. Furukawa T, Ogura T, Katayama Y, Hiraoka M. 1998. Characteristics of rabbit ClC-2 current expressed in Xenopus oocytes and its contribution to volume regulation. *Am. J. Physiol. Cell Physiol.* 274:C500–C12

48. Xiong H, Li C, Garami E, Wang Y, Ramjeesingh M, et al. 1999. ClC-2 activation modulates regulatory volume decrease. *J. Membr. Biol.* 167:215–21

49. Roman RM, Smith RL, Feranchak AP, Clayton GH, Doctor RB, Fitz JG. 2001. ClC-2 chloride channels contribute to HTC cell volume homeostasis. *Am. J. Physiol. Gastrointest. Liver Physiol.* 280:G344–G53

50. Nordeen MH, Jones SM, Howell KE, Caldwell JH. 2000. GOLAC: an endogenous anion channel of the Golgi complex. *Biophys. J.* 78:2918–28

51. Buyse G, Voets T, Tytgat J, De Greef C, Droogmans G, et al. 1997. Expression of human pICln and ClC-6 in Xenopus oocytes induces an identical endogenous chloride conductance. *J. Biol. Chem.* 272:3615–21

52. Koumi S, Sato R, Aramaki T. 1994. Characterization of the calcium-activated chloride channel in isolated guinea-pig hepatocytes. *J. Gen. Physiol.* 104:357–73

53. Weinman SA, Meng X-J. 1998. Intracellular calcium is required for cAMP-activated but not swelling-activated chlorided conductances in rat hepatocytes. *Hepatology* 28:532A (Abstr.)

54. Kidd JF, Thorn P. 2000. Intracellular Ca^{2+} and Cl^- channel activation in secretory cells. *Annu. Rev. Physiol.* 62:493–513

55. Capiod T, Ogden DC. 1989. Properties of membrane ion conductances evoked by hormonal stimulation of guinea-pig and rabbit isolated hepatocytes. *Proc. R. Soc. London Ser. B* 236:187–201

56. Pauli BU, Abdel-Ghany M, Cheng HC, Gruber AD, Archibald HA, Elble RC. 2000. Molecular characteristics and functional diversity of CLCA family members. *Clin. Exp. Pharmacol. Physiol.* 27:901–5

57. Fuller CM, Benos DJ. 2000. Electrophysiological characteristics of the Ca^{2+}-activated Cl^- channel family of anion transport proteins. *Clin. Exp. Pharmacol. Physiol.* 27:906–10

58. Gruber AD, Pauli BU. 1999. Molecular cloning and biochemical characterization of a truncated, secreted member of the human family of Ca^{2+}-activated Cl^- channels. *Biochim. Biophys. Acta* 1444:418–23

59. Gruber AD, Schreur KD, Ji HL, Fuller CM, Pauli BU. 1999. Molecular cloning and transmembrane structure of hCLCA2 from human lung, trachea, and mammary gland. *Am. J. Physiol. Cell Physiol.* 276:C1261–C70

60. Gandhi R, Elble RC, Gruber AD, Schreur KD, Ji HL, et al. 1998. Molecular and functional characterization of a calcium-sensitive chloride channel from mouse lung. *J. Biol. Chem.* 273:32096–101

61. Gruber AD, Elble RC, Ji HL, Schreur KD, Fuller CM, Pauli BU. 1998. Genomic cloning, molecular characterization, and functional analysis of human CLCA1, the first human member of the family of Ca^{2+}-activated Cl^- channel proteins. *Genomics* 54:200–14

62. Cohn JA, Strong TV, Picciotto MR, Nairn AC, Collins FS, Fitz JG. 1993. Localization of the cystic fibrosis transmembrane conductance regulator in human bile duct epithelial cells. *Gastroenterology* 105:1857–64

63. Kinnman N, Lindblad A, Housset C, Buentke E, Scheynius A, et al. 2000. Expression of cystic fibrosis transmembrane conductance regulator in liver tissue from patients with cystic fibrosis. *Hepatology* 32:334–40

64. Kim JA, Kang YS, Lee SH, Lee EH, Yoo BH, Lee YS. 1999. Glibenclamide induces apoptosis through inhibition of cystic fibrosis transmembrane conductance regulator (CFTR) Cl$^-$ channels and intracellular Ca^{2+} release in HepG2 human hepatoblastoma cells. *Biochem. Biophys. Res. Commun.* 261:682–88

65. Bortner CD, Hughes FMJ, Cidlowski JA. 1997. A primary role for K$^+$ and Na$^+$ efflux in the activation of apoptosis. *J. Biol. Chem.* 272:32436–42

66. Hughes FMJ, Bortner CD, Purdy GD, Cidlowski JA. 1997. Intracellular K$^+$ suppresses the activation of apoptosis in lymphocytes. *J. Biol. Chem.* 272:30567–76

67. Hughes FMJ, Cidlowski JA. 1999. Potassium is a critical regulator of apoptotic enzymes in vitro and in vivo. *Adv. Enzyme Regul.* 39:157–71

68. Yu SP, Yeh C, Strasser U, Tian M, Choi DW. 1999. NMDA receptor-mediated K$^+$ efflux and neuronal apoptosis. *Science* 284:336–39

69. Maeno E, Ishizaki Y, Kanaseki T, Hazama A, Okada Y. 2000. Normotonic cell shrinkage because of disordered volume regulation is an early prerequisite to apoptosis. *Proc. Natl. Acad. Sci. USA* 97:9487–92

70. Szabo I, Lepple-Wienhues A, Kaba KN, Zoratti M, Gulbins E, Lang F. 1998. Tyrosine kinase-dependent activation of a chloride channel in CD95-induced apoptosis in T lymphocytes. *Proc. Natl. Acad. Sci. USA* 95:6169–74

71. Meng X-J, Carruth MW, Weinman SA. 1997. Leukotriene D$_4$ activates a chloride conductance in hepatocytes from lipopolysaccharide-treated rats. *J. Clin. Invest.* 99:2915–22

72. Nietsch HH, Roe MW, Fiekers JF, Moore AL, Lidofsky SD. 2000. Activation of potassium and chloride channels by tumor necrosis factor alpha. Role in liver cell death. *J. Biol. Chem.* 275:20556–61

73. Li X, Wang T, Weinman SA. 2000. MRP-2 modulates volume regulation by speeding channel activation and changing chloride channel distribution in hepatocytes. *Hepatology* 32:316A (Abstr.)

74. Dickson RC, Bronk SF, Gores GJ. 1992. Glycine cytoprotection during lethal hepatocellular injury from adenosine triphosphate depletion. *Gastroenterology* 102:2098–107

75. Marsh DC, Vreugdenhil PK, Mack VE, Belzer FO, Southard JH. 1993. Glycine protects hepatocytes from injury caused by anoxia, cold ischemia and mitochondrial inhibitors, but not injury caused by calcium ionophores or oxidative stress. *Hepatology* 17:91–98

76. Waters SL, Schnellmann RG. 1996. Extracellular acidosis and chloride channel inhibitors act in the late phase of cellular injury to prevent death. *J. Pharmacol. Exp. Ther.* 278:1012–17

77. Carini R, Bellomo G, Grazia DC, Albano E. 1997. Glycine protects against hepatocyte killing by KCN or hypoxia by preventing intracellular Na$^+$ overload in the rat. *Hepatology* 26:107–12

78. Frank A, Rauen U, De Groot H. 2000. Protection by glycine against hypoxic injury of rat hepatocytes: inhibition of ion fluxes through nonspecific leaks. *J. Hepatol.* 32:58–66

79. Wondergem R, Gong W, Monen SH, Dooley SN, Gonce JL, et al. 2001. Blocking swelling-activated chloride current inhibits mouse liver cell proliferation. *J. Physiol.* 532:661–72

80. Van Dyke RW. 1996. Acidification of lysosomes and endosomes. *Subcell. Biochem.* 27:331–60

81. Pollock NS, Kargacin ME, Kargacin GJ. 1998. Chloride channel blockers inhibit Ca^{2+} uptake by the smooth

muscle sarcoplasmic reticulum. *Biophys. J.* 75:1759–66

82. Gaxiola RA, Yuan DS, Klausner RD, Fink GR. 1998. The yeast CLC chloride channel functions in cation homeostasis. *Proc. Natl. Acad. Sci. USA* 95:4046–50

83. Grabe M, Oster G. 2001. Regulation of organelle acidity. *J. Gen. Physiol.* 117:329–44

84. Van Dyke RW. 1993. Acidification of rat liver lysosomes: quantitation and comparison with endosomes. *Am. J. Physiol. Cell Physiol.* 265:C901–C17

85. Tilly BC, Mancini GM, Bijman J, van Gageldonk PG, Beerens CE, et al. 1992. Nucleotide-activated chloride channels in lysosomal membranes. *Biochem. Biophys. Res. Commun.* 187:254–60

86. Morier N, Sauvé R. 1994. Analysis of a novel double-barreled anion channel from rat liver rough endoplasmic reticulum. *Biophys. J.* 67:590–602

87. Eliassi A, Garneau L, Roy G, Sauvé R. 1997. Characterization of a chloride-selective channel from rough endoplasmic reticulum membranes of rat hepatocytes: evidence for a block by phosphate. *J. Membr. Biol.* 159:219–29

88. Davis LI. 1995. The nuclear pore complex. *Annu. Rev. Biochem.* 64:865–96

89. Tabares L, Mazzanti M, Clapham DE. 1991. Chloride channels in the nuclear membrane. *J. Membr Biol.* 123:49–54

90. Guihard G, Proteau S, Payet MD, Escande D, Rousseau E. 2000. Patch-clamp study of liver nuclear ionic channels reconstituted into giant proteoliposomes. *FEBS Lett.* 476:234–39

91. Jentsch TJ, Friedrich T, Schriever A, Yamada H. 1999. The CLC chloride channel family. *Pflugers Arch.* 437:783–95

92. Wills NK, Fong P. 2001. ClC chloride channels in epitheilia: recent progress and remaining puzzles. *NIPS* 16:161 (Abstr.)

93. Waldegger S, Jentsch TJ. 2000. From tonus to tonicity: physiology of CLC chloride channels. *J. Am. Soc. Nephrol.* 11:1331–39

94. Maduke M, Miller C, Mindell JA. 2000. A decade of CLC chloride channels: structure, mechanism, and many unsettled questions. *Annu. Rev. Biophys. Biomol. Struct.* 29:411–38

95. Uchida S. 2000. In vivo role of CLC chloride channels in the kidney. *Am J. Physiol. Renal Physiol.* 279:F802–F8

96. Piwon N, Gunther W, Schwake M, Bosl MR, Jentsch TJ. 2000. ClC-5 Cl⁻-channel disruption impairs endocytosis in a mouse model for Dent's disease. *Nature* 408:369–73

97. Kida Y, Uchida S, Miyazaki H, Sasaki S, Marumo F. 2001. Localization of mouse CLC-6 and CLC-7 mRNA and their functional complementation of yeast CLC gene mutant. *Histochem. Cell Biol.* 115:189–94

98. Shimada K, Weinman SA. 1998. Chloride channel mRNA expression in hepatocytes is altered by osmotic stress. *Hepatology* 28:532A (Abstr.)

99. Roman RM, Smith RL, Clayton GH, Fitz JG. 1998. ClC Cl⁻ channels are broadly expressed in liver epithelial cells. *Hepatology* 28:530A (Abstr.)

100. Steinmeyer K, Schwappach B, Bens M, Vandewalle A, Jentsch TJ. 1995. Cloning and functional expression of rat CLC-5, a chloride channel related to kidney disease. *J. Biol. Chem.* 270:31172–77

101. Gunther W, Luchow A, Cluzeaud F, Vandewalle A, Jentsch TJ. 1998. ClC-5, the chloride channel mutated in Dent's disease, colocalizes with the proton pump in endocytotically active kidney cells. *Proc. Natl. Acad. Sci. USA* 95:8075–80

102. Sakamoto H, Sado Y, Naito I, Kwon TH, Inoue S, et al. 1999. Cellular and subcellular immunolocalization of ClC-5 channel in mouse kidney: colocalization with H⁺-ATPase. *Am. J. Physiol. Renal Physiol.* 277:F957–F65

103. Devuyst O, Christie PT, Courtoy PJ, Beauwens R, Thakker RV. 1999. Intrarenal and subcellular distribution of the human chloride channel, CLC-5, reveals

a pathophysiological basis for Dent's disease. *Hum. Mol. Genet.* 8:247–57

104. Landry D, Sullivan S, Nicolaides M, Redhead C, Edelman A, et al. 1993. Molecular cloning and characterization of p64, a chloride channel protein from kidney microsomes. *J. Biol. Chem.* 268:14948–55

105. Landry DW, Akabas MH, Redhead C, Edelman A, Cragoe EJ, Al-Awqati Q. 1989. Purification and reconstitution of chloride channels from kidney and trachea. *Science* 244:1469–72

106. Edwards JC. 1999. A novel p64-related Cl⁻ channel: subcellular distribution and nephron segment-specific expression. *Am. J. Physiol. Renal Physiol.* 276:F398–F408

107. Berryman M, Bretscher A. 2000. Identification of a novel member of the chloride intracellular channel gene family (CLIC5) that associates with the actin cytoskeleton of placental microvilli. *Mol. Biol. Cell.* 11:1509–21

108. Edwards JC, Kapadia S. 2000. Regulation of the bovine kidney microsomal chloride channel p64 by p59fyn, a Src family tyrosine kinase. *J. Biol. Chem.* 275:31826–32

109. Redhead CR, Edelman AE, Brown D, Landry DW, Al-Awqati Q. 1992. A ubiquitous 64-kDa protein is a component of a chloride channel of plasma and intracellular membranes. *Proc. Natl. Acad. Sci. USA* 89:3716–20

110. Valenzuela SM, Martin DK, Por SB, Robbins JM, Warton K, et al. 1997. Molecular cloning and expression of a chloride ion channel of cell nuclei. *J. Biol. Chem.* 272:12575–82

111. Tonini R, Ferroni A, Valenzuela SM, Warton K, Campbell TJ, et al. 2000. Functional characterization of the NCC27 nuclear protein in stable transfected CHO-K1 cells. *FASEB J.* 14:1171–78

112. Tulk BM, Edwards JC. 1998. NCC27, a homolog of intracellular Cl⁻ channel p64, is expressed in brush border of renal prox-

imal tubule. *Am. J. Physiol. Renal Physiol.* 274:F1140–F49

113. Valenzuela SM, Mazzanti M, Tonini R, Qiu MR, Warton K, et al. 2000. The nuclear chloride ion channel NCC27 is involved in regulation of the cell cycle. *J. Physiol.* 529(3):541–52

114. Tulk BM, Schlesinger PH, Kapadia SA, Edwards JC. 2000. CLIC-1 functions as a chloride channel when expressed and purified from bacteria. *J. Biol. Chem.* 275:26986–93

115. Sellinger M, Weinman SA, Henderson RM, Zweifach A, Boyer JL, Graf J. 1992. Anion channels in rat liver canalicular plasma membranes reconstituted into planar lipid bilayers. *Am. J. Physiol. Gastrointest. Liver Physiol.* 262:G1027–G32

116. Heiss NS, Poustka A. 1997. Genomic structure of a novel chloride channel gene, CLIC2, in Xq28. *Genomics* 45:224–28

117. Qian Z, Okuhara D, Abe MK, Rosner MR. 1999. Molecular cloning and characterization of a mitogen-activated protein kinase-associated intracellular chloride channel. *J. Biol. Chem.* 274:1621–27

118. Howell S, Duncan RR, Ashley RH. 1996. Identification and characterisation of a homologue of p64 in rat tissues. *FEBS Lett.* 390:207–10

119. Fernandez-Salas E, Sagar M, Cheng C, Yuspa SH, Weinberg WC. 1999. p53 and tumor necrosis factor alpha regulate the expression of a mitochondrial chloride channel protein. *J. Biol. Chem.* 274:36488–97

120. Chuang JZ, Milner TA, Zhu M, Sung CH. 1999. A 29 kDa intracellular chloride channel p64H1 is associated with large dense-core vesicles in rat hippocampal neurons. *J. Neurosci.* 19:2919–28

121. Duncan RR, Westwood PK, Boyd A, Ashley RH. 1997. Rat brain p64H1, expression of a new member of the p64 chloride channel protein family in endoplasmic reticulum. *J. Biol. Chem.* 272:23880–86

122. Nishizawa T, Nagao T, Iwatsubo T, Forte JG, Urushidani T. 2000. Molecular cloning and characterization of a novel chloride intracellular channel-related protein, parchorin, expressed in water-secreting cells. *J. Biol. Chem.* 275:11164–73

123. Urushidani T, Chow D, Forte JG. 1999. Redistribution of a 120 kDa phosphoprotein in the parietal cell associated with stimulation. *J. Membr. Biol.* 168:209–20

124. Tsujimoto Y, Shimizu S. 2000. Bcl-2 family: life-or-death switch. *FEBS Lett.* 466:6–10

125. Crompton M. 1999. The mitochondrial permeability transition pore and its role in cell death. *Biochem. J.* 341(2):233–49

126. Roos N, Benz R, Brdiczka D. 1982. Identification and characterization of the pore-forming protein in the outer membrane of rat liver mitochondria. *Biochim. Biophys. Acta* 686:204–14

127. Colombini M. 1989. Voltage gating in the mitochondrial channel, VDAC. *J. Membr. Biol.* 111:103–11

128. Yu WH, Forte M. 1996. Is there VDAC in cell compartments other than the mitochondria? *J. Bioenerg. Biomembr.* 28:93–100

129. Nagasawa M, Kanzaki M, Iino Y, Morishita Y, Kojima I. 2001. Identification of a novel chloride channel expressed in the endoplasmic reticulum, Golgi apparatus and nucleus. *J. Biol. Chem.* 276:20413–18

130. Ballatori N, Truong AT, Jackson PS, Strange K, Boyer JL. 1995. ATP depletion and inactivation of an ATP-sensitive taurine channel by classic ion channel blockers. *Mol. Pharmacol.* 48:472–76

131. Brandt S, Jentsch TJ. 1995. ClC-6 and ClC-7 are two novel broadly expressed members of the CLC chloride channel family. *FEBS Lett.* 377:15–20

132. Edwards JC, Tulk B, Schlesinger PH. 1998. Functional expression of p64, an intracellular chloride channel protein. *J. Membr. Biol.* 163:119–27

Annu. Rev. Physiol. 2002. 64:635–61

BILE SALT TRANSPORTERS

Peter J. Meier and B. Stieger

*Division of Clinical Pharmacology and Toxicology, Department of Medicine, University
Hospital, 8091 Zurich, Switzerland; e-mail: meierabt@kpt.unizh.ch;
bstieger@kpt.unizh.ch*

Key Words   liver, small intestine, enterohepatic circulation, transport proteins

■ **Abstract**   Bile salts are the major organic solutes in bile and undergo extensive enterohepatic circulation. Hepatocellular bile salt uptake is mediated predominantly by the Na^+-taurocholate cotransport proteins Ntcp (rodents) and NTCP
(humans) and by the Na^+-independent organic anion-transporting polypeptides Oatp1,
Oatp2, and Oatp4 (rodents) and OATP-C (humans). After diffusion (bound by intracellular bile salt–binding proteins) to the canalicular membrane, monoanionic bile
salts are secreted into bile canaliculi by the bile salt export pump Bsep (rodents)
or BSEP (humans). Both belong to the ATP-binding cassette (ABC) transporter superfamily. Dianionic conjugated bile salts are secreted into bile by the multidrug-
resistance-associated proteins Mrp2/MRP2. In bile ductules, a minor portion of
protonated bile acids and monomeric bile salts are reabsorbed by non-ionic diffusion and the apical sodium-dependent bile salt transporter Asbt/ASBT, transported
back into the periductular capillary plexus by Mrp3/MRP3 [and/or a truncated form
of Asbt (tAsbt)], and subjected to cholehepatic shunting. The major portion of biliary bile salts is aggregated into mixed micelles and transported into the intestine,
where they are reabsorbed by apical Oatp3, the apical sodium-dependent bile salt
transporter (ASBT), cytosolic intestinal bile acid-binding protein (IBABP), and basolateral Mrp3/MRP3 and tAsbt. Transcriptional and posttranscriptional regulation of
these enterohepatic bile salt transporters is closely related to the regulation of lipid
and cholesterol homeostasis. Furthermore, defective expression and function of bile
salt transporters have been recognized as important causes for various cholestatic liver
diseases.

ENTEROHEPATIC CIRCULATION OF BILE SALTS

Bile acids are amphipathic steroidal compounds derived from the enzymatic catabolism of cholesterol in the liver (1). The predominant bile acids in humans are
dihydroxylated ($3\alpha OH,7\alpha OH$) chenodeoxycholic acid and the more hydrophilic
trihydroxylated ($3\alpha OH,7\alpha OH,12\alpha OH$) cholic acid. These primary bile acids are
synthesized within hepatocytes and conjugated at the terminal (C_{24}) carboxyl
group with the amino acids taurine and glycine (2). This amidation increases the

amphipathic character and decreases the ionization constants of bile acids, rendering them more hydrophilic and more readily excretable into bile. The glycoconjugates, which predominate in humans, and tauroconjugates, which predominate in rodents, have pK_a values of approximately 4 and 2, respectively, and exist predominantly in their anionic salt form at physiological pH (3). Therefore, they are called bile salts. Bile salts are the major organic solutes in bile, and their vectorial secretion from blood into bile represents the major driving force for hepatic bile formation. Although bile is isoosmotic to plasma, bile salts are concentrated up to 1000-fold in bile, necessitating active transport by hepatocytes against a concentration gradient. The detergent   properties of bile salts aid in the solubilization of biliary phospholipids and cholesterol into mixed micelles. A minority of bile salts and unconjugated protonated bile acids travel as monomers in bile and are subject to the cholehepatic shunt pathway, i.e., they are reabsorbed by biliary ductular cells (cholangiocytes), returned to the hepatocytes via the periductular capillary plexus, and resecreted across hepatocytes into bile (Figure 1). The majority of biliary bile salts associated with mixed micelles are secreted via the bile ducts and gallbladder into the lumen of the small intestine where they act as detergents to emulsify dietary fats and lipid-soluble vitamins. In addition, intraduodenal bile salts regulate pancreatic secretions and the release of gastrointestinal peptides (4, 5). The vast majority of bile salts are efficiently reabsorbed from the small intestine through a combination of sodium-independent absorption in the proximal small intestine and active sodium-dependent absorption in the distal ileum (6, 7). The absorbed bile salts are then returned to the liver in the portal circulation and resecreted into bile. This efficient enterohepatic circulation ensures that from the total bile salt pool of adult humans (3–4 g), which circulates 6 to 10 times per 24 h through the enterohepatic pathway, only about 0.5 g bile salts are lost per day through fecal excretion. This loss is compensated for by de novo hepatic bile salt synthesis from cholesterol, which contributes less than 3% of bile salts secreted with hepatic bile (8). The intrinsic links between intestinal bile salt absorption, hepatic bile salt synthesis, and hepatic cholesterol degradation have recently been delineated by the discovery that hydrophobic bile salts can upregulate the ileal bile acid-binding protein and down-regulate hepatic cholesterol 7α-hydroxylase (CYP7A1) through the action of the nuclear bile salt

---→

Figure 1   Bile salt transporters within enterohepatic circulation pathway. See text for explanation. Abbreviations: ASBT, apical sodium bile salt transporter; BSEP, bile salt export pump; BS, bile salts; BAH, uncharged (protonated) bile acids; CA, carbonic anhydrase; Chol, cholesterol; IBABP, intestinal bile acid-binding protein; IBAT, intestinal bile acid transporter; MDR1, P-glycoprotein; Mdr2/MDR3, phosphatidylcholine flippase; MRPs, multidrug-resistance-associated proteins; NTCP, Na^+-taurocholate cotransport protein; OA, organic anions; OATP, organic anion-transporting protein; PL, phosphatidylcholine; tAsbt, truncated ASBT; XH, protonated solutes; O, lipid vesicles; ⊗, mixed micelles.

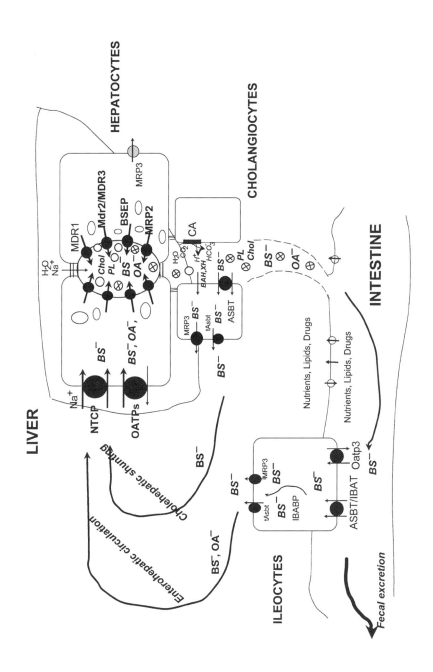

receptor FXR (farnesoid X receptor) or BAR (bile acid receptor) on gene transcription (9–12). By this mechanism, bile salts can regulate cholesterol homeostasis, as well as their own enterohepatic circulation. This review summarizes the transport systems involved in the movement of bile salts across the enterohepatic membrane barriers (Figure 1) and highlights the regulatory principles that control the physiological interactions between the enterohepatic circulation of bile salts and the elimination of cholesterol from the body. Several complementary reviews have recently been published that provide further in-depth insights into the molecular regulation of enterohepatic bile salt circulation and its consequences for lipid biology under various physiological and pathophysiological conditions (1, 9, 13–20).

BILE SALT TRANSPORT IN HEPATOCYTES

Vectorial hepatocellular secretion of bile salts from blood into bile is a major mechanism for both hepatic bile formation and ongoing enterohepatic bile salt circulation. This concentrative transport process is governed by the polarized expression of distinct transport systems at the basolateral (sinusoidal) and apical (canalicular) surface domains of hepatocytes. Furthermore, intracellular cytosolic bile salt-binding proteins and, although to a lesser extent, vesicle-associated transport processes are involved in the transcellular delivery of bile salts from the basolateral to the canalicular plasma membrane of hepatocytes.

Basolateral Bile Salt Uptake into Hepatocytes

The specific architecture of the liver sinusoids allows the passage of protein-bound compounds through endothelial fenestrae into the space of Disse, where the basolateral uptake systems of hepatocytes can extract bile salts and other organic molecules from albumin (21). The first-pass extraction of conjugated bile salts from sinusoidal blood ranges from 75 to 90% depending on bile salt structure and remains constant irrespective of systemic bile salt concentrations (22, 23). Under physiological conditions, bile salt uptake occurs predominantly in periportal hepatocytes, resulting in a lobular concentration gradient between zone 1 (periportal) and zone 3 (perivenous) hepatocytes (24). Uptake involves two processes: cotransport with sodium coupled to the electrochemical sodium gradient (maintained by Na^+/K^+-ATPase) and sodium-independent bile salt/organic anion exchange (1, 15).

Na^+-TAUROCHOLATE COTRANSPORTING PROTEIN (NTCP/NTCP)   Sodium-dependent uptake of taurine and glycine conjugated bile salts is mediated by the Na^+-taurocholate cotransporting polypeptide that has been isolated from rat (Ntcp, *gene symbol Slc10a1*) (25), mouse (Ntcp1/2, *Slc10a1*), (26), rabbit (27), and human (NTCP, *SLC10A1*) (28) liver (Table 1). Rat Ntcp and human NTCP consist of 362

TABLE 1 Enterohepatic bile salt transporters in rat, mouse, and human

Species	Cell type	Localization	Carrier	Function	References
Human	Hepatocyte	Basolateral	NTCP (*SLC10A1*)	Na$^+$-dependent bile salt uptake	(28, 32)
Rat	Hepatocyte	Basolateral	Ntcp (*Slc10a1*)	Na$^+$-dependent bile salt uptake	(25, 30)
Mouse	Liver		Ntcp1/2 (*Slc10a1*)	Na$^+$-dependent bile salt uptake	(26)
Rabbit	Liver		Ntcp (*Slc10a1*)	Na$^+$-dependent bile salt uptake	(27)
Human	Liver		OATP-A (OATP, OATP-1) (*SLC21A3*)	Na$^+$-independent bile salt uptake; multispecific	(103)
Human	Hepatocyte	Basolateral	OATP-C (OATP2,LST-1) (*SLC21A6*)	Na$^+$-dependent bile salt uptake; multispecific	(90, 95–98)
Human	Hepatocyte	Basolateral	OATP8 (LST-2) (*SLC21A8*)	Na$^+$-independent bile salt uptake, multispecific	(90, 99)
Rat	Hepatocyte	Basolateral	Oatp1 (*Slc21a1*)	Na$^+$-independent bile salt uptake, multispecific	(66, 68)
Rat	Hepatocyte	Basolateral	Oatp2 (*Slc21a5*)	Na$^+$-independent bile salt uptake, multispecific	(80, 81)
Rat	Enterocyte	Apical	Oatp3 (*Slc21a7*)	Na$^+$-independent bile salt uptake, multispecific	(7, 62)
Rat	Hepatocyte	Basolateral	Oatp4 (*Slc21a10*)	Na$^+$-independent bile salt uptake, multispecific	(87–89)
Mouse	Liver		Oatp1 (*Slc21a1*)	Na$^+$-independent bile salt uptake, multispecific	(69)
Human	Enterocyte, cholangiocyte		ASBT (*SLC10A2*)	Na$^+$-dependent bile salt uptake	(161)
Rat	Enterocyte Cholangiocyte Enterocyte Cholangiocyte	Apical Basolateral	Asbt (*Slc10a2*) tAsbt (*Slc10a2*)	Na$^+$-dependent bile salt uptake bile salt efflux	(153, 162) (153)
Mouse	Enterocyte		Asbt (*Slc10a2*)	Na$^+$-depdendent bile salt uptake	(163)
Rabbit	Enterocyte		Asbt (*Slc10a2*)	Na$^+$-dependent bile salt uptake	(27)
Hamster	Enterocyte		Asbt (*Slc10a2*)	Na$^+$-dependent bile salt uptake	(160)
Human	Hepatocyte	Canalicular	BSEP (*ABCB11*)	ATP-dependent bile salt transport	(132, 133)
Rat	Hepatocyte	Canalicular	Bsep (*Abcb11*)	ATP-dependent bile salt transport	(127)

(Continued)

TABLE 1   (*Continued*)

Species	Cell type	Localization	Carrier	Function	References
Mouse	Hepatocyte		Bsep (*Abcb11*)	ATP-dependent bile salt transport	(128, 129)
Human	Hepatocyte	Canalicular	MRP2 (*ABCC2*)	ATP-dependent organic anion (dianionic bile salt) transport, multispecific	(185)
Rat	Hepatocyte	Canaliclar	Mrp2 (*Abcc2*)	ATP-dependent organic anion (dianionic bile salt) transport, multispecific	(186, 187)
Human	Hepatocyte Cholangiocyte	Basolateral Basolateral	MRP3 (*ABCC3*)	ATP-dependent organic anion (including bile-salts) transport, multispecific	(106, 108) (152)
Rat	Hepatocyte Cholangiocyte	Basolateral	Mrp3 (*Abcc3*)	ATP-dependent organic anion (including bile-salts) transport, multispecific	(105, 106)

and 349 amino acids, respectively, and show a 77% amino acid sequence identity (29). These transporters are expressed exclusively in hepatocytes and are localized strictly to the basolateral plasma membrane (Figure 1) (30–32). They possess seven or even nine transmembrane spanning domains with an exoplasmic N terminus and a cytoplasmic C terminus (25, 28, 33). Ntcp/NTCP mediate sodium-coupled uptake of taurocholate and other bile salts with the same stoichiometry (Na^+:bile salt $= 2$:1) and similar K_m values, as previously published for hepatocellular bile salt uptake in isolated hepatocytes (15, 34, 35). During development of rat liver, Na^+-dependent bile salt uptake and expression of Ntcp appear in parallel on days 18–20 of gestation (36, 37). Also, parallel decreases of Na^+-taurocholate cotransport and Ntcp expression were observed in primary cultures of rat hepatocytes (38, 39) and during in situ regeneration of rat liver (40). These studies, together with a 95% reduction of Na^+-dependent taurocholate uptake after specific inhibition of Ntcp expression in *Xenopus laevis* oocytes by antisense oligonucleotides (41), indicate that Ntcp is the predominant Na^+-dependent bile salt uptake system in rat hepatocytes. The exact physiological role of sinusoidal plasma membrane associated "microsomal epoxide hydrolase" in overall Na^+-dependent bile salt uptake is still unclear (15). The *Ntcp/NTCP*-genes are localized on chromosomes 6q24, 12, and 14 in rat, mouse, and humans, respectively (28, 42, 43).

Ntcp/NTCP function is regulated by changes in gene expression and consequently protein synthesis and by protein trafficking and the rate of insertion into the plasma membrane. *Ntcp* gene transcription is controlled by a hierarchical network of hepatocyte-enriched transcription factors and ligand-activated nuclear receptors (44). First, the *Ntcp* promoter has been identified as the primary target of the divergent homeobox gene *Hex* (45). The Hex protein represents an essential transcription factor for hepatocyte development and differentiation (46).

TABLE 2   Transcription factors and ligand-activated nuclear receptors involved in the regulation of bile salt transporter gene expression

Nuclear factors	Experimental model	bile salt transporter expression	Reference
HNF3β	Overexpression in transgenic mice	↓Ntcp, ↔Oatp1, ↔Bsep, ↓Mdr2	(47)
HNF4α	Conditional knockout mouse	↓ Ntcp, ↓Oatp1, ↑ Bsep, ↓ Mdr2	(50)
HNF1α	Knockout mouse	↓ Ntcp, ↓ Oatp1, ↓ Oatp2, ↔ Bsep, ↔ Mdr2, ↓ ASBT	(49)
	Promoter expression studies	↑ OATP-C, ↑OATP8, ↑ Oatp4	(94)
FXR/BAR	Wild-type mouse, bile acid feeding	↓ Ntcp, ↔ Oatp1, ↑ Bsep, ↑ Mrp2, ↔ ASBT	(52)
	Promoter expression studies	↑ BSEP	(136)
	Knockout mouse, bile acid feeding	↔ Ntcp, ↑ Oatp1, ↔ Bsep, ↑ Mdr2, ↔ ASBT	(52)
PXR	Rat treatment, promoter expression studies	↔ Ntcp, ↔ Oatp1, ↑ Oatp2	(76, 77, 85)

↑, upregulation; ↔, no change; ↓, down-regulation.

Second, overexpression of hepatocyte nuclear factor 3β (HNF3β) is associated with down-regulation of *Ntcp* gene expression (Table 2) (47). Third, downregulation of *Ntcp* gene expression is preceded by decreased activity of HNF1α in endotoxin-induced cholestasis (48), and HNF1α and HNF4α null mice have decreased Ntcp expression levels (Table 2) (49, 50). Fourth, the retinoid X receptor (RXR) and retinoic acid receptor (RAR) heterodimer RXR:RAR is an important transactivator of rat *Ntcp* gene expression (51). Fifth, the farnesoid X receptor (FXR) or bile acid receptor (BAR) is involved in the down-regulation of *Ntcp* expression during cholate feeding (Table 2) (52) and probably also in cholestasis. Similar to the bile acid-induced down-regulation of *CYP7A1* (53, 54), the negative feedback regulation of Ntcp by bile acid-activated FXR/BAR occurs via inhibition of RXR:RAR by the FXR/BAR-dependent small heterodimer partner 1 (SHP-1) (55). This coordinated down-regulation of both bile acid synthesis and bile salt import provides a protective mechanism against hepatocellular bile acid damage in cholestasis. Finally, prolactin-dependent up-regulation of Ntcp in the postpartum period is mediated by Stat5, a member of the signal transducers and activation of transcription family of transcription factors (56). On the posttranscriptional level, cAMP has been shown to stimulate sodium-dependent taurocholate uptake into hepatocytes by translocating Ntcp to the plasma membrane (57, 58). cAMP causes rapid translocation of Ntcp from a preformed vesicular pool to the plasma membrane via the actin cytoskeleton, as convincingly demonstrated with a functional Ntcp-green fluorescent protein (GFP) conjugate in transfected HepG2 cells (59). Hence, cAMP-dependent targeting and membrane insertion of Ntcp represent important mechanisms for short-term regulation of basolateral bile salt uptake into hepatocytes in response to physiological stimuli.

ORGANIC ANION-TRANSPORTING POLYPEPTIDES (Oatps/OATPs)   Sodium-independent hepatocellular uptake of bile salts is mediated by several members of the Oatp/OATP family of membrane transporters, which are classified in the solute carrier gene family $Slc21a/SLC21A$ (Table 1). Oatps/OATPs represent multispecific transporters that mediate, in addition to conjugated and unconjugated bile salts, hepatocellular uptake of a vast variety of other amphipathic organic compounds, including bromosulphophthalein (BSP), bilirubin, steroids and steroid conjugates, thyroid hormones, peptides, mycotoxins, and numerous drugs (15). Certain Oatps/OATPs are predominantly or even exclusively expressed in extrahepatic tissues such as brain (60–63), kidney (64), lung (65), and intestine (7).

In the rat, there are now three Oatps localized in the liver for which bile salts are substrates (Table 1). Oatp1 ($Slc21a1$) consists of 670 amino acids and represents a glycoprotein with 12 predicted transmembrane-spanning domains (66). Its native apparent molecular mass is approximately 80 kDa in rat liver plasma membrane vesicles (67, 68). It is selectively expressed at the basolateral plasma membrane of rat hepatocytes and transports conjugated and unconjugated bile salts with K_m values between 7 and 54 μM (15). The Oatp1 of mouse liver has also been isolated and shown to have the same substrate specificity as rat Oatp1 (69). Oatp1 probably works as an exchanger and accepts bicarbonate (70) and/or glutathione (71) as physiological counter-anions. In developing rat liver, Oatp1 is not expressed prior to postnatal day 15, indicating that it does not play a major role in the early development of the enterohepatic circulation (72, 73). Oatp1 is down-regulated after cholate feeding (74) in cholestatic liver and following partial hepatectomy (75) and in primary cultured rat hepatocytes (38). Similar to Ntcp, $Oatp1$ gene expression is also down-regulated in HNF1α and HNF4α knockout mice (Table 2) (49, 50). In FXR/BAR knockout mice, cholate feeding was associated with increased Oatp1 expression (52). These observations indicate that $Oatp1$ gene expression is, at least in part, controlled by transcription factors similar to those that up-regulate Ntcp [e.g., HNF1α, FXR/BAR, and SHP-1(?)], although SHP-1 dependent down-regulation of Oatp1 remains to be determined. Because Oatp1 is induced neither by phenobarbital nor by pregnenolone-16α-carbonitrile (76, 77), the promiscuous xenobiotic pregnane X receptor (PXR) is probably not involved in the transcriptional regulation of the $Oatp1$ gene (Table 2) (78). Functional down-regulation of Oatp1 on the protein level occurs via serine phosphorylation by extracellular ATP (79). By this mechanism, Oatp1 can lose its transport activity without leaving the cell surface, thus indicating that the phosphorylation state of membrane-associated Oatp1 protein must be considered when assessing alterations of its functional expression in pathobiological states.

Oatp2 (*Slc21a5*), originally cloned from rat brain, is 77% identical with Oatp1 and is also expressed basolaterally in hepatocytes (Table 1) (80, 81). Further comparisons of Oatp1 and Oatp2 showed overlapping substrate specificities with respect to bile salts, but with differences in their acinar localization along the liver sinusoids (68, 81). Whereas Oatp1 showed a homogeneous lobular distribution, Oatp2 was predominantly expressed in perivenous hepatocytes, excluding the innermost first and second cell layers surrounding the central vein (81, 82). Because the major uptake of bile salts occurs in periportal hepatocytes (24, 83), Oatp1 is implicated in the sodium-independent uptake of bile salts under normal conditions. However, in models of cholestasis such as treatment with estrogens, an event associated with down-regulation of both Ntcp and Oatp1 (75), pericentral hepatocytes in zone 3 of the sinusoids are recruited for bile salt transport (83). Therefore, Oatp2 may assume a more important role in situations in which the expression or activity of Ntcp and/or Oatp1 is compromised (84). Similar to Ntcp and Oatp1, *Oatp2* gene expression is down-regulated in primary cultured rat hepatocytes (38) and in HNF1α knockout mice (Table 2) (49). However, unlike Ntcp and Oatp1, *Oatp2* gene expression is induced by phenobarbital and pregnenolone-16α-carbonitrile treatment (76, 77) through a PXR-dependent mechanism (Table 2) (85). Because lithocholic acid, a cholestatic secondary bile acid formed in the intestine by bacterial 7α-dehydroxylation of chenodeoxycholic acid, is an endogenous ligand of PXR, concomitant PXR-dependent up-regulation of Oatp2 (uptake) and CYP3A (hydroxylation) represents an important constitutive response in the hepatic detoxification of both cholestatic bile salts and xenobiotic chemicals (85, 86).

The third rat liver Oatp is Oatp4 (*Slc21a10*) (Table 1) (87), a full-length isoform of the so-called liver-specific transporter 1 or rlst-1 (88). It is exclusively expressed at the basolateral membrane of hepatocytes (89) and shares 43–44% amino acid identity with Oatp1 and Oatp2 (90). Although three rlst-1 splice-variants have been detected (91), full-length Oatp4 represents quantitatively and functionally the most important protein in rat liver (87). Oatp4 mediates similar sodium-independent bile salt and organic anion transport as do Oatp1 and Oatp2, although the affinities for some substrates can vary (89). Oatp4 belongs to the same subfamily as human OATP-C (64% amino acid identity) and human OATP8 (66%) and appears to be especially involved in the hepatic clearance of certain anionic peptides including the exogenous heptapeptide microcystin (92) and the gastrointestinal peptide hormone cholecystokinin 8 (CCK-8) (93). Although the regulation of *Oatp4* gene expression has not been studied in detail, recent findings indicate that HNF1α might be positively involved (Table 2) (94), thus further supporting the concept that HNF1α is essential for bile salt uptake across the basolateral hepatocyte plasma membrane (49). Collectively, a comparison of the data on Oatp-mediated bile salt transport with the data of sodium-independent bile salt uptake in the isolated perfused rat liver and in isolated hepatocytes indicates that the Oatp1, 2, and 4 transporters can account for the bulk of sodium-independent bile salt uptake in rat liver (15).

Human liver expresses four OATPs (OATP-A, OATP-B, OATP-C, and OATP8), although all with different expression levels and significance for bile salt uptake

(Table 1) (90). The most important sodium-independent bile salt transporter of human liver is OATP-C (*SLC21A6*), which has also been called LST-1, OATP2, or OATP6 (Table 1) (90, 95–97). It is selectively expressed at the basolateral membrane of human hepatocytes and transports taurocholate with a slightly lower affinity compared with human NTCP (15). OATP-C exhibits large overlapping substrate specificities with the other OATPs of human liver (90), but a unique property of OATP-C is transport of unconjugated bilirubin (98). Although OATP-C shares many transport substrates with rat Oatp4, it remains uncertain whether the two proteins represent truly orthologous gene products. OATP8 (*SLC21A8*) is 80% identical with OATP-C and shares the same selective basolateral expression in human hepatocytes (99). In contrast to OATP-C, bile salts were not transported when OATP8 was expressed in mammalian cells (99) but were identified as low-affinity substrates in the oocyte expression system (90). However, and in keeping with the nature of Oatps/OATPs as multispecific organic substrate carriers, OATP8 transports numerous organic anions, as well as digoxin and various peptides (90, 92, 93). Interestingly, the *OATP-C* and *OATP8* genes are both localized on chromosome 12p12, and their expression is under similar control by HNF1α (Table 2) (94). These observations, together with the similar genomic organization of human *OATP-A*, *OATP-C*, and *OATP8* genes (99, 100) and the mouse *Oatp2* and *Oatp4* genes (101, 102), lends further support to the hypothesis that HNF1α is a global regulator of hepatic uptake of bile salts and numerous other amphipathic albumin-bound compounds.

Although OATP-A (*SLC21A3*) was originally cloned from human liver (103), its hepatic expression level is low and, therefore, its overall contribution to hepatic bile salt uptake probably minor (97). Its predominant expression is in the brain, where it is involved in the transport of various drugs and opioid peptides across the blood-brain barrier (104). Finally, OATP-B (*SLC21A9*) is expressed at the basolateral membrane of human hepatocytes, but its spectrum of transport substrates does not include bile salts (90).

BASOLATERAL BILE SALT EFFLUX   Under normal physiological conditions, basolateral efflux of bile salts into portal blood plasma is negligible. However, under cholestatic conditions, basolateral bile salt efflux can compensate, at least in part, for the disrupted canalicular bile salt secretory pathway. This is indicated by an up-regulation of the expression of the basolateral ABC transporter Mrp3 (*Abcc3*) (105), which is capable of pumping bile salts out of hepatocytes (Table 1) (106). Basolateral Mrp3(rat)/MRP3(human) (*ABCC3*) expression is also up-regulated in Mrp2-deficient transport mutant rat strains (107) and in the liver of patients with the Dubin-Johnson syndrome (108). Whether Oatps/OATPs contribute to basolateral bile salt efflux via bile salt/organic anion exchange remains to be investigated.

Bile Salt Transport Across Hepatocytes

After basolateral uptake, bile salts are rapidly transferred across hepatocytes for canalicular secretion. Transcellular bile salt passage can occur within seconds, and an intravenous bolus is almost totally recovered in bile within 10 min (109). Essentially, two processes have been proposed to be involved, and they might be used to different degrees by hydrophilic and hydrophobic bile salt species. First, under physiological bile salt load conditions, the majority of intracellular bile salts bind to cytosolic proteins and diffuse to the canalicular membrane along the prevailing basolateral > canalicular bile salt concentration gradient. Hepatocellular cytosolic bile salt-binding proteins thus far identified include the 3α-hydroxysteroid dehydrogenase (3α-HSD), glutathione S-transferases and liver fatty acid-binding protein (L-FABP) in rat liver, and a 36-kDa bile acid-binding protein in human liver (13, 110). The latter binds bile acids with a higher affinity than the rat binding proteins, but its exact function in bile acid transport is not known. In rat hepatocytes, the functional role of cytosolic bile acid-binding proteins has been supported by the demonstration that inhibition of bile acid binding to 3α-HSD by indomethacin resulted in the redistribution of bile acids out of the cell and into the media (111). 3α-HSD and other cytoplasmic dehydrogenases with a molecular mass between 30 and 37 kDa bind lithocholic acid with a greater affinity than chenodeoxycholic acid (112). The same is also true for L-FABP, which represents the most prominent cytoplasmic protein component of rat hepatocytes that is labeled by photoreactive bile acid probes (112). Expression of 3α-HSD and L-FABP is controlled by HNF-1α and/or by the bile acid sensor FXR/BAR (49, 52). Loss of HNF-1α leads to down-regulation of both L-FABP and 3α-HSD. After cholate feeding, FXR/BAR null mice completely lose L-FABP expression. Bile salts also interact with the cytosolic phospholipid transfer protein, by which process they may modulate intracellular phospholipid transport (113).

Second, under high bile salt loads, hydrophobic bile salts partition into intracellular membranes and may accumulate within membrane-bound compartments (110, 114, 115). Although early morphological studies have indicated vesicle-associated intracellular bile salt transport (109), several more recent observations do not support the occurrence of significant vesicular transport of bile salts within cells under physiological conditions (for reviews, see 13, 116). For example, although the microtubule inhibitor colchicine decreased the canalicular accumulation of fluorescent derivatives of lithocholate and ursodeoxycholate, it had little effect on the cytoplasmic distribution and canalicular accumulation of fluorescent derivatives of cholate and chenodeoxycholate in hepatocyte couplets (117, 118). Thus whereas transcellular transport of lithocholate and ursodeoxycholate might be dependent on vesicle transport, the primary bile salts cholate and chenodeoxycholate are probably not. However, bile salts have been shown to regulate microtubule function, which is critical for the regulated insertion and retrieval of transport proteins into and from the canalicular membrane (118, 119). Such bile salt-dependent vesicular recycling of canalicular transporters can explain the correlation between pericanalicular vesicle fusion with the canalicular membrane and increased maximum taurocholate transport during bile salt infusion in the perfused rat liver (120–122). The degree of regulation by bile salts depends in part on their hydrophobicity, a determinant for activation of second messenger pathways such as cytosolic free calcium and protein kinase C (116).

Canalicular Bile Salt Secretion from Hepatocytes

Canalicular secretion of bile salts is an ATP-dependent process (123–126). The canalicular bile salt export pump (Bsep, *Abcb11*) was cloned from rat (127) and mouse (128–130) liver and is a member of the ABC transporter gene superfamily (Table 1). Its transport function was demonstrated in vesicles isolated from Sf9 cells transfected with Bsep. Rat Bsep has 12 putative transmembrane-spanning domains and is composed of 1321 amino acids with a molecular mass of 160 kDa in rat liver. Its amino acid sequence is 49 and 48 % identical with rat Mdr1b (*Abcb1b*) and rat Mdr2 (*Abcb4*), respectively. The K_m value for monoanionic bile salts of rat Bsep ranges from 2 to 5 μM for bile salts, which is in agreement with the K_m values of 2 to 6 μM for ATP-dependent monoanionic bile salt transport in isolated canalicular plasma membrane vesicles (131). Together with additional *cis*-inhibition experiments, the data indicate that Bsep is the major bile salt export pump in rat liver (131). This view is supported by the identification of the human *BSEP* gene by positional cloning in individuals with progressive familial intrahepatic cholestasis type 2 (PFIC2), an inherited progressive liver disease characterized by high serum bile salt concentrations in conjunction with low γ-glutamyltransferase serum levels (132). In such patients, several mutations in the *BSEP* gene have been identified that result in the absence of canalicular BSEP expression and a decrease of biliary bile salts to less than 1% of normal (133). Interestingly, however, Bsep knockout mice, which exhibit growth retardation and liver steatosis with mild cholestasis, show only minimally reduced bile flow and residual bile salt secretion of about 30% compared with wild-type mice (134). Although canalicular secretion of taurocholate is completely blocked, Bsep null mice continue to secrete muricholic acid together with a novel tetrahydroxylated bile acid. Furthermore, Bsep knockout mice exhibit increased biliary cholesterol excretion, which may be attributed to repression of bile salt biosynthesis by bile salts accumulating within hepatocytes. These exciting new observations might indicate the presence of an additional canalicular bile salt transporter in mouse liver. Alternatively, it might be possible that the secreted bile salt pool in Bsep knockout animals may represent sulfated and/or glucuronidated dianionic bile salts, which are known to be transported by the canalicular ABC transporter Mrp2 (*Abcc2*) (Table 1) (135).

Canalicular secretion of bile salts represents the rate-limiting step in the overall bile salt transport from blood into bile. The canalicular bile salt transport maximum can be adapted to the actual metabolic needs by transcriptional and posttranscriptional regulation. On the transcriptional level, canalicular *Bsep* gene expression is up-regulated by cholate feeding through an FXR/BAR-dependent mechanism (52). An FXR/BAR-dependent transactivation has also been demonstrated for the human *BSEP* promoter (136). In contrast to the basolateral bile salt uptake systems (see above), canalicular expression of Bsep is relatively well preserved in cholestasis and primary cultured hepatocytes (38, 137), which might be related to the fact that canalicular expression of Bsep (and of Mrp2) is not under the control of HNF-1α (Table 2) (49). On the posttranscriptional level, acute adaptation of the canalicular bile salt transport maximum can occur via rapid insertion of Bsep into, or its retrieval from, the canalicular plasma membrane. Bsep insertion into the canalicular membrane is stimulated by hypoosmotic swelling (138) and infusion of taurocholate (139). Furthermore, recruitment and direct stimulation of Bsep is dependent on phosphoinositide 3-kinase (PI3-kinase) (139) and PI3-kinase-generated lipid products (140). Also, Bsep can be directly phosphorylated in vitro by protein kinase C (130). Newly synthesized Bsep is directly targeted from the Golgi apparatus to the canalicular membrane via one or more subcanalicular vesicular pools (141, 142). Finally, canalicular Bsep function can be inhibited by a variety of cholestatic drugs and drug metabolites, including cyclosporine A, rifampicine, glibenclamide, and estradiol-17β-glucuronide (131, 143), which results in increased bile salt concentrations in serum and eventually in cholestatic liver injury. These observations highlight the importance of Bsep in canalicular bile salt secretion and its important role in their enterohepatic circulation.

Once secreted into bile canaliculi, bile salts stimulate the release of phosphatidylcholine and probably also of cholesterol from the outer leaflet of the canalicular membrane (144). Continuous supply of phosphatidylcholine to the outer leaflet of the canalicular membrane is provided by the phospholipid translocator Mdr2 (*Abcb4*) and its human orthologue MDR3 (*ABCB4*) (Figure 1) (144). Biliary lipids and bile salts are aggregated within vesicles and mixed micelles to protect the biliary epithelium from the detergent properties of bile salts. Lack of Mdr2 results in progressive bile ductular damage and cholestasis in knockout mice (145). In humans, defective MDR3 expression is associated with progressive familial intrahepatic cholestasis type 3 (PFIC3), an inherited liver disease characterized by high serum bile salt concentrations in conjunction with high γ-glutamyltransferase serum levels (146). These observations demonstrate that a coordinated function of Bsep/BSEP and Mdr2/MDR3 is important for ongoing bile formation and normal lipid homeostasis.

BILE SALT TRANSPORT IN BILE DUCT EPITHELIAL CELLS

Within biliary ductules, preferential absorption of monomeric bile acids results in bicarbonate-rich choleresis (Figure 1) (3), a process that is markedly stimulated by the infusion of the dihydroxy bile acid, ursodeoxycholic acid, at rates that exceed the hepatocyte's conjugation capacity (147, 148) or by the synthetic C_{23} nor- and C_{22}-dinordihydroxy bile acids that hardly undergo conjugation in hepatocytes (149). Although these unconjugated bile acids may passively enter cholangiocytes, conjugated bile salts are reabsorbed in a sodium-dependent manner across the apical plasma membrane of cholangiocytes by the bile salt transport ASBT (150, 151). ASBT is identical to the ileal bile acid transporter (IBAT) (Table 1). As in cholestatic hepatocytes, basolateral export of bile salts from cholangiocytes may also be mediated by the ABC-cassette transporter Mrp3/MRP3 in both rats (Figure 1) (105) and humans (152). Another candiate transporter is an alternatively spliced and truncated form of ASBT (so-called tAsbt), which contains 154 instead of 348 amino acids encoding a protein of 19 kDa (Table 1). This tAsbt can mediate taurocholate efflux and has been selectively localized to the basolateral plasma membrane domain of cholangiocytes (153). tAsbt-mediated taurocholate efflux might represent the sodium-independent anion-exchange mechanism, which was previously proposed to be involved in basolateral bile salt efflux from cholangiocytes (154).

BILE SALT TRANSPORT IN THE INTESTINE

Efficient reabsorption of bile salts in the intestine and delivery to the portal circulation is the second major mechanism for enterohepatic circulation of bile salts (Figure 1). Whereas conjugated bile salts are, owing to their low pKa values, predominantly trapped in the lumen of the gut and hence require protein-mediated reabsorption, unconjugated bile salts with pKa values of 5 or higher may be absorbed in significant portions by passive diffusion (110). Bile salt absorption in the jejunum is a sodium-independent and non-saturable process indicating uptake by passive diffusion or by low-affinity transport systems. In contrast, bile salt absorption in the ileum is sodium dependent and saturable demonstrating the presence of high-affinity protein-mediated uptake.

Apical Bile Salt Uptake

The sodium-independent bile salt transporter Oatp3 (*Slc21a7*), cloned from rat ileum, was demonstrated to be expressed in the brush border membrane of jejunal enterocytes (Table 1) (7). Oatp3 was originally cloned from retina (62) and is 80% identical with rat Oatp1. It is a high-affinity bile salt transporter with K_m values ranging from 4 to 21 μM for conjugated and unconjugated bile salts (7). Its K_m value for taurocholate is 21 μM, which is compatible with the K_m value for taurocholate (54 μM), determined in isolated rat jejunal brush border membrane vesicles (155), but is much lower than the value of 10 mM reported for perfused guinea pig jejunum (156). Whether this discrepancy is because of experimental differences (e.g., the presence of an unstirred water layer in the perfused intestine versus virtual absence of an unstirred layer in isolated brush border membrane vesicles) and/or to species differences remains to be elucidated. Uptake of bile salts into jejunual brush border membrane vesicles is stimulated by an in-to-out bicarbonate gradient, which is reminiscent of the *trans*-stimulation of Oatp1 by bicarbonate (70). However, whether Oatp3 can also mediate bile salt/bicarbonate exchange has not been established. Furthermore, it remains to be proven whether functional expression of bile salt-transporting OATPs also occurs at the brush border membrane of human intestine, although preliminary evidence for intestinal expression of OATP-A has been provided by a PCR approach (157). Nevertheless, the identification of Oatp3 as a sodium-independent bile salt uptake system along the major parts of the rat intestine suggests that the previously thought significance of mere passive diffusion for overall intestinal bile salt absorption must be reconsidered.

The distal ileum is the site of expression of IBAT or ASBT (*Slc10a2/SLC10A2*) (158, 159). ASBT was cloned from hamster (160), human (161), rat (162), rabbit (27), and mouse (163) ileum. It is a protein of 348 amino acids and has seven or possibly nine (33) transmembrane-spanning domains. It is about 35% identical with and has a size (48 kDa) similar to rat Ntcp and belongs to the same transporter family (*SLC10A*). In accordance with data from ileal brush border membrane vesicles (164), ASBT is electrogenic with a sodium-to-bile salt stoichiometry of 2:1 (165). ASBT is expressed biphasically during rat development, with first expression on day 22 of gestation, a transient decrease, and then a sharp increase of expression at postnatal days 17 and 18 (162, 166). The K_m values of the hamster and human ASBTs are 33 and 17 μM, respectively (160, 161). Extensive characterization of the substrate specificity of rabbit ASBT led to the development of a three-dimensional model of a pharmacophore for ASBT (167). The model predicts that ring D, the methyl group at position 18, and the α-oriented hydroxyl groups at positions 7 and 12 are essential for binding, whereas the hydroxyl group at position 3 allows for a certain flexibility.

Transcriptional regulation of *ASBT* gene expression is under control of HNF1α because HNF1α null mice do not express ASBT (Table 2) (49). However, and in contrast to the liver transporter Ntcp, expression of ASBT is not regulated in an FXR/BAR-dependent manner (52). This finding is supported by the lack of effect of intestinal bile salt depletion on the expression of ASBT in rat ileum (168). However, other studies indicate that altering the bile salt pool might affect the intestinal expression of ASBT, although both cholate feeding and feeding of bile salt-binding fibers increased intestinal ASBT expression (169, 170), whereas bile duct ligation was associated with down-regulation of dimeric, but not monomeric, ASBT in rat ileum (171). Also, an increased bile salt pool induced by cholesterol feeding has been reported to stimulate ASBT expression in rabbit but not in rat ileum (172). Hence, the effects of manipulating the bile salt pool on ASBT expression are controversial and probably also species specific. In addition to the bile salt pool, expression of ASBT is also regulated by hormones such as glucocorticoids, which lead to an up-regulation of ASBT (173). ASBT expression is restricted to the terminal ileum, and resection of the terminal ileum does not induce its expression in more proximal areas of the intestine (159). In humans, type IV hypertriglyceridemia is associated with abnormal absorption of bile salts with reduced expression levels of ASBT (174). The physiological importance of ASBT is highlighted by the fact that patients with primary bile acid malabsorption suffer from congenital diarrhea and steatorrhea concomitant with interrupted enterohepatic circulation of bile salts and reduced plasma cholesterol levels. These patients have mutations in the *ASBT* gene, some of which render the protein nonfunctional (175). Although considerable more work is required to definitively establish the transcriptional and posttranscriptional regulation of intestinal ASBT expression, the findings so far reported underline the physiological significance of ASBT-mediated intestinal bile salt uptake for the enterohepatic circulation and overall lipid homeostasis.

Cytosolic Bile Salt-Binding Protein(s)

Radiation inactivation analysis of the rabbit ileal bile salt uptake system suggests a molecular mass of 450 kDa, compatible with a multimeric transport system (176). Photoaffinity labeling studies identified a 14-kDa protein cytoplasmatically attached to ASBT. This ileal bile acid-binding protein (I-BABP) is closely related in sequence and function to the L-FABP (177). According to Kramer et al. (176), the functional ileal bile salt uptake complex contains four ASBT dimers and four I-BABPs. I-BABP contains a deep bile acid-binding pocket (178) and represents the most important protein for transcellular movement of bile salts across intestinal epithelial cells. This view is supported by the observations that bile salts increase their own binding to I-BABP (179). Also, bile acids up-regulate the expression of I-BABP (180) via the transcription factor FXR/BAR (11, 181). During postnatal development of the rat, I-BABP and ASBT expression levels rise in parallel (162, 182). Furthermore, similar to ASBT (see above), dexamethasone treatment also leads to an up-regulation of I-BABP (182). Thus the majority of observations indicate that intestinal ASBT and I-BABP expression might be under the control of similar, if not identical, transcription factors, although this assumption remains to be validated in further physiological and pathophysiological experimental models.

Basolateral Bile Salt Efflux

Basolateral efflux of bile salts from enterocytes is the least understood transport step. Functionally, the presence of an anion exchange mechanism(s) in the basolateral plasma membrane has been documented in membrane vesicles isolated from intestinal epithelial cells (183). Recently, the ASBT splice variant tAsbt, which can function as an anion exchanger (see above), was reported to be expressed twofold higher than full-length ASBT in the ileum (153). Another candidate protein for basolateral efflux of bile salts is Mrp3/MRP3, which is expressed in both rat and human small intestine (107, 184). Because Mrp3 transports bile salts (106) and is expressed in the basolateral plasma membrane of cholangiocytes (105, 152), it may also act as a basolateral bile salt export pump in the small intestine.

CONCLUSIONS AND PERSPECTIVES

The progress made in the identification and characterization of bile salt transport systems has greatly increased our understanding of enterohepatic physiology and pathophysiology. Carrier cloning and the rapidly increasing understanding of the transcriptional regulation of gene expression have allowed for a more complete delineation of the relationship between enterohepatic bile salt transport and lipid and cholesterol homeostasis. However, as evidenced by the Bsep knockout mice, it remains possible that additional bile salt carriers wait to be identified. This is especially true for the intestine, where the identity of the basolateral bile salt efflux system(s) has still not been defined. Further major challenges for the near future include the detailed molecular characterization of the interactions of bile salt transporters with intracellular bile salt-binding proteins and with the mechanisms of transcellular lipid and cholesterol transport, as well as the complete elucidation of the regulatory transporter network and its significance for hepatobiliary disease processes.

Visit the Annual Reviews home page at www.AnnualReviews.org

LITERATURE CITED

1. St-Pierre MV, Kullak-Ublick GA, Hagenbuch B, Meier PJ. 2001. Transport of bile acids in hepatic and non-hepatic tissues. *J. Exp. Biol.* 204:1673–86

2. Falany CN, Johnson MR, Barnes S, Diasio RB. 1994. Glycine and taurine conjugation of bile acids by a single enzyme. Molecular cloning and expression of human liver bile acid CoA:amino acid *N*-acyltransferase. *J. Biol. Chem.* 269:19375–79

3. Hofmann A. 1994. Bile acids. In *The Liver: Biology and Pathobiology*, ed. IM Arias, JL Boyer, N Fausto, WB Jakoby, DA Schacter, DA Shafritz, pp. 678–710. New York: Raven

4. Riepl RL, Fiedler F, Ernstberger M, Teufel J, Lehnert P. 1996. Effect of intraduodenal taurodeoxycholate and L-phenylalanine on pancreatic secretion and on gastroenteropancreatic peptide release in man. *Eur. J. Med. Res.* 1:499–505

5. Koop I, Schindler M, Bosshammer A, Scheibner J, Stange E, Koop H. 1996. Physiologic control of cholecystokinin release and pancreatic enzyme secretion by intraduodenal bile acids. *Gut* 39:661–67

6. Love M, Dawson PA. 1998. New insights into bile acid transport. *Curr. Opin. Lipidol.* 9:225–29

7. Walters HC, Craddock AL, Fusegawa H, Willingham MC, Dawson PA. 2000. Expression, transport properties, and chromosomal location of organic anion transporter subtype 3. *Am. J. Physiol. Gastrointest. Liver Physiol.* 279:G1188–G200

8. Hofmann AF. 1999. Bile acids: the good, the bad, and the ugly. *News Physiol. Sci.* 14:24–29

9. Russell DW. 1999. Nuclear orphan receptors control cholesterol catabolism. *Cell* 97:539–42

10. Parks DJ, Blanchard SG, Bledsoe RK, Chandra G, Consler TG, et al. 1999. Bile acids: natural ligands for an orphan nuclear receptor. *Science* 284:1365–68

11. Makishima M, Okamoto AY, Repa JJ, Tu H, Learned RM, et al. 1999. Identification of a nuclear receptor for bile acids. *Science* 284:1362–65

12. Wang H, Chen J, Hollister K, Sowers LC, Forman BM. 1999. Endogenous bile acids are ligands for the nuclear receptor FXR/BAR. *Mol. Cell* 3:543–53

13. Agellon LB, Torchia EC. 2000. Intracellular transport of bile acids. *Biochim. Biophys. Acta* 1486:198–209

14. Chawla A, Saez E, Evans RM. 2000. "Don't know much bile-ology." *Cell* 103: 1–4

15. Kullak-Ublick GA, Stieger B, Hagenbuch B, Meier PJ. 2000. Hepatic transport of bile salts. *Semin. Liver Dis.* 20:273–92

16. Thompson R, Strautnieks S. 2000. Inherited disorders of transport in the liver. *Curr. Opin. Genet. Dev.* 10:310–13

17. Müller M. 2000. Transcriptional control

of hepatocanicular transporter gene expression. *Semin. Liver Dis.* 20:323–37

18. Repa JJ, Mangelsdorf DJ. 2000. The role of orphan nuclear receptors in the regulation of cholesterol homeostasis. *Annu. Rev. Cell Dev. Biol.* 16:459–81

19. Schoonjans K, Brendel C, Mangelsdorf DJ, Auwerx J. 2000. Sterols and gene expression: control of affluence. *Biochim. Biophys. Acta* 1529:114–25

20. Trauner M, Boyer JL. 2001. Cholestatic syndromes. *Curr. Opin. Gastroenterol.* 17:242–56

21. Reichen J. 1999. The role of the sinusoidal endothelium in liver function. *News Physiol. Sci.* 14:117–21

22. Hofmann AF. 1998. Bile secretion and the enterohepatic circulation of bile acids. In *Gastrointestinal and Liver Disease*, ed. M Feldman, BF Scharschmidt, MH Sleisenger, pp. 937–48. Philadelphia/London: Saunders

23. Meier PJ. 1995. Molecular mechanisms of hepatic bile salt transport from sinusoidal blood into bile. *Am. J. Physiol. Gastrointest. Liver Physiol.* 269:G801–G12

24. Groothuis GM, Hardonk MJ, Keulemans KP, Nieuwenhuis P, Meijer DFF. 1982. Autoradiographic and kinetic demonstration of acinar heterogeneity of taurocholate transport. *Am. J. Physiol. Gastrointest. Liver Physiol.* 243:G455–G62

25. Hagenbuch B, Stieger B, Foguet M, Lübbert H, Meier PJ. 1991. Functional expression cloning and characterization of the hepatocyte Na$^+$/bile acid cotransport system. *Proc. Natl. Acad. Sci. USA* 88:10629–33

26. Cattori V, Eckhardt U, Hagenbuch B. 1999. Molecular cloning and functional characterization of two alternatively spliced Ntcp isoforms from mouse liver. *Biochim. Biophys. Acta* 1445:154–59

27. Kramer W, Stengelin S, Baringhaus KH, Enhsen A, Heuer H, et al. 1999. Substrate specificity of the ileal and the hepatic Na$^+$/bile acid cotransporters of the rabbit. I. Transport studies with membrane vesicles and cell lines expressing the cloned transporters. *J. Lipid Res.* 40:1604–17

28. Hagenbuch B, Meier PJ. 1994. Molecular cloning, chromosomal localization and functional characterization of a human liver Na$^+$/bile acid cotransporter. *J. Clin. Invest.* 93:1326–31

29. Meier PJ, Eckhardt U, Schroeder A, Hagenbuch B, Stieger B. 1997. Substrate specificity of sinusoidal bile acid and organic anion uptake systems in rat and human liver. *Hepatology* 26:1667–77

30. Stieger B, Hagenbuch B, Landmann L, Höchli M, Schroeder A, Meier PJ. 1994. In situ localization of the hepatocytic Na$^+$/taurocholate cotransporting polypeptide in rat liver. *Gastroenterology* 107:1781–87

31. Ananthanarayanan M, Ng OC, Boyer JL, Suchy FJ. 1994. Characterization of cloned rat liver Na$^+$-bile acid cotransporter using peptide and fusion protein antibodies. *Am. J. Physiol. Gastrointest. Liver Physiol.* 267:G637–G43

32. Kullak-Ublick GA, Glasa J, Böker C, Oswald M, Grützner U, et al. 1997. Chlorambucil-taurocholate is transported by bile acid carriers expressed in human hepatocellular carcinomas. *Gastroenterology* 113:1295–305

33. Hallén S, Bränden M, Dawson PA, Sachs G. 1999. Membrane insertion scanning of the human ileal sodium/bile acid cotransporter. *Biochemistry* 38:11379–88

34. Weinman SA. 1997. Electrogenicity of Na$^+$-coupled bile acid transporters. *Yale J. Biol. Med.* 70:331–40

35. Hagenbuch B, Meier PJ. 1996. Sinusoidal (basolateral) bile salt uptake system of hepatocytes. *Semin. Liver Dis.* 16:129–36

36. Hardikar W, Ananthanarayanan M, Suchy FJ. 1995. Differential ontogenic regulation of basolateral and canalicular bile acid transport proteins in rat liver. *J. Biol. Chem.* 270:20841–46

37. Suchy FJ, Bucuvalas JC, Goodrich AL,

Moyer MS, Blitzer BL. 1986. Taurocholate transport and Na$^+$-K$^+$-ATPase activity in fetal and neonatal rat liver plasma membrane vesicles. *Am. J. Physiol. Gastrointest. Liver Physiol.* 251:G665–G73

38. Rippin SJ, Hagenbuch B, Meier PJ, Stieger B. 2001. Cholestatic expression pattern of sinusoidal and canalicular organic anion transport systems in primary cultured rat hepatocytes. *Hepatology* 33:776–82

39. Liang D, Hagenbuch B, Stieger B, Meier PJ. 1993. Parallel decrease of Na$^+$-taurocholate cotransport and its encoding mRNA in primary cultures of rat hepatocytes. *Hepatology* 18:1162–66

40. Green RM, Gollan JL, Hagenbuch B Meier PJ, Beier DR. 1997. Regulation of hepatocyte bile salt transporters during hepatic regeneration. *Am. J. Physiol. Gastrointest. Liver Physiol.* 273:G621–G27

41. Hagenbuch B, Scharschmidt BF, Meier PJ. 1996. Effect of antisense oligonucleotides on the expression of hepatocellular bile acid and organic anion uptake systems in *Xenopus laevis* oocytes. *Biochem. J.* 316:901–4

42. Cohn MA, Rounds DJ, Karpen SJ, Ananthanarayanan M, Suchy FJ. 1995. Assignment of a rat liver Na$^+$/bile acid cotransporter gene to chromosome 6q24. *Mamm. Genome* 6:60–61

43. Green RM, Ananthanarayanan M, Suchy FJ, Beier DR. 1998. Genetic mapping of the Na$^+$-taurocholate cotransporting polypeptide to mouse chromosome 12. *Mamm. Genome* 9:598

44. Karpen SJ, Sun AQ, Kudish B, Hagenbuch B, Meier PJ, et al. 1996. Multiple factors regulate the rat liver basolateral sodium-dependent bile acid cotransporter gene promoter. *J. Biol. Chem.* 271:15211–21

45. Denson LA, Karpen SJ, Bogue CW, Jacobs HC. 2000. Divergent homeobox gene *Hex* regulates promoter of the Na$^+$-dependent bile acid cotransporter. *Am.*

J. Physiol. Gastrointest. Liver Physiol. 279:G347–G55

46. Keng VW, Yagi H, Ikawa M, Nagano T, Myint Z, et al. 2000. Homeobox gene *Hex* is essential for onset of mouse embryonic liver development and differentiation of the monocyte lineage. *Biochem. Biophys. Res. Commun.* 276:1155–61

47. Rausa FM, Tan Y, Zhou H, Yoo KW, Stolz DB, et al. 2000. Elevated levels of hepatocyte nuclear factor 3β in mouse hepatocytes influence expression of genes involved in bile acid and glucose homeostasis. *Mol. Cell. Biol.* 20:8264–82

48. Trauner M, Arrese M, Lee YH, Boyer JL, Karpen SJ. 1998. Endotoxin downregulates rat hepatic *Ntcp* gene expression via decreased activity of critical transcription factors. *J. Clin. Invest.* 101:2092–100

49. Shih DQ, Bussen M, Sehayek E, Ananthanarayanan M, Shneider BL, et al. 2001. Hepatocyte nuclear factor-1α is an essential regulator of bile acid and plasma cholesterol metabolism. *Nat. Genet.* 27:375–82

50. Hayhurst GP, Lee YH, Lambert G, Ward JM, Gonzalez FJ. 2001. Hepatocyte nuclear factor 4α (nuclear factor 2A1) is essential for maintenance of hepatic gene expression and lipid homeostasis. *Mol. Cell. Biol.* 21:1393–403

51. Denson LA, Auld KL, Schiek DS, McClure MH, Mangelsdorf DJ, Karpen SJ. 2000. Interleukin-1β suppresses retinoid transactivation of two hepatic transporter genes involved in bile formation. *J. Biol. Chem.* 275:8835–43

52. Sinal CJ, Tohkin M, Miyata M, Ward JM, Lambert G, Gonzalez FJ. 2000. Targeted disruption of the nuclear receptor FXR/BAR impairs bile acid and lipid homeostasis. *Cell* 102:731–44

53. Goodwin B, Jones SA, Price RR, Watson MA, McKee DD, et al. 2000. A regulatory cascade of the nuclear receptors FXR, SHP-1, and LRH-1 represses bile acid biosynthesis. *Mol. Cell* 6:517–26

54. Lu TT, Makishima M, Repa JJ, Schoon-jans K, Kerr TA, et al. 2000. Molecular basis for feedback regulation of bile acid synthesis by nuclear receptors. *Mol. Cell* 6:507–15

55. Denson LA, Sturm E, Echevarria W, Zimmermann TL, Makishima M, et al. 2001. The orphan nuclear receptor, shp, mediates bile acid-induced feedback inhibition of the bile acid transporter, ntcp. *Gastroenterology* 121:140–47

56. Ganguly TC, O'Brien ML, Karpen SJ, Hyde JF, Suchy FJ, Vore M. 1997. Regulation of the rat liver sodium-dependent bile acid cotransporter gene by prolactin. *J. Clin. Invest.* 99:2906–14

57. Grüne S, Engelkind LR, Anwer MS. 1993. Role of intracellular calcium and protein kinase in the activation of hepatic Na^+/taurocholate cotransport by cyclic AMP. *J. Biol. Chem.* 268:17734–41

58. Mukhopadhayay S, Ananthanarayanan M, Stieger B, Meier PJ, Suchy FJ, Anwer MS. 1997. cAMP increases liver Na^+-taurocholate cotransport by translocating transporter to plasma membrane. *Am. J. Physiol. Gastrointest. Liver Physiol.* 273:G842–G48

59. Dranoff JA, McClure M, Burgstahler AD, Denson LA, Crawford AR, et al. 1999. Short-term regulation of bile acid uptake by microfilament-dependent translocation of rat Ntcp to the plasma membrane. *Hepatology* 30:223–29

60. Gao B, Stieger B, Noé B, Fritschy JM, Meier PJ. 1999. Localization of the organic anion transporting polypeptide 2 (Oatp2) in capillary endothelium and choroid plexus epithelium of rat brain. *J. Histochem. Cytochem.* 47:1255–63

61. Gao B, Meier PJ. 2001. Organic anion transport across the choroid plexus. *Microsc. Res. Tech.* 52:60–64

62. Abe T, Kakyo M, Sakagami H, Tokui T, Nishio T, et al. 1998. Molecular characterization and tissue distribution of a new organic anion transporter subtype (oatp3) that transports thyroid hormones and tau-

rocholate and comparison with oatp2. *J. Biol. Chem.* 273:22395–401

63. Angeletti RH, Novikoff PM, Juvvadi SR, Fritschy JM, Meier PJ, Wolkoff AW. 1997. The choroid plexus epithelium is the site of the organic anion transport protein in the brain. *Proc. Natl. Acad. Sci. USA* 94:283–86

64. Choudhuri S, Ogura K, Klassen CD. 2001. Cloning, expression, and ontogeny of mouse organic anion-transporting polypeptide-5, a kidney-specific organic anion transporter. *Biochem. Biophys. Res. Commun.* 280:92–98

65. Tamai I, Nezu J, Uchino H, Sai Y, Oku A, et al. 2000. Molecular identification and characterization of novel members of the human organic anion transporter (OATP) family. *Biochem. Biophys. Res. Commun.* 273:251–60

66. Jacquemin E, Hagenbuch B, Stieger B, Wolkoff AW, Meier PJ. 1994. Expression cloning of a rat liver Na^+-independent organic anion transporter. *Proc. Natl. Acad. Sci. USA* 91:133–37

67. Bergwerk AJ, Shi X, Ford AC, Kanai N, Jacquemin E, et al. 1996. Immunologic distribution of an organic anion transport protein in rat liver and kidney. *Am. J. Physiol. Gastrointest. Liver Physiol.* 271:G231–G28

68. Eckhardt U, Schroeder A, Stieger B, Höchli M, Landmann L, et al. 1999. Polyspecific substrate uptake by the hepatic organic anion transporter Oatp1 in stably transfected CHO cells. *Am. J. Physiol. Gastrointest. Liver Physiol.* 276:G1037–G42

69. Hagenbuch B, Adler ID, Schmid TE. 2000. Molecular cloning and functional characterization of the mouse organic-anion-transporting polypeptide 1 (Oatp1) and mapping the gene to chromosome X. *Biochem. J.* 345:115–20

70. Satlin LM, Amin V, Wolkoff AW. 1997. Organic anion transporting polypeptide mediates organic anion/HCO_3 exchange. *J. Biol. Chem.* 272:26340–45

71. Li L, Lee TK, Meier PJ, Ballatori N. 1998. Identification of glutathione as a driving force and leukotriene C4 as a substrate for Oatp1, the sinusoidal organic solute transporter. *J. Biol. Chem.* 273:16184–91

72. Angeletti RH, Bergwerk AJ, Novikoff PM, Wolkoff AW. 1998. Dichotomous development of the organic anion transport protein and choroid plexus. *Am. J. Physiol. Cell Physiol.* 275:C882–C87

73. Dubuisson C, Cresteil D, Desrochers M, Decimo D, Hadchouel M, Jacquemin E. 1996. Ontogenic expression of the Na^+-independent organic anion transporting polypeptide (oatp) in rat liver and kidney. *J. Hepatol.* 25:932–40

74. Fickert P, Zollner G, Fuchsbichler A, Pojer C, Zenz R, et al. 2001. Effects of ursodeoxycholic and cholic acid feeding on hepatocellular transporter expression in mouse liver. *Gastroenterology* 121:170–83

75. Lee JL, Boyer JL. 2000. Molecular alterations in hepatocyte transport mechanisms in acquired cholestatic liver disorders. *Semin. Liver Dis.* 20:373–84

76. Hagenbuch N, Reichel C, Stieger B, Cattori V, Fattinger KE, et al. 2001. Effect of phenobarbital on the expression of bile salt and organic anion transporters of rat liver. *J. Hepatol.* 34:881–87

77. Rausch-Derra LC, Hartley DP, Meier PJ, Klassen CD. 2001. Differential effects of microsomal enzyme inducing chemicals on the hepatic expression of rat organic anion transporters Oatp1 and Oatp2. *Hepatology* 33:1469–78

78. Jones SA, Moore LB, Shenk JL, Wisely GB, Hamilton GA, et al. 2000. The pregnane X receptor: a promiscuous xenobiotic receptor that has diverged during evolution. *Mol. Endocrinol.* 14:27–39

79. Glavy JS, Wu SM, Wang PJ, Orr GA, Wolkoff AW. 2000. Down-regulation by extracellular ATP of rat hepatocyte organic anion transport is mediated by serine phosphorylation of Oatp1. *J. Biol. Chem.* 275:1479–84

80. Noé B, Hagenbuch B, Stieger B, Meier PJ. 1997. Isolation of a multispecific organic anion and cardiac glycoside transporter from rat brain. *Proc. Natl. Acad. Sci. USA* 94:10346–50

81. Reichel C, Gao B, van Montfoort J, Cattori V, Rahner C, et al. 1999. Localization and function of the organic anion-transporting polypeptide Oatp2 in rat liver. *Gastroenterology* 117:688–95

82. Kakyo M, Sakagami H, Nishio T, Nakai D, Nakagomi R, et al. 1999. Immunohistochemical distribution and functional characterization of an organic anion polypeptide 2 (oatp2). *FEBS Lett.* 445:343–46

83. Buscher HP, Meder I, MacNelly S, Gerok W. 1993. Zonal changes of hepatobiliary taurocholate transport in intra-hepatic cholestasis induced by 17 alpha-ethinyl estradiol:a histoautoradiographic study in rats. *Hepatology* 17:494–99

84. Aiso M, Takikawa H, Yamanaka M. 2000. Biliary excretion of bile acids and organic anions in zone 1- and zone 3-injured rats. *Liver* 20:38–44

85. Staudinger JL, Goodwin B, Jones SA, Hawkins-Brown D, MacKenzie KI, et al. 2001. The nuclear receptor PXR is a lithocholic acid sensor that protects against liver toxicity. *Proc. Natl. Acad. Sci. USA* 98:3369–74

86. Xie W, Radominska-Pandya A, Shi Y, Simon CM, Nelson MC, et al. 2001. An essential role for nuclear receptors SXR/PXR in detoxification of cholestatic bile acids. *Proc. Natl. Acad. Sci. USA* 98:3375–80

87. Cattori V, Hagenbuch B, Hagenbuch N, Stieger B, Ha R, et al. 2000. Identification of organic anion transporting poylpeptide 4 (Oatp4) as a major full-length isoform of the liver-specific transporter-1 (rlst-1) in rat liver. *FEBS Lett.* 474:242–45

88. Kakyo M, Unno M, Tokui T, Nakagomi

R, Nishio T, et al. 1999. Molecular characterization and functional regulation of a novel rat liver-specific organic anion transporter rlst-1. *Gastroenterology* 117:770–75

89. Cattori V, van Montfoort J, Stieger B, Landmann L, Meijer DKF, et al. 2001. Localization of organic anion transporting polypeptide 4 (Oatp4) in rat liver and comparison of its substrate specificity with Oatp1, Oatp2 and Oatp3. *Pflügers Arch.* In press

90. Kullak-Ublick GA, Ismair MG, Stieger B, Landmann L, Huber R, et al. 2001. Organic anion-transporting polypeptide B (OATP-B) and its functional comparison with three other OATPs of human liver. *Gastroenterology* 120:525–33

91. Choudhuri S, Ogura K, Klaassen CD. 2000. Cloning of the full-length coding sequence of rat liver-specific organic anion transporter-1 (rlst-1) and a splice variant and partial characterization of the rat lst-1 gene. *Biochem. Biophys. Res. Commun.* 274:79–86

92. Fischer WJ, Cattori V, Meier PJ, Dietrich DR, Hagenbuch B. 2001. Organic anion transporting polypeptides (OATPs) mediate uptake of microcystin into brain and liver. *J. Appl. Physiol.* In press

93. Ismair MG, Stieger B, Cattori V, Hagenbuch B, Fried M, et al. 2001. Hepatic uptake of cholecystokinin octapeptide (CCK-8) by organic anion transporting polypeptides Oatp4 (Slc21a6) and OATP8 (SLC21A8) of rat and human liver. *Gastroenterology* 121: In press

94. Jung D, Hagenbuch B, Gresh L, Pontoglio M, Meier PJ, Kullak-Ublick GA. 2001. Characterization of the human OATP-C (*SLC21A6*) gene promoter and regulation of liver-specific *OATP* genes by hepatocyte nuclear factor 1α. *J. Biol. Chem.* 276:37204–14

95. König J, Yunhai C, Nies AT, Keppler D. 2000. A novel human organic anion transporting polypeptide localized to the basolateral hepatocyte membrane.

Am. J. Physiol. Gastrointest. Liver Physiol. 278:G156–G64

96. Hsiang B, Zhu Y, Wang Z, Wu Y, Sasseville V, et al. 1999. A novel human hepatic organic anion transporting polypeptide (OATP2). Identification of a liver-specific human organic anion transporting polypeptide and identification of rat and human hydroxymethylglutaryl-CoA reductase inhibitor transporters. *J.Biol. Chem* 274:37161–68

97. Abe T, Kakyo M, Tokui T, Nakagomi R, Nishio T, et al. 1999. Identification of a novel gene family encoding human liver-specific organic anion transporter LST-1. *J. Biol. Chem.* 274:17159–63

98. Cui Y, König J, Leier I, Buchholz U, Keppler D. 2001. Hepatic uptake of bilirubin and its conjugates by the human organic anion-transporting polypeptide SLC21A6. *J. Biol. Chem.* 276:9626–30

99. König J, Cui Y, Nies T, Keppler D. 2000. Localization and genomic organization of a new hepatocellular organic anion transporting polypeptide. *J. Biol. Chem.* 275:23161–68

100. Kullak-Ublick GA, Beuers U, Fahney C, Hagenbuch B, Meier PJ, Paumgartner G. 1997. Identification and functional characterization of the promoter region of the human organic anion transporting polypeptide gene. *Hepatology* 26:991–97

101. Ogura K, Choudhuri S, Klaassen CD. 2000. Full-length cDNA cloning and genomic organization of the mouse liver-specific organic anion transporter-1 (lst-1). *Biochem. Biophys. Res. Commun.* 272:563–70

102. Ogura K, Choudhuri S, Klaassen CD. 2001. Genomic organization and tissue-specific expression of splice variants of mouse organic anion transporting polypeptide 2. *Biochem. Biophys. Res. Commun.* 281:431–39

103. Kullak-Ublick GA, Hagenbuch B, Stieger B, Schteingart CD, Hofmann AF, et al. 1995. Molecular and functional

characterization of an organic anion trans-porting polypeptide cloned from human liver. *Gastroenterology* 109:1274–82

104. Gao B, Hagenbuch B, Kullak-Ublick GA, Benke D, Aguzzi A, Meier PJ. 2000. Organic anion-transporting polypeptides mediate transport of opioid peptides across the blood-brain barrier. *J. Pharm. Exp. Ther.* 294:73–79

105. Soroka CJ, Lee JL, Azzaroli F, Boyer JL. 2001. Cellular localization and up-regulation of multidrug resistance-associated protein 3 in hepatocytes and cholangiocytes during obstructive cholestasis in rat liver. *Hepatology* 33:783–91

106. Hirohashi T, Suzuki H, Takikawa H, Sugiyama Y. 2000. ATP dependent transport of bile salts by rat multidrug resistance-associated protein 3 (Mrp3). *J. Biol. Chem.* 275:2905–10

107. Hirohashi T, Suzuki H, Ito K, Ogawa K, Kume K, et al. 1998. Hepatic expression of multidrug resistance-associated protein-like proteins maintained in Eisai hyperbilirubinemic rats. *Mol. Pharmacol* 53:1068–75

108. König J, Rost D, Cui Y, Keppler D. 1999. Characterization of the human multidrug resistance protein isoform MRP3 localized to the basolateral hepatocyte membrane. *Hepatology* 29:1156–63

109. Crawford JM, Berken CA, Gollan JL. 1988. Role of the hepatocyte microtubular system in the excretion of bile salts and biliary lipid: implications for intracellular vesicular transport. *J. Lipid Res.* 29:144–56

110. Bahar RJ, Stolz AS. 1999. Bile acid transport. *Gastroenterol. Clin. N. Am.* 28:27–58

111. Takikawa H, Stolz A, Kaplowitz N. 1987. Cyclical oxidation-reduction of the C3 position on bile acids catalyzed by rat hepatic 3 alpha-hydroxysteroid dehydrogenase. 1. Studies with the purified enzyme, isolated rat hepatocytes, and inhibition

by indomethacin. *J. Clin. Invest.* 80:852–60

112. Stolz A, Takikawa H, Ookhtens M, Kaplowitz N. 1989. The role of cytoplasmic proteins in hepatic bile acid transport. *Annu. Rev. Physiol.* 51:161–76

113. Cohen DE, Leonard MR, Carey MC. 1994. In vitro evidence that phospholipid secretion into bile may be coordinated intracellularly by the combined actions of bile salts and the specific phosphatidylcholine transfer protein of liver. *Biochemistry* 33:9975–80

114. Lamri Y, Roda A, Dumont M, Feldmann G, Erlinger S. 1988. Immunoperoxidase localization of bile salts in rat liver cells. Evidence for a role of the Golgi apparatus in bile salt transport. *J. Clin. Invest.* 82:1173–82

115. Kullak-Ublick GA, Stieger B, Hagenbuch B, Meier PJ. 2000. Hepatic transport of bile salts. *Semin. Liver Dis.* 20:273–92

116. Crawford JM. 1996. Role of vesicle-mediated transport pathways in hepatocellular bile secretion. *Semin. Liver Dis.* 16:169–89

117. Wilton JC, Matthews GM, Burgoyne RD, Mills CO, Chipman JK, Coleman R. 1994. Fluorescent choleretic and cholestatic bile salts take different paths across the hepatocyte: transcytosis of glycolithocholate leads to an extensive redistribution of annexin II. *J. Cell Biol.* 127:401–10

118. El Seaidy AZ, Mills CO, Elias E, Crawford JM. 1997. Lack of evidence for vesicle trafficking of fluorescent bile salts in rat hepatocyte couplets. *Am. J. Physiol. Gastrointest. Liver Physiol.* 272:G298–G309

119. Marks DL, LaRusso NF, McNiven MA. 1995. Isolation of the microtubule-vesicle motor kinesin from rat liver: selective inhibition by cholestatic bile acids. *Gastroenterology* 108:824–33

120. Häussinger D, Saha N, Hallbrucker C, Lang F, Gerok W. 1993. Involvement of microtubules in the swelling-induced

stimulation of transcellular taurocholate transport in perfused rat liver. *Biochem. J.* 291:355–60

121. Bruck R, Haddad P, Graf J, Boyer JL. 1992. Regulatory volume decrease stimulates bile flow, bile acid excretion, and exocytosis in isolated perfused rat liver. *Am. J. Physiol. Gastrointest. Liver Physiol.* 262:G806–G12

122. Gatmaitan ZC, Nies AT, Arias IM. 1997. Regulation and translocation of ATP-dependent apical membrane proteins in rat liver. *Am. J. Physiol. Gastrointest. Liver Physiol.* 272:G1041–G49

123. Nishida T, Gatmaitan Z, Che M, Arias IM. 1991. Rat liver canalicular membrane vesicles contain an ATP-dependent bile acid transport system. *Proc. Natl. Acad. Sci. USA* 88:6590–94

124. Müller M, Ishikawa T, Berger U Klunemann C, Lucka L, et al. 1991. ATP-dependent transport of taurocholate across the hepatocyte canalicular membrane mediated by a 110-kDa glycoprotein binding ATP and bile salt. *J. Biol. Chem.* 266:18920–26

125. Adachi Y, Kobayashi H, Kurumi Y, Shouji M, Kitano M, Yamamoto T. 1991. ATP-dependent taurocholate transport by rat liver canalicular membrane vesicles. *Hepatology* 14:655–59

126. Stieger B, O'Neill B, Meier PJ. 1992. ATP-dependent bile salt transport in canalicular rat liver plasma-membrane vesicles. *Biochem. J.* 284:67–74

127. Gerloff T, Stieger B, Hagenbuch B, Madon J, Landmann L, et al. 1998. The sister of P-glycoprotein represents the canalicular bile salt export pump of mammalian liver. *J. Biol. Chem.* 273:10046–50

128. Green RM, Hoda F, Ward KL. 2000. Molecular cloning and characterization of the murine bile salt export pump. *Gene* 241:117–23

129. Lecureur V, Sun D, Hargrove P, Schuetz EG, Kim RB, et al. 2000. Cloning and expression of murine sister of P-glycoprotein reveals a more dis-

criminating transporter than *MDR1*/P-glycoprotein. *Mol. Pharmacol.* 57:24–35

130. Noe J, Hagenbuch B, Meier PJ, St-Pierre MV. 2001. Characterization of the mouse bile salt export pump overexpressed in the baculovirus system. *Hepatology* 33:1223–31

131. Stieger B, Fattinger K, Madon J, Kullak-Ublick GA, Meier PJ. 2000. Drug- and estrogen-induced cholestasis through inhibition of the hepatocellular bile salt export pump of rat liver. *Gastroenterology* 118:422–30

132. Strautnieks SS, Bull LN, Knisely AS, Kocoshis SA, Dahl N, et al. 1998. A gene encoding a liver-specific ABC transporter is mutated in progressive familial intrahepatic cholestasis. *Nat. Genet.* 20:233–38

133. Jansen PL, Strautnieks SS, Jacquemin E, Hadchouel M, Sokal EM, et al. 1999. Hepatocanalicular bile salt export pump deficiency in patients with progressive familial intrahepatic cholestasis. *Gastroenterology* 117:1370–79

134. Wang R, Salem M, Yousef IM, Tuchweber B, Lam P, et al. 2001. Targeted inactivation of sister of P-glycoprotein gene (spgp) in mice results in nonprogressive but persistent intrahepatic cholestasis. *Proc. Natl. Acad. Sci. USA* 98:2011–16

135. Keppler D, König J. 2000. Hepatic secretion of conjugated drugs and endogenous substances. *Semin. Liver Dis.* 20:265–72

136. Ananthanarayanan M, Balasubramanian N, Makishima M, Suchy FJ. 2001. Human bile salt export bump (BSEP) promoter is transactivated by the farnesoid X receptor/bile acid receptor (FXR/BAR). *J. Biol. Chem.* 276:28857–65

137. Lee JM, Trauner M, Soroka CJ, Stieger B, Meier PJ, Boyer JL. 2000. Expression of the bile salt export pump is maintained after chronic cholestasis in the rat. *Gastroenterology* 118:163–72

138. Schmitt M, Kubitz R, Lizun S, Wettstein M, Häussinger D. 2001. Regulation of the

dynamic localization of the rat Bsep gene-encoded bile salt export pump by aniso-osmolarity. *Hepatology* 33:509–18

139. Misra S, Ujhazy P, Gatmaitan Z, Varticovski L, Arias IM. 1998. The role of phosphoinositide 3-kinase in taurocholate-induced trafficking of ATP-dependent canalicular transporters in rat liver. *J. Biol. Chem.* 273:26638–44

140. Misra S, Ujhazy P, Varticovski L, Arias IM. 1999. Phosphoinositide 3-kinase lipid products regulate ATP-dependent transport by sister of P-glycoprotein and multidrug resistance-associated protein 2 in bile canalicular membrane vesicles. *Proc. Natl. Acad. Sci. USA* 96:5814–19

141. Kipp H, Arias IM. 2000. Newly synthesized canalicular ABC transporters are directly targeted from the Golgi to the hepatocyte apical domain in rat liver. *J. Biol. Chem.* 275:15917–25

142. Kipp H, Pichetshote N, Arias IM. 2001. Transporters on demand. Intrahepatic pools of canalicular ATP binding cassette transporters in rat liver. *J. Biol. Chem.* 276:7218–24

143. Böhme M, Müller M, Leier I, Jedlitschky G, Keppler D. 1994. Cholestasis caused by inhibition of the adenosine triphosphate-dependent bile salt transport in rat liver. *Gastroenterology* 107:255–65

144. Oude Elferink RP, Groen AK. 2000. Mechanisms of biliary lipid secretion and their role in lipid homeostasis. *Semin. Liver Dis.* 20:293–305

145. Smit JJM, Schinkel AH, Oude Elferink RP, Groen AK, Wagenaar E, et al. 1993. Homozygous disruption of the murine mdr2 P-glycoprotein gene leads to a complete absence of phospholipid from bile and to liver disease. *Cell* 75:451–62

146. de Vree JM, Jacquemin E, Sturm E, Cresteil D, Bosma PJ, et al. 1998. Mutations in the MDR3 gene cause progressive familial intrahepatic cholestasis. *Proc. Natl. Acad. Sci. USA* 95:282–87

147. Dumont M, Uchman S, Erlinger S. 1980. Hypercholeresis induced by ursodeoxy-cholic acid and 7-ketolithocholic acid in the rat. *Gastroenterology* 79:82–89

148. Kitani K, Kanai S. 1982. Effect of ursodeoxycholate on the bile flow in the rat. *Life Sci.* 31:1973–85

149. Kirkpatrick RB, Green MD, Hagey LR, Hofmann AF, Tephly TR. 1988. Effect of side chain length on bile acid conjugation: glucuronidation, sulfation and coenzyme A formation of nor-bile acids and their natural C24 homologs by human and rat liver fractions. *Hepatology* 8:353–57

150. Alpini G, Glaser SS, Rodgers R, Phinizy JL, Robertson WE, et al. 1997. Functional expression of the apical Na^+-dependent bile acid transporter in large but not small rat cholangiocytes. *Gastroenterology* 113:1734–40

151. Lazaridis KN, Pham L, Tietz P, Marinelli RA, de Groen PC, et al. 1997. Rat cholangiocytes absorb bile acids at their apical domain via the ileal sodium-dependent bile acid transporter. *J .Clin. Invest.* 100:2714–21

152. Kool M, van der Linden M, de Haas M, Scheffer GL, de Vree JML, et al. 1999. MRP3, an organic anion transporter able to transport anti-cancer drugs. *Proc. Natl. Acad. Sci. USA* 96:6914–19

153. Lazaridis KN, Tietz P, Wu T, Kip S, Dawson PA, LaRusso NF. 2000. Alternative splicing of the rat sodium/bile acid transporter changes its cellular localization and transport properties. *Proc. Natl. Acad. Sci. USA* 97:11092–97

154. Benedetti A, Di Sario A, Marucci L, Svegliati-Baroni G, Schteingart CD, et al. 1997. Carrier-mediated transport of conjugated bile acids across the basolateral membrane of biliary epithelial cells. *Am. J. Physiol. Gastrointest. Liver Physiol.* 272:G1416–G24

155. Amelsberg A, Jochims C, Richter CP, Nitsche R, Fölsch UR. 1999. Evidence for an anion exchange mechanism for uptake of conjugated bile acid from the rat jejunum. *Am. J. Physiol. Gastrointest. Liver Physiol.* 276:G737–G42

156. Amelsberg A, Schteingart CD, Ton-Nu H-T, Hofmann AF. 1996. Carrier-mediated jejunal absorption of conjugated bile acids in the guinea pig. *Gastroenterology* 110:1098–106

157. Richter CP, Amelsberg A, Nitsche R. 1999. Localization of the organic anion transport polypeptide (OATP) at the brush border side in human jejunum. *Gastroenterology* 116:L0376

158. Stelzner M, Hoagland V, Somasundaram S. 2000. Distribution of bile acid absorption and bile acid transporter gene message in the hamster ileum. *Pflügers Arch.* 440:157–62

159. Coppola CP, Gosche JR, Arrese M, Ancowitz B, Madsen J, et al. 1998. Molecular analysis of the adaptive response of intestinal bile acid transport after ileal resection in the rat. *Gastroenterology* 115:1172–78

160. Wong MH, Oelkers P, Craddock AL, Dawson PA. 1994. Expression cloning and characterization of the hamster ileal sodium-dependent bile acid transporter. *J. Biol. Chem.* 270:1340–47

161. Wong MH, Oelkers P, Dawson PA. 1995. Identification of a mutation in the ileal sodium-dependent bile acid transporter gene that abolishes transport activity. *J. Biol. Chem.* 270:27228–34

162. Shneider BL, Dawson PA, Christie DM, Hardikar W, Wong MH, Suchy FJ. 1995. Cloning and molecular characterization of the ontogeny of a rat ileal sodium-dependent bile acid transporter. *J. Clin. Invest.* 95:745–54

163. Saeki T, Matoba K, Furukawa H, Kirifuji K, Kanamoto R, Iwami K. 1999. Characterization, cDNA cloning, and functional expression of mouse ileal sodium-dependent bile acid transporter. *J. Biochem.* 125:846–51

164. Lücke H, Stange G, Kinne R, Murer H. 1978. Taurocholate-sodium co-transport by brush-border membrane vesicles isolated from rat ileum. *Biochem. J.* 174:951–58

165. Weinman SA, Carruth MW, Dawson PA. 1998. Bile acid uptake via the human apical sodium-bile acid cotransporter is electrogenic. *J. Biol. Chem.* 273:34691–95

166. Shneider BL, Setchell KDR, Crossman MW. 1997. Fetal and neonatal expression of the apical sodium-dependent bile acid transporter in the rat ileum and kidney. *Ped. Res.* 42:189–94

167. Baringhaus K-H, Matter H, Stengelin S, Kramer W. 1999. Substrate specificity of the ileal and the hepatic Na$^+$/bile aicd cotransporters of the rabbit. II. A reliable 3D QSAR pharmacophore model for the ileal Na$^+$/bile acid cotransporter. *J. Lipid Res.* 40:2158–68

168. Arrese M, Trauner M, Sacchiero RJ, Crossman MW, Shneider BL. 1998. Neither intestinal sequestration of bile acids nor common bile duct ligation modulate the expression and function of the rat ileal bile acid transporter. *Hepatology* 28:1081–87

169. Stravitz RT, Sanyal AJ, Pandak WM, Vlahcevic ZR, Beets JW, Dawson PA. 1997. Induction of sodium-dependent bile acid transporter messenger RNA, protein and activity in rat ileum by cholic acid. *Gastroenterology* 113:1599–608

170. Buhman KK, Furumoto EJ, Donkin SS, Story JA. 2000. Dietary psyllium increases expression of ileal apical sodium-dependent bile acid transporter mRNA coordinately with dose-responsive changes in bile acid metabolism. *J. Nutr.* 130:2137–42

171. Sauer P, Stiehl A, Fitscher BA, Riedel H-D, Benz C, et al. 2000. Downregulation of ileal bile acid absorption in bile-duct-ligated rats. *J. Hepatol.* 33:2–8

172. Xu G, Shneider BL, Shefer S, Nguyen LB, Batta AK, et al. 2000. Ileal bile acid transport regulates bile acid pool, synthesis and plasma cholesterol levels differently in cholesterol-fed rats and rabbits. *J. Lipid Res.* 41:298–304

173. Nowicki MJ, Shneider BL, Paul JM,

Heubi JE. 1997. Glucocorticoids upregulate taurocholate transport by ileal brush-border membrane. *Am. J. Physiol. Gastrointest. Liver Physiol.* 273:G197–G203

174. Duane WC, Hartich LA, Bartman AE, Ho SB. 2000. Diminished gene expression of ileal apical sodium bile acid transporter explains impaired absorption of bile acid in patients with hypertriglyceridemia. *J. Lipid Res.* 41:1384–89

175. Oelkers P, Kirby LC, Heubi JE, Dawson PA. 1997. Primary bile acid malabsorption caused by mutations in the ileal sodium-dependent bile acid transporter gene (*SLC10A2*). *J. Clin. Invest.* 99:1880–87

176. Kramer W, Girbig F, Gutjahr U, Kowaleski S. 1995. Radiation inactivation analysis of the Na⁺/bile acid co-transport system from rabbit ileum. *Biochem. J.* 306:241–46

177. Gong Y-Z, Everett ET, Schwartz DA, Norris JS, Wilson FA. 1994. Molecular cloning, tissue distribution, and expression of a 14-kDa bile acid-binding protein from rat ileal cytosol. *Proc. Natl. Acad. Sci. USA* 91:4741–45

178. Kramer W, Sauber K, Baringhaus K-H, Kurz M, Stengelin S, et al. 2001. Identification of the bile acid-binding site of the ileal lipid-binding protein by photoaffinity labeling matrix-assisted laser desorption ionization-mass spectrometry, and NMR structure. *J. Biol. Chem.* 276:7291–301

179. Kramer W, Corsiero D, Friedrich M, Girbig F, Stengelin S, Weyland C. 1998. Intestinal absorption of bile acids: paradoxical behaviour of the 14-kDa ileal lipid-binding protein in differential photoaffinity labelling. *Biochem. J.* 333:335–41

180. Kanda T, Foucand L, Nakamura Y, Niot I, Besnard P, et al. 1998. Regulation of expression of human intestinal bile acid-binding protein in CaCo-2 cells. *Biochem. J.* 330:261–65

181. Grober J, Zaghini I, Fujii H, Jones SA, Kliewer SA, et al. 1999. Identification of a bile acid-responsive element in the human ileal bile acid-binding protein gene. Involvement of the farnesoid X receptor/9-*cis*-retinoic acid receptor heterodimer. *J. Biol. Chem.* 274:29749–54

182. Hwang ST, Henning SJ. 2000. Hormonal regulation of expression of ileal bile acid binding protein in suckling rats. *Am. J. Physiol. Regulatory Integrative Comp. Physiol.* 278:R1555–R63

183. Weinberg SL, Burckhardt G, Wilson FA. 1986. Taurocholate transport by rat intestinal basolateral plasma membrane vesicles. Evidence for the presence of an anion exchange transport system. *J. Clin. Invest.* 78:44–50

184. Kiuchi Y, Suzuki H, Hirohashi T, Tyson CA, Sugiyama Y. 1998. cDNA cloning and inducible expression of human multidrug resistance associated protein 3 (MRP3). *FEBS Lett.* 433:149–52

185. Paulusma CC, Kool M, Bosma PJ, Scheffer GL, ter Borg F, et al. 1997. A mutation in the human canalicular multispecific organic anion transporter gene causes the Dubin-Johnson syndrome. *Hepatology* 25:1539–42

186. Paulusma CC, Bosma PJ, Zaman GJ, Bakker CT, Otter M, et al. 1996. Congenital jaundice in rats with a mutation in a multidrug resistance-associated protein gene. *Science* 271:1126–28

187. Büchler M, König J, Brom M, Kartenbeck J, Spring H, et al. 1996. cDNA cloning of the hepatocyte canalicular isoform of the multidrug resistance protein, cMrp, reveals a novel conjugate export pump deficient in hyperbilirubinemic mutant rats. *J. Biol. Chem.* 271:15091–98

Annu. Rev. Physiol. 2002. 64:663–80

MECHANISMS OF IRON ACCUMULATION IN HEREDITARY HEMOCHROMATOSIS

Robert E. Fleming[1] and William S. Sly[2]

[1]Department of Pediatrics, and the [2]Edward A. Doisy Department of Biochemistry and Molecular Biology, Saint Louis University School of Medicine, 1402 South Grand Boulevard, St. Louis, Missouri 63104; e-mail: flemingr@slu.edu; slyws@slu.edu

Key Words   iron absorption, HFE, β_2-microglobulin, transferrin receptor

■ **Abstract**   Hereditary hemochromatosis (HH) is a common inborn error of iron metabolism characterized by excess dietary iron absorption and iron deposition in several tissues. Clinical consequences include hepatic failure, hepatocellular carcinoma, diabetes, cardiac failure, impotence, and arthritis. Despite the discovery of the mutation underlying most cases of HH, considerable uncertainty exists in the mechanism by which the normal gene product, HFE, regulates iron homeostasis. Knockout of the *HFE* gene clearly confers the HH phenotype on mice. However, studies on HFE expressed in cultured cells have not yet clarified the mechanism by which *HFE* mutations lead to increased dietary iron absorption. Recent discoveries suggest other genes, including a second transferrin receptor and the circulating peptide hepcidin, participate in a shared pathway with HFE in regulation of iron absorption. This review summarizes our current understanding of the relationship between iron stores and absorption and presents models to explain the dysregulated iron homeostasis in HH.

INTRODUCTION

Hereditary hemochromatosis (HH) is a common inherited disorder of iron metabolism in which intestinal iron absorption is excessive relative to body iron stores. Over time the excess absorbed iron leads to saturation of serum transferrin, deposition of iron in parenchymal cells of several tissues, and cellular toxicity (1). The autosomal-recessive inheritance pattern of HH has long been recognized, but the underlying gene defect was identified only recently. Surprisingly, the responsible gene did not encode a protein with metal-transport characteristics, but rather a major histocompatibility complex (MHC) class I–like molecule (2). The cloning of the *HFE* gene has established a foundation for a better understanding of the molecular and cell biology of iron absorption and its altered regulation in HH. The molecular mechanism by which the HFE mutation increases intestinal iron absorption is still being determined. This chapter reviews the current understanding of the biology of iron uptake and the role of the *HFE* and presents working models

0066-4278/02/0315-0663$14.00

to explain the mechanisms by which the HFE mutation leads to dysregulation of intestinal iron absorption.

REGULATION OF IRON HOMEOSTASIS

Maintaining normal iron homeostasis is essential for the organism, as both iron deficiency and iron excess are associated with cellular dysfunction (3). Because there are no significant physiological mechanisms to modulate iron loss, homeostasis is dependent upon tightly linking body iron requirements with intestinal iron absorption. Intestinal iron absorption is tied to body iron requirements by at least two functionally defined regulators. One is the erythropoietic regulator, which adjusts intestinal iron absorption in response to the demands of erythropoiesis, independent of body iron stores (4). The other is the stores regulator, which acts on a pathway that facilitates a slow accumulation of dietary iron. The stores regulator is of great physiological importance because it prevents iron overload after ensuring iron needs are met (5). Several lines of evidence suggest that the set point for the stores regulator is altered in patients with HH, such that absorption of dietary iron is excessive relative to body iron stores. An understanding of the physiology of intestinal iron absorption is necessary for understanding the dysregulation that occurs in HH.

Iron Absorption

Nearly all absorption of dietary iron occurs in the proximal small intestine (primarily duodenum), where it may be taken up either as free iron or as heme. Absorption of iron across the enterocytes occurs in two stages: uptake across the apical membrane and transfer across the basolateral membrane. Prior to uptake, dietary free iron requires reduction from the ferric (Fe^{3+}) to the ferrous (Fe^{2+}) state. This is accomplished by the ferric reductase Dcytb, expressed on the luminal surface of the proximal small intestine (6). The ferrous ion is taken up by the apical transporter DMT1 (also known as DCT1, Nramp2) (7). Uptake of heme occurs by a yet unidentified transporter, possibly involving the molecule mobilferrin (8). Iron is then released from heme within the enterocyte by heme oxygenase. The iron may be stored within the cell as ferritin and lost with the sloughed senescent enterocyte or transferred across the basolateral membrane to the plasma. This latter process occurs via the transporter ferroportin1 (9) [other names Ireg1 (10), MTP1 (11)] and requires oxidation of iron to the ferric state by the molecule hephaestin (12) (Figure 1, step 1; see color insert). Patients with HH demonstrate increased luminal reductase activity (13), increased DMT1 expression (14), decreased enterocyte ferritin stores (15), and increased ferroportin1 expression (10, 16). Each of these changes would be expected to contribute to the increased iron absorption seen in HH.

Hepatic Iron Storage

Absorbed iron is bound to transferrin and circulates initially through the portal system of the liver, which is the major site of iron storage. Hepatocytes take up transferrin-bound iron via the classical transferrin receptor (TfR) and possibly by the more highly expressed homologous protein, TfR2 (17, 18) (Figure 1, step 2). Hepatocytes may also take up transferrin-bound iron by non-receptor-mediated mechanisms (19). Non-transferrin-bound iron, found in the circulation only when transferrin becomes highly saturated, may be taken up by hepatocytes as well (20). This latter mechanism appears to contribute to the hepatic iron deposition in HH (21). The means by which iron is released from hepatocytes is poorly understood; however, the iron export protein ferroportin1 is expressed in hepatocytes (9, 11).

Reticuloendothelial Iron Storage

Reticuloendothelial (RE) macrophages acquire iron from surface transferrin receptors (22) or from phagocytosis of senescent erythrocytes (23). Heme oxygenase within the cells releases the iron from the erythrocyte heme. The iron is either retained (stored as ferritin) or released into the plasma via the iron export protein ferroportin1. The released iron is oxidized to the ferric state in the plasma by ceruloplasmin and is bound to circulating transferrin. HH patients have paradoxical sparing of iron loading in the RE system (24) owing to increased release of RE iron (25) and/or decreased uptake of transferrin-bound iron (26).

Communication Between Sites of Iron Storage and Uptake

The means by which the sites of iron storage (hepatocytes and the RE system) and uptake (duodenum) communicate with each other to maintain normal iron homeostasis is uncertain. This stores regulator has been proposed to involve soluble factors such as transferrin-bound iron, serum ferritin, or serum TfR (27, 28). Recent evidence suggests that circulating levels of the liver peptide hepcidin convey information on liver iron stores to reticuloendothelial cells and intestinal epithelial cells (Figure 1, step 3). In mice, hepatic iron loading is associated with increased hepcidin expression in the liver (29). Loss of hepcidin expression in mice is associated with excess iron in periportal hepatocytes and iron sparing of RE cells, a phenotype resembling HH (30). *HFE*, the gene that is mutated in HH, is normally expressed in duodenal crypt cells (31) and in RE cells, suggesting that the HFE is a participant in the hepcidin-mediated signaling between these cells and liver iron stores (Figure 1, step 4).

Tight linkage of dietary iron absorption with body stores occurs in the proximal small intestine. Here duodenal crypt cells, the precursor cells for the absorptive enterocytes, sense the iron needs of the body and are programmed as they mature into absorptive enterocytes (32, 33) to express appropriate levels of the previously described iron transport proteins (4, 16, 34) (Figure 1, step 5). This programming is thought to involve changes in the labile iron pool of the crypt cells, which in turn changes the mRNA-binding activities of the iron regulatory proteins (IRPs) (35, 36). Transcripts encoding ferritin, DMT1, and ferroportin1 each contain iron responsive elements (IREs) and are possibly modulated in their expression by changes in IRP-binding activities.

The importance of the stores regulator is highlighted by the findings in patients with HH. These individuals absorb dietary iron excessively relative to body stores, suggesting that the set point for the stores regulator is altered. The excess iron accumulates over time, leading to tissue damage and organ failure (37). *HFE*, the gene defective in HH, encodes an MHC class I integral membrane protein found in a physical complex with β_2-microglobulin (β_2M) (2). In the duodenum the HFE/β_2M complex is confined to the crypt cells, where it is physically associated with TfR (38). HFE and TfR are also expressed in RE cells. The observation in patients with HH of increased dietary iron absorption and decreased RE iron stores suggests that HFE disruption leads to disordered uptake (or release) of plasma-derived iron in these cells. The potential mechanisms by which this may occur are discussed in detail after consideration of the molecular genetics and cell biology of HFE.

MOLECULAR GENETICS OF THE *HFE* GENE

HFE Gene and Transcript

The genomic structure of *HFE* is similar to other MHC class I–like molecules. Each of the first six exons of the *HFE* gene encodes one of the six distinct domains of the HFE protein (Figure 2; see color insert). The mutation found to be associated with HH was a single base change in exon 4, resulting in the substitution of tyrosine for cysteine at amino acid 282 of the deduced amino acid sequence of the unprocessed protein (C282Y). This corresponds to amino acid 260 of the mature protein after removal of the 22-residue signal sequence. (Most other MHC class I proteins are numbered from the first amino acid of the mature sequence.) Proof that *HFE* is the gene defective in HH was provided when knockout of the mouse gene resulted in iron overload (39).

The predominant *HFE* transcript is ~4.2 kb, although additional minor transcripts, both longer and shorter, have been reported (2, 40, 41). At least some of the smaller transcripts are attributable to alternative splicing events and differential use of polyadenylation signals in exon 7 (40, 41). However, the levels of translated protein from the *HFE* splice-variant transcripts and their physiological significance are unknown. *HFE* mRNA is expressed at low levels in most human tissues (2) and has been detected in multiple cell lines of epithelial or fibroblastic origin (although, unlike other MHC class I genes, not in lymphohematopoietic cells).

The promoter region of the *HFE* gene contains sequences homologous to several *cis*-acting elements, including GATA, NF-IL6, AP1, AP2, CREB, PEA3, γ-IRE, GFI1, HNF-3β, and HFH2 (42); however, functional characterization has not been reported. In contrast to other MHC molecules, *HFE* expression was not induced in response to various cytokines in cell culture (43). Sequences conferring transcriptional regulation by metal ions have not been identified in the *HFE* gene. The effect of iron on *HFE* expression is unclear. One study found that *HFE* mRNA and protein levels increased with increased cellular iron in CaCo2 cells (a human intestinal cell line) (44); however, another study found that neither iron chelation nor iron replacement affected *HFE* mRNA levels in these cells (45). IREs have not been identified in either the 3′ or 5′ untranslated region of the *HFE* transcript.

HFE Gene Product

The *HFE* gene encodes a 343–amino acid protein consisting of six distinct domains: a 22–amino acid signal peptide, three extracellular domains (designated the $\alpha 1$, $\alpha 2$, and $\alpha 3$ loops), a transmembrane region, and a short intracellular region (2, 46) (Figures 2, 3; see color insert). Intramolecular disulfide bridges exist within the $\alpha 2$ and $\alpha 3$ domains. Crystallographic studies demonstrate that the $\alpha 1$ and $\alpha 2$ loops together form a superdomain that includes two antiparallel α helices (46). Although the groove between the antiparallel α helices is analogous to the peptide-binding groove in antigen-presenting MHC class I proteins, studies suggest that HFE does not have a functional peptide-binding groove (46, 47).

In the duodenum, the primary site of dietary iron absorption, HFE is confined to the crypt cells (Figure 4; see color insert) (31, 38). HFE co-localizes in these cells with TfR in a distribution consistent with the recycling endosome (38). It has been postulated that HFE in duodenal crypt cells may play a key role in modulating dietary iron absorption (see below). HFE protein has been detected in epithelial cell populations of other parts of the intestine as well, but the crypt cell specificity is seen only in the proximal small intestine (38).

HFE has been also detected immunohistochemically in a number of other tissues, including brain (especially capillary endothelium), liver (sinusoidal lining cells, bile duct epithelial cells, and Kupffer cells) (48), placenta (syncytiotrophoblasts) (49), tissue macrophages (50), and circulating monocytes and granulocytes (50). It is likely that HFE expression in Kupffer cells and tissue macrophages play a role in modulating iron uptake and storage by the RE system.

HFE MUTATIONS

C282Y

The majority of patients with classical HH have a single base change in exon 4 of the *HFE* gene, leading to the substitution of tyrosine for cysteine at amino acid 282 of the unprocessed protein (C282Y). This mutation was also proven to cause murine HH when knockin of the C282Y mutation (51) produced the HH phenotype in the mouse. The impaired function of the C282Y mutant protein in HH appears to be caused by loss of interaction of the protein with $\beta_2 M$, with secondary block in intracellular transport of HFE leading to accelerated turnover, and decreased presentation on the cell surface. Several immunohistochemical studies have examined HFE expression in tissue samples from HH patients and normal controls. Decreased expression of the C282Y protein was observed in Kupffer cells (52) and macrophages (50). Within duodenal enterocytes, granular cytoplasmic staining was observed in C282Y homozygotes but not in controls (53). It was concluded that the granular staining likely represents abnormally processed HFE protein.

The proportion of HH patients homozygous for C282Y varies in different populations, ranging from 100% of HH cases in a study from Australia to only 64% in a study from Italy (54, 55). In the United States, Britain, and France, 82–90% of patients with a clinical diagnosis of HH are homozygous for C282Y (2, 56–58). Prevalence of the C282Y mutation is greatest in Caucasian subjects of European ancestry, in which the carrier frequency is ∼10–15% (59, 60). In other ethnic populations, the C282Y mutation is less common and is always associated with the ancestral Caucasian haplotype (61, 62). These observations suggest that the C282Y mutation occurred once on an ancestral (possibly Celtic) haplotype that spread from northern Europe to other regions of the world (54, 57). The haplotype containing the C282Y mutation extends ∼7 Mb, suggesting that the mutation arose during the past 2000 years (64). It has been proposed that the C282Y mutation, by leading to increased iron absorption and accumulation of body iron stores, provided a selective advantage to a population with limited dietary iron availability. Not all individuals homozygous for the C282Y mutation develop clinical findings of HH (60, 65). This observation suggests incomplete penetrance of the C282Y mutation and raises the possibility that other genes involved in iron transport act as modifiers of the HH phenotype.

H63D

More common than the C282Y mutation in the general population is a missense mutation at nucleotide 187 of the *HFE* open reading frame, which results in the substitution of histidine for aspartate at amino acid 63 of the unprocessed protein (H63D) (2). The H63D mutation is found in 15–40% of Caucasians (66), a frequency too high to implicate this allele in most cases of HH. In fact, H63D homozygosity appears to lead to only a slight (∼4-fold) increased risk for iron loading (67). Compound heterozygosity for the H63D mutation with C282Y is found with a higher frequency in patients with iron overload than predicted from the general population frequencies of these two alleles (2). The risk for iron loading in the C282Y/H63D compound heterozygote, however, is nearly 200-fold lower than in the C282Y homozygote (67). Interestingly, H63D appears to form a salt bridge with a residue in the α2 loop of HFE that binds to TfR, possibly providing a molecular basis for its effect on iron homeostasis.

The H63D mutation has been found on many haplotypes, suggesting it may have arisen in different populations at historically different times. H63D is found at the highest allele frequencies in Europe, in countries bordering the Mediterranean, in the Middle East, and in the Indian subcontinent. Because the H63D haplotype is shorter (700 kb or fewer) than the C282Y haplotype, it is thought to be evolutionarily older (64).

Other *HFE* Mutations

HFE mutations other than C282Y and H63D have been found in isolated patients with iron overload (68). These mutations include S65C, G93R, I105T (69), Q127H, and R330M (Figure 5; see color insert). A causal relationship between these mutations and disordered iron homeostasis, however, remains to be established. Of these, the S65C substitution is most common, accounting for 7.8% of chromosomes carrying neither C282Y nor H63D in probands with an HH phenotype (70). Interestingly, G93R and I105T are located in a region of the $\alpha 1$ domain of HFE that binds TfR. In addition to the above described missense mutations, two *HFE* frameshift mutations (P160ΔC and V68ΔT) (68), two nonsense mutations (71), and one splice site mutation (72) have been reported. Other than P160ΔC, each of these truncating mutations occurred in HH patients who were compound heterozygotes for the C282Y mutation.

CELL BIOLOGY OF THE HFE PROTEIN

Association of HFE with β_2-Microglobulin

The homology between HFE and other MHC class I molecules led to the prediction that HFE is physically associated with β_2-microglobulin (β_2M). HFE-β_2M association was demonstrated in human duodenum (38) and placenta (49), as well as cultured cells (73–75). The HH-like phenotype of β_2M-knockout mice provides functional evidence of the importance of the HFE-β_2M association in the whole organism (76, 77). C282Y mutant protein (expressed in cell culture) binds poorly with β_2M and is less abundantly presented at the cell surface (74, 78). In cell culture, much of the expressed mutant C282Y protein is retained in the endoplasmic reticulum and middle Golgi compartments, fails to undergo late Golgi processing, and is subject to accelerated degradation (74). HH patients likewise demonstrate reduced cell surface expression of the C282Y mutant protein (50). Furthermore, the association between HFE and β_2M appears to be necessary for the interaction of HFE with TfR (79) (see below).

Association of HFE with Transferrin Receptor

The first mechanistic link between HFE and cellular iron metabolism was provided by the observation that HFE forms a complex with TfR. The physical association of HFE with TfR has been demonstrated in placental syncytiotrophoblasts (49), the site of maternal-fetal iron transport, and in duodenal crypt enterocytes, the site of regulation of dietary iron absorption (38). Studies in transfected cell lines suggest that the HFE-TfR complex forms shortly after the biosynthesis of each protein (75). While HFE complexed with TfR was found to be stable, uncomplexed HFE protein was rapidly degraded. HFE and TfR were found to remain associated at the cell surface and in intracellular vesicles. These observations suggest that HFE, TfR, and diferric transferrin (FeTf) undergo endocytosis as a complex. The TfR endosomal compartment becomes acidified upon endocytosis (pH 5.5–6.0), a process that leads to the release of iron from FeTf. Experiments using soluble HFE and soluble TfR demonstrate that the HFE-TfR complex dissociates upon lowering the pH from 7.5 to 6.0 (46), suggesting that HFE protein dissociates from TfR in the acidified endosome. The effect of dissociation of HFE from TfR on iron release and export from the endosome is unknown. The fate of the HFE protein in the endosome after dissociation from TfR is likewise undetermined.

The HFE-TfR complex has been crystallized and the amino acids participating in their interaction identified. Interestingly, HFE and FeTf appear to bind a similar region of the transferrin receptor. This observation has made the stoichiometric relationship between TfR and HFE of particular interest. A 2:2 (HFE monomer:TfR monomer) relationship was demonstrated in crystallography studies using high (millimolar) concentrations of soluble HFE and soluble TfR (80). If relevant in vivo, this observation suggests that HFE-TfR complex would be unable to bind FeTf. However, a ternary complex of HFE-TfR-FeTf with a stoichiometry of 1:2:1 has been observed in experiments using micromolar concentrations of soluble HFE and soluble TfR (81). Likewise, HFE overexpressed in HeLa cells appears to have a 1:2 stoichiometry with TfR during intracellular trafficking (79). Although the stoichiometry of the HFE-TfR complex at the cell membrane is unknown, it appears unlikely that all cell-surface TfR is complexed with two HFE molecules, as overexpression of HFE in transfected cells does not completely prevent FeTf binding to TfR. The effect of HFE on Tf-mediated iron uptake is discussed below.

Effect of *HFE* Mutations on HFE-TfR Interaction

Several observations suggest that the changes in iron homeostasis from *HFE* mutations are the result of alterations in the normal interaction between HFE and TfR. C282Y mutant HFE protein expressed in cell culture does not associate with TfR (82). H63D, although remote from residues that interact with TfR (Figure 5), appears to form a salt bridge with the TfR-binding region of the α2 loop of HFE (46). Overexpressed H63D protein is capable, nonetheless, of forming a complex with TfR (82). As mentioned above, isolated patients with iron overload carry mutations in the region of HFE that binds TfR, suggesting that alteration in HFE-TfR interaction may be responsible. Compound mutant mice hemizygous for TfR and lacking HFE accumulate more liver iron than mice lacking HFE alone (51), supporting the concept that interaction between these two proteins is necessary in the normal control of iron absorption. However, no patients with iron overload carrying mutations in TfR have been identified (83).

Effect of HFE on TfR-Mediated Iron Uptake in Cultured Cells

The effect of HFE on TfR-mediated iron uptake has been investigated in several studies. In HeLa cells, overexpression of HFE leads to a decrease in Tf-mediated iron uptake (43, 84–86). The decreased iron uptake is, in turn, associated with an increase in cellular iron regulator protein (IRP) binding activity (85, 86), a decrease in intracellular ferritin levels (75, 84–86), and an increase in TfR expression (75, 85). These observations are all consistent with an iron-deficient phenotype.

The basis for the decreased Tf-bound iron uptake in HeLa cells overexpressing HFE is unclear, and data have been conflicting. While overexpression of HFE in HeLa cells was reported to decrease affinity of FeTf for TfR (82), the observed effect would not be expected to have functional consequences under physiological conditions, where FeTf levels are well above that required to saturate TfR. Furthermore, no change in the affinity of FeTf for TfR was noted upon overexpressing HFE in HeLa cells in another study (43). This study instead found that overexpressed HFE reduced the number of functional FeTf-binding sites and also impaired TfR endocytosis, leading to an accumulation of nonfunctional TfR on the cell surface. The impaired endocytosis was subsequently attributed to HFE-induced phosphorylation of TfR (79). In contrast, however, a third group (84) found that overexpression of HFE in HeLa cells had no effect on the rate of TfR endocytosis (or exocytosis). They instead proposed that normal HFE protein decreases the cellular acquisition of iron from endocytic FeTf.

The effect of HFE overexpression has also been examined in the human intestinal epithelial cell line CaCo2 (87). As seen in HeLa cells, HFE overexpression produced a relatively iron-deficient state in these cells. Surprisingly, however, apical (presumably DMT1-mediated) iron uptake was decreased despite an increase in DMT1 protein. These results probably do not have implications for the DMT1-mediated uptake of dietary iron across the apical enterocyte, as HFE does not appear to be expressed in these cells. If, however, HFE does affect DMT1-mediated iron transport, there are possible implications for the transfer of iron from the endocytic compartment to the cytoplasm because both HFE and DMT are present within the TfR endosome.

It should be noted, however, that each of the studies on the effect of HFE overexpression (in HeLa or CaCo2 cells) was performed without overexpression of β_2M. It is thus possible that abnormally processed HFE (unassociated with β_2M) influenced the iron homeostasis in these cells (perhaps analogous to the situation in the β_2M knockout mouse). It has been conjectured that because overexpressed HFE leads to relative cellular iron deficiency, loss of functional HFE might lead to cellular iron excess and account for the iron deposition in HH. However, such a mechanism is unlikely to explain iron loading of the liver, as hepatocytes appear to express little (if any) HFE (48). Furthermore, RE cells (which do express HFE) demonstrate relative iron sparing rather than loading in HH patients (24). This iron sparing is apparently, at least in part, from decreased uptake of Tf-bound iron (26). Transfection of HH macrophages with wild-type HFE was found to increase accumulation of Fe delivered by FeTf and increased the Fe-ferritin pool within the transfected cells. These results suggest that the *HFE* mutation directly leads to the paradoxical iron-deficient phenotype of RE cells in HH.

The iron stores regulator appears to influence dietary iron absorption by altering the iron status of the duodenal crypt cells, the precursor cells for the absorptive villus enterocytes. The daughter enterocytes become programmed by this signal to an appropriate level of expression of the previously described iron transport proteins (Figure 1). Similar to RE cells, duodenal enterocytes in HH demonstrate a relative iron-deficient phenotype (88–90). Specifically, duodenal mucosal samples from HH patients demonstrate increased ferric reductase activity (13), increased DMT1 mRNA content (14), relatively decreased ferritin levels (15), and increased ferroportin1 mRNA content (10, 16). All are consistent with changes observed with dietary iron deficiency. It has thus been proposed that loss of functional HFE in duodenal crypt cells leads to relative iron deficiency, a situation similar to that seen in RE cells (38, 91). Such an effect in the crypt cells could alter expression of iron transport proteins in daughter enterocytes, and lead to an inappropriately high absorption of dietary iron.

Models for Effect of *HFE* Mutation on Dietary Iron Absorption

As stated above, the intestinal villus cells in HH patients have characteristics more consistent with relative iron deficiency than excess (88–90). How does loss of HFE lead to relatively low iron in these cells? Two hypothesized models have been put forward. In one model, loss of HFE leads to increased transferrin-mediated iron uptake by the crypt cells (33). In the other model, loss of HFE leads to decreased transferrin-mediated iron uptake by these cells (38, 91).

The first model is based on the observation that overexpression of HFE in HeLa cells leads to decreased Tf-mediated iron uptake and an iron-deficient phenotype. Conversely, it is reasoned, loss of HFE would lead to increased Tf-mediated iron uptake and iron excess. The increased iron uptake would then lead (at least transiently) to an increase in the intracellular labile iron pool, which decreases IRP binding activity. The decreased IRP activity increases translation of the iron efflux transporter (by the interaction of IRP with an IRE located in the 3' untranslated region of the mRNA). The increased expression of ferroportin1 increases iron efflux from daughter enterocytes, ultimately leading to an iron-deficient phenotype. This model assumes a long half-life for the ferroportin1 protein (maintaining ferroportin1 levels even though the cells have become relatively iron deficient). In addition to this caveat, there are other potential problems with this model. Ferroportin1 mRNA is highly expressed in villus enterocytes, but not crypt cells, making regulation by the crypt cell regulatory iron pool unlikely. Furthermore, iron deficiency leads to an increase in enterocyte ferroportin1 (a direction opposite of that expected) (9–11). This latter observation has raised questions as to the role of the 5' IRE in the regulation of ferroportin1 expression in the intestine (11).

The second model to explain the iron-deficient phenotype of the HH intestine is perhaps better supported. In this model, loss of HFE leads to decreased transferrin-mediated iron uptake in intestinal crypt cells (consistent with observations in circulating macrophages of HFE patients), leading to a decrease in the labile iron pool. This stabilizes the mRNA for the iron transport protein DMT1 (via the interaction of IRPs with an IRE in the 3′ untranslated region of the DMT1 transcript). The DMT1 protein is translated in daughter enterocytes and leads to increased uptake of dietary iron. The increased dietary iron uptake possibly then increases the regulatory iron pool in the villus enterocytes (35). IRP binding activity would then be decreased, possibly leading to increased translation of ferroportin1 mRNA in the villus cells (via the IRE in the 5′ untranslated region of the ferroportin1 mRNA). Both apical and basolateral iron transport would thereby be increased: apical via the effect of IRP binding activity on the DMT1 mRNA expressed in the crypt cells and basolateral by the effect of IRP binding activity on ferroportin1 mRNA expressed in the villus cells.

This latter model is supported by the observations that DMT1 mRNA is predominantly expressed toward the crypts (7), whereas ferroportin1 mRNA is predominantly expressed at the villus tip (11). Increased DMT1 mRNA levels have been reported in HH patients (14) and *Hfe−/−* mice (92) in association with increased DMT1-mediated iron transport (93). Furthermore, compound mutant mice deficient in both HFE and DMT1 do not manifest iron loading (94). However, not all studies have demonstrated increased duodenal DMT1 mRNA levels in *Hfe−/−* mice (95). Furthermore, DMT1 protein was undetectable by immunoblot and immunohistochemistry in *Hfe* knockout (and control) mice, but not in iron-deficient mice (95). The level of sensitivity to detect DMT1 may have been an issue, however, in this analysis. Thus the basis for the effect of *HFE* mutation on dietary iron absorption remains uncertain. Ultimately, testing each of the proposed mechanisms requires understanding of the consequences of *HFE* mutation on the labile iron pool of the duodenal crypt and villus cells separately.

Murine *Hfe* and Experimental Models of *Hfe* Gene Disruption

The murine *Hfe* gene is structurally similar to the human gene (96). However, rather than being telomeric to the murine MHC complex (chromosome 17), the murine *Hfe* gene has been evolutionarily translocated to chromosome 13 (97, 98). The mouse transcript is ~1.5–2.0 kb and found in multiple tissues (96, 97). The deduced murine Hfe amino acid sequence is 66% similar to the human (97). Both C282 and H63 in the human HFE sequence are conserved in the mouse (C294 and H67).

Four different *Hfe* gene disruptions have been reported in the mouse: an exon 4 knockout (39), an exon 3 disruption/exon 4 knockout (51), an exon 2–3 knockout (99), and a C282Y knockin (51). In each model, the mice manifested increased hepatic iron levels (39, 51, 99). Elevated transferrin saturations (39) and increased intestinal iron absorption (99) have been reported as well. Like HH patients, these mice demonstrate relative sparing of iron loading in RE cells (39). *Hfe* disruption caused no immunologic manifestations (99). Interestingly, mice that are homozygous for the C282Y mutation have less severe iron loading than *Hfe−/−* mice, indicating that the C282Y mutation is not a null allele (51). *Hfe* knockout mice have been bred to mice carrying mutations in other genes involved in normal iron homeostasis (94). Studies using these animals indicate that DMT1, hephaestin, β_2M, and TfR can modify the HH phenotype. A role for naturally occurring strain-dependent gene modifiers was demonstrated by the variation in iron loading seen upon breeding the *Hfe* knockout allele onto genetically defined mouse strains (100).

OTHER FORMS OF HEMOCHROMATOSIS

In addition to *HFE*, disruptions in several other genes have been demonstrated to result in an HH phenotype in human patients or animal models. These include the genes for β_2M (76, 77), TfR2 (101, 102), and hepcidin (30). Mutations in several additional genes have been shown to cause iron loading, but not the characteristic periportal iron deposition in hepatocytes, and relative sparing of iron deposition in RE macrophages. Mutations in ferroportin1, although leading to periportal iron loading in hepatocytes (103, 104), lead to loading rather than sparing of iron in RE macrophages. The common phenotype of β_2M, TfR2, and hepcidin mutations implicates each as a participant in the stores regulator pathway. A speculative model for feedback between hepatocyte iron stores and intestinal iron absorption, which includes a role for each of these molecules, is presented in Figure 1 (see color insert).

SUMMARY

The recent identification and cloning of *HFE* has greatly expanded our understanding of many aspects of HH. The introduction of genetic tests for the C282Y and H63D mutations has allowed presymptomatic diagnosis and added precision to studies of the population genetics of HH. Mouse models of HH and cell culture studies have increased our understanding of the normal physiology of HFE protein and the consequences of its disruption. Current data suggest that HFE exerts its effects on cellular iron homeostasis by complexing with TfR and influencing the level of Tf-mediated cellular iron uptake. The mechanism by which *HFE* mutation leads to increased intestinal iron uptake is incompletely understood; however, it appears to involve the inappropriate sensing of body iron status by HFE-expressing duodenal crypt cells and an increase in expression of genes involved in dietary iron absorption. The pace of discovery of genes influencing iron absorption makes it likely that additional insights will soon add to our understanding of the mechanisms by which *HFE* mutation contributes to the pathogenesis of hereditary hemochromatosis.

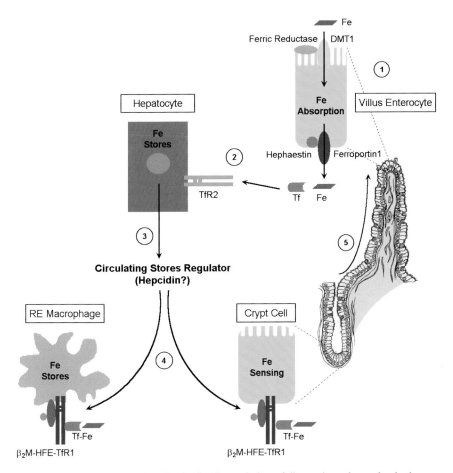

Figure 1 Proposed mechanism for feedback regulation of dietary iron absorption by hepatic iron stores. (1) Uptake of dietary iron occurs in duodenal villus enterocytes via DMT1 after reduction to the ferrous state by the ferric reductase. The iron is transported out of the enterocytes by ferroportin1 and oxidized to the ferric state by hephaestin. The iron binds to apotransferrin and enters the portal circulation. (2) Hepatocytes take up the iron via several mechanisms including TfR2. (3) In response to increased iron stores, hepatocytes release hepcidin. (4) Hepcidin interacts with RE macrophages and duodenal enterocytes to increase intracellular iron in these cells by a mechanism involving the HFE-β_2M-TfR complex. (5) The increased iron in crypt cells decreases expression of DMT1, ferroportin1, and ferric reductase in daughter enterocytes. Villus figure (step 5) adapted with permission from *Gastroenterology and Hepatology Vol. 7, Small Intestine.* 1997. Philadelphia: Curr. Med. p. 1.4

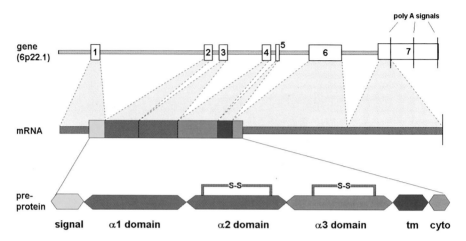

Figure 2 Human *HFE* gene product. The *HFE* gene consists of 7 (known) exons (*top*) that contribute to the major *HFE* mRNA transcript (*middle*). Each of the first 6 exons encodes a distinct domain of HFE protein (*bottom*). CDS, coding sequence; UT, untranslated sequence; tm, transmembrane domain; cyto, cytoplasmic domain.

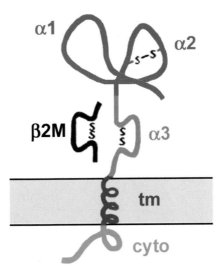

Figure 3 HFE protein in association with β_2 microglobulin at the cell surface. The three extracellular domains of HFE are designated α1, α2, and α3. β_2M is shown associated with the α3 domain. Adapted and modified from Reference 2.

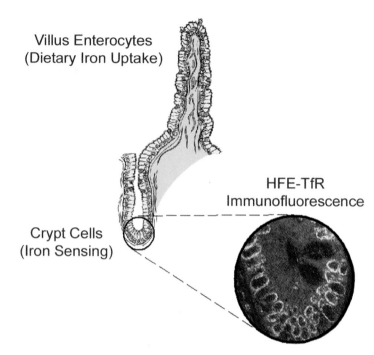

Figure 4 HFE co-localization with TfR in human duodenal crypt cells. Double immuno fluorescence of HFE protein and TfR demonstrates that both proteins are expressed in the same crypt enterocytes. Co-localization of HFE and TfR is indicated by the bright yellow perinuclear signal. Adapted and modified from Reference 38. Villus figure adapted with permission from *Gastroenterology and Hepatology Vol. 7, Small Intestine.* 1997. Philadelphia: Curr. Med. p. 1.4

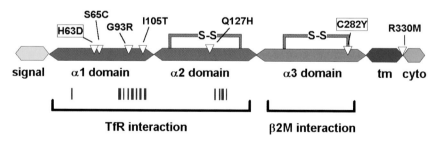

Figure 5 Identified mutations in HFE protein. Positions of known missense mutations in the HFE protein are identified with triangles. Positions of amino acid residues thought to participate in the interaction of HFE with TfR (80) are designated with vertical lines: tm, transmembrane domain; cyto, cytoplasmic domain.

Visit the Annual Reviews home page at www.AnnualReviews.org

LITERATURE CITED

1. Nichols GM, Bacon BR. 1989. Hereditary hemochromatosis: pathogenesis and clinical features of a common disease. *Am. J. Gastroenterol.* 84:851–62

2. Feder JN, Gnirke A, Thomas W, Tsuchihashi Z, Ruddy DA, et al. 1996. A novel MHC class I-like gene is mutated in patients with hereditary haemochromatosis. *Nat. Genet.* 13:399–408

3. Andrews NC. 1999. Disorders of iron metabolism. *N. Engl. J. Med.* 341:1986–95

4. Finch C. 1994. Regulators of iron balance in humans. *Blood* 84:1697–702

5. Gavin MW, McCarthy DM, Garry PJ. 1994. Evidence that iron stores regulate iron absorption–a setpoint theory. *Am. J. Clin. Nutr.* 59:1376–80

6. McKie AT, Barrow D, Latunde-Dada GO, Rolfs A, Sager G, et al. 2001. An iron-regulated ferric reductase associated with the absorption of dietary iron. *Science* 291:1755–59

7. Gunshin H, Mackenzie B, Berger UV, Gunshin Y, Romero MF, et al. 1997. Cloning and characterization of a mammalian proton-coupled metal-ion transporter. *Nature* 388:482–88

8. Wolf G, Wessling-Resnick M. 1994. An integrin-mobilferrin iron transport pathway in intestine and hematopoietic cells. *Nutr. Rev.* 52:387–89

9. Donovan A, Brownlie A, Zhou Y, Shepard J, Pratt SJ, et al. 2000. Positional cloning of zebrafish ferroportin1 identifies a conserved vertebrate iron exporter. *Nature* 403:776–81

10. McKie AT, Marciani P, Rolfs A, Brennan K, Wehr K, et al. 2000. A novel duodenal iron-regulated transporter, IREG1, implicated in the basolateral transfer of iron to the circulation. *Mol. Cell* 5:299–309

11. Abboud S, Haile DJ. 2000. A novel mammalian iron-regulated protein involved in intracellular iron metabolism. *J. Biol. Chem.* 275:19906–12

12. Vulpe CD, Kuo YM, Murphy TL, Cowley L, Askwith C, et al. 1999. Hephaestin, a ceruloplasmin homologue implicated in intestinal iron transport, is defective in the sla mouse. *Nat. Genet.* 21:195–99

13. Raja KB, Pountney D, Bomford A, Przemioslo R, Sherman D, et al. 1996. A duodenal mucosal abnormality in the reduction of Fe(III) in patients with genetic haemochromatosis. *Gut* 38:765–69

14. Zoller H, Pietrangelo A, Vogel W, Weiss G. 1999. Duodenal metal-transporter (DMT-1, NRAMP-2) expression in patients with hereditary haemochromatosis. *Lancet* 353:2120–23

15. Whittaker P, Skikne BS, Covell AM, Flowers C, Cooke A, et al. 1989. Duodenal iron proteins in idiopathic hemochromatosis. *J. Clin. Invest.* 83:261–67

16. Zoller H, Koch RO, Theurl I, Obrist P, Pietrangelo A, et al. 2001. Expression of the duodenal iron transporters divalent-metal transporter 1 and ferroportin 1 in iron deficiency and iron overload. *Gastroenterology* 120:1412–19

17. Kawabata H, Yang R, Hirama T, Vuong PT, Kawano S, et al. 1999. Molecular cloning of transferrin receptor 2. A new member of the transferrin receptor-like family. *J. Biol. Chem.* 274:20826–32

18. Fleming RE, Migas MC, Holden CC, Waheed A, Britton RS, et al. 2000. Transferrin receptor 2: continued expression in mouse liver in the face of iron overload and in hereditary hemochromatosis. *Proc. Natl. Acad. Sci. USA* 97:2214–19

19. Bonkovsky HL. 1991. Iron and the liver. *Am. J. Med. Sci.* 301:32–43

20. Brissot P, Wright TL, Ma WL, Weisiger RA. 1985. Efficient clearance of nontransferrin-bound iron by rat liver. Implications for hepatic iron loading in iron

overload states. *J. Clin. Invest.* 76:1463–70

21. Batey RG, Lai Chung Fong P, Shamir S, Sherlock S. 1980. A non-transferrin-bound serum iron in idiopathic hemochromatosis. *Digest. Dis. Sci.* 25:340–46

22. Testa U, Kuhn L, Petrini M, Quaranta MT, Pelosi E, Peschle C. 1991. Differential regulation of iron regulatory element-binding protein(s) in cell extracts of activated lymphocytes versus monocytes-macrophages. *J. Biol. Chem.* 266:13925–30

23. Deiss A. 1983. Iron metabolism in reticuloendothelial cells. *Semin. Hematol.* 20:81–90

24. McLaren GD. 1989. Reticuloendothelial iron stores and hereditary hemochromatosis: a paradox. *J. Lab. Clin. Med.* 113:137–38

25. Moura E, Noordermeer MA, Verhoeven N, Verheul AF, Marx JJ. 1998. Iron release from human monocytes after erythrophagocytosis in vitro: an investigation in normal subjects and hereditary hemochromatosis patients. *Blood* 92:2511–19

26. Montosi G, Paglia P, Garuti C, Guzman CA, Bastin JM, et al. 2000. Wild-type HFE protein normalizes transferrin iron accumulation in macrophages from subjects with hereditary hemochromatosis. *Blood* 96:1125–29

27. Taylor P, Martinez-Torres C, Leets I, Ramirez J, Garcia-Casal MN, Layrisse M. 1988. Relationships among iron absorption, percent saturation of plasma transferrin and serum ferritin concentration in humans. *J. Nutr.* 118:1110–15

28. Cook JD. 1990. Adaptation in iron metabolism. *Am. J. Clin. Nutr.* 51:301–8

29. Pigeon C, Ilyin G, Courselaud B, Leroyer P, Turlin B, et al. 2001. A new mouse liver-specific gene, encoding a protein homologous to human antimicrobial peptide hepcidin, is overexpressed during iron overload. *J. Biol. Chem.* 276:7811–19

30. Nicolas G, Bennoun M, Devaux I, Beaumont C, Grandchamp B, et al. 2001. Lack of hepcidin gene expression and severe tissue iron overload in Upstream Stimulator Factor 2 (USF2) knockout mice. *Proc. Natl. Acad. Sci. USA* 98:8780–85

31. Parkkila S, Waheed A, Britton RS, Feder JN, Tsuchihashi Z, et al. 1997. Immunohistochemistry of HLA-H, the protein defective in patients with hereditary hemochromatosis, reveals unique pattern of expression in gastrointestinal tract. *Proc. Natl. Acad. Sci. USA* 94:2534–39

32. Anderson GJ. 1996. Control of iron absorption. *J. Gastroenterol. Hepatol.* 11:1030–32

33. Roy CN, Enns CA. 2000. Iron homeostasis: new tales from the crypt. *Blood* 96:4020–27

34. Sayers MH, English G, Finch C. 1994. Capacity of the store-regulator in maintaining iron balance. *Am. J. Hematol.* 47:194–97

35. Schumann K, Moret R, Kunzle H, Kuhn LC. 1999. Iron regulatory protein as an endogenous sensor of iron in rat intestinal mucosa. Possible implications for the regulation of iron absorption. *Eur. J. Biochem.* 260:362–72

36. Eisenstein RS, Blemings KP. 1998. Iron regulatory proteins, iron responsive elements and iron homeostasis. *J. Nutr.* 128:2295–98

37. Bacon BR. 2001. Hemochromatosis: diagnosis and management. *Gastroenterology* 120:718–25

38. Waheed A, Parkkila S, Saarnio J, Fleming RE, Zhou XY, et al. 1999. Association of HFE protein with transferrin receptor in crypt enterocytes of human duodenum. *Proc. Natl. Acad. Sci. USA* 96:1579–84

39. Zhou XY, Tomatsu S, Fleming RE, Parkkila S, Waheed A, et al. 1998. HFE gene knockout produces mouse model of hereditary hemochromatosis. *Proc. Natl. Acad. Sci. USA* 95:2492–97

40. Thenie A, Orhant M, Gicquel I, Fergelot P, Le Gall JY, et al. 2000. The HFE gene undergoes alternate splicing

processes. *Blood Cells Mol. Dis.* 26:155–62

41. Sanchez M, Bruguera M, Rodes J, Oliva R. 2001. Complete characterization of the 3′ region of the human and mouse hereditary hemochromatosis HFE gene and detection of novel splicing forms. *Blood Cells Mol. Dis.* 27:35–43

42. Sanchez M, Queralt R, Bruguera M, Rodes J, Oliva R. 1998. Cloning, sequencing and characterization of the rat hereditary hemochromatosis promoter: comparison of the human, mouse and rat HFE promoter regions. *Gene* 225:77–87

43. Salter-Cid L, Brunmark A, Li Y, Leturcq D, Peterson PA, et al. 1999. Transferrin receptor is negatively modulated by the hemochromatosis protein HFE: implications for cellular iron homeostasis. *Proc. Natl. Acad. Sci. USA* 96:5434–39

44. Han O, Fleet JC, Wood RJ. 1999. Reciprocal regulation of HFE and Namp2 gene expression by iron in human intestinal cells. *J. Nutr.* 129:98–104

45. Feder JN, Penny DM, Irrinki A, Mintier GA, Lebrón JA, et al. 1999. The hereditary hemochromatosis gene and iron homeostasis. In *Molecular Biology of Hematopoiesis*, ed. NG Abraham, A Tabilio, M Martelli, S Asanao, A Donfrancesco. New York: Plenum. 418 pp.

46. Lebrón JA, Bennett MJ, Vaughn DE, Chirino AJ, Snow PM, et al. 1998. Crystal structure of the hemochromatosis protein HFE and characterization of its interaction with transferrin receptor. *Cell* 93:111–23

47. Madden DR, Gorga JC, Strominger JL, Wiley DC. 1992. The three-dimensional structure of HLA-B27 at 2.1 Å resolution suggests a general mechanism for tight peptide binding to MHC. *Cell* 70:1035–48

48. Bastin JM, Jones M, O'Callaghan CA, Schimanski L, Mason DY, Townsend AR. 1998. Kupffer cell staining by an HFE-specific monoclonal antibody: implications for hereditary haemochromatosis. *Br. J. Haematol.* 103:931–41

49. Parkkila S, Waheed A, Britton RS, Bacon BR, Zhou XY, et al. 1997. Association of the transferrin receptor in human placenta with HFE, the protein defective in hereditary hemochromatosis. *Proc. Natl. Acad. Sci. USA* 94:13198–202

50. Parkkila S, Parkkila AK, Waheed A, Britton RS, Zhou XY, et al. 2000. Cell surface expression of HFE protein in epithelial cells, macrophages, and monocytes. *Haematologica* 85:340–45

51. Levy JE, Montross LK, Cohen DE, Fleming MD, Andrews NC. 1999. The C282Y mutation causing hereditary hemochromatosis does not produce a null allele. *Blood* 94:9–11

52. Byrnes V, Ryan E, O'Keane C, Crowe J. 2000. Immunohistochemistry of the HFE protein in patients with hereditary hemochromatosis, iron deficiency anemia, and normal controls. *Blood Cells Mol. Dis.* 26:2–8

53. Zuccon L, Corsi B, Levi S, Mattioli M, Fracanzani AL, et al. 2000. Immunohistochemistry of HFE in the duodenum of C282Y homozygotes with antisera for recombinant HFE protein. *Haematologica* 85:346–51

54. Jazwinska EC, Pyper WR, Burt MJ, Francis JL, Goldwurm S, et al. 1995. Haplotype analysis in Australian hemochromatosis patients: evidence for a predominant ancestral haplotype exclusively associated with hemochromatosis. *Am. J. Hum. Genet.* 56:428–33

55. Carella M, D'Ambrosio L, Totaro A, Grifa A, Valentino MA, et al. 1997. Mutation analysis of the HLA-H gene in Italian hemochromatosis patients. *Am. J. Hum. Genet.* 60:828–32

56. Beutler E, Gelbart T, West C, Lee P, Adams M, et al. 1996. Mutation analysis in hereditary hemochromatosis. *Blood Cells Mol. Dis.* 22:187–94

57. Merryweather-Clarke AT, Pointon JJ, Shearman JD, Robson KJ. 1997. Global prevalence of putative haemochromatosis mutations. *J. Med. Genet.* 34:275–78

58. Jouanolle AM, Fergelot P, Gandon G, Yaouanq J, Le Gall JY, David V. 1997. A candidate gene for hemochromatosis: frequency of the C282Y and H63D mutations. *Hum. Genet.* 100:544–47

59. Beckman LE, Saha N, Spitsyn V, Van Landeghem G, Beckman L. 1997. Ethnic differences in the HFE codon 282 (Cys/Tyr) polymorphism. *Hum. Hered.* 47:263–67

60. Olynyk JK, Cullen DJ, Aquilia S, Rossi E, Summerville L, Powell LW. 1999. A population-based study of the clinical expression of the hemochromatosis gene. *N. Engl. J. Med.* 341:718–24

61. Chang JG, Liu TC, Lin SF. 1997. Rapid diagnosis of the HLA-H gene Cys 282 Tyr mutation in hemochromatosis by polymerase chain reaction–a very rare mutation in the Chinese population. *Blood* 89:3492–93

62. Cullen LM, Gao X, Easteal S, Jazwinska EC. 1998. The hemochromatosis 845 G → A and 187 C → G mutations: prevalence in non-Caucasian populations. *Am. J. Hum. Genet.* 62:1403–7

63. Deleted in proof

64. Rochette J, Pointon JJ, Fisher CA, Perera G, Arambepola M, et al. 1999. Multicentric origin of hemochromatosis gene (HFE) mutations. *Am. J. Hum. Genet.* 64:1056–62

65. Jouanolle AM, Fergelot P, Raoul ML, Gandon G, Roussey M, et al. 1998. Prevalence of the C282Y mutation in Brittany: penetrance of genetic hemochromatosis? *Ann. Genet.* 41:195–98

66. Bacon BR, Powell LW, Adams PC, Kresina TF, Hoofnagle JH. 1999. Molecular medicine and hemochromatosis: at the crossroads. *Gastroenterology* 116:193–207

67. Risch N. 1997. Haemochromatosis, HFE and genetic complexity. *Nat. Genet.* 17:375–76

68. Pointon JJ, Wallace D, Merryweather-Clarke AT, Robson KJ. 2000. Uncommon mutations and polymorphisms in the hemochromatosis gene. *Genet. Test.* 4:151–61

69. Barton JC, Sawada-Hirai R, Rothenberg BE, Acton RT. 1999. Two novel missense mutations of the HFE gene (I105T and G93R) and identification of the S65C mutation in Alabama hemochromatosis probands. *Blood Cells Mol. Dis.* 25:147–55

70. Mura C, Raguenes O, Ferec C. 1999. HFE mutations analysis in 711 hemochromatosis probands: evidence for S65C implication in mild form of hemochromatosis. *Blood* 93:2502–5

71. Piperno A, Arosio C, Fossati L, Vigano M, Trombini P, et al. 2000. Two novel nonsense mutations of HFE gene in five unrelated Italian patients with hemochromatosis. *Gastroenterology* 119:441–45

72. Wallace DF, Dooley JS, Walker AP. 1999. A novel mutation of HFE explains the classical phenotype of genetic hemochromatosis in a C282Y heterozygote. *Gastroenterology* 116:1409–12

73. Feder JN, Tsuchihashi Z, Irrinki A, Lee VK, Mapa FA, et al. 1997. The hemochromatosis founder mutation in HLA-H disrupts beta2-microglobulin interaction and cell surface expression. *J. Biol. Chem.* 272:14025–28

74. Waheed A, Parkkila S, Zhou XY, Tomatsu S, Tsuchihashi Z, et al. 1997. Hereditary hemochromatosis: effects of C282Y and H63D mutations on association with beta2-microglobulin, intracellular processing, and cell surface expression of the HFE protein in COS-7 cells. *Proc. Natl. Acad. Sci. USA* 94:12384–89

75. Gross CN, Irrinki A, Feder JN, Enns CA. 1998. Co-trafficking of HFE, a nonclassical major histocompatibility complex class I protein, with the transferrin receptor implies a role in intracellular iron regulation. *J. Biol. Chem.* 273:22068–74

76. de Sousa M, Reimao R, Lacerda R, Hugo P, Kaufmann SH, Porto G. 1994.

Iron overload in beta 2-microglobulin-deficient mice. *Immunol. Lett.* 39:105–11

77. Santos M, Schilham MW, Rademakers LH, Marx JJ, de Sousa M, Clevers H. 1996. Defective iron homeostasis in beta 2-microglobulin knockout mice recapitulates hereditary hemochromatosis in man. *J. Exp. Med.* 184:1975–85

78. Feder JN, Tsuchihashi Z, Irrinki A, Lee VK, Mapa FA, et al. 1997. The hemochromatosis founder mutation in HLA-H disrupts beta2-microglobulin interaction and cell surface expression. *J. Biol. Chem.* 272:14025–28

79. Salter-Cid L, Brunmark A, Peterson PA, Yang Y. 2000. The major histocompatibility complex-encoded class I-like HFE abrogates endocytosis of transferrin receptor by inducing receptor phosphorylation. *Genes Immun.* 1:409–17

80. Bennett MJ, Lebrón JA, Bjorkman PJ. 2000. Crystal structure of the hereditary haemochromatosis protein HFE complexed with transferrin receptor. *Nature* 403:46–53

81. Lebrón JA, West AP Jr., Bjorkman PJ. 1999. The hemochromatosis protein HFE competes with transferrin for binding to the transferrin receptor. *J. Mol. Biol.* 294:239–45

82. Feder JN, Penny DM, Irrinki A, Lee VK, Lebrón A, et al. 1998. The hemochromatosis gene product complexes with the transferrin receptor and lowers its affinity for ligand binding. *Proc. Natl. Acad. Sci. USA* 95:1472–77

83. Tsuchihashi Z, Hansen SL, Quintana L, Kronmal GS, Mapa FA, et al. 1998. Transferrin receptor mutation analysis in hereditary hemochromatosis patients. *Blood Cells Mol. Dis.* 24:317–21

84. Roy CN, Penny DM, Feder JN, Enns CA. 1999. The hereditary hemochromatosis protein, HFE, specifically regulates transferrin-mediated iron uptake in HeLa cells. *J. Biol. Chem.* 274:9022–28

85. Corsi B, Levi S, Cozzi A, Corti A, Alti-mare D, et al. 1999. Overexpression of the hereditary hemochromatosis protein, HFE, in HeLa cells induces an iron-deficient phenotype. *FEBS Lett.* 460:149–52

86. Riedel HD, Muckenthaler MU, Gehrke SG, Mohr I, Brennan K, et al. 1999. HFE downregulates iron uptake from transferrin and induces iron-regulatory protein activity in stably transfected cells. *Blood* 94:3915–21

87. Arredondo M, Munoz P, Mura C, Nunez MT. 2001. HFE inhibits apical iron uptake by intestinal epithelial (Caco-2) cells. *FASEB J.* 15:1276–78

88. Powell LW, Campbell CB, Wilson E. 1970. Intestinal mucosal uptake of iron and iron retention in idiopathic haemochromatosis as evidence for a mucosal abnormality. *Gut* 11:727–31

89. Whittaker P, Skikne BS, Covell AM, Flowers C, Cooke A, et al. 1989. Duodenal iron proteins in idiopathic hemochromatosis. *J. Clin. Invest.* 83:261–67

90. Pietrangelo A, Casalgrandi G, Quaglino D, Gualdi R, Conte D, et al. 1995. Duodenal ferritin synthesis in genetic hemochromatosis. *Gastroenterology* 108:208–17. Erratum. *Gastroenterology* 108(6):1963

91. Kuhn LC. 1999. Iron overload: molecular clues to its cause. *Trends Biochem. Sci.* 24:164–66

92. Fleming RE, Migas MC, Zhou X, Jiang J, Britton RS, et al. 1999. Mechanism of increased iron absorption in murine model of hereditary hemochromatosis: increased duodenal expression of the iron transporter DMT1. *Proc. Natl. Acad. Sci. USA* 96:3143–48

93. Griffiths WJ, Sly WS, Cox TM. 2001. Intestinal iron uptake determined by divalent metal transporter is enhanced in HFE-deficient mice with hemochromatosis. *Gastroenterology* 120:1420–29

94. Levy JE, Montross LK, Andrews NC. 2000. Genes that modify the hemochromatosis phenotype in mice. *J. Clin. Invest.* 105:1209–16

95. Canonne-Hergaux F, Levy JE, Fleming MD, Montross LK, Andrews NC, Gros P. 2001. Expression of the DMT1 (NRAMP2/DCT1) iron transporter in mice with genetic iron overload disorders. *Blood* 97:1138–40

96. Riegert P, Gilfillan S, Nanda I, Schmid M, Bahram S. 1998. The mouse HFE gene. *Immunogenetics* 47:174–77

97. Hashimoto K, Hirai M, Kurosawa Y. 1997. Identification of a mouse homolog for the human hereditary haemochromatosis candidate gene. *Biochem. Biophys. Res. Commun.* 230:35–39

98. Albig W, Drabent B, Burmester N, Bode C, Doenecke D. 1998. The haemochromatosis candidate gene HFE (HLA-H) of man and mouse is located in syntenic regions within the histone gene cluster. *J. Cell Biochem.* 69:117–26

99. Bahram S, Gilfillan S, Kuhn LC, Moret R, Schulze JB, et al. 1999. Experimental hemochromatosis due to MHC class I HFE deficiency: immune status and iron metabolism. *Proc. Natl. Acad. Sci. USA* 96:13312–17

100. Fleming RE, Holden CC, Tomatsu S, Waheed A, Brunt EM, et al. 2001. Mouse strain differences determine severity of iron accumulation in Hfe knockout model of hereditary hemochromatosis. *Proc. Natl. Acad. Sci. USA* 98:2707–11

101. Camaschella C, Roetto A, Cali A, De Gobbi M, Garozzo G, et al. 2000. The gene TFR2 is mutated in a new type of haemochromatosis mapping to 7q22. *Nat. Genet.* 25:14–15

102. Roetto A, Totaro A, Piperno A, Piga A, Longo F, et al. 2001. New mutations inactivating transferrin receptor 2 in hemochromatosis type 3. *Blood* 97:2555–60

103. Njajou OT, Vaessen N, Joosse M, Berghuis B, van Dongen JW, et al. 2001. A mutation in SLC11A3 is associated with autosomal dominant hemochromatosis. *Nat. Genet.* 28:213–14

104. Montosi G, Donovan A, Totaro A, Garuti C, Pignatti E, et al. 2001. Dominant hemochromatosis is associated with a mutation in the ferroportin (SCL11A3) gene. *J. Clin. Invest.* 108:619–23

Annu. Rev. Physiol. 2002. 64:681–708

MOLECULAR PATHOGENESIS OF LUNG CANCER

Sabine Zöchbauer-Müller[1], Adi F. Gazdar[1,2], and John D. Minna[1,3,4]

[1]Hamon Center for Therapeutic Oncology Research, Departments of [2]Pathology, [3]Internal Medicine, and [4]Pharmacology, The University of Texas Southwestern Medical Center, Dallas, Texas, 75390-8593; Sabine.Zoechbauer@akh-wien.ac.at; Adi.Gazdar@UTSouthwestern.edu; John.Minna@UTSouthwestern.edu

Key Words   tumor suppressor gene, allele loss, methylation, angiogenesis, preneoplasia

■ **Abstract**   Lung cancer is the most common cause of cancer death in the United States, killing more than 156,000 people every year. In the past two decades, significant progress has been made in understanding the molecular and cellular pathogenesis of lung cancer. Abnormalities of proto-oncogenes, genetic and epigenetic changes of tumor suppressor genes, the role of angiogenesis in the multistage development of lung cancer, as well as detection of molecular abnormalities in preinvasive respiratory lesions, have recently come into focus. Efforts are ongoing to translate these findings into new clinical strategies for risk assessment, chemoprevention, early diagnosis, treatment selection, and prognosis and to provide new targets and methods of treatment for lung cancer patients. All these strategies should aid in reducing the number of newly diagnosed lung cancer cases and in increasing the survival and quality of life of patients with lung cancer.

INTRODUCTION

Lung cancer is one of the most prevalent and lethal cancers in the world. In the United States about 156,000 people died because of lung cancer last year, which represents 28% of all cancer deaths (1). Although the rate of lung cancer deaths for males is decreasing in the United States, the mortality associated with lung cancer among women continues to increase (2). Cigarette smoking is the main risk factor for lung cancer, accounting for about 90% of the cases in men and 70% of the cases in women (3, 4). Exposures to other environmental and occupational respiratory carcinogens may be interactive with cigarette smoking and thus also influence lung cancer incidence. Nevertheless, prevention of smoking initiation and increased smoking cessation remain the best long-term methods to prevent lung cancer development. The major histologic types of lung cancer are non-small cell lung cancers

(NSCLC),[1] which represent about 80% of lung cancers and are divided into squamous cell carcinoma, adenocarcinoma (including bronchiolo-alveolar carcinoma), large cell carcinoma and mixed types, and small cell lung cancers (SCLC), which represent about 20% of lung cancers. Development of metastases when the primary tumors are still small, coupled with lack of methods for early diagnosis and of systemic therapies with great efficacy to deal with micrometastatic disease are the main reasons why the prognosis of lung cancer patients is still poor, with overall 5-year survival rates of about 14% (5). Thus new methods for early detection and identification of smokers at greatest risk for developing lung cancer, such as spiral-computed tomography screening for early lung cancers, biomarkers for lung cancer risk assessment, new approaches for lung cancer prevention (chemoprevention), and new drugs based on rational targets, are necessary and need to be developed. In this review we summarize the advances that have been made in understanding the molecular and cellular biology of lung cancer over the recent years (major molecular abnormalities summarized in Table 1). It is hoped that these new findings will result in novel approaches for prevention and early detection of lung cancer and new effective treatment strategies for lung cancer patients.

AUTOCRINE AND PARACRINE SYSTEMS: GROWTH FACTORS AND THEIR RECEPTORS

Many growth factors and their receptors are expressed by lung cancer cells or adjacent normal cells leading to the development of several autocrine or paracrine growth stimulatory loops (6). One of the best-characterized autocrine systems involves gastrin-releasing peptide and other bombesin-like peptides (GRP/BN). In normal tissues, bombesin-like peptides stimulate growth of bronchial epithelial cells and are important regulators of human lung development (7, 8). The effects of GRP are mediated through the family of G protein-coupled bombesin receptors, which includes the gastrin-releasing peptide receptor (GRPR), the neuromedin B receptor (NMBR), and the bombesin receptor subtype 3 (BRS-3). NMBR

---

[1]**Abbreviations:** NSCLC, non-small cell lung cancer; SCLC, small cell lung cancer; GRP, gastrin-releasing peptide; GRPR, gastrin-releasing peptide receptor; NMB, neuromedin B; NMBR, neuromedin B receptor; BRS-3, bombesin receptor subtype 3; IGF, insulin-like growth factor; PDGR, platelet-derived growth factor; EGF, epidermal growth factor; EGFR, epidermal growth factor receptor; HGF, hepatocyte growth factor; TSG, tumor suppressor gene; LOH, loss of heterozygosity; FHIT gene, fragile histidine triad gene; RARβ, retinoic acid receptor β; p14, p14^{ARF}; RB, retinoblastoma gene; CDK, cyclin-dependent protein kinase; APC, adenomatous polyposis coli; CDH13, H-cadherin; TIMP-3, tissue inhibitor of metalloproteinase-3; p16, p16^{INK4a}; MGMT, O^6-methylguanine-DNA methyltransferase; DAPK, death-associated protein kinase; ECAD, E-cadherin; GSTP1, glutathione-S-transferase P1; p15, p15^{INK4b}; VEGF, vascular epithelial growth factor; RER, replication error repair

TABLE 1   Major molecular abnormalities in the pathogenesis of lung cancer

Growth factors, receptors, and activation of proto-oncogenes
  GRP/BN and their receptors, *IGF, HGF* and its receptor *MET, EGFR,
  HER2/neu*
  RAS mutations
  MYC amplification and deregulated expression
  Abnormal *BCL-2* expression
  Cyclin D1 expression

Loss of function of tumor suppressor genes (*p53, RB, p16, FHIT, RASSF1A, APC*)[a]

Aberrant methylation resulting in loss of gene expression (*APC, CDH13, RARβ, FHIT,
  RASSF1A, TIMP-3, p16, MGMT, DAPK*)

Activation of tumor-stimulated angiogenesis for tumor growth

Expression of telomerase activity and cellular immortality

Loss of components of apoptosis pathways

Potentially loss of DNA repair mechanisms

[a]Frequent allele loss at chromosomal sites 1p13, 1p36, 3p14-cen, 3p21.3–22, 3p25–26, 4p15.1–15.3, 4q25–26, 4q33–34, 5q21–22, 6p21.3, 6q25, 8p21–22, 9p21, 11p13, 13q14, 17p13.1, 18q21.1, 22q13-qter indicate the presence of multiple other TSGs.

expression is common in NSCLCs, whereas the expression of GRPR and BRS-3 is less frequent (9). Also the ligands for these receptors are often expressed in lung cancer cells leading to potential self-stimulatory (autocrine) loops. The NMB gene is expressed in 100% of SCLCs and NSCLCs (10). Approximately 20–60% of SCLCs express GRP, whereas NSCLCs express GRP less frequently (11). The expression of this loop most likely represents a reactivation of a program required for embryonal lung development because mutations of either GRP/BN or its receptor have not been found in lung cancers. However, how and why this reactivation occurs is unknown. Activation of GRPR in human airways has been associated with a proliferative response of bronchial cells to GRP and with long-term tobacco use. Recently, Shriver et al. (12) reported that the *GRPR* mRNA is expressed more frequently in women than in men in the absence of smoking and that expression of this gene is activated earlier in women in response to tobacco exposure. Additionally, the presence of two expressed copies of the X-linked GRPR gene in females may be a factor in the increased susceptibility of women to tobacco-induced lung cancer. Blocking this system with antibodies or receptor antagonists results in inhibition of tumor growth in model systems; this concept is being explored as a new therapeutic avenue (13). Other peptides with potential autocrine growth function in lung cancer are insulin-like growth factors (IGF) I and II (14). Wu et al. (15) tested the hypothesis that accumulation of genetic damage is dependent on an individual's intrinsic carcinogen sensitivity and on various humoral factors such as IGFs. The mean blood levels of IGF-I and the molar ratio of IGF-I/IGF-binding protein-3 were higher in patients with advanced or poorly differentiated cancer than in patients with early or well-differentiated cancers.

The *c-erbB-1* proto-oncogene encodes the receptor for the epidermal growth factor (EGFR), which regulates epithelial proliferation and differentiation and is activated in lung cancer cells by overexpression (16, 17). The *EGFR* gene was found to be overexpressed in 13% of NSCLCs (18), and EGFR protein expression by tumors seems to be a poor prognosis risk factor in NSCLC patients (19). *c-erbB-2* (also known as *HER2/neu*), a related proto-oncogene that encodes a protein product of 185 kDa (p185neu), is another growth factor receptor and is frequently expressed in NSCLCs, but not in SCLCs (20, 21). Using an anti-p185HER2 antibody, inhibition of human lung cancer cell line growth has been demonstrated (22), whereas anti-EGFR antibodies can inhibit the growth of human tumor cell lines overexpressing EGFR. The use of humanized monoclonal anti-HER2 antibody (Herceptin^{TM}) has been tested in breast cancer treatment with promising results and is also being evaluated clinically in lung cancer. Additionally, several new drugs have been developed that block the tyrosine kinase activity of these receptors, which leads to tumor growth inhibition in preclinical models; these drugs have also recently gone into clinical trials.

The hepatocyte growth factor (HGF) and its receptor comprise another growth factor/receptor complex that may play a role in lung cancer development. HGF, which stimulates epithelial cells to proliferate, move, and carry out differentiation programs, is expressed in many NSCLCs (23, 24) where it is associated with impaired survival (25). The receptor for HGF is encoded by the oncogene *MET*, which is expressed in normal lung epithelium, SCLCs, and NSCLCs (23, 24, 26). Thus these tumor cells have both the receptor and the ligand for this growth factor.

Data in the literature concerning the expression of estrogen and progesterone receptors in human lung cancer are discordant. Di Nunno et al. (27) did not find estrogen or progesterone receptor expression in a large series of NSCLC samples, whereas Su et al. (28) reported that ∼40% of NSCLCs investigated had either a postive estrogen or progesterone receptor status. Expression of the estrogen receptor-related protein p29, but not of the estrogen receptor itself, was found frequently in NSCLCs and was correlated with shorter survival time in women (29).

DOMINANT ONCOGENES

Dominant oncogenes or proto-oncogenes play an important role in the regulation of normal cellular growth and development, and physiologically their expression is tightly regulated (30). Activation of these genes can occur by several genetic mechanisms such as DNA amplification, translocation, and point mutations.

RAS

The *RAS* genes (*HRAS*, *KRAS*, *NRAS*) play an essential role in signal transduction pathways and code for guanosine triphosphate (GTP)-binding proteins. RAS-GTP generates a growth-promoting signal, whereas hydrolysis of RAS-bound

GTP to guanosine diphosphate (GDP) inactivates this RAS signal. In the presence of oncogenic *RAS* mutations, GTP cannot be hydrolyzed to GDP, resulting in constitutively active RAS-GTP growth-promoting activity. The *RAS* oncogenes usually acquire their transforming capacity by point mutations, and they have been detected in about 20–30% of lung adenocarcinomas and in 15–20% of all NSCLCs at *KRAS* codons 12, 13, or 61 (11, 31). However, *RAS* point mutations are never found in SCLCs. Mutations occur preferentially in *KRAS* and account for approximately 90% of *RAS* mutations in lung adenocarcinomas. Almost all *KRAS* mutations affect codon 12, and most mutations are G-T transversions of the type that are associated with cigarette smoke carcinogens (31). A correlation between *RAS* mutations and poor prognosis of NSCLC patients has been reported by several authors (32–34). Farnesyltransferase inhibitors, which block a lipid modification of *RAS* essential for its activity, are currently being investigated in clinical trials.

MYC

RAS signaling activates nuclear proto-oncogene products like MYC, which on heterodimerization transcriptionally activates downstream genes, thereby stimulating cell growth. The *MYC* proto-oncogene family includes *MYC*, *NMYC*, and *LMYC* and encodes nuclear DNA-binding proteins, which are involved in transcriptional regulation (35). Activation of the *MYC* genes occurs by gene amplification or transcriptional dysregulation and results in protein overexpression (36). Abnormal *MYC* expression is frequently observed in SCLCs but is less common in NSCLCs (11).

BCL-2

One key member of the normal pathway for programmed cell death (apoptosis) is the *BCL-2* proto-oncogene. It protects cells from apoptosis and is negatively regulated by p53. Forced overexpression of oncogenic *MYC* or *RAS* in fibroblasts can lead to apoptosis in the face of nutrient deprivation. It is likely that tumor cells overexpress *BCL-2* to overcome apoptotic signals from *MYC* and *RAS* expression. Immunohistochemical studies have shown that BCL-2 protein is expressed frequently in SCLCs and NSCLCs (37, 38), where the expression is higher in squamous cell carcinomas than in adenocarcinomas (37). Of interest, some studies demonstrate a survival benefit for patients with BCL-2-positive tumors (37–40) while others do not confirm this finding (41). A higher response rate to chemotherapy was observed for BCL-2-positive tumors compared with BCL-2-negative tumors, suggesting that BCL-2 expression reflects a higher susceptibility to cytotoxic treatment (38). Antisense *BCL-2* drugs, which block translation of BCL-2 protein in tumor cells, will soon be tested in clinical trials. These drugs may work either alone or by increasing a tumor cell response to standard chemotherapy and radiotherapy.

Notch-3

Notch-3 is located on chromosome 19p in a region that was found to be translocated to chromosome 15q in a metastatic lung carcinoma (42) and is one of the

Notch family members involved in differentiation and neoplasia. *Notch-3* overexpression is frequently found in NSCLCs and is associated with chromosome 19p translocation (42). This finding is of particular interest because it demonstrates that a specific chromosome translocation occurs in a common epithelial cancer and activates a gene not previously implicated in lung cancer.

AIS

AIS (amplified in squamous cell carcinoma) is a *p53* homologue located on the distal long arm of chromosome 3 with multiple protein products (p40, p51, p63, p73L). Fluorescent in situ hybridization analysis revealed frequent amplification of this gene locus in primary lung squamous cell carcinomas, and protein overexpression was observed in lung squamous cell carcinomas and tumors known to harbor a high frequency of *p53* mutations, suggesting that AIS plays an oncogenic role in lung squamous cell carcinomas (43). Circulating anti-p40 (AIS) antibodies have been detected in the sera of respiratory tract cancer patients, but the presence or absence of AIS antibodies were independent of other clinicopathological characteristics of these patients (44).

CHROMOSOMAL SITES OF FREQUENT ALLELE LOSS AND TUMOR SUPPRESSOR GENES (TSGs)

According to Knudson's two-hit hypothesis (45), loss of function of TSGs requires that both alleles have to be inactivated. One allele is inactivated by mutation, methylation (epigenetic) changes, or other changes that target the individual TSG, whereas the other allele is usually inactivated as part of loss of many genetic markers by deletion, nonreciprocal translocation, or mitotic recombination in the chromosomal region referred to as allele loss or loss of heterozygosity (LOH) (30, 46). Thus consistent LOH for genetic markers at a given locus in many tumors is strong evidence for the presence of one or more TSGs in that region. Recently, a genome-wide high-resolution search of LOH was performed on SCLCs and NSCLCs in order to detect new loci that may harbor TSGs (47). In total, 399 microsatellite markers separated by an approximate distance of 10 cM (centimorgans) were used to detect sites of frequent LOH. Overall, 22 different regions with more than 60% LOH were identified. Thirteen regions showed a preference for SCLC, 7 regions a preference for NSCLC, and 2 regions affected both SCLC and NSCLC. The chromosomal arms with the most frequent LOH were 1p, 3p, 4p, 4q, 5q, 8p, 9p (*p16*), 9q, 10p, 10q, 13q (*Rb*), 15q, 17p (*p53*), 18q, 19p, Xp, and Xq. Interestingly, SCLC and NSCLC had different regions of frequent LOH, and NSCLC had more of these regions than SCLC, suggesting that SCLC and NSCLC frequently undergo different genetic alterations. The occurrence of LOH in primary SCLCs compared with primary squamous cell carcinomas and primary adenocarcinomas was also investigated in the study by Wistuba et al. (48) using

19 polymorphic microsatellite markers at 12 chromosomal regions. Each tumor type had a characteristic pattern of allelic loss, and the bronchial epithelium accompanying SCLCs showed a much higher frequency of LOH compared with squamous cell carcinomas and adenocarcinomas. The study by Sanchez-Cespedes et al. (49) reported a much higher frequency of widespread chromosomal abnormalities in lung adenocarcinomas from smokers compared with infrequent changes in tumors arising in nonsmokers.

So far, 3p allele loss has been shown to be the most frequent molecular alteration in lung cancers (47, 50, 51). However, 3p allele loss occurs not only in tumors but also in the normal epithelium of smokers without lung cancer, and hyperplasias, dysplasias, and carcinoma in situ in the respiratory epithelium accompanying lung cancers, suggesting that it is an early change in the multistep pathogenesis of lung cancer (51–53). A high-resolution 3p LOH study in primary lung tumors and preneoplastic/preinvasive lesions using a panel of 28 microsatellite markers demonstrated a progressive increase in the frequency and size of 3p allelic loss regions with increasing severity of histopathological preneoplastic/preinvasive changes (51).

LOH involving markers on 3p is found in more than 90% of SCLCs and more than 80% of NSCLCs (30). Several distinct 3p regions have been identified that show frequent allele loss, including 3p25–26, 3p21.3–22, 3p14, and 3p12 and suggest that several TSGs are likely located on chromosome 3p (30, 54). Additionally, several independent homozygous deletions at 3p21.3, 3p14.2, and 3p12 were found in lung cancers, which represents strong evidence for the presence of TSGs (30). A 630-kb homozygous deletion on 3p21.3 has recently been defined, and 25 genes have been cloned in this region, some of which are strong candidate TSGs (55). One of these is the *RASSF1* gene, which has a predicted Ras association domain and homology to the Ras effector Nore 1 (56, 57). The *RASSF1* gene encodes several major transcripts that are produced by alternative promoter selection and alternative mRNA splicing. mRNA expression of one of these transcripts, *RASSF1A*, is frequently lost in lung cancer. The major mechanism for inactivating *RASSF1A* is by aberrant methylation of its promoter region turning off its expression; inactivation of *RASSF1A* by mutation is rare (56–58). Additionally, the study by Burbee et al. (57) shows that patients whose tumors are methylated for *RASSF1A* have a shorter overall survival rate than patients whose tumors are not methylated for *RASSF1A*. Thus these data strongly support the candidacy of *RASSF1A* as a TSG that plays a major role in the pathogenesis of lung cancer. The *FHIT* (fragile histidine triad) gene, a candidate TSG that spans the FRA3B common fragile site at 3p14.2, was found to be frequently abnormal in lung cancer (59, 60). Aberrant *FHIT* transcripts were detected in 80% of SCLC and 40% of NSCLC specimens (59, 60), and absent FHIT protein expression was found in ~50% of all lung cancers (61, 62). LOH at the *FHIT* gene locus was seen more frequently in smokers than in nonsmokers, suggesting that *FHIT* is a molecular target of tobacco smoke carcinogens (61). Recently, aberrant methylation of the 5′ CpG island of the *FHIT* gene was shown to be an important mechanism for silencing this gene in lung cancer (63). Transfection of a wild-type copy of *FHIT* into lung cancer cells can reverse the malignant phenotype and induce tumor cell apoptosis (64, 65); this suggests that FHIT overexpression could serve as a future therapeutic approach.

The retinoic acid receptor β-2 (*RARβ*) gene located at 3p24 has been intensively studied in lung cancer and found to have defective function, thus making it a candidate TSG. This is particularly important given the interest in using retinoids as chemoprevention agents for lung cancer. *RARβ* is a key retinoid receptor that mediates growth control responses, and considerable evidence suggests that *RARβ* abnormalities exist in lung cancers (66–69). A recent study suggested that the suppressive effect of RARβ2 is isoform specific because expression of RARβ1 and also RARα1 could not mimic the growth-suppressive effect of RARβ2 in the lung cancer cell line Calu-1 (70). Frequent loss of *RARβ* mRNA expression has been described in both primary NSCLCs and bronchial biopsy specimens from heavy smokers (69, 71, 72). In addition, diminished or absent RARβ protein expression was seen in ~50% of resected NSCLCs (73). The mechanism underlying this loss of expression was shown in the study by Virmani et al. (74) who reported that aberrant methylation of *RARβ* is a frequent event (~50%) in both SCLCs and NSCLCs.

Several other 3p genes including the BRCA1-binding protein *BAP1* (3p21), the base excision repair gene *hOGG1* (3p25), the DNA mismatch repair gene *hMLH1* (3p21), and other genes such as the semaphorins (*SEMA3B* and *SEMA3F*), the calcium channel $\alpha_2\delta_2$ auxiliary subunit gene (*CACNA2D2*), and the human *RON* gene also may function as TSGs (55, 75–83).

By using an approach for genome-wide screening for homozygous deletions, a homozygously deleted region on chromosome 2q has been identified (84). This region harbors the lipoprotein receptor-related protein-deleted in tumors (*LRP-DIT*) gene. Homozygous deletions in *LRP-DIT* were detected in 17% of NSCLC cell lines, and expression of only abnormal transcripts missing parts of the *LRP-DIT* sequence was observed additionally in 30% of NSCLC cell lines. However, no *LRP-DIT* alterations were found in SCLC cell lines. Recently, the *TSLC1* gene has been identified at chromosome 11q23.2 (85). This region is of particular interest because LOH occurs frequently and, in addition, tumorigenicity of A549 lung cancer cells can be suppressed by this region. Moreover, loss of *TSLC1* expression was observed frequently in NSCLCs, and aberrant methylation of the promoter region has been identified as a major mechanism for inactivating this gene. The *PPP2R1B* gene, located on chromosome 11q22-24, which encodes the beta isoform of the A subunit of the serine/threonine protein phosphatase 2A (PP2A), was found to be altered by mutations in lung, colon, and breast cancer, thus suggesting it as a putative TSG (86).

p53

The *p53* gene, located at chromosome region 17p13.1, encodes a 53-kDa nuclear protein. This protein acts as a transcription factor to turn on the expression of a DNA damage response program, blocks the progression of cells through the cell cycle in the late G1 phase, and triggers apoptosis (87). DNA damage is the major signal for p53 phosphorylation, which is catalyzed by the kinase encoded by the ataxia telangiectasia gene *ATM*. Phosphorylated p53 acts as a specific DNA-binding transcription factor for several genes including *p21/WAF1/CIP1*, *MDM2*, *GADD45*, *BAX*, and cyclin G (30). Activation of these genes results in apoptosis, cell cycle arrest, and DNA repair. Mutations of the *p53* gene comprise some of the most common genetic changes associated with cancer and cause loss of tumor suppressor function and loss of ability to induce apoptosis. The prevalent type of point mutations is a GC to TA transversion causing missense mutations. This type of mutation appears to be related to benzo(*a*)pyrene-induced damage from cigarette smoking (88). A significantly higher risk for *p53* mutations was observed in smokers compared with nonsmokers (89). However, alcohol use seems to have an impact on the occurrence of *p53* mutations, which were found more often in alcohol drinkers who smoke compared with nondrinkers who smoke or in nonsmokers (90). *p53* missense mutations lead to increased protein half-life, resulting in higher levels of p53 protein, which can easily be detected by immunohistochemistry. Abnormal p53 expression by immunostaining was reported in 40–70% of SCLCs and 40–60% of NSCLCs (91, 92).

There are numerous reports that investigated the association of *p53* abnormalities with the prognosis of NSCLC patients; however, the results were discordant as to whether *p53* abnormalities influence prognosis (93–98). To resolve this, a recent meta-analysis was performed to determine whether alterations of *p53* adversely affect survival of NSCLC patients by investigating 43 published articles (99). The negative prognostic effect of *p53* alterations was highly significant in patients with adenocarcinomas but not in patients with squamous cell carcinomas. *p53* alterations were detected either as overexpression in protein studies or as mutation in DNA studies. The incidence of *p53* alterations in DNA studies was lower than that in protein studies, and the incidence of p53 overexpression and mutations in adenocarcinomas was significantly lower than that in squamous cell carcinomas.

As a new treatment approach, *p53* has been introduced into clinical trials with retroviral and adenoviral gene therapy delivered directly into tumors with initially promising antitumor responses (100). A recent study investigated the additional benefit from adenoviral *p53* gene therapy directly injected into tumors in patients undergoing first-line chemotherapy for NSCLC (101). However, no differences in response rates or survival were observed between the group treated with additional *p53* gene therapy and the group treated with chemotherapy alone. The successful systemic delivery of *p53* by liposomes has been shown recently in lung cancer (102) and needs to be investigated further for treatment of primary and disseminated lung cancer. In addition, vaccine trials using mutant p53 peptides have been completed (103).

The *p16^INK4*-Cyclin D1-CDK4-*RB* Pathway

p16^INK4 (*p16*) was mapped to the critical region at chromosome 9p21, which frequently undergoes allele loss and mutations (104). However, *p16* is also frequently inactivated by aberrant methylation of its promoter region (105–108). p16 functions in the pathway by binding to cyclin-dependent protein kinase 4 (CDK4) and inhibiting the ability of CDK4 to interact with cyclin D1 (109). CDKs, particularly the cyclin D1–associated CDK4 and CDK6, phosphorylate the product of the retinoblastoma (*RB*) gene and thus release cells from its growth inhibitory effects at G0/G1 (109). Mechanisms that affect the balance between *p16* and cyclin D1 result in abnormal cell cycling and growth (104). *p16* abnormalities are frequently found in NSCLCs but are rare in SCLCs (30). Adenoviral vector-based gene therapy with *p16* is being developed in preclinical studies.

p14^{ARF} (p14) is the second alternative reading frame protein encoded by the *p16* locus and appears to be important in growth regulation. By binding to the MDM2-p53 complex, it prevents p53 degradation, thereby leading to p53 activation. Loss of p14 expression was more frequently found in lung tumors with neuroendocrine features (110). However, aberrant methylation of the *p14* promoter region did not occur frequently in NSCLCs (108).

The other key component in this pathway is the *RB* gene (109). *RB* was originally identified as being inactivated in childhood retinoblastomas (111). It is located in chromosome region 13q14.11 and encodes a 105-kDa protein that is important in regulating the cell cycle during G0/G1 phase (112). Inactivation of either p16 or RB expression inactivates the pathway, which is found in ∼90% of all lung cancers. However, it is uncommon to have both *RB* and *p16* inactivated in the same tumor. Loss of RB function can occur by deletions, mutations, or splicing abnormalities. Abnormalities of the RB protein are found in more than 90% of SCLCs and about 15–30% of NSCLCs (92, 113–115). Absence of RB protein expression was associated with poor prognosis in NSCLC patients in one study (116) but was not confirmed by others (115, 117). Other members of the *RB* gene family are *p107* and *pRB2/p130*. *RB2/p130* is thought to be a TSG, and loss of expression was found to be correlated with worse histological grading and development of metastasis in lung cancers (118). Recently, Claudio et al. (119) reported a high frequency of *RB2/p130* mutations in primary lung tumors. Retrovirus-mediated delivery of wild-type *RB2/p130* to a lung tumor cell line potently inhibited tumorigenesis, suggesting that *RB2/p130* may be a candidate for gene therapy trials for lung cancer.

APC (Adenomatous Polyposis Coli) Gene/wnt Pathway

The *APC* gene encodes a large protein with multiple cellular functions and interactions, including roles in signal transduction in the wnt-signaling pathway, mediation of intercellular adhesion, stabilization of the cytoskeleton, and possibly regulation of the cell cycle and apoptosis (120). Loss of expression of APC has been frequently observed in primary lung cancers, and aberrant methylation is the most important mechanism for inactivating expression of this gene in lung cancer (121). With loss of APC expression, the wnt-signaling pathway is constitutively turned on, resulting in accumulation of β-catenin as the result of wnt-signal, which in turn activates transcription of genes such as the c-*myc* oncogene and *cyclin D1*, both of which regulate cell cycle progression. Although this could also occur by β-catenin mutations, these mutations are rare in lung cancer.

The occurrence of mutations in the *PTEN/MMAC1* gene, which is located at the chromosomal region 10q23.3, has been investigated in a large number of lung cancers (122). However, it seems that genetic abnormalities of this gene are only involved in a relatively small subset of lung cancers.

Aberrant Promoter Methylation

Aberrant methylation of normally unmethylated CpG-rich areas, also known as CpG islands that are located in or near the promoter region of many genes, has been associated with transcriptional inactivation of TSGs in human cancer (105, 123). Methylation serves as an alternative to the genetic loss of a TSG function by deletion or mutation. As discussed above, several genes are frequently methylated in primary lung tumors including the genes adenomatous polyposis coli (*APC*), retinoic acid receptor β-2 (*RARβ*), *CDH13* (H-cadherin), fragile histidine triad (*FHIT*), *RASSF1A*, tissue inhibitor of metalloproteinase-3 (*TIMP-3*), *p16*, and death-associated protein kinase (*DAPK*) (56–58, 63, 74, 105, 108, 121, 124–128) (Table 2). A significantly shorter disease-free survival for patients whose tumors were methylated for *DAPK* was reported by Tang et al. (128), and Burbee et al. (57) found a shorter overall survival for patients whose tumors were methylated for *RASSF1A*. Methylated DNA sequences can be detected in primary lung cancers, circulating in serum DNA from lung cancer patients, in sputum samples prior to the onset of invasive lung cancer, as well as in precursor lesions for lung carcinomas (106, 129, 130). These findings indicate that aberrant methylation can develop during the preneoplastic process and thus may serve as a potential biomarker for early diagnosis of lung cancer, as well as in following disease load. Determining the methylation status of certain genes in bronchial biopsies, bronchioloalveolar washings, and sputum samples from high-risk individuals such as heavy smokers is being tested as a marker for lung cancer risk assessment. Aberrant methylation can be reversed in vitro by drugs that block methylation such as 5-aza-2'-deoxycytidine, which results in gene re-expression and tumor growth inhibition (63, 74). Histone deacetylase inhibitors also can reverse the methylation status of genes and frequently are additive or synergistic with 5-aza-2'-deoxycytidine. Because of the frequency of tumor- acquired methylation, clinical trials with demethylating drugs such as 5-aza-2'-deoxycytidine, with or without histone deacetylase inhibitors, are being developed (131).

Tumor Angiogenesis

Angiogenesis is important in neoplastic development and progression because both tumor growth and metastatic dissemination of tumor cells depend on vascular support (132). An increasing number of angiogenic factors, i.e., inducers and inhibitors regulating endothelial cell proliferation and migration, have been identified (132–134). Angiogenic factors affect vasculature formation, growth patterns,

TABLE 2   Frequencies of genes methylated in lung cancers

Gene	NSCLCs (%)	SCLCs[a] (%)
APC	46	26
CDH13	43–45	20
RARβ	40–43	76
FHIT	37	64
RASSF1A	30–40	100
TIMP-3	19–26	ND
p16	25–41	ND
MGMT	21–27	ND
DAPK	16–44	ND
ECAD	18	ND
p14	6–8	ND
GSTP1	7–9	ND
BRCA1	4	ND
p73	0	ND
hMLH1	0	ND
p15	0	ND

[a]Most SCLC data are from tumor cell lines because of the clinical difficulty in getting pretreatment tumor samples for study. ND, not done; NSCLC, non-small cell lung cancer; SCLC, small cell lung cancer; APC, adenomatous polyposis coli; CDH13, H-cadherin; RARβ, retinoic acid receptor β-2; FHIT, fragile histidine triad; TIMP-3, tissue inhibitor of metalloproteinase-3; p16, p16INK4a; MGMT, O^6-methylguanine-DNA methyltransferase; DAPK, death-associated protein kinase; ECAD, E-cadherin; GSTP1, glutathione-S-transferase P1; p14, p14ARF; p15, p15INK4b. Data extracted from references 56–58, 63, 74, 105–107, 120, 122–126.

and vascular permeability, modulate host response, and influence tumor invasion, metastasis, and prognosis. Tumor cells and their precursor cells are able to secrete angiogenic substances that depend on certain factors including hypoxia and alterations in dominant and recessive oncogenes such as p53 and RAS (135, 136). Among these, the vascular endothelial growth factor (VEGF) appears to play a crucial role and is frequently expressed by lung cancers (137). These growth factors and their receptors are prime regulators of both physiological and pathological angiogenesis (134). So far, two receptors for VEGF, which are selectively expressed in endothelium, have been characterized, and antibodies have been developed that can block the interaction between VEGF and its receptors (138). In addition, VEGF was identified as one factor responsible for inhibition of the functional maturation of dentritic cells (139). This fact is important because dendritic cells are important for antigen presentation and suggest that inadequate function may be responsible for the escape of tumors from the host immune system. VEGF expression in NSCLCs was significantly associated with new vessel formation and was an adverse prognostic factor in these patients (140). Koukourakis et al. (141) investigated the activated microvessel density in early operable NSCLCs and found it significantly higher in the invading front of the tumors and in the normal lung adjacent to the tumors compared with normal lung distal to the tumor or the inner tumor areas. These results suggest that activated microvessel density serves as an independent prognostic factor in NSCLC patients. Upregulation of platelet-derived endothelial cell growth factor may be associated with a worse prognosis in patients with node-negative NSCLCs (142). Additionally, Fontanini et al. (143) reported that the microvessel count was a highly significant adverse predictor of both overall and disease-free survival in patients with NSCLC, suggesting that the evaluation of tumor angiogenesis may be useful in the postsurgical staging of NSCLC patients to identify subsets of patients who may benefit from adjuvant treatment studies. New treatment approaches directed against angiogenic factors or their receptors are being investigated in clinical trials in lung cancer. These include humanized monoclonal anti-VEGF antibodies, anti-VEGF receptor antibodies, and drugs blocking the VEGF receptors' tyrosine kinase activity essential for their function.

TOBACCO SMOKE CARCINOGENS

Tobacco smoke is responsible for about 90% of all cases of lung cancer. The three major classes of carcinogens in tobacco smoke are the polycyclic hydrocarbons [such as benzo(*a*)pyrene], nitrosamines, and aromatic amines (144). The carcinogenic effects of tobacco smoke in the lung involve the induction of carcinogen-activating and inactivating enzymes, as well as covalent DNA adduct formation, which may result in DNA misreplication and mutation. A significant association between the level of benzo(*a*)pyrene-induced DNA adducts and risk for lung cancer has been reported and suggests that subjects sensitive to benzo(*a*)pyrene-induced DNA damage may have a suboptimal ability to remove benzo(*a*)pyrene-DNA adducts. These subjects are thus susceptible to tobacco carcinogen exposure and may be at increased risk of developing lung cancer (145).

Alterations in Smoke-Damaged Respiratory Epithelium

Lung cancers are believed to arise after a series of progressive pathological changes in the respiratory epithelium, and many of them can already be found in preneoplastic/preinvasive lesions or even in the cytologically normal bronchial epithelium from smokers. Bennett et al. (146) have observed mutations in the *p53* gene in dysplastic bronchial epithelium, and *RAS* mutations were described in areas of dysplasia and in atypical alveolar hyperplasia (147). LOH at chromosomal regions 8p and 9p occurs early in the multistage development of invasive lung cancer; however, LOH at 3p is the earliest and most frequent event (51, 52, 148, 149). Allele loss at 8p21–23, commenced at the hyperplasia/metaplasia stage, was seen in 65% of smokers without cancer and persisted for up to 48 years after smoking cessation. Similar to LOH found at chromosome 3p, there was also a progressive increase in the LOH frequency and in the size of allele loss with increasing severity of histopathologic preneoplastic changes. LOH was also detected in plasma DNA from individuals at high risk of lung cancer (150). Recently, aberrant methylation of certain genes has been linked to early stages of respiratory carcinogenesis. Belinsky et al. (129) reported that aberrant methylation of *p16* can be detected in lung pre-cursor lesions. The same authors found aberrant methylation of the genes *p16* and O^6-methylguanine-DNA-methyltransferase (*MGMT*) in sputum samples from pa-tients with squamous cell lung carcinomas up to 3 years before clinical diagnosis (130). *p16* methylation and *p53* mutations, but not *KRAS* mutations, were found in the bronchial epithelium from chronic smokers before any clinical evidence of neoplasia (151), and p16 methylation, *p53* mutations, *KRAS* mutations, and mi-crosatellite instability were detected in bronchoalveolar lavage fluid from patients with early-stage lung cancer (152). Additionally, aberrant methylation of *FHIT* has been found in the smoking-damaged bronchial epithelium from heavy smokers without cancer (63). A high frequency of mitochondrial DNA (mtDNA) mutations have been described in various malignant tumors including lung cancer (153). The mutated mtDNA was detectable in bronchoalveolar lavage fluids, suggesting that it may serve as a powerful molecular marker for detection of lung cancer. The functional significance of such mitochondrial changes is currently unknown.

The fact that specific alterations can be detected in preneoplastic/preinvasive le-sions suggests that these abnormalities may be useful as biomarkers for lung cancer. These biomarkers could be used to identify individuals at high risk for developing lung cancer, monitor the efficacy of lung cancer chemoprevention trials, diagnose lung cancer in early stages, and monitor the efficacy of lung cancer therapies. Additionally, study of biomarkers in tumors could identify patients with different prognoses and allow tailoring of therapy. Samples used to test for biomarkers for risk assessment have to be obtainable in an easy, noninvasive and inexpensive way. Sputum samples fulfill these criteria and therefore are attractive material for biomarker studies.

Microsatellite Alterations

Microsatellite alterations, which involve a single shift of individual allelic bands, were seen in about 35% of SCLCs and 22% of NSCLCs (30). Thus microsatellite alterations have been tested as molecular biomarkers for early detection of cancer cells in sputum and bronchial washings (152, 154). In other human tumors such as colon cancer, the replication error repair (RER) phenotype results in "laddering" of short tandem repeat sequences associated with inherited or acquired mutations in DNA mismatch repair genes such as *MSH2* and *hMLH1* (30). However, this RER phenotype or mutations in these genes have not been found in lung cancers. Lung tumors with microsatellite alterations at selected tetranucleotide repeats have a high frequency of *p53* mutations and do not display a phenotype consistent with defects in mismatch repair (155). The molecular abnormalities underlying such microsatellite alterations in lung cancer are unknown.

Telomerase Activity

The ends of human chromosomes (telomeres) contain the hexameric TTAGGG tandem repeats. During normal cell division, the absence of telomerase activity is associated with progressive telomere shortening, leading to cell senescence and normal cell mortality (30). On the contrary, germ cells, some stem cells, and most cancer cells have telomerase activity that results in replacing the hexameric repeats, therefore leading to potential cellular immortality (30). The majority of SCLCs and about 80% of NSCLCs (156, 157) show high levels of telomerase activity, which is associated with increased cell proliferation rates and advanced stage in NSCLCs (157). Because of this, anti-telomerase drugs are being developed as new therapeutics for a variety of cancers including lung cancer. Telomerase components are activated in the latent preneoplastic stages of lung cancer. The mechanism for re-expression of the catalytic component hTERT or the RNA component of telomerase in tumors is currently unknown. Normal epithelial cells can be immortalized and transformed to malignancy by the combination of hTERT, SV40 T antigen, and a mutated *RAS* gene (158).

Neuroendocrine Phenotype of Lung Tumors

The classification of neuroendocrine (NE) lung tumors includes carcinoids and SCLCs and has been enlarged with a new entity, the large cell NE carcinomas (LCNEC) (159). NE lung tumors share certain morphological, ultrastructural, immunohistochemical, and other molecular characteristics that sustain their NE phenotype (e.g., NE secretory granules at electron microscopy, NE markers at immunohistochemistry) (159). Specific NE markers include chromogranin, synaptophysin, and neural cell adhesion molecule (NCAM). NE lung tumors appear to be epithelial tumors characterized by their preferential NE differentiation but retain their propensity to follow multidirectional differentiation pathway. The derivation of all histologic types of lung cancer from a common endodermal stem cell is likely to be responsible for the frequent multidirectional differentiation in lung tumors. However, this stem cell has not yet been identified.

Viral Factors in the Pathogenesis of Lung Cancer

There is no evidence yet that HIV infection leads to an increased incidence of lung cancer. However, lung tumors arising in HIV-positive patients with or without the acquired immunodeficiency syndrome (AIDS) have a severalfold increase in the frequency of microsatellite alterations, which indicates increased genetic instability (160).

Lung cancer is the leading cause of cancer death in Taiwanese women, although less than 10% of female lung cancer patients are smokers, which suggests that other factors are important for developing lung cancer (161). A recent study indicated that human papillomavirus (HPV) oncogenic subtypes 16/18 may be involved in the pathogenesis of lung cancer of these Taiwanese women (161). Fifty-five percent of lung tumor patients had HPV 16/18 DNA compared with 27% of noncancer control subjects, which had undergone thoracic surgery for lung diseases other than cancer. Also the odds ratio ($\sim$10-fold) of HPV16/18 infection of nonsmoking female lung cancer patients was much higher compared with nonsmoking male lung cancer patients (odds ratio of $\sim$2). Additionally, HPV 16/18 DNA was uniformly located in lung tumor cells, but not in the adjacent noninvolved lung. These results strongly suggest that HPV infection with virus subtypes known to be oncogenic for cervical cancer is associated with lung cancer development of nonsmoking Taiwanese female lung cancer patients. Because oncogenic HPV subtypes encode E6 and E7 viral oncoproteins, which inactivate p53 and Rb protein, respectively, HPV infection provides several key mutations at an early stage (162).

The impact of the well-known oncogenic DNA virus simian virus 40 (SV40) to the development of malignant mesotheliomas was until recently controversial (163, 164). Shivapurkar et al. (165) reported that SV40 sequences are frequently (56%) present in malignant mesotheliomas and are absent in adjacent lung and normal mesothelial tissues and in lung carcinomas, which strongly implicates their role in the pathogenesis of malignant mesotheliomas. SV40 encodes viral oncoprotein large T antigen, which binds to and inactivates p53 and Rb proteins, and small t antigen, which inactivates PP2A function by displacing regulatory (B) subunits from the catalytic (C) and structural subunits (A). Thus the oncogenic virus again provides inactivation of key TSG proteins (166). Although SV40 is definitely involved in the pathogenesis of mesotheliomas, it does not appear to be involved in lung cancer pathogenesis.

Jaagsiekte sheep retrovirus (JSRV) can induce rapid, multifocal lung cancer in sheep, but JSRV is a simple retrovirus having no known oncogenes so the mechanism of oncogenesis is still unknown. Recently, HYAL2, a glycosylphosphatidylinositol (GPI)-anchored cell-surface protein, has been identified as the receptor for JSRV (167). Of great interest is the fact that the *HYAL2* gene resides in the 600-kb 3p21.3 TSG homozygous deletion region (55). Lung cancer induced by JSRV closely resembles human bronchiolo-alveolar carcinoma. Further studies are necessary to investigate the relationship of JSRV oncogenesis to human bronchiolo-alveolar carcinoma.

Second Primary Lung Cancers

The risk of developing a second lung cancer in patients who survived resection of NSCLC is approximately 1–2% per patient per year (168). The average risk of developing a second lung cancer in patients who survived SCLC is approximately 6% per patient per year (168). Because of the high risk of developing a second lung cancer, these patients need to be followed carefully for many years. This increased risk probably represents a persistent field defect in the respiratory epithelium of patients cured of one lung cancer. This defect probably involves the multiple somatically acquired genetic changes detected in respiratory epithelium described previously that predispose these individuals to lung cancer development. In a related scenario, molecular changes including *p53* and *KRAS* mutations and analysis of LOH and microsatellite alterations at nine chromosomal regions were investigated in second primary lung cancers that followed therapy (usually chemo- or radiotherapy) for Hodgkin's disease (169). The overall frequency of microsatellite alterations was substantially greater in the second primary lung cancers than in the respective sporadic cancers. However, no differences in the occurence of LOH or in the pattern of *p53* and *KRAS* mutations were observed. These results suggest that microsatellite alterations, which reflect widespread genomic instability, occur frequently in second primary lung cancers, although future research is necessary to address the possible contribution of these alterations to the pathogenesis of second primary lung cancers.

Prognostic Markers in Lung Cancer

A recent study (170) investigated a panel of nine molecular markers including p53, Bcl-2, ErbB-2 (HER-2/neu), KI-67, RB, EGFR, factor VIII (as a marker of angiogenesis), sialyl-Tn antigen, and CD-44 in more than 400 stage I NSCLCs by immunostaining. RB and sialyl-Tn antigen were scored abnormal if loss of staining was seen, whereas the others were scored abnormal with increased expression. Among men, the only marker associated with decreased survival was ErbB2, whereas among women, p53, RB, CD-44, and factor VIII were of prognostic significance. Sabel et al. (171) reported a negative impact of CD40 expression on survival of lung cancer patients. Cyclin E is a G1 cyclin and one of the key regulators of the G1-S transition. The expression of cyclin E was investigated by immunohistochemistry in a large series of NSCLCs (172). High-cylin E expression was found more frequently in tumors from smokers than from nonsmokers, in squamous cell carcinomas than in nonsquamous carcinomas, and in later-stage (pT2-4) tumors than in early-stage (pT1) tumors. Additionally, patients whose tumors showed high-level cyclin E expression survived a significantly shorter time than patients with tumors having low-level expression. However, other CDK2-associated cyclins, including cyclin E2, cyclin A1, and cyclin A2, do not have a prognostic role in NSCLC (173).

CONCLUSIONS

The understanding of lung cancer pathogenesis has grown rapidly over the recent years, but our knowledge will grow even more in the next decade with the information from the Human Genome Project. The use of techniques such as microarrays for testing expression of nearly all human genes and their isoforms at the same time in lung cancer, or other genome-wide strategies involving proteinomics, will provide large amounts of information that need to be translated into clinical practice and integrated into our understanding of lung cancer pathogenesis. The main goals for future studies should focus on how this information can be used in terms of risk assessment, prevention, early detection of lung cancer, and development of new therapeutic targets. New treatment strategies including drugs that block oncogene functions such as tyrosine kinase inhibitors, gene therapy, monoclonal antibodies against growth factors and receptors, angiogenesis inhibitors, vaccines, apoptosis modulators, demethylating agents, and new drugs targeted at abnormal pathways are being developed and introduced into clinical trials. These approaches should help to decrease the number of lung cancer deaths through early detection and treatment and increase the cure and prolong the survival of patients with lung cancer.

ACKNOWLEDGMENT

This work was supported by grants from the Austrian Science Foundation (J1658-MED, J1860-MED), by a National Cancer Institute Lung Cancer SPORE grant (P50 CA70907), and The G. Harold and Leila Y. Mathers Charitable Foundation.

Visit the Annual Reviews home page at www.AnnualReviews.org

LITERATURE CITED

1. Jemal A, Chu KC, Tarone RE. 2001. Recent trends in lung cancer mortality in the United States. *J. Natl. Cancer Inst.* 93:277–83

2. Greenlee RT, Murray T, Bolden S, Wingo PA. 2000. Cancer statistics, 2000. *CA Cancer J. Clin.* 50:7–33

3. Doll R, Peto R. 1981. The causes of cancer: quantitative estimates of avoidable risks of cancer in the United States today. *J. Natl. Cancer Inst.* 66:1191–2308

4. Shopland DR. 1995. Tobacco use and its contribution to early cancer mortality with a special emphasis on cigarette smoking. *Environ. Health Perspect.* 103:131–42

5. Travis WD, Travis LB, Devesa SS. 1995. Lung cancer. *Cancer* 75:191–202

6. Viallet J, Sausville EA. 1996. Involvement of signal transduction pathways in lung cancer biology. *J. Cell Biochem. Suppl.* 24:228–36

7. Sunday ME, Hua J, Torday JS, Reyes B, Shipp MA. 1992. CD10/neutral endopeptidase 24.11 in developing human fetal lung. Patterns of expression and modulation of peptide-mediated proliferation. *J. Clin. Invest.* 90:2517–25

8. Spindel ER. 1996. Roles of bombesin-like peptides in lung development and lung injury. *Am. J. Respir. Cell Mol. Biol.* 14:407–8

9. DeMichele MA, Davis AL, Hunt JD, Landreneau RJ, Siegfried JM. 1994. Expression of mRNA for three bombesin receptor subtypes in human bronchial epithelial cells. *Am. J. Respir. Cell Mol. Biol.* 11:66–74

10. Cardona C, Rabbitts PH, Spindel ER, Ghatei MA, Bleehen NM, et al. 1991. Production of neuromedin B and neuromedin B gene expression in human lung tumor cell lines. *Cancer Res.* 51:5205–11

11. Richardson GE, Johnson BE. 1993. The biology of lung cancer. *Semin. Oncol.* 20:105–27

12. Shriver SP, Bourdeau HA, Gubish CT, Tirpak DL, Davis AL, et al. 2000. Sex-specific expression of gastrin-releasing

peptide receptor: relationship to smoking history and risk of lung cancer. *J. Natl. Cancer Inst.* 92:24–33

13. Cuttitta F, Carney DN, Mulshine J, Moody TW, Fedorko J, et al. 1985. Bombesin-like peptides can function as autocrine growth factors in human small-cell lung cancer. *Nature* 316:823–26

14. Quinn KA, Treston AM, Unsworth EJ, Miller MJ, Vos M, et al. 1996. Insulin-like growth factor expression in human cancer cell lines. *J. Biol. Chem.* 271:11477–83

15. Wu X, Yu H, Amos CI, Hong WK, Spitz MR. 2000. Joint effect of insulin-like growth factors and mutagen sensitivity in lung cancer risk. *Growth Horm. IGF Res.* 10:S26–27

16. Rusch V, Baselga J, Cordon-Cardo C, Orazem J, Zaman M, et al. 1993. Differential expression of the epidermal growth factor receptor and its ligands in primary non-small cell lung cancers and adjacent benign lung. *Cancer Res.* 53:2379–85

17. Rusch V, Klimstra D, Venkatraman E, Pisters PWT, Langenfeld J, Dmitrovsky E. 1997. Overexpression of the epidermal growth factor receptor and its ligand transforming growth factor alpha is frequent in resectable non-small cell lung cancer but does not predict tumor progression. *Clin. Cancer Res.* 3:515–22

18. Reissmann PT, Koga H, Figlin RA, Holmes EC, Slamon DJ. 1999. Amplification and overexpression of the cyClin. D1 and epidermal growth factor receptor genes in non-small-cell lung cancer. Lung Cancer Study Group. *J. Cancer Res. Clin. Oncol.* 125:61–70

19. Ohsaki Y, Tanno S, Fujita Y, Toyoshima E, Fujiuchi S, et al. 2000. Epidermal growth factor receptor expression correlates with poor prognosis in non-small cell lung cancer patients with p53 overexpression. *Oncol. Rep.* 7:603–7

20. Weiner DB, Nordberg J, Robinson R, Nowell PC, Gazdar A, et al. 1990. Expression of the neu gene-encoded protein (P185neu) in human non-small cell carcinomas of the lung. *Cancer Res.* 50:421–25

21. Rachwal WJ, Bongiorno PF, Orringer MB, Whyte RI, Ethier SP, Beer DG. 1995. Expression and activation of erbB-2 and epidermal growth factor receptor in lung adenocarcinomas. *Br. J. Cancer* 72:56–64

22. Kern JA, Torney L, Weiner D, Gazdar A, Shepard HM, Fendly B. 1993. Inhibition of human lung cancer cell line growth by an anti-p185HER2 antibody. *Am. J. Respir. Cell Mol. Biol.* 9:448–54

23. Harvey P, Warn A, Newman P, Perry LJ, Ball RY, Warn RM. 1996. Immunoreactivity for hepatocyte growth factor/scatter factor and its receptor, met, in human lung carcinomas and malignant mesotheliomas. *J. Pathol.* 180:389–94

24. Olivero M, Rizzo M, Madeddu R, Casadio C, Pennacchietti S, et al. 1996. Overexpression and activation of hepatocyte growth factor/scatter factor in human non-small-cell lung carcinomas. *Br. J. Cancer* 74:1862–68

25. Siegfried JM, Weissfeld LA, Singh-Kaw P, Weyant RJ, Testa JR, Landreneau RJ. 1997. Association of immunoreactive hepatocyte growth factor with poor survival in resectable non-small cell lung cancer. *Cancer Res.* 57:433–39

26. Singh-Kaw P, Zarnegar R, Siegfried JM. 1995. Stimulatory effects of hepatocyte growth factor on normal and neoplastic human bronchial epithelial cells. *Am. J. Physiol. Lung Cell Mol. Physiol.* 268: L1012–L20

27. Di Nunno L, Larsson LG, Rinehart JJ, Beissner RS. 2000. Estrogen and progesterone receptors in non-small cell lung cancer in 248 consecutive patients who underwent surgical resection. *Arch. Pathol. Lab. Med.* 124:1467–70

28. Su JM, Hsu HK, Chang H, Lin SL, Chang HC, et al. 1996. Expression of estrogen and progesterone receptors in

non-small-cell lung cancer: immunohisto-chemical study. *AntiCancer Res.* 16: 3803–6

29. Vargas SO, Leslie KO, Vacek PM, Socinski MA, Weaver DL. 1998. Estrogen-receptor-related protein p29 in primary nonsmall cell lung carcinoma: pathologic and prognostic correlations. *Cancer* 82: 1495–500

30. Sekido Y, Fong KM, Minna JD. 1998. Progress in understanding the molecular pathogenesis of human lung cancer. *Biochim. Biophys. Acta* 1378:F21–59

31. Rodenhuis S, Slebos RJ. 1990. The ras oncogenes in human lung cancer. *Am. Rev. Respir. Dis.* 142:S27–30

32. Slebos RJ, Kibbelaar RE, Dalesio O, Kooistra A, Stam J, et al. 1990. K-ras oncogene activation as a prognostic marker in adenocarcinoma of the lung. *N. Engl. J. Med.* 323:561–65

33. Mitsudomi T, Steinberg SM, Oie HK, Mulshine JL, Phelps R, et al. 1991. ras gene mutations in non-small cell lung cancers are associated with shortened survival irrespective of treatment intent. *Cancer Res.* 51:4999–5002

34. Rosell R, Li S, Skacel Z, Mate JL, Maestre J, et al. 1993. Prognostic impact of mutated K-ras gene in surgically resected non-small cell lung cancer patients. *Oncogene* 8:2407–12

35. Grandori C, Eisenman RN. 1997. Myc target genes. *Trends Biochem. Sci.* 22: 177–81

36. Krystal G, Birrer M, Way J, Nau M, Sausville E, et al. 1988. Multiple mechanisms for transcriptional regulation of the myc gene family in small-cell lung cancer. *Mol. Cell. Biol.* 8:3373–81

37. Pezzella F, Turley H, Kuzu I, Tungekar MF, Dunnill MS, et al. 1993. bcl-2 protein in non-small-cell lung carcinoma. *N. Engl. J. Med.* 329:690–94

38. Kaiser U, Schilli M, Haag U, Neumann K, Kreipe H, et al. 1996. Expression of bcl-2-protein in small cell lung cancer. *Lung Cancer* 15:31–40

39. Fontanini G, Vignati S, Bigini D, Mussi A, Lucchi M, et al. 1995. Bcl-2 protein: a prognostic factor inversely correlated to p53 in non-small-cell lung cancer. *Br. J. Cancer* 71:1003–7

40. Higashiyama M, Doi O, Kodama K, Yokouchi H, Nakamori S, Tateishi R. 1997. bcl-2 oncoprotein in surgically resected non-small cell lung cancer: possibly favorable prognostic factor in association with low incidence of distant metastasis. *J. Surg. Oncol.* 64:48–54

41. Anton RC, Brown RW, Younes M, Gondo MM, Stephenson MA, Cagle PT. 1997. Absence of prognostic significance of bcl-2 immunopositivity in non-small cell lung cancer: analysis of 427 cases. *Hum. Pathol.* 28:1079–82

42. Dang TP, Gazdar AF, Virmani AK, Sepetavec T, Hande KR, et al. 2000. Chromosome 19 translocation, overexpression of Notch3, and human lung cancer. *J. Natl. Cancer Inst.* 92:1355–57

43. Hibi K, Trink B, Patturajan M, Westra WH, Caballero OL, et al. 2000. AIS is an oncogene amplified in squamous cell carcinoma. *Proc. Natl. Acad. Sci. USA* 97:5462–67

44. Yamaguchi K, Patturajan M, Trink B, Usadel H, Koch W, et al. 2000. Circulating antibodies to p40 (AIS) in the sera of respiratory tract cancer patients. *Int. J. Cancer* 89:524–28

45. Knudson AG Jr. 1989. Hereditary cancers disclose a class of cancer genes. *Cancer* 63:1888–91

46. Zöchbauer-Müller S, Minna JD. 2000. The biology of lung cancer including potential clinical applications. *Chest Surg. Clin. N. Am.* 10:691–708

47. Girard L, Zöchbauer-Müller S, Virmani AK, Gazdar AF, Minna JD. 2000. Genome-wide allelotyping of lung cancer identifies new regions of allelic loss, differences between small cell lung cancer and non-small cell lung cancer, and loci clustering. *Cancer Res.* 60:4894–906

48. Wistuba II, Berry J, Behrens C, Maitra A,

Shivapurkar N, et al. 2000. Molecular changes in the bronchial epithelium of patients with small cell lung cancer. *Clin. Cancer Res.* 6:2604–10

49. Sanchez-Cespedes M, Ahrendt SA, Piantadosi S, Rosell R, Monzo M, et al. 2001. Chromosomal alterations in lung adenocarcinoma from smokers and nonsmokers. *Cancer Res.* 61:1309–13

50. Virmani AK, Fong KM, Kodagoda D, McIntire D, Hung J, et al. 1998. Allelotyping demonstrates common and distinct patterns of chromosomal loss in human lung cancer types. *Genes Chromosomes Cancer* 21:308–19

51. Wistuba II, Behrens C, Virmani AK, Mele G, Milchgrub S, et al. 2000. High resolution chromosome 3p allelotyping of human lung cancer and preneoplastic/preinvasive bronchial epithelium reveals multiple, discontinuous sites of 3p allele loss and three regions of frequent breakpoints. *Cancer Res.* 60:1949–60

52. Hung J, Kishimoto Y, Sugio K, Virmani A, McIntire DD, et al. 1995. Allele-specific chromosome 3p deletions occur at an early stage in the pathogenesis of lung carcinoma. *J. Am. Med. Assoc.* 273:558–63

53. Wistuba II, Lam S, Behrens C, Virmani AK, Fong KM, et al. 1997. Molecular damage in the bronchial epithelium of current and former smokers. *J. Natl. Cancer Inst.* 89:1366–73

54. Hibi K, Takahashi T, Yamakawa K, Ueda R, Sekido Y, et al. 1992. Three distinct regions involved in 3p deletion in human lung cancer. *Oncogene* 7:445–49

55. Lerman MI, Minna JD. 2000. The 630-kb lung cancer homozygous deletion region on human chromosome 3p21.3: identification and evaluation of the resident candidate tumor suppressor genes. The International Lung Cancer Chromosome 3p21.3 Tumor Suppressor Gene Consortium. *Cancer Res.* 60:6116–33

56. Dammann R, Li C, Yoon JH, Chin PL,

Bates S, Pfeifer GP. 2000. Epigenetic inactivation of a RAS association domain family protein from the lung tumour suppressor locus 3p21.3. *Nat. Genet.* 25:315–19

57. Burbee DG, Forgacs E, Zöchbauer-Müller S, Shivakumar L, Fong KM, et al. 2001. Epigenetic inactivation of RASSF1A in lung and breast cancers and malignant phenotype suppression. *J. Natl. Cancer Inst.* 93:691–99

58. Agathanggelou A, Honorio S, Macartney DP, Martinez A, Dallol A, et al. 2001. Methylation associated inactivation of RASSF1A from region 3p21.3 in lung, breast and ovarian tumours. *Oncogene* 20:1509–18

59. Sozzi G, Veronese ML, Negrini M, Baffa R, Cotticelli MG, et al. 1996. The FHIT gene 3p14.2 is abnormal in lung cancer. *Cell* 85:17–26

60. Fong KM, Biesterveld EJ, Virmani A, Wistuba I, Sekido Y, et al. 1997. FHIT and FRA3B 3p14.2 allele loss are common in lung cancer and preneoplastic bronchial lesions and are associated with cancer-related FHIT cDNA splicing aberrations. *Cancer Res.* 57:2256–67

61. Sozzi G, Sard L, De Gregorio L, Marchetti A, Musso K, et al. 1997. Association between cigarette smoking and FHIT gene alterations in lung cancer. *Cancer Res.* 57:2121–23

62. Geradts J, Fong KM, Zimmerman PV, Minna JD. 2000. Loss of Fhit expression in non-small-cell lung cancer: correlation with molecular genetic abnormalities and clinicopathological features. *Br. J. Cancer* 82:1191–97

63. Zöchbauer-Müller S, Fong KM, Maitra A, Lam S, Geradts J, et al. 2001. 5′ CpG island methylation of the *FHIT* gene is correlated with loss of gene expression in lung and breast cancer. *Cancer Res.* 61:3581–85

64. Siprashvili Z, Sozzi G, Barnes LD, McCue P, Robinson AK, et al. 1997. Replacement of Fhit in cancer cells suppresses

tumorigenicity. *Proc. Natl. Acad. Sci. USA* 94:13771–76

65. Ji L, Fang B, Yen N, Fong K, Minna JD, Roth JA. 1999. Induction of apoptosis and inhibition of tumorigenicity and tumor growth by adenovirus vector-mediated fragile histidine triad (FHIT) gene over-expression. *Cancer Res.* 59:3333–39

66. Gebert JF, Moghal N, Frangioni JV, Sugarbaker DJ, Neel BG. 1991. High frequency of retinoic acid receptor beta abnormalities in human lung cancer. *Oncogene* 6:1859–68

67. Geradts J, Chen JY, Russell EK, Yankas-kas JR, Nieves L, Minna JD. 1993. Human lung cancer cell lines exhibit resistance to retinoic acid treatment. *Cell Growth Diff.* 4:799–809

68. Lu XP, Fanjul A, Picard N, Pfahl M, Rungta D, et al. 1997. Novel retinoid-related molecules as apoptosis inducers and effective inhibitors of human lung cancer cells in vivo. *Nat. Med.* 3:686–90

69. Xu XC, Sozzi G, Lee JS, Lee JJ, Pasto-rino U, et al. 1997. Suppression of retinoic acid receptor beta in non-small-cell lung cancer in vivo: implications for lung cancer development. *J. Natl. Cancer Inst.* 89:624–29

70. Toulouse A, Morin J, Dion PA, Houle B, Bradley WE. 2000. RARbeta2 specificity in mediating RA inhibition of growth of lung cancer-derived cells. *Lung Cancer* 28:127–37

71. Xu XC, Lee JS, Lee JJ, Morice RC, Liu X, et al. 1999. Nuclear retinoid acid receptor beta in bronchial epithelium of smokers before and during chemoprevention. *J. Natl. Cancer Inst.* 91:1317–21

72. Ayoub J, Jean-Francois R, Cormier Y, Meyer D, Ying Y, et al. 1999. Placebo-controlled trial of 13-*cis*-retinoic acid activity on retinoic acid receptor-beta expression in a population at high risk: implications for chemoprevention of lung cancer. *J. Clin. Oncol.* 17:3546–52

73. Picard E, Seguin C, Monhoven N, Roch-ette-Egly C, Siat J, et al. 1999. Expres-sion of retinoid receptor genes and pro-teins in non-small-cell lung cancer. *J. Natl. Cancer Inst.* 91:1059–66

74. Virmani AK, Rathi A, Zöchbauer-Müller S, Sacchi N, Fukuyama Y, et al. 2000. Promoter methylation and silencing of the retinoic acid receptor-beta gene in lung carcinomas. *J. Natl. Cancer Inst.* 92:1303–7

75. Jensen DE, Proctor M, Marquis ST, Gardner HP, Ha SI, et al. 1998. BAP1: a novel ubiquitin hydrolase which binds to the BRCA1 RING finger and enhances BRCA1-mediated cell growth suppres-sion. *Oncogene* 16:1097–112

76. Radicella JP, Dherin C, Desmaze C, Fox MS, Boiteux S. 1997. Cloning and char-acterization of hOGG1, a human homolog of the OGG1 gene of *Saccharomyces cerevisiae*. *Proc. Natl. Acad. Sci. USA* 94:8010–15

77. Papadopoulos N, Nicolaides NC, Wei YF, Ruben SM, Carter KC, et al. 1994. Mutation of a mutL homolog in hereditary colon cancer. *Science* 263:1625–29

78. Wei MH, Latif F, Bader S, Kashuba V, Chen JY, et al. 1996. Construction of a 600-kilobase cosmid clone contig and generation of a transcriptional map surrounding the lung cancer tumor sup-pressor gene (TSG) locus on human chro-mosome 3p21.3: progress toward the iso-lation of a lung cancer TSG. *Cancer Res.* 56:1487–92

79. Roche J, Boldog F, Robinson M, Robin-son L, Varella-Garcia M, et al. 1996. Dis-tinct 3p21.3 deletions in lung cancer and identification of a new human sema-phorin. *Oncogene* 12:1289–97

80. Sekido Y, Bader S, Latif F, Chen JY, Duh FM, et al. 1996. Human semaphorins A(V) and IV reside in the 3p21.3 small cell lung cancer deletion region and demon-strate distinct expression patterns. *Proc. Natl. Acad. Sci. USA* 93:4120–25

81. Xiang RH, Hensel CH, Garcia DK, Carl-son HC, Kok K, et al. 1996. Isolation of the human semaphorin III/F gene

(SEMA3F) at chromosome 3p21, a region deleted in lung cancer. *Genomics* 32:39–48

82. Gao B, Sekido Y, Maximov A, Saad M, Forgacs E, et al. 2000. Functional properties of a new voltage-dependent calcium channel alpha(2)delta auxiliary subunit gene (CACNA2D2). *J. Biol. Chem.* 275:12237–42

83. Angeloni D, Danilkovitch-Miagkova A, Ivanov SV, Breathnach R, Johnson BE, et al. 2000. Gene structure of the human receptor tyrosine kinase RON and mutation analysis in lung cancer samples. *Genes Chromosomes Cancer* 29:147–56

84. Liu CX, Musco S, Lisitsina NM, Forgacs E, Minna JD, Lisitsyn NA. 2000. LRP-DIT, a putative endocytic receptor gene, is frequently inactivated in non-small cell lung cancer cell lines. *Cancer Res.* 60:1961–67

85. Kuramochi M, Fukuhara H, Nobukuni T, Kanbe T, Maruyama T, et al. 2001. TSLC1 is a tumor-suppressor gene in human non-small-cell lung cancer. *Nat. Genet.* 27:427–30

86. Wang SS, Esplin ED, Li JL, Huang L, Gazdar A, et al. 1998. Alterations of the PPP2R1B gene in human lung and colon cancer. *Science* 282:284–87

87. Sidransky D, Hollstein M. 1996. Clinical implications of the p53 gene. *Annu. Rev. Med.* 47:285–301

88. Greenblatt MS, Bennett WP, Hollstein M, Harris CC. 1994. Mutations in the p53 tumor suppressor gene: clues to cancer etiology and molecular pathogenesis. *Cancer Res.* 54:4855–78

89. Husgafvel-Pursiainen K, Boffetta P, Kannio A, Nyberg F, Pershagen G, et al. 2000. p53 mutations and exposure to environmental tobacco smoke in a multicenter study on lung cancer. *Cancer Res.* 60:2906–11

90. Ahrendt SA, Chow JT, Yang SC, Wu L, Zhang MJ, et al. 2000. Alcohol consumption and cigarette smoking increase the frequency of p53 mutations in non-small cell lung cancer. *Cancer Res.* 60:3155–59

91. Nishio M, Koshikawa T, Kuroishi T, Suyama M, Uchida K, et al. 1996. Prognostic significance of abnormal p53 accumulation in primary, resected non-small-cell lung cancers. *J. Clin. Oncol.* 14:497–502

92. Geradts J, Fong KM, Zimmerman PV, Maynard R, Minna JD. 1999. Correlation of abnormal RB, p16ink4a, and p53 expression with 3p loss of heterozygosity, other genetic abnormalities, and clinical features in 103 primary non-small cell lung cancers. *Clin. Cancer Res.* 5:791–800

93. Mitsudomi T, Oyama T, Kusano T, Osaki T, Nakanishi R, Shirakusa T. 1993. Mutations of the p53 gene as a predictor of poor prognosis in patients with non-small-cell lung cancer. *J. Natl. Cancer Inst.* 85:2018–23

94. Kawasaki M, Nakanishi Y, Kuwano K, Yatsunami J, Takayama K, Hara N. 1997. The utility of p53 immunostaining of transbronchial biopsy specimens of lung cancer: p53 overexpression predicts poor prognosis and chemoresistance in advanced non-small cell lung cancer. *Clin. Cancer Res.* 3:1195–200

95. Tomizawa Y, Kohno T, Fujita T, Kiyama M, Saito R, et al. 1999. Correlation between the status of the p53 gene and survival in patients with stage I non-small cell lung carcinoma. *Oncogene* 18:1007–14

96. Lee JS, Yoon A, Kalapurakal SK, Ro JY, Lee JJ, et al. 1995. Expression of p53 oncoprotein in non-small-cell lung cancer: a favorable prognostic factor. *J. Clin. Oncol.* 13:1893–903

97. Apolinario RM, van der Valk P, de Jong JS, Deville W, van Ark-Otte J, et al. 1997. Prognostic value of the expression of p53, bcl-2, and bax oncoproteins, and neovascularization in patients with radically resected non-small- cell lung cancer. *J. Clin. Oncol.* 15:2456–66

98. Hashimoto T, Tokuchi Y, Hayashi M,

Kobayashi Y, Nishida K, et al. 1999. p53 null mutations undetected by immunohistochemical staining predict a poor outcome with early-stage non-small cell lung carcinomas. *Cancer Res.* 59:5572–77

99. Mitsudomi T, Hamajima N, Ogawa M, Takahashi T. 2000. Prognostic significance of p53 alterations in patients with non-small cell lung cancer: a meta-analysis. *Clin. Cancer Res.* 6:4055–63

100. Roth JA, Swisher SG, Merritt JA, Lawrence DD, Kemp BL, et al. 1998. Gene therapy for non-small cell lung cancer: a preliminary report of a phase I trial of adenoviral p53 gene replacement. *Semin. Oncol.* 25:33–37

101. Schuler M, Herrmann R, De Greve JL, Stewart AK, Gatzemeier U, et al. 2001. Adenovirus-mediated wild-type p53 gene transfer in patients receiving chemotherapy for advanced non-small-cell lung cancer: results of a multicenter phase II study. *J. Clin. Oncol.* 19:1750–58

102. Ramesh R, Saeki T, Smyth Templeton N, Ji L, Stephens LC, et al. 2001. Successful treatment of primary and disseminated human lung cancers by systemic delivery of tumor suppressor genes using an improved liposome vector. *Mol. Ther.* 3:337–50

103. DeLeo AB. 1998. p53-based immunotherapy of cancer. *Crit. Rev. Immunol.* 18: 29–35

104. Serrano M, Lee H, Chin L, Cordon-Cardo C, Beach D, DePinho RA. 1996. Role of the INK4a locus in tumor suppression and cell mortality. *Cell* 85:27–37

105. Merlo A, Herman JG, Mao L, Lee DJ, Gabrielson E, et al. 1995. 5′ CpG island methylation is associated with transcriptional silencing of the tumour suppressor p16/CDKN2/MTS1 in human cancers. *Nat. Med.* 1:686–92

106. Esteller M, Sanchez-Cespedes M, Rosell R, Sidransky D, Baylin SB, Herman JG. 1999. Detection of aberrant promoter hypermethylation of tumor suppressor genes in serum DNA from non-small cell lung cancer patients. *Cancer Res.* 59:67–70

107. Kashiwabara K, Oyama T, Sano T, Fukuda T, Nakajima T. 1998. Correlation between methylation status of the p16/CDKN2 gene and the expression of p16 and Rb proteins in primary non-small cell lung cancers. *Int. J. Cancer* 79:215–20

108. Zöchbauer-Müller S, Fong KM, Virmani AK, Geradts J, Gazdar AF, Minna JD. 2001. Aberrant promoter methylation of multiple genes in non-small cell lung cancers. *Cancer Res.* 61:249–55

109. Sherr CJ. 1996. Cancer cell cycles. *Science* 274:1672–77

110. Gazzeri S, Della Valle V, Chaussade L, Brambilla C, Larsen CJ, Brambilla E. 1998. The human p19ARF protein encoded by the beta transcript of the p16INK4a gene is frequently lost in small cell lung cancer. *Cancer Res.* 58:3926–31

111. Yunis JJ, Ramsay N. 1978. Retinoblastoma and sub-band deletion of chromosome 13. *Am. J. Dis. Child.* 132:161–63

112. Ewen ME. 1994. The cell cycle and the retinoblastoma protein family. *Cancer Metastasis Rev.* 13:45–66

113. Reissmann PT, Koga H, Takahashi R, Figlin RA, Holmes EC, et al. 1993. Inactivation of the retinoblastoma susceptibility gene in non-small-cell lung cancer. The Lung Cancer Study Group. *Oncogene* 8:1913–19

114. Cagle PT, el-Naggar AK, Xu HJ, Hu SX, Benedict WF. 1997. Differential retinoblastoma protein expression in neuroendocrine tumors of the lung. Potential diagnostic implications. *Am. J. Pathol.* 150:393–400

115. Dosaka-Akita H, Hu SX, Fujino M, Harada M, Kinoshita I, et al. 1997. Altered retinoblastoma protein expression in non-small cell lung cancer: its synergistic effects with altered ras and p53 protein status on prognosis. *Cancer* 79:1329–37

116. Xu HJ, Quinlan DC, Davidson AG, Hu SX, Summers CL, et al. 1994. Altered retinoblastoma protein expression and

prognosis in early-stage non-small-cell lung carcinoma. *J. Natl. Cancer Inst.* 86: 695–99

117. Shimizu E, Coxon A, Otterson GA, Steinberg SM, Kratzke RA, et al. 1994. RB protein status and clinical correlation from 171 cell lines representing lung cancer, extrapulmonary small cell carcinoma, and mesothelioma. *Oncogene* 9:2441–48

118. Baldi A, Esposito V, De Luca A, Fu Y, Meoli I, et al. 1997. Differential expression of Rb2/p130 and p107 in normal human tissues and in primary lung cancer. *Clin. Cancer Res.* 3:1691–97

119. Claudio PP, Stiegler P, Howard CM, Bellan C, Minimo C, et al. 2001. RB2/p130 gene-enhanced expression down-regulates vascular endothelial growth factor expression and inhibits angiogenesis in vivo. *Cancer Res.* 61:462–68

120. Fearnhead NS, Britton MP, Bodmer WF. 2001. The abc of apc. *Hum. Mol. Genet.* 10:721–33

121. Virmani AK, Rathi A, Sathyanarayana UG, Padar A, Huang CX, et al. 2001. Aberrant methylation of the adenomatous polyposis coli (APC) gene promoter 1A in breast and lung carcinomas. *Clin. Cancer Res.* 7:1998–2004

122. Forgacs E, Biesterveld EJ, Sekido Y, Fong KM, Muneer S, et al. 1998. Mutation analysis of the PTEN/MMAC1 gene in lung cancer. *Oncogene* 17:1557–65

123. Baylin SB, Herman JG, Graff JR, Vertino PM, Issa JP. 1998. Alterations in DNA methylation: a fundamental aspect of neoplasia. *Adv. Cancer Res.* 72:141–96

124. Sato M, Mori Y, Sakurada A, Fujimura S, Horii A. 1998. The H-cadherin (CDH13) gene is inactivated in human lung cancer. *Hum. Genet.* 103:96–101

125. Toyooka KO, Toyooka S, Virmani AK, Sathyanarayana UG, Euhus DM, et al. 2001. Loss of expression and aberrant methylation of the CDH13 (H-cadherin) gene in breast and lung carcinomas. *Cancer Res.* 61:4556–60

126. Bachman KE, Herman JG, Corn PG, Merlo A, Costello JF, et al. 1999. Methylation-associated silencing of the tissue inhibitor of metalloproteinase-3 gene suggests a suppressor role in kidney, brain, and other human cancers. *Cancer Res.* 59:798–802

127. Esteller M, Corn PG, Baylin SB, Herman JG. 2001. A gene hypermethylation profile of human cancer. *Cancer Res.* 61: 3225–29

128. Tang X, Khuri FR, Lee JJ, Kemp BL, Liu D, et al. 2000. Hypermethylation of the death-associated protein (DAP) kinase promoter and aggressiveness in stage I non-small-cell lung cancer. *J. Natl. Cancer Inst.* 92:1511–16

129. Belinsky SA, Nikula KJ, Palmisano WA, Michels R, Saccomanno G, et al. 1998. Aberrant methylation of p16 (INK4a) is an early event in lung cancer and a potential biomarker for early diagnosis. *Proc. Natl. Acad. Sci. USA* 95:11891–96

130. Palmisano WA, Divine KK, Saccomanno G, Gilliland FD, Baylin SB, et al. 2000. Predicting lung cancer by detecting aberrant promoter methylation in sputum. *Cancer Res.* 60:5954–58

131. Momparler RL, Eliopoulos N, Ayoub J. 2000. Evaluation of an inhibitor of DNA methylation, 5-aza-2'-deoxycytidine, for the treatment of lung cancer and the future role of gene therapy. *Adv. Exp. Med. Biol.* 465:433–46

132. Hanahan D, Folkman J. 1996. Patterns and emerging mechanisms of the angiogenic switch during tumorigenesis. *Cell* 86:353–64

133. Folkman J. 1997. Angiogenesis and angiogenesis inhibition: an overview. *Exs* 79:1–8

134. Veikkola T, Alitalo K. 1999. VEGFs, receptors and angiogenesis. *Semin. Cancer Biol.* 9:211–20

135. Rak J, Filmus J, Finkenzeller G, Grugel S, Marme D, Kerbel RS. 1995. Oncogenes as inducers of tumor angiogenesis. *Cancer Metastasis Rev.* 14:263–77

136. Chiarugi V, Magnelli L, Gallo O. 1998.

Cox-2, iNOS and p53 as play-makers of tumor angiogenesis. *Int. J. Mol. Med.* 2:715–19

137. O'Byrne KJ, Koukourakis MI, Giatromanolaki A, Cox G, Turley H, et al. 2000. Vascular endothelial growth factor, platelet-derived endothelial cell growth factor and angiogenesis in non-small-cell lung cancer. *Br. J. Cancer* 82:1427–32

138. Brekken RA, Overholser JP, Stastny VA, Waltenberger J, Minna JD, Thorpe PE. 2000. Selective inhibition of vascular endothelial growth factor (VEGF) receptor 2 (KDR/Flk-1) activity by a monoclonal anti-VEGF antibody blocks tumor growth in mice. *Cancer Res.* 60:5117–24

139. Gabrilovich DI, Chen HL, Girgis KR, Cunningham HT, Meny GM, et al. 1996. Production of vascular endothelial growth factor by human tumors inhibits the functional maturation of dendritic cells. *Nat. Med.* 2:1096–103

140. Fontanini G, Vignati S, Boldrini L, Chin S, Silvestri V, et al. 1997. Vascular endothelial growth factor is associated with neovascularization and influences progression of non-small cell lung carcinoma. *Clin. Cancer Res.* 3:861–65

141. Koukourakis MI, Giatromanolaki A, Thorpe PE, Brekken RA, Sivridis E, et al. 2000. Vascular endothelial growth factor/KDR activated microvessel density versus CD31 standard microvessel density in non-small cell lung cancer. *Cancer Res.* 60:3088–95

142. Koukourakis MI, Giatromanolaki A, O'Byrne KJ, Comley M, Whitehouse RM, et al. 1997. Platelet-derived endothelial cell growth factor expression correlates with tumour angiogenesis and prognosis in non-small-cell lung cancer. *Br. J. Cancer* 75:477–81

143. Fontanini G, Lucchi M, Vignati S, Mussi A, Ciardiello F, et al. 1997. Angiogenesis as a prognostic indicator of survival in non-small-cell lung carcinoma: a prospective study. *J. Natl. Cancer Inst.* 89:881–86

144. Gazdar AF, Minna JD. 1997. Cigarettes, sex, and lung adenocarcinoma. *J. Natl. Cancer Inst.* 89:1563–65

145. Li D, Firozi PF, Wang LE, Bosken CH, Spitz MR, et al. 2001. Sensitivity to DNA damage induced by benzo(a)pyrene diol epoxide and risk of lung cancer: a case-control analysis. *Cancer Res.* 61:1445–50

146. Bennett WP, Colby TV, Travis WD, Borkowski A, Jones RT, et al. 1993. p53 protein accumulates frequently in early bronchial neoplasia. *Cancer Res.* 53: 4817–22

147. Westra WH, Baas IO, Hruban RH, Askin FB, Wilson K, et al. 1996. K-ras oncogene activation in atypical alveolar hyperplasias of the human lung. *Cancer Res.* 56:2224–28

148. Kishimoto Y, Sugio K, Hung JY, Virmani AK, McIntire DD, et al. 1995. Allele-specific loss in chromosome 9p loci in preneoplastic lesions accompanying non-small-cell lung cancers. *J. Natl. Cancer Inst.* 87:1224–29

149. Wistuba II, Behrens C, Virmani AK, Milchgrub S, Syed S, et al. 1999. Allelic losses at chromosome 8p21–23 are early and frequent events in the pathogenesis of lung cancer. *Cancer Res.* 59:1973–79

150. Allan JM, Hardie LJ, Briggs JA, Davidson LA, Watson JP, et al. 2001. Genetic alterations in bronchial mucosa and plasma DNA from individuals at high risk of lung cancer. *Int. J. Cancer* 91:359–65

151. Kersting M, Friedl C, Kraus A, Behn M, Pankow W, Schuermann M. 2000. Differential frequencies of p16(INK4a) promoter hypermethylation, p53 mutation, and K-ras mutation in exfoliative material mark the development of lung cancer in symptomatic chronic smokers. *J. Clin. Oncol.* 18:3221–29

152. Ahrendt SA, Chow JT, Xu LH, Yang SC, Eisenberger CF, et al. 1999. Molecular detection of tumor cells in bronchoalveolar lavage fluid from patients with early stage lung cancer. *J. Natl. Cancer Inst.* 91:332–39

153. Fliss MS, Usadel H, Caballero OL, Wu L,

Buta MR, et al. 2000. Facile detection of mitochondrial DNA mutations in tumors and bodily fluids. *Science* 287:2017–19

154. Liloglou T, Maloney P, Xinarianos G, Hulbert M, Walshaw MJ, et al. 2001. Cancer-specific genomic instability in bronchial lavage: a molecular tool for lung cancer detection. *Cancer Res.* 61:1624–28

155. Ahrendt SA, Decker PA, Doffek K, Wang B, Xu L, et al. 2000. Microsatellite instability at selected tetranucleotide repeats is associated with p53 mutations in non-small cell lung cancer. *Cancer Res.* 60: 2488–91

156. Hiyama K, Hiyama E, Ishioka S, Yamakido M, Inai K, et al. 1995. Telomerase activity in small-cell and non-small-cell lung cancers. *J. Natl. Cancer Inst.* 87: 895–902

157. Albanell J, Lonardo F, Rusch V, Engelhardt M, Langenfeld J, et al. 1997. High telomerase activity in primary lung cancers: association with increased cell proliferation rates and advanced pathologic stage. *J. Natl. Cancer Inst.* 89:1609–15

158. Hahn WC, Meyerson M. 2001. Telomerase activation, cellular immortalization and cancer. *Ann. Med.* 33:123–29

159. Brambilla EM, Lantuejoul S, Sturm N. 2000. Divergent differentiation in neuroendocrine lung tumors. *Semin. Diagn. Pathol.* 17:138–48

160. Wistuba II, Behrens C, Milchgrub S, Virmani AK, Jagirdar J, et al. 1998. Comparison of molecular changes in lung cancers in HIV-positive and HIV-indeterminate subjects. *J. Am. Med. Assoc.* 279:1554–59

161. Cheng YW, Chiou HL, Sheu GT, Hsieh LL, Chen JT, et al. 2001. The association of human papillomavirus 16/18 infection with lung cancer among nonsmoking Taiwanese women. *Cancer Res.* 61:2799–803

162. McGlennen RC. 2000. Human papillomavirus oncogenesis. *Clin. Lab. Med.* 20:383–406

163. Mulatero C, Surentheran T, Breuer J, Rudd RM. 1999. Simian virus 40 and human pleural mesothelioma. *Thorax* 54: 60–61

164. Pepper C, Jasani B, Navabi H, Wynford-Thomas D, Gibbs AR. 1996. Simian virus 40 large T antigen (SV40LTAg) primer specific DNA amplification in human pleural mesothelioma tissue. *Thorax* 51:1074–76

165. Shivapurkar N, Wiethege T, Wistuba II, Milchgrub S, Muller KM, Gazdar AF. 2000. Presence of simian virus 40 sequences in malignant pleural, peritoneal and noninvasive mesotheliomas. *Int. J. Cancer* 85:743–45

166. Weiss R, Giordano A, Furth P, DeCaprio J, Pipas J, et al. 1998. SV40 as an oncogenic virus and possible human pathogen. *Dev. Biol. Stand.* 94:355–60, 69–82

167. Rai SK, Duh FM, Vigdorovich V, Danilkovitch-Miagkova A, Lerman MI, Miller AD. 2001. Candidate tumor suppressor HYAL2 is a glycosylphosphatidylinositol (GPI)-anchored cell-surface receptor for jaagsiekte sheep retrovirus, the envelope protein of which mediates oncogenic transformation. *Proc. Natl. Acad. Sci. USA* 98:4443–48

168. Johnson BE. 1998. Second lung cancers in patients after treatment for an initial lung cancer. *J. Natl. Cancer Inst.* 90:1335–45

169. Behrens C, Travis LB, Wistuba II, Davis S, Maitra A, et al. 2000. Molecular changes in second primary lung and breast cancers after therapy for Hodgkin's disease. *Cancer Epidemiol. Biomarkers Prev.* 9:1027–35

170. D'Amico TA, Aloia TA, Moore MB, Herndon JE, Brooks KR, et al. 2000. Molecular biologic substaging of stage I lung cancer according to gender and histology. *Ann. Thoracic Surg.* 69:882–86

171. Sabel MS, Yamada M, Kawaguchi Y, Chen FA, Takita H, Bankert RB. 2000. CD40 expression on human lung cancer correlates with metastatic spread. *Cancer Immunol. Immunother.* 49:101–8

172. Mishina T, Dosaka-Akita H, Hommura F, Nishi M, Kojima T, et al. 2000. Cyclin E expression, a potential prognostic marker for non-small cell lung cancers. *Clin. Cancer Res.* 6:11–16

173. Muller-Tidow C, Metzger R, Kugler K, Diederichs S, Idos G, et al. 2001. Cyclin E is the only cyclin-dependent kinase 2-associated cyclin that predicts metastasis and survival in early stage non-small cell lung cancer. *Cancer Res.* 61:647–53

Annu. Rev. Physiol. 2002. 64:709–48

β-Defensins in Lung Host Defense

Brian C. Schutte and Paul B. McCray, Jr.

Department of Pediatrics, Genetics Ph.D. Program, University of Iowa College of Medicine, Iowa City, Iowa; e-mail: paul-mccray@uiowa.edu

Key Words   innate defense, epithelia, antimicrobial peptides, mucosa

■ **Abstract**   Host defenses at the mucosal surface of the airways evolved to present many layers of protection against inhaled microbes. Normally, the intrapulmonary airways are sterile. Airway secretions contain numerous factors with antimicrobial activity that contribute to innate defenses. Many protein and peptide components exert bacteriostatic or bacteriocidal effects against a wide variety of organisms and may act in synergistic or additive combinations. The β-defensins are a relatively recently described family of peptide antimicrobials that are widely expressed at mucosal surfaces, including airway and submucosal gland epithelia. These small cationic peptides are products of individual genes that exhibit broad-spectrum activity against bacteria, fungi, and some enveloped viruses. Their expression in airway epithelia may be constitutive or inducible by bacterial products or pro-inflammatory cytokines. β-defensins also act as chemokines for adaptive immune cells, including immature dendritic cells and T cells via the CCR6 receptor, and provide a link between innate and adaptive immunity. Alterations in the function of the β-defensins may contribute to disease states. Here we review much of the biology of the β-defensins, including gene discovery, genomic organization, molecular structure, regulation of expression, and function.

INTRODUCTION

Alexander Fleming was the first to make the observation that respiratory secretions possess endogenous antibacterial activity (1, 2). In the years since his discovery of lysozyme, a large body of work has revealed the complexity and intricacies of pulmonary host defense. By virtue of the lung's large surface area and daily encounters with microbes and other inhaled and aspirated particulates, many strategies evolved to counter infectious challenges, avoid chronic inflammatory states, and preserve the sterility of the intrapulmonary airways and alveoli. Among the front lines of these defenses in epithelia and secretions are the β-defensins. These β-sheet antimicrobial peptides comprise an important facet of mucosal host defense (3–5). Their place in innate immunity is an evolving story, as genes encoding the human β-defensins continue to be discovered, and their role in lung host defense continues to be investigated. In this chapter, we review several aspects of the β-defensin gene family, emphasizing the human gene family, their peptide products, and their functions in pulmonary mucosal immunity.

0066-4278/02/0315-0709$14.00

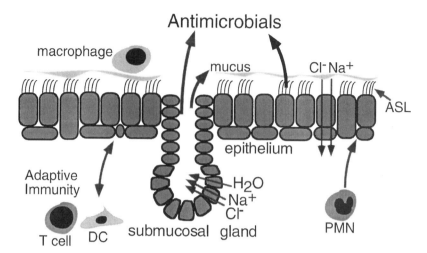

Figure 1   Schematic representation of host defense at the epithelial surface of the lung. Components of the innate and adaptive immune systems are indicated diagramatically. In addition to secreting and modifying the composition of airway surface liquid (ASL), surface and submucosal gland epithelia secrete mucus and many antimicrobials. DC, dendritic cell; PMN, polymorphonuclear neutrophil.

Innate Host Defense in the Lung

The airway epithelium directly interacts with the environment in a dynamic fashion. Because of this interplay between the host and environment, many systems evolved to clear or inactivate the pathogens encountered (see Figure 1). These defense systems are numerous and include components of both innate and acquired immunity. The acquired, specific, or adaptive immune system is made up of elements of the cellular immune system, predominantly T and B lymphocytes. This arm of host defense arose later in evolution and is characterized by a requirement for previous exposure and a time lag of days to weeks before it is maximally activated (6). The clonal expansion of T and B cells helps develop and maintain antigen-specific, cell-mediated, and humoral responses. These responses are features of immunologic memory, differentiating self from non-self, and prevent the unwanted and possibly damaging consequences of autoimmune and other chronic inflammatory diseases that may have profound consequences in the lung. Remarkably, these are uncommon pulmonary problems, a testimony to the overall effectiveness of the system.

Innate immunity at mucosal surfaces is a more primitive defense system, thought to have arisen more than half a billion years ago (6–9). Found in all multicellular organisms from plants to insects to vertebrates, this system complements adaptive immunity by its "ever ready" nature, as many of its components are continuously present and require no previous encounter or memory for their activity. The production of innate immune factors may also be induced by an infectious or inflammatory stimulus (6, 10, 11). In the lung, the innate immune system consists of many components with overlapping, and in some cases redundant, functions. The mucocilary apparatus forms one important line of defense by trapping inhaled particulate matter that is then swept toward the mouth by cilia. The airways and alveoli are also inhabited by professional phagocytic cells, most notably the alveolar macrophage, which can engulf inhaled or aspirated microbes (12). Other immune effector cells such as dendritic cells or neutrophils are present nearby in the submucosal space or vascular compartments (Figure 1).

The epithelial cells lining the conducting airways, submucosal glands, and alveoli also play a vital role in lung health. In addition to serving as an important physical barrier, two major functions of these cells are the production and modification of airway surface liquid and the manufacture and secretion of a number of host defense factors. Airway surface liquid (ASL) composition is complex and as yet not fully characterized (13). Two-dimensional gel electrophoresis analysis of bronchoalveolar lavage (BAL) and nasal secretions reveals that ASL is a complex mixture of many hundred proteins, peptides, and fragments (14–16) (also see http://www.umh.ac.be/~biochim/proteomic.htm). It is secreted and modified by submucosal gland and surface epithelial cells of the lung and also contains plasma transudates (17). The secretion of ASL is osmotically coupled to the active transport of ions across the epithelium, and modifications of its composition occur in response to specific neuro-humoral signals (18). Submucosal glands also contribute to the production of macromolecules including mucins. ASL thus provides an environment for ciliary function and hydrates the mucins contributing to the mucus blanket. Presumably, the normal ASL composition provides an optimal microenvironment for the function of the microbicidal compounds it also contains.

Antimicrobial Factors of Airway Surface Liquid

In addition to transporting liquid and electrolytes, airway epithelia produce several proteins and peptides with antimicrobial activity (listed below in Table 1). It is difficult to accurately measure the concentrations of all ASL components in vivo owing to the inherent problems in sampling this microenvironment. Many of these factors are produced in more abundance by submucosal gland epithelia, but some are also produced by the surface epithelia. These factors act in a broad-spectrum fashion, in some cases exerting antimicrobial effects against bacteria, fungi, and viruses. The proteins of highest concentration in ASL are lysozyme, lactoferrin, and secretory leukocyte proteinase inhibitor (SLPI) (19–21). An interesting feature of most of the antimicrobial components of ASL is the dependence of their activity on the ionic strength of the solution in which they are assayed. As the NaCl or divalent cation concentrations increase, antimicrobial activity decreases dramatically (20).

Airway secretions contain many of the same bacteriostatic and microbicidal products found in mucosal secretions throughout the body. The cationic,

TABLE 1   Antimicrobial products in ASL (partial listing[a])

Product	Relative concentration	Source
Lysozyme	μg-mg/ml	epithelia, neutrophils
SLPI	μg/ml	epithelia, macrophages
Lactoferrin	μg/ml	epithelia, neutrophils
IgA secretory component	μg/ml	epithelia
Phospholipase A2	μg-mg/ml	epithelia, neutrophils
SP-A, SP-D	ng-μg/ml	epithelia
Defensins ($\alpha-$, $\beta-$)	ng-mg/ml	neutrophils, epithelia
Cathelicidins (LL37)	?	neutrophils, epithelia
BPI	?	neutrophils
Peroxidases	?	neutrophils, epithelia
Anionic peptide	0.8–1.3 mM	epithelia

[a]Note: in some cases these values are derived from BAL specimens where concentrations are diluted ~100–1000-fold.

bacteriolytic protein lysozyme is a specific product of serous cells, neutrophils, and other mononuclear cells and is found in substantial quantities in airway secretions (21, 22). Lysozyme exerts its antimicrobial activity by enzymatically cleaving glycosidic bonds of the bacterial membrane peptidoglycans. SLPI, a low-molecular-weight, cationic proteinase inhibitor, is also produced by epithelia and macrophages and contains most of the anti-neutrophil proteinase activity in the airways (23, 24). Interestingly, the N-terminal domain of the protein also exhibits antimicrobial activity. Lactoferrin is found in airway secretions and is also produced by epithelia and neutrophils in a pattern similar to lysozyme (25–27). It is a cationic, iron-binding protein that may act by inhibiting growth of iron-requiring bacteria. Lactoferrin also has direct microbicidal properties through its N-terminal cationic fragment (28). Secretory type II phospholipase A2 (PLA2) is produced by both epithelia and blood-derived cells. PLA2 enzymatically targets the degradation of phospholipids in bacterial cell walls (29). There is evidence that the expression of lysozyme, lactoferrin, and PLA2 are all increased with airway inflammation. The epithelium also secretes the secretory component of IgA which is required for the translocation of IgA from the basolateral to apical membrane where it is released (30). Bacterial permeability-inducing protein (BPI) is a neutrophil product that acts by disrupting gram-negative bacteria outer membranes (31). It can also potentiate the activity of proteins such as PLA2. Other epithelial products with antimicrobial properties include surfactant proteins A (SP-A) (32, 33) and D (SP-D) (34, 35), anionic peptides (36), peroxidases, and proline-rich proteins (37). In addition to these protein components, peptide antimicrobials are also produced, including defensins and cathelicidins.

Antimicrobial Peptides and Lung Host Defense

The peptide components of pulmonary host defense include members of the defensin and cathelicidin classes of antimicrobials. Although the focus of the review is the β-defensins, mention of the cathelicidins and α- and θ-defensins is warranted. The cathelicidin family of antimicrobial peptides is widely distributed in animals (38, 39). They contain an N-terminal domain called cathelin, which contains an SH3 domain involved in protein binding (40), and a C-terminal domain with antimicrobial activity (39).

Although the cathelin domains are highly conserved across species lines, the C-terminal antimicrobial domains are structurally diverse. The first cathelicidin described was cow bactenecin (38). The mature peptide has broad-spectrum bactericidal activity. Cathelicidins have since been identified in other species including humans (FALL39/LL37, hCAP18) (41), mice (mCRAMP) (42), rats (rCRAMP), rabbits (38), horses (43), and sheep (SMAP29 and SMAP34) (44–46). These antimicrobial peptides uniformly adopt an amphipathic α-helical structure on binding to polyanionic moieties (lipopolysaccharides or teichoic acids) of the bacterial cell wall. Incorporation of the hydrophobic face of the helix in the lipid bilayer is generally believed to promote distortion and ultimately disruption of the membrane, thus leading to cell death. In contrast to the defensins, the antimicrobial activity of the cathelicidins is generally preserved over a wide range of ionic strengths. In humans and chimpanzees there is evidence supporting the expression of LL37 in neutrophils and in epithelia, including the airways (47–49).

Three families of defensin peptides have been isolated from vertebrates. The α-defensins are two- or three-exon, single-gene products of neutrophils, macrophages, and Paneth cells of the intestine (31). In humans, the α-defensins are expressed in neutrophils (DEFA1, DEFA2, DEFA3, DEFA4) and Paneth cells and other epithelia including the epithelium of the vagina, endocervix, and fallopian tubes (DEFA5, DEFA6). These cationic peptides are distinct from the β-defensins based on the pattern of pairing of their three disulfide bonds (31). They make up as much as 50% of the protein content in neutrophil azurophilic granules. Following engulfment of microbes by neutrophils, α-defensins are released into the phagolysosome where they exert their antimicrobial effects. In Paneth cells, DEFA5 and DEFA6 are present in cytoplasmic granules (50). The β-defensins are found in mammalian granulocytes, epithelia, and mucosal secretions (5). A third family of peptides, termed θ-defensins, has only been identified in leukocytes of rhesus macaques (51).

Epithelial β-Defensins—Pulmonary Expression and Function

β-defensins are cationic peptides with broad-spectrum antimicrobial activity that are products of epithelia and leukocytes (31). These two-exon, single-gene products are expressed at epithelial surfaces and secreted at sites including the skin (52), cornea (53), tongue (54, 55), gingiva (54, 56), salivary glands (54), esophagus (55), intestine (57), kidney (58, 59), urogenital tract (58), and respiratory epithelium

(60–62). To date, three β-defensins of epithelial origin, HBD-1 (63), HBD-2 (52), and HBD-3 (64, 65), have been identified and characterized in humans. There is evidence that all these gene products are expressed in airway epithelia. The properties of vertebrate β-defensins are compiled in Table 2.

DISCOVERY OF β-DEFENSIN GENES AND GENE PRODUCTS

Bovine tracheal antimicrobial peptide (TAP) was the first pulmonary epithelial β-defensin discovered and investigated in detail (66, 67). In what became a paradigm for β-defensin discovery, TAP was isolated using a combination of acid extraction, gel filtration, and reverse-phase HPLC. These purification steps took full advantage of the characteristically small (<5 kDa) and cationic defensin proteins. Amino acid and cDNA sequence analysis showed that this gene product contained the familiar six-cysteine motif of the previously identified mammalian defensins (66) but that the spacing between the cysteine residues was different. Using a similar strategy (68), Selsted and collaborators discovered a group of 13 defensin-like proteins from bovine neutrophils (BNBD). Although these highly cationic peptides shared the six-cysteine motif of defensins, these authors suggested that the BNBD peptides were members of a new family of defensins. They showed that the tridisulfide motif of the BNBD peptides was different from the α-defensins and named this new family of antimicrobial peptides, which includes TAP, the β-defensins.

Since those initial discoveries, the family of vertebrate β-defensins has grown to at least 60 members, 43 from mammals, 6 from birds, and 11 β-defensin-like proteins from marsupials and reptiles (Table 2). Although half of these proteins were discovered through protein purification approaches, new paradigms for defensin gene discovery have emerged that rely on nucleic acid homology. These new approaches include PCR amplification, nucleic acid hybridization, and similarity searches of cDNA and genomic DNA sequence databases. Because of their small size, β-defensin genes provide excellent targets for PCR amplification. Diamond and colleagues not only led the way in isolating TAP, the first β-defensin gene product, but in the same study, they were the first to use PCR to clone a β-defensin gene (66). Subsequent studies used PCR to discover 10 vertebrate β-defensin genes derived from chicken (69), goat (70), mouse (55), pig (72), rat (73), rhesus monkey (49), sheep (74), and turkey (69). The diversity of this group of species implies that PCR is a robust approach for identifying β-defensin homologs. A gene ortholog provides a common function in different species; consequently, the structure of the ortholog is more conserved than gene paralogs, structurally similar genes that provide a related function within a species. In 10 of 12 cases, the primers used to amplify the β-defensin gene were derived from the cDNA sequence of a different species. Zhao and coworkers (69) provide the only report of a β-defensin gene discovered using PCR primers designed from a different β-defensin gene from the same species. A similar trend has been observed from studies that use direct nucleic

TABLE 2 Properties of vertebrate β-defensin genes and their gene products

Species	Gene	Method of discovery[a]	Swiss-Prot	Genbank	Cysteine spacing[b]	H+K+R[c]	Exons (intron size kb)	Chromosome (method)	Expression inducible (transcription binding sites)[d]	References
Bovine	BNDB4[e]	Protein	P46162	U36200	6 4 9 6	8	2 (1.5)	27 (F, SCH)	N, M	(168, 105, 190, 191)
	EBD	Genomic clone	O02775	AF000362	6 4 9 6	7	2 (1.5)	27 (F, SCH)	EE I (NF-IL-6, no NFκB)	(105, 75)
	LAP	Protein	Q28880	S76279	6 4 9 6	10	2		To I (NFκB)	(114, 115)
	TAP	Protein	P25068	L13373	6 4 9 6	9	2 (1.5)	27 (F, SCH)	Tr I (NFκB)	(105, 190, 115, 66, 67, 116)
Chicken	GAL1	Protein	AAC36051	AF033335	6 4 9 5	9			M, L	(192, 193)
	GAL2	Protein	AAC36052	AF033336	4 4 9 5	8			M, L	(192, 193)
	GAL3	PCR	AAG09213	AF181952	6 4 9 5	7			To, Tr	(69)
Chimp	BD1	cDNA	AAF04110	AF188607	6 4 9 6	8			AE, S	(49)
	BD2	cDNA	Q9TT12	AF209855	6 4 9 6	8				
	EP2C	cDNA	AAF87720	AF283553	6 3 9 6	7			Ep	(78)
	EP2D	cDNA	AAF87721	AF263554	6 4 9 6	8			Ep	(78)
	EP2E	cDNA	AAF87722	AF263555	6 4 9 6	8			Ep	(78)
Goat	BD1	PCR	O97946	Y17679	6 4 9 6	13			AE	(70)
	BD2	PCR	O97942	AJ009877	6 4 9 6	12			EE	(70)

(*Continued*)

TABLE 2 (*Continued*)

Species	Gene	Method of discovery[a]	Swiss-Prot	Genbank	Cysteine spacing[b]	H+K+R[c]	Exons (intron size kb)	Chromosome (method)	Expression inducible (transcription binding sites)[d]	References
Human	DEFB1	Protein	Q09753	U50930-1	6 4 9 6	8	2 (7 kb)	8p22-p23 (F)	K, AE not I	(63, 61, 106, 62, 58, 109)
	DEFB2	Protein	O15263	Z71389	6 4 9 6	8	2 (1.6 kb)	8p22-p23 (F, RH)	S, L, Tr I (NFκB)	(52, 107, 108, 98, 54, 119, 194)
	DEFB3	HTGS protein	AAF73853	AF217245	6 4 9 6	13	2 (0.9 kb)	8p22-p23 (HGP)	S, Tr I (NF-IL-6, but no NFκB)	(65, 64, 77)
	EP2C	cDNA	AAG21880	AY005129	6 3 9 6	6	3 (0.6, 1.2)	8p22-p23 (HGP)	Ep I (androgens)	(94)
	EP2E	cDNA	AAG21881	AY005129	6 4 9 6	11	2 (1.4)	8p22-p23 (HGP)	Ep I (androgens)	(94)
	EP2D HE2B1	cDNA	AAF37187 AAG21882	AF168617 AY005129	6 4 9 6	8	3 (0.6, 12)	8p22-p23 (HGP)	Ep I (androgens)	(65, 94, 95)
Mouse	Defb1	EST	P56386	AF003524	6 4 9 6	9	2 (15 kb)	8 (L, F)	K	(84, 97, 99)
	Defb2	Genomic	P82020	AJ011800	6 4 9 6	8	2 (4 kb)	8 (RH)	K, AE I	(55, 76)
	Defb3	PCR	Q9WTL0	AF092929	6 3 9 6	10	2 (1.7 kb)	8A4 (F, RH)	To, Es, Tr I (NFκB)	(55, 100)
	Defb4	PCR	P82019	AF155882	6 3 9 6	6	2 (2.4 kb)	8 (RH)	To, Tr, Es not I No NFκB	(55)
	Defb5	EST	AAG49340	AF318068	6 4 9 6	5	2			
Pig	BD1	PCR	O62697	AF031666	6 4 9 6	9	2 (1.5)	15q14-q15.1 (F)	To, Tr not I (no NFκB or NF-IL-6)	(72, 101, 111)
Platypus	DLP2[f]	Protein	1D6BA		6 7 7 6	7			VG	(90)
Rat	RBD1	EST	O89117	X89820	6 4 9 6	9		16 (RH)	K	(86)
	RBD2	EST	O88514		6 3 9 6	8		16 (RH)	L	(86)
	Bin1b	PCR		AF217088	6 4 9 6	9	2 (1.3 kb)		Ep I	(73)

Species	Name	cDNA	Protein	Genomic	Cys spacing[b]	[c]		VG[d]	Ref
Rattlesnake, trop	CRO1[g]		P24331		6 6 11 5	14			(195)
Rhesus	BD1	PCR	O18794	AF014016	6 4 9 6	9		Tr	(49)
	BD2	PCR	AAK26259	AF288286	6 4 9 6	8		Tr	(49)
Sheep	SBD1	PCR	O19038	U75250	6 4 9 6	14	2 (1.4)	Tr	(74, 110)
	SBD2	PCR	O19039	U75251	6 4 9 6	8	2	GI	(74, 110)
Turkey	THP1	Protein	AAC36053	AF033337	6 4 9 5	8			(192, 193, 196)
	THP2	Protein	AAC36054	AF033338	4 4 9 5	10	26 (CMP)		(192, 193, 196)
	THP3	PCR	AAG09213	AF181953	6 4 9 5	6	26	Tr	(69)

[a] β-defensins were discovered by protein purification (Protein), identification and sequencing of cDNA (cDNA) or genomic (Genomic) clones, PCR amplification (PCR), or sequence similarity searches of expressed sequence tag (EST) or high-throughput genomic sequence (HTGS) databases.

[b] The number of amino acids that separate the cysteine residues (C1-C2, C2-C3, C3-C4, C4-C5) in the six-cysteine β-defensin domain. A missing number indicates the absence of a cysteine residue at one of the characteristic positions.

[c] Total number of positively charged residues in the known or predicted mature β-defensin peptide. For this calculation, the only residues counted were between seven amino acids before the first cysteine (C1) and seven amino acids beyond the last cysteine (C6).

[d] Tissues or cell types that significantly express the indicated β-defensin genes are neutrophils (N), macrophages (M), lung (L), tongue (To), trachea (Tr), enteric epithelia (EE), airway epithelia (AE), kidney (K), venom gland (VG), intestine (GI), esophogus (Es), skin (S), epididymis (Ep). The list of tissues for each gene is not inclusive. β-defensin genes whose expression has been demonstrated to be inducible, and whether they contain the indicated transcription-binding site, are indicated (I).

[e] The BNDB4 gene is representative of the 14 known defensin proteins isolated from bovine neutrophils. The related proteins and accession numbers are BNDB1 (P46159), BNDB2 (P46160), BNDB3 (P46161), BNDB5 (P46163), BNDB6 (P46164), BNDB7 (P46165), BNDB8 (P46166), BNDB9 (P46167), BNDB10 (P46168), BNDB11 (P46169), BNDB12 (P46170), BNDB13 (P46171), and BDC7 (O18815).

[f] The DLP2 gene is representative of the 4 known defensin-like proteins isolated from male platypus venom gland. The other proteins and accession numbers are DLP1 (1B8WA), DLP3 (Torres et al. 1999) and DLP4 (Torres et al. 1999).

[g] The CRO1 gene is representative of the 7 known defensin-like proteins isolated from snake venom. The related proteins and accession numbers are CRO2 (P24332), CRO3 (P24333), CRO (AAC02995), CRO (AAC06241), an unnamed protein (I00927) and myotoxin A (JC5324).

acid hybridization screens for discovering β-defensin genes. Although two groups successfully used nucleic acid hybridization at low stringency to identify novel β-defensin paralogs in cow (75) and mouse (76), more recent studies identified chimp β-defensin genes using cDNA probes derived from their human orthologs (77, 78).

Another approach for identifying novel β-defensin genes has been to perform sequence similarity searches against sequence databases for proteins, cDNA, and genomic DNA. This approach is gaining momentum rapidly as new sequence information from many species pours into the genetic repositories from large genome centers (79–83). Searches of the expressed sequence tag database (dbEST), a database composed of partial cDNA sequences, led to the discovery of two mouse β-defensin genes (84, 85) and two rat β-defensin genes (86). This is certainly a robust approach for finding β-defensin genes that have moderate sequence similarity and are expressed constitutively in a species that is being studied by large genomic-based approaches. However, many β-defensin genes are expressed only as a result of induction by infection or inflammation (see below). A genomic DNA-based approach is an alternative, nucleic acid method that does not rely on expression. As discussed below, the defensin genes are clustered in a narrow region of the genome. By determining the sequence of this region and then analyzing it, it should be possible to identify novel defensin genes. Jia and collaborators (65) used this approach to identify a new human β-defensin gene *DEFB3*. This gene is not represented in dbEST, and expression of the *DEFB3* gene can only be detected in lung tissue following induction with IL-1. This gene was also discovered by Schroder and colleagues (64) by isolating the gene product from psoriatic scales that contained 10–30 times more DEFB3 protein than normal skin. The *DEFB3* gene is representative of many β-defensin genes that would be difficult to discover using methods that rely on expression from normal tissues.

Are additional airway defensin genes waiting to be discovered, and if so, how will they be found? It is certain that the list of genes in Table 2 will grow longer if only through the identification of homologs through PCR-based and genomic studies of new species. For example, in mice, Southern blot analysis suggests that there is a large β-defensin gene family (55), and the mouse genome is currently being sequenced (81). However, as discussed above, these nucleic acid–based methods are not suited for discovering new classes of defensin genes or new subclasses of β-defensin genes. They rely heavily on sequence similarity, but a hallmark of the β-defensin gene family is the diversity of their primary structures. If a genomics-based approach is to be successful in identifying new β-defensin genes, additional analysis tools must be employed that rely less on the primary structure, e.g., the nucleotide or amino acid sequence, and more on the secondary or tertiary structure of the predicted protein. The solution and crystal structure for several β-defensin gene products were recently determined and indicate common secondary and tertiary structures despite the diversity of the primary sequence (87–90). If a new wave of discovery occurs in the future, it may arise from computer-based approaches that can test entire genomes for sequences that encode proteins predicted to fold and create the characteristic tridisulfide linkages of β-defensin gene products.

STRUCTURE OF THE VERTEBRATE β-DEFENSINS

The primary structure of each β-defensin gene product is characterized by small size, a six-cysteine motif, high cationic charge, and exquisite diversity beyond these features. The most characteristic feature of defensin proteins is their six-cysteine motif. Both α-defensin and β-defensin proteins have a six-cysteine motif that forms a network of three disulfide bonds. However, the spacing between the cysteines and the connections between the three disulfide bonds differ between the α- and β-defensin proteins (68). The three disulfide bonds in the β-defensin proteins are between C1-C5, C2-C4, and C3-C6 (Figure 2, see color insert), whereas those in the α-defensin proteins are between C1-C6, C2-C4, and C3-C5. The most common spacing between adjacent cysteine residues is 6, 4, 9, 6, 0 (Table 2) and 1, 4, 9, 9, 0, respectively. The spacing between the cysteines in the β-defensin proteins can vary by one or two amino acids except for C5 and C6, located nearest the C terminus. In all known vertebrate β-defensin genes, these two cysteine residues are adjacent to each other.

A second feature of the β-defensin proteins is their small size. Each β-defensin gene encodes a preproprotein that ranges in size from 59 to 80 amino acids with an average size of 65 amino acids. This gene product is then cleaved by an unknown mechanism to create the mature peptide that ranges in size from 36 to 47 amino acids, with an average size of 45 amino acids. The exceptions to these ranges are the EP2/HE2 gene products that contain the β-defensin motif and are expressed in the epididymis (78, 91–95). Using alternative splicing and a secondary promoter, the human *HE2/EP2* gene produces three isoforms that carry the β-defensin motif, EP2C, EP2D, and EP2E (94). The size of the preproproteins is 113, 133, and 80, respectively. The additional length of isoforms EP2C and EP2D is from the insertion of an extra exon between the leader sequence and the six-cysteine motif, and *EP2D* encodes 20 additional amino acids after the adjacent C5 and C6 residues. Although the structure of the EP2E isoform is most similar to other β-defensin genes, it encodes the same 20–amino acid tail as isoform EP2D after the adjacent C5 and C6 residues. The function of these additional amino acids at the C terminus is unknown. However, it is interesting to note that the rat homolog for the *EP2E* gene, *Bin1b* (73), encodes only six amino acids beyond the C5 and C6 residues. This gene possesses antimicrobial activity that is typical of β-defensin proteins (73), suggesting that the additional amino acids of its human and chimpanzee orthologs are not necessary for antibiotic function.

A third feature of β-defensin proteins is the high concentration of cationic residues. The number of positively charged residues (arginine, lysine, histidine) in the mature peptide ranges from 6 to 14 with an average of 9 (Table 2). It has been proposed that the high positive charge density allows the β-defensin peptides to bind and insert into the cellular membrane, where they kill the cell either by forming a pore (96) or by simply permeabilizing the cell through an electrostatic interaction without forming a pore (89). The relationship between the killing activity and the charge density of the β-defensin proteins is supported by the observations that the antimicrobial activity of many β-defensin proteins is salt-sensitive (58, 60, 61, 86, 97–101), possibly because mono- and divalent cations interfere with the binding to the negatively charged bacterial surface. An exception to this rule is the protein encoded by *DEFB3* whose bactericidal activity against *Staphylococcus aureus* is not salt-sensitive at physiologic salt concentrations (64). As noted previously (65), the *DEFB3* gene encodes six more positively charged amino acids than the other two human β-defensin genes, *DEFB1* and *DEFB2*. Future experiments will likely test whether these additional positive charges are related to the salt-insensitive *Staphylococcus* killing activity of this protein.

The final feature of the β-defensin gene products is their diverse primary structure but apparent conservation of tertiary structure. Figure 2 shows an alignment of the amino acid sequence of the six-cysteine domain for a subset of the β-defensin proteins. These few β-defensin proteins were chosen to exemplify the diversity of amino acid sequences of this family. Each protein shown represents a group of related β-defensin proteins, and each group was identified by the multiple sequence alignment of all of the vertebrate β-defensin proteins (Figure 3). Beyond the six cysteines, no single amino acid at a given position is conserved in all 60 members of this protein family. However, some positions are conserved that appear to be important for secondary and tertiary structures and function (see below).

A similar theme of amino acid diversity is observed for the leader domain for the β-defensin proteins (Figure 4, see color insert). As in Figure 2, this subset of β-defensin protein sequences was chosen as representative of the groups identified by the multiple sequence alignment of all the available protein sequences (Figure 5) to highlight their diversity. Again, no single residue is completely conserved, although two general patterns are noticeable. First, like other secreted proteins (insulin, erythropoetin), the leader domain for the β-defensin proteins is highly hydrophobic. Second, a positively charged amino acid is located at position two in all available β-defensin protein sequences. This is remarkable given the hydrophobic nature of leader domains and the paucity of conservation of amino acids in β-defensin proteins. The function of this amino acid is unknown; however, it may provide an excellent marker for identifying the first exon of novel β-defensin genes.

Despite the great diversity of the primary amino acid sequence of the β-defensin proteins, the limited data suggest that the tertiary structure of this protein family is conserved and provides a unifying theme for antimicrobial activity. The solution structure has been determined for the proteins encoded by *BNBD-12* from cow (87), *DEFB2* from human (88), and *DPL1* from platypus (90). The structural core for each of these proteins is a triple-stranded, antiparallel β-sheet, as exemplified for the proteins encoded by *BNBD-12* and *DEFB2* (Figure 6*A,B*, see color insert). The three β-strands are connected by a β-turn and a β-hairpin loop, and the second β-strand also contains a β-bulge. When these structures are folded

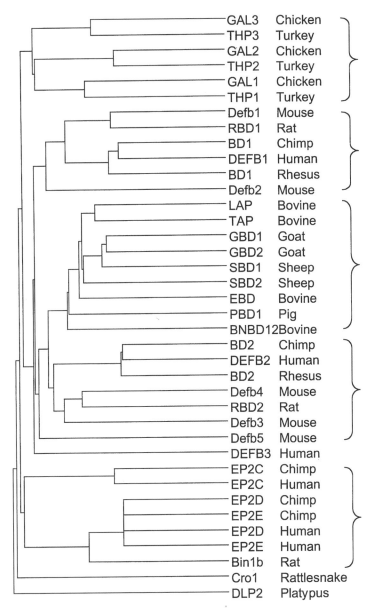

Figure 3 Dendrogram based on the six-cysteine domain of vertebrate β-defensin proteins. Dendrogram was constructed with the computer program PileUp (Genetics Computer Group); all known β-defensin proteins are included, except for 13 proteins from bovine (BNBD1-11, -13, BDC7) with high similarity to BNBD12, three proteins from platypus (DLP2-4) with high similarity to DLP2, and six proteins from snake venom (Cro2-5, myotoxin A and an unnamed protein). Brackets at right indicate most related groups.

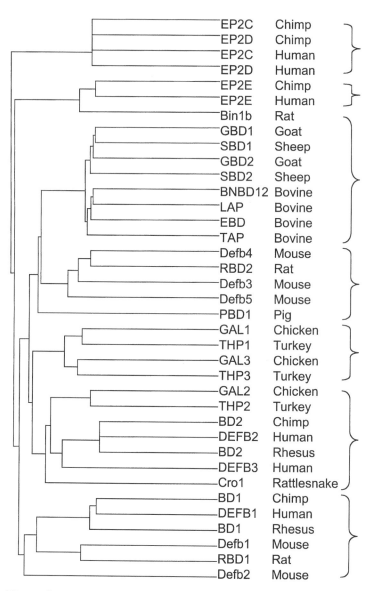

Figure 5 Dendrogram based on the leader domain of vertebrate β-defensin proteins. Dendrogram was constructed as in Figure 3.

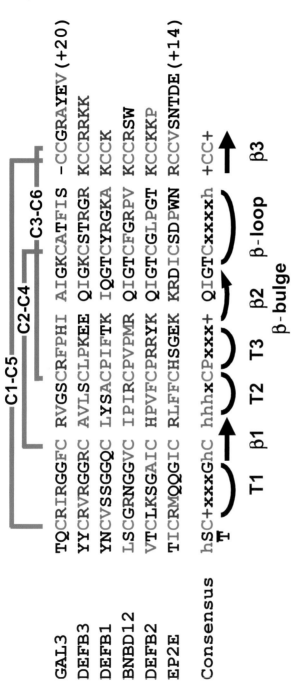

Figure 2 Amino acid alignment of the six-cysteine domains of β-defensin proteins. The β-defensin proteins shown were chosen to highlight the diversity of this protein family. Green brackets represent disulfide linkages between the cysteine residues indicated. Residues are color coded, cysteines (*green*), positively charged (*blue*), and hydrophobic (*red*). The C terminus is truncated by the indicated number of amino acids for GAL3 and EP2E. The consensus sequence was derived from a multiple sequence alignment of all 60 known β-defensin proteins. Specific residues, hydrophobic residues (*red h*) and positively charged residues (*blue +*) are listed if found at that position in more than half of the proteins. The residues involved in the secondary structures are indicated. Secondary structures found in β-defensin proteins include three β-turns (T1, T2, T3), three antiparallel β-sheets (β1, β2, β3), a β-loop; a β-bulge is present in the β2 sheet.

```
EP2C       MRQRLLPSVTSLLLVALLFPGSS
EP2E       MKVFFLFAVLFCLVQTNSGDVPP
TAP        MRLHHLLLALLFLVLSAWSGFTQ
Defb4      MRIHYLLFTFLLVLLSPLAAFTQ
GAL1       MRIVYLLLPFILLLAQGAAGSSQ
DEFB2      MRVLYLLFSFLFIFLMPLPGVFG
DEFB1      MRTSYLLLFTLCLLLSEMASGGN

Consensus  MRhxxLLhhhhhhhhxxxxxx
```

Figure 4 Amino acid alignment of the leader domain of β-defensin proteins. The β-defensin proteins shown were chosen to highlight the diversity of this protein family. Residues are color coded and the consensus sequence was constructed as in Figure 2.

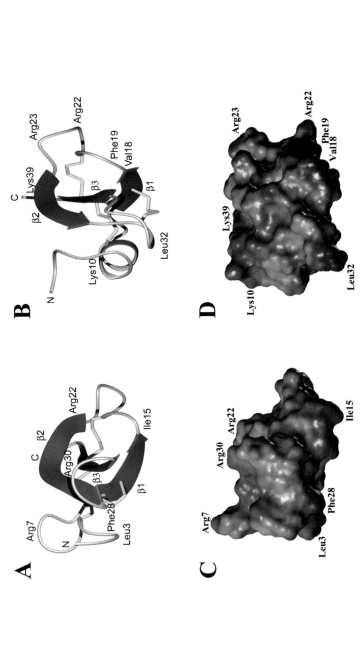

Figure 6 Tertiary structure of BNBD-12 and HBD–2 proteins. Ribbon representation of the lowest energy conformers for BNBD-12 (*A*) and HBD-2 (*B*) proteins. The structures shown are the first from the Protein Data Bank and were generated with the program Molecule Analysis and Molecule Display (MOLMOL) (197). The N and C termini are labeled. The three β-sheets are indicated by blue ribbons and the disulfide bonds by yellow bars. Hydrophobicity map for the lowest energy conformers for the BNBD-12 (*C*) and HBD-2 (*D*) proteins. The hydrophilic areas are shown in blue and the hydrophobic areas are shown in reddish brown (SYBYL 6.5, Tripos, Inc., St Louis, MO). The locations of key hydrophilic and hydrophobic residues are indicated for comparison between the primary sequence (Figure 2) and each of the models in this figure.

into their proper tertiary structure, the apparently random sequence of cationic and hydrophobic residues is concentrated into two faces of a globular protein. One face is hydrophilic and contains many of the positively charged side chains and the other is hydrophobic (Figure 6*C*,*D*). In solution, HBD-2 exhibited a α-helical segment near the N terminus that was not previously ascribed to solution structures of α-defensins or to the β-defensin BNBD-12. The authors speculate that this novel structural element might contribute to the specific microbicidal or chemokine-like properties of HBD-2 (88). Presumably, it is the amphipathic nature of these proteins that allows them to be effective antimicrobial agents. Initially, an electrostatic interaction occurs between the cationic surface of the defensin protein and the polyanionic surface of the bacterial membrane. Then, the hydrophobic surface gains entry into the membrane and ultimately leads to disruption of the membrane. The amino acids whose side chains are directed toward the surface of the protein are less conserved between β-defensin proteins and may partly explain the difference in specificity for antimicrobial activity, whereas the amino acid residues in the three β-strands of the core β-sheet are more highly conserved (Figure 2). For example, the β-hairpin loop in BNBD-12 has an arginine residue (Arg30) that contributes to the hydrophilic face, whereas the analogous position in the DEFB2 gene product contains a leucine residue (Leu32) that contributes to the hydrophobic face (Figure 2 and Figure 6). Recently, the crystalline structure of HBD-2 was reported. Hoover and coworkers determined two high-resolution X-ray structures of HBD-2 from crystals formed by 30 mg/ml solutions (89). Crystals of HBD-2 formed dimers topologically distinct from that of HDEFA3, and a quaternary octameric arrangement of HBD-2 was conserved in two crystal forms. These structures were the first to demonstrate dimerization of β-defensins. From these structural and electrostatic properties of the HBD-2 octamer, the authors suggest an electrostatic charge-based mechanism of membrane permeabilization by β-defensins rather than a mechanism based on formation of bilayer-spanning pores (89).

Finally, we note that the overall topology of these proteins is strikingly similar to the α-defensin proteins (102–104), even though the primary structures of the α- and β-defensins are even more diverse and the three disulfide linkages of these two protein families are different. These observations suggest that the core triple-stranded β-sheet structure of the α- and β-defensins is an extremely accommodating platform from which organisms can evolve diverse antimicrobial peptides to respond to a plethora of invading pathogens.

STRUCTURE AND ORGANIZATION OF THE β-DEFENSIN GENES

From the available data (Table 2), the typical β-defensin gene is comprised of two exons separated by an intron that is usually 1.5 kb, but can be as large as 16 kb. The processed transcript varies from 300 to 400 nt in length with a 5′ UTR ∼35 nt, an open reading frame of ~200 nt, and a 3′ UTR of ~100 nt. The first exon includes the 5′ UTR and encodes the leader domain of the preproprotein; the second exon encodes the mature peptide with the six-cysteine domain. The *EP2/HE2* gene is a notable exception to these rules. This gene has two promoters and eight exons that produce at least nine transcripts. Three of these transcripts, EP2C, EP2D, and EP2E, encode proteins with the β-defensin-like motif (65, 78, 94). As is described below, these three transcripts appear to be derived from the exons of two adjacent β-defensin genes. Both putative β-defensin genes contain an exon that encodes a leader domain and an exon that encodes the six-cysteine domain.

As initially recognized in the cow genome (105), the β-defensin genes are organized as a cluster in mammalian genomes. Subsequently, the α-defensin genes in human, mouse, and rat were shown to co-localize with the β-defensin genes on human chromosome 8p23-p22 (106), mouse chromosome 8 (84), and rat chromosome 16 (86), suggesting a single gene cluster for both defensin gene families. As predicted, all mapped β-defensin genes in human (64, 65, 94, 107–109) and mouse (84, 85, 97, 99, 100) co-localize in their respective genomes. In addition, the β-defensin genes in sheep (110) and pig (111) map to regions that are syntenic to the human and mouse defensin clusters (Table 2).

To date, the order and orientation of the β-defensin genes is only available for the human genome. Initially, Linzmeier and collaborators (109) showed by multicolor FISH that the *DEFB1* gene was more telomeric than the *DEFB2* gene. They estimated that the distance between these two genes was about 400 kb. Jia and collaborators (65) then analyzed the genomic sequence from the BAC clones that were used as FISH probes in the previous study. They found that the *EP2*, *DEFB3*, and *DEFB2* genes co-localized to a single BAC clone within a 50 kb region (Figure 7). These three genes are transcribed from telomere to centromere. As alluded to above, *DEFB3* and *DEFB2* have the typical two-exon structure for a β-defensin gene. However, the *EP2* gene has eight exons and two promoters that produce at least nine different transcripts. Frohlich and coworkers (94) found that exons one and four encode the leader domain and exons three and six encode the six-cysteine

$\longrightarrow$

Figure 7   Physical map of the defensin cluster on human chromosome 8p23-p22. The map was constructed by analysis of two large DNA sequence contigs, NT_008268 and NT_019483, downloaded from GenBank (www.ncbi.nlm.nih.gov). Nucleotide distance (Mb) was obtained from the length of the sequence contigs; the genetic (cM) and radiation hybrid (cR) positions were obtained from the Human Physical Map (http://carbon.wi.mit.edu:8000/cgi-bin/contig/phys_map) (198). Vertical lines connect STSs to the clones that contain them as determined by BLAST (112). Horizontal bars represent the indicated BAC clone. Transcript map of the region contains the genes *EP2* (also known as *HE2*), *DEFB3*, and *DEFB2*. Exons that encode the six-cysteine β-defensin motif are indicated (β). Arrows indicate the direction of transcription for those transcripts that encode β-defensin proteins.

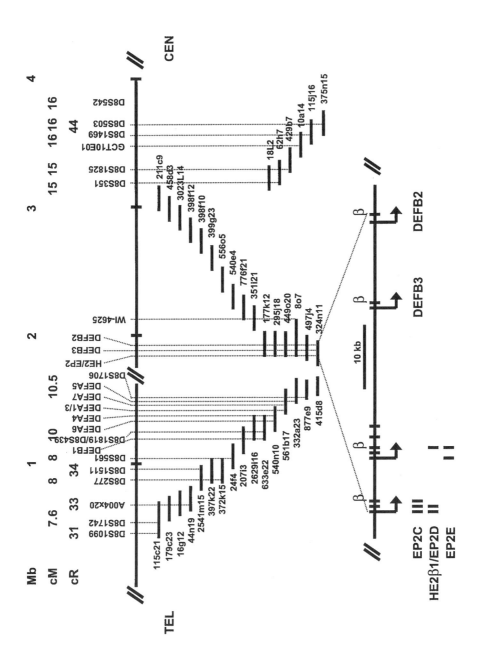

motif. As shown in Figure 7, three EP2 transcripts, EP2C, EP2D, and EP2E, include both a leader domain and the six-cysteine motif (65, 78, 94). The EP2C transcript encodes the first leader domain and the first six-cysteine domain; the EP2D transcript encodes the first leader domain but the second six-cysteine domain; the EP2E transcript encodes the second leader and both six-cysteine domains. The structure of the *EP2E* gene product is most similar to the other β-defensin proteins. Based on high sequence similarity, the *Bin1b* gene product appears to be the rat ortholog for the human *EP2E* gene product, and it was shown to have antimicrobial activity (73). The transcripts EP2C and EP2D are unique. They include a third exon that encodes 53 additional amino acids between the leader and the six-cysteine domains. It is not known whether these two EP2 isoforms have antimicrobial activity and if the additional amino acids alter that activity.

Since the time the genomic analysis of the BAC clones that contain the *DEFB1* and *DEFB2* genes was completed, an additional genomic sequence from 8p22-p23 was added to the public databases. The sequence for these BAC clones now makes up a small portion of two large sequence contigs (Figure 7). The *DEFB1* gene is contained in a contig of 16 clones that spans about 1.5 Mb sequence (Genbank #NT_008268), and the *EP2/DEFB3/DEFB2* genes are contained in a contig of 22 clones that spans about 2 Mb sequence (Genbank #NT_019483). We performed tBLASTn (112) searches for sequences that encode β-defensin-like proteins in this 3.5 Mb region, using the amino acid sequences of all the known human β-defensin proteins as the subject sequence. No additional defensin-like sequences were detected. We also searched these sequences for sequence tagged sites (STSs) and placed them on the map according to their position in the clones. These STSs were used for constructing genetic and physical maps. The linear order of these STSs in both the genetic and physical maps is identical to the order deduced from the sequence contig, indicating that the sequence contigs are a reliable representation of the region at chromosome 8p22-p23.

We note however that there are several unresolved issues concerning the genomic structure of the defensin locus. First, the gene order, deduced from this sequence contig, is not consistent with previous studies. The sequence contig suggests the following gene order, *TEL-DEFB1-DEFA1-DEFB2-CEN*, but based on multi-color FISH analysis, Linzmeier and collaborators (109) observed a different gene order, *TEL-DEFA1-DEFB1-DEFB2-CEN*. Also Harder and colleagues (107) showed that the *DEFB1* and *DEFB2* genes were located on a single YAC clone (733g4) that was derived from the YAC contig WC8.1 (www.genome.wi.mit.edu). However, this YAC clone also contains the genetic markers D8S1695 and D8S1759. These markers are located at 22 and 24 cM, respectively, on the genetic map. The sequence contig indicates that the *DEFB1* and *DEFB2* genes are located near the 10 cM region. In the human genome, 1 cM on a genetic map is roughly equivalent to 1 Mb of sequence (79). Therefore, it is difficult to reconcile how a 1.5 Mb YAC can contain two genes and two markers that are at least 12 cM apart on a genetic map. One possible explanation for these results is that the region surrounding the defensin locus is genetically unstable. We note that partial 8p deletion is a common chromosomal abnormality (113). Additional genomic analysis must be performed to determine whether the DNA sequence instability of genomic clones is related to this in vivo genetic instability.

The second unresolved issue concerning the genomic structure of the defensin locus is the presence of a gap in the sequence contig that separates the *DEFB1* and α-defensin genes from the other β-defensin genes. A gap also exists in the YAC physical map constructed by the Whitehead Institute (www.genome.wi.mit.edu). The STSs on the telomeric side of the sequence gap map to the YAC contig WC8.0, whereas the STSs on the centromeric side of the sequence gap map to the YAC contig WC8.1. There are no known YAC clones that bridge these two contigs. These gaps suggest that there is a region inside the defensin locus that is unclonable or, at least, not maintained in large genomic clones such as BACs or YACs. How large is this gap? The only estimate is from the multi-color FISH data (109) and suggests that the distance between the *DEFB1* and *DEFB2* genes is around 400 kb. However, the reliability of this estimate is in doubt because the order of the defensin genes in that study is not consistent with the sequence contigs. It is interesting to speculate whether other defensin genes are located in this sequence gap and whether similar problems will arise when this region is sequenced in other species.

EXPRESSION OF β-DEFENSIN GENE PRODUCTS IN THE LUNGS OF ANIMAL SPECIES

Bovine tracheal antimicrobial peptide (TAP) was the first pulmonary epithelial β-defensin investigated in detail (66). TAP expression was restricted to the epithelium of the conducting airways and was not present in submucosal glands, alveoli, or other pulmonary cell types (67). Expression of TAP mRNA in the lung was developmentally regulated, as the message was undetectable in trachea from fetuses of five or six months gestation (67). A second bovine epithelial β-defensin, termed lingual antimicrobial peptide (LAP), was later cloned and shown to be present in respiratory epithelia (114). The regulation of TAP and LAP mRNA expression in the lung was studied in detail using cultured bovine tracheal epithelia (115). TAP and LAP expression are inducible and their induction is regulated at the level of transcription by agents including bacterial LPS and TNF-α (115, 116). Bovine tracheal epithelia also express CD14, and this protein was shown to be integrally involved in the binding of LPS to cow airway epithelia to activate TAP expression (116). Furthermore, genomic analysis of the 5′ flanking region of the TAP gene reveals consensus sequence sites for NF-IL-6 and NF-κB, implying a role for inflammatory mediators in the regulation of airway β-defensin expression (116). These seminal studies of TAP and LAP have proven to be a very useful paradigm for the identification of pulmonary β-defensins and investigation of several aspects of their regulation and function. In addition, the observation that respiratory epithelia may express more than a single β-defensin gene product, a principle first shown in cows, appears to hold true in most mammalian species studied to date (Table 2).

Expression studies of β-defensins have also been performed in several other experimental models, including mouse (Table 2). In the mouse, six β-defensins (*Defb1-6*) have been discovered and there is evidence that at least five of these gene products are expressed in the lung from mRNA analysis. *Defb1* is constitutively expressed in lung and other tissues and the peptide product shows salt-sensitive antimicrobial activity against gram-positive and gram-negative bacteria (84, 97, 99, 117). *Defb1* shares sequence homology and a tissue distribution similar to human DEFB1. *Defb2* was identified by Morrison and collaborators (76). The expression of *Defb2* mRNA in murine airway epithelia was undetectable without LPS treatment of the cells (76). *Defb3* is an inducible gene product expressed in lung, small intestine, and liver (55, 100). The *Defb4* mRNA expression was restricted to the esophagus, tongue, and trachea and was not induced in the lung by challenging animals with intratracheal *Pseudomonas aeruginosa* (55). The *Defb5* sequence is deposited in Genbank (AF318068), but expression studies have not yet been reported. *Defb6* was recently reported by Yamaguchi and colleagues as inducible and predominantly expressed in skeletal muscle in addition to trachea, esophagus, and tongue (118). Identification and characterization of all murine β-defensins will provide an opportunity to understand further the role of these peptides in host defense through animal model studies and the generation of β-defensin-deficient animals by gene targeting.

EXPRESSION AND REGULATION OF HUMAN β-DEFENSINS IN THE RESPIRATORY EPITHELIUM

The expression of DEFB1, DEFB2, and DEFB3 has been investigated in some detail in human pulmonary epithelia (52, 60–62, 64, 65, 98, 119). Similar to results reported in kidney, urogenital, and other epithelia (58, 120), DEFB1 expression in the lung is constitutive (61, 62, 98). Goldman and colleagues found that DEFB1 mRNA was expressed throughout the respiratory epithelia of non–cystic fibrosis (CF) and CF lungs (61). In a human bronchial xenograft model, antisense oligonucleotides to DEFB1 abolished ASL microbicidal activity in non-CF grafts, suggesting a role for the peptide in pulmonary innate immunity (61). Singh and colleagues investigated the expression and function of DEFB1 and DEFB2 in airway epithelia (98). The DEFB1 and DEFB2 mRNAs were expressed in excised surface and submucosal gland epithelia from non-CF and CF specimens (98). The pro-inflammatory cytokine IL-1β stimulated the expression of DEFB2 mRNA and peptide but not DEFB1 mRNA in primary cultures of well-differentiated airway epithelia. Secreted HBD-2 peptide was detected on the apical surface of well-differentiated primary cultures of airway epithelia grown at the air-liquid interface, and there was a clear increase in peptide abundance following 24 h of IL-1β stimulation (98). Using Western blot analysis, HBD-2 was identified in bronchoalveolar lavage (BAL) (121) fluid from patients with CF or inflammatory lung diseases, but not in normal volunteers, whereas HBD-1 was found at similar levels in all conditions (98, 122, 123). Both HBD-1 and HBD-2 were found in BAL fluid in concentrations of several ng/ml (98, 122, 123). These data suggest that in the lung HBD-2 may be important in the response to inflammation, whereas HBD-1 may serve as a defense in the absence of inflammation.

A third human β-defensin was discovered independently by two laboratories, one using a genomics-based approach (65), the other isolating the peptide from skin (64). Jia and coworkers detected DEFB3 mRNA in skin, esophagus, gingival keratinocytes, placenta, and trachea (65). Furthermore, in fetal lung explants and gingival keratinocytes, DEFB3 mRNA expression was induced by treatment with IL-1β for 24 h (65). Harder and colleagues found that *P. aeruginosa* and TNF-α both induced expression of DEFB3 mRNA in cultured A549 cells (64). Thus DEFB3 expression in airway epithelia is also regulated by infectious and inflammatory stimuli. In addition to expressing three classic β-defensins, there is evidence for expression of HE2β1 mRNA in bronchial epithelia by RT-PCR analysis (65). The rat gene product *Bin1b* is the homolog for the EP2e isoform and exhibits antimicrobial activity against *Escherichia coli* (73). However, further work is needed to establish whether there is a role for products of the *HE2/EP2* gene in pulmonary host defense.

CONSTITUTIVE AND INDUCIBLE EXPRESSION OF β-DEFENSINS IN AIRWAY EPITHELIA

It is well established that β-defensin expression in airway epithelia may be constitutive or inducible. Studies in a number of epithelial models have failed to demonstrate evidence for the transcriptional regulation of *DEFB1* gene expression by stimuli including pro-inflammatory cytokines and regulators of second messenger pathways (98, 120). DEFB1 mRNA abundance does not increase in cultured airway epithelia in response to a variety of pro-inflammatory stimuli (98). Interestingly, HBD-1 protein expression increases in urine and breast tissue during pregnancy (58, 124), suggesting that certain hormonal conditions may regulate its expression.

In contrast to *DEFB1*, there is considerable evidence from studies of airway epithelia in vitro and from plasma and BAL analysis in vivo demonstrating that *DEFB2* and *DEFB3* respond to pro-inflammatory stimuli with increases in mRNA or peptide abundance. In cultured airway epithelia the DEFB2 mRNA abundance increased markedly in response to stimulation with *P. aeruginosa* (119), TNF-α (119), or IL-1β (98, 119). Similarly, DEFB3 mRNA abundance increased in response to the same stimuli in cultured epithelia (64) or cultured fetal lung explants (65). The HBD-2 peptide is present in higher levels in BAL or plasma specimens from patients with inflammatory lung diseases such as cystic fibrosis (98), interstitial lung disease (98), pneumonia (122), and atypical mycobacterial infections (123). In contrast, HBD-1 concentrations in BAL and plasma do not appear to change in the inflammatory disease states (98, 123). In studies of the composition and antimicrobial activity of human nasal secretions, Cole et al. detected HBD-2 peptide in concentrations of 0.3–4 μg/ml in three of five specimens assayed (19). These studies indicate that β-defensin expression in the airway epithelium and respiratory secretions is responsive to a variety of environmental stimuli.

The finding that the expression of the *DEFB2* and *DEFB3* genes in airway epithelia is induced in response to bacteria and pro-inflammatory cytokines raises several interesting questions regarding the relevant in vivo stimuli and regulatory pathways underlying these responses. The discovery of the mammalian homologs of the *Drosophila* Toll receptor has fueled tremendous interest in signal transduction in adaptive immune cells and epithelia via these Toll-like receptors (TLRs) (8, 125). In the fruit fly, Toll receptors on the surface cells of the fat body are activated by the ligand Spaetzle, leading to signal transduction via the NF-κB homologs Cactus/Dorsal and the expression of inducible antimicrobial peptides such as drosomycin and others (126). A growing body of evidence demonstrates that the TLR family of receptors in mammals plays an important role as pattern recognition receptors on the surface of adaptive immune cells and epithelia, allowing the host to differentiate self from non-self (6, 8). By analogy to the *Drosophila* system, in which Toll signaling leads to the expression of antimicrobial peptides, humans have at least two inducible β-defensins in airway epithelia.

What are the relevant in vivo receptors and ligands regulating the expression of inducible antimicrobial peptides in human airway epithelia? This is a question for which the answers are only beginning to emerge. The IL-1 receptor is expressed on airway epithelia and is a member of the TLR family of receptors. Signaling via the IL-1 receptor is linked to signal transduction via NF-κB-dependent pathways. Expression of TLRs 1–6 and CD14 was recently documented in human airway epithelia (127). Furthermore, LPS was shown to induce *DEFB2* gene expression in a CD14-dependent fashion and stimulate signaling via NF-κB (128). The implication of CD14 in this signaling pathway suggests that TLR4 is a relevant receptor for recognition of LPS in airway epithelia (127). Mutations in the human TLR4 gene that co-segregated with a blunted response to inhaled LPS were recently identified (128). These TLR4 missense mutations (Asp299Gly and Thr399Ile) affect the extracellular domain of the TLR4 receptor and demonstrate that these sequence changes alter the ability of the host to respond to environmental stress (128). There is increasing evidence for specificity within the TLR family (6). For example, TLR2 signals in response to peptidoglycan from gram-positive bacterial cell walls (129, 130), whereas TLR4 recognizes gram-negative LPS (131–133). The TLR5 is a receptor for bacterial flagellae (134), and TLR9 binds and signals in response to bacterial DNA CpG motifs (135).

It remains to be determined how the TLR family of receptors is involved in the responses in airway epithelia that lead to the induction of β-defensin expression. Harder and colleagues reported that a mucoid strain of *P. aeruginosa*, TNF-α, IL-1β, and high concentrations (>10 μg/ml) of *Pseudomonas* LPS, all induced DEFB2 mRNA in a dose-dependent fashion in cultured normal bronchial and tracheal as well as normal and CF-derived nasal epithelial cells (119). Interestingly, two nonmucoid strains and IL-6 had no effect on *DEFB2* gene expression (119). It is possible that the presence of TLRs on important immune effector cells in the lung such as macrophages and dendritic cells may lead to secondary signaling that stimulates airway epithelia. For example, in a preliminary study, Singh and coworkers found that the induction of the *DEFB2* gene expression in well-differentiated primary cultures of human airway epithelia was much more responsive to LPS when the epithelia were exposed to the bacterial product in the presence of alveolar macrophages (136). This suggests that airway epithelia may wait for a second response signal such as IL-1 release from macrophages before responding by inducing expression and secretion of HBD-2. Additional study is needed to determine the relative importance of multiple bacterial products and cytokines in the direct regulation of β-defensin expression in airway epithelia.

Analysis of the putative promoter regions of the *DEFB2* and *DEFB3* genes reveals interesting contrasts. The 5′ flanking sequence of *DEFB2* contains numerous consensus sequence sites for regulators of transcription including NF-κB, NF-IL-6 and AP-1 (52, 108, 127). In contrast, there are no NF-κB response elements present in the sequence upstream of the transcription start site of the *DEFB3* gene, although numerous AP-1 consensus sites are present (65). This observation suggests that the expression of *DEFB2* and *DEFB3* genes in airway epithelia is likely to be regulated differently.

ANTIMICROBIAL ACTIVITY OF β-DEFENSIN PEPTIDES

The most widely studied aspect of β-defensin function is their antimicrobial properties (3). β-defensin peptides are produced as pre-propeptides and then cleaved to release a C-terminal active peptide fragment; however, the pathways for the intracellular processing, storage, and release of the human β-defensin peptides in airway epithelia are unknown. Although it is well-documented that β-defensin peptides are present in mucosal secretions, it is also possible that important antibacterial functions of the peptides may be related to their presence intracellularly or when attached to cell surfaces or secreted mucins. In general, β-defensin activity is microbicidal rather than bacteriostatic and requires micromolar concentrations. Broad-spectrum antimicrobial activity against gram-positive and gram-negative bacteria, fungi, and enveloped viruses has been reported, but most studies have focused on the antibacterial activity (3, 137). Characteristically, the antimicrobial activity of the β-defensin peptides is salt sensitive and their killing is markedly reduced as the ionic strength of the solutions increases (i.e., NaCl > 50 mM).

Studies of the 38–amino acid native bovine TAP revealed activity in vitro against *E. coli, Staphylococcus aureus, Klebsiella pneumonia, P. aeruginosa,* and *Candida albicans,* consistent with a broad spectrum of activity (66, 138). The secreted form of the HBD-1 peptide exists in five forms, each representing progressive N-terminal truncations from a 47–amino acid mature peptide to the smallest 36–amino acid form (58). Respiratory secretions contain at least three forms of HBD-1 (98, 123). Studies by Valore and colleagues showed that HBD-1 peptide activity was not changed appreciably by low pH but was inhibited by high salt conditions (58). Interestingly, of the HBD-1 peptides detected in urine, only the 36–amino acid peptide retained its activity against *E. coli* in physiologic salt concentrations (58). HBD-1 has also been shown to exhibit activity against *P. aeruginosa* and *Listeria monocytogenes* (58, 61, 98). HBD-2 is present as a 41–amino acid peptide and is microbicidal in micromolar concentrations against *P. aeruginosa, E. coli,* and *C. albicans* (52, 60, 98). HBD-2 shows little activity against gram-positive *S. aureus* (52). Harder and colleagues isolated the 45–amino acid HBD-3 peptide from psoriatic scales by specifically focusing on peptides that showed activity against *S. aureus* (64). HBD-3 showed microbicidal activity against *S. aureus,* vancomycin-resistant *Enterococcus faecium, P. aeruginosa, C. albicans,* and *Streptococcus pyogenes* (64). Interestingly, the *S. aureus* killing was retained in solutions containing NaCl concentrations in excess of 150 mM (64). A striking feature of the HBD-3 peptide is its greater density of cationic residues when compared with HBD-1 and HBD-2, especially at its C terminus (64, 65). Perhaps the greater charge density of HBD-3 facilitates greater interactions and therefore activity with the gram-positive bacteria cell wall. Thus from ongoing investigations of the human β-defensins, patterns of unique antimicrobial spectrums are beginning to emerge. It will be interesting to learn how the variable spectrums of activity may relate to different pathways for inducing peptide expression in response to infection or inflammation.

The defensin peptides are thought to initiate their interactions with bacteria through simple electrostatic interactions with bacterial cell walls. It is this property that confers the characteristic salt sensitivity to the peptides. Their amphiphilic design allows the peptides to interact with membranes such that the charged regions bind to anionic phospholipid head groups (i.e., LPS, techoic acid) and water and the nonpolar surface is buried in the lipid phase. While there is compelling evidence that defensins permeabilize bacterial cell membranes, the mechanism of the effect is not known. It is well documented that α-defensins sequentially permeabilize the outer and inner membranes of *E. coli* (139). Defensins may act by forming oligomeric membrane spanning pores, by disrupting lipid membranes, or through a combination of such effects (3, 88, 89). The lower anionic lipid content of the cell membranes of multicellular organisms is thought to provide a degree of specificity and protection against damage to host cells.

SYNERGY AMONG COMPONENTS OF AIRWAY SURFACE LIQUID—ANTIMICROBIAL SOUP

Why does ASL contain such a large number of substances with antimicrobial activity? This complex "soup" provides the host an array of innate immune factors to present to the potential pathogens. Interestingly, the mechanisms of action of the various protein and peptide antimicrobials vary. Thus this complex system presents many layers of protection, some bacteriostatic, some bactericidal, and in many ways is highly redundant. Sorting out how the "meat and potatoes" of this antimicrobial soup work with less abundant "spices" is a somewhat daunting task. It is likely that these multiple components of the innate mucosal immune system in the lung act additively or synergistically to prevent infection or colonization of the lung from a variety of inhaled or aspirated pathogens. Indeed this supposition is supported by several observations. In studies of nasal secretions, Cole et al. found that the antimicrobial activity was reduced by boiling the samples (19). Boiling effectively eliminates the activity of the abundant antimicrobial proteins lysozyme and lactoferrin. However, the microbicidal activity of the secretions was not restored by simple supplementation with lysozyme and lactoferrin, suggesting that other less abundant factors contribute to the net activity (19). Bals and coworkers reported that the in vitro antimicrobial activity of HBD-2 was synergistic with lysozyme and lactoferrin (60). The combinations of lysozyme-lactoferrin, lysozyme-SLPI, and lactoferrin-SLPI were found to be synergistic in vitro by Singh and colleagues (140). Furthermore, the triple combination of lysozyme, lactoferrin, and SLPI showed enhanced synergy (140). The combined peptides HBD-1 and LL-37 were additive in their effects on killing (140). These data suggest that the antibacterial potency of ASL may be significantly increased by synergistic and additive interactions between antimicrobial factors.

ROLE OF β-DEFENSIN PEPTIDES IN COMMUNICATION WITH THE ADAPTIVE IMMUNE SYSTEM

In addition to their broad-spectrum antimicrobial properties, there is evidence that the β-defensins may act as chemokines for immature dendritic cells and memory T cells and thus serve as a bridge between the innate and adaptive immune systems (141, 142). Studies by Yang and colleagues revealed that HBD-1 and HBD-2 were selectively chemotactic for cells expressing human CCR6, a chemokine receptor preferentially expressed by immature dendritic cells and memory T cells (141, 143, 144). In contrast to the micromolar concentrations needed to kill bacteria, the β-defensin chemokine activities were present at nanomolar concentrations (141). The HBD-1 and -2-induced chemotaxis was sensitive to pertussis toxin and was inhibited by antibodies to CCR6. The binding of iodinated CCL20 (also termed LARC or MIP-3α), the only reported chemokine ligand for CCR6, to CCR6-transfected cells was competitively inhibited by the β-defensins. The chemokine activity of CCL20 was approximately 10-fold greater than that of HBD-1 and HBD-2 (141). Thus β-defensins may also promote adaptive immune responses in the lung by recruiting dendritic and T cells to the site of microbial invasion through interaction with CCR6. HBD-2 also stimulates mast cells to release histamine whereas HBD-1 has no such effect (145).

These studies have implications for pulmonary mucosal immunity and suggest several areas for future studies. The sub-epithelial space in the airways contains a dense meshwork of dendritic cells (146, 147). Studies by Holt and others have carefully defined many aspects of the dynamic life cycle and responses of pulmonary dendritic cells to infectious and immunologic stimuli in the airways (148–151). It had previously been demonstrated that the lung expresses CCL20 mRNA abundantly, suggesting that respiratory epithelia may be one source for the production of this chemokine (152–154). If airway epithelia are sources for both β-defensins and CCL20, the regulation of their production and relative abundances may indicate possible differential functions in mucosal responses. Recent creation of CCR6 knockout mice may provide tools to investigate this further (155, 156). Detailed investigation of the vectoral secretion and local concentrations of the β-defensins and CCL20 in pulmonary epithelia and secretions may help clarify how immature dendritic cells and T cells of the airway are stimulated to move in response to infectious and inflammatory stimuli.

OTHER PROPERTIES OF β-DEFENSIN PEPTIDES

In addition to their direct antimicrobial effects and chemokine functions, defensins have been reported to have several other activities. Murphy and coworkers demonstrated that α-defensins can act synergistically with insulin to promote proliferation of epithelial cells and fibroblasts (157). Such properties may be important to promote healing in areas of epithelial damage in the airways. Additional functions of the α-defensins include inhibition of fibrinolysis (157a) and inhibition of ACTH-stimulated cortisol production (158). The α-defensins also display anti-inflammatory activities, including regulation of complement activation (159) and proteinase inhibitor secretion (160). Lencer and coworkers reported that the mouse intestinal α-defensins cryptdins 2 and 3 selectively permeabilized the apical cell membrane of epithelial cells in culture to elicit a physiologic Cl^- secretory response (161). It is not known whether the β-defensin peptides are capable of affecting the bioelectric properties of the airway or alveolar epithelium.

ORIGINS OF THE DEFENSIN GENES

Is an ancestral gene at the root or is it a forest of smaller trees with a different gene at each root? Several clues can be used to address this question: conservation of sequence, expression, function, and genomic location. As described above, elements of the primary sequence of the defensin genes are conserved, especially the six-cysteine motif. However, as a whole the primary sequence of the defensin genes is not conserved. Even the spacing for the six-cysteine motif of the β-defensin genes is variable (Table 2). Similar diversity is found for gene expression and gene function. How can such diversity be explained if these genes did arise from a common ancestor? Hughes suggests positive selection (162). He compared the rate of synonymous mutations (d_S) with the rate of nonsynonymous mutations

(d_N). Synonymous mutations do not alter the amino acid sequence and most will not affect the gene function (163), whereas nonsynonymous mutations alter the amino acid sequence and the majority of nonsynonymous mutations will be deleterious to the gene function. Consequently, for most genes, d_S should exceed d_N. To the contrary, Hughes found that for many of the α-defensin genes and for the β-defensin genes from cow and sheep, d_N exceeded d_S, suggesting that positive selection was promoting diversification of these genes following duplication (162). Analogous arguments could also be made to explain the diversity of expression patterns, gene regulation, and gene function for the defensin gene family. Therefore, despite the great diversity of these properties for the defensin gene family, it is possible that these genes originate from a single, common ancestral gene. In addition, the β-defensin genes cluster at a single locus along with the α-defensin genes in the human (106), mouse (99), and rat (86) genomes. Although this observation is consistent with the hypothesis that the defensin genes originated from a single ancestral gene, Hughes cautions that, as with other genes (164), linkage may reflect selection for a group of genes with a common pattern of expression or function (162). Which, if any, of the defensin genes is the progenitor? At this time, the genetic record for the defensin gene family is still too sparse to answer this question. Comparative genomic analysis of many additional vertebrate and multicellular organisms must be performed to determine whether it is a single tree and which gene lies at its base.

POSSIBLE LINKS BETWEEN ASL PEPTIDE ACTIVITY AND PULMONARY DISEASE

A striking feature of a number of the protein and peptide components of ASL, including the β-defensins, is the observation that their activity is inhibited by high NaCl concentrations (20, 60, 61, 98). Furthermore, the antimicrobial activity of some ASL components is very sensitive to divalent cation concentrations (165, 166) and may be optimal at neutral pH (165, 167). These observations raise the possibility that changes in ASL composition, such as elevated NaCl concentrations or a fall in pH (168), could have a profound effect on the function of innate immune factors, including β-defensins. Several investigators have suggested that such phenomena may contribute to disease pathogenesis in cystic fibrosis (61, 169), but this hypothesis remains controversial, in part because of the difficulties inherent in accurately measuring ASL composition in vivo (170–173). Because the airway surface liquid salt concentration may be elevated in cystic fibrosis airways in vivo, Singh et al. examined the effect of salt on the synergistic combinations (140). They found that as the ionic strength increased, synergistic interactions among ASL components were lost (140). These results suggest that the increased salt concentrations that may exist in cystic fibrosis could inhibit airway defenses by diminishing these synergistic interactions. Further detailed studies are needed to understand the composition of normal airway surface liquid, including salt concentration and pH, and how it may change in disease states such as cystic fibrosis, asthma, or chronic bronchitis.

The expression of β-defensins in the human lung is developmentally regulated, a finding that may have clinical implications. The relative immune-deficient state of the newborn reflects immaturity of both innate and adaptive responses (174). DEFB1 mRNA expression was quantified using RNase protection assays and was detected in human postnatal and adult lung but not in 15- and 22-week gestation fetal lung (62). In preliminary studies, it was noted that the DEFB2 mRNA is in low abundance prenatally and increases at postnatal time points. Similar results were found for the *DEFB3* gene (T. Starner & P. B. McCray, unpublished data). In cultured mid-gestation human fetal lung explants, IL-1β treatment stimulated increases in the abundance of both DEFB2 and DEFB3 mRNAs, whereas DEFB1 did not change (T. Starner & P. B. McCray, unpublished data) (65). Many clinical studies support a protective role for breast milk feeding in decreasing the incidence and/or lessening the severity of infectious diseases in infants (175). It is interesting to note that human breast milk contains HBD-1 peptide in μg/ml quantities (124, 176). Perhaps the peptide exerts microbicidal effects that influence colonization or infection of the nasopharynx or upper gastrointestinal tract. The finding that β-defensin expression is reduced in the fetal lung raises the possibility that immaturity of β-defensin expression in the pulmonary epithelium of preterm infants contributes to their increased risk of infection (177). In addition, it is possible that immaturity of the ion transport properties of the respiratory epithelium in infants could alter the airway surface liquid composition in a fashion unfavorable to optimal antimicrobial activity (178, 179).

There has been little study of how commonly used pulmonary medications may impact the expression or function of antimicrobial peptides in the lung. Duits and coworkers investigated the effects of dexamethasone on the expression of the *DEFB1, 2,* and *3* genes in cultured bronchial epithelial cells (77). Using a RT-PCR-based assay, they reported that dexamethasone inhibited the inducible expression of DEFB3 mRNA, but not the expression of the *DEFB1* and *DEFB2* genes, in cultured airway epithelial cells. They also investigated the expression of the human β-defensins in mononuclear phagocytes and found detectable mRNA for DEFB1 and DEFB2, but not DEFB3, in this cell type. Dexamethasone had no effect on the expression of DEFB1 and DEFB2 mRNAs in mononuclear phagocytes (77). Investigation of the impact of therapeutic agents on the innate immune functions of the respiratory epithelium is an area deserving of further study.

Although the defensin peptides are an important component of neutrophil and epithelial microbicidal function, some evidence suggests that if they are present in pulmonary secretions in high concentrations, they may have harmful effects on the host (160, 180, 181). It is well documented that defensin peptides can cause dose-dependent changes in the barrier properties of epithelia. Nygaard reported that human neutrophil α-defensins 1–3 increased the mannitol permeability and decreased transepithelial resistance of Madin-Darby canine kidney (MDCK) monolayers (180). However, the peptides were not cytolytic to MDCK cells as measured by lactate dehydrogenase release. Because the α-defensins may accumulate in airway secretions of patients with chronic inflammatory lung disorders, including cystic fibrosis and asthma, it is possible that the peptides contribute to disease pathogenesis (182, 183). For example, in cystic fibrosis, concentrations of α-defensins in secretions may reach mg/ml amounts (184). Slutsky and coworkers reported that human α-defensins caused dose-dependent changes in lung function, influx of neutrophils, and release of pro-inflammatory cytokines in mice following intratracheal instillation (181). It is currently not known whether the human β-defensins may have cytotoxic effects on epithelia in the concentrations present in vivo.

Islam and colleagues recently reported that infections by *Shigella* species may reduce the expression of the antibacterial peptides LL-37 and HBD-1 in intestinal epithelia from patients with bacillary dysenteries and in *Shigella*-infected cell cultures of epithelial and monocyte origin (185). The ability to modulate expression of host defense factors may be an important virulence strategy for some bacteria. Such downregulation of innate immune factors could contribute to bacterial adherence or infection in the respiratory epithelium.

CLINICAL APPLICATIONS OF ANTIMICROBIAL PEPTIDES

Antimicrobial peptides, including β-defensins and cathelicidins, are interesting candidates for investigation and development as therapeutic agents for topical or systemic administration (3, 186–189). Their broad spectrum of activity and the low incidence of the development of bacterial resistance with repeated exposure are attractive features. Another area for potential therapeutic development is the approach of augmenting or supplementing endogenous antimicrobial activity by stimulating expression pharmacologically. Developments in these areas will occur as we advance our understanding of the complexity of airway surface liquid composition and the antimicrobial components it contains.

ACKNOWLEDGMENTS

We acknowledge the support of the National Institutes of Health HL-61234 (P. B. M. and B. C. S.), P30-HD27748 (Frank Morriss and B. C. S.), and the Children's Miracle Network Telethon. We thank our colleagues Chuck Bevins, Brian Tack, Hong Peng Jia, Tim Starner, Pradeep Singh, Mike Welsh, and Joe Zabner for their helpful comments and suggestions. We thank Bill Kearney, Monali Sawai, and the University of Iowa NMR Facility for preparing the BNBD-12 and HBD-2 structure figure. We thank Margaret Malik, Andrea Penisten, Autumn Bradley, and Joe Mitros for technical assistance. We acknowledge the support of the Cell Culture Core, partially supported by the Cystic Fibrosis Foundation, NHLBI (PPG HL-51670), and the Center for Gene Therapy for Cystic Fibrosis (NIH P30 DK54759).

Visit the Annual Reviews home page at www.AnnualReviews.org

LITERATURE CITED

1. Fleming A. 1922. On a remarkable bacteriolytic element found in tissues and secretions. *Proc. R. Soc. London Ser. B* 93:306–19
2. Fleming A, Allison VD. 1922. Observations on a bacteriolytic substance ("lysozyme") found in secretions and tissues. *Br. J. Exp. Pathol.* 3:252–60
3. Ganz T, Lehrer RI. 1999. Antibiotic peptides from higher eukaryotes: biology and applications. *Mol. Med. Today* 5:292–97
4. Lehrer RI, Bevins CL, Ganz T. 1998. Defensins and other antimicrobial peptides. In *Mucosal Immunology*, ed. PL Ogra, J Mestecky, ME Lamm, WM Strober, J Bienstock, pp. 89–99. New York: Academic
5. Huttner KM, Bevins CL. 1999. Antimicrobial peptides as mediators of epithelial host defense. *Pediatr. Res.* 45:785–94
6. Kimbrell DA, Beutler B. 2001. The evolution and genetics of innate immunity. *Nat. Rev. Genet.* 2:256–67
7. Medzhitov R, Janeway CA. 1998. Commentary. Self-defense: the fruit fly style. *Proc. Natl. Acad. Sci. USA* 95:429–30
8. Hoffmann JA, Kafatos FC, Janeway CA, Jr, Ezekowitz RAB. 1999. Phylogenetic perspectives in innate immunity. *Science* 284:1313–18
9. Boman HG. 2000. Innate immunity and the normal microflora. *Immunol. Rev.* 173:5–16
10. Zhang G, Ghosh S. 2001. Toll-like receptor-mediated NF-kappaB activation: a phylogenetically conserved paradigm in innate immunity. *J. Clin. Invest.* 107:13–19
11. Travis SM, Singh PK, Welsh MJ. 2001. Antimicrobial peptides and proteins in the innate defense of the airway surface. *Curr. Opin. Immunol.* 13:89–95
12. Bezdíček P, Crystal R. 1997. Pulmonary

macrophages. In *The Lung: Scientific Foundations*, ed. RG Crystal, pp. 859–69. Philadelphia: Raven. 2nd ed.
13. Lindahl M, Stahlbom B, Tagesson C. 1999. Newly identified proteins in human nasal and bronchoalveolar lavage fluids: potential biomedical and clinical applications. *Electrophoresis* 20:3670–76
14. Lindahl M, Stahlbom B, Tagesson C. 1995. Two-dimensional gel electrophoresis of nasal and bronchoalveolar lavage fluids after occupational exposure. *Electrophoresis* 16:1199–204
15. Lindahl M, Svartz J, Tagesson C. 1999. Demonstration of different forms of the anti-inflammatory proteins lipocortin-1 and Clara cell protein-16 in human nasal and bronchoalveolar lavage fluids. *Electrophoresis* 20:881–90
16. Wattiez R, Hermans C, Cruyt C, Bernard A, Falmagne P. 2000. Human bronchoalveolar lavage fluid protein two-dimensional database: study of interstitial lung diseases. *Electrophoresis* 21:2703–12
17. McCray PB Jr, Welsh MJ. 1998. Transport function of airway epithelia and submucosal glands. In *Fishman's Pulmonary Diseases and Disorders*, ed. AP Fishman, JA Elias, JA Fishman, MA Grippi, LR Kaiser, RM Senior, pp. 129–37. New York: McGraw-Hill
18. Boucher RC. 1994. Human airway ion transport Part II. *Am. J. Respir. Crit. Care Med.* 150:581–93
19. Cole AM, Dewan P, Ganz T. 1999. Innate antimicrobial activity of nasal secretions. *Infect. Immun.* 67:3267–75
20. Travis SM, Conway BA, Zabner J, Smith JJ, Anderson NN, et al. 1999. Activity of abundant antimicrobials of the human airway. *Am. J. Respir. Cell Mol. Biol.* 20:872–79

21. Schnapp D, Harris A. 1998. Antibacterial peptides in bronchoalveolar lavage fluid. *Am. J. Respir. Cell Mol. Biol.* 19:352–56

22. Konstan MW, Chen PW, Sherman JM, Thomassen MJ, Wood RE, Boat TF. 1981. Human lung lysozyme: sources and properties. *Am. Rev. Respir. Dis.* 123:120–24

23. Franken C, Meijer CJ, Dijkman JH. 1989. Tissue distribution of antileukoprotease and lysozyme in humans. *J. Histochem. Cytochem.* 37:493–98

24. Hiemstra PS, Maassen RJ, Stolk J, Heinzel-Weiland R, Steffens GJ, Dijkman JH. 1996. Antibacterial activity of antileukiprotease. *Infect. Immun.* 64:4520–24

25. Ellison RT, Giehl TJ, LaForce FM. 1988. Damage of the outer membrane of enteric gram-negative bacteria by lactoferrin and transferrin. *Infect. Immun.* 56:2774–81

26. Masson PL, Heremans JF, Prignot JJ, Wauters G. 1966. Immunohistochemical localization and bacteriostatic properties of an iron-binding protein from bronchial mucus. *Thorax* 21:538–44

27. Thompson AB, Bohling T, Payvandi F, Rennard SI. 1990. Lower respiratory tract lactoferrin and lysozyme arise primarily in the airways and are elevated in association with chronic bronchitis. *J. Lab. Clin. Med.* 115:148–58

28. Nibbering PH, Ravensbergen E, Welling MM, van Berkel LA, van Berkel PH, et al. 2001. Human lactoferrin and peptides derived from its N terminus are highly effective against infections with antibiotic-resistant bacteria. *Infect. Immun.* 69:1469–76

29. Weinrauch Y, Elsbach P, Madsen LM, Foreman A, Weiss J. 1996. The potent anti-*Staphylococcus aureus* activity of a sterile rabbit inflammatory fluid is due to a 14-kD phospholipase A2. *J. Clin. Invest.* 97:250–57

30. Reynolds HY. 1997. Integrated host defense against infections. In *The Lung,* ed. RG Crystal, JB West, ER Weibel, PJ Barnes, pp. 2353–65. New York: Raven

31. Ganz T, Weiss J. 1997. Antimicrobial peptides of phagocytes and epithelia. *Semin. Hematol.* 34:343–54

32. LeVine AM, Bruno MD, Huelsman KM, Ross GF, Whitsett JA, Korfhagen TR. 1997. Surfactant protein A-deficient mice are susceptible to group B streptococcal infection. *J. Immunol.* 158:4336–40

33. LeVine AM, Kurak KE, Bruno MD, Stark JM, Whitsett JA, Korfhagen TR. 1998. Surfactant protein-A-deficient mice are susceptible to *Pseudomonas aeruginosa* infection. *Am. J. Respir. Cell Mol. Biol.* 19:700–8

34. Crouch EC. 1998. Minireview: collectins and pulmonary host defense. *Am. J. Respir. Cell Mol. Biol.* 19:177–201

35. Mason RJ, Greene K, Voelker DR. 1998. Surfactant protein A and surfactant protein D in health and disease. *Am. J. Physiol. Lung Cell Mol. Physiol.* 275:L1–L13

36. Brogden KA, De Lucca AJ, Bland J, Elliott S. 1996. Isolation of an ovine pulmonary surfactant-associated anionic peptide bactericidal for *Pasteurella haemolytica. Proc. Natl. Acad. Sci. USA* 93:412–16

37. Salathe M, Holderby M, Forteza R, Abraham WM, Wanner A, Conner GE. 1997. Isolation and characterization of a peroxidase from the airway. *Am. J. Respir. Cell Mol. Biol.* 17:97–105

38. Zanetti M, Gennaro R, Romeo D. 1995. Cathelicidins: a novel protein family with a common proregion and a variable C-terminal antimicrobial domain. *FEBS Lett.* 174:1–5

39. Gennaro R, Zanetti M. 2000. Structural features and biological activities of the cathelicidin-derived antimicrobial peptides. *Biopolymers* 55:31–49

40. Chan YR, Zanetti M, Gennaro R, Gallo RL. 2001. Anti-microbial activity and cell binding are controlled by sequence determinants in the anti-microbial peptide PR-39. *J. Invest. Dermatol.* 116:230–35

41. Agerberth A, Gunne H, Odeberg J, Kogner P, Boman HG, Gudmundsson GH. 1995. FALL-39, a putative human peptide antibiotic, is cysteine-free and expressed in bone marrow and testis. *Proc. Natl. Acad. Sci. USA* 92:195–99

42. Gallo RL, Kim KJ, Bernfield M, Kozak CA, Zanetti M, et al. 1997. Identification of CRAMP, a cathelin-related antimicrobial peptide expressed in the embryonic and adult mouse. *J. Biol. Chem.* 272:13088–93

43. Skerlavaj B, Scocchi M, Gennaro R, Risso A, Zanetti M. 2001. Structural and functional analysis of horse cathelicidin peptides. *Antimicrobial Agents Chemother.* 45:715–22

44. Mahoney MM, Lee AY, Brezinski-Caliguri DJ, Huttner KM. 1995. Molecular analysis of the sheep cathelin family reveals a novel antimicrobial peptide. *FEBS Lett.* 377:519–22

45. Bagella L, Scocchi M, Zanetti M. 1995. cDNA sequences of three sheep myeloid cathelicidins. *FEBS Lett.* 376:225–28

46. Skerlavaj B, Benincasa M, Risso A, Zanetti M, Gennaro R. 1999. SMAP-29: a potent antibacterial and antifungal peptide from sheep leukocytes. *FEBS Lett.* 463:58–62

47. Agerberth B, Grunewald J, Castanos VE, Olsson B, Jornvall H, et al. 1999. Antibacterial components in bronchoalveolar lavage fluid from healthy individuals and sarcoidosis patients. *Am. J. Respir. Crit. Care Med.* 160:283–90

48. Bals R, Wang X, Zasloff M, Wilson JM. 1998. The peptide antibiotic LL-37/hCAP-18 is expressed in epithelia of the human lung where it has broad antimicrobial activity at the airway surface. *Proc. Natl. Acad. Sci. USA* 95:9541–46

49. Bals R, Lang C, Weiner DJ, Vogelmeier C, Welsch U, Wilson JM. 2001. Rhesus monkey (*Macaca mulatta*) mucosal antimicrobial peptides are close homologues of human molecules. *Clin. Diagn. Lab. Immunol.* 8:370–75

50. Mallow EB, Harris A, Salzman N, Russell JP, Deberardinis RJ, et al. 1996. Human enteric defensins—Gene structure and developmental expression. *J. Biol. Chem.* 271:4038–45

51. Tang YQ, Yuan J, Osapay G, Osapay K, Tran D, et al. 1999. A cyclic antimicrobial peptide produced in primate leukocytes by the ligation of two truncated alpha-defensins. *Science* 286:498–502

52. Harder J, Bartels J, Christophers E, Schroder J-M. 1997. A peptide antibiotic from human skin. *Nature* 387:861–62

53. McNamara N, Van R, Tuchin OS, Fleiszig SM. 1999. Ocular surface epithelia express mRNA for human beta defensin-2. *Exp. Eye Res.* 69:483–90

54. Mathews MS, Jia HP, Guthmiller JM, Losh G, Graham S, et al. 1999. Production of beta-defensin antimicrobial peptides by the oral mucosa and salivary glands. *Infect. Immun.* 67:2740–45

55. Jia HP, Wowk SA, Schutte BC, Lee SK, Vivado A, et al. 2000. A novel murine beta-defensin expressed in tongue, esophagus, and trachea. *J. Biol. Chem.* 275:33314–20

56. Krisanaprakornkit S, Weinberg A, Perez CN, Dale BA. 1998. Expression of the peptide antibiotic human β-defensin 1 in cultured gingival epithelial cells and gingival tissue. *Infect. Immun.* 66:4222–28

57. O'Neil DA, Porter EM, Elewaut D, Anderson GM, Eckmann L, et al. 1999. Expression and regulation of the human beta-defensins hBD-1 and hBD-2 in intestinal epithelium. *J. Immunol.* 163:6718–24

58. Valore EV, Park CH, Quayle AJ, Wiles KR, McCray PB Jr, Ganz T. 1998. Human β-defensin-1, an antimicrobial peptide of urogenital tissues. *J. Clin. Invest.* 101:1633–42

59. Zucht HD, Grabowsky J, Schrader M, Liepke C, Jurgens M, et al. 1998. Human β-defensin-1: a urinary peptide present in variant molecular forms and its putative functional implication. *Eur. J. Med. Res.* 3:315–23

60. Bals R, Wang X, Wu Z, Freeman T, Bafna V, et al. 1998. Human β-defensin 2 is a salt-sensitive peptide antibiotic expressed in human lung. *J. Clin. Invest.* 102:874–80

61. Goldman MJ, Anderson MG, Stolzenberg ED, Kari PU, Zasloff M, Wilson JM. 1997. Human β-defensin-1 is a salt-sensitive antibiotic in lung that is inactivated in cystic fibrosis. *Cell* 88:1–9

62. McCray PB Jr, Bentley L. 1997. Human airway epithelia express a β-defensin. *Am. J. Respir. Cell. Mol. Biol.* 16:343–49

63. Bensch KW, Raida M, Magert HJ, Schulz-Knappe P, Forssmann WG. 1995. hBD-1: a novel beta-defensin from human plasma. *FEBS Lett.* 368:331–35

64. Harder J, Bartels J, Christophers E, Schroder JM. 2001. Isolation and characterization of human beta-defensin-3, a novel human inducible peptide antibiotic. *J. Biol. Chem.* 276:5707–13

65. Jia HP, Schutte BC, Schudy A, Linzmeier R, Guthmiller JM, et al. 2001. Discovery of new human beta-defensins using a genomics-based approach. *Gene* 263:211–18

66. Diamond G, Zasloff M, Eck H, Brasseur M, Maloy WL, Bevins CL. 1991. Tracheal antimicrobial peptide, a cysteine-rich peptide from mammalian tracheal mucosa: peptide isolation and cloning of a cDNA. *Proc. Natl. Acad. Sci. USA* 88:3952–56

67. Diamond G, Jones DE, Bevins CL. 1993. Airway epithelial cells are the site of expression of a mammalian antimicrobial peptide gene. *Proc. Natl. Acad. Sci. USA* 90:4596–600

68. Selsted ME, Tang YQ, Morris WL, McGuire PA, Novotny MJ, et al. 1993. Purification, primary structures, and antibacterial activities of beta-defensins, a new family of antimicrobial peptides from bovine neutrophils. *J. Biol. Chem.* 268:6641–48

69. Zhao C, Nguyen T, Liu L, Sacco RE, Brogden KA, Lehrer RI. 2001. Gallinacin-3, an inducible epithelial beta-defensin in the chicken. *Infect. Immun.* 69:2684–91

70. Zhao C, Nguyen T, Liu L, Shamova O, Brogden K, Lehrer RI. 1999. Differential expression of caprine beta-defensins in digestive and respiratory tissues. *Infect. Immun.* 67:6221–24

71. Deleted in proof

72. Zhang G, Wu H, Shi J, Ganz T, Ross CR, Blecha F. 1998. Molecular cloning and tissue expression of porcine β-defensin-1. *FEBS Lett.* 424:37–40

73. Li P, Chan HC, He B, So SC, Chung YW, et al. 2001. An antimicrobial peptide gene found in the male reproductive system of rats. *Science* 291:1783–85

74. Huttner KM, Brezinski-Caliguri DJ, Mahoney MM, Diamond G. 1998. Antimicrobial peptide expression is developmentally regulated in the ovine gastrointestinal tract. *J. Nutr.* 128:297S–99S

75. Tarver AP, Clark DP, Diamond G, Russell JP, E-Bromage H, et al. 1998. Enteric β-defensin: molecular cloning and characterization of a gene with inducible intestinal epithelial cell expression associated with cryptosporidium parvum infection. *Infect. Immun.* 66:1045–56

76. Morrison GM, Davidson DJ, Dorin JR. 1999. A novel mouse beta defensin, Defb2, which is upregulated in the airways by lipopolysaccharide. *FEBS Lett.* 442:112–16

77. Duits LA, Langermans JA. Paltsansing S, van der Straaten T, Vervebbe RA, et al. 2000. Expression of beta-defensin-1 in chimpanzee (*Pan troglodytes*) airways. *J. Med. Primatol.* 29:318–23

78. Frohlich O, Po C, Murphy T, Young LG. 2000. Multiple promoter and splicing mRNA variants of the epididymis-specific gene EP2. *J. Androl.* 21:421–30

79. Lander ES, Linton LM, Birren B, Nusbaum C, Zody MC, et al. 2001. Initial sequencing and analysis of the human genome. *Nature* 409:860–921

80. Venter JC, Adams MD, Myers EW, Li

PW, Mural RJ, et al. 2001. The sequence of the human genome. *Science* 291:1304–51

81. Marshall E. 2000. Genomics. Public-private project to deliver mouse genome in 6 months. *Science* 290:242–43

82. Scheetz TE, Raymond MR, Nishimura DY, McClain A, Roberts C, et al. 2001. Generation of a high-density rat EST map. *Genome Res.* 11:497–502

83. Elgar G, Clark MS, Meek S, Smith S, Warner S, et al. 1999. Generation and analysis of 25 Mb of genomic DNA from the pufferfish Fugu rubripes by sequence scanning. *Genome Res.* 9:960–71

84. Huttner KM, Kozak CA, Bevins CL. 1997. The mouse genome encodes a single homolog of the antimicrobial peptide human β-defensin 1. *FEBS Lett.* 413:45–49

85. Jia HP, Wowk SA, Schutte BC, Lee SK, Vivado A, et al. 2000. A novel murine beta-defensin expressed in tongue, esophagus, and trachea. *J. Biol. Chem.* 275:33314–20

86. Jia HP, Mills JN, Barahmand-Pour F, Nishimura D, Mallampali RK, et al. 1999. Molecular cloning and characterization of rat genes encoding homologues of human beta-defensins. *Infect. Immun.* 67:4827–33

87. Zimmermann GR, Legault P, Selsted ME, Pardi A. 1995. Solution structure of bovine neutrophil β-defensin-12: The peptide fold of the β-defensins is identical to that of the classical defensins. *Biochemistry* 34:13663–71

88. Sawai MV, Jia HP, Liu L, Aseyev V, Wiencek JM, et al. 2001. The NMR structure of human beta-defensin-2 reveals a novel alpha- helical segment. *Biochemistry* 40:3810–16

89. Hoover DM, Rajashankar KR, Blumenthal R, Puri A, Oppenheim JJ, et al. 2000. The structure of human beta-defensin-2 shows evidence of higher order oligomerization. *J. Biol. Chem.* 275:32911–18

90. Torres AM, Wang X, Fletcher JI, Alewood D, Alewood PF, et al. 1999. Solution structure of a defensin-like peptide from platypus venom. *J. Biochem.* 341:785–94

91. Kirchhoff C, Osterhoff C, Habben I, Ivell R, Kirchloff C. 1990. Cloning and analysis of mRNAs expressed specifically in the human epididymis. *Int. J. Androl.* 13:155–67. Erratum *Int. J. Androl.* 1990 13(4):327

92. Krull N, Ivell R, Osterhoff C, Kirchhoff C. 1993. Region-specific variation of gene expression in the human epididymis as revealed by in situ hybridization with tissue-specific cDNAs. *Mol. Reprod. Dev.* 34:16–24

93. Osterhoff C, Kirchhoff C, Krull N, Ivell R. 1994. Molecular cloning and characterization of a novel human sperm antigen (HE2) specifically expressed in the proximal epididymis. *Biol. Reprod.* 50:516–25

94. Frohlich O, Po C, Young LG. 2001. Organization of the human gene encoding the epididymis-specific ep2 protein variants and its relationship to defensin genes. *Biol. Reprod.* 64:1072–79

95. Hamil KG, Sivashanmugam P, Richardson RT, Grossman G, Ruben SM, et al. 2000. HE2beta and HE2gamma, new members of an epididymis-specific family of androgen-regulated proteins in the human. *Endocrinology* 141:1245–53

96. White SH, Wimley WC, Selsted ME. 1995. Structure, function, and membrane integration of defensins. *Curr. Opin. Struct. Biol.* 5:521–27

97. Bals R, Goldman MJ, Wilson JM. 1998. Mouse β-defensin 1 is a salt-sensitive antimicrobial peptide present in epithelia of the lung and urogenital tract. *Infect. Immun.* 66:1225–32

98. Singh PK, Jia HP, Wiles K, Hesselberth J, Liu L, et al. 1998. Production of β-defensins by human airway epithelia. *Proc. Natl. Acad. Sci. USA* 95:14961–66

99. Morrison GM, Davidson DJ, Kilanowski FM, Borthwick DW, Crook K, et al. 1998. Mouse beta defensin-1 is a functional

homolog of human beta defensin-1. *Mamm. Genome* 9:453–57

100. Bals R, Wang X, Meegalla RL, Wattler S, Weiner D, et al. 1999. Mouse beta-defensin 3 is an inducible antimicrobial peptide expressed in the epithelia of multiple organs. *Infect. Immun.* 67:3542–47

101. Shi J, Zhang G, Wu H, Ross C, Blecha F, Ganz T. 1999. Porcine epithelial beta-defensin 1 is expressed in the dorsal tongue at antimicrobial concentrations. *Infect. Immun.* 67:3121–27

102. Hill CP, Yee J, Selsted ME, Eisenberg D. 1991. Crystal structure of defensin HNP-3, an amphiphilic dimer: mechanisms of membrane permeabilization. *Science* 251:1481–85

103. Pardi A, Hare DR, Selsted ME, Morrison RD, Bassolino DA, Bach AC 2nd. 1988. Solution structures of the rabbit neutrophil defensin NP-5. *J. Mol. Biol.* 201:625–36

104. Pardi A, Zhang XL, Selsted ME, Skalicky JJ, Yip PF. 1992. NMR studies of defensin antimicrobial peptides. 2. Three-dimensional structures of rabbit NP-2 and human HNP-1. *Biochemistry* 31:11357–64

105. Gallagher DSJ, Ryan AM, Diamond G, Bevins CL, Womack JE. 1995. Somatic cell mapping of β-defensin genes to cattle syntenic group U25 and fluorescence in situ localization to chromosome 27. *Mamm. Genet.* 6:554–56

106. Liu L, Zhao C, Heng HHQ, Ganz T. 1997. The human-beta-defensin-1 and alpha-defensins are encoded by adjacent genes: two peptide families with differing disulfide topology share a common ancestry. *Genomics* 43:316–20

107. Harder J, Siebert R, Zhang Y, Matthiesen P, Christophers E, et al. 1997. Mapping of the gene encoding human β-defensin-2 (DEFB2) to chromosome region 8p22–p23.1. *Genomics* 46:472–75

108. Liu L, Wang L, Jia HP, Zhao C, Heng HHQ, et al. 1998. Structure and mapping of the human β-defensin HBD-2 gene and its expression at sites of inflammation. *Gene* 222:237–44

109. Linzmeier R, Ho CH, Hoang BV, Ganz T. 1999. A 450–kb contig of defensin genes on human chromosome 8p23. *Gene* 233:205–11

110. Huttner KM, Lambeth MR, Burkin HR, Burkin DJ, Broad TE. 1998. Localization and genomic organization of sheep antimicrobial peptide genes. *Gene* 206:85–91

111. Zhang G, Hiraiwa H, Yasue H, Wu H, Ross CR, et al. 1999. Cloning and characterization of the gene for a new epithelial beta-defensin. Genomic structure, chromosomal localization, and evidence for its constitutive expression. *J. Biol. Chem.* 274:24031–37

112. Altschul SF, Gish W, Miller W, Myers EW, Lipman DJ. 1990. Basic local alignment search tool. *J. Mol. Biol.* 215:403–10

113. Digilio MC, Marino B, Guccione P, Giannotti A, Mingarelli R, Dallapiccola B. 1998. Deletion 8p syndrome. *Am. J. Med. Genet.* 75:534–36

114. Schonwetter BS, Stolzenberg ED, Zasloff MA. 1995. Epithelial antibiotics induced at sites of inflammation. *Science* 267:1645–48

115. Russell JP, Diamond G, Tarver AP, Scanlin TF, Bevins CL. 1996. Coordinate induction of two antibiotic genes in tracheal epithelial cells exposed to the inflammatory mediators lipopolysaccharide and tumor necrosis factor alpha. *Infect. Immun.* 64:1565–68

116. Diamond G, Russell JP, Bevins CL. 1996. Inducible expression of an antibiotic peptide gene in lipopolysaccharide-challenged tracheal epithelial cells. *Proc. Natl. Acad. Sci. USA* 93:5156–60

117. McCray PB Jr, Zabner J, Jia HP, Welsh MJ, Thorne PS. 1999. Efficient killing of inhaled bacteria in deltaF508 mice: role of airway surface liquid composition. *Am. J. Physiol. Lung Cell Mol. Physiol.* 277:L183–L90

118. Yamaguchi Y, Fukuhara S, Nagase T,

Tomita T, Hitomi S, et al. 2001. A novel mouse β-defensin, mBD-6, predominantly expressed in skeletal muscle. *J. Biol. Chem.* 14:14

119. Harder J, Meyer-Hoffert U, Teran LM, Schwichtenberg L, Bartels J, et al. 2000. Mucoid *Pseudomonas aeruginosa*, TNF-alpha, and IL-1beta, but not IL-6, induce human beta-defensin-2 in respiratory epithelia. *Am. J. Respir. Cell Mol. Biol.* 22:714–21

120. Zhao C, Wang I, Lehrer RI. 1996. Widespread expression of beta-defensin hBD-1 in human secretory glands and epithelial cells. *FEBS Lett.* 396:319–22

121. Zielenski J, Corey M, Rozmahel R, Markiewicz D, Aznarez I, et al. 1999. Detection of a cystic fibrosis modifier locus for meconium ileus on human chromosome 19q13. *Nat. Genet.* 22:128–29

122. Hiratsuka T, Nakazato M, Date Y, Ashitani J, Minematsu T, et al. 1998. Identification of human β-defensin-2 in respiratory tract and plasma and its increase in bacterial pneumonia. *Biochem. Biophys. Res. Commun.* 249:943–47

123. Ashitani J, Mukae H, Hiratsuka T, Nakazato M, Kumamoto K, Matsukura S. 2001. Plasma and BAL fluid concentrations of antimicrobial peptides in patients with *Mycobacterium avium*-intracellular infection. *Chest* 119:1131–37

124. Jia HP, Starner T, Ackermann M, Kirby P, Tack BF, McCray PB Jr. 2001. Abundant human beta-defensin-1 expression in milk and mammary gland epithelium. *J. Pediatr.* 138:109–12

125. Medzhitov R, Preston-Hurlburt P, Janeway CA. 1997. A human homologue of the *Drosophila* Toll protein signals activation of adaptive immunity. *Nature* 388:394–97

126. Tauszig S, Jouanguy E, Hoffmann JA, Imler JL. 2000. From the cover: toll-related receptors and the control of antimicrobial peptide expression in *Drosophila*. *Proc. Natl. Acad. Sci. USA* 97:10520–25

127. Becker MN, Diamond G, Verghese MW, Randell SH. 2000. CD14-dependent lipopolysaccharide-induced beta-defensin-2 expression in human tracheobronchial epithelium. *J. Biol. Chem.* 275:29731–36

128. Arbour NC, Lorenz E, Schutte BC, Zabner J, Kline JN, et al. 2000. TLR4 mutations are associated with endotoxin hyporesponsiveness in humans. *Nat. Genet.* 25:187–91

129. Ozinsky A, Underhill DM, Fontenot JD, Hajjar AM, Smith KD, et al. 2000. The repertoire for pattern recognition of pathogens by the innate immune system is defined by cooperation between toll-like receptors. *Proc. Natl. Acad. Sci. USA* 97:13766–71

130. Schwandner R, Dziarski R, Wesche H, Rothe M, Kirschning CJ. 1999. Peptidoglycan- and lipoteichoic acid-induced cell activation is mediated by toll-like receptor 2. *J. Biol. Chem.* 274:17406–9

131. Brightbill HD, Libraty DH, Krutzik SR, Yang RB, Belisle JT, et al. 1999. Host defense mechanisms triggered by microbial lipoproteins through toll-like receptors. *Science* 285:732–36

132. Aliprantis AO, Yang RB, Mark MR, Suggett S, Devaux B, et al. 1999. Cell activation and apoptosis by bacterial lipoproteins through toll-like receptor-2. *Science* 285:736–39

133. Lien E, Means TK, Heine H, Yoshimura A, Kusumoto S, et al. 2000. Toll-like receptor 4 imparts ligand-specific recognition of bacterial lipopolysaccharide. *J. Clin. Invest.* 105:497–504

134. Hayashi F, Smith KD, Ozinsky A, Hawn TR, Yi EC, et al. 2001. The innate immune response to bacterial flagellin is mediated by Toll-like receptor 5. *Nature* 410:1099–103

135. Hemmi H, Takeuchi O, Kawai T, Kaisho T, Sato S, et al. 2000. A Toll-like receptor recognizes bacterial DNA. *Nature* 408:740–45

136. Singh P, Travis S, Smith JJ, Jia HP, Tack B, et al. 1998. Expression and activity of β-defensins-1 and 2 (hBD-1 and hBD-2) in airway epithelia. *Am. J. Respir. Crit. Care Med.* 157:A58 (Abstr.)

137. Daher KA, Selsted ME, Lehrer RI. 1986. Direct inactivation of viruses by human granulocyte defensins. *J. Virol.* 60:1068–74

138. Lawyer C, Pai S, Watabe M, Bakir H, Eagleton L, Watabe K. 1996. Effects of synthetic form of tracheal antimicrobial peptide on respiratory pathogens. *J. Antimicrob. Chemo.* 37:599–604

139. Lehrer RI, Barton A, Daher KA, Harwig SS, Ganz T, Selsted ME. 1989. Interaction of human defensins with *Escherichia coli.* Mechanism of bactericidal activity. *J. Clin. Invest.* 84:553–61

140. Singh PK, Tack BF, McCray PB Jr, Welsh MJ. 2000. Synergistic and additive killing by antimicrobial factors found in human airway surface liquid. *Am. J. Physiol. Lung Cell Mol. Physiol.* 279:L799–L805

141. Yang D, Chertov O, Bykovskaia SN, Chen Q, Buffo MJ, et al. 1999. Beta-defensins: linking innate and adaptive immunity through dendritic and T cell CCR6. *Science* 286:525–28

142. Ganz T. 1999. Defensins and host defense. *Science* 286:420–21

143. Liao F, Rabin RL, Smith CS, Sharma G, Nutman TB, Farber JM. 1999. CC-chemokine receptor 6 is expressed on diverse memory subsets of T cells and determines responsiveness to macrophage inflammatory protein 3 alpha. *J. Immunol.* 162:186–94

144. Baba M, Imai T, Nishimura M, Kakizaki M, Takagi S, et al. 1997. Identification of CCR6, the specific receptor for a novel lymphocyte-directed CC chemokine LARC. *J. Biol. Chem.* 272:14893–98

145. Niyonsaba F, Someya A, Hirata M, Ogawa H, Nagaoka I. 2001. Evaluation of the effects of peptide antibiotics hu-

man beta-defensins-1/-2 and LL-37 on histamine release and prostaglandin D(2) production from mast cells. *Eur. J. Immunol.* 31:1066–75

146. Holt PG. 2000. Antigen presentation in the lung. *Am. J. Respir. Crit. Care Med.* 162:S151–56

147. Dieu MC, Vanbervliet B, Vicari A, Bridon JM, Oldham E, et al. 1998. Selective recruitment of immature and mature dendritic cells by distinct chemokines expressed in different anatomic sites. *J. Exp. Med.* 188:373–86

148. Holt PG, Schon-Hegrad MA, Oliver J. 1988. MHC class II antigen-bearing dendritic cells in pulmonary tissues of the rat. Regulation of antigen presentation activity by endogenous macrophage populations. *J. Exp. Med.* 167:262–74

149. McWilliam AS, Nelson D, Thomas JA, Holt PG. 1994. Rapid dendritic cell recruitment is a hallmark of the acute inflammatory response at mucosal surfaces. *J. Exp. Med.* 179:1331–36

150. McWilliam AS, Napoli S, Marsh AM, Pemper FL, Nelson DJ, et al. 1996. Dendritic cells are recruited into the airway epithelium during the inflammatory response to a broad spectrum of stimuli. *J. Exp. Med.* 184:2429–32

151. McWilliam AS, Marsh AM, Holt PG. 1997. Inflammatory infiltration of the upper airway epithelium during Sendai virus infection: involvement of epithelial dendritic cells. *J. Virol.* 71:226–36

152. Hieshima K, Imai T, Opdenakker G, Van Damme J, Kusuda J, et al. 1997. Molecular cloning of a novel human CC chemokine liver and activation-regulated chemokine (LARC) expressed in liver. Chemotactic activity for lymphocytes and gene localization on chromosome 2. *J. Biol. Chem.* 272:5846–53

153. Hromas R, Gray PW, Chantry D, Godiska R, Krathwohl M, et al. 1997. Cloning and characterization of exodus, a novel beta-chemokine. *Blood* 89:3315–22

154. Dieu-Nosjean MC, Massacrier C, Homey B, Vanbervliet B, Pin JJ, et al. 2000. Macrophage inflammatory protein 3alpha is expressed at inflamed epithelial surfaces and is the most potent chemokine known in attracting Langerhans cell precursors. *J. Exp. Med.* 192:705–18

155. Cook DN, Prosser DM, Forster R, Zhang J, Kuklin NA, et al. 2000. CCR6 mediates dendritic cell localization, lymphocyte homeostasis, and immune responses in mucosal tissue. *Immunity* 12:495–503

156. Varona R, Villares R, Carramolino L, Goya I, Zaballos A, et al. 2001. CCR6-deficient mice have impaired leukocyte homeostasis and altered contact hypersensitivity and delayed-type hypersensitivity responses. *J. Clin. Invest.* 107:R37–45

157. Murphy CJ, Foster BA, Mannis MJ, Selsted ME, Reid TW. 1993. Defensins are mitogenic for epithelial cells and fibroblasts. *J. Cell. Physiol.* 155:408–13

157a. Higazi AA, Ganz T, Kariko K, Cines DB. 1996. Defensin modulates tissue-type plasminogen activator and plasminogen binding to fibrin and endothelial cells. *J. Biol. Chem.* 271:17650–55

158. Zhu QZ, Hu J, Mulay S, Esch F, Shimasaki S, Solomon S. 1988. Isolation and structure of corticostatin peptides from rabbit fetal and adult lung. *Proc. Natl. Acad. Sci. USA* 85:592–96

159. Prohaszka Z, Nemet K, Csermely P, Hudecz F, Mezo G, Fust G. 1997. Defensins purified from human granulocytes bind C1q and activate the classical complement pathway like the transmembrane glycoprotein gp41 of HIV-1. *Mol. Immunol.* 34:809–16

160. Van Wetering S, Mannesse-Lazeroms SPG, Dijkman JH, Hiemstra PS. 1997. Effect of neutrophil serine proteinases and defensins on lung epithelial cells: modulation of cytotoxicity and IL-8 production. *J. Leukoc. Biol.* 62:217–26

161. Lencer WI, Cheung G, Strohmeier GR, Currie MG, Ouellette AJ, et al. 1997. Induction of epithelial chloride secretion by channel-forming cryptdins 2 and 3. *Proc. Natl. Acad. Sci. USA* 94:8585–89

162. Hughes AL. 1999. Evolutionary diversification of the mammalian defensins. *Cell Mol. Life Sci.* 56:94–103

163. Kimura M. 1977. Preponderance of synonymous changes as evidence for the neutral theory of molecular evolution. *Nature* 267:275–76

164. Hughes AL. 1998. Phylogenetic tests of the hypothesis of block duplication of homologous genes on human chromosomes 6, 9, and 1. *Mol. Biol. Evol.* 15:854–70

165. Turner J, Cho Y, Dinh N, Waring AJ, Lehrer RI. 1998. Activities of LL-37, a cathelin-associated antimicrobial peptide of human neutrophils. *Antimicrob. Agents Chemother.* 9:2206–14

166. Travis SM, Anderson NN, Forsyth WR, Espiritu C, Conway BD, et al. 2000. Bactericidal activity of mammalian cathelicidin-derived peptides. *Infect. Immun.* 68:2748–55

167. Lehrer RI, Selsted ME, Szklarek D, Fleischmann J. 1983. Antibacterial activity of microbicidal cationic proteins 1 and 2, natural peptide antibiotics of rabbit lung macrophages. *Infect. Immun.* 42:10–14

168. Choi JY, Muallem D, Kiselyov K, Lee MG, Thomas S. 2001. Aberrant CFTR-dependent HCO_3-transport in mutations associated with cystic fibrosis. *Nature* 410:94–97

169. Smith JJ, Travis SM, Greenberg EP, Welsh MJ. 1996. Cystic fibrosis airway epithelia fail to kill bacteria because of abnormal airway surface fluid. *Cell* 85:229–36

170. Guggino WB. 1999. Minireview. Cystic fibrosis and the salt controversy. *Cell* 96:607–10

171. Wine JJ. 1997. A sensitive defense: salt and cystic fibrosis. *Nat. Med.* 3:1–2

172. Knowles MR, Robinson JM, Wood RE, Pue CA, Mentz WM, et al. 1997. Ion composition of airway surface liquid of patients with cystic fibrosis as compared with normal and disease-control subjects. *J. Clin. Invest.* 100:2588–95

173. Boucher RC. 1999. Topical review—molecular insights into the physiology of the 'thin film' of airway surface liquid. *J. Physiol.* 516:631–38

174. Lawton AR, Cooper MD. 1996. Ontogeny of immunity. In *Immunologic Disorders: In Infants and Children*, ed. ER Stiehm, pp. 1–13. Philadelphia: Saunders

175. Scariati PD, Grummer-Strawn LM, Fein SB. 1997. A longitudinal analysis of infant morbidity and the extent of breast-feeding in the United States. *Pediatrics* 99:E5

176. Tunzi CR, Harper PA, Bar-Oz B, Valore EV, Semple JL, et al. 2000. Beta-defensin expression in human mammary gland epithelia. *Pediatr. Res.* 48:30–35

177. Barton L, Hodgman JE, Pavlova Z. 1999. Causes of death in the extremely low birth weight infant. *Pediatrics* 103:446–51

178. Barker PM, Gowen CW, Lawson EE, Knowles MR. 1997. Decreased sodium ion absorption across nasal epithelium of very premature infants with respiratory distress syndrome. *J. Pediatr.* 130:373–77

179. Smith DE, Otulakowski G, Yeger H, Post M, Cutz E, O'Brodovich HM. 2000. Epithelial Na(+) channel (ENaC) expression in the developing normal and abnormal human perinatal lung. *Am. J. Respir. Crit. Care Med.* 161:1322–31

180. Nygaard SD, Ganz T, Peterson MW. 1993. Defensins reduce the barrier integrity of a cultured epithelial monolayer without cytotoxicity. *Am. J. Respir. Cell. Mol. Biol.* 8:193–200

181. Zhang H, Porro G, Orzech N, Mullen B, Liu M, Slutsky AS. 2001. Neutrophil defensins mediate acute inflammatory response and lung dysfunction in dose-related fashion. *Am. J. Physiol. Lung Cell Mol. Physiol.* 280:L947–L54

182. Hiemstra PS, Van Wetering S, Stolk J. 1998. Neutrophil serine proteinases and defensins in chronic obstructive pulmonary disease: effects on pulmonary epithelium. *Eur. Respir. J.* 12:1200–8

183. van Wetering S, Sterk PJ, Rabe KF, Hiemstra PS. 1999. Defensins: key players or bystanders in infection, injury, and repair in the lung? *J. Allergy Clin. Immunol.* 104:1131–38

184. Soong LB, Ganz T, Ellison A, Coughey GH. 1997. Purification and characterization of defensins from cystic fibrosis sputum. *Inflamm. Res.* 46:98–102

185. Islam D, Bandholtz L, Nilsson J, Wigzell H, Christensson B, et al. 2001. Downregulation of bactericidal peptides in enteric infections: a novel immune escape mechanism with bacterial DNA as a potential regulator. *Nat. Med.* 7:180–85

186. Breithaupt H. 1999. The new antibiotics: Can novel antibacterial treatments combat the rising tide of drug-resistant infections? *Nat. Biotech.* 17:1165–69

187. Yu Q, Lehrer RI, Tam JP. 2000. Engineered salt-insensitive alpha-defensins with end-to-end circularized structures. *J. Biol. Chem.* 275:3943–49

188. Hancock REW, Lehrer R. 1998. Cationic peptides: a new source of antibiotics. *Trends Biotech.* 16:82–88

189. Kelley KJ. 1996. Using host defenses to fight infectious diseases. *Nat. Biotech.* 14:587

190. Ryan LK, Rhodes J, Bhat M, Diamond G. 1998. Expression of beta-defensin genes in bovine alveolar macrophages. *Infect. Immun.* 66:878–81

191. Yount NY, Yuan J, Tarver A, Castro T, Diamond G, et al. 1999. Cloning and expression of bovine neutrophil

beta-defensins. Biosynthetic profile during neutrophilic maturation and localization of mature peptide to novel cytoplasmic dense granules. *J. Biol. Chem.* 274:26249–58

192. Harwig SS, Swiderek KM, Kokryakov VN, Tan L, Lee TD, et al. 1994. Gallinacins: cysteine-rich antimicrobial peptides of chicken leukocytes. *FEBS Lett.* 342:281–85

193. Brockus CW, Jackwood MW, Harmon BG. 1998. Characterization of beta-defensin prepropeptide mRNA from chicken and turkey bone marrow. *Animal Genet.* 29:283–89

194. Diamond G, Kaiser V, Rhodes J, Russell JP, Bevins CL. 2000. Transcriptional regulation of beta-defensin gene expression in tracheal epithelial cells. *Infect. Immun.* 68:113–19

195. Smith LA, Schmidt JJ. 1990. Cloning and nucleotide sequences of crotamine genes. *Toxicon* 28:575–85

196. Evans EW, Beach GG, Wunderlich J, Harmon BG. 1994. Isolation of antimicrobial peptides from avian heterophils. *J. Leukoc. Biol.* 56:661–65

197. Koradi R, Billeter M, Wuthrich K. 1996. MOLMOL: a program for display and analysis of macromolecular structures. *J. Mol. Graph.* 14:51–55, 29–32

198. Dib C, Faure S, Fizames C, Samson D, Drouot N, et al. 1996. A comprehensive genetic map of the human genome based on 5,264 microsatellites. *Nature* 380:152–54

Annu. Rev. Physiol. 2002. 64:749–74

REGULATION OF ENDOTHELIAL NITRIC OXIDE SYNTHASE: Location, Location, Location

Philip W. Shaul

Department of Pediatrics University of Texas Southwestern Medical Center, 5323 Harry Hines Boulevard, Dallas, Texas 75390-9063; e-mail: pshaul@mednet.swmed.edu

Key Words caveolae, epithelium, estrogen, high density lipoprotein, low density lipoprotein

■ **Abstract** Endothelial nitric oxide synthase (eNOS) is expressed in vascular endothelium, airway epithelium, and certain other cell types where it generates the key signaling molecule nitric oxide (NO). Diminished NO availability contributes to systemic and pulmonary hypertension, atherosclerosis, and airway dysfunction. Complex mechanisms underly the cell specificity of eNOS expression, and co- and post-translational processing leads to trafficking of the enzyme to plasma membrane caveolae. Within caveolae, eNOS is the downstream target member of a signaling complex in which it is functionally linked to both typical G protein-coupled receptors and less typical receptors such as estrogen receptor (ER) α and the high-density lipoprotein receptor SR-BI displaying novel actions. This compartmentalization facilitates dynamic protein-protein interactions and calcium- and phosphorylation-dependent signal transduction events that modify eNOS activity. Further understanding of these mechanisms will enable us to take preventive and therapeutic advantage of the powerful actions of NO in multiple cell types.

INTRODUCTION

The endothelial isoform of nitric oxide synthase (eNOS) is one of three isoenzymes that converts L-arginine to L-citrulline plus the key signaling molecule nitric oxide (NO). eNOS is acutely activated by agonists of diverse G protein-coupled cell surface receptors and by physical stimuli such as hemodynamic shear stress, and varying oxygenation. As the name implies, eNOS was originally cloned from vascular endothelium, and NO generated by eNOS regulates blood pressure, platelet aggregation, leukocyte adherence, and vascular smooth muscle cell mitogenesis. Diminished endothelial NO production or availability has been implicated in the pathogenesis of a variety of forms of systemic and pulmonary hypertension and in other vascular disorders including atherosclerosis (1–6).

In addition to its role in the vascular wall, there is evidence that eNOS is expressed in diverse epithelial cell types including in the airway. NO is present in

0066-4278/02/0315-0749$14.00

expired gas (7), and studies in animal models as well as in humans indicate that the principal source of expired NO is the airway rather than the pulmonary vasculature (8, 9). The functions of epithelium-derived NO in the airway include smooth muscle relaxation and bacteriostasis, as well as the modulation of ciliary motility, mucin secretion, and plasma exudation (7, 10). Studies in the perinatal period indicate that airway NO is also critically involved in the regulation of lung liquid production and tissue resistance (11–14). There is also evidence that airway epithelial NOS expression is attenuated during inflammatory conditions, potentially contributing to airway dysfunction (15). Epithelial eNOS expression has also been demonstrated in the kidney, where it plays a role in tubular function (16, 17), and in epithelium of both the male and female reproductive organs, with differential expression during the estrus cycle, suggesting that it is involved in modulating reproduction (18, 19). eNOS has also been discovered in cardiac myocytes, platelets, and the hippocampus (20).

Soon after its initial isolation and characterization in endothelial cells, functional eNOS was found to be primarily associated with the plasma membrane (21). Because eNOS activity is acutely regulated by multiple extracellular stimuli and the NO produced is a labile, cytotoxic messenger molecule with primarily paracrine function (1, 3), both the cell specificity of eNOS expression and the intracellular site of NO synthesis have a major influence on the biological activity of the molecule. Borrowing the phrase from the real estate industry, which emphasizes the importance of location, location, location, this review focuses on the cellular and molecular mechanisms underlying eNOS cell-specific and subcellular localization. For information regarding the regulation of eNOS abundance and the biochemistry of eNOS enzymatic activity, including the process of eNOS uncoupling and superoxide generation, the reader is referred to other reviews (22–24). Paralleling the number of times that the term location is stated in the title, three major topics are covered in the current review. First, the mechanisms underlying the unique capacity for eNOS gene expression in endothelium and epithelium are examined. The processes regulating eNOS trafficking and localization to caveolae, which are specialized microdomains involved in the compartmentalization of signaling molecules on the plasma membrane, are then addressed. Finally, the mechanisms regulating eNOS activity in caveolae are presented. It is anticipated that issues of eNOS localization will continue to warrant strong consideration in our efforts to understand attenuated NO availability as a pathogenetic mechanism in multiple disease states and our attempts to supercede NO deficiency through gene therapy and the modulation of the activity of the enzyme.

CELL-SPECIFICITY OF eNOS EXPRESSION

eNOS Expression in Endothelium and Epithelium

The cell specificity of eNOS expression is exemplified by the immunohistochemical analyses of the distribution of the enzyme and the two other major NOS isoforms, inducible NOS (iNOS) and neuronal NOS (nNOS), in the lung. Studies in ovine lung indicate that eNOS protein is readily detectable in the endothelium at all levels of the pulmonary vasculature, and it is also found in bronchial and proximal bronchiolar epithelium. iNOS is constitutively coexpressed with eNOS in the bronchial and proximal bronchiolar epithelium, and it is also detected in the epithelium of terminal and respiratory bronchioles. nNOS protein is present in the epithelium at all levels including the alveolar wall. These immunohistochemical findings were confirmed by isoform-specific reverse transcription polymerase chain reaction assays and NADPH diaphorase histochemistry (25). Studies in human and rat lung have yielded similar findings for epithelial eNOS (26, 27). Thus there is both a high degree of cell specificity of eNOS expression and redundancy of NOS isoform expression in certain pulmonary epithelial cell types.

Molecular Basis of Cell-Specific eNOS Expression

The mechanisms underlying cell-specific eNOS expression in airway epithelium have been evaluated in transient transfection studies in NCI-H441 human bronchiolar epithelial cells, which are of Clara cell lineage, using the human eNOS promoter fused to a luciferase reporter gene (Luc) (28). eNOS expression in airway epithelium was first demonstrated in H441 cells (29). Transfection with 1624 bp of the eNOS promoter sequence 5′ to the initiation ATG (−1624eNOS-Luc) yielded a 19-fold increase in promoter activity relative to vector alone. Similarly, −1624eNOS-Luc yielded a 35-fold increase in promoter activity in cultured pulmonary artery endothelial cells. However, there was no detectable promoter activity with −1624eNOS-Luc transfected into CCD-18Lu lung fibroblasts, consistent with the exclusive cell-specific expression of pulmonary eNOS in the proximal airway epithelium and vascular endothelium. There was 6-fold greater promoter activity in H441 cells compared with the non-Clara cell human airway epithelial lines BEAS-2B and NHBE, suggesting further selectivity of relative eNOS gene expression in different epithelial cell types. 5′ deletion of the eNOS promoter from −1624 to −994, −318, and −279 did not alter basal eNOS promoter activity in H441 cells. However, further deletion from −279 to −248 reduced basal promoter activity by 65%, and activity was completely lost with deletion to −79. Point mutations of −1624eNOS-Luc revealed that the positive regulatory element between −279 and −248 is the consensus GATA-binding motif at −254 and that the positive regulatory element between −248 and −79 is the Sp1 binding motif at −125 (Figure 1*a*). Parallel studies in endothelial cells revealed identical findings (Figure 1*b*). Electrophoretic mobility shift assays (EMSA) yielded two epithelial nuclear protein-DNA complexes with the GATA site and four complexes with the Sp1 site. Immunodepletion with antisera to GATA-2 prevented formation of both GATA-nuclear protein complexes, and antisera to Sp1 supershifted the slowest migrating Sp1-nuclear protein complex. Whereas EMSA with epithelial and lung fibroblast nuclei showed identical complexes with the Sp1 site, the slower-migrating GATA-nuclear protein complex was unique to epithelial cells. These findings indicate that in airway epithelium the interaction of Sp1 nuclear protein with the Sp1 site at −125 is required for basal activation of the eNOS gene, and

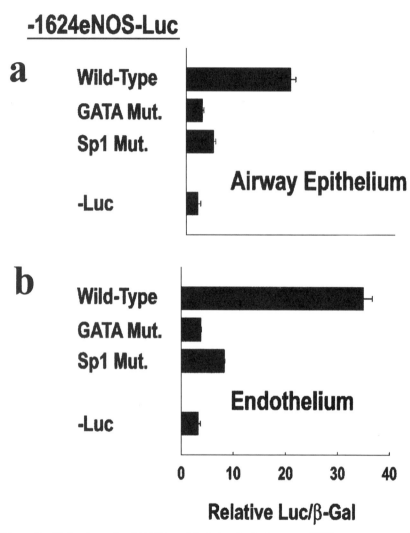

Figure 1 Role of proximal GATA and Sp1 sites in the human eNOS gene promoter in H441 airway epithelial cells (*a*) and pulmonary endothelial cells (*b*). The −1624eNOS-Luc wild-type construct or site-directed mutants (Mut.) of the GATA consensus site at 254 or the Sp1-binding motif at 125 or vector alone (Luc) were cotransfected with SV40-driven β-galactosidase plasmid (β-gal), and relative activities (Luc/β-gal) were determined in cell lysates 72 h later. Values are mean $\pm$ SEM ($n = 3$), and the results are representative of the findings of three independent experiments (reprinted with permission from Reference 30).

that GATA-2 binding to the GATA site at -254 is critically involved in the cell specificity of eNOS expression (30). When considered along with work by other investigators (31), it is apparent that identical mechanisms most likely underlie cell-specific eNOS expression in epithelium and endothelium (Figure 2, see color insert). Furthermore, there is evidence of potential selectivity of eNOS expression in microvascular beds that may be related to a platelet-derived growth factor response element in the eNOS promoter (32). Thus multiple mechanisms at the level of eNOS gene transcription regulate the cell specificity of expression of the enzyme.

eNOS LOCALIZATION IN CAVEOLAE

Trafficking of eNOS to Caveolae

Caveolae are specialized plasma membrane microdomains originally studied in numerous cell types for their involvement in the transcytosis of macromolecules. This earlier work included the landmark observation of receptor-mediated uptake of folate by caveolae (33). Caveolae are enriched in glycosphingolipids, cholesterol, sphingomyelin, and lipid-anchored membrane proteins, and they are characterized by a light buoyant density and resistance to solubilization by Triton X-100 at $4°C$. Once the caveola coat protein caveolin was identified, it was possible to purify this specialized membrane domain, and it was discovered that caveolae also contain a variety of signal transduction molecules. The list of resident signaling molecules includes G protein-coupled receptors such as the muscarinic acetylcholine receptor, G-proteins, and molecules involved in the regulation of intracellular calcium homeostasis such as a plasma membrane calcium pump, and protein kinase C (34–36). Knowledge that eNOS activity is acutely regulated by extracellular factors and that the protein is primarily associated with the plasma membrane prompted the initial examinations of eNOS in caveolae.

In studies of cultured endothelial cells, eNOS localization was first evaluated in subcellular fractions including the caveolae and non-caveolae portions of the plasma membrane that were isolated by a detergent-free method which takes advantage of the unique buoyant density of caveolae membranes (37) (Figure 3a). Within the plasma membrane, the marker protein caveolin was detected exclusively in the caveolae membrane fraction, and eNOS protein was also found to be highly enriched in caveolae membranes. It was then ascertained as to whether the localization of eNOS to caveolae correlates with NOS enzymatic activity (Figure 3b). NOS activity was 7-fold greater in the plasma membrane fraction than in cytosol, and within the plasma membrane, NOS activity was undetectable in the non-caveolae fraction, whereas it was 9- to 10-fold greater in caveolae membranes compared with the whole plasma membrane. In repeat experiments, 51–86% of the total enzymatic activity in the postnuclear supernatant was recovered in the plasma membrane, and 57–100% of the total activity in the plasma membrane was recovered in the caveolae fraction, indicating that in quiescent endothelial cells the majority of functional enzyme is localized to caveolae. To confirm the

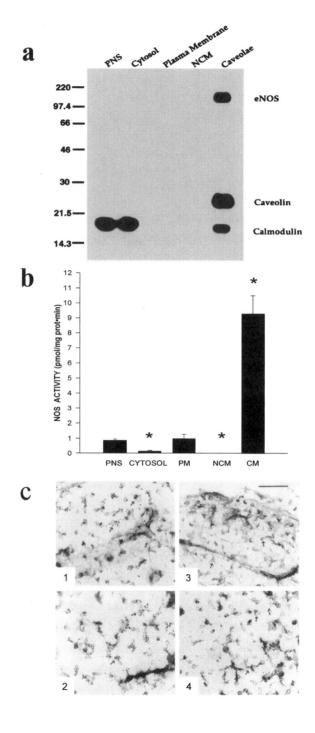

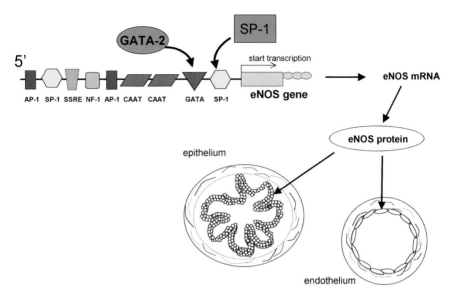

Figure 2 Mechanisms underlying cell-specific eNOS gene expression in epithelium and endothelium. The interaction of Sp1 nuclear protein with the Sp1 site in the core eNOS promoter is required for basal activation of the gene. The cell specificity of eNOS expression entails GATA-2 binding to the GATA site in the core promoter.

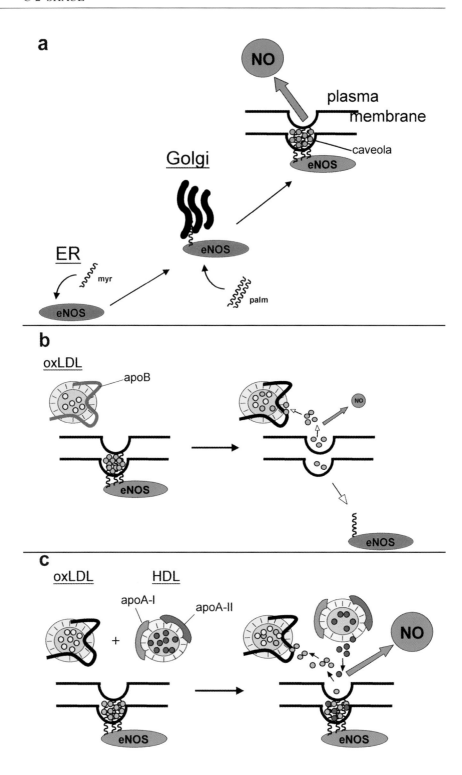

Figure 6 (opposite) Basis for eNOS targeting to caveolae, and the effects of oxidized LDL (oxLDL) and HDL on the process. (*a*) Normal eNOS targeting to caveolae. Myristoylation of the eNOS protein results in eNOS trafficking to the Golgi apparatus, where it is palmitoylated. Myristoylated palmitoylated eNOS is targeted to caveolae, which are enriched in cholesterol (*orange circles*). Localization to caveolae optimizes the capacity for eNOS activation (*change from red to green*). (*b*) Effect of oxLDL on eNOS localization and function in caveolae. OxLDL serves as a cholesterol acceptor, thereby disrupting the specialized lipid environment in caveolae. eNOS is displaced from caveolae to intracellular domains, where the capacity to activate the enzyme is attenuated (*change from green to red*). (*c*) Effect of HDL on eNOS redistribution by oxLDL. At the same time that oxLDL is removing cholesterol from caveolae, HDL is maintaining the total cholesterol content of caveolae by the provision of cholesterol ester (*blue circles*). eNOS is retained in caveolae where it can be effectively activated (*green*).

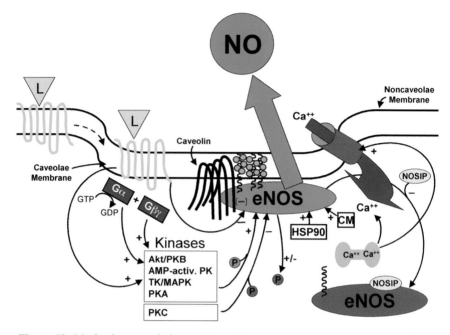

Figure 10 Mechanisms regulating eNOS activity in endothelial cell caveolae. eNOS is localized to cholesterol-enriched (*orange circles*) caveolae by myristoylation and palmitoylation, where interaction with caveolin attenuates the activity of the enzyme. Typical eNOS stimulation is initiated by ligand (L) binding to plasma membrane receptors, which may reside in caveolae or are mobilized to caveolae upon activation. Receptor activation leads to the stimulation of multiple potential kinase pathways through processes that may be G protein-mediated. Phosphorylation events result in eNOS stimulation (+) or eNOS inactivation (-) depending on the kinase involved and the site of eNOS phosphorylation. Dephosphorylation events are also stimulatory or inhibitory. eNOS activity requires the binding of calcium-calmodulin; the calcium is derived from either intracellular stores, which are likely to be in close proximity to caveolae membranes, or from calcium influx. eNOS activity is enhanced by HSP90 binding, and it is diminished by association with the C-terminal domain of certain G protein-coupled receptors. eNOS interacting protein (NOSIP) binds to eNOS to cause translocation of the enzyme from caveolae to intracellular sites, resulting in diminished NOS activity (*change from green to red*).

findings made by subfractionation using an entirely independent method, eNOS was localized by immunoelectron microscopy (Figure 3*c*). Control experiments entailed localization of caveolin, which exists in three major isoforms: caveolin-1, caveolin-2, and caveolin-3. Caveolin-1 and caveolin-2 are ubiquitously expressed, whereas caveolin-3 is found exclusively in muscle cells (38). Using whole-mount plasma membrane preparations of fibroblasts and endothelial cells processed for the localization of caveolin-1, gold particles were found over small invaginations on the membrane surface with the characteristic appearance of caveolae (Figure 3*c*, *Panels 1 and 2*). In contrast, coated pits and smooth membrane areas were not decorated with gold. When identically processed membranes from fibroblasts were labeled with anti-eNOS antiserum, gold particles were not found over caveolae structures or elsewhere (Figure 3*c*, *Panel 3*). However, when endothelial membranes were labeled with anti-eNOS antiserum, caveolae structures were specifically decorated with gold, but the coated pits and smooth membrane were not (Figure 3*c*, *Panel 4*). Thus the detection of eNOS in isolated endothelial caveolae fractions is not the result of nonspecific association of the enzyme with caveolae membranes during the purification process; alternatively, it accurately reflects enzyme localization to the microdomain.

The mechanisms underlying eNOS localization to caveolae were then explored, first by determining if the process is faithfully reconstituted by transient transfection of wild-type eNOS cDNA into COS-7 cells, which possess caveolae but do not express NOS constitutively. The distribution of wild-type eNOS in transfected COS-7 cells mimicked that in endothelial cells, with NOS activity in caveolae enriched 7- to 12-fold relative to whole plasma membrane (Table 1). These findings indicate that the eNOS protein and caveolae membranes collectively possess all of the properties required for post-translational targeting of the enzyme to caveolae. The basis for trafficking to the microdomain was then determined using the recently obtained knowledge that eNOS is both N-terminally myristoylated at the glycine residue in position 2 and palmitoylated at the cysteine residues at positions 15 and 26 (39, 40). cDNA encoding either a myristoylation-deficient mutant of

---

Figure 3   eNOS is localized to endothelial cell caveolae. (*a*) Immunoblot analysis for eNOS, caveolin-1, and calmodulin in subcellular fractions from endothelial cells. Samples of postnuclear supernatant (PNS), cytosol, plasma membrane (PM), noncaveolae membrane (NCM), and caveolae membrane (CM) were evaluated. (*b*) NOS enzymatic activity in subcellular fractions from endothelial cells. L-[^{3}H]arginine conversion to L-[^{3}H]citrulline was measured in the presence of excess substrate, cofactors, calcium, and calmodulin. Enzymatic activity was undetectable in NCM. Results are mean $\pm$ SEM ($n = 4$–6), *p < 0.05 versus plasma membrane. (*c*) Localization of caveolin and eNOS in plasma membranes by immunoelectron microscopy. Immunogold labeling was performed using antibody to caveolin-1 in fibroblasts (*Panel 1*) and endothelial cells (*Panel 2*), and antibody to eNOS in fibroblasts (*Panel 3*) and endothelial cells (*Panel 4*). Caveolae (*arrows*) are evident in both cell types. Bar $= 0.45$ μm (reprinted with permission from Reference 37).

TABLE 1   Nitric oxide synthase activity in subcellular fractions of COS-7 cells transfected with wild-type, *myr*⁻ mutant, or *palm*⁻ mutant eNOS[a]

PNS	CYTO	PM	(PM/CYTO)	NCM	CM	(CM/NCM)	(CM/PM)
Wild-Type							
A. 2.2 ± 0.1	0.11 ± 0.01^b	1.7 ± 0.2	(15.5)	0.41 ± 0.04^b	15.4 ± 2.4^b	(37.6)	(9.1)
B. 8.8 ± 0.2^b	12.9 ± 0.2^b	16.7 ± 0.1	(5.8)	4.5 ± 0.3^b	205.2 ± 3.4^b	(45.6)	(12.3)
C. 31.7 ± 0.3^b	4.3 ± 0.1^b	63.2 ± 2.0	(14.7)	15.3 ± 0.5^b	417.8 ± 10.7^b	(27.3)	(6.6)
Myr⁻ Mutant							
A. 16.2 ± 0.7^b	9.0 ± 0.3^b	4.3 ± 0.2	(0.48)	3.3 ± 0.3	1.7 ± 0.2^b	(0.52)	(0.40)
B. 18.6 ± 0.3^b	11.3 ± 0.2^b	3.1 ± 0.2	(0.27)	1.9 ± 0.04^b	1.2 ± 0.2^b	(0.63)	(0.39)
C. 35.8 ± 1.6^b	11.3 ± 0.9	10.5 ± 0.9	(0.93)	5.1 ± 0.9^b	2.7 ± 0.4^b	(0.26)	(0.26)
Palm⁻ Mutant							
A. 17.1 ± 0.7^b	7.1 ± 0.1	9.6 ± 0.2	(1.4)	3.8 ± 0.2^b	10.6 ± 0.3	(2.8)	(1.10)
B. 7.0 ± 0.1^b	2.0 ± 0.03	3.0 ± 0.6	(1.5)	0.6 ± 0.04^b	3.8 ± 0.6	(6.3)	(1.27)
C. 53.8 ± 1.2^b	15.2 ± 0.2^b	36.2 ± 0.9	(2.4)	13.2 ± 0.6^b	38.1 ± 1.0	(2.9)	(1.05)

[a]Activity (pmol citrulline formed/mg protein/min) was measured in postnuclear supernatant (PNS), cytosol (CYTO), plasma membrane (PM), non-caveolar membrane (NCM), and caveolar membrane (CM). Results from three independent transfection experiments (A, B, and C) are shown for each construct. The ratios of NOS activity in different fractions are shown in parentheses. Mean $\pm$ SEM, $n = 4$–6, [b]$p < 0.05$ versus PM (reprinted with permission 37).

eNOS (*myr*⁻), which is incapable of both myristoylation and palmitoylation or cDNA for wild-type eNOS was transiently transfected into COS-7 cells, and NOS activity was evaluated in subcellular fractions 72 h later (Table 1). In multiple studies of *myr*⁻ mutant cells, the ratio of NOS activity in plasma membrane versus cytosol was less than one, contrasting with 6- to 16-fold greater activity in plasma membrane versus cytosol in wild-type eNOS cells. A ratio of one or less indicates a lack of enrichment, whereas a ratio of greater than one indicates enrichment. In the myr⁻ mutant cells, the ratio of NOS activity in caveolae versus non-caveolae plasma membranes was also less than one; this differed markedly with the 27- to 46-fold greater NOS activity in caveolae versus non-caveolae plasma membranes noted in wild-type eNOS cells. These observations indicate that acylation targets eNOS to caveolae. Experiments were then performed to elucidate the specific role of palmitoylation in targeting the enzyme to caveolae. Robinson & Michel showed that the substitition of serines for the cysteines at positions 15 and 26 completely prevents palmitoylation of the protein, whereas myristoylation is unaffected (*palm*⁻ mutant) (40). In repeated studies of COS-7 cells transfected with the *palm*⁻ mutant eNOS cDNA, the relative abundance of NOS activity in plasma membrane versus cytosol was diminished compared with that in wild-type cells (1.4- to 2.4-fold greater activity in plasma membrane compared with cytosol, versus 6- to 16-fold difference, respectively), but it was not as severely attenuated as in *myr*⁻ cells. In addition, in *palm*⁻ mutant (myristoylation-competent) cells, there was a decrease in relative NOS activity in caveolae versus non-caveolae plasma membrane (3- to 6-fold difference) compared with wild-type cells (27- to 46-fold difference), but the decrease was less than that found in *myr*⁻ mutant cells (0.3- to 0.6-fold difference). As such, there is approximately a 10-fold enhancement in the targeting of eNOS to caveolae owing to myristoylation alone, and the targeting is augmented an additional 10-fold by palmitoylation. These results indicate that both acylation processes are necessary for optimal targeting of eNOS to caveolae. Importantly, studies of caveolae isolated from rat lung yielded results that were in agreement with those made in the cultured endothelial cells (41), indicating that caveolae localization of eNOS does indeed occur in intact endothelium. Furthermore, characterization of the localization of *palm⁻* mutant eNOS has revealed that the mutant protein is found primarily in the Golgi apparatus, whereas the *myr⁻* mutant is distributed throughout the cell, and that normal dual acylation is required for optimal function of eNOS in a physiologic context (42–44). Thus eNOS myristoylation targets the protein to the Golgi apparatus where it is palmitoylated, and the myristoylated and palmitoylated protein is targeted to the caveolae membrane (Figure 6A, see color insert).

Role of the Lipid Environment in eNOS Localization in Caveolae

In addition to the modifications of the eNOS protein described above, the specialized lipid environment within caveolae is critical to the targeting and the function of the enzyme within the microdomain. Such considerations are particularly relevant to our understanding of hypercholesterolemia-induced vascular disease and atherosclerosis because these disorders are characterized by an early and selective impairment of endothelium-derived relaxation. In the early phase of the disease process, there is attenuated responsiveness to receptor-dependent stimuli such as acetylcholine, whereas responsiveness to receptor-independent stimuli such as the calcium ionophore A23187 is not altered. Compounded by the inactivation of NO by superoxide anions that soon ensues, there is diminished NO bioavailability that results in increased neutrophil adherence to the endothelium, and the pathogenesis of atherosclerosis is initiated (4–6, 45). Studies have further shown that oxidized LDL (oxLDL) inhibits NO-mediated responses and that antioxidants attenuate these effects by reducing both the formation of free radicals and the oxidative modification of LDL that lead to impaired NO-related responses (46).

Because membrane cholesterol is essential for normal caveolae function (34), and the initiating events in atherogenesis are characterized by attenuated endothelial NO production in response to extracellular stimuli, the effects of oxLDL on the subcellular location of eNOS and on eNOS function were investigated (47). In cells briefly exposed to lipoprotein-deficient serum (LPDS), HDL or native LDL (nLDL), both eNOS and caveolin remained highly enriched in the caveolae fraction of the plasma membrane (Figure 4a, *Panels 1 and 2*). In contrast, brief exposure to oxLDL induced both eNOS and caveolin to move from caveolae and the plasma membrane to an internal membrane fraction containing endoplasmic reticulum, Golgi apparatus, mitochondria, and numerous other intracellular organelles. However, PKCα and GM1, two additional resident proteins in caveolae, did not translocate from caveolae with oxLDL exposure (Figure 4a, *Panel 3*). Identical findings were obtained in cells treated with the protein synthesis inhibitor cycloheximide, and the removal of oxLDL and continued treatment with cycloheximide (recovery) permitted eNOS and caveolin to return to caveolae membranes

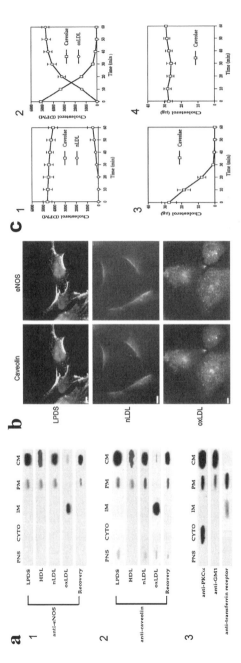

Figure 4 Oxidized LDL displaces eNOS from endothelial cell caveolae through depletion of caveolae cholesterol. (*a*) Oxidized LDL (oxLDL) but not lipoprotein-deficient serum (LPDS), HDL or nLDL exposure (60 min) alters the subcellular distribution of eNOS and caveolin. Additional studies were performed in oxLDL-treated cells that were washed and incubated for an additional 120 min in LPDS only (recovery). Samples of postnuclear supernatant (PNS), cytosol (CYTO), intracellular membranes (IM), plasma membrane (PM), non-caveolae membrane (NCM), and caveolae membrane (CM) were isolated, and immunoblot analysis was performed for eNOS (*Panel 1*) and caveolin-1 (*Panel 2*). Control experiments in oxLDL-treated cells included immunoblot analyses for the caveolae resident proteins PKCα and GM1, and for transferrin receptors that are found in coated pits (*Panel 3*). (*b*) OxLDL but not LPDS or nLDL induces eNOS and caveolin to co-localize in an internal membrane compartment. Cells were incubated for 60 min, fixed, and processed for double-label immunofluorescence. (*c*) oxLDL depletes caveolae of cholesterol. For Panels 1 and 2, endothelial cells were labeled with [³H]acetate for 18 h at 37°C, washed and incubated with nLDL (*Panel 1*) or oxLDL (*Panel 2*), for 0–60 min at 37°C. The medium was collected, and the nLDL or oxLDL was isolated by centrifugation. The cells were washed, processed to isolate caveolae, and [³H]cholesterol was extracted and measured. For Panels 3 and 4, unlabeled cells were incubated with oxLDL (*Panel 3*) or HDL (*Panel 4*) for 0–60 min at 37°C. The cells were washed, caveolae were isolated, and the amount of total cholesterol associated with caveolae was determined. Results are mean ±SEM (*n* = 8) (reprinted with permission from Reference 47).

(Figure 4*a*, *Panels 1 and 2*). The observations made by subcellular fractionation were confirmed using indirect immunofluorescence (Figure 4*b*).

Having demonstrated that oxLDL causes eNOS redistribution in endothelial cells, the effects of the lipoprotein on the activation of the enzyme by acetylcholine were evaluated. OxLDL attenuated the activation of eNOS at all concentrations of acetylcholine tested, yielding a shift in the dose-response curve to acetylcholine to the right by 100-fold. Nonspecific disruption of signal transduction was excluded by the finding that oxLDL treatment did not alter acetylcholine or bradykinin-stimulated prostacyclin production. The potential effects of oxLDL on eNOS modification were assessed, and the palmitoylation and myristoylation of eNOS were not altered. In addition, eNOS phosphorylation was unaffected by oxLDL (47). Thus there is marked subcellular redistribution of eNOS provoked by oxLDL, and this process is not related to changes in the co- and post-translation modification of the enzyme protein.

The next series of studies determined if the effects of oxLDL on eNOS distribution are alternatively explained by modifications in the caveolae lipid environment. Endothelial cells were labeled with ^{3}H-acetate and exposed to nLDL or oxLDL, the media were collected and caveolae were isolated, and the amounts of sterol in the extracellular media and caveolae were determined (Figure 4*c*). nLDL did not alter the amount of radiolabeled sterol in caveolae, and negligible amounts accumulated in the extracellular media (Figure 4*c*, *Panel 1*). In contrast, radiolabeled sterol was readily transferred from caveolae to oxLDL in the media, such that the caveolae were essentially completely lacking labeled sterol by 30 min (Figure 4*c*, *Panel 2*). In parallel, the total cholesterol content of caveolae was depleted by oxLDL, whereas HDL serving as a control had no effect (Figure 4*c*, *Panels 3 and 4*). Thus by acting as an acceptor of cholesterol, oxLDL causes depletion of caveolae cholesterol. To determine if this process leads to eNOS displacement, the effect of cyclodextrin, which extracts cholesterol from caveolae, was determined. Similar to oxLDL, cyclodextrin caused redistribution of both eNOS and caveolin to the intracellular membrane fraction, whereas PKCα localization to caveolae was unaltered (47). Thus oxLDL causes a depletion of caveolae cholesterol, and the perturbation in the lipid environment leads to eNOS redistribution and an attenuated capacity to activate the enzyme (Figure 6*B*). This process may play a critical role in the pathogenesis of hypercholesterolemia-induced vascular disease and atherosclerosis.

Because there is a strong negative correlation between HDL levels and the risk for atherosclerosis (48, 49), and HDL mediates cholesterol trafficking (50, 51), the hypothesis was tested that HDL modifies the effects of oxLDL on eNOS localization and activation in caveolae (52). The addition of HDL to medium containing oxLDL prevented eNOS and caveolin displacement from the caveolae fraction (Figure 5*a*), and it also restored acetylcholine-induced stimulation of the enzyme (Figure 5*b*). To delineate the basis of the effects of HDL on eNOS targeting, further studies of caveolae cholesterol homeostasis were performed (Figure 5*c*). Whereas oxLDL alone caused a marked reduction in caveolae sterol content (Figure 5*c*, *Panel 1*), cotreatment with HDL completely prevented this effect

(Figure 5c, *Panel 3*). Further experiments demonstrated that the capacity of HDL to maintain the concentration of caveolae-associated cholesterol in the face of oxLDL is not through the inhibition of cholesterol transport out of caveolae by oxLDL (Figure 5c, *Panel 4*); it results from the uptake of cholesterol esters supplied by HDL (Figure 5c, *Panel 5*). Thus in the presence of oxLDL the unique lipid environment within caveolae is preserved by HDL, thereby maintaining the normal subcellular localization and function of eNOS and potentially explaining at least a portion of the potent antiatherogenic properties of HDL (Figure 6C).

eNOS REGULATION IN CAVEOLAE

Protein-Protein Interactions Regulating eNOS

Considerable attention has been paid to the mechanisms by which the localization of eNOS to caveolae affects the function of the enzyme, including the interaction of

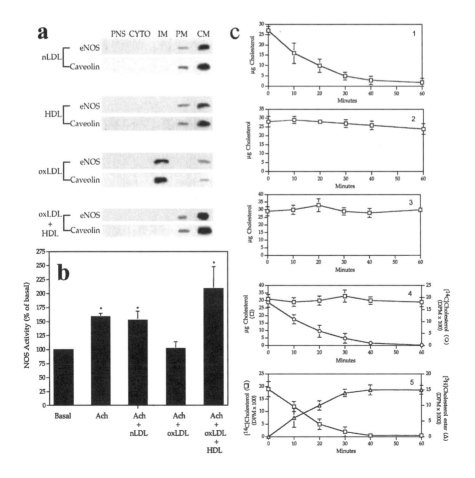

eNOS with other resident proteins such as the caveola coat protein caveolin. First, it was demonstrated in studies of endothelial and cardiac myocyte cell lysates that eNOS coimmunoprecipitates with caveolin-1 and caveolin-3, respectively (53, 54). Then, in vitro studies and experiments with eNOS and caveolin-1 overexpression in COS-7 cells revealed that both N- and C-terminal domains of caveolin interact directly with the eNOS oxygenase domain and inhibit eNOS catalytic activity (55–58). In vitro manipulations further indicated that calcium-calmodulin may disrupt the interaction between eNOS and caveolin, leading to enhanced enzymatic activity (57). Experiments using particulate and soluble cellular fractions suggest that this may be a cyclic phenomenon, with depalmitoylation and dissociation of eNOS and caveolin leading to mobilization of the enzyme from the particulate fraction upon agonist stimulation, followed by reassociation of eNOS and caveolin in the particulate fraction (59). Further work indicates that the 20-amino acid scaffolding domain (amino acids 82–101) of caveolin binds and modifies the activity of eNOS, as well as that of other resident signaling molecules including protein kinase C (PKC) and G protein α subunits (58, 60, 61). More recently, Bucci et al. demonstrated that a chimeric peptide with a cellular internalization sequence fused to the caveolin-1 scaffolding domain was efficiently incorporated into blood vessels and endothelial cells, resulting in selective inhibition of acetylcholine-induced vasodilation and NO production (62). Thus there is accumulating evidence that eNOS-caveolin interactions have a dramatic impact on the activity of the enzyme.

←——

Figure 5   HDL prevents oxLDL-induced inhibition of eNOS localization and activation in caveolae. (*a*) Endothelial cells were treated for 60 min with nLDL, HDL, or oxLDL, or with oxLDL followed by HDL for an additional 15 min. Samples of postnuclear supernatant (PNS), cytosol (CYTO), intracellular membranes (IM), plasma membrane (PM), noncaveolae membrane (NCM), and caveolae membrane (CM) were isolated, and immunoblot analysis was performed for eNOS and caveolin-1. (*b*) After treating endothelial cells as described above, NOS activity was evaluated in intact cells by measuring L-[^{3}H]arginine conversion to L-[^{3}H]citrulline over 15 min in the absence of exogenous stimulation (basal) or in the presence of acetylcholine (Ach, 10^{-6} M). Values are mean $\pm$ SEM ($n = 4$), [*]p < 0.05 versus basal. (*c*) HDL maintains the sterol content of caveolae. Endothelial cells were incubated with oxLDL (*Panel 1*) or HDL (*Panel 2*) for 0–60 min at 37°C. For Panel 3, cells were pretreated with oxLDL and then HDL was added for an additional 15 min. The cells were fractionated to isolate caveolae and the mass of cholesterol associated with caveolae was determined. In Panel 4, cells were radiolabeled with [^{14}C]acetate for 18 h, and HDL and oxLDL were added simultaneously for 0–60 min. The cells were processed to measure the amount of [^{14}C]cholesterol associated with caveolae and the mass of cholesterol associated with caveolae. In Panel 5, cellular cholesterol pools were radiolabeled as described in Panel 4, and HDL was labeled with [^{3}H]cholesterol ester. [^{3}H]HDL and oxLDL were incubated with cells as described in Panel 4, and the amount of [^{14}C]cholesterol and [^{3}H]cholesterol associated with caveolae was determined. Values are mean $\pm$ SEM ($n = 3$) (reprinted with permission from Reference 52).

There is also evidence of other protein-protein interactions that modify eNOS function. Ju and colleagues (64) have demonstrated that the activity of the enzyme is negatively regulated by its association with the C-terminal domain of G protein-coupled receptors such as the bradykinin B2 receptor, which is mobilized to caveolae upon agonist stimulation (64). In addition, stimulation of endothelial cells by histamine, vascular endothelial growth factor (VEGF), or shear stress leads to the binding of heat shock protein 90 to eNOS, which causes allosteric activation of the enzyme (65). Furthermore, yeast two-hybrid screening has recently identified a novel 34-kDa protein designated NOSIP (eNOS interacting protein) that binds to the C-terminal region of the eNOS oxygenase domain and promotes translocation of the enzyme from caveolae to intracellular sites, resulting in the attenuation of NO production (66).

Signal Transduction Pathways Regulating eNOS

In addition to modulation by protein-protein interaction, multiple signal transduction pathways converge to regulate eNOS. The activation of the enzyme in response to multiple hormone agonists such as estradiol, bradykinin, and VEGF occurs in association with elevations in cytosolic calcium concentrations (67–69). In contrast, eNOS activation by shear stress and isometric vessel contraction occurs independently of changes in intracellular calcium levels (70, 71). Shear stress-induced enzyme activation is regulated by potassium channels, and it is prevented by tyrosine kinase inhibition, indicating that the process also entails tyrosine phosphorylation (70, 72, 73). However, dependence on tyrosine phosphorylation is probably not limited to calcium-independent mechanisms such as shear stress because pharmacologic approaches indicate that it is also necessary for eNOS activation by estradiol and VEGF (69, 74). There is also evidence for a role of MAP kinases in the modulation of eNOS activity by calcium-dependent agonists, with MAP kinase activation interestingly yielding eNOS stimulation by estradiol and eNOS inactivation by bradykinin (74, 75). In most paradigms, it remains uncertain whether the relevant tyrosine kinase or MAP kinase-mediated phosphorylation events involve residues on eNOS or residues on signaling molecules proximal to eNOS.

In addition to regulation by calcium and via tyrosine phosphorylation, multiple protein kinases modify eNOS activity through effects on serine phosphorylation at position 1177. These kinases include AMP-activated protein kinase, PKC, cAMP-dependent protein kinase (PKA), and the serine/threonine kinase Akt, which is also known as protein kinase B (PKB). Factors that activate eNOS through PKB/Akt-mediated phosphorylation of Ser-1177 include estradiol (76, 77), shear stress (78, 79), VEGF (80, 81), and insulin-like growth factor-1 (80). In contrast to the activation that occurs with phosphorylation of Ser-1177, phosphorylation of the threonine at position 497 yields attenuated eNOS activity (82). There is evidence of coordinated regulation of eNOS activity by agonists such as VEGF, which cause both phosphorylation of Ser-1177 and dephosphorylation of Thr-497

(83). Recent work has further demonstrated that PKA signaling leads to eNOS phosphorylation at Ser-1177 and dephosphorylation of Thr-497, thereby enhancing enzymatic activity, whereas PKC promotes both the dephosphorylation of Ser-1177 and the phosphorylation at Thr-497, resulting in attenuated enzyme activity. Furthermore, the dephosphorylation processes are mediated by phosphatases PP2A and PP1 acting selectively at Ser-1177 and Thr-497, respectively (83). Thus eNOS activity is regulated by a complex combination of protein-protein interactions and signal transduction cascades involving calcium mobilization and phosphorylation events.

eNOS Signaling Module in Caveolae

Over the last several years, numerous observations indicate that many of the signal transduction molecules described above are colocalized with eNOS in caveolae (84). Such colocalization has suggested the existence of a functional signaling module that compartmentalizes the multiple events regulating the level of NO production. However, direct evidence for such a module was unavailable until recent studies were performed to further understand how eNOS is stimulated by estradiol via nongenomic actions of estrogen receptor α (ERα), which was previously known to serve solely as a nuclear transcription factor (74). These mechanisms were pursued because estradiol is an important agonist for endothelial eNOS in both physiologic and pathophysiologic circumstances, affording atheroprotection to premenopausal women compared with men and to postmenopausal women receiving estrogen replacement, and also modifying responses to vascular injury (85). Estradiol was also one of the first agents demonstrated to activate eNOS in airway epithelium, where the hormone may play a role in modifying airway responsiveness (86).

The subcellular site of interaction between ERα and eNOS was initially determined in experiments employing isolated endothelial cell plasma membranes. Estradiol caused an increase in eNOS activity in the isolated membranes in the absence of added calcium, calmodulin, or eNOS cofactors, the activation was blocked by the ER antagonist ICI 182,780 and ERα antibody, and immuno-identification experiments detected ERα protein in the plasma membrane fraction. Plasma membranes from COS-7 cells expressing eNOS and ERα also displayed ER-mediated eNOS stimulation, whereas membranes from cells expressing eNOS alone or ERα plus a myristoylation-deficient mutant eNOS were insensitive. Further fractionation of endothelial cell plasma membranes revealed ERα protein in caveolae, and estradiol caused stimulation of eNOS in isolated caveolae that was ER dependent; in contrast, non-caveolae membranes were insensitive. Responses to estradiol and the more classical eNOS agonists acetylcholine and bradykinin were also compared in isolated caveolae, and equivalent, robust eNOS activation was observed with all three agents (Figure 7). In addition, calcium chelation prevented estradiol-stimulated eNOS activity in both isolated whole plasma membranes and caveolae. These cumulative findings reveal that a subpopulation of ERα is localized to

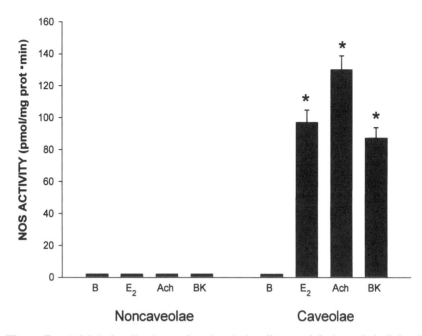

Figure 7 eNOS is localized to a functional signaling module in endothelial cell caveolae. L-[³H]arginine conversion to L-[³H]citrulline was measured in isolated noncaveolae and caveolae fractions of endothelial cell plasma membranes, in the absence (basal, B) or presence of 10^{-8} M estradiol (E_2), 10^{-6} M acetylcholine (Ach), or 10^{-6} M bradykinin (BK). Values are mean $\pm$ SEM, ($n = 4$–6), [*]$p < 0.05$ versus basal (reprinted with permission from Reference 87).

endothelial cell caveolae where they are coupled to eNOS in a functional signaling module that regulates the local calcium environment (87). Mechanisms occurring within this signaling module are most likely critical to the atheroprotective properties of estradiol (85).

More recent work has ascertained whether the dramatic antiatherogenic features of HDL are further explained by caveolae-related mechanisms other than those involving changes in membrane cholesterol content (see above). The severity of atheroma formation in both humans and animal models is inversely related to serum concentrations of HDL cholesterol and the major HDL apolipoprotein apoA-I (50, 51, 88). However, the mechanisms by which HDL and apoA-I are antiatherogenic are complex and not well understood (50), and although they are involved in reverse cholesterol transport from peripheral tissues to the liver, serum concentrations of HDL and apoA-I do not control the degree of reverse cholesterol transport (89). Studies of the direct effects of HDL on endothelial function were therefore undertaken, and they demonstrated that HDL causes rapid and dramatic stimulation of eNOS in cultured endothelial cells in the absence of changes in membrane cholesterol. In contrast, eNOS was not activated by purified forms of apoA-I or apoA-II, or by native LDL. Heterologous expression experiments in Chinese hamster ovary cells revealed that the class B scavenger receptor, known as scavenger receptor-BI (SR-BI), mediates the effects of HDL on the enzyme (90). In addition, HDL caused enhanced endothelium- and nitric oxide (NO)-dependent relaxation in aortae from wild-type mice, but not in aortae from homozygous null SR-BI knockout mice (91). Endothelial cell subfractionation experiments revealed that HDL activation of eNOS is demonstrable in isolated caveolae, but not in non-caveolae membranes (Figure 8a), and that SR-BI and eNOS are colocalized in caveolae (Figure 8b). An initial assessment of the constituents of HDL and SR-BI required for the process was performed by antibody blockade in isolated endothelial cell plasma membranes. HDL stimulation of eNOS in the membranes was attenuated by antibodies to apoA-I and to the C-terminal cytoplasmic tail of SR-BI, but not by antibody to apoA-II. Thus HDL activates eNOS via SR-BI in caveolae through a process that requires apoA-I binding. The resulting increase in NO production may be critical to the atheroprotective features of HDL and apoA-I (92). When these observations are considered along with the evidence for estradiol-, acetylcholine- and bradykinin-induced stimulation of eNOS in isolated caveolae (87), it is apparent that key processes regulating the activity of the enzyme are compartmentalized in this unique microdomain. It is anticipated that isolated endothelial cell caveolae will serve as a helpful model system in future studies of eNOS regulation at the plasma membrane.

The role of localized calcium homeostasis in the modulation of eNOS activity in caveolae has been further delineated in intact cells employing a chimera containing the full-length eNOS cDNA and an HA-tagged aequorin sequence (EHA), as well as a myristoylation-deficient EHA (MHA) containing the myr^- eNOS mutant, which cannot be myristoylated or palmitoylated (93). These investigations were prompted by previous work with potassium oxalate precipitation showing that calcium is highly localized to the cell surface and, in particular, to caveolae in certain cell types (94), and the more recent observation that calcium waves in endothelial cells induced by ATP and other agonists preferentially originate at caveolin-rich cell edges (95). As anticipated, the chimeric EHA and MHA proteins displayed targeting in transfected COS-7 cells that was similar to wild-type eNOS and myr^- eNOS, respectively, with EHA localized primarily to plasma membrane and MHA primarily found in cytosolic and intracellular membrane fractions. Both constructs retained enzymatic eNOS activity and aequorin-mediated calcium sensitivity. Cytosolic-directed aequorin alone (cytoAEQ) and plasma membrane-associated EHA and intracellular MHA were compared in their ability to sense changes in local calcium concentration in transfected COS-7 cells using histamine, which stimulates intracellular calcium release via the generation of inositol 1,4,5-triphosphate (96) (Figure 9a). Both cytosolic aequorin and MHA displayed preferential sensitivity to calcium originating from intracellular pools, whereas capacitative calcium entry (CCE), which occurs following the release of intracellular calcium, was sensed preferentially by EHA. In parallel, measurements

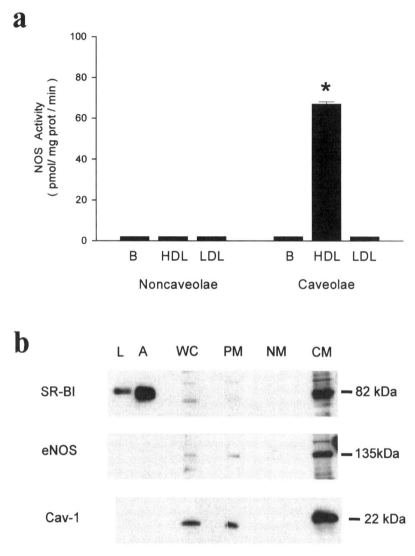

Figure 8 HDL activates eNOS in isolated endothelial cell caveolae, which are enriched in SR-BI. (*a*) L-[³H]arginine conversion to L-[³H]citrulline was measured in isolated non-caveolae and caveolae fractions of endothelial cell plasma membranes in the
absence (basal, B) or presence of HDL or nLDL. Values are mean ± SEM ($n = 4$–6),
*p < 0.05 versus basal. (*b*) Immunoblot analyses for SR-BI, eNOS and caveolin-1(Cav-
1) in endothelial whole cell lysate (WC), intact plasma membrane (PM), noncaveolae
membrane (NM) and caveolae membrane (CM). Liver (L) and adrenal (A) were used
as positive controls for SR-BI (reprinted with permission from Reference 92).

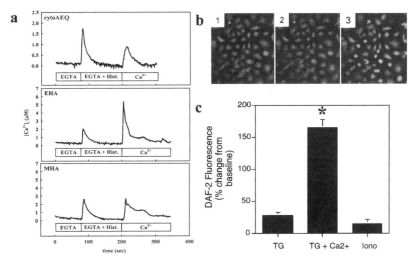

Figure 9   eNOS activation by capacitative calcium entry (CCE) occurs in endothelial cell caveolae. (*a*) Effects of histamine-induced Ca^{2+} mobilization and capacitative entry on the intracellular calcium concentration $[Ca^{2+}]_i$ of cells transfected with EHA or MHA construct. COS-7 cells transfected with cytosolic aequorin (cytoAEQ), EHA, or MHA were stimulated with 100 μM histamine in nominally Ca^{2+}-free buffer (EGTA) at 60 s, followed by the addition of 4 mM Ca^{2+} at 200 s. The resulting luminescence was measured and transformed into $[Ca^{2+}]_i$. CCE upon the addition of extracellular Ca^{2+} was sensed preferentially by EHA. (*b*) NO production in endothelial cells mediated by thapsigargin (TG)-induced CCE. Cells loaded with the fluorescent NO indicator DAF-2 diacetate were equilibrated for 1 min in Krebs-Ringer phosphate buffer containing 0.54 mM Ca^{2+} (*Panel 1*). At the end of equilibration, cells were treated for 1 min with 100 nM TG (*Panel 2*) followed by the addition of 4 mM Ca^{2+} (*Panel 3*). Fluorescence was recorded and photographed with a 20 objective. (*c*) Comparison of eNOS activation stimulated by TG-induced CCE versus ionomycin (Iono) treatment performed in a manner to yield equivalent increases in $[Ca^{2+}]_i$. DAF-2 diacetate fluorescence was measured in 10 cells after addition of 100 nM TG either in the absence (TG) or presence of 4 mM extracellular Ca^{2+} (TG + Ca^{2+}), or with 10 μM ionomycin in the presence of 30 μM extracellular $[Ca^{2+}]$. Values are mean $\pm$ SEM, *p < 0.05 versus TG alone (reprinted with permission from Reference 93).

of eNOS activation in intact, transfected COS-7 cells revealed that the eNOS enzymatic activity of EHA was more sensitive to calcium influx via CCE than intracellular release, whereas MHA eNOS activity was more responsive to intracellular calcium release. When DAF-2 flourescence was employed to assess endogenous eNOS activation in endothelial cells (Figure 9*b*), a 10-fold greater increase in NO production was found with thapsigargin-induced CCE compared with stimulation with the calcium ionophore ionomycin done in a manner to yield comparable rises in intracellular calcium (Figure 9c) (93). These findings indicate that acylation-mediated targeting of eNOS to caveolae places the enzyme in close proximity to sources of calcium entry via CCE. Thus multiple signal transduction events initiated by numerous stimuli converge on eNOS within caveolae to regulate the activity of the enzyme.

SUMMARY AND FUTURE DIRECTIONS

Since the identification of endothelium-derived relaxing factor as NO in 1987 and the ensuing cloning of eNOS in 1992 (39, 97–100), exhaustive efforts have been put forth to understand the processes mediating eNOS function in vascular endothelium, in certain types of epithelium, and in other cells such as cardiac myocytes. Considerable work has focused on regulatory mechanisms related to changes in eNOS abundance and the biology of eNOS enzymatic activity (22–24). However, it has also been appreciated that complex mechanisms underly the cell specificity of eNOS expression and that co- and post-translational processing leads to trafficking of the enzyme to plasma membrane caveolae. Within caveolae, eNOS is the downstream target member of a signaling complex in which it is functionally linked to both typical G protein-coupled receptors and less typical receptors such as ERα and SR-BI that display novel actions. This compartmentalization facilitates dynamic protein-protein interactions and calcium- and phosphorylation-dependent signal transduction events that modify the response of the enzyme to extracellular stimuli (Figure 10, see color insert). The combination of cell-specific expression, subcellular trafficking, and local signaling events provides both precise fine-tuning of NO-mediated responses under physiologic circumstances and multiple mechanisms whereby NO availability may be attenuated in numerous disease states. The current and future challenges in this area of research are many. The processes responsible for the subcellular movement of eNOS and its modifiers and the mechanisms underlying their interactions warrant further dissection in simplified model systems. New knowledge is also required about the processes whereby differential eNOS regulation occurs in various vascular sites and in nonendothelial cell types. Additionally, the functional impacts of many of the features of eNOS subcellular localization and regulation that have been rigorously defined in cell culture must still be delineated in intact model systems. It is only through such focused efforts that new approaches will ultimately be developed to take preventative and therapeutic advantage of the powerful actions of NO in multiple cell types.

ACKNOWLEDGMENTS

The author thanks many colleagues and collaborators, past and present, who have contributed to the data summarized in this review. These include Richard G.W. Anderson, Urs A. Arnet, Alison Blair, Ken L. Chambliss, Patricia A. Conrad, Blair E.

Cox, Zohre German, Lisa D. Hahner, Helen H. Hobbs, Richard E. Karas, Pingsheng Liu, Charles J. Lowenstein, Ping Lu, Yves L. Marcel, Michael E. Mendelsohn, Thomas Michel, Chieko Mineo, Sherri Osborne-Lawrence, Margaret C. Pace, Lisa J. Robinson, Todd S. Sherman, Eric J. Smart, Annette Uittenbogaard, Yunshu Ying, Ivan S. Yuhanna, and Yan Zhu. The author also thanks Marilyn Dixon for preparing this manuscript. This work was supported by National Institutes of Health grants HL58888, HL53546, and HD30276; the Lowe Foundation; and the Crystal Charity Ball.

Visit the Annual Reviews home page at www.AnnualReviews.org

LITERATURE CITED

1. Moncada S, Higgs A. 1993. The L-arginine-nitric oxide pathway. *N. Engl. J. Med.* 329:2002–12

2. Bredt DS, Snyder SH. 1994. Nitric oxide: a physiologic messenger molecule. *Annu. Rev. Biochem.* 63:175–95

3. Nathan C, Xie QW. 1994. Regulation of biosynthesis of nitric oxide. *J. Biol. Chem.* 269:13725–28

4. Harrison DG. 1994. Endothelial dysfunction in atherosclerosis. *Basic Res. Cardiol.* 89:87–102

5. Cohen RA. 1995. The role of nitric oxide and other endothelium-derived vasoactive substances in vascular disease. *Prog. Cardiovasc. Dis.* 38:105–28

6. Flavahan NA. 1992. Atherosclerosis or lipoprotein-induced endothelial dysfunction: potential mechanisms underlying reduction in EDRF/nitric oxide activity. *Circulation* 85:1927–38

7. Gaston B, Drazen JM, Loscalco J, Stamler JS. 1994. The biology of nitrogen oxides in the airways. *Am. J. Respir. Crit. Care Med.* 149:538–51

8. Gustaffson LE, Leone AM, Persson MG, Wiklund NP, Moncada S. 1991. Endogenous nitric oxide is present in the exhaled air of rabbits, guinea pigs and humans. *Biochem. Biophys. Res. Commun.* 181:852–57

9. Dweik RA, Laskowski D, Abu-Soud HM, Kaneko FT, Hutte R, et al. 1998. Nitric oxide synthesis in the lung. Regulation by oxygen through a kinetic mechanism. *J. Clin. Invest.* 101:660–66

10. Barnes PJ. 1995. Nitric oxide and airway disease. *Ann. Med.* 27:389–93

11. Cummings JJ. 1995. Pulmonary vasodilator drugs decrease lung liquid production in fetal sheep. *J. Appl. Physiol.* 79:1212–18

12. Cummings JJ. 1997. Nitric oxide decreases lung liquid production in fetal lambs. *J. Appl. Physiol.* 83:1538–44

13. Jakupaj M, Martin RJ, Dreshaj IA, Potter CF, Haxhiu MA, et al. 1997. Role of endogenous NO in modulating airway contraction mediated by muscarinic receptors during development. *Am. J. Physiol. Lung Cell. Mol. Physiol.* 273:L531–L36

14. Potter CF, Dreshaj IA, Haxhiu MA, Stork EK, Chatburn RL, et al. 1997. Effect of exogenous and endogenous nitric oxide on the airway and tissue components of lung resistance in the newborn piglet. *Pediatr. Res.* 41:886–91

15. Folkerts G, van der Linde HJ, Nijkamp FP. 1995. Virus-induced airway hyperresponsiveness in guinea pigs is related to a deficiency in nitric oxide. *J. Clin. Invest.* 95:26–30

16. Tracey WR, Pollock JS, Murad F, Nakane M, Förstermann U. 1994. Identification of an endothelial-like type III NO synthase in LLC-PK1 kidney epithelial cells. *Am. J. Physiol. Cell Physiol.* 266:C22–C28

17. Ujiie K, Yuen J, Hogarth L, Danzier R, Star RA. 1994. Localization and regulation of endothelial NO synthase mRNA expression in rat kidney. *Am. J. Physiol. Renal Physiol.* 267:F296–F302

18. Zini A, O'Bryan MK, Magid MS, Schlegel PN. 1996. Immunohistochemical localization of endothelial nitric oxide synthase in human testis, epididymis, and vas deferens suggests a possible role for nitric oxide in spermatogenesis, sperm maturation, and programmed cell death. *Biol. Reprod.* 55:935–41

19. Chatterjee S, Gangula PR, Dong YL, Yallampalli C. 1996. Immunocytochemical localization of nitric oxide synthase-III in reproductive organs of female rats during the oestrous cycle. *Histochem. J.* 28:715–23

20. Michel T, Feron O. 1997. Nitric oxide synthases: which, where, how, and why. *J. Clin. Invest.* 100:2146–52

21. Hecker M, Mulsch A, Bassenge E, Förstermann U, Busse R. 1994. Subcellular localization and characterization of nitric oxide synthase(s) in endothelial cells: physiological implications. *Biochem. J.* 299:247–52

22. Papapetropoulos A, Rudic RD, Sessa WC. 1999. Molecular control of nitric oxide synthases in the cardiovascular system. *Cardiovasc. Res.* 43:509–20

23. Andrew PJ, Mayer B. 1999. Enzymatic function of nitric oxide synthases. *Cardiovasc. Res.* 43:521–31

24. Govers R, Rabelink TJ. 2001. Cellular regulation of endothelial nitric oxide synthase. *Am. J. Physiol. Renal Physiol.* 280:F193–F206

25. Sherman TS, Chen Z, Yuhanna IS, Lau KS, Margraf LR, et al. 1999. Nitric oxide synthase isoform expression in the developing lung epithelium. *Am. J. Physiol. Lung Cell. Mol. Physiol.* 276:L383–L90

26. Giaid A, Saleh D. 1995. Reduced expression of endothelial nitric oxide synthase in the lungs of patients with pulmonary hypertension. *N. Engl. J. Med.* 333:214–21

27. Xue C, Botkin SJ, Johns RA. 1996. Localization of endothelial NOS at the basal microtubule membrane in ciliated epithelium of rat lung. *J. Histochem. Cytochem.* 44:463–71

28. Robinson LJ, Weremowicz S, Morton CC, Michel T. 1994. Isolation and chromosomal localization of the human endothelial nitric oxide synthase (NOS3) gene. *Genomics* 19:350–57

29. Shaul PW, North AJ, Wu LC, Wells LB, Brannon TS, et al. 1994. Endothelial nitric oxide synthase is expressed in cultured human bronchiolar epithelium. *J. Clin. Invest.* 94:2231–36

30. German Z, Chambliss KL, Pace MC, Arnet UA, Lowenstein CJ, Shaul PW. 2000. Molecular basis of cell-specific endothelial nitric-oxide synthase expression in airway epithelium. *J. Biol. Chem.* 275:8183–89

31. Zhang R, Min W, Sessa WC. 1995. Functional analysis of the human eNOS promoter. *J. Biol. Chem.* 270:15320–26

32. Guillot PV, Guan J, Liu L, Kuivenhoven JA, Rosenberg RD, et al. 1999. A vascular bed-specific pathway regulates cardiac expression of endothelial nitric oxide synthase. *J. Clin. Invest.* 103:799–805

33. Anderson RGW, Kamen BA, Rothberg KG, Lacey SW. 1992. Potocytosis: sequestration and transport of small molecules by caveolae. *Science* 225:410–11

34. Chang W-J, Rothberg KG, Kamen BA, Anderson RGW. 1992. Lowering the cholesterol content of MA104 cells inhibits receptor-mediated transport of folate. *J. Cell Biol.* 118:63–69

35. Chang WJ, Ying YS, Rothberg KG, Hooper NM, Turner AJ, et al. 1994. Purification and characterization of smooth muscle cell caveolae. *J. Cell Biol.* 126:127–38

36. Conrad PA, Smart EJ, Ying Y, Anderson RGW, Bloom GS. 1995. Caveolin cycles between plasma membrane caveolae

and the Golgi complex by microtubule-dependent and microtubule-independent steps. *J. Cell Biol.* 131:1421–33

37. Shaul PW, Smart EJ, Robinson LJ, German Z, Yuhanna IS, et al. 1996. Acylation targets endothelial nitric oxide synthase to plasmalemmal caveolae. *J. Biol. Chem.* 271:6518–22

38. Schlegel A, Lisanti MP. 2001. Caveolae and their coat proteins, the caveolins: from electron microscopic novelty to biological launching pad. *J. Cell. Physiol.* 186:329–37

39. Lamas S, Marsden PA, Li GK, Tempst P, Michel T. 1992. Endothelial nitric oxide synthase: molecular cloning and characterization of a distinct constitutive enzyme iosform. *Proc. Natl. Acad. Sci. USA* 89:6348–52

40. Robinson LJ, Michel T. 1995. Mutagenesis of palmitoylation sites in endothelial nitric oxide synthase identifies a motif for dual acylation and subcellular targeting. *Proc. Natl. Acad. Sci. USA* 92:11776–80

41. Garcia-Cardena G, Oh P, Liu J, Schnitzer JE, Sessa WC. 1996. Targeting of nitric oxide synthase to endothelial cell caveolae via palmitoylation: implications for nitric oxide signaling. *Proc. Natl. Acad. Sci. USA* 3:6448–53

42. Liu J, Garcia-Cardena G, Sessa WC. 1996. Palmitoylation of endothelial nitric oxide synthase is necessary for optimal stimulated release of nitric oxide: implications for caveolae localization. *Biochemistry* 35:13277–81

43. Liu J, Hughes TE, Sessa WC. 1997. The first 35 amino acids and fatty acylation sites determine the molecular targeting of endothelial nitric oxide synthase into the Golgi region of cells: a green fluorescent protein study. *J. Cell Biol.* 137:1525–35

44. Sowa G, Liu J, Papapetropoulos A, Rex-Haffner M, Hughes TE, et al. 1999. Trafficking of endothelial nitric-oxide synthase in living cells. Quantitative evidence supporting the role of palmitoylation as a kinetic trapping mechanism limiting membrane diffusion. *J. Biol. Chem.* 274:22524–31

45. Lefer AM, Ma XL. 1993. Decreased basal nitric oxide release in hypercholesterolemia increases neutrophil adherence to rabbit coronary artery endothelium. *Arterioscler. Thromb.* 13:771–76

46. Plane F, Jacobs M, McManus D, Bruckdorfer KR. 1993. Probucol and other antioxidants prevent the inhibition of endothelium-dependent relaxation by low density lipoproteins. *Atherosclerosis* 103:73–79

47. Blair A, Shaul PW, Yuhanna IS, Conrad PA, Smart EJ. 1999. Oxidized low-density lipoprotein displaces eNOS from plasmalemmal caveolae and impairs eNOS activation. *J. Biol. Chem.* 274:32512–19

48. Grundy SM. 1986. Cholesterol and coronary heart disease. A new era. *J. Am. Med. Assoc.* 256:2849–58

49. Tall AR. 1990. Plasma high density lipoproteins. Metabolism and relationship to atherogenesis. *J. Clin. Invest.* 86:379–84

50. Krieger M. 1998. The "best" of cholesterols, the "worst" of cholesterols: a tale of two receptors. *Proc. Soc. Exp. Biol. Med.* 95:4077–80

51. Fidge NH. 1999. High density lipoprotein receptors, binding proteins, and ligands. *Lipid Res.* 40:187–201

52. Uittenbogaard A, Shaul PW, Yuhanna IS, Blair A, Smart EJ. 2000. High density lipoprotein prevents oxidized low density lipoprotein-induced inhibition of endothelial nitric-oxide synthase localization and activation in caveolae. *J. Biol. Chem.* 275:11278–83

53. Feron O, Belhassen L, Kobzik L, Smith TW, Kelly RA, et al. 1996. Endothelial nitric oxide synthase targeting to caveolae. Specific interactions with caveolin isoforms in cardiac myocytes and endothelial cells. *J. Biol. Chem.* 271:22810–14

54. Garcia-Cardena G, Fan R, Stern DF, Liu J, Sessa WC. 1996. Endothelial nitric

oxide synthase is regulated by tyrosine phosphorylation and interacts with caveolin-1. *J. Biol. Chem.* 271:27237–40

55. Garcia-Cardena G, Martasek P, Masters BSS, Skidd PM Couet J, et al. 1997. Dissecting the interaction between nitric oxide synthase (NOS) and caveolin. *J. Biol. Chem.* 272:25437–40

56. Ju H, Zou R, Venema VJ, Venema RC. 1997. Direct interaction of endothelial nitric-oxide synthase and caveolin-1 inhibits synthase activity. *J. Biol. Chem.* 272:18522–25

57. Michel JB, Feron O, Sacks D, Michel T. 1997. Reciprocal regulation of endothelial nitric-oxide synthase by Ca^{2+}-calmodulin and caveolin. *J. Biol. Chem.* 272:15583–86

58. Michel JB, Feron O, Sase K, Prabhakar P, Michel T. 1997. Caveolin versus calmodulin. Counterbalancing allosteric modulators of endothelial nitric oxide synthase. *J. Biol. Chem.* 272:25907–12

59. Feron O, Saldana F, Michel JB, Michel T. 1998. The endothelial nitric-oxide synthase-caveolin regulatory cycle. *J. Biol. Chem.* 273:3125–28

60. Couet J, Li S, Okamoto T, Ikezu T, Lisanti MP. 1997. Identification of peptide and protein ligands for the caveolin-scaffolding domain. Implications for the interaction of caveolin with caveolae-associated proteins. *J. Biol. Chem.* 272: 6525–33

61. Oka N, Yamamoto M, Schwencke C, Kawabe J, Ebina T, et al. 1997. Caveolin interaction with protein kinase C. Isoenzyme-dependent regulation of kinase activity by the caveolin scaffolding domain peptide. *J. Biol. Chem.* 272:33416–21

62. Bucci M, Gratton JP, Rudic RD, Acevedo L, Roviezzo F, et al. 2000. In vivo delivery of the caveolin-1 scaffolding domain inhibits nitric oxide synthesis and reduces inflammation. *Nat. Med.* 6:1362–67

63. DeWeerd WF, Leeb-Lundberg LM. 1997. Bradykinin sequesters B2 bradykinin receptors and the receptor-coupled Galpha subunits Galphaq and Galphai in caveolae in DDT1 MF-2 smooth muscle cells. *J. Biol. Chem.* 272:17858–66

64. Ju H, Venema VJ, Marrero MB, Venema RC. 1998. Inhibitory interactions of the bradykinin B2 receptor with endothelial nitric-oxide synthase. *J. Biol. Chem.* 273:24025–29

65. Garcia-Cardena C, Fan R, Shah V, Sorrentino R, Cirino G, et al. 1998. Dynamic activation of endothelial nitric oxide synthase by Hsp90. *Nature* 392:821–24

66. Dedio J, Konig P, Wohlfart P, Schroeder C, Kummer W, et al. 2001. NOSIP, a novel modulator of endothelial nitric oxide synthase activity. *FASEB J.* 15:79–89

67. Goetz RM, Thatte HS, Prabhakar P, Cho MR, Michel T, et al. 1999. Estradiol induces the calcium-dependent translocation of endothelial nitric oxide synthase. *Proc. Natl. Acad. Sci. USA* 96:2788–93

68. Gosink EC, Forsberg EJ. 1993. Effects of ATP and bradykinin on endothelial cell Ca^{2+} homeostasis and formation of cGMP and prostacyclin. *Am. J. Physiol. Cell Physiol.* 265:C1620–C29

69. Papapetropoulos A, Garcia-Cardena G, Madri JA, Sessa WC. 1997. Nitric oxide production contributes to the angiogenic properties of vascular endothelial growth factor in human endothelial cells. *J. Clin. Invest.* 100:3131–39

70. Ayajiki K, Kindermann M, Hecker M, Fleming I, Busse R. 1996. Intracellular pH and tyrosine phosphorylation but not calcium determine shear stress-induced nitric oxide production in native endothelial cells. *Circ. Res.* 78:750–58

71. Fleming I, Bauersachs J, Schafer A, Scholz D, Aldershvile J, et al. 1999. Isometric contraction induces the Ca^{2+}-independent activation of the endothelial nitric oxide synthase. *Proc. Natl. Acad. Sci. USA* 96:1123–28

72. Ohno M, Gibbons GH, Dzau VJ, Cooke

JP. 1993. Shear stress elevates endothelial cGMP. Role of a potassium channel and G protein coupling. *Circulation* 88:193–97

73. Corson MA, James NL, Latta SE, Nerem RM,   Berk BC, et al. 1996. Phosphorylation of endothelial nitric oxide synthase in response to fluid shear stress. *Circ. Res.* 79:984–91

74. Chen Z, Yuhanna IS, Galcheva-Gargova Z,   Karas RH,   Mendelsohn ME, et al. 1999. Estrogen receptor alpha mediates nongenomic activation of eNOS by estrogen. *J. Clin. Invest.* 103:401–6

75. Bernier SG,   Haldar S,   Michel T. 2000. Bradykinin-regulated interactions of the mitogen-activated protein kinase pathway with the endothelial nitric-oxide synthase. *J. Biol. Chem.* 275:30707–15

76. Hisamoto K,   Ohmichi M,   Kurachi H, Hayakawa J,   Kanda Y, et al. 2001. Estrogen induces the Akt-dependent activation of endothelial nitric-oxide synthase in vascular endothelial cells. *J. Biol. Chem.* 276:3459–67

77. Haynes MP, Sinha D, Russell KS, Collinge M,   Fulton D, et al. 2000. Membrane estrogen receptor engagement activates endothelial nitric oxide synthase via the PI3-kinase-Akt pathway in human endothelial cells. *Circ. Res.* 87:677–82

78. Dimmeler S,   Fleming I,   Fisslthaler B, Hermann C, Busse R, Zeiher AM. 1999. Activation of nitric oxide synthase in endothelial cells by Akt-dependent phosphorylation. *Nature* 399:601–5

79. Gallis B,   Corthals GL,   Goodlett DR, Ueba H,   Kim F, et al. 1999. Identification of flow-dependent endothelial nitric-oxide synthase phosphorylation sites by mass spectrometry and regulation of phosphorylation and nitric oxide production by the phosphatidylinositol 3-kinase inhibitor LY294002. *J. Biol. Chem.* 274:30101–8

80. Michell BJ,   Griffiths JE,   Mitchelhill KI,   Rodriguez-Crespo I,   Tiganis T, et al. 1999. The Akt kinase signals directly to endothelial nitric oxide synthase. *Curr. Biol.* 9:845–48

81. Fulton D,   Gratton JP,   McCabe TJ, Fontana J,   Fujio Y, et al. 1999. Regulation of endothelium-derived nitric oxide production by the protein kinase Akt. *Nature* 399:597–601

82. Chen ZP,   Mitchelhill KI,   Michell BJ, Stapleton D,   Rodriguez-Crespo I,   Witters LA. 1999. AMP-activated protein kinase phosphorylation of endothelial NO synthase. *FEBS Lett.* 443:285–89

83. Michell BJ,   Chen ZZ,   Tiganis T,   Stapleton D,   Katsis F, et al. 2001. Coordinated control of endothelial nitric-oxide synthase phosphorylation by protein kinase C and the cAMP-dependent protein kinase. *J. Biol. Chem.* 276:17625–28

84. Shaul PW,   Anderson RGW. 1998. Role of plasmalemmal caveolae in signal transduction. *Am. J. Physiol. Lung Cell. Mol. Physiol.* 275:L843–L51

85. Mendelsohn ME,   Karas RH. 1999. The protective effects of estrogen on the cardiovascular system. *N. Engl. J. Med.* 340:1801–11

86. Kirsch EA, Yuhanna IS, Chen Z, German Z,   Sherman TS,   Shaul PW. 1999. Estrogen acutely stimulates endothelial nitric oxide synthase in H441 human airway epithelium. *Am. J. Respir. Cell Mol. Biol.* 20:658–66

87. Chambliss KL, Yuhanna IS, Liu P, German Z,   Sherman TS, et al. 2000. ERà and eNOS are organized into a functional signaling module in caveolae. *Circ. Res.* 87:E44–52

88. Gordon DJ,   Rifkind BM. 1989. High-density lipoprotein—the clinical implications of recent studies. *N. Engl. J. Med.* 321:1311–16

89. Jolley CD, Woollett LA, Turley SD, Dietschy JM. 1998. Centripetal cholesterol flux to the liver is dictated by events in the peripheral organs and not by the plasma high density lipoprotein or apolipoprotein A-I concentration. *J. Lipid Res.* 39:2143–49

90. Acton SL, Scherer PE, Lodish HF, Krieger M. 1994. Expression cloning of SR-BI, a CD36-related class B scavenger receptor. *J. Biol. Chem.* 269:21003–9

91. Rigotti A, Trigatti BL, Penman M, Rayburn H, Herz J, et al. 1997. A targeted mutation in the murine gene encoding the high density lipoprotein (HDL) receptor scavenger receptor class B type I reveals its key role in HDL metabolism. *Proc. Natl. Acad. Sci. USA* 94:12610–15

92. Yuhanna IS, Zhu Y, Cox BE, Hahner LD, Osborne-Lawrence S, et al. 2001. High-density lipoprotein binding to scavenger receptor-BI activates endothelial nitric oxide synthase. *Nat. Med.* 7:853–57

93. Lin S, Fagan KA, Shaul PW, Cooper DMF, Rodman DM. 2000. Sustained endothelial nitric-oxide synthase activation requires capacitative Ca^{2+} entry. *J. Biol. Chem.* 275:17979–85

94. Isshiki M, Anderson RGW. 1999. Calcium signal transduction from caveolae. *Cell Calcium* 26:201–8

95. Isshiki M, Ando J, Korenaga R, Kogo H, Fujimoto T, et al. 1998. Endothelial Ca^{2+} waves preferentially originate at specific loci in caveolin-rich cell edges. *Proc. Natl. Acad. Sci. USA* 95:5009–14

96. Berridge MJ. 1993. Inositol trisphosphate and calcium signaling. *Nature* 361:315–25

97. Palmer RM, Ferrige AG, Moncada S. 1987. Nitric oxide release accounts for the biological activity of endothelium-derived relaxing factor. *Nature* 327:524–26

98. Ignarro LJ, Buga GM, Wood KS, Byrns RE, Chaudhuri G. 1987. Endothelium-derived relaxing factor produced and released from artery and vein is nitric oxide. *Proc. Natl. Acad. Sci. USA* 84:9265–69

99. Sessa WC, Harrison JK, Barber CM, Zeng D, Durieux ME, et al. 1992. Molecular cloning and expression of a cDNA encoding endothelial cell nitric oxide synthase. *J. Biol. Chem.* 267:15274–76

100. Nishida K, Harrison DG, Navas JP, Fisher AA, Dockery SP, et al. 1992. Molecular cloning and characterization of the constitutive bovine aortic endothelial cell nitric oxide synthase. *J. Clin. Invest.* 90:2092–96

Annu. Rev. Physiol. 2002. 64:775–802

GM-CSF Regulates Pulmonary Surfactant Homeostasis and Alveolar Macrophage-Mediated Innate Host Defense

Bruce C. Trapnell and Jeffrey A. Whitsett

Division of Pulmonary Biology, Children's Hospital Medical Center, Cincinnati, Ohio 45229-3039; e-mail: bruce.trapnell@chmcc.org, jeff.whitsett@chmcc.org

Key Words   GM-CSF receptor, signaling, transcriptional control, PU.1, differentiation

■ **Abstract**   Recent studies in transgenic mice have revealed important insights into the roles of GM-CSF in regulation of surfactant homeostasis and lung host defense. Interruption of the GM-CSF signaling pathway by targeted ablation of the GM-CSF gene or its receptor ($GM^{-/-}$ or $GM R_{\beta c}^{-/-}$ mice, respectively) resulted in pulmonary alveolar proteinosis (PAP) but no hematologic abnormalities. Alveolar macrophages from $GM^{-/-}$ mice have reduced capacity for surfactant catabolism, cell adhesion, phagocytosis, bacterial killing, Toll-receptor signaling, and expression of various pathogen-associated molecular pattern recognition receptors, suggesting arrest at an early stage of differentiation. PAP and abnormalities of alveolar macrophage function were corrected by local expression of GM-CSF in the lung, and expression of the transcription factor PU.1 in alveolar macrophages of $GM^{-/-}$ mice rescued most defects. Recently, a strong association of auto-antibodies to GM-CSF or GM-CSF receptor gene mutations with PAP has implicated GM-CSF signaling abnormalities in the pathogenesis of PAP in humans. Together, these observations demonstrate that GM-CSF has a critical role in regulation of surfactant homeostasis and alveolar macrophage innate immune functions in the lung.

INTRODUCTION

Granulocyte-macrophage colony-stimulating factor (GM-CSF) was initially identified as an activity present in lung cell–conditioned medium, capable of stimulating growth of granulocytes and macrophages from cultured hematopoietic progenitors (1). Purification of natural GM-CSF and production of the recombinant protein from cloned cDNA permitted extensive in vitro and in vivo studies that defined a number of biologic activities for GM-CSF (1, 2; reviewed in 3, 4). GM-CSF functions as a growth factor predominantly affecting cells of the phagocytic lineage, but also stimulates production of eosinophils, erythrocytes, megakaryocytes, and dendritic cells (3, 5–8). Owing to its stimulatory effect on bone marrow, GM-CSF is

0066-4278/02/0315-0775$14.00

now widely used therapeutically to ameliorate chemotherapy-induced neutropenia and to stimulate hematopoietic recovery after bone-marrow transplantation (9). Separate from its effects on progenitor cell proliferation, GM-CSF stimulates a number of functions of mature hematopoietic cells (3, 4), including alveolar macrophages (10–13), and also influences the growth of nonhematopoietic cells such as the alveolar epithelium of the lung (14). Recently, a critical role for GM-CSF in the lung was demonstrated by studies in several transgenic mouse models wherein GM-CSF function was altered by targeted ablation of genes for either GM-CSF (GM$^{-/-}$ mice) or its receptor (GM R$_{\beta c}^{-/-}$ mice) or by overexpression of GM-CSF in various tissues (Table 1). This review summarizes data from these models and recent advances in our knowledge regarding the critical role of GM-CSF in the regulation of alveolar macrophage functions in surfactant homeostasis and innate immune lung host defense.

MOLECULAR BIOLOGY OF GM-CSF AND ITS RECEPTOR

GM-CSF is a 23-kDa, glycosylated, monomeric secreted polypeptide encoded by a 2.5-kb gene comprised of 4 exons located near genes for other hematologic growth factor family members in both humans (5q22-31) and mice (11q30-31) (15–17; see 3 for a review of GM-CSF biology). Murine and human GM-CSF share modest structural homology at the level of nucleotide (70%) and protein (56%) sequence (3, 18) and do not exhibit cross-species receptor binding or biological activity. Mature murine GM-CSF comprises 124 residues, human comprises 127 residues, and both are derived from a precursor containing a 25–amino acid signal peptide. GM-CSF is produced by multiple cell types, is present in serum and most tissues, albeit at low concentrations, and is also found in association with the extracellular matrix and as an integral membrane protein.

The effects of GM-CSF are mediated through heteromeric cell-surface receptors expressed on monocytes, macrophages, granulocytes, and other cells (19), including type II alveolar epithelial cells (14) (Figure 1). The GM-CSF receptor (GM-CSF R) is composed of α (CDw116; GM-CSF R$_\alpha$) and β (GM-CSF R$_{\beta c}$) chains; the latter is common to receptors for GM-CSF, IL-3, and IL-5 (19–21). Neither GM-CSF R$_\alpha$ nor β_c chains contain a tyrosine kinase catalytic domain, but the β_c chain constitutively associates with JAK2, which is a tyrosine kinase (22). GM-CSF binds with low affinity to the α chain, which then associates with the β chain, thus increasing α chain binding affinity, and initiates JAK2 autophosphorylation and postreceptor signaling. Signal transmission occurs through multiple pathways, each requiring distinct regions of the α (23) and β_c (24, 25) receptor chains. JAK2 activates STAT, MAPK, and a third distinct, albeit less well-defined pathway (26, 27). Non-JAK2 pathways have also been implicated in GM-CSF R signaling (28). The complex postreceptor signal transduction pathways modulated by GM-CSF are currently the subject of intense investigation (see recent reviews 26, 27, 29, 30 for further details).

The biological importance of GM-CSF receptor signaling in the mouse lung is supported by data demonstrating expression of GM-CSF R_α mRNA in both alveolar macrophages and type II alveolar epithelial cells, immunohistochemical staining of both cell types with anti-GM-CSF R_α chain antibodies, and enhanced proliferation of both cell types in transgenic mice overexpressing GM-CSF only in the lung (SPC-GM$^{+/+}$/GM$^{-/-}$) (Table 1) (14). The consequences of GM-CSF signaling for surfactant homeostasis and alveolar macrophage innate immune functions in lung host defense form the remainder of this review.

ROLE OF GM-CSF SIGNALING IN SURFACTANT HOMEOSTASIS

Studies with GM$^{-/-}$ and GM $R_{\beta c}^{-/-}$ mice demonstrated that GM-CSF signaling is required for homeostasis of pulmonary surfactant in the alveoli of the lung. Genetic ablation of GM-CSF or GM-CSF $R_{\beta c}$ in mice is caused the accumulation of surfactant phospholipids and proteins in the lungs. Pathological changes in the lungs of the GM$^{-/-}$ or GM $R_{\beta c}^{-/-}$ mice resemble those in human pulmonary alveolar proteinosis (PAP). These gene targeting experiments revealed unexpected links between GM-CSF signaling, alveolar macrophage function, and surfactant homeostasis. Recent clinical studies demonstrated that auto-antibodies against GM-CSF are likely involved in the pathogenesis of human PAP in some patients, thus providing the foundation for the development of novel strategies for diagnosis and treatment of this rare, but often debilitating, pulmonary disorder.

Pulmonary Surfactant Composition and Function

Surfactant is comprised of ~90% lipids, 10% proteins, and lesser amounts of carbohydrates (31). Approximately 80–90% of the surfactant lipids are phospholipids, phosphatidylcholine (PC) accounting for 70–80% of the surfactant phospholipids recovered in bronchoalveolar lavage (BAL) fluid from mammals (32). Di-saturated PC, principally dipalmitoyl-PC (DPPC), is the major surface-active component; ~60% of PC in pulmonary surfactant is DPPC. Surfactant-associated proteins SP-A, SP-B, SP-C, and SP-D contribute to the surface-active properties and structural forms of intraalveolar surfactant and may be involved in intracellular trafficking of phospholipid components (31, 33). The surfactant-associated proteins SP-B and SP-C are extremely hydrophobic proteins of approximately 8 and 4 kDa, respectively (31). SP-B and SP-C likely interact with phospholipids to enhance the surface-active properties of rapid spreading and stability during compression. SP-A and SP-D are collagenous glycoproteins with monomeric molecular ratios of 28–36 and 43 kDa, respectively (31). SP-A and SP-D bind some bacteria and viruses that enter the lung. SP-A and SP-D bind and activate alveolar macrophages, influencing chemotaxis, production of oxygen radicals, secretion of proteolytic enzymes, and inflammatory cytokines (31, 34–36) and influence both structure and

TABLE 1   Mice with genetic modifications of GM-CSF signaling

Mouse line[a]	Endogenous gene modification	Transgene[b]	Phenotype	Reference[c]
RV-GM[+]	None	MoMLV-GM-CSF	Elevation of GM-CSF levels in multiple tissues; massive accumulation of macrophages in tissues throughout the body; generalized myositis; blindness; early death	134
GM[−/−]	Homozygous GM-CSF gene ablation	None	Absence of GM-CSF; pulmonary alveolar proteinosis; multiple abnormalities in alveolar macrophages; impaired innate immunity in lung; no gross hematological abnormalities	59, 60
SPC-GM[+/+]	None	SP-C-GM-CSF	Marked elevation of GM-CSF levels only in the lungs with normal levels in blood; progressive alveolar macrophage accumulation; alveolar type II cell hyperplasia; no gross hematological abnormalities	14
SPC-GM[+/+]/GM[−/−]	Homozygous GM-CSF gene ablation	SP-C-GM-CSF	Marked elevation of GM-CSF in lung but not in blood; pulmonary alveolar proteinosis and alveolar macrophage abnormalities of GM[−/−] mice corrected; progressive alveolar macrophage accumulation; alveolar type II cell hyperplasia; no gross hematological abnormalities	14

Murine line	Genetic manipulation	Transgene	Phenotype	Citation
GM $R_{\beta c}^{-/-}$	Homozygous GM-CSF receptor β_c chain gene ablation	None	Absence of GM-CSF receptor β_c chain expression; pulmonary alveolar proteinosis; multiple abnormalities in alveolar macrophages; no gross hematological abnormalities	73, 78
SP-D$^{-/-}$/GM$^{-/-}$	Homozygous SP-D gene and homozygous GM-CSF gene ablation	None	Absence of GM-CSF and SP-D; pulmonary alveolar proteinosis characterized by surfactant lipid elevation out of proportion (higher) compared to elevation of surfactant proteins; no gross hematological abnormalities	76
SPC-GM$^{+/+}$/SP-D$^{-/-}$/GM$^{-/-}$	Homozygous SP-D gene and homozygous GM-CSF gene ablation	SP-C-GM-CSF	Marked elevation of GM-CSF in lung but not in blood; absence of SP-D; pulmonary alveolar proteinosis partly corrected; surfactant phospholipids elevated in lung; progressive alveolar macrophage accumulation; alveolar type II cell hyperplasia; no gross hematological abnormalities	77
GM$^{-/-}$/Csfm$^{Op/Op}$	Homozygous GM-CSF gene ablation and homozygous M-CSF gene null point mutations	None	Absence of GM-CSF and M-CSF; pulmonary alveolar proteinosis; generalized reduction in tissue macrophages and blood monocytes; susceptibility to pulmonary infections; osteopetrosis; absence of tooth eruption	135

[a]The names of murine lines harboring various and sometimes multiple genetic manipulations to alter GM-CSF signaling have been shortened to improve readability. Abbreviations include: RV, retroviral; GM, GM-CSF; SPC, surfactant protein C; GM $R_{\beta c}$, GM-CSF receptor β_c chain.

[b]Transgenes are composed of the following genetic elements: MoMLV-GM-CSF, Moloney murine leukemia virus vector expressing the murine GM-CSF cDNA from the promoter within the upstream viral long-terminal repeat; SPC-GM, human 3.7-kb surfactant protein C promoter expressing the murine GM-CSF cDNA. The viral promoter of the MoMLV-GM-GMCSF transgene results in widespread expression of the transgene in multiple tissues. In contrast, the lung tissue-specific SPC promoter of the SP-C-GM-CSF transgene results in expression only in the respiratory epithelium of the lung with confinement of GM-CSF in the lung.

[c]Citations refer to the initial study describing creation of the indicated murine models. See text for further details of the pulmonary and other phenotypes more fully defined in studies subsequent to the initial report of the model.

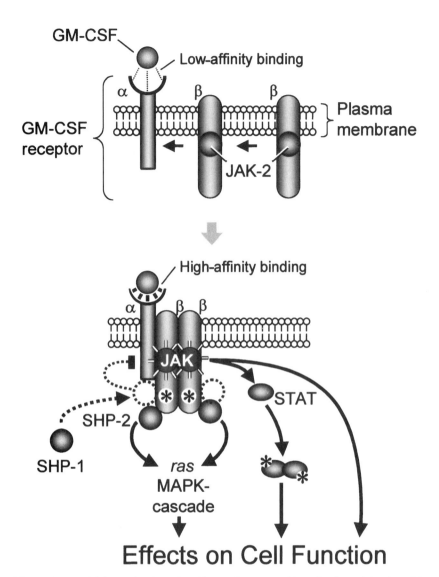

Figure 1 Model for activation, signaling pathways, and inactivation of the GM-CSF receptor. GM-CSF initially binds with low affinity to the GM-CSF R_α chain (*top*), which then associates with the affinity-converting GM-CSF $R_{\beta c}$ chain to form a six polypeptide complex consisting of two α chains, two β chains, and two JAK2 chains (*bottom*). JAK2, which is constitutively bound to GM-CSF $R_{\beta c}$, is then activated and phosphorylates tyrosine residues of the receptor (*asterisks*). These phosphorylated regions bind Src-homology-2 (SH2)-containing proteins including Stat5 and SHP-2, which are themselves phosphorylated and activate at least three cascades that mediate nuclear signaling. These signaling pathways converge on c-fos, c-myc, and numerous other target genes, resulting in multiple effects on cell function. Receptor inactivation occurs when SHP-1, another SH2-containing protein, binds to the receptor and extinguishes JAK2 activation.

function of pulmonary surfactant. SP-D plays a critical role in the regulation of surfactant phospholipid but not protein metabolism in the lung in vivo (37, 38).

Overview of Surfactant Metabolism

Surfactant phospholipids and proteins are synthesized, stored, secreted, and recycled or catabolized by alveolar type II epithelial cells (Figure 2) (37). Surfactant is stored in type II cells in large inclusion bodies known as lamellar bodies. When lamellar bodies are exocytosed into the alveolar lumen, interactions among lipids,

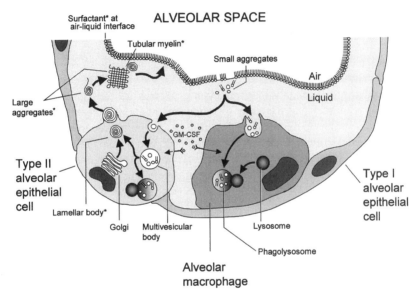

Figure 2 Surfactant metabolism. This diagram illustrates current understanding of surfactant protein and phospholipid synthetic, secretory, recycling, and catabolic pathways in the lung. Synthesis occurs in alveolar type II epithelial cells, followed by processing in the Golgi network. Surfactant proteins and phospholipids are stored in lamellar bodies (*asterisks*). Lamellar bodies are exocytosed into the liquid covering the alveolar surface of the alveolar space, where they assemble into the structures known as tubular myelin. The surfactant large-aggregate forms, extracellular lamellar bodies, and tubular myelin all have surface-active properties. Phospholipids from large-aggregate forms move to form continuous monolayers/multilayers of phospholipids that line the alveolar spaces and airways, with polar heads oriented toward the liquid and acyl chains toward the air. Surfactant is inactivated by mechanical and biological processes and converted into the surface-inactive, small-aggregate forms. Approximately 70–80% of surfactant small-aggregate forms are taken up by type II cells, sorted to lysosomes, and catabolized or reutilized. The remaining surfactant pool is phagocytosed and catabolized by alveolar macrophages.

SP-A, SP-B, and SP-C organize the material into lattice-like structures known as tubular myelin. Lipids from tubular myelin subsequently form monolayers and multilayers at the air-liquid interface to reduce surface tension and prevent alveolar collapse. Surfactant can be separated by centrifugation into heavy and light fractions, containing large and small aggregates. The heavy fraction contains surfactant with surface-active properties; the light fraction is an inactive form. The large aggregate surfactant form is changed to small aggregate vesicles through mechanical or biological actions, and these small aggregate particles are taken up by type II epithelial cells and alveolar macrophages, thus clearing inactivated surfactant from the alveolar space (39). In the adult rabbit lung ~70–80% of surfactant is taken up and reutilized by type II cells; most of the remaining 20–30% of surfactant is phagocytosed and catabolized by alveolar macrophages. A small portion of surfactant phospholipid is removed from the respiratory tract by the mucociliary escalator. The cellular and biochemical mechanisms mediating surfactant clearance, recycling, and catabolism remain poorly understood.

In healthy individuals, alveolar and lung tissue surfactant pool sizes are tightly regulated. At birth, alveolar surfactant pool sizes are determined by the degree of fetal maturation and secretion following the initiation of breathing (39). Surfactant concentrations increase in late gestation and are influenced by glucocorticoids and other hormones and factors. Surfactant secretion accompanies the birth process, initiating a cycle of secretion and recycling that maintains alveolar surfactant concentration in the postnatal lung. During the neonatal period, surfactant recycling is highly efficient; ~85–90% of surfactant taken up by type II cells from the alveolar spaces is recycled (39). In adult mice and rabbits, recycling efficiency has been estimated to be ~50% (39, 40). After intratracheal administration, exogenous surfactant or surfactant components are cleared from the alveoli of adult animals, with a half life of ~6–10 hours; however, radio-labeled surfactant components can be detected for up to 6–10 days because of recycling (31, 41). The alveolar pools of surfactant lipids and proteins are determined by the concerted effects of surfactant secretion, clearance, and recycling.

Pulmonary Alveolar Proteinosis

Pulmonary alveolar proteinosis (PAP) was first described in humans in 1958 (42) as a pulmonary disorder associated with the accumulation of surfactant lipids and proteins in the airspaces. Lungs of PAP patients are filled with surfactant-like material that impairs gas exchange causing dyspnea, fatigue, and exercise intolerance. Alveolar proteinosis material is rich in phospholipids and proteins. Ultrastructurally, PAP material is rich in tubular myelin, membranous vesicles, and lamellated structures resembling lamellar bodies. The biochemical composition of PAP material is similar to that of normal surfactant, consistent with the conclusion that the alveolar material in PAP represents an aberrant accumulation of pulmonary surfactant (43, 44). Except for the accumulation of surfactant, lung histology in

PAP is generally unchanged unless the condition is complicated by pneumonia. Patients with PAP are at increased risk for viral, bacterial, and fungal infections (45). Histologic abnormalities in PAP include the presence of enlarged alveolar macrophages that contain numerous phospholipid inclusions (Figure 3, see color insert), supporting the concept that abnormalities of surfactant catabolism by the alveolar macrophages may contribute to the disorder (46, 47).

PAP is classified as primary or secondary depending on the presence or absence of nonpulmonary pathology. In primary PAP, pathology is confined to the lung. In secondary PAP, typical abnormalities in the lung are also associated with hematologic malignancies, lymphomas, and immunosuppression (48–51). Secondary PAP also occurs with lung injury from inhalation of industrial particles, such as silica, aluminum, or titanium dust, or infection by *Pneumocystis carinii* (52–55). In some cases of secondary PAP, pathological findings include thickened pulmonary interstitium and fibrosis, supporting the concept that lung injury contributes to abnormalities in surfactant homeostasis.

ANIMAL MODELS OF PAP

Several animal models with pulmonary abnormalities similar to those in human PAP have been described, including naturally occurring variants of the beige mouse and a severe-combined immunodeficient (SCID) mouse designated as CB.17scid/scid (40, 56). Surfactant accumulations with similarity to PAP occur in rats exposed to inhaled silica or other particles (54, 57, 58). However, the cellular and molecular determinants of alveolar proteinosis in these models are not known.

Ablation of GM-CSF by Gene Targeting Caused PAP in Mice

$GM^{-/-}$ mice were generated by gene targeting in two independent laboratories (59, 60). Surprisingly, the only organ pathology identified in these mice was the marked accumulation of pulmonary surfactant phospholipids and proteins in the lungs (Figure 3). Microscopic evaluation of lung sections from $GM^{-/-}$ mice revealed an abundance of inclusions and enlarged foamy alveolar macrophages. The alveolar spaces contained amorphous eosinophilic material typical of PAP in humans. Mononuclear cell infiltrates were routinely observed in perivascular and peribronchiolar regions of the lungs from $GM^{-/-}$ mice. These characteristic mononuclear cell infiltrates increased with age. Under vivarium conditions, no pulmonary inflammation or infection in the $GM^{-/-}$ mice was detected; however, $GM^{-/-}$ mice are susceptible to opportunistic bacterial and fungal organisms (59, 61, 62). The histologic and biochemical characteristics of PAP were similar in both $GM^{-/-}$ and $GM\ R_{\beta c}^{-/-}$ mice (Figure 3).

Ultrastructural studies revealed that tubular myelin and lamellar bodies accumulated in the alveoli of $GM^{-/-}$ mice (59, 60). SP-A and SP-B levels measured by

ELISA were ≥10-fold higher in BAL fluid from GM$^{-/-}$ mice than from wild-type (i.e., GM-CSF$^{+/+}$) mice (60). The increased concentrations of surfactant proteins in the lungs of GM$^{-/-}$ mice were not directly related to changes in surfactant protein mRNA transcription or accumulation because SP-A, SP-B, and SP-C mRNAs, as assessed by S1 nuclease protection assays, were not altered in lung tissue from the GM$^{-/-}$ mice (60).

Immunohistochemistry confirmed the marked increase in alveolar SP-B in lungs of GM$^{-/-}$ mice (Figure 3). Alveolar phospholipid pool size was increased 10-fold, and lung tissue phospholipid was increased 2–4-fold in GM$^{-/-}$ mice compared with wild-type control mice (63).

Decreased Surfactant Phospholipid and Protein Clearance in GM$^{-/-}$ Mice

Metabolic studies were conducted to determine whether surfactant accumulation in GM$^{-/-}$ mice was caused by alterations in synthesis, secretion, recycling, and/or catabolism. Synthesis and secretion of saturated phosphatidylcholine (Sat PC) in GM$^{-/-}$ mice were similar to that measured in wild-type mice, despite the increased Sat PC concentrations already present in the lung tissues and airways of GM$^{-/-}$ mice. Likewise, mRNA concentrations for the surfactant proteins SP-A, SP-B, and SP-C were unaltered (60). However, clearance of Sat PC and SP-A from the airways of GM$^{-/-}$ mice was markedly impaired. Thus metabolic studies demonstrated that lack of GM-CSF signaling impaired recycling and/or catabolism but did not directly alter synthesis or secretion of surfactant phospholipids and proteins (Figure 4).

GM-CSF Is Expressed in Lung Cells

Several lines of evidence support the concept that local signaling pathways for GM-CSF are important in the lung. GM-CSF was first isolated from mouse lung cell–conditioned media (1), and GM-CSF is more easily detected in the lung than in blood (61, 62). GM-CSF mRNA was detected in various cell types, including myeloid, endothelial, fibroblasts, and alveolar macrophages in mice and primates (64). GM-CSF expression was also observed in nonhematopoietic cells, including lung fibroblasts, bronchial, and tracheal cells, and in alveolar type II cells (64–67). In situ hybridization was used to identify GM-CSF transcripts in bronchiolar cells and in hyperplastic alveolar cells of inflamed pulmonary tissue in human lung sections (68). GM-CSF activity was detected in conditioned media obtained from rat type II cells in vitro (65). Thus, GM-CSF is expressed in multiple cell types of the lung and, therefore, was a candidate for modulation of surfactant homeostasis.

Because ablation of GM-CSF caused PAP, reconstitution of GM-CSF in GM$^{-/-}$ mice was tested for correction of pulmonary abnormalities. To investigate the spatial requirement for GM-CSF in correction of PAP, GM-CSF was selectively expressed in respiratory epithelial cells of GM$^{-/-}$ mice, using the human SP-C promoter (SPC-GM$^{+/+}$/GM$^{-/-}$ mice) (Table 1). Expression of GM-CSF in the lung

A. Normal surfactant clearance

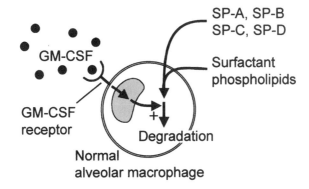

B. Pulmonary alveolar proteinosis

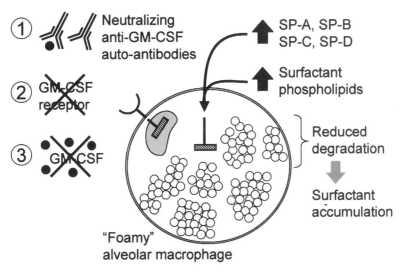

Figure 4 Model illustrating the current concepts of surfactant accumulation in the lungs in human and murine PAP. (*A*) Normally, GM-CSF stimulates the terminal differentiation of alveolar macrophages, enhancing their capacity for uptake and catabolism of surfactant proteins (SP-A, SP-B, SP-C, SP-D) and surfactant phospholipids. (*B*) Similar pulmonary histopathological features (see Figure 3) are produced by interruption of GM-CSF signaling through a variety of mechanisms, including (1) the presence of neutralizing anti-GM-CSF auto-antibodies in acquired idiopathic PAP in humans, (2) mutations of the GM-CSF receptor in both humans and mice, and (3) disruption of the gene encoding GM-CSF in mice. Interruption of the GM-CSF signaling pathway causes decreased catabolism of surfactant protein and lipids by alveolar macrophages leading to both intra- and extracellular accumulation.

completely corrected the alveolar proteinosis resulting from the targeted ablation of the endogenous GM-CSF gene (69). Likewise, inhalation but not systemic administration of GM-CSF corrected PAP in $GM^{-/-}$ mice (70), demonstrating that the presence of GM-CSF locally in the lung was necessary and sufficient to restore pulmonary surfactant homeostasis.

Lung-Specific GM-CSF Expression Increased Lung Size and Caused Type II Cell Hyperplasia and Alveolar Macrophage Accumulation

Local overexpression of GM-CSF in the lungs of $SPC\text{-}GM^{+/+}$ or $SPC\text{-}GM^{+/+}/GM^{-/-}$ mice caused unexpected changes in lung size and in the numbers of type II cells and alveolar macrophages (14). Morphometric analysis confirmed that lungs from mice expressing the SPC-GM transgene in either $GM^{-/-}$ or wild-type genetic background were 30–40% larger and that the numbers of type II cells increased approximately fourfold compared with lungs of mice without this transgene. Increased expression of proliferating cell nuclear antigen (PCNA) and proSP-C indicated increased proliferation of both alveolar macrophages and type II cells in the lungs of $SPC\text{-}GM^{+/+}$ or $SPC\text{-}GM^{+/+}/GM^{-/-}$ mice.

Surfactant Metabolism in $GM^{-/-}$ and $SPC\text{-}GM^{+/+}/GM^{-/-}$ Mice

Local expression of GM-CSF in the lungs of GM-CSF-targeted mice $(SPC\text{-}GM^{+/+}/GM^{-/-})$ restored surfactant phospholipid and protein pool sizes to normal. Two important changes in surfactant metabolism were detected in these mice: (a) Incorporation of radio-labeled surfactant phospholipid precursors was increased in lung tissues of $SPC\text{-}GM^{-/-}$ mice compared with that in wild-type mice, consistent with the increased number of type II cells in $SPC\text{-}GM^{+/+}/GM^{-/-}$ mice, and (b) alveolar clearance of exogenous radio-labeled DPPC or SP-B was enhanced in $SPC\text{-}GM^{+/+}/GM^{-/-}$ mice compared with wild-type controls, consistent with the increased numbers and activity of type II cells and alveolar macrophages. Thus overexpression of GM-CSF in the lungs of $SPC\text{-}GM^{+/+}/GM^{-/-}$ mice increased alveolar clearance of surfactant phospholipid and protein, supporting the hypothesis that GM-CSF influences surfactant catabolism within the local milieu.

Most of the pulmonary and alveolar macrophage abnormalities observed in $GM^{-/-}$ mice were corrected by reconstitution of GM-CSF in the lung through protein replacement, somatic gene transfer, or transgenic approaches (69–72). Because GM-CSF regulates functions in both alveolar macrophages and type II alveolar epithelial cells, these studies lacked clarity with regard to the target of the "therapeutic" effects of GM-CSF replacement. However, correction of PAP in $GM\ R_{\beta c}^{-/-}$ mice by transplantation of bone marrow from wild-type mice suggests that effects of GM-CSF on surfactant metabolism are mediated by changes in alveolar macrophage function rather than in type II alveolar epithelial cells (73).

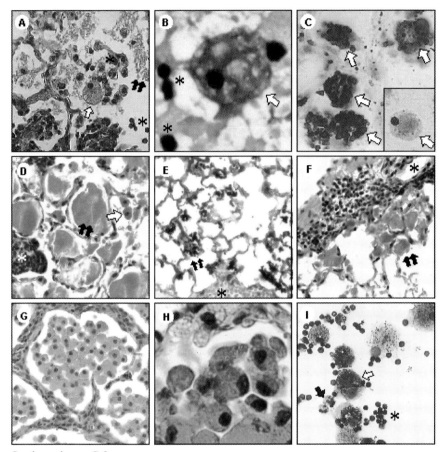

See legend page C-2

Figure 3 Lung and alveolar macrophage histopathology in mice with altered GM-CSF expression and human acquired, idiopathic PAP. (*A*) Lung histopathology in GM$^{-/-}$ mice stained with hematoxylin and eosin (H&E) showing foamy alveolar macrophages (*white arrow*), alveolar eosinophilic proteinosis material (*double black arrows*) and the typical lymphocytosis (*asterisk*). (*B*) Appearance of a typical foamy macrophage and proteinaceous material in GM$^{-/-}$ mice. Lymphocytes (*asterisks*) of normal size illustrate the markedly enlarged size of foamy macrophages. (*C*) Foam cells in GM$^{-/-}$ mice contain large amounts of lipid as demonstrated by oil-red-O staining (*white arrows*). Note the minimal accumulation in alveolar macrophages of age-matched wild-type mice (*inset*). (*D*) Both extracellular (*double black arrows*) and intracytoplasmic material within alveolar macrophages (*white arrow*) in GM$^{-/-}$ mice stains with periodic-acid Schiff reagent. (*E*) Immunohistochemical staining for surfactant protein B demonstrates its marked accumulation in the extracellular material (*double black arrows*) and within alveolar macrophages (not well-seen in this photomicrograph). Note the presence of lymphocyte accumulation (*asterisk*). (*F*) Lung histopathology after H&E staining in GM R$_{\beta c}^{-/-}$ mice. The extracellular proteinaceous material is similar in appearance to that of GM$^{-/-}$ mice. Pulmonary lymphocytosis also occurs in GM R$_{\beta c}^{-/-}$ mice. (*G*) Histologic appearance of the lung from a SPC-GM$^{+/+}$/GM$^{-/-}$ mouse stained with H&E, showing marked accumulation of alveolar macrophages with a normal morphologic appearance and lack of PAP. (*H*) Normal morphologic appearance of alveolar macrophages in SPC-GM$^{+/+}$/GM$^{-/-}$ mice. (*I*) Abnormal morphology of human alveolar macrophages (*white arrow*) from an individual with acquired idiopathic PAP. Also present are numerous lymphocytes (*asterisk*), neutrophils (*black arrow*), and enlarged foamy alveolar macrophages.

Catabolic Defect in Alveolar Macrophages from $GM^{-/-}$ Mice

Whole-animal metabolic studies demonstrated that the defect in surfactant lipid and protein metabolism in the $GM^{-/-}$ mice was not related to changes in surfactant synthesis or secretion but to decreased clearance (63). In vitro studies utilizing freshly isolated alveolar macrophages from $GM^{-/-}$ mice demonstrated marked defects in catabolism of both SP-A and surfactant phospholipids (74). Degradation of both SP-A and DPPC was markedly decreased in the alveolar macrophages from $GM^{-/-}$ mice. Furthermore, rates of surfactant catabolism were increased in alveolar macrophage from mice expressing increased amounts of GM-CSF in the lung. Uptake and binding of surfactant components by alveolar macrophages from $GM^{-/-}$ mice were not deficient. These findings, and the observations that bone marrow transplantation of $GM\ R_{\beta c}^{+/+}$ precursor cells substantially corrected alveolar proteinosis in $GM\ R_{\beta c}^{-/-}$ mice, supports the conclusion that the primary defect in surfactant homeostasis in the absence of GM-CSF signaling was caused by deficiency in catabolism of surfactant proteins and lipids by alveolar macrophages, likely related to the requirement for GM-CSF-dependent pathways for differentiation of alveolar macrophages in the lung.

Role of SP-D in Surfactant Lipid Homeostasis

The surfactant proteins SP-A, SP-B, SP-C, and SP-D accumulate in the lungs of patients and mice with PAP and, in turn, may influence surfactant structure and homeostasis. Furthermore, recent studies demonstrated that SP-D itself plays an important role in surfactant lipid organization and metabolism. Surfactant phospholipid, but not proteins, were increased five- to sixfold in the alveoli and lungs of $SP-D^{-/-}$ mice. In contrast to findings in the $GM^{-/-}$ mice, surfactant catabolism by the alveolar macrophages from $SP-D^{-/-}$ mice was not deficient; the abnormalities in surfactant content were mediated by abnormalities in the size of both intracellular and extracellular surfactant phospholipid pools (75). Experiments in double transgenic $SP-D^{-/-}/GM^{-/-}$ or triple transgenic $SPC-GM^{+/+}/GM^{-/-}/SP-D^{-/-}$ mice demonstrated that effects of SP-D and GM-CSF were additive, reflecting distinct, noncompensating regulatory pathways by which surfactant protein levels influence surfactant homeostasis (76).

Ablation of GM-CSF Receptors Caused PAP

Ablation of the murine gene encoding GM-CSF $R_{\beta c}$ produced alveolar proteinosis in mice (Figure 4) (77, 78). Wild-type bone marrow (i.e., from $GM\ R_{\beta c}^{+/+}$ mice) transplanted into lethally irradiated $GM\ R_{\beta c}^{-/-}$ substantially corrected alveolar proteinosis in $GM\ R_{\beta c}^{-/-}$ mice. Wild-type macrophages were found in the lungs of the $GM\ R_{\beta c}^{-/-}$ mice, suggesting that the engrafted wild-type alveolar macrophages had corrected the PAP. Because GM-CSF receptors are present on alveolar macrophages, alveolar type II epithelial cells, and bronchiolar epithelial cells, these cells are potential targets of GM-CSF (14, 64). Likewise, GM-CSF signaling between human bronchiolar epithelial and dendritic/Langerhans cells has been reported (79). The identification of GM-CSF receptors in various pulmonary cells supports the hypothesis that GM-CSF directly or indirectly influences alveolar macrophages and other pulmonary cells to modulate surfactant homeostasis.

CLINICAL IMPLICATIONS FOR DIAGNOSIS AND TREATMENT OF PAP

Prior to the generation of GM-CSF gene ablated mice (59, 60), the molecular basis of PAP was unknown. However, the spontaneous development of alveolar proteinosis in both $GM^{-/-}$ and $GM\,R_{\beta c}^{-/-}$ mice and correction of PAP by bone marrow transplantation in $GM\,R_{\beta c}^{-/-}$ mice (73) suggested that mutations or alterations in function of GM-CSF, the GM-CSF receptor, or its signal transduction pathways may cause abnormalities in alveolar macrophages resulting in clinical PAP (Figure 4). Recent studies demonstrated decreased GM-CSF receptor activity in four of eight patients with severe PAP, and in one case a point mutation was identified in the gene encoding the GM-CSF $R_{\beta c}$ (80). Although GM-CSF mutations have not been identified in humans with PAP, interruption of GM-CSF signaling by neutralizing antibodies directed against GM-CSF has been implicated in the pathogenesis of idiopathic PAP (81).

Autoimmunity Against GM-CSF and PAP

Analysis of serum and BAL fluid from adult patients with idiopathic PAP revealed the presence of neutralizing antibody to GM-CSF in most adult patients with the disorder (81–83). Whereas GM-CSF was readily detected in BAL fluid from the PAP patients, it was associated with binding proteins that were subsequently found to be antibodies of both IgG and IgM classes, supporting the concept that acquired idiopathic PAP is primarily an autoimmune disease caused by inhibition of GM-CSF activity (Figure 4). Several clinical trials evaluating GM-CSF for therapy of PAP have demonstrated improvement in some, but not all, PAP patients after systemic treatment with recombinant GM-CSF (84, 85). Thus, therapeutic strategies to increase GM-CSF activity or inhibit immune responses represent novel clinical approaches for PAP. Clinically, the success of lung lavage for treatment of PAP may depend on removal of both surfactant and anti-GM-CSF antibody from pulmonary tissues.

ROLE OF THE ALVEOLAR MACROPHAGE IN LUNG HOST DEFENSE

The extensive, yet fragile, surface of the lung is protected by multiple structural, innate immune, and adaptive immune defense mechanisms. These defenses are composed of physical, cellular, and molecular components and include epithelial

barriers, mucociliary clearance, antimicrobial polypeptides, professional phagocytes, inflammatory and other cytokines, natural killer (NK) cells, and cells and
products of the acquired immune system. Abundant and increasing evidence indicates that alveolar macrophages play critical roles in initial defenses against
pulmonary pathogens (86, 87) and that macrophage innate immune functions significantly influence subsequent acquired immune responses (88). Lung macrophages are bone marrow–derived mononuclear phagocytes found on the epithelial
surfaces of airways and alveoli in approximate proportion to the surface area, as
well as the interstitium of the lung (89). As the resident professional phagocyte,
alveolar macrophages occupy a central position in lung host defense, provide a
first line of defense against inhaled pathogens, and interact with numerous components of both the innate and acquired immune systems. These cells phagocytose
inhaled particles and pathogens and play a role in the catabolism of protein, lipid,
and carbohydrate components of the lung. Alveolar macrophages selectively respond to noxious ingestants (i.e., stimulation of inflammation during bacterial
phagocytosis), whereas responses to nonnoxious ingestants are generally molli-
fied (i.e., antiinflammatory responses during phagocytosis of apoptotic cells) (90).
An extensive literature describes the modulation of multiple alveolar macrophage
functions by various endogenous factors including cytokine growth factors such
as GM-CSF (3, 4). Notwithstanding, the mechanisms regulating the commitment,
proliferation, differentiation, and function of alveolar macrophages in vivo remain
poorly understood.

The Role of GM-CSF in Alveolar Macrophage Ontogeny

The alveolar macrophage is generally believed to be a relatively long-lived cell
derived from hematologic stem cell progenitors in the bone marrow (87, 91, 92),
although a second, marrow-independent embryonic lineage has been described
(93). Local proliferation of alveolar macrophage does occur ($\sim$1% in humans)
(94), but this does not appear to be the principal mechanism of cellular replenishment. Rather, monocytes constitutively enter the lung (95) and differentiate
into morphologically, histochemically, and functionally distinct macrophage populations (96, 97). GM-CSF promotes monocytic and granulocytic progenitor cell
growth, differentiation, and activation (9, 98) and enhances the proliferation and
accumulation of alveolar macrophages in vivo (11, 13, 14). Whereas GM-CSF
was initially considered a primary regulator of myelopoiesis, myeloid progenitor survival, lineage commitment, differentiation, and proliferation are also influenced by other hematologic cytokines including macrophage-colony stimulating
factor (M-CSF), interleukin-3 (IL-3), and IL-6 (99). Because the activities of GM-
CSF overlap with those of other cytokines, the singular role played by GM-CSF
in myelopoiesis has been difficult to assess (100). Surprisingly, disturbances in
hematopoiesis or myelopoiesis were not observed in $GM^{-/-}$ or $GM\ R_{\beta c}^{-/-}$ mice,
demonstrating that GM-CSF signaling is not critical to hematopoiesis in mice
(59, 60, 78, 101). The observation that both $GM^{-/-}$ and $GM\ R_{\beta c}^{-/-}$ mice developed

abnormalities in the lung (i.e., alveolar macrophage dysfunction and PAP) demonstrated a nonredundant role for GM-CSF in alveolar macrophage function and pulmonary homeostasis (59, 60). Whereas most aspects of hematopoiesis, including erythropoiesis, myelopoiesis, and extra-pulmonary immune function, were normal in both $GM^{-/-}$ and $GM\,R_{\beta c}^{-/-}$ mice, the abnormalities in lung histology and alveolar macrophage morphology and function were similar in both models. Alveolar macrophages in $GM^{-/-}$ and $GM\,R_{\beta c}^{-/-}$ mice acquire an abnormal foamy appearance with increasing age. This morphologic abnormality in $GM^{-/-}$ mice was corrected by expression of GM-CSF in the lungs of SPC-$GM^{+/+}/GM^{-/-}$ mice (Table 1) (69) or by pulmonary aerosolization of GM-CSF (70). Together, these data demonstrate that GM-CSF has a critical role in the lung but is not vital for early myelopoiesis or for the constitutive accumulation of alveolar macrophages in vivo. Functional overlap of GM-CSF with other family members (100) appears to explain the normal hematopoietic and systemic immune functions in nonpulmonary tissues in the $GM^{-/-}$ and $GM\,R_{\beta c}^{-/-}$ mice.

GM-CSF Regulates Multiple Functions in Alveolar Macrophages

Extensive in vitro and in vivo studies have demonstrated that GM-CSF enhances multiple functions in myeloid cells, including cell adhesion and chemotaxis (102, 103); Fc receptor expression (104); complement- and antibody-mediated phagocytosis (104, 105); oxidative metabolism (106, 107); intracellular killing of bacteria, fungi, protozoa, and viruses (108–110); cytokine signaling (111); and antigen presentation (4, 112). GM-CSF stimulates differentiation and functions of alveolar macrophages in vitro (10, 12, 113). Similarly, stimulation of monocytes with GM-CSF alters the expression of numerous genes as determined by serial analysis of gene expression (SAGE) (114), a finding consistent with the diverse effects of GM-CSF on myeloid cells.

Pulmonary Host Defense in Human PAP

Patients with PAP have increased susceptibility to microbial lung infections, suggesting that pulmonary host defenses are defective (42, 45, 115). The presence of histologically abnormal alveolar macrophages in PAP (e.g., large, foamy cells) suggests the possibility that alveolar macrophage host defense functions may be abnormal (47, 116). Alveolar macrophages obtained by segmental lavage of patients with PAP demonstrated giant lysosomes, poor survival in tissue culture, impaired chemotactic activity, decreased cell adhesion, and decreased intracellular killing of *Candida pseudotropicalis* (47). Other studies revealed defects in phagocytosis and phagolysosome fusion in alveolar macrophages from PAP patients (117). These defects could be recapitulated in normal alveolar macrophages incubated with cell-free fractions of PAP lavage fluid isolated at 20,000 × g and

250 $\times$ g, respectively (117). Whereas peripheral blood monocytes from a patient with PAP were morphologically normal, incubation of normal monocyte-derived macrophages with PAP lavage material recapitulated morphological abnormalities in vitro (47). Interestingly, defective alveolar macrophage function could be improved by therapeutic whole-lung lavage (118, 119). Recently, a 40,000 $\times$ g cell-free fraction of lung lavage from PAP patients inhibited proliferation of normal monocytes grown in the presence of GM-CSF (83). This fraction also inhibited the growth of and blocked the binding of ^{125}I-GM-CSF to a GM-CSF-dependent cell line in vitro. Subsequent studies demonstrated that this inhibitory activity was caused by neutralizing anti-GM-CSF antibodies. Such auto-antibodies were present in all of 24 individuals with idiopathic PAP from five countries but not in individuals with other secondary or congenital forms of PAP, other lung diseases, or normal controls (82). Together, these studies suggest that host defense defects in individuals with acquired idiopathic PAP may be related to the inhibition of GM-CSF and its effects on normal alveolar macrophage functions.

Host Defense Abnormalities in GM$^{-/-}$ and GM R$_{\beta c}^{-/-}$ Mice

The critical nature of GM-CSF stimulation of innate immunity in the lung in vivo was initially demonstrated in GM$^{-/-}$ mice wherein decreased pulmonary clearance of bacteria and increased susceptibility to intratracheal group B *Streptococcus* was observed (61). Superoxide production by alveolar macrophages from GM$^{-/-}$ mice was markedly decreased following exposure to group B *Streptococcus* or phorbol myristate acetate. GM$^{-/-}$ mice were also susceptible to *Pneumocystis carinii* pneumonia in vivo (62). Exaggerated inflammatory responses were observed following bacterial (61), parasitic (62), or viral lung infection (120) in the GM$^{-/-}$ mice. Transgenic expression of GM-CSF in lungs of GM$^{-/-}$ mice (e.g., SPC-GM$^{+/+}$/GM$^{-/-}$ mice) (Table 1) corrected defects in bacteria and fungal pulmonary clearance and inhibited exuberant inflammatory responses observed following pulmonary infection in the GM$^{-/-}$ mice (61). Alveolar macrophages from GM$^{-/-}$ mice showed decreased cell adhesion, decreased phagocytosis of gram negative or positive bacteria, zymosan, latex beads (121), and decreased internalization of adenovirus (120). Independent of the phagocytic abnormality, intracellular killing of both gram-positive and gram-negative bacteria was reduced in alveolar macrophages of GM$^{-/-}$ mice (122).

Paradoxically, despite exaggerated proinflammatory cytokine levels following microbial infection of the lungs, alveolar macrophages from GM$^{-/-}$ mice failed to release TNFα following exposure to lipopolysaccharide (121, 122). Consistent with that observation, reduced expression of multiple components of the Toll-like receptor (TLR) signaling pathway, including TLR4, TLR2, and CD14, was observed in alveolar macrophages from GM$^{-/-}$ mice (123). Expression of other pathogen-associated molecular pattern receptors, such as the mannose receptor, was reduced in alveolar macrophages from GM$^{-/-}$ mice (123). Interestingly, expression of all these receptors on alveolar macrophages was rescued by specific expression of GM-CSF in the lung (i.e., in SPC-GM$^{+/+}$/GM$^{-/-}$ mice) (121).

Together, these observations suggest that GM-CSF regulates multiple, diverse functions in alveolar macrophages and are consistent with the concept that GM-CSF has a critical role in regulating the terminal differentiation of alveolar macrophages in the lung.

Transcriptional Control of Alveolar Macrophage Accumulation and Function

Knowledge of the mechanisms regulating macrophage gene expression and differentiation has advanced considerably in recent years (reviewed in 124–126). Myelopoiesis is regulated by transcription factors that can be grouped into those necessary for macrophage development and those important for activation or repression of crucial genes but not yet demonstrated to be essential (124). These include factors (*a*) necessary for survival of stem and/or pluripotent myeloid cells (e.g., GATA-2, SCL, and C-myb), or (*b*) necessary for myeloid differentiation (e.g., PU.1, AML1), (*c*) important in regulating intermediate stages of myeloid cell differentiation (e.g., C/EBPβ, HOXB7, c-Myc), or (*d*) important in regulating macrophage maturation (e.g., C/EBPα, EGR-1, IRF-1, NF-Y, Jun/Fos, C-Maf, and STAT proteins) (126).

PU.1 Mediates Action of GM-CSF in the Alveolar Macrophage

PU.1 is an ets-family transcription factor that regulates myeloid and B-cell lineage development (127, 128). Targeted ablation of the PU.1 locus in mice blocked macrophage and B lymphocyte development and delayed neutrophil and T lymphocyte development (128, 129). PU.1 was not required for myeloid lineage commitment (130) but promoted both proliferation (131) and differentiation (132) of myeloid progenitors. PU.1 is regarded as a "master" transcription factor because it modulates a large number of functionally diverse genes in early myeloid cells, consistent with a significant role in regulation of myeloid differentiation (125). Some of the many genes regulated by PU.1 include those encoding cell surface molecules (immunoglobulin receptors FcγRI and FcγRIII; integrin complement receptors CR3 (CD11b/CD18), CR4 (CD11c/CD18); receptors for GM-CSF, M-CSF, and G-CSF; macrosialin; scavenger receptors I and II; mannose receptor MHC II I-Aβ; and intracellular enzymes (lysozyme, NADPH oxidase subunits pg91^{phox} and p47^{phox}) among others (reviewed in 125). PU.1 also autoregulates its own expression at the level of transcription (133).

PU.1 is expressed in alveolar macrophages in the lungs of wild-type but not GM$^{-/-}$ mice, and expression is restored in alveolar macrophages in SPC-GM$^{+/+}$/GM$^{-/-}$ mice (121). Thus in mice, the presence and level of local GM-CSF expression in the lung correlates with the presence and level of PU.1 in alveolar

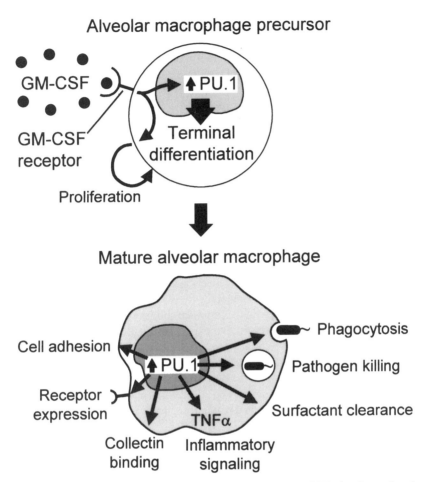

Figure 5   Model illustrating the mechanism by which GM-CSF stimulates alveolar macrophage innate immune functions. GM-CSF regulates expression of PU.1 in alveolar macrophages in vivo. Retroviral vector-mediated, constitutive expression of PU.1 in alveolar macrophages of $GM^{-/-}$ mice rescued the defects in cell adhesion, pathogen-associated molecular pattern receptors (e.g., Toll-like receptors, mannose receptor); surfactant protein and lipid uptake and degradation; Toll-like receptor signaling; phagocytosis of bacteria, fungi, and virus; and, independent of uptake, intracellular killing of bacteria. These observations suggest that GM-CSF stimulates terminal differentiation of alveolar macrophages through the global transcription factor PU.1.

macrophages. In the context of the strong effects of PU.1 as a master regulator of macrophage differentiation, this suggests that PU.1 might mediate the effects of GM-CSF on alveolar macrophage innate immune function by stimulating the terminal differentiation of alveolar macrophage precursors in the lung (Figure 5). This hypothesis is supported by the recent demonstration that constitutive, retroviral-mediated expression of PU.1 in alveolar macrophages from $GM^{-/-}$ mice rescued alveolar macrophage cell adhesion, surfactant protein and degradation, expression of pathogen pattern recognition receptors, phagocytosis, bacterial killing, and Toll-receptor signaling (121).

SUMMARY

Ablation of GM-CSF signaling in the mouse results in profound alterations in pulmonary surfactant homeostasis, causing accumulation of surfactant lipids and proteins in the alveolar spaces and lung tissues. Metabolic studies in $GM^{-/-}$ mice demonstrated impaired surfactant catabolism by alveolar macrophages. Local expression of GM-CSF in lung epithelial cells corrected the aberrant morphology of alveolar macrophages and pulmonary alveolar proteinosis in $GM^{-/-}$ mice, supporting the concept that GM-CSF signaling pathways within the lung influence surfactant catabolism. PAP in $GM\ R_{\beta c}^{-/-}$ mice was substantially corrected by bone marrow transplantation, showing that alveolar macrophages play an important role in surfactant homeostasis. Expression of the transcription factor PU.1 in murine alveolar macrophages correlates with the presence and level of expression of GM-CSF locally in the lung, and constitutive expression of PU.1 in alveolar macrophages from $GM^{-/-}$ mice substantially rescues the abnormalities of innate immune function and surfactant catabolism. These observations, which link GM-CSF signaling, PU.1, and alveolar macrophage function, support a role for GM-CSF/PU.1 in regulating the terminal differentiation of alveolar macrophages and distinguish the unique importance of PU.1 in the lung from its role early in monocyte-macrophage differentiation.

The observation that the presence of neutralizing, anti-GM-CSF antibodies is strongly correlated with PAP in individuals with the acquired, idiopathic form of the disease supports a role for GM-CSF in lung surfactant homeostasis in humans and the concept that PAP patients with disorders of GM-CSF receptor structure or function may benefit from bone marrow transplantation to restore alveolar macrophage function. The finding that many PAP patients with abnormalities in alveolar macrophage function similar to those of $GM^{-/-}$ mice have auto-antibodies against GM-CSF offers therapeutic opportunities to improve clinical outcomes. Thus studies in the $GM^{-/-}$ and $GM\ R_{\beta c}^{-/-}$ transgenic mice have provided unexpected insights into the mechanisms by which GM-CSF controls alveolar macrophage differentiation and surfactant catabolism that have important implications for diagnosis and therapy of PAP.

ACKNOWLEDGMENTS

This work was supported in part by National Institutes of Health grants HL28623 (J.A.W.), HL69549 (B.C.T.), and National Institutes of Health SCOR grant HL56387.

Visit the Annual Reviews home page at www.AnnualReviews.org

LITERATURE CITED

1. Burgess AW, Camakaris J, Metcalf D. 1977. Purification and properties of colony-stimulating factor from mouse lung-conditioned medium. *J. Biol. Chem.* 252:1998–2003

2. Gough NM, Gough J, Metcalf D, Kelso A, Grail D, et al. 1984. Molecular cloning of cDNA encoding a murine haematopoietic growth regulator, granulocyte-macrophage colony stimulating factor. *Nature* 309:763–67

3. Rasko JE. 1994. Granulocyte-macrophage colony stimulating factor. In *The Cytokine Handbook*, ed. R Derynck, pp. 343–69. London: Academic

4. Tarr PE. 1996. Granulocyte-macrophage colony-stimulating factor and the immune system. *Med. Oncol.* 13:133–40

5. Metcalf D. 1980. Clonal analysis of proliferation and differentiation of paired daughter cells: action of granulocyte-macrophage colony-stimulating factor on granulocyte-macrophage precursors. *Proc. Natl. Acad. Sci. USA* 77:5327–30

6. Metcalf D, Johnson GR, Burgess AW. 1980. Direct stimulation by purified GM-CSF of the proliferation of multipotential and erythroid precursor cells. *Blood* 55:138–47

7. Johnson GR, Metcalf D. 1980. Detection of a new type of mouse eosinophil colony by Luxol-fast-blue staining. *Exp. Hematol.* 8:549–61

8. Inaba K, Inaba M, Deguchi M, Hagi K, Yasumizu R, et al. 1993. Granulocytes, macrophages, and dendritic cells arise from a common major histocompatibility complex class II-negative progenitor in mouse bone marrow. *Proc. Natl. Acad. Sci. USA* 90:3038–42

9. Metcalf D. 1999. Cellular hematopoiesis in the twentieth century. *Semin. Hematol.* 36:5–12

10. Akagawa KS, Kamoshita K, Tokunaga T.

1988. Effects of granulocyte-macrophage colony-stimulating factor and colony-stimulating factor-1 on the proliferation and differentiation of murine alveolar macrophages. *J. Immunol.* 141:3383–90

11. Worgall S, Singh R, Leopold PL, Kaner RJ, Hackett NR, et al. 1999. Selective expansion of alveolar macrophages in vivo by adenovirus-mediated transfer of the murine granulocyte-macrophage colony-stimulating factor cDNA. *Blood* 93:655–66

12. Lemaire I, Yang H, Lauzon W, Gendron N. 1996. M-CSF and GM-CSF promote alveolar macrophage differentiation into multinucleated giant cells with distinct phenotypes. *J. Leukocyte Biol.* 60:509–18

13. Nakata K, Akagawa KS, Fukayama M, Hayashi Y, Kadokura M, Tokunaga T. 1991. Granulocyte-macrophage colony-stimulating factor promotes the proliferation of human alveolar macrophages in vitro. *J. Immunol.* 147:1266–72

14. Huffman Reed JA, Rice WR, Zsengeller ZK, Wert SE, Dranoff G, Whitsett JA. 1997. GM-CSF enhances lung growth and causes alveolar type II epithelial cell hyperplasia in transgenic mice. *Am. J. Physiol. Lung Cell Mol. Physiol.* 273:L715–L25

15. Le Beau MM, Pettenati MJ, Lemons RS, Diaz MO, Westbrook CA, et al. 1986. Assignment of the GM-CSF, CSF-1, and FMS genes to human chromosome 5 provides evidence for linkage of a family of genes regulating hematopoiesis and for their involvement in the deletion (5q) in myeloid disorders. *Cold Spring Harb. Symp. Quant. Biol.* 51:899–909

16. Barlow DP, Bucan M, Lehrach H, Hogan BL, Gough NM. 1987. Close genetic and physical linkage between the murine haemopoietic growth factor genes GM-CSF

and multi-CSF (IL3). *EMBO J.* 6:617–23

17. Lee JS, Young IG. 1989. Fine-structure mapping of the murine IL-3 and GM-CSF genes by pulsed-field gel electrophoresis and molecular cloning. *Genomics* 5:359–62

18. Miyatake S, Otsuka T, Yokota T, Lee F, Arai K. 1985. Structure of the chromosomal gene for granulocyte-macrophage colony stimulating factor: comparison of the mouse and human genes. *EMBO J.* 4:2561–68

19. Goodall GJ, Bagley CJ, Vadas MA, Lopez AF. 1993. A model for the interaction of the GM-CSF, IL-3 and IL-5 receptors with their ligands. *Growth Factors* 8:87–97

20. Gearing DP, King JA, Gough NM, Nicola NA. 1989. Expression cloning of a receptor for human granulocyte-macrophage colony-stimulating factor. *EMBO J.* 8:3667–76

21. Hayashida K, Kitamura T, Gorman DM, Arai K, Yokota T, Miyajima A. 1990. Molecular cloning of a second subunit of the receptor for human granulocyte-macrophage colony-stimulating factor (GM-CSF): reconstitution of a high-affinity GM-CSF receptor. *Proc. Natl. Acad. Sci. USA* 87:9655–59

22. Quelle FW, Sato N, Witthuhn BA, Inhorn RC, Eder M, et al. 1994. JAK2 associates with the beta c chain of the receptor for granulocyte-macrophage colony-stimulating factor, and its activation requires the membrane-proximal region. *Mol. Cell. Biol.* 14:4335–41

23. Matsuguchi T, Zhao Y, Lilly MB, Kraft AS. 1997. The cytoplasmic domain of granulocyte-macrophage colony-stimulating factor (GM-CSF) receptor alpha subunit is essential for both GM-CSF-mediated growth and differentiation. *J. Biol. Chem.* 272:17450–59

24. Duronio V, Clark-Lewis I, Federsppiel B, Wieler JS, Schrader JW. 1992. Tyrosine phosphorylation of receptor beta sub-units and common substrates in response to interleukin-3 and granulocyte-macrophage colony-stimulating factor. *J. Biol. Chem.* 267:21856–63

25. Sakamaki K, Miyajima I, Kitamura T, Miyajima A. 1992. Critical cytoplasmic domains of the common beta subunit of the human GM-CSF, IL-3 and IL-5 receptors for growth signal transduction and tyrosine phosphorylation. *EMBO J.* 11:3541–49

26. Watanabe S, Itoh T, Arai K. 1996. Roles of JAK kinases in human GM-CSF receptor signal transduction. *J. Allergy Clin. Immunol.* 98:S183–91

27. D'Andrea RJ, Gonda TJ. 2000. A model for assembly and activation of the GM-CSF, IL-3 and IL-5 receptors: insights from activated mutants of the common beta subunit. *Exp. Hematol.* 28:231–43

28. Hackenmiller R, Kim J, Feldman RA, Simon MC. 2000. Abnormal Stat activation, hematopoietic homeostasis, and innate immunity in c-fes−/− mice. *Immunity* 13:397–407

29. Ward AC, Oomen SP, Smith L, Gits J, van Leeuwen D, et al. 2000. The SH2 domain-containing protein tyrosine phosphatase SHP-1 is induced by granulocyte colony-stimulating factor (G-CSF) and modulates signaling from the G-CSF receptor. *Leukemia* 14:1284–91

30. Woodcock JM, Bagley CJ, Lopez AF. 1999. The functional basis of granulocyte-macrophage colony stimulating factor, interleukin-3 and interleukin-5 receptor activation, basic and clinical implications. *Int. J. Biochem. Cell Biol.* 31:1017–25

31. Weaver TE, Whitsett JA. 1991. Function and regulation of expression of pulmonary surfactant-associated proteins. *Biochem. J.* 273:249–64

32. Batenburg JJ. 1992. Surfactant phospholipids: synthesis and storage. *Am. J. Physiol. Lung Cell Mol. Physiol.* 262:L367–L85

33. Wright JR. 1990. Clearance and recycling

of pulmonary surfactant. *Am. J. Physiol. Lung Cell Mol. Physiol.* 259:L1–L12

34. Pison U, Wright JR, Hawgood S. 1992. Specific binding of surfactant apoprotein SP-A to rat alveolar macrophages. *Am. J. Physiol. Lung Cell Mol. Physiol.* 262:L412–L17

35. Wintergerst E, Manz-Keinke H, Plattner H, Schlepper-Schafer J. 1989. The interaction of a lung surfactant protein (SP-A) with macrophages is mannose dependent. *Eur. J. Cell. Biol.* 50:291–98

36. Wright JR, Youmans DC. 1993. Pulmonary surfactant protein A stimulates chemotaxis of alveolar macrophage. *Am. J. Physiol. Lung Cell Mol. Physiol.* 264: L338–L44

37. Fisher JH, Sheftelyevich V, Ho YS, Fligiel S, McCormack FX, et al. 2000. Pulmonary-specific expression of SP-D corrects pulmonary lipid accumulation in SP-D gene-targeted mice. *Am. J. Physiol. Lung Cell Mol. Physiol.* 278:L365–L73

38. Botas C, Poulain F, Akiyama J, Brown C, Allen L, et al. 1998. Altered surfactant homeostasis and alveolar type II cell morphology in mice lacking surfactant protein D. *Proc. Natl. Acad. Sci. USA* 95:11869–74

39. Jobe AH, Ikegami M. 1993. Surfactant metabolism. *Clin. Perinatol.* 20:683–96

40. Gross NJ, Barnes E, Narine KR. 1988. Recycling of surfactant in black and beige mice: pool sizes and kinetics. *J. Appl. Physiol.* 64:2017–25

41. Chander A, Fisher AB. 1990. Regulation of lung surfactant secretion. *Am. J. Physiol. Lung Cell Mol. Physiol.* 258:L241–L53

42. Rosen SG, Castleman B, Liebow AA. 1958. Pulmonary alveolar proteinosis. *N. Engl. J. Med.* 258:1123–42

43. Onodera T, Nakamura M, Sato T, Akino T. 1983. Biochemical characterization of pulmonary washings of patients with alveolar proteinosis, interstitial pneumonitis and alveolar cell carcinoma. *Tohoku J. Exp. Med.* 139:245–63

44. Crouch E, Persson A, Chang D. 1993. Accumulation of surfactant protein D in human pulmonary alveolar proteinosis. *Am. J. Pathol.* 142:241–48

45. Persson A. 1988. Pulmonary alveolar proteinosis. In *Fishman's Pulmonary Diseases and Disorders*, ed. AP Fishman, pp. 1225–30. New York: McGraw-Hill

46. Golde DW. 1979. Alveolar proteinosis and the overfed macrophage [editorial]. *Chest* 76:119–20

47. Golde DW, Territo M, Finley TN, Cline MJ. 1976. Defective lung macrophages in pulmonary alveolar proteinosis. *Ann. Intern. Med.* 85:304–9

48. Carnovale R, Zornoza J, Goldman AM, Luna M. 1977. Pulmonary alveolar proteinosis: its association with hematologic malignancy and lymphoma. *Radiology* 122:303–6

49. Green D, Dighe P, Ali NO, Katele GV. 1980. Pulmonary alveolar proteinosis complicating chronic myelogenous leukemia. *Cancer* 46:1763–66

50. Hildebrand FL Jr, Rosenow EC 3rd, Habermann TM, Tazelaar HD. 1990. Pulmonary complications of leukemia. *Chest* 98:1233–39

51. Steens RD, Summers QA, Tarala RA. 1992. Pulmonary alveolar proteinosis in association with Fanconi's anemia and psoriasis. A possible common pathogenetic mechanism. *Chest* 102:637–38

52. Keller CA, Frost A, Cagle PT, Abraham JL. 1995. Pulmonary alveolar proteinosis in a painter with elevated pulmonary concentrations of titanium. *Chest* 108:277–80

53. Hook GE. 1991. Alveolar proteinosis and phospholipidoses of the lungs. *Toxicol. Pathol.* 19:482–513

54. Dethloff LA, Gilmore LB, Brody AR, Hook GE. 1986. Induction of intra- and extra-cellular phospholipids in the lungs of rats exposed to silica. *Biochem. J.* 233:111–18

55. Ruben FL, Talamo TS. 1986. Secondary pulmonary alveolar proteinosis occurring in two patients with acquired immune

deficiency syndrome. *Am. J. Med.* 80: 1187–90

56. Jennings VM, Dillehay DL, Webb SK, Brown LA. 1995. Pulmonary alveolar proteinosis in SCID mice. *Am. J. Respir. Cell Mol. Biol.* 13:297–306

57. Heppleston AG, Fletcher K, Wyatt I. 1974. Changes in the composition of lung lipids and the "turnover" of dipalmitoyl lecithin in experimental alveolar lipoproteinosis induced by inhaled quartz. *Br. J. Exp. Pathol.* 55:384–95

58. Ballantyne B. 1994. Pulmonary alveolar phospholipoproteinosis induced by Orasol Navy Blue dust. *Hum. Exp. Toxicol.* 13:694–99

59. Stanley E, Lieschke GJ, Grail D, Metcalf D, Hodgson G, et al. 1994. Granulocyte/macrophage colony-stimulating factor-deficient mice show no major perturbation of hematopoiesis but develop a characteristic pulmonary pathology. *Proc. Natl. Acad. Sci. USA* 91:5592–96

60. Dranoff G, Crawford AD, Sadelain M, Ream B, Rashid A, et al. 1994. Involvement of granulocyte-macrophage colony-stimulating factor in pulmonary homeostasis. *Science* 264:713–16

61. LeVine AM, Reed JA, Kurak KE, Cianciolo E, Whitsett JA. 1999. GM-CSF-deficient mice are susceptible to pulmonary group B streptococcal infection. *J. Clin. Invest.* 103:563–69

62. Paine R 3rd, Preston AM, Wilcoxen S, Jin H, Siu BB, et al. 2000. Granulocyte-macrophage colony-stimulating factor in the innate immune response to Pneumocystis carinii pneumonia in mice. *J. Immunol.* 164:2602–9

63. Ikegami M, Ueda T, Hull W, Whitsett JA, Mulligan RC, et al. 1996. Surfactant metabolism in transgenic mice after granulocyte macrophage-colony stimulating factor ablation. *Am. J. Physiol. Lung Cell Mol. Physiol.* 270:L650–L58

64. Gasson JC. 1991. Molecular physiology of granulocyte-macrophage colony-stimulating factor. *Blood* 77:1131–45

65. Blau H, Riklis S, Kravtsov V, Kalina M. 1994. Secretion of cytokines by rat alveolar epithelial cells: possible regulatory role for SP-A. *Am. J. Physiol. Lung Cell Mol. Physiol.* 266:L148–L55

66. Churchill L, Friedman B, Schleimer RP, Proud D. 1992. Production of granulocyte-macrophage colony-stimulating factor by cultured human tracheal epithelial cells. *Immunology* 75:189–95

67. Smith SM, Lee DK, Lacy J, Coleman DL. 1990. Rat tracheal epithelial cells produce granulocyte/macrophage colony-stimulating factor. *Am. J. Respir. Cell Mol. Biol.* 2:59–68

68. Kato M, Schleimer RP. 1994. Antiinflammatory steroids inhibit granulocyte/macrophage colony-stimulating factor production by human lung tissue. *Lung* 172:113–24

69. Huffman JA, Hull WM, Dranoff G, Mulligan RC, Whitsett JA. 1996. Pulmonary epithelial cell expression of GM-CSF corrects the alveolar proteinosis in GM-CSF-deficient mice [see comments]. *J. Clin. Invest.* 97:649–55

70. Reed JA, Ikegami M, Cianciolo ER, Lu W, Cho PS, et al. 1999. Aerosolized GM-CSF ameliorates pulmonary alveolar proteinosis in GM-CSF-deficient mice. *Am. J. Physiol. Lung Cell Mol. Physiol.* 276:L556–L63

71. Zsengeller ZK, Reed JA, Bachurski CJ, LeVine AM, Forry-Schaudies S, et al. 1998. Adenovirus-mediated granulocyte-macrophage colony-stimulating factor improves lung pathology of pulmonary alveolar proteinosis in granulocyte-macrophage colony-stimulating factor-deficient mice. *Hum. Gene Ther.* 9: 2101–9

72. Ikegami M, Jobe AH, Huffman Reed JA, Whitsett JA. 1997. Surfactant metabolic consequences of overexpression of GM-CSF in the epithelium of GM-CSF-deficient mice. *Am. J. Physiol. Lung Cell Mol. Physiol.* 273:L709–L14

73. Nishinakamura R, Wiler R, Dirksen U,

Morikawa Y, Arai K, et al. 1996. The pulmonary alveolar proteinosis in granulocyte macrophage colony-stimulating factor/interleukins 3/5 beta c receptor-deficient mice is reversed by bone marrow transplantation. *J. Exp. Med.* 183:2657–62

74. Yoshida M, Ikegami M, Reed JA, Chroneos ZC, Whitsett JA. 2001. GM-CSF regulates surfacant Protein-A and lipid catabolism by alveolar macrohpages. *Am. J. Physiol. Lung Cell Mol. Physiol.* 280:L379–L86

75. Ikegami M, Whitsett JA, Jobe A, Ross G, Fisher J, Korfhagen T. 2000. Surfactant metabolism in SP-D gene-targeted mice. *Am. J. Physiol. Lung Cell Mol. Physiol.* 279:L468–L76

76. Ikegami M, Hull W, Yoshida H, Wert S, Whitsett JA. 2001. SP-D and GM-CSF regulate surfactant homeostasis via distinct mechanisms. *Am. J. Physiol. Lung Cell Mol. Physiol.* 281:L697–L703

77. Nishinakamura R, Nakayama N, Hirabayashi Y, Inoue T, Aud D, et al. 1995. Mice deficient for the IL-3/GM-CSF/IL-5 beta c receptor exhibit lung pathology and impaired immune response, while beta IL3 receptor-deficient mice are normal. *Immunity* 2:211–22

78. Robb L, Drinkwater CC, Metcalf D, Li R, Kontgen F, et al. 1995. Hematopoietic and lung abnormalities in mice with a null mutation of the common beta subunit of the receptors for granulocyte-macrophage colony-stimulating factor and interleukins 3 and 5. *Proc. Natl. Acad. Sci. USA* 92:9565–69

79. Tazi A, Bouchonnet F, Grandsaigne M, Boumsell L, Hance AJ, Soler P. 1993. Evidence that granulocyte macrophage colony-stimulating factor regulates the distribution and differentiated state of dendritic cells/Langerhans cells in human lung and lung cancers. *J. Clin. Invest.* 91:566–76

80. Dirksen U, Nishinakamura R, Groneck P, Hattenhorst U, Nogee L, et al. 1997. Human pulmonary alveolar proteinosis associated with a defect in GM-CSF/IL-3/IL-5 receptor common beta chain expression. *J. Clin. Invest.* 100:2211–17

81. Kitamura T, Tanaka N, Watanabe J, Watanabe U, Kanegasaki S, et al. 1999. Idiopathic pulmonary alveolar proteinosis as an autoimmune disease with neutralizing antibody against granulocyte/macrophage colony-stimulating factor. *J. Exp. Med.* 190:875–80

82. Kitamura T, Uchida K, Tanaka N, Tsuchiya T, Watanabe J, et al. 2000. Serological diagnosis of idiopathic pulmonary alveolar proteinosis. *Am. J. Respir. Crit. Care Med.* 162:658–62

83. Tanaka N, Watanabe J, Kitamura T, Yamada Y, Kanegasaki S, Nakata K. 1999. Lungs of patients with idiopathic pulmonary alveolar proteinosis express a factor which neutralizes granulocyte-macrophage colony stimulating factor. *FEBS Lett.* 442:246–50

84. Seymour JF, Dunn AR, Vincent JM, Presneill JJ, Pain MC. 1996. Efficacy of granulocyte-macrophage colony-stimulating factor in acquired alveolar proteinosis. *N. Engl. J. Med.* 335:1924–52

85. Kavuru MS, Sullivan EJ, Piccin R, Thomassen MJ, Stoller JK. 2000. Exogenous granulocyte-macrophage colony-stimulating factor administration for pulmonary alveolar proteinosis. *Am. J. Respir. Crit. Care Med.* 161:1143–48

86. Fels AO, Cohn ZA. 1986. The alveolar macrophage. *J. Appl. Physiol.* 60:353–69

87. Bezdicek P, Crystal RG. 1997. Pulmonary macrophages. In *The Lung: Scientific Foundations*, ed. RG Crystal, PJ Barnes, JB West, ER Weibel, pp. 859–75. Philadelphia: Lippincott-Raven

88. Medzhitov R, Janeway C Jr. 2000. Innate immunity. *N. Engl. J. Med.* 343:338–44

89. Crapo JD, Barry BE, Gehr P, Bachofen M, Weibel ER. 1982. Cell number and cell characteristics of the normal human lung. *Am. Rev. Respir. Dis.* 126:332–37

90. Aderem A, Underhill DM. 1999. Mechanisms of phagocytosis in macrophages. *Annu. Rev. Immunol.* 17:593–623

91. van Furth R, Cohn ZA. 1968. The origin and kinetics of mononuclear phagocytes. *J. Exp. Med.* 128:415–35

92. Thomas ED, Ramberg RE, Sale GE, Sparkes RS, Golde DW. 1976. Direct evidence for a bone marrow origin of the alveolar macrophage in man. *Science* 192:1016–18

93. Lichanska AM, Browne CM, Henkel GW, Murphy KM, Ostrowski MC, et al. 1999. Differentiation of the mononuclear phagocyte system during mouse embryogenesis: the role of transcription factor PU.1. *Blood* 94:127–38

94. Bitterman PB, Saltzman LE, Adelberg S, Ferrans VJ, Crystal RG. 1984. Alveolar macrophage replication. One mechanism for the expansion of the mononuclear phagocyte population in the chronically inflamed lung. *J. Clin. Invest.* 74:460–69

95. Kennedy DW, Abkowitz JL. 1998. Mature monocytic cells enter tissues and engraft. *Proc. Natl. Acad. Sci. USA* 95:14944–49

96. Van Furth R. 1992. Production and migration of monocytes and macrophages. In *Mononuclear Phagocytes: Biology of Monocytes and Macrophages*, ed. R Van Furth. Dordrecht: Kluwer

97. Hume DA, Robinson AP, MacPherson GG, Gordon S. 1983. The mononuclear phagocyte system of the mouse defined by immunohistochemical localization of antigen F4/80. Relationship between macrophages, Langerhans cells, reticular cells, and dendritic cells in lymphoid and hematopoietic organs. *J. Exp. Med.* 158:1522–36

98. Nicola NA, Metcalf D. 1986. Specificity of action of colony-stimulating factors in the differentiation of granulocytes and macrophages. *Ciba Found. Symp.* 118:7–28

99. Clark SC, Kamen R. 1987. The human hematopoietic colony-stimulating factors. *Science* 236:1229–37

100. Krumwieh D, Weinmann E, Siebold B, Seiler FR. 1990. Preclinical studies on synergistic effects of IL-1, IL-3, G-CSF and GM-CSF in cynomolgus monkeys. *Int. J. Cell Cloning* 8:229–47; Discussion 47–48

101. Nishinakamura R, Miyajima A, Mee PJ, Tybulewicz VL, Murray R. 1996. Hematopoiesis in mice lacking the entire granulocyte-macrophage colony-stimulating factor/interleukin-3/interleukin-5 functions. *Blood* 88:2458–64

102. Weisbart RH, Golde DW, Gasson JC. 1986. Biosynthetic human GM-CSF modulates the number and affinity of neutrophil f-Met-Leu-Phe receptors. *J. Immunol.* 137:3584–87

103. Arnaout MA, Wang EA, Clark SC, Sieff CA. 1986. Human recombinant granulocyte-macrophage colony-stimulating factor increases cell-to-cell adhesion and surface expression of adhesion-promoting surface glycoproteins on mature granulocytes. *J. Clin. Invest.* 78:597–601

104. Weisbart RH, Kacena A, Schuh A, Golde DW. 1988. GM-CSF induces human neutrophil IgA-mediated phagocytosis by an IgA Fc receptor activation mechanism. *Nature* 332:647–48

105. Collins HL, Bancroft GJ. 1992. Cytokine enhancement of complement-dependent phagocytosis by macrophages: synergy of tumor necrosis factor-alpha and granulocyte-macrophage colony-stimulating factor for phagocytosis of Cryptococcus neoformans. *Eur. J. Immunol.* 22:1447–54

106. Wing EJ, Ampel NM, Waheed A, Shadduck RK. 1985. Macrophage colony-stimulating factor (M-CSF) enhances the capacity of murine macrophages to secrete oxygen reduction products. *J. Immunol.* 135:2052–56

107. Coleman DL, Chodakewitz JA, Bartiss AH, Mellors JW. 1988. Granulocyte-macrophage colony-stimulating factor enhances selective effector functions

of tissue-derived macrophages. *Blood* 72:573–78

108. Fleischmann J, Golde DW, Weisbart RH, Gasson JC. 1986. Granulocyte-macrophage colony-stimulating factor enhances phagocytosis of bacteria by human neutrophils. *Blood* 68:708–11

109. Ruef C, Coleman DL. 1990. Granulocyte-macrophage colony-stimulating factor: pleiotropic cytokine with potential clinical usefulness. *Rev. Infect. Dis.* 12:41–62

110. Weiser WY, Van Niel A, Clark SC, David JR, Remold HG. 1987. Recombinant human granulocyte/macrophage colony-stimulating factor activates intracellular killing of *Leishmania donovani* by human monocyte-derived macrophages. *J. Exp. Med.* 166:1436–46

111. Gennari R, Alexander JW, Gianotti L, Eaves-Pyles T, Hartmann S. 1994. Granulocyte macrophage colony-stimulating factor improves survival in two models of gut-derived sepsis by improving gut barrier function and modulating bacterial clearance. *Ann. Surg.* 220:68–76

112. Morrissey PJ, Bressler L, Park LS, Alpert A, Gillis S. 1987. Granulocyte-macrophage colony-stimulating factor augments the primary antibody response by enhancing the function of antigen-presenting cells. *J. Immunol.* 139:1113–19

113. Chen BD, Mueller M, Chou TH. 1988. Role of granulocyte/macrophage colony-stimulating factor in the regulation of murine alveolar macrophage proliferation and differentiation. *J. Immunol.* 141:139–44

114. Hashimoto S, Suzuki T, Dong HY, Yamazaki N, Matsushima K. 1999. Serial analysis of gene expression in human monocytes and macrophages. *Blood* 94:837–44

115. Davidson JM, Macleod WM. 1969. Pulmonary alveolar proteinosis. *Br. J. Dis. Chest* 63:13–28

116. Harris JO. 1979. Pulmonary alveolar proteinosis: abnormal in vitro function of alveolar macrophages. *Chest* 76:156–59

117. Gonzalez-Rothi RJ, Harris JO. 1986. Pulmonary alveolar proteinosis. Further evaluation of abnormal alveolar macrophages. *Chest* 90:656–61

118. Okano A, Sato A, Chida K, Iwata M, Yasuda K, et al. 1990. Improvement in alveolar macrophage function after therapeutic lung lavage in pulmonary alveolar proteinosis. *Nihon Kyobu Shikkan Gakkai Zasshi* 28:723–28

119. Bury T, Corhay JL, Saint-Remy P, Radermecker M. 1989. Alveolar proteinosis: restoration of the function of the alveolar macrophages after therapeutic lavage. *Rev. Mal. Respir.* 6:373–75

120. Zsengeller Z, Halvinas H, Otake K, Whitsett JA, Trapnell BC. 2000. Alveolar macrophage-mediated adenovirus uptake and clearance from the lung is modulated by granulocyte-macrophage colony stimulating factor. *Respir. Crit. Care Med.* 163:A901

121. Shibata Y, Berclaz P-Y, Chroneos Z, Yoshida H, Whitsett JA, Trapnell BC. 2001. GM-CSF regulates alveolar macrophage differentiation and innate immunity in the lung through PU.1. *Immunity* 15:557–67

122. Shibata Y, Berclaz P-Y, Whitsett JA, Trapnell B. 2001. GM-CSF regulates innate immunity in the lung by coordinate, marked stimulation of phagocytosis, pathogen killing and toll receptor signaling in alveolar macrophages through the transcription factor PU.1. *Pediatr. Pulmonol.* Suppl. 269

123. Shibata Y, Berclaz P-Y, Trapnell BC. 2001. Arrest of alveolar macrophage differentiation, in vivo, in GM-CSF-deficient mice is associated with altered expression of transcription factors PU.1 and ICSBP. *Respir. Crit. Care Med.* 163:A790 (Abstr.)

124. Tenen DG, Hromas R, Licht JD, Zhang DE. 1997. Transcription factors, normal myeloid development, and leukemia. *Blood* 90:489–519

125. Lloberas J, Soler C, Celada A. 1999. The key role of PU.1/SPI-1 in B cells, myeloid cells and macrophages. *Immunol. Today* 20:184–89

126. Valledor AF, Borras FE, Cullell-Young M, Celada A. 1998. Transcription factors that regulate monocyte/macrophage differentiation. *J. Leukocyte Biol.* 63:405–17

127. Klemsz MJ, McKercher SR, Celada A, Van Beveren C, Maki RA. 1990. The macrophage and B cell-specific transcription factor PU.1 is related to the ets oncogene. *Cell* 61:113–24

128. Scott EW, Simon MC, Anastasi J, Singh H. 1994. Requirement of transcription factor PU.1 in the development of multiple hematopoietic lineages. *Science* 265:1573–77

129. McKercher SR, Torbett BE, Anderson KL, Henkel GW, Vestal DJ, et al. 1996. Targeted disruption of the PU.1 gene results in multiple hematopoietic abnormalities. *EMBO J.* 15:5647–58

130. Olson MC, Scott EW, Hack AA, Su GH, Tenen DG, et al. 1995. PU.1 is not essential for early myeloid gene expression but is required for terminal myeloid differentiation. *Immunity* 3:703–14

131. Celada A, Borras FE, Soler C, Lloberas J, Klemsz M, et al. 1996. The transcription factor PU.1 is involved in macrophage proliferation. *J. Exp. Med.* 184:61–69

132. DeKoter RP, Walsh JC, Singh H. 1998. PU.1 regulates both cytokine-dependent proliferation and differentiation of granulocyte/macrophage progenitors. *EMBO J.* 17:4456–68

133. Chen H, Ray-Gallet D, Zhang P, Hetherington CJ, Gonzalez DA, et al. 1995. PU.1 (Spi-1) autoregulates its expression in myeloid cells. *Oncogene* 11:1549–60

134. Lang RA, Metcalf D, Cuthbertson RA, Lyons I, Stanley E, et al. 1987. Transgenic mice expressing a hemopoietic growth factor gene (GM-CSF) develop accumulations of macrophages, blindness, and a fatal syndrome of tissue damage. *Cell* 51:675–86

135. Liesche GJ, Stanley E, Grail D, Hodgson G, Sinickas V, et al. 1994. Mice lacking both macrophage- and granulocyte-macrophage-colony stimulating factor have macrphages and coexistent osteopetrosis and severe lung disease. *Blood* 84:27–35

Annu. Rev. Physiol. 2002. 64:803–43

HUMAN AND MURINE PHENOTYPES ASSOCIATED WITH DEFECTS IN CATION-CHLORIDE COTRANSPORT

Eric Delpire[1,2,3] and David B. Mount[4]

[1]Departments of Anesthesiology, [2]Molecular Physiology and Biophysics, and [3]Center for Molecular Neuroscience; [4]Department of Medicine, Division of Nephrology, Nashville VA Medical Center and Vanderbilt University Medical Center, Nashville, Tennessee 37232; e-mail: eric.delpire@mcmail.vanderbilt.edu; david.mount@mcmail.vanderbilt.edu

Key Words Na-K-2Cl cotransporter, K-Cl cotransporter, Na-Cl cotransporter, brain, kidney

■ **Abstract** The diuretic-sensitive cotransport of cations with chloride is mediated by the cation-chloride cotransporters, a large gene family encompassing a total of seven Na-Cl, Na-K-2Cl, and K-Cl cotransporters, in addition to two related transporters of unknown function. The cation-chloride cotransporters perform a wide variety of physiological roles and differ dramatically in patterns of tissue expression and cellular localization. The renal-specific Na-Cl cotransporter (NCC) and Na-K-2Cl cotransporter (NKCC2) are involved in Gitelman and Bartter syndrome, respectively, autosomal recessive forms of metabolic alkalosis. The associated phenotypes due to loss-of-function mutations in NCC and NKCC2 are consistent, in part, with their functional roles in the distal convoluted tubule and thick ascending limb, respectively. Other cation-chloride cotransporters are positional candidates for Mendelian human disorders, and the K-Cl cotransporter KCC3, in particular, may be involved in degenerative peripheral neuropathies linked to chromosome 15q14. The characterization of mice with both spontaneous and targeted mutations of several cation-chloride cotransporters has also yielded significant insight into the physiological and pathophysiological roles of several members of the gene family. These studies implicate the Na-K-2Cl cotransporter NKCC1 in hearing, salivation, pain perception, spermatogenesis, and the control of extracellular fluid volume. Targeted deletion of the neuronal-specific K-Cl cotransporter KCC2 generates mice with a profound seizure disorder and confirms the central role of this transporter in modulating neuronal excitability. Finally, the comparison of human and murine phenotypes associated with loss-of-function mutations in cation-chloride cotransporters indicates important differences in physiology of the two species and provides an important opportunity for detailed physiological and morphological analysis of the tissues involved.

0066-4278/02/0315-0803$14.00

INTRODUCTION

With the final deciphering of the human genome, the functional characterization of newly identified proteins is proceeding at an accelerated pace, and a more complete picture now emerges for each family of proteins. Simultaneously, genetic linkage rapidly identifies potential candidates for human diseases; mutations in proteins are found to cause diseases (monogenic diseases) or to predispose individuals to more complex disorders (multigene or multifactorial disorders). Animal models are also being created with the hope of understanding the function of these proteins and the pathophysiology associated with their dysfunction. This review examines the significant contribution to human diseases of a relatively small family of membrane transporters, the electroneutral cation-chloride cotransporter family.

Electroneutral inorganic cation-chloride cotransporters are encoded by a small number of genes that share some homology with eukaryote cationic amino acid transporters and prokaryote and eukaryote amino acid permeases (Figure 1). All members from this superfamily of proteins are integral membrane proteins consisting of 10 to 12 transmembrane segments (Figure 2). For most of the cation-chloride cotransporters, the basic membrane topology has been predicted by Kyte-Doolittle hydropathy analysis and other methods. In the case of NKCC1, however, it is clear that the amino- and carboxyl-terminal domains are subject to phosphorylation by protein kinases, indicating an intracellular location. The basic topology model of NKCC1 was recently confirmed experimentally using in vitro translation, with TM1, TM3, TM5, TM7, TM9, and TM11 being inserted into the membrane as signal anchors and TM2, TM4, TM6, TM8, TM10, and TM12 being inserted as the corresponding stop transfers (69). Most of the core region of the cation-chloride cotransporters, from transmembrane domains TM3 to TM12, have homology to the large amino acid permease domain (pfam00324). This leaves the putative amino- and carboxyl-terminal tails and transmembrane segments 1 and 2 to define the cation-chloride cotransporters. The two first transmembrane segments TM1 and TM2 and the carboxyl terminus show high degrees of homology between members, whereas the amino terminus is poorly conserved (83, 87, 104, 159). There is strong evidence that the intracellular amino- and carboxyl-terminal tails of the cotransporters are involved in the regulation of cotransport activity (11, 142, 143). It is worth stressing the significance of TM2, which possesses three spliced variants in NKCC2 (93, 94, 158, 161, 177). This transmembrane segment is implicated in ion affinity (95, 96). Chimeric approaches and site-directed mutagenesis experiments reveal the importance of TM2 in defining the affinity of NKCC1 for both Na^+ and K^+ (97). Finally, a recent study using cross-linking reagents provides evidence that the structural unit of the secretory Na-K-2Cl cotransporter KCC1 is a homodimer (155). This intriguing finding needs further confirmation and warrants a functional analysis of monomers and homodimers. The possibility of heterodimer formation is of further interest, particularly given the co-expression of two or more K-Cl cotransporters in multiple cell types.

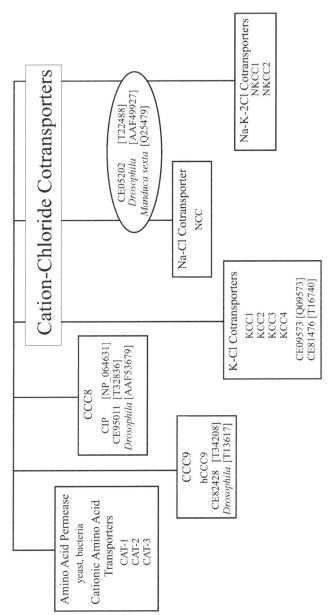

Figure 1 Diagram representing the functionally characterized cation-chloride cotransporters and their relationship with yeast amino acid permeases and vertebrate cationic amino acid transporters. Electroneutral coupled movement of Na^+, K^+, and Cl^- occurs under three different ion combinations: K-Cl cotransport, Na-Cl cotransport, and Na-K-2Cl cotransport. New related proteins (CCC8 and CCC9) of unknown function are distantly related to the cation-chloride cotransporters. *C. elegans* and *Drosophila* orthologs of these proteins have been included. Accession numbers are indicated in brackets.

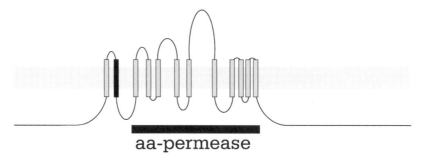

aa-permease

Figure 2   Members of this superfamily of membrane transporters have 12 transmembrane domains and large amino- and carboxyl-terminal tails. Transmembrane domains TM3–TM12 have homology to amino acid permease domain (pfam00324). Transmembrane domain TM2 (highlighted) is involved in ion binding.

Cation-chloride cotransporters can be divided into subfamilies, two of which are relatively well defined: the Na^+-dependent cation-chloride cotransporters, which include one Na-Cl cotransporter (NCC, also known as TSC) and two Na-K-2Cl cotransporters (NKCC1 and NKCC2, also known as BSC2 and BSC1, respectively), and the Na^+-independent K-Cl cotransporters (KCCs), of which there are four. Two additional genes have recently been identified in the human genome: one protein (CIP, or CCC8) with unknown transport function, which has been proposed to regulate NKCC1 function (26), and CCC9 (D. B. Mount, unpublished results), which is as yet uncharacterized. An expressed pseudogene with particular homology to NCC is also found on human chromosome 1 (D. B. Mount, unpublished results).

This review mostly focuses on the role that these cotransporters play in human disorders and discusses the consequences of gene manipulation in mice. We have chosen to exclude KCC4 because this gene is not implicated in human disease and has not been inactivated in mice. Several extensive reviews on the function and regulation of the major cation-chloride cotransporters have been published in the past few years; we refer the interested reader to this substantive background information (41, 81–83, 104, 129, 130, 159, 160, 196).

With the complete sequencing of the human genome, all cation-chloride cotransporters have now been mapped to human chromosomes (Table 1). We have also included in the table the location of some of the genes in the mouse genome.

Na^+-DEPENDENT COTRANSPORTERS

Secretory Na-K-2Cl Cotransporter: NKCC1

Na-K-2Cl cotransporters mediate the electroneutral transport of 1 Na^+, 1 K^+, and 2 Cl^- ions across cell membranes. There are two closely related genes encoding Na-K-2Cl cotransporters: NKCC1, often referred to as the secretory Na-K-2Cl cotransporter because of its expression in many secretory epithelia, and NKCC2,

TABLE 1 Chromosomal localization of cotransporters in human and mouse chromosomes

Protein	Gene	Human	Reference	Mouse	Reference
NCC	Slc12a3	16q13	(205)	8 cM 45	(175)
NKCC1	Slc12a2	5q23.2	(179)	18 cM 31	(45)
NKCC2	Slc12a1	15q15-q21.1	(203)	2 cM 69.5	(188)
KCC1	Slc12a4	16q22.1	(128)	8 cM 53	(181)
KCC2	Slc12a5	20			
KCC3	Slc12a6	15q13-q14	(88, 162)		
KCC4	Slc12a7	5p15.3	(162)		
CCC8(CIP)		7q22	(26)		
CCC9		3q21-22	(D. B. Mount, unpublished)		

the absorptive Na-K-2Cl contransporter, the apical cotransporter that reabsorbs Na-K-Cl from the glomerular ultrafiltrate as it passes through the renal thick ascending limb (TAL).

EXPRESSION OF NKCC1   The cloning of NKCC1 from shark rectal gland (244), mouse mIMCD-3 cells (45), and human T84 cells (179) and the development of monoclonal and polyclonal antibodies facilitated the identification of NKCC1 in multiple tissues. It is in the nervous system (185, 187) and the kidney (74, 105) that the most extensive characterization of NKCC1 expression has been reported. Immunolocalization studies revealed that whereas the cotransporter is expressed in many different tissues, it is found, in fact, in specific cell types. In the brain, for instance, the cotransporter is highly expressed on the apical membrane of choroid plexus epithelial cells (184, 243) and is also found in young neurons and oligodendrocytes, yet is absent from astrocytes. In the kidney, the cotransporter is expressed in the juxtaglomerular apparatus, including the glomerular and extraglomerular mesangial cells, and the renin-positive juxtaglomerular cells in the afferent arteriole. Within the distal nephron, NKCC1 is also found on the basolateral membrane of murine inner medullary collecting duct cells, whereas in rat it is found at the basolateral membrane of type A intercalated cells of the outer medullary collecting duct. The rest of the nephron is devoid of NKCC1 expression. It is worth noting that antibodies might have a limited detection threshold and might not detect cells with low expression levels. This could account for the failure to detect expression in cells that are known to demonstrate functional characteristics of the cotransporter. For instance, NKCC1 antibodies do not label astrocytes in vivo, whereas the cotransporter was functionally demonstrated in isolated astrocytes (223, 238). Induction of NKCC1 expression by isolation and culture of the cells was demonstrated by Raat et al. (189). Whether these cells originally expressed NKCC1 at levels below detection is still a matter of debate.

DISRUPTION OF THE NKCC1 GENE IN MOUSE   Significant insight on the function of NKCC1has been gained from the development of knockout mice (42, 60, 174) and from the molecular characterization of an existing mouse mutant (48) in the colonies of the Jackson Laboratory (Bar Harbor, Maine). Before examining the phenotypes, we address some interesting functional clues gained from these targeting experiments. First, the gene encoding NKCC1 (Slc12a2) was disrupted by targeting multiple sites: exon 9 (transmembrane segment TM7 (42), exon 6 (TM4) (60), exons 9 to 11 (TM7–8), and exons 24 and 25 (carboxyl terminus) (174). All these deletions led to similar phenotypes. None of these targeting experiments involves the amino terminus of the cotransporter. This is worth noting because there is always the possibility for expressing a nonfunctional fragment of the cotransporter, which might rise to an unexpected phenotype. Fortunately, we demonstrated with our mouse model that the modified NKCC1 transcript was very unstable (42). Second, it is of importance that disruption of the carboxyl terminus of the cotransporter results in a nonfunctional protein (174). This observation is consistent with prior reports showing that mutants of various cation-chloride cotransporters involving deletion of carboxyl-terminal segments are nonfunctional in heterologous expression systems (95, 215). A similar conclusion can be drawn from the *shaker* mouse, which carries a mutation in the NKCC1 gene that results in a frame shift within exon 21 and disruption of the carboxyl-terminal tail of the cotransporter (48). This mouse model is interesting in another respect; the frame shift mutation occurs in an alternatively spliced exon (191). The *shaker* mouse must therefore express a cotransporter isoform lacking exon 21. Based on RT-PCR data, the cotransporter lacking exon 21 is primarily expressed in the brain (191). Western blot analysis of *shaker* tissues using antibodies directed against epitopes from other exons in the gene should indicate expression of this alternative isoform; however, this has not been reported. This interesting aspect aside, the reported *shaker* mouse phenotype does not differ significantly from the phenotype observed in the other NKCC1 mouse models.

Sensorineural deafness   The most striking phenotype of the NKCC1 *null* mice is the *shaker/waltzer* phenotype (42, 48, 60), consisting of circling and head bobbing. This behavior is indicative of inner ear dysfunction. Tests of auditory function using startle response under acoustic stimulus (42) or auditory brain stem responses to sound stimuli (60) revealed that homozygous NKCC1 mutants are deaf. Histological analysis of the cochlea revealed the collapse of the endolymphatic cavity, indicative of impaired epithelial secretion. The Reissner's membrane, separating the scala media from scala vestibuli, ordinarily under tension owing to the pressure of the secreted liquid, is found lying on top of the stria vascularis, the secreting epithelium. A similar behavioral phenotype and inner ear defect can be found in the knockout of *minK* (*isk*), a subunit of a K^+ channel, expressed on the apical membrane of the stria vascularis (233). Several studies have localized NKCC1 on the basolateral membrane of this secreting epithelium (35, 42, 78, 154). Thus the basolateral Na-K-2Cl cotransporter carries K^+ ions from blood into the epithelial cells, whence K^+ channels on the apical membrane secrete this cation into the cochlear chamber. Fluid is being secreted together with K^+, and the disruption of either mechanism results in reduction or absence of endolymph secretion. Hearing is affected through at least two mechanisms: (*a*) absence of liquid impairs transmission of sound vibration and movement of the tectonic membrane above the sensory hair cells, and (*b*) absence of the K^+-rich endolymph leads to degeneration of these hair cells. The *shaker/waltzer* phenotype indicates a vestibular dysfunction, and although not reported, NKCC1 homozygous mice are likely to display the irreversible collapse of the vestibular wall observed in the *minK* (*isk*) mutant mice (233).

Intestinal phenotype   Flagella and coworkers reported increased incidence of death among NKCC1 homozygous mice around weaning (60). This phenomenon, however, was not reported in other knockout models of the cotransporter (42, 174). Flagella et al. suggested that the periweaning mortality could be related to bleeding in the intestine. They also observed a worm-like appearance of the cecum in sick homozygous mutants, similar to that described in CFTR-deficient mice (60). However, Grubb and colleagues reported that, in contrast to CFTR-deficient mice, NKCC1 null mice exhibit only very mild intestinal abnormalities in the form of scattered dilated crypts (79). Comparison between the CFTR-deficient and the NKCC-deficient mouse models suggests that sufficient liquid secretion exists in the intestine of NKCC1 mice to circumvent the intestinal problems that plague the cystic fibrosis mice. In support of this view, injection of *Escherichia coli* enterotoxins into the stomach of 4–5-day-old NKCC1 homozygous mice induced normal fluid accumulation in the intestine, suggesting no significant impairment in fluid transport (60). Short-circuit conductance measurements in wild-type and homozygous mice revealed a significant role for the Na-K-2Cl cotransporter in fluid secretion. These transepithelial conductance studies also revealed the adaptability of the tissue, which uses HCO_3 secretion to a greater extent in the absence of Cl^- secretion (79).

Salivary gland phenotype   Basolateral Na-K-2Cl cotransport participates in the production and regulation of fluid secretion by salivary glands (33). Saliva production involves the isotonic secretion of a plasma-like primary fluid from the acinar cells, followed by NaCl reabsorption in the absence of water reabsorption. In vivo functional studies performed in wild-type and homozygous NKCC1 mice have demonstrated that NKCC1-deficient mice show a significant reduction in the total volume of saliva produced (57). This deficit correlates with a loss of bumetanide-sensitive Cl^- uptake mechanism in acinar cells. Of interest was the observation that disruption of NKCC1 also resulted in the production of a hyperosmotic fluid with higher NaCl content. The authors suggest that NaCl reabsorption is decreased because of a down-regulation of epithelial Na^+ channels (57).

Blood pressure, renin phenotype   The renal-specific apical cotransporters NCC and NKCC2 play significant roles in controlling blood pressure. Numerous studies (summarized in 171) have shown increased erythrocyte and vascular smooth muscle Na-K-2Cl cotransporter in primary hypertension. This increase in NKCC1 activity, however, likely represents an epiphenomenon rather than a direct association. Based on the confined expression of NKCC1 in the afferent arteriole of the glomerulus (renin-containing cells) and the extraglomerular mesangium, we suggest a possible role for NKCC1 in urinary Cl⁻ sensing, a mechanism involved in modulating both the renin/angiotensin system and tubuloglomerular feedback (see below for NKCC2). Intrarenal renin levels are significantly increased in NKCC1 knockout mice, consistent with a significant reduction in blood pressure and/or an impaired sensing process (E. Delpire, unpublished data). Measurements of blood pressure in two NKCC1 knockout models yielded conflicting data (60, 174). Flagella and coworkers reported a significant reduction in mean arterial blood pressure in NKCC1 knockout mice (68.6 ± 2.8 mm Hg versus 91.8 ± 3.9 mm Hg in control mice). These data were not confirmed by Pace et al. (174) who measured a systolic blood pressure of 100 ± 7 mm Hg in NKCC1-deficient mice versus 108 ± 8 mm Hg in control mice. Furthermore, varying dietary sodium content from 6% to less than 0.02% did not significantly affect blood pressure in either genotype. Whether these opposite results come from strain differences remains to be determined.

Male infertility   Before focusing on the male sterility phenotype, we should mention that only one third of NKCC1−/− females become pregnant and carry to term (43, 174). Successful pregnancy is followed by significantly increased pup mortality, presumably due to mother's neglect or to a lactation deficit. Homozygote NKCC1−/− males demonstrate complete sterility due to a deficit in spermatocyte production (43, 174). The size of the homogygous testis is about 25% of control testis. Histological analysis of homozygote testis revealed an accumulation of spermatogonia, the presence of leptotene and zygotene spermatocytes, a reduced number of late pachytene and diplotene spermatocytes, and an absence of round and elongated spermatids (43, 174). Tunel staining revealed increased apoptotic activity at the center of the seminiferous tubes but not the base, suggesting cell death of maturing spermatocytes but not spermatogonia (43). Blood testosterone levels (43) and luteinizing hormone (LH) levels (E. Delpire, unpublished data) were significantly reduced in NKCC1-deficient animals, suggesting a deficit in the hypothalamus/pituitary axis pathway.

Sensory perception phenotype   In contrast to most central neurons, which respond to GABA by hyperpolarizing their membranes, terminals and cell body of primary sensory neurons elicit large depolarizing currents in response to the neurotransmitter (135). This depolarization is mediated by an efflux of Cl⁻ ions through GABA$_A$ receptors, which is made possible by the expression of an active inward Cl⁻ transport mechanism in these cells (46, 168). Cation-chloride cotransporters were investigated early on, based on the blocking effect of furosemide but not

4,4′-diisothiocyanatostilbene-2,2′-disulfonic acid (DIDS) on the inward Cl⁻ pumping mechanism (9). It was the involvement of a K-Cl cotransporter mechanism that was first proposed, irrespective of the thermodynamical constraints associated with K-Cl cotransport. In 1988, Alvarez-Leefmans et al. proposed the participation of another cation-chloride cotransporter, the Na-K-2Cl cotransporter, for active accumulation of Cl⁻ in frog dorsal root ganglion neurons (3). The involvement of the cotransporter was based on the observations that intracellular Cl⁻ accumulation was dependent on extracellular Cl⁻ activity and the simultaneous presence of extracellular Na⁺ and K⁺. Furthermore, Cl⁻ accumulation was more sensitive to bumetanide than furosemide, indicative of Na-K-2Cl cotransport (3, 4). A similar conclusion was reached while examining Cl⁻ in developing *Xenopus laevis* larvae Rohon-Beard neurons (195). The molecular identification of Na-K-2Cl cotransporters and the development of isoform-specific polyclonal antibodies allowed us in 1997 to identify NKCC1 as the primary cation-chloride cotransporter expressed in rodent DRG neurons.

Using the gramicidin-perforated patch method, we recently demonstrated that NKCC1 alone can account for accumulation of intracellular Cl⁻ above electrochemical potential equilibrium in wild-type mouse DRG neurons. High intracellular Cl⁻ concentrations in turn promote GABA depolarization. Absence of the cotransporter in NKCC1 knockout mice results in a collapse of the Cl⁻ gradient and leads to significantly decreased depolarizing GABA responses, or even, in some neurons, small hyperpolarizing GABA responses at resting membrane potential (219). Partial depolarization of primary afferent terminals implies that fewer Ca^{2+} channels are activated upon arrival of an action potential, which reduces the amount of transmitter released and thus decreases the probability of the postsynaptic membrane to reach the threshold for creation of an action potential; this action is called presynaptic inhibition. Presynaptic inhibition is believed to be critical in filtering sensory noise coming from the periphery (242). Abnormal GABA responses in DRG neurons are likely to affect presynaptic inhibition and result in a sensory perception phenotype. Several types of sensory fibers include large fibers connected to muscle that are involved in proprioception and thinner fibers coming from various receptors that transmit temperature and pain sensations (Figure 3). Deficit in proprioception is difficult to assess in the NKCC1 knockout mice because of the strong locomotor phenotype that results from the inner ear defect. In contrast, we were able to demonstrate a nociception (pain) phenotype in homozygous NKCC1 mice, consistent with abnormal GABA neurotransmission in the terminals of nociceptive fibers (219).

A ROLE FOR NKCC1 IN HUMAN DISEASE?   In contrast to NKCC2, a renal-specific gene involved in the renal tubular disorder Bartter syndrome (see below), the widely expressed NKCC1 has not been yet linked to any human disorders. This certainly does not diminish the functional importance of the cotransporter. First, a complete absence of cotransporter expression might be embryonic lethal in humans. Second, along with other factors, a change in NKCC1 expression or activity might

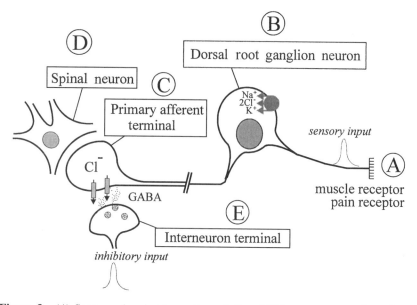

Figure 3 (*A*) Sensory signals trigger the activity of different receptors. Muscle receptors are important for the fine control of gait and movement. These proprioception receptors are connected to the spinal cord through large fibers. Smaller afferent fibers transmit pain (nociception), joint sensation (kinesthesia), and temperature and touch perception. (*B*) All sensory fibers have their nuclei located in the dorsal root ganglia. The sensory neurons express high amounts of NKCC1 (Na-K-2Cl contransporter) protein. (*C*) In the spinal cord, the terminals of these primary afferent (sensory) fibers contact spinal neurons. The GABA$_A$ receptor is expressed in primary afferent terminals where the intracellular Cl$^-$ concentration is high. Thus GABA produces depolarizing responses. (*D*) In the spinal cord, neurons in the dorsal columns receive processes from the primary afferent fibers. These neurons can be either sensory or motor neurons. They signal back to the muscles or sensory receptors through the ventral columns or make connections to the brain. (*E*) Interneurons come into contact with synaptic boutons of primary sensory fibers. They release GABA and induce depolarization of the terminals causing presynaptic inhibition.

contribute to complex disorders. For instance, based on the prominent role that the cotransporter plays in hearing (as indicated by the phenotype of the knockout mouse), the demonstration that NKCC1 expression decreases in aging gerbils, and the observation that NKCC1 heterozygote mouse hearing deteriorates with age, it is likely that the cotransporter plays a role in age-related hearing loss in humans. Involvement of NKCC1 has also been invoked in hypertension (2, 22, 65, 136). The demonstration of low blood pressure and high renin levels in NKCC1-deficient mice, as well as the well-known effect of furosemide on the vasculature, strengthens the view that NKCC1 activity might affect blood pressure.

Absorptive Na-K-2Cl Cotransporter: NKCC2

NKCC2 is a pivotal protein in renal physiology, with prominent roles in the reabsorption of cations (218), acid secretion (76), and countercurrent multiplication (115). This important transporter was cloned in 1994 (63), using a probe from flounder NCC to isolate full-length cDNAs from a rat kidney library. The functional expression of full-length native NKCC2 in HEK293 cells has been problematic (95–97); however, several splice-forms (see below) are functional when expressed in *Xenopus* oocytes (64, 73, 182, 183).

PHYSIOLOGICAL LOCALIZATION OF NKCC2   In contrast to NKCC1, the expression of NKCC2 is confined to the kidney (63, 93, 158, 177). Both in situ hybridization (63, 93, 170) and single nephron RT-PCR (246) have been utilized to localize NKCC2 transcripts to the cortical and medullary thick ascending limb (TAL). As mentioned above, much of TM2 in NKCC2 is encoded by three 96-nucleotide cassette exons, denoted A, B, and F by Payne & Forbush (177). The significant $3'$ homology between these three exons poses a technical problem for localization studies. However, the inclusion of cassette exons in NKCC2 differs along the TAL (93); the inner stripe appears to express mostly the F cassette, the outer stripe the A cassette, and the cortical TAL and macula densa the B cassette (246). Full-length mouse NKCC2 isoforms containing all three cassettes are functional in *Xenopus* oocytes (183), and preliminary data indicate that the three rabbit isoforms differ in cation affinity (73).

Immunolocalization using antibodies to NKCC2 of variable specificity have detected expression of this transporter at the apical membrane of the TAL (54, 106, 144). Immunoelectron microscopy reveals expression at both the apical membrane and within a population of sub-apical vesicles, suggesting a role for membrane trafficking in the regulation of NKCC2 (167). The TAL of both rat (1) and hamster (226) contains two morphological subtypes, a rough-surfaced cell type (R cells) with prominent apical microvilli and a smooth-surfaced cell type (S cells). Both S and R cell types in rat TAL express NKCC2, and S cells in particular have an abundance of NKCC2-positive sub-apical vesicles (167). In the hamster TAL, cells can also be separated functionally into those with high apical and low basolateral K^+ conductance and a weak basolateral Cl^- conductance (LBC cells), and a second population with low apical and high basolateral K^+ conductance, combined with a prominent basolateral Cl^- conductance (HBC cells) (226, 247). The relative frequency of the morphological and functional subtypes in CTAL and MTAL suggests that HBC cells correspond to S cells and LBC cells to R cells (226). Of note, the staining of rat TAL with antibodies to renal outer medullary K^+ channel (ROMK), which encodes at least a component of the apical K^+ conductance in the TAL, is also heterogeneous (245). However, it is unknown which morphological subtype of the TAL, S or R, expresses ROMK.

NKCC2 is also expressed in macula densa cells (106, 167, 170), which have been shown to possess apical Na-K-2Cl cotransport activity (127). This localization is of central significance in renal physiology, given the role of the macula densa in tubular regulation of renin release and tubuloglomerular feedback. Both processes are dependent on the activity of NKCC2 in that luminal loop diuretics block both tubuloglomerular feedback (98) and the suppression of renin release by luminal chloride (85). Moreover, there is a pathway of cation-chloride cotransporters between the lumen of the macula densa and the rest of the juxtaglomerular apparatus. Thus NKCC2 is expressed at the apical membrane of the macula densa, KCC4 at the basolateral membrane of the macula densa (163), and NKCC1 in mesangial cells (both extraglomerular and glomerular) and within the juxtaglomerular cells that synthesize and secrete renin (105).

Apical Na-K-2Cl cotransport plays a crucial role in the transepithelial absorption of cations by the TAL and macula densa. In particular, the K^+ dependence of NKCC2 is required to ensure adequate intracellular K^+ for recycling by ROMK and other apical K^+ channels (86). This K^+ recycling generates a lumen-positive potential difference, which drives paracellular reabsorption of Na^+, Ca^{2+}, and Mg^{2+}. Sun et al. have demonstrated that the K^+ dependence of mouse NKCC2 is under hormonal control, such that in the absence of vasopressin, apical bumetanide-sensitive Na^+ transport is K^+-independent, switching to a K^+-dependent Na-K-2Cl mode in tubules exposed to vasopressin (218). The molecular mechanism for this switch is not entirely clear but may involve interactions between full-length NKCC2 isoforms and a carboxyl-terminal truncated form (158, 183), which can mediate K^+-independent Na-Cl cotransport under specific conditions (182). Activation of NKCC2 by vasopressin is consistent with the role of this transporter in countercurrent multiplication and urinary concentration (115), and recent publications have demonstrated significant effects of vasopressin on the abundance of both NKCC2 (110) and ROMK (53) in the TAL. NKCC2 is also under opposing negative control by prostaglandin E_2, which reduces expression of the transporter protein (58).

Another important function of the TAL involving NKCC2 is the absorption of ammonium (NH_4^+) from the tubular lumen, followed by countercurrent multiplication (76). The NH_4^+ ion has the same ionic radius as K^+ and can be transported in lieu of K^+ by both NKCC2 (7, 127) and NKCC1 (237), in addition to a number of other pathways. As is the case for other cations, countercurrent multiplication of NH_4^+ by the TAL greatly increases the concentration of NH_4^+/NH_3 available for secretion in the collecting duct. The NH_4^+ produced by the proximal tubule in response to acidosis is thus reabsorbed across the TAL, concentrated by countercurrent multiplication in the medullary interstitium, and secreted in the collecting duct. The capacity of the TAL to reabsorb NH_4^+ is increased during acidosis, at least in part owing to an increase in NKCC2 expression (7). Hyperkalemia in turn appears to inhibit renal acid excretion by competing with NH_4^+ for reabsorption by the TAL (51); this may be a major factor in the acidosis associated with various defects in K^+ excretion (50).

NKCC2 IN BARTTER SYNDROME   Bartter syndrome, an autosomal-recessive form of hypokalemic hypochloremic metabolic alkalosis, provides a vivid demonstration of the physiological role of both the TAL and NKCC2. The clinical presentation of the more common forms of familial metabolic alkalosis can be separated into three separate phenotypic subtypes (80, 125, 194): classic or adult Bartter syndrome, antenatal Bartter syndrome (also known as hyperprostaglandin E syndrome), and Gitelman syndrome. Gitelman syndrome, almost exclusively the result of mutations in NCC, is described in more detail below. Other classifications are dictated by the specific disease genes involved (201, 203, 204), and thus avoid the issue of phenotypic variability and overlap (see below).

Patients with classic Bartter syndrome typically suffer from polyuria and polydipsia, and manifest a hypokalemic, hypochloremic alkalosis. They may have an increase in urinary calcium excretion, and 20% are hypomagnesemic (80). Other features include marked elevation of serum angiotensin-II, serum aldosterone, and plasma renin. Patients with antenatal Bartter syndrome present earlier in life with a severe systemic disorder characterized by marked electrolyte wasting, polyhydramnios, and significant hypercalciuria with nephrocalcinosis. Prostaglandin synthesis and excretion are significantly increased and may account for many of the systemic symptoms. As in some forms of nephrogenic diabetes insipidus, decreasing prostaglandin synthesis with cyclo-oxygenase inhibition with indomethacin can improve polyuria in Bartter syndrome by reducing the amplifying inhibition of urinary concentrating mechanisms by prostaglandins. Indomethacin also increases serum potassium and decreases plasma renin activity but does not correct the basic tubular defect. Of interest, COX-2 immunoreactivity is increased in the TAL and macula densa of patients with Bartter syndrome (118), and a recent report suggests a clinical benefit of newer cyclo-oxygenase inhibitors specific for COX-2 in this syndrome (114). This also fits with the induction of COX-2 in macula densa and cortical TAL by low luminal chloride, via a pathway that includes NKCC2 and p38 MAP kinase (30).

Early clearance studies in Bartter syndrome suggested that these patients had a defect in the function of the TAL (71). Many of the clinical features are mimicked by the administration of loop diuretics, and at least a subset of patients with antenatal Bartter syndrome do not show a diuretic response to furosemide (117). NKCC2 was thus an early candidate gene for this disorder. As expected, soon after the initial cloning of NKCC2 (63, 177), disease-associated mutations were found in the human NKCC2 gene in four kindreds with antenatal Bartter syndrome (203); in the genetic classification of Bartter syndrome, these patients are considered to have Bartter syndrome type I. Although the functional consequence of disease-associated NKCC2 mutations has not been studied, the first (203) and subsequent reports (15, 230) include patients with frameshift mutations and premature stop codons that predict the absence of a functional NKCC2 protein.

Bartter syndrome is a genetically heterogeneous disease (see detailed reviews in 80, 125, 194). Given the role of apical K^+ permeability in the TAL, encoded at least in part by ROMK (245), this K^+ channel was another early candidate gene. Multiple disease-associated mutations in ROMK have been reported in patients with Bartter syndrome type II, most of whom exhibit the antenatal phenotype

(204). Finally, mutations in Bartter syndrome type III have been reported in the chloride channel CLC-NKB (201), which is expressed at the basolateral membrane of at least the TAL and distal convoluted tubule (DCT) (227). Those patients with mutations in CLC-NKB typically have the classic Bartter phenotype, with a relative absence of nephrocalcinosis. In a significant fraction of patients with Bartter syndrome, the NKCC2, ROMK, and CLC-NKB genes are not involved (201). For example, a subset of patients with associated sensorineural deafness exhibit linkage to chromosome 1p31 (21, 235). No doubt the positional cloning of this and other Bartter syndrome genes will have a considerable impact on the understanding of the TAL and DCT.

Despite the reasonable correlation between the disease gene involved and the associated subtype of familial alkalosis, there is significant phenotypic overlap and phenotypic variability. For example, patients with mutations in CLC-NKB most frequently exhibit classic Bartter syndrome, but can present with a more severe antenatal Bartter phenotype or even with a phenotype similar to Gitelman syndrome (119). With respect to Bartter syndrome secondary to mutations in NKCC2, a number of patients have been described with variant presentations, including an absence of hypokalemia (15). Of particular relevance to the role of NKCC2 in NH_4^+ recycling and acid excretion (7, 76), a number of patients have also been described with renal tubular acidosis (15). In addition, a decreased ability to appropriately acidify the urine after an oral ammonium chloride load is observed in Bartter syndrome (194) and has recently been confirmed in selected patients with mutations in NKCC2 (15).

DISRUPTION OF THE NKCC2 GENE IN MOUSE   To further understand the pathophysiology of Bartter syndrome, Takahashi et al. created a mouse strain with a targeted deletion in NKCC2 (220). A null allele of NKCC2 was successfully created by deleting 12 kb of the promoter and exons 1–3. Unlike antenatal forms of Bartter syndrome, polyhydramnios did not develop in the $-/-$ fetuses. By day 1, however, $-/-$ mice were clearly volume depleted, as judged by an increase in hematocrit compared with that of wild-type mice. All untreated homozyogous mice died within 2 weeks, with a progressive weight loss and failure to thrive, linked in part to a failure to nurse and obtain adequate milk (220). Biochemical studies revealed a severe anion-gap metabolic acidosis, renal insufficiency, hypernatremia, and hyperkalemia. Both the profound volume depletion and the abrogation of chloride sensing by the macula densa via NKCC2 were expected to lead to activation of renin synthesis and release by the juxtaglomerular apparatus. This was indeed the case, with a 200-fold elevation of plasma renin in the $-/-$ pups and a striking recruitment of renin-secreting cells in the juxtaglomerular apparatus and renal arterioles.

Attempts to rescue the NKCC2 $-/-$ mice by subcutaneous injection of saline, as reported in mice with a targeted deletion of the mineralocorticoid receptor (19), were not successful. However, treatment with indomethacin for the first 3 weeks of life succeeded in rescuing 10% of $-/-$ mice beyond weaning. Serum chemistries of 3-month-old −/− mice (in the absence of indomethacin) were notable for mild hypokalemia, increased bicarbonate, and mild alkalosis; serum chloride was not reported. Plasma renin was elevated 23-fold in the adult −/− mice, and the excretion of PGE2 was increased 2-fold. Urine calcium excretion was also increased, with normal serum calcium, and nephrocalcinosis was observed.

An unexpected abnormality in the NKCC2 −/− mice was a significant increase in the excretion of mouse major urinary protein (MUP), a 20-kDa protein produced in the liver. Although this proteinuria was attributed to a proximal tubular defect (220), the authors did not examine the urinary excretion of other proteins known to undergo reabsorption by the proximal tubule, nor did they examine whether the NKCC2 −/− mice had a defect in function or expression of megalin, which coordinates the reabsorption of multiple low-molecular-weight proteins by the proximal tubule (133).

The adult NKCC2 −/− mice exhibited a profound polyuria, equivalent to a urine output of 30 liters in a 60-kg human (220). As reviewed above, this was an expected result of targeting NKCC2, given the importance of this transporter in countercurrent multiplication and urinary concentration (115). Urine osmolality was 371 ± 20 mOsm versus 2314 ± 74 mOsm in +/+ mice, and failed to increase in response to water restriction or vasopressin administration. The defects in urinary concentration and calcium excretion were also unresponsive to the re-administration of indomethacin. As a result of the sustained polyuria, all the −/− adult mice developed progressive hydronephrosis. To examine the relative role of this hydronephrosis and NKCC2 deletion in the metabolic syndrome of the −/− mice, the authors treated normal 4-week-old mice with furosemide for 4 weeks. These mice developed only mild hydronephrosis but exhibited a urine output comparable to that of the −/− mice, with an attenuated response to vasopressin. Of interest, furosemide-treated mice also had an increase in the excretion of MUP, indicating that the tubular reabsorption of this protein is linked in some way to the function of NKCC2.

To summarize, untreated homozygotes NKCC2 −/− mice, died within 7 days with profound volume depletion, renal insufficiency, failure to thrive, and metabolic acidosis. This phenotype was rescued by indomethacin until weaning at 3 weeks of age. Adult −/− mice had a similar phenotype to human Bartter syndrome, with a hypokalemic metabolic alkalosis, increased plasma renin, hypercalciuria, nephrocalcinosis, and polyuria. An unexpected difference was the presence of MUP proteinuria in the NKCC2 −/− mice, which does not have a human counterpart. As emphasized by Takahashi et al., differences between the human and mouse phenotypes are mostly quantitative. Thus the −/− mice had milder hypokalemia, but exhibited profound polyuria and marked hydronephrosis. Similar quantitative differences in the murine and human phenotypes generated by loss-of-function mutations in other transport proteins have been reported. For example, targeted deletion of the α subunit of the mouse amiloride-sensitive sodium channel (ENaC) results in neonatal pulmonary edema and death within the first 2 days of life (92). In contrast, humans with systemic pseudohypoaldosteronism due to loss-of-function mutations in α-ENaC suffer from a comparatively mild pulmonary phenotype (109).

Na-Cl Cotransporter: NCC

NCC from the winter flounder (*Pseudopleuronectes americanus*) was the first cation-chloride cotransporter to be cloned, using expression cloning in *Xenopus* oocytes (64). This impressive feat was followed closely by the cloning of rat NCC (63). Rat and flounder NCC have the expected transport characteristics of Na-Cl cotransporters, i.e., they mediate Cl^--dependent, K^+-independent uptake of $^{22}Na^+$ that is highly sensitive to thiazide diuretics and resistant to loop diuretics.

PHYSIOLOGICAL LOCALIZATION OF NCC   The thiazide-sensitive Na-Cl cotransporter NCC is essentially a renal-specific gene, although there are some indications of extra-renal expression in osteoblasts (10), placenta (28), and endocrine pancreas (13). Both in situ hybridization (8, 63, 169) and single nephron RT-PCR (246) localize NCC to the DCT in rat, rabbit, and human. Immunolocalization with NCC-specific antibodies indicates expression at the apical membrane of the DCT (111, 186). Immunogold labeling reveals expression in a population of sub-apical vesicles, suggesting a role for membrane trafficking in the regulation of NCC (186).

The DCT reabsorbs 5–7% of the filtered sodium chloride load, primarily through NCC. However, there is clear evidence for an alternate pathway for Na-Cl absorption at the apical membrane of the DCT that consists of parallel Na-H exchange and chloride-formate or chloride-oxalate exchange. As in the proximal tubule (108), perfusion of the DCT with solutions containing formate or oxalate increases transepithelial salt transport, a phenomenon that is blocked by DIDS and [*N*-ethyl-*N*-isopropyl]-2′-4′ amiloride (EIPA) (239). DCTs express the Na-H exchanger NHE-2 at the apical membrane; although the molecular identity of the apical chloride-base exchanger in the DCT is not known, it is likely a member of the SLC26 gene family of anion exchangers (200).

The DCT is an aldosterone-sensitive epithelium, positive for both the mineralocorticoid receptor and the enzyme 11β-hydroxysteroid dehydrogenase-2 (11β-HSD2) (20, 231). NCC expression increases dramatically in rats on a low-salt diet, and similar increases can be induced by aldosterone and fludrocortisone (111), indicating that this transporter is under transcriptional control by mineralocorticoids. Aldosterone has additional effects on the cellular architecture of the DCT. Thus there is striking hypertrophy and hyperplasia of the DCT in mice with a targeted deletion of 11β-HSD2 (120). In this mouse model of the human disorder syndrome of apparent mineralocorticoid excess (SAME) (240), cortisol has an unprotected mineralocorticoid effect in aldosterone-responsive epithelia such as the DCT. The role of NCC in the hypertension and nephrocalcinosis of patients with SAME (157, 240) is an intriguing issue for future study.

The DCT also appears to be an estrogen-responsive epithelium. Estrogen affects renal sodium handling, and ovariectomized female rats have increased sodium excretion compared with controls. The expression of NCC is also reduced in ovariectomized rats, and treatment with estrogen dramatically increases apical labeling of NCC in the DCT (232). As is the case with mineralocorticoid, the overall ultrastructure of DCT cells is also affected by estrogen, with an increase in the number of apical microprojections.

In addition to these hormonal influences, evidence suggests that the DCT undergoes remodeling in response to the load of filtered sodium chloride (86, 102, 137). Increases in NaCl delivery to the DCT, induced by either increased dietary salt or furosemide treatment, are generally associated with cell hypertrophy, increased binding of thiazides, and increases in tubular transport capacity. Treatment of rats with furosemide also appears to increase the amount of NCC transcript in the DCT (169). Thiazide treatment and dietary NaCl restriction are associated in turn with decreases in DCT transport capacity. Thiazide treatment of rats has been shown to induce dose-dependent apoptosis of the DCT, sparing the connecting tubule (CNT) (137). These morphological changes are accompanied by marked decreases in the amount of NCC protein and transcript in the DCT. Remodeling of the DCT, and potentially other nephron segments, by hypertrophy or apoptosis may occur in response to changes in luminal NaCl delivery (211). Clearly, however, there may be important hormonal mediators in both processes (116, 120, 232).

It has been known for some time that luminal thiazide in the early DCT is hypocalciuric (34), suggesting that NCC is also involved in calcium homeostasis. The role of NCC in calcium transport has been examined using an immortalized cell line from mouse DCT in which apical thiazide increases intracellular calcium $[Ca^{2+}]_i$ (70). The blockade of apical Na-Cl cotransport by thiazide decreases internal cell Cl^- through the continued efflux of Cl^- through basolateral Cl^- channels. Cells treated with thiazide hyperpolarize toward the K^+ equilibrium potential; this hyperpolarization increases $[Ca^{2+}]_i$, which enters through dihydropyridine-sensitive apical Ca^{2+} channels. However, evidence of expression of NCC in this cell line has not been published, although similar findings have been published in an osteoblast cell line that expresses NCC (10).

The phenotypic transition between the NKCC2-positive cells of the cortical TAL and the NCC-positive cells of the DCT is dramatic, occurring at a variable distance after the macula densa (169, 186). In contrast, in the rat and human kidney, the distal border of the DCT is much less abrupt; the CNT is composed of a mixture of NCC-positive CNT cells, NCC-negative CNT cells, and NCC-negative intercalated cells (169, 186). The transition between the DCT and the CNT is more distinct in the rabbit (8). Of relevance to the role of NCC in calcium homeostasis is the presence of calbindin D28, an intracellular calcium-binding protein implicated in transcellular calcium transport in all cells that express NCC (186). In addition, within the late DCT and early CNT, some cells with apical NCC also express the basolateral Na/Ca exchanger and Ca-ATPase (169, 186). More recently, the epithelial calcium channel (ECaC) has been found to colocalize with NCC in the CNT but not DCT of rabbit kidney (89). ECaC has many of the properties expected of the apical absorptive Ca^{2+} channel of the distal nephron (90).

NCC IN GITELMAN SYNDROME   A major advance in the understanding of hereditary Bartter-like syndromes was the observation that a large subset of patients can be distinguished by the presence of hypocalciuria (14), in contrast to the frequent occurrence of hypercalciuria and nephrocalcinosis in patients with classic and/or antenatal Bartter syndrome. These hypocalciuric patients are now classified as having Gitelman syndrome (75). Gitelman syndrome is overall a much milder disorder than Bartter syndrome and is further characterized by hypokalemia, alkalosis, and hypomagnesemia. Although plasma renin activity may be increased, renal prostaglandin excretion is not elevated (141), another distinguishing feature.

As with Bartter syndrome and furosemide (117), the absence of a diuretic response to thiazide (225) in Gitelman syndrome was an early indication that the thiazide-sensitive Na-Cl cotransporter NCC was the disease gene. In addition, hypocalciuria is an expected consequence of inactivating NCC (34). Soon after the cloning of rat NCC (63), inactivating mutations in the human gene were described in 12 kindreds with Gitelman syndrome (205). This was subsequently confirmed in a number of other kindreds (134, 147, 222). The functional consequence of eight coding sequence Gitelman mutations was studied in the orthologous mouse NCC, using heterologous expression in *Xenopus* oocytes (124). Mutant NCC constructs did not mediate thiazide-sensitive NaCl uptake, confirming that the mutations in Gitelman syndrome lead to a loss of function. Oocytes expressing mutant NCC proteins did not glycosylate the core NCC protein, which suggests abnormalities in trafficking and processing of the disease-associated mutants. Whereas wild-type NCC reached the cell membrane, oocytes expressing mutant NCC did not exhibit membrane staining with NCC-specific antibodies. Similar trafficking defects have been described in genetic disorders of other transporters, including the sodium-dependent glucose cotransporter (146).

The hypocalciuria in Gitelman syndrome is not accompanied by changes in serum calcium, phosphate, vitamin D, or PTH (36). However, there are clear differences in bone density between affected and unaffected members of specific Gitelman kindreds. Homozygote affecteds have much higher bone densities than unaffected wild-type family members, whereas heterozygotes have intermediate values for both bone density and calcium excretion (36). An interesting association has recently been described between chondrocalcinosis, the abnormal deposition of calcium pyrophosphate dihydrate (CPPD) in joint cartilage, and Gitelman syndrome (32). Other subjects also exhibit ocular choroidal calcification (234). The role of NCC in determining bone density, blood pressure, and other clinical parameters in the general population is an intriguing issue for future study, particularly given the high percentage of compound heterozygotes in kindreds with Gitelman syndrome (205).

DISRUPTION OF THE NCC GENE IN MOUSE   Mice with an inactivation of the NCC gene were recently generated by Schultheis et al. (198). These animals exhibit some but not all of the features of Gitelman syndrome. Serum chemistries were notable for an absence of both hypokalemia and alkalosis, in marked contrast to the human disorder. This is reminiscent of the surprisingly mild hypokalemia detected in the NKCC2 $-/-$ mice, which has been attributed to a much higher ratio of dietary potassium to sodium in mice versus humans (220). Renal renin mRNA was increased in NCC $-/-$ mice compared with controls, on both a sodium-depleted and sodium-replete diet; however, serum aldosterone was not increased. Blood pressure in $-/-$ mice was only slightly lower than that of controls on a sodium-replete diet but was much lower than that of wild-type mice when dietary sodium was restricted. This confirms the suggestion that inactivating mutations in NCC may have an anti-hypertensive effect (205) and increases interest in the phenotypic consequence of heterozygous inactivation of NCC in humans.

The NCC $-/-$ mice did exhibit significant hypocalciuria and hypomagnesemia compared with wild-type controls, as expected. The potential mechanism of the hypocalciuria is discussed above. The mechanism of the hypomagnesemia remains obscure and awaits molecular characterization of the transepithelial magnesium transport pathways in the DCT.

The availability of NCC knockout mice affords the opportunity to characterize the morphological effects of NCC inactivation in the DCT, which has not been studied in human subjects with Gitelman syndrome. Consistent with the observation that thiazide causes apoptosis and remodeling of the DCT (137), detailed morphological analysis reveals a reduction in DCT cells in the renal cortex, from $6.1 \pm 1.2\%$ in $+/+$ mice to $1.9 \pm 0.7\%$ in $-/-$ mice. Those DCT cells that were present in $-/-$ mice were morphologically abnormal, with a decreased cellular height and less elongated mitochondria that were not closely associated with the basolateral membrane folds. The morphological abnormalities in the DCT accentuate the possibility that hypertrophy and hyperplasia of this epithelium are determined by apical Na-Cl absorption (210).

A GAIN-OF-FUNCTION OF NCC IN OTHER HUMAN DISORDERS?   A frequent theme in the renal tubular disorders that affect salt excretion is that genetic disorders in these pathways can result from either a loss or a gain of function. For example, in the case of the mineralocorticoid receptor, inactivating mutations cause pseudohypoaldosteronism type I (68), whereas an activating mutation can lead to early-onset hypertension (67). Inactivating mutations in the subunits of the renal amiloride-sensitive sodium channel also lead to pseudohypoaldosteronism type I (29), whereas activating mutations result in Liddle's syndrome, another form of hypertension (84).

Pseudohypoaldosteronism type II (PHA2), also known as Gordon's syndrome, would appear to represent a gain-of-function in NCC and the DCT. PHA2, which has been described as the mirror image of Bartter syndrome (77), consists of autosomal-dominant hypertension and hyperkalemia, with suppression of the renin-angiotensin axis and a hyperchloremic acidosis. The full phenotype can be reversed by aggressive salt restriction (112). Further characterization has demonstrated hyperabsorption of NaCl, but not Na^+ accompanied by other anions (221). This observation and the almost universal therapeutic effect of thiazides (77) suggests increased activity of NCC in PHA2. However, linkage analysis has excluded NCC as the causative gene (202) but has resulted in localization of three different PHA2 loci on chromosomes 1, 17, and 12 (47, 145). Although some patients with PHA2 may in fact have increased reabsorption of NaCl within the proximal tubule (113), it is expected that the characterization of the causative genes for PHA2 will impact on the understanding of the biology of the DCT and the regulation of NCC. As an interesting aside, patients with both PHA2 (216) and SAME (157, 240) have been described with hypercalciuria and nephrolithiasis, which suggests that overactive NCC (PHA2) and/or hyperplasia (SAME) of the DCT (120) does increase urinary calcium excretion, as opposed to the effect in Gitelman syndrome.

Na^+-INDEPENDENT K-Cl COTRANSPORTERS

KCC1

KCC1 was first cloned by Gillen and coworkers (72). These authors demonstrated widespread expression of KCC1, with abundant transcript in brain, colon, heart, kidney, liver, lung, spleen, stomach, pancreas, and muscle. This pattern of expression is consistent with a housekeeping role for KCC1 in cell volume maintenance and regulation. The physiology of K-Cl cotransporter is undoubtedly best studied in red blood cells from various vertebrate species (for reviews see 129, 130). Although erythroid K-Cl cotransport is generally considered to be mediated by KCC1 (181, 217), recent evidence suggests that additional isoforms are expressed (132). In fact, some biophysical characteristics of red cell K-Cl cotransport are a better fit for KCC3 than for KCC1. For instance, the anion series, i.e., $Br > Cl > I > SCN = NO_3$, coincides with the order of K^+ transport through KCC3, as demonstrated in heterologous expression systems (150).

Consensus exists highlighting the importance of phosphorylation/dephosphorylation mechanisms in regulating K-Cl cotransport. Experiments pioneered by Jennings in rabbit (100, 212) and extended in human, mouse, dog, and sheep red blood cells (16, 103, 122, 172) demonstrated that okadaic acid and calyculin A, inhibitors of the serine-threonine protein phosphatase PP-1, inhibit K-Cl cotransport activation by a number of stimuli. The heterologous expression of various cloned KCCs reveals that hypotonic activation is also abrogated by phosphatase inhibition (152). Recent data from human red cells suggest an important additional role for the phosphatase PP-2A (17, 18). The simplest model posits that activation of the KCCs is related to dephosphorylation by PP-1 and/or PP-2A, whereas phosphorylation by still-unidentified kinases causes tonic inactivation under isotonic conditions. The upstream phosphatases are in turn under negative control by staurosporine-sensitive kinases, which are likely members of the Src family of cytoplasmic tyrosine kinases (38).

THE ROLE OF K-Cl COTRANSPORT IN HEMOGLOBINOPATHIES   Erythroid K-Cl cotransporter mediates the efflux of K^+ and Cl^- ions with an obligatory exit of water such that the activation of K-Cl cotransport results in a decrease in cell volume. The importance of K-Cl cotransport in red cell volume regulation is illustrated by the BXD-31 mouse strain, in which increased red cell K-Cl cotransport has been linked to a resistance to osmotic lysis (6). The relationships between K-Cl cotransport and red cell hydration and red cell abnormalities in patients with hemoglobin C, hemoglobin S, and β-thalasemia have been active areas of investigation (for reviews, see 23, 24). The simplest scheme is that abnormal activation of K-Cl cotransporter leads to increased dehydration of the erythrocyte, which initiates or aggravates the hemolytic and/or sickling process. The evidence is unequivocal that red cells from patients with sickle cell disease exhibit activation of both K-Cl cotransport and a Ca^{2+}-activated K^+ channel (Gardos channel, IK1) (23, 24). The relative role of K-Cl cotransport and the Gardos K^+ channel in sickle cell disease is, however, a matter of controversy. Thus K-Cl cotransport may be responsible for the formation of intermediate-density sickle cells, whereas the Gardos channel functions primarily in the generation of "hyperdense" cells (199). Alternatively, activation of the Gardos channel may be the primary event (reviewed in 149). What does seem clear, however, is that both pathways participate in the sickling process because the pharmacological inhibition of both K-Cl cotransport (37) and the Gardos K^+ channel (25) are of benefit in patients with sickle cell disease. The characterization of mice with targeted deletions of KCC1 (217), KCC3, and the Gardos channel (IK1/SK4) (228) promises to yield important tools for the study of cation transport in the sickling process. Moreover, variation in the corresponding human genes may have modifying effects on the clinical course of sickle cell anemia and other hemoglobinopathies.

KCC2

Evidence for a second gene encoding a K-Cl cotransporter was found during an EST database search in 1995, and the corresponding full-length cDNA clone was subsequently isolated from a rat brain cDNA library (178). Although no function was reported in the original publication, this new cDNA was found closely related to the functionally characterized K-Cl cotransporter, KCC1. Its transport properties were later demonstrated in HEK-293 cells (176) and in *Xenopus laevis* oocytes (215).

EXPRESSION OF KCC2   The first publication reporting the cloning of KCC2 demonstrated the restricted expression pattern of this K-Cl cotransporter; the cotransporter was heavily expressed in brain, and no transcript was detected in colon, heart, kidney, liver, lung, spleen, or stomach (178). Using in situ hybridization, KCC2 was further confined to neurons: No signal could be found in white matter tracts or in other non-neuronal structures within the brain. This first observation was later confirmed by immunolocalization studies using carboxyl-terminal antibodies (139, 241). Neuronal expression appears to be limited to central neurons, including retinal neurons (229, 236), with the possible exception of dorsal root ganglion (sensory) neurons, where the cotransporter was detected by PCR (139, 193) and immunostaining (139).

The neuron-restricted expression pattern of KCC2 is generated in part by the presence of a neuronal-restrictive silencing element in the mouse KCC2 gene (107). This element is recognized by a transcription factor (NRSF) that silences transcription of the gene in non-neuronal cells (121, 156). The transcription factor is expressed in neuron precursors and is down-regulated during neuronal maturation (197).

DISRUPTION OF THE KCC2 GENE IN MOUSE   The importance of KCC2 in the regulation of CNS excitability has been established by the characterization of mice with a targeted deletion of the KCC2 gene (90a, 138, 242a). Complete deletion of KCC2 results in a severe motor deficit leading to respiratory failure at birth and mortality (90a). By targeting exon 1, we created a mouse model with 95% reduction in KCC2 expression. Mice with a trace amount of KCC2 protein demonstrated handling-induced seizure behavior that was evident as early as the day after birth (138, 242a). Seizures were naturally triggered by movement in the cage, and the frequent seizure activity produced significant injury in the brain. It is likely that the accumulated injury is responsible for the short life span of KCC2 homozygous mice; by postnatal age P17, all homozygous KCC2 mice had died. The study of heterozygous animals provides important insights into the role of the cotransporter: Whereas no overt phenotype or brain injury could be observed in these animals carrying one normal allele, electrophysiological measurements in the hippocampus showed some degree of hyperexcitability, which suggests that heterozygote mice, expressing half the amount of KCC2 protein, might demonstrate a lower threshold for epileptic seizures. This was confirmed by examining susceptibility of pentylenetetrazole (PTZ)-induced seizures in both adult wild-type and heterozygous mice; the heterozygous animals demonstrated a twofold increase in epileptic seizures (242a). Thus a reduction in KCC2 expression results in an increased susceptibility to the development of seizures.

Using Long-Evans rats exposed to one episode of global ischemia, Reid and coworkers demonstrated that the loop diuretics furosemide, bumetanide, and ethacrynic acid prevented sound-triggered seizures (192). In this complex model, several factors could account for this inhibitory effect. First, the expression of cation-chloride NKCC1 or KCC2 could be affected by ischemia, modifying the Cl^- driving force toward more depolarizing potentials. Second, in the presence of continuous sensory input, neural activity could raise the extracellular K^+ concentration, thus reducing or reversing the gradient for K-Cl cotransport. In combination, these factors could facilitate the development of seizure activity. Inhibition of the cotransporter(s) by loop diuretics could then reduce or prevent these sound-triggered seizures.

A ROLE FOR KCC2 IN HUMAN EPILEPSY?   The interesting seizure phenotype observed in KCC2 heterozygous and homozygous mice prompts us to examine the potential role of KCC2 in human epilepsy. Human KCC2 is located on chromosome 20q between markers D20S119 and D20S197 (207). Although this arm of chromosome 20 contains two epilepsy genes, there is no linkage with the segment containing the KCC2 gene. This, however, does not exclude the possibility that KCC2 constitutes a risk factor for epilepsy. Based on the observation that KCC2 homozygous mice have a higher susceptibility for epileptic seizures, it is likely that variations in KCC2 expression or function in humans will result in a lower threshold for epileptic seizures. Variation in KCC2 expression or function can result from polymorphisms in the promoter of the gene or mild mutations in the coding sequence. For instance, the human KCC2 gene contains a polymorphic $(CA)_n$ repeat just 5′ of a conserved binding site for NRSF (207).

NEURONAL FUNCTION OF KCC2   The precise role of KCC2 in CNS function is still not completely understood. Several hypotheses can account for the phenotype observed in the KCC2 knockout mice. First, it is likely that KCC2 actively participates in the development of inhibitory GABA responses. Thus absence of KCC2 would greatly affect the efficiency of the inhibitory pathway. Second, KCC2 may actively regulate ion gradients during synaptic activity. Absence of KCC2 could then lead to the inability to conserve Cl^- and/or K^+ gradients, especially during high rates of synaptic activity. Third, the early function of KCC2 might be critical during brain development and the formation of brain circuitry.

During early postnatal brain development, the excitatory and inhibitory pathways, which are present in the adult brain, are incompletely developed. GABAergic synapses develop early and GABA first produces most of the excitatory drive in the immature brain. During the first two weeks after birth, GABA gradually becomes inhibitory (12, 62, 123, 140, 164). This shift in GABA response from depolarizing to hyperpolarizing comes from a decrease in intracellular Cl^- in neurons (39, 173). The glycinergic pathway follows a similar maturation pattern (55, 206). It is during that time frame that the expression of two major cation-chloride cotransporters changes significantly: expression of the inward Na-K-2Cl cotransporter, NKCC1, decreases (187), whereas expression of KCC2 is induced (31, 139, 193). This K-Cl cotransporter likely represents the long-sought mechanism that in mature neurons accounts for a Cl^- concentration below electrochemical equilibrium (153, 224).

Several studies have directly examined the participation of a K-Cl cotransporter (presumably KCC2) in the regulation of intracellular Cl^- and the effect of GABA (39, 61, 64a, 99). Using cultured rat midbrain neurons, whole-cell patch recording, and furosemide, Jarolimek and coworkers (99) demonstrated that the GABA reversal potential was 20 mM more hyperpolarized in dendrites than in soma. This difference was attributed to the activity of a furosemide-sensitive and K^+-dependent transport mechanism: a K-Cl cotransporter. Similar conclusions were reached by examining the effect of furosemide on GABA receptor activation in neocortical slices (somatosensory cortex) (39, 61). Using KCC2-specific antisense oligonucleotides, Rivera et al. demonstrated more directly the participation of KCC2 in the development of GABA hyperpolarizing responses during neuronal maturation (193).

The second hypothesis evolved from the observation that during sustained activation of dendritic $GABA_A$ receptors, the postsynaptic membrane response reverses back to depolarization (208, 209). These activity-dependent depolarizing responses can be found at the dendrites but not at the cell soma. This hypothesis assumes that, during prolonged synaptic activity, the Cl^- gradient collapses, allowing bicarbonate ions to play a greater role. Indeed, $GABA_A$ receptors also permeate HCO_3^-; the $HCO_3^-:Cl^-$ ratio for the receptor is 5:1. Because the HCO_3^- reversal potential is 50 mV more positive than the resting membrane potential (101, 209) in the absence of a Cl^- gradient, the outward movement of bicarbonate leads to membrane depolarization.

Experimental evidence for the role of a K-Cl cotransporter in regulating intracellular Cl^- during repetitive synaptic activity comes from the work of Ehrlich and coworkers in lateral superior olive (LSO) neurons (55). Using voltage step protocols (increasing or decreasing) and repetitive application of glycine, they demonstrated that under sufficiently short time intervals, the glycine reversal potential ($E_{glycine}$) was shifted in positive or negative directions, depending on whether inward or outward Cl^- currents were activated. This observation was similar to the shift in E_{GABA} seen in adult neurons following prolonged activation of $GABA_A$ receptors (91). Of interest was the observation that the shift in $E_{glycine}$ was significantly lower in more mature neurons, indicating the presence of a better Cl^- regulation mechanism in mature cells. Based on a furosemide-induced depolarizing shift in $E_{glycine}$ in maturing LSO neurons and the ontogeny of KCC2 in CNS neurons, it is likely that this Cl^- regulating mechanism is the K-Cl cotransporter KCC2. Thus it can be proposed that in mature neurons, KCC2 participates in the regulation of intracellular Cl^- during synaptic transmission. The cotransporter likely has the turnover capacity to deal with some level of repetitive synaptic activity. However, once the activity reaches a certain level, the cotransporter becomes rate-limiting and the Cl^- gradient cannot be maintained.

The role of the K-Cl cotransporter might not be limited to the regulation of Cl^- during high synaptic activity; K^+ ions are released in the extracellular space in amounts large enough to reverse the driving force for K-Cl cotransporter (176).

KCC3

KCC3 was independently cloned in 1999 by three groups (88, 162, 190). Comparison of the protein sequences reported by Hiki et al. (88) and Mount et al. (162) suggested the presence of alternative start sites and 5′ ends. Analyses of the human and mouse genes identified two alternative first exons, denoted exons 1a and 1b, and 5′-RACE-PRC of the more 3′ exon 1b indicates that the two KCC3 proteins (KCC3a and KCC3b) are generated by transcriptional initiation at alternative promoters (D. B. Mount, unpublished results).

EXPRESSION OF KCC3   All three groups reported expression of KCC3 by Northern blot analysis in multiple tissues, including kidney and brain. Using exon-specific probes and Northern blot analysis, KCC3a was identified in brain, muscle, and kidney, whereas KCC3b was most abundant in kidney (180). Differential expression of the two KCC3 isoforms in brain and kidney was confirmed by Western blot analysis: The protein band is 10–20 kDa larger in brain than kidney, consistent with the size difference between KCC3a and KCC3b (88, 162, 190).

Nervous system   Using a specific polyclonal amino-terminal antibody, KCC3 was detected at the base of choroid plexus epithelium; no signal could be seen on the apical membrane (180). Immunolocalization of KCC3 on the basolateral membrane of choroid plexus epithelial cells suggests a role in K^+ reabsorption. Although no functional evidence exists for a K-Cl cotransporter on the basolateral membrane of choroid plexus epithelium, an apical K-Cl cotransporter was functionally demonstrated in *Necturus* choroid plexus epithelial cells (248). This suggests expression of yet another K-Cl cotransporter (KCC1 and/or KCC4) in this epithelium. Following stringent epitope retrieval manipulations, KCC3 could be detected in white matter tracts throughout the brain, as well as in large, far-reaching neurons. Double immunostaining revealed co-expression of KCC3 with myelin basic protein and $2',3'$-cyclic nucleotide-$3'$-phosphodiesterase indicating expression of the K-Cl cotransporter in oligodendroglial cells. Large neurons included CA1 hippocampal neurons, cortical pyramidal neurons, and cerebellar Purkinje neurons. Because KCC2 is expressed in CA1 hippocampal pyramidal neurons, staining of KCC3 in these cells indicates expression of more than one K-Cl cotransporter. Whether these neurons also express KCC1 is not known. The KCC3 cotransporter is expressed at significant levels in spinal cord and in white matter–rich dorsal and ventral columns, yet is barely detectable in peripheral nerves (180).

Kidney   Although KCC1, KCC3, and KCC4 are expressed in the kidney, the physiological role of renal KCCs is not known with certainty. As reviewed in detail elsewhere, there is evidence for K-Cl cotransport at the basolateral membrane of multiple nephron segments, in particular the proximal tubule (160). Within the kidney, KCC3-specific antibodies exclusively label the basolateral membrane of the proximal tubule, from the S1 segment to S3 (163). Of note, KCC4 is co-expressed with KCC3 at the basolateral membrane of the proximal tubule, although expression of this KCC extends into the distal nephron as well. KCC3 and KCC4 are likely to function as swelling-activated exit mechanisms for K^+ and Cl^-, and as such may play an important role in transepithelial salt transport by the kidney (160).

DISRUPTION OF THE KCC3 GENE IN MOUSE   As is the case with KCC2, it is a mouse model that gives important clues on the importance of the KCC3 cotransporter in the nervous system. A mouse lacking KCC3 was created by homologous recombination in embryonic stem cells (44). To completely disrupt KCC3 expression, disruption was targeted downstream of the two alternative first exons (1a and 1b). Thus exon 3, which encodes a 31–amino acid stretch of the amino-terminal tail of KCC3, was targeted for deletion. Heterozygous mice were undistinguishable from wild-type mice. In contrast, homozygous mice developed a posture/gait phenotype evident prior to weaning. The mice have difficulties in keeping their bodies upright, and they walk with uncoordinated movements, including sudden limb extensions (slippings). Whereas wild-type and heterozygous mice performed equally on Rotorod, wire hang, and beam task tests, homozygous mice failed these three tests, demonstrating a strong motor coordination and strength phenotype. In addition to this phenotype, both heterozygous and homozygous animals demonstrated decreased exploratory behavior in the activity chamber, indicated by distance traveled as well as rearing, suggesting additional CNS phenotypes. Examination of peripheral nerves showed evidence of demyelination: presence of axons only partially myelinated as well as myelin deposits. Together, these data suggest sensorimotor neuropathy. In contrast to most animal models of demyelination, this KCC3 model is unique in its very early onset.

AGENESIS OF CORPUS CALLOSUM AND PERIPHERAL NEUROPATHY (ACCPN)   The human gene encoding KCC3 was mapped by radiation hybrid analysis to chromosome 15q14-15 between markers D15S1040 and D15S118 (162). This region of chromosome 15 is genetically linked to two idiopathic epilepsies, juvenile myoclonic epilepsy (56) and familial rolandic epilepsy (166), a peripheral neuropathy associated both with or without agenesis of the corpus callosum (ACCPN) (27) and with a potentially allelic disorder, recessive familial spastic paraparesis (FSP) (165). Based on the early onset of the locomotor phenotype observed in the knockout mouse, there is a good probability that the KCC3 gene is involved in ACCPN and/or FSP. However, coding mutations in the 26 exons of KCC3 (151) have already been excluded in kindreds with juvenile myoclonic epilepsy and rolandic epilepsy (213).

The first description of an autosomal recessive syndrome that associates agenesis of the corpus callosum (ACC), mental deficiency, and peripheral motor deficit was provided by Andermann and coworkers in 1972 (5). Although originally described from a single county in the Province of Quebec, Canada, the disorder may also be found in other populations with high rates of consanguinity (40). This progressive neuropathy is significant in its early onset and severity. Signs of impaired motor development can be noticed as early as 4–6 months of age, and delays in motor milestones become more obvious thereafter. ACCPN patients demonstrate both muscle weakness and decreased nerve velocities. Ultrastructural analyses of nerve and muscle reveals demyelination with axonal degeneration and neurogenic atrophy of the muscle (40, 126).

ACC is a common brain malformation observed in humans, with an incidence possibly as high as 230 per 10,000 children with developmental disabilities. Primary ACC is caused by defects in the formation of commisural axons or failure of properly formed axons to cross the midline. Because of the peripheral neuropathy, Dobyns speculated that the corpus callosum abnormality in ACCPN patients might represent a degeneration of the commisural axons rather than true agenesis (49). It is intriguing that affected individuals can present a complete agenesis of the corpus callosum, whereas affected siblings have seemingly normal formation of the corpus callosum (148). ACCPN is associated with mild to moderate mental retardation and frequent occurrence of a psychotic syndrome. In a study of 62 ACCPN patients, no significant relationship between corpus callosum agenesis and psychosis could be found (59). In contrast, a significant association between posterior fossa atrophy and psychosis was established in the same study, suggesting an association between cerebellar abnormalities and a schizophrenia-like syndrome. The association of psychotic features with ACCPN is of particular interest because the region of 15q14 containing KCC3 was recently linked to periodic catatonia (214), a subset of schizophrenia.

CONCLUSION

It has been more than 20 years since the first reports describing the basic functional characteristics of Na-K-2Cl cotransport (66) and K-Cl cotransport (52, 131), and a scant 9 years since the published cloning of the first cation-chloride cotransporter (64). Viewed in this light, the recent progress in understanding the physiological and pathophysiological roles of this gene family has been most gratifying. Future issues in human genetics include the further investigation of KCC3 in neurodegenerative disorders such as FSP and ACCPN. The positional cloning and characterization of the three PHA2 genes (47, 145) and novel Bartter syndrome genes (21, 235) will be important events in renal physiology, and promise to yield new insight into the function and regulation of NCC and NKCC2. The role of more subtle variation in cation-chloride cotransporters in human polygenic disease is an important issue for future study—one wonders first and foremost about the role of NCC, NKCC1, and NKCC2 in human hypertension, of KCC2 in epilepsy, and of the red cell KCCs (KCC1 and KCC3) as modifier genes in hemoglobinopathies. In mouse genetics, the role of the KCCs and associated regulatory proteins in the osmotic resistance of the BXD-31 mouse strain is an important question (6). The first generation of mice with targeted deletion of the cation-chloride cotransporters has yielded important new information. Clearly, however, understanding of the precise role of individual pathways in specific tissues will need a combination of both multiple knockouts and tissue-specific knockouts. For example, KCC3 and KCC4 are both heavily expressed in the renal proximal tubule (163), suggesting that mice with inactivation of both these KCCs, either global or tissue-specific, will be required to understand the role of K-Cl cotransport in transepithelial salt transport (160). Likewise, inducible and brain-region-specific knockouts will provide important additional information on the role that these cation-chloride cotransporters play in controlling CNS excitability.

Visit the Annual Reviews home page at www.AnnualReviews.org

LITERATURE CITED

1. Allen F, Tisher CC. 1976. Morphology of the ascending thick limb of Henle. *Kidney Int.* 9:8–22
2. Alvarez-Guerra M, Nazaret C, Garay RP. 1998. The erythrocyte Na,K,Cl cotransporter and its circulating inhibitor in Dahl salt-sensitive rats. *J. Hypertens.* 16:1499–504

3. Alvarez-Leefmans FJ, Gamiño SM, Giraldez F, Nogueron I. 1988. Intracellular chloride regulation in amphibian dorsal root ganglion neurons studied with ion-selective microelectrodes. *J. Physiol.* 406:225–46

4. Alvarez-Leefmans FJ, Nani A, Marquez S. 1998. Chloride transport, osmotic balance, and presynaptic inhibition. In *Presynaptic Inhibition and Neural Control*, ed. P Rudomin, R Romo, LM Mendell, pp. 50–79. New York: Oxford Univ. Press

5. Andermann F, Andermann E, Joubert M, Karpati G, Carpenter S, Melancon D. 1972. Familial agenesis of the corpus callosum with anterior horn cell disease: a syndrome of mental retardation, areflexia and paraparesis. *Trans. Am. Neurol. Assoc.* 97:242–44

6. Armsby CC, Stuart-Tilley AK, Alper SL, Brugnara C. 1996. Resistance to osmotic lysis in BXD-31 mouse erythrocytes: association with upregulated K-Cl cotransport. *Am. J. Physiol. Cell Physiol.* 270:C866–C77

7. Attmane-Elakeb A, Mount DB, Sibella V, Vernimmen C, Hebert SC, Bichara M. 1998. Stimulation by in vivo and in vitro metabolic acidosis of expression of rBSC1, the Na-K(NH4)-2Cl cotransporter of the rat medullary thick ascending limb. *J. Biol. Chem.* 273:33681–91

8. Bachmann S, Velazquez H, Obermüller N, Reilly RF, Moser D, Ellison DH. 1995. Expression of the thiazide-sensitive Na-Cl cotransporter by rabbit distal convoluted tubule cells. *J. Clin. Invest.* 96:2510–14

9. Ballanyi K, Grafe P. 1985. An intracellular analysis of gamma-aminobutyric-acid-associated ion movements in rat sympathetic neurones. *J. Physiol.* 365:41–58

10. Barry ELR, Gesek FA, Kaplan MR, Hebert SC, Friedman PA. 1997. Expression of the Na-Cl cotransporter in osteoblast-like cells: effect of thiazide diuretics. *Am. J. Physiol. Cell Physiol.* 272(1 Pt 1):C109–C16

11. Behnke R, Bieswal F, Forbush BI. 1999. A protein phosphatase binding site in the N-terminus of the Na-K-2Cl cotransporter protein (NKCC1) is important in determining the activity of the transporter. *FASEB J.* 13:A397

12. Ben-Ari Y, Cherubini E, Corradetti R, Gaiarsa JL. 1989. Giant synaptic potentials in immature rat CA3 hippocampal neurones. *J. Physiol.* 416:303–25

13. Bernstein PI, Zawalach W, Bartiss A, Reilly R, Palcso M, Ellison DH. 1995. The thiazide-sensitive Na-Cl cotransporter is expressed in rat endocrine pancreas. *J. Am. Soc. Nephrol.* 6:732

14. Bettinelli A, Bianchetti MG, Girardin E, Caringella A, Cecconi M, et al. 1992. Use of calcium excretion values to distinguish two forms of primary renal tubular hypokalemic alkalosis: Bartter and Gitelman syndromes. *J. Pediatr.* 120:38–43

15. Bettinelli A, Ciarmatori S, Cesareo L, Tedeschi S, Ruffa G, et al. 2000. Phenotypic variability in Bartter syndrome type I. *Pediatr. Nephrol.* 14:940–45

16. Bize I, Dunham PB. 1994. Staurosporine, a protein kinase inhibitor, activates K-Cl cotransport in LK sheep erythrocytes. *Am. J. Physiol. Cell Physiol.* 266:C759–C70

17. Bize I, Guvenc B, Buchbinder G, Brugnara C. 2000. Stimulation of human erythrocyte K-Cl cotransport and protein phosphatase type 2A by *N*-ethylmaleimide: role of intracellular Mg^{2+}. *J. Membr. Biol.* 177:159–68

18. Bize I, Guvenc B, Robb A, Buchbinder G, Brugnara C. 1999. Serine/threonine protein phosphatases and regulation of K-Cl cotransport in human erythrocytes. *Am. J. Physiol. Cell Physiol.* 277:C926–C36

19. Bleich M, Warth R, Schmidt-Hieber M, Schulz-Baldes A, Hasselblatt P, et al. 1999. Rescue of the mineralocorticoid receptor knock-out mouse. *Pflügers Arch.* 438:245–54

20. Bostanjoglo M, Reeves WB, Reilly RF,

Velazquez H, Robertson N, et al. 1998. 11Beta-hydroxysteroid dehydrogenase, mineralocorticoid receptor, and thiazide-sensitive Na-Cl cotransporter expression by distal tubules. *J. Am. Soc. Nephrol.* 9:1347–58. Erratum. *J. Am. Soc. Nephrol.* 9(11):2179

21. Brennan TM, Landau D, Shalev H, Lamb F, Schutte BC, et al. 1998. Linkage of infantile Bartter syndrome with sensorineural deafness to chromosome 1p. *Am. J. Hum. Genet.* 62:355–61

22. Brown RA, Chipperfield AR, Davis JP, Harper AA. 1999. Increased (Na+K+Cl) cotransport in rat arterial smooth muscle in deoxycorticosterone (DOCA)/salt-induced hypertension. *J. Vasc. Res.* 36: 492–501

23. Brugnara C. 1995. Erythrocyte dehydration in pathophysiology and treatment of sickle cell disease. *Curr. Opin. Hematol.* 2:132–38

24. Brugnara C. 1997. Erythrocyte membrane transport physiology. *Curr. Opin. Hematol.* 4:122–27

25. Brugnara C, Gee B, Armsby CC, Kurth S, Sakamoto M, et al. 1996. Therapy with oral clotrimazole induces inhibition of the Gardos channel and reduction of erythrocyte dehydration in patients with sickle cell disease. *J. Clin. Invest.* 97:1227–34

26. Caron L, Rousseau F, Gagnon E, Isenring P. 2000. Cloning and functional characterization of a cation-Cl-cotransporter-interacting protein. *J. Biol. Chem.* 275: 32027–36

27. Casaubon LK, Melanson M, Lopes-Cendes I, Marineau C, Andermann E, et al. 1996. The gene responsible for a severe form of peripheral neuropathy and agenesis of the corpus callosum maps to chromosome 15q. *Am. J. Hum. Genet.* 58: 28–34

28. Chang H, Tashiro K, Hirai M, Ikeda K, Kurokawa K, Fujita T. 1996. Identification of a cDNA encoding a thiazide-sensitive sodium-chloride cotransporter from the human and its mRNA expression in various tissues. *Biochem. Biophys. Res. Comm.* 223:324–28

29. Chang SS, Grunder S, Hanukoglu A, Rosler A, Mathew PM, et al. 1996. Mutations in subunits of the epithelial sodium channel cause salt wasting with hyperkalaemic acidosis, pseudohypoaldosteronism type 1. *Nat. Genet.* 12:248–53

30. Cheng HF, Wang JL, Zhang MZ, McKanna JA, Harris RC. 2000. Role of p38 in the regulation of renal cortical cyclooxygenase-2 expression by extracellular chloride. *J. Clin. Invest.* 106:681–88

31. Clayton GH, Owens GC, Wolf JS, Smith RL. 1998. Ontogeny of cation-Cl− cotransporter expression in rat neocortex. *Brain Res. Dev. Brain Res.* 109:281–92

32. Cobeta-Garcia JC, Gascon A, Iglesias E, Estopinan V. 1998. Chondrocalcinosis and Gitelman's syndrome. A new association? *Ann. Rheum. Dis.* 57:748–49

33. Cook DI, Young JA. 1989. Fluid and electrolyte secretion by salivary glands. In *Handbook of Physiology. The Gastrointestinal System. Salivary, Pancreatic, Gastric and Hepatobiliary Secretion.* IV: 1–23. Bethesda, MD: Am. Physiol. Soc. Section editor: SG Schultz

34. Costanzo LS. 1985. Localization of diuretic action in microperfused rat distal tubules: Ca and Na transport. *Am. J. Physiol. Renal Physiol.* 248:F527–F35

35. Crouch JJ, Sakaguchi N, Lytle C, Schulte BA. 1997. Immunohistochemical localization of the Na-K-Cl co-transporter (NKCC1) in the gerbil inner ear. *J. Histochem.* 45:773–78

36. Cruz D, Simon D, Lifton RP. 1999. Inactivating mutations in the Na-Cl cotransporter is associated with high bone density. *J. Am. Soc. Nephrol.* 10:597A

37. De Franceschi L, Bachir D, Galacteros F, Tchernia G, Cynober T, et al. 1997. Oral magnesium supplements reduce erythrocyte dehydration in patients with sickle cell disease. *J. Clin. Invest.* 100: 1847–52

38. De Franceschi L, Fumagalli L, Olivieri O, Corrocher R, Lowell CA, Berton G. 1997. Deficiency of Src family kinases Fgr and Hck results in activation of erythrocyte K/Cl cotransport. *J. Clin. Invest.* 99:220–27

39. DeFazio RA, Keros S, Quick MW, Hablitz JJ. 2000. Potassium-coupled chloride cotransport controls intracellular chloride in rat neocortical pyramidal neurons. *J. Neurosci.* 20:8069–76

40. Deleu D, Bamanikar SA, Muirhead D, Louon A. 1997. Familial progressive sensorimotor neuropathy with agenesis of the corpus callosum (Andermann syndrome): a clinical, neuroradiological and histopathological study. *Eur. Neurol.* 37:104–9

41. Delpire E, Kaplan MR, Plotkin MD, Hebert SC. 1996. The Na-(K)-Cl cotransporter family in the mammalian kidney: molecular identification and function(s). *Nephrol. Dial. Transplant.* 11:1967–73

42. Delpire E, Lu J, England R, Dull C, Thorne T. 1999. Deafness and imbalance associated with inactivation of the secretory Na-K-2Cl co-transporter. *Nat. Genet.* 22:192–95

43. Delpire E, Lu J, England R, Orgebin-Crist MC. 1999. Phenotypic characterization of the NKCC1/BSC2 mouse knockout. *J. Am. Soc. Nephrol.* 10:30A

44. Delpire E, Mount DB, Lu J, England R, Kirby M, McDonald MP. 2001. Locomotor defects associated with disruption of the K-Cl cotransporter KCC3 gene. *FASEB J.* 15:A440

45. Delpire E, Rauchman MI, Beier DR, Hebert SC, Gullans SR. 1994. Molecular cloning and chromosome localization of a putative basolateral Na-K-2Cl cotransporter from mouse inner medullary collecting duct (mIMCD-3) cells. *J. Biol. Chem.* 269:25677–83

46. Deschenes M, Feltz P, Lamour Y. 1976. A model for an estimate in vivo of the ionic basis of presynaptic inhibition: an intracellular analysis of the GABA-induced depolarization in rat dorsal root ganglia. *Brain Res.* 118:486–93

47. Disse-Nicodeme S, Achard JM, Desitter I, Houot AM, Fournier A, et al. 2000. A new locus on chromosome 12p13.3 for pseudohypoaldosteronism type II, an autosomal dominant form of hypertension. *Am. J. Hum. Genet.* 67:302–10

48. Dixon MJ, Gazzard J, Chaudhry SS, Sampson N, Schulte BA, Steel KP. 1999. Mutation of the Na-K-Cl co-transporter gene Slc12a2 results in deafness in mice. *Hum. Mol. Genet.* 8:1579–84

49. Dobyns WB. 1996. Absence makes the search grow longer. *Am. J. Hum. Genet.* 58:7–16

50. DuBose TD. 1997. Hyperkalemic hyperchloremic metabolic acidosis: pathophysiologic insights. *Kidney Int.* 51:591–602

51. DuBose TD Jr, Good DW. 1992. Chronic hyperkalemia impairs ammonium transport and accumulation in the inner medulla of the rat. *J. Clin. Invest.* 90:1443–49

52. Dunham PB, Ellory JC. 1981. Passive potassium transport in low potassium sheep red cells: dependence upon cell volume and chloride. *J. Physiol.* 318:511–30

53. Ecelbarger CA, Kim GH, Knepper MA, Liu J, Tate M, et al. 2001. Regulation of potassium channel Kir 1.1 (ROMK) abundance in the thick ascending limb of Henle's loop. *J. Am. Soc. Nephrol.* 12:10–18

54. Ecelbarger CA, Terris J, Hoyer JR, Nielsen S, Wade JB, Knepper MA. 1996. Localization and regulation of the rat renal Na(+)-K(+)-2Cl$^-$ cotransporter, BSC-1. *Am. J. Physiol. Renal Physiol.* 271:F619–F28

55. Ehrlich I, Lohrke S, Friauf E. 1999. Shift from depolarizing to hyperpolarizing glycine action in rat auditory neurones is due to age-dependent Cl$^-$ regulation. *J. Physiol.* 520:121–37

56. Elmslie FV, Rees M, Williamson MP, Kerr M, Kjeldsen MJ, et al. 1998. Genetic mapping of a major susceptibility locus

for juvenile myoclonic epilepsy on chromosome 15q. *Hum. Mol. Genet.* 6:1329–34

57. Evans RL, Park K, Turner RJ, Watson GE, Nguyen H-V, et al. 2000. Severe impairment of salivation in $Na^+/K^+/2Cl^-$ cotransporter (NKCC1)-deficient mice. *J. Biol. Chem.* 275:26720–26

58. Fernandez-Llama P, Ecelbarger CA, Ware JA, Andrews P, Lee AJ, et al. 1999. Cyclooxygenase inhibitors increase Na-K-2Cl cotransporter abundance in thick ascending limb of Henle's loop. *Am. J. Physiol. Renal Physiol.* 277:F219–F26

59. Filteau M-J, Pourcher E, Bouchard RH, Baruch P, Mathieu J, et al. 1991. Corpus callosum agenesis and psychosis in Andermann syndrome. *Arch. Neurol.* 48:1275–80

60. Flagella M, Clarke LL, Miller ML, Erway LC, Giannella RA, et al. 1999. Mice lacking the basolateral Na-K-2Cl cotransporter have impaired epithelial chloride secretion and are profoundly deaf. *J. Biol. Chem.* 274:26946–55

61. Fukuda A, Muramatsu K, Okabe A, Shimano Y, Hida H, et al. 1998. Changes in intracellular Ca^{2+} induced by 4 $GABA_A$ receptor activation and reduction in Cl^- gradient in neonatal rat neocortex. *J. Neurophysiol.* 79:439–46

62. Gaiarsa J-L, McLean H, Congar P, Leinekugel X, Khazipov R, et al. 1995. Postnatal maturation of gamma-aminobutyric $acid_{A and B}$-mediated inhibition in the CA3 hippocampal region of the rat. *J. Neurobiol.* 26:339–49

63. Gamba G, Miyanoshita A, Lombardi M, Lytton J, Lee W-S, et al. 1994. Molecular cloning, primary structure, and characterization of two members of the mammalian electroneutral sodium-(potassium)-chloride cotransporter family expressed in kidney. *J. Biol. Chem.* 269:17713–22

64. Gamba G, Saltzberg SN, Lombardi M, Miyanoshita A, Lytton J, et al. 1993. Primary structure and functional expression of a cDNA encoding the thiazide-sensitive, electroneutral sodium-chloride cotransporter. *Proc. Natl. Acad. Sci. USA* 90:2749–53

64a. Ganguly K, Schinder AF, Wong ST, Poo M. 2001. GABA itself promotes the developmental switch of neuronal GABAergic responses from excitation to inhibition. *Cell* 105:521–32

65. Garay RP, Alvarez-Guerra M. 1999. Na-K-Cl cotransporters and 'salt-sensitive' arterial hypertension. *Arch. Mal. Coeur Vaiss.* 92:1033–38

66. Geck P, Pietrzyk C, Burckhardt B-C, Pfeiffer B, Heinz E. 1980. Electrically silent cotransport of Na^+, K^+ and Cl^- in Ehrlich cells. *Biochim. Biophys. Acta* 600:432–47

67. Geller DS, Farhi A, Pinkerton N, Fradley M, Moritz M, et al. 2000. Activating mineralocorticoid receptor mutation in hypertension exacerbated by pregnancy. *Science* 289:119–23

68. Geller DS, Rodriguez-Soriano J, Vallo Boado A, Schifter S, Bayer M, et al. 1998. Mutations in the mineralocorticoid receptor gene cause autosomal dominant pseudohypoaldosteronism type I. *Nat. Genet.* 19:279–81

69. Gerelsaikhan T, Turner RJ. 2000. Transmembrane topology of the secretory Na^+-K^+-$2Cl^-$ cotransporter NKCC1 studied by in vitro translation. *J. Biol. Chem.* 275:40471–77

70. Gesek FA, Friedman PA. 1992. Mechanism of calcium transport stimulated by chlorothiazide in mouse distal convoluted tubule cells. *J. Clin. Invest.* 90:429–38

71. Gill JR Jr, Bartter FC. 1978. Evidence for a prostaglandin-independent defect in chloride reabsorption in the loop of Henle as a proximal cause of Bartter's syndrome. *Am. J. Med.* 65:766–72

72. Gillen CM, Brill S, Payne JA, Forbush B III. 1996. Molecular cloning and functional expression of the K-Cl cotransporter from rabbit, rat, and human. A new

member of the cation-chloride cotransporter family. *J. Biol. Chem.* 271:16237–44

73. Giménez I, Isenring P, Forbush B III. 1999. The three splice variants of the renal Na-K-Cl cotransporter (NKCC2) differ in their affinity for Na⁺. *FASEB J.* 13: A64

74. Ginns SM, Knepper MA, Ecelbarger CA, Terris J, He X, et al. 1996. Immunolocalization of the secretory isoform of Na-K-Cl cotransporter in rat renal intercalated cells. *J. Am. Soc. Nephrol.* 7:2533–42

75. Gitelman HJ, Graham JB, Welt LG. 1966. A new familial disorder characterized by hypokalemia and hypomagnesemia. *Trans. Assoc. Am. Phys.* 79:221–35

76. Good DW. 1994. Ammonium transport by the thick ascending limb of Henle's loop. *Annu. Rev. Physiol.* 56:623–47

77. Gordon RD. 1986. Syndrome of hypertension and hyperkalemia with normal glomerular filtration rate. *Hypertension* 8:93–102

78. Goto S, Oshima T, Ikeda K, Ueda N, Takasaka T. 1997. Expression and localization of the Na-K-2Cl cotransporter in the rat cochlea. *Brain Res.* 765:324–26

79. Grubb BR, Lee E, Pace AJ, Koller BH, Boucher RC. 2000. Intestinal ion transport in NKCC1-deficient mice. *Am. J. Physiol. Gastrointest. Liver Physiol.* 279: G707–G18

80. Guay-Woodford LM. 1998. Bartter syndrome: unraveling the pathophysiologic enigma. *Am. J. Med.* 105:151–61

81. Haas M. 1994. The Na-K-Cl cotransporters. *Am. J. Physiol. Cell Physiol.* 267:C869–C85

82. Haas M, Forbush B III. 2000. The Na-K-Cl cotransporter of secretory epithelia. *Annu. Rev. Physiol.* 62:515–34

83. Haas M, Forbush B III. 1998. The Na-K-Cl cotransporters. *J. Bioenerg. Biomembr.* 30:161–72

84. Hansson JH, Nelson-Williams C, Suzuki H, Schild L, Shimkets R, et al. 1995. Hypertension caused by a truncated epithelial sodium channel gamma subunit: genetic heterogeneity of Liddle syndrome. *Nat. Genet.* 11:76–82

85. He X-R, Greenberg SG, Briggs JP, Schermann J. 1995. Effects of furosemide and verapamil on the NaCl dependency of macula densa-mediated renin secretion. *Hypertension* 26:137–42

86. Hebert SC. 1992. Nephron heterogeneity. In *Handbook of Physiology—Renal Physiology.* 1:875–925. New York: Oxford Univ. Press

87. Hebert SC, Gullans SR. 1995. The electroneutral sodium-(potassium)-chloride co-transporter family: a journey from fish to the renal co-transporters. *Curr. Opin. Nephrol. Hypert.* 4:389–91

88. Hiki K, D'Andrea RJ, Furze J, Crawford J, Woollatt E, et al. 1999. Cloning, characterization, and chromosomal location of a novel human K⁺-Cl⁻ cotransporter. *J. Biol. Chem.* 274:10661–67

89. Hoenderop JG, Hartog A, Stuiver M, Doucet A, Willems PH, Bindels RJ. 2000. Localization of the epithelial Ca(2+) channel in rabbit kidney and intestine. *J. Am. Soc. Nephrol.* 11:1171–78

90. Hoenderop JG, van der Kemp AW, Hartog A, van Os CH, Willems PH, Bindels RJ. 1999. The epithelial calcium channel, ECaC, is activated by hyperpolarization and regulated by cytosolic calcium. *Biochem. Biophys. Res. Commun.* 261:488–92

90a. Hubner CA, Stein V, Hermans-Borgmeyer I, Meyer T, Ballanyi K, et al. 2001. Disruption of KCC2 reveals an essential role of K-Cl cotransport already in early synaptic inhibition. *Neuron* 30:515–24

91. Huguenard JR, Alger BE. 1986. Whole-cell voltage-clamp study of the fading of GABA-activated currents in acutely dissociated hippocampal neurons. *J. Neurophysiol.* 56:1–18

92. Hummler E, Barker P, Gatzy J, Beermann F, Verdumo C, et al. 1996. Early death due to defective neonatal lung liquid

clearance in alpha-ENaC-deficient mice. *Nat. Genet.* 12:325–28

93. Igarashi P, Vanden Heuvel GB, Payne JA, Forbush III B. 1995. Cloning, embryonic expression and alternative splicing of a murine kidney specific Na-K-Cl cotransporter. *Am. J. Physiol. Renal Physiol.* 269:F405–F18

94. Igarashi P, Vanden Heuvel GB, Quaggin SE, Payne JA, Forbush B III. 1994. Cloning, embryonic expression, and chromosomal localization of murine renal Na-K-Cl cotransporter (NKCC2). *J. Am. Soc. Nephrol.* 5:288

95. Isenring P, Forbush B III. 1997. Ion and bumetanide binding by the Na-K-2Cl cotransporter: importance of transmembrane domains. *J. Biol. Chem.* 272:24556–62

96. Isenring P, Jacoby SC, Chang J, Forbush B III. 1998. Mutagenic mapping of the Na-K-Cl cotransporter for domains involved in ion transport and bumetanide binding. *J. Gen. Physiol.* 112:549–58

97. Isenring P, Jacoby SC, Forbush B III. 1998. The role of transmembrane domain 2 in cation transport by the Na-K-Cl cotransporter. *Proc. Natl. Acad. Sci. USA* 95:7179–84

98. Ito S, Carretero O. 1990. An in vitro approach to the study of macula densa-mediated glomerular hemodynamics. *Kidney Int.* 328:1206–10

99. Jarolimek W, Lewen A, Misgeld U. 1999. A furosemide-sensitive K^+-Cl^- cotransporter counteracts intracellular Cl^- accumulation and depletion in cultured rat midbrain neurons. *J. Neurosci.* 19:4695–704

100. Jennings ML, Schultz RK. 1991. Okadaic acid inhibition of KCl cotransport. Evidence that protein dephosphorylation is necessary for activation of transport by either swelling or *N*-ethylmaleimide. *J. Gen. Physiol.* 97:799–817

101. Kaila K. 1994. Ionic basis of GABA$_A$ receptor channel function in the nervous system. *Prog. Neurosci.* 42:489–537

102. Kaissling B, Bachmann S, Kriz W. 1985. Structural adaptation of the distal convoluted tubule to prolonged furosemide treatment. *Am. J. Physiol. Renal Physiol.* 248:F374–F81

103. Kaji D, Tsukitani Y. 1991. Role of protein phosphatase in activation of KCl cotransport in human erythrocytes. *Am. J. Physiol. Cell Physiol.* 260:C176–C82

104. Kaplan MR, Mount DB, Delpire E, Gamba G, Hebert SC. 1996. Molecular mechanisms of NaCl cotransport. *Annu. Rev. Physiol.* 58:649–68

105. Kaplan MR, Plotkin MD, Brown D, Hebert SC, Delpire E. 1996. Expression of the mouse Na-K-2Cl cotransporter, mBSC2, in the terminal IMCD, the glomerular and extraglomerular mesangium and the glomerular afferent arteriole. *J. Clin. Invest.* 98:723–30

106. Kaplan MR, Plotkin MD, Lee W-S, Xu Z-C, Lytton H, Hebert SC. 1996. Apical localization of the Na-K-2Cl cotransporter, rBSC1, on rat thick ascending limbs. *Kidney Int.* 49:40–47

107. Karadsheh MF, Delpire E. 2001. A neuronal restrictive silencing element is found in the KCC2 gene: molecular basis for KCC2 specific expression in neurons. *J. Neurophysiol.* 85:995–97

108. Karniski LP, Aronson PS. 1985. Chloride/formate exchange with formic acid recycling: a mechanism of active chloride transport across epithelial membranes. *Proc. Natl. Acad. Sci. USA* 82:6362–65

109. Kerem E, Bistritzer T, Hanukoglu A, Hofmann T, Zhou Z, et al. 1999. Pulmonary epithelial sodium-channel dysfunction and excess airway liquid in pseudohypoaldosteronism. *N. Engl. J. Med.* 341:156–62

110. Kim GH, Ecelbarger CA, Mitchell C, Packer RK, Wade JB, Knepper MA. 1999. Vasopressin increases Na-K-2Cl cotransporter expression in thick ascending limb of Henle's loop. *Am. J. Physiol. Renal Physiol.* 276:F96–F103

111. Kim GH, Masilamani S, Turner R, Mitchell C, Wade JB, Knepper MA. 1998. The thiazide-sensitive Na-Cl cotransporter is an aldosterone-induced protein. *Proc. Natl. Acad. Sci. USA* 95:14552–57

112. Klemm SA, Gordon RD, Tunny TJ, Finn WL. 1990. Biochemical correction in the syndrome of hypertension and hyperkalaemia by severe dietary salt restriction suggests renin-aldosterone suppression critical in pathophysiology. *Clin. Exp. Pharmacol. Physiol.* 17:191–95

113. Klemm SA, Gordon RD, Tunny TJ, Thompson RE. 1991. The syndrome of hypertension and hyperkalemia with normal GFR (Gordon's syndrome): Is there increased proximal sodium reabsorption? *Clin. Invest. Med.* 14:551–58

114. Kleta R, Basoglu C, Kuwertz-Broking E. 2000. New treatment options for Bartter's syndrome. *N. Engl. J. Med.* 343:661–62

115. Knepper MA, Kim GH, Fernandez-Llama P, Ecelbarger CA. 1999. Regulation of thick ascending limb transport by vasopressin. *J. Am. Soc. Nephrol.* 10:628–34

116. Kobayashi S, Clemmons DR, Nogami H, Roy AK, Venkatachalam MA. 1995. Tubular hypertrophy due to work load induced by furosemide is associated with increases of IGF-1 and IGFBP-1. *Kidney Int.* 47:818–28

117. Kockerling A, Reinalter SC, Seyberth HW. 1996. Impaired response to furosemide in hyperprostaglandin E syndrome: evidence for a tubular defect in the loop of Henle. *J. Pediatr.* 129:519–28

118. Komhoff M, Jeck ND, Seyberth HW, Grone HJ, Nusing RM, Breyer MD. 2000. Cyclooxygenase-2 expression is associated with the renal macula densa of patients with Bartter-like syndrome. *Kidney Int.* 58:2420–24

119. Konrad M, Vollmer M, Lemmink HH, Vanden Heuvel LP, Jeck N, et al. 2000. Mutations in the chloride channel gene CLCNKB as a cause of classic Bartter

syndrome. *J. Am. Soc. Nephrol.* 11:1449–59

120. Kotelevtsev Y, Brown RW, Fleming S, Kenyon C, Edwards CR, et al. 1999. Hypertension in mice lacking 11beta-hydroxysteroid dehydrogenase type 2. *J. Clin. Invest.* 103:683–89

121. Kraner SD, Chong JA, Tsay HJ, Mandel G. 1992. Silencing the type II sodium channel gene: a model for neural-specific gene regulation. *Neuron* 9:37–44

122. Krarup T, Dunham PB. 1996. Reconstitution of calyculin-inhibited K-Cl cotransport in dog erythrocyte ghosts by exogenous PP-1. *Am. J. Physiol. Cell Physiol.* 270:C898–902

123. Kriegstein AR, Suppes T, Prince DA. 1987. Cellular and synaptic physiology and epileptogenesis of developing rat neocortical neurons in vitro. *Dev. Brain Res.* 34:161–71

124. Kunchaparty S, Bernstein PL, Bartiss A, Desir GV, Reilly R, Ellison DH. 1995. Evidence for alternative splicing of the thiazide-sensitive Na-Cl transporter in mouse kidney. *FASEB J.* 9:A586

125. Kurtz I. 1998. Molecular pathogenesis of Bartter's and Gitelman's syndromes. *Kidney Int.* 54:1396–410

126. Labrisseau A, Vanasse M, Brochu P, Jasmin G. 1984. The Andermann syndrome: agenesis of the corpus callosum associated with mental retardation and progressive sensorimotor neuronopathy. *Can. J. Neurol. Sci.* 11:257–61

127. Lapointe JY, Laamarti A, Bell PD. 1998. Ionic transport in macula densa cells. *Kidney Int. (Suppl.)* 67:S58–S64

128. Larsen F, Solheim J, Kristensen T, Kolsto AB, Prydz H. 1993. A tight cluster of five unrelated human genes on chromosome 16q22.1. *Hum. Mol. Genet.* 2:1589–95

129. Lauf PK, Adragna NC. 2000. K-Cl cotransport: properties and molecular mechanism. *Cell. Physiol. Biochem.* 10:341–54

130. Lauf PK, Bauer J, Adragna NC, Fujise H,

Zade-Oppen AAM, et al. 1992. Erythrocyte K-Cl cotransport: properties and regulation. *Am. J. Physiol. Cell Physiol.* 263:C917–C32

131. Lauf PK, Theg BE. 1980. A chloride dependent K^+ flux induced by *N*-ethylmaleimide in genetically low K^+ sheep and goat erythrocytes. *Biochem. Biophys. Res. Comm.* 70:221–42

132. Lauf PK, Zhang J, Delpire E, Fyffe REW, Mount DB, Adragna NC. 2001. Erythrocyte K-Cl cotransport: immunocytochemical and functional evidence for more than one KCC isoform in HK and LK sheep red blood cells. *Comp. Biochem. Physiol. A* 130:499–509

133. Leheste JR, Rolinski B, Vorum H, Hilpert J, Nykjaer A, et al. 1999. Megalin knockout mice as an animal model of low molecular weight proteinuria. *Am. J. Pathol.* 155:1361–70

134. Lemmink HH, Vanden Heuvel LPWJ, van Dijk A, Merkx GFM, Smilde TJ, et al. 1996. Linkage of Gitelman syndrome to the human thiazide-sensitive sodium-chloride cotransporter (hTSC) gene with identification of mutations in three Dutch families. *Ped. Nephrol.* 10:403–7

135. Levy RA. 1977. The role of GABA in primary afferent depolarization. *Prog. Neurobiol.* 9:211–67

136. Lluch MM, de la Sierra A, Poch E, Coca A, Aguilera MT, et al. 1996. Erythrocyte sodium transport, intraplatelet pH, and calcium concentration in salt-sensitive hypertension. *Hypertension* 27:919–25

137. Loffing J, Loffing-Cueni D, Hegyi I, Kaplan MR, Hebert SC, et al. 1996. Thiazide treatment of rats provokes apoptosis in distal tubule cells. *Kidney Int.* 50:1180–90

138. Lovinger DM, Delpire E. 2000. Frequent seizures and early lethality associated with disruption of the mouse KCC2 gene. *J. Neurosci.* 26:1148

139. Lu J, Karadsheh M, Delpire E. 1999. Developmental regulation of the neuronal-specific isoform of K-Cl cotransporter KCC2 in postnatal rat brains. *J. Neurobiol.* 39:558–68

140. Luhmann HJ, Prince DA. 1991. Postnatal maturation of the GABAergic system in rat neocortex. *J. Neurophysiol.* 65:247–63

141. Luthy C, Bettinelli A, Iselin S, Metta MG, Basilico E, et al. 1995. Normal prostaglandinuria E2 in Gitelman's syndrome, the hypocalciuric variant of Bartter's syndrome. *Am. J. Kidney Dis.* 25:824–28

142. Lytle C, Forbush B III. 1990. The [Na-K-2Cl] cotransport protein is activated and phosphorylated by cell shrinkage in a secretory epithelium. *J. Cell Biol.* 111:312A

143. Lytle C, Forbush B III. 1996. Regulatory phosphorylation of the secretory Na-K-Cl cotransporter: modulation by cytoplasmic Cl. *Am. J. Physiol. Cell Physiol.* 270:C437–C48

144. Lytle C, Xu J-C, Biemesderfer D, Forbush B III. 1995. Distribution and diversity of Na-K-Cl cotransport proteins: a study with monoclonal antibodies. *Am. J. Physiol. Cell Physiol.* 269:C1496–C505

145. Mansfield TA, Simon DB, Farfel Z, Bia M, Tucci JR, et al. 1997. Multilocus linkage of familial hyperkalaemia and hypertension, pseudohypoaldosteronism type II, to chromosomes 1q31–42 and 17p11–q21. *Nat. Genet.* 16:202–5

146. Martin MG, Turk E, Lostao P, Kerner C, Wright EM. 1996. Defects in Na^+/glucose cotransporter (SGLT1) trafficking and function cause glucose-galactose malabsorption. *Nat. Genet.* 12:216–20

147. Mastroianni N, Bettinelli A, Bianchetti M, Colussi G, De Fusco M, et al. 1996. Novel molecular variants of the Na-Cl cotransporter gene are responsible for Gitelman syndrome. *Am. J. Hum. Genet.* 59:1019–26

148. Mathieu J, Bédard F, Prévost C, Langevin P. 1990. Neuropathie sensitivomotrice héréditaire avec ou sans agénésie du corps calleux: étude radiologique et

clinique de 64 cas. *Can. J. Neurol. Sci.* 17:103–8

149. McGoron AJ, Joiner CH, Palascak MB, Claussen WJ, Franco RS. 2000. Dehydration of mature and immature sickle red blood cells during fast oxygenation/deoxygenation cycles: role of KCl cotransport and extracellular calcium. *Blood* 95:2164–68

150. Mercado A, Mount DB, Vazquez N, Song L, Gamba G. 2000. Functional characteristics of the renal KCCs. *FASEB J.* 14:A341

151. Mercado A, Song L, George AJ, Delpire E, Mount DB. 1999. Molecular, functional, and genomic characterization of KCC3 and KCC4. *J. Am. Soc. Nephrol.* 10:38A

152. Mercado A, Song L, Vazquez N, Mount DB, Gamba G. 2000. Functional comparison of the K^+-Cl^- cotransporters KCC1 and KCC4. *J. Biol. Chem.* 275:30326–34

153. Misgeld U, Deisz RA, Dodt HU, Lux HD. 1986. The role of chloride transport in postsynaptic inhibition of hippocampal neurons. *Science* 232:1413–15

154. Mizuta K, Adachi M, Isawa KH. 1997. Ultrastructural localization of the Na-K-2Cl cotransporter in the lateral wall of the rabbit cochlear duct. *Hearing Res.* 106:154–62

155. Moore-Hoon ML, Turner RJ. 2000. The structural unit of the secretory Na^+-K^+-$2Cl^-$ cotransporter (NKCC1) is a homodimer. *Biochemistry* 39:3718–24

156. Mori N, Stein R, Sigmund O, Anderson DJ. 1990. A cell type-preferred silencer element that controls the neural-specific expression of the SCG10 gene. *Neuron* 4:583–94

157. Moudgil A, Rodich G, Jordan SC, Kamil ES. 2000. Nephrocalcinosis and renal cysts associated with apparent mineralocorticoid excess syndrome. *Pediatr. Nephrol.* 15:60–62

158. Mount DB, Baekgaard A, Hall AE, Plata C, Xu JZ, et al. 1999. Isoforms of the apical Na-K-2Cl transporter in murine thick

ascending limb. I: Molecular characterization and intra-renal localization. *Am. J. Physiol. Renal Physiol.* 276:F347–F58

159. Mount DB, Delpire E, Gamba G, Hall AE, Poch E, et al. 1998. The electroneutral cation-chloride cotransporters. *J. Exp. Biol.* 201:2091–102

160. Mount DB, Gamba G. 2001. Renal K-Cl cotransporters. *Curr. Opin. Nephrol. Hypertens.* 10:685–91

161. Mount DB, Hall AE, Plata C, Villaneuva Y, Kaplan MR, et al. 1995. Characterization of alternatively spliced transcripts of the murine apical bumetanide-sensitive Na-(K)-Cl cotransporter gene. *J. Am. Soc. Nephrol.* 6:347

162. Mount DB, Mercado A, Song L, Xu J, George JAL, et al. 1999. Cloning and characterization of KCC3 and KCC4, new members of the cation-chloride cotransporter gene family. *J. Biol. Chem.* 274:16355–62

163. Mount DB, Song L, Mercado A, Gamba G, Delpire E. 2000. Basolateral localization of renal tubular K-Cl cotransporters. *J. Am. Soc. Nephrol.* 11:35A

164. Muller D, Oliver M, Lynch G. 1989. Developmental changes in synaptic properties in hippocampus of neonatal rats. *Dev. Brain Res.* 49:105–14

165. Murillo FM, Kobayashi H, Pegoraro E, Galluzzi G, Creel G, et al. 1999. Genetic localization of a new locus for recessive familial spastic paraparesis to 15q13–15. *Neurology* 53:5–7

166. Neubauer BA, Fiedler B, Himmelein B, Kampfer F, Lassker U, et al. 1998. Centrotemporal spikes in families with rolandic epilepsy: linkage to chromosome 15q14. *Neurology* 51:1608–12

167. Nielsen S, Maunsbach AB, Ecelbarger CA, Knepper MA. 1998. Ultrastructural localization of Na-K-2Cl cotransporter in thick ascending limb and macula densa of rat kidney. *Am. J. Physiol. Renal Physiol.* 275:F885–F93

168. Nishi S, Minota S, Karczmar AG. 1974. Primary afferent neurones: the ionic

mechanism of GABA-mediated depolarization. *Neuropharmacology* 13:215–19

169. Obermüller N, Bernstein P, Velazquez H, Reilly R, Moser D, et al. 1995. Expression of the thiazide-sensitive Na-Cl cotransporter in rat and human kidney. *Am. J. Physiol. Renal Physiol.* 269:F900–F10

170. Obermüller N, Kunchaparty S, Ellison DH, Bachmann S. 1996. Expression of the Na-K-2Cl cotransporter by macula densa and thick ascending limb cells of rat and rabbit nephron. *J. Clin. Invest.* 98:635–40

171. Orlov SN, Adragna NC, Adarichev VA, Hamet P. 1999. Genetic and biochemical determinants of abnormal monovalent ion transport in primary hypertension. *Am. J. Physiol. Cell Physiol.* 276:C511–C36

172. Orringer EP, Brochenbrough JS, Whitney JA, Glosson PS, Parker JC. 1991. Okadaic acid inhibits activation of K-Cl cotransport in red cells containing hemoglobin S and C. *Am. J. Physiol. Cell Physiol.* 261:C591–C93

173. Owens DF, Boyce LH, Davis MBE, Kriegstein AR. 1996. Excitatory GABA responses in embryonic and neonatal cortical slices demonstrated by gramicidin perforated-patch recordings and calcium imaging. *J. Neurosci.* 16:6414–23

174. Pace AJ, Lee E, Athirakul K, Coffman TM, O'Brien DA, Koller BH. 2000. Failure of spermatogenesis in mouse lines deficient in the Na^+-K^+-$2Cl^-$ cotransporter. *J. Clin. Invest.* 105:441–50

175. Pathak BG, Shaughnessy JDJ, Meneton P, Greeb J, Shull GE, et al. 1996. Mouse chromosomal location of three epithelial sodium channel subunit genes and an apical sodium chloride cotransporter gene. *Genomics* 33:124–27

176. Payne JA. 1997. Functional characterization of the neuronal-specific K-Cl cotransporter: implications for $[K^+]_o$ regulation. *Am. J. Physiol. Cell Physiol.* 273:C1516–C25

177. Payne JA, Forbush B III. 1994. Alternatively spliced isoforms of the putative

renal Na-K-Cl cotransporter are differentially distributed within the rabbit kidney. *Proc. Natl. Acad. Sci. USA* 91:4544–48

178. Payne JA, Stevenson TJ, Donaldson LF. 1996. Molecular characterization of a putative K-Cl cotransporter in rat brain. A neuronal-specific isoform. *J. Biol. Chem.* 271:16245–52

179. Payne JA, Xu J-C, Haas M, Lytle CY, Ward D, Forbush B III. 1995. Primary structure, functional expression, and chromosome localization of the bumetanide sensitive Na-K-Cl cotransporter in human colon. *J. Biol. Chem.* 270:17977–85

180. Pearson M, Lu J, Mount DB, Delpire E. 2001. Localization of the K-Cl cotransporter, KCC3, in the central and peripheral nervous systems: expression in choroid plexus, large neurons, and white matter tracts. *Neuroscience* 103:483–93

181. Pellegrino CM, Rybicki AC, Musto S, Nagel RL, Schwartz RS. 1998. Molecular identification and expression of erythroid K:Cl cotransporter in human and mouse erythroleukemic cells. *Blood Cell. Mol. Dis.* 24:31–40

182. Plata C, Meade P, Hall A, Welch RC, Vazquez N, et al. 2001. Alternatively spliced isoform of apical Na(+)-K(+)-Cl(−) cotransporter gene encodes a furosemide-sensitive Na(+)-Cl(−) cotransporter. *Am. J. Physiol. Renal Physiol.* 280:F574–F82

183. Plata C, Mount DB, Rubio V, Hebert SC, Gamba G. 1999. Isoforms of the apical Na-K-2Cl transporter in murine thick ascending limb. II: Functional characterization and mechanism of activation by cyclic-AMP. *Am. J. Physiol. Renal Physiol.* 276:F359–F66

184. Plotkin MD, Kaplan MR, Peterson L, Hebert SC, Delpire E. 1996. Expression of the Na-K-2Cl cotransporter, BSC2, in the rat central nervous system. *FASEB J.* 10:A145

185. Plotkin MD, Kaplan MR, Peterson LN, Gullans SR, Hebert SC, Delpire E. 1997.

Expression of the Na$^+$-K$^+$-2Cl$^-$ cotransporter BSC2 in the nervous system. *Am. J. Physiol. Cell Physiol.* 272:C173–C83

186. Plotkin MD, Kaplan MR, Verlander JW, Lee W-S, Brown D, et al. 1996. Localization of the thiazide sensitive Na-Cl cotransporter, rTSC, in the rat kidney. *Kidney Int.* 50:174–83

187. Plotkin MD, Snyder EY, Hebert SC, Delpire E. 1997. Expression of the Na-K-2Cl cotransporter is developmentally regulated in postnatal rat brains: a possible mechanism underlying GABA's excitatory role in immature brain. *J. Neurobiol.* 33:781–95

188. Quaggin SE, Payne JA, Forbush B III, Igarashi P. 1995. Localization of the renal Na-K-Cl cotransporter gene (*Slc12a1*) on mouse chromosome 2. *Mamm. Genome* 6:557–61

189. Raat NJH, Delpire E, van Os CH, Bindels RJM. 1996. Culturing induced expression of basolateral Na$^+$-K$^+$-2Cl$^-$ cotransporter BSC2 in proximal tubule, aortic endothelium, and vascular smooth muscle. *Pflügers Arch.* 431:458–60

190. Race JE, Makhlouf FN, Logue PJ, Wilson FH, Dunham PB, Holtzman EJ. 1999. Molecular cloning and functional characterization of KCC3, a new K-Cl cotransporter. *Am. J. Physiol. Cell Physiol.* 277:C1210–C19

191. Randall J, Thorne T, Delpire E. 1997. Partial cloning and characterization of *Slc12a2*: the gene encoding the secretory Na$^+$-K$^+$-2Cl$^-$ cotransporter. *Am. J. Physiol. Cell Physiol.* 273:C1267–C77

192. Reid KH, Guo SZ, Iyer VG. 2000. Agents which block potassium-chloride cotransport prevent sound-triggered seizures in post-ischemic audiogenic seizure-prone rats. *Brain Res.* 864:134–37

193. Rivera C, Voipio J, Payne JA, Ruusuvuori E, Lahtinen H, et al. 1999. The K$^+$/Cl$^-$ co-transporter KCC2 renders GABA hyperpolarizing during neuronal maturation. *Nature* 397:251–55

194. Rodriguez-Soriano J. 1998. Bartter and related syndromes: the puzzle is almost solved. *Pediatr. Nephrol.* 12:315–27

195. Rohrbough J, Spitzer NC. 1996. Regulation of intracellular Cl$^-$ levels by Na$^+$-dependent Cl$^-$ cotransport distinguishes depolarizing from hyperpolarizing GABA$_A$ receptor-mediated responses in spinal neurons. *J. Neurosci.* 16:82–91

196. Russell JM. 2000. Sodium-potassium-chloride cotransport. *Physiol. Rev.* 80:211–76

197. Schoenherr CJ, Anderson DJ. 1995. The neuron-restrictive silencer factor (NRSF): a coordinate repressor of multiple neuron-specific genes. *Science* 267:1360–63

198. Schultheis PJ, Lorenz JN, Meneton P, Nieman ML, Riddle TM, et al. 1998. Phenotype resembling Gitelman's syndrome in mice lacking the apical Na$^+$-Cl$^-$ cotransporter of the distal convoluted tubule. *J. Biol. Chem.* 273:29150–55

199. Schwartz RS, Musto S, Fabry ME, Nagel RL. 1998. Two distinct pathways mediate the formation of intermediate density cells and hyperdense cells from normal density sickle red blood cells. *Blood* 92:4844–55

200. Scott DA, Karniski LP. 2000. Human pendrin expressed in *Xenopus laevis* oocytes mediates chloride/formate exchange. *Am. J. Physiol. Cell Physiol.* 278: C207–C11

201. Simon DB, Bindra RS, Mansfield TA, Nelson-Williams C, Mendonca E, et al. 1997. Mutations in the chloride channel gene, CLCNKB, cause Bartter's syndrome type III. *Nat. Genet.* 17:171–78

202. Simon DB, Farfel Z, Ellison D, Bia M, Tucci J, Lifton RP. 1995. Examination of the thiazide-sensitive Na-Cl cotransporter as a candidate gene in Gordon's syndrome. *J. Am. Soc. Nephrol.* 6:632

203. Simon DB, Karet FE, Hamdan JM, Di Pietro A, Sanjad SA, Lifton RP. 1996. Bartter's syndrome, hypokalaemic alkalosis with hypercalciuria, is caused by mutations in the Na-K-2Cl cotransporter NKCC2. *Nat. Genet.* 13:183–88

204. Simon DB, Karet FE, Rodriguez-Soriano J, Hamdan JH, DiPietro A, et al. 1996. Genetic heterogeneity of Bartter's syndrome revealed by mutations in the K$^+$ channel, ROMK. *Nat. Genet.* 14:152–56

205. Simon DB, Nelson-Williams C, Johnson Bia M, Ellison D, Karet FE, et al. 1996. Gitelman's variant of Bartter's syndrome, inherited hypokalaemic alkalosis, is caused by mutations in the thiazide-sensitive Na-Cl cotransporter. *Nat. Genet.* 12:24–30

206. Singer JH, Talley EM, Bayliss D, Berger AJ. 1998. Development of glycinergic synaptic transmission to rat brain stem motoneurons. *J. Neurophysiol.* 80:2608–20

207. Song L, Mercado A, Desai R, George AL, Gamba G, Mount DB. 2001. Characterization of hKCC2, the human neuronal-specific K-Cl cotransporter. *FASEB J.* 15: A440

208. Staley KJ, Proctor WR. 1999. Modulation of mammalian dendritic GABA(A) receptor function by the kinetics of Cl$^-$ and HCO$_3$$^-$-transport. *J. Physiol.* 519:693–712

209. Staley KJ, Soldo BL, Proctor WR. 1995. Ionic mechanisms of neuronal excitation by inhibitory GABA$_A$ receptors. *Science* 269:977–81

210. Stanton BA. 1989. Renal potassium transport: morphological and functional adaptations. *Am. J. Physiol. Regulatory Integrative Comp. Physiol.* 257:R989–R97

211. Stanton BA, Kaissling B. 1989. Regulation of renal ion transport and cell growth by sodium. *Am. J. Physiol. Renal Physiol.* 257:F1–F10

212. Starke LC, Jennings ML. 1993. K-Cl cotransport in rabbit red cells: further evidence for regulation by protein phosphatase. *Am. J. Physiol. Cell Physiol.* 264:C118–C24

213. Steinlein OK, Neubauer B, Sander T, Song L, Stoodt J, Mount DB. 2001. Mutation analysis of the potassium chloride cotransporter KCC3 (SLC12A6) in rolandic and idiopathic generalized epilepsy. *Epilepsy Res.* 44:191–95

214. Stober G, Saar K, Ruschendorf F, Meyer J, Nurnberg G, et al. 2000. Splitting schizophrenia: periodic catatonia-susceptibility locus on chromosome 15q15. *Am. J. Hum. Genet.* 67:1201–7

215. Strange K, Singer TD, Morrison R, Delpire E. 2000. Dependence of KCC2 K-Cl cotransporter activity on a conserved carboxy terminus tyrosine residue. *Am. J. Physiol. Cell Physiol.* 279:C860–C67

216. Stratton JD, McNicholas TA, Farrington K. 1998. Recurrent calcium stones in Gordon's syndrome. *Br. J. Urol.* 82:925

217. Su W, Shmukler BE, Chernova MN, Stuart-Tilley AK, De Franceschi L, et al. 1999. Mouse K-Cl cotransporter KCC1: cloning, mapping, pathological expression, and functional regulation. *Am. J. Physiol. Cell Physiol.* 277:C899–C912

218. Sun A, Grossman EB, Lombardi M, Hebert SC. 1991. Vasopressin alters the mechanism of apical Cl$^-$ entry from Na$^+$:Cl$^-$ to Na$^+$:K$^+$:2Cl$^-$ cotransport in mouse medullary thick ascending limb. *J. Membr. Biol.* 120:83–94

219. Sung K-W, Kirby M, McDonald MP, Lovinger DM, Delpire E. 2000. Abnormal GABA$_A$-receptor mediated currents in dorsal root ganlion neurons isolated from Na-K-2Cl cotransporter null mice. *J. Neurosci.* 20:7531–38

220. Takahashi N, Chernavvsky DR, Gomez RA, Igarashi P, Gitelman HJ, Smithies O. 2000. Uncompensated polyuria in a mouse model of Bartter's syndrome. *Proc. Natl. Acad. Sci. USA* 97:5434–39

221. Take C, Ikeda K, Kurasawa T, Kurokawa K. 1991. Increased chloride reabsorption as an inherited renal tubular defect in familial type II pseudohypoaldosteronism. *N. Engl. J. Med.* 324:472–76

222. Takeuchi K, Kure S, Kato T, Taniyama Y, Takahashi N, et al. 1996. Association of a mutation in thiazide-sensitive Na-Cl cotransporter with familial Gitelman's

syndrome. *J. Clin. Endocrinol. Metab.* 81:4496–99

223. Tas PWL, Massa PT, Kress HG, Koschel K. 1987. Characterization of an Na$^+$/K$^+$/Cl$^-$ cotransport in primary cultures of rat astrocytes. *Biochim. Biophys. Acta* 903:411–16

224. Thompson SM, Deisz RA, Prince DA. 1988. Outward chloride/cation co-transport in mammalian cortical neurons. *Neurosci. Lett.* 89:49–54

225. Tsukamoto T, Kobayashi T, Kawamoto K, Fukase M, Chihara K. 1995. Possible discrimination of Gitelman's syndrome from Bartter's syndrome by renal clearance study: report of two cases. *Am. J. Kidney Dis.* 25:637–41

226. Tsuruoka S, Koseki C, Muto S, Tabei K, Imai M. 1994. Axial heterogeneity of potassium transport across hamster thick ascending limb of Henle's loop. *Am. J. Physiol. Renal Physiol.* 267:F121–F29

227. Vandewalle A, Cluzeaud F, Bens M, Kieferle S, Steinmeyer K, Jentsch TJ. 1997. Localization and induction by dehydration of ClC-K chloride channels in the rat kidney. *Am. J. Physiol. Renal Physiol.* 272:F678–F88

228. Vandorpe DH, Shmukler BE, Jiang L, Lim B, Maylie J, et al. 1998. cDNA cloning and functional characterization of the mouse Ca^{2+}-gated K$^+$ channel, mIK1. Roles in regulatory volume decrease and erythroid differentiation. *J. Biol. Chem.* 273:21542–53

229. Vardi N, Zhang LL, Payne JA, Sterling P. 2000. Evidence that different cation chloride cotransporters in retinal neurons allow opposite responses to GABA. *J. Neurosci.* 20:7657–63

230. Vargas-Poussou R, Feldmann D, Vollmer M, Konrad M, Kelly L, et al. 1998. Novel molecular variants of the Na-K-2Cl cotransporter gene are responsible for antenatal Bartter syndrome. *Am. J. Hum. Genet.* 62:1332–40

231. Velazquez H, Naray-Fejes-Toth A, Silva T, Andujar E, Reilly RF, et al. 1998.

Rabbit distal convoluted tubule coexpresses NaCl contransporter and 11 beta-hydroxysteroid dehydrogenase II mRNA. *Kidney Int.* 54:464–72

232. Verlander JW, Tran TM, Zhang L, Kaplan MR, Hebert SC. 1996. Estrogen enhances thiazide-sensitive NaCl cotransporter (TSC) density in the apical plasma membrane of distal convoluted tubule (DCT) in ovariectomized rats. *J. Am. Soc. Nephrol.* 7:1293

233. Vetter DE, Mann JR, Wangemann P, Liu J, McLaughlin KJ, et al. 1996. Inner ear defects induced by null mutation of the *isk* gene. *Neuron* 17:1251–64

234. Vezzoli G, Soldati L, Jansen A, Pierro L. 2000. Choroidal calcifications in patients with Gitelman's syndrome. *Am. J. Kidney Dis.* 36:855–58

235. Vollmer M, Jeck N, Lemmink HH, Vargas R, Feldmann D, et al. 2000. Antenatal Bartter syndrome with sensorineural deafness: refinement of the locus on chromosome 1p31. *Nephrol. Dial. Transplant.* 15:970–74

236. Vu TQ, Payne JA, Copenhagen DR. 2000. Localization and developmental expression patterns of the neuronal K-Cl cotransporter (KCC2) in the rat retina. *J. Neurosci.* 20:1414–23

237. Wall SM, Trinh HN, Woodward KE. 1995. Heterogeneity of NH$_4^+$ transport in mouse inner medullary collecting duct cells. *Am. J. Physiol. Renal Physiol.* 38:F536–F44

238. Walz W, Hertz L. 1984. Sodium transport in astrocytes. *J. Neurosci. Res.* 11:231–39

239. Wang T, Agulian SK, Giebisch G, Aronson PS. 1993. Effects of formate and oxalate on chloride absorption in rat distal tubule. *Am. J. Physiol. Renal Physiol.* 264:F730–F36

240. White PC, Mune T, Agarwal AK. 1997. 11 beta-hydroxysteroid dehydrogenase and the syndrome of apparent mineralocorticoid excess. *Endocr. Rev.* 18:135–56

241. Williams JR, Sharp JW, Kumari VG,

Wilson M, Payne JA. 1999. The neuron-specific K-Cl cotransporter, KCC2. Antibody development and initial characterization of the protein. *J. Biol. Chem.* 274:12656–64

242. Willis WD. 1999. Dorsal root potentials and dorsal root reflexes: a double-edged sword. *Exp. Brain Res.* 124:395–421

242a. Woo N-S, Lu J, England R, McClellan R, Dufour S, et al. 2002. Hyperexcitability and epilepsy associated with disruption of the mouse neuronal-specific K-Cl cotransporter gene. *Hippocampus.* In press

243. Wu Q, Delpire E, Hebert SC, Strange K. 1998. Functional demonstration of Na-K-2Cl cotransporter activity in isolated, polarized choroid plexus cells. *Am. J. Physiol. Cell Physiol.* 275:C1565–C72

244. Xu J-C, Lytle C, Zhu TT, Payne JA, Benz EJ, Forbush B III. 1994. Molecular cloning and functional expression of the bumetanide-sensitive Na-K-2Cl co-transporter. *Proc. Natl. Acad. Sci. USA* 91:2201–5

245. Xu JZ, Hall AE, Peterson LN, Bienkowski MJ, Eessalu TE, Hebert SC. 1997. Localization of the ROMK protein on apical membranes of rat kidney nephron segments. *Am. J. Physiol. Renal Physiol.* 273:F739–F48

246. Yang T, Huang YG, Singh I, Schnermann J, Briggs JP. 1996. Localization of bumetanide- and thiazide-sensitive Na-K-Cl cotransporters along the rat nephron. *Am. J. Physiol. Renal Physiol.* 271:F931–F39

247. Yoshitomi K, Koseki C, Taniguchi J, Imai M. 1987. Functional heterogeneity in the hamster medullary thick ascending limb of Henle's loop. *Pflügers Arch.* 408:600–8

248. Zeuthen T. 1994. Cotransport of K^+, Cl^- and H_2O by membrane proteins from choroid plexus epithelium of *Necturus maculosus*. *J. Physiol.* 478:203–19

Annu. Rev. Physiol. 2002. 64:845–76

Renal Genetic Disorders Related to K^+ and Mg^{2+}

David G. Warnock

Division of Nephrology, Departments of Medicine and Physiology, and Nephrology Research and Training Center, University of Alabama at Birmingham, Birmingham, Alabama 35294; e-mail: dwarnock@nrtc.uab.edu

Key Words   epithelia transport, epithelial sodium channel, Liddle's syndrome, Bartter's syndrome, Gitelman's syndrome

■ **Abstract**   The recent knowledge of the renal epithelial transport systems has exploded with the identification, cloning, and characterization of a large number of membrane transport proteins. The fundamental aspects of these transporters are beginning to emerge at the molecular level and are summarized in the accompanying contributions in this volume of the *Annual Review of Physiology*. The aim of my review is to integrate this body of knowledge with the understanding of the clinical disorders of human mineral homeostasis that accompany gain, loss, or dysregulation of function of these transport systems. The specific focus is on the best defined human clinical syndromes in which there are derangements in K^+ and Mg^{2+} homeostasis.

INTRODUCTION

A variety of inherited disorders alter specific renal epithelial transport functions. This review addresses those hereditary diseases in which the transport of electrolytes by the renal tubular epithelium is deranged and the defect that has been attributed to a specific transport protein. Of central importance are the primary Na^+ transport systems in the thick ascending limb (TAL), distal convoluted (DCT) and connecting tubules (CT), and the collecting tubules (CCT) (Figure 1). This review does not include those disorders in which the principal disturbance involves transport of bicarbonate, protons, or calcium; rather the focus is on those renal genetic disorders that affect K^+ and Mg^{2+} balance (1).

RENAL K^+ HANDLING

Nearly all of the K^+ in the glomerular ultrafiltrate is reabsorbed along the nephron, and the final urinary excretion very nearly matches the daily dietary intake. Because most of the filtered load is reabsorbed, the urinary excretory component is secreted

0066-4278/02/0315-0845$14.00

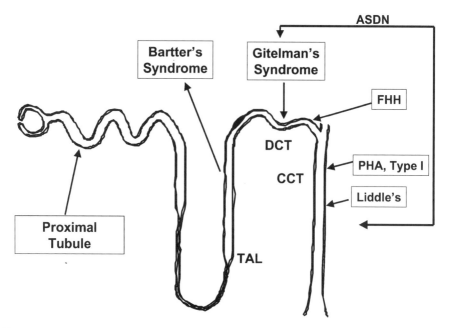

Figure 1   Localization of ion transporters and disease loci along the nephron. ASDN, aldosterone-sensitive distal nephron; TAL, thick ascending limb; DCT, distal convoluted tubule, CCT, cortical collecting tubule. Genetic syndromes include FHH, familial hyperkalemia with hypertension; PHA, Type I, pseudohypoaldosteronism, Type I.

into the luminal fluid compartment in the distal nephron (2, 3), with the predominate fraction contributed in a regulated fashion by the aldosterone-sensitive segments of the distal nephron (ASDN). Two major factors directly affect the rate of K^+ secretion: mineralocorticoid tone and Na^+ delivery to the ASDN (Figure 2). K^+ secretion increases in proportion to the delivery of Na^+. The primary process is the electrogenic reabsorption of Na^+ through the epithelial Na^+ channel (ENaC) in the apical membrane; the lumen-negative transepithelial electrical potential drives K^+ secretion.

Mineralocorticoid hormones, usually aldosterone, enhance ENaC activity and thereby drive K^+ secretion. In addition, in so far as increased Na^+ delivery is accompanied by increased volume flow along the nephron, the concentration of K^+ in the lumen is reduced, which also facilitates the K^+ secretory process. As shown in Figure 3, the K^+ secretory channel [e.g., the renal outer medullary K^+ channel (ROMK1)] is physically distinct from ENaC, thus there is no direct coupling or 1:1 exchange of cations (4). Furthermore, the regulation of each channel can be independent. Mineralocorticoids have an important direct effect on ENaC regulation (5). However, hyperkalemia is an important stimulus for aldosterone secretion, providing an important linkage between the rates of Na^+ reabsorption

K+ Secretion

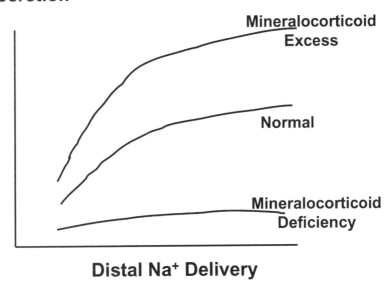

Distal Na+ Delivery

Figure 2 Dependence of K$^+$ secretion on Na$^+$ delivery in the aldosterone-sensitive segments of the distal nephron.

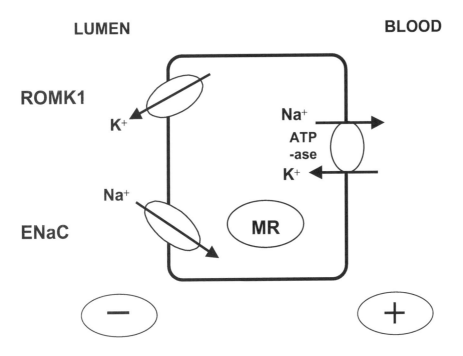

Figure 3 Principal cell of CCT. ENaC, epithelial Na$^+$ channel; ROMK1, renal outer medulla K$^+$ channel, 1; MR, mineralocorticoid receptor.

and K^+ secretion in the ASDN (2–4). Independently of mineralocorticoids, protein kinase A, phosphatidyl-*bis*-phosphate, and serum K^+, per se (6, 7), are important regulatory factors of the apical membrane ROMK1 activity in the ASDN (7–9).

RENAL K^+ WASTING DISORDERS WITH LOW PLASMA RENIN ACTIVITY

Without important exception, all of the currently defined Mendelian forms of human hypertension can be attributed to derangements in the function and/or regulation of ENaC activity in the ASDN and of the processes that control the final adjustments of NaCl balance (10). Concomitant with the inappropriate ENaC activity due to constitutive or applied activation, renal K^+ wasting occurs as predicted from the schema in Figure 2.

Liddle's Syndrome

Mutations of the ENaC subunits can result in gain of function (Liddle's syndrome), with the predictable clinical phenotype of low-renin hypertension and suppressed aldosterone secretion (1). They were originally described as truncations or frameshifts in the β or γ subunits of ENaC (11). These mutations delete a critical proline-rich region of the cytosolic tail, which interacts with a cytoskeletal protein called Nedd4 (12). Several pedigrees with mutations in this proline-rich region are associated with gain of ENaC function and low-renin hypertension with metabolic alkalosis and renal potassium wasting (11, 13). De novo ENaC mutations have been reported in sporadic cases of Liddle's syndrome (14, 15). An increasing number of mutations have been described, but to date, only the β and γ subunits have been involved (16). With wider appreciation of this rare form of genetic hypertension, the diagnosis is more frequently being made in young (17), middle-aged (18), and even elderly patients (19). Despite the excitement of defining the genetic causes of human hypertension, numerous screening efforts of various ethnic populations have not changed the original view that Liddle's syndrome is a rare cause of human hypertension.

Although activating mutations of ENaC subunits are uncommon in patients with essential hypertension, there may be polymorphisms in these genes that have important effects on the regulation of ENaC activity. None of the ENaC polymorphisms, which have potential importance at the epidemiologic level, have been shown to have any demonstrable effects in the in vitro expression systems used to examine ENaC activity (20).

Liddle's syndrome is the phenotypic extreme of low-renin hypertension (21). In the original studies of the index case, aldosterone secretion was markedly suppressed, accounting for the descriptor pseudoaldosteronism (22). The studies following renal transplantation of the proband (23) suggested that the suppression of the plasma renin activity was the result of ENaC gain-of-function in Liddle's syndrome. Continued suppression of aldosterone excretion, despite rigorous dietary salt restriction, is best explained by chronic suppression of 18-hydroxylase (adrenal aldosterone synthase) as a consequence of chronic volume expansion (21).

Low-renin hypertension is a common clinical disorder, especially among subjects of African ancestry (20). Despite suppressed plasma renin activity, aldosterone is not suppressed. Aldosterone secretion remains under control of angiotensin II in African-American subjects although the plasma aldosterone response to angiotensin II infusion is blunted compared to Caucasian controls (24). Recent insights into these issues have been obtained from studies of the original Liddle's pedigree (21). Gas chromatograpic-mass spectrometry was used to examine the urinary steroid metabolite profile in these patients. Marked suppression of aldosterone metabolite excretion, as well as all of the other adrenal 18-hydroxylase urinary steroid metabolites, was observed, denoting a chronic, global suppression of aldosterone synthase activity in the adrenal zona glomerulosa in Liddle's syndrome (21). Of note, in the original proband studied two years after successful renal transplantation, the plasma renin activity and aldosterone excretion were restored to normal (23), confirming that suppression of aldosterone synthase activity in Liddle's syndrome and the suppressed plasma renin activity are simply indices of chronic sustained volume expansion (21). The lack of downregulation of ENaC activity in the face of persistent volume expansion underlies the pathophysiology of Liddle's syndrome. A similar lack of downregulation of ENaC activity, especially in the face of dietary salt excess could underlie more common forms of low-renin hypertension (20).

Renal K$^+$ wasting and hypokalemia are generally, but not universally, observed in most patients with Liddle's syndrome (22, 23, 25). The reasons for variations in the serum K$^+$ concentration are not understood, but renal K$^+$ wasting follows from the schema presented in Figure 2 and would be worsened by increased dietary Na$^+$ intake and delivery to the ASDN. The primary defect in Liddle's syndrome is constitutive activation of ENaC in the apical membrane of the principal cells of the CCT (Figure 3). As a result, it appears as if there is mineralocorticoid effect despite suppression of aldosterone synthesis and secretion because of chronic volume expansion (20, 26). Due to this apparent mineralocorticoid effect and suppression of aldosterone secretion, with ongoing renal K$^+$ excretion, the ratio of urinary aldosterone to K$^+$ appears to be a sensitive index of ENaC activation resulting from gain-of-function mutations or the effects of mineralocorticoids other than aldosterone (20, 26).

S810L Mutation in the Mineralocorticoid Receptor

A second form of pseudoaldosteronism recently described it the result of a mutation in the hormone-binding domain of the mineralocorticoid receptor (27). This syndrome is referred to herein as Type II pseudoaldosteronism, whereas Liddle's syndrome is Type I pseudoaldosteronism (Table 1). Although the phenotype of this autosomal-dominant disorder strongly resembles Liddle's syndrome with

TABLE 1   Suggested nomenclature for human genetic syndromes in which blood pressure is altered inversely to aldosterone secretion: pseudoaldosteronism and pseudohypoaldosteronism

Descriptor	Inheritance	Mutations	Phenotypes	OMIM
Pseudoaldosteronism				
PA1	AD	γ and β ENaC subunits	$\Uparrow$ BP, $\Downarrow$ K$^+$, $\Downarrow$ pH	177200
PA2	AD	S810L MR	$\Uparrow$ BP, $\Downarrow$ K$^+$, $\Downarrow$ pH	605115
Pseudohypoaldosteronism				
PHA1	AR	α, γ and β ENaC subunits	$\Uparrow$ K$^+$, $\Uparrow$ pH, $\Downarrow$ BP (neonatal)	264350
PHA2	AD	MR	$\Uparrow$ K$^+$, $\Uparrow$ pH, $\Downarrow$ BP (mild)	177735

Abbreviations: AD, autosomal-dominant; AR, autosomal recessive; BP, blood pressure; ENaC, epithelial Na$^+$ channel; K$^+$, serum potassium; MR, mineralocorticoid receptor; OMIM, online Mendelian inheritance in man; PA, pseudoaldosteronism; pH, systemic pH; PHA, pseudohypoaldosteronism; S820L, serine to leucine mutation at position 820 of the MR.

hypertension, hypokalemia, suppressed peripheral renin activity, and aldosterone secretion, there is also an especially severe presentation during pregnancy (27). Geller et al. (27) described a serine-to-leucine mutation at position 810 in the hormone-binding domain of the mineralocorticoid receptor that changes the affinity of the receptor for a number of steroids. Progesterone in particular has an exceptionally high affinity for the receptor, thus accounting for the especially severe presentation during pregnancy. Spironolactone is also an agonist for the mutated receptor; therefore spironolactone is ineffective in Liddle's syndrome (22) and contraindicated in treatment of this second form of pseudoaldosteronism (27). It appears that the mutated receptor has constitutively activated basal activity or may be responding to other endogenous steroids (e.g., 19-norprogesterone or 17-OH progesterone) to account for the clinical phenotype in males and nonpregnant females with type 2 pseudoaldosteronism.

Glucocorticoid Remediable Aldosteronism (GRA, Familial Hyperaldosteronism, Type I)

GRA results from a chimeric gene product that places 18 hydroxylase (aldosterone synthase) under the control of an ACTH promoter, which regulates its level and cellular site of expression. As a consequence, ACTH-regulated 18 hydroxylase activity is aberrantly expressed in the zona fasciculata and acts on cortisol, which is normally produced in this zone, to form 18-hydroxycortisol (18-OH F) and 18-oxocortisol (28). GRA is an important autosomal-dominant cause of human genetic hypertension because it is relatively common, and the syndrome represents the first example of successful application of the înewî genetics to define an inherited human hypertensive syndrome (29). The continuing stimulation of adrenal steroidogenesis by ACTH, and the suppressive effects of modest doses of dexamethasone are usual features of the syndrome. Recent studies have used genetic testing to prospectively ascertain affected family members and further define the clinical phenotype. Although pregnant women with GRA do not appear to be more prone to pre-eclampsia, they do have chronic hypertension and are at increased risk for exacerbation of their hypertension during pregnancy; they have a Cesarean section rate twice that of other general, or even hypertensive, obstetrical populations (30). Males appear to be more severely affected than females, and the severity of hypertension correlates with the level of aldosterone secretion; angiotensin II has very little effect on aldosterone secretion, but glucocorticoids suppress it, indicating that a substantial amount of aldosterone is synthesized by the hybrid gene product in the severe forms of GRA (31–34). The fundamental mechanism of renal K^+ wasting and hypokalemia is similar to Liddle's syndrome.

Familial Hyperaldosteronism, Type II

A similar presentation as GRA has been described by Stowasser & Gordon and termed familial hyperaldosteronism, Type II; Type II is associated with adrenal hyperplasia and aldosteronomas. This distinct form of familial hyperaldosteronism is not suppressible by glucocorticoids and has been mapped to chromosome 7p22 (38). Stowasser & Gordon term GRA as familial hyperaldosteronism Type I (35). Both forms are autosomal-dominant forms of low-renin, mineralocorticoid excess hypertension (Table 1).

Apparent Mineralocorticoid Excess (AME) Syndrome

We now have much more precise knowledge of the underlying pathophysiology of the AME syndrome (1). AME results from inactivation of 11β hydroxysteroid dehydrogenase, Type II (11β-OH SDH II), causing reduced metabolism of cortisol to cortisone, and local cortisol excess and mineralocorticoid response in aldosterone-target cells, which coexpress 11β-OH SDH II and Type I mineralocorticoid receptors. The balance between cortisol and cortisone metabolism determines whether cortisol will exert any mineralocorticoid effect at the target tissue level. Even though plasma cortisol levels are not necessarily elevated, the metabolic clearance of cortisol is prolonged in AME, and there is an excess urinary excretion of the reduced metabolites of cortisol compared with cortisone (THF/THE ratio).

Dietary salt excess does not appear to affect 11β OH SDH II activity (39), but some hypertensive patients may have mild 11β OH SDH II deficiency and are therefore especially susceptible to the effects of exogenous inhibitors (40). Salt-sensitive hypertension has recently been associated with reduced 11β-OH SDH II activity and with a specific polymorphism in the 11β-OH SDH II gene (41). The allelic frequency of 11β OH SDH II polymorphisms needs to be more fully described in human populations because there could be genetically determined differences in the levels of 11β OH SDH II activity or in the response to exogenous inhibitors of its activity. More information is needed about the levels of its intrinsic activity and the regulation of 11β OH SDH II activity in human hypertensive populations (40, 42). The underlying mechanism of renal K^+ wasting is similar to Liddle's syndrome; in this instance, the apparent mineralocorticoid effect (Figure 2) is exerted by cortisol at the level of the mineralocorticoid receptor in the ASDN.

RENAL K$^+$ WASTING DISORDERS WITH HIGH PLASMA RENIN ACTIVITY

In addition to the hypertensive disorders discussed above, there are functional defects in the transporters of the TAL and DCT that are associated with volume depletion and consequent elevated plasma renin activity and aldosterone excretion (29). The mechanistic implications of these human genetic disorders have been confirmed with transgenic mouse models of the human disease (11). In these disorders, there is an absolute increase in the aldosterone secretion and, because of the proximal defect(s), an ongoing delivery of Na$^+$ to the downstream ASDN. Renal K$^+$ wasting and hypokalemia are nearly universal findings and entirely consistent with the schema presented in Figure 2. In addition to activation of the K$^+$ secretory processes in the principal cells of the CCT (Figure 3), functional defects in the K$^+$ absorption via the Na-K-Cl cotransporter in the TAL may also occur (Figure 4).

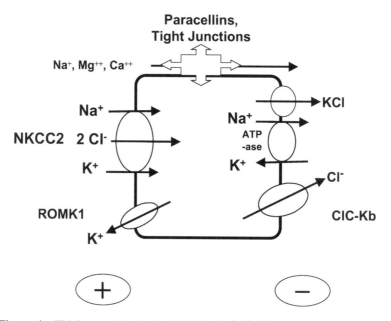

Figure 4   Thick ascending limb. NKCC2, Na$^+$K$^+$2Cl$^-$ cotransporter; ROMK1, renal outer medulla K$^+$ channel, 1; KCl, KCl transporter; ClC-Kb, basolateral chloride channel.

Bartter's Syndrome

Bartter and colleagues in 1962 described hypokalemic, hypochloremic metabolic alkalosis in two children and a man; this distinctive syndrome now bears his eponym. Other characteristics of Bartter's syndrome are increased urinary excretion of potassium and prostaglandins, normal or low blood pressure despite elevated plasma renin activity and aldosterone concentrations, a relative vascular resistance to the pressor effects of exogenous angiotensin-II, and hyperplasia of the juxtaglomerular apparatus (43). Most patients are diagnosed in infancy, childhood, or early adolescence, and there is no racial, ethnic, or sex predilection. Although many cases appear to be sporadic, Bartter's syndrome does occur in families, with an autosomal-recessive mode of inheritance (44).

Bartter's syndrome is not a single disease but rather a set of closely related renal tubular disorders. At least three phenotypic subsets have been distinguished: an antenatal hypercalciuric variant, also termed hyperprostaglandin E syndrome, which presents at birth with polyhydramnios and dehydration; classic Bartter's syndrome, which presents in young children, often as failure to thrive; and a distinct form of the antenatal syndrome associated with sensorineural deafness (45, 46). This heterogeneity and the diverse array of physiologic derangements have long confounded efforts to understand the fundamental defects in the Bartter-like syndromes. The proposed primary defects include juxtaglomerular hyperplasia, insensitivity to angiotensin-II, prostaglandin-kallikrein-kinin overproduction, a defect in K$^+$ transport that results in excessive K$^+$ excretion, and a defect in NaCl transport in the TAL or DCT (44, 47). Elevated plasma renin and absence of hypertension excludes a primary adrenal steroid excess syndrome, and the finding of high urinary Cl$^-$ excretion rules out secondary hyperaldosteronism from extrarenal fluid loss. These findings are consistent with only two conditions, Bartter's syndrome and chronic diuretic therapy (44, 47).

These and other clinical data suggest that Bartter's syndrome results from defective transepithelial transport of NaCl in the TAL. The genes encoding several thick ascending limb proteins have been cloned, including the bumetanide-sensitive Na-K-2Cl cotransporter (NKCC2, alternatively designated BSC1), the apical, ATP-regulated K$^+$ channel (ROMK1), and the kidney-specific Cl$^-$ channel, ClC-Kb (Figure 4). Loss-of-function mutations in each of these transport proteins have been documented in some but not all patients with antenatal Bartter's syndrome or the more classic phenotype (48–52). Mutations in these transporters appear to account for nearly all of the currently recognized cases of Bartter's syndrome. The antenatal form, with sensorineural deafness, has been linked to a different chromosomal locus than any of the three transporters described above (45, 46). In addition, other possible candidate loci to consider include the extracellular Ca^{2+}-sensing receptor that regulates NKCC2 and ROMK1 function and the basolateral KCl cotransporter (53–55).

The antenatal form of Bartter's syndrome is usually linked to mutations in the NKCC2 or ROMK1 and presents with severe hypokalemia and metabolic alkalosis. This form can be life threatening in utero when associated with marked fetal polyuria, polyhydramnios, and premature delivery (56, 57). Occasionally, neonatal Bartter's syndrome can be associated initially with hypernatremia and hyperchloremia, suggesting nephrogenic diabetes insipidus, but the usual presentation quickly becomes apparent (58). Hypercalciuria, secondary nephrocalcinosis, and early onset osteopenia are also observed in these patients. A knockout mouse model with a NKCC2 deletion has been developed and is marked by polyuria and hypercalciuria (59).

Type II Bartter's syndrome has been linked to mutations in the apical membrane ROMK1 channel in the TAL (Figure 4). This form is also associated with the antenatal presentation, increased PGE_2 excretion, severe salt wasting and nephrocalcinosis (60, 61). The mutations in ROMK1 include complete gene deletions, truncations, and missense mutations. When examined in various expression systems, these mutations have provided novel insights into the regulation and functional domains of ROMK1 (62–65).

The third defined locus in Bartter's syndrome is the basolateral Cl^- channel referred to as ClC-Kb (Figure 4). These mutations are less often, but occasionally, associated with the antenatal form of Bartter's syndrome. The more usual presentation is classical Bartter's syndrome, which manifests during childhood rather than in the antenatal or neonatal period (44). There can be some marked phenotypic variability, with transition to a hypomagnesemic variant (66); nephrocalcinosis is not a prominent feature of this disorder (56). Thakker has noted the similarities in renal Ca^{2+} wasting observed with mutations in ClC-Kb in Bartter's syndrome, Type III and mutations in another Cl^- channel (ClC-5) expressed in proximal tubule endocytic vesicles, and mutated in Dent's disease (1, 67).

Impaired transport of NaCl in the TAL is associated with reduction in the lumen-positive electrical transport potential that normally drives the paracellular reabsorption of Ca^{2+} and Mg^{2+} (68) and causes increased urinary loss of these ions. Hypercalciuria is a common feature of Bartter's syndrome, and in antenatal Bartter's syndrome it often leads to nephrocalcinosis. Hypomagnesemia, in contrast, is relatively uncommon (69). A striking feature of Bartter's syndrome is the depressed vascular reactivity, which may reflect the chronic effects of angiotensin II generation from the high renin and aldosterone levels or enhanced nitric oxide synthase activity, as marked by urinary excretion of nitrates (70, 71). The antenatal form of Bartter's syndrome is often associated with increased renal PGE_2 excretion and is treated with nonsteroidal anti-inflammatory agents, which, unfortunately can have adverse long-term effects on renal function (72, 73). In one such case, successful renal transplantation has been reported with a "cure" of the Bartter's syndrome (73).

Gitelman's Syndrome

This syndrome was first recognized as a variant of Bartter's syndrome in which patients have hypomagnesemia and hypocalciuria (74). Thiazide diuretics inhibit the NaCl cotransporter [NCCT, also known as the thiazide-sensitive cotransporter

LUMEN   **BLOOD**

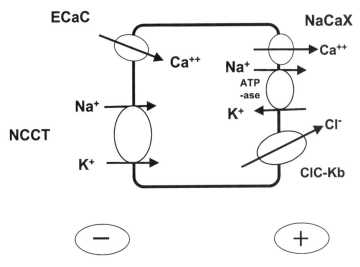

Figure 5 Distal convoluted tubule. NCCT, NaCl cotransporter; ECaC, epithelial Ca^{2+} channel; ROMK1, renal outer medulla K$^+$ channel, 1; NaCaX, basolateral Na$^+$/Ca^{2+} exchanger; ClC-Kb, basolateral chloride channel.

(TSC)] in the DCT (Figure 5), and patients with Gitelman's syndrome have a subnormal natriuretic response to intravenous chlorothiazide but a prompt natriuresis after furosemide administration (75). Other authors have also noted the marked response to loop diuretics in Gitelman's syndrome without any response to thiazide agents (76, 77). These studies suggest that the NCCT is an obvious candidate gene in Gitelman's syndrome. Loss of apparent thiazide-sensitive NCCT activity in peripheral mononuclear cells from patients with Gitelman's syndrome is also consistent with this thesis (78). A knockout mouse model developed in 1998 manifested very mild volume depletion, did not tolerate a low-salt diet, and had hypocalciuria and hypomagnesemia (79); the response to furosemide was greater than thiazides, similar to Gitelman's syndrome. Another interpretation of these responses to diuretics implicates a functional hypertrophy of the TAL associated with deletion of the NCCT protein gene product in Gitelman's syndrome or in the knockout mouse model (13). Ellison (69) has noted that thiazides do not usually cause renal Mg^{2+} wasting, but in contrast to thiazide therapy, which is episodic and not continuous since the time of conception, the knockout mouse model and patients with Gitelman's syndrome have much more profound derangements of DCT function and at least functional evidence for hypertrophy of the TAL (13).

Mutations associated with putative loss of NCCT function have been identified in patients with this syndrome (80–82). The initial efforts mapped the causative locus to chromosome 16q (82, 83), and NCCT mutations were then defined (80, 81). A number of mutations have now been defined, including truncations and missense mutations. Most appear to involve the intracellular and C-terminal domain of NCCT, and in expression systems cause decreased functional expression with processing disturbances in delivering the matured protein to the cell surface (84–87).

Melander et al. (88) studied four Swedish patients who were found to be compound heterozygotes for mutations in NCCT. These same mutations were then searched for in a population of patients with primary hypertension, but none were found to be heterozygotes for any of the mutations that had been associated with Gitelman's syndrome (88). Of even greater interest, a series of NCCT polymorphisms were defined in the hypertensive patients and compared with normotensive controls and the patients with Gitelman's syndrome. A novel Arg904Gln polymorphism was identified in the hypertensive subjects; Gln904 homozygotes were over-represented in the hypertensive subjects (5 of 292), compared with controls (0 of 264), suggesting that this polymorphism may functionally activate NCCT and be causally related to or at least increase the risk of developing hypertension in this defined subset. Loss of NCCT function causes defective reabsorption of NaCl in the DCT, which normally reabsorbs about 7% of the filtered load of NaCl. This defect would increase solute delivery to the collecting tubule, with consequent mild volume contraction and aldosterone-stimulated K^+ and H^+ ion secretion, which would result in mild hypokalemic metabolic alkalosis. The extent of the volume contraction, the degree of stimulation of vasopressin and the renin-angiotensin-aldosterone axis, and the extent of K^+ depletion appear to be less marked than in Bartter's syndrome and are not sufficient to increase renal and systemic prostaglandin E_2 production substantially. Urinary prostaglandin excretion remains normal in Gitelman's syndrome (44). Bettinelli et al. (89) emphasized the importance of hypocalciuria and hypomagnesemia in Gitelman's syndrome compared with Bartter's syndrome; the molar urinary Ca^{2+}/creatinine ratio was <0.2, and the serum Mg^{2+} <0.75 mM in Gitelman's syndrome. With this distinction, Gitelman's could no longer be simply viewed as a variant of Bartter's syndrome, but in fact a distinct entity, setting the stage for the family studies and linkage efforts in both syndromes. Since then, the possibility of some phenotypic variation has been appreciated in Gitelman's syndrome; a childhood form has been described that appears to be autosomal-recessive and is associated with severe Mg^{2+} wasting, weakness, and even tetanic episodes during childhood (90). Complete sequence information is not currently available to further define this clinical phenotype.

Other clinical features of Gitelman's syndrome has been recently recognized, including short stature (91, 92), chondrocalcinosis with calcium pyrophosphate dihydrate crystal deposition disease, and sclerochoroidal calcifications, all of which appear to be influenced by the degree of hypomagnesemia (93–97). Pregnancy is usually successful in Gitelman's syndrome, and several cases have been diagnosed in patients who develop severe hypokalemia and hypomagnesemia during their pregnancies; in contrast to Bartter's syndrome, Gitelman's syndrome is usually accompanied with oligohydramnios during pregnancy (98).

Pharmacologic inhibition of NCCT function by thiazides is known to stimulate calcium reabsorption by the DCT and, presumably, the mutations that inactivate NCCT in Gitelman's syndrome cause hypocalciuria in the same manner. Thiazides block entry of NaCl across the luminal membrane of the cells of the DCT. The resulting decrease in Cl$^-$ entry and fall in the intracellular Cl$^-$ concentration hyperpolarizes the basolateral membrane. Ca^{2+} entry via the apical voltage-activated Ca^{2+} channels appears to be stimulated. This channel has been termed the epithelial calcium channel (ECaC) (99–102) and CaT2 (103, 104) and has been defined by expression cloning in the oocyte system. The limitation to Na$^+$ entry also lowers the cell Na$^+$ concentration and facilitates exchange of Na$^+$ for Ca^{2+} across the basolateral membrane (Figure 5). Thus thiazides cause changes in both luminal Ca^{2+} entry and basolateral Ca^{2+} exit that increase reabsorption of Ca^{2+} in the DCT, with resultant hypocalciuria. The cause of the persisting renal Mg^{2+} wasting has not been completely defined and is discussed below. Mg^{2+} depletion is an important part of the syndrome and renal Mg^{2+} wasting is clearly the cause (105). Mg^{2+} wasting accounts for many of the symptoms (97) and for relative PTH resistance with low serum Ca^{2+} levels. In contrast, the plasma 1,25-(OH)2-D$_3$ levels and bone density appear to be normal (106).

In view of the importance of Mg^{2+} wasting and depletion in Gitelman's syndrome, this provides the focal point for therapeutic approaches to the patients. Mg^{2+} repletion will assist in K$^+$ repletion and reduce the risk of tetanic crises (107). Elevated aldosterone levels are common and seem to worsen the renal Mg^{2+} wasting so that spironolactone and/or amiloride can be useful agents in treating the Mg^{2+} depletion (108).

RENAL K$^+$ RETENTION SYNDROMES

Pseudohypoaldosteronism, Type I

Pseudohypoaldosteronism Type I is associated with renal salt wasting; elevated NaCl in sweat, stool, and saliva; hyperkalemia; and elevated plasma renin activity and aldosterone concentrations (109). It is an autosomal-recessive disorder that involves multiple organ systems and is especially marked in the neonatal period with vomiting, hyponatremia, failure to thrive, and occasionally the respiratory distress syndrome. These children do not respond to exogenous mineralocorticoids, but with aggressive salt replacement and control of hyperkalemia, they can survive and the severity of the disorder appears to lessen as they mature (109, 110).

The importance of respiratory tract infections has been emphasized (111), and prematurity can be associated with the respiratory distress syndrome (112). Similar to cystic fibrosis, there may be an increased incidence of bronchopneumonia due to *Pseudomonas* (113), but progressive pulmonary functional decline is not observed (114). Excessive airway liquid with chest congestion, cough, and wheezing can be seen in children with this syndrome, and in adulthood there appears to be increased mucociliary clearance (115).

The three ENaC subunits are obvious candidate genes for this disorder (PHA1), and mutations have been described in each subunit that cause loss-of-function of ENaC (116, 117). A relatively large number of mutations and splice variants have been described in all three subunits and are reviewed in Reference (16).

A knockout mouse model of pseudohypoaldosteronism, Type I has been generated by deletion of the α subunit; early neonatal death from respiratory distress occurs, presumably owing to the failure of alveolar fluid clearance at the time of birth (118). If these mice are rescued from the respiratory failure by engineered expression of the α subunit of the ENaC in lung, they survive and have salt wasting and hyperkalemia (119). A β subunit knockout model has also been generated that manifests hyperkalemia with consequent neonatal death at approximately 38 h (120); the hyperkalemia is due to renal K^+ retention because urinary K^+ excretion is low even during the postnatal diuretic phase. [Of note, such hyperkalemia has been commonly observed in low- birth-weight human neonates, which resolves with postnatal Na^+ diuresis (121) and with prenatal glucocorticoid administration (122).] If these β knockout mice survive to adulthood, they are asymptomatic unless put on a low-salt diet when they become hyperkalemic, hypotensive, lose weight, and develop metabolic acidosis (123).

There was initially some confusion in the descriptions of pseudohypoaldosteronism because of uncertainty about the inheritance and reports that there were defects in the mineralocorticoid receptor in some kindreds. In 1991, a distinction was made by Hanukoglu between a severe, multisystem syndrome, which appeared to be autosomal-recessive, and a dominant form that was primarily associated with hyperkalemia (124). Other reports had described such cases of childhood hyperkalemia without the severe salt wasting that is characteristic of the recessive form of the syndrome (125, 126). In retrospect, it appears that some degree of confusion initially resulted from studies of Type I mineralocorticoid receptor abundance and binding characteristics in peripheral blood monocytes (127, 128), and the apparent lack of mutation in the mineralocorticoid receptor (129, 130).

This issue has been clarified with the description of ENaC mutations in the recessive form, as described above, and of a number of mutations in the mineralocorticoid receptor in patients with the autosomal-dominant form of pseudohypoaldosteronism, Type I (131–133). These patients present with salt wasting and hyperkalemia but do not have pulmonary or other organ system involvement. This result was anticipated by the finding that carbenoxolone, which inhibits 11-beta hydroxysteroid dehydrogenase Type II (the enzyme that converts cortisol to cortisone), can partially correct the apparent mineralocorticoid resistance in these patients (111). By slowing the conversion of cortisol to cortisone, carbenoxolone raises the intracellular cortisol concentration sufficiently to maintain high activation of the wild-type receptor and thus overcome the functional defect in the mutant receptor. This is consistent with haplo-insufficiency of the wild-type mineralocorticoid receptor in this autosomal-dominant form of pseudohypoaldosteronism, whereas the multiorgan defects in the recessive form are explained by loss-of-function mutations in the subunits of the epithelial sodium channel.

Now that the distinction between the dominant and recessive forms of pseudohypoaldosteronism can be made on a genetic basis, there are still some dominant forms that cannot be readily explained by mutations in the mineralocorticoid receptor (131), which suggests some unidentified sources of genetic heterogenity remain. There are also other less-appreciated clinical features that may provide clues to such heterogenity. Hypercalciuria has been described in a few instances, along with nephrocalcinosis and increased urinary PGE$_2$ excretion (134), and a response to thiazide diuretics with reduction of the hypercalciuria and improvement of the hyperkalemia (135). Such findings suggest that there could be regional differences in the expression of the defective channel complexes along the distal nephron. Finally, it may be possible to discern relatively mild symptoms or findings in the heterozygote parents and relatives of children affected with the recessive form of the syndrome. Two severely affected sibs with high Na$^+$/K$^+$ excretion ratios in urine, sweat, saliva, and stool were described who had salt wasting and severe hyperkalemia during upper respiratory infections (136). Of note, the father and his sister were not symptomatic but had high sweat Na$^+$ levels and chronically increased aldosterone secretion, as measured by increased urinary tetrahydro-aldosterone levels, perhaps manifesting the previously undetected heterozygotic carrier state of the disorder (136).

Familial Hyperkalemia with Hypertension

This is an autosomal-dominant disorder(s) with persistent hyperkalemia with normal glomerular filtration rate, reduced renal K$^+$ secretion, hyperchloremic metabolic acidosis and, especially in adults, low-renin hypertension.

The first two descriptions of this syndrome mentioned hypertension as a prominent feature (137, 138), but subsequent reports of pediatric cases laid greater emphasis on the hyperkalemia with normal glomerular filtration rate (139, 140). Hyperchloremic metabolic acidosis is also a feature of the syndrome, but this can be relatively mild and is clearly affected in its severity by the hyperkalemia (141). Farfel et al. (142) and Nahum et al. (143) have suggested that a proximal acidification defect could be superimposed on the basic distal renal tubular acidosis; hyperkalemia can suppress renal ammoniagenesis and proximal bicarbonate reabsorption and appears to play a major role in the acidification defect that causes the hyperchloremic metabolic acidosis in this syndrome.

Subsequent descriptions detailed the development of low-renin hypertension in this syndrome with low or normal levels of plasma aldosterone or aldosterone secretion (141, 142, 144–151). Nahum et al. (143) reported that the serum aldosterone was high and plasma renin completely suppressed during severe hyperkalemia, but the serum aldosterone level was normalized when the hyperkalemia was corrected. Throckmorton & Bia (152) reported a 41-year-old male with hyperkalemia, mild hypertension, normal aldosterone levels, and no metabolic acidosis. Careful studies of the largest pedigree yet reported demonstrate that hypertension appears to become more severe as the subjects age, and initially the blood pressure may, in fact, be normal (153, 154).

The low-renin state, as an index of volume expansion, is a manifestation of altered chloride reabsorption in the distal nephron (150, 151). There is an important relation between Cl^- delivery to the ASDN and K^+ secretion. The normal relationship between NaCl delivery and K^+ secretion depicted in Figure 1 is flattened and moved to the right. With this apparent increase in Cl^- avidity, net NaCl reabsorption is favored and K^+ secretion (and to some extent, H^+ secretion) is reduced. If other anions are substituted, as during Na_2SO_4 or $NaHCO_3$ infusions, K^+ secretion will be enhanced and the hyperkalemia will improve (150, 151). A similar effect is seen with thiazide diuretics in this syndrome (139, 140, 146, 148–151, 155–157); hyperkalemia is improved, presumably by inhibition of NaCl reabsorption in the DCT. Gordon et al. reported that dietary Na^+ restriction can also completely reverse the suppressed renin and aldosterone and hyperkalemia (144). Schambelan et al. proposed that the underlying pathophysiologic mechanism, an inward Cl^- absorptive rate in the ASDN, might shunt the electrogenic current generated by Na^+ transport and thereby reduce K^+ secretion (150). Although this shunt pathway was explicitly viewed as a paracellular route for Cl^- permeation, and recent studies of the determinants of paracellular ion permeability in the TAL support this interpretation (158, 159), it is also worth noting that other recent studies have demonstrated apical expression of a CFTR-related protein (160) that could provide a transcellular route for transepithelial Cl^- transport in the DCT, CT, and CCT. This transporter may provide another candidate gene for linkage efforts and studies of its regulation.

The beneficial effects of thiazides could be explained by localizing the disease process to the CT, which is the transition between the DCT and the CCT, and where both NCCT and ENaC are expressed (Figure 6). Gordon et al. (147) initially suggested that distal NaCl absorption was stimulated proximal to the distal K^+ secretory sites so that the delivery of NaCl and thus K^+ secretion in the more distal segments of the nephron was decreased. With reference to Figure 1, the Na^+ delivery is moved to the left and K^+ secretion is reduced.

A third explanation for the syndrome focuses on a primary derangement of K^+ secretion in the ASDN. This hypothesis accounts for the initial hyperkalemic presentation in childhood (139, 140, 153, 154), with subsequent development of systemic hypertension owing to enhanced renal NaCl reabsorption and volume expansion. The longitudinal studies (153, 154) support this view. Nahum et al. (143) have reported that dDAVP appears to stimulate renal K^+ secretion and can correct the hyperkalemia, thus supporting the view of an impairment of K^+ secretion. Gordon et al. (147) reported that stopping therapy in a patient treated with thiazide for 23 years was promptly followed by recurrence of all of the biochemical derangements, but the blood pressure remained normal.

LUMEN   **BLOOD**

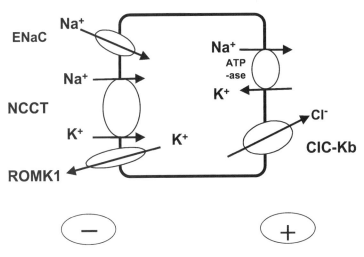

Figure 6   Connecting tubule. NCCT, NaCl cotransporter; ENaC, epithelial Na$^+$ channel; ROMK1, renal outer medulla K$^+$ channel, 1; ClC-Kb, basolateral chloride channel.

It may not be coincidental that three distinct genetic loci have been defined by linkage studies (153, 161, 162). It is possible that each of these distinct sites on chromosome 1q31–42 (161), chromosome 12p13.3 (153), and chromosome 17p11–q21 (161, 162), codes for a specific transport process corresponding to the three pathophysiologic processes, but the nature of these genes is simply not defined at this time. Although gain-of-function mutations of the NCCT in the CT could explain the entire syndrome, all three linkage studies have definitively ruled out this gene, which is located on chromosome 16q13 (82). With respect to the chromosome 17 locus, Soubrier (163) has suggested that this may be a locus of potential importance for hypertension in the general population, consistent with a later report of genome scanning efforts with the Framingham cohort (164).

   In addition to these three suggested pathophysiologies, there are additional phenotypic variations in this syndrome which will have to be fit into any unifying explanation. Although the deranged K$^+$ secretion probably reflects a functional limitation on the apical ROMK1 channel in the CT, this channel still appears to be responsive to hyperpolarization of the transmembrane electrical potential when SO$_4^=$ or HCO$_3^-$ are substituted by Cl$^-$ (150, 151), and the channel can be stimulated by dDAVP (143). Increasing distal Na^+ delivery with furosemide after ROMK1 stimulation with dDAVP (165) improves hyperkalemia and appears to restore the normal relationship depicted in Figure 1. High-dose exogenous steroids can improve the hyperkalemia and metabolic acidosis (166, 167). It is possible that additional genetic heterogeneity may emerge as the currently defined loci are identified and applied to the known pedigrees.[1]

SUGGESTED NOMENCLATURE FOR PSEUDOALDOSTERONISM AND PSEUDOHYPOALDOSTERONISM, TYPES I AND II

Now that the human phenotypes associated with mutational derangements of ENaC and the mineralocorticoid receptor have been described and can be directly attributed to specific mutations of the relevant genes, it may be useful to systematize the nomenclature of these disorders. In the genetically defined pseudoaldosteronism disorders, Type I could be assigned to the gain-of-functions mutations of ENaC (168), and Type II assigned to the S810L gain-of-function mutation of the mineralocorticoid receptor (27). In parallel, the loss-of-function ENaC mutations would continue to be described as pseudohypoaldosteronism Type I, but the term pseudohypoaldosteronism Type II would be used to describe the syndromes associated with loss-of-function mutations of the mineralocorticoid receptor (133). This nomenclature is presented in Table 1. Previously, pseudohypoaldosteronism Type II and a variety of eponymic descriptors have been used to describe familial hyperkalemia with hypertension, but as these syndromes have become better defined, neither pseudohypoaldosteronism nor an eponymic designation appear to be appropriate descriptors. Familial hyperkalemia with hypertension describes the fundamental disorders in this syndrome and is the most appropriate nomenclature (149, 154).

MAGNESIUM DISORDERS

This section describes the genetic disorders for which the genetic basis has been established, or at least strongly linked, with a specific chromosomal location and a pathophysiologic basis that directly relates to the current understanding of renal Mg^{2+} handling. Great progress has been made, but a number of variant syndromes need to be better defined (169) now that at least three of the familial

---

[1]Since the submission of this review, mutations have been identified in members of a family of novel serine/threonine kinases (Wilson et al. 2001. *Science* 293:1107–12). Human hypertension caused by mutations in WNK kinases, which are limited to the familial hyperkalemia with hypertension loci on chromosomes 1 and 2, confirm the genetic heterogeneity of this syndrome.

TABLE 2   Familial disorders of magnesium homeostasis

Descriptor	Inheritance	Mutations	Phenotypes	OMIM
Hypomagnesemia, hypercalciuria and nephrocalcinosis	AR	Paracellin I Claudin 16 (3q27)	⇓ Serum Mg^{2+}, ⇑ Urine Mg^{2+} ⇑ Urine Ca^{2+}, ⇑ PTH, stones, nephrocalcinosis, polyuria, ocular problems, ESRD ⇓	603959 248250
Hypomagnesemia with hypocalciuria	AD	γNa$^+$-K$^+$-ATPase FXYD2 (11q23)	⇓ Serum Mg^{2+}, ⇑ Urine Mg^{2+} ⇓ Urine Ca^{2+}	154020 601814
Hypomagnesemia with secondary hypocalcemia	AR	? (9q12-22.2)	⇓ Serum Mg^{2+}, ⇓ Urine Mg^{2+} ⇓ Serum Ca^{2+}, ⇑ PTH	602014

Abbreviations: AD, autosomal-dominant; AR, autosomal-recessive; OMIM, online Mendelian inheritance in man.

hypomagnesemic disorders have been well described (Table 2). Disorders of the divalent cation (Ca^{2+}/Mg^{2+}) sensing receptor are not addressed (see 170, 171 for recent reviews).

Familial Hypomagnesemia with Hypercalciuria and Nephrocalcinosis

This is an autosomal-recessive disorder that has been mapped by positional cloning to chromosome 3q27. It consists of severe renal Mg^{2+} wasting with hypercalciuria and early onset nephrocalcinosis and nephrolithiasis (172–175). Progressive renal insufficiency during childhood has been observed (174, 175); resolution of the syndrome following successful renal transplantation has been reported (174). Other features of this disorder included elevated PTH and resistance to PTH due to hypomagnesemia, polyuria, ocular difficulties, sensorineural deafness, and urinary tract infections. Serum Ca^{2+}, K$^+$, phosphate, and pH are normal (174, 176). The heterozygotes appear to be normal, although a high frequency of hypercalciuria has been noted (174).

The cause of this renal Mg^{2+} wasting syndrome was shown to be putative loss-of-function mutations in Paracellin-I (159). This membrane protein is a member of the claudin family of proteins that are integral constitutes of tight junctions between epithelial cells (158). Claudin 16 (Paracellin I) was localized to the tight junctions of the TAL with confocal microscopy and to the TAL and DCT with in situ hybridization (159). The N terminus is intracellular, and the first extracellular membrane loop is highly negatively charged and is presumed to play an important role in determining the divalent cation permeability of the paracellular pathway of the TAL. As previously described, this pathway is the predominant route for divalent cation reabsorption, driven by the lumen-positive potential difference in the TAL (Figure 3). Mutations in this integral member of the tight junction reduce Mg^{2+} permeability and are the proximate cause of the renal Mg^{2+} wasting that underlies this disorder.

Familial Hypomagnesemia with Hypocalciuria

This is an autosomal-dominant form of renal Mg^{2+} wasting that also exhibits hypocalciuria. Mg^{2+} infusion studies demonstrate a reduction in the renal Mg^{2+} absorptive capacity (177). The two original families were studied with linkage analysis; the defect was mapped to chromosome 11q23 and subsequently shown to be attributable to a point mutation in the γ subunit of the Na^+-K^+-ATPase complex (178). A transmembrane lysine was converted to arginine by the single-base pair change. Localization of the mutant subunit, which appears to reduce the overall expression of Na^+-K^+-ATPase in the plasma membrane, needs to be carried out to determine if the renal Mg^{2+} wasting can be localized to a defined nephron segment. An autosomal-recessive form has also been reported (179) but has not yet been assigned to a chromosomal region or putative gene. The associated hypocalciuria indicates that the DCT may be a region of interest in the dominant syndrome.

Hypomagnesemia with Secondary Hypocalemia

This autosomal-recessive disorder presents shortly after birth with severe hypomagnesemia and hypocalcemia (180). Tetanic convulsions can occur, and severe neurologic impairment can result if aggressive replacement therapy is not instituted (181, 182). The primary disorder appears to be caused by a defect in gastrointestinal Mg^{2+} absorption, so parenteral replacement therapy may be necessary. Although the putative intestinal Mg^{2+} has not been identified, the syndrome has been linked to chromosome 9q12–22.2 (183). Whereas the primary defect is clearly in the intestinal Mg^{2+} absorptive pathways, Cole & Quamme have speculated that there may also be a renal leak of Mg^{2+} in this syndrome owing to dysfunction of an apical Mg^{2+} channel in the DCT (169). Expression cloning of the intestinal and renal apical membrane Mg^{2+} channels would be a worthwhile effort.

Magnesium Wasting in Bartter's and Gitelman's Syndromes

There are similarities between the effects of diuretics and inactivating mutations in the transport proteins with regard to the excretion of both NaCl and Ca^{2+}, but the differences in Mg^{2+} excretion are not consistent with the diuretic paradigm. Loop diuretics increase urinary Mg^{2+} excretion, whereas thiazides have little effect on it. Most patients with Gitelman's syndrome have increased urinary Mg^{2+} excretion and marked hypomagnesemia, but hypomagnesemia is uncommon and, when present, mild in patients with Bartter's syndrome. The physiologic basis underlying these differences in Mg^{2+} excretion is not known (169).

In mammals about 60% of filtered Mg^{2+} is normally reabsorbed in the tTAL, and 5 to 10% is reabsorbed in the DCT; there is very little reabsorption of Mg^{2+} in the collecting duct (184, 185). The mechanisms for Mg^{2+} reabsorption in the DCT are similar to those for Ca^{2+} and include a luminal Mg^{2+} channel and a basolateral Na^+/Mg^{2+} exchanger. (185) Volume depletion, metabolic alkalosis, and vasopressin stimulate Mg^{2+} transport in the DCT, and aldosterone can potentiate this effect of vasopressin (186). These hormonal actions appear to be counterbalanced by the effects of K$^+$ depletion, which inhibits Mg^{2+} reabsorption in the DCT (187). The effects of K$^+$ depletion are profound and will even decrease Mg^{2+} absorption that has been stimulated by (1,25)-OH$_2$-vitamin D3 (187, 188).

Based on these physiologic observations, it appears that net Mg^{2+} excretion is determined by the balance of hormonal effects and intracellular K$^+$ stores in the DCT. Thus the marked salt wasting and aldosterone stimulation in patients with antenatal and classic Bartter's syndromes may lead to stimulation of Mg^{2+} reabsorption in the DCT, substantially mitigating the Mg^{2+} wasting caused by the transport defect in the TAL. Because loop diuretics are short-acting, there may be relatively less stimulation of Mg^{2+} reabsorption in the DCT in those patients treated with these drugs compared with patients who have loss-of-function mutations in the relevant transport proteins (1, 169). In contrast to the Bartter syndromes, the volume-depletion, metabolic alkalosis, and stimulation of aldosterone secretion are all less severe in Gitelman's syndrome. The effect of hypokalemia may predominate in this disorder and thus may explain the profound Mg^{2+} wasting that is a cardinal feature of Gitelman syndrome.

Very recently, the apical membrane Ca^{2+} channel (ECaC, CaT2) has been defined for the DCT and CT (99, 100, 102, 104, 189). A striking and unique feature of this transporter compared with other Ca^{2+} channels is its strong inward rectification and negative voltage dependence; it is activated by hyperpolarization rather than depolarization of the apical membrane (99, 102). Although there is functional evidence for a similar Mg^{2+} entry pathway (187), this channel has not been cloned and expressed so its biophysical properties have yet to be defined. If it is hypothesized that this putative Mg^{2+} channel (EMagC?) shows voltage dependence that is less steep than ECaC, then the preferential Mg^{2+} wasting in Gitelman's could be explained. In the setting of chronic K$^+$ depletion, the cell membrane potentials in the DCT and CT are likely to be depolarized. If the voltage-gating of EMagC is even steeper than ECaC, then there would be a much greater effect on Mg^{2+} entry than Ca^{2+} entry, and selective renal Mg^{2+} wasting could occur. Although this explanation is entirely speculative at this point, there is no doubt about the evident need to clone, express, and characterize the relevant Mg^{2+} transport pathways.

CONCLUSIONS

Each of the syndromes reviewed herein demonstrates the power of molecular and genetic techniques in defining the underlying pathophysiology of human diseases. The candidate gene approach was directly applied in the example of Liddle's syndrome and pseudohypoaldosteronism Type I. Familial hypomagnesemia with hypercalciuria and nephrocalcinosis provides a contrasting example in which genetic linkage studies identified a previously unknown candidate gene and thereby illuminated the physiologic role of newly described protein. Bartter's and Gitelman's syndromes demonstrate that phenotypic variations may be attributed to different genetic loci, all of which express proteins that participate in an integrated physiologic process (i.e., NaCl reabsorption in the TAL and DCT). Further advances may include identification of additional genes to explain genetic heterogeneity, for example in Bartter's syndrome, as well as the identification of modifying genes that modulate phenotypic heterogeneity. Polymorphisms in these transporters that cause lesser functional consequences may explain more common electrolyte and transport disorders such as idiopathic hypercalciuria, diuretic-induced renal potassium wasting, and even some forms of low-renin hypertension.

ACKNOWLEDGMENTS

The author is grateful to Drs. Jean-Michel Achard and Xavier Jeunemaître for helpful comments during the preparation of this manuscript.

Visit the Annual Reviews home page at www.AnnualReviews.org

LITERATURE CITED

1. Scheinman SJ, Guay-Woodford LM, Thakker RV, Warnock DG. 1999. Genetic disorders of renal electrolyte transport. *N. Engl. J. Med.* 340:177–87
2. Giebisch G, Wang W. 1996. Potassium transport: from clearance to channels and pumps. *Kidney Int.* 49:1624–31
3. Giebisch G. 1998. Renal potassium transport: mechanisms and regulation. *Am. J. Physiol. Renal Physiol.* 274:F817–F33
4. Palmer LG. 1999. Potassium secretion and the regulation of distal nephron K channels. *Am. J. Physiol. Renal Physiol.* 277:F821–F25
5. Rossier BC. 1997. Cum grano salis—the epithelial sodium channel and the control of blood pressure. *J. Am. Soc. Nephrol.* 8:980–92
6. Wald H, Garty H, Palmer LG, Popovtzer MM. 1998. Differential regulation of ROMK expression in kidney cortex and medulla by aldosterone and potassium. *Am. J. Physiol. Renal Physiol.* 275:F239–F45
7. Palmer LG, Frindt G. 2000. Aldosterone and potassium secretion by the cortical collecting duct. *Kidney Int.* 57:1324–28

8. Leung YM, Zeng WZ, Liou HH, Solaro CR, Huang CL. 2000. Phosphatidylinositol 4,5-bisphosphate and intracellular pH regulate the ROMK1 potassium channel via separate but interrelated mechanisms. *J. Biol. Chem.* 275:10182–89
9. Liou HH, Zhou SS, Huang CL. 1999. Regulation of ROMK1 channel by protein kinase A via a phosphatidylinositol 4,5-bisphosphate-dependent mechanism. *Proc. Natl. Acad. Sci. USA* 96:5820–25
10. Lifton RP, Gharavi AG, Geller DS. 2001. Molecular mechanisms of human hypertension. *Cell* 104:545–56
11. Meneton P, Oh YS, Warnock DG. 2001. Genetic renal tubular disorders of renal ion channels and transporters. *Semin. Nephrol.* 21:81–93
12. Abriel H, Loffing J, Rebhun JF, Pratt JH, Schild L, et al. 1999. Defective regulation of the epithelial Na-channel by Nedd4 in Liddle's syndrome. *J. Clin. Invest.* 103:667–73
13. Meneton P, Warnock DG. 2001. Involvement of renal apical Na transport systems in the control of blood pressure. *Am. J. Kidney Dis.* 37:S39–S47

14. Uehara Y, Sasaguri M, Kinoshita A, Tsuji E, Kiyose H, et al. 1998. Genetic analysis of the epithelial sodium channel in Liddle's syndrome. *J. Hypertens.* 16: 1131–35

15. Yamashita Y, Koga M, Takeda Y, Enomoto N, Uchida S, et al. 2001. Two sporadic cases of Liddle's syndrome caused by de novo ENaC mutations. *Am. J. Kidney Dis.* 37:499–504

16. Oh Y, Warnock DG. 2000. Disorders of the epithelial Na$^+$ channel in Liddle's syndrome and autosomal recessive pseudohypoaldosteronism, type 1. *Exp. Nephrol.* 8:320–25

17. Vania A, Tucciarone L, Mazzeo D, Capodaglio PF, Cugini P. 1997. Liddle's syndrome: a 14-year follow-up of the youngest diagnosed case. *Pediatr. Nephrol.* 11:7–11

18. Oh J, Kwon KH. 2000. Liddle's syndrome: a report in a middle-aged woman. *Yonsei Med. J.* 41:276–80

19. Matsushita T, Miyahara Y, Matsushita M, Yakabe K, Yamaguchi K, et al. 1998. Liddle's syndrome in an elderly woman. *Intern. Med.* 37:391–95

20. Warnock DG. 2001. Genetic forms of human hypertension. *Curr. Opin. Nephrol. Hyperten.* 10:493–99

21. Warnock DG. 1999. The epithelial sodium channel in hypertension. *Curr. Hypertens. Reports* 1:158–63

22. Liddle GW, Bledsoe T, Coppage WSJ. 1963. A familial renal disorder simulating primary aldosteronism but with negligible aldosterone secretion. *Trans. Assoc. Am. Physicians* 76:199–213

23. Botero-Velez M, Curtis JJ, Warnock DG. 1994. Brief report: Liddle's syndrome revisited–a disorder of sodium reabsorption in the distal tubule. *N. Engl. J. Med.* 330:178–81

24. Fisher ND, Gleason RE, Moore TJ, Williams GH, Hollenberg NK. 1994. Regulation of aldosterone secretion in hypertensive blacks. *Hypertension* 23: 179–84

25. Findling JW, Raff H, Hansson JH, Lifton RP. 1997. Liddle's syndrome: prospective genetic screening and suppressed aldosterone secretion in an extended kindred. *J. Clin. Endocrinol. Metab.* 82:1071–74

26. Warnock DG. 1998. Liddle syndrome: an autosomal dominant form of human hypertension. *Kidney Int.* 53:18–24

27. Geller DS, Farhi A, Pinkerton N, Fradley M, Moritz M, et al. 2000. Activating mineralocorticoid receptor mutation in hypertension exacerbated by pregnancy. *Science* 289:119–23

28. Lifton RP, Dluhy RG, Powers M, Rich G, Cook S, et al. 1992. A chimaeric 11 β-hydroxylase/aldosterone synthase gene causes glucocorticoid-remediable aldosteronism and human hypertension. *Nature* 355:262–65

29. Lifton RP. 1996. Molecular genetics of human blood pressure variation. *Science* 272:676–80

30. Wycoff JA, Seely EW, Hurwitz S, Anderson BF, Lifton RP, Dluhy RG. 2000. Glucocorticoid-remediable aldosteronism and pregnancy. *Hypertension* 35:668–72

31. Stowasser M, Huggard PR, Rossetti TR, Bachmann AW, Gordon RD. 1999. Biochemical evidence of aldosterone overproduction and abnormal regulation in normotensive individuals with familial hyperaldosteronism type I. *J. Clin. Endocrinol. Metab.* 84:4031–36

32. Stowasser M, Bachmann AW, Huggard PR, Rossetti TR, Gordon RD. 2000. Treatment of familial hyperaldosteronism type I: only partial suppression of adrenocorticotropin required to correct hypertension. *J. Clin. Endocrinol. Metab.* 85:3313–18

33. Stowasser M, Bachmann AW, Huggard PR, Rossetti TR, Gordon RD. 2000. Severity of hypertension in familial hyperaldosteronism type I: relationship to gender and degree of biochemical disturbance. *J. Clin. Endocrinol. Metab.* 85:2160–66

34. Stowasser M, Gartside MG, Taylor WL, Tunny TJ, Gordon RD. 1997. In familial hyperaldosteronism type I, hybrid gene-induced aldosterone production dominates that induced by wild-type genes. *J. Clin. Endocrinol. Metab.* 82:3670–76

35. Stowasser M, Gordon RD. 2000. Primary aldosteronism: learning from the study of familial varieties. *J. Hypertens.* 18:1165–76

36. Torpy DJ, Gordon RD, Lin JP, Huggard PR, Taymans SE, et al. 1998. Familial hyperaldosteronism type II: description of a large kindred and exclusion of the aldosterone synthase (CYP11B2) gene. *J. Clin. Endocrinol. Metab.* 83:3214–18

37. Torpy DJ, Stratakis CA, Gordon RD. 1998. Linkage analysis of familial hyperaldosteronism type II–absence of linkage to the gene encoding the angiotensin II receptor type 1. *J. Clin. Endocrinol. Metab.* 83:1046

38. Lafferty AR, Torpy DJ, Stowasser M, Taymans SE, Lin JP, et al. 2000. A novel genetic locus for low renin hypertension: familial hyperaldosteronism type II maps to chromosome 7 (7p22). *J. Med. Genet.* 37:831–35

39. Ingram MC, Wallace AM, Collier A, Fraser R, Connell JMC. 1996. Sodium status, corticosteroid metabolism and blood pressure in normal human subjects and in a patient with abnormal salt appetite. *Clin. Exp. Pharmacol. Physiol.* 23:375–78

40. Warnock DG. 2000. Low renin hypertension in the next millennium. *Semin. Nephrol.* 20:40–46

41. Lovati E, Ferrari P, Dick B, Jostarndt K, Frey BM, et al. 1999. Molecular basis of human salt sensitivity: the role of the 11beta-hydroxysteroid dehydrogenase type 2. *J. Clin. Endocrinol. Metab.* 84:3745–49

42. Ferrari P, Krozowski ZS. 2000. Role of 11 beta-hydroxysteroid dehydrogenase type 2 in blood pressure regulation. *Kidney Int.* 57:1374–81

43. Bartter FC, Pronove P, Gill J, MacCardle RC. 1962. Hyperplasia of the juxtaglomerular complex with hyperaldosteronism and hypokalemic alkalosis: a new syndrome. *Am. J. Med.* 33:811–28

44. Guay-Woodford L. 1998. Bartter syndrome: unraveling the pathophysiologic enigma. *Am. J. Med.* 105:151–61

45. Vollmer M, Jeck N, Lemmink HH, Vargas R, Feldmann D, et al. 2000. Antenatal Bartter syndrome with sensorineural deafness: refinement of the locus on chromosome 1p31. *Nephrol. Dialysis Transplant* 15:970–74

46. Brennan TM, Landau D, Shalev H, Lamb F, Schutte BC, et al. 1998. Linkage of infantile Bartter syndrome with sensorineural deafness to chromosome 1p. *Am. J. Hum. Genet.* 62:355–61

47. Schwartz ID, Alon US. 1996. Bartter syndrome revisited. *J. Nephrol.* 9:81–87

48. International Collaborative Study Group for Barter-Like Syndromes. 1997. Mutations in the gene encoding the inwardly-rectifiying renal potassium channel, ROMK, cause the antenatal variant of Bartter syndrome: evidence for genetic heterogeneity. *Hum. Mol. Genet.* 6:17–26

49. Vollmer M, Koehrer M, Topaloglu R, Strahm B, Omran H, Hildebrandt F. 1998. Two novel mutations of the gene for Kir 1.1 (ROMK) in neonatal Bartter syndrome. *Pediatr. Nephrol.* 12:69–71

50. Simon DB, Bindra RS, Mansfield TA, Nelson-Williams C, Mendonca E, et al. 1997. Mutations in the chloride channel gene, CLCNKB, cause Bartter's syndrome type III. *Nat. Genet.* 17:171–78

51. Simon DB, Karet FE, Rodriguez-Soriano J, Hamdan JH, DiPietro A, et al. 1996. Genetic heterogeneity of Bartter's syndrome revealed by mutations in the K$^+$ channel, ROMK. *Nat. Genet.* 14:152–56

52. Simon DB, Karet FE, Hamdan JM, DiPietro A, Sanjad SA, Lifton RP. 1996. Bartter's syndrome, hypokalaemic alkalosis with hypercalciuria, is caused by

mutations in the Na-K-2Cl cotransporter NKCC2. *Nat. Genet.* 13:183–88

53. Vargas R, Dechaux M, Hue H, Niaudet P, Antignac C. 1998. Association of pseudo-Bartter syndrome and autosomal dominant hypocalcemia due to a mutation in the calcium-sensing receptor gene. *J. Am. Soc. Nephrol.* 9:395A (Abstr.)

54. Gillen CM, Brill S, Payne JA, Forbush B 3rd. 1996. Molecular cloning and functional expression of the K-Cl cotransporter from rabbit, rat, and human. A new member of the cation-chloride cotransporter family. *J. Biol. Chem.* 271:16237–44

55. Gillen CM, Forbush B 3rd. 1999. Functional interaction of the K-Cl cotransporter (KCC1) with the Na-K-Cl cotransporter in HEK-293 cells. *Am. J. Physiol. Cell Physiol.* 276:C328–C36

56. Konrad M, Vollmer M, Lemmink HH, van den Heuvel LP, Jeck N, et al. 2000. Mutations in the chloride channel gene CLCNKB as a cause of classic Bartter syndrome. *J. Am. Soc. Nephrol.* 11:1449–59

57. Vargas-Poussou R, Feldmann D, Vollmer M, Konrad M, Kelly L, et al. 1998. Novel molecular variants of the Na-K-2Cl cotransporter gene are responsible for antenatal Bartter syndrome. *Am. J. Hum. Genet.* 62:1332–40

58. Bettinelli A, Ciarmatori S, Cesareo L, Tedeschi S, Ruffa G, et al. 2000. Phenotypic variability in Bartter syndrome type I. *Pediatr. Nephrol.* 14:940–45

59. Takahashi N, Chernavvsky DR, Gomez RA, Igarashi P, Gitelman HJ, Smithies O. 2000. Uncompensated polyuria in a mouse model of Bartter's syndrome. *Proc. Natl. Acad. Sci. USA* 97:5434–39

60. Feldmann D, Alessandri JL, Deschenes G. 1998. Large deletion of the 5′ end of the ROMK1 gene causes antenatal Bartter syndrome. *J. Am. Soc. Nephrol.* 9:2357–59

61. Jeck N, Derst C, Wischmeyer E, Ott H, Weber S, et al. 2001. Functional het-erogeneity of ROMK mutations linked to hyperprostaglandin E syndrome. *Kidney Int.* 59:1803–11

62. Ortega B, Millar ID, Beesley AH, Robson L, White SJ. 2000. Stable, polarised, functional expression of Kir1.1b channel protein in Madin-Darby canine kidney cell line. *J. Physiol.* 528(Pt 1):5–13

63. Schulte U, Hahn H, Konrad M, Jeck N, Derst C, et al. 1999. pH gating of ROMK (K(ir)1.1) channels: control by an Arg-Lys-Arg triad disrupted in antenatal Bartter syndrome. *Proc. Natl. Acad. Sci. USA* 96:15298–303

64. Schwalbe RA, Bianchi L, Accili EA, Brown AM. 1998. Functional consequences of ROMK mutants linked to antenatal Bartter's syndrome and implications for treatment. *Hum. Mol. Genet.* 7:975–80

65. Flagg TP, Tate M, Merot J, Welling PA. 1999. A mutation linked with Bartter's syndrome locks Kir 1.1a (ROMK1) channels in a closed state. *J. Gen. Physiol.* 114:685–700

66. Jeck N, Konrad M, Peters M, Weber S, Bonzel KE, Seyberth HW. 2000. Mutations in the chloride channel gene, CLCNKB, leading to a mixed Bartter-Gitelman phenotype. *Pediatr. Res.* 48:754–58

67. Thakker RV. 2000. Molecular pathology of renal chloride channels in Dent's disease and Bartter's syndrome. *Exp. Nephrol.* 8:351–60

68. Greger R. 1985. Ion transport mechanisms in thick ascending limb of Henle's loop of mammalian nephron. *Physiol. Rev.* 65:760–97

69. Ellison DH. 2000. Divalent cation transport by the distal nephron: insights from Bartter's and Gitelman's syndromes. *Am. J. Physiol. Renal Physiol.* 279:F616–F25

70. Calo L, Davis PA, Milani M, Cantaro S, Antonello A, et al. 1999. Increased endothelial nitric oxide synthase mRNA level in Bartter's and Gitelman's syndrome. Relationship to vascular reactivity. *Clin. Nephrol.* 51:12–17

71. Balat A, Cekmen M, Yurekli M, Kutlu O, Islek I, et al. 2000. Adrenomedullin and nitrite levels in children with Bartter syndrome. *Pediatr. Nephrol.* 15:266–70

72. Schachter AD, Arbus GS, Alexander RJ, Balfe JW. 1998. Non-steroidal anti-inflammatory drug-associated nephrotoxicity in Bartter syndrome. *Pediatr. Nephrol.* 12:775–77

73. Kim JY, Kim GA, Song JH, Lee SW, Han JY, et al. 2000. A case of living-related kidney transplantation in Bartter's syndrome. *Yonsei Med. J.* 41:662–65

74. Gitelman HJ, Graham JB, Welt LG. 1966. A new familial disorder characterized by hypokalemia and hypomagnesemia. *Trans. Assoc. Am. Physicians* 79:221–35

75. Sutton RA, Mavichak V, Halabe A, Wilkins GE. 1992. Bartter's syndrome: evidence suggesting a distal tubular defect in a hypocalciuric variant of the syndrome. *Miner. Electrolyte Metab.* 18:43–51

76. Tsukamoto T, Kobayashi T, Kawamoto K, Fukase M, Chihara K. 1995. Possible discrimination of Gitelman's syndrome from Bartter's syndrome by renal clearance study: report of two cases. *Am. J. Kidney Dis.* 25:637–41

77. Colussi G, Rombola G, Brunati C, De Ferrari ME. 1997. Abnormal reabsorption of Na^+/Cl^- by the thiazide-inhibitable transporter of the distal convoluted tubule in Gitelman's syndrome. *Am J. Nephrol.* 17:103–11

78. Abuladze N, Yanagawa N, Lee I, Jo OD, Newman D, et al. 1998. Peripheral blood mononuclear cells express mutated NCCT mRNA in Gitelman's syndrome: evidence for abnormal thiazide-sensitive NaCl cotransport. *J. Am. Soc. Nephrol.* 9:819–26

79. Schultheis PJ, Lorenz JN, Meneton P, Nieman ML, Riddle TM, et al. 1998. Phenotype resembling Gitelman's syndrome in mice lacking the apical Na^+-Cl^- cotransporter of the distal convoluted tubule. *J. Biol. Chem.* 273:29150–55

80. Simon DB, Nelson-Williams C, Bia MJ, Ellison D, Karet FE, et al. 1996. Gitelman's variant of Bartter's syndrome, inherited hypokalaemic alkalosis, is caused by mutations in the thiazide-sensitive Na-Cl cotransporter. *Nat. Genet.* 12:24–30

81. Lemmink HH, van den Heuvel LP, van Dijk HA, Merkx GF, Smilde TJ, et al. 1996. Linkage of Gitelman syndrome to the thiazide-sensitive sodium-chloride cotransporter gene with identification of mutations in Dutch families. *Pediatr. Nephrol.* 10:403–7

82. Pollak MR, Delaney VB, Graham RM, Hebert SC. 1996. Gitelman's syndrome (Bartter's variant) maps to the thiazide-sensitive cotransporter gene locus on chromosome 16q13 in a large kindred. *J. Am. Soc. Nephrol.* 7:2244–48

83. Karolyi L, Ziegler A, Pollak M, Fischbach M, Grzeschik KH, et al. 1996. Gitelman's syndrome is genetically distinct from other forms of Bartter's syndrome. *Pediatr. Nephrol.* 10:551–54

84. Lemmink HH, Knoers NV, Karolyi L, van Dijk H, Niaudet P, et al. 1998. Novel mutations in the thiazide-sensitive NaCl cotransporter gene in patients with Gitelman syndrome with predominant localization to the C-terminal domain. *Kidney Int.* 54:720–30

85. Kunchaparty S, Palcso M, Berkman J, Velazquez H, Desir GV, et al. 1999. Defective processing and expression of thiazide-sensitive Na-Cl cotransporter as a cause of Gitelman's syndrome. *Am. J. Physiol. Renal Physiol.* 277:F643–F49

86. Mastroianni N, Bettinelli A, Bianchetti M, Colussi G, De Fusco M, et al. 1996. Novel molecular variants of the Na-Cl cotransporter gene are responsible for Gitelman syndrome. *Am. J. Hum. Genet.* 59:1019–26

87. Monkawa T, Kurihara I, Kobayashi K,

Hayashi M, Saruta T. 2000. Novel mutations in thiazide-sensitive Na-Cl cotransporter gene of patients with Gitelman's syndrome. *J. Am. Soc. Nephrol.* 11:65–70

88. Melander O, Orho-Melander M, Bengtsson K, Lindblad U, Rastam L, et al. 2000. Genetic variants of thiazide-sensitive NaCl-cotransporter in Gitelman's syndrome and primary hypertension. *Hypertension* 36:389–94

89. Bettinelli A, Bianchetti MG, Girardin E, Caringella A, Cecconi M, et al. 1992. Use of calcium excretion values to distinguish two forms of primary renal tubular hypokalemic alkalosis: Bartter and Gitelman syndromes. *J. Pediatr.* 120:38–43

90. Bettinelli A, Bianchetti MG, Borella P, Volpini E, Metta MG, et al. 1995. Genetic heterogeneity in tubular hypomagnesemia-hypokalemia with hypocalciuria (Gitelman's syndrome). *Kidney Int.* 47:547–51

91. Ko CW, Koo JH. 1999. Recombinant human growth hormone and Gitelman's syndrome. *Am. J. Kidney Dis.* 33:778–81

92. Bettinelli A, Rusconi R, Ciarmatori S, Righini V, Zammarchi E, et al. 1999. Gitelman disease associated with growth hormone deficiency, disturbances in vasopressin secretion and empty sella: a new hereditary renal tubular-pituitary syndrome? *Pediatr. Res.* 46:232–38

93. Bourcier T, Blain P, Massin P, Grunfeld JP, Gaudric A. 1999. Sclerochoroidal calcification associated with Gitelman syndrome. *Am. J. Ophthalmol.* 128:767–68

94. Hisakawa N, Yasuoka N, Itoh H, Takao T, Jinnouchi C, et al. 1998. A case of Gitelman's syndrome with chondrocalcinosis. *Endocr. J.* 45:261–67

95. Punzi L, Calo L, Schiavon F, Pianon M, Rosada M, Todesco S. 1998. Chondrocalcinosis is a feature of Gitelman's variant of Bartter's syndrome. A new look at the hypomagnesemia associated with calcium pyrophosphate dihydrate crystal deposition disease. *Rev. Rheum. Engl. Ed.* 65:571–74

96. Vezzoli G, Soldati L, Jansen A, Pierro L. 2000. Choroidal calcifications in patients with Gitelman's syndrome. *Am. J. Kidney Dis.* 36:855–58

97. Cruz DN, Shaer AJ, Bia MJ, Lifton RP, Simon DB. 2001. Gitelman's syndrome revisited: an evaluation of symptoms and health-related quality of life. *Kidney Int.* 59:710–17

98. Jones JM, Dorrell S. 1998. Outcome of two pregnancies in a patient with Gitelman's syndrome—a case report. *J. Maternal-Fetal Invest.* 8:147–48

99. Nilius B, Vennekens R, Prenen J, Hoenderop JG, Bindels RJ, Droogmans G. 2000. Whole-cell and single channel monovalent cation currents through the novel rabbit epithelial Ca^{2+} channel ECaC. *J. Physiol.* 527(Pt 2):239–48

100. Muller D, Hoenderop JG, Merkx GF, van Os CH, Bindels RJ. 2000. Gene structure and chromosomal mapping of human epithelial calcium channel. *Biochem. Biophys. Res. Commun.* 275:47–52

101. Muller D, Hoenderop JG, Meij IC, van den Heuvel LP, Knoers NV, et al. 2000. Molecular cloning, tissue distribution, and chromosomal mapping of the human epithelial Ca^{2+} channel (ECaC1). *Genomics* 67:48–53

102. Hoenderop JG, van der Kemp AW, Hartog A, van Os CH, Willems PH, Bindels RJ. 1999. The epithelial calcium channel, ECaC, is activated by hyperpolarization and regulated by cytosolic calcium. *Biochem. Biophys. Res. Commun.* 261:488–92

103. Yue L, Peng JB, Hediger MA, Clapham DE. 2001. CaT1 manifests the pore properties of the calcium-release-activated calcium channel. *Nature* 410:705–9

104. Peng JB, Chen XZ, Berger UV, Vassilev PM, Brown EM, Hediger MA. 2000. A rat kidney-specific calcium transporter in the distal nephron. *J. Biol. Chem.* 275:28186–94

105. Peters N, Bettinelli A, Spicher I, Basilico E, Metta MG, Bianchetti MG. 1995.

Renal tubular function in children and adolescents with Gitelman's syndrome, the hypocalciuric variant of Bartter's syndrome. *Nephrol. Dialysis Transplant* 10:1313–19

106. Bianchetti MG, Bettinelli A, Casez JP, Basilico E, Metta MG, et al. 1995. Evidence for disturbed regulation of calciotropic hormone metabolism in Gitelman syndrome. *J. Clin. Endocrinol. Metab.* 80:224–28

107. Bettinelli A, Basilico E, Metta MG, Borella P, Jaeger P, Bianchetti MG. 1999. Magnesium supplementation in Gitelman syndrome. *Pediatr. Nephrol.* 13:311–14

108. Colussi G, Rombola G, De Ferrari ME. 1994. Distal nephron function in familial hypokalemia-hypomagnesemia (Gitelman's syndrome). *Nephron* 66:122–23

109. Oberfield SE, Levine LS, Carey RM, Bejar R, New MI. 1979. Pseudohypoaldosteronism: multiple target organ unresponsiveness to mineralocorticoid hormones. *J. Clin. Endocrinol. Metab.* 48:228–34

110. Petersen S, Giese J, Kappelgaard AM, Lund HT, Lund JO, et al. 1978. Pseudohypoaldosteronism. Clinical, biochemical and morphological studies in a long-term follow-up. *Acta Paediatr. Scand.* 67:255–61

111. Hanukoglu A, Bistritzer T, Rakover Y, Mandelberg A. 1994. Pseudohypoaldosteronism with increased sweat and saliva electrolyte values and frequent lower respiratory tract infections mimicking cystic fibrosis. *J. Pediatr.* 125:752–55

112. Malagon-Rogers M. 1999. A patient with pseudohypoaldosteronism type 1 and respiratory distress syndrome. *Pediatr. Nephrol.* 13:484–86

113. Marthinsen L, Kornfalt R, Aili M, Andersson D, Westgren U, Schaedel C. 1998. Recurrent Pseudomonas bronchopneumonia and other symptoms as in cystic fibrosis in a child with type I pseudohypoaldosteronism. *Acta Paediatr.* 87:472–74

114. Schaedel C, Marthinsen L, Kristoffersson AC, Kornfalt R, Nilsson KO, et al. 1999. Lung symptoms in pseudohypoaldosteronism type 1 are associated with deficiency of the alpha-subunit of the epithelial sodium channel. *J. Pediatr.* 135:739–45

115. Kerem E, Bistritzer T, Hanukoglu A, Hofmann T, Zhou Z, et al. 1999. Pulmonary epithelial sodium-channel dysfunction and excess airway liquid in pseudohypoaldosteronism. *N. Engl. J. Med.* 341:156–62

116. Chang SS, Grunder S, Hanukoglu A, Rosler A, Mathew PM, et al. 1996. Mutations in subunits of the epithelial sodium channel cause salt wasting with hyperkalaemic acidosis, pseudohypoaldosteronism type 1. *Nat. Genet.* 12:248–53

117. Strautnieks SS, Thompson RJ, Gardiner RM, Chung E. 1996. A novel splice-site mutation in the gamma subunit of the epithelial sodium channel gene in three pseudohypoaldosteronism type 1 families. *Nat. Genet.* 13:248–50

118. Hummler E, Barker P, Gatzy J, Beermann F, Verdumo D, et al. 1996. Early death due to defective neonatal lung liquid clearance in αENaC-deficient mice. *Nat. Genet.* 12:325–28

119. Hummler E, Barker P, Talbot C, Wang Q, Verdumo C, et al. 1997. A mouse model for the renal salt-wasting syndrome pseudohypoaldosteronism. *Proc. Nat. Acad. Sci. USA* 94:11710–15

120. McDonald FJ, Yang B, Hrstka RF, Drummond HA, Tarr DE, et al. 1999. Disruption of the beta subunit of the epithelial Na^+ channel in mice: hyperkalemia and neonatal death associated with a pseudohypoaldosteronism phenotype. *Proc. Natl. Acad. Sci. USA* 96:1727–31

121. Lorenz JM, Kleinman LI, Markarian K. 1997. Potassium metabolism in extremely low birth weight infants in the first week of life. *J. Pediatr.* 131:81–86

122. Omar SA, DeCristofaro JD, Agarwal BI, LaGamma EF. 2000. Effect of prenatal steroids on potassium balance in

extremely low birth weight neonates. *Pediatrics* 106:561–67

123. Pradervand S, Barker PM, Wang Q, Ernst SA, Beermann F, et al. 1999. Salt restriction induces pseudohypoaldosteronism type 1 in mice expressing low levels of the beta-subunit of the amiloridesensitive epithelial sodium channel. *Proc. Natl. Acad. Sci. USA* 96:1732–37

124. Hanukoglu A. 1991. Type I pseudohypoaldosteronism includes two clinically and genetically distinct entities with either renal or multiple target organ defects. *J. Clin. Endocrinol. Metab.* 73:936–44

125. Ballauff A, Wendel U, Kupke I, Kuhnle U. 1994. A partial form of pseudohypoaldosteronism type I without renal sodium wasting. *J. Pediatr. Endocrinol.* 7:57–60

126. Claris Appiani A, Marra G, Tirelli SA, Goj V, Romeo L, et al. 1986. Early childhood hyperkalemia: variety of pseudohypoaldosteronism. *Acta Paediatr. Scand.* 75:970–74

127. Komesaroff PA, Verity K, Fuller PJ. 1994. Pseudohypoaldosteronism: molecular characterization of the mineralocorticoid receptor. *J. Clin. Endocrinol. Metab.* 79:27–31

128. Kuhnle U, Hinkel GK, Akkurt HI, Krozowski Z. 1995. Familial pseudohypoaldosteronism: a review on the heterogeneity of the syndrome. *Steroids* 60:157–60

129. Zennaro MC, Borensztein P, Jeunemaitre X, Armanini D, Corvol P, Soubrier F. 1995. Molecular characterization of the mineralocorticoid receptor in pseudohypoaldosteronism. *Steroids* 60:164–67

130. Zennaro MC, Borensztein P, Jeunemaitre X, Armanini D, Soubrier F. 1994. No alteration in the primary structure of the mineralocorticoid receptor in a family with pseudohypoaldosteronism. *J. Clin. Endocrinol. Metab.* 79:32–38

131. Viemann M, Peter M, Lopez-Siguero JP, Simic-Schleicher G, Sippell WG. 2001. Evidence for genetic heterogeneity of pseudohypoaldosteronism Type 1: identification of a novel mutation in the human mineralocorticoid receptor in one sporadic case and no mutations in two autosomal dominant kindreds. *J. Clin. Endocrinol. Metab.* 86:2056–59

132. Tajima T, Kitagawa H, Yokoya S, Tachibana K, Adachi M, et al. 2000. A novel missense mutation of mineralocorticoid receptor gene in one Japanese family with a renal form of pseudohypoaldosteronism type 1. *J. Clin. Endocrinol. Metab.* 85:4690–94

133. Geller DS, Rodriquez-Soriano J, Boado AV, Schifter S, Bayer M, et al. 1998. Mutations in the mineralocorticoid receptor gene cause autosomal dominant pseudohypoaldosteronism, type I. *Nat. Genet.* 19:279–81

134. Shalev H, Ohali M, Abramson O. 1994. Nephrocalcinosis in pseudohypoaldosteronism and the effect of indomethacin therapy. *J. Pediatr.* 125:246–48

135. Stone RC, Vale P, Rosa FC. 1996. Effect of hydrochlorothiazide in pseudohypoaldosteronism with hypercalciuria and severe hyperkalemia. *Pediatr. Nephrol.* 10:501–3

136. Popow C, Pollak A, Herkner K, Scheibenreiter S, Swoboda W. 1988. Familial pseudohypoaldosteronism. *Acta Paediatr. Scand.* 77:136–41

137. Paver WKA, Pauline GJ. 1964. Hypertension and hyperpotassemia without renal disease in a young male. *Med. J. Australia* 2:305–6

138. Arnold JE, Healy JK. 1969. Hyperkalemia, hypertension and systemic acidosis without renal failure associated with a tubular defect in potassium excretion. *Am. J. Med.* 47:461–72

139. Weinstein SF, Allan DM, Mendoza SA. 1974. Hyperkalemia, acidosis, and short stature associated with a defect in renal potassium excretion. *J. Pediatr.* 85:355–58

140. Spitzer A, Edelmann CM Jr, Goldberg LD, Henneman PH. 1973. Short stature, hyperkalemia and acidosis: a defect in

renal transport of potassium. *Kidney Int.* 3:251–57

141. Brautbar N, Levi J, Rosler A, Leitesdorf E, Djaldeti M, et al. 1978. Familial hyperkalemia, hypertension, and hyporeninemia with normal aldosterone levels. A tubular defect in potassium handling. *Arch. Intern. Med.* 138:607–10

142. Farfel Z, Iaina A, Levi J, Gafni J. 1978. Proximal renal tubular acidosis: association with familial normaldosteronemic hyperpotassemia and hypertension. *Arch. Intern. Med.* 138:1837–40

143. Nahum H, Paillard M, Prigent A, Leviel F, Bichara M, et al. 1986. Pseudohypoaldosteronism type II: proximal renal tubular acidosis and dDAVP-sensitive renal hyperkalemia. *Am J. Nephrol.* 6:253–62

144. Gordon RD, Geddes RA, Pawsey CG, O'Halloran MW. 1970. Hypertension and severe hyperkalaemia associated with suppression of renin and aldosterone and completely reversed by dietary sodium restriction. *Australas Ann. Med.* 19:287–94

145. Gordon RD. 1986. The syndrome of hypertension and hyperkalemia with normal glomerular filtration rate: Gordon's syndrome. *Aust. NZ J. Med.* 16:183–84

146. Gordon RD, Hodsman GP. 1986. The syndrome of hypertension and hyperkalaemia without renal failure: long term correction by thiazide diuretic. *Scott/ Med/ J.* 31:43–44

147. Gordon RD. 1986. The syndrome of hypertension and hyperkalaemia with normal GFR. A unique pathophysiological mechanism for hypertension? *Clin. Exp. Pharmacol. Physiol.* 13:329–33

148. Farfel Z, Iaina A, Rosenthal T, Waks U, Shibolet S, Gafni J. 1978. Familial hyperpotassemia and hypertension accompanied by normal plasma aldosterone levels: possible hereditary cell membrane defect. *Arch. Intern. Med.* 138:1828–32

149. Isenring P, Lebel M, Grose JH. 1992. Endocrine sodium and volume regulation in familial hyperkalemia with hypertension. *Hypertension* 19:371–77

150. Schambelan M, Sebastian A, Rector FC Jr. 1981. Mineralocorticoid-resistant renal hyperkalemia without salt wasting (type II pseudohypoaldosteronism): role of increased renal chloride reabsorption. *Kidney Int.* 19:716–27

151. Take C, Ikeda K, Kurasawa T, Kurokawa K. 1991. Increased chloride reabsorption as an inherited renal tubular defect in familial type II pseudohypoaldosteronism. *N. Engl. J. Med.* 324:472–76

152. Throckmorton DC, Bia MJ. 1991. Pseudohypoaldosteronism: case report and discussion of the syndrome. *Yale J. Biol. Med.* 64:247–54

153. Disse-Nicodème S, Achard JM, Desitter I, Houot AM, Fournier A, et al. 2000. A new locus on chromosome 12p13.3 for pseudohypoaldosteronism type II, an autosomal dominant form of hypertension. *Am. J. Hum. Genet.* 67:302–10

154. Disse-Nicodème S, Achard J-M Fiquet-Kempf B, Potier J, Delahousse M, et al. 2001. Hypertension familiale hyperkaliemique: analyse de la variabilité phenotypique par l'étude de 7 familles. In *Actualités Néphrologiques de l'Hôpital Necker*, ed. JP Grunfield, pp. 60–71. Paris: Medicine-Sciences Flammarion

155. Semmekrot B, Monnens L, Theelen BG, Rascher W, Gabreels F, Willems J. 1987. The syndrome of hypertension and hyperkalaemia with normal glomerular function (Gordon's syndrome). A pathophysiological study. *Pediatr. Nephrol.* 1:473–78

156. Pasman JW, Gabreels FJ, Semmekrot B, Renier WO, Monnens LA. 1989. Hyperkalemic periodic paralysis in Gordon's syndrome: a possible defect in atrial natriuretic peptide function. *Ann. Neurol.* 26:392–95

157. Wayne VS, Stockigt JR, Jennings GL. 1986. Treatment of mineralocorticoid-resistant renal hyperkalemia with hypertension (type II pseudohypoaldosteronism). *Aust. NZ J. Med.* 16:221–23

158. Yu AS. 2000. Paracellular solute transport: more than just a leak? *Curr. Opin. Nephrol. Hypertens.* 9:513–15

159. Simon DB, Lu Y, Choate KA, Velazquez H, Al-Sabban E, et al. 1999. Paracellin-1, a renal tight junction protein required for paracellular Mg^{2+} resorption. *Science* 285:103–6

160. Van Huyen JP, Bens M, Teulon J, Vandewalle A. 2001. Vasopressin-stimulated chloride transport in trans-immortalized mouse cell lines derived from the distal convoluted tubule and cortical and inner medullary collecting ducts. *Nephrol. Dialysis Transplant* 16:238–45

161. Mansfield TA, Simon DB, Farfel Z, Bia M, Tucci JR, et al. 1997. Multilocus linkage of familial hyperkalaemia and hypertension, pseudohypoaldosteronism type II, to chromosomes 1q31–42 and 17p11-q21. *Nat. Genet.* 16:202–5

162. O'Shaughnessy KM, Fu B, Johnson A, Gordon RD. 1998. Linkage of Gordon's syndrome to the long arm of chromosome 17 in a region recently linked to familial essential hypertension. *J. Hum. Hypertens.* 12:675–78

163. Soubrier F. 1998. Gordon's syndrome, renal tubule, chromosome 17 and essential hypertension: a credible link? *J. Hum. Hypertens.* 12:663–64

164. Levy D, DeStefano AL, Larson MG, O'Donnell CJ, Lifton RP, et al. 2000. Evidence for a gene influencing blood pressure on chromosome 17. Genome scan linkage results for longitudinal blood pressure phenotypes in subjects from the Framingham heart study. *Hypertension* 36:477–83

165. Erdogan G, Corapcioglu D, Erdogan MF, Hallioglu J, Uysal AR. 1997. Furosemide and dDAVP for the treatment of pseudohypoaldosteronism type II. *J Endocrinol. Invest.* 20:681–84

166. Shoker A, Morris G, Skomro R, Laxdal V. 1996. Pseudohypoaldosteronism with normal blood pressure. *Clin. Nephrol.* 46:105–11

167. Travis PS, Cushner HM. 1986. Mineralocorticoid-induced kaliuresis in type-II pseudohypoaldosteronism. *Am. J. Med. Sci.* 292:235–40

168. Shimkets RA, Warnock DG, Bositis CM, Nelson-Williams C, Hansson JH, et al. 1994. Liddle's syndrome: heritable human hypertension caused by mutations in the beta subunit of the epithelial sodium channel. *Cell* 79:407–14

169. Cole DE, Quamme GA. 2000. Inherited disorders of renal magnesium handling. *J. Am. Soc. Nephrol.* 11:1937–47

170. Brown EM, MacLeod RJ. 2001. Extracellular calcium sensing and extracellular calcium signaling. *Physiol. Rev.* 81:239–97

171. Brown EM. 2000. Familial hypocalciuric hypercalcemia and other disorders with resistance to extracellular calcium. *Endocrinol. Metab. Clin. North Am.* 29:503–22

172. Milazzo SC, Ahern MJ, Cleland LG, Henderson DR. 1981. Calcium pyrophosphate dihydrate deposition disease and familial hypomagnesemia. *J. Rheumatol.* 8:767–71

173. Manz F, Scharer K, Janka P, Lombeck J. 1978. Renal magnesium wasting, incomplete tubular acidosis, hypercalciuria and nephrocalcinosis in siblings. *Eur. J. Pediatr.* 128:67–79

174. Praga M, Vara J, Gonzalez-Parra E, Andres A, Alamo C, et al. 1995. Familial hypomagnesemia with hypercalciuria and nephrocalcinosis. *Kidney Int.* 47:1419–25

175. Nicholson JC, Jones CL, Powell HR, Walker RG, McCredie DA. 1995. Familial hypomagnesaemia–hypercalciuria leading to end-stage renal failure. *Pediatr. Nephrol.* 9:74–76

176. Benigno V, Canonica CS, Bettinelli A, von Vigier RO, Truttmann AC, Bianchetti MG. 2000. Hypomagnesaemia-hypercalciuria-nephrocalcinosis: a report of nine cases and a review. *Nephrol. Dialysis Transplant* 15:605–10

177. Geven WB, Monnens LA, Willems HL,

Buijs WC, ter Haar BG. 1987. Renal magnesium wasting in two families with autosomal dominant inheritance. *Kidney Int.* 31:1140–44

178. Meij IC, Koenderink JB, van Bokhoven H, Assink KF, Groenestege WT, et al. 2000. Dominant isolated renal magnesium loss is caused by misrouting of the Na⁺, K⁺-ATPase gamma-subunit. *Nat. Genet.* 26:265–66

179. Geven WB, Monnens LA, Willems JL, Buijs W, Hamel CJ. 1987. Isolated autosomal recessive renal magnesium loss in two sisters. *Clin. Genet.* 32:398–402

180. Paunier L, Radde IC, Kooh SW, Conen PE, Fraser D. 1968. Primary hypomagnesemia with secondary hypocalcemia in an infant. *Pediatrics* 41:385–402

181. Shalev H, Phillip M, Galil A, Carmi R, Landau D. 1998. Clinical presentation and outcome in primary familial hypomagnesaemia. *Arch. Dis. Child* 78:127–30

182. Abdulrazzaq YM, Smigura FC, Wettrell G. 1989. Primary infantile hypomagnesaemia; report of two cases and review of literature. *Eur. J. Pediatr.* 148:459–61

183. Walder RY, Shalev H, Brennan TM, Carmi R, Elbedour K, et al. 1997. Familial hypomagnesemia maps to chromosome 9q, not to the X chromosome: genetic linkage mapping and analysis of a balanced translocation breakpoint. *Hum. Mol. Genet.* 6:1491–97

184. Quamme GA, de Rouffignac C. 2000. Epithelial magnesium transport and regulation by the kidney. *Front. Biosci.* 5:D694–711

185. Quamme GA. 1997. Renal magnesium handling: new insights in understanding old problems. *Kidney Int.* 52:1180–95

186. Dai LJ, Ritchie G, Kerstan D, Kang HS, Cole DE, Quamme GA. 2001. Magnesium transport in the renal distal convoluted tubule. *Physiol. Rev.* 81:51–84

187. Dai LJ, Friedman PA, Quamme GA. 1997. Cellular mechanisms of chlorothiazide and cellular potassium depletion on Mg²⁺ uptake in mouse distal convoluted tubule cells. *Kidney Int.* 51:1008–17

188. Ritchie G, Kerstan D, Dai LJ, Kang HS, Canaff L, et al. 2001. 1,25(OH)₂D₃ stimulates Mg²⁺ uptake into MDCT cells: modulation by extracellular Ca²⁺ and Mg²⁺. *Am. J. Physiol. Renal Physiol.* 280:F868–F78

189. Vennekens R, Hoenderop JG, Prenen J, Stuiver M, Willems PH, et al. 2000. Permeation and gating properties of the novel epithelial Ca²⁺ channel. *J. Biol. Chem.* 275:3963–69

Annu. Rev. Physiol. 2002. 64:877–97

EPITHELIAL SODIUM CHANNEL AND THE CONTROL OF SODIUM BALANCE: Interaction Between Genetic and Environmental Factors

Bernard C. Rossier, Sylvain Pradervand, Laurent Schild, and Edith Hummler
Institute of Pharmacology and Toxicology, University of Lausanne, Rue du Bugnon 27, CH-1005 Lausanne, Switzerland; e-mail: Bernard.Rossier@ipharm.unil.ch

Key Words   ENaC, salt intake, hypertension, lung fluid clearance, aldosterone

■ **Abstract**   The epithelial sodium channel (ENaC) expressed in aldosterone-responsive epithelial cells of the kidney and colon plays a critical role in the control of sodium balance, blood volume, and blood pressure. In lung, ENaC has a distinct role in controlling the ionic composition of the air-liquid interface and thus the rate of mucociliary transport. Loss-of-function mutations in ENaC cause a severe salt-wasting syndrome in human pseudohypoaldosteronism type 1 (PHA-1). Gain-of-function mutations in ENaC β and γ subunits cause pseudoaldosteronism (Liddle's syndrome), a severe form of salt-sensitive hypertension. This review discusses genetically defined forms of a salt sensitivity and salt resistance in human monogenic diseases and in animal models mimicking PHA-1 or Liddle's syndrome. The complex interaction between genetic factors (ENaC mutations) and the risk factor (salt intake) can now be studied experimentally. The role of single-nucleotide polymorphisms (SNPs) in determining salt sensitivity or salt resistance in general populations is one of the main challenges of the post-genomic era.

INTRODUCTION

In aldosterone-responsive epithelial cells (kidney, colon), the epithelial sodium channel (ENaC) plays a pivotal role in the control of sodium balance, blood volume, and blood pressure (1). In glucocorticoid-responsive epithelia, as in distal lung airways, ENaC may have a distinct role in controlling fluid reabsorption at the air-liquid interface, thereby determining the rate of mucociliary transport (2–4). In human and in animal models, imbalance of ENaC activity may lead to numerous pathologies, for example, an abnormal increase of ENaC activity may contribute to hypertension or to decreased mucociliary transport, as observed in cystic fibrosis. An abnormal decrease of ENaC activity may lead to (*a*) severe renal salt-losing syndromes with a hypotensive phenotype, (*b*) a respiratory distress syndrome in premature newborns (5), and (*c*) high-altitude pulmonary edema in adult (6).

0066-4278/02/0315-0877$14.00   **877**

In all instances, the intensity of the clinical phenotype depends on a complex interaction between genetic and environmental factors. Monogenic diseases offer a unique opportunity to study this interaction in simple terms and define the relative importance of a given gene in a pathophysiological cascade.

In this short review, we discuss genetically defined forms of salt sensitivity and salt resistance and focus on the role of sodium transport in the renal collecting duct for the final adjustment of sodium balance, blood volume, and blood pressure. For understanding the pathophysiology of many human diseases, a general strategy has been defined (Figure 1). First, starting from clinical observations of heritable diseases (7, 8), a pathophysiological hypothesis is made and then tested in vivo. Two monogenic diseases, pseudohypoaldosteronism Type 1 (PHA-1) and pseudoaldosteronism (Liddle's syndrome), have aided in defining the critical role of these genes in the control of sodium balance and blood pressure. Second (Figure 1), the molecular mechanisms by which the mutations cause the phenotypes are studied ex vivo and in vitro. However, the molecular mechanisms established in vitro are generally not helpful in elucidating the pathophysiology of the diseases in vivo, which requires the establishment of animal models, mainly transgenic mice models, that will mimic the human diseases (9). With these models, the pathophysiology of the diseases can be studied and the modifying factors (modifier or interacting genes modulating the phenotypic expression of the mutated genes) identified in order to develop new therapeutic strategies and test new drugs.

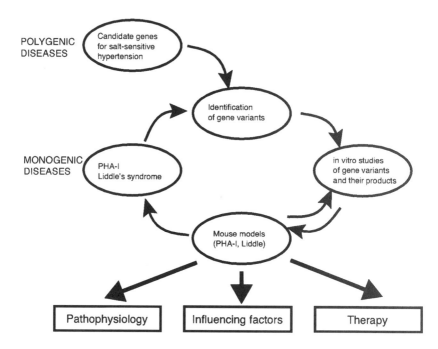

Figure 1   Strategy: from monogenic to polygenic diseases through animal models.

Monogenic diseases are typically severe but rare (allelic frequency less than 1% in human populations). They define new physiological paradigms and signaling cascades and help to define candidate genes that could play a role in the more complex and more frequent polygenic diseases (essential hypertension, diabetes, obesity, etc). Genetic factors are thought to be allelic polymorphisms that are frequent (over 1%) in the human genome. Most of the allelic polymorphisms are short repetitive sequences (microsatellites) or single-nucleotide polymorphisms (SNPs), which are thought to be present in about 1 out of 1000 nucleotides in the human genome, which consists of 3 billion basepairs. As discussed below, the functional significance of SNPs in the coding or non-coding regions of candidate genes can be studied according to the same flow chart (Figure 1), and animal models mimicking polygenic diseases can be engineered.

Numerous reviews have recently been published on many aspects of ENaC and related gene families and can be found in the reference list (4, 10–18).

HYPERTENSIVE AND HYPOTENSIVE PHENOTYPE: SALT-SENSITIVITY VERSUS SALT-RESISTANCE

Hypertension is the most frequently found disease in human populations. Genetic and nongenetic factors are involved. High-salt intake has been proposed as a major risk factor. Short-term experiments (one week exposure to high-salt intake) in healthy volunteers and hypertensive patients showed, according to an operational definition, an abnormal blood pressure response in 30% of normal individuals and up to 50% of hypertensive patients, which defined both groups as salt sensitive (19). Controlled studies in which the salt intake was precisely controlled over a relatively long period of time (30 days) have indicated that salt intake contributes directly to the measured blood pressure (20). Over 400 participants were randomly assigned to eat either a controlled diet, typical of intake in the United States or the so-called dietary approach to stop hypertension (DASH) diet, which contained, in random order, high, intermediate, or low levels of sodium, for 30 consecutive days. The DASH diet was associated with a significantly lower systolic blood pressure at each sodium level, and the difference was greater with high sodium concentrations than with lower ones. The effect was statistically significant in both groups of participants with or without hypertension, but the effect was more striking (−11 mm versus −7 mm of mercury) in participants with hypertension (20).

Long-term effects of salt intake as a risk factor have been extensively examined in large epidemiological studies and have led to controversial conclusions (21, 22). In some studies, there was a good correlation between the 24-h urinary excretion of sodium and the measurement of blood pressure (22). However, on the other hand, reduction of salt intake did not lead to significant decreases in blood pressure (21).

How can we explain the significant effect of salt intake in short-term (1 to 4 weeks), well-controlled, double-blind studies, whereas long-term epidemiological studies on large populations did not reach a consensus? These contradictory results may be explained by two kinds of confounding factors (1). First, long-term studies of low- or high-salt intake as a single experimental variable over years in human populations are obviously difficult if not impossible. Second, epidemiological studies have not taken into account the possibility that susceptibility genes could confer salt sensitivity or salt resistance in a large proportion of the population. Regarding the first factor, Denton et al. (23) were able to demonstrate in chimpanzees (the species genetically closest to humans) that the addition of salt within the human dietetic range causes a highly significant rise in their blood pressure. Interestingly, the effect of salt differed between individuals, and only 60% of the cohort developed high blood pressure, defining two sub-populations of salt-sensitive or salt-resistant animals. Regarding the second factor, the identification of mutations in the epithelial sodium channel (ENaC) β subunit as a cause of a monogenic form of salt-sensitive hypertension in humans (Liddle's syndrome) (24) has highlighted the importance of a single mutated gene as sufficient to induce large changes in blood pressure.

According to the hypothesis put forward by Guyton over 10 years ago (25), control of blood pressure at steady state and on a long-term basis is critically dependent on renal mechanisms. During the last 6 years, a number of genes expressed in various parts of the nephron have been shown to be directly involved in the control of blood pressure (26, 27). The identification of mutations in monogenic diseases such as the Bartter's or the Gitelman's syndrome clearly indicates that defects in ion transporters expressed in the thick ascending limb (TAL) or in the distal convoluted tubule (DCT) may lead to severe salt-wasting syndromes with a hypotensive phenotype. Thus far, no gain-of-function mutations in these transporters (NaK2CL or NaCl cotransporters, Cl or K channels) have been identified in TAL or DCT. In DCT a gain-of-function mutation in the NaCl cotransporter (or its regulatory cascade) has been postulated to explain the Type-2 pseudohypoaldosteronism (Gordon's syndrome) (28, 29). In the aldosterone-sensitive distal nephron (ASDN), i.e., the connecting tubule (CNT), the cortical collecting duct (CCD), and, to some extent, the outer medullary collecting duct (OMCD) and inner medullary collecting duct (IMCD) (30–32), the final control of sodium reabsorption is achieved through an amiloride-sensitive electrogenic sodium reabsorption, which is under tight hormonal control (33). The main limiting factor in sodium reabsorption in this part of the nephron is the apically located amiloride-sensitive epithelial sodium channel (ENaC), whose activity is controlled by aldosterone, vasopressin, and insulin (33).

ENaC AS A LIMITING FACTOR OF AMILORIDE-SENSITIVE ELECTROGENIC SODIUM TRANSPORT

Aldosterone-Dependent Sodium Transport

The most important conserved function of aldosterone is to promote sodium reabsorption with potassium and hydrogen secretion across the so-called tight epithelia that display a high transepithelial electrical resistance and an amiloride-sensitive electrogenic sodium transport (33). According to a well-accepted model of epithelial sodium transport, the two major steps in sodium reabsorption by renal epithelia are facilitated transport driven by an electrochemical potential difference across the apical membrane from urine to cell and active transport driven by metabolic energy across the basolateral membrane, from cell to interstitium. The apical membrane entry step is mediated by the sodium-selective, amiloride-sensitive ion channel ENaC, and the exit step is catalyzed by Na,K-ATPase, the ouabain-sensitive sodium pump (33). These two in-series mechanisms are the rate-limiting steps for sodium transport and thus the most likely final effectors of action for aldosterone or for any other hormone that regulates the overall reabsorption process. The classic model of the mechanism of aldosterone action in tight epithelia (33) proposes the following steps: (*a*) Aldosterone crosses the plasma membrane and binds to its cytosolic receptor, either the mineralocorticoid (MR) or glucocorticoid (GR) receptor. MR and GR are protected from illicit occupation by high levels of plasma glucocorticoids (cortisol or corticosterone) through the metabolizing action of 11-β-HSD2, which transforms the active cortisol into cortisone, an inactive metabolite, unable to bind either to MR or GR. (*b*) The receptor-hormone complex is translocated to the nucleus where it interacts with the promoter region of target genes, activating or repressing their transcriptional activity. (*c*) Aldosterone-induced proteins (AIPs) or aldosterone-repressed proteins (ARPs) mediate an increase in transepithelial sodium transport. Early effects are produced by the activation of pre-existing transport proteins (ENaC, Na, K-ATPase) via yet uncharacterized mediators. The late effect is characterized by an accumulation of additional transport proteins and other elements of the sodium transport machinery (33). Three monogenic forms of salt-sensitive hypertension in humans map to genes expressed in principal cells: ENaC (Liddle's syndrome) (27), syndrome of apparent mineralocorticoid excess (AME) owing to gene inactivation of 11-β-HSD2 (34–36), and gain-of-function mutations in the mineralocorticoid binding site of MR (37). It is therefore remarkable that mutations of genes involved in the two main entry steps in the aldosterone signaling pathway (MR and 11-β-HSD2) and in the main effector (ENaC) are indeed able to induce severe salt-sensitive hypertension.

Biophysical and Physiological Properties of ENaC

The epithelial sodium channel is characterized by its remarkable ion cationic selectivity ($P_{NA}/P_K > 20$), a low urinary conductance ($\sim$4–5 pS in the presence of sodium), gating kinetics characterized by long closing and opening times (3–5 s) and, finally, a high sensitivity to amiloride (K_i 0.1 μM) (38). The epithelial sodium channel is a heteromultimeric protein made of three homologous subunits (α, β, γ). Coexpression of all three subunits in the *Xenopus* oocyte constitutes the channel with all the physiological and pharmacological properties of the native channel described in the apical membrane of cortical collecting duct principal cells. The subunits of the sodium channels are proteins of 650–700 amino acids. At the protein level, the identity between each subunit is $\sim$35%, suggesting that duplication of the ancestor gene is ancient but appears only in metazoans from worm to vertebrate (39). Each subunit passes the membrane only twice, exposing a large 50-kDa ectodomain outside of the cell, whereas the two amino and carboxy termini ($\sim$8–10 kDa) are cytoplasmic. Each subunit is glycosylated. Quantitative analysis of cell surface expression of ENaC α, β, and γ subunits shows that they assemble according to a fixed stoichiometry with α ENaC as the most abundant subunit (40, 41) (Figure 2, see color insert). Functional assays based on differential sensitivity to channel blockers, elicited by mutations tagging each α, β, and γ subunit, are consistent with a four-subunit stoichiometry composed of two α, one β, and one γ subunit. Expression of concatameric cDNA constructs made of different combinations of ENaC subunits confirms the four-subunit channel stoichiometry and shows that the arrangements of the subunits on the channel pore consist of two α subunits separated by β and γ subunits (41) (Figure 2, see color insert). Similar architectures have been proposed by Kleyman and colleagues (42) for ENaC and by Barbry and colleagues for FaNaC (43), a homotetrameric channel involved in synaptic transmission in snails. At variance with these data, an octameric and/or nonameric structure has been proposed by two other laboratories (44, 45). The discrepancies between the two sets of data will not be easily reconciled until structural data provide a more definitive answer to this important question.

Tissue-specific expression of ENaC is now well documented in sodium-transporting epithelia sensitive to aldosterone. ENaC subunits are found in ASDN (30), in the surface epithelia of colon (46), and in the duct cells of exocrine glands (sweat glands, salivary glands) (46). ENaC is expressed in epithelial cells of non-transporting epithelia (keratinocytes and hair follicles), which are mineralocorticoidresponsive but not involved in overall sodium balance (47–55), and it is proposed to play a role in skin differentiation and hair growth (56). Aldosterone-dependent ENaC expression in taste buds suggests an important role in salt tasting (57). ENaC is also expressed in distal and proximal airways, as well as in nasal mucosa (48–51). In this tissue, ENaC participates in the control of the extracellular fluid at the air-cell interface by controlling the ionic composition of the alveolar surface liquid (ASL). This process is not under mineralocorticoid control but rather glucocorticoid. In other tissues or organs, ENaC may have additional specific functions, e.g., in hearing by controlling endolymph cationic composition in inner ear (52, 53), but ENaC is apparently not involved in the mechano-electrical transducer apparatus (54).

Gene Inactivation of ENaC Subunits in Mice

Gene inactivation experiments in mice are considered of prime importance in order to understand the function of any given gene in vivo. A number of simple predictions can be made. If gene expression is of critical importance during early development or embryogenesis, it should lead to the absence of $-/-$ animals

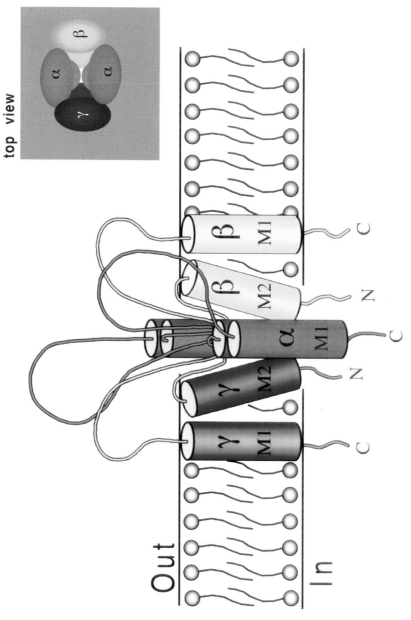

Figure 2 ENaC: structure membrane topology and stochiometry model.

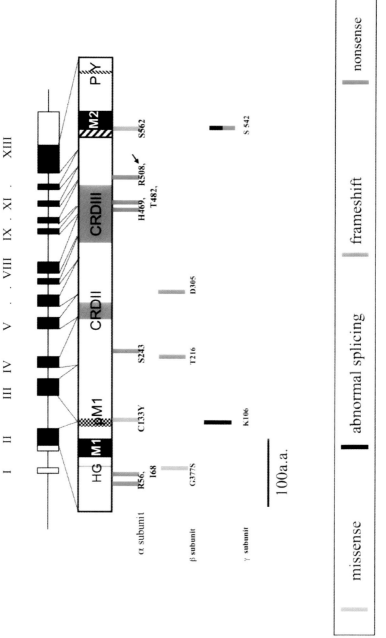

Figure 3 Human PHA-1 mutations: genomic organization and linear model of ENaC subunits.

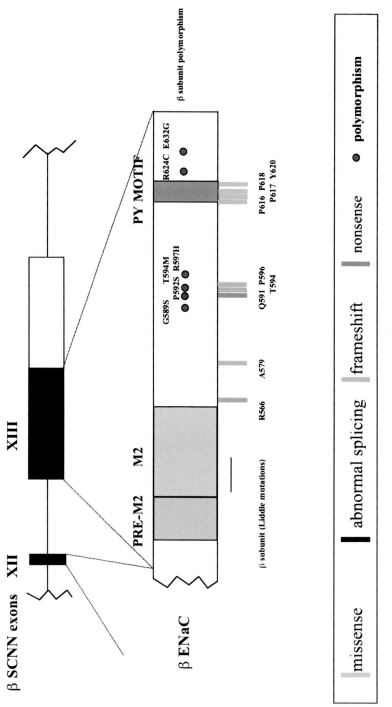

Figure 4 Human Liddle mutations: genomic organization of exon XII and XIII and the C terminus of the β and γ subunits.

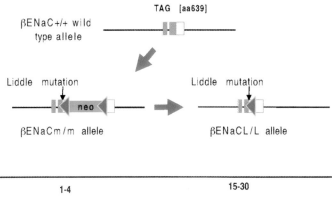

mRNA expression (% of wt)	1-4	15-30
protein expression (% of wt)	ND	~10
normal salt diet	compensated salt-wasting phenotype **(chronic PHA-1)**	compensated salt-retention phenotype: hypervolemia but normal blood pressure and electrolytes
low salt diet	induced salt-wasting phenotype **(acute PHA-1)**	
high salt diet		salt-induced hypertension, hypokalaemia, metabolic alcalosis **(Liddle's syndrome)**

Figure 5 Mouse models mimicking salt-induced hypertension and inducible PHA by salt restriction.

at birth or a severe shift in the expected Mendelian distribution when two heterozygotes are bred: (25% $-/-$ versus 50% $+/-$ versus 25% $+/+$ animals). At the other extreme, no obvious phenotype is observed. Null mutant animals develop normally, are fertile, and have no main pathophysiological anomaly. This can be interpreted either that the clinical phenotype can be observed only under stress conditions or, alternatively, that the gene of interest is redundant so that other pathways can take over the deficit of the missing protein. Each subunit of ENaC has been now separately inactivated (Table 1). The genotypic analysis of newborn animals indicates a Mendelian ratio (roughly 25% $+/+$, 50% $+/-$, and 25% $-/-$), suggesting that embryonic and fetal development in $-/-$ animals for α, β, or γ was not impaired. A striking feature of inactivation of any subunit was, however, 100% lethality within 2 to 4 days after birth. As summarized in Table 1, the clinical phenotype of each gene inactivation was different. For the α knockout (58), neonates developed respiratory distress and died within 40 h after birth from failure to clear their lungs of liquid, showing that ENaC plays a critical role in the adaptation of the newborn lung to air breathing (58). The β ENaC-deficient mice showed normal prenatal development but died within 2 days after birth, most likely of hyperkaliemia (59). The β subunit is apparently required for ENaC function in the renal collecting duct, but in contrast to the α subunit, the β subunit is not required for the transition from a liquid-filled to an air-filled lung (59). Relative to controls, γ ENaC $-/-$ pups exhibited low urinary potassium and high urinary sodium concentrations and died between 24 and 36 h after birth, probably from hyperkaliemia (60). Newborn γ ENaC $-/-$ mice cleared their lung liquid more slowly than control littermates, but lung water at 12 h was nearly normal. It was concluded that γ ENaC facilitates neonatal lung liquid clearance and is critical for renal sodium and potassium transport (60).

TABLE 1   Summary of mouse mutations mimicking human diseases

Line	mRNA (%wt)			ENaC activity (% wt)		Phenotype	References
	Kidney	Lung	Colon	Colon	Lung		
Null alleles							
αENaC$^{-/-}$	—	—	—	ND	0	RDS ($+/-$ renal phenotype)	
βENaC$^{-/-}$	—	—	ND	ND	ND	Acute PHA	(59)
γENaC$^{-/-}$	—	ND	ND	ND	~15	Acute PHA ($+/-$ lung phenotype)	(60)
Mutant alleles							
βENaC$^{m/m}$	~1	~1	~4	~20	~15	Inducible PHA	(66)
βENaC$^{L/L}$	~13	ND	~28	~190	ND	Salt sensitive HTA	(66)
Transgenic rescue							
αENaC$^{-/-Tg}$	~10	~10	-150	~15	30–50	Chronic PHA	(65)

ENaC: MONOGENIC HUMAN DISEASES
AND THEIR ANIMAL MODELS

PHA-1

HUMAN DISEASE   Pseudohypoaldosteronism type 1 (PHA-1) is an inherited disease characterized by severe neonatal salt-wasting, hyperkaliemia, metabolic acidosis, and is unresponsiveness to mineralocorticoid hormones (7, 61). An autosomal-dominant form is caused by a loss-of-function mutation in the mineralocorticoid receptor (62), whereas the more severe recessive form is caused by loss-of-function mutations in the ENaC subunits (63) (summary in Table 2). Figure 3 (see color insert) illustrates the genomic organization of the SCNN genes encoding the ENaC subunits and presents a linear representation of the primary structure of the protein. ENaC subunits share a number of conserved domains important for channel function. A highly conserved histidine-glycine (HG) motif in the cytoplasmic N-terminal end (pre M1) is important for channel opening and closing. The transmembrane spanning segments M1 and M2 are likely involved in the 1 on the permeation pathway. A post-M1 domain, as well as two cysteine-rich domains in the extracellular loop, are characteristics of the members of the ENaC gene family. A short sequence preceding the M2 transmembrane segment forms the extracellular channel pore with the amiloride-binding site and the ion-selectivity filter. The mutations causing PHA-1 are distributed all along the ENaC sequence. It is conceivable that nonsense frameshift mutations, as well as mutations leading to abnormal splicing in the α, β, and γ ENaC subunits, cause a channel decrease/loss of function. Missense mutations causing PHA-1 are found in critically important domains of the protein. The βG37S is found in the HG motif involved in channel gating, the αC133Y in the pre-M1 domain may impair intracellular trafficking of the channel, and the αS562L and γS542 mutations occur in the heart of the channel pore, the ion-selectivity filter. The mechanism by which this mutation leads to a hypoactive channel in the apical membrane of epithelial cells has been recently reviewed extensively (38, 64).

ANIMAL MODELS   The first mouse models mimicking PHA-1 (see above β or γ knockouts) led to a lethal phenotype within 40 to 50 h after birth, preventing detailed analysis of the disease. In humans, PHA-1 can be adequately treated by preventing hyperkaliemia (use of potassium resins) and by an adequate intake of salt (up to 30 g/day) for an adult patient suffering from PHA-1. Therefore, it will be interesting to study the long-term effect of a severe salt-wasting syndrome in animal models.

As shown in Table 1, two animal models fulfill these criteria. Transgenic expression of α ENaC driven by a cytomegalovirus promoter in α ENaC $-/-$ knockout mice (α ENaC $-/-$ Tg) rescued the perinatal lethal pulmonary phenotype and partially restored sodium transport in renal, colonic, and pulmonary epithelia (65). At days 5 to 9, however, 50% mortality was observed, owing to severe PHA-1

TABLE 2   Summary of human mutations causing PHA-1 and Liddle

PHA-1		
Codon	**Nucleotide**	**Reference**
αR56stop	α166C → T	(72)
αI68 frame shift	α203delTC	(63)
αC133Y		(109, 110)
? frame shift	α604delAC	(72)
Abnormal sequence after codon Ser 243 and truncation after Ala 247	α729 del A	(74)
H469 frameshift	α1404delC	(72)
Abnormal sequence after codon 482 and premature stop after 495	α1449 del C	(74)
αR508stop	1522C → T	(63)
α S562L (+ W493R)	1685C → T	(74)
β G37S	236G → A	(63)
βT216 frame shift	β647insA	(72)
β305 frame shift	β915delC	(72)
γ KYS106-108 to N/exon deletion	γ318GTACAGdel	(111)
γS542 frame shift/abnormal splicing	γ1627delG/1570 G-A	(112)

Liddle Syndrome

Codon	**Nucleotide**	**Reference**
βR566stop	1696C → T	(24)
βA579fr	1735del1736	(113)
βQ591stop	1770C → T	(24)
βT594fr	1907insC	(24)
βP596fr	1913delC	(24)
βR597 fr	1916insC	(114)
βP617S	1976C → T	(114)
βP618L	1980C → T	(115)
βY620H	1985T → C	(116)
βW573stop	1718G → A	(117)
γW575stop	1724G → A	(118)

with metabolic acidosis, urinary salt-wasting, and growth retardation. The adult α ENaC $-/-$ Tg survivors exhibited a compensated PHA-1 with normal acid-base and electrolyte values, but a sixfold elevation of plasma aldosterone compared with wild-type littermate controls (65). In a second model, the β subunit locus gene was disrupted by the introduction of a 1-kb neomycin-resistance gene cassette in the 3' untranslated region of the gene (β_m), which was required to introduce the Liddle mutation, a stop codon at position 639 of the mouse β subunit (66) (Figure 3). These mice showed low levels of β ENaC mRNA expression in kidney and lung ($\beta_{m/m}$: ~1% and colon ~4% of wild-type). In homozygous mutant β ENaC mice, no significant β_m ENaC protein could be detected by immunofluorescence staining. At birth, there was a small delay in lung liquid clearance that paralleled diminished amiloride-sensitive sodium absorption in tracheal mucosa. Under normal salt intake, the mice showed a normal growth rate and looked phenotypically sound. However, adult $\beta_{m/m}$ ENaC mice exhibited a significantly reduced ENaC activity in colon and elevated plasma aldosterone levels, suggesting hypovolemia and a compensated PHA-type 1 phenotype. The phenotype was clinically silent, as $\beta_{m/m}$ ENaC mice showed no weight loss, normal plasma sodium and potassium concentration, normal blood pressure, and a compensated metabolic acidosis. On a 1-week low-salt diet, however, β ENaC mutant mice developed clinical symptoms of acute PHA, leading to weight loss, hyperkaliemia, decreased blood pressure and, ultimately, death (66). In this specific model, salt restrictions thus induce PHA, whereas low levels of β subunits (undetectable by immunostaining) are enough to insure normal life, providing that salt intake is sufficient. It is important to note, however, that in this model the mutated β subunit ENaC has the Liddle mutation and thus should be retained at the cell surface.

In summary, both the β and γ ENaC knockouts exhibit a phenotype that resembles the clinical manifestations of human neonatal PHA-1. The phenotypes of the β and γ subunit knockout experiments were expected to lead to the PHA-1 phenotype (7, 61), because loss-of-function mutations in the β t (63, 67) or γ subunit (68) led to a classical PHA-1 phenotype in humans. The severe lethal lung phenotype observed in the α knockout was, however, not observed in humans. In patients suffering from the more severe (recessive) form of PHA-1, a lung phenotype manifests later (a few month after birth) and is never as dramatic as in the mouse. This discrepancy could be explained by the following not mutually exclusive hypothesis:

Species differences: (a) the presence of species-specific channels and subunits [a δ subunit that so far has been identified only in human tissues (69)] or regulatory proteins expressed in the human versus mouse lung; (b) fetal maturation of the lung: Lung maturation in mice does not appear to be as advanced as it is for humans at birth. The anatomical immaturity (70) of the mouse lung at birth could explain the more severe phenotype in this species.

Mutation differences: Frameshift, missense mutations and premature stop codons have been described in patients suffering from PHA-1 (Figure 3; Table 2).

In humans, the truncation of the α subunit is predicted to abolish ENaC activity in all tissues because it deletes the pore-forming region of the subunit (preM2/M2 in Figure 3) present in at least two copies in the channel complex (see Figure 2). In mice, gene targeting of the α subunit is predicted to fully abolish ENaC activity in all tissues because it deletes one third of the protein, including the first transmembrane domain and part of the extracellular loop (58).

Two types of studies have been conducted in order to understand this apparent discrepancy. For example, it was shown that the human PHA-1 stop mutation (αR508stop) of the α ENaC subunit induced a measurable sodium current in *Xenopus* oocytes that coexpress β and γ subunits (71). The αR508stop mutant variant was co-assembled with β and γ subunits and was present at the cell surface at a lower density, consistent with a lower sodium current. The data suggest a novel role for the α subunit in the assembly and targeting of an active channel to the cell surface and suggest that the channel pore, consisting of only β and γ subunits ($\beta 2 \gamma 2$ heterotetramer), can provide significant residual activity. It was postulated that this activity may be sufficient to explain the absence of a severe pulmonary phenotype in patients with PHA-1 (71). The striking lung phenotype observed in the mouse also suggested that human patients suffering from PHA-1 may have a yet undetected clinical lung phenotype. Patients with systemic PHA-1 fail to absorb liquid from airway surfaces, resulting in an increased volume of liquid in the airways. The results demonstrated that sodium transport also has an important role in regulating the volume of liquid on human airway surfaces (72–74). Along the same line of reasoning, Barker and colleagues have also shown that in severe prematurity, neonates may develop a respiratory distress syndrome (75). This can be explained in part by a low level of sodium transport in airways, as observed by measuring the amiloride-sensitive electrical potential difference (PD) in nasal mucosa, assuming that this electrophysiological measurement reflects the activity of ENaC in distal airways. The role of ENaC in different forms of pulmonary edema is now under intense investigation. A pathophysiological condition that may be relevant is high-altitude pulmonary edema (HAPE), which can be triggered by low pO_2 at high altitude (6). In adult transgenic mice in which the expression and/or the regulation of α ENaC has been altered, lung pulmonary edema may develop when exposed to hypoxia (76).

Liddle's syndrome

HUMAN DISEASE   Liddle's syndrome is an autosomal-dominant form of salt-sensitive hypertension, characterized by early onset of severe hypertension, normo- or hypokaliemia, metabolic alkalosis, and repressed renin and aldosterone secretion (8, 77). Liddle's syndrome results from mutations in the cytoplasmic C terminus of either the β or γ subunit of ENaC, which leads to constitutively increased channel activity (Figure 4, see color insert; Table 2). The mutations result in the elimination of 45 to 75 normal amino acids from the C terminus of either the

β or γ subunit. Missense mutations mapped more precisely the affected region, which was defined as a proline-rich domain (PPxY) (78, 79). Under physiological conditions, the PPxY motif binds to Nedd-4, a ubiquitin ligase protein that promotes internalization and presumably degradation of ENaC molecules (80). The lack of this repressor activity leads to an increased cell surface expression of ENaC molecules and, therefore, explains a constitutive reabsorption of sodium with concomitant consequences on the control of blood pressure. Recently, physiological ENaC partners in kidney cells (Nedd4-2 or *Xenopus* Nedd4) have been identified (81). The molecular mechanisms by which Nedd-4 interacts with ENaC have also been recently reviewed in detail (82, 83) and are not be discussed in this review.

Despite a good understanding of the molecular mechanisms by which the Liddle mutation leads to a hyperactive channel, there is still little information about how salt-sensitive hypertension develops in vivo. The study of Denton and colleagues in chimpanzees (23) shows that chronic and long-term exposure to high-salt intake (almost 2 years), which led to a severe hypertensive phenotype, was not sufficient to make this phenotype irreversible. Upon removal of the high-salt diet, the blood pressure of the hypertensive animals quickly returned to normal. Again, development of animal models that would allow time-course studies of dose-dependence and reversibility of salt-sensitive hypertension would be important for the development of prophylactic or therapeutic measures.

ANIMAL MODEL   A mouse model for the Liddle syndrome has recently been generated by Cre-loxP-mediated recombination (Table 1 and Figure 5, see color insert) (84). Under a normal salt diet, mice heterozygous (L/+) and homozygous (L/L) for the Liddle mutation (L) develop normally during the first 3 months of life. In these mice, blood pressure is not different from wild-type, despite evidence for increased sodium reabsorption in distal colon and low plasma aldosterone, suggesting chronic hypervolemia. Under high-salt intake, the Liddle mice develop high blood pressure, metabolic alkalosis, and hypokaliemia, accompanied by cardiac and renal hypertrophy. This animal model recapitulates, to a large extent, the clinical symptoms of the human form of salt-sensitive hypertension and establishes a causal relationship between dietary salt, a gene expressed in the kidney, and hypertension.

As shown in Figure 5, the introduction of the mutated allele in the mouse genome led to a relatively low expression of mRNA in this animal ($\sim$50% of wild-type in (L/+) animals and only 14% in (L/L) animals). Despite this low abundance of mutated β mRNA, it was sufficient to generate a hypertensive phenotype in the presence of a high-salt diet. It is remarkable that a change in expression of the same mutated allele (from 1% to 14%) can induce two opposite phenotypes, i.e., a severe salt-wasting and salt-gaining (hypertensive) phenotype. This points to the critical role of ENaC in achieving sodium balance in vivo. Precise regulatory mechanisms are required in order to achieve sodium homeostasis, but still little is known about this regulatory mechanism.

FROM MONO- TO POLYGENIC DISEASES

Monogenic diseases are characterized by mutations with rare allelic frequency in human populations (less than 0.1%), often leading to very severe phenotypes. More frequent mutations (allelic polymorphisms), often SNPs, have an allelic frequency of >1 or 2% in human populations. What can be the functional consequence of such polymorphisms? It is presently a key issue in understanding the genetic factors of polygenic diseases. For the common forms of hypertension, which are multifactorial, genetic factors account for about 30% of the variation of blood pressure in human populations. The development of the disease depends therefore on the interaction between genetic factors, on one hand, and risk factors, on the other hand. Many genes have been now associated with blood pressure control (85) or essential hypertension: ENaC (86), 11-β-HSD2 (87), angiotensinogen (88, 89), and genes on the renin angiotensin aldosterone axis. See recent reviews in References 90–92, α subunit of adducin (93), glucagon receptor (94), $G_{\alpha 3}$ (95), $G_{\alpha s}$ (96), endothelin 2 (97), and $\alpha 2$ adrenergic receptor (98). Most of these genes are expressed in the kidney or in the principal cells. They all may participate as candidate genes in the salt-sensitive phenotype. In most cases, however, for the studies demonstrating positive associations, there are an equal number of studies showing lack of association with a specific polymorphism (89, 92). In a recent review, Corvol and colleagues summarized (89) the current opinion about the candidate gene approach to the understanding of the genetics of human essential hypertension. Taking the ENaC and angiotensinogen genes as examples, the main pitfalls of this approach are the following:

1. Many linkage or association studies have limited statistical confirmation.
2. The genetic findings may vary greatly according to the populations studied.
3. There is a need for better phenotyping of the hypertensive population.
4. The causal relationship between molecular variance and hypertension is and will be difficult to establish firmly.
5. The contribution of the genetics studied in rodents to the molecular genetics of human hypertension must be re-examined.
6. Most molecular variance leads to a low attributable risk in the population or a low individual effect at individual levels.
7. It is too early to propose dietary recommendation and specific drug treatment according to patients' phenotypes.

PERSPECTIVES

The complexity of salt-sensitive hypertension requires a number of new developments along the lines listed below.

- Identification of new candidate genes expressed in the principal cells of the cortical collecting duct. The motivation for searching out new candidate genes comes from clinical observations that a large number of families suffering from hypertension with a Liddle phenotype (low plasma aldosterone, low renin, hypokaliemia, metabolic acidosis) have not mapped to any of the known monogenic diseases so far described (MR receptor, 11-β-HSD2 or ENaC). This can be achieved, for example, by differential hybridization (99), RT-PCR differential display (100–102), DNA chips or SAGE (103), protein-protein interaction assay (80), and complementation functional cloning (104, 105). A number of new interesting candidate genes have recently been identified: (*a*) corticosteroid-induced genes: CHIF (106), K-Ras (100), sgk (101, 102), GILZ (103); (*b*) genes regulating ENaC activity: Nedd4-2 (107), mCAP-1 (104); and (*c*) genes yet to be cloned and identified in the transcriptome of the principal cell (103).

- There is a need to improve clinical phenotyping and especially to find better ways to assess salt sensitivity and salt resistance in humans. When the clinical phenotyping is done carefully, the associated studies may become a powerful tool to analyze the importance of genetic polymorphisms, as shown recently by Ferrari et al., studying polymorphisms of the 11-β-HSD2 genes (87) or by Bianchi and colleagues for α adducin (93). The study of isolated human populations in a restricted area showing a more homogenous genetic background may also help to solve the problem.

- An enormous amount of data will be generated by the identification of over 3 million SNPs in the human genome. The real challenge, however, is the functional and physiological understanding of a combination of a large number of SNPs mapping many different genes. This requires a much better understanding of animal and human physiology and intermediate phenotypes. For instance, a new and provocative (but testable) signaling cascade inside the kidney for the renin-angiotensin system has been proposed (91). Gene targeting in mice is in the foreseeable future and is the only practical approach for studying the interaction between genetic and environmental factors in vivo (9, 108). Renal physiology in mice must be considerably extended and methods improved to validate such an experimental model for its relevance to human disease. In the case of the Liddle mutation, the mouse model appears to be valid for salt-sensitive hypertension (84). Digenic and polygenic models of salt-sensitive hypertension can now be engineered.

- Finally, the identification of the structure of ENaC and a better understanding of the three-dimensional structure of major binding sites for sodium and amiloride will allow the development of new, more selective drugs: for instance, antagonists that, unlike amiloride, will not be antagonized by high urinary sodium. Such novel compounds would qualify as good candidates for an anti-salt pill. On the other hand, the need for channel openers or activators is important for the treatment of respiratory distress syndrome in neonates, in premature newborns, and in some forms of pulmonary edema, i.e., HAPE.

Visit the Annual Reviews home page at www.AnnualReviews.org

LITERATURE CITED

1. Rossier BC. 1997. 1996 Homer Smith Award Lecture. Cum grano salis: the epithelial sodium channel and the control of blood pressure. *J. Am. Soc. Nephrol.* 8:980–92

2. Boucher RC. 1994. Human airway ion transport. *Am. J. Resp. Crit. Care* 150: 271–81

3. Guggino WB. 1999. Cystic fibrosis and the salt controversy. *Cell* 96:607–10

4. Matalon S, O'Brodovich H. 1999. Sodium channels in alveolar epithelial cells: molecular characterization, biophysical properties, and physiological significance. *Annu. Rev. Physiol.* 61:627–61

5. Barker PM, Gowen CW, Lawson EE, Knowles MR. 1997. Decreased sodium ion absorption across nasal epithelium of very premature infants with respiratory distress syndrome. *J. Pediatr.* 130:373–77

6. Scherrer U, Sartori C, Lepori M, Allemann Y, Duplain H, et al. 1999. High-altitude pulmonary edema: from exaggerated pulmonary hypertension to a defect in transepithelial sodium transport. *Adv. Exp. Med. Biol.* 474:93–107

7. Hanukoglu A. 1991. Type I pseudohypoaldosteronism includes two clinically and genetically distinct entities with either renal or multiple target organ defects. *J. Clin. Endocr. Metab.* 73:936–44

8. Liddle GW, Bledsoe T, Coppage WS. 1963. A familial renal disorder simulating primary aldosteronism but with negligible aldosterone secretion. *Trans. Assoc. Am. Physicians* 76:199–213

9. Smithies O. 1993. Animal models of human genetic diseases. *Trends Genet.* 9:112–16

10. Voilley N, Galibert A, Bassilana F, Renard S, Lingueglia E, et al. 1997. The amiloride-sensitive Na$^+$ channel: from primary structure to function. *Comp. Biochem. Phys. A* 118:193–200

11. Garty H, Palmer LG. 1997. Epithelial sodium channels: function, structure, and regulation. *Physiol. Rev.* 77:359–96

12. Stokes JB. 1999. Disorders of the epithelial sodium channel: insights into the regulation of extracellular volume and blood pressure. *Kidney Int.* 56:2318–33

13. Kleyman TR, Sheng S, Kosari F, Kieber-Emmons T. 1999. Mechanism of action of amiloride: a molecular prospective. *Semin. Nephrol.* 19:524–32

14. Benos DJ, Stanton BA. 1999. Functional domains within the degenerin/epithelial sodium channel (Deg/ENaC) superfamily of ion channels. *J. Physiol.* 520:631–44

15. Oh YS, Warnock DG. 2000. Disorders of the epithelial Na(+) channel in Liddle's syndrome and autosomal recessive pseudohypoaldosteronism type 1. *Exp. Nephrol.* 8:320–25

16. Warnock DG. 2000. Aldosterone-related genetic effects in hypertension. *Curr. Hypertens. Rep.* 2:295–301

17. Mano I, Driscoll M. 1999. DEG ENaC channels: a touchy superfamily that watches its salt. *BioEssays.* 21:568–78

18. Bachmann S, Bostanjoglo M, Schmitt R, Ellison DH. 1999. Sodium transport-related proteins in the mammalian distal nephron—distribution, ontogeny and functional aspects. *Anat. Embryol.* 200:447–68

19. Luft FC, Weinberger MH. 1997. Heterogeneous responses to changes in dietary salt intake—the salt-sensitivity paradigm. *Am. J. Clin. Nutr.* 65:S612–17

20. Sacks FM, Svetkey LP, Vollmer WM, Appel LJ, Bray GA, et al. 2001. Effects on blood pressure of reduced dietary sodium and the dietary approaches to stop hypertension (DASH) diet. *New Engl. J. Med.* 344:3–10

21. Midgley JP, Matthew AG, Greenwood CM, Logan AG. 1996. Effect of reduced

dietary sodium on blood pressure: a meta-analysis of randomized controlled trials. *J. Am. Med. Assoc.* 275:1590–97

22. Dyer AR, Elliott P, Shipley M. 1994. Urinary electrolyte excretion in 24 hours and blood pressure in the Intersalt study. *Am. J. Epidemiol.* 139:940–51

23. Denton D, Weisinger R, Mundy NI, Wickings EJ, Dixson A, et al. 1995. The effect of increased salt intake on blood pressure of chimpanzees. *Nat. Med.* 1:1009–16

24. Shimkets RA, Warnock DG, Bositis CM, Nelson-Williams C, Hansson JH, et al. 1994. Liddle's syndrome: heritable human hypertension caused by mutations in the beta subunit of the epithelial sodium channel. *Cell* 79:407–14

25. Guyton AC. 1991. Blood pressure control—special role of the kidneys and body fluid. *Science* 252:1813–16

26. Lifton RP. 1996. Molecular genetics of human blood pressure variation. *Science* 272:676–80

27. Lifton RP, Gharavi AG, Geller DS. 2001. Molecular mechanisms of human hypertension. *Cell* 104:545–56

28. Disse-Nicodeme S, Achard JM, Desitter I, Houot AM, Fournier A, et al. 2000. A new locus on chromosome 12p13.3 for pseudohypoaldosteronism type II, an autosomal dominant form of hypertension. *Am. J. Hum. Genet.* 67:302–10

29. Stiefel P, Garcia-Morillo S, Miranda ML, Garcia-Donas MA, Pamies E, et al. 2000. Gordon's syndrome: increased maximal rate of the Na-K-Cl cotransport and erythrocyte membrane replacement of sphingomyelin by phosphatidylethanolamine. *J. Hypertens.* 18:1327–30

30. Loffing J, Pietri L, Aregger F, Bloch-Faure M, Ziegler U, et al. 2000. Differential subcellular localization of ENaC subunits in mouse kidney in response to high- and low-Na diets. *Am. J. Physiol. Renal Physiol.* 279:F252–F58

31. Loffing J, Zecevic M, Feraille E, Kaisling B, Asher C, et al. 2001. Aldosterone in-duces rapid apical translocation of ENaC in early portion of renal collecting system: possible role of SGK. *Am. J. Physiol. Renal Physiol* 280:F675–F82

32. Hager H, Kwon T-H, Vinnikova AK, Masilamani S, Brooks HL, et al. 2001. Immunocytochemical and immunoelectron microscopic localization of alpha-, beta-,gamma-ENaC in rat kidney. *Am. J. Physiol. Renal Physiol.* 280:F1093–F106

33. Verrey F, Hummler E, Schild L, Rossier BC, Seldin DW, Giebisch G, eds. 2000. Control of Na transport by aldosterone. In *The Kidney: Physiology and Pathophysiology.* Chpt. 53:1441–71. Philadelphia: Lippincott. Vol. 1. 3rd. ed.

34. Wilson RC, Krozowski ZS, Li K, Obeyesekere VR, Razzaghyazar M, et al. 1995. A mutation in the HSD11B2 gene in a family with apparent mineralocorticoid excess. *J. Clin. Endocr. Metab.* 80:2263–66

35. Li A, Tedde R, Krozowski ZS, Pala A, Li KXZ, et al. 1998. Molecular basis for hypertension in the "type II variant" of apparent mineralocorticoid excess. *Am. J. Hum. Genet.* 63:370–79

36. Morineau G, Marc JM, Boudi A, Galons H, Gourmelen M, et al. 1999. Genetic, biochemical, and clinical studies of patients with A328V or R213C mutations in 11 beta HSD2 causing apparent mineralocorticoid excess. *Hypertension* 34:435–41

37. Geller DS, Farhi A, Pinkerton N, Fradley M, Moritz M, et al. 2000. Activating mineralocorticoid receptor mutation in hypertension exacerbated by pregnancy. *Science* 289:119–23

38. Kellenberger S, Schild L. 2001. Structure and function of the ENaC/Degenerin ion channels. *Physiol. Rev.* Submitted

39. Tavernarakis N, Driscoll M. 1997. Molecular modeling of mechanotransduction in the nematode *Caenorhabditis elegans*. *Annu. Rev. Physiol.* 59:659–89

40. Firsov D, Schild L, Gautschi I, Merillat AM, Schneeberger E, Rossier BC. 1996. Cell surface expression of the epithelial

Na channel and a mutant causing Liddle syndrome: a quantitative approach. *Proc. Natl. Acad. Sci. USA* 93:15370–75

41. Firsov D, Gautschi I, Merillat AM, Rossier BC, Schild L. 1998. The heterotetrameric architecture of the epithelial sodium channel (ENaC). *EMBO J.* 17:344–52

42. Kosari F, Sheng S, Li J, Mak DO, Foskett JK, Kleyman TR. 1998. Subunit stoichiometry of the epithelial sodium channel. *J. Biol. Chem.* 273:13469–74

43. Coscoy S, Lingueglia E, Lazdunski M, Barbry P. 1998. The FMRFamide activated sodium channel is a tetramer. *J. Biol. Chem.* 273:8317–22

44. Eskandari S, Snyder PM, Kreman M, Zampighi GA, Welsh MJ, Wright EM. 1999. Number of subunits comprising the epithelial sodium channel. *J. Biol. Chem.* 274:27281–86

45. Snyder PM, Cheng C, Prince LS, Rogers JC, Welsh MJ. 1998. Electrophysiological and biochemical evidence that DEG/ENaC cation channels are composed of nine subunits. *J. Biol. Chem.* 273:681–84

46. Duc C, Farman N, Canessa CM, Bonvalet JP, Rossier BC. 1994. Cell-specific expression of epithelial sodium channel alpha, beta, and gamma subunits in aldosterone-responsive epithelia from the rat: localization by in situ hybridization and immunocytochemistry. *J. Cell Biol.* 127:1907–21

47. Brouard M, Casado M, Djelidi S, Barrandon Y, Farman N. 1999. Epithelial sodium channel in human epidermal keratinocytes: expression of its subunits and relation to sodium transport and differentiation. *J. Cell Sci.* 112:3343–52

48. Farman N, Talbot CR, Boucher R, Fay M, Canessa C, et al. 1997. Noncoordinated expression of alpha-, beta-, and gamma-subunit mRNAs of epithelial Na$^+$ channel along rat respiratory tract. *Am. J. Physiol. Cell Physiol.* 272:C131–C41

49. Rochelle LG, Li DC, Ye H, Lee E, Talbot

CR, Boucher RC. 2000. Distribution of ion transport mRNAs throughout murine nose and lung. *Am. J. Physiol. Lung Cell Mol. Physiol.* 279:L14–L24

50. Talbot CL, Bosworth DG, Briley EL, Fenstermacher DA, Boucher RC, et al. 1999. Quantitation and localization of ENaC subunit expression in fetal, newborn, and adult mouse lung. *Am. J. Resp. Cell Mol. Biol.* 20:398–406

51. Tchepichev S, Ueda J, Canessa C, Rossier BC, O'Brodovich H. 1995. Lung epithelial Na channel subunits are differentially regulated during development and by steroids. *Am. J. Physiol. Cell Physiol.* 269:C805–C12

52. Gründer S, Muller A, Ruppersberg JP. 2001. Developmental and cellular expression pattern of epithelial sodium channel alpha, beta and gamma subunits in the inner ear of the rat. *Eur. J. Neurosci.* 13:641–48

53. Couloigner V, Fay M, Djelidi S, Farman N, Escoubet B, et al. 2001. Location and function of the epithelial Na channel in the cochlea. *Am. J. Physiol. Renal Physiol.* 280:F214–F22

54. Rüsch A, Hummler E. 1999. Mechanoelectrical transduction in mice lacking the alpha-subunit of the epithelial sodium channel. *Hearing Res.* 131:170–76

55. Oda Y, Imanzahrai A, Kwong A, Komuves L, Elias PM, et al. 1999. Epithelial sodium channels are upregulated during epidermal differentiation. *J. Invest. Dermatol.* 113:796–801

56. Mauro T, Guitard M, Oda Y, Crumrine D, Komuves L, et al. 2001. The ENaC channel is required for normal epidermal differentiation. *J. Invest. Dermatol.* Submitted

57. Lin WH, Finger TE, Rossier BC, Kinnamon SC. 1999. Epithelial Na$^+$ channel subunits in rat taste cells: localization and regulation by aldosterone. *J. Comp. Neurol.* 405:406–20

58. Hummler E, Barker P, Gatzy J, Beermann F, Verdumo C, et al. 1996. Early

death due to defective neonatal lung liquid clearance in alpha-ENaC-deficient mice. *Nat. Genet.* 12:325–28

59. McDonald FJ, Yang B, Hrstka RF, Drummond HA, Tarr DE, et al. 1999. Disruption of the beta subunit of the epithelial Na⁺ channel in mice: hyperkaliemia and neonatal death associated with a pseudohypoaldosteronism phenotype. *Proc. Natl. Acad. Sci. USA* 96:1727–31

60. Barker PM, Nguyen MS, Gatzy JT, Grubb B, Norman H, et al. 1998. Role of gamma ENaC subunit in lung liquid clearance and electrolyte balance in newborn mice. *J. Clin. Invest.* 102:1634–40

61. Kuhnle U, Nielsen MD, Tietze HU, Schroeter CH, Schlamp D, et al. 1990. Pseudohypoaldosteronism in eight families: different forms of inheritance are evidence for various genetic defects. *J. Clin. Endocr. Metab.* 70:638–41

62. Geller DS, Rodriguez-Soriano J, Vallo Boado A, Schifter S, Bayer M, et al. 1998. Mutations in the mineralocorticoid receptor gene cause autosomal dominant pseudohypoaldosteronism type I. *Nat. Genet.* 19:279–81

63. Chang SS, Gründer S, Hanukoglu A, Rosler A, Mathew PM, et al. 1996. Mutations in subunits of the epithelial sodium channel cause salt wasting with hyperkalaemic acidosis, pseudohypoaldosteronism type 1. *Nat. Genet.* 12:248–53

64. Bonny O, Hummler E. 2000. Dysfunction of epithelial sodium transport: from human to mouse. *Kidney Int.* 57:1313–18

65. Hummler E, Barker P, Talbot C, Wang Q, Verdumo C, et al. 1997. A mouse model for the renal salt-wasting syndrome pseudohypoaldosteronism (PHA-1). *Proc. Natl. Acad. Sci. USA* 94:11710–15

66. Pradervand S, Barker PM, Wang Q, Ernst SA, Beermann F, et al. 1999. Salt restriction induces pseudohypoaldosteronism type 1 in mice expressing low levels of the beta-subunit of the amiloride-sensitive epithelial sodium channel. *Proc. Natl. Acad. Sci. USA* 96:1732–37

67. Gründer S, Firsov D, Chang SS, Fowler Jaeger N, Gautschi I, et al. 1997. A mutation causing pseudohypoaldosteronism type 1 identifies a conserved glycine that is involved in the gating of the epithelial sodium channel. *EMBO J.* 16:899–907

68. Strautnieks SS, Thompson RJ, Gardiner RM, Chung E. 1996. A novel splice-site mutation in the gamma subunit of the epithelial sodium channel gene in three pseudohypoaldosteronism type 1 families. *Nat. Genet.* 13:248–50

69. Waldmann R, Champigny G, Bassilana F, Voilley N, Lazdunski M. 1995. Molecular cloning and functional expression of a novel amiloride-sensitive Na⁺ channel. *J. Biol. Chem.* 270:27411–14

70. Ballard PL. 1977. *Lung Development. Hormones and Lung Maturation*, pp. 1–23. Berlin: Springer-Verlag

71. Bonny O, Chraibi A, Loffing J, Jaeger NF, Gründer S, et al. 1999. Functional expression of a pseudohypoaldosteronism type I mutated epithelial Na⁺ channel lacking the pore-forming region of its alpha subunit. *J. Clin. Invest.* 104:967–74

72. Kerem E, Bistritzer T, Hanukoglu A, Hofmann T, Zhou ZQ, et al. 1999. Pulmonary epithelial sodium-channel dysfunction and excess airway liquid in pseudohypoaldosteronism. *New Engl. J. Med.* 341:156–62

73. Prince LS, Launspach JL, Geller DS, Lifton RP, Pratt JH, et al. 1999. Absence of amiloride-sensitive sodium absorption in the airway of an infant with pseudohypoaldosteronism. *J. Pediatr.* 135:786–89

74. Schaedel C, Marthinsen L, Kristoffersson AC, Kornfalt R, Nilsson KO, et al. 1999. Lung symptoms in pseudohypoaldosteronism type 1 are associated with deficiency of the alpha-subunit of the epithelial sodium channel. *J. Pediatr.* 135:739–45

75. Barker PM, Gowen CW, Lawson EE, Knowles MR. 1997. Decreased sodium

ion absorption across nasal epithelium of very premature infants with respiratory distress syndrome. *J. Pediatr.* 130:373–77

76. Hummler E, Barker P, Beermann F, Gatzy J, Verdumo C, et al. 1997. Role of the epithelial sodium channel in lung liquid clearance. *Chest* 111:113S

77. Botero Velez M, Curtis JJ, Warnock DG. 1994. Liddle's syndrome revisited—a disorder of sodium reabsorption in the distal tubule. *New Engl. J. Med.* 330:178–81

78. Schild L, Canessa CM, Shimkets RA, Gautschi I, Lifton RP, Rossier BC. 1995. A mutation in the epithelial sodium channel causing Liddle disease increases channel activity in the *Xenopus laevis* oocyte expression system. *Proc. Natl. Acad. Sci. USA* 92:5699–703

79. Snyder PM, Price MP, McDonald FJ, Adams CM, Volk KA, et al. 1995. Mechanism by which Liddle's syndrome mutations increase activity of a human epithelial sodium channel. *Cell* 83:969–78

80. Staub O, Dho S, Henry P, Correa J, Ishikawa T, et al. 1996. WW domains of Nedd4 bind to the proline-rich PY motifs in the epithelial Na$^+$ channel deleted in Liddle's syndrome. *EMBO J.* 15:2371–80

81. Abriel H, Loffing J, Rebhun JF, Pratt JH, Schild L, et al. 1999. Defective regulation of the epithelial Na$^+$ channel by Nedd4 in Liddle's syndrome. *J. Clin. Invest.* 103:667–73

82. Staub O, Plant P, Ishikawa T, Schild L, Rotin D. Benos DJ, ed. 1999. Regulation of ENaC by interacting proteins and by ubiquitination. In *Amiloride-Sensitive Sodium Channels: Physiology and Functional Diversity. Current Topics in Membranes*, 47:65–86. San Diego: Academic

83. Staub O, Abriel H, Plant P, Ishikawa T, Kanelis V, et al. 2000. Regulation of the epithelial Na$^+$ channel by Nedd4 and ubiquitination. *Kidney Int.* 57:809–15

84. Pradervand S, Wang Q, Burnier M, Beermann F, Horisberger J-D, et al. 1999. A mouse model for Liddle's syndrome. *J. Am. Soc. Nephrol.* 10:2527–33

85. Nagy Z, Busjahn A, Bahring S, Faulhaber HD, Gohlke HR, et al. 1999. Quantitative trait loci for blood pressure exist near the IGF-1, the Liddle syndrome, the angiotensin II-receptor gene and the renin loci in man. *J. Am. Soc. Nephrol.* 10:1709–16

86. Baker EH, Dong YB, Sagnella GA, Rothwell M, Onipinla AK, et al. 1998. Association of hypertension with T594M mutation in beta subunit of epithelial sodium channels in black people resident in London. *Lancet* 351:1388–92

87. Lovati E, Ferrari P, Dick B, Jostarndt K, Frey BM, et al. 1999. Molecular basis of human salt sensitivity: the role of 11 beta-hydroxysteroid dehydrogenase Type 2. *J. Clin. Endocr. Metab.* 84:3745–49

88. Jeunemaitre X, Soubrier F, Kotelevtsev Y, Lifton R, Williams CS, et al. 1992. Molecular basis of human hypertension: role of angiotensinogen. *Cell* 71:169–80

89. Corvol P, Persu A, Gimenez-Roqueplo AP, Jeunemaitre X. 1999. Seven lessons from two candidate genes in human essential hypertension: angiotensinogen and epithelial sodium channel. *Hypertension* 33:1324–31

90. Siffert W. 1998. G proteins and hypertension: an alternative candidate gene approach. *Kidney Int.* 53:1466–70

91. Lalouel JM, Rohrwasser A, Terreros D, Morgan T, Ward K. 2001. Angiotensinogen in essential hypertension: from genetics to nephrology. *J. Am. Soc. Nephrol.* 12:606–15

92. Luft FC. 2001. Molecular genetics of salt-sensitivity and hypertension. *Drug Metab. Dispos.* 29:500–4

93. Manunta P, Cusi D, Barlassina C, Righetti M, Lanzani C, et al. 1998. Alpha-adducin polymorphisms and renal sodium handling in essential hypertensive patients. *Kidney Int.* 53:1471–78

94. Brand E, Bankir L, Plouin P-F, Soubrier F. 1999. Glucagon receptor gene mutation (Gly40Ser) in human essential hypertension. *Hypertension* 34:15–17

95. Siffert W, Rosskopf D, Siffert G, Busch S, Moritz A, et al. 1998. Association of a human G-protein beta3 subunit variant with hypertension. *Nat. Genet.* 18:45–48

96. Jia H, Hingorani AD, Sharma P, Hopper R, Dickerson C, et al. 1999. Association of the G(s)alpha gene with essential hypertension and response to beta-blockade. *Hypertension* 34:8–14

97. Sharma P, Hingorani A, Jia H, Hopper R, Brown MJ. 1999. Quantitative association between a newly identified molecular variant in the endothelin-2 gene and human essential hypertension. *J. Hypertens.* 17:1281–87

98. Svetkey LP, Chen YT, McKeown SP, Preis L, Wilson AF. 1997. Preliminary evidence of linkage of salt sensitivity in black Americans at the beta 2–adrenergic receptor locus. *Hypertension* 29:918–22

99. Attali B, Latter H, Rachamim N, Garty H. 1995. A corticosteroid-induced gene expressing an "IsK-like" K^+ channel activity in *Xenopus* oocytes. *Proc. Natl. Acad. Sci. USA* 92:6092–96

100. Spindler B, Mastroberardino L, Custer M, Verrey F. 1997. Characterization of early aldosterone-induced RNAs identified in A6 kidney epithelia. *Pflügers Arch.* 434:323–31

101. Chen SY, Bhargava A, Mastroberardino L, Meijer OC, Wang J, et al. 1999. Epithelial sodium channel regulated by aldosterone-induced protein sgk. *Proc. Natl. Acad. Sci. USA* 96:2514–19

102. Naray-Fejes-Toth A, Canessa C, Cleaveland ES, Aldrich G, Fejes-Toth G. 1999. sgk is an aldosterone-induced kinase in the renal collecting duct—effects on epithelial Na^+ channels. *J. Biol. Chem.* 274:16973–78

103. Robert-Nicoud M, Flahaut M, Elalouf JM, Nicod M, Salinas M, et al. 2001. Transcriptome of a mouse kidney cortical collecting duct cell line: effects of aldosterone and vasopressin. *Proc. Natl. Acad. Sci. USA* 98:2712–16

104. Vallet V, Chraibi A, Gaeggeler HP, Horisberger JD, Rossier BC. 1997. An epithelial serine protease activates the amiloride-sensitive sodium channel. *Nature* 389:607–10

105. Vallet V, Horisberger JD, Rossier BC. 1998. Epithelial sodium channel regulatory proteins identified by functional expression cloning. *Kidney Int.* 54:S109–14

106. Deleted in proof

107. Kamynina E, Debonneville C, Bens M, Vandewalle A, Staub O. 2001. A novel mouse Nedd4 protein suppresses the activity of the epithelial Na^+ channel. *FASEB J.* 15:204–14

108. Meneton P, Ichikawa I, Inagami T, Schnermann J. 2000. Renal physiology of the mouse. *Am. J. Physiol. Renal Physiol* 278:F339–F51

109. Grunder S, Chang SS, Lifton R, Rossier BC. 1998. PHA-1: a novel thermosensitive mutation in the ectodomain of alphaENaC. *J. Am. Soc. Nephrol.* 9:35A (Abstr.)

110. Firsov D, Robert-Nicoud M, Gruender S, Schild L, Rossier BC. 1999. Mutational analysis of cysteine-rich domains of the epithelium sodium channel (ENaC). Identification of cysteines essential for channel expression at the cell surface. *J. Biol. Chem.* 274:2743–49

111. Strautnieks SS, Thompson RJ, Hanukoglu A, Dillon MJ, Hanukoglu I, et al. 1996. Localisation of pseudohypoaldosteronism genes to chromosome 16p12.2–13.11 and 12p13.1-pter by homozygosity mapping. *Hum. Mol. Genet.* 5:293–99

112. Adachi M, Tachibana K, Asakura Y, Abe S, Nakae J, et al. 2001. Clinical case seminar—compound heterozygous mutations in the gamma subunit gene of ENaC (1627delG and 1570–1G → A) in one sporadic Japanese patient with a systemic form of pseudohypoaldosteronism type 1. *J. Clin. Endocr. Metab.* 86:9–12

113. Jeunemaitre X, Bassilana F, Persu A, Dumont C, Champigny G, et al. 1997. Genotype-phenotype analysis of a newly

discovered family with Liddle's-syndrome. *J. Hypertens.* 15:1091–100

114. Inoue J, Iwaoka T, Tokunaga H, Takamune K, Naomi S, et al. 1998. A family with Liddle's syndrome caused by a new missense mutation in the β subunit of the epithelial sodium channel. *J. Clin. Endocr. Metab.* 83:2210–13

115. Hansson JH, Schild L, Lu Y, Wilson TA, Gautschi I, et al. 1995. A de novo missense mutation of the beta subunit of the epithelial sodium channel causes hypertension and Liddle syndrome, identifying a proline-rich segment critical for regulation of channel activity. *Proc. Natl. Acad. Sci. USA* 92:11495–99

116. Tamura H, Schild L, Enomoto N, Matsui N, Marumo F, Rossier BC. 1996. Liddle disease caused by a missense mutation of

beta subunit of the epithelial sodium channel gene. *J. Clin. Invest.* 97:1780–84

117. Hansson JH, Nelson-Williams C, Suzuki H, Schild L, Shimkets R, et al. 1995. Hypertension caused by a truncated epithelial sodium channel γ subunit: genetic heterogeneity of Liddle syndrome. *Nat. Genet.* 11:76–82

118. Yamashita Y, Koga M, Takeda Y, Enomoto N, Uchida S, et al. 2001. Two sporadic cases of Liddle's syndrome caused by de novo ENaC mutations. *Am. J. Kidney Dis.* 37:499–504

119. Ambrosius WT, Bloem LJ, Zhou L, Rebhun JF, Snyder PM, et al. 1999. Genetic variants in the epithelial sodium channel in relation to aldosterone and potassium excretion and risk for hypertension. *Hypertension* 34:631–37

Annu. Rev. Physiol. 2002. 64:899–923

GENETIC DISEASES OF ACID-BASE TRANSPORTERS

Seth L. Alper

Molecular Medicine and Renal Units, Beth Israel Deaconess Medical Center,
Department of Medicine and Cell Biology, Harvard Medical School Boston,
Massachusetts 02215; e-mail: salper@caregroup.harvard.edu

Key Words renal tubular acidosis, AE1 anion exchanger, vacuolar H^+-ATPase, carbonic anhydrase, sodium bicarbonate cotransporter, chloride channel.

■ **Abstract** Genetic disorders of acid-base transporters involve plasmalemmal and organellar transporters of H^+, HCO_3^-, and Cl^-. Autosomal-dominant and -recessive forms of distal renal tubular acidosis (dRTA) are caused by mutations in ion transporters of the acid-secreting Type A intercalated cell of the renal collecting duct. These include the AE1 Cl^-/HCO_3^- exchanger of the basolateral membrane and at least two subunits of the apical membrane vacuolar (v)H^+-ATPase, the V_1 subunit B1 (associated with deafness) and the V_0 subunit *a4*. Recessive proximal RTA with ocular disease arises from mutations in the electrogenic Na^+-bicarbonate cotransporter NBC1 of the proximal tubular cell basolateral membrane. Recessive mixed proximal-distal RTA accompanied by osteopetrosis and mental retardation is associated with mutations in cytoplasmic carbonic anhydrase II. The metabolic alkalosis of congenital chloride-losing diarrhea is caused by mutations in the DRA Cl^-/HCO_3^- exchanger of the ileocolonic apical membrane. Recessive osteopetrosis is caused by deficient osteoclast acid secretion across the ruffled border lacunar membrane, the result of mutations in the vH^+-ATPase V_0 subunit or in the CLC-7 Cl^- channel. X-linked nephrolithiasis and engineered deficiencies in some other CLC Cl^- channels are thought to represent defects of organellar acidification. Study of acid-base transport disease-associated mutations should enhance our understanding of protein structure-function relationships and their impact on the physiology of cell, tissue, and organism.

INTRODUCTION

Optimal function of physiological processes requires control of intracellular and extracellular pH. All cell types must maintain cytoplasmic pH within a narrow range, while generating and responding to metabolic acid and base loads, and while maintaining acidic and alkaline environments within distinct subcellular organellar compartments. The pH of blood and extracellular fluid must be maintained within a narrow range around pH 7.4, whereas specialized cell types can secrete large amounts of acid or base into delimited lumenal spaces. Segregation and regulation of these cellular and extracellular compartments of different pH requires control of acid-base transporter protein activity, biosynthesis and degradation, trafficking,

0066-4278/02/0315-0899$14.00

899

anchoring, and interaction with other polypeptides. Systemic pH homeostasis is achieved through integrated function of the intracellular and plasmalemmal acid-base transporters of kidneys, lungs, and gut, coordinated and regulated by endocrine and nervous systems.

This review summarizes the molecular bases for genetic diseases of plasmalemmal and organellar acid-base transporters. These diseases include (in varying combinations) renal tubular acidoses, deafness, osteopetrosis, mental retardation, alkalotic chloride-losing diarrhea, nephrolithiasis, ocular disease, and brain disease. The associated mutant acid-base transporters include the SLC4A and SLC26A bicarbonate transporters, vH$^+$-ATPase subunits, and carbonic anhydrase II. Acid-base disorders of Cl$^-$ channel diseases are also discussed. Diseases of renal Na$^+$ reabsorption cause secondary acid-base disturbances. Inborn errors of metabolism associated with acid-base disturbances are not addressed.

PHYSIOLOGY OF URINARY ACIDIFICATION

The human body fueled by a western diet daily generates $\sim$1 mmol of mineral acid per kg of body weight. This acid load must be excreted by the kidney. In addition, the daily glomerular filtration of the human kidneys includes in its $\sim$180 liters about 4.5 mol of filtered bicarbonate that must be recovered (1–3). The polarized epithelial cells of the proximal tubule reclaim 80–90% of filtered bicarbonate in a complex process requiring coordinated actions of transport and enzymatic activities in the apical and basolateral membranes and in the cytoplasm (Figure 1). Proximal tubular epithelial cells secrete H$^+$ across their apical membranes into the tubular lumen via the NHE3 Na$^+$/H$^+$ exchanger ($\sim$2/3 of the total) and the vH$^+$-ATPase ($\sim$1/3 of the total). H$^+$ is made available for secretion by intracellular generation of H$^+$ and HCO$_3^-$ from cytoplasmic carbonic anhydrase II-catalyzed hydration of CO$_2$. As H$^+$ is secreted in the lumen, intracellular HCO$_3^-$ is translocated across the basolateral membrane into the interstitium by the electrogenic kNBC1 Na$^+$-bicarbonate cotransporter. Rapid equilibration between extracellular CO$_2$ and HCO$_3^-$ is aided by lumenally disposed glycosyl-phosphatidylinositol-anchored carbonic anhydrase IV at the apical surface, as well as by basolateral surface carbonic anhydrase activity (4).

The small fraction of lumenal bicarbonate not reclaimed by the proximal tubule is reabsorbed across thick ascending limb cells by a process requiring lumenal membrane Na$^+$/H$^+$ exchange (NHE2 and/or NHE3). The nonvolatile acid load is secreted by the Type A intercalated cells of the connecting segment and collecting duct via apical vH$^+$-ATPase, cytoplasmic carbonic anhydrase CAII, and the kAE1 Cl$^-$/HCO$_3^-$ exchanger of the basolateral membrane (Figure 1). Membrane-bound carbonic anhydrase likely also contributes to intercalated cell H$^+$ secretion (5). Lumenal H$^+$ is trapped by urinary buffers. These include ammonium, generated in the proximal tubule and subject to countercurrent recycling, and phosphate, subject to its regulated proximal tubular reabsorption. Loss-of-function mutations

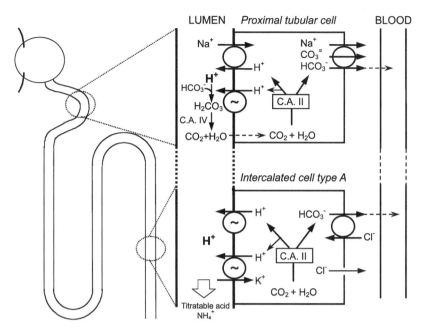

Figure 1 Schematic of major sites of nephron acid-base handling: bicarbonate reabsorption by the proximal tubular cell and acid secretion by the collecting duct Type A intercalated cell. Reproduced from (9) with permission.

in the genes encoding polypeptides that mediate some of these functions in the proximal and distal nephron underlie multiple heritable forms of renal tubular acidosis. Mutations in other contributing gene products are hypothesized to act as genetic modifiers.

Familial Distal Renal Tubular Acidosis (dRTA)

Primary familial dRTA is diagnosed by failure of the kidney to generate sufficiently acidic urine in the setting of spontaneous systemic metabolic acidosis (complete dRTA) or after an imposed acid load (incomplete dRTA). The inadequate urinary acidification is usually accompanied by hypocitraturia, hypercalciuria, and nephrocalcinosis, leading to interstitial disease, urinary concentration defects, nephrolithiasis, and ultimately to chronic renal insufficiency. Affected children present with failure to thrive and growth retardation. The variably severe hypokalemia can present as muscle weakness, periodic paralysis, or cardiac arrhythmia. Decreased ammonium excretion is the result of reduced NH_4^+ trapping in the less acid urine, in the setting of normal renal proximal ammoniagenesis. Chronic interstitial disease secondary to nephrocalcinosis contributes to the electrolyte abnormalities and concentration defects by disruption of the medullary countercurrent gradient (3, 6–9).

Familial dRTA is inherited in both dominant and recessive patterns. De novo mutations have also been documented. Recessive cases are more likely than dominant cases to be of severe and early onset. Only the recessive forms of dRTA are in some cohorts accompanied by sensorineural hearing loss. Bicarbonate supplementation reliably resolves acidosis. If started sufficiently early in the course of the disease, bicarbonate supplementation can restore normal growth and bone density, reverse hypocitraturia and hypercalciuria, and halt renal calcification processes. Bicarbonate treatment does not, however, ameliorate hearing loss (6).

Familial defects in distal urinary acidification might arise from any primary defect in acid secretion by the collecting duct Type A intercalated cell. Such defects in apical H^+ translocation from cell to lumen could reflect mutations in a subunit of the heteromultimeric vacuolar H^+- ATPase, mutations in a subunit of the heterodimeric H^+/K^+ ATPases, or mutations in one of several candidate apical Cl^- conductances. Defects could also be predicted in carbonic anhydrase II or in basolateral HCO_3^- translocation from cell into interstitium, reflecting mutations in the kAE1 Cl^-/HCO_3^- exchanger. In addition, secondary defects in urinary acidification due to impaired ENaC-mediated Na^+ reabsorption across collecting duct principal cells could lead to a more lumen-positive apical membrane potential unfavorable for electrogenic H^+ secretion. Constitutive upregulation of bicarbonate secretion from Type B intercalated cells might also result in mild dRTA. Lastly, mutations in genes encoding proteins that regulate activity, biosynthesis, targeting, retention, or degradation of any of these acid/base transporters could cause dRTA.

AE1 mutations in dominant and recessive forms of dRTA have been found through the candidate gene approach (9). Mutations in vH+-ATPase subunits have been found in recessive dRTA through whole-genome linkage mapping, followed by candidate screening within the linked region (6).

AE1 Mutations in Distal Renal Tubular Acidosis

The 911 amino acid human erythroid AE1 (eAE1) Cl^-/HCO_3^- exchanger SLC4A1 is the major intrinsic membrane protein of the red cell and serves to increase the total CO_2 carrying capacity of the blood. The *AE1* gene on human chromosome 17q21-22 is transcribed in erythroid precursors under control of an erythroid-specific promoter upstream of exon 1. AE1 is also expressed at high level in the Type A intercalated cell of the collecting duct. Renal transcription arises from a distinct promoter within intron 3 of the 20 exon *AE1* gene (10). The resultant kidney transcript encodes a kidney-specific kAE1 polypeptide that lacks the N-terminal 65 (human) (11) or 79 (mouse) amino acids present in eAE1 (12).

The ~400 amino acid N-terminal cytoplasmic domain of eAE1 (Figure 2, see color insert) anchors the membrane protein to the underlying cytoskeleton via interactions with ankyrin-1 and proteins 4.1 and 4.2. The N-terminal region of this domain also binds glycolytic enzymes and hemoglobin (13, 14). With the exception of the little understood protein kanadaptin (15), the identity of kAE1-binding proteins in the intercalated cell remains unknown. The ~500 amino acid C-terminal domain includes 12 to 14 transmembrane spans and a short C-terminal cytoplasmic tail (Figure 2). The topographical disposition of the first 9 and (more provisionally) of the final 2 transmembrane spans is generally accepted. Several models have been proposed for the structure of the possibly flexible intervening region (16–19). Although residues have been identified whose mutation modifies anion permeation (20–22), the anion translocation pathway remains to be identified. The short C-terminal tail includes a binding site for carbonic anhydrase II (23). Transfection and mutagenesis experiments suggest that the resulting local channeling of transported substrate is important to maintain high rates of Cl^-/HCO_3^- exchange (24) and that the model of the AE1/carbonic anhydrase "metabolon" (23, 24) is likely to be replicated by other bicarbonate transporters. The AE1 transmembrane domain reconstituted into two-dimensional lipid crystalline arrays has been visualized by cryo-electron microscopy and image reconstruction to 20 Å resolution (25).

The majority of AE1 mutations reported to date apparently cause only erythroid abnormalities without renal phenotype (Figure 2) (26–28). Most cause autosomal-dominant forms of hereditary spherocytic anemia (HS) and are not encountered in homozygous form, suggesting embryonic lethality. HS mutations are usually associated with moderately decreased total eAE1 polypeptide in the erythrocyte membrane, and apparent absence of mutant polypeptide. The mutants studied to date reveal abnormal trafficking in *Xenopus* oocytes (28) and in 293 human embryonic kidney (HEK) cells (30). The mutants fail to reach the medial Golgi and are misfolded as evidenced by failure to bind to SITS-affigel (30). Both the functionally inactive mutant AE1 Δ400-408 (Southeast Asian Ovalocytosis, SAO) and the gain-of-function mutant AE1 P868L (hereditary acanthocytosis) (Figure 2) also change red cell shape, but without symptomatic phenotype.

Dominant HS-associated AE1 mutations are generally not associated with dRTA. Conversely, dRTA-associated AE1 mutations are generally not associated with HS. Whereas HS missense mutations are distributed throughout AE1 cytoplasmic and transmembrane domains, dRTA mutations are restricted to AE1's transmembrane domain (Figure 2). The almost complete segregation between mutations associated with HS and with dRTA is not understood. In only two instances have homozygous recessive AE1 mutations been associated with the combination of neonatal life-threatening anemia and apparent distal renal tubular acidosis. Both of these mutations were associated with complete absence of red cell membrane eAE1: V488M in TM4 of human AE1 (31), and the equivalent in bovine AE1 of human R646X (32), terminating after TM7. [Two additional families with dominant HS due to AE1 truncation mutants exhibited deficient urinary acidification accompanied by proximal-type bicarbonaturia (33, 34); thus the renal defect could not be attributed simply to dRTA.]

Autosomal dominant dRTA was first associated with exon 14 nucleotide substitutions encoding missense mutations in residue 589, in which the wild-type Arg is converted to His (27, 28, 35), Ser (27, 35), or Cys (27). Red cells of these individuals displayed normal AE1 polypeptide abundance, ~25% reduction in magnitude of AE1-mediated sulfate transport with normal pH-dependence, and minimally altered AE1 affinity for the antagonist DIDS. Heterologous expression of kAE1 R589H in *Xenopus* oocytes revealed $\sim$50% of wild-type function, whether measured as Cl^-/Cl^- exchange or as Cl^-/HCO_3^- exchange (27, 28) and without abnormal inhibition by outer medullary values of pH or osmolarity (28). Enhancement of function and surface expression by co-expressed glycophorin A (36) was retained. Coexpression of mutant and wild-type AE1 forms indicated no dominant-negative phenotype that would explain defective urinary acidification. kAE1 R589C exhibited wild-type Cl^- transport activity (27). Interestingly, R589 is adjacent to K590, the covalent binding site for the inhibitor phenyl isothiocyanate, and this pair of basic residues is conserved in all AE anion exchangers.

Other dominant dRTA-associated AE1 mutations exhibited similarly modest decrements in function. These findings pose a central question about dominant dRTA. How do partial loss-of-function mutations (which in *Xenopus* oocytes and red cells do not exhibit a dominant-negative behavior) produce an impaired urinary acidification phenotype, whereas more severe loss-of-function mutants associated with HS leave urinary acidification apparently unaltered. Oocyte expression experiments cannot detect some temperature-sensitive defects in stability or trafficking or defects in polarized sorting. Preliminary cytological results in transiently transfected polarized MDCK cells grown on permeable filter supports suggest that kAE1 R589H polypeptide exhibits at steady state a wild-type level of accumulation and restriction to the basolateral membrane (D. Prabakaran & S. L. Alper, unpublished data). However, renal cortical tissue sections from one individual with AE1 R589H revealed intercalated cells with decreased vH^+-ATPase immunostaining and apparently absent kAE1 (interpretation was complicated by the coincident scarring of chronic pyelonephritis) (37).

Other dominant dRTA AE1 mutations have minimal or no red cell phenotype. Heterozygosity for AE1 S613F within putative TM7 is associated with increased AE1-mediated sulfate uptake, normal iodide uptake in red cells, and near-normal Cl^- uptake in *Xenopus* oocytes (27). Heterozygosity for AE1 A858D in or near putative TM14 is associated with slight reductions in red cell AE1 polypeptide content and AE1-mediated sulfate flux. AE1 A858D expressed in *Xenopus* oocytes exhibited only 28% of wild-type Cl^- uptake but was moderately stimulated by glycophorin A coexpression to 43% of the wild-type stimulated value, with equivalent enhancement of surface polypeptide expression (38).

Studies of red cells heterozygous for AE1 R901X resulting from a 13-nucleotide duplication in exon 20 (35) have not been reported. This mutation truncates C-terminal cytoplasmic residues that include a putative Type II PDZ recognition domain capable (as a fusion protein) of interaction with the PDZ domain-containing protein PICK1 (39), but retains residues 886–890 that constitute the AE1-binding site for carbonic anhydrase II (23, 24). Preliminary expression data in *Xenopus* oocytes (N. Dahl, L. Jiang & S. L. Alper, unpublished data.) show Cl^-/Cl^- and Cl^-/HCO_3^- exchange activities of magnitude equivalent to or greater than those exhibited by the dominant dRTA mutant AE1 R589H (28) and without dominant-negative phenotype.

Among individuals with dominant dRTA, correlation between genotype and phenotype has not been apparent. Incomplete and complete dRTA are present within individual families. Extent of nephrocalcinosis and degree of hypokalemia vary among family members carrying the same mutation. Within one R589C family, the father presented with severe nephrocalcinosis/lithiasis and isosthenuria, but without metabolic acidosis (incomplete dRTA). In contrast, his daughter was acidotic, hypokalemic, and hypercalciuric, but without nephrocalcinosis (40). Thus clinical phenotype can be altered by (still undefined) modifier genes, some of which may themselves be RTA genes. Deafness has not been documented in association with dominant dRTA.

Within the intercalated cell, dominant dRTA-associated AE1 mutations likely do exert a dominant-negative effect. Additional functional studies, including those contrasting erythroid and intercalated cell types, will be required to understand AE1-associated dRTA in the absence of erythroid disease.

AE1 Mutations in Recessive Distal Renal Tubular Acidosis

In Caucasian populations, AE1 has not been associated with classical recessive dRTA (35). However, AE1 mutations are a major cause of recessive dRTA in populations in Thailand (41, 42), Malaysia, and Papua-New Guinea (38). The first recessive dRTA mutation found was *AE1 G701D*. In this first family described, patient red cell AE1 content and AE1-mediated sulfate flux were normal, but in *Xenopus* oocytes, AE1 G701D exhibited complete loss-of-function owing to lack of surface accumulation and intracellular retention (41). Since red cell function appeared normal, glycophorin A or another erythroid-specific AE1-binding protein were candidates to explain the difference. Indeed, glycophorin A coexpression completely rescued AE1 G701D function and surface expression (41). AE1 G701D C-terminally linked to GFP was also retained inside polarized MDCK cells and was rescued to the basolateral surface by coexpressed glycophorin A (D. Prabakaran & S. L. Alper, unpublished data). Thus the G701D mutation produces a conditional trafficking phenotype. It folds and traffics normally in the presence of glycophorin A (the situation in erythroid precursor cells). However, intercalated cells lack detectable glycophorin A (41), and in its absence, AE1 G701D cannot accumulate at the cell surface. A functional equivalent of glycophorin A in intercalated cells has not been defined.

In some Southeast Asian patients, dRTA is associated with ovalocytosis, and early case reports of this association (43) fueled interest in AE1 as a dRTA candidate gene. These patients are compound heterozygotes of AE1 dRTA mutations with the Δ400–408 in-frame deletion ovalocytosis mutation SAO (Figure 2). dRTA with ovalocytosis has been found with the AE1 mutations G701D, A858D, and ΔV850 in *trans* with AE1 SAO. Compound heterozygosity with the alleles *A858D* and *ΔV850* has also been reported (38). These more C-terminal mutants accumulated to lower levels in the red cell and exhibited reduced transport function (only 29 and 14% of wild-type activity in *Xenopus* oocytes), but retained responsiveness to glycophorin A coexpression. Interestingly, although AE1 SAO is not dominant-negative in the red cell (44), glycophorin A enhancement of A858D function in oocytes was strongly inhibited in the presence of coexpressed SAO (38). This inhibitory effect of coexpressed AE1 SAO was also evident, though more modest, in the presence of other recessive mutants (38, 45). The ability of glycophorin A to effect rescue of the recessive dRTA mutation AE1 G701D did not require normally homodimeric glycophorin A. Four mutants with impaired homodimerization retained the ability to enhance wild-type AE1 translocation to the oocyte surface, as well as to rescue AE1 G701D function to normal levels (46).

The two most C terminal of the recessive dRTA mutations in AE1 reside in a region implicated in anion transport function. Covalent modification of A858C or of S852C in an otherwise Cys-less AE1 inhibited Cl^-/HCO_3^- exchange by 40% in 293 HEK cells (17). Thus the integrity of the ecto-loop between putative TMs 13 and 14 appears important to AE1 function, especially in the absence of glycophorin A in intercalated cells.

Recessive and compound heterozygous AE1 mutations show functional defects that account for dRTA without anemia, due to loss-of-function or hypofunction in the Type A intercalated cell to a greater degree than in the erythrocyte. However, a role for altered targeting of kAE1 remains possible. One patient with SAO and dRTA (genetically yet undefined) responded to oral bicarbonate loading with a normally elevated value for the "urine minus blood carbon dioxide tension difference" (U-B pCO_2) in an alkaline urine, whereas patients with dRTA usually fail to elevate urine pCO_2 in this setting (47). Although consistent with accumulation of a mis-targeted mutant AE1 in the apical membrane of the intercalated cell, other explanations are possible for the result of this highly sensitive but less specific test (48).

The AE1 $(-/-)$ mouse (49, 50) has runting, and severe anemia with very fragile red cell membranes. These AE1 $(-/-)$ membranes are 50% depleted of ankyrin-1 and devoid of the AE1-binding proteins 4.2 and glycophorin A (51), despite retention of apparently normal spectrin-actin cytoskeletal ultrastructure. Although AE1 expression was absent in AE1 $(-/-)$ mouse kidney (49), vH^+-ATPase-expressing intercalated cells remained present (A. K. Stuart-Tilley, L. L. Peters & S. L. Alper, unpublished data). The targeted disruption of AE1 exon 3 in one of the mouse lines, despite its intended preservation of the kAE1 open reading frame (50), affected the kidney-specific promoter of the *AE1* gene as evidenced by loss of *kAE1* mRNA from kidney (B. E. Shmukler & S. L. Alper, unpublished data).

ACID-BASE DISORDERS IN DISEASES OF SLC26A Cl^-/BASE EXCHANGERS

Alkali-secreting Type B intercalated cells express Cl^-/HCO_3^- exchange activity in their apical membranes. In the mouse, this activity is encoded by the SLC26 anion exchanger gene family member, pendrin (SLC26A4) on chromosome 7q22 (52, 53). In humans, pendrin mutations cause the Pendred syndrome of deafness

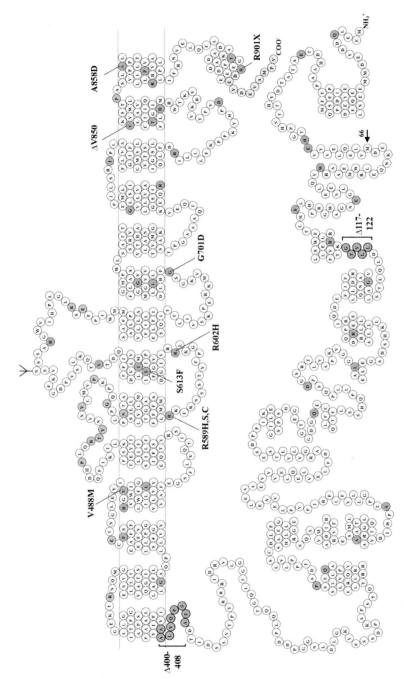

See text page C-2

Figure 2 Model of human AE1 secondary structure highlighting sites of missense or in-frame deletion mutations associated with dRTA (*green*); frameshift, nonsense, in-frame deletion, and missense mutations associated with HS and other erythroid dyscrasias (*orange*); and sites of blood group polymorphisms (*blue*). Consecutive green residues at the C terminus represent the dRTA mutation *R901X*. Consecutive orange residues at Δ117–22 and Δ400–408 represent in-frame deletions. Met 66 (*arrow*) is the first residue of the kidney AE1 isoform kAE1. Secondary structure of residues 55–356 in the N-terminal cytoplasmic domain is based on the crystal structure (131). Arrangement of the transmembrane helices is adapted from Wood (132) and Popov et al. (16). Topographical disposition of TMs 1–9 is supported by covalent labeling or proteolytic cleavage data. The uncertain disposition of TMs 10–12 may include a re-entrant loop (16) or an elongated, flexible transmembrane span (17), either of which has been postulated to contribute to the anion translocation pathway. Approximation of the C- terminal tail to N-terminal residues of AE1 is not meant to imply their interaction.

and variably penetrant hypothyroid goiter. With its broad anion selectivity (54, 55), pendrin likely mediates Cl^-/I^- exchange in thyrocytes and perhaps Cl^-/HCO_3^- or Cl^-/formate exchange in the cochlea. Deaf pendrin $(-/-)$ mice are not spontaneously alkalotic, as would be predicted if pendrin-mediated HCO_3^- secretion by Type B intercalated cells were essential under normal dietary conditions. However, perfusion of cortical collecting ducts isolated from deoxycorticosterone-treated pendrin $(-/-)$ mice demonstrated absence of the normally activated bicarbonate secretion (53). An upregulatory mutation in pendrin might conceivably cause a mild or incomplete dRTA.

The AE4 Cl^-/HCO_3^- exchanger has also been proposed as an apical Cl^-/HCO_3^- exchanger of Type B intercalated cells (56). Conflicting reports on rat AE4 localization place it in either the Type B intercalated cell apical membrane (57) or the Type A cell basolateral membrane (58). Whatever the localization of AE4, genetics in human and mouse suggest that AE4 is central neither to Type A cell-mediated acid secretion nor to Type B cell-mediated bicarbonate secretion.

The pendrin gene is immediately adjacent to the DRA (downregulated in adenoma) gene (*SLC26A3*), encoding a transporter polypeptide of nearly 50% amino acid identity. Wild-type DRA mediates Cl^-/HCO_3^- exchange (59) in the apical membrane of the enterocyte. Mutations in the DRA gene cause recessive congenital chloride-losing diarrhea (60). This syndrome is associated with prenatal-onset watery diarrhea, metabolic alkalosis, dehydration, and severe electrolyte disturbances. Untreated, the volume contraction can lead to renal failure. After correction of serum electrolytes, fecal $[Cl^-]$ is always greater >90 mM (61).

The relative importance of bicarbonate transport in the physiological function of two other disease-associated SLC26 anion exchangers remains uncertain. The DTD (*SLC26A2*) gene on chromosome 5q32 encodes a sulfate/chloride/oxalate exchanger (62), and mutations in the gene cause recessive diastrophic chondrodysplasia. The disease has been attributed to inadequate sulfate uptake into chondrocytes, with consequent inadequate sulfation of cartilage glycosaminoglycans. CFEX/PAT (SLC26A6) (64) is an apical Cl^-/formate/bicarbonate exchanger of the proximal tubule apical membrane (65), which is also expressed in pancreas. Although CFEX likely contributes to proximal tubular Cl^- reabsorption, roles for CFEX in proximal nephron regulation of acid-base balance and in pancreatic bicarbonate secretion remain undefined.

vH^+-ATPase Subunit Mutations in Distal Renal Tubular Acidosis

Although the genes encoding the many subunits of the vH^+-ATPase were long considered as candidate genes for dRTA, mutations in genes encoding so widely expressed an activity were thought unlikely to generate the restricted phenotype of primary dRTA. However, the combined results of whole-genome screening of large recessive dRTA pedigrees with the sequencing of the human genome have led to rejection of this initial concern. Some subunits of the vH^+-ATPase are members of gene families with tissue-specific expression of particular members. Mutations in two vH$^+$-ATPase gene members of such multigene families have been shown to cause recessive dRTA. Their expression is restricted entirely or nearly entirely to the kidney (66, 67).

The vH$^+$-ATPases are a family of multisubunit ATP-dependent proton pumps responsible for acidification of intracellular organelles and acidification of lumena or interstitial spaces adjacent to cell plasma membranes (68, 69). The vH$^+$-ATPases are hetero-oligomeric complexes composed of 13 polypeptide types. These can be fractionated into a soluble cytoplasmically disposed V_1 domain of $\sim$570 kDa, which encompasses the ATPase activity, and a membrane-associated V_o domain of $\sim$260 kDa that includes the proton translocation pathway (Figure 3). The X-ray crystal structure of the V_1-related F_1 domain from the structurally related mitochondrial F-ATPase has been solved (70), and the electron microscopic image analysis of the F_0 domain from bovine brain clathrin-coated pits is available at 21 Å resolution (71).

The V_1 domain is composed of 8 subunit types A-H with molecular masses (in kDa) of 70 (A), 60 (B), 40 (C), 34 (D), 33 (E), 14 (F), and 16 (G), with a stoichiometry estimated at $A_3B_3C_1D_1E_1F_1G_2H_1$. The V_0 domain is composed of 5 subunit types a (100–116 kDa), c and c' (each 17 kDa), c'' (19 kDa), and d (38 kDa), with a stoichiometry of $a_1d_1c_1''c_6c_6'$. The proteolipid subunits c, c', and c'', each

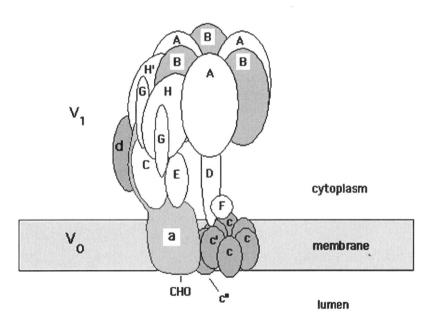

Figure 3 Schematic of the hetero-oligomeric vH$^+$-ATPase. dRTA mutations have been defined in the ATP6B1 isoform of the B subunit in the V_1 cytoplasmic ATPase complex and in the ATP6N1B isoform of the *a* subunit of the V_0 transmembrane H$^+$ pore complex (*shaded*). Modified from (72) with permission.

predicted to encode four transmembrane α-helices, are thought to assemble into a ring structure within the lipid bilayer (72). The V_1 domain likely is attached to the V_0 domain by a central stalk made up of subunit D. By analogy with mechanistic studies on the F-ATPase, hydrolysis of ATP by the catalytic site of subunit A is thought to force rotation of the stalk like a crankshaft inside the A_3B_3 hexamer. Rotation of the D subunit crankshaft in turn drives rotation of the proteolipid ring (73). Rotation of the proteolipid ring with respect to the immobile a and associated subunits is thought to open the H^+ permeability pathway at or near the interface between the a subunit and some of the proteolipid subunits.

Whole-genome linkage analysis by Karet et al. mapped a gene for recessive dRTA to chromosome 2p13 with a multipoint lod score of 7.34 (66). Although the B1 subunit of the vH^+-ATPase (ATP6B1) had been previously assigned to chromosome 2q, its candidacy was pursued because in kidney its expression is restricted to intercalated cells, whereas the B2 subunit is expressed in both the brush border crypts of proximal tubule and widely in other tissues (74). The B1 gene was indeed remapped within the dRTA maximum likelihood interval, cloned, and its 14 exons sequenced. Fifteen mutations distributed throughout the 513 amino acid B1 sequence were identified in 19 of 62 evaluated kindreds, which included 7 missense mutations in addition to premature terminations, frameshifts, and splice-site mutations (66).

Bilateral sensorineural hearing loss was documented in 87% of evaluated families. Immunocytochemical analysis confirmed expression of B1 subunit in the mouse cochlea and endolymphatic sac (66), including the interdental cells and endolymphatic sac cells whose basolateral membranes contain abundant AE Cl^-/HCO_3^- exchanger (75). The failure of bicarbonate therapy to correct deafness while successfully treating systemic symptoms of dRTA suggests that transepithelial acid secretion is required for normal cochlear development and hair cell survival. The restricted sites of expression of ATP6B1 explain the limited phenotype of mutant-associated disease because ATP6B2 is thought to serve the same function in tissues other than intercalated cells and cochlea.

The mouse B1 gene encodes a B1 polypeptide 93% identical to human B1 gene. B1 subunit $(-/-)$ mice are born in Mendelian ratios, but unlike humans with B1 mutations, grow at normal rates. However, the mice do fail to acidify their urine in response to oral acid loading. Thus absence of expression of the vH^+-ATPase gene in the mouse has produced a model of recessive incomplete dRTA (76).

These findings strongly suggested other tissue-specific subunits of vH^+-ATPase are candidate dRTA genes. Indeed, in a second group of families, recessive dRTA with preserved hearing mapped to chromosome 7q33-34 (77) in which a maximum likelihood interval of 4 cM was defined (67). Within a BAC previously mapped to this interval, an EST was found to encode a sequence homologous to the widely transcribed ATP6N1A a subunit of the vH^+-ATPase V_0 subunit. The new gene, *ATP6N1B* (now called *a4*), encoded in 23 exons a protein of 840 amino acids 61% identical to *ATP6N1A*, and with 50% and 40% respective identities to homologs in *Drosophila melanogaster* and *Saccharomyces cerevisiae*. The cytoplasmic N-terminal 250 amino acids preceded 6 to 9 predicted transmembrane helices, several potential N-glycosylation sites, and a short C-terminal tail. Initial Northern analysis revealed expression only in fetal and adult kidney. Eight mutations in as many kindreds included the three missense mutations *P524L*, *M580T* (invariant among all eukaryotic orthologs), and the charge-altering *G820R*, predicted to lie within a transmembrane span. Initial immunocytochemical studies localized ATP6N1B polypeptide to the apical membrane of Type A intercalated cells of human kidney cortex, but not to proximal tubular cells (67). Thus its distribution in human kidney parallels that of ATP6B1.

The syntenic mouse *Atp6n1b* gene has recently been cloned and encodes an *a4* polypeptide 85% identical to the human ortholog. RT-PCR analysis confirms kidney-specific expression. Unlike the human *a4* 116-kDa subunit, the mouse *a4* 105-kDa V_0 subunit of vH$^+$-ATPase is expressed in apical membrane of proximal tubule and in occasional cells of epidydimis and vas deferens, and perhaps also in basolateral membrane of Type B intercalated cells, as well as in apical membrane of Type A cells (78).

Additional recessive dRTA mutations in both *ATP6B1* and *ATP6N1B* genes have recently been found, including new missense mutations. In addition, families have been identified in which recessive dRTA maps no previously identified loci (79), suggesting that further recessive dRTA genes will be identified.

vH$^+$-ATPase Mutations in Osteopetrosis

ATP6N1B is only one of four *a* subunits in human and mouse. The *S. cerevisiae* genome encodes two *a* subunits that differ in their proton pump energy-coupling efficiency, subcellular localization, and in their subunit association regulation by cellular metabolic status (80). The *C. elegans* genome encodes four *a* subunits that differ in their cell type localization and developmental stage of expression (81). In human and mouse, the bone disease osteopetrosis is associated with mutations in a different *a* subunit isoform.

Autosomal-recessive osteopetrosis is a rare early-onset disorder of abnormally dense bones caused by failure of osteoclast bone resorption and remodeling (82). Clinical findings include osteosclerosis, pancytopenia due to secondary hematopoietic dysfunction, and hepatosplenomegaly. Associated bony compression of cranial nerves often leads to progressive loss of hearing and vision (but primary degeneration may also contribute). Life expectancy rarely exceeds 20 years, but bone marrow transplantation can be curative. In this complex of diseases, impaired osteoclast-mediated acidification of the bony lacunar space leads to decreased dissolution of calcium apatite at the bone front.

The apical vH$^+$-ATPase of the osteoclast is pharmacologically distinct from the renal vH$^+$-ATPase and contains in its V_0 complex the *a3* subunit OC116 (83). OC116 was mapped to chromosome 11q13, syntenic with the murine (*oc*) osteosclerosis mutation (84). The *oc* gene was found to be the mouse ortholog of OC116/*a3*. The *oc* mutation in *a3* is a 1.6-kb deletion that includes the translation initiation site (85). An engineered $a3$ knockout mouse also exhibited severe osteopetrosis. Osteoclasts isolated from these animals failed to generate resorption pits on bone chips, but blood and urine pH, renal vH^+-ATPase activity, and hepatic lysosomal vH^+-ATPase activity were normal (86). In humans, mutations in the $a3$ gene account for $\sim$50% of kindreds with recessive osteopetrosis. Nearly all mutations discovered are frameshift or termination mutations (82, 83). The murine osteopetroses resulting from genetic disruption of the mouse genes encoding the c-src tyrosine kinase and the cellular differentiation factor macrophage-colony stimulating factor have not yet been encountered among human osteopetroses.

An alternatively spliced transcript of OC116/$a3$ encodes the lymphocyte $a3$ protein TIRC7, an N-terminally truncated version beginning at $a3$ amino acid 217. Anti-TIRC7 antibodies prevent T cell proliferation triggered by alloantigen or mitogen. Because a soluble fragment of TIRC7 can produce T cell unresponsiveness that can be rescued by interleukin-2, TIRC7 is considered a lymphocyte costimulatory molecule that binds to a surface receptor. Anti-TIRC7 antibody administration in a rat model of acute renal allograft rejection prolonged graft survival (87). However, TIRC7 remains to be implicated in lymphocyte vH^+-ATPase function or to be linked to disease.

ACID-BASE TRANSPORT DISORDERS
OF Cl^- CHANNEL DISEASES

The ability of the vH^+-ATPase to acidify any lumenal or organellar space can be limited by the associated membrane's conductances for H^+, Cl^-, and K^+. H^+ conductance appears to vary progressively along the secretory pathway (88), but Cl^- conductance is likely also regulated (89). In the absence of these conductances, the vH^+-ATPase can generate a sizable transmembrane electrical potential. However, in order to generate the $\sim$1000-fold pH gradient of which it is capable, the H^+ pump-generated electrical gradient must be shunted by Cl^- conductance, which allows movement of Cl^- ion in parallel with H^+. Thus Cl^- channels are candidate disease genes for disorders of acid transport.

One such acid transport disease is another form of recessive osteopetrosis. Kornak et al. reported that knockout of the mouse ClC-7 gene leads to early-onset osteopetrosis and retinal degeneration (90). ClC-7 was detected in the ruffled lacunar membrane of the osteoclast, the retina, and other eye tissues, in addition to expression elsewhere. The developmental time course of retinal degeneration preceeded apparent bony compression of the optic nerve, suggesting an intrinsic retinal defect. Isolated ClC-7 $(-/-)$ osteoclasts had poorly developed ruffled borders that failed to resorb bone in vitro despite the presence of vH^+-ATPase at the osteoclast surface. However, these osteoclasts retained the ability to acidify intracellular vesicles that likely correspond to lysosomes or prelysosomal compartments.

The clinical similarity of the mouse disease with human recessive osteopetrosis led to analysis of the CLCN7 gene in 12 cases of the human disease. In one of these

12 cases, compound heterozygosity for CLCN7 mutations Q555X and R762Q was exhibited. Only the missense mRNA was present in patient tissue, but its ClC7 translation product was undetectable in the patient's fibroblasts (90).

The ClC-5 chloride channel was identified as the gene mutated in Dent's disease (also referred to as X-linked recessive nephrolithiasis and X-linked recessive hypophosphatemic rickets). Positional cloning was aided by identification of a microdeletion in one patient (91). Dent's disease is characterized by renal proximal tubular low-molecular-weight proteinuria, hyperphosphaturia, and hypercalciuria that often leads to nephrocalcinosis and nephrolithiasis. Untreated cases sometimes progress to renal osteomalacia and to renal insufficiency, occasionally associated with secondary incomplete dRTA (92).

In kidney, ClC-5 polypeptide is expressed in proximal tubule at the base of the brush border and in Type A intercalated cells of cortex, as well as in thick ascending limb (93). Its absence in ClC-5 ($-/-$) mice (94, 95) and its 90% reduction in a ClC-5 antisense ribozyme transgenic mouse (96) were associated with microproteinuria and with impaired in vivo proximal tubular endocytosis of several types of tagged filtered proteins. However, the several mouse models differed in the penetrance of hypercalciuria, nephrocalcinosis, and bone disease. Endocytosis regulated by CLC-5 likely depends on a proline-based internalization signal in the C-terminal cytoplasmic domain, which can interact with WW-domain-containing proteins such as ubiquitin ligases (97). Direct demonstration of ClC-5's requirement for endosomal acidification is still lacking. However, H^+-ATPase-mediated lumenal acidification was not reduced in isolated perfused CCD or proximal tubule from the antisense ribozyme mouse (98), suggesting that CLC-5 is not required for proximal or distal H^+ secretion in the mouse, or alternatively that its complete absence is required to detect its contribution to lumenal acid secretion. In interesting contrast, expression of CFTR appears to be required for vH^+-ATPase function in mouse intercalated cells through a mechanism independent of Cl^- channel activity (99). Levels of the intercalated cell Cl^- channel AQP6 (100, 101) are regulated by acid-base loading protocols (102), but a role for AQP6 in vH^+-ATPase-mediated lumenal or vesicular acidification remains uncertain.

Absence of ClC-3 chloride channel in the ClC-3 ($-/-$) mouse led to reduced body size and to postnatal degeneration of retina and hippocampus. Although swelling-activated Cl^- conductance was unchanged in primary hepatocytes and pancreatic acinar cells of ClC3 ($-/-$) mice, the rate and extent of synaptic vesicle acidification was reduced compared with that of wild-type littermates (103). This constitutes the first direct evidence of involvement of a mammalian ClC Cl^- channel in organellar acidification. It seems likely, although unproven, that the pathology of the CLC-3 ($-/-$) mouse derives entirely from a reduction of organellar Cl^- conductance. Absence of the ClC-2 chloride channel in the ClC-2 ($-/-$) mouse led to degeneration of male germ cells, postnatal degeneration of retinal photoreceptor cells (which never developed outer segments), and reduced transepithelial currents across retinal pigment epithelium (104). However, data on organellar acidification were not presented.

Recent data on CFTR-regulated bicarbonate transport has drawn increasing interest. In a recombinant overexpression system, CFTR-associated bicarbonate transport (but not chloride currents) correlated with clinical state of pancreatic sufficiency, an index of ductal HCO_3^- secretion (105). However, the correlation was less good for high sweat $[Cl^-]$ (106). CFTR may increase HCO_3^- secretion in pancreatic duct (and perhaps also in bronchial submucous gland and other tissues) either as an intrinsic HCO_3^- channel/transporter or as a regulator of apical Cl^-/HCO_3^- exchangers (likely DRA, CFEX/PAT and/or other SLC26 anion exchangers).

NBC1 Mutations in Recessive Proximal Renal Tubular Acidosis

Permanent, isolated proximal RTA is a rare autosomal-recessive disease characterized by impaired proximal tubular bicarbonate reabsorption with maintenance of other proximal reabsorption functions, such as those for glucose, amino acids, phosphate, and citrate. This type of proximal RTA can present with growth retardation alone, or together with ocular abnormalities such as band keratopathy, cataracts, and glaucoma, and with mental retardation (107). Dietary bicarbonate supplementation is difficult because the proximal tubular capacity for proximal bicarbonate reabsorption is greatly reduced, and compensatory upregulation of bicarbonate reabsorption in more distal nephron segments is limited. However, very-high-dose bicarbonate supplementation can enhance growth even if correction of metabolic acidosis is incomplete (108).

Na^+-bicarbonate cotransport across the basolateral membrane of proximal tubular cells is required for proximal bicarbonate reabsorption and is almost certainly mediated by the kidney NBC1 isoform of the electrogenic Na^+-bicarbonate cotransporter, kNBC1 (109, 110). Therefore, the human NBC1 (*SLC4A4*) gene (111) was evaluated as a candidate gene for isolated proximal RTA in two families, and two missense mutations were found (112). R510H lies within the transmembrane domain of NBC1 at or near the exofacial end of putative TM4, and the R298S mutation is in the putative N-terminal cytoplasmic domain. Initial measurement of Cl^--independent, amiloride-resistant pH_i recovery in transfected ECV304 cells acid-loaded by nigericin pretreatment indicated that both mutant polypeptides functioned at $\sim$55% of the wild-type rate (112).

Two additional NBC1 mutations have been found in families with proximal RTA. The *Q29X* mutation (107) severely truncates the 1007 amino acid knBC1 protein of kidney, but leaves intact the pNBC1 polypeptide of pancreatic duct and other tissues, whose variant N-terminal sequence (85 unique amino acids in place of the 41 N-terminal residues of kAE1) is generated by alternate promoter usage (111). The missense mutation *S427L* has been found in an Israeli family (113). Preliminary data in *Xenopus* oocytes suggest that the mutant NBC1 polypeptide R510H traffics abnormally (114) and that R298S and S427L mutant polypeptides also exhibit impaired transport activities (M. F. Romero, personal communication).

No pancreatic abnormalities have been reported to date in these patients. However, the ocular abnormalities that accompany isolated proximal RTA in these patients are consistent with the finding that both knBC1 and pnBC1 are expressed in multiple ocular tissues of human and rat (115). In addition, human lens epithelial cell Na^+-bicarbonate cotransport and total NBC1 polypeptide are reduced in parallel by recombinant adenoviral transduction of NBC1 antisense ribozyme. Interestingly, the proximal RTA patient with the *Q29X* NBC1 mutation predicted to preserve pnBC1 biosynthesis has, in addition to mental retardation, bilateral glaucoma without band keratopathy or cataract (107).

The stoichiometry of Na^+-bicarbonate cotransport in the intact proximal tubule is 3 HCO_3^-:1 Na^+, but for knBC1 expressed in intact *Xenopus* oocytes (116) or macropatches (117) it is 2:1. However, a 2:1 stoichiometry of transport would not support observed rates of proximal tubular bicarbonate reabsorption. Stoichiometry estimates in transfected monolayers of polarized epithelial cells suggest that whereas knBC1 and pnBC1 display equivalent stoichiometries of Na^+-HCO_3^- transport, cell type–specific factors can determine that transport stoichiometry (118). The ability of submicromolar $[Ca^{2+}]$ addition to the cytosolic face of oocyte macropatches to convert knBC1 stoichiometry from 2:1 to 3:1 (116) suggests a possible mechanism by which missense mutations might impair NBC1 transport function. Reduction of NBC1 transport stoichiometry from 3:1 to 2:1 by protein phosphorylation (119), perhaps mediated by NHERF (120), might also be altered in some mutants. The shift in stoichiometry from 2:1 to 3:1 may represent a change in one of the anion-binding sites from a preference for HCO_3^- to a preference for CO_3^{2-} (116).

The NHE3 $(-/-)$ mouse exhibits a mild metabolic acidosis, with greatly reduced proximal tubular HCO_3^- reabsorption and distal compensation (121) and lacks proximal tubular Cl^-/formate exchange (122). Although, human *NHE3* mutations have not yet been reported in proximal RTA, *NHE3* remains a candidate or a modifier gene for the disease.

Carbonic Anhydrase II Mutations in Mixed Renal Tubular Acidosis

The earliest example of a heritable renal tubular acidosis for which a cause was found was carbonic anhydrase II deficiency (123). This soluble, cytosolic enzyme is widely expressed. Patients with this deficiency exhibit osteopetrosis and cerebral calcification, as well as a mixed RTA with proximal and distal components. In different kindreds, mild or severe mental retardation has been described. Three of the 13 mutations described to date account for ~90% of patients. Bone marrow or stem cell transplantation can correct the osteopetrosis and halt progression of cerebral calcification, but cannot correct the mixed RTA and attendant growth retardation (124).

A carbonic anhydrase II-null mouse was created during an ethylnitrosourea mutagenesis screening program (125). This mouse had renal tubular acidosis with respiratory compensation (126) but lacked osteopetrosis (125). Intercalated cells developed normally in this mouse, but disappeared from the outer medulla and were depleted from cortex during aging (127). The carbonic anhydrase II-null mouse provided the first example of transient therapeutic correction of a genetic renal defect by gene therapy (128). Retrograde injection into the renal pelvis of a liposome suspension containing plasmid DNA encoding CAII led to renal parenchymal expression of CAII protein and partial, transient correction of the urinary acidification defect. However, because precise cellular localization of CAII polypeptide was not performed, the developmental consequences to collecting duct cell phenotype of restored CAII expression were not determined (125).

Acid-Base Disturbances Secondary to Genetic Disorders of Renal Na^+ Reabsorption

Genetic impairment of Na^+ reabsorption by the ENaC Na^+ channel of the collecting duct principal cell apical membrane diminishes the electrical driving force favoring intercalated cell H^+ secretion, as well as principal cell K^+ secretion. Thus hyperkalemic metabolic acidosis usually accompanies hypotensive salt-wasting syndromes. In addition to recessive deficiencies of aldosterone synthase and steroid-21-hydroxylase, these syndromes include dominant and recessive forms of pseudohypoaldosteronism Type I, due, respectively, to heterozygous loss-of-function mutations in the mineralocorticoid receptor and to homozygous loss-of-function mutations in the α, β, or γ subunits of the ENaC Na^+ channel (129).

Conversely, genetic enhancement of principal cell ENaC-mediated Na^+ reabsorption is accompanied by increased collecting duct H^+ and K^+ secretion, leading to hypokalemic metabolic alkalosis in the setting of either hypertensive Na^+ retention or hypotensive salt wasting. Examples include hypertension-associated upregulatory mutations of aldosterone action such as glucocorticoid-remediable aldosteronism, 11β-hydroxysteroid dehydrogenase deficiency (apparent mineralocorticoid excess), and an activating mutation in the mineralocorticoid receptor rendering it sensitive to further activation by steroids lacking 21-hydroxyl groups such as progesterone. Hypokalemic metabolic alkalosis can also accompany upregulatory mutations in the β and γ subunits of ENaC in the hypertensive Liddle's syndrome (129).

Hypokalemic alkalosis in the setting of salt-wasting accompanies loss-of-function mutations of the distal convoluted tubule thiazide receptor NCC1 in Gitelman's syndrome, as well as loss-of-function mutations in any of three components required for NaCl reabsorption by the thick ascending limb of Henle. These include the bumetanide receptor NKCC2 of the apical membrane, the apical membrane K^+ channel ROMK, and the basolateral Cl^- channel CLCKNB (129). All of these salt-wasting disorders increase lumenal Na^+ presentation to ENaC in the principal cell apical membrane, with consequently enhanced H^+ and K^+ secretion.

A novel class of mutation is represented by the rare autosomal-dominant pseudohypoaldosteronism Type II (PHAII), a thiazide-responsive, chloride-dependent, hyperkalemic hypertension sometimes associated with metabolic acidosis. Two genes recently identified in PHAII kindreds encode two isoforms of WNK serine-threonine kinases. WNK1 polypeptide is present throughout epithelial cells of the distal convoluted tubule and collecting duct. WNK4 is restricted to tight junctions of distal convoluted tubule cells but appears both cytoplasmic and junctional in collecting duct cells (130). The presumed ion transporter substrates or downstream targets of the WNK kinases, and their relationship to urinary H^+ secretion, remain to be determined.

CONCLUSION

Acid-base transport is essential to all cells and tissues. Some acid-base transporter polypeptides or subunits are expressed in so wide a range of cell types, or perform functions with so little available redundancy, that loss-of-function mutations might be embyronic lethal. Those genetic disorders of acid-base transport that allow birth and postnatal development tend to have restricted spectra of tissue involvement. This may reflect restriction of mutant gene product expression to the tissues phenotypically affected or indicate that the mutant gene product is normally rate-limiting only in the affected tissues. Such disorders include those arising from mutations in the AE1, DRA, and pendrin Cl^-/HCO_3^- exchangers, select subunits of the vH^+-ATPase, the NBC1 Na^+-HCO_3^- cotransporter, carbonic anhydrase II, certain organellar Cl^- channels, and gene products contributing to renal Na^+ reabsorption.

The growing number of mouse models of acid-base transport(er) disease reveals that orthologous mutations in mouse and human often exhibit phenotypes that, though similar, differ in important ways. This survey of genetic diseases of acid-base transporters highlights the importance of parallel investigation in multiple experimental systems. Studies of these acid-base transporters must be carried out at molecular, cellular, genetic, and organismic levels. At the organismic level, studies are required in genetically defined model organisms, in mouse, and in humans. Each type of investigation will yield its particular insights.

Visit the Annual Reviews home page at www.AnnualReviews.org

LITERATURE CITED

1. Seldin D, Giebisch G, eds. 2001. *The Kidney.* Philadelphia: Lippincott, Williams & Wilkins. 2942 pp. 3rd ed.
1a. Bevensee M, Aronson PS, Alper SL, Boron WF. 2001. Intracellular pH regulation. See Ref. 1, pp. 391–444
2. Hamm LL, Alpern RJ. 2001. Cellular mechanisms of renal tubular acidification. See Ref. 1, pp. 1935–80
3. Rodriguez-Soriano J. 2000. New insights into the pathogenesis of renal tubular

acidosis—from functional to molecular studies. *Pediatr. Nephrol.* 14:1121–36
4. Tsuruoka S, Swenson ER, Petrovic S, Fujimura A, Schwartz GJ. 2001. Role of basolateral carbonic anhydrase in proximal tubular fluid and bicarbonate absorption. *Am. J. Physiol. Renal Physiol.* 280:F146–F54
5. Schwartz GJ, Kittelberger AM, Barnhart DA, Vijayakumar S. 2000. Carbonic anhydrase IV is expressed in H^+-secreting

cells of rabbit kidney. *Am. J. Physiol. Renal Physiol.* 278:F894–F904

6. Karet FE. 2000. Inherited renal tubular acidosis. *Adv. Nephrol. (Necker Hosp.)* 30:147–62

7. Battle D, Ghanekar H, Jain S, Mitra A. 2001. Hereditary distal renal tubular acidosis: new understandings. *Annu. Rev. Med.* 52:471–84

8. Sabatini S, Kurtzman NA. 2001. Biochemical and genetic advances in distal renal tubular acidosis. *Semin. Nephrol.* 21:94–106

9. Shayakul C, Alper SL. 2000. Inherited renal tubular acidosis. *Curr. Opin. Nephrol. Hypertens.* 9:541–46

10. Sahr KE, Taylor WM, Daniels BP, Rubin HL, Jarolim P. 1994. The structure and organization of the human erythroid anion exchanger (AE1) gene. *Genomics* 24:491–501

11. Kollert-Jons A, Wagner S, Hubner S, Appelhans H, Drenckhahn D. 1993. Anion exchanger 1 in human kidney and oncocytoma differs from erythroid AE1 in its NH$_2$ terminus. *Am. J. Physiol. Renal Physiol.* 265:F813–F21

12. Brosius FC, Alper SL, Garcia AM, Lodish HF. 1989. The major kidney band 3 gene transcript predicts an amino-terminal truncated band 3 polypeptide. *J. Biol. Chem.* 264:7784–87

13. Alper SL. 1994. The band 3-related AE anion exchanger gene family. *Cell. Physiol. Biochem.* 4:265–81

14. Tanner M.J. 1997. The structure and function of band 3 (AE1): recent developments. *Mol. Membr. Biol.* 14:155–65

15. Chen J, Vijayakumar S, Li X, Al-Awqati Q. 1998. Kanadaptin is a protein that interacts with the kidney but not the erythroid form of band 3. *J. Biol. Chem.* 273:1038–43

16. Popov M, Li J, Reithmeier RAF. 1999. Transmembrane folding of the human erythrocyte anion exchanger (AE1, Band 3) determined scanning and insertional N-glycosylation mutagenesis. *Biochem. J.* 339:269–79

17. Fujinaga J, Tang X-B, Casey JR. 1999. Topology of the membrane domain of the human erythocyte anion exchange protein AE1. *J. Biol. Chem.* 274:6626–33

18. Groves JD, Tanner MJ. 1999. Structural model for the organization of the transmembrane spans of the human red-cell anion exchanger (band 3; AE1). *Biochem. J.* 344:699–711

19. Ota K, Sakaguchi M, Hamasaki N, Mihara K. 1998. Assessment of topogenic functions of anticipated transmembrane segments of human band 3. *J. Biol. Chem.* 273:28286–91

20. Chernova MN, Jiang L, Vandorpe DH, Hand M, Crest M, et al. 1997. Electrogenic sulfate/chloride exchange in *Xenopus* oocytes mediated by murine AE1 E699Q. *J. Gen. Physiol.* 109:345–60

21. Muller-Berger S, Karbach D, Konig J, Lepke S, Wood PG. 1995. Inhibition of mouse erythroid band 3-mediated chloride transport by site-directed mutagenesis of histidine residues and its reversal by second site mutation of Lys 558, the locus of covalent H$_2$DIDS binding. *Biochemistry* 34:9315–24

22. Tang XB, Kovacs M, Sterling D, Casey JR. 1999. Identification of residues lining the translocation pore of human AE1 plasma membrane anion exchange protein. *J. Biol. Chem.* 274:3557–64

23. Vince JW, Reithmeier RA. 2000. Identification of the carbonic anhydrase II binding site in the Cl$^-$/HCO$_3$$^-$-anion exchanger AE1. *Biochemistry* 39:5527–33

24. Sterling D, Reithmeier RA, Casey JR. 2001. A transport metabolism: functional interaction of carbonic anhydrase II and chloride/bicarbonate exchangers. *J. Biol. Chem.* In press

25. Wang DN, Sarabia VE, Reithmeier RA, Kuhlbrandt W. 1994. Three-dimensional map of the dimeric membrane domain of the human erythrocyte anion exchanger, Band 3. *EMBO J.* 14:3230–35

26. Jarolim P, Murray JL, Rubin HL, Taylor

WM, Prchal JT, et al. 1996. Characterization of 13 novel band 3 gene defects in hereditary spherocytosis with band 3 deficiency. *Blood* 88:4366–74

27. Bruce LJ, Cope DL, Jones GK, Schofield AE, Burley M, et al. 1997. Familial distal renal tubular acidosis is associated with mutations in the red cell anion exchanger (band 3, AE1) gene. *J. Clin. Invest.* 100:1693–707

28. Jarolim P, Shayakul C, Prabakaran D, Jiang L, Stuart-Tilley AK, et al. 1998. Autosomal dominant distal renal tubular acidosis is associated in three families with heterozygosity for the R589H mutation in the AE1 (band 3) Cl⁻/HCO3⁻-exchanger. *J. Biol. Chem.* 273:6380–88

29. Chernova MN, Humphreys BD, Robinson DH, Garcia A-M, Brosius FC, Alper SL. 1997. Functional consequences of mutations in the transmembrane domain and the carboxy-terminus of the murine AE1 anion exchanger. *Biochim. Biophys. Acta* 1329:111–23

30. Quilty JA, Reithmeier RA. 2000. Trafficking and folding defects in hereditary spherocytosis mutants of the human red cell anion exchanger. *Traffic* 1:987–98

31. Ribeiro ML, Alloisio N, Almeida H, Gomes C, Texier P, et al. 2000 Severe hereditary spherocytosis and distal renal tubular acidosis associated with the total absence of band 3. *Blood* 96:1602–4

32. Inaba M, Yawata A, Koshino I, Sato K, Takeuchi M, et al. 1996. Defective anion transport and marked spherocytosis with membrane instability caused by hereditary total deficiency of red cell band 3 in cattle due to a nonsense mutation. *J. Clin. Invest.* 97:1804–17

33. Lima PR, Gontijo JA. Lopes de Faria JB, Costa FF, Saad ST. 1997. Band 3 Campinas: a novel splicing mutation in the band 3 (AE1) gene associated with hereditary spherocytosis, hyperactivity of Na⁺/Li⁺ countertransport and an abnormal renal bicarbonate handling. *Blood* 90:2810–18

34. Rysava R, Tesar V, Jirsa M, Brabec V,

Jarolim P. 1997. Incomplete distal renal tubular acidosis coinherited with a mutation in the band 3 (AE1) gene. *Nephrol. Dial. Transplant* 12:1869–73

35. Karet FE, Gainza FJ, Gyory AZ, Unwin RJ, Wrong O, et al. 1998. Mutations in the chloride-bicarbonate exchanger gene AE1 cause autosomal dominant but not autosomal recessive distal renal tubular acidosis. *Proc. Natl. Acad. Sci. USA* 95:6337–42

36. Groves JD, Tanner MJ. 1992. Glycophorin A facilitates the expression of human band 3-mediated anion transport in *Xenopus* oocytes. *J. Biol. Chem.* 267:22163–70

37. Shayakul C, Jarolim P, Ideguchi H, Prabakaran D, Cortez D, et al. 1999. Use of a CA repeat polymorphism physically linked to the human AE1 (band 3) anion exchanger gene to diagnose autosomal dominant distal renal tubular acidosis associated with the AE1 mutation R589H, and immunocytochemical study of kidney from one patient. *J. Am. Soc. Nephrol.* 10:442A (Abstr.)

38. Bruce LJ, Wrong O Toye AM, Young MT, Ogle G, et al. 2000. Band 3 mutations, renal tubular acidosis and Southeast Asian ovalocytosis in Malaysia and Papua New Guinea: loss of up to 95% band 3 transport in red cells. *Biochem. J.* 350:41–51

39. Cowan CA, Yokoyama N, Bianchi LM, Henkemeyer M, Fritzsch B. 2000. EphB2 guides axons at the midline and is necessary for normal vestibular function. *Neuron* 26:417–30

40. Weber S, Soergel M, Jeck N, Konrad M. 2000. Atypical distal renal tubular acidosis confirmed by mutation analysis. *Pediatr. Nephrol.* 15:201–4

41. Tanphaichitr VS, Sumboonnanonda A, Ideguchi H, Shayakul C, Brugnara C, et al. 1998. Novel AE1 mutations in recessive distal renal tubular acidosis: rescue of loss-of-function by glycophorin A. *J. Clin. Invest.* 102:2173–79

42. Vasuvattakul S, Yenchitsomanus P, Vachuanichsanong P, Thuwajit P, Kaitwatcharachai C, et al. Autosomal recessive distal renal tubular acidosis associated with

Southeast Asian ovalocytosis. *Kidney Int.* 56:1674–82

43. Baehner RL, Gilchrist GS, Anderson EJ. 1968. Hereditary elliptocytosis and primary renal tubular acidosis in a single family. *Am. J. Dis. Child.* 115:414–19

44. Jennings ML, Gosselink PG. 1995. Anion exchange protein in Southeast Asian ovalocytes: heterodimer formation between normal and variant subunits. *Biochemistry* 34:3588–95

45. Shayakul C, Jariyawat S, Kaewkaukul N, Sophasan S. Functional rescue of anion exchanger 1 (AE1) G701D by glycophorin A is attenuated by co-expression of AE1 Δ400–408: a basis for transport defect in autosomal recessive distal renal tubular acidosis (dRTA). *J. Am. Soc. Nephrol.* 12:10A (Abstr.)

46. Young MT, Beckmann R, Toye AM, Tanner MJA. 2000. Red cell glycophorin A-band 3 interactions associated with the movement of band 3 to the cell surface. *Biochem. J.* 350:53–60

47. Kaitwatcharachai C, Vasuvattakul S, Yenchitsomanus P, Thuwajit P, Malasit P, et al. 1999. Distal renal tubular acidosis and high urine carbon dioxide tension in a patient with Southeast Asian ovalocytosis. *Am. J. Kidney Dis* 33:1147–52

48. Kurtzman N. 1990. Disorders of distal acidification. *Kidney Int.* 38:720–27

49. Peters LL, Shivdasani RA, Liu S-C, Hanspal M, John KM, et al. 1996. Anion exchanger 1 (band 3) is required to prevent erythrocyte membrane surface loss but not to form the membrane skeleton. *Cell* 86:917–27

50. Southgate CD, Chishti AH, Mitchell B, Yi SJ, Palek J. 1996. Targeted disruption of the murine erythroid band 3 gene results in spherocytosis and severe haemolytic anaemia despite a normal membrane skeleton. *Nat. Genet.* 14:227–30

51. Hassoun H, Hanada T, Lutchman M, Sahr KE, Palek J, et al. 1998. Complete deficiency of glycophorin A in red blood cells from mice with targeted inactivation of the band 3 (AE1) gene. *Blood* 91:2146–51

52. Everett LA, Glaser B, Beck JC, Idol JR, Buchs A, et al. 1997. Pendred syndrome is caused by mutations in a putative sulphate transporter gene (PDS). *Nat. Genet.* 17:411–22

53. Royaux IE, Wall SM, Karniski LP, Everett LA, Suzuki K, et al. 2001. Pendrin, encoded by the Pendred syndrome gene, resides in the apical region of renal intercalated cells and mediates bicarbonate secretion. *Proc. Natl. Acad. Sci. USA* 98:4221–26

54. Scott DA, Wang R, Kreman TM, Sheffield VC, Karniski LP. 1999. The Pendred syndrome gene encodes a chloride-iodide transporter. *Nat. Genet.* 21:440–43

55. Scott DA, Karniski LP. 2000. Human pendrin expressed in *Xenopus laevis* oocytes mediates chloride/formate exchange. *Am. J. Physiol. Cell Physiol.* 278:C207–C11

56. Tsuganezawa H, Kobayashi K, Iyori M, Araki T, Koizumi A, et al. 2001. A new member of the HCO3(−) transporter superfamily is an apical anion exchanger of beta-intercalated cells in the kidney. *J. Biol. Chem.* 276:8180–89

57. Petrovic S, Wang Z, Greeley T, Amlal H, Soleimani M. 2001. Apical Cl⁻/HCO3⁻-exchangers pendrin and AE4 in the kidney: localization and regulation in metabolic acidosis. *J. Am. Soc. Nephrol.* 12:7A (Abstr.)

58. Elkjaer ML, Hager H, Ishibashi K, Muallem S, Frokier J, Neilsen S. 2001. Laser confocal microscopical and immunoelectron micrscopical localization of anion exchanger AE4 in rat kidney. *J. Am. Soc. Nephrol.* 12:3A (Abstr.)

59. Melvin JE, Park K, Richardson L, Schlutheis PJ, Shull GE. 1999. Mouse down-regulated in adenoma (DRA) is an intestinal Cl⁻/HCO₃⁻-exchanger and is up-regulated in colon of mice lacking the NHE3 Na⁺/H⁺ exchanger. *J. Biol. Chem.* 274:22855–61

60. Hoglund P, Haila S, Socha J, Tomaszewski L, Saarialho-Kere U, et al. 1996. Mutations

of the Down-regulated in adenoma (DRA) gene cause congenital chloride diarrhoea. *Nat. Genet.* 14:316–19

61. Kere J, Lohi H, Hoglund P. 1999. Genetic disorders of membrane transport: congenital chloride diarrhea. *Am. J. Physiol. Gastrointest. Liver Physiol.* 276:G7–G13

62. Satoh H, Susaki M, Shukunami C, Iyama K, Negoro T, Hiraki Y. 1999. Functional analysis of diastrophic dysplasia sulfate transporter. *J. Biol. Chem.* 273:12307–15

63. Hastbacka J, de la Chapelle A, Mahtani MM, Clines G, Reeve et al. 1994. The diastrophic dysplasia gene encodes a novel sulfate transporter: positional cloning by fine-structure linkage disequilibrium mapping. *Cell* 78:1073–87

64. Lohi H, Kujala M, Kerkela E, Saarialho-Kere Kestila M, Kere J. 2000. Mapping of five new putative anion transporter genes in human and characterization of SLC26A6, a candidate gene for pancreatic anion exchanger. *Genomics* 70:102–12

65. Knauf F, Yang CL, Thomson RB, Mentone SA, Giebisch G, Aronson PS. 2001. Identification of a chloride-formate exchanger on the brush border membrane of renal proximal tubule cells. *Proc. Natl. Acad. Sci. USA* 98:9425–30

66. Karet FE, Finberg KE, Nelson RD, Nayir A, Mocan H, et al. 1999. Mutations in the gene encoding B1 subunit of the H^+-ATPase cause renal tubular acidosis with sensorineural deafness. *Nat. Genet.* 21:84–90

67. Smith AN, Skaug J, Choate KA, Nayir A, Bakkaloglu A, et al. 2000. Mutations in ATP6N1B, encoding a new kidney vacuolar proton pump 116-kD subunit, cause recessive distal renal tubular acidosis with preserved hearing. *Nat. Genet.* 26:71–75

68. Forgac M. 1999. Structure and properties of the vacuolar (H^+)-ATPases. *J. Biol. Chem.* 274:12951–54

69. Schoonderwoert VTG, Martens GJM. 2001. Proton pumping in the secretory pathway. *J. Membr. Biol.* 182:159–69

70. Abrahams JP, Leslie AG, Lutter R, Walker JE. 1994. Structure at 2,8 Å resolution of F-1 ATPase from bovine heart mitochondria. *Nature* 370:621–28

71. Wilkens S, Forgac M. 2001. Three-dimensional structure of the vacuolar proton ATPase proton channel by electron microscopy. *J. Biol. Chem.* 276:44064–68

72. Xu T, Vasilyeva E, Forgac M. 1999. Subunit interactions in the clathrin-coated vesicle vacuolar $(H+)$-ATPase complex. *J. Biol. Chem.* 274:28909–15

73. Panke O, Gumbiowski K, Junge W, Engelbrecht S. 2000. F-ATPase: specific observation of the rotating c-subunit oligomer of EF(0)EF(1). *FEBS Lett.* 472:34–38

74. Nelson R, Guo XL, Masood K, Brown D, Kalkbrenner M, Gluck S. 1992. Selectively amplified expression of an isoform of the vacuolar H(+)-ATPase 56 kilodalton subunit in renal intercalated cells. *Proc. Natl. Acad. Sci. USA* 89:3541–45

75. Stankovic KM, Brown D, Alper SL, Adams JC. 1997. Localization of pH regulating proteins H^+ ATPase and Cl^-/$HCO3^-$-exchanger in the guinea pig inner ear. *Hearing Res.* 114:21–34. Erratum. 1998. *Hearing Res.* 124:91–92

76. Finberg KE, Wang T, Wagner C, Geibel JP, Dou H, Lifton RP. 2001. Generation and characterization of H^+ ATPase B1 subunit-deficient mice. *J. Am. Soc. Nephrol.* 12:3A–4A (Abstr.)

77. Karet FE, Finberg KE, Nayir A, Bakkaloglu A, Ozen S, et al. 1999. Localization of a gene for autosomal recessive distal renal tubular acidosis with normal hearing (rdRTA2) to 7q33–34. *Am. J. Hum. Genet.* 65:1656–65

78. Smith AN, Finberg KE, Wagner CA, Lifton RP, Devonald MAJ, et al. 2001. Molecular cloning and characterization of Atp6n1b: a novel fourth murine vacuolar H^+-ATPase α-subunit gene. *J. Biol. Chem.* 276:42382–88

79. Stover EH, Bavalia C, Eady N, Borthwick KJ, Smith AN, Karet FE. 2001. Novel ATP6B1 and ATP6N1B mutations in autosomal recessive distal renal tubular

acidosis. *J. Am. Soc. Nephrol.* 12:560A (Abstr.)

80. Kawasaki-Nishi S, Nishi T, Forgac M. 2001. Yeast V-ATPase complexes containing different isoforms of the 100-kDa α subunit differ in coupling efficiency and in vivo dissociation. *J. Biol. Chem.* 276:17941–48

81. Oka T, Toyomura T, Hongo K, Wada Y, Futai M. 2001. Four subunit a isoforms of *C. elegans* vacuolar H$^+$-ATPase: cell-specific expression during development. *J. Biol. Chem.* In press

82. Sobacchi C, Frattini A, Orchard P, Porras O, Tezcan I, et al. 2001. The mutational spectrum of human malignant autosomal recessive osteopetrosis. *Hum. Mol. Genet.* 10:1767–73

83. Kornak U, Schulz A, Friedrich W, Uhlhaas S, Kremens B, et al. 2000. Mutations in the α3 subunit of the vacuolar H$^+$-ATPase cause infantile malignant osteopetrosis. *Hum. Mol. Genet.* 9:2059–63

84. Heaney C, Shalev H, Elbedour K, Carmi R, Staack JB, et al. 1998. Human autosomal recessive osteopetrosis maps to 11q13, a position predicted by comparative mapping of the murine osteosclerosis (oc) mutation. *Hum. Mol. Genet.* 7:1407–10

85. Scimeca J-C, Franchi A, Trojani C, Parrinello H, Grosgeorge J, et al. 2000. The gene encoding the mouse homolog of the human osteoclast-specific 116 kDa V-ATPase subunit bears a deletion in osteosclerotic (oc/oc) mutants. *Bone* 26:207–13

86. Li Y-P, Chen W, Liang Y, Li E, Stashenko P. 1999. *Atp6i*-deficient mice exhibit severe osteopetrosis due to loss of osteoclast-mediated extracellular acidification. *Nat. Genet.* 23:447–52

87. Heinemann T, Bulwin G-C, Randall J, Schnieders B, Sandhoff K, et al. 1999. Genomic organization of the gene coding for TIRC7, a novel membrane protein essential for T cell activation. *Genomics* 57:398–406

88. Wu MM, Grabe M, Adams S, Tsien RY, Moore HP, Machen TE. 2001. Mechanisms of pH regulation in the secretory pathway. *J. Biol. Chem.* 276:33027–35

89. Grabe M, Oster G. 2001. Regulation of organelle acidity. *J. Gen. Physiol.* 117:329–44

90. Kornak U, Kasper D, Bosl MR, Kaiser E, Schweizer M, et al. 2001. Loss of the ClC-7 chloride channel leads to osteopetrosis in mice and man. *Cell* 104:205–15

91. Lloyd SE, Pearce SH, Fisher SE, Steinmeyer K, Schwappach B, et al. 1996. A common molecular basis for three inherited kidney stone diseases. *Nature* 379:445–49

92. Kelleher CL, Buckalew VM, Frederickson ED, Rhodes DJ, Conner DA, et al. 1998. CLCN5 mutation Ser244Leu is associated with X-linked renal failure without X-linked recessive hypophosphatemic rickets. *Kidney Int.* 53:31–37

93. Gunther W, Luchow A, Cluzeaud F, Vandewalle A, Jentsch TJ. 1998. ClC-5, the chloride channel mutated in Dent's disease, colocalizes with proton pump in endocytically active kidney cells. *Proc. Natl. Acad. Sci. USA* 95:8075–80

94. Piwon N, Gunther W, Schwake M, Bosl M, Jentsch TJ. 2000. ClC-5 Cl$^-$ channel disruption impairs endocytosis in a mouse model for Dent's disease. *Nature* 408:369–73

95. Wang SS, Devuyst O, Courtoy PJ, Wang XT, Wang H, et al. 2000. Mice lacking renal chloride channel, CLC-5, are a model for Dent's disease, a nephrolithiasis disorder associated with defective receptor-mediated endocytosis. *Hum. Mol. Genet.* 9:2937–45

96. Luyckx VA, Leclercq B, Dowland LK, Yu AS. 1999. Diet-dependent hypercalciuria in transgenic mice with reduced CLC5 chloride channel expression. *Proc. Natl. Acad. Sci. USA* 96:12174–19

97. Schwake M, Friedrich T, Jentsch TJ. 2001. An internalization signal in ClC-5, an endosomal Cl$^-$ channel mutated in Dent's disease. *J. Biol. Chem.* 276:12049–54

98. Wagner CA, Enck A, Giebisch G, Yu AS, Geibel JP. 2001. Normal plasma membrane H⁺-ATPase function in the proximal tubule and collecting duct in a mouse model for Dent's disease. *J. Am. Soc. Nephrol.* 12:11A (Abstr.)

99. Wagner CA, Lukewille U, Breton S, Brown D, Grubb BR, et al. 2001. Complete loss of vacuolar H⁺-ATPase function in intercalated cells in cortical collecting ducts from CFTR knock-out mice. *J. Am. Soc. Nephrol.* 12:11A (Abstr.)

100. Yasui M, Kwon TH, Knepper MA, Neilsen S, Agre P. 1999. Aquaporin-6: an intracellular vesicle water channel protein in renal epithelia. *Proc. Natl. Acad. Sci. USA* 96:5808–13

101. Yasui M, Hazawa A, Kwon TH, Nielsen S, Guggino WB, Agre P. 1999. Rapid gating and anion permeability of an intracellular aquaporin. *Nature* 402:184–87

102. Promeneur D, Kwon TH, Yasui M, Kim GH, Frokiaer J, et al. 2000. Regulation of AQP6 mRNA and protein expression in rats in response to altered acid-base or water balance. *Am. J. Physiol. Renal Physiol.* 279:F1014–F26

103. Stobrawa SM, Breiderhoff T, Takamori S, Engel D, Schweizer M, et al. 2001. Disruption of ClC-3, a chloride channel expressed on synaptic vesicles, leads to a loss of the hippocampus. *Neuron* 29:185–96

104. Bosl RM, Stein V, Hubner C, Zdebik AA, Jordt SE, et al. 2001. Male germ cells and photoreceptors, both dependent on close cell-cell interactions, degenerate upon ClC-2 Cl⁻ channel disruption. *EMBO J.* 20:1289–99

105. Choi JY, Muallem D, Kiselyov K, Lee MG, Thomas PJ, Muallem S. 2001. Aberrant CFTR-dependent HCO₃⁻-transport in mutations associated with cystic fibrosis. *Nature* 410:94–97

106. Wine JJ. 2001. Cystic fibrosis: the 'bicarbonate before chloride' hypothesis. *Curr. Biol.* 11:R463–66

107. Igarashi T, Inatomi J, Sekine T, Seki G,

Shimadzu M, et al. 2001. Novel nonsense mutation in the Na/HCO₃ cotransporter gene (SLC4A4) in a patient with permanent isolated proximal renal tubular acidosis and bilateral glaucoma. *J. Am. Soc. Nephrol.* 12:713–18

108. Shiohara M, Igarashi T, Mori T, Komiyama A. 2000. Genetic and long-term data on a patient with permanent isolated proximal renal tubular acidosis. *Eur. J. Pediatr.* 159:892–94

109. Romero MF, Hediger MA, Boulpaep EL, Boron WF. 1997. Expression cloning and characterization of an electrogenic Na⁺/HCO3⁻-cotransporter. *Nature* 387:409–13

110. Soleimani M, Burnham CE. 2001. Na⁺-HCO₃⁻-cotransporters (nbc): cloning and characterization. *J. Membr. Biol.* 183:71–84

111. Abuladze N, Song M, Pushkin A, Newman D, Lee I, et al. 2000. Structural organization of the human NBC1 gene: kNBC1 is transcribed from an alternative promoter in intron 3. *Gene* 251:109–22

112. Igarashi T, Inatomi J, Sekine T, Cha SH, Kanai Y, et al. 1999. Mutations in SLC4A4 cause permanent isolated proximal renal tubular acidosis with ocular abnormalities. *Nat. Genet.* 23:264–66

113. Dinour D, Knecht A, Serban I, Holtzman EJ. 2000. A novel missense mutation in the sodium bicarbonate cotransporter (NBC-1) causes congenital proximal renal tubular acidosis with ocular defects. *J. Am. Soc. Nephrol.* 11:3A (Abstr.)

114. Sciortino CM, Romero MF. 2001. Functional characterization of a Na⁺/HCO3 cotransporter point mutation that results in Type II renal tubular acidosis. *FASEB J.* 15:A502 (Abstr.)

115. Usui T, Hara M, Satoh H, Moriyama N, Kagaya H, et al. 2001. Molecular basis of ocular abnormalities associated with proximal renal tubular acidosis. *J. Clin. Invest.* 108:107–15

116. Sciortino CM, Romero MF. 1999. Cation and voltage dependence of rat kidney

electrogenic Na$^+$-HCO$_3^-$-cotransporter, rkNBC, expressed in oocytes. *Am. J. Physiol. Renal Physiol.* 277:F611–F23

117. Muller-Berger S, Ducoudret O, Diakov A, Fromter E. 2001. The renal-HCO$_3^-$-cotransporter expressed in *Xenopus laevis* oocytes: change in stoichiometry in response to elevation of cytosolic Ca^{2+} concentration. *Pflügers Arch.* 442:718–28

118. Gross E, Hawkins K, Abuladze N, Pushkin A, Cotton CU, et al. 2001. The stoichiometry of the electrogenic sodium bicarbonate cotransporter NBC1 is cell-type dependent. *J. Physiol.* 531:597–603

119. Gross E, Abuladze N, Pushkin A, Sassani P, Kukkipati R, et al. 2001. Phosphorylation of Ser^{982} in knBC1 shifts the HCO$_3^-$:Na$^+$ stoichiometry from 3:1 to 2:1. *J. Am. Soc. Nephrol.* 12:4A (Abstr.)

120. Weinman EJ, Evangelista CM, Steplock D, Liu MZ, Shenolikar S, Bernardo A. 2001. Essential role for NHERF in cAMP-mediated inhibition of Na$^+$-HCO$_3^-$ Cotransporter in BSC-1 cells. *J. Biol. Chem.* In press

121. Schultheis PJ, Clarke LL, Meneton P, Miller ML, Soleimani M, et al. 1998. Renal and intestinal absorptive defects in mice lacking the NHE3 Na$^+$/H$^+$ exchanger. *Nat. Genet.* 19:282–85

122. Wang T, Yang CL, Abbiati T, Shull GE, Giebisch G, Aronson PS. 2001. Essential role of NHE3 in facilitating formate-dependent NaCl absorption in the proximal tubule. *Am. J. Physiol. Renal Physiol.* 281:F288–F92

123. Sly WS, Whyte MP, Sundaram V, Tashian RE, Hewett-Emmett D, et al. 1985. Carbonic anhydrase II deficiency in 12 families with the autosomal recessive syndrome of osteopetrosis with renal tubular acidosis and cerebral calcification. *N. Engl. J. Med.* 313:139–45

124. McMahon C, Will A, Hu P, Shah GN, Sly WS, Smith OP. 2001. Bone marrow transplantation corrects osteopetrosis in the carbonic anhydrase II deficiency syndrome. *Blood* 97:1947–50

125. Lewis SE, Erickson RP, Barnett LB, Venta PJ, Tashian RE. 1988. *N*-ethylnitrosourea-induced null mutation at the mouse Car-2 locus: an animal model for human carbonic anhydrase II deficiency syndrome. *Proc. Natl. Acad. Sci. USA* 85:1962–66

126. Lien YH, Lai LW. 1998. Respiratory acidosis in carbonic anhydrase II-deficient mice. *Am. J. Physiol. Lung Cell Mol. Physiol.* 274:L301–L4

127. Breton S, Alper SL, Gluck SL, Sly WS, Barker JE, Brown D. 1995. Depletion of intercalated cells from collecting ducts of carbonic anhydrase II-deficient (CAR2 null) mouse. *Am. J. Physiol. Renal Physiol.* 269:F761–F74

128. Lai LW, Chan DM, Erickson RP, Hsu SJ, Lien YH. 1998. Correction of renal tubular acidosis in carbonic anhydrase II-deficient mice with gene therapy. *J. Clin. Invest.* 101:1320–25

129. Lifton RP, Gharavi AG, Geller DS. 2001. Molecular mechanisms of human hypertension. *Cell* 104:545–56

130. Wilson FH, Disse-Nicodeme S, Choate KA, Ishikawa K, Nelson-Williams C, et al. 2001. Human hypertension caused by mutations in WNK kinases. *Science* 293:1107–12

131. Zhang D, Kiyatkin A, Bolin JT, Low PS. 2000. Crystallographic structure and functional interpretation of the cytoplasmic domain of erythrocyte membrane band 3. *Blood* 96:2925–33

132. Wood PG. 1992. The anion exchange proteins: homology and secondary structure. *Prog. Cell Res.* 2:325–52

SUBJECT INDEX

CUMULATIVE INDEXES

CONTRIBUTING AUTHORS, VOLUMES 60–64

CHAPTER TITLES, VOLUMES 60–64

Cardiovascular Physiology

Endocrinology

Gastrointestinal Physiology

Neurophysiology

Perspectives

Renal and Electrolyte Physiology

Respiratory Physiology

Special Topics

Apoptosis

Circadian Rhythms